13th International Conference on Micro and Nanotechnology for Power Generation and Energy Conversion Applications

AA003149

(PowerMEMS 2013)

Journal of Physics: Conference Series Volume 476

London, United Kingdom
3-6 December 2013

ISBN: 978-1-62993-452-5
ISSN: 1742-6588

Printed from e-media with permission by:

Curran Associates, Inc.
57 Morehouse Lane
Red Hook, NY 12571

Some format issues inherent in the e-media version may also appear in this print version.

Printed by Curran Associates, Inc. (2014)

For permission requests, please contact the Institute of Physics
at the address below.

Institute of Physics
Dirac House, Temple Back
Bristol BS1 6BE UK

Phone: 44 1 17 929 7481
Fax: 44 1 17 920 0979

techtracking@iop.org

Additional copies of this publication are available from:

Curran Associates, Inc.
57 Morehouse Lane
Red Hook, NY 12571 USA
Phone: 845-758-0400
Fax: 845-758-2634
Email: curran@proceedings.com
Web: www.proceedings.com

TABLE OF CONTENTS

INVITED TALK B

THERMAL HARVESTERS AND DEVICES

HARVESTER THEORY AND MODELLING

POSTER SESSION A

Journal of Physics: Conference Series **476** (2013) 011001 doi:10.1088/1742-6596/476/1/011001

2013

PowerMEMS

The 13th International Conference on Micro- and Nano-Technology for Power Generation and Energy Conversion Applications

Conference Proceedings

London

3rd – 6th December 2013

The 13th International Conference on Micro- and Nano-Technology for Power Generation and Energy Conversion Applications

Credits:
PowerMEMS 2013 logo and website by www.jumpingstone.com
Frontcover image by Worldoflard (http://www.flickr.com/photos/worldoflard/)

Journal of Physics: Conference Series **476** (2013) 011001 doi:10.1088/1742-6596/476/1/011001

WELCOME TO POWERMEMS 2013

It is a pleasure to welcome you to The Royal Society in London and the 13th International Conference on Micro- and Nano-Technology for Power Generation and Energy Conversion Applications, or PowerMEMS 2013.

The objective of PowerMEMS 2013 is to catalyse innovation in miniature, micro- and nano-scale technologies for power generation and energy conversion. The conference aims to stimulate the exchange of insights and information, and the development of new ideas in the Power MEMS/NEMS field as well as at the meso-scale. It will allow the attendees to interact and network within our multidisciplinary community that includes professionals from many branches of science and engineering.

The technical program is led by four invited speakers covering inductive power transfer, chip scale power sources, thermal energy harvesting and implantable biofuel cells. We received 177 abstracts and following a careful reviewing process by the Technical Program Committee a total of 137 papers were selected for presentation. These have been organised into 16 Oral sessions in two parallel streams and two poster sessions that have been augmented by 10 late news papers. The oral and regular poster papers are, for the first time, being published by the Institute of Physics.

We have made every effort to make PowerMEMS 2013 the busiest yet and have included for the first time the PowerMEMS School. This two-day school held at Imperial College London covered a wide range of power-MEMS topics including technologies for power generation, power transmission, energy storage, power electronics interfaces and metrology. Registrations for the School exceeded our expectations and it was full by early November. We hope this, and other activities such as the Discussion Panel and the inclusion of late news papers, will make PowerMEMS 2013 a memorable success. We have also reached out to new communities, such as those working in wireless power transfer and RF harvesting to broaden the technology remit of our conference.

The social program is an important aspect of any conference and the PowerMEMS 2013 banquet will be held in the Science Museum. This provides a fantastic opportunity to network whilst viewing some of the fundamental engineering innovations that have ultimately bought us all here today.

There is a long list of individuals we would like to thank for their support in organising PowerMEMS 2013. Once again the TPC, chaired by Eric Yeatman and Douglas Paul, have given us their valuable time and effort in reviewing abstracts. The PowerMEMS School chairs Einar Halvorsen and Shad Roundy and the expert speakers made the School possible. The local organising committee, led by Alwyn Elliott, have provided us with invaluable assistance in making PowerMEMS 2013 happen. The financial support from Imperial College London, the University of Southampton and conference sponsors has also been gratefully appreciated. Finally, we would like to thank you all for attending and helping in making PowerMEMS 2013 a success. We wish you a productive and enjoyable conference and a wonderful stay in London.

Sincerely

P. D. Mitcheson

Paul Mitcheson

Steve Beeby

Journal of Physics: Conference Series **476** (2013) 011001 doi:10.1088/1742-6596/476/1/011001

CONFERENCE OFFICIALS

Conference Co-Chairs

Stephen Beeby, *University of Southampton, UK*
Paul Mitcheson, *Imperial College London, UK*

Technical Program Committee Co-Chairs

Douglas Paul, *University of Glasgow, UK*
Eric Yeatman, *Imperial College London, UK*

PowerMEMS School Co Chairs

Einar Halvorsen, *Vestfold University College, Norway*
Shad Roundy, *University of Utah, USA*

Local Organising Committee Chair

Alwyn Elliott, *Imperial College London, UK*

International Steering Committee

Mark Allen, *Georgia Institute of Technology, USA*
Steve Beeby, *University of Southampton, UK*
Young-Ho Cho, *KAIST, South Korea*
Alan Epstein, *Massachusetts Institute of Technology, USA*
Masayoshi Esashi, *Tohoku University, Japan*
Luc Fréchette, *Université de Sherbrooke, Canada*
Reza Ghodssi, *University of Maryland, USA*
Hiroki Kuwano, *Tohoku University, Japan*
Jeff Lang, *Massachusetts Institute of Technology, USA*
Ryutaro Maeda, *National Institute of Advanced Industrial Science and Technology, Japan*
Kazusuke Maenaka, *University of Hyogo, Japan*
Paul Mitcheson, *Imperial College London, UK*
Albert Pisano, *UC San Diego, USA*
Susumu Sugiyama, *Ritsumeikan University, Japan*
Yuju Suzuki, *University of Tokyo, Japan*
Shuji Tanaka, *Tohoku University, Japan*
Miwako Waga, *Susano Berkeley LLC., Japan*
Peter Woias, *University Freiburg IMTEK, Germany*

Technical Program Committee

David Arnold, *University of Florida, USA*
Seiji Aoyagi, *Kansai University, Japan*
Skandar Basrour, *Tima Laboratory, France*
Philippe Basset, *Université Paris Est / ESIEE Paris, France*
Mustafa Ilker Beyaz, *Antalya International University, Turkey*
Danick Briand, *École Polytechnique Fédérale de Lausanne (EPFL), Switzerland*
Steve Burrow, *University of Bristol, UK*
Luc Fréchette, *University of Sherbrooke, Canada*
Takayuki Fujita, *University of Hyogo, Japan*
Florian Herrault, *Georgia Institute of Technology, USA*
Andrew Holmes, *Imperial College London, UK*

Journal of Physics: Conference Series **476** (2013) 011001 doi:10.1088/1742-6596/476/1/011001

Hanseup Kim, *University of Utah, USA*
Seong-Hyok Kim, *Georgia Institute of Technology, USA*
Sejin Kwon, *KAIST, South Korea*
Carol Livermore, *Northeastern University, USA*
Matthew McCarthy, *Drexel University, USA*
Jae Park, *Kwangwoon University, South Korea*
Paul Ronney, *University of Southern California, USA*
Nico de Rooij, *École Polytechnique Fédérale de Lausanne (EPFL), Switzerland*
Yuji Suzuki, *University of Tokyo, Japan*
Shuji Tanaka, *Tohoku University, Japan*
Luis Velasque-Garcia, *Massachusetts Institute of Technology, USA*
Huib Visser, *Imec Netherlands*
Ruud Vullers, *Imec, Netherlands*
C. Mike Waits, *US Army Research Laboratory, USA*
Xiaohong Wang, T*singhua University, China*
David Yates, *Imperial College London, UK*

Local Organising Committee

Dibin Zhu, *University of Southampton, UK*
Alex Weddell, *University of Southampton, UK*
Tzern Toh, *Imperial College London, UK*
Michail Kiziroglou, *Imperial College London, UK*
Christopher Kwan, *Imperial College London, UK*

Peer review statement

All papers published in this volume of *Journal of Physics: Conference Series* have been peer reviewed through processes administered by the proceedings Editors. Reviews were conducted by expert referees to the professional and scientific standards expected of a proceedings journal published by IOP Publishing.

Inductive power transfer: Powering our future

Grant A. Covic[1]
The University of Auckland, New Zealand

E-mail: ga.covic@auckland.ac.nz

Abstract. The ability to provide power without wires was imagined over a century ago, but assumed commercially impractical and impossible to realise. However for more than two decades the University of Auckland has been at the forefront of developing and commercialising this technology alongside its industrial partners. This research has proven that significant wireless power can be transferred over relatively large air-gaps efficiently and robustly. Early solutions were applied in industrial applications to power moving vehicles in clean room systems, industrial plants, and in theme parks, but more recently this research has helped develop technology that has the ability to impact us directly at home. The seminar will describe some of the early motivations behind this research, and introduce some of the solutions which have been developed by the team of researchers at Auckland over two decades, many of which have found their way into the market. It will also describe how the technology has recently been re-developed to enable battery charging of electric vehicles without the need to plug in, and alongside this how it has the potential to change the way we drive in the future.

1. About The Speaker

Prof Grant A. Covic (Senior Member IEEE) received his BE (Hons), and PhD degrees in Electrical and Electronic Engineering from The University of Auckland (UoA), New Zealand in 1986 and 1993 respectively. He was appointed as a full time Lecturer in 1992, a Senior lecturer in 2000, an Associate Professor in 2007 and to Professor in 2013 within the Electrical and Computer Engineering Department at the UoA, New Zealand. In 2010 he co-founded (with Prof. John Boys) a new global start-up company "HaloIPT" focusing on electric vehicle (EV) wireless charging infrastructure, which was sold in late 2011. Presently he heads power electronics research at the UoA and co-leads the interoperability sub-team within the SAE J2954 wireless charging standard for EVs.

Today his research and consulting interests include power electronics, electric vehicle battery charging and inductive (contact-less) power transfer (IPT) from which he has published more than 100 refereed papers in international journals and conferences. He also holds a number of US patents with many more pending, from which licenses in specialized application areas of IPT have been granted around the world.

A 3-DOF SOI MEMS ultrasonic energy harvester for implanted devices

A G Fowler[1], S O R Moheimani[1] and S Behrens[2]

[1] School of Electrical Engineering and Computer Science, University of Newcastle, Callaghan, NSW 2308, Australia

[2] CSIRO Energy Technology, PO Box 330, Newcastle, NSW 2300, Australia

E-mail: Reza.Moheimani@newcastle.edu.au

Abstract. This paper reports the design and testing of a microelectromechanical systems (MEMS) energy harvester that is designed to harvest electrical energy from an external source of ultrasonic waves. This mechanism is potentially suited to applications including the powering of implanted devices for biomedical applications. The harvester employs a novel 3-degree of freedom design, with electrical energy being generated from displacements of a proof mass via electrostatic transducers. A silicon-on-insulator MEMS process was used to fabricate the device, with experimental characterization showing that the harvester can generate 24.7 nW, 19.8 nW, and 14.5 nW of electrical power respectively through its x-, y-, and z-axis vibrational modes.

1. Introduction

Energy harvesting for microscale applications remains a major focus of microelectromechanical systems (MEMS)-based research. Applications such as wearable electronics, wireless network nodes, and structural monitoring systems stand to gain a significant benefit by having their running time increased through the integration of MEMS-based energy harvesters [1–3]. In the majority of cases, such harvesting devices are designed to use environmental vibration sources to produce electrical power [4–6].

Implanted electronic systems in biomedical applications such as patient monitoring and drug delivery continue to use batteries as a primary power source, meaning that a surgical procedure is often required for battery replacement [7]. As a result, there is a continued interest in the development of methods to safely and reliably transfer electrical energy to electronic devices within the body. The use of ultrasonic waves as part of a MEMS-based energy harvesting system is a potential solution to this problem, with ultrasound having been demonstrated as being a viable means of transferring energy to a remote receiver [8, 9].

A MEMS-based 2-degree of freedom (DOF) ultrasonic energy harvester was reported in [10–12]. The device featured mechanical resonance modes in the in-plane x and y directions, and was used to harvest electrical energy from an external source of ultrasonic waves via electrostatic comb-finger transducers.

2. Design of a Novel 3-DOF Energy Harvester

This paper presents a MEMS energy harvester that is designed to utilize ultrasonic waves from an external transmitter as a source of mechanical excitation. In contrast to the 2-DOF system

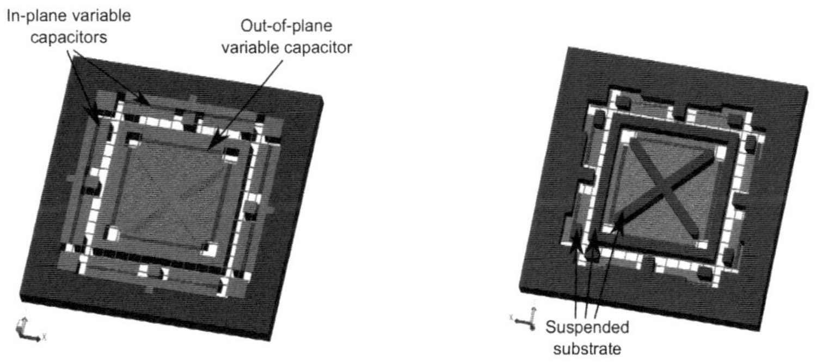

(a) (b)

Figure 1: a) Schematic diagram of 3-DOF ultrasonic energy harvester. b) Underside of harvester, with suspended substrate elements shown.

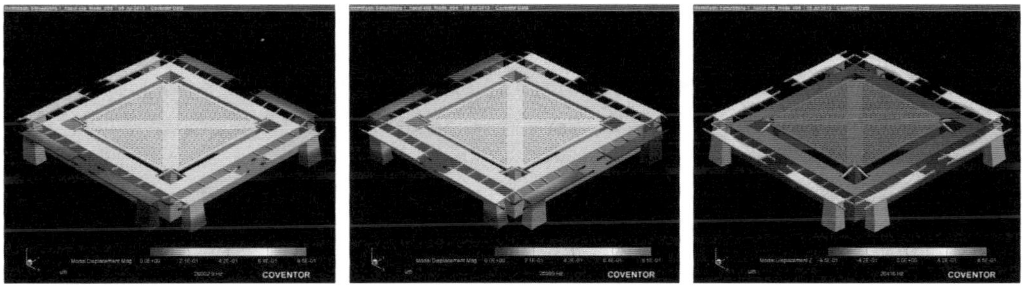

(a) (b) (c)

Figure 2: Simulated resonance modes. a) X axis. b) Y axis. c) Z axis.

reported in [12], the harvester uses a novel 3-DOF design that features mechanical resonance modes in the in-plane (x and y) and out-of-plane (z) directions. This configuration allows the system to harvest electrical energy in any orientation with respect to the external ultrasonic transmitter. Electrostatic comb-finger transducers are used to generate electrical energy from the mechanical vibrations of the harvester's proof mass.

2.1. In-Plane Harvesting Mechanism

The MEMS harvester utilizes a parallel kinematic design for its in-plane harvesting mechanism, which is an extension of the 2-DOF structure presented in [12]. A schematic diagram of the device is shown in Fig. 1a. The harvester contains a proof mass in the center of the device, which is anchored to the substrate via a number of beam flexures. In addition to decoupling the motions of the mass in the x and y directions, the flexures are designed to create in-plane mechanical resonance modes at the desired frequency.

The in-plane oscillations of the proof mass are converted into electrical energy through the use of in-plane, overlap-varying comb finger electrodes placed around the edges of the device. Due to the decoupled motion of the mass, displacements in the x and y directions are harvested via independent electrodes.

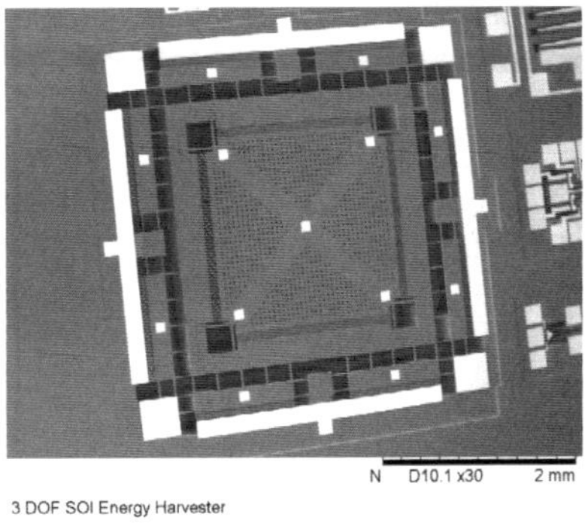

(a)

(b)

Figure 3: a) Image of fabricated MEMS harvester obtained by an SEM. b) Closer view of electrically isolating cuts in SOI layer.

2.2. Out-of-Plane Harvesting Mechanism

The MEMS harvester is also designed to feature an out-of-plane harvesting mechanism that enables the system to harvest energy from ultrasonic waves in the out-of-plane (z) direction. This mechanism is created by selectively etching the central structure to create a smaller 'sub-mass', which remains connected to the remainder of the proof mass by a set of secondary beam flexures. This creates a nested structure, with the secondary flexures being designed to allow this sub-mass to possess a mechanical resonance mode in the out-of-plane direction. The entire mass structure continues to move in unison as a single unit for in-plane displacements.

The out-of-plane vibrations of the sub-mass are harvested via a separate electrostatic transducer, which takes the form of additional comb finger electrodes attached to both the sub-mass and the surrounding structure. This creates a variable capacitor whose capacitance is dependent on the out-of-plane displacement of the sub-mass. The silicon-on-insulator (SOI) layer is etched in strategic locations to electrically isolate the sub-mass from the rest of the moving structure. Elements of the die's substrate are retained in these areas to maintain a mechanical connection between the electrically-isolated components, as shown in Fig. 1b.

2.3. Modal Simulations

The harvester is designed to be mechanically excited by ultrasonic waves from an external transmitter. To maximize the amplitude of the resulting vibrations and thus the harvested electrical power, the device's resonance mode frequencies should be matched to the frequency of the transmitter. A target frequency of 25 kHz was used in the design of the harvester, as this is a common frequency for off-the-shelf ultrasonic transmitters.

The harvester was designed using CoventorWare, and a simulation of the device's mechanical resonance modes are shown in Fig. 2. Both the in-plane and out-of-plane modes are located close to 25 kHz.

3. MEMS Harvester Fabrication

The harvester was fabricated via MEMSCAP's SOIMUMPs prototype process [13], with 25 μm-thick doped silicon comprising the main device layer. The 400μm-thick silicon substrate is largely etched away except in certain areas as described in Section 2.2, where it remains connected to the SOI layer via a 2μm-thick layer of buried oxide.

Scanning electron microscope (SEM) images of the fabricated MEMS harvester are shown in Fig. 3.

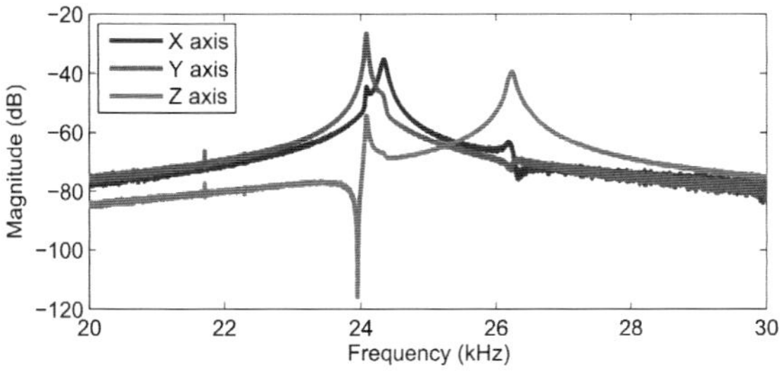

(a)

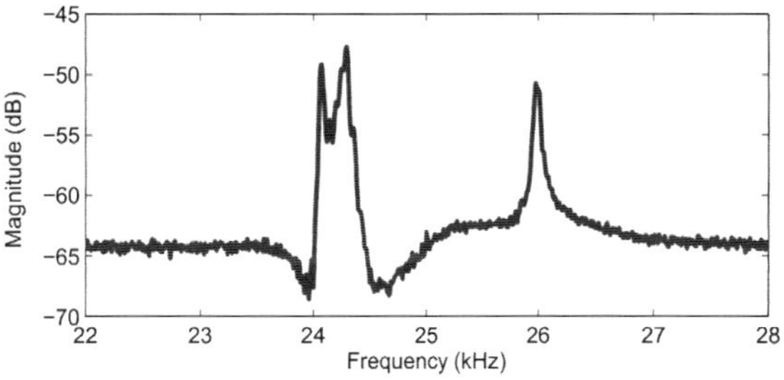

(b)

Figure 4: a) Magnitude plot of frequency response obtained using 3D micro system analyzer. b) Harvester frequency response obtained using a signal analyzer while all three modes are excited by an ultrasonic transmitter.

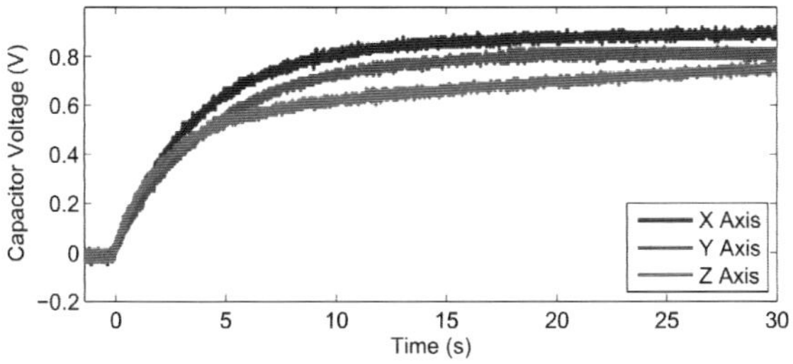

Figure 5: Rising voltage on load capacitor due to harvesting from ultrasonic waves. Separate tests were performed with the transmitter aligned along the harvester's x, y, and z axes.

4. Experimental Characterization

4.1. Identification of Resonance Modes

The resonance frequencies of the fabricated device were measured using a Polytec MSA-050-3D Micro System Analyzer (MSA). The MSA uses three laser vibrometers to measure the vibrations of a test subject, with the vibrations being resolved into their x-, y-, and z-axis components. The harvester was mechanically excited by applying a small-amplitude chirp signal with a 60 V DC bias to the device's capacitive electrodes, with the resulting frequency response being shown in Fig. 4a. This magnitude plot indicates three clear resonance frequencies at 24.30 kHz, 24.08 kHz, and 26.23 kHz, corresponding with the resonance modes of the x, y, and z axes respectively. These are close to the target frequency of 25 kHz, with the variations being attributed to tolerances in the fabrication process.

The operation of the harvester's electrostatic transducers was verified by performing a second frequency response measurement using the device's electrical output. Each of the harvester's transducers were biased with a 60 V DC voltage, and were connected in parallel to the input of a signal analyzer. The harvester was mechanically excited by a 25 kHz Prowave 250SR160 ultrasonic transmitter, which has a center frequency that is close to the frequencies of the harvester's resonant modes. The transmitter was driven by a 20 V_{rms} wideband chirp signal and placed near the MEMS device at a 45° angle to the harvester's three axes. The three expected resonance modes are clearly evident in the obtained frequency response (shown in Fig. 4b), with the frequency of each mode being consistent with those obtained using the MSA.

4.2. Measurement of Harvested Power

The device's power harvesting ability was measured by using the converted electrical energy to charge a load capacitor. The electrostatic transducers were each connected to a full-wave diode rectifier and a 1 μF electrolytic capacitor, with the voltage on the capacitor being monitored by an oscilloscope connected via a low-noise voltage preamplifier.

Testing was performed by placing the ultrasonic transmitter approximately 5 cm away from the MEMS harvester while aligned along one of the device's primary axes. The transmitter was driven by a 20 V_{rms} sine wave at the identified frequency of the corresponding resonance mode, while the electrostatic transducers were again biased with 60 V DC.

The evolution of the voltages on the load capacitor resulting from the harvesting of the ultrasonic waves is shown in Fig. 5. Based on the calculated change in energy stored by the capacitor over the initial 15 second period, the average power harvested by the device was determined to be 24.7 nW, 19.8 nW, and 14.5 nW for the x, y, and z axes respectively.

5. Conclusion

The design and characterization of a MEMS ultrasonic energy harvester has been presented, which features a novel 3-DOF design that allows it to harvest energy regardless of its position with respect to an external source of ultrasonic waves.

The harvester was fabricated using a SOI MEMS process, with experimental testing showing that its resonance modes are close to the targeted frequency of 25 kHz. Its energy harvesting performance was demonstrated by exciting the device using an ultrasonic transmitter, with a load capacitor being used to collect the harvested electrical power.

This ultrasonic energy harvesting mechanism is particularly suited to applications such as the powering of implanted electronic devices for biomedical applications. The additional considerations that must be met to create a practical implentation of such an implantable system are the subject of future work.

Acknowledgments

This research was supported by the Australian Research Council (ARC) and the Commonwealth Scientific and Industrial Research Organisation (CSIRO). Experimental work was performed at the Laboratory for Dynamics and Control of Nanosystems at the University of Newcastle, Australia.

References

[1] Williams C B and Yates R B 1995 Analysis of a micro-electric generator for microsystems *The 8th International Conference on Solid-State Sensors and Actuators, 1995 and Eurosensors IX (Transducers '95)* vol 1 pp 369–372

[2] Torres E O and Rincón-Mora G A 2006 Electrostatic energy harvester and Li-ion charger circuit for micro-scale applications *49th IEEE International Midwest Symposium on Circuits and Systems, 2006* pp 65–69

[3] Harb A 2011 *Renewable Energy* **36** 2641–2654

[4] Mitcheson P D, Green T C, Yeatman E M and Holmes A S 2004 *Journal of Microelectromechanical Systems* **13** 429–440

[5] Beeby S P, Tudor M J and White N M 2006 *Measurement Science and Technology* **17** 175–195

[6] Chao P C P 2011 *IEEE Sensors Journal* **11** 3106–3121

[7] Mhetre M R, Nagdeo N S and Abhyankar H K 2011 Micro energy harvesting for biomedical applications: A review *2011 3rd International Conference on Electronics Computer Technology* pp 1–5

[8] Arra S, Leskinen J, Heikkila J and Vanhala J 2007 Ultrasonic power and data link for wireless implantable applications *2007 2nd International Symposium on Wireless Pervasive Computing* pp 567–571

[9] Ozeri S, Shmilovitz D, Singer S and Wang C C 2010 *Ultrasonics* **50** 666–74

[10] Zhu Y, Moheimani S O R and Yuce M R 2009 A 2-DOF wideband electrostatic transducer for energy harvesting and implantable applications *2009 IEEE Sensors* pp 1542 – 1545

[11] Zhu Y, Moheimani S O R and Yuce M R 2010 *IEEE Electron Device Letters* **31** 374–376

[12] Zhu Y, Moheimani S O R and Yuce M R 2011 *IEEE Sensors Journal* **11** 155–161

[13] Cowen A, Hames G, Monk D, Wilcenski S and Hardy B 2011 *SOIMUMPs Design Handbook*

Helmholtz Resonator for Lead Zirconate Titanate Acoustic Energy Harvester

Tomohiro Matsuda, Kazuki Tomii, Saori Hagiwara, Shuntaro Miyake, Yuichi Hasegawa, Takamitsu Sato, Yuta Kaneko, and Yasushiro Nishioka

Department of Precision Machinery Engineering, College of Science and Technology, Nihon University, 7-24-1 Narashinodai, Funabashi, Chiba 274-8501, Japan

E-mail: nishioka@eme.cst.nihon-u.ac.jp

Abstract. Acoustic energy harvesters that function in environments where sound pressure is extremely high (~150 dB), such as in engine rooms of aircrafts, are expected to be capable of powering wireless health monitoring systems. This paper presents the power generation performances of a lead-zirconate-titanate (PZT) acoustic energy harvester with a vibrating PZT diaphragm. The diaphragm had a diameter of 2 mm, consisting of Al(0.1 μm)/PZT(1 μm)/Pt(0.1 μm)/Ti(0.1 μm)/SiO2(1.5 μm). The harvester generated a power of 1.7x10-13 W under a sound pressure level of 110 dB at the first resonance frequency of 6.28 kHz. It was found that the generated power was increased to 6.8×10-13 W using a sound-collecting Helmholtz resonator cone with the height of 60 mm. The cone provided a Helmholtz resonance at 5.8 kHz, and the generated power increased from 3.4×10-14 W to 1.4×10-13 W at this frequency. The cone was also effective in increasing the bandwidth of the energy harvester.

1. Introduction

Energy harvesters have attracted considerable attention for possible application to independently operating intelligence systems [1,2]. PZT [$Pb(Zr,Ti)O_3$] is an excellent candidate for harvesting power from ambient vibration sources because it can efficiently convert mechanical strain to electrical charge [3,4]. Several bulk PZT piezoelectric energy harvesters have been reported [5-9]. In addition, currently, energy harvesters using thin PZT films are being extensively studied because they were expected to be integrated into silicon integrated circuits (IC) [10-13].

Acoustic energy harvesters that function in an environment where sound pressure is extremely high (~150 dB) such as in an engine room of an aircraft, are expected to be capable of powering wireless health monitoring systems. Horowitz *et al.* have recently reported on a microelectromechanical systems (MEMS) acoustic energy harvester that contains a PZT piezoelectric film as a diaphragm [14]. They claimed that their energy harvester with a PZT (0.27-μm thick) diaphragm with a diameter of 2.4 mm should exhibit a power density of 0.34 $\mu W/cm^2$ at a sound pressure level (SPL) of 149 dB under the first resonance vibration mode. Shinoda *et al.* [15] fabricated PZT energy harvesters with similar sizes to those of Horowitz *et al.*, compared the performances of their energy harvesters with those of Horowitz *et al.* at SPL of 100 dB, and reported that higher generated power could be realized by vibrating PZT diaphragms at the third-resonance frequency. Kimura *et al.* reported that PZT acoustic harvesters with non-wet electrode fabrication processes exhibited further improved power density at the first resonance [16]. However, these previous reports on the PZT acoustic energy harvesters did not utilize the different polarity charges in the central area and the peripheral area of the diaphragms. Tomioka *et el.* reported on similar PZT acoustic energy harvesters considering to utilization of different polarizations of generated charges on the vibrating diaphragm with a PZT piezoelectric capacitor [17]. Iizumi *et al.* reported on further improved performances of PZT acoustic energy harvesters that utilized different charge polarizations at the peripheral area and the central area of a PZT diaphragm with a diameter of 2 mm, using a combined device consisting of a peripheral harvester and central harvester [18].

Horowitz *et al.* also discussed the influences of Helmholtz sound resonator placed in front of the energy harvesters, and suggested an increase of the frequency bandwidth of the harvesters [14]. However, the influences of the Helmholtz sound collecting cones with a trumpet shape have not been reported. In this paper, the power generation of a similar PZT energy harvester equipped with a Helmholtz sound collecting cone with a trumpet shape was investigated.

2. Device Structure

Figure 1(a) shows the top-view design of the PZT energy harvester investigated here. The two top Al electrodes cover the peripheral area and the central area of the PZT diaphragm, respectively. A schematic of a cross-section of the energy harvester is shown in Figure 1(b). A piezoelectric capacitor with the diaphragm structure of Al (0.1 μm)/PZT (1.0 μm)/Pt (0.1 μm)/Ti (0.1 μm) was formed on 1.5-μm-thick SiO_2 substrate. The cavity with a diameter of 2.0 mm was fabricated on a 300-μm-thick silicon wafer. The processing details have been already reported previously [15]-[17].

When sound pressure is applied from above the device, the diaphragm vibrates, resulting in the deflection of the diaphragm. The deflected PZT capacitor generates a voltage difference between the Al electrodes and the Pt/Ti common electrode owing to the polarization of PZT. In the case of exciting the first resonance, in which the entire diaphragm was vibrated in the same phase, the generated charge polarizations were different between the peripheral area and the central area as schematically shown in Figure 1(c). The two energy harvesters were defined between the peripheral electrode and the common electrode. However, only the peripheral harvester was investigated in this work.

The resonance frequencies of the diaphragm under a sound pressure level of 110 dB were measured between 1 kHz and 17 kHz using the characterization system described in previous reports [15], [16]. In order to measure the resonance frequency of the energy harvester, an amplifier with a gain of 10,000 was connected to amplify the output voltage signals from the energy harvester. The measured lowest resonance peak appeared at 6.28 kHz. This suggested that the first resonance occurred at 6.28 kHz. The simulated first resonance frequency obtained using the infinite element method well coincided with the value of 6.28 kHz as previously reported [15]-[18].

In this study, a sound collector cone made of thin paper was attached in front of a speaker (Figure 2). Two sound collector cones with height H of 60 and 70 mm were used here. To ensure that the cone covered the speaker, the diameter of the bottom of the cone was 80 mm. Figure 2(c) shows the sound pressure measured at the top end of the cone as a function of the sound frequency. The sound pressure varied as a function of the sound frequency. When using the cone, the increase of the sound pressure of approximately 15 dB was realized, as seen in Figure 2(c). It should be noted that the resonance peak frequencies can be adjusted by modifying H. The shape of the cone was designed to have a maximum sound pressure at a frequency of approximately 5-7 kHz, to enhance the harvested energy near the first resonance of 6.28 kHz. Frequency positions fn at the peak sound pressure can be explained using a simple Helmholtz resonance theory [14]. The resonance frequency can be obtained by assuming that n-times half of the wavelength of the sound is equal to H. Thus, fn is expressed as shown in Eq. 1;

$$f_n = \frac{n}{2H} c \quad \text{Eq. (1).}$$

where n is an integer, and c is the velocity of sound at room temperature (345 m/s). The calculated resonance frequencies for the Helmholtz resonance of the cones with H of 60 and 70 mm are shown in Table 1.

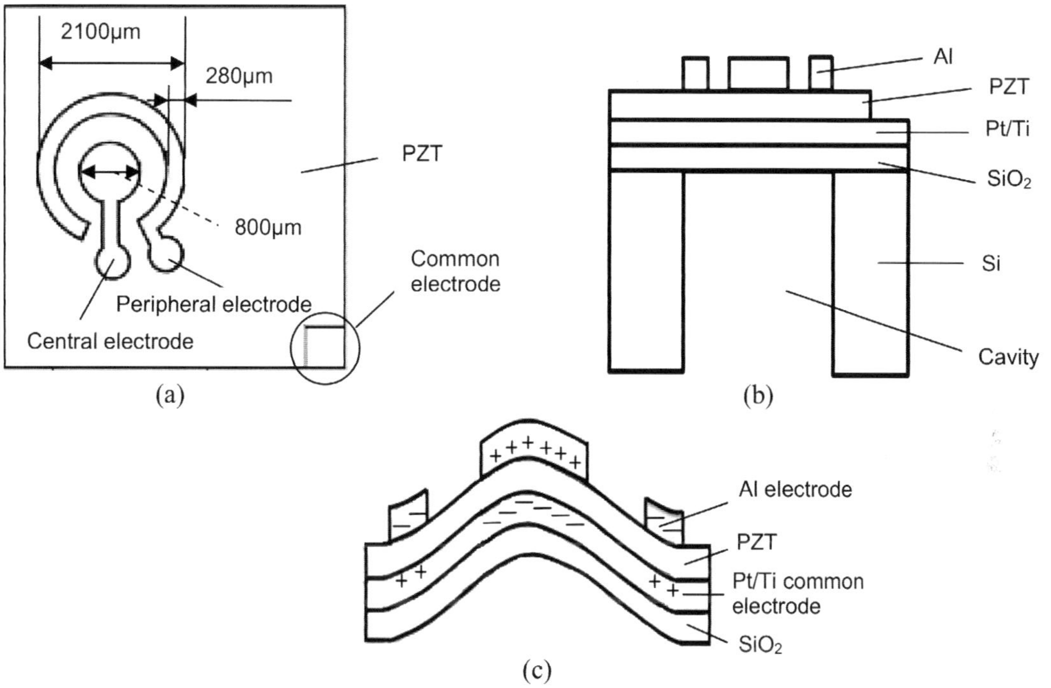

Figure 1. Structure of PZT energy harvester; (a) top view, (b) Schematic cross-section, and (c) concept of connected energy harvester.

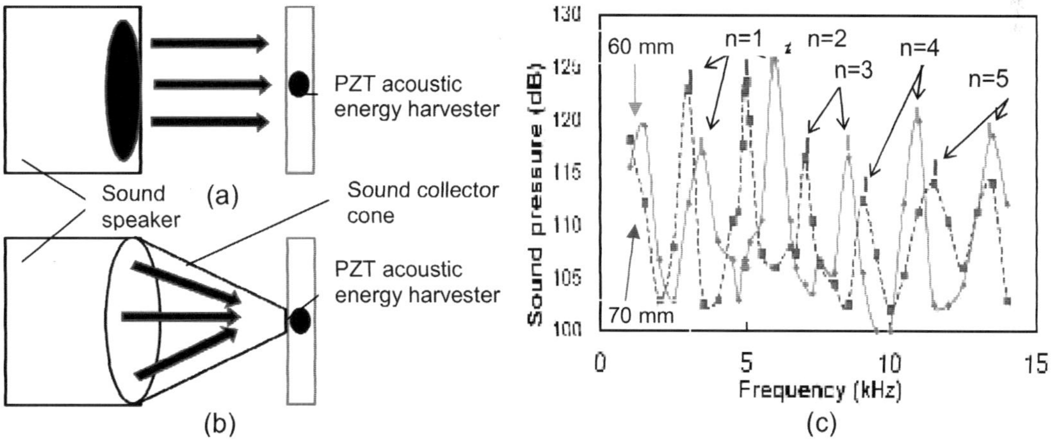

Figure 2 Sound pressure measurement setups; (a) without using sound collecting cone, (b) with using sound correcting cone and (c) measured sound pressure at the end of paper cone as function of sound frequency.

Table 1. Calculated resonance frequencies for the Helmholtz resonance with the diameter D of 80 mm and the height H of 60 mm and 70 mm.

	H=60mm	H=70mm
n=1	2.87 kHz	2.46 kHz
n=2	5.75 kHz	4.92 kHz
n=3	8.62 kHz	7.39 kHz
n=4	11.5 kHz	9.85 kHz
n=5	14.3 kHz	12.3 kHz
n=6	17.2 kHz	14.7 kHz

3. Results and Discussion

Figure 3(a) compares the frequency dependencies of the open-circuit output voltages for the peripheral device with and without the sound collector cone with height of 60 mm. The sound pressure was controlled to be 110 dB within the measured frequency range. Sharp peaks for the first resonance for the peripheral device measured with or without the cone were observed at 6.28 kHz. The output voltages for the device measured using the cone showed distinct peaks at 5.8 kHz and 6.28 kHz. The peak at 5.8 kHz corresponds to the Helmholtz resonance (see Table 1, n=2 with H=60 mm). The peak at 6.28 kHz increased by approximately two times. This output voltage seems to increase because of the wide Helmholtz resonance peak at 5.8 kHz. The peak voltage value at 5.8 kHz was increased from 0.0034 to 0.0292 mV, which is an increase in a factor of approximately 8.6. Thus, a significant increase in the generated power can be expected using the sound collector cone. In addition, the bandwidth of the output voltage was increased. Table 2 summarizes these results.

Figure 3(b) shows the relationships between the generated power of the energy harvester and the sound pressure in different cases. The power was measured at a load resistance of 56.8 Ω that was determined by impedance matching between the harvester and the load resistance. The harvester generated the power of 1.7×10^{-13} W under the sound pressure level of 110 dB at the first resonance frequency of 6.28 kHz, without using the cone. The harvester generated the power of approximately 3.4×10^{-14} W at 5.8 kHz. The measured generated powers for the harvester using the cone with H of 60 mm are also compared in Figure 3(b). A significant increase in the generated power was realized using the cone at the both frequencies. These values of the generated power were 1.4×10^{-13} W at 5.8 kHz and 6.8×10^{-13} W at 6.28 kHz, respectively.

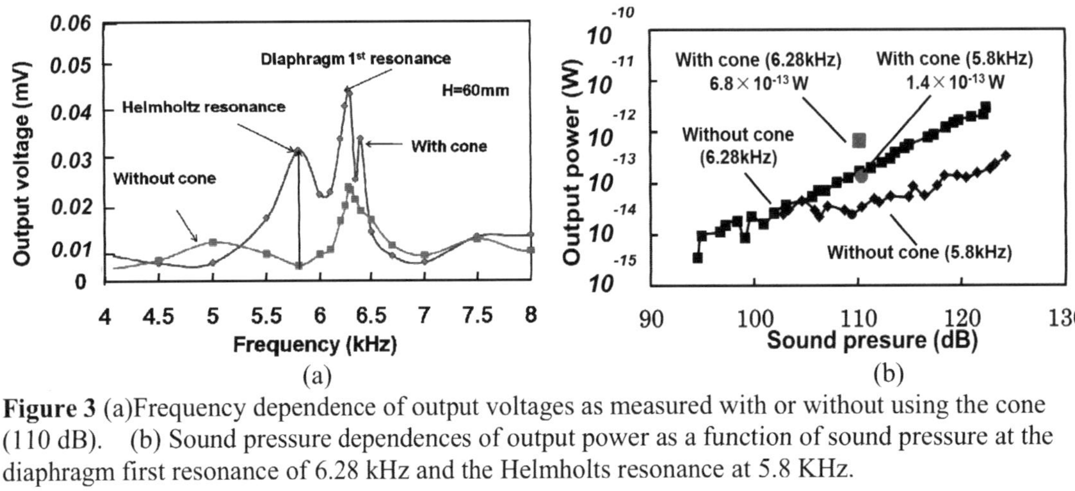

Figure 3 (a)Frequency dependence of output voltages as measured with or without using the cone (110 dB). (b) Sound pressure dependences of output power as a function of sound pressure at the diaphragm first resonance of 6.28 kHz and the Helmholts resonance at 5.8 KHz.

Table 2. Comparison of resonance frequencies and open-circuit output voltage using and without using the sound collector cone.

	First resonance (kHz)	Output voltage (mV)	Helmholtz resonance *n*=2 (kHz)	Output Voltage (mV)
Without cone	6.28	0.0204	5.8	0.0034
With cone	6.28	0.0410	5.8	0.0292

4. Conclusions
This paper presented an energy harvester with a vibrating PZT diaphragm. The diaphragm had a diameter of 2 mm consisting of Al(0.1 μm)/PZT(1 μm)/Pt(0.1 μm)/Ti(0.1 μm)/SiO_2(1.5 μm). The harvester generated a power of 1.7×10^{-13} W under a sound pressure level of 110 dB at the first resonance frequency of 6.28 kHz. It was found that the generated power was increased to 6.8×10^{-13} W using a sound-collecting Helmholtz resonator cone with the height of 60 mm. The cone provided a Helmholtz resonance at 5.8 kHz, and the generated power increased from 3.4×10^{-14} W to 1.4×10^{-13} W at this frequency. The cone was also effective to increase the bandwidth of the energy harvester.

5. References
[1] Shu Y C and Lien I C 2006 *Smart Mater. Struct.* **15** 1499
[2] Beeby S P, Tudor M J and White N M 2007 *Meas. Sci. Technol.* **17** R175
[3] Priya S 2007 *J. Electroceram.* **19** 165
[4] Togo M, Takamura A, Asai T, Kaji H and Nishizawa M 2007 *Electrochim. Acta* **52** 4669
[5] Sodano H A, Park G and Inman D J 2004 *Strain* **40** 49
[6] Ericka M, Vasic D, Costa F, Poulin G, and Tliba S 2005 *J. Phys. IV* **128** 187-193
[7] Shen D, Wikle III H C. Choe S-Y, and Kim D J 2008 *Appl. Phys. Express* **1** 098002
[8] Harigai T, Adachi H, and Fujii E 2010 *J. Appl. Phys.* **107** 096101
[9] Morimoto K, Kanno I, Wada K, and Kotera H 2010 *Sens. Actuators A* **163** 428
[10] Barton S.C, Gallaway J and Atanassov P 2004 *Chem. Rev.* **104** 4867
[11] Rakbamrung P, Lallart M, Guyomar D, Muensit N, Thanachayanont C, Lucat C, Guiffard B, Petit L and Sukwisut P 2010 *Sens. Actuators A*, **163** 493
[12] Bedekar V, Oliver J, Zhang S, and Priya S 2009 *Jpn. J. Appl. Phys.* **48** 091406
[13] Renaud M, Karakaya K, Sterken T, Fiorini P, Van Hoof P, and Puers R 2008 *Sens. Actuators A*, **145** 380
[14] Horowitz S B, Sheplak M, Cattafesta N, and Nishida T, 2006 *Jpn. J. Micromech. Microeng.* **16** S174
[15] Shinoda S, Tai T, Itoh H, Sugou H, Ichioka H, Kimura S, and Nishioka Y 2010 *Jpn. J. Appl. Phys.* **49** 04DL21
[16] Kimura S, Tomioka S, Iizumi S, Tsujimoto K, Sugou T, and Nishioka Y 2011 *Jpn. J. Appl. Phys.* **50** 06GM14
[17] Tomioka S, Kimura S, Iizumi S, Tsujimoto K, Uchida Y, Tomii K, Matsuda T, and Nishioka Y 2011 *Jpn. J. Appl. Phys.* **50** 09ND16
[18] Iizumi S, Kimura S, Tomioka S, Tsujimoto K, Uchida Y, Tomii K, Matsuda T, and Nishioka Y 2012 *Procedia Engineering* **25**, 187

Electrodynamic wireless power transmission to a torsional receiver

K M McEachern[1] **and D P Arnold**
Interdisciplinary Microsystems Group, Department of Electrical and Computer Engineering, University of Florida, Gainesville, FL 32611, USA

E-mail: kmceachern@ufl.edu

Abstract. This paper presents a wireless power transmission (WPT) concept that uses electrodynamic coupling and torsional motion of a permanent magnet in the receiver. The system is shown to transfer an average power of 3.09 mW (power density equal to 143 μW/cm^3) over a distance of 1 cm, an average power of 1.98 mW over a distance of 2 cm, and an average power of 126 μW over a distance of 7 cm. We also demonstrate unaltered power transmission through conductive media, including a human hand and an aluminum plate, highlighting a key advantage of the electrodynamic wireless power transmission approach.

1. Introduction

Many groups currently research the use of inductively coupled coils to achieve wireless power transmission (WPT) [1-5]. Inductively coupled WPT utilizes an alternating magnetic field from a transmitter coil to induce a current in a receiver coil some distance, *d*, away via electromagnetic (inductive) coupling. Electrodynamic WPT is an alternative method, in which the magnetic field induces mechanical motion in a permanent magnet situated above a receiver coil [6-7]. The motion of the magnet induces a current through the receiver coil, and if a load is connected to the receiver coil, WPT from the transmitter coil to the load is realized. Inductively coupled WPT solutions typically operate in the MHz range [1-4], with the exception of [5], which uses frequency control circuitry to transmit at a frequency around 72 kHz. Conversely, the electrodynamically coupled approach allows for much lower frequencies in the range of 10's of Hz [6-7]. This lower operation frequency facilitates WPT through electrically conductive media, such as the human body or metal walls, which have been shown to attenuate signals in the MHz range [1]. Additionally, the limit placed by IEEE on the magnetic field a person's head or torso may be exposed to for frequencies of 20-759 Hz is 0.904 mT, whereas for frequencies 3 kHz to 5 MHz, the allowable magnetic field exposure is limited to 0.205-0.229 mT [8-9]. The use of lower frequency range may therefore be advantageous for biomedical and consumer applications, since higher fields can be safely used to transmit power within a populated environment (e.g. a room), or potentially even to receivers located inside a patient's body (e.g. for implanted medical devices).

Previously, our group has experimented with a prototype electrodynamically coupled WPT system that yielded power transmission of 0.15 mW over a distance of 2.2 cm [6], and 0.9 mW over a distance of 1 cm [7]. The system employed a cantilever beam with a magnet as a tip mass and had two operating modes: force mode and torque mode in which the field from the transmitter induced a force

[1] To whom any correspondence should be addressed.

or torque, respectively, on the magnet. Both modes induced vibration of the cantilever beam, but efficiency and maximum power transmitted were both higher in the torque mode of operation. This prompted the question: what if the receiver was built with an architecture to better capitalize on torsional motion of the magnet? Hence, the system presented in this paper was designed with torsion as the desired vibrational mode, with hopes of increasing the power output and/or efficiency.

2. WPT System Design

The torsional electrodynamic WPT system has two distinct parts: a transmitter coil and a receiver, as illustrated in figure 1. The transmitter coil has an outer diameter of 63 mm, an inner diameter of 49 mm, and a height of 12.3 mm. It consists of approximately 435 turns of AWG 28 copper wire, estimated from the resistance of the wire. The receiver comprises a disc-shaped permanent magnet suspended over a compact receiver coil by rubber bands that function as torsion springs (figure 2). The magnet used is an N45 grade NdFeB diametrically magnetized ring magnet with an outer diameter of 6.35 mm, an inner diameter of 3.17 mm, and a thickness of 3.17 mm. The receiver coil's outer diameter is 26.5 mm, inner diameter is 13 mm, and height is 8.4 mm and comprises approximately 187 turns of AWG 25 copper wire, estimated from the resistance of the wire. The distance d between the transmitter and receiver is defined as the distance between the center of the transmit coil and the center of the receiver coil.

To transmit power, an alternating electric current is input to the transmitter coil, which induces a time-varying magnetic field. When in range of the receiver, the magnetic field from the transmitter applies an alternating torque on the receiver magnet, causing the magnet to rotate back and forth over the receiver coil. The resulting time-varying magnetic flux from the magnet motion induces a current in the load, and hence power delivery. Compared to a translational motion, the torsional motion of this architecture has the potential to induce a higher rate of change in magnetic flux through the coil, since the magnet alternates between completely in-plane (zero flux) and out-of-plane (maximum flux) with respect to the receiver coil in just one quarter of a period. When the transmitter frequency equals the resonant frequency of the torsional resonator, a very large response is observed (mechanical quality factor of 47).

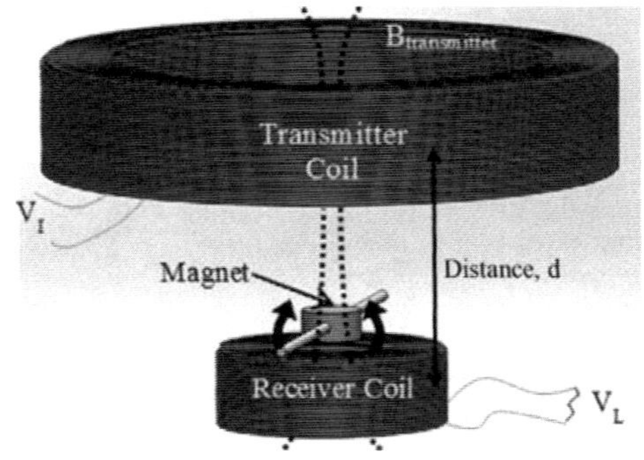

Figure 1. Schematic of the electrodynamic torsional wireless power transmission system.

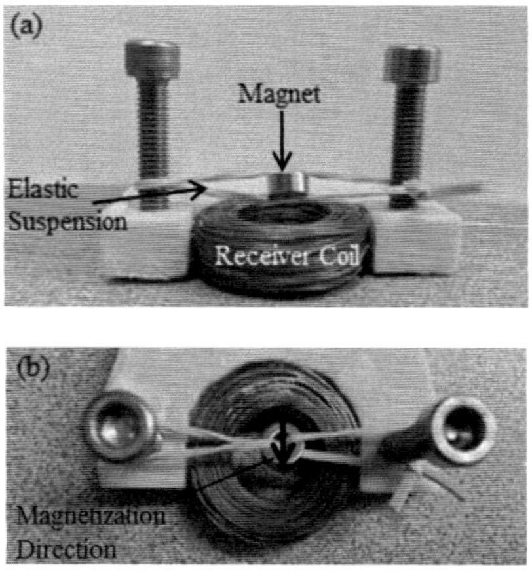

Figure 2. Front (a) and top (b) views of the wireless power receiver. The arrow indicates the magnetization direction.

3. Test Setup

The test setup used in the characterization of the wireless power transmission system is as follows. An Agilent 33120A function generator was used to supply the AC input power to the transmitter coil. The rms voltage supplied was measured via an Agilent DSOX2004A 4-channel oscilloscope, while the current supplied was measured with a Tektronix TCPA312 current probe. The probe was connected to a Tektronix TCPA 300 current probe amplifier, which was in turn connected to the oscilloscope to display the rms current measurement. Additionally, the rms output voltage was measured across the leads of the receiver coil via the oscilloscope. A 100 Ω potentiometer was used as a resistive load. A Keithley 2400 SourceMeter was used to measure the resistance of the load potentiometer.

4. Results

Using the test setup described in the previous section, first the open-circuit voltage was measured over varying input current and frequency, at a fixed transmission distance of 1 cm. Additionally, the power transmitted was measured over varying load resistance, frequency, and distance. Finally, the output power was measured under the same input conditions while transmitting through three different media.

4.1. Open-circuit voltage

At a transmission distance of 1 cm, for constant-amplitude current inputs of 1 mA_{rms}, 5 mA_{rms}, and 10 mA_{rms}, the open-circuit voltage was measured across the output of the receiver coil for different excitation frequencies. For each excitation level, the current amplitude was held constant to ensure a constant amplitude magnetic field from the transmitter, since the impedance of the transmitter was frequency dependent. As a consequence, changes in the receiver output voltage were due only to changes in the frequency of the alternating magnetic field, and not changes in the magnetic field amplitude. As can be seen in figure 3, a resonant dynamic response was observed with peak output at 53-56 Hz. However, the peak voltage amplitude did not scale proportionally with the input current, and a "spring softening" type behavior was noted, both indicating non-linear behavior.

To prove the output from the receiver was caused by the magnet's torsional motion relative to the receiver coil and not simply due to inductive coupling between the transmitter coil and receiver coil, the magnet was removed from the receiver, and the open-circuit rms output voltage was measured for the same distance, frequencies, and input excitation as before. From figure 4, it is clear that the contribution from inductive coupling between the coils (~1 mV) is negligible when operating at resonance.

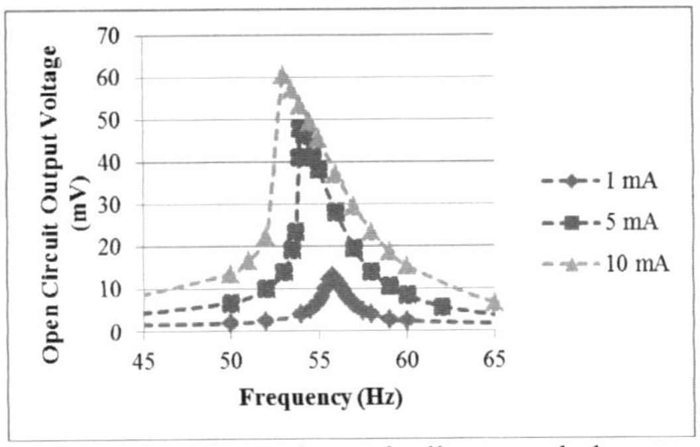

Figure 3. Electrodynamically coupled case. Open-circuit rms output voltage vs. frequency and vs. input current.

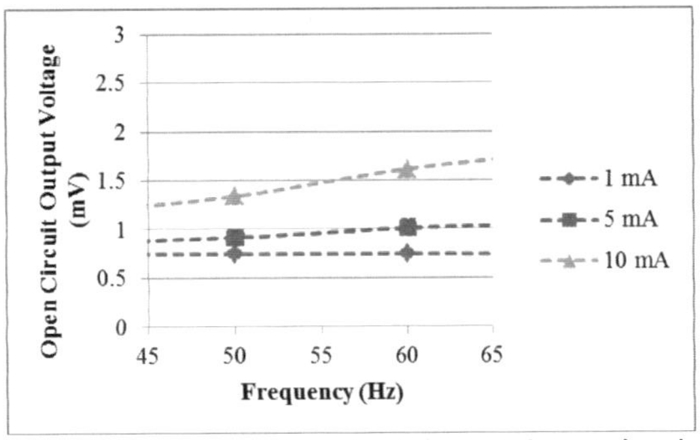

Figure 4. Inductively coupled case. Open-circuit rms output voltage vs. frequency and vs. input current. Note: the y-axis scale differs from the electrodynamically coupled case in figure 3.

4.2. Maximizing output power

For the same transmission distance of 1 cm, the output power was then measured across a potentiometer connected with the receiver coil's terminals. The optimal load resistance was measured by tuning the resistance until maximum time-average output power was achieved. As shown in figure 5 below, for a 5 mA_{rms} input current and a fixed 1 cm transmission distance, operating at a frequency of 54 Hz, the optimal load resistance was found to be 4.65 Ω, yielding an output power of 141.8 μW. However, it must be noted that this load resistance was only optimal at this distance and for this input current. The same process was repeated to find the optimal load resistance for 20 mA_{rms} and 100 mA_{rms} input currents at the same fixed distance of 1 cm, which were found to be 2.63 Ω and 1.20 Ω, respectively. The system's output power was also measured over varying frequency, in order to determine the optimal operating frequency of the system. For a 5 mA_{rms} input and a load resistance of 4.65 Ω, the system transmitted a maximum power of 161.6 μW at an optimal frequency of 55 Hz, as can be seen in figure 6.

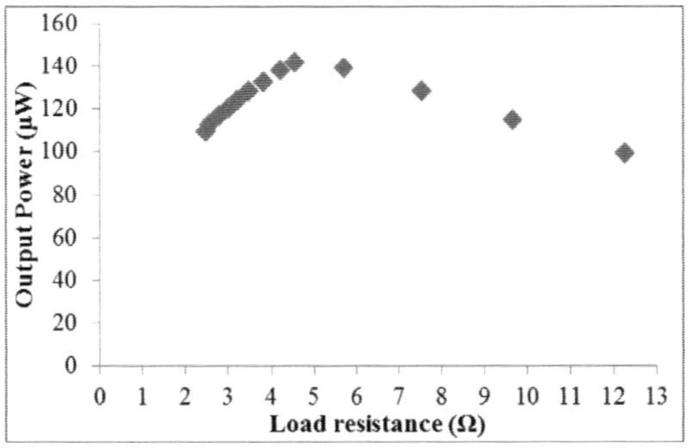

Figure 5. Average output power vs. load resistance for transmitter current of 5 mA_{rms},, frequency of 54 Hz, and distance of 1 cm.

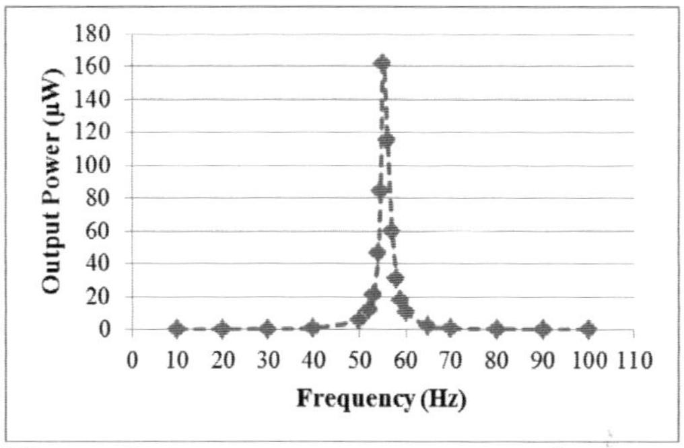

Figure 6. Average output power vs. frequency for transmitter current of 5 mA_{rms}, load resistance of 4.65 Ω and distance of 1 cm.

4.3. Power vs. distance

The maximum transmission distance between the transmitter and receiver is paramount in many practical wireless transmission implementations. Hence, the system was characterized for transmitter currents of both 20 mA_{rms} and 100 mA_{rms} using the optimal load resistances found in the previous section. The transmitter and receiver were set a certain distance apart, and then the excitation frequency was tuned to achieve the maximum output for that distance. This process was repeated for distances ranging from 1 cm to 10 cm. As expected, the power decays with distance as shown in figure 7. At a distance of 1 cm, the transmitted power was 1.04 mW for a 20 mA_{rms} input current and 3.09 mW for an input current of 100 mA_{rms}. For many low-power sensing applications, 100 μW is sufficient power. With this benchmark in mind, an output power of 100 μW was achieved at a distance of 4.3 cm for an input current of 20 mA_{rms} and at a distance of 7.4 cm for an input current of 100 mA_{rms}.

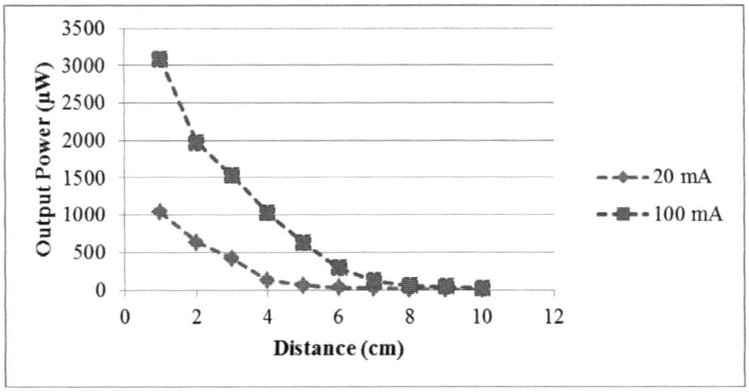

Figure 7. Average output power vs. transmission distance for transmitter currents of 20 mA_{rms} (2.63 Ω load) and 100 mA_{rms} (1.20 Ω load).

4.4. Transmission through solid media

To demonstrate this system's ability to transmit power through conductive media, measurements were taken both with and without solid objects obstructing the path between the transmitter and receiver. The solid media tested are two media likely to come in between a transmitter and receiver if used in commercial or industrial applications: a human hand and a 6.75 mm thick plate of aluminum.

As seen in figure 8 below, the power transmission hardly differs when compared with transmission through air. For a $100mA_{rms}$ current in the transmitter coil and a transmit distance of 8 cm, output powers of 94.1 μW, 93.6 μW, and 93.6 μW were measured for transmission through air, a human hand, and the aluminum, respectively. The variation between these three measurements is small enough to be attributed to experimental error.

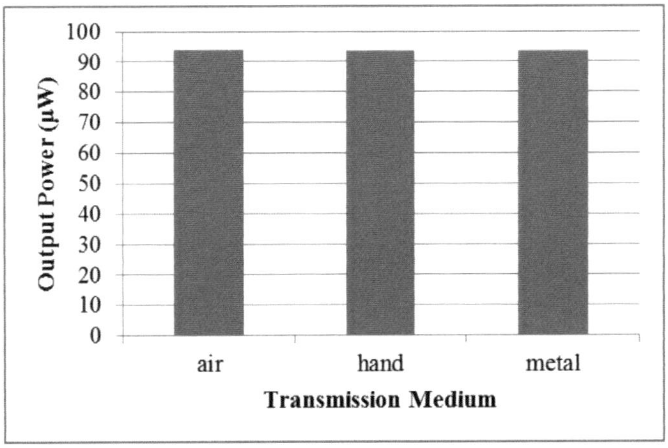

Figure 8. Average output power at a transmission distance of 8 cm for a 100 mA_{rms} input through three media: air, a human hand, and aluminum.

5. Conclusions

The results from this initial prototype are promising, with a maximum observed output of over 3 mW. Furthermore, an output power of 126 μW was achieved over a distance of 7 cm. It was demonstrated that power transmission with our device is virtually unimpeded through both a human hand and a plate of aluminum, illustrating a key potential advantage of the electrodynamic WPT approach for real-world applications. This is especially true for bio-applications in which the transmitted power needs to pass through the skin to reach the receiver(s) inside the body.

Ongoing work is focused on shrinking the receiver's size and increasing the maximum transmission distance; all this while keeping the maximum transmittable power at a practical, usable level. Additional work is also being done on experimenting with different suspension materials, with the hopes of achieving higher efficiency, since rubber elastic produces a highly damped response.

Acknowledgements
This work was supported in part by a National Science Foundation Graduate Research Fellowship for Kelly McEachern.

References
[1] Kurs A, Karalis A, Moffatt R, Joannopoulos J D, Fisher P and Soljačić M 2007 *Science* **317** 83-6
[2] Lee S and Lorenz R D 2011 *IEEE Trans. Ind. Appl.* **47** pp 2495-504
[3] Sample A P, Meyer D A, and Smith J R 2011 *IEEE Trans. Ind. Electron.* **58** pp 544-54
[4] Zhang F, Hackworth S A, Fu W, Li C, Mao Z, and Sun M 2011 *IEEE Trans. Magn.* **47** pp 1478-81
[5] Si P, Hu A P, Malpas S and Budgett D 2008 *IEEE Trans. Biomed. Circuit. Syst.* **2** 22
[6] Arnold D P, Mur-Miranda J O and Challa V R 2012 *Smart Mater. Struct.* **21** 115-7
[7] Challa V R, Divakar V, Mur-Miranda J O and Arnold D P 2011 *Proc. PowerMEMS* pp 335–8
[8] 2005 IEEE Standard for Safety Levels with Respect to Human Exposure to Radio Frequency Electromagnetic Fields, 3 kHz to 300 GHz *IEEE Std C95.1*
[9] 2002 IEEE Standard for Safety Levels With Respect to Human Exposure to Electromagnetic Fields, 0-3 kHz *IEEE Std C95.6*

Long range inductive power transfer system

James Lawson, Manuel Pinuela, David C Yates, Stepan Lucyszyn, and Paul D Mitcheson

James Lawson, Electronic and Electrical Engineering Department, Imperial College London, South Kensington Campus, London, SW7 2AZ, UK

E-mail: `james.lawson08@imperial.ac.uk`

Abstract. We report upon a recently developed long range inductive power transfer system (IPT) designed to power remote sensors with mW level power consumption at distances up to 7 m. In this paper an inductive link is established between a large planar (1 x 1 m) transmit coil (Tx) and a small planer (170 x 170 mm) receiver coil (Rx), demonstrating the viability of highly asymmetrical coil configurations that real-world applications such as sensor networks impose. High Q factor Tx and Rx coils required for viable power transfer efficiencies over such distances are measured using a resonant method. The applicability of the Class-E amplifier in very low magnetic coupling scenarios and at the high frequencies of operation required for high Q operation is demonstrated by its usage as the Tx coil driver.

1. Introduction

Attempts to transmit power using electromagnetic fields beyond 1 m have tended to use the far field, rather than near field region [1, 2]. However such operation involves radiating energy and is intrinsically inefficient, unless the receiver intercepts a significant portion of the radiated energy. This limits such systems to the use of antennas with narrow beam pattens. The technology presented here uses loop antennas which have very broad magnetic field distributions allowing power transfer to receivers over a wide area [3]. Coils are kept electrically small, to keep radiation losses to a minimum, avoiding significant radio emission [4]. Ambient RF systems are capable of tens of μW with cm scale antennas [5] and although they are able to operate in many urban environments without an explicit power source many applications are not achievable at this power level. Hubregt J. Visser has developed a directional radiative system, capable of delivering hundreds of μW to the load at a distance of 1 m [2]. However, inductive coupling has been recently demonstrated with watt level received power at distances over 0.6 m [6]. Another IPT system has been demonstrated transmitting power with 15% efficiency over 2 m [7], using a 1 x 1 m Tx and 1 x 1 m Rx coil at a frequency of 513 kHz.

When optimising the coils in an inductive power transfer system within a given dimensional constraint, the most important parameter is Q factor. The link efficiency η_{link} is given by [8]:

$$\eta_{link} = \frac{k^2 Q_{Tx} Q_{Rx}}{\left(1 + \sqrt{1 + k^2 Q_{Tx} Q_{Rx}}\right)^2} \tag{1}$$

where the coupling factor k is given by:

$$k = \frac{M_{TxRx}}{\sqrt{L_{Tx} L_{Rx}}} \tag{2}$$

Coil's self inductance, L, is proportional to the number of turns squared, while the mutual inductance between the Tx and Rx coils, M_{TxRx}, is proportional to the turns product of the two coils. Therefore k is effectively set by the coil diameter and varies little with changes in the number of turns. As such, maximising Q factor by selection of appropriate operating frequency and minimising the coil's loss resistance are key to obtaining high link efficiency [6].

2. Design

2.1. Miniaturised receiver coils

Direct measurement of coil parameters via the reflection coefficient using a traditional network analyser is not generally possible for Qs greater than 30 [9]. Therefore measurement of coil parameters was performed using a transmission-type resonant measurement [10] with 2 pF coupling capacitors to minimally load the resonators. An estimate of resonator loading was obtained using the peak transmission coefficient [11].

$$Q_L(f_0) = \frac{f_0}{f_2 - f_1} \tag{3}$$

$$Q_0(f_0) = \frac{Q_L(f_0)}{1 - |S_{21}(f_0)|} \tag{4}$$

Where Q_L is the loaded Q factor obtained from the resonator bandwidth, $|S_{21}(f_0)|$ the magnitude of the resonant peak, f_2 is the upper $-3dB$ point and f_1 the lower $-3dB$ point. Inductance of the coils varies little with frequency and therefore was found by measurement of the resonant frequency with a large resonant capacitor to minimise the effects of the coil self capacitance. The capacitors used to resonate the coils (American Technical Ceramics) have Qs in excess of 10,000 at the frequencies of interest and were therefore assumed to be lossless. A series of 20 mm outside diameter Rx coils were constructed, aiming to maximise the Q factor at 3 MHz, which is the target frequency of operation of the IPT system. This frequency of operation was primarily dictated by the need to avoid spurious radio emission from the Tx coil, caused by the transmission wavelength approaching its electrical length. Coil parameters (turns, turn spacing) were optimised in an EM simulator (CST Microwave Studio) before construction and via empirical models [12, 13, 14]. The number of turns and conductor diameter for the Rx coil was constrained by manufacturing capability. It was found that PCB coils, due to the small conductor cross-sectional areas, had poor Q compared to copper wire coils. The wire coils were constructed by threading enameled wire through a former, composed of thin 100 (μm) flexible PCB material, with holes drilled at each wire corner; a thin ferrite, TDK IRJ04 ($\mu = 40$, depth $= 0.5$mm) was then stuck to the reverse side of the coil to reduce the effect of metallic planes behind the coil. The ferrite had little measurable effect on the coils Q factor.

The results shown in figure 1 compare favourably with published research, achieving almost 2x the Q factor of optimised PCB coils at 3 MHz of 40 mm diameter [12]. Litz wire was found to offer a small advantage in Q factor over equivalent diameter solid wire; the lack of dramatic improvement is likely due to the reduced conductor area available in such finely stranded insulated wire. However these miniaturised 20 mm coils only harvest 1 mW within a meter of the transmitting coil and are of limited use for a long range system. A 170 mm diameter Rx coil was therefore developed for the long range wireless power transfer system using a 3D printed ABS former loaded with 3 overlayed parallel connected litz wires shown in figure 2. Due to its greater Q factor (280 at 3 MHz) and 72x greater area than the miniaturised coils, it was found to be able to harvest greater than a mW up to 7 m from the Tx coil. To obtain an estimate of the power transferred, not taking into account the environment the coil is placed in, the mutual inductance has been estimated via the loop areas of the antennas using [14]:

$$M_{TxRx} = \frac{\mu_0 A_{Tx} A_{Rx}}{2r^3} \tag{5}$$

Where A_{TX}, A_{RX} are the sum of the areas of the turns for the Tx and Rx coils respectively and r is the on axis displacement of the coils.

Table 1. Miniaturised 20 mm diameter receiver coils

Coil	Spacing (mm)	Conductor	Turns	Inductance (μH)	Q(3 MHz)
Litz wire	0.5	Litz wire 0.025 mm x 160	11	1.11	97
Solid wire	0.5	0.321 mm diameter copper wire	11	1.25	68
PCB	0.25	0.4 x 0.035 mm Cu track with HASL	15	2.07	30

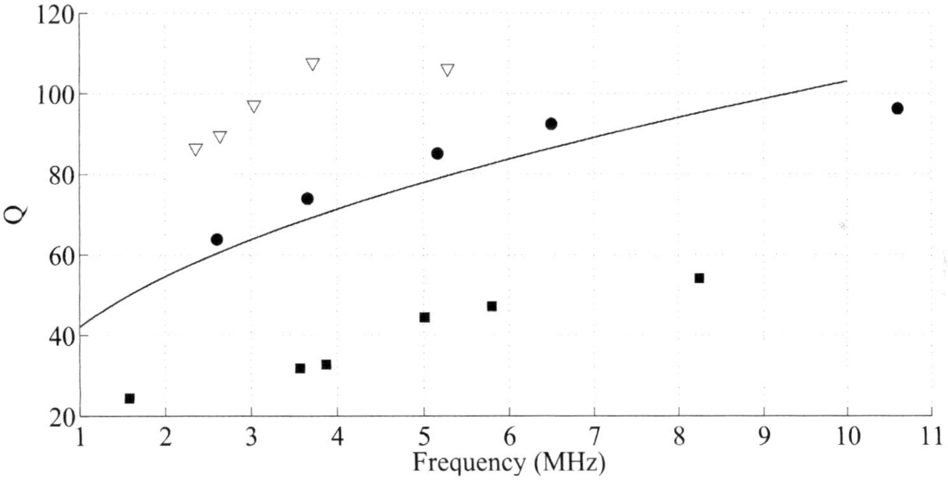

Figure 1. Q factor measurements and simulation results for 2cm coils. Simulation of solid wire coil ——, Litz wire coil ▽, Solid wire ●, PCB coil ■.

Table 2. Long range IPT system coils. [1]Simulation result.

Coil	Diameter (m)	Spacing (mm)	Conductor	Turns	Inductance (μH)	Q(3 MHz)
Rx	0.17	1.5	3 strands of 1.3 mm litz	6	9.70	280
Tx	1	30	28 mm Cu pipe	2	7.63	2890[1]

Figure 2. 170 mm and 20 mm Rx coils

Figure 3. 1 m Tx coil and class-E amplifier

2.2. Long range IPT approximations
In a conventional IPT system, where efficiency is > 50%, the Rx coil reflected impedance makes up the majority of the primary tank impedance formed between the Tx coil and its tuning capacitor [6]. As this varies with position and Rx load, the Class-E amplfier must be retuned for optimal efficiancy. However, with a long-range system, the Rx coil is so weakly coupled that it has negligable effect on the Tx coil impedance, simplifying both design and operation. Furthermore, the impedance match for the Rx coil can be simplified to the conjugate match of the Rx tank at the operating frequency:

$$Z_{tank} = \frac{1}{\frac{1}{R_{Rx}+j\omega L_{Rx}} + j\omega C} \tag{6}$$

where R_{Rx} is the series loss resistance of the Rx coil, L_{Rx} the inductance of the Rx coil and C the turning capacitance used to resonate the LC tank at the systems operating frequency. As this model makes use of the coil's loss resistance, the Q factor measurements of the coils can be verified by checking that the calculated impedance match is optimal. The match should be resistive at resonance, since the tank presents a real impedance. M_{TxRx} can be found using the measured magnetic field generated by the circulating current from a free wheeling primary I_{Tx} and the area A of flux that the Rx coil encloses (i.e. sum of areas of all turns). The free wheeling coil current can be found using the coil voltage and the impedance of the Tx coil (from inductance and Q factor measurements) or by approximating the coil as line segments of current and measuring the magnetic field at a known close location to the Tx coil (e.g. 1 m away on axis).

$$M_{TxRx} = \frac{A_{rx}B}{I_{Tx}} \tag{7}$$

This methodology allows for prediction of system efficiency in real-world environments where the magnetic field strength is modified by conductive objects.

3. Results

The 1 x 1 m Tx coil and 170 x 170 mm Rx coil were displaced from 2 to 10 m, on axis, in a large laboratory with the coils located far away from any boundary and with no objects placed between the coils. The laboratory contained a number of non-removable conductive objects, such as the flooring (steel backed) and the bench supports (steel bar). This is reflected in the magnetic

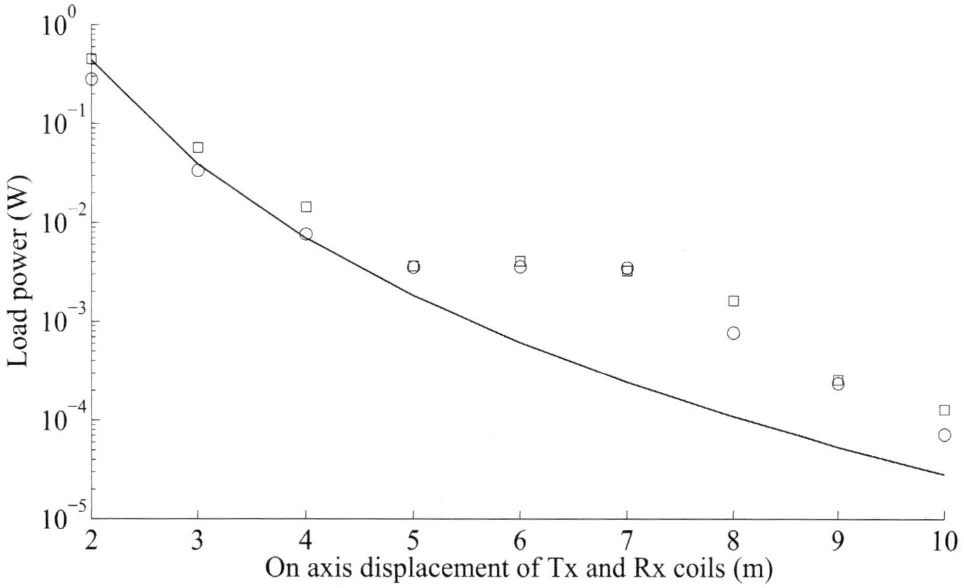

Figure 4. Load power transferred to 170 x 170mm Rx coil from 1 x1 m Tx coil with 98W DC power input to Tx amplifier. Simulation using round loops approximation ——, Prediction using local magnetic field strength □, Measured power at Rx coil ○ .

field strength readings in table 3 which are greater than would be expected at 6 m and beyond. This shows that in real-world environments, range may be greater than expected for long-range systems. Magnetic field readings were taken using a 50 mm Aaronia PBS1 magnetic field probe and Agilent FieldFox Handheld spectrum analyser, orientated such that the component of the field in the direction of displacement was measured. To measure the received power for the system, the Rx coil was first was tuned to the excitation frequency of the Tx coil and then impedance matched with a 51 kΩ 1206 surface mount resistor. The Rx coil was then positioned at the locations the magnetic field probe readings had previously been taken with the system operated at the same 98 W DC power input to the Class-E amplifier. The voltage at the load was then measured to determine the power transferred to the load from the Tx coil, using a Fluke 190 Scopemeter connected to the tuning capacitors, via a short stub of coaxial cable to avoid adding additional loop area to the Rx coil. To test the maximum power input to the system, a 246 W DC power input test was run where 10.9 mW was transferred to the receiver coil located 6 m away on axis; the limiting factor was found to be the 1,000 V rating of the MOSFET used in the Class-E amplifier, due to the high impedance of the free-wheeling Tx coil. The litz wire 20 x 20 mm coil was found to be able to harvest 832 μW at 1 m for 98 W of DC power input to the Tx amplifier.

4. Conclusion

A practical long-range IPT system has been demonstrated. While showing low efficiency, useful amounts of power (> 10 mW) at distances up to 6 m from the Tx coil to compact 170 x 170 mm receivers was recovered. This system makes use of recent advances in IPT technology, using a quasi-resonant Tx coil with a Class-E driver, MHz transmission frequency, quasi-resonant primary and resonant secondary coils [6]. Novel coil constructions, with the aim of maximising Q factor, have been developed for the link. We have demonstrated the systems performance

Table 3. long-range IPT system results

Distance (m)	Pin (W)	Pout (mW)	Tx voltage (kVpp)	B field (μT)	Clock (MHz)
1	97.9	NA	4.06	2.91	2.87126
2	97.9	282	4.06	0.331	2.87126
3	97.9	33.6	4.06	0.117	2.87126
4	97.9	7.53	4.06	0.0588	2.87126
5	97.9	3.52	4.06	0.0294	2.87126
6	97.9(246)	3.57(10.9)	4.06(8.19)	0.0322(NA)	2.87126(2.87721)
7	97.9	3.47	4.06	0.0278	2.87126
8	97.9	0.771	4.06	0.0197	2.87126
9	97.9	0.235	4.06	0.0078	2.87126
10	97.9	0.072	4.06	0.0056	2.87126

in a real-world environment, showing the validity of using magnetic field measurements for performance prediction. In addition we have demonstrated with a 246 W power input test that greater Tx coil power levels are possible, given the main constraint of the maximum drain-source voltage the MOSFET can withstand in the Class-E amplifier. Although efficiency is very low, many sensors can be powered at the same time from the same transmitter and distributed around a room, increasing overall system efficiency and providing an alternative solution to energy harvesters for power supplies in remote applications.

Acknowledgments
Funding is gratefully acknowledged from Her Majesty's Government.

References
[1] Strassner B and Chang K 2013 *Proceedings of the IEEE* **101** 1379–1396 ISSN 0018-9219
[2] Visser H 2012 Aspects of far-field rf energy transport *Microwave Conference (EuMC), 2012 42nd European* pp 317–320
[3] Finkenzeller K 2003 *RFID Handbook* 2nd ed (New York: John Wiley & Sons) ISBN 0-470-84402-7
[4] Yates D, Holmes A and Burdett A 2004 *Circuits and Systems I: Regular Papers, IEEE Transactions on* **51** 1405–1413 ISSN 1549-8328
[5] Pinuela M, Mitcheson P and Lucyszyn S 2013 *Microwave Theory and Techniques, IEEE Transactions on* **61** 2715–2726 ISSN 0018-9480
[6] Pinuela M, Yates D, Lucyszyn S and Mitcheson P 2013 *Power Electronics, IEEE Transactions on* **28** 2437–2447 ISSN 0885-8993
[7] Garnica J, Chinga R and Lin J 2013 *Proceedings of the IEEE* **101** 1321–1331 ISSN 0018-9219
[8] Vandevoorde G and Puers R 2001 *Sensors and Actuators A: Physical* **92** 305 – 311 ISSN 0924-4247 URL `http://www.sciencedirect.com/science/article/pii/S092442470100588X`
[9] Kuhn W and Boutz A 2010 *Microwave Theory and Techniques, IEEE Transactions on* **58** 1046–1055 ISSN 0018-9480
[10] Kajfez D 1999 Q factor measurements, analog and digital University University of Mississippi URL `http://www.ee.olemiss.edu/darko/rfqmeas2b.pdf`
[11] Kajfez D 2011 *Q factor measurements using MATLAB* 1st ed (London: Artech House)
[12] Jow U M and Ghovanloo M 2007 *Biomedical Circuits and Systems, IEEE Transactions on* **1** 193–202 ISSN 1932-4545
[13] Terman F E 1947 *Radio Engineers' Handbook* 2nd ed (New York and London: McGraw-Hill)
[14] Ramo S, Whinnery J R and Duzer T V 1994 *Fields and Waves in Communication Electronics* 3rd ed (New York: John Wiley & Sons) ISBN 978-0471585510

Evaluation of Vertical Integrated Nanogenerator Performances in Flexion

R Tao, R Hinchet, G Ardila and M Mouis

IMEP-LAHC, Grenoble-INP, Univ. J. Fourier, Univ. Savoie, CNRS , 3 Parvis L. Néel BP 257, 38016, Grenoble, France

E-mail: taoran07@gmail.com

Abstract. Piezoelectric nanowires have attracted great interest as new building blocks of mechanical energy harvesting systems. This paper presents the design improvements of mechanical energy harvesters integrating vertical ZnO piezoelectric nanowires onto a Silicon or plastic membrane. We have calculated the energy generation and conversion performance of ZnO nanowires based vertical integrated nanogenerators in flexion mode. We show that in flexion mode ZnO nanowires are superior to bulk ZnO layer. Both mechanical and electrical effects of matrix materials on the potential generation and energy conversion are discussed, in the aim of guiding further improvement of nanogenerator performance.

1. Introduction

Recently, piezoelectric nanowires (NWs) have attracted great interest as the fundamental building blocks for future electronic autonomous devices. Their unique electrical and mechanical properties can be exploited advantageously for applications such as sensors [1], piezoelectric diodes [2] and energy harvesting devices [3]. Among the latter, nanogenerators (NGs) have received much attention, where piezoelectricity is employed to convert mechanical energy to electrical energy for the operation of low-power electronics [4]. Given that the size-induced enhancement of electromechanical properties has been primarily reported for piezoelectric NWs [5], the development of new experimental, computational, and theoretical approaches is required to meet challenges at nanoscale. In this work, finite element method (FEM) simulations are used to calculate the energy generation and conversion performance of ZnO NWs based energy harvesters. The design of the devices is based on the Vertically Integrated NanoGenerator (VING) as proposed in [6].

2. Nanogenerator structure

The energy harvesting device (Figure 1) consists of a VING structure sandwiched between two aluminium electrodes on a substrate. The 1cm × 1cm substrate is made of silicon or plastic membrane: the former is compatible with current electric circuits and devices, while the latter is more flexible and adaptable to future independent self-powered nanosystems. As a continuation of previous work [7], Si membrane was used to calculate the strain distribution under given pressure. The functional core structure of the NG is the piezoelectric NWs integrated into a thin insulating PMMA layer. The NWs grow along their c-axis on a thin ZnO seed layer deposited using spattering. Si_3N_4 insulating layers with no internal strain avoid the short circuit and increase NG robustness. We have previously evaluated NGs under compression [7]. Here we extend the study to a flexion mode using a thin membrane as a mechanical transducer. The hydrostatic pressure generates a force bending the membrane, compressing the NW from the sidewalls.

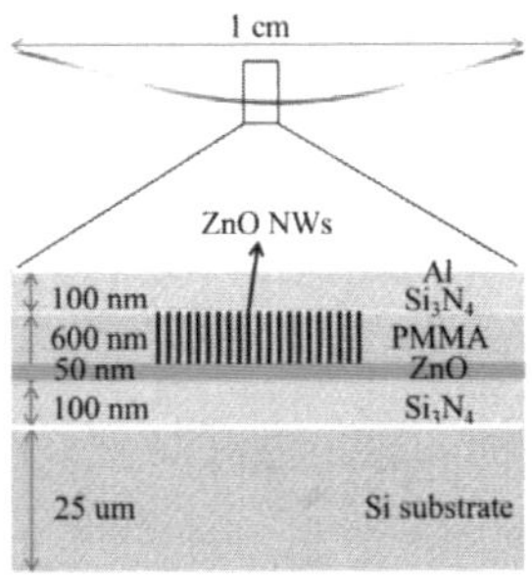

Figure 1. Cross-section view of the device with a zoom where the VING is located.

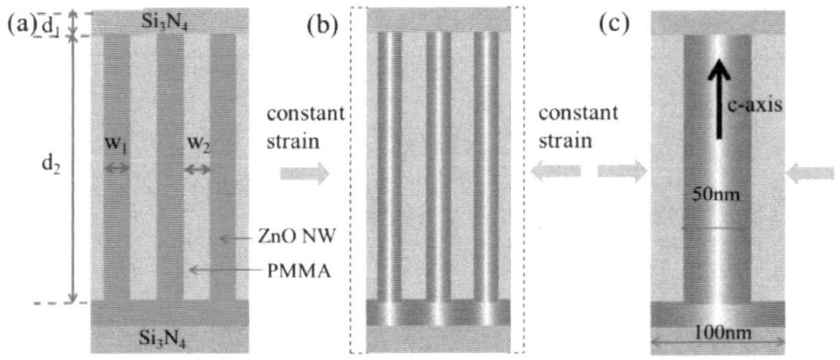

Figure 2. Schematic of nanogenerator cells structure. The radius of the NW is 25 nm and the width of the cell is 100 nm. When the membrane is bent, the single nanogenerator cell is compressed laterally.

3. Nanogenerator operation and modelling assumptions

Under strain, the NW active layer of the VING structure generates a potential difference which can be used to power an external circuit by means of a capacitive displacement current.

The hypothesis of an applied hydrostatic pressure is applied over the membrane structure (Figure 1). The main issue compared to previews studies [7] is how the constraint is transferred to the NW. For this reason, although the active part of the device is made of millions of NWs integrated into a polymer matrix, in this study only a single core cell of the device was considered (Figure 2c) with appropriate boundary conditions for fast and reliable FEM modeling.

When the device membrane is under flexion, the NG core cell is compressed laterally and extended along the c-axis (Figure 2). In the core cell, the energy conversion mechanism has been divided into 3 steps [7]: mechanical energy transfer, mechanical to electrical energy conversion and, finally, electrical energy transfer to the output circuit. In the first step, the total input mechanical energy ξ is considered to be composed of two parts. One (ξ_1) is the energy stored in the Si_3N_4 layers, and the other (ξ_2) is the energy stored in the NW/PMMA compound. The energy that reaches the core NW is used in the piezoelectric transfer process (ξ_3). Thus, the mechanical energy transfer efficiency is expressed as $\eta_m=\xi_3/(\xi_1+\xi_2)$, where ξ_1 is supposed to be a constant. Thus the major influencing factor is the ratio of ξ_3 and ξ_2, whose reciprocal increases with $(w_2E_{NW})/(w_1E_{PMMA})$ where w_1 and w_2 are geometrical parameters (see Figure 2) and E is the Young Modulus. The energy conversion efficiency η_p of the second step depends on the NW's electromechanical properties and is proportional to the square of piezoelectric strain constant e_{33}^2 [7]. The third step brings dielectric losses to our device with the efficiency (η_p) defined as $1/(1+d_1\varepsilon_{2eq}/d_2\varepsilon_1)$, where d is a geometrical parameter (see Figure 2), ε_{2eq} and ε_1 are the relative permittivity of the NW/matrix compound and the Si_3N_4 layer respectively. Finally the total conversion efficiency is the product of the three individual step efficiencies.

4. Simulation

FEM simulations of the device were performed. The coupling between the mechanical, electrical and piezoelectric equations was considered for a core NG cell. The input impact described by the strain in the cell was calculated from a 2D model of the membrane. Because the NW layer was far from neutral plane, the hydrostatic pressure generated constant strain at sidewalls of each cell along c-axis. We calculated that the NW/PMMA layer had a vertically homogeneous strain of -9.36×10^{-4} (within linear deformation) when the device is bent by a hydrostatic pressure of 1.8 kPa, which is the limit of the linear deformation regime. The cell size was 100nm × 100nm × 850nm, with NW radius r=25nm and length L=600nm. Young's modulus, Poisson's ratio (υ) and relative permittivity of the matrix material were varied around parameters of PMMA in order to investigate optimization trends. Several real insulating compounds were used as the matrix material. Their functions on potential generation and energy conversion were compared with each other, as well as with a ZnO layer (the thickness is 118 nm) which has the same volume as ZnO NW.

Analysis is focused on the displacement, the strain tensor S_{xx} and the electric potential distributions. The hardness of the matrix material results in two different types of distribution of the displacement and the strain. When a soft matrix material (PMMA) was used, the matrix was more compressed than the NW as indicated by the displacement distribution (Figure 3a). The strain tensor S_{xx} distribution (Figure 3b) shows that strain was concentrated in the matrix material instead of the NW. The opposite resulted when a hard matrix (Al_2O_3) was used (Figure 3c) and strain is concentrated in the NW (Figure 3d). This is consistent with the mechanical energy transfer efficiency (η_m) defined in section 3. As the Young's modulus of the matrix material increases, the efficiency increases. More mechanical energy is transferred into the core NW, influencing the final potential and energy generation.

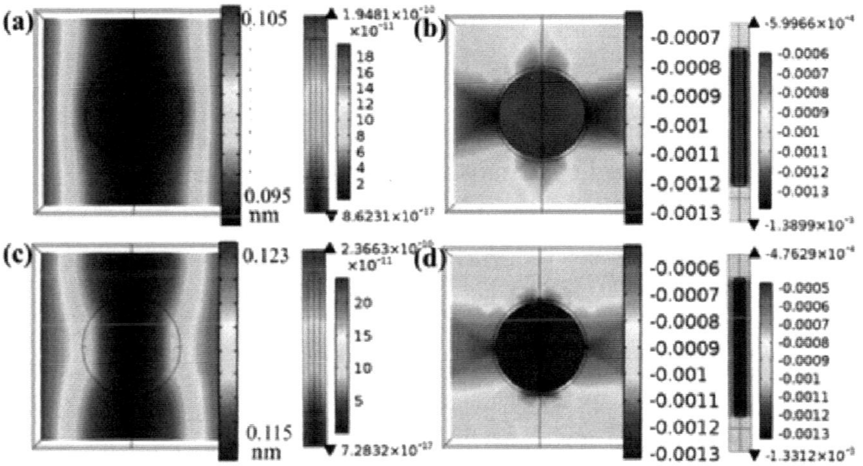

Figure 3. Topview and sideview mapping of (a) displacement and (b) strain when PMMA is used; (c) displacement and (d) strain when Al_2O_3 is used.

Several insulating matrix materials were considered to improve the NG performances. In the simulations, the total volume of ZnO was kept constant. The potential (Figure 4) was calculated along the central axis of the NG cell. From simulations, the potential generation ability of bulk ZnO (yellow line) is weaker than that of NW based NG cells. Zones (1) and (4) correspond to the Si_3N_4 insulating layers, where dielectric losses are unavoidable. Zone (2) corresponds to the thin seed ZnO layer, while zone (3) is corresponding to the NW layer. Globally, the NG cell with harder matrix materials (Al_2O_3, Si_3N_4) generated higher potential.

To clarify the energy transfer process, Young's modulus, Poisson's ratio and relative permittivity of matrix materials were varied independently. The former two parameters mainly influence the mechanical energy transfer and slightly affect the piezoelectric energy transfer. The relative permittivity affects the electrical energy transfer to the output circuit. Mechanical and electrical parameters of real matrix materials are listed in Table 1. First, the effects of Young's modulus and Poisson's ratio on the potential generation were compared (Figure 5). The curves were plotted against one parameter while the two others were kept equal to that of PMMA. As the value of Young's modulus spans from 3GPa (PMMA) to 400GPa (Al_2O_3), the potential increases from 0.8V to 4.4V. In contrast, varying from 0.17 to 0.40, the change of Poisson's ratio only brings a potential difference of 1V. The relationship between the relative permittivity and the electric potential is more complex. In fact, the relative permittivity of the matrix material is not the direct factor that influences the electrical energy transfer process. Efficiency η_e decreases with increasing ε_{eq} of the matrix/NW compound. Here

ε_{eq} is not a linear combination of the permittivity of the NW and the matrix. As a result, the potential curve reaches a maximum value when the permittivity is close to 2.5 (Figure 6). Considering that the relative permittivity of real matrix materials varies from 2.09 (SiO_2) to 9.7 (Si_3N_4), the potential variation is less than 0.5V. In the simulation, the effect of Young's modulus is more significant than the effects of Poisson's ratio and relative permittivity.

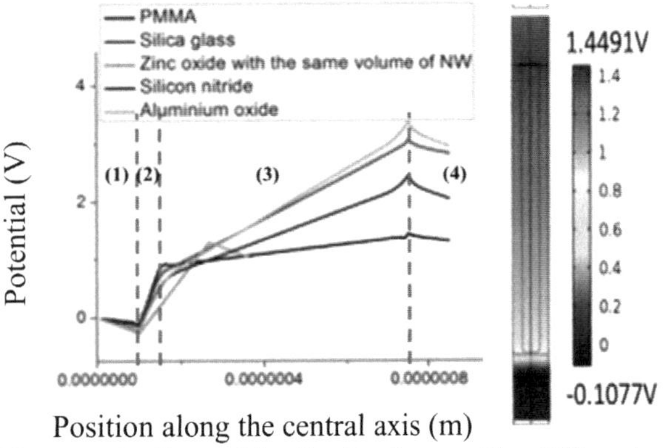

Figure 4. Potential distribution with different matrix materials along central axis (left). Sideview of potential distribution with PMMA matrix (right).

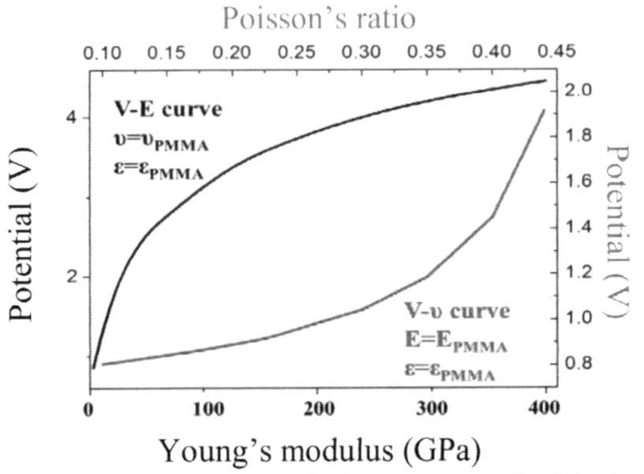

Figure 5. The potential changing trend with the increasing Young's modulus and Poisson's ratio of matrix materials.

Table 1. Mechanical and electrical parameters of different matrix materials.

	PMMA	SiO_2	ZnO	Si_3N_4	Al_2O_3
Young's modulus (GPa)	3	73.1	111	250	400
Poisson's ratio	0.40	0.17	0.35	0.23	0.22
Relative permittivity	3.0	2.09	8.9	9.7	5.7

The relationship between Young's modulus and the energy transfer process is discussed in Figure 7, 8 and 9. The solid line represents the simulation results of a material with fixed Poisson ratio (0.4) and relative permittivity (3.0) changing with increasing Young's Modulus, while the symbols represent simulation results for real matrix materials. At constant input strain, the strain energy is proportional to the equivalent Young's modulus of the matrix/NW compound, which means that more strain energy is needed to compress the NG cell with harder matrix material. With constant Poisson's ratio and relative permittivity, the total electric energy mainly benefits from the high mechanical energy transfer efficiency at high E. For hard materials, large relative permittivity reduces the electrical energy transfer to the output circuit. For PMMA, the low relative permittivity decreased the dielectric losses. The total energy conversion efficiency η is calculated by:

$$\eta = \frac{total\ electric\ energy}{total\ strain\ energy} = \eta_m \times \eta_p \times \eta_e \qquad (1)$$

Within the different materials investigated, SiO_2 achieved the highest conversion efficiency with low relative permittivity and medium hardness.

The design has to be a compromise between mechanical energy transfer and electrical energy transfer. Low mechanical energy transfer obtained with soft materials can be translated into a low sensitivity of the device to external forces. On the other hand, low electrical energy transfer obtained with hard material (they usually have high relative permittivity) can increase the dielectric losses.

5. Conclusion

In this article, we have evaluated a VING structure working in flexion mode for the first time. The effect of matrix materials on NG cells was analyzed using FEM simulations, in order to improve the device performance. Comparing with PMMA, a NG cell using an aluminum oxide matrix improves the potential by two times and the electric energy by six times which is essential for energy harvesters. On the other hand, silica glass will increase the energy conversion efficiency and could be better used in sensors applications.

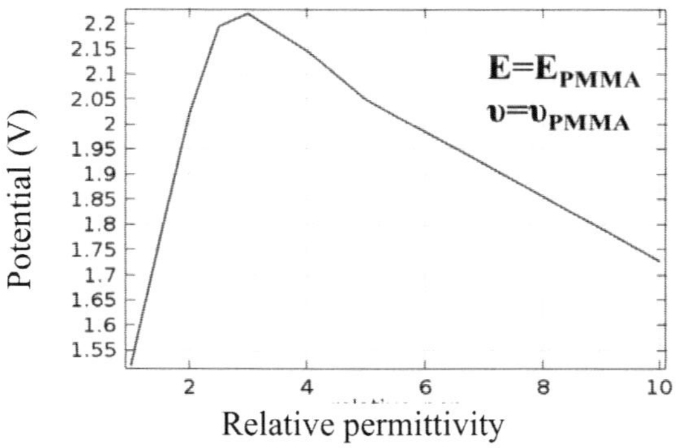

Figure 6. Potential changing with the increasing relative permittivity of matrix materials

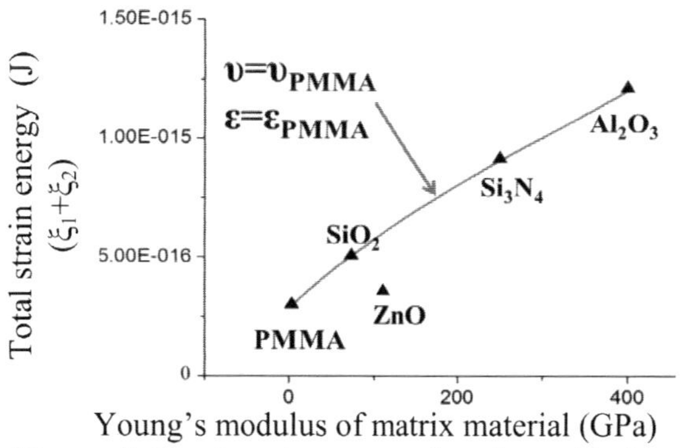

Figure 7. Total strain energy changing with the Young's modulus of matrix materials (per cell and per mechanical stimulus).

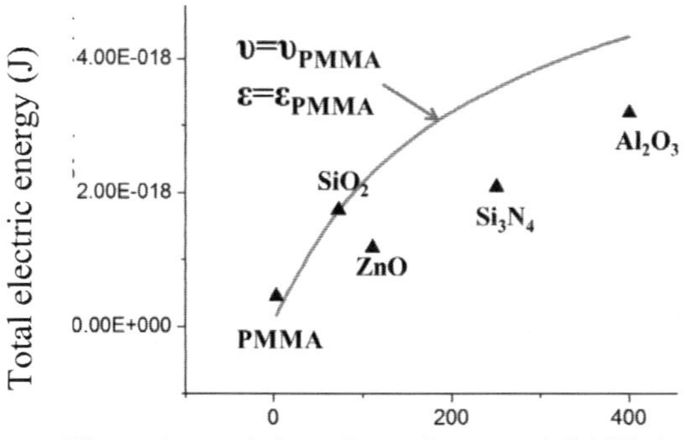

Figure 8. Total electric energy changing with the Young's modulus of matrix materials (per cell and per mechanical stimulus).

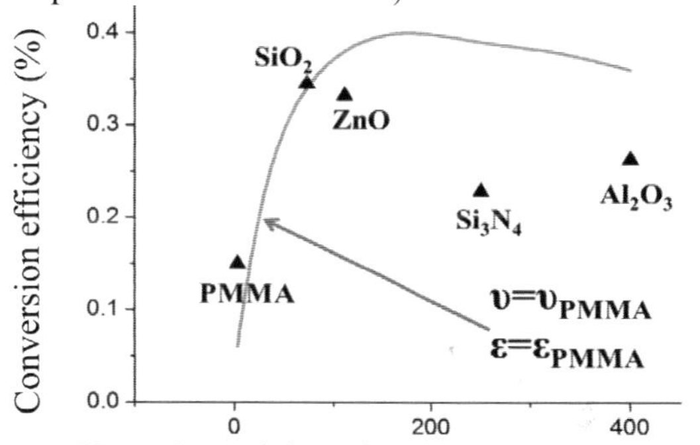

Figure 9. Energy conversion efficiency changing with the Young's modulus of matrix materials.

6. Acknowledgement
This work has been supported by CNRS INSIS Institute within the framework of Exploratory Projects on energy.

References

[1] Zhou J, Fei P, Gao Y F, Liu J, Bao G and Wang Z L 2008 *Nano Lett.* **8** 2725
[2] He J H, Hsin C L, Liu J, Chen L J and Wang Z L 2007 *Adv. Mater.* **19** 781
[3] Wang Z L and Song J 2006 *Science* **312** 242
[4] Hu Y, Zhang Y, Xu C, Zhu G and Wang Z L 2010 *Nano Lett.* **10** 5025
[5] Hinchet R, Ferreira J, Keraudy J, Ardilal G, Pauliac-Vaujour E, Mouis M and Montès L 2012 Scaling rules of piezoelectric nanowires in view of sensor and energy harvester integration *IEDM, 2012 IEEE Int. (San Francisco, CA, 10-13 December 2012)* p 6.2.1 - 6.2.4
[6] Xu S, Qin Y, Xu C, Wei Y, Yang R and Wang Z L 2010 *nature nanotech.* **5** 367
[7] Hinchet R, Lee S, Ardila G, Montes L, Mouis M and Wang Z L Design and guideline rules for the performance improvement of vertically integrated nanogenerator *PowerMEMS 2012 (Atlanta, USA, 2-5 December 2012)*

Development of Piezoelectric MEMS Vibration Energy Harvester Using (100) Oriented $BiFeO_3$ Ferroelectric Film

S Murakami[1], T Yoshimura[2], K Satoh[1], K Wakazono[2], K Kariya[2] and N Fujimura[2]
[1]Technology Research Institute of Osaka Prefecture,
2-7-1 Ayumino, Izumi, Osaka 594-1157, Japan
[2]Osaka Prefecture University,
1-1 Gakuen-cho, Naka-ku, Sakai, Osaka 599-8531, Japan
E-mail: sh-murakami@tri-osaka.jp

Abstract. Piezoelectric vibration energy harvesters (VEHs) with unimorph structure have been developed using Si micro-electrical mechanical systems (MEMS) technology. Since we revealed that (100) epitaxial $BiFeO_3$ (BFO) piezoelectric films have high figure-of-merit on energy conversion, (100)-oriented BFO films have been prepared on (100)-oriented $LaNiO_3$ bottom electrodes by the sol-gel method. We fabricated the piezoelectric VEHs using BFO films with resonance frequencies of ~100 Hz. The maximum output power density of these VEHs was determined to be 10.5 $\mu Wmm^{-3}G^{-2}$ (G=9.8 ms^{-2}) at a load resistance of 1 MΩ, which exceeds or is comparable to those of the best-performing VEHs using other piezoelectric films.

1. Introduction

Recently, considerable attention has been directed toward vibration energy harvesters (VEHs). This arises from the fact that wasted mechanical energy from vibrations is present in various environments and can be converted into electrical power for microelectronic systems, such as wireless sensor remote systems. Several methods to date have been reported to be useful in obtaining electrical energy from vibrational energy including electromagnetic induction, electrostatic generation, and piezoelectric effect. Although each of methods seems to provide a useful amount of energy, the piezoelectric effect has advantages of high energy density and miniaturization potential [1–5]. Therefore, we have focused on piezoelectric VEHs using micro-electrical mechanical systems (MEMS) technology [6,7]. We have already demonstrated that VEHs using polycrystalline $BiFeO_3$ (BFO) piezoelectric film can produce output voltages of 1.5 VG^{-1} (G=9.8 m/s^2) and electrical power of 2.8 $\mu Wmm^{-3}G^{-2}$ at a load resistance of 1 MΩ with a resonance frequency of ~98 Hz [8]. The characteristics are comparable to those of MEMS VEHs using other piezoelectric film materials such as $Pb(Zr,Ti)O_3$ (PZT) AlN and $(K,Na)NbO_3$ (KNN) [5].

In this study, we fabricated MEMS VEHs using (100)-oriented BFO films on (100)-oriented $LaNiO_3$ bottom electrodes, which have larger piezoelectric responses than polycrystalline BFO films [8]. The energy harvesting performance of MEMS VEHs was characterized at resonance frequencies of ~100 Hz. These VEHs showed higher performance than the VEHs using polycrystalline BFO films.

2. Fabrication process and experiment

The structure of the fabricated piezoelectric MEMS VEHs of the present study is shown schematically in figure 1. It consists of a single cantilever structure, which is composed of a supporting silicon membrane, $LaNiO_3$/Pt/Ti bottom electrode, BFO piezoelectric film, Pt upper electrode, and Cu proof mass. A schematic of the MEMS fabrication process of VEHs is shown in figure 2. The starting material was a silicon-on-insulator (SOI) wafer with 5-µm-thick Si, 1-µm-thick SiO_2 (insulating layer), and 650-µm-thick Si-bulk. The structures of the VEHs were fabricated as follows;

1) An SOI wafer was oxidized on both sides with a thickness of 0.5 µm.
2) Thin diaphragm structures were formed using the anisotropic etching technique with TMAH (25 wt%, 90 °C) etchant. An important point here is to stop etching, saving a final diaphragm with about 100-µm-thick Si-bulk, so as to take care of process-handling in the post-process flow.
3) We used $LaNiO_3$/Pt/Ti films as bottom electrodes. The thicknesses of $LaNiO_3$, Pt, and Ti were 150, 200, and 20 nm, respectively. The Pt/Ti film was deposited on the front surface by sputtering, and patterned using a lift-off technique to form the bottom electrode. The $LaNiO_3$ film was deposited by sputtering, and patterned by wet-etching with HF and HNO_3 etchant.
4) A piezoelectric BFO thin film was prepared by the sol-gel method. The film was crystallized by annealing at 500 °C for 10 min in a rapid thermal annealing furnace. The thickness of the film was 250 nm. The BFO film was patterned by etching with HF and HNO_3 etchant.
5) The Pt upper electrode with a thickness of 200 nm was formed by sputtering and a lift-off technique.
6) To form the cantilevers, the SiO_2 and Si layers on the front surface around the cantilevers were removed completely by BHF etchant and by reactive ion etching (RIE) with SF_6 reactive gas, respectively.
7) The Si-bulk and insulator layer below the cantilever were all etched out by RIE with SF_6 reactive gas and by BHF etchant, respectively.
8) A proof mass made of Cu was attached at the free end of the cantilever. The proof mass was about 0.4 mg and had dimensions of $0.40\times0.45\times0.25$ mm^3.

We could finally fabricate the piezoelectric VEHs with unimorph structure, as shown in figure 3. The width, length, and thickness of the cantilever were 0.4 mm, 3 mm, and 6 µm, respectively. Therefore, the effective volume of the fabricated VEH was $0.40\times3.0\times0.256$ mm^3.

The experimental setup with a vibrator for measuring the output voltage and electrical power of the fabricated piezoelectric MEMS VEHs is shown in figure 4. Vibrations were applied to the VEH in the perpendicular direction using a vibrator. A load resistance was connected to the bottom and upper electrodes. Output electrical power was determined by measuring the voltage of the load resistance with a lock-in amplifier.

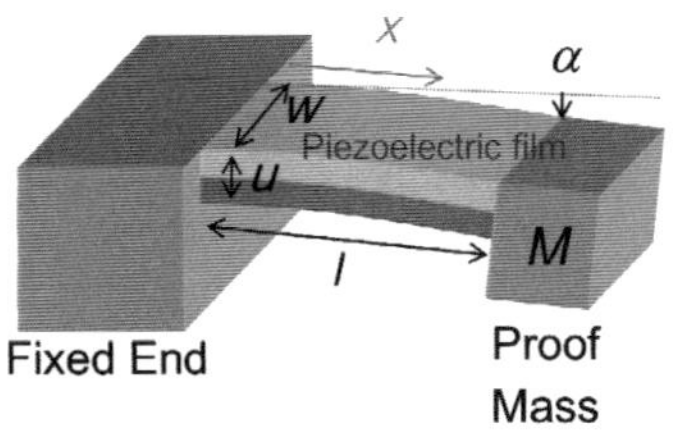

Figure 1. Structure of vibration energy harvester using a piezoelectric film.

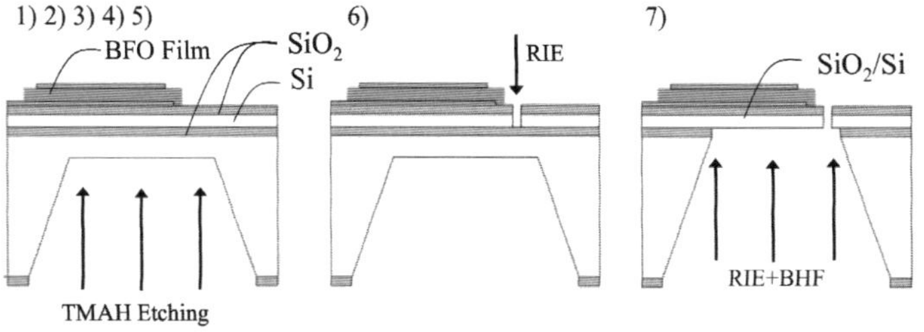

Figure 2. Fabrication of a piezoelectric vibration energy harvester using the MEMS technique.

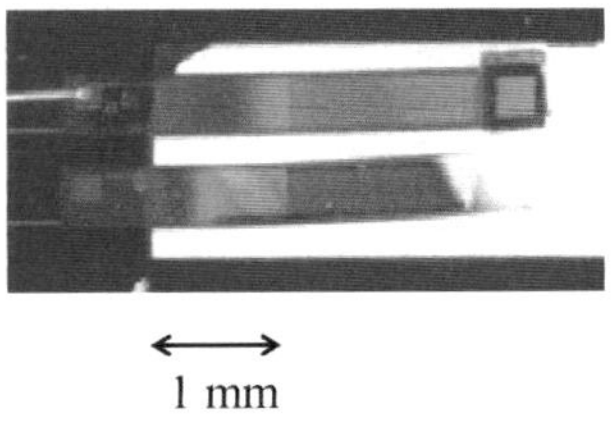

Figure 3. Photo of a piezoelectric energy harvesting device with a $BiFeO_3$ films.

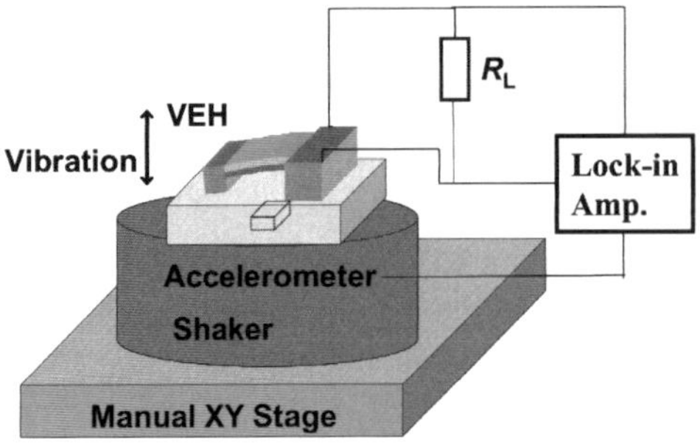

Figure 4. Experimental setup with a shaker for measuring the output voltage and electrical power of the fabricated piezoelectric vibration energy harvester.

3. Piezoelectric thin films for vibration energy harvester

When the VEH device is accelerated in the perpendicular direction, the proof mass receives an inertial force, which bends the cantilever, thereby inducing a strain. Consequently, the piezoelectric film converts the strain to electrical power in the d_{31} mode. The charge density D_3 induced by the piezoelectric effect is given by

$$D_3 = e_{31,f}(x_1 + x_2), \quad (1)$$

where $e_{31,f}$ is the effective transverse piezoelectric coefficient and x_i the in-plain strain. $e_{31,f}$ is defined as [9]

$$e_{31,f} = \frac{d_{31}}{s_{11}^E + s_{12}^E}, \quad (2)$$

where d_{31} is the transverse piezoelectric coefficient and s_{ij}^E the elastic compliance coefficient.

$e_{31,f}$ is a useful parameter rather than d_{31} to discuss the piezoelectric property of the piezoelectric film for the VEHs, because $e_{31,f}$ can be determined directly by the method based on substrate bending and collecting accumulating charges [6,9]. The output power of the VEHs is also proportional to $e_{31,f}^2$, as described below. Although d_{31} might be more familiar than $e_{31,f}$ for the characterization of the piezoelectricity, using d_{31} is not practical in modelling MEMS devices using piezoelectric films. This is because s_{ij}^E of the films are different from those of the bulk, and depend significantly on preparation conditions of the films.

Accounting for the distribution of strain along the longitudinal direction of the cantilever, the maximum electrical power dissipated in the load resistor, R_L, which is connected to the piezoelectric film on the cantilever at the impedance matching point, is given by the following equation: [8]

$$P_{R\max} = \frac{MQ^2A^2}{\omega_r}\frac{9d}{4l^3u}\left(\frac{1}{4}x^3 - lx^2 + l^2x\right)\frac{(1-\nu)^2}{E}\frac{e_{31,f}^{\;2}}{\varepsilon_0\varepsilon_r}, \quad (3)$$

where M is the mass of the proof mass; Q mechanical quality factor of the cantilever; A acceleration of the applied vibration; ω_r angular velocity; d thickness of the piezoelectric film; l and u length and thickness of the cantilever, respectively; ν Poisson's ratio; E Young's modulus of the cantilever; and ε_0 and ε_r are the permittivity of the vacuum and the relative permittivity of the piezoelectric film, respectively. As can be seen in Eq. (3), the maximum electrical power, $P_{R\max}$, is proportional to M, Q^2, A^2, and $1/\omega_r$. In addition, from the last term of Eq. (3),

$$FOM = \frac{e_{31,f}^{2}}{\varepsilon_0 \varepsilon_r} \quad (4)$$

can be regarded as the figure-of-merit (FOM) of the piezoelectric film for vibrational energy harvesting. Thus, the development of piezoelectric films with large $e_{31,f}$ and low ε_r is important to increase the output energy of the VEHs.

In general, $Pb(Zr,Ti)O_3$ (PZT) films are often used as piezoelectric films in energy harvesting. Because ε_r of the BFO films is an order of magnitude lower than that of the PZT films, BFO films are expected to have high FOM. Therefore, we have so far demonstrated the piezoelectric VEHs with polycrystalline BFO films that were prepared on the Pt/Ti bottom electrodes. The VEHs, fabricated using the MEMS process, produced output energies comparable to those of VEHs using other piezoelectric films, as described above [8].

Growth of the (100)-oriented BFO films is one of the most important factors to obtain larger $e_{31,f}$ and consequently higher FOM [6,7,10]. In the present study, the BFO films, which were prepared by the sol-gel method on the (100)-oriented $LaNiO_3$ bottom electrodes, showed (100) orientation, and exhibited high $e_{31,f}$ of –2.8 C/m^2, which is close to that of the (100) epitaxial BFO film [7]. The FOMs of the (100)-oriented BFO film and the polycrystalline BFO film were determined to be 8.5 and 1.1 GPa, respectively. Therefore, VEHs using the (100)-oriented BFO film can be expected to show higher performance than those using the polycrystalline BFO film.

4. Characteristics of the vibration energy harvester using (100) BFO films

The vibration frequency dependence of the normalized output voltage of the VEH using the (100)-oriented BFO film is shown in figure 5. The accelerations of the applied vibrations were 0.009, 0.018 and 0.054 G_{rms}. It can be seen that the VEH has a resonance frequency around 92–95 Hz and that the normalized maximum output voltage at the resonance frequency is independent of the acceleration. In contrast, hysteresis behaviours were observed near the resonance frequency and the non-linear resonance increased with increasing acceleration. These behaviours were observed also with the VEHs using the polycrystalline BFO films. It appears that the nonlinear resonance of the VEH mainly originates from the dumping effect of air and the nonlinear elasticity of the cantilever [8].

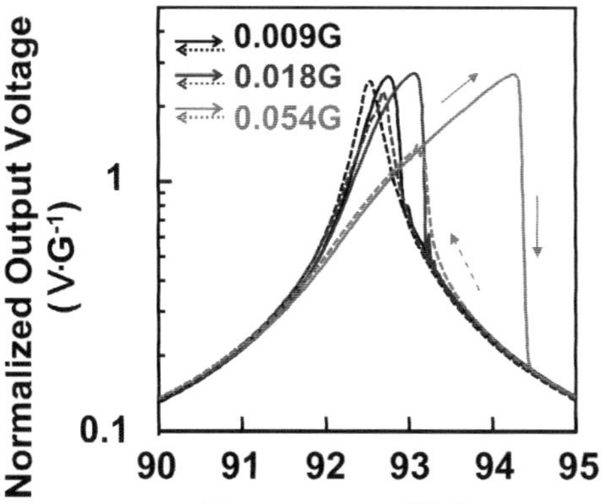

Figure 5. Normalized output voltage as a function of vibration frequency of the harvester using the (100)-oriented $BiFeO_3$ film.

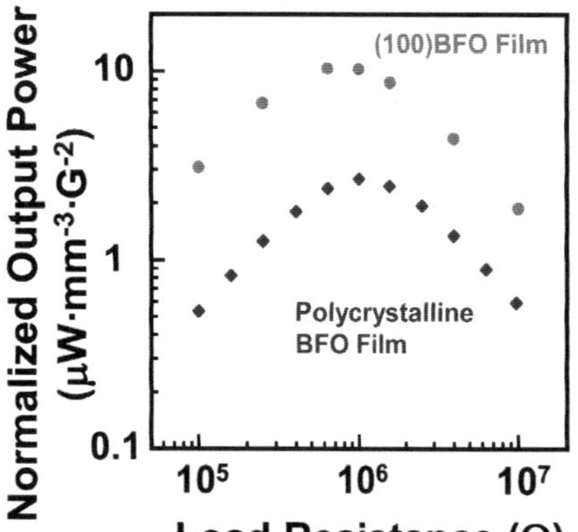

Figure 6. Load resistance dependences of the normalized output power of the harvesters with (100)-oriented or polycrystalline $BiFeO_3$ films.

The load resistance dependence of the output power density of the VEH with the (100)-oriented BFO film was characterized at the acceleration of 0.018 G_{rms} and the resonance frequency of 93 Hz as shown in figure 6. The output power is the electrical power that is dissipated in the load resistance, and normalized using G^2 and the effective volume of the VEH. The output power density of the VEH using the polycrystalline BFO film [8] is also shown in figure 6. The maximum normalized output power of the VEH using the (100)-oriented BFO films is 10.5 $\mu Wmm^{-3}G^{-2}$ at a load resistance of 1 MΩ, which is about three times larger than that of the VEH using the polycrystalline BFO films. This result is comparable to or larger than those of the best-performing VEHs using other piezoelectric films.

5. Conclusions

We fabricated piezoelectric VEHs using MEMS technology. To obtain better performance, we prepared (100)-oriented BFO films on the (100)-oriented $LaNiO_3$ bottom electrode by the sol-gel method. The VEH with the low resonance frequency of ~100 Hz showed a high output power density of 10.5 $\mu Wmm^{-3}G^{-2}$ at a 1-MΩ load resistance. It exceeded or was comparable to those of the best-performing VEHs using other piezoelectric films. These results indicated that the (100)-oriented BFO film is a good candidate material for piezoelectric VEHs.

Acknowledgments

This work was supported by the Industrial Technology Research Grant Program 2011 from the New Energy and Industrial Technology Development Organization (NEDO) of Japan.

References

[1] Roundy S, Wright P K and Rabaey J 2003 Comput. Commun. **26** 1131
[2] Roundy S and Wright P K 2004 Smart Mater. Struct. **13** 1131
[3] Mitcheson P D, Reilly E K, Toh T, Wright P K and Yeatman E M 2007 J. Micromech. Microeng. **17** S211
[4] Priya S 2007 J. Electroceram. **19** 167
[5] Kim S G, Priya S and Kanno I 2012 MRS Bulletin **37** 1039
[6] Yoshimura T, Sakiyama H, Oshio T, Ashida A and Fujimura N 2010 Jpn. J. Appl. Phys. **49** 021501
[7] Ujimoto K, Yoshimura T, Ashida A and Fujimura N 2012 Appl. Phys. Lett. **100** 102901
[8] Yoshimura T, Murakami S, Wakazono K, Kariya K and Fujimura N 2013 Appl. Phys. Exp. **6** 051501
[9] Dubois M A and Muralt P 1999 Sens. Actuators A **77** 106
[10] Yoshimura T and Trolier-McKinstry S 2002 J. Appl. Phys. **92** 3979

Perovskite enhanced solid state ZnO solar cells

L Loh[1,2], J Briscoe[2] and S Dunn[2]
[1]School of Engineering, Nanyang Polytechnic, 180 Ang Mo Kio Ave 8, S568930, Singapore
[2]School of Engineering and Materials Science, Queen Mary, University of London, Mile End Road, London, E1 4NS, UK

E-mail: s.c.dunn@qmul.ac.uk

Abstract. This paper will report on the design, fabrication and testing of a solid-state perovskite enhanced ZnO solar cell. The p-type perovskite material used is bismuth ferrite (BFO) which has an absorption range within the blue range of the visible light spectrum. The solid state solar cell, was sensitized with N719 dye and used a CuSCN hole conductor. A disadvantage of ZnO is its poor chemical stability in acidic and corrosive environments. As chemical solution techniques were used in depositing BFO, a buffer method using an aminosilane ((3-aminopropyltrimethoxysilane or $H_2N(CH_2)_3Si(OCH_3)_3$)) coating was used to provide a protective coating on the ZnO nanorods before the BFO film was spin coated onto the ZnO nanorods. The photovoltaic performance of the solar cells were tested using a Keithley 2400 source meter under 100mW/cm^2, AM 1.5G simulated sunlight, where improvements in J_{sc} and efficiency were observed. The BFO was able to harness more electrons and also acted as a buffer from electron recombination.

Keywords: solar cell, perovskite, ZnO, BFO, CuSCN

INTRODUCTION

Since the development of the dye sensitised solar cell by Michael Gratzel[1,2], there has been a significant amount of research in solar cells using low cost materials and techniques. ZnO has been investigated as an alternative n-type material with its similar band structure and physical properties. ZnO has the advantage of higher electronic mobility and electron density (1 - 5 cm^{2}V^{-1}S^{-1} and 1 – 5 x 10^{18} cm^{-3} respectively)[3] and has been studied using different nanostructures, doping of the semiconductor material, combining with composites, as well as using different processing technologies to improve the performance of the solar cell, which peaked at about 7%[4]. Liquid electrolytes used in redox systems, suffer from leakage, instability, dye desorption and corrosion of the electrodes[5]. Studies on solid-state solar cells where a solid hole conducting material is employed to replace the liquid redox electrolyte, have shown better stability. The use of the p-type semiconductor, copper thiocyanate (CuSCN) with a bandgap of 3.6 eV, hole conductivity of greater than 5×10^{-4} Scm^{-1}, chemical stability, transparency in the visible light spectrum range and high carrier mobility[6] has been reported with ZnO solar cells using various sensitizers. ZnO combined with sintered CdSe gave a stabilized efficiency of 1.5%. 0.1% efficiency was achieved using ZnO nanorods sensitized by *(cis-bis bis(2,2'-bipyridyl-4,4'-dicarboxylato) isothiocyanato ruthenium(II) bis-tetrabutylammonium) or* N719 dye for 1.8 μm length ZnO nanorods [7] and 1.7% efficiency[8] using 11-12 μm long ZnO nanorods.

The use of perovskite materials to display a PV effect was first investigated by Brady et al[9] in 1939 on tartaric acids. Since the 1970s, perovskite materials such as barium titanate ($BaTiO_3$)[10] and lead zirconate titanate (PZT)[11,12] have been studied for their ferroelectric properties which were attributed to the non-centrosymmetry of the crystal structure. The use of perovskite-like materials like $CH_3NH_3PbI_3$[13–15] and $CsSnI_3$[16] recently received significant interest when they were shown to produce high performance TiO_2-based dye sensitized solar cells.

This paper will report the enhancements in performance on ZnO-based dye sensitized solid state solar cells using perovskite, Bismuth Ferrite ($BiFeO_3$ or BFO). The interest in BFO was initiated in 2003[17,18] where the photoconductivity was studied. BFO possesses an R3c space group, which exhibits weak ferromagnetism at room temperature, ferroelectric and piezoelectric properties below the Curie temperature of 1100 K and anti-ferromagnetic ordering below the Néel temperature of 640 K. BFO has been applied in solar cell applications in a few instances, and this includes a biotemplated synthesized BFO mesoporous-iodide electrolyte-based DSSC[19], in a CNT/BFO/Pt structure with CdSe quantum dots [20,21] as an electron conductor and in an inorganic-organic BFO/P3HT/Au hybrid solid state solar cell where BFO is classified by the authors as an n-type semiconductor to P3HT's p-type material[22].

EXPERIMENTS

2 mm thick glass with fluorine tin oxide (FTO) coating of 15Ω were used as substrates. A seed layer of ZnO was deposited using an aqueous method with 5mM of absolute ethanolic zinc acetate ($Zn(CH_3CO_2)_2$) by dropping on the conductive face of the substrate. The substrates were rinsed with ethanol and dried with nitrogen. The process was repeated 5 times before going through an annealing cycle on a hot plate at 350°C. The whole process was repeated 3 times. The growth of the ZnO nanorods were synthesized by an aqueous method. The seeded substrates were suspended face down in a fixture in 25mM of zinc nitrate hexahydrate ($Zn(NO_3)_2.6H_2O$) and 25mM of hexamethylenetetramine or HMT ($C_6H_{12}N_4$) in deionised (DI) water[23] at 90°C for 2.5 hours. This process was repeated with a change in precursor solution for 6 times for a total of 15 hours of growth. An aminosilane buffer layer was deposited on the ZnO nanorods using 3-aminopropyltriethoxysilane or $H_2N(CH_2)_3Si(OC_2H_5)_3$. The ZnO nanorods were soaked in an aqueous solution of 4% by volume 3-aminopropyltriethoxysilane in toluene at 60°C for 30 minutes. The samples were then rinsed with toluene and dried at 90°C for an hour. The BFO sol was formulated from bismuth nitrate pentahydrate ($Bi(NO_3)_3.5H_2O$) and iron nitrate nonahydrate ($Fe(NO_3)_3.9H_2O$) mixed in 2-methoxyethanol ($CH_3OCH_2CH_2OH$), with 20% and 0.2% by volume of the dehydrating agent acetic anhydride ($(CH_3CO)_2O$) and ethanolamine respectively. The BFO sol was then spin coated at 5000 rpm onto the aminosilane coated ZnO nanorods and preannealed at 350°C for 3 minutes on a hotplate. Samples were annealed at 600°C for 1 hour. The samples were next soaked in 0.5mM of N719 dye for 15 hours, rinsed with absolute ethanol and dried with nitrogen. Dupont's 25 μm thick Surlyn barrier film was then laid around the active area to act as a spacer. The sample was then spray coated with CuSCN at 90°C using 0.2 M CuSCN in dipropyl sulfide. Au counter electrodes were sputtered using a Denton Vacuum Explorer 14 thermal evaporator using a power of 125 W, base pressure of 4 x 10^{-2} mTorr and working pressure of 100 mTorr at ambient temperature in 100% Argon (Ar) atmosphere.

RESULTS AND DISCUSSIONS

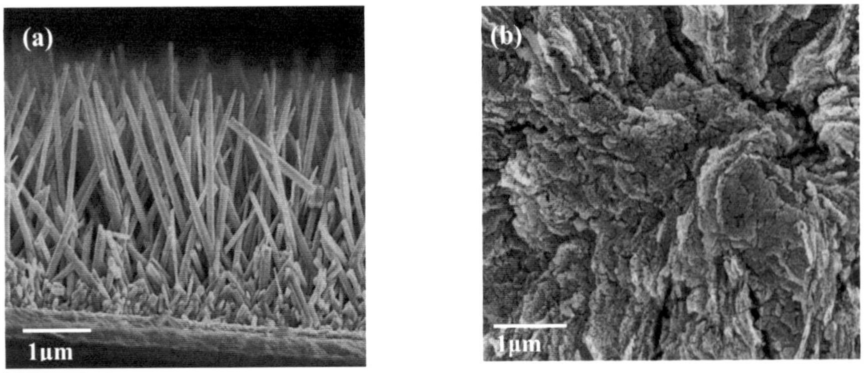

Figure 1. (a) Cross Sectional SEM Image of ZnO nanorods, (b) Dissolution of ZnO by BFO

The ZnO nanorods synthesized for 15 hours achieved about 2.3 μm long ZnO nanorods with (0001) crystallinity (Figure 1(a)). ZnO is chemically unstable in corrosive environments. The use of aminosilane buffer before the deposition of BFO was an important step as the BFO sol had a pH of less than 1. Direct deposition of BFO on the ZnO nanorods caused the dissolution of the ZnO nanorods(Figure 1(b)). The aminosilane was eventually removed during the annealing process of the ZnO/BFO nanorod structures. Figure 2 shows SEM images of the BFO coated ZnO nanorods. The TEM images show that the BFO had very thin coating which were particulate in nature, forming an almost hierarchical structure of nanorods coated with particles. Areas of even coating showed coatings of 1.4 nm – 2 nm thickness.

Figure 2. BFO coated ZnO nanorods: (a) SEM topographical view, (b) TEM image of evenly coated ZnO nanorod, (c) TEM image of 1.4nm thick BFO coating, (d) CuSCN infiltrated ZnO/BFO nanorods

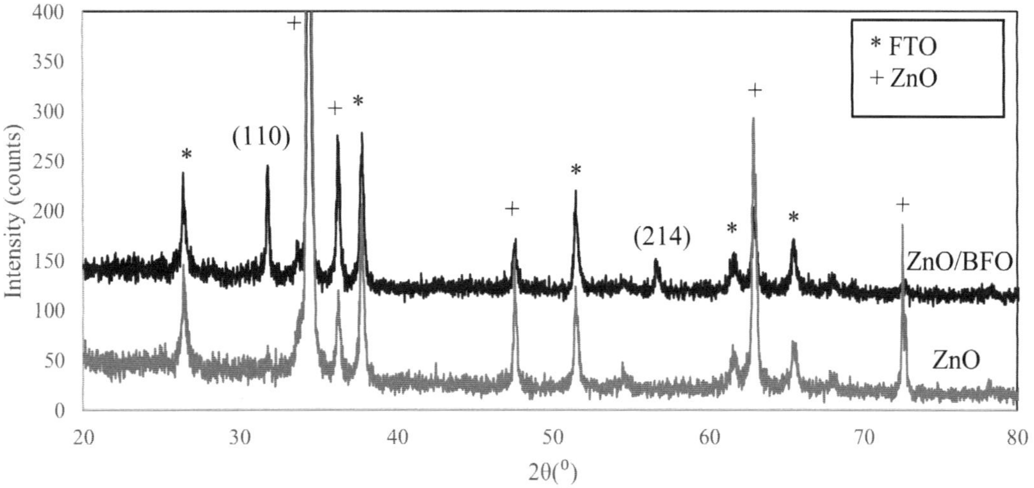

Figure 3. XRD comparison of ZnO nanorods and BFO coated ZnO nanorods

XRD scans of the samples in Figure 3, using the Panalytical X'pert Pro X-ray diffraction (XRD) confirm the presence of rhombohedral, R3c BFO crystallinity with a strong (110) crystal plane and a smaller (214) secondary BFO plane. The optical absorbance spectra in Figure 4, measured using the Perkin Elmer Lambda 950 UV visible spectroscopy showed the increased absorbance when BFO was added. There is a minor absorbance peak at around 517 nm indicating the presence of BFO.

The solar cells were tested with the IVT Solar VS 6820 solar simulator system with a Keithley 2400 source meter under 100mW/cm^2, AM 1.5G simulated sunlight. The J-V curves in Figure 5 showed a significant improvement in the performance of the solar cells with the addition of the BFO coating. There was a 100% increase in the J_{sc} and more than double the improvement in the efficiency.

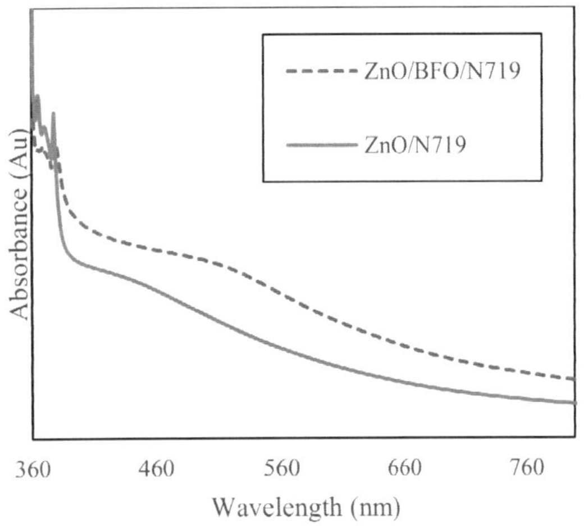

Figure 4. Comparison of absorbance data

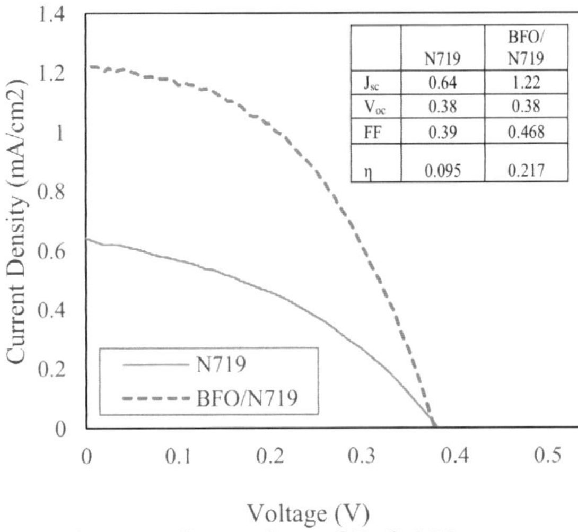

Figure 5. Comparison of ZnO J-V curves

The indicative energy band diagram of the ZnO/BFO/N719/CuSCN is shown in Figure 6. The conduction band of N719 at 3.76 eV is lower than the 3.3 eV for BFO. On photoexcitation, the electrons tunnel through the BFO to the ZnO conduction band where the electrons are collected and flow through and external load to the counter electrode. The oxidised N719 and BFO via N719 is regenerated back to the ground state by capturing an electron from the valence band of the CuSCN via hole transfer. Recombination of the electrons occur with the N719 or with the CuSCN. However, with the lower N719 conduction band to the BFO's, the BFO acts as an electron blocking layer which helps suppress the electron recombination, thus resulting in the increased J_{sc} for the ZnO/BFO/N719 system.

CONCLUSION

In conclusion, we have shown that the addition of a BFO shell on ZnO nanorods was able to enhance the performance of a solid state solar cell with an increase in J_{sc} and having more than 2 times improvements in efficiency. BFO mainly acts as an electron blocker which prevents the electrons from recombining with the N719 due to the locations of their conduction bands. We have also shown that lower cost aqueous techniques can be used to deposit coatings using corrosive sols onto ZnO using a temporary buffer method. As a non-optimised construction of the solar cell was used, there is strong potential for improving the results further through optimised construction techniques.

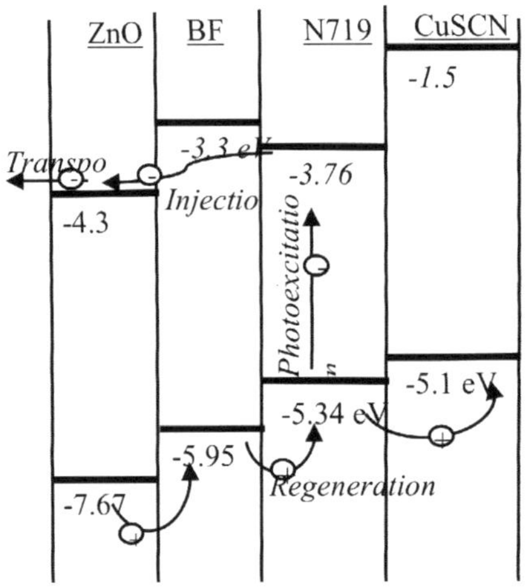

Figure 6. Representation of ZnO/BFO/ N719/CuSCN Energy

REFERENCES

1. Gratzel M 2005 *Inorg. Chem.* **44** 6841–51.
2. Gratzel M 2003 *J. Photochem. Photobio. C-Photochem.Rev.* **4** 145–53
3. Law M, Greene LE, Johnson JC, Saykally R and Yang PD 2005 *Nature Mat.* **4** 455–9
4. Xu CK, Wu JM, Desai U V and Gao D *J. Amer. Chem. Soc.* 2011 **133** 8122–5
5. Boschloo G and Hagfeldt A 2009 Accts. Chem. Res. **42** 1819–26
6. Zhang QB, Guo HH, Feng ZF, Lin LL, Zhou JZ and Lin ZH 2010 *Electrochimica Acta.* **55** 4889–94
7. Tena-Zaera R, Ryan MA, Katty A, Hodes G, Bastide S and Levy-Clement C 2006 *Comptes Rendus Chimie* **9** 717–29
8. Edri E, Rabinovich E, Niitsoo O, Cohen H, Bendikov T and Hodes G 2010 *J. Phys. Chem. C.* **114** 13092–7
9. Dittrich T, Kieven D, Rusu M, Belaidi A, Tornow J, Schwarzburg K, et al. 2008 *Appl. Phys. Lett.* **93**
10. Postels B, Kasprzak A, Buergel T, Bakin A, Schlenker E, Wehmann HH, et al 2008 *J. Kor. Phys. Soc.* **53** 115–8
11. Desai U V, Xu CK, Wu JM and Gao D 2012 *Nanotechnology* **23**
12. Etgar L, Gao P, Xue ZS, Peng Q, Chandiran AK, Liu B, et al. 2012 *J. Amer. Chem. Soc.* **134** 17396–9
13. Kim H-S, Lee J-W, Yantara N, Boix PP, Kulkarni SA, Mhaisalkar S, et al 2013 *Nano Lett.* **13** 2412–7
14. Kojima A, Teshima K, Shirai Y and Miyasaka T 2009 *J. Amer. Chem. Soc.* **131** 6050
15. Chung I, Lee B, He JQ, Chang RPH and Kanatzidis MG 2012 *Nature* **485** 486–494
16. Brody PS and Crowne F 1975 J. Elect. Mat. **4** 955–71
17. Glass AM, Linde DVD and Negran TJ 1974 *Appl. Phys. Lett.* **25** 233–5
18. Basu SR, Martin LW, Chu YH, Gajek M, Ramesh R, Rai RC, et al. 2008 *Appl. Phys. Lett.* **92** 91905
19. Nuraje N, Dang XN, Qi JF, Allen MA, Lei Y and Belcher AM. 2012 *Adv. Mat.* **24** 2885–9
20. Zang YY, Xie D, Chen Y, Wu X, Ren TL, Zhu HW, et al. 2012 *J. .Appl. Phys.* **112**
21. Zang YY, Xie D, Chen Y, Wu X, Ren TL, Wei JQ, et al. 2012 *Nanoscale* **4** 2926–30
22. Liu ZK and Yan F 2011 *Physica Status Solidi-Rapid Res. Lett.* **5** 367–9
23. Zhang QF, Dandeneau CS, Zhou XY and Cao GZ 2009 *Adv. Mat.* **21** 4087–108

Piezoelectric Enhancement of Hybrid Organic/Inorganic Photovoltaic Device

Joe Briscoe,[1] Safa Shoaee,[2] James R. Durrant[2] and Steve Dunn[1]

1 Centre for Materials Research, School of Engineering and Materials, Queen Mary University of London, E1 4NS, UK.

2 Centre for Plastic Electronics, Department of Chemistry, Imperial College London, London SW7 2AZ, UK.

j.briscoe@qmul.ac.uk

Abstract. Solar cells are produced using solution processing that combine ZnO nanorods with the conjugated polymer poly(3-hexylthiophene) (P3HT). ZnO nanorods have an average length and diameter of 2.7 μm and 81 nm, and are well coated with P3HT. The solar cells have a power conversion efficiency of 1.24 %, which increases to 1.78 % when 10 kHz acoustic vibrations are applied to the device at 75 dB using a loudspeaker. Transient absorption studies demonstrate that the efficiency increase originates from a decrease in the non-geminate recombination rate in the system. It is proposed that electric fields at the ZnO:P3HT interface arising from the piezoelectric effect in ZnO increase the charge-carrier separation, producing this reduction in recombination and associated efficiency increase.

1. Introduction

Solar cells which combine nanostructured wide band gap inorganic semiconductors with conjugated polymers have the potential to utilize both the strong light absorption of the polymer and the superior charge transport properties and stability of the inorganic material. Metal-oxide semiconductors are generally used in such structures, the most common being TiO_2 and ZnO [1]. ZnO nanorods or wires are attractive for this application because their morphology allows a direct charge transport pathway from the charge separation interface to the cathode [2], and they have good carrier mobility [3].

Poly(3-hexylthiophene) (P3HT) is a well-established conjugated polymer photovoltaic material, which has been studied in combination with ZnO nanostructures [4–9]. The highest reported efficiency for this type of device is 0.76 % [7]. Higher efficiencies are obtained when P3HT is combined with the fullerene derivative [6,6]-phenyl-C_{61}-butyric acid methyl ester (PCBM) in bulk heterojunction devices, which have achieved efficiencies over 5 % [10]. There is therefore a desire to find strategies to improve the efficiency of inorganic-organic hybrid devices to make use of the improved stability that they may offer. The ZnO:P3HT system is therefore a useful standard system in which to study novel techniques for such improvements.

In addition to photovoltaic applications, ZnO is also a piezoelectric material; when a force (e.g. pressure or vibration) is applied to ZnO it develops an internal polarization and associated electric field due to asymmetric displacement of the anions and cations in the lattice [11]. This has led to ZnO nanorods being used for kinetic energy harvesting where the electric field is used to generate a voltage and/or current [12–15]. Some attempts have also been made to utilize this effect in a photovoltaic structure to combine the two energy harvesting methods [16–18]. However, since the voltage and current output of the piezoelectric energy harvester was much smaller than the photovoltaic output, the addition of vibration only led to slight increases in photovoltage [16,17] unless the illumination intensity was reduced significantly to make to two outputs commensurate [18]. Therefore alternative approaches may be required to optimally combine piezoelectric and photovoltaic outputs.

2. Experimental Methods

To grow ZnO nanorods indium-tin oxide (ITO)-coated glass substrates seeded with a sputtered ZnO film (100 nm) were suspended in an aqueous solution of 15 mM zinc nitrate and 25 mM hexamethylenetetramine and heated to 90 °C for 4 hours [19] a total of 8 times in fresh solutions followed by annealing in air at 400 °C for 1 hour. ZnO nanorod samples were immersed overnight in a solution of P3HT in chlorobenzene (2 g/l) and then dried by N_2 gas. Subsequently a P3HT layer was spin coated from chlorobenzene (45 g/l) at 1100 rpm. Gold contacts were deposited by evaporation.

Photovoltaic performance of the solar cells was measured by recording the current-voltage characteristics of the device with a Keithley 2400 SMU while illuminating with a solar simulator fitted with an AM 1.5 filter at 1 sun (100 $mWcm^{-2}$) illumination. Transient absorption decays were measured by exciting the sample film under a nitrogen atmosphere using a commercially available optical parametric oscillator (Oppolette) pumped by Nd:YAG laser (Lambda Photometrics) with an excitation wavelength of 500 nm, a pump intensity of 0.4 - 20 $\mu J.cm^{-2}$ and a repetition frequency of 20 Hz. Absorption (980 nm) was probed using a 100 W quartz halogen lamp (Bentham, IL 1) with a stabilised power supply (Bentham, 605). The signal from the photodiode was pre-amplified and sent to the main amplification system with an electronic band-pass filter (Costronics Electronics and was collected with a digital oscilloscope (Tektronics, TDS220), triggered by the signal of the pump laser pulse from a photodiode (Thorlabs Inc., DET210). Two monochromators and appropriate optical cut-off filters were placed before and after the sample to reduce stray light, scattered light and sample emission.

For all tests, the external vibration was applied at a fixed distance through a loud speaker at 75 dB, with frequencies ranging between 1 – 50 kHz.

3. Results

The as-produced ZnO nanorods were on average 81 nm in diameter and 2.7 µm long, giving an aspect ratio of 33:1 (Figure 1a). Using high aspect ratio nanorods maximizes the interface between the ZnO and P3HT, which has been shown previously to lead to increased device photocurrent [7]. This also maximizes the possible response to applied vibration, as longer rods have been shown to produce higher output voltages in energy harvesting devices using similar hybrid structures [14]. The cross-section micrograph of the ZnO structure after coating with P3HT confirms that this interface could be fully utilized, as the P3HT coats the ZnO conformally and penetrates to the base of the rods to the extent that the nanorods are barely visible.

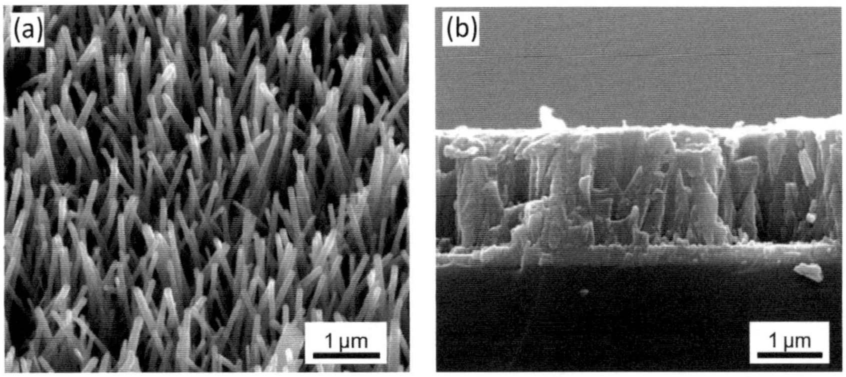

Figure 1. SEM micrographs of the as-grown ZnO nanorods at 30° tilt (a) and cross-section of the ZnO nanorods coated with P3HT (b).

Under illumination the P3HT:ZnO nanorod device produces a power converstion efficiency (PCE) of 1.24 % under 1 sun illumination (Figure 2a). When the same test was performed while applying acoustic vibration to the device using a loudspeaker, the device efficiency increased by 44 % to 1.78 %. This increase in efficiency resulted from an increased open-circuit voltage (V_{oc}) of around 0.1 V and short-circuit current density (J_{sc}) of around 1 mA/cm^2. An increased V_{oc} of 18 mV has been reported previously for ZnO-nanorod-based solar cells, which was attributed to the addition of the piezoelectrically-generated voltage to the photovoltage [16,18]. However, the significant increase in V_{oc} and accompanying increase in J_{sc} reported here has not been observed previously.

To investigate the origin of the increase in efficiency transient absorption spectroscopy (TAS) was performed on the device to monitor the photoinduced absorption of the P3HT cations (Figure 2b). It has been shown previously that such measurements can indicate the yield of dissociated charge carriers and their recombination dynamics in ZnO:P3HT [5]. For ZnO:P3HT the decays show approximately exponential dynamics and the lifetimes in the μs–ms timescale increase significantly with the application of applied vibration. These decays have previously been assigned to non-geminate recombination of dissociated charge carriers [5]. Such an increase in lifetime therefore indicates that the application of external vibration significantly reduces the rate of non-geminate recombination in the ZnO:P3HT system.

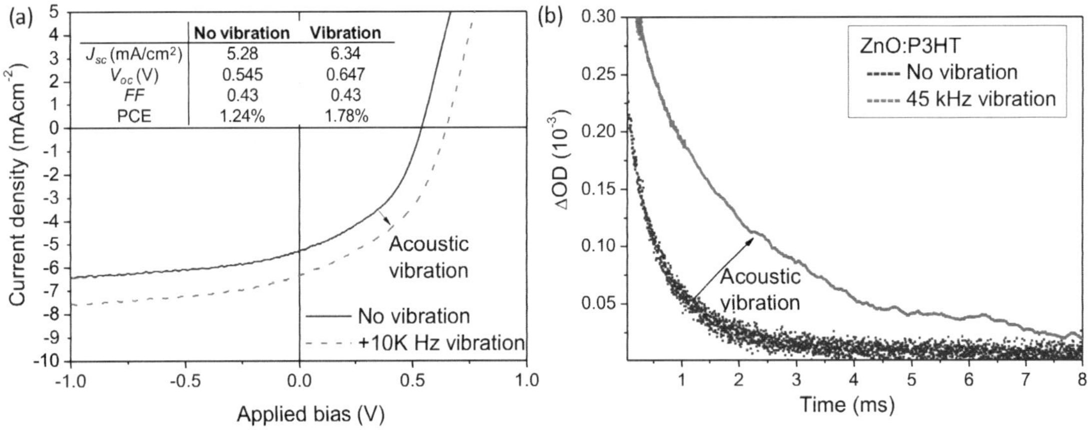

Figure 2. (a) Current density-voltage measurements of ZnO:P3HT solar cells under 1 sun AM 1.5 illumination without and with applied vibration from a loudspeaker at 10 kHz. Acoustic vibration leads to an increase in both open-circuit voltage (V_{oc}) and short-circuit current density (J_{sc}) leading to an overall efficiency increase of 44 %. Key device parameters are shown. (b) Transient absorption signals of the ZnO:P3HT system without and with applied vibration. The lifetime of the P3HT$^+$ polaron increases significantly with the application of applied vibration.

4. Discussion

Although other effects resulting from vibrations such as local heating, improved interfacial contact and structural reorganisation cannot be fully disregarded, the possibility that the piezoelectricity of the ZnO in this system could influence the recombination dynamics should be considered. It has recently been shown that the application of vibrations to a very similar polymer:ZnO nanorod system can lead to tens of mV being measured in an external circuit [14]. For this to be measured a much higher voltage and therefore electric field must exist in the nanorods, as screening effects by free carriers mean that a large portion of the generated voltage is not measured externally. It is also notable that this effect is much more significant in the ZnO:P3HT system than in the similar ZnO:PCBM:P3HT system reported in the literature [18]. One difference between these two systems is the location of the exciton separation and charge recombination interface. In the ZnO:P3HT system it is at the ZnO surface, whereas in the ZnO:PCBM:P3HT system it is between the P3HT and PCBM, therefore separated from the ZnO interface. This gives further support to the hypothesis that the origin of the enhancement lies with the ZnO, or a similar effect would also have been observed with ZnO:PCBM:P3HT.

Considering the evidence from the TAS measurements that the vibration-induced enhancement originates from a reduction in non-geminate recombination (Figure 2a), a model is proposed to explain the origin of this effect. As shown in Figure 3a, in the non-vibration case non-geminate recombination occurs at the ZnO:P3HT interface when an electron from ZnO recombines with a positive polaron from P3HT. When vibrations are applied to the ZnO it will bend and compress, leading to a polarization gradient in each nanorod. This results in an electric field in the nanorod which shifts the bands as shown in Figure 3b. As the electric field oscillates either the electron in ZnO or positive polaron in P3HT drift away from the interface, and the carrier moving towards the interface cannot cross it due to the high energy barrier without an opposite carrier with which to recombine. Therefore the net effect is to reduce the non-geminate recombination in the system, and therefore PCE, as observed. This also explains why the effect is not observed in the ZnO:PCBM:P3HT system, as here the majority of non-geminate recombination occurs at the PCBM:P3HT interface and therefore is not influenced by the polarization in the ZnO.

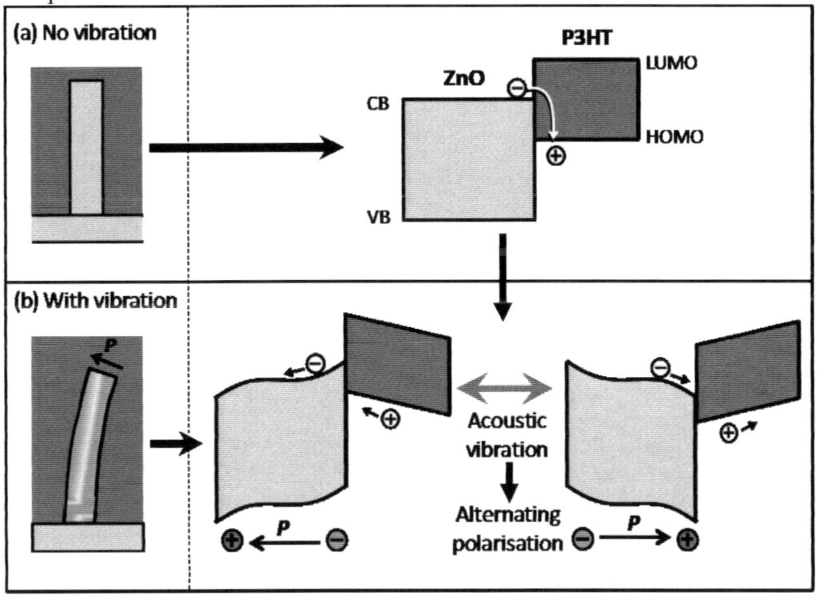

Figure 3. Schematic of charge carrier dynamics at the ZnO:P3HT interface without (a) and with (b) acoustic vibration. Flat bands are shown at equilibrium for clarity. In (a) non-geminate recombination is represented, which is suppressed in (b) as the electric field associated with the polarization, P, in the ZnO causes band bending that withdraws one type of carrier away from the barrier.

5. Conclusions

We have shown that the efficiency of a ZnO nanorod:P3HT solar cell increases from 1.24 to 1.78 % with the application of external acoustic vibrations. Transient absorption studies demonstrate that this effect arises largely from a significant decrease in the non-geminate recombination rate. We have proposed a model whereby the vibration-induced piezoelectric polarization creates band bending at the ZnO:P3HT interface which increases the spatial separation of charge carriers and reduces the recombination as observed. It is possible that other effects arising from vibration may also influence the charge carrier dynamics in the solar cell, and more studies are therefore needed to understand the mechanisms in more detail. However, the observation that such significant improvements in solar cell efficiency can be achieved with the application of acoustic vibrations could lead to a new avenue for solar cell optimization where devices are placed in areas of high vibration to utilize this effect.

References

[1] Gershon T 2011 Metal oxide applications in organic-based photovoltaics *Mater. Sci. Technol.* **27** 1357–71
[2] Law M, Greene L E, Johnson J C, Saykally R and Yang P 2005 Nanowire dye-sensitized solar cells *Nat. Mater.* **4** 455–9
[3] Huynh W U, Dittmer J J and Alivisatos A P 2002 Hybrid nanorod-polymer solar cells *Science* **295** 2425–7
[4] Peiro A M, Ravirajan P, Govender K, Boyle D S, O'Brien P, Bradley D D C, Nelson J and Durrant J R 2006 Hybrid polymer/metal oxide solar cells based on ZnO columnar structures *J. Mater. Chem.* **16** 2088–96
[5] Ravirajan P, Peiró A M, Nazeeruddin M K, Graetzel M, Bradley D D C, Durrant J R and Nelson J 2006 Hybrid Polymer/Zinc Oxide Photovoltaic Devices with Vertically Oriented ZnO Nanorods and an Amphiphilic Molecular Interface Layer *J. Phys. Chem. B* **110** 7635–9
[6] Oosterhout S D, Wienk M M, van Bavel S S, Thiedmann R, Koster L J A, Gilot J, Loos J, Schmidt V and Janssen R A J 2009 The effect of three-dimensional morphology on the efficiency of hybrid polymer solar cells *Nat. Mater.* **8** 818–24
[7] Baeten L, Conings B, Boyen H-G, D'Haen J, Hardy A, D'Olieslaeger M, Manca J V and Van Bael M K 2011 Towards Efficient Hybrid Solar Cells Based on Fully Polymer Infiltrated ZnO Nanorod Arrays *Adv. Mater.* **23** 2802–5
[8] Li Y, Lu P, Jiang M, Dhakal R, Thapaliya P, Peng Z, Jha B and Yan X 2012 Femtosecond Time-Resolved Fluorescence Study of TiO2-Coated ZnO Nanorods/P3HT Photovoltaic Films *J. Phys. Chem. C* **116** 25248–56
[9] Sung Y-H, Liao W-P, Chen D-W, Wu C-T, Chang G-J and Wu J-J 2012 Room-Temperature Tailoring of Vertical ZnO Nanoarchitecture Morphology for Efficient Hybrid Polymer Solar Cells *Adv. Funct. Mater.* **22** 3808–14
[10] Lee S-H, Kim J-H, Shim T-H and Park J-G 2009 Effect of interface thickness on power conversion efficiency of polymer photovoltaic cells *Electron. Mater. Lett.* **5** 47–50
[11] Jaffe B, Cook J M and Jaffe H 1971 *Piezoelectric Ceramics* (London and New York: Academic Press)
[12] Choi M-Y, Choi D, Jin M-J, Kim I, Kim S-H, Choi J-Y, Lee S Y, Kim J M and Kim S-W 2009 Mechanically Powered Transparent Flexible Charge-Generating Nanodevices with Piezoelectric ZnO Nanorods *Adv. Mater.* **21** 2185–9
[13] Xu S, Qin Y, Xu C, Wei Y, Yang R and Wang Z L 2010 Self-powered nanowire devices *Nat. Nano.* **5** 366–73
[14] Briscoe J, Stewart M, Vopson M, Cain M, Weaver P M and Dunn S 2012 Nanostructured p-n Junctions for Kinetic-to-Electrical Energy Conversion *Adv. Energy Mater.* **2** 1261–8
[15] Briscoe J, Jalali N, Woolliams P, Stewart M, Weaver P M, Cain M and Dunn S 2013 Measurement techniques for piezoelectric nanogenerators *Energy Environ. Sci.* **6** 3035–45
[16] Xu C, Wang X and Wang Z L 2009 Nanowire Structured Hybrid Cell for Concurrently Scavenging Solar and Mechanical Energies *J. Am. Chem. Soc.* **131** 5866–72
[17] Xu C and Wang Z L 2011 Compact Hybrid Cell Based on a Convoluted Nanowire Structure for Harvesting Solar and Mechanical Energy *Adv. Mater.* **23** 873–7
[18] Choi D, Lee K Y, Jin M-J, Ihn S-G, Yun S, Bulliard X, Choi W, Lee S Y, Kim S-W, Choi J-Y, Kim J M and Wang Z L 2011 Control of naturally coupled piezoelectric and photovoltaic properties for multi-type energy scavengers *Energy Environ. Sci.* **4** 4607–13
[19] Leschkies K S, Divakar R, Basu J, Enache-Pommer E, Boercker J E, Carter C B, Kortshagen U R, Norris D J and Aydil E S 2007 Photosensitization of ZnO Nanowires with CdSe Quantum Dots for Photovoltaic Devices *Nano Lett.* **7** 1793–8

PowerMEMS 2013 IOP Publishing
Journal of Physics: Conference Series **476** (2013) 012010 doi:10.1088/1742-6596/476/1/012010

Real World Testing Of A Piezoelectric Rotational Energy Harvester For Human Motion

P Pillatsch, E M Yeatman and A S Holmes

Imperial College London, Optical and Semiconductor Devices Group, Electrical and Electronic Engineering, SW7 2AZ, London, UK

E-mail: p.pillatsch10@imperial.ac.uk

Abstract.
Harvesting energy from human motion is challenging because the frequencies are generally low and random compared to industrial machinery that vibrates at much higher frequencies. One of the most promising and popular strategies to overcome this is frequency up-conversion. The transducing element is actuated at its optimal frequency of operation, higher than the source excitation frequency, through some kind of catch and release mechanism. This is beneficial for efficient power generation. Such devices have now been investigated for a few years and this paper takes a previously introduced piezoelectric rotational harvester, relying on beam plucking for the energy conversion, to the next step by testing the device during a half marathon race. The prototype and data acquisition system are described in detail and the experimental results presented. A comparison of the input excitation, based on an accelerometer readout, and the output voltage of the piezoelectric beam, recorded at the same time, confirm the successful implementation of the system. For a device functional volume of 1.85 cm^3, a maximum power output of 7 μW was achieved when the system was worn on the upper arm. However, degradation of the piezoelectric material meant that the performance dropped rapidly from this initial level; this requires further research. Furthermore, the need for intermediate energy storage solutions is discussed, as human motion harvesters only generate power as long as the wearer is actually moving.

1. Introduction

Energy harvesting from surrounding sources has been a popular area of research in recent years. Some industrially viable solutions have been presented for waste heat and machine vibration harvesting. A general overview of the field can be found in [1], whereas [2] gives a more specific introduction into piezoelectric devices.

Generating energy from human motion is of particular interest to power medical sensors and implants as well as sports devices such as heart rate monitors. Ideally, such systems should operate maintenance free and go unnoticed by the user for the majority of the time and therefore batteries are impractical. However, the main inhibitors are the low and random frequencies that we tend to move in. Frequency up-conversion offers a promising solution – the human excitation acts on a proof mass which then actuates a transducer in a way that it can freely oscillate at its natural frequency. The overall efficiency of conversion is increased because the transducer always operates at its ideal frequency, regardless of external excitation. Piezoelectric devices have been presented in [3] with two piezoelectric springs, in [4] for a direct-force knee joint harvester and in [5] for a linear beam plucking device with magnetic coupling. The device described in this article has most recently been discussed in [6], where an experimentally verified simulation model was introduced.

The goal of this paper is to show real world results of a miniaturized prototype worn during a half marathon running race. Testing in such an event gives the opportunity to analyse the performance in a tough environment and to collect data over a long period of time. It is also a significant step in confirming the feasibility of human energy harvesting outside a well controlled laboratory environment.

2. Prototype And Experimental Set-Up

The operation principle can be understood from figure 1. An eccentric proof mass is free to swing around its axis of rotation. The conversion of mechanical to electrical energy is achieved via a piezoelectric beam that is actuated at each pass of the rotor via magnetic coupling and then left to ring down. A photograph of the actual prototype with a £1 coin for size comparison can be seen in figure 2. The dimensions of the rotor match those of a Seiko Kinetic wristwatch and the reader will appreciate the simplicity and small number of parts necessary for this device.

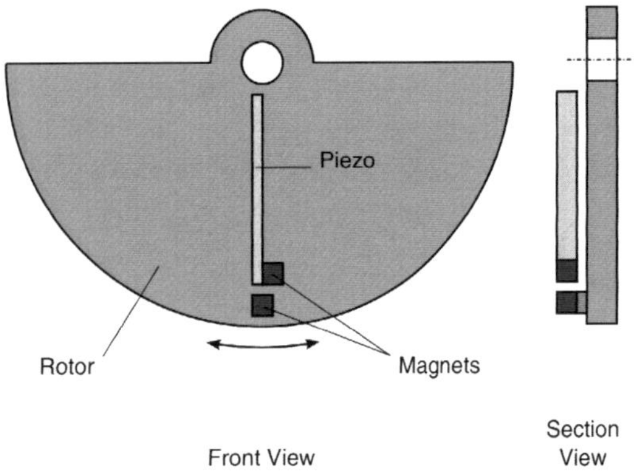

Figure 1: Functional principle of the rotational harvester

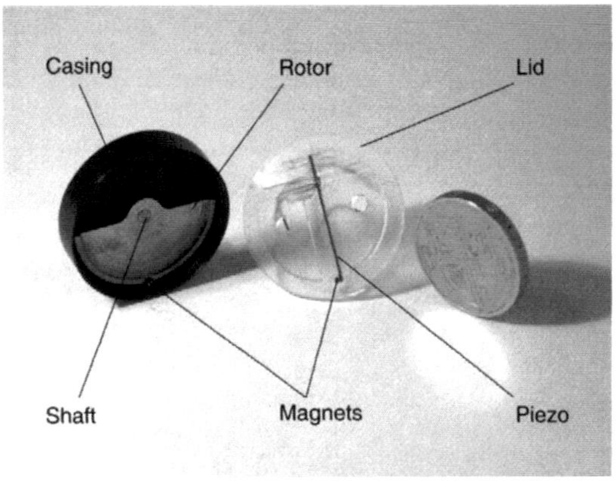

Figure 2: Photograph of the tested prototype with a £1 coin for size comparison

The illustrations in isometric and section view, given in figures 3 and 4 respectively, further clarify the mechanical arrangement of the different components and the device size. The bearings taking the rotor shaft are both housed on the same side of the casing. As a result, the piezoelectric beam can occupy almost the entire diameter on the side of the perspex lid to give a free length of 19.5 mm. With a layer thickness of only 130 µm for each of the piezoceramic sides of the series connected bimorph, and a carbon fibre structural centre of 110 µm, the resulting natural frequency of the beam is around 400 Hz. Based on a measured capacitance of 3.1 nF a load of 150 kΩ was chosen as an impedance match for the power output measurements. The Q-factor of the beam is around 40 and the NdFeB permanent magnets ($1 \times 1 \times 1$ mm, strength N45) are in an advantageous repelling arrangement, based on findings in [6]. The total device volume is 5 cm^3, at a weight of 10.5 g. However, the functional volume of the components involved in power generation, i.e. without the casing, is only 1.85 cm^3.

The measurement system, shown in figure 5, comprises a three axis ADXL335 accelerometer and a Logomatic data logger that records the voltage output of the piezo at 2000 Hz sampling rate. Since, the accelerometer data does not need to be sampled at such a high rate, a separate Ardulog datalogger saves the three channels to a microSD card. This programmable board also controls the Logomatic to ensure that both sets of data are on the same timescale. After packaging and installing the harvester, this set-up was worn in a cell-phone sleeve on the

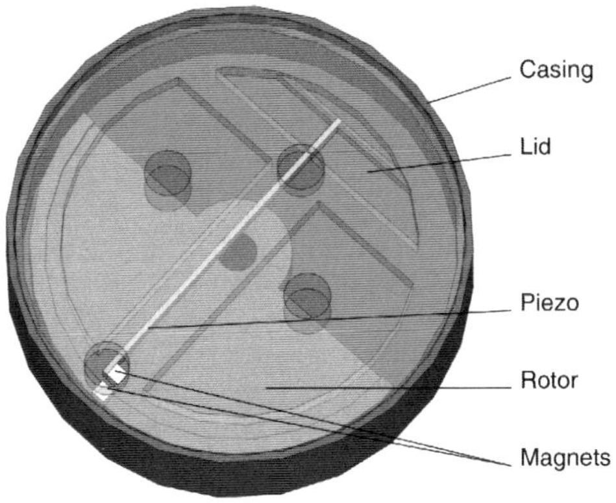

Figure 3: Rendering of the rotational harvester prototype in isometric view

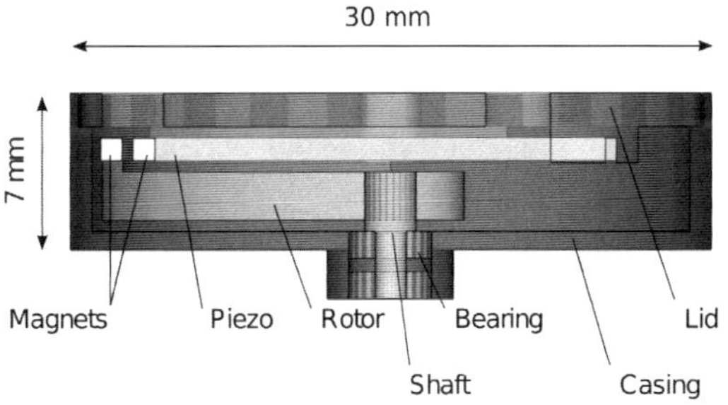

Figure 4: Rendering of the rotational harvester prototype in section view

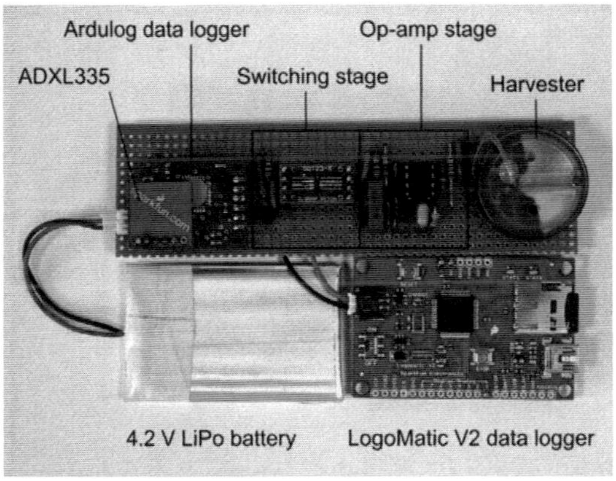

Figure 5: Photograph of the data logging system

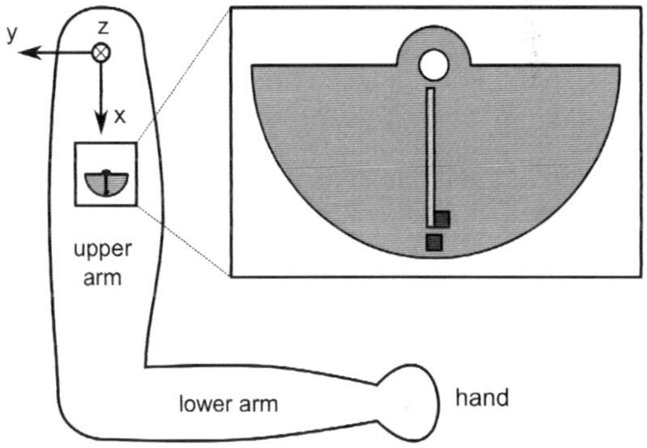

Figure 6: Orientation of the harvester and accelerometer coordinate system when mounted on the upper arm

right upper arm during the race. The device orientation and the coordinate system of the accelerometer are shown in figure 6.

3. Results

Figure 7 shows a sample output of the accelerometer data. The x-axis clearly shows the influence of gravity, whereas the z-axis, being close to the rotational axis of the shoulder, sees very little excitation in comparison. The corresponding FFT in figure 8 confirms that the majority of source motion happens in the x- and y-directions at frequencies around 1 Hz and 2.5 Hz.

Figure 9 depicts the recorded piezo voltage starting at the same time as the accelerometer data but zoomed in further. Three consecutive actuations of the beam with the subsequent ring-down can clearly be recognized. The FFT in figure 10, taken over the middle one of those three actuations, proves that the low frequency excitation, as seen in figure 8, has successfully been converted to a much higher transducer frequency of about 400 Hz.

Finally, two challenges for the feasibility of human motion energy harvesting can also be distinguished from the gathered experimental data. First, the piezo voltage for an entire five minute measurement file of a pre race test is given in figure 11. The data logging system was

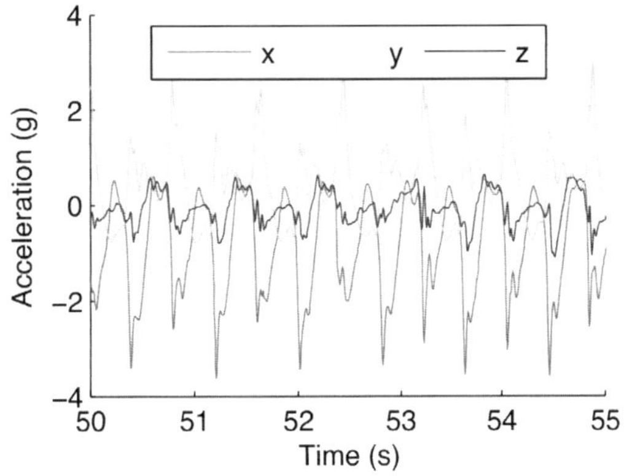

Figure 7: Sample of the accelerometer output for the device when worn on the upper arm while running

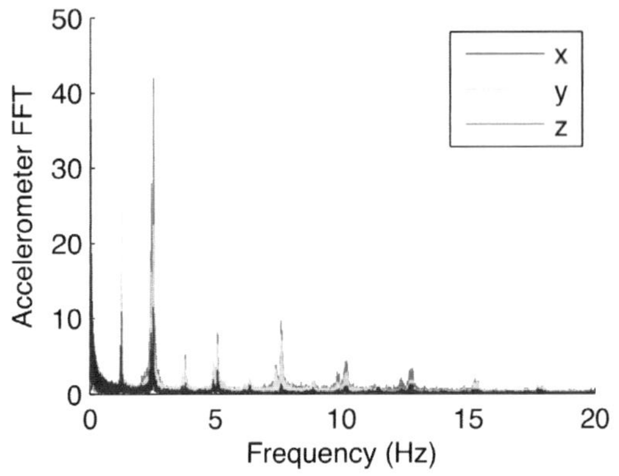

Figure 8: FFT of the accelerometer output showing the dominant excitation frequencies

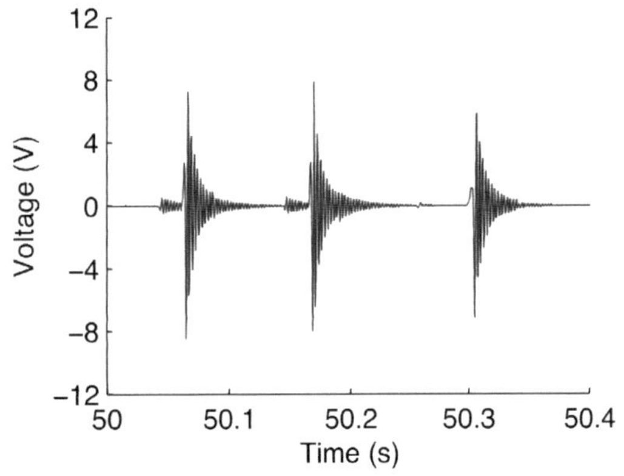

Figure 9: Sample of the measured voltage across the piezo beam with attached resistive load

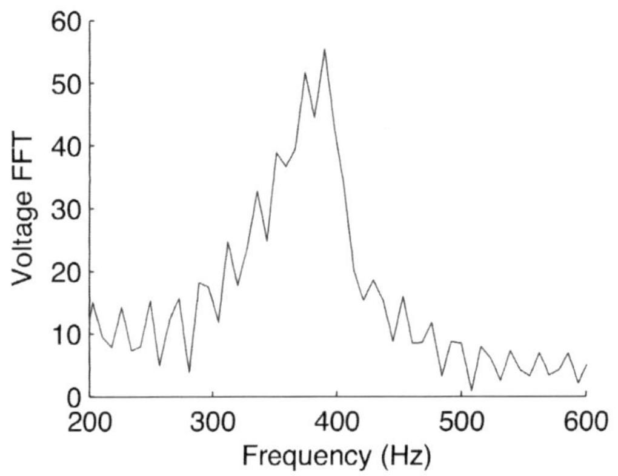

Figure 10: FFT of the voltage output of the piezo beam

designed to make use of the entire ADC range, which causes a slight truncation of data at the top. However, this only affects a negligible portion of the first peak of the decaying waveform after each actuation. More importantly, there are large gaps in the voltage around a time between 100 s and 150 s. This is not a malfunction of the device; the gaps occurred while stopping at a red traffic light to cross the street. Obviously, if the wearer is not moving, no energy can be harvested and this graph clearly illustrates the need for intermediate energy storage solutions to overcome those downtimes, if sensors are required to operate continuously. Second, the average power output for each of the data files is plotted over time in figure 12. Starting from a decent power output of 7 μW, the performance dropped very rapidly to then level out at about 0.5 μW. Examination of the prototype did not show any signs of wear on the mechanics of the device and thus the decrease in power output is a result of degradation of the piezoelectric material. More recent findings suggest that this is due to an excessive initial deflection of the beams before release which can be alleviated by further separating the magnets and consequently weakening the coupling. However, further investigation on this topic is required.

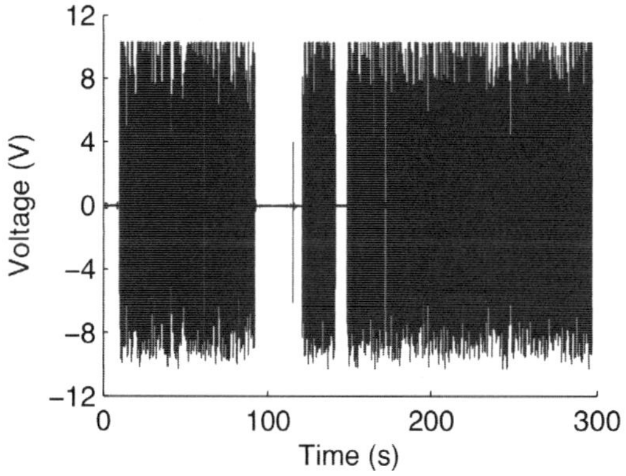

Figure 11: Sample of an entire five minute data set for the voltage output of the piezoelectric beam while running

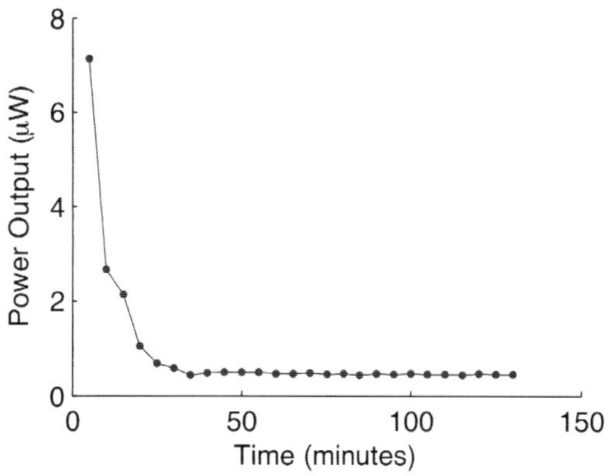

Figure 12: Degradation of the power output over the pre race test and actual race

4. Conclusions

This paper reports the first real world experimental data of a rotational piezoelectric energy harvester for human motion. The mechanics of the prototype and the portable data collection system are described. The device uses the principle of frequency up-conversion to transform the random low frequency excitation of a runner into a higher transducer frequency by magnetically plucking a piezoelectric beam. This has the potential for increased efficiency of power generation.

The successful operation of the up-conversion is verified through comparison of an FFT analysis on the output voltage waveform of the harvester and three axis accelerometer data recorded simultaneously.

Identified challenges for successful implementation of human motion energy harvesting in the future are the degradation of device performance over time and the need for intermediate energy storage, e.g. rechargeable batteries or capacitors, to overcome downtimes of the power generation when the wearer is not moving.

Nevertheless, these results are encouraging because the mechanics of the device with its contact-less coupling did not show any signs of wear after the testing and, with a functional volume exactly the same as that of a Seiko Kinetic harvester, a peak power output of 7 μW was achieved – enough to power a wristwatch or other novel sensors for health monitoring.

Acknowledgments

The authors would like to thank the Engineering and Physical Sciences Research Council (EPSRC) for financial support as well as Michael Riedel and Klaus van der Linden from Johnson Matthey Catalysts (Germany) GmbH for providing helpful discussions during the design of the piezoelectric bending beams.

References

[1] Mitcheson P D, Yeatman E M, Rao G K, Holmes A S and Green T C 2008 *Proc. of the IEEE* **96** 1457–1486
[2] Kim S G, Priya S and Kanno I 2012 *MRS Bulletin* **37** 1039–1050 ISSN 0883-7694
[3] Galchev T, Kim H and Najafi K 2011 *Journal of Microelectromechanical Systems* **20** 852–866
[4] Pozzi M, Aung M S H, Zhu M, Jones R K and Goulermas J Y 2012 *Smart Materials and Structures* **21** 075023 (8pp) ISSN 0964-1726
[5] Tang Q C, Yang Y L and Li X 2011 *Smart Materials and Structures* **20** 125011 (6pp) ISSN 0964-1726
[6] Pillatsch P, Yeatman E M and Holmes A S 2013 *Transducers* pp 1364–1367 ISBN 9781467359832

Journal of Physics: Conference Series **476** (2013) 012011 doi:10.1088/1742-6596/476/1/012011

A compact human-powered energy harvesting system

Yuan Rao, Kelly M. McEachern, and David P. Arnold
Interdisciplinary Microsystems Group, Dept. Electrical and Computer Engineering, University of Florida, Gainesville, FL 32611, USA

E-mail: yrao1@ufl.edu

Abstract. This paper presents a fully functional, self-sufficient body-worn energy harvesting system for passively capturing energy from human motion, with the long-term vision of supplying power to portable, wearable, or even implanted electronic devices. The system requires no external power supplies and can bootstrap from zero-state-of-charge to generate electrical energy from walking, jogging and cycling; convert the induced ac voltage to a dc voltage; and then boost and regulate the dc voltage to charge a Li-ion-polymer battery. Tested under normal human activities (walking, jogging, cycling) when worn on different parts of the body, the 70 cm^3 system is shown to charge a 3.7 V rechargeable battery at charge rates ranging from 33 μW to 234 μW.

1. Introduction

Vibrational energy harvesting systems that convert ambient mechanical energy in the environment to usable electrical energy represent a promising emerging technology to achieve autonomous, self-renewable, and maintenance-free operation of wireless electronic devices and systems [1-3]. A complete energy harvesting system comprises three main components: an energy harvester that converts the mechanical vibrations into electrical energy, an energy harvesting interface circuit that conditions and regulates the energy, and an energy storage element that stores the intermittent harvested energy.

In this work, a fully self-contained energy harvesting system is demonstrated using an input-powered interface circuit and a non-resonant electrodynamic harvester, designed specifically for harvesting energy from human movements. Compared with a previous design [4], both the harvester and the interface circuit are redesigned and improved. The current system achieves 10% higher net power density (3.3 μW/cm^3) with 30% smaller volume (70 cm^3).

2. System Design

2.1. Energy Harvester

Compared with mechanical vibration sources, human-induced motions are challenging for energy harvesting design because of their low-frequency (1–10 Hz), aperiodic, and time-varying characteristics. The commonly used high-Q resonant-type energy harvesters, which are based on an under-damped, single-degree-of-freedom, mass-spring-damper system are generally not well suited for human movement energy harvesting. The reason is that these resonant systems are optimized to achieve maximum output power within a small frequency range under oscillatory accelerations. Moreover, it is difficult to tune the resonant frequency to the low frequencies of human motions (1–10 Hz) and maintain high quality factor, especially all while maintaining compact device dimensions. Another major restriction of conventional resonant harvesters is that they are typically designed for only one rectilinear degree of freedom, while normal human movements occur in three dimensions and involve a high degree of rotational, rather than oscillatory, motions.

The above challenges motivate the use of a non-resonant harvester architecture, which can respond over a broad range of vibration frequencies and amplitudes, and the use of a multi-directional architecture that can respond to motions in multiple axes. Our research group has previously reported a unique, omnidirectional electrodynamic energy harvester design [5], which is replicated with modifications in the system reported here.

As shown in figure 1, the harvester structure is fabricated in two symmetric hemispheres using a Nylon plastic material from a 3D printer (Shapeways, Inc.). The two halves are fit together to form a spherical cavity (diameter=3 cm) with a permanent magnet ball (diameter=1.27 cm, Grade N40 NdFeB) inside. The upper and lower sections are then each wrapped with ~1400 turns of 34 AWG copper wire forming two independent coils with a resistance of 110 Ω each. The harvester has a total volume of about 39 cm^3 and weighs 68 g.

In operation, the magnet ball moves chaotically within this spherical housing when subjected to external vibrations/motions. The motion of the magnet ball induces a time-varying magnetic flux in each of the surrounding coils, thus generating a voltage according to Faraday's Law. When an electrical load is connected, a current will flow through the coil, thus converting mechanical energy into electrical energy. In the experiments, the two coils are counter-wound and connected in series to improve power generation.

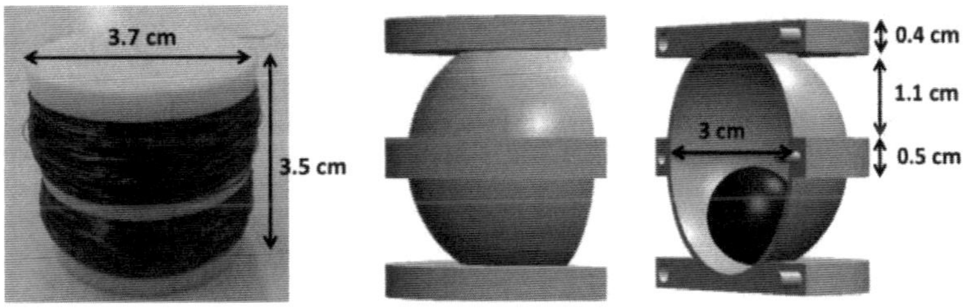

Figure 1. Photograph (left) and 3-D schematic (right) of the energy harvester.

2.2. Energy Harvesting Circuits

The pseudo-random output voltage of the harvester is conditioned by an input-powered energy harvesting circuit, which itself is powered by this time-varying voltage. The block diagram of the circuit is shown in figure 2. It consists of an ac/dc stage and a dc/dc stage, converting the energy from the harvester to the load (battery). The harvester is represented by a voltage source V_S and an output impedance of R_S. The battery is modeled by a constant-voltage (CV) load.

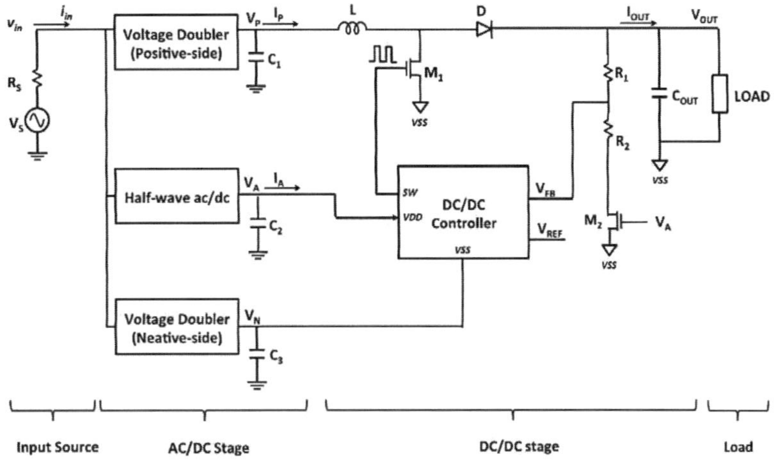

Figure 2. Block diagram of the interface circuit

Two ac/dc converters are used in the ac/dc stage: a voltage doubler, which serves as the primary rectifier in the power path and an auxiliary half-wave ac/dc converter to provide the supply voltage of the dc/dc controller. The dc/dc stage is a boost converter with on-chip dc/dc controller to provide a regulated output voltage for battery charging. Compared with the previous interface circuit design, which used a half-wave rectifier architecture [4], the voltage doubler front-end used here has the benefit of higher output power, and also lower voltage conversion burden to the following dc/dc stage. The auxiliary half-wave ac/dc converter is used to lower the minimum operational voltage of the dc/dc converter (currently 1.2 V_{pk} at 3.7 V CV load).

The circuits are fabricated in silicon using an On Semi 3M-2P 0.5-µm CMOS process. Figure 3 shows the voltage doubler chipset, with a size of 1.6 mm by 1.6 mm, packaged into one SOIC16 package. figure 4 shows the dc/dc controller die, with a size of 1.6 mm by 1.6 mm, packaged in a SOIC 24 package. Both chips are covered by glass lids. Besides the on-chip circuit, there are several external components. The inductor (L) is 22 µH, the capacitors (C_1, C_2, C_3 and C_{out}) are 220 µF, and the feedback resistors (R_1 and R_2) are 3 MΩ and 1 MΩ.

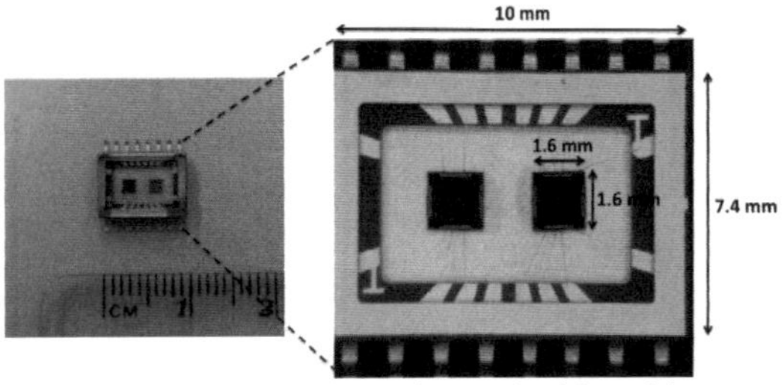

Figure 3. Photo of the voltage doubler chipset

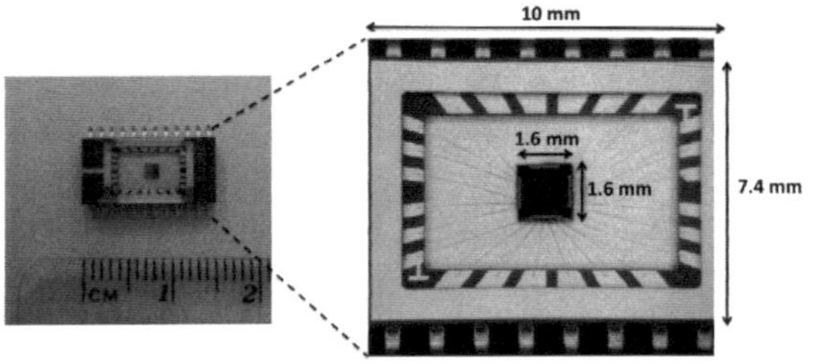

Figure 4. Photo of the dc/dc controller chip

Measurements show that at open-load the system turns on when the input is above 1 V_{pk} and turns off when the input drops below about 600 mV_{pk}. When the system enters sleep mode, there is no measurable standby power consumption. It's important to acknowledge that a commercial button cell battery with nominal voltage of 1.5 V is used to provide the reference voltage (V_{REF}) for the dc/dc controller, because there is no on-chip reference circuit. Note that this battery is connected to the gate of a MOSFET, and therefore the power consumption from this battery is negligible. In future designs, this battery can be replaced by an on-chip bandgap reference circuit.

2.3. Energy Storage

In this design where the system size is strictly limited, a rechargeable lithium-ion (Li-ion) battery is used as the energy storage element. A Li-ion battery is chosen here not only because of its small size, light weight, and good energy density, but also because it has no memory effect and slow self-discharge rate. However, Li-ion batteries should be handled with caution, especially in high temperatures, because they can easily ignite or explode. A commercial Li-ion polymer rechargeable battery with nominal voltage of 3.7 V and maximum capacity of 65 mAh is used [7]. The battery has a smaller size (23mm × 12mm × 4mm) and a longer life time (up to 500 cycles charge/discharge) than conventional rechargeable cells. It has a built-in self-protection circuit to avoid over-charge or over-discharge.

2.4. System Prototype

The interface chips, capacitors, diodes, and inductor are soldered on the top PCB1 as indicated in figure 5, whereas the button battery and the rechargeable battery are mounted on the bottom PCB2. These two round PCBs with 3.2 cm diameters are mounted on top and bottom surfaces of the harvester to form the complete harvesting system. Figure 6 depicts the prototype system, which has a cylinder structure with 6.3 cm height and 3.8 cm diameter, leading to a total volume of about 70 cm^3 and a total weight of 81 g.

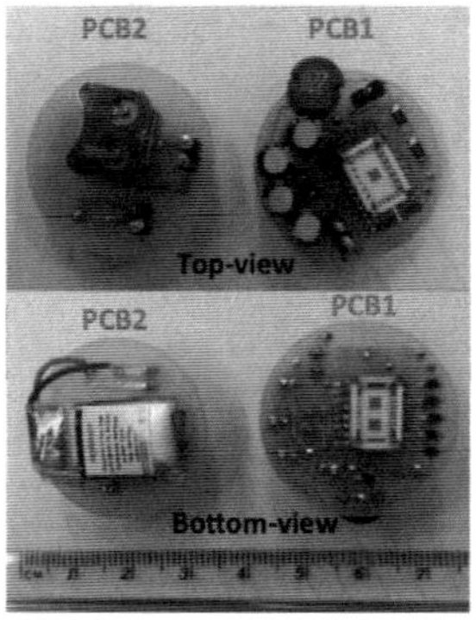

Figure 5. Photograph of the circuit boards.

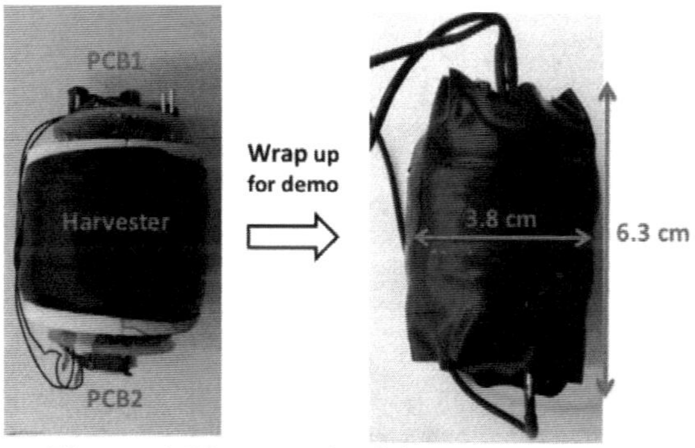

Figure 6. Photograph of the system.

3. System Measurement

3.1. Measurement Method

The energy reclamation performance of the complete energy harvesting system is then measured when subjected to real human activities. In the experiments, the system is attached to a person's ankle, wrist, or upper arm. These locations are chosen for two reasons. The first reason is that more motion/vibrations are expected at these locations, and thus more energy can be harvested [8]. Another reason is that to attach the system on these locations won't cause too much discomfort in daily human activities, which is important for future applications. Because different people may have different gaits in daily activities, the generated energy will vary from person to person. Therefore, in the experiment, the same person is tested for all the activities, so that the result is consistent among different locations and movements.

The system energy reclamation is measured for three types of movements: walking, jogging and cycling. Walking and jogging are conducted on a treadmill, while cycling is performed on a stationary cycling machine. By doing so, the speed of each activity is accurately controlled and replicated. Each activity type is tested for 10-minute duration and is repeated by attaching the system to the different parts of the body. To quantify the harvested energy, the battery voltage is measured at 1-minute intervals during the activity. Comparing the voltages to the separately measured battery charging curve at 100 μA charging current, the total energy delivered from the system to the battery is estimated.

3.2. Delivered Energy

Figure 7 (a) shows the estimated energy delivered to the battery versus time when the system is mounted on the ankle for jogging, walking and cycling. Comparing these three activities, the most accumulated energy of 142 mJ is delivered after 10 minutes of jogging. This result is within our expectation, because jogging generates larger vibration accelerations than walking and cycling.

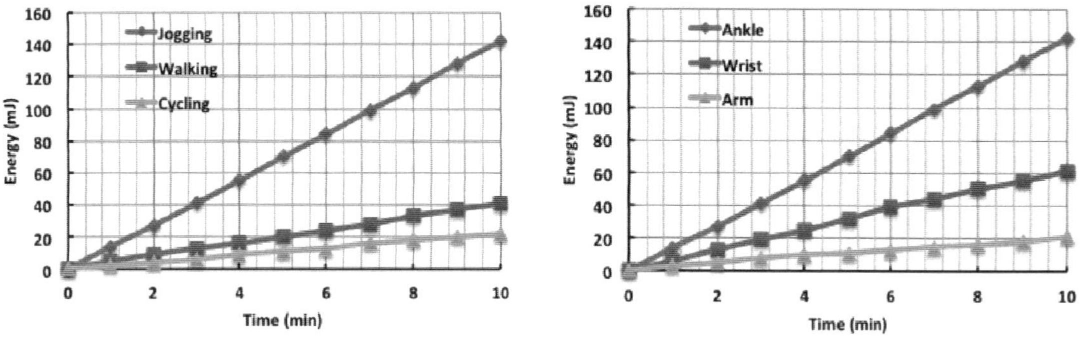

(a) From the human ankle (b) From jogging

Figure 7. Energy delivered to the battery

Similarly, figure 7 (b) plots the energy delivered during jogging when the system is mounted at the ankle, arm and wrist. Here, cycling delivers more energy than walking and cycling for the same period of time. Note that in both figures, the energy delivered increases almost linearly with the time of human movements, because the speed of each human activity remains constant.

3.3. Output Power

The average harvested power can be estimated by calculating the slope of the lines in figure 7. Table 1 summarizes the measurement results. The average power delivered to the battery ranges from 33 μW to 234 μW for the different configurations tested. When mounted on the ankle during jogging, a maximum power delivery of 234 μW is measured. Due to the relatively small upper-body vibrations, no significant power is measured at arm during walking and cycling, and the same for the wrist during cycling. This is attributed to the energy harvester not generating sufficient ac voltage amplitude for the interface circuitry to function.

Table 1. Average power delivered to the battery

	Ankle	**Wrist**	**Arm**
Jogging	234 μW	100 μW	210 μW
Walking	67 μW	33 μW	N/A
Cycling	34 μW	N/A	N/A

4. Conclusion

A complete self-sufficient energy harvesting system—including a magnetically based harvester, a self-powered interface circuit, and a rechargeable battery—is demonstrated and characterized. The harvester employs spherical magnetic harvester structure, favorable for harvesting multi-directional vibrations from human movements. The system successfully scavenges and converts mechanical energy from ordinary human movements to electrical energy for charging a battery. The measurement result shows that a sustained average power of 234 μW is delivered to the battery during jogging when the system is mounted on the ankle. The total volume of the system is 70 cm^3, and therefore the net power density is about 3.34 $\mu W/cm^3$. Compared with a previous design [5], the current system performance improves with a 10% higher net power density (3.3 $\mu W/cm^3$) and a 30% smaller volume (70 cm^3).

Acknowledgement
This work was supported in part by Texas Instruments via a fellowship for Yuan Rao.

References

[1] S. Roundy 2003 *Energy Scavenging for Wireless Sensor Nodes with a Focus on Vibration to Electricity Conversion* (PhD Dissertation. Berkeley)
[2] S. Beeby, M. Tudor and N. White 2006 *Meas. Sci. Technol.,* ***17*** 175
[3] R. J. M. Vullers, R. v. Schaijk, H. J. Visser, J. Penders and C. V. Hoof 2010 *IEEE Solid-State Circuits magazine*
[4] Y. Rao, S. Cheng and D. P. Arnold 2012 *PowerMEMS* 101.
[5] B. J. Bowers and D. P. Arnold 2009 *J. Micromech. Microeng* **19** 094008
[6] http://www.onsemi.com/PowerSolutions/product.do?id=NSR0320MW2T1G
[7] http://www.tenergy.com/30117
[8] N. G. Elvin and A. A. Elvin 2012 *IEEE Trans. on Mechatronics* **18** 637

Design, fabrication and characterization of an inductive human motion energy harvester for application in shoes

K Ylli[1,3], D Hoffmann[1], B Folkmer[1] and Y Manoli[1,2]

[1] HSG-IMIT – Institute of Micromachining and Information Technology, Wilhelm-Schickard-Str.10, 78052 Villingen-Schwenningen, Germany
[2] Fritz Huettinger Chair of Microelectronics, Department of Microsystems Engineering – IMTEK, Georges-Koehler-Allee 102, 79110 Freiburg, Germany

E-mail: klevis.ylli@hsg-imit.de

Abstract. The concept of energy harvesting has been in the focus of research for more than two decades now and with the continuous device miniaturization and reduction in power consumption of the electronics it has become a viable power source for mobile systems. The increasing desire for mobility and longevity in terms of battery life has eventually led to wearable systems, i.e. electronic circuits with their power supply which are being integrated into textiles and everyday life. This paper reports the development of a cylindrical inductive energy harvesting device which exploits the accelerations available in the plane of the foot during walking. The modeling and characterization of the system is based upon real-world acceleration data recorded during treadmill runs. Although a wider range of test subjects would be required to increase the statistical relevance of the measured data, it is concluded that the energy provided by this system is sufficient to power low energy circuits at comparatively slow walking velocities. Additionally, the obtained knowledge can be used to develop a smaller, parallelized system.

1. Introduction

In the past two decades the field of energy harvesting has received significant attention from researchers around the globe. The development of modern materials, the optimization of known principles as well as the ongoing miniaturization of electronic components and the reduction of their power consumption has made energy harvesting from ambient vibrations and motion a feasible alternative to conventional battery powered systems. Correspondingly, the development of smart textiles and wearable systems has been driven forward resulting in ever increasing mobility and the desire to be energy independent. Kinetic energy harvesters have thus been developed which exploit the energy available in human motion and in particular the energy expended in the human gait. Due to the low frequency of the human gait of approximately 1 Hz [1], conventional resonant energy harvesters are not a feasible solution.

Human motion harvesters developed to date include devices that mechanically translate the downward motion of the heel into a horizontal motion and drive an electromagnetic generator [2],

[3] To whom any correspondence should be addressed.

devices that exploit the material properties of electro active polymers to generate a power output upon applying mechanical stress [2] and shock induced harvesters which vibrate at their resonant frequency once deflected out of the equilibrium position upon heel strike [3].

In this work an energy harvester is presented, that exploits the acceleration of the foot with only one passive moving component and hence no risk of wire breaks or other mechanical failure due to its straightforward construction. An approach different from others seen in literature [4] was taken in the development process by using acceleration data recorded during a treadmill run as the input source for the differential equation that models the system instead of simplifying the signal.

2. Overview of the available excitations and the developed system model

2.1. Human Gait: An Overview

There are several different types of feasible non-resonant mechanical excitation sources for energy harvesting devices that can be observed in the human gait. The first possible excitation is provided by the weight of the test subject. Researchers have come to the conclusion that the dynamic force acting upon the sole during walking can reach up to 130% of the human weight and therefore easily surpass a force of 1000 N which can be exploited by energy harvesters [5].

A second excitation source can be found in the shocks, which occur upon heel-strike as mentioned previously [3]. The third type of excitation, which is also the basis for the harvester presented in this work, is the acceleration of the foot during walking. Treadmill measurements were performed at constant running velocities of 4 km/h, 6 km/h, 8 km/h and 10 km/h to record the accelerations. Figure 1 shows the accelerometer attachment to the shoe sole. Figure 2 shows the recorded acceleration and the integrated velocity for the exemplary running speed of 10 km/h.

Figure 1. Accelerometer placement and recorded axes.

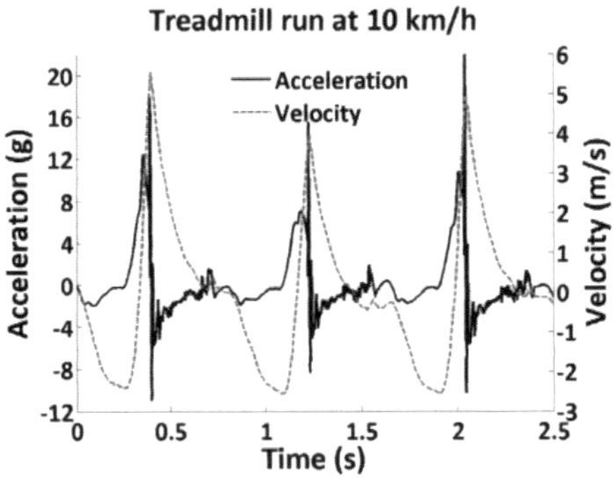

Figure 2. Recorded y-axis (horizontal) acceleration and integrated velocity for a running speed of 10 km/h.

Table 1. Average acceleration peaks recorded in y-direction during treadmill runs.

Running velocity	Positive peak average (g)	Negative peak average (g)
4 km/h	5	-4.28
6 km/h	8.7	-5.45
8 km/h	15	-4.16
10 km/h	23	-3.26

Large accelerations and velocities are readily available during walking as can been seen in Table 1, which motivated the development of an energy harvesting device that exploits these low-frequency, large amplitude accelerations. In order to be as accurate as possible, the recorded acceleration curve is used as input into the software model.

2.2. System Modeling

The developed energy harvester is based on the principle of induction and employs two copper wire coils through which a generator magnet can move freely (see figure 3). The change in the magnetic field induces a voltage in the coil windings which provides the harvester output power across a suitable load. The main aim of this device was to gain a better understanding of the processes involved: from the motion of the generator magnet (GM) to the fabrication of a suitable housing which protects the harvester and allows the integration of the device in a shoe sole.

a) b)

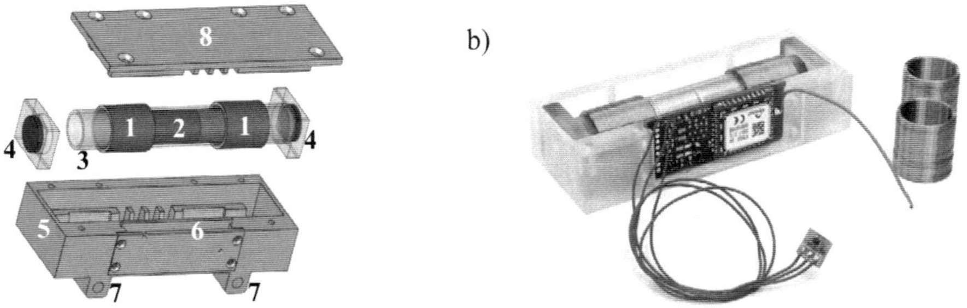

Figure 3. a) Schematic of the designed harvester prototype. Labeled parts: 1. Coils, 2. Generator magnet 3. PVC channel, 4. Stopper magnets (magnetic springs), 5. Housing, 6. Electronics compartment, 7. Boreholes for external attachment to shoe, 8. Lid. **b)** Fabricated device and spare coils; test electronics include a rectifier, a temperature sensor and an EnOcean wireless transmitter.

In order to model the energy harvesting device, a differential equation was developed which considers the relevant forces acting upon the GM which is the only moving harvester part. The considered forces include the friction force between the GM and the PVC channel, the electromotive force caused by the induced voltage in the coil and the repulsive magnetic force between the generator magnet and the stopper magnets (SM). The SMs are placed as magnetic springs at either end of the PVC channel to avoid collisions with the housing. The air compression due to the air in the channel being pushed by the GM was not considered for this first prototype.

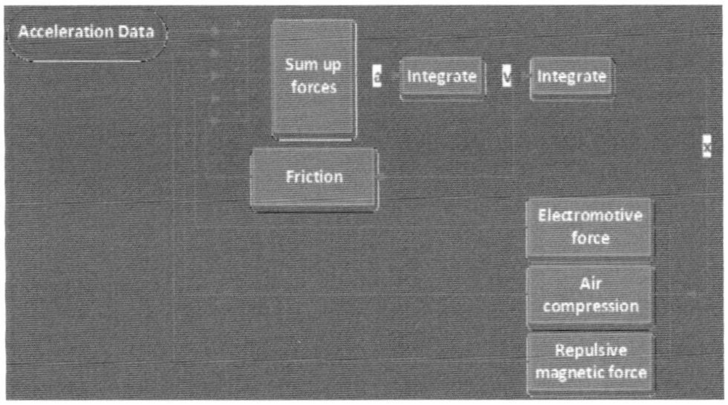

Figure 4. Block diagram showing the forces acting in the differential equation of the system model.

The simulation was performed using 2D FEA-software and rotational symmetry to compute the magnetic field distribution and subsequently the coupling coefficient and the induced voltage according to (1) and (2). A more detailed description of the analytical calculation of the induced voltage can be found in [6].

$$k = -\frac{d\varphi}{dx} \tag{1}$$

$$U_{ind} = -\frac{d\varphi}{dt} = -\left(\frac{dA}{dt}B + \frac{dB}{dt}A\right) = -\frac{d\varphi}{dx}\cdot\frac{dx}{dt} = k\cdot\dot{x} \tag{2}$$

3. Prototype development

3.1. Optimization and Fabrication

A cylindrical setup was chosen for this device because of the availability of the components as well as the fact that rotational symmetry can be exploited to greatly facilitate the calculations performed by the simulation software. A genetic algorithm was used to perform the optimization of the following set of parameters within feasible boundaries: the GM radius, the GM height, the length of the coil, the wall thickness (outer radius minus inner radius) of the coil, the SM height and the location of the coil. The values of the key parameters are listed in table 2.

Table 2. Optimized key geometrical parameters of prototype device.

GM diameter	GM height	PVC Channel: inner diameter	PVC Channel: length	Coil: inner diameter	Coil: outer diameter
13.5 mm	50 mm	13.6 mm	100 mm	16 mm	19 mm

The housing was 3D-printed in plastic and contains grooves that accommodate the coils at precisely the optimum position as obtained by the optimization procedure. The coils were placed to allow the generator magnet to enter the first coil while it leaves the second coil and the coils were reverse connected in series to achieve a larger voltage output.

3.2. Treadmill measurements

In order to test the device under realistic conditions, it was attached externally to a running shoe and the output voltage across the optimum load resistance was measured on a data logger. The resulting power output for different coil-magnet configurations with an optimal load is shown in figures 5 and 6. As was the case in the simulations, it is essential that the actual excitation of the foot is used for the characterization of the device instead of a simplified representation, e.g. a sine signal.

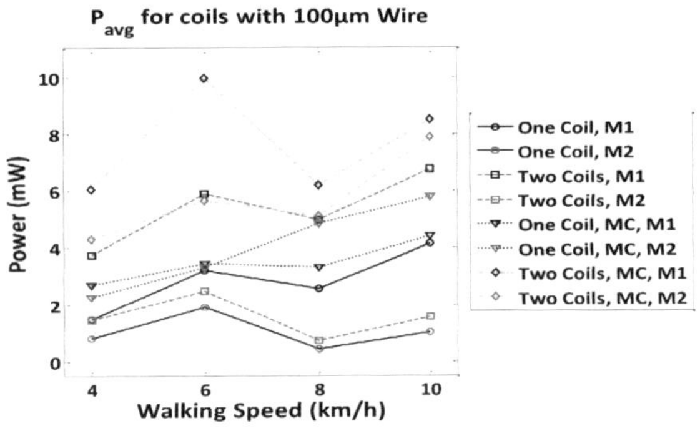

Figure 5. Power output for different device configurations with 100 µm copper wire coils. M1 and M2 are "weak" and "strong" spring magnets respectively. MC denotes a modified channel with a larger inner diameter of 14 mm.

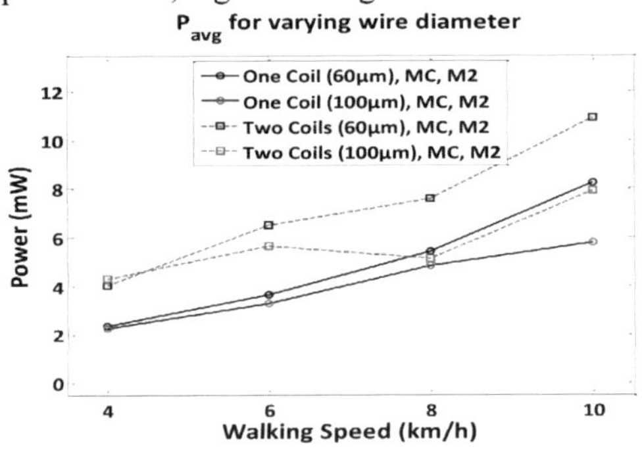

Figure 6. Power output for configurations of varying coil wire diameter. The resistance of the coils is 230 Ω and 1640 Ω respectively for the 60 µm wire and 100 µm wire.

3.3. Analysis

It can be seen from figures 5 and 6 that a system with two coils in series performs significantly better in terms of output power than a single coil system. It is also observed that measurements with a slightly enlarged channel (marked "MC" in the figures) show an increased power output, as the GM can move more freely and a certain amount of air can pass around it instead of being pushed in front of it. Comparatively weak SMs are shown to lead to a larger power output, as the magnetic field of the stronger SMs has a greater range and inhibits the movement of the GM to a certain degree.

In figure 6 it is shown that a configuration with 60 μm coil wire produces a larger power output than the 100 μm configuration, although the copper fill factor, i.e. the percentage of the coil cross-section actually filled with copper, is smaller (see norm IEC 60317) . This result is unexpected and has to be investigated further by using a variety of test subjects of whom everyone has an own unique style of walking.

A significant observation from figure 5 concerns the transition from fast walking at 6 km/h to slow running at 8 km/h. As can be seen, there is no continuous increase in output power as the treadmill velocity increases. This can be attributed to the change in the walking style from the abrupt movements of fast walking to the smooth movements of jogging. Table 3 illustrates this observation by listing the peak-to-peak velocities as the result of the integration of the acceleration data. As shown, there is no linear increase in the peak-to-peak velocity due to the described effect.

Table 3. Peak-to-peak velocity change of the foot during running.

Treadmill speed	peak-to-peak foot velocity
4 km/h	3.9101 m/s
6 km/h	4.9175 m/s
8 km/h	5.1595 m/s
10 km/h	8.1262 m/s

4. Conclusion

While this first prototype is simple and rather large in size, it shows a promising power output of up to approximately 10 mW with RMS-voltages between one and six volts depending on the configuration and the walking speed (voltage peaks are significantly higher). The mentioned power output sufficed to supply an experimental circuit which measured the temperature at the foot of the subject and transmitted the data up to seven times per second using an enOcean transmitter. Additionally it was shown that slower velocities do not necessarily translate into a low power output of the system.

The experience gained in the development of this device has led to a number of improvements of the system model, e.g. the calculation of the coupling factor depending on the position of the GM relative to the SMs or the consideration of the damping due to the air pushed in front of the GM.

Future work will include a more detailed analysis on the effect of changes in geometrical parameters and tolerances on the device output as well as the development of a smaller, parallelized system based on the knowledge obtained in this work. In order to characterize the devices and their behavior more accurately and increase the statistical significance of the obtained data, the number of test subjects will be increased.

Acknowledgments

This research (IGF-Project: 17742N) was supported by the "Programm zur Förderung der industriellen Gemeinschaftsforschung und -entwicklung (IGF)", Germany.

References

[1] Kadaba M P, Ramakrishnan H K and Wootten M E 1990 *J. Orthop. Res* **8** no. 3 383–92
[2] Paradiso J A and Starner T 2005 *IEEE Pervasive Computing* **4** no. 1 18–27
[3] Moro L and Benasciutti D 2010 *Smart Mater. Struct* **19** no. 11 115011
[4] Shen J X, Wang C F, Luk P C K, Miao D M, Shi D and Xu C 2013 *IEEE Trans. on Ind. Applicat.* **49** no. 2 990–6
[5] Niu P, Chapman P, Riemer R, and Zhang X 2004 *IEEE Power Electronics Specialists Conference, 2004. PESC 04.* (*Aachen*) **6** 2100–6.
[6] Spreemann D and Manoli Y 2012 *Electromagnetic vibration energy harvesting devices: Architectures, design, modeling and optimization* (Dordrecht: Springer) 2012

Electromagnetic Vibration Energy Harvester Using Springless Proof Mass and Ferrofluid as a Lubricant

S H Chae[1], S Ju[1], Y Choi[1], S Jun[1], S M Park[1], S Lee[1], H W Lee[2], and C-H Ji[1]
[1]Department of Electronics Engineering, Ewha Womans University, Seoul, Korea
[2]Department of Neurology, Ewha Womans University School of Medicine, Seoul, Korea

E-mail:cji@ewha.ac.kr

Abstract. This paper presents an electromagnetic energy harvester using an array of rectangular permanent magnets as springless proof mass and ferrofluid as a lubricating material. Lateral motion of the multi-pole magnet array generates voltage across an array of copper windings formed under the aluminum channel in response to low frequency external vibrations such as human-body-induced motion. A proof-of-concept device has been fabricated and output voltage has been measured at various input frequencies and accelerations provided by a vibration exciter. Device with ferrofluid lubrication generated maximum open-circuit voltage of 0.47V at 3g vibration at 12Hz, which is 8% higher than that of the device without lubricant. Maximum output power of 71.26μW has been obtained at 40.8Ω with the device with ferrofluid lubrication.

1. Introduction

In a variety of areas including wireless sensor nodes and medical implements, energy harvesting has attracted research interest as an alternative to conventional internal energy sources which are primary batteries in general. Harvesting energy from vibration is drawing notable attention due to rather straightforward power generation mechanism and availability of vibration sources in the environment.

Electromagnetic induction based rotary power generators have been widely used and researched in macro and miniature scale applications up to the present [1]. Due to the complexity of these systems and difficulties related with translation of motion and providing support for the moving part, utilization of spring-supported linear motion has prevailed in most of the electromagnetic energy harvesters [2]. In general, harvesters based on spring-mass-damper systems require relatively high frequency, especially in miniature devices, and frequency matching which limit the operation frequency band. Alternative techniques are required to harvest energy from low frequency vibrations such as human-body-induced motion. To overcome the issues of conventional spring-mass-damper-based devices, wide variety of harvesters using springless proof mass have been researched [3-5].

To realize an electromagnetic harvester with springless proof mass, means to provide stable relative motion between the permanent magnet and windings while maintaining a narrow gap and magnetization direction are crucial. Rolling magnets can be utilized at the expense of rather random variation of magnetization direction [5]. Ball bearing has proven to be effective in some applications, but the wear issue has to be taken into account [6]. Utilization of cylindrical magnet has also been demonstrated [4]. In this research, we have utilized ferrofluid as a lubricant for a multi-pole rectangular magnet assembly which undergoes a linear sliding motion. Ferrofluid is a colloidal suspension containing ferromagnetic nanoparticles with friction-reducing capabilities on a magnet surface. A proof-of-concept electromagnetic vibration energy harvester with springless sliding magnet assembly and ferrofluid as a lubricant has been fabricated and tested.

2. Harvester design

A schematic diagram of the proposed energy harvester is shown in figure 1. Proof-of-concept harvester design utilizes the sliding motion of four NdFeB bar magnets covered with a pure iron plate oscillating in aluminum housing. When an external lateral vibration is applied to the device, multi-pole magnets oscillate horizontally and voltage is induced across an array of copper windings connected in series. Pure iron plate covering the top side of the magnets enhances the magnetic flux across the windings. Ferrofluid reduces the friction between the magnet surface and aluminum housing. A set of five coils is formed by winding a 0.1mm-diameter copper wire around the bobbins attached to the bottom of the housing. Coil pitch and height of coil are 3.07mm and 2.1mm, respectively. Gap between the magnet assembly and coil is 0.2mm.

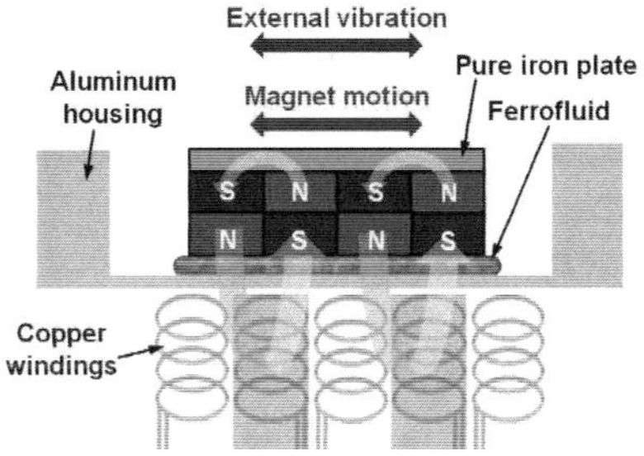

Figure 1. Schematic of the proposed energy harvester using springless multi-pole planar magnets with ferrofluid lubrication.

3. Fabrication

Figure 2 shows the fabricated electromagnetic energy harvester. Copper windings are formed with 0.1mm-diameter copper wire and bobbins machined with acrylic plate. Number of turns for each winding is 50. An array of four NdFeB bar magnets, which measure 12.7x3.18x1.59mm^3 each, are attached and covered with a pure iron plate (12.7x12.7x0.2mm^3). Aluminum housing has been fabricated by precision CNC (computer numerical control) milling process. Channel width and length are 14.7mm and 19.05mm, respectively. Assembled device, which measures 23.05x18.7x4.5mm^3, has been covered with an acrylic plate and tested with and without the ferrofluid. Figure 3 shows the array of magnets without pure iron plate before and after and ferrofluid application. As shown in figure 3(b), dispensed droplet of ferrofluid tends to gather around the points of higher magnetic flux density. When the magnet is flipped and placed inside the housing, squeezed ferrofluid droplet forms a thin layer between the magnets and aluminum housing which enables the magnet assembly to smoothly glide on the aluminum surface.

(a) (b)

Figure 2. Proof-of-concept energy harvester after assembly: (a) bottom side, (b) top side

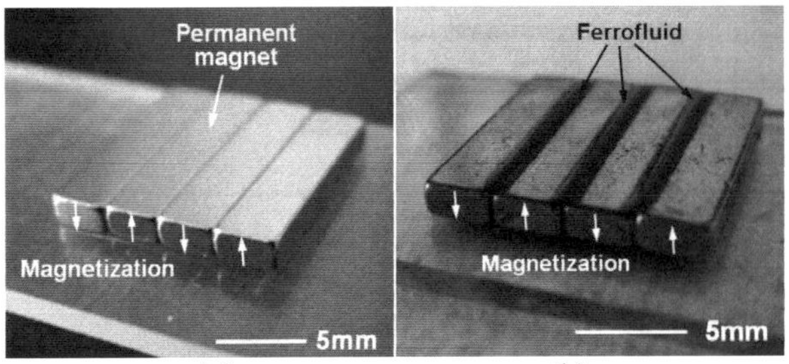

(a) (b)

Figure 3. Magnet array: (a) in pristine condition, (b) after ferrofluid dispense.

4. Results and discussion

Assembled energy harvester has been tested with a vibration exciter, which provides various input frequencies and accelerations. Figure 4 shows the experimental setup for the vibration exciter test. Sinusoidal vibration has been applied laterally to the harvester with frequency ranging from 7Hz to 20Hz with 1Hz step. Acceleration has been varied from 0.5g to 3g with 0.5g step at each frequency. Because of the limitation of the vibration exciter, frequencies under 7Hz and the accelerations above 3g could not be tested.

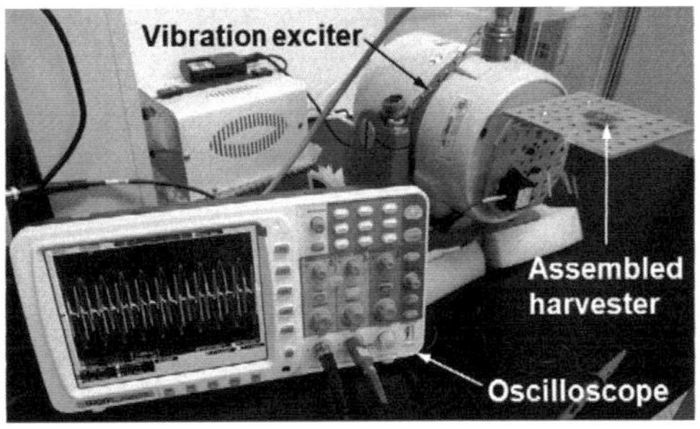

Figure 4. Experimental setup for the vibration exciter test

Open-circuit voltages of the device with and without ferrofluid has been measured and compared at various input frequencies and accelerations (figure 5). Both devices showed relative flat frequency response with clear increasing trend as input acceleration was increased. Also, output voltage decrease at higher frequency range was observed from both devices. At 3g vibration, maximum open-circuit voltages of 0.47V at 12Hz and 0.44V at 13Hz have been obtained with devices with and without ferrofluid, respectively. As shown in figure 6, application of ferrofluid increases the open-circuit voltage of the device throughout entire frequency and acceleration ranges.

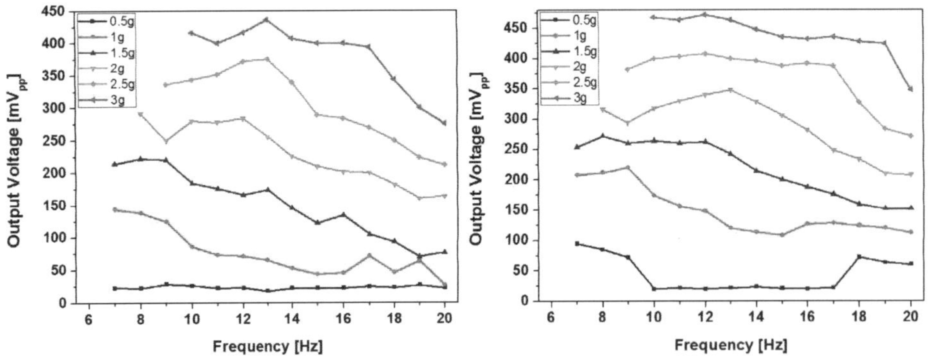

(a) (b)

Figure 5. Peak-to-peak open-circuit voltage at various input frequencies and accelerations: (a) device without ferrofluid, (b) device with ferrofluid.

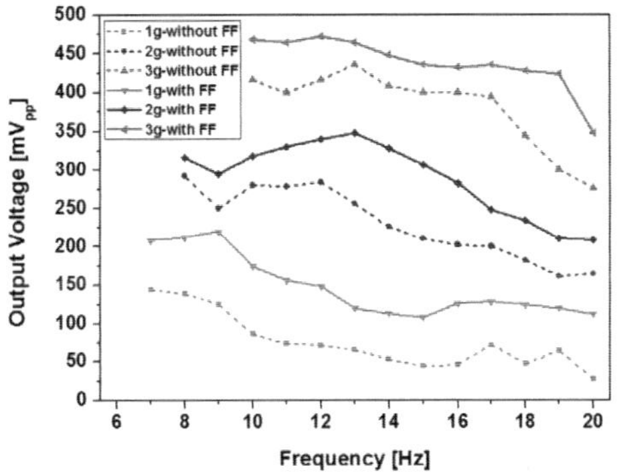

Figure 6. Comparison of the open-circuit voltage with and without ferrofluid.

Figure 7 shows the typical open-circuit voltage waveforms of the fabricated devices. Due to reduced friction between magnet assembly and underlying aluminum surface, travel range of the magnet per cycle has been increased due to ferrofluid lubrication. Moreover, velocity of the magnet has been improved which can be evidenced by higher output voltage amplitude. Increase in velocity was large enough to overcome the enlarged gap between the magnet and coil as a result of magnet flotation on ferrofluid.

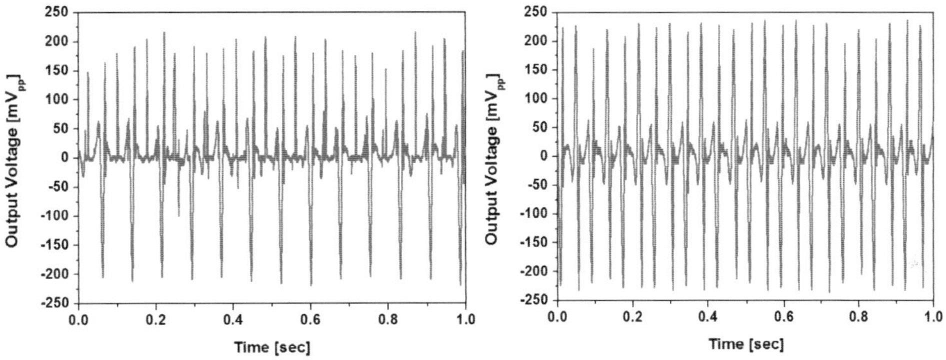

(a) (b)

Figure 7. Typical open-circuit voltage waveform: (a) device without ferrofluid (3g acceleration at 13Hz), (b) device with ferrofluid (3g acceleration at 12Hz)

Output power and root-mean-square output voltage has been obtained at load resistances ranging from 11.6Ω to 51.5Ω for both devices (figure 8). Input frequencies and accelerations for maximum open-circuit voltages have been used in the experiment. Maximum output power of 71.26μW has been obtained at 40.8Ω using the device with ferrofluid. For the device without ferrofluid, however, maximum output power of 74.10μW has been achieved at 24.6Ω. Output power of the device without ferrofluid was higher than the device with ferrofluid in 23.6-33.5Ω range, which require further investigation.

Although still a lot of room for performance improvement exists, the concept of utilizing ferrofluid as a lubricant for springless electromagnetic vibration energy harvester has been tested successfully. In addition to the improvement in output characteristics of the harvesting device, ferrofluid lubrication is expected to protect the device from wear and provide enhanced long term reliability. For further improvement of the overall device performance, geometries of the channel and magnet assembly need to be optimized to reduce the planar rotation of the magnet. A thorough quantitative analysis of dispensed ferrofluid volume on magnet flotation and modification of the contact regions to hydrophobic surfaces could further improve the output.

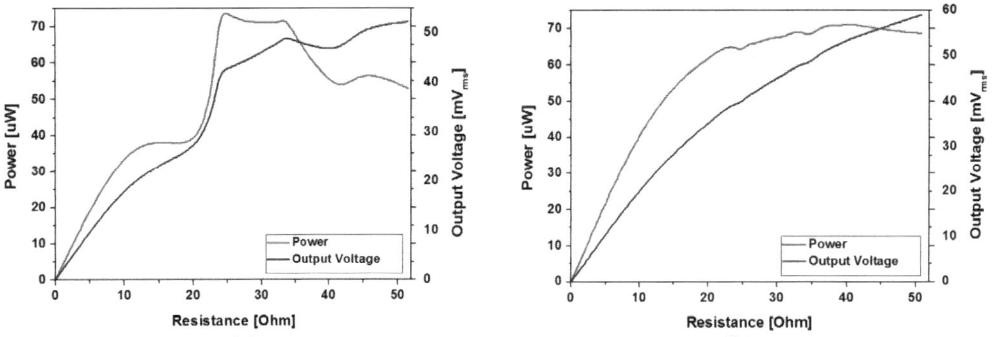

(a) (b)

Figure 8. Output power and voltage at various load resistances: (a) device without ferrofluid (13Hz at 3g), (b) device with ferrofluid (12Hz at 3g)

5. Conclusion

We have proposed an electromagnetic vibration energy harvester with springless proof mass and ferrofluid as a lubricant. A proof-of-concept energy harvester consisting of multi-pole magnets and array of copper windings has been fabricated and experimented. Effect of ferrofluid as a lubricant has been verified by comparing the harvester output with and without ferrofluid. Maximum open-circuit voltage of 0.47V has been generated with the device with ferrofluid at 12Hz at 3g, which was 8% higher than that without ferrofluid. Maximum output power 71.26μW has been obtained from the device with ferrofluid while the device without ferrofluid has generated 74.10μW of maximum power. Although output power characteristics require further investigation, we have successfully demonstrated the feasibility of utilizing ferrofluid as a lubricating material for springless electromagnetic vibration energy harvester for low frequency vibrations.

6. Acknowledgement

This work was supported by the Ewha Global Top 5 Grant 2011 of Ewha Womans University and by the Pioneer Research Center Program through the National Research Foundation of Korea funded by the Ministry of Science, ICT & Future Planning (2010-0019453).

References

[1] Herrault F *et al.* 2008 Ultraminiaturized high-speed permanent-magnet generators for milliwatt-level power generation *J. Microelectromech. Syst.* **17** 1376-1387

[2] Roundy S and Takahashi E 2012 A cost-effective planar electromagnetic energy harvesting transducer *Proc. PowerMEMS 2012* (Atlanta, 2-5 December 2012) 10-13

[3] Galchev T *et al.* 2012 A new multi-dimensional low-frequency springless proximity inertial harvester for converting human and environmental motion *Proc. PowerMEMS 2012* (Atlanta, 2-5 December 2012) 117-120

[4] Pillatsch P *et al.* 2012 A scalable piezoelectric impulse-excited energy harvester for human body excitation *Smart Mater. Struct.* **21** 115018

[5] Bowers B J and Arnold D P 2009 Spherical, rolling magnet generators for passive energy harvesting from human motion *J. Micromech. Microeng.* **19** 094008

[6] Hergert R J *et al.* 2013 Performance of integrated retainer rings in silicon micro-turbines with thrust style micro-ball bearings *J. Micromech. Microeng.* **23** 065033

Resilient, Nanostructured, High-Current, and Low-Voltage Neutralizers for Electric Propulsion of Small Spacecraft in Low Earth Orbit

A A Fomani, A I Akinwande, and L F Velásquez-García

Microsystems Technology Laboratories, Massachusetts Institute of Technology
77 Massachusetts Ave., Cambridge, Massachusetts 02139, USA

aafomani@mit.edu

Abstract. We report propellantless neutralizers resilient to oxygen and low-vacuum environments based on arrays of Pt-coated, self-aligned, and gated Si field emitters. These devices emit currents in excess of 1 mA at bias voltages of less than 120 V, adequate for neutralizing the plume of a small spacecraft's electric propulsion system. The reported devices produce similar currents at fivefold less voltage and emitting area than state-of-the-art CNT neutralizers. Long-term (3 hours) continuous emission in a 1 μTorr oxygen partial pressure environment was demonstrated, confirming the compatibility of these neutralizers with low Earth orbit (LEO) conditions. A robust processing sequence was developed that could be employed for high-yield fabrication of large-area field emission neutralizers with active areas larger than 10 cm^2 for current emission higher than 100 mA.

1. Introduction

Development of robust, low-power, and high-current neutralizers that do not consume propellant is necessary to advance the state of the art of electric propulsion (EP) systems for small spacecraft. EP systems are particularly attractive for small spacecraft because they deliver thrust at higher specific impulses than chemical rockets, resulting in important propellant mass savings [1]. In EP systems such as field emission electric propulsion thrusters (FEEPs), ion engines, and hall thrusters, a beam of positive ions, i.e., plume, is ejected at high speed to produce thrust [1] - [4]. If the plume is not neutralized, the operation of the electric rocket will negatively charge the spacecraft over time. This built-up charge reduces the propulsion efficiency and eventually stops the thruster. Moreover, it can damage the onboard instruments through arcing and back-ion bombardment. Consequently, neutralizers, i.e. electron sources that maintain the overall charge neutrality of the spacecraft, are an essential part of many EP systems [5].

Currently, neutralizers based on hollow cathodes and hot filaments are employed in state-of-the-art EP systems [6], [7]. Hollow cathodes generate plasma for efficient electron emission at current levels of 100 mA or more with specific currents as high as 10 mA/W [7]. However, these neutralizers are not compatible with miniaturized spacecraft because they consume propellant at rates comparable to the mass flow rate of the plume [8] and they deliver mA-level currents at low specific currents (~0.1 mA/W) [7]. On the other hand, filament cathodes exploit thermionic emission where electrons are excited to vacuum by thermal energy. Hence, these cathodes do not consume propellant but their specific current is typically as low as 1 mA/W [9]. Furthermore, filament cathodes are not suitable for operation at low Earth orbit (LEO) as they quickly degrade when exposed to residual gasses [10].

Field emission neutralizers (FENs) are an attractive option for EP systems of small spacecraft because of their low power consumption, high specific current, small size, and lack of propellant consumption [11]. In field emission (figure 1(a)), electrons tunnel out of the material due to the application of a high electric field that bends the vacuum level and decreases the width of the energy barrier. Consequently, FENs are in principle more energy efficient than thermionic-based neutralizers because of the huge difference in average energy required to emit an electron. The energy required for thermionic emission depends on the thermal capacitance of the cathode which is mostly wasted. The previously reported FENs are based on Spindt-type Mo field emitter arrays (FEAs) [9], [11], carbon nanotube FEAs [12] - [14], Si FEAs with AlN or diamond-like carbon coatings [15], and BN thin-film emitters [16]. For operation in LEO, neutralizers must withstand long-term operation in environments with oxygen partial pressures of $\sim 5\times 10^{-7}$ Torr [17]. Carbon nanotube-based FENs could satisfy these requirements; however, they require biases higher than 600 V for 1 mA current emission [14]. Operation at low voltages (<100 V) has been demonstrated using Mo FEAs [9] and AlN-coated FEAs [15]; however, these devices should quickly degrade when exposed to oxygen ($\sim 7\times 10^{-7}$ Torr), similar to HfC-coated Si [18]. In this work, we report FENs with continuous electron emission of 1 mA with an extraction gate voltage less than 200 V and long-term operation in low vacuum environment with oxygen partial pressures as high as 1×10^{-6} Torr. The FENs are based on nanometer-sharp (< 10 nm radius), Pt-coated, and self-aligned gated Si FEAs (320,000 tips in 0.32 cm^2).

2. Pt-coated Si FEN

The schematic of a Pt-coated Si FEN is illustrated in figure 1(b). The device has a self-aligned configuration to maximize the extraction efficiency of electrons. Each FEN has an array of Pt-coated Si tips individually surrounded by a gate electrode with 3 μm-wide aperture. The tips are located 0.7 μm below the gate plane and a 2.5 μm-thick SiO_x/SiN_x stack layer is used as the gate dielectric. The selected device geometry ensures a field factor (β) of $\sim 10^6$ cm^{-1} for a 5-10 nm radius tip, which results in field emission at voltages as low as 60 V. A thick gate dielectric is necessary for reliable long-term operation so the field in the dielectric is a small fraction of the breakdown field (E_C ~ 800−1000 V/μm) of the dielectric. The tips are coated with Pt to improve their resistance against corrosive gasses and back-streaming ions. A metal mesh, mounted 2 mm above the FEN, is used as a transparent anode to extract electrons while retarding the positive ions approaching the tip array.

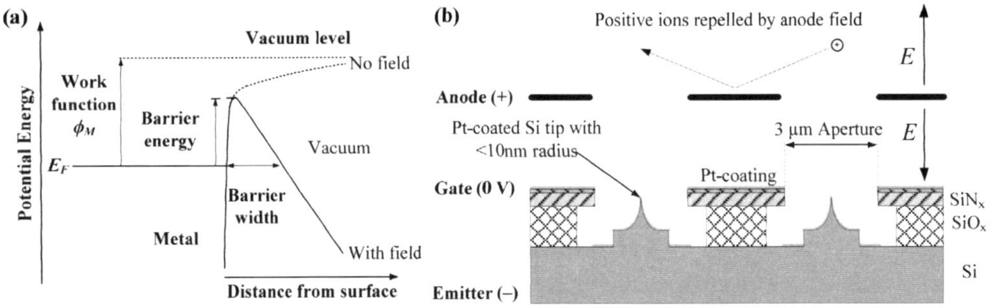

Figure 1. (a) Potential energy diagram of electrons at the vicinity of a metal surface w/wo normal electrostatic incident field. (b) Cross-section schematic of field emission neutralizer; the electric field direction is shown with a metal mesh used as anode.

We developed a high-yield process flow to fabricate large-area FEAs that is compatible with array areas larger than 10 cm^2. First, a thick gate dielectric is incorporated in the device by etching a matrix of Si pillars (3 μm-tall) buried under a 3 μm-thick SiO_x film. The sample is then planarized until the SiO_x film is removed from the top of the pillars, and a 200 nm-thick SiO_2 layer is thermally grown over the exposed Si. After this, the gate stack consisting of SiN_x (300 nm) and n-Poly Si (200 nm) is deposited and protected by a 200 nm-thick SiO_x film. Next, the gate aperture and SiO_2 disks over Si pillars are defined with a concentric mask in a single lithography step to form a self-aligned device structure. Subsequently, the tips are produced over the pillars by isotropic dry etching and oxidation sharpening using the SiO_2 disks as an etch/oxidation mask. Finally, the oxide coating the tips is stripped using HF, and a 10 nm layer of Pt is e-beam deposited over the array. Figure 2(a) shows the optical image of a 0.6 cm × 0.6 cm array with 320,000 tips and a 2 mm × 2 mm gate contact. SEM images of the gated tips with 3 µm aperture are shown in figure 2(b) and 2(c).

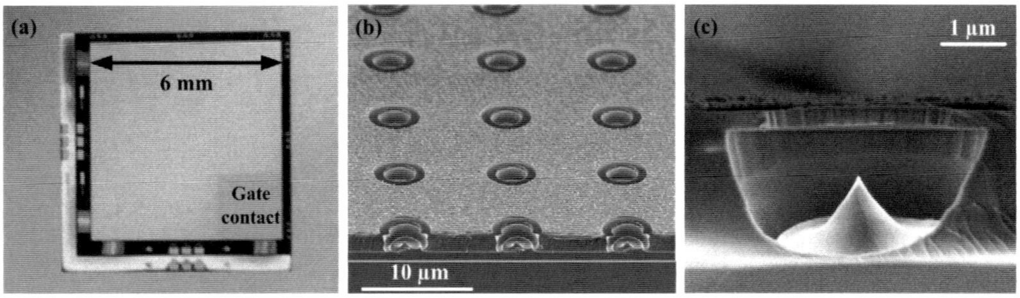

Figure 2. (a) Optical image of a field emitter neutralizer; (b) angle-view and (c) cross-section SEM images of self-aligned gated tips.

3. Neutralizer performance in high vacuum

The field emission characteristics of the device in high vacuum were measured by applying a negative voltage to the emitter while the gate and anode terminals were biased at 0 and 1100 V, respectively. The *I-V* characteristics of the device as a function of the gate-emitter voltage (V_{GE}) are shown in figure 3. At pressures below 5×10^{-8} Torr, currents as high as 1 mA are emitted at V_{GE} of less than 120 V with more than 95% transmission through the gate. As expected, the emission current exhibits a linear Fowler-Nordheim behavior at current levels above 50 µA, where the field emission current is larger than the leakage current through the gate insulator (see inset of figure 3). The extracted field factor β is $> 10^6$ cm^{-1}, in good agreement with electric field simulations.

Figure 4 illustrates the variation in gate and anode currents, as well as the gate-emitter voltage, for a device emitting 1 mA for 20 hrs. The gate current remained below 50 µA during the experiment and variation in V_{GE} was lower than 15 V. The changes in V_{GE} at high vacuum that occur sharply (see data around 11 hours) could be due to mechanical vibrations and poor gate contact. Moreover, the electrical field in the gate insulator of the device is estimated to be less than 50 V/µm, i.e., substantially lower than the critical field of the gate insulator. Consequently, a long-term operation well beyond 20 hours is expected for the device.

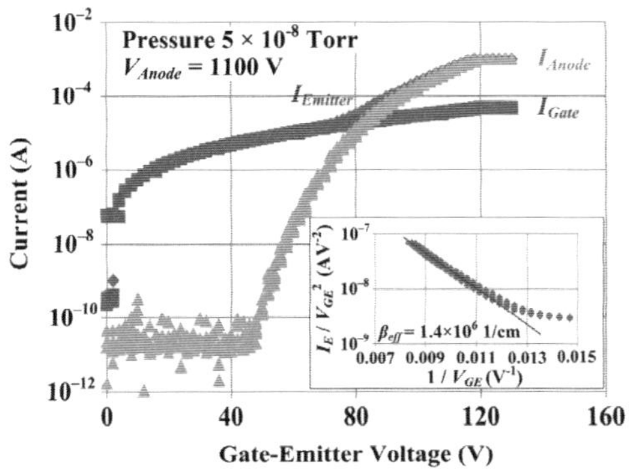

Figure 3. Transfer characteristics of FEN; inset shows linear FN behavior with field factor > 10^6.

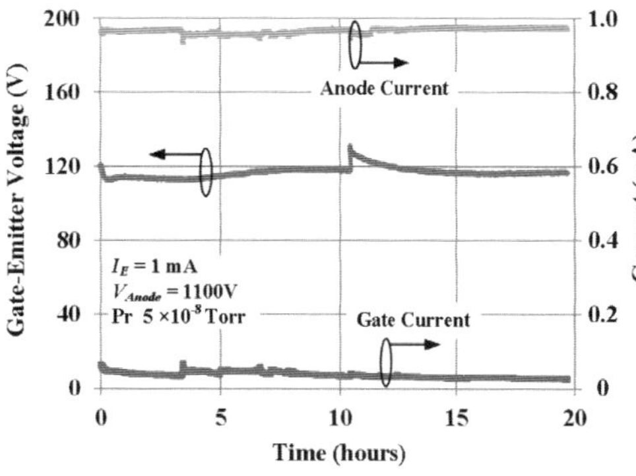

Figure 4. Long-term FEN performance: gate-emitter voltage, anode current, and gate current over time for a device emitting 1 mA.

4. Neutralizer performance in poor vacuum and oxygen residual gas

The performance of the FENs was studied in N_2 and air at pressures of 5×10^{-6} and 5×10^{-7} Torr. For these experiments, gas was admitted into the chamber through a precision leak valve to set the chamber pressure. The devices were biased at a constant current of 1 mA with anode and gate voltages at 1100 V and 0 V, respectively. After 3 hours of continuous operation, the gas valve was closed to reduce the pressure below 5×10^{-8} Torr (the pressure dropped below 5×10^{-8} Torr in less than 30 s), and the experiment continued for one more hour. Figure 5 shows the variation of V_{GE} over time for devices operated at 5×10^{-7} and 5×10^{-6} Torr in both N_2 and air atmospheres. The FENs had initial gate-emitter voltages of 110-120 V for emission currents of 1 mA at high vacuum ($<5\times10^{-8}$ Torr). Less than 10 V increase in V_{GE} was sufficient to maintain 1 mA current for 3 hours in N_2 at 5×10^{-7} Torr, and the initial device characteristics were restored by operation of the device in high vacuum. In contrast, degradation in air or at higher pressures in N_2 was faster and not fully reversible. This could be due to passivation of the tip surface by N_2 and O_2 that increases the work function of the emitting surface. This is consistent with the higher impact of O_2 on the device characteristics. Another mechanism for device degradation can be tip damage by ion bombardment, which could be mitigated by reducing the mesh voltage. Interestingly, the gate current of the devices operated in N_2 had higher fluctuations than those exposed to air as illustrated in figure 6. This could be explained by passivation of gate dielectric sidewalls by O_2 that reduces the surface leakage current.

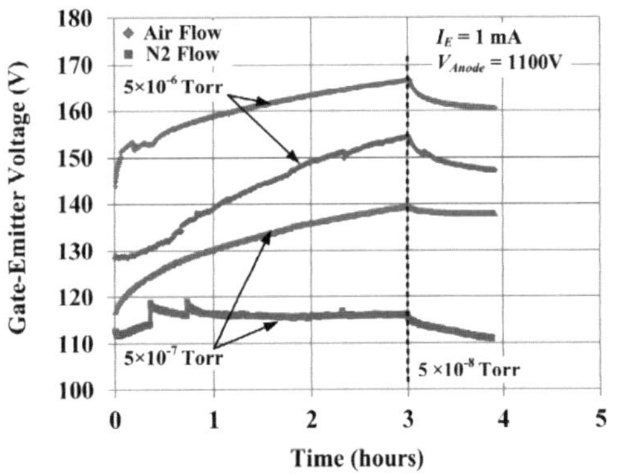

Figure 5. Field emission at 5×10^{-6} and 5×10^{-7} Torr in both N_2 and air atmospheres. Device degradation was reversible after operation in N_2 at 5×10^{-7} Torr.

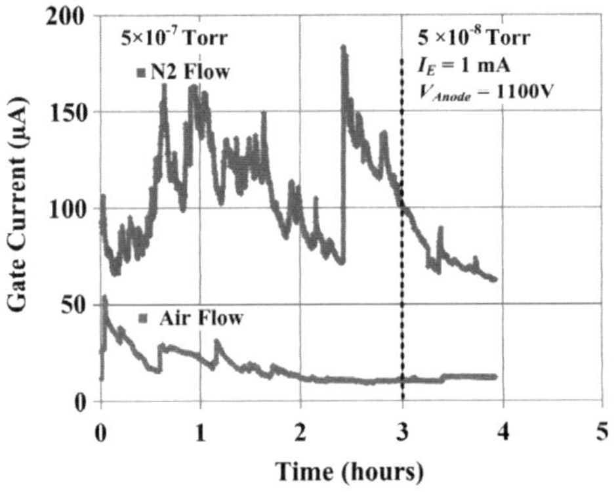

Figure 6. Variation of gate current over time during operation of FEN in N_2 and air at 5×10^{-6} Torr. The gate current was lower and more stable in air than in pure nitrogen.

5. Conclusions

Arrays of Pt-coated, self-aligned and gated tips are reported as field emission neutralizers for electric propulsion of small spacecraft in low Earth orbit. The neutralizers consist of 320,000 tips with 10 μm pitch and 5-10 nm tip radii, and have an integrated self-aligned gate electrode with 3 μm apertures. The devices emit currents in excess of 1 mA at bias voltages below 120 V, i.e., similar currents at five-fold less bias voltage and emission area than state-of-the-art CNT neutralizers. The devices have a 2.5 μm-thick gate dielectric to prevent device failure due to dielectric breakdown; the tips are coated with a 10 nm-thick Pt film to improve the tip resistance against ion bombardment and reactive gasses. Continuous emission for 3 hours at pressures of 5×10^{-6} Torr in air was demonstrated. Less than 60 V increase in the gate-emitter voltage was sufficient to maintain the emission current at 1 mA. As expected, less device degradation was observed at lower pressures and in N_2 compared to air. Nevertheless, degradation rates seem to saturate, suggesting that our FENs could achieve long-term operation at gate-emitter bias voltages below 250 V, which could be easily accommodated as the electric field intensity in the gate dielectric would remain significantly less than the breakdown field of insulator (800–1000 V/μm). Furthermore, FENs with tens of cm^2 active area and 5-10 μm-thick gate dielectrics could be fabricated using the same process flow to increase the emission current and device reliability and hence be able to satisfy the requirements of larger electric propulsion systems.

Acknowledgment
The devices were made at the Microsystems Technology Laboratories of the Massachusetts Institute of Technology. This work was funded in part by the Defense Advanced Research Projects Agency / Microsystems Technology Office, contract W31P4Q-10-1-0005 (Program Manager A. Shkel). Any opinions, findings, and conclusions or recommendations expressed in this publication are those of the authors and do not necessarily reflect the views of the US Government and therefore, no official endorsement of the US Government should be inferred.

References

[1] Sutton G. P. 2001 *Rocket propulsion elements* (New York: John Wiley & Sons)
[2] Tajmar M, Genovese A and Steiger W 2004 Indium field emission electric propulsion microthruster experimental characterization *J. Propul. Power* **20** 211
[3] Brophy J R and Noca M 1998 Electric propulsion for solar system exploration *J. Propul. Power* **14** 700
[4] Cardiff E H, Jamiesont B G, Norgaard P C and Chepko A B 2004 The NASA GSFC MEMS colloidal thruster *in Proc. AIAA/ASME/SAE/ASEE Joint Conf. and Exhibit* 3592
[5] Tajmar M 2003 *Advanced space propulsion systems* (Vienna: Springer)
[6] Sengupta A, Brophy J R and Goodfellow K D 2003 Status of the extended life test of the deep space 1 flight spare ion engine after 30,352 hours of operation *in Proc. AIAA* 4558
[7] Genovese A, Marcuccio S, Petracchi P and Andrenucci M 1996 Neutralization Tests of a mN FEEP Thruster *in Proc. AIAA* 2725
[8] Hargus W A and Reed G 2002 The air force clustered hall thruster program *in Proc. AIAA/ASME/SAE/ASEE Joint Propul. Conf. Exhibit* 3678
[9] Marrese C M and Polk J E 2000 Molybdenum field emission array cathode performance in primarisy xenon environments: Preliminary experimental results at low operating voltages *in Proc. AIAA/ASME/SAE/ASEE Joint Propul. Conf. Exhibit* 3266
[10] Gross J H 2006 *Mass spectrometry: A textbook* (Heidelberg, Germany: Springer-Verlag)
[11] Marrese C M 2000 A review of field emission cathode technologies for electric propulsion systems and instruments *in Proc. IEEE Aerospace Conf.* **4** 85
[12] Aplin K L, Kent B J, Song W and Castelli C 2009 Field emission performance of multiwalled carbon nanotubes for a low-power spacecraft neutraliser *Acta Astronautica* **64** 875
[13] Williams L T, Kumsomboone V S, Ready W J and Walker M L R 2010 Lifetime and failure mechanisms of an arrayed carbon nanotube field emission cathode *IEEE Trans Electron Dev.* **57** 3163
[14] Gasdaska C J, Falkos P, Hruby V, Robin M, Demmons N, McCormick R, Spence D and Young J 2004 Testing of carbon nanotube field emission cathodes *in Proc. AIAA/ASME/SAE/ASEE Joint Propul. Conf. Exhibit* 3427
[15] Aplin K L, Kent B J, Collingwood C M, Wang L, Stevens R, Huq S E and Malik A 2011 Use of coated silicon field emitters as neutralisers for fundamental physics space missions *Adv. Space Res.* **48** 1265
[16] Goldberg H, Encarnación P, Morris D, Gilchrist B and Clarke R 2004 Cold-cathode electron field emission of boron nitride thin film with a MEMS-based gate for space applications *in Proc. AIAA/ASME/SAE/ASEE Joint Propul. Conf. Exhibit* 3499
[17] Ceccanti F, Marcuccio S and Andrenucci M 2001 FEEP thruster survivability in the LEO atomic oxygen environment *in Proc. International Electric Propulsion Conf. IPEC* 295
[18] Nicolaescu D, Nagao M, Sato T, Filip V, Kanemaru S and Itoh J 2005 Emission statistics for Si and HfC emitter arrays after residual gas exposure *J. Vac. Sci. Technol. B* **23** 707

Development and Characterization of Piezoelectric Artificial Cochlear with micro Actuator mimicking Human Cochlear

Y Jung[1], S Kim[1], J Kwak[1], H Kang[1], Y H Lee[1], S Park[1], W Kim[1] and S Hur[1]
[1] Department of Nature-Inspired Nanoconvergence System, Korea Institute of Machinery and Materials, 156 Gajeongbuk-ro, Yuseong-gu, Daejeon, 305-343, Korea

E-mail: yjung@kimm.re.kr

Abstract. This paper presents the development of piezoelectric artificial cochlear (P-AC) capable of analyzing incoming acoustic or mechanical signals without external power source. The P-AC consists of membrane part and package part. The package part provides liquid environment through which the incoming signal is transmitted to membrane part. The membrane part responds to the transmitted signal and local area of the membrane part vibrates differently depending on its local resonant frequency. Previously in our group, we have demonstrated the feasibility of the P-AC with trapezoidal membrane part as sound analyzer by using mouth simulator as a sound input. In this research, we modified the P-AC to have the membrane part of logarithmically varying width. Also by incorporating mico-actuator into the package part that mimic the function of stapes bone in middle ear, we created similar environment to cochlear where human basilar membrane vibrates. The fabricated P-AC successfully demonstrates frequency separation of incoming mechanical signal from micro-actuator into several frequency bands within human hearing range.

1. Introduction

Human ear is a miniaturized acoustic transducer of great sensitivity (20 μPa ~ 60 Pa) and wide dynamic frequency range (20 Hz ~ 20 kHz). Scientists have tried to understand the mechanosensory system in human ear and found that the phenomenon of frequency separation of acoustic signal and its energy conversion by studying the structure of basilar membrane within the cochlear – a snail shell-shaped organ in the inner ear [1]. Basilar membrane has a structure of trapezoidal shape with varying thickness, which is thick and narrow near the base (close to oval window) and becomes thinner and wider near the apex. Thus the base area in the basilar membrane has a high resonant frequency, but the apex area has a low resonant frequency. This unique structure provides a frequency separation of incoming signal. As a basilar membrane vibrates upon incoming acoustic stimulation, the stereocilia on the hair cells near the vibrating basilar membrane become mechanically affected and generate bioelectrical signal. This bioelectrical signal is transmitted to the auditory nerves. The conversion of mechanical vibration input to electrical signal output is similar to that of piezoelectric membrane or cantilever.

Several researchers have studied to develop a sound analyzer inspired by human cochlear. Xu demonstrated a frequency filter function of acoustic sensor by using an epoxy cantilever structure [2]. Chen et al. developed beam type acoustic sensor and characterized its vibration and frequency selectivity in the water [3]. An acoustic sensor with polyimide membrane with Si_3N_4 beams was developed by White and Grosh and its vibrational behaviour was measured by using capacitive sensors

[4]. Physical cochlear model made of polyimide membrane with Al beams was developed by Puria et al. and characterized with Laser vibrometer [5]. These researches have presented promising results of artificial cochlear as a sound analyzer but energy conversion of vibration to electrical signal was not possible. Recently, Ito et al. developed a biomimetic acoustic sensor with piezoelectric membrane and presented its capability of frequency separation and energy conversion of acoustic signal input from a speaker [6].

The objective of this research is to develop biomimetic piezoelectric artificial cochlear (P-AC) having functions of frequency separation and corresponding electric signal generation and to characterize the developed acoustic sensor in a similar fashion as human basilar membrane vibrates upon the mechanical stimulation by stapes bone in the middle ear.

2. Design and Fabrication

The P-AC consists of two components; membrane part and package part (figure 1). The membrane part has piezoelectric thin film as a membrane, stainless use steel (SUS) frame defining the shape of membrane and patterned electrical pads. The schematic of the membrane part is shown in figure 2. The length in y axis of opening in the SUS frame is 28 mm and the width in x axis varies logarithmically from 0.97 mm (y = 0 mm) to 8.0 mm (y = 28 mm). The thickness of piezoelectric polymer film is 25.4 μm.

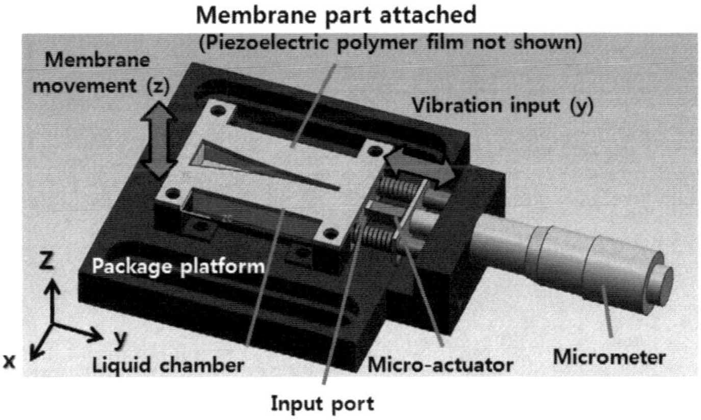

Figure 1. 3D drawing of package part of P-AC

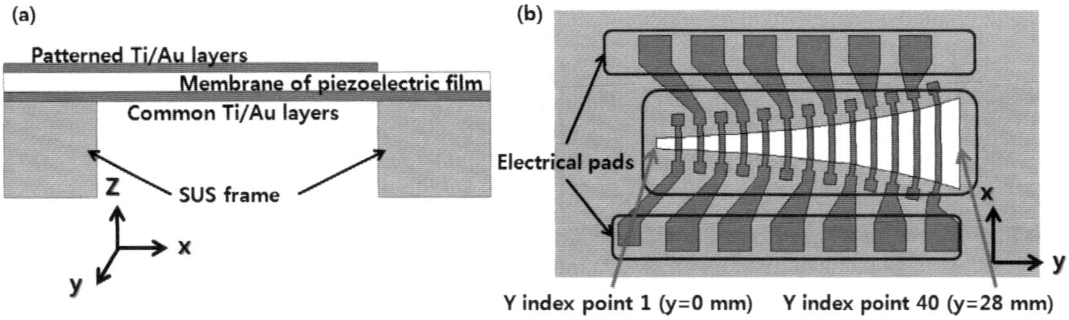

Figure 2. 2D schematic of membrane part design of P-AC (a) from side view and (b) from top view.

A common electrical pad is on the bottom side of piezoelectric polymer film and 13 patterned electrical pads are designed on the top side. The package part consists of liquid chamber, platform and micro actuator. The membrane part is attached to liquid chamber and liquid chamber is fixed on package platform. A micrometer assembled with package platform can adjust the position of micro actuator such that the input port of liquid chamber is in precise contact with the tip of micro actuator (figure 1).

The fabrication process involves microfabrication process to make a patterned electrical pads and assembly process of membrane part and package part. The piezoelectric polymer film used was polyvinylidene difluoride (PVDF) film of thickness 25.4 μm (Kynar® Film, Professional Plastics, Singapore). 20 nm / 200 nm thick titanium (Ti) / gold (Au) layers were sputtered with a shadow mask on the top side of PVDF film. The shadow mask has openings for patterned electrical pads. Additionally Ti/Au layers were sputtered on the bottom side without shadow mask for common electrical pad for harvesting piezoelectric signal. The SUS frame was prepared with precision machining with a line saw process and attached with processed PVDF film. For improving piezoelectricity of PVDF film, corona poling process was introduced before bonding process. The package platform and liquid chamber was prepared with conventional machining process. The micro actuator used was PICMA® Stack Multilayer Piezoactuaotr (P-883.11, PI-Korea, Korea) and assembled on the platform with micrometer. Finally, the membrane part was assembled over the liquid chamber and fixed together with the platform. The Fabricated P-AC is shown in figure 3.

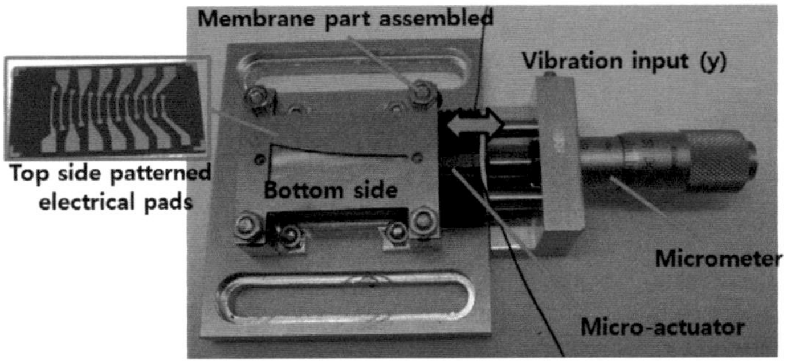

Figure 3. Photograph of fabricated and assembled P-AC

3. Experimental Results

The fabricated P-AC was characterized mechanically by measuring its vibrational property upon external mechanical stimulation from micro actuator. The micro actuator was operated by electrical signal from function generator which can control the magnitude and frequency of mechanical stimulation. The membrane vibration or displacement in z axis was measured with scanning laser Doppler vibrometer system (PSV-I-400 LR and OFV-505, Polytec, Germany) (figure 4). Artificial body fluid was introduced into the liquid chamber until the fluid becomes fully contacted with the piezoelectric polymer film. Along the central part of membrane in y axis, the vibration data were collected on 40 different points (Y index point 1 (y = 0 mm) ~ Y index point 40 (y = 28 mm)). The frequencies of applied signals were 100 Hz ~ 4,000 Hz in 100 Hz step. The upper limit of experimental frequency was decided based on the preliminary test results of frequency response of micro actuator which has a 3-dB cut-off frequency near 4,000 Hz.

The measured vibrational data of the P-AC membrane showed the clear frequency separation into four different frequency bands. First frequency band was 100 Hz ~ 400 Hz and has a displacement peak near Y index point 35. Second frequency band 700 Hz ~ 1400 Hz (z-displacement peak near Y index point 22). Third and fourth frequency bands were 2200 Hz ~ 2700 Hz and 3200 Hz ~ 3700 Hz (figure 5). The displacement peak was above 1 μm in all frequency bands. The vibration envelope at the frequencies between two neighbouring frequency bands showed mixed shape of multiple peaks. The piezoelectric signal output from membrane vibration was measured as a preliminary in air environment with sound pressure input of 94 dBSPL (= 1.0 Pa) and the maximum pieozoelectric signal was 6.3 mV_{PP} at 1,200 Hz while the displacement is below 0.1 μm.

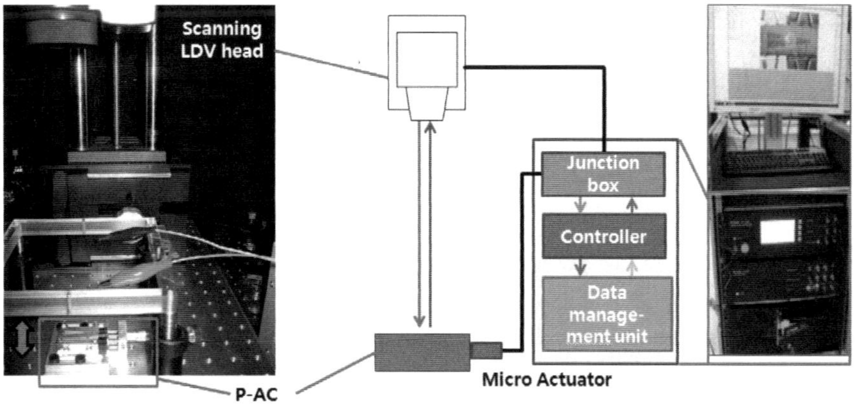

Figure 4. Schematic and photograph of experimental setup for measurement of vibration properties of P-AC

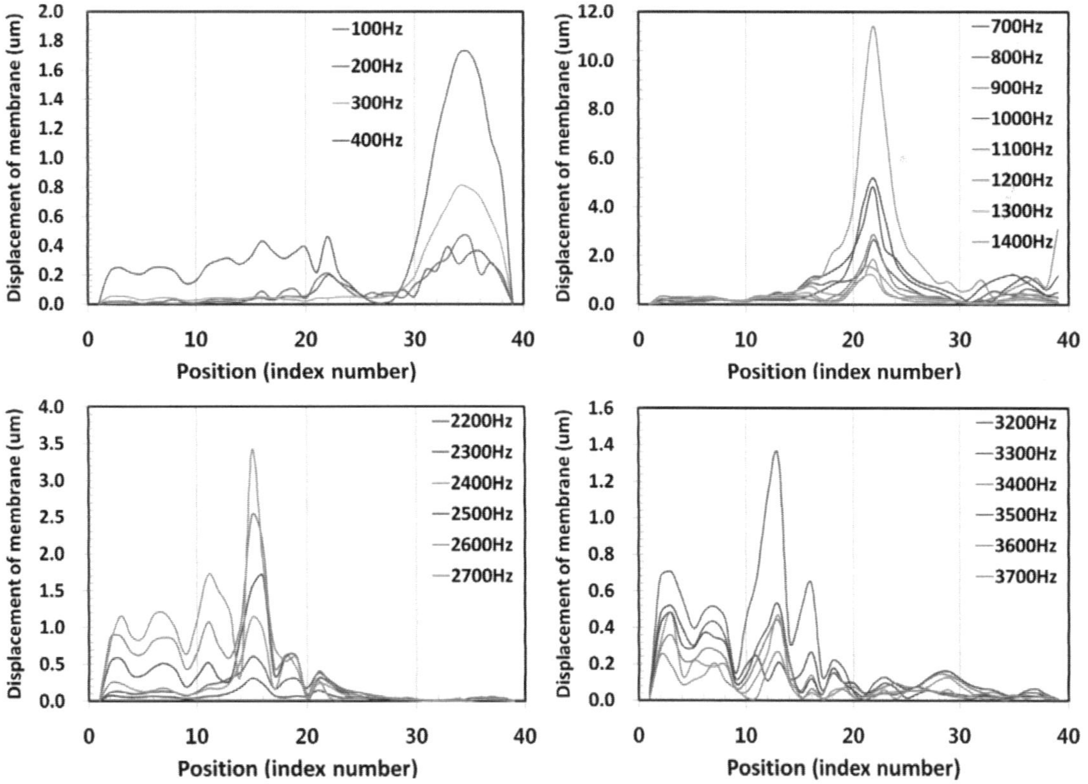

Figure 5. Vibration envelope of P-AC upon mechanical input from artificial stapes (micro actuator) within four different frequency bands of 100 Hz ~ 400 Hz, 700 ~ 1,400 Hz, 2,200 Hz ~ 2,700 Hz and 3,200 Hz ~ 3,700 Hz.

4. Discussion and Conclusion

In this study, we presented the fabrication and characterization of a piezoelectric artificial cochlear made of piezoelectric polymer film as a membrane. The piezoelectric signal output from each electrical pad was measured successfully in preliminary test condition, but measurement of piezoelectric signal output in liquid condition turn out to require further package design and experimental setup change. Additional tests will be followed. The characterization results of vibration show clear frequency separation of incoming signal from micro actuator – artificial stapes into four different frequency bands within 100 Hz ~ 4,000 Hz range. With further size reduction and enhanced frequency separation with piezoelectric signal output, the P-AC is expected to play an important role in the development of a total implantable cochlear system that operates with much less power and maintenance compared to conventional cochlear implant.

Acknowledgement

This research was supported by the Pioneer Research Center Program through the National Research Foundation of Korea funded by the Ministry of Science, ICT & Future Planning (2009-0082960, 2010-0018347).

References

[1] von Bekesy G 1974 *Annu Rev Physiol.* **36** 1-16
[2] Xu T, Bachman M, Zeng F and Li G 2004 *Sens Actuators A Phys.* **114(2-3)** 176-82
[3] Chen F, Cohen HI, Bifano TG, Castle J, Fortin J, Kapusta C, Mountain DC, Zosuls A and Hubbard AE 2006 *J. Acoust Soc Am.* **119(1)** 394-405
[4] White RD and Grosh K 2005 *Proc Natl Acad Sci USA* **102(5)** 1296-301
[5] Wittbrodt MJ, Steele CR and Puria S 2005 *Audiology and Neurotology* **11(2)** 104-12
[6] Shintaku H, Nakagawa T, Kitagawa D, Tanujava H, Kawano S and Ito J 2010 *Sens Actuators A Phys.* **158(2)** 183-92

A Compact X-ray Generator Using a Nanostructured Field Emission Cathode and a Microstructured Transmission Anode

S Cheng, F A Hill, E V Heubel, and L F Velásquez-García

Microsystems Technology Laboratories, *Massachusetts Institute of Technology*
77 Massachusetts Ave., Cambridge, Massachusetts 02139, USA

shcheng@mit.edu

Abstract. We report the design, fabrication, and preliminary characterization of a compact X-ray generator for improved X-ray absorption imaging that uses a nanostructured field emission cathode (FEC) as the electron source and a microstructured transmission anode as the X-ray generating element. FECs consume less power, respond faster, and tolerate lower vacuum than thermionic cathodes used in conventional X-ray generators. The use of a transmission anode, instead of a conventional reflection anode, allows filtering of the background radiation (bremsstrahlung) while allowing efficient generation of X-rays at lower voltages by exciting atomic shell transitions, resulting in emission of X-rays with narrow spectral linewidth for sharper imaging of biological tissue. The fabricated FEC contains arrays of self-aligned, gated field emitters that turn on at bias voltages under 30 V and transmit 99.5% of the electrons to the anode. The FEC emits a maximum current of 1.2 μA per field emitter (588 μA total array current) at a bias voltage of 85 V. A facility is built and tested to generate X-rays with an FEC and a transmission anode. Using the facility, we obtained an X-ray absorption image of an ex-vivo sample that clearly shows soft tissue and fine bone structures.

1. Introduction

A conventional X-ray generator consists of a thermionic cathode and a reflection anode inside of a vacuum chamber that has an X-ray transmission window. The cathode generates a beam of electrons that is accelerated towards the anode, which is biased at tens of kilovolts above the cathode voltage. Some of the electrons collide with the anode and convert their kinetic energy into radiation, a fraction of which escapes the vacuum chamber through a transmission window made of a suitable material, such as beryllium. The X-ray emission is a mix of bremsstrahlung radiation (broad, continuous spectrum) and fluorescence (emission at specific peaks corresponding to atomic shell transitions). Conventional X-ray technology requires high vacuum to operate, does not efficiently produce X-rays, and has overall low power efficiency [1].

In a thermionic cathode, electrons are "boiled" off the surface of a hot filament (heated to a temperature greater than 1100 K) when the internal energy of the electrons is sufficient to overcome the potential barrier holding the electrons within the material. Thermionic cathodes have a number of issues including low power efficiency due to the filament heating, low brightness, and quick degradation of the filament when exposed to residual gasses. Field emission cathodes (FECs) are an attractive alternative to thermionic cathodes in X-ray systems because FECs consume less power, respond faster, and tolerate lower vacuum than thermionic cathodes [2]. FECs use high surface electric fields (> 3×10^7 V/cm) to decrease and narrow the potential barrier that traps electrons within the emitting material, causing electrons to tunnel to vacuum. FECs use sharp emitters to generate high electric fields on the surface of the emitter tips. To generate high electric fields at low voltage, a gate electrode with an aperture is built in proximity to a nanometer-sharp tip, and a bias voltage is applied between the gate and the emitter so that the gate is positive with respect to the emitter. Arrays of field emitters can be implemented to increase the total emission current, emission uniformity, and reliability of the cathode.

Carbon nanotube (CNT)-based FECs have been recently used in X-ray generators [3-5]. The cathodes are composed of a substrate with a CNT forest and an electrode gate bonded to the substrate. It is relatively simple to fabricate CNT forests; however, the CNTs are tightly packed, the location of the CNTs is random, the height and diameter of the CNTs vary with a wide distribution, and a cluster of CNTs share the same gate aperture, resulting in high operating voltages (> 1000 V) and non-uniform current emission across the CNT forest. Metal tips can also be used in FECs. Cathodes composed of arrays of batch-microfabricated gated metal emitters are possible thanks to the Spindt technology [6], where metal tips are formed by evaporating metal at an angle onto a rotating substrate with pre-fabricated gate apertures. A cone structure can be made because of the shadowing effect of the aperture. Low-voltage operation (< 100 V) is possible because of the small gate-to-tip distance and the small tip diameter; in addition, the Spindt technology produces self-aligned gate apertures so that the electron sources have low emittance. Metal field emitters have lifetime issues as they quickly degrade when exposed to even low partial pressures of oxygen (~5×10^{-7} Torr) [7].

This paper reports efficient generation of X-rays using a batch-microfabricated field emission cathode composed of arrays of self-aligned, gated, and nanometer-sharp n-silicon tips, and a microstructured transmission anode. The nanostructured silicon FEC operates at low voltage and reliably achieves high-current emission with high transmission. The transmission anode efficiently generates X-rays while reducing the background radiation, resulting in emission of X-rays with narrow spectral linewidth for sharp imaging of biological tissue.

2. Device design and fabrication

2.1. Cathode design and fabrication

Figure 1(a) shows a scanning electron microscope (SEM) image of the cross section of a gated field emitter. The emitter consists of an n-silicon tip 2.5 µm tall with a 5 nm tip radius, a 1.5 µm thick gate oxide layer, and a 500 nm thick n-doped polysilicon gate with a 0.6 µm diameter aperture. The separation between the gate aperture and the tip is about 0.5 µm. Figure 1(b) shows an optical microscope image of the whole FEC chip. The chip consists of a 10×10 array of electron guns with square packing and a pitch of 300 µm. Each electron gun contains a 7×7 array of field emitters arranged in square packing with a 10 µm pitch. Each column of 10 electron guns can be independently turned on, such that activation of a column results in the operation of 490 field emitters at once.

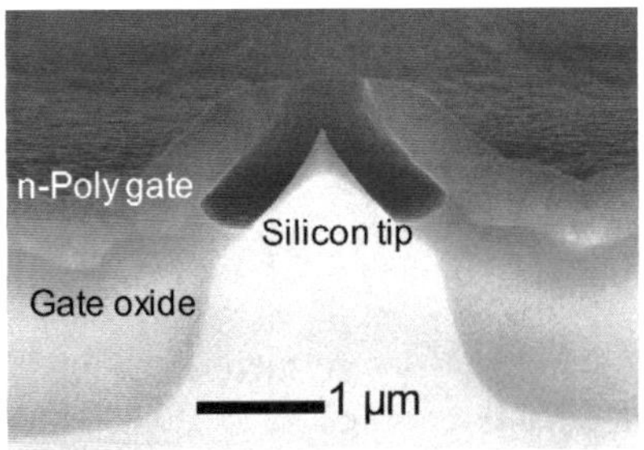

(a)

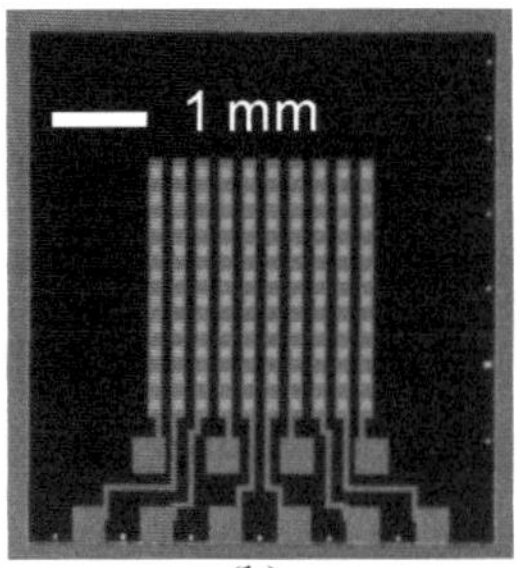

(b)

Figure 1. (a) Cross section of a field emitter, (b) picture of a full FEC chip.

Figure 2 illustrates the process flow to fabricate the FEC chip. The process starts with a 6" n-type silicon wafer (100 Ω-cm resistivity) with a film stack consisting of a 100 nm-thick thermal oxide film, a 100 nm-thick silicon-rich silicon nitride film and a 500 nm-thick chemical vapour deposition (CVD) oxide film. The film stack is dry etched to form 2.2 μm diameter discs (Figure 2(a)). Each disc functions as the etch mask for one emitter tip. The silicon is etched using an isotropic dry etch to undercut the discs and form tips with 200-300 nm radii, and a second anisotropic dry etch further etches the exposed silicon to increase the height of the emitters (Figure 2(b)). The wafer is oxidized to sharpen the tips, resulting in 5 nm tip radii (Figure 2(c)). The nitride and oxide films are then striped using diluted hydrofluoric acid (Figure 2(d)), completing the fabrication of the emitter tips. The next step is to add the gate. First, 4 μm of CVD oxide is deposited (Figure 2(e)). This oxide layer is planarized using chemical mechanical polishing (CMP) (Figure 2(f)). A buffered oxide etch (BOE) is used to etch back the oxide layer just until the tips are exposed (Figure 2(g)), and then the emitters are coated with 0.5 μm-thick CVD oxide layer. The different oxide thicknesses over the different areas of the emitters are chosen to achieve both high breakdown voltage as well as low turn-on voltage. Next, a 500 nm-thick n-doped polysilicon film (Figure 2(h)) is deposited, and the substrate is polished using CMP to expose the gate oxide (Figure 2(i)). The polysilicon film is patterned to form the gate arrays, contact pads and to electrically isolate emitters that are not part of the same column. Finally, a BOE dip is performed to etch the oxide immediately surrounding the emitter tips (Figure 2(j)).

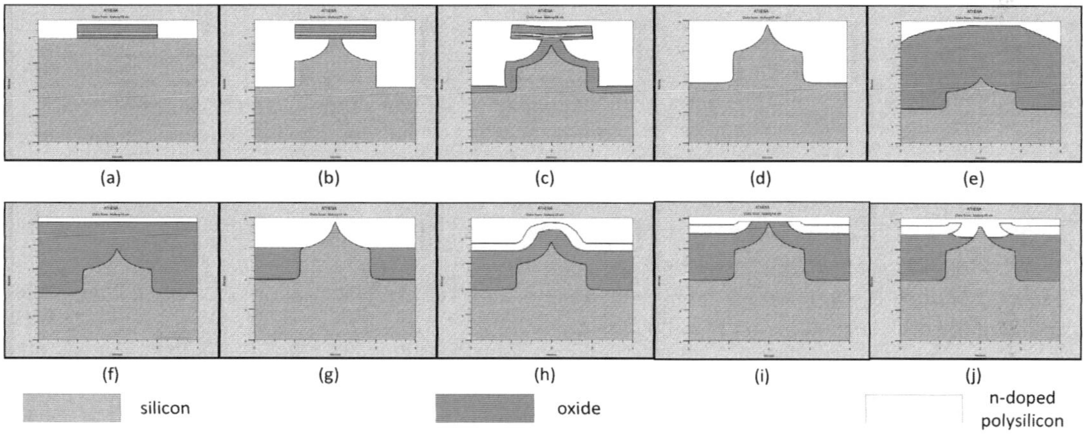

Figure 2. (a)-(j) Silvaco simulation results of the fabrication process flow of the gated field emitters.

2.2. Anode design and fabrication

The anode is a 1" diameter 300 μm-thick beryllium wafer that has a 2.5 μm-thick evaporated gold film on one side. In a transmission anode, electrons impact an X-ray-generating film that is thick enough to absorb most of the electrons, but thin enough to avoid reabsorbing the majority of the X-rays produced (transmission anodes are designed for a particular anode voltage). A transmission anode helps reduce the background radiation and also improves the clarity of the images [1]. If the anode voltage is close to the first ionization potential of the atomic shell transitions of the X-ray generating material, the radiation spectra will be mainly composed of the atomic shell transition peak that is also the high-energy part of the spectra, resulting in emission of X-rays with narrow spectral linewidth for sharper imaging of low-Z materials.

3. Device characterization

Current-voltage (I-V) characterization of the FECs is conducted inside an ultra-high vacuum chamber at a pressure of less than 5×10^{-10} Torr. The field emitters are negatively biased with a Keithley 237 source measurement unit (SMU) by making contact to the backside of the chip. Probes are used to make contact to the polysilicon contact pads that are connected to the gate electrode of each column; the gate electrode is grounded using a picoammeter Keithley 6485 that also measures the gate leakage current. A phosphor screen is mounted 2 cm above the wafer surface, serving as a collector and an electron spot imager. The phosphor screen is positively biased at 1100 V with a second Keithley 237 SMU. During the I-V characterization, the emitter voltage is swept between 0 V and -85 V while measuring the emitter current, the collector current and the gate leakage current. A typical I-V curve (Figure 3(a)) shows that the FEC turns on at ~30 V, and that almost all the current emitted from the tips is transmitted to the collector. Figure 3(b) shows the Fowler-Nordheim (F-N) plot; from its slope, a field factor, β, equal to 8×10^5 cm^{-1}, is estimated. We estimate that the emitter tip radius is 5 nm, which is in agreement with measurements of the tip radii using SEM imaging, verifying that the current is field emitted. The linearity between the emitter and collector currents (Figure 5(c)) further indicates that the currents are field emitted, and the slope of the plot demonstrates that over 99.5% of the electrons are transmitted to the collector.

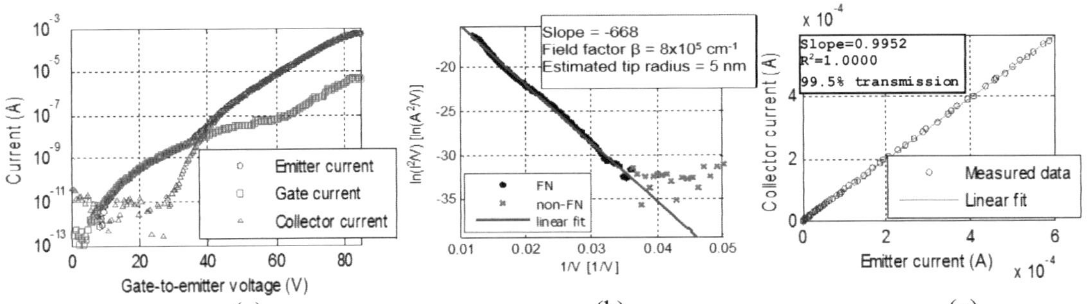

(a) (b) (c)

Figure 3. Results of field emission tests on a column of field emitters: (a) current-voltage characteristics, (b) Fowler-Nordheim plot, and (c) collector current plotted as a function of the emitted current indicating that the transmission is greater than 99.5%

The compact X-ray source is shown in Figure 4. The system consists of an 8" cubic vacuum chamber that contains the FEC and the transmission anode. The chamber is maintained at a pressure of 2×10^{-8} Torr using a turbomolecular pump and a diaphragm pump. The cathode voltage is applied using a flange with a plurality of SHV-5 feedthroughs for individual addressability of the FEC columns. The cathode holder is mounted onto the same flange, which is connected to a linear positioning stage that is used to adjust the axial position of the cathode. The chamber has a high voltage feedthrough that allows a maximum voltage of 60 kV to be applied to the anode. A beryllium window is located behind the anode holder to achieve a maximum amount of X-ray transmission. Using an FEC and a microstructured transmission anode biased at 30 kV, we have obtained a high-quality X-ray absorption image of an ex-vivo human wrist that clearly shows soft tissue and fine bone structures (Figure 5(a)). This image exhibits superior quality compared to the images from miniaturized FEC reflection anode X-ray sources previously reported [5]. The spectrum of the X-ray emission (measured using an Amptek X-123 X-ray spectrometer) (Figure 5(b)) shows a relatively small level of background radiation and emission peaks that correspond to the first two L-shell transitions of gold.

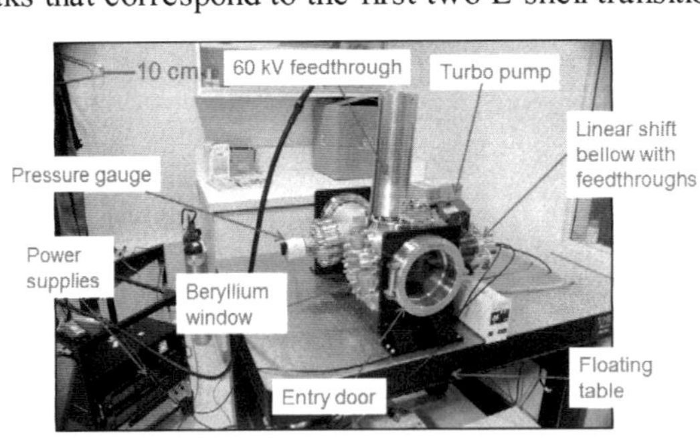

Figure 4. Picture of the x-ray source

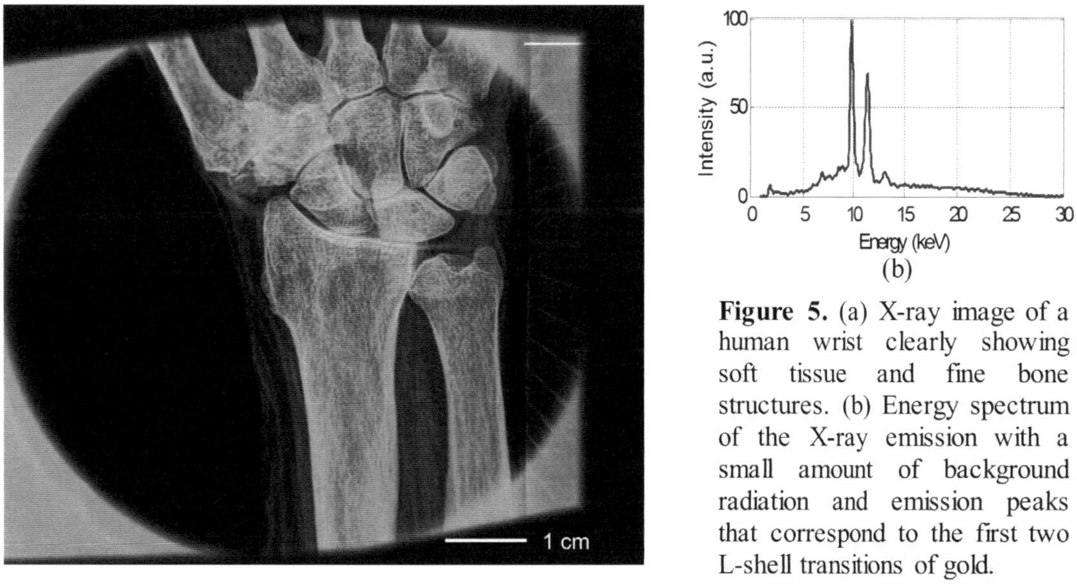

(a) (b)

Figure 5. (a) X-ray image of a human wrist clearly showing soft tissue and fine bone structures. (b) Energy spectrum of the X-ray emission with a small amount of background radiation and emission peaks that correspond to the first two L-shell transitions of gold.

4. Conclusions

A nanostructured silicon-tip field emission cathode capable of low-voltage operation and high current emission with high gate transmission has been demonstrated. A compact X-ray source has been built with the field emission cathode and a microstructured transmission anode. The X-ray source generates radiation that allows clear imaging of low-Z materials, in particular biological samples.

Acknowledgments

The devices were made at the Microsystems Technology Laboratories of the Massachusetts Institute of Technology. The authors would like to thank R Gupta from the Massachusetts General Hospital for providing the space to install the X-ray generator, supplying the ex-vivo sample, and for helping acquire the X-ray absorption image. This work was funded in part by the Defense Advanced Research Projects Agency / Microsystems Technology Office, grant N66001-11-4204 (Program Managers A I Akinwande and D Palmer, Principal Investigator R Gupta), subaward 219877 through the Massachusetts General Hospital. Any opinions, findings, and conclusions or recommendations expressed in this publication are those of the authors and do not necessarily reflect the views of the US Government and therefore, no official endorsement of the US Government should be inferred.

References

[1] Beckhoff B, Kanngießer B, Langhoff N, Wedell R, and Wolff H 2006 *Handbook of Practical X-ray Fluorescence Analysis* (Berlin: Springer)

[2] Slack C M and Ehrke L F 1941 Field emission X-ray tube *J. Appl. Phys.* **12** 165

[3] Kita S, Watanabe Y, Ogawa A, Ogura K, Sakai Y, Matsumoto Y, Isokane Y, Okuyama F, Nakazato T, and Otsuka T 2008 Field-emission-type X-ray source using carbon-nanofibers *J. Appl. Phys.* **103** 064505

[4] Wang S, Calderon X, Peng R, Schreiber E C, Zhou O, and Chang S 2011 A carbon nanotube field emission multipixel X-ray array source for microradiotherapy application Appl. Phys. Lett. **98** 213701

[5] Choi H Y, Kim J U, and Lee C J Development of carbon nanotube based reflection type X-ray source 2009 *Acta Physica Polonica A* **115** 1078

[6] Spindt C A 1968 A thin-film field-emission cathode *J. Appl. Phys.* **39** 3504

[7] Ceccanti F, Marcuccio S and Andrenucci M 2001 FEEP thruster survivability in the LEO atomic oxygen environment in *Proc. International Electric Propulsion Conf. IPEC* 295

An all-metallic microburner for a millimeter-scale thermophotovoltaic generator

Walker R Chan[1], Benjamin A Wilhite[2], Jay J Senkevich[1], Marin Soljacic[1], John Joannopoulos[1] and Ivan Celanovic[1]

[1] Institute for Soldier Nanotechnology, Massachusetts Institute of Technology, 77 Massachusetts Avenue, Cambridge, MA 02139, USA

[2] Artie McFerrin Department of Chemical Engineering, Texas A&M University, 200 University Drive, College Station, TX 77843, USA

E-mail: wrchan@mit.edu

Abstract.
Thermophotovoltaics (TPVs) is the conversion of heat to electricity via the thermal emission of photons and their subsequent absorption and conversion to electricity by infrared photovoltaic cells. One of the key challenges is designing a robust microburner with an integrated selective emitter—and having the system operate at 1000°C for thousands of hours. Previous attempts at TPV system demonstrations tend to be large, inefficient, and have limited lifetimes. Here we present a novel all metallic microburner design and experimental results in the context of a proposed small, robust, and efficient TPV generator. Fabricated entirely by machining and welding, the microburner is comprised of a serpentine channel in a 20 × 20 mm slab of Inconel with inlet and outlet capillaries that double as mechanical supports. The microburner has a thermal power input of 50 to 100 W and reaches temperatures of 700 to 1100°C. The metallic microburner is robust under high temperature operation and a 2D tantalum photonic crystal can be attached by welding for high fuel-to-electricity conversion efficiency. We characterized the microburner's exhaust composition and temperature distribution which matched well with CFD simulations. We operated the microburner at 60 W of input power (reaching about 1000°C) for 135 hours before it failed.

1. Introduction

The ongoing proliferation of power-intensive mobile devices has driven the development of increasingly compact, energy-dense power sources. Batteries represent a mature technology capable of providing energy densities of up to 1 MJ/kg, whereas hydrocarbon fuels have energy densities approaching 45 MJ/kg while promising rapid refueling. To this end, researchers have explored several possible energy conversion routes for providing next-generation portable power including mechanical heat engines, fuel cells, thermoelectrics, and thermophotovoltaics [1].

Thermophotovoltaics (TPVs) is the conversion of heat to electricity via the thermal emission of photons and their subsequent absorption and conversion to electricity by infrared photovoltaic (PV) cells. TPVs present an extremely appealing approach for small-scale power sources due to the combination of high power density limited ultimately by Planck blackbody emission, multifuel operation due to the ease of generating heat, and a fully static conversion process. Small-scale TPVs have yet to be demonstrated and are particularly challenging because of the need to develop strong synergistic interactions between chemical, thermal, optical, and

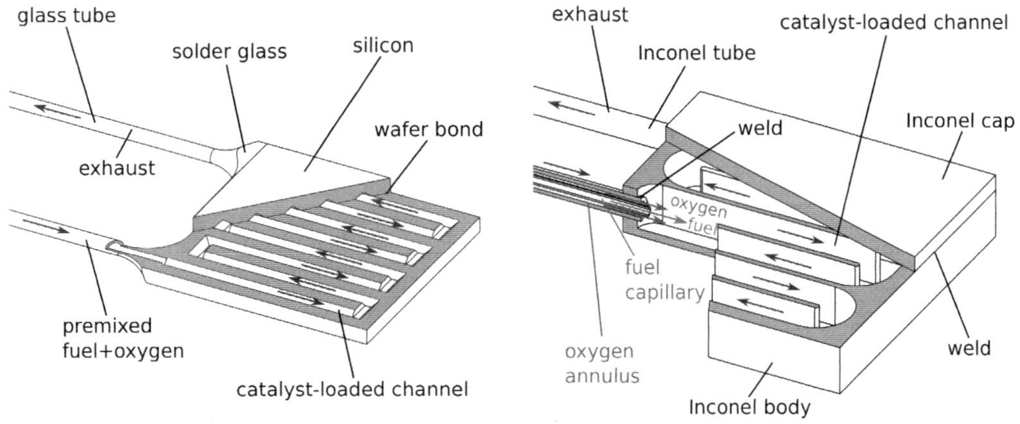

(a) Old silicon microburner. (b) Novel metallic microburner.

Figure 1: Schematics of the silicon and metal microburners.

	Silicon Microburner	Metallic Microburner
Active area	10 × 10 × 1.4 mm thick	20 × 20 × 3.2 mm thick
Channels	0.4 × 0.4 mm by KOH etching	2.3 × 2.7 mm deep by machining
Tubes	Single premixed inlet/single exhaust, 0.55 OD × 0.4 mm ID borosilicate glass tubes attached with solder glass	Single unmixed tube-in-tube inlet/dual exhaust, 1.6 OD × 1.1 mm ID Inconel tubes attached by welding.
Emitter	Five layer Si/SiO_2 stack deposited by chemical vapor deposition directly on the microburner	Bare Inconel (this work), 2D tantalum photonic crystal attached by welding (in progress)
Typical operation	<1 hour at 13 W and 800°C with catalytic combustion	135 hours at 60 W and 1000°C with homogeneous combustion
Fuel to electricity	2.7% (measured)	5% (simulated)

Table 1: Parameters of the silicon and metallic microburners.

electrical domains, which in turn give rise to requirements for extreme materials performance and subsystems synchronization [2–5].

In this work we propose an all metallic, robust, high temperature microburner which overcomes limitations associated with previous silicon-based microburner designs [5–8]. A comparison of the presently reported metallic microburner with our previously reported silicon microburner is presented in Fig. 1 and Table 1. The metallic microburner addresses the three primary shortcomings of the silicon microburner: (i) cracking due to a thermal expansion mismatch between the solder glass and silicon, (ii) the inherent difficulties of integrating a metallic photonic crystal (described in Ref. [9]) with silicon for improved fuel-to-electricity conversion efficiency and (iii) the high costs and long lead times associated with silicon micromachining.

2. Experiment

The metallic mircoburner was assembled of machined components as shown in Fig. 1b by electron beam welding and mounted in a vacuum chamber to prevent convective heat loss. A 5% platinum on porous alumina catalyst was loaded into the microburner by washcoating [5–8]. To ignite

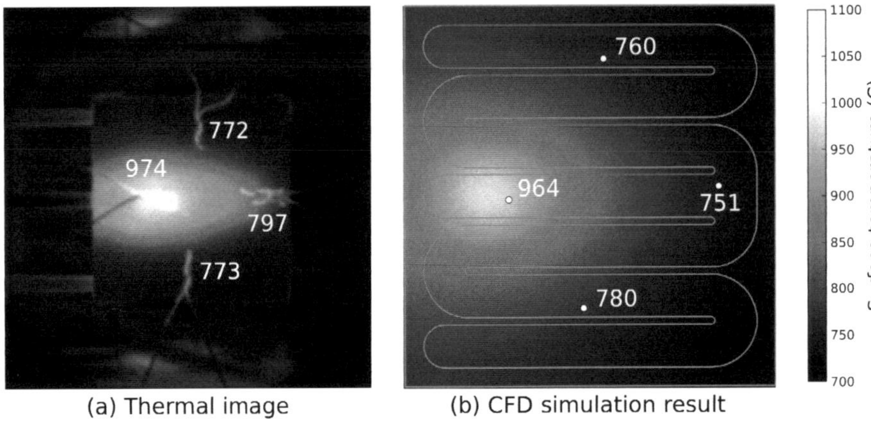

Figure 2: A thermal image (a) and CFD simulation result (b) for the microburner operating at $Q = 60$ W and $\phi = 1.5$.

the microburner, it was heated to approximately 400°C with a halogen lamp. Above that temperature, the propane kinetics over the catalyst were sufficient for autothermal operation, and the halogen lamp was shut off.

Instumentation consisted of four thermocouples, a thermal imager, and a mass spectrometer (MS). The K-type thermocouples were spot welded to the body of the microburner and are visible in Fig. 2(a). The FLIR SC660 thermal imager looked through a NaCl window at the microburner. The thermal images were calibrated from thermocouple readings. The Pfeiffer GSD 301 MS sniffed the exhaust gases after they passed through a condenser chilled to 0°C to remove water vapor. Exhaust composition was found using a least squares method to fit a combination of NIST mass spectra and our own calibrations to the raw data.

Steady state temperature versus total propane flow was measured by increasing propane flow in increments from 30 to 60 standard cubic centimeters per minute (sccm), corresponding to a total latent heat input of $Q = 60$–91 W, while maintaining an oxygen flow of 7.5 times that of propane (an equivalence ratio of $\phi = 1.5$). Surface temperatures were allowed to stabilize at each set of flow conditions, typically 5 to 10 minutes. Flowrates were then decreased to 30 sccm propane to investigate possible hysteresis. Experimental data is presented in Figure 3a.

At a constant propane feed of 40 sccm, the equivalence ratio was varied between 1 and 2 by a similar process. For all equivalence ratios $\phi \geq 1.25$, only carbon dioxide and unreacted oxygen were present in the exhaust, indicating complete combustion.

Once the above data was obtained, we conducted a longterm study summarized in Fig. 3b. The microburner was ignited at $Q = 60$ W and $\phi = 1.5$ (point a). The periodic dips in temperature and MS data (points b and c) are caused by the periodic removal and de-watering of the condenser. The occasional changes in steady state temperature (point d and e) are due to a transition between homogeneous and catalytic combustion. The MS data begins to indicate incomplete combustion around 135 hours (point f), evidenced by reduction in CO_2 signal and increased O_2 and propane signals. Work is currently underway to ascertain the cause of device failure.

3. Model

Initially, a first order analytical model was used to predict microburner temperature. Assuming complete combustion and a uniform temperature, the heat of combustion was equated to heat loss by radiation, conduction down the tubes, and heat carried out in the exhaust. The analytical model accurately captures the trend versus propane flow in Fig. 3a but obviously cannot predict

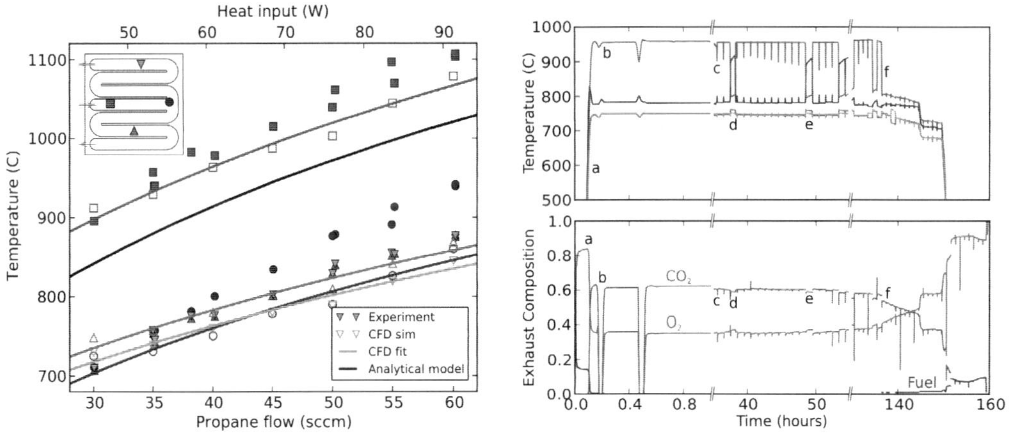

(a) Steady state temperatures.

(b) Long term study.

Figure 3: Experimental and modeling results. (a) summarizes the steady state temperature as a function of propane flow with $\phi = 1.5$. Individual CFD simulations do not fall exactly on a smooth curve because of the mesh size used. (b) is a long term study of the microburner running at $Q = 60$ W and $\phi = 1.5$.

the observed 200°C temperature distribution. A uniform temperature is important for efficient operation of the PV cells and reactor robustness.

In order to model the temperature distribution, we conducted a computational fluid dynamics (CFD) simulation of the microburner in COMSOL using the Chemical Engineering Module and a combination of weakly compressible flow, convection and conduction and Maxwell-Stefan multicomponent diffusion physics packages. The homogeneous combustion of propane was described using a single-expression kinetic model [10]. Gas-phase dynamic viscosity, thermal conductivity, heat capacity, and diffusivity were calculated according to the methods in Ref. [11–13] from literature data. Inlet flow was assumed to have a uniform velocity across the channel cross-section, have a uniform temperature of ambient, and consist of uniformly premixed propane and oxygen with no tube-in-tube segregated feed to facilitate convergence. Outlet boundary conditions assumed constant pressure of 1 bar and were open to convective heat and mass transport. A combination of no-slip, no mass flux (impermeable wall) and continuity of heat flux boundary conditions were used to describe fluid-wall interfaces. Radiative heat losses from all external surfaces of the solid-phase were described using a simple Stefan-Boltzman expression assuming an emissivity of 0.5 for all surfaces. Finite-element meshes consisted of approximately 10^5 individual elements, corresponding to 10^6 degrees of freedom. Typical solution times were an hour.

Simulation predictions of burner surface temperature at $Q = 60$ W and $\phi = 1.5$ are presented alongside thermal images obtained experimentally in Fig. 2. A summary of measured and simulated thermocouple temperatures over the span of propane flow rates studied is presented in Fig 3, alongside the analytical model. Simulations predict a more symmetric and focused hot-spot than experimentally observed; this is attributed to a combination of (i) assumption of pre-mixed fuel and oxidant leading to a pre-mixed flame shape, as opposed to a diffusion-flame shape expected with segregated feed, and (ii) known limitations of the single-expression kinetic model for accurately predicting flame location [14]. Even with these assumptions, we obtained a good overall temperature agreement.

As noted above, two distinct operating regimes were observed within the microburner, corresponding to homogeneous and catalytic combustion. In the homogeneous regime (shown in Fig. 2), a stable flame is present near the reactor inlet, as evidenced by the presence of an intense hot-spot located at the reactor inlet. In contrast, the catalytic regime is characterized by a cooler and more distributed hot-spot along the entire central channel length. The microburner would oftentimes begin in the catalytic regime but would transition the the homogeneous regime. Once in the homogeneous regime, it would not transition back to the catalytic regime without some external disturbance, such as a brief decrease in fuel flow. Given that homogeneous combustion was predominant in the present study, simulations reported herein neglected catalytic combustion. In future work, the combustion regime can be controlled using thermal and radical quenching by altering the channel geometry [15].

4. Conclusion

In this work, we presented a metallic microburner that will serve as a platform for a portable TPV system. The metallic microburner is more robust than the previous silicon one, allowing for higher temperature operation. Furthermore, the 2D tantalum photonic crystal (described in Ref. [9]) can be welded to the metallic microburner for a twofold increase fuel-to-electricity conversion efficiency as compared to previous silicon-based devices. Experimental and simulation results confirmed stable operation at flow rates corresponding to $Q = 45$–90 W with homogeneous flame corresponding to a hot-spot temperature of up to 1100°C.

Acknowledgments

This work was partially supported by the Army Research Office through the Institute for Soldier Nanotechnologies under Contract Nos. DAAD-19-02-D0002 and W911NF-07-D0004. WC and MS were partially supported by the MIT S3TEC Energy Research Frontier Center of the Department of Energy under Grant No. DE-SC0001299. BW acknowledges the support of the National Science Foundation under Award No. 1311818, administered by ENG-CBET-PRE.

References

[1] Mitsos A 2009 *Microfabricated power generation devices : design and technology* (Weinheim: Wiley-VCH) ISBN 978-3527320813
[2] Doyle E, Shukla K and Metcalfe C 2001 Development and demonstration of a 25 watt thermophotovoltaic power source for a hybrid power system Tech. Rep. TR04-2001 National Aeronautics and Space Administration
[3] Fraas L M, Avery J E and Huang H X 2003 *Semiconductor Science and Technology* **18** S247
[4] Wenming Y, Siawkiang C, Chang S, Hong X and Zhiwang L 2005 *Journal of Micromechanics and Microengineering* **15** S239
[5] Chan W R, Bermel P, Pilawa-Podgurski R C N, Marton C H, Jensen K F, Senkevich J J, Joannopoulos J D, Soljacic M and Celanovic I 2013 *Proceedings of the National Academy of Sciences* **110** 5309–5314
[6] Blackwell B S 2008 *Design, fabrication, and characterization of a micro fuel processor* Ph.D. thesis Massachusetts Institute of Technology
[7] Arana L, Schaevitz S, Franz A, Schmidt M and Jensen K 2003 *Microelectromechanical Systems, Journal of* **12** 600–612
[8] Nielsen O M, Arana L R, Baertsch C D, Jensen K F and Schmidt M A 2003 *Transducers, Solid-State Sensors, Actuators and Microsystems, 12th International Conference on, 2003* vol 1 pp 714–717
[9] Rinnerbauer V, Ndao S, Xiang Yeng Y, Senkevich J J, Jensen K F, Joannopoulos J D, Soljacic M, Celanovic I and Geil R D 2013 *Journal of Vacuum Science Technology B: Microelectronics and Nanometer Structures* **31** 011802–011802–7
[10] Westbrook C K and Dryer F L 1981 *Combustion Science and Technology* **27** 31–43
[11] Wilke C R 1950 *The Journal of Chemical Physics* **18** 517–519
[12] Bird R 2007 *Transport phenomena* (New York: J. Wiley)
[13] Fuller E N, Schettler P D and Giddings J C 1966 *Industrial & Engineering Chemistry* **58** 18–27
[14] Vlachos D, Schmidt L and Aris R 1993 *Combustion and Flame* **95** 313 – 335
[15] Miesse C M, Masel R I, Jensen C D, Shannon M A and Short M 2004 *AIChE Journal* **50** 3206–3214 ISSN 1547-5905

Innovative thermal energy harvesting for future autonomous applications

Stephane Monfray [1]
STMicroelectronics, France

E-mail: stephane.monfray@st.com

Abstract. As communicating autonomous systems market is booming, the role of energy harvesting will be a key enabler. As example, heat is one of the most abundant energy sources that can be converted into electricity in order to power circuits. Harvesting systems that use wasted heat open new ways to power autonomous sensors when the energy consumption is low, or to create systems of power generators when the conversion efficiency is high. The combination of different technologies (low power μ-processors, μ-batteries, radio, sensors…) with new energy harvesters compatible with large varieties of use-cases with allow to address this booming market. Thanks to the conjunction of ultra-low power electronic development, 3D technologies & Systems in Package approaches, the integration of autonomous sensors and electronics with ambient energy harvesting will be achievable. The applications are very wide, from environment and industrial sensors to medical portable applications, and the Internet of things may also represent in the future a several billions units market.

1. Introduction

In this presentation, we will address an innovative way of harvesting heat, and present the fabrication of thin modules that work without a heat sink at temperatures close to ambient. The key point of the integration of this technology is the ability to keep a large thermal gradient on a device by intelligent control over the thermal flow. To provide an innovative system of conversion of heat into electricity, the concept is based on two key principles: 1-the conversion of a continuous stream of heat into mechanical impulses, 2-the conversion of these mechanical pulses into stored electrical energy. The generated electric power density can be increased by miniaturizing each device, but an optimal working point should be found in order to have high power due to scaling and enough signal so as to overcome the threshold of the harvesting circuit. The concept is unique in the sense that it is based on a matrix structure of micro or mini energy nodes which will work together and which do not require the use of radiators, thanks to the controlled thermal resistance. This opens the door to new properties and features of the object, with better performance for low power autonomous applications.

[1] To whom any correspondence should be addressed.

Figure 1. Innovative product features planned: Flexibility, low cost, large area.

2. About The Speaker

Dr Stephane Monfray is the Principal Scientist in the Advanced Devices Technology group at STMicroelectronics in Crolles. Since 2008, he is a member of the STMicroelectronics experts staff. He was involved in FP6 and FP7 projects (Nanocmos, Pullnano, Duallogic) and leads since 2012 a national project (FUI) on energy harvesting. He is the author and co-author of more than 70 publications in major conferences and journals, of more than 30 patents, of a book chapter. He was instructor at the SOI conference shortcourse in 2010, and at MIGAS in 2011, and he had multiple participations and paper presentations at the IEDM in 2001, 2002, 2004, 2007 and 2010. He was the co-recipient of the Paul Rappaport Award in 2000, and he is the co-recipient of the 2012 French Electronic Grand Prize "General Ferrié" for his work on thin film devices.

Journal of Physics: Conference Series **476** (2013) 012019 doi:10.1088/1742-6596/476/1/012019

Micro Thermal Diode with Glass Thermal Insulation Structure Embedded in Vapor Chamber

T Hirayanagi[1], T Tsukamoto[2], M Esashi[3] and S Tanaka[1]
[1] Department of Bioengineering and Robotics, Tohoku University, Sendai, Japan
[2] Micro System Integration Center, Tohoku University, Sendai, Japan
[3] Advanced Institute for Material Research, Tohoku University, Sendai, Japan

E-mail: hirayanagi@mems.mech.tohoku.ac.jp

Abstract. This paper reports a novel micro thermal diode with an embedded micro glass thermal insulation structure. The diodicity is given by wettability-driven one-way fluid circulation mechanism without any external force. The water circulation only occurs in "forward" mode, resulting in low thermal resistance. The glass thermal insulation structure enhances thermal insulation in "reverse" mode. The thermal insulation in forward and reverse mode, R_f and R_r, were measured as 2.06±0.34 K/W and 3.04±0.38 K/W, respectively. Therefore, the performance index, R_r/R_f, was 1.47±0.14.

1. Introduction

A thermal diode is one of important components of a micro thermal device such as a micro refrigerator and a thermal energy harvester [1]. A thermosyphon, in which one-way working fluid circulation is driven by gravity force, can be used as a thermal diode. However, the heat transfer property strongly depends on the gravity direction. In addition, the gravity force becomes quite small compared with surface tension in a microscale region, which limits the miniaturization of the thermosyphon. Therefore, surface-tension-driven fluid circulation mechanism is more suitable in a micro thermal diode. In this paper, we propose a novel micro thermal diode, in which the one-way working fluid circulation is driven by wettability difference between condensation and evaporation parts and thermal insulation in the reverse direction is realized by an embedded thick glass microstructure.

2. Structure and principle of the micro thermal diode

Figure 1 shows the structure and working mechanism of the micro thermal diode. The device consists of two plates: an outlet-side plate with a superhydrophobic surface and an inlet-side plate with microchannels with a hydrophilic surface. The microchannels are filled with working fluid. A glass microstructure, which is embedded in the inlet-side channel plate, forms a thermal insulation chamber wall.

In forward mode, the inlet-side plate with the microchannels is heated, while the outlet-side plate with the superhydrophobic surface is cooled (Fig. 1 (b)). The working fluid in the microchannels evaporates and condenses on the superhydrophobic surface, which makes small droplets on the surface. The droplet glows by coalescing, and then returns to the microchannel when it touches to the hydrophilic microchannel wall. Thus, the evaporation-condensation cycle continues like a heat pipe. The evaporation rate from microchannels is higher than that of pool boiling [2]. In addition, the heat transfer coefficient of drop-wise condensation is much higher than that of film-wise condensation [3].

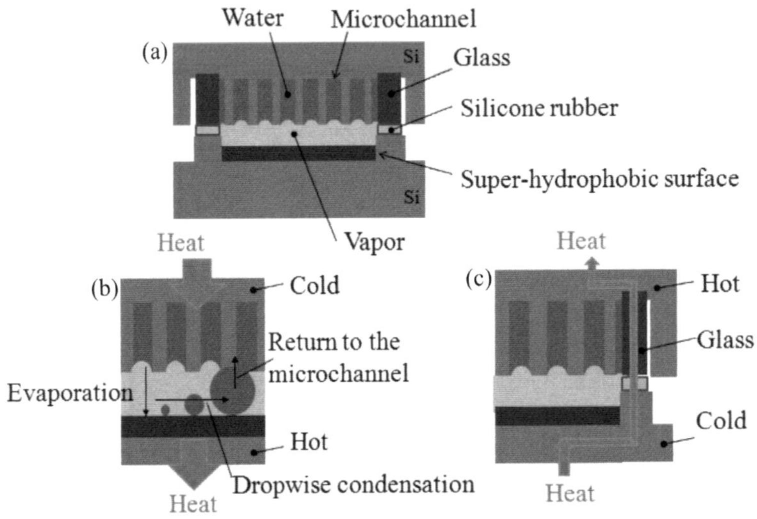

Fig. 1 Structure and working mechanism of the micro thermal diode. (a) Cross sectional schematic of the device. (b) Water circulation mechanism in forward mode and (c) heat flow path in reverse mode.

From these mechanisms, the thermal resistance in forward mode is quite low.

On the contrary, the outlet-side plate with the superhydrophobic surface is heated in reverse mode (Fig. 1 (c)). The working fluid on the superhydrophobic surface evaporates and condensed on the microchannel wall. However, the condensed liquid remains in the microchannels because of their hydrophilicity, and does not return to the heating area [4]. Therefore, the heat mainly flows thorough the chamber wall, which is made of thick glass for high thermal insulation.

3. Device fabrication

Figure 2 shows the fabrication process. First, micromolds for the glass thermal insulation structure were formed on a Si substrate (Fig. 2 (a.1)). A borosilicate glass plate was anodically bonded to it under vacuum condition at an applied voltage of 600 V and a substrate temperature of 400°C (Fig 2 (a.2)). The glass was reflowed at 900°C under atmospheric pressure to fill the vacuum-sealed molds (Fig. 2 (a.3)). The unnecessary glass overlapping the surface was removed by grinding and polishing (Fig. 2 (a.4)), and the microchannels were formed by deep reactive ion etching (Fig. 2 (a.5)). Finally, the surface was covered with SiO_2 to increase the wettability (Fig. 2 (a.6)). Figure 3 shows the fabricated microchannels and the embedded glass thermal insulation structure on the inlet-side plate.

For the outlet-side plate, another Si substrate was etched to form mesa structures, and a thin film of Au/Cr was deposited (Fig. 2 (b.2)). Then, a Ni/PTFE film was electroplated using a solution consisting of 1.2 M of $Ni(NH_2SO_3)$, 0.5 M of H_3BO_3, and 40 g/L of PTFE particles with an average diameter of 220 nm [5]. A pulsed electroplating method was used, where "on" time was fixed at 10 ms and "off" time was varied from 2 ms to 10 ms. The temperature of the solution was kept at 55±1°C, and the current density in "on" period was set at 0.2 mA/dm^2. The film was annealed at 300°C in a vacuum furnace to form a network of PTFE, and Ni was etched by nitric acid to make a porous PTFE surface. Figure 4 shows the superhydrophobic surface made of porous PTFE and a water droplet on it. Figure 5 shows the contact angle of the water droplet on the superhydrophobic surface. The contact angle increased with the etching time due to lotus leaf effect. When the etching time was too long, however, the contact angle again reduced, because the PTFE particles were partially removed. The maximum contact angle obtained was about 160°.

4. Experimental result

4 μL of water was sealed between the two plates with a silicone rubber gasket. Figure 6 shows the experimental setup. The test device was placed between two heat conduction blocks, one of which was

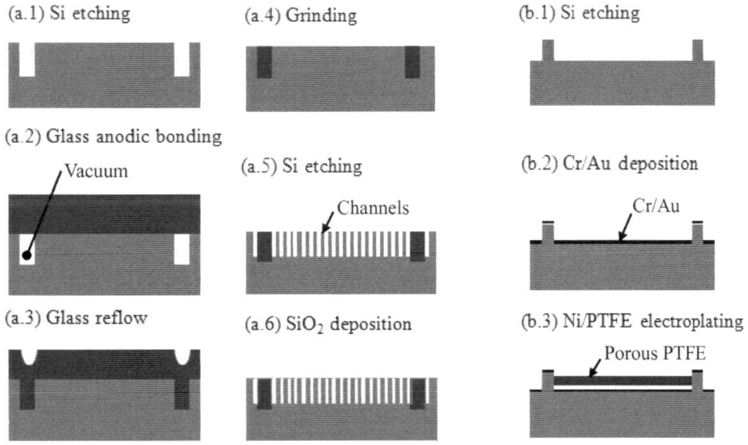

Fig. 2 Fabrication process of (a) the inlet-side plate with microchannels and (b) the outlet-side plate with a superhydrophobic surface.

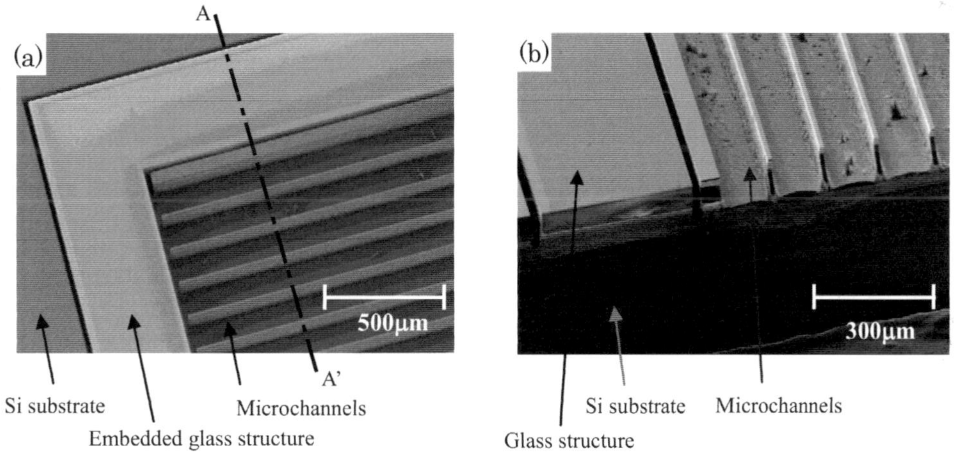

Fig. 3 SEM of the inlet-side plate with microchannels. (a)Top view and (b) A-A' cross sectional view.

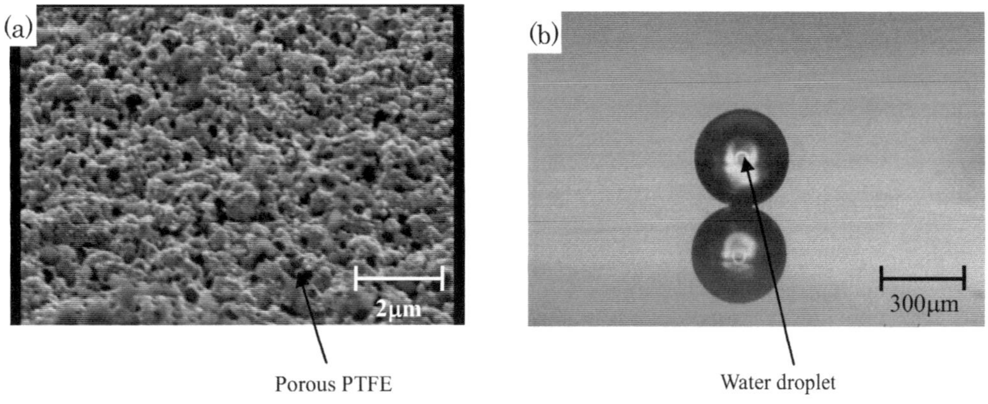

Fig. 4 Superhydrophobic surface. (a) Porous PTFE and (b) a water droplet on it.

heated by an electric heater and the other was connected to a heat sink for cooling. Heat flow through the device, Q, was measured by a heat flux sensor (FMR-200-K, Concept Engineering Gmbh, Germany), and temperature difference across the device, ΔT, was measured by thermocouples attached on both sides of the device. The thermal resistance, R_{th}, was then calculated as

$$R_{\mathrm{th}} = \Delta T / Q.$$

The performance index of the thermal diode was defined as

$$R = R_{\mathrm{r}} / R_{\mathrm{f}},$$

where R_{f} and R_{r} are the thermal resistance in forward and reverse mode, respectively.

Figure 7 shows measured thermal resistance in forward and reverse mode. When the heat flow was small, the S/N ratio of the data was low due to small temperature difference across the device. Therefore, we use the experimental results where the heat flow was larger than 1 W for the following discussion. The conductive thermal resistance of the glass wall was measured using the device without water. The measured thermal resistance was about 4.09±0.33 K/W, which is roughly consistent with a theoretically estimated value of 3.0 K/W. The reverse mode thermal resistance was measured as 3.04±0.38 K/W, and thus heat mainly flows through the glass wall in reverse mode. On the other hand, the forward mode thermal resistance was about 2.06±0.34 K/W, which was much smaller than that of reverse mode. This result suggests that water droplets returned to the microchannels against the gravity as intended. The performance index, $R_{\mathrm{r}}/R_{\mathrm{f}}$, was 1.47±0.14.

5. Conclusion

We designed and fabricated a micro thermal diode with a glass thermal insulation structure embedded in a vapor chamber. The fabrication processes of the glass micro structure and the porous PTFE superhydrophobic surface were developed. The thermal resistances in both forward and reverse mode were measured. The result confirmed the flow direction dependency of the thermal resistance, and suggested that the water circulation mechanism worked against the gravity as intended. The performance index, $R_{\mathrm{r}}/R_{\mathrm{f}}$, was as high as about 1.47±0.14.

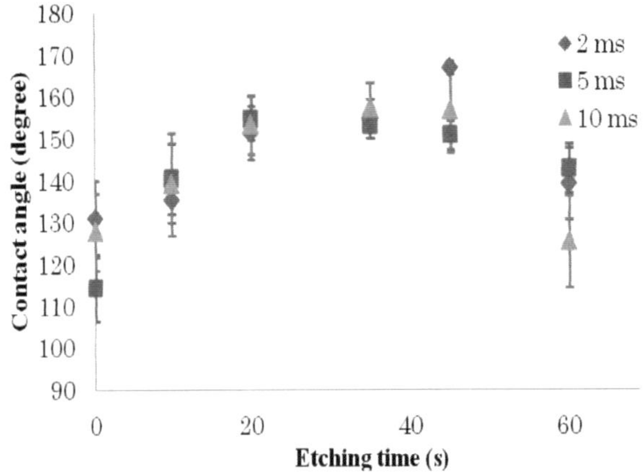

Fig. 5 Contact angle of water on the superhydrophobic surface with respect to etching time. Different conditions of "off" time in the pulsed electroplating exhibited similar results.

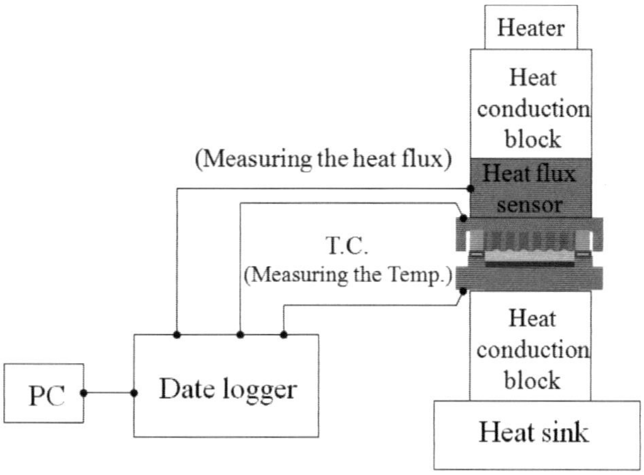

Fig. 6 Experimental setup to measure thermal resistance.

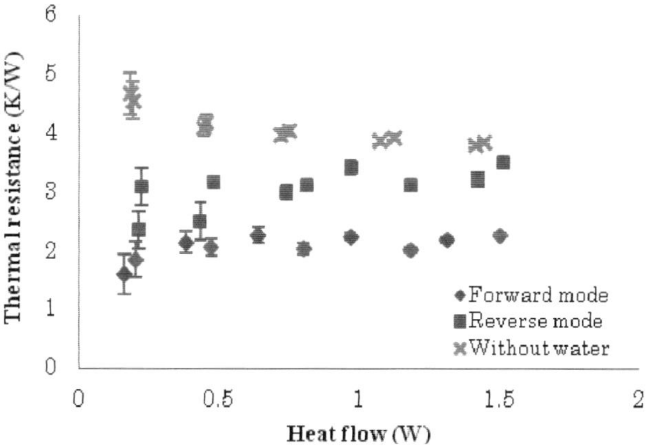

Fig. 7 Measured thermal resistance in forward and reverse mode compared with that without water.

6. Reference

[1] Mohamad A 1997 Integrated Solar Collector-Storage Tank System with Thermal Diode *Sol. Energy* **61** 221
[2] Tsukmoto T and Imai R 2006 Thermal characteristics of a high heat flux micro-evaporator *Exp. Thermo Fluid Sci*. **30** 837-842
[3] Rose J 2002 Dropwise condensation theory and experiment: a review *Inst. Mech. Eng.* **216** 115
[4] Boreyko J, Zhao Y and Chen C 2011 Planar jumping-droplet thermal diodes *Appl. Phys. Lett.* **99** 234105
[5] Abe T and Esashi M 2001 Ultrahydrophobic Surface Selectively Modified by Pulse Electrodeposition from Aqueous Dispersion of PTFE Particles *Sens. Actuators* **3B** 1044

Acknowledgement
This research was partly supported by Creation of Innovation Centers for Advanced Interdisciplinary Research Areas Program.

Performance evaluation of a thermoelectric energy harvesting device using various phase change materials

A Elefsiniotis[1], M E Kiziroglou[2], S W Wright[2], T T Toh[2], P D Mitcheson[2], T Becker[1], E M Yeatman[2] and U Schmid[3]
[1]EADS Innovation Works, Communication & Sensor, 81663 Munich, Germany
[2]Department of Electrical and Electronic Engineering, Imperial College London, SW7 2AZ, U.K.
[3]Institute of Sensor and Actuator Systems, Vienna University of Technology, Floragasse 7, 1040, Austria

E-mail: alexandros.elefsiniotis@eads.net

Abstract. This paper compares the performance of a group of organic and inorganic phase change materials for a heat storage thermoelectric energy harvesting device. The device consists of thermoelectric generators and a closed container filled with a phase change material. One side of the generators is mounted on the aircraft fuselage and the other to the thermal mass. The group of inorganic and organic phase change materials was tested across two temperature ranges. These ranges are defined as "positive" and "negative", with the former being a sweep from +35°C to -5°C and the latter being a sweep from +5°C to -35°C. The performance in terms of electrical energy output and power produced is examined in detail for each group of materials.

1. Introduction

In an effort to reduce the maintenance cost of an aircraft over its lifetime, wireless health monitoring sensors (WHMS) have been under consideration. The sensors may vary in function, from crack-wires, to multi-axial strain gauges, to acousto-ultrasonic, which usually require different amounts of energy. The energy that such sensors require traditionally is provided by either cables or rechargeable batteries. However, neither of these two solutions is really optimal, because on the one hand cables introduce design complexity, weight and installation cost and on the other hand batteries require regular recharging and/or replacement during the lifetime of an aircraft (≈30 years).

A way to create a truly autonomous WHMS is to convert energy from an otherwise unused, localised source and convert it to useable electrical energy. This concept is known as energy harvesting and a study of different possible harvesting methods for aircraft applications has been reported in [1]. Thermoelectric energy harvesting has been shown to be a promising solution for powering completely autonomous WHMS in aircraft [2].

A possible realization of a thermoelectric energy harvesting device uses an aircraft's fuselage as a heat source or sink and phase change materials (PCMs) as a dynamic thermal mass in order to increase the temperature difference across a thermoelectric generator (TEG). PCMs, while undergoing a phase change, absorb or release energy additional to their specific heat, known as latent heat, without changing their temperature. This effect enhances the temperature difference across the TEG, effectively increasing its energy conversion potential. In general, the latent heat is the most important physical property of a PCM in terms of thermal energy storage capabilities, while the phase change temperature determines the effective operating temperature range of the device.

The performance of such devices has been studied during a flight test campaign, demonstrating average electrical energy output densities as high as 2.4 J/g of PCM [2], and can be theoretically as high as 10 J/g [3]. The scope of this paper is to investigate the performance of two different PCM groups, one consisting of organic and one of inorganic materials, in terms of power and total electrical energy output.

2. Design Aspects

The harvester described in this paper consists of a closed aluminium container, a polyurethane insulating sheath, two TEGs (Marlow TG 12-2.5L [4]) electrically connected in series, and a 10 Ω resistor as a matched load. A photograph of the prototype device is shown in Figure 1.

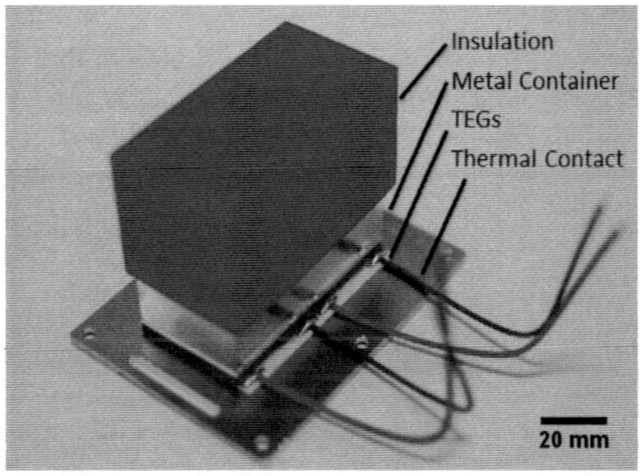

Figure 1. Photograph of the thermoelectric energy harvester.

For each test, the container was filled with 23 ml of PCM, which is 7 ml less than the internal volume of the container to allow material expansion during phase change. Separate experiments were conducted for materials exhibiting positive and negative phase change temperatures. The entire device, which would be exposed to the interior fuselage temperature, is subjected to a 40°C temperature change: +5°C to -35°C for the "negative" group and +30°C to -10°C for the "positive" group.

Previous versions of such devices were tested with distilled water or inorganic solutions as PCMs [5, 6]. Table 1 summarizes the physical properties of all the materials used to determine the performance of the proposed device. Distilled water is included in Table 1 as a reference for comparing the physical properties of the other PCMs.

Table 1. Physical properties of the inorganic and organic PCMs used to evaluate the performance of the device. E-series PCMs are supplied by PCM Products Ltd. and the RT series PCMs by Rubitherm.

Physical properties	Distilled water	Inorganic materials		Organic materials		
		E-15	E-11	RT10-HCG	RT10-HC	RT-9
T_m (°C)	0	-15	-11.6	+10 – +9	+10 – +4	-9 – -10
C_p (kJ/kgK)	4.2	3.87	3.55	2	2	2
ρ (kg/m^3)	1000	1060	1090	770	770	770
ΔH (kJ/kg)	330	303	301	152	195	260

3. Theoretical Background

The PCMs in this application act as thermal batteries, which are "charged" or "discharged" during ascent and descent respectively. The maximum thermal energy that can theoretically be stored is

$$Q = \int_{T_m}^{T_h} mC_{p,l}(T)dT + m\Delta H + \int_{T_c}^{T_m} mC_{p,s}(T)dT, \qquad (1)$$

where $C_{p,l}$ and $C_{p,s}$ are the specific heat capacity in the liquid and solid phase, respectively. The mass m, and the temperature on the hot and cold sides and the phase change temperature are represented by T_h, T_c and T_m respectively. ΔH signifies a material's latent heat.

The thermal energy that the PCMs can store also indicates the heat flux that goes through the TEGs. The overall efficiency of the TEGs, can be simplified, due to the relatively small ΔT [3], to

$$\eta_{TEG}(\Delta T) = \frac{\Delta T}{T_h} \cdot \frac{ZT_h}{4} \qquad (2)$$

where Z is the figure of merit.

To calculate the maximum TEG efficiency, a ΔT of 20°C, occurring during phase change, was used. This ΔT was assumed to be a representative average temperature difference, a choice that was backed by the experimental data. The theoretical values for the thermal energy stored in the PCMs as well as the harvested electrical energy output are shown in Table 2.

The open-circuit voltage developed by the TEGs is proportional to the temperature difference across them. It is calculated by the Seebeck equation, $V_{TEG} = \alpha(T)\Delta T$ where α is the Seebeck coefficient. Combining this with (2) and assuming an optimal load resistance, R_l, equal to the total internal resistance, R_i, the power output is maximized [3] and given by:

$$P_{out} = \frac{V_{TEG}^2}{R_l} = \frac{\alpha^2}{4R_i}\Delta T^2. \qquad (3)$$

4. Experimental Results

A climate chamber (CTS T-70/200) was used to approximate the fuselage temperature profiles found on an Airbus test aircraft. The fuselage temperature change rate is about 0.050 K/s [7] while the climate chamber can achieve 0.067 K/s, and this rate is used for the experiments.

The "positive" PCM group consists of two materials, which have a phase change temperature of +10°C. The cooling and heating rates used by the climate chamber simulates the ascent and descent phases, which each take approximately 17 min. Assuming that the PCM is in thermal equilibrium with the fuselage temperature at the beginning of the flight, the latent heat is released immediately after the aircraft has reached cruising altitude, as Figure 2a illustrates. During this time, the largest possible temperature difference across the thermoelectric generator is achieved and thus the peak electrical power is achieved as seen in Figure 2b. Thermal equilibrium between the fuselage and the PCM is achieved after 45 min. and thus no more electrical energy can be produced. A similar but reversed behaviour is observed during descent. The electrical power and energy outputs are shown in Figure 2b.

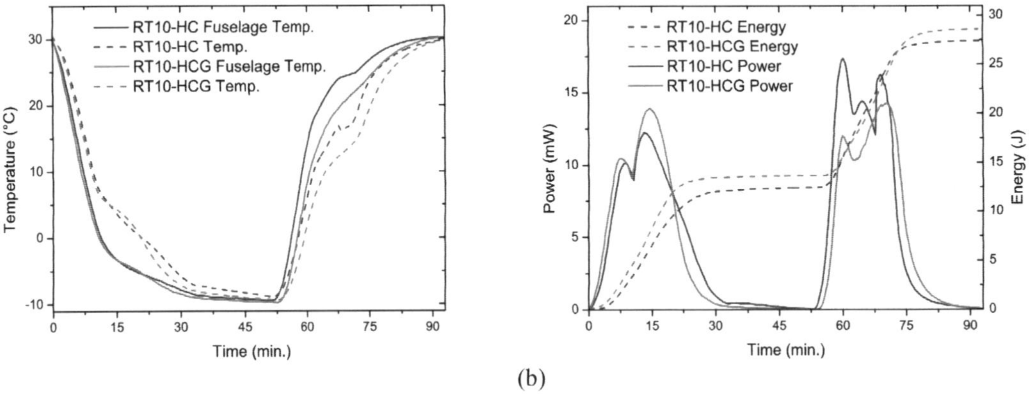

(a) (b)

Figure 2. Temperature, power and energy experimental results of the "positive" group. The total harvested electrical energy is 27.4 J for RT10-HC and 28.6 J for RT10-HCG.

Equation (1), when considered with the PCM properties, suggests that in general most of the thermal energy stored in a PCM comes from its latent heat. The latent heat is usually an order of magnitude greater than the heat capacity over a 40° C temperature change.

Inorganic materials exhibit a significant peak in output electrical power during solidification compared to organic materials. Any material can undergo a phenomenon called super-cooling, during which its temperature can be lower than its phase change temperature without any solidification taking place. This phenomenon is generally more prominent in pure materials and weaker in solutions or organic materials [8]. The super-cooling effect significantly affects the electrical power output of a device and imposes difficulties in numerical simulations (due to its discontinuous nature and non-deterministic occurrence).

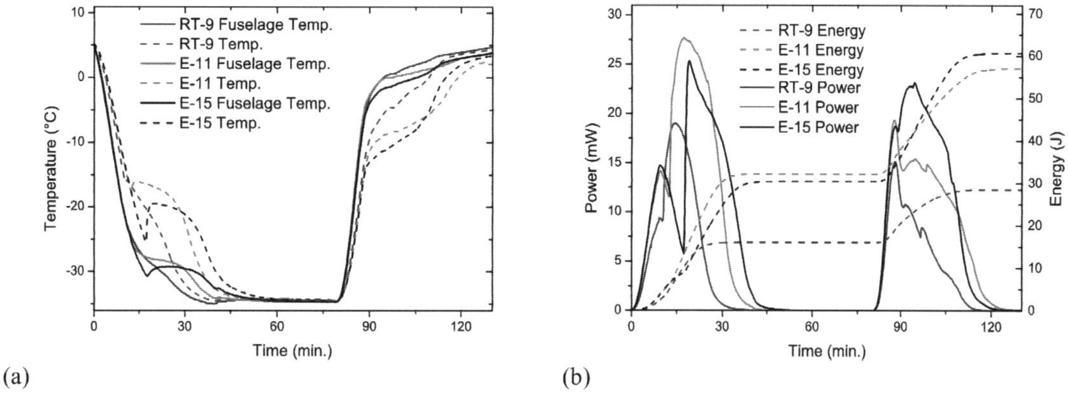

Figure 3. Temperature, power and energy experimental results of the "negative" group. The total harvested electrical energy is 28.5 J, 58.8 J and 60.6 J for RT-9, E-11 and E-15 respectively.

The electrical energy outputs of the PCMs in the negative group correspond to their latent heat. This observation shows that the theoretical predictions correlate well with the experimental results. However, the measured electrical energy output is less than the calculated value because heat loss to the environment is not accounted for. The overall efficiency of the device using the E-15 is 24%. The electrical energy outputs of the "negative" and "positive" groups are summarized in Table 2.

Table 2. Theoretical calculations and experimental data for the thermoelectric harvesting device. The TEG conversion efficiency η is 1.2%.

Calculated values	"Positive" group		"Negative" group		
	RT10-HCG	RT10-HC	RT-9	E-11	E-15
Energy stored, Q (J)	10700	12700	15600	20400	21100
Theoretical energy output (J)	128.1	151.8	187.7	244.5	252.7
Experimental energy output (J)	28.6	27.4	28.5	58.8	60.6

5. Conclusions and Outlook

The thermoelectric harvesting device described in this paper has shown great potential. For the "negative" temperature range, the inorganic PCMs are capable of generating almost twice the electrical energy output compared to their organic counterparts. This is largely due to their greater latent heat and specific heat values. They are therefore the preferred choice for the relevant temperature profile. Furthermore, inorganic salt-solution and organic paraffin based PCMs are compared in terms of energy performance for heat storage harvesting applications. A total electrical energy output of more than 60 J from 23 ml of a salt-based PCM (E-15) is acquired, close to the reference performance of water [7]. However, inorganic PCMs are more corrosive to the container material than organic materials. Consequently, alternative container designs are under consideration, in order to achieve the highest possible efficiency with minimum device deterioration over time. Further geometrical aspects are under investigation as well as the use of multiple PCMs in isolated cavities, in order to extend the operating temperature range of the device [9].

6. Acknowledgments

The authors would like to thank Mr. Nikolaos Kokorakis for his help in conducting the experiments and overall support.

7. References

[1] T. Becker, M. Kluge, J. Schalk, K. Tiplady, Ch. Paget, U. Hilleringmann and T. Otterpohl, Autonomous sensor nodes for aircraft structural health monitoring, IEEE Sensors Journal, Vol. 9, No. 11, pp. 1589-1595, 2009.

[2] Elefsiniotis, A., Samson, D., Becker, T., and Schmid, U. (2013). Investigation of the performance of thermoelectric energy harvesters under real flight conditions. Journal of Electronic Materials, 42(7), 2301-2305.

[3] Kiziroglou, M. E., Wright, S. W., Toh, T. T., Mitcheson, P. D., Becker, T., & Yeatman, E. M. (2013). Design and fabrication of heat storage thermoelectric harvesting devices. IEEE Transactions on Industrial Electronics, 61(1), DOI: 10.1109/TIE.2013.2257140

[4] Marlow Industries, Technical datasheet, Thermoelectric Generator TG12-2.5, DOC #102-0341REV H

[5] Samson, D., Kluge, M., Becker, T., and Schmid, U. (2011). Wireless sensor node powered by aircraft specific thermoelectric energy harvesting. Sensors and Actuators A: Physical, 172(1), 240-244.

[6] Elefsiniotis, A., Kokorakis, N., Becker, T., and Schmid, U. (2013, May). Design and material aspects for thermoelectric energy harvesting devices in aircrafts. In SPIE Microtechnologies (pp. 87631N-87631N). International Society for Optics and Photonics.

[7] D. Samson, T. Otterpohl, M. Kluge, U. Schmid and T. Becker, Aircraft-Specific Thermoelectric Generator Module, Journal of Electronic Materials, Vol. 39, No. 9, pp. 2092-2095, 2010.

[8] Feltham, D. L., & Worster, M. G. (2000). Similarity solutions describing the melting of a mushy layer. Journal of Crystal Growth, 208(1), 746-756.

[9] A. Elefsiniotis, N. Kokorakis, Th. Becker, and U. Schmid, Performance of a low temperature energy harvesting device for powering wireless sensor nodes in aircrafts applications, Proc. of the 17th Intern. Conf. on Solid State Sensors and Actuators, Transducers2013, pp. 2276-2279, Barcelona, 2013.

Combined Pyroelectric, Piezoelectric and Shape Memory Effects for Thermal Energy Harvesting

D Zakharov[1,2], B Gusarov[1,2], E Gusarova[3], B Viala[3], O Cugat[1,2], J Delamare[1,2] and L Gimeno[1,2]

[1]G2Elab, Univ. Grenoble Alpes, F38000 Grenoble, France
[2]CNRS, Univ. Grenoble Alpes, F38000 Grenoble, France
[3]CEA, LETI, MINATEC Campus, F38000 Grenoble, France

dmitry.zakharov@g2elab.grenoble-inp.fr

Abstract. This work proposes an enhanced method for thermal energy harvesting exploiting combined pyroelectric, piezoelectric and shape memory (SME) effects, and presents its experimental validation. A material which is pyroelectric is also piezoelectric. If it is combined with a material with SME, which generates large strain and stress in a rather narrow temperature range, the resulting composite material would generate voltage from temperature variations using two different energy conversion principles at once: (1) pyroelectric effect, (2) piezoelectric effect driven by SME. A Macro Fiber Composite piezoelectric was shown here to exhibit significant pyroelectric effect (~4 V/°C). When combining it with a SME Ti-Ni-Cu alloy into a laminated structure, this effect increased by 50%. This increase may be an order of magnitude higher for an optimized system. Such composites open an opportunity to harvest thermal energy from natural sources, since this method can increase the rather low efficiency of current pyroelectric materials especially for small temperature variations.

1. Introduction

Besides macro-power systems such as solar cells, wind and watermills, unconventional small-scale power systems are emerging from labs. They aim to bring energy autonomy to ultra-low-power (ULP) electronics. This gives rise to substantial research activities in the fields of micro or nano-energy harvesting and ULP devices. As a result, devices are being optimized to require less and less energy, down to picojoules for some applications [1], while harvesting systems are now able to generate energies up to the millijoule range [2]. Micro- and nano- harvesting systems are starting to offer a real alternative over battery-powered solutions with virtually inexhaustible sources, no maintenance and little adverse environmental effects.

Considering thermal sources which are widely available in our environment, energy can be directly converted into electricity by means of thermoelectricity (Seebeck effect) or pyroelectricity. Thermoelectric and pyroelectric power generators are already in common use. However, for acceptable efficiency, applications require significant spatial- and time-dependent temperature variations respectively. This in turn requires optimized cold source management systems. The cold source usually consists in a radiator whose size and shape can become problematic at small scales.

Another strong limitation stems from the low voltage or current delivered by these generators, which require complex and power-consuming buck-boost converters.

An alternative scheme is to indirectly convert heat into electricity through mechanical transformations using smart composite materials. Such composite requires two materials being mechanically coupled. This principle is discussed in detail in the next section.

2. Principle

Recently, we proposed and studied an original hybrid laminated composite material for harvesting quasi-static temperature variations [3]. It consisted of Macro Fiber Composite (MFC) piezoelectric and Ti-Ni-Cu shape memory alloy (SMA). During temperature variations, large thermally induced strain generated by the SMA layer is transmitted to the piezoelectric layer, which in turn generates a voltage. In this way thermally induced strain is converted into useful voltage with neither complex electronics nor cold source management system (no radiator).

In this previous work, pyroelectric effect was not considered for MFC, relying on literature data [4]; however, here we give first experimental evidence that pyroelectricity is indeed significant in MFC. As a consequence, the SMA/MFC composite material can generate voltage from temperature variations using two different energy conversion principles at once: (1) pyroelectric effect, and (2) piezoelectric effect driven by shape memory effect (SME). The principle is illustrated in figure 1. Since all pyroelectrics are also piezoelectrics, this approach can be applied to enhance the electric response of any pyroelectric material in a certain temperature range, where the thermoelastic straining of the SMA is pre-determined to take place. When designing such a functional material, arrangements should be made to ensure that both effects will induce voltage of the same sign, otherwise they risk canceling each other.

This approach is different from that using the difference of thermal expansion coefficients of the pyroelectric and its substrate to drive piezoelectricity [5]: since SME generates much larger strains in a narrow temperature ranges. This is especially advantageous for harvesting small temperature variations.

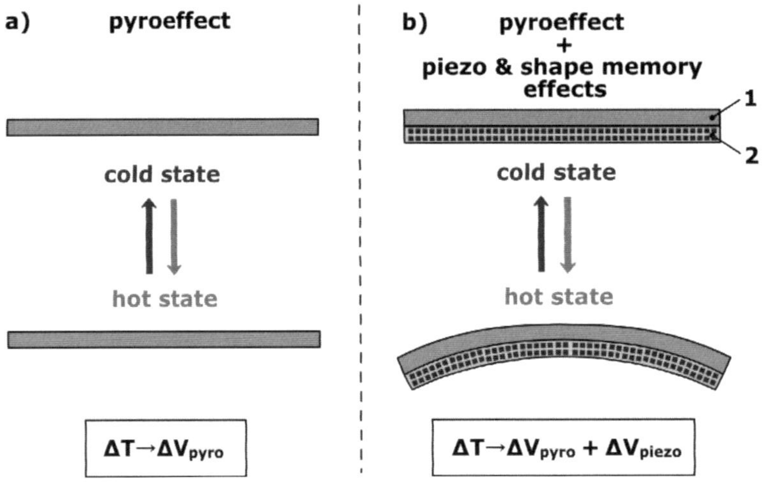

Figure 1. Principle of proposed enhancement of pyroelectric material performance. 1 – pyro/piezo-electric layer, 2 – SMA layer with SME pre-determined at the temperature range of interest. When temperature changes by ΔT, (a) pyroelectric material alone generates voltage ΔV_{pyro} due to pyroeffect, (b) hybrid composite generates voltage ΔV_{pyro} due to pyroeffect plus ΔV_{piezo} due to piezoelectric effect driven by SME.

3. Results and discussion

3.1. Pyroelectric effect for MFC

Figure 2 shows absolute values of voltage generated by MFC alone when dipping in hot and cold water, measured by electrostatic non-contact voltmeter. The use of electrostatic voltmeter allows observing time-dependent direct voltage generation with no losses [6]. In this experiment, the MFC was firstly moved from the air (room temperature) into the hot water for about a minute (curve 1). After registering the curve but before taking the sample out, the generated voltage was discharged into a resistor, thereby bringing the MFC back to the uncharged state. Then it was quickly taken out from the hot water and put into the cold water (curve 2, the first slope between 0 and ~2 s corresponds to the travel in air before immersion in cold water). As observed on the plot, for both cases the voltage firstly increased due to pyroelectricity and then decreased due to the intrinsic charge leakage. Leakage at heating was significantly higher (the decreasing time $t_{1/2}$ from maximum to half -maximum was 3 s at heating versus 30 s at cooling). The leakage is known to increase with temperature [7].

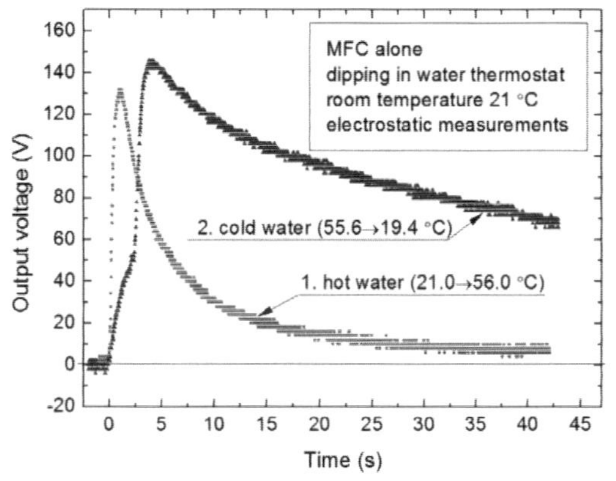

Figure 2. Absolute electric response of MFC alone when heated (1) and cooled (2) by dipping in hot and cold water (opposite voltage signs). Voltage firstly increases due to pyroelectricity, and then decreases due to intrinsic charge leakage. Leakage is faster at high temperatures.

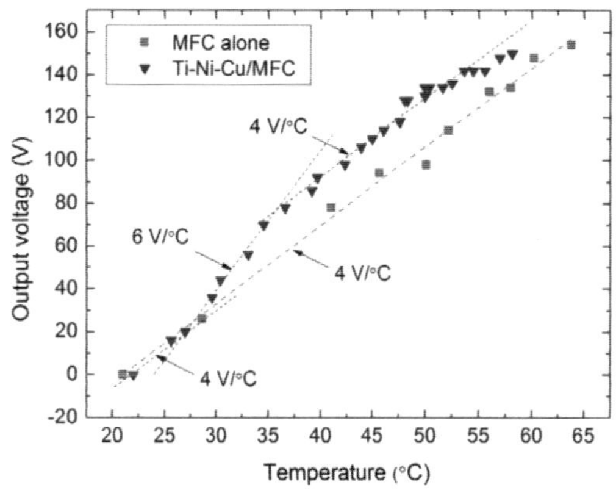

Figure 3. Peak voltage generated by MFC alone and by Ti-Ni-Cu/MFC composite, when heated to various temperatures. In the region ~30-40 °C the composite shows an increase of "pyro"-electric voltage coefficient due to thermoelastic straining of Ti-Ni-Cu SMA, and then resumes the "pyro" slope.

The peak voltage generated by MFC alone was measured at various temperatures of hot water. This dependence was linear and characterized by a remarkable high pyroelectric voltage coefficient ρ_V ~4 V/°C (figure 3, red filled squares). In fact, MFC has been used mainly as an actuator up to now and its pyroelectric properties were not studied before. Our work gives evidence of significant pyroelectric effect for MFC in the working conditions described above. This result is rather expected, since MFC contains PZT (PZT5A1 type) [8], which is known to have pyroelectric properties [9]. The pyroelectric voltage coefficient ρ_V can be evaluated knowing its pyroelectric charge coefficient ρ_Q, relative dielectric permittivity ε_r and thickness h using the following expression [10]:

$$\rho_V = \frac{\rho_Q}{\varepsilon_r \cdot \varepsilon_0} \cdot h ,$$

where ε_0 is the vacuum permittivity ($8.85 \cdot 10^{-12}$ Fm^{-1}). Using the following values: $\rho_Q = 2.38 \cdot 10^{-4}$ $Cm^{-2}K^{-1}$ [9], $\varepsilon_r = 1800$ [11], $h = 127$ μm [4], one obtains $\rho_V = 1.9$ V/°C. The measured value is of the same order of magnitude which is consistent with the theoretical calculation.

3.2. Ti-Ni-Cu/MFC composite

The fabrication details of our Ti-Ni-Cu/MFC composite were reported in [3]. Its electric response is also plotted on figure 3 (filled blue triangles). The pyroelectric voltage coefficient ρ_V is the same as that of MFC alone below ~30 °C. But in the region ~30-40 °C it increases by ~50% (6 V/°C) due to the thermoelastic straining of Ti-Ni-Cu SMA. Thereby, above ~40 °C, after the transformation in the SMA has been completed, the output voltage of the composite stays higher than that of MFC alone, since the piezoelectric voltage contributes in addition to the pyroelectric one. In fact, this increase was expected to be much higher since the SMA was pre-strained to about 1-3%. Part of the strain is believed to be lost at the interface (glue), which is critical for the performance of such functional composites. Also, the sizes of the SMA ribbon and the MFC used in this work were not perfectly adapted: the SMA layer covered only 15% of the MFC layer. As a consequence, a part of PZT fibers were probably not experiencing the deformation.

3.3. Sign of the effects

It is important here to deal with the question of the sign of the different effects. When the MFC was subjected to heating, the observed voltage was positive. As it was explained in [3], the Ti-Ni-Cu ribbon has been pre-deformed to shorten at heating. Considering the Ti-Ni-Cu/MFC bilayer, we assume that it bends and fibers become tensile. Thus since PZT piezoelectric voltage coefficient g_{33} is positive [11], the generated voltage is positive. For such type of composites mechanical engineering (planar or flexural deformation) must be considered with respect to the sign of piezo- and pyro-coefficients.

3.4. Optimization

These first experiments were aimed at proving the proposed concept of enhanced thermal energy harvester, and the system was not optimized. Optimization of the composite can be realized by taking an SMA element of the same area as the MFC in order to provide homogeneous deformation of all the PZT fibers. The thickness of the SMA element and its preset strain should be optimized in order to apply adequate strain and stress to the MFC. The optimized glue connection should allow transferring the mechanical work of the SMA with minimal losses. All these arrangements should allow us to obtain thermally-induced stain in MFC layer close to 0.1% and corresponding contribution of piezoelectric effect of more than 300 V [12] in a narrow temperature range (ΔT ~10-20 °C for Ti-Ni-Cu SMA) resulting in significant enhancement of the thermoelectric-like response.

4. Conclusion

The proposed method of combining pyroelectric material with a shape memory material provides an enhanced electric response compared to MFC pyroelectricity alone. This opens an opportunity to harvest thermal energy from natural sources, since the new method can increase the rather low efficiency of current pyroelectric materials, especially for small temperature variations. Since the exploited physical effects do not deteriorate at microscale, the proposed concept is suitable for MEMS.

Acknowledgements

This work was supported by the French National Research Agency (ANR) under the Carnot Institute Program and by the Nanosciences Foundation of Grenoble under RTRA program. The research leading to these results has received funding from the European Community's Seventh Framework Programme (FP7/2007-2013) under grant agreement NANOFUNCTION n°257375.

References

[1] Mark M, Chen Y, Sutardja C, Tang C, Gowda S, Wagner M, Werthimer D and Rabaey J 2011 *Symp. on VLSI circuits (Kyoto, Japan, 15–17 June)* pp 168-171
[2] Radousky H B and Liang H 2012 *Nanotechnology* **23** 502001

[3] Zakharov D, Lebedev G, Cugat O, Delamare J, Viala B, Lafont T, Gimeno L and Shelyakov A 2012 *J. Micromech. Microeng.* **22** 094005
[4] Williams R B, Inman D J and Wilkie W K 2004 *Journal of Thermal Stresses* **27** 903
[5] Chang H H S and Huang Z 2010 *Smart Mater. Struct.* **19** 065018
[6] Gusarova E, Gusarov B, Zakharov D, Bousquet M, Viala B, Cugat O, Delamare J and Gimeno L 2013 An improved method for piezoelectric characterization of polymers for energy harvesting applications *Proc. 13th Intal Conf. on Micro and Nanotechnology for Power Generation and Energy Conversion Applications (London, UK, December 3-6, 2013)*
[7] Lipscomb I P, Weaver P M, Swingler J and McBride J W 2009 *Sensors Actuators A* **151** 179–86
[8] Macro Fiber Composite (MFC) brochure at http://www.smart-material.com/Datasheets.html
[9] Xie J, Mane X P, Green C W, Mossi K M and Leang K K 2010 *Journal of Intelligent Material Systems and Structures* **21**(3) 243
[10] Pintilie L and Pintilie I 1997 *Ferroelectrics* **200** 219-235
[11] Piezoelectric Brochure at http://www.morganelectroceramics.com/products/piezoelectric/
[12] Lafont T, Gimeno L, Delamare J, Lebedev G A, Zakharov D I, Viala B, Cugat O, Galopin N, Garbuio L and Geoffroy O 2012 *J. Micromech. Microeng.* **22** 094009

Thermoelectric energy harvesting system for demonstrating autonomous operation of a wireless sensor node enabled by a multipurpose interface

Joachim Leicht[1], Peter Heilmann[1], Thorsten Hehn[2], Xiaoming Li[3], Dominic Maurath[1], Christian Moranz[1], Marcell Thewes[4], Gerd Scholl[4] and Yiannos Manoli[1,2]

[1] Fritz Huettinger Chair of Microelectronics,
Department of Microsystems Engineering — IMTEK, University of Freiburg, Germany
[2] HSG-IMIT — Institute of Micromachining and Information Technology, Villingen-Schwenningen, Germany
[3] Daimler AG, Research & Development, Sindelfingen, Germany
[4] Electrical Measurement Engineering, Helmut-Schmidt-University, Hamburg, Germany

E-mail: joachim.leicht@imtek.uni-freiburg.de

Abstract. This paper demonstrates the autonomous operation of a wireless sensor node exclusively powered by thermoelectric energy harvesting. Active operation of a wireless sensor system is demonstrated successfully by means of an on-line programmable emulation kit that enables various thermoelectric energy harvesting scenarios. Moreover, this emulation kit accomplishes autonomous wireless sensor node operation by interfacing a small-scaled thermogenerator via a CMOS integrated autonomous multipurpose energy harvesting interface circuit performing maximum power point tracking.

1. Introduction

The extremely large number of sensors in todays means of transportation such as automobiles, aircrafts, or trains require the reduction of the size of these sensor systems as well as the power drawn from the battery. Therefore, autonomous wireless sensor networks composed of self-sufficient tiny sensor nodes need to be applied [1].

In order to increase the sensor network energy efficiency, low power radio transmission protocols are necessary which ensure reliable and proper operation with respect to established specifications such as given by the automotive industry [1].

Thermoelectric energy harvesting is an attractive candidate for successfully meeting the powering challenge since various temperature gradients arise during transportation. Moreover, thermoelectric energy harvesting is suitable for real world applications because very small scaled off-the-shelf devices which provide high power densities such as Micropelt® thermogenerators (TEGs) allow efficient and miniaturized wireless sensor network solutions [2–4].

The reliable operation of wireless sensor nodes by means of energy harvesting requires interface circuits which optimize the ambient energy extraction and which manage the wireless sensor node powering. Furthermore, the storage of excessive harvested energy has to be controlled by these interfaces [4–7].

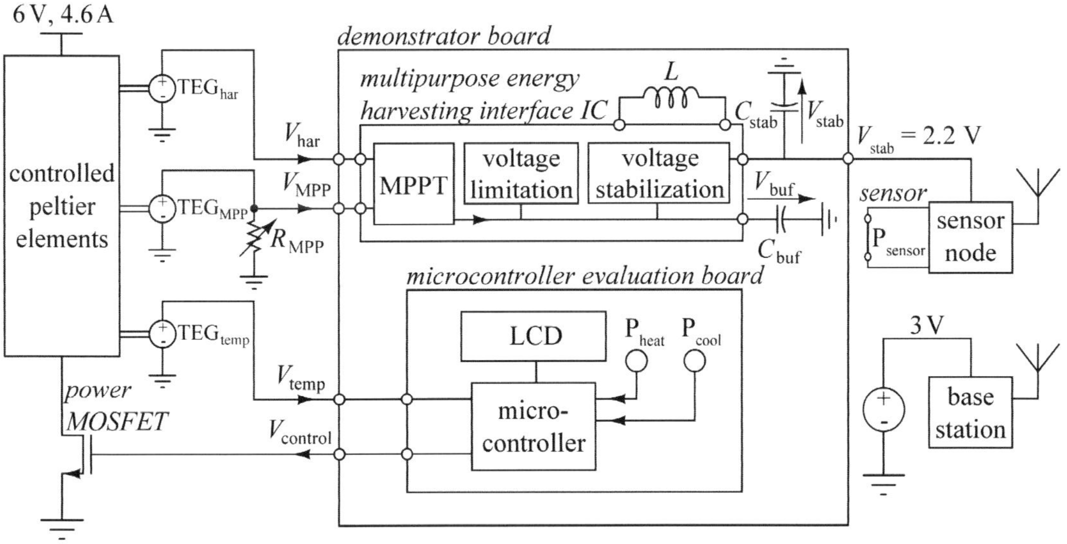

Figure 1. Overview of the programmable thermoelectric energy harvesting emulation kit that enables various thermoelectric energy harvesting scenarios and the autonomous operation of a wireless sensor node by means of a multipurpose interface circuit.

This paper presents the programmable thermoelectric energy harvesting emulation kit outlined in Figure 1 which is composed of a thermoelectric energy harvester, a CMOS integrated circuit (IC) for efficiently interfacing the energy harvester, and a wireless sensor application designed for automobile safety systems [1, 3, 6]. This emulation kit allows setting of a defined temperature difference across the thermoelectric energy harvester and triggering of wireless data transmissions. Autonomous sensor node operation is established by the CMOS integrated energy harvesting interface circuit that enables maximum power point (MPP) tracking (MPPT) [6].

Section 2 describes the applied thermogenerator while the interface is covered in Section 3. Section 4 details the emulation kit and autonomous wireless sensor node operation is verified in Section 5. Finally, in Section 6 the paper draws a conclusion.

2. Thermoelectric energy harvester

The used thermoelectric energy harvester is the TGP-751 thermogenerator from Micropelt® [3]. This energy harvester can be modeled by a DC voltage source in series to the internal resistance R_{TEG} as shown in Figure 2 [4]. The open circuit voltage V_{TEGoc} is generated because of the Seebeck effect and R_{TEG} is the total series resistance of the thermoelectric legs.

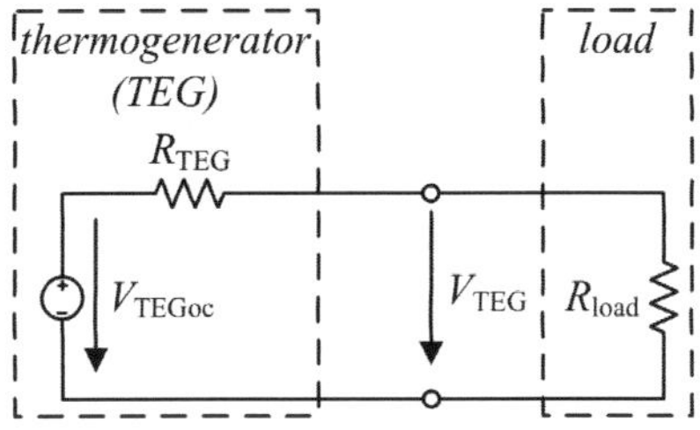

Figure 2. Equivalent circuit model of the thermoelectric energy harvester.

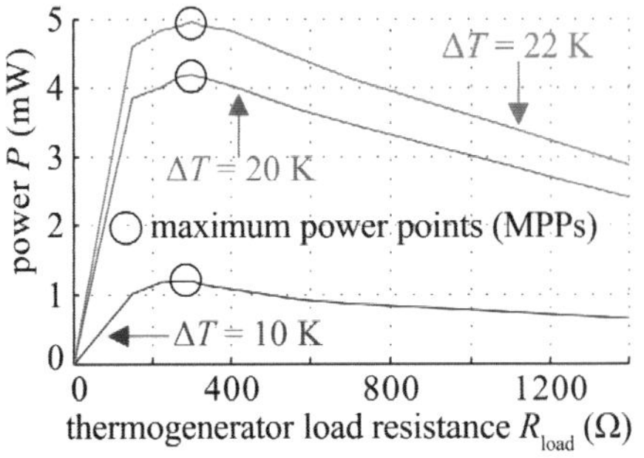

Figure 3. Measured power characteristics of the Micropelt® thermogenerator TGP-751 at several temperature differences ΔT.

The thermogenerator delivers maximum power $P_{\max}$ when the load resistance R_{load} matches the internal resistance R_{TEG}. The terminal voltage V_{TEG} at the maximum power point (V_{MPP}) is therefore 50 % of V_{TEGoc}. Hence, the maximum electrical load power $P_{\max}$ can be given as

$$P_{\max} = \frac{{V_{\mathrm{MPP}}}^2}{R_{\mathrm{TEG}}} = \frac{{V_{\mathrm{TEGoc}}}^2}{4R_{\mathrm{TEG}}} = \frac{(\alpha\Delta T)^2}{4R_{\mathrm{TEG}}}, \tag{1}$$

where α is the Seebeck coefficient and ΔT is the temperature difference across the harvester [4].

The Seebeck coefficient provided by the TGP-751 is specified to be 110 mV/K and the value of the internal resistance is specified to be 300 Ω. Both characteristics are declared by the datasheet for an ambient temperature of 25 °C [3]. Measured load power characteristics for different temperature differences across the applied thermogenerator are shown in Figure 3 and obviously the maximum power point occurs at a load resistance of around 300 Ω.

3. Multipurpose energy harvesting interface

Figure 1 illustrates the applied self-sufficient multipurpose CMOS integrated energy harvesting interface circuit which is detailed in [6] and Figure 4 shows the prototype chip micrograph. This interface allows optimal electrical energy output of the thermogenerator by means of maximum power point tracking. In [6] this interface is applied for highly efficient energy harvesting with electromagnetic kinetic energy transducers that deliver AC voltages.

In contrast to that, this paper proposes the use of the same interface for DC sources such as thermogenerators. Thus, a multipurpose interface is introduced enabling the use of different physical principles with a single interface chip and the storage of the harvested energy in the energy buffer C_{buf} as depicted in Figure 1. Using the harvested energy, a wireless sensor node connected to the interface as outlined in Figure 1 can be operated autonomously and reliably by the stabilized output voltage V_{stab} regulated by the multipurpose energy harvesting interface.

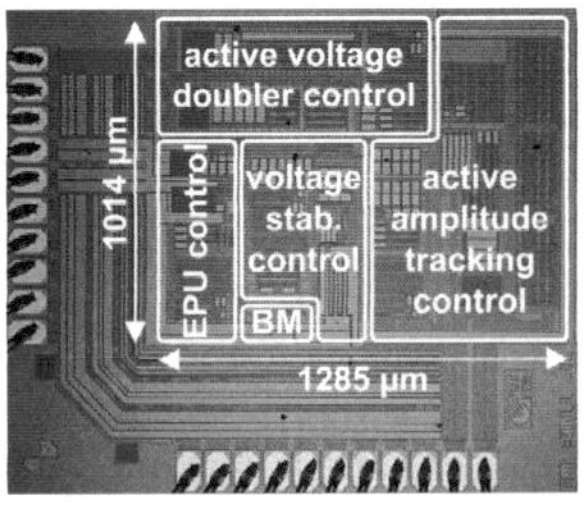

Figure 4. Micrograph of the prototype interface chip fabricated in a 0.35 µm CMOS process [6].

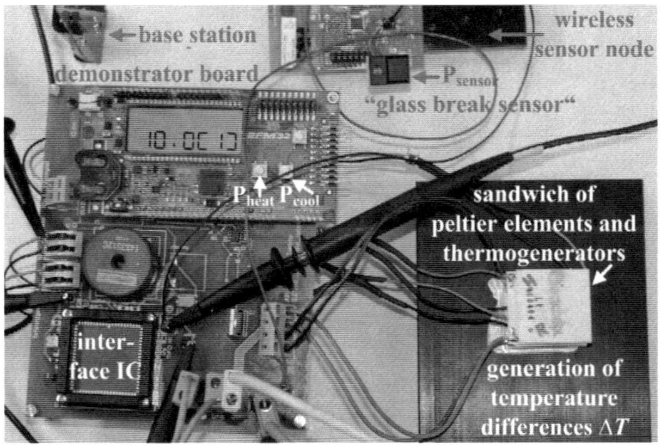

Figure 5. Photograph of the experimental thermoelectric energy harvesting emulation kit.

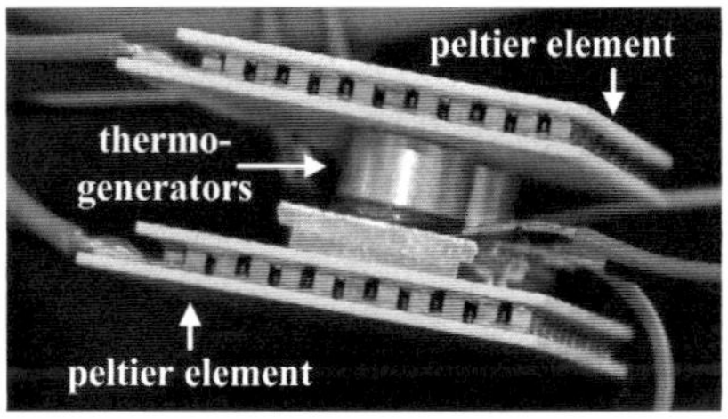

Figure 6. Sandwich of peltier elements and thermogenerators to enable a defined temperature difference ΔT across the thermoelectric energy harvesting device.

4. Thermoelectric energy harvesting emulation kit

Figure 5 shows a photograph of the experimental emulation kit that is schematically shown in Figure 1. It is composed of a demonstrator board, peltier elements, thermogenerators, a wireless sensor node, and a radio base station.

The demonstrator board comprises the multipurpose energy harvesting interface IC and a microcontroller evaluation board (EFM32TG-STK3300 [8]). Two peltier elements and three identical thermogenerators (Micropelt® TGP-751 [3]) are sandwiched as shown in Figure 6. This sandwich allows defined temperature differences by cooling and heating the peltier elements accordingly via an on-off control. This control consists of a microcontroller implementation, the thermogenerator $\mathrm{TEG_{temp}}$ (Figure 1) for sensing the current temperature difference, pushbuttons (P_{heat} and P_{cool}) for setpoint programming and a power MOSFET used as actuating element.

The thermogenerator $\mathrm{TEG_{har}}$ is the actual energy harvesting device and $\mathrm{TEG_{MPP}}$ in conjunction with R_{MPP} is used to track the maximum power point dynamically (Figure 1) [6, 7].

A sensor network for automobile safety systems as described in [1] is exemplified by the radio base station and the wireless sensor node. The pushbutton P_{sensor} emulates the glass break sensor needed for the intrusion detection of this safety system and triggers radio transmissions.

Figure 7 presents the final emulation kit comprising the experimental kit that cooperates with a controller area network (CAN) [1]. The shown car window is monitored by the above mentioned automobile safety system and an intrusion indicated by the emulated glass break sensor is displayed on the computer screen completing a remote intrusion monitoring system.

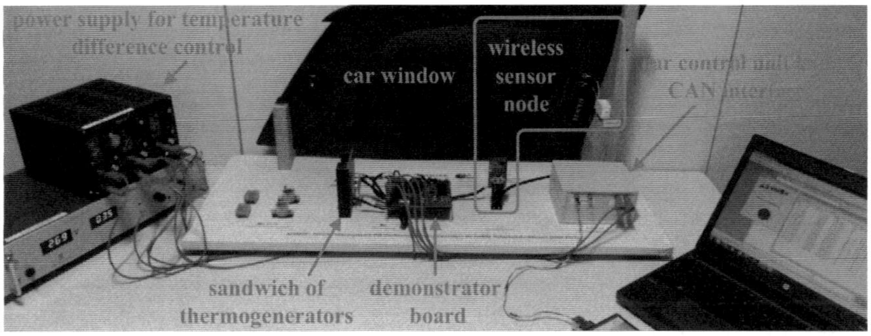

Figure 7. Photograph of the final thermoelectric energy harvesting emulation kit.

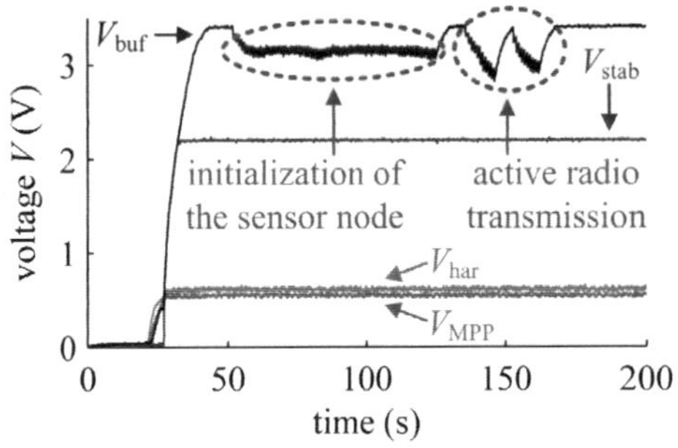

Figure 8. Measured system cold start-up and wireless sensor node operation at a ΔT of 9 K.

5. Autonomous wireless sensor node operation

Figure 8 shows a measured cold start-up from a fully discharged energy buffer capacitor C_{buf} for a temperature difference ΔT across the harvester of 9 K. The harvester output voltage V_{har} is controlled up to the maximum power point voltage V_{MPP} by the autonomous multipurpose interface in order to operate the harvester TEG_{har} close to its maximum power point.

The stabilized output voltage V_{stab} is controlled to 2.2 V in order to supply the wireless sensor node. The first voltage drop interval of V_{buf} encircled in Figure 8 indicates the initialization of the wireless sensor node. Active radio transmission of the sensor node is shown by the next encircled voltage drops. The stabilized voltage V_{stab} shows negligible influence due to these voltage drops and therefore reliable autonomous operation of the wireless sensor node is demonstrated.

Energy not used by the wireless sensor node or by the interface circuitry is stored in the energy buffer C_{buf}. In order to keep the system within the maximum allowed electrical operation conditions the interface circuit limits the buffer voltage V_{buf} as shown in Figure 8 [1, 6].

6. Conclusion

Autonomous operation of a wireless sensor node without needing a pre-charged energy storage element is successfully demonstrated by means of the presented on-line programmable thermoelectric energy harvesting emulation kit. The self-sufficient CMOS integrated multipurpose energy harvesting interface circuitry implemented in the emulation kit operates the very small scaled off-the-shelf thermogenerator close to its maximum power point in order to optimize the ambient energy extraction. Moreover, this interface manages the powering of the wireless sensor node reliably and stores excessive harvested energy in an energy buffer.

Acknowledgments

This work is part of the GRK1322 funded by the German Research Foundation (DFG), and it is supported by ASYMOF funded by the Federal Ministry of Education and Research (BMBF).

References

[1] Thewes M, Li X and Scholl G 2012 *Proc. IEEE Int. Multi-Conf. on Systems, Signals and Devices (Chemnitz)* pp 1–5
[2] Böttner H, Nurnus J, Schubert A and Volker F 2007 *Proc. IEEE Int. Conf. on Thermoelectrics (Jeju)* pp 306–309
[3] Micropelt GmbH *http://www.micropelt.com/products/thermogenerator_package.php* access date: 2013–09–27
[4] Kluge M, Samson D, Becker T, Gavrikov A, Bennemann B and Nurnus J 2010 *Proc. PowerMEMS (Leuven)* pp 2–3 2
[5] Manoli Y 2010 *Proc. IEEE Eur. Solid-State Circuits Conf. (Seville)* pp 27–36
[6] Leicht J, Maurath D and Manoli Y 2012 *Proc. IEEE Eur. Solid-State Circuits Conf. (Bordeaux)* pp 101–104
[7] Brunelli D, Moser C, Thiele L and Benini L 2009 *IEEE Trans. Circuits Syst. I, Reg. Papers* **56** 2519–2528
[8] Energy Micro AS *http://www.energymicro.com/downloads/tools-documents* access date: 2013–09–27

Physical insight into electromagnetic kinetic energy transducers and appropriate energy conditioning for enhanced micro energy harvesting

Joachim Leicht[1], Thorsten Hehn[2], Dominic Maurath[1], Christian Moranz[1] and Yiannos Manoli[1,2]

[1] Fritz Huettinger Chair of Microelectronics,
Department of Microsystems Engineering - IMTEK,
University of Freiburg, Germany
[2] HSG-IMIT - Institute of Micromachining and Information Technology,
Villingen-Schwenningen, Germany

E-mail: `joachim.leicht@imtek.uni-freiburg.de`

Abstract. This paper proposes a new method for modeling electromagnetic kinetic energy transducers and gives analytical expressions that enable the design of efficient energy conditioning circuitry. The introduced transducer modeling approach achieves high accuracy without requiring a large set of parameters. The presented transducer characterization allows physical insight into fully assembled and packaged transducers in order to extract the required transducer model parameters without knowledge of the individual components. Moreover, the electromagnetic coupling, the parasitic damping, and the optimal load can be modeled with a dependence on the external excitation. Precise co-simulation with CMOS integrated energy conditioning circuitry is possible implementing this model in a circuit simulator.

1. Introduction

Kinetic micro energy harvesting enables the use of ambient vibrations in order to supply low power electronic systems such as wireless sensor nodes [1, 2]. In order to guarantee autonomous and reliable system operation, the ambient energy extraction has to be optimized. Therefore, a well-designed kinetic energy harvesting system composed of a kinetic micro power transducer and an efficient energy conditioning such as active rectification is needed (Figure 1).

Efficient system design requires an accurate transducer model and precise definitions of the optimal operation conditions of the entire system. In order to design an efficient energy harvester highly accurate transducer modeling is mandatory and a large set of model parameters is needed [3, 4]. These parameters are difficult to extract from a fabricated harvester without deep transducer design knowledge which usually applies to circuit designers. However, for the design of efficient energy conditioning circuits a reduced parameter set is sufficient that describes the optimum transducer operation precisely. Harvester disassembly allows to extract this set [5].

A fully packaged small-scaled electromagnetic kinetic energy transducer prototyped by the HSG-IMIT (Figure 1) is characterized and modeled accurately in this work without disassembly. Moreover, definitions are proposed that detail the optimum operation of energy conditioning based on ultra-low power CMOS integrated active rectification with low voltage drop [6].

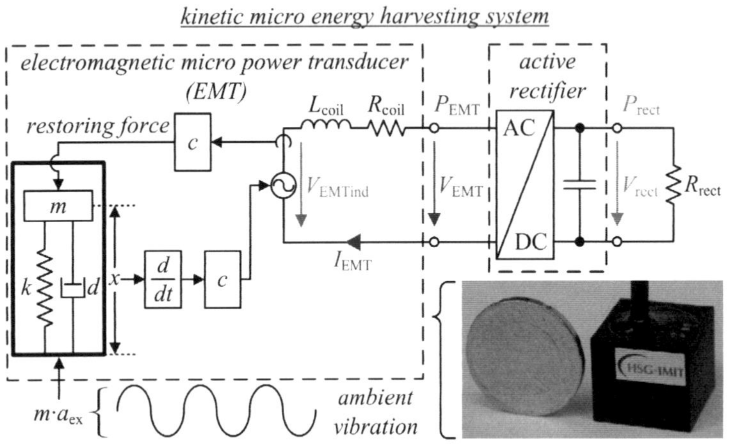

a_{ex}:	excitation acceleration
m:	seismic mass
k:	spring stiffness
d:	parasitic damping
x:	relative mass displacement
c:	electromagnetic coupling
L_{coil}:	inductance of the coil
R_{coil}:	resistance of the coil
V_{EMTind}:	induced voltage of the EMT
I_{EMT}:	EMT current
V_{EMT}:	EMT output voltage
P_{EMT}:	EMT output power
V_{rect}:	rectified voltage
P_{rect}:	rectifier load power
R_{rect}:	rectifier load resistance

Figure 1. Kinetic micro energy harvesting system and nomenclature of the system parameters.

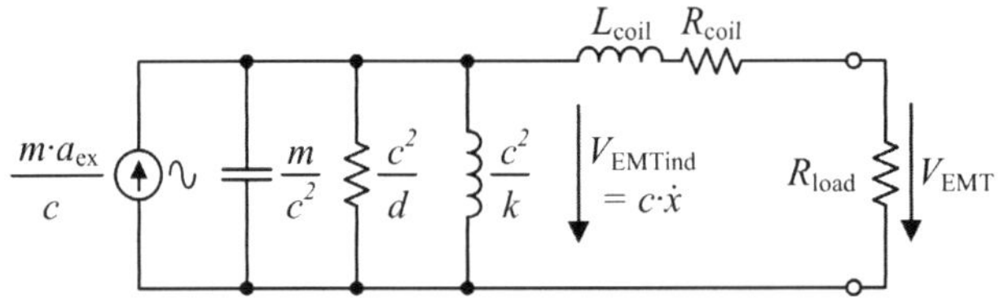

Figure 2. Equivalent circuit model of the electromagnetic kinetic energy transducer connected to a load resistance.

2. Electromagnetic kinetic energy transducer modeling and characterization

Figure 2 details the applied equivalent circuit model comprising the needed model parameters [7]. Maximum power P_{max} is harvested at resonance when the transducer load resistance R_{load} matches its optimum R_{opt} [3–6, 8]. By means of a defined transducer excitation ensured by a controlled vibration table [6] the maximum power, the optimal load resistance, and the resonance frequency f_{res} can be measured by performing sweeps over frequency and over the load resistance.

The characterized electromagnetic transducer is a low damped electromagnetic spring-mass-damper system and so the resonance frequency almost equals the natural frequency f_0 [8]. Thus, for all presented expressions f_{res} is assumed to be f_0 and the spring stiffness k can be given as

$$k = (2\pi f_{\mathrm{res}})^2 m = (2\pi f_0)^2 m, \tag{1}$$

where m is the seismic mass.

Measurement results of P_{max} and R_{opt} at resonance are presented in Figure 3 for different transducer peak excitation accelerations $\hat{a}_{\mathrm{ex}}$. By means of these results and the easily measurable coil series resistance R_{coil} the expressions

$$d_{\mathrm{m}} = \frac{d}{m^2} = \frac{\hat{a}_{\mathrm{ex}}{}^2 (R_{\mathrm{opt}} - R_{\mathrm{coil}})}{8 R_{\mathrm{opt}} P_{\mathrm{max}}}, \tag{2}$$

$$c_{\mathrm{m}} = \frac{c}{m} = \frac{\hat{a}_{\mathrm{ex}} (R_{\mathrm{opt}} - R_{\mathrm{coil}})}{(8 R_{\mathrm{opt}} P_{\mathrm{max}})^{1/2}} = \sqrt{d (R_{\mathrm{opt}} - R_{\mathrm{coil}})} \tag{3}$$

can be derived, where d_{m} embodies the parasitic damping d normalized to the square of the seismic mass m^2 and c_{m} the electromagnetic coupling normalized to m.

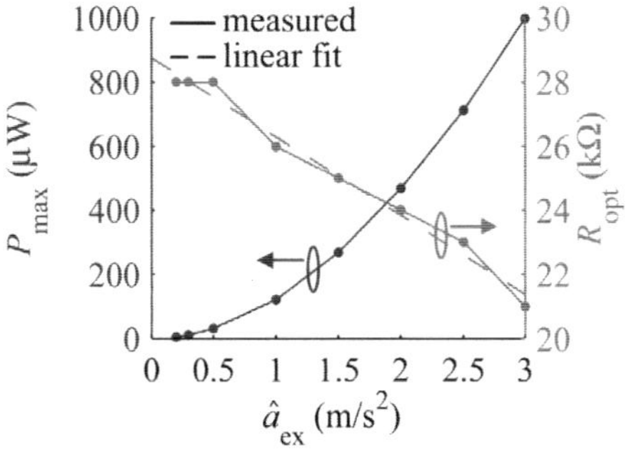

Figure 3. Measured maximum power P_{max} and measured optimal load resistance R_{opt} of the electromagnetic kinetic energy transducer.

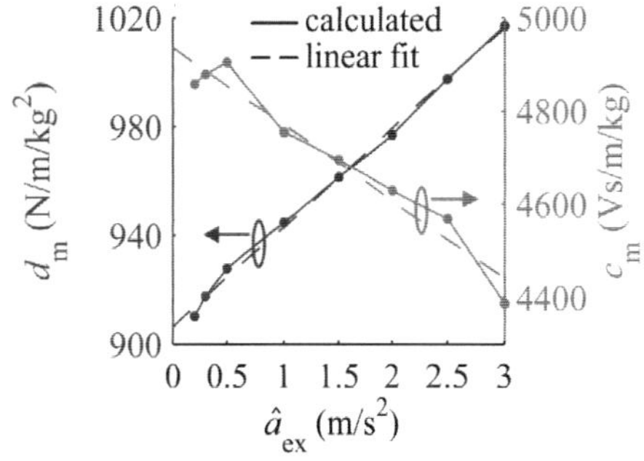

Figure 4. Normalized parasitic damping d_m and corresponding normalized electromagnetic coupling c_m.

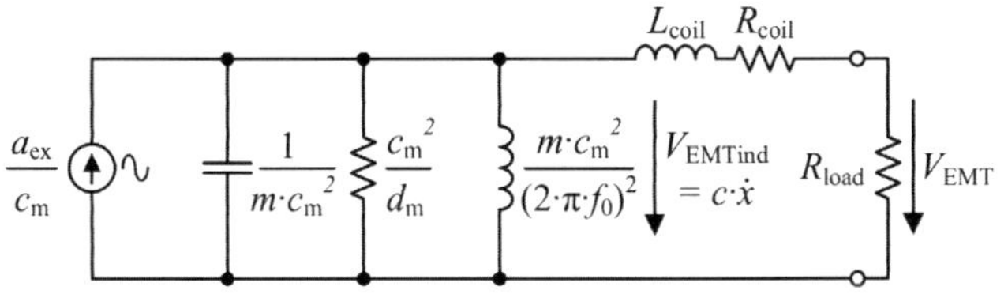

Figure 5. Normalized equivalent circuit model of the electromagnetic kinetic energy transducer connected to a load resistance.

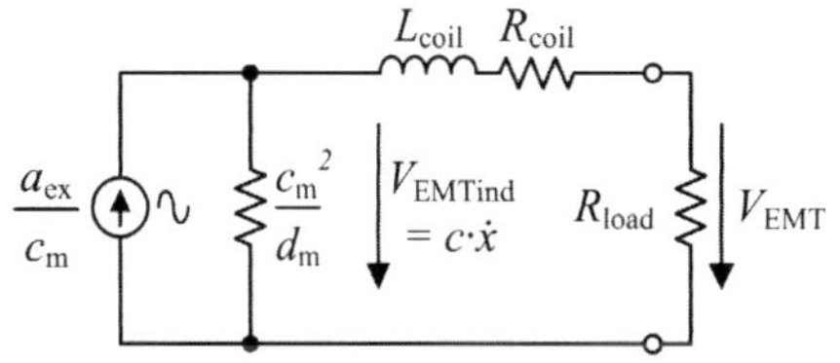

Figure 6. Normalized equivalent circuit model of the transducer connected to a resistive load at resonance.

Figure 4 illustrates calculation results of d_m and c_m as well as the corresponding linear fits which are used in circuit simulations. This allows accurate simulations even though the optimal load resistance, the electromagnetic coupling as well as the parasitic damping depend on the external excitation as demonstrated in Figure 3 and Figure 4 [4, 9].

The normalized parameters d_m and c_m result in the normalized equivalent circuit model of the electromagnetic kinetic energy harvester shown in Figure 5. In Figure 6 the corresponding reduced normalized model valid at resonance is given. It is demonstrated that the seismic mass is completely incorporated in the normalized parameters at resonance for this linear load scenario.

3. Guidelines for optimal energy conditioning circuitry design

In order to harvest the highest possible energy by means of active rectification the transducer must be excited at resonance and the active rectifier (Figure 1) must be operated at the maximum power point (MPP) (Figure 7) [6]. Maximum power point operation can be accomplished via a MPP-harvesting interface [10]. The design of such an interface requires knowledge about the optimum transducer load and the optimum rectifier load that enables MPP operation.

The normalized model given in Figure 6 allow to express the optimal load resistance R_{opt} and the maximum load power P_{max} of the electromagnetic kinetic energy transducer at resonance using the original model parameters [3–5, 8] as well as the extractable normalized parameters:

$$R_{opt} = \frac{c^2}{d} + R_{coil} = \frac{{c_m}^2}{d_m} + R_{coil}, \tag{4}$$

$$P_{max} = \frac{{\hat{V}_{opt}}^2}{2R_{opt}} = \frac{(m\hat{a}_{ex}c)^2}{8d(c^2 + R_{coil}d)} = \frac{(\hat{a}_{ex}c_m)^2}{8d_m({c_m}^2 + R_{coil}d_m)} = \frac{{\hat{a}_{ex}}^2}{8d_m}\frac{R_{opt} - R_{coil}}{R_{opt}}, \tag{5}$$

where $\hat{V}_{\mathrm{opt}}$ is the optimal peak voltage across R_{opt} delivered by the transducer.

Assuming an ideal full wave bridge rectifier composed of ideal diodes allows to give approximate MPP-harvesting interface design expressions by means of R_{opt}, P_{max}, and R_{coil}:

$$P_{\mathrm{MPP}} \approx P_{\mathrm{max}} \frac{R_{\mathrm{opt}} - R_{\mathrm{coil}}}{R_{\mathrm{opt}}}, \tag{6}$$

$$V_{\mathrm{MPP}} \approx \sqrt{2P_{\mathrm{max}} R_{\mathrm{opt}}} \frac{R_{\mathrm{opt}} - R_{\mathrm{coil}}}{R_{\mathrm{opt}}} = \hat{V}_{\mathrm{opt}} \frac{R_{\mathrm{opt}} - R_{\mathrm{coil}}}{R_{\mathrm{opt}}}, \tag{7}$$

$$R_{\mathrm{MPP}} \approx 2(R_{\mathrm{opt}} - R_{\mathrm{coil}}), \tag{8}$$

$$\alpha_{\mathrm{MPP}} \approx 2 \arccos \left(\frac{R_{\mathrm{opt}} - R_{\mathrm{coil}}}{R_{\mathrm{opt}} + R_{\mathrm{coil}}} \right), \tag{9}$$

where P_{MPP} is the maximum rectifier load power enabled by the optimal rectifier load resistance R_{MPP}. At the MPP the rectified voltage V_{rect} (Figure 1) is driven to V_{MPP}. This voltage level induces the conduction angle α_{MPP} of the ideal diodes. The given expressions are also applicable for a low voltage drop active rectifier with low power loss as verified in Section 4.

Excitations below the mechanical seismic mass displacement limit of the packaged harvester shown in Figure 1 when loaded with the optimal resistance at resonance are considered [3].

4. Results and verifications

Figure 7 shows simulation results (dashed) and measurement results (solid) of P_{max} for an excitation of $\hat{a}_{\mathrm{ex}} = 1\,\mathrm{m/s^2}$ at resonance. Moreover, measured (solid) and simulated (dashed) characteristics of the active rectifier load power P_{rect} are shown. These simulation results are achieved with model parameters valid for the actual excitation ($\hat{a}_{\mathrm{ex}} = 1\,\mathrm{m/s^2}$). Simulation and measurement results correspond almost perfectly. The dotted characteristics show simulation results achieved with model parameters determined at another excitation ($\hat{a}_{\mathrm{ex}} = 3\,\mathrm{m/s^2}$, Figure 4) different from the actual excitation ($\hat{a}_{\mathrm{ex}} = 1\,\mathrm{m/s^2}$). The mismatch of these dotted characteristics compared to the measurement results demonstrates that it is insufficient to use only a single set of constant parameters. The presented simulations are performed with the original model shown in Figure 2 implemented in the circuit simulator applying the real weight of the seismic mass.

In Figure 8 the seismic mass m is swept up to the weight of the packaged transducer which is the possible maximum weight by applying the normalized model shown in Figure 5 with the

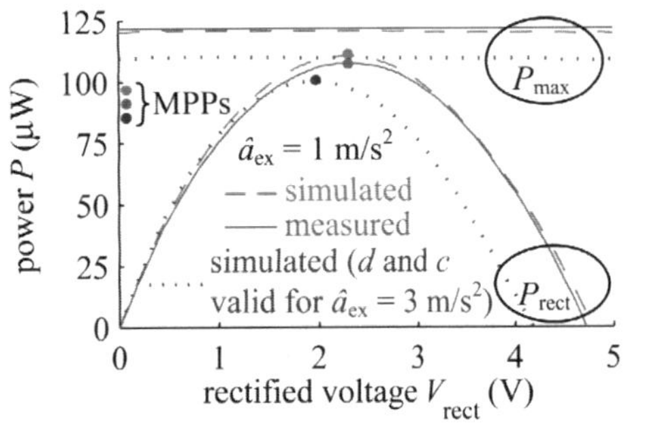

Figure 7. Comparison of measurement and simulation results of the maximum transducer power P_{max} as well as the active rectifier load power P_{rect} at resonance.

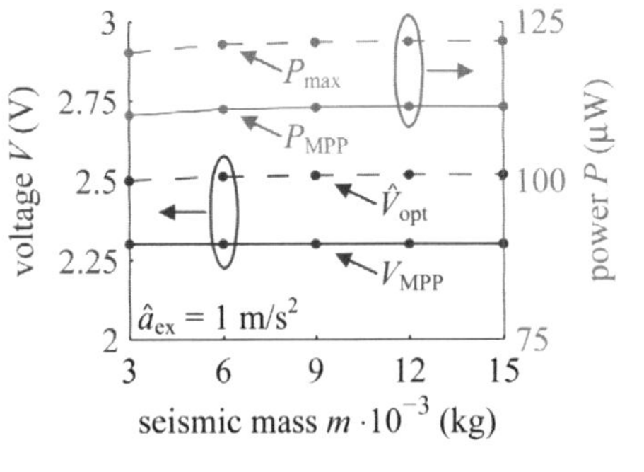

Figure 8. Simulated sweep of m at resonance performed for the transducer connected to its optimal load resistance and to the active rectifier operated at its maximum power point.

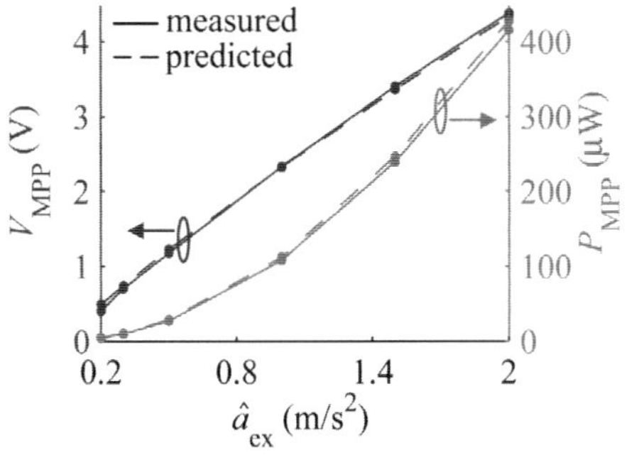

Figure 9. Measured and predicted maximum rectifier load power P_{MPP} and related V_{MPP}.

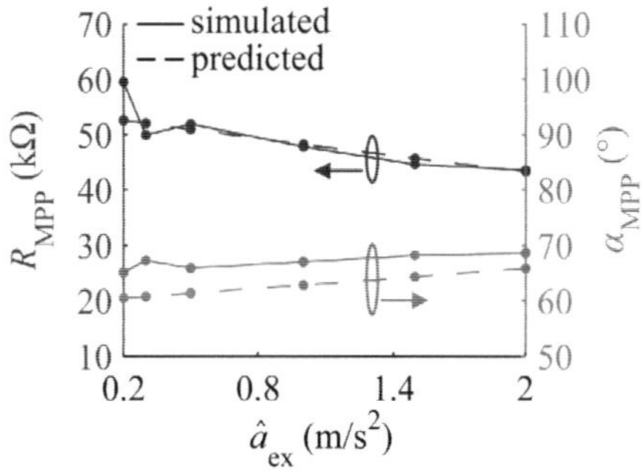

Figure 10. Simulated as well as predicted R_{MPP} and α_{MPP} which enable MPP operation.

linear fits of the calculated parameters (Figure 4) implemented in the circuit simulator. Figure 8 shows the maximum transducer load power P_{max}, the optimal transducer peak voltage $\hat{V}_{opt}$, the maximum active rectifier load power P_{MPP}, and the optimal rectified voltage V_{MPP} for an excitation of $\hat{a}_{ex} = 1\,m/s^2$ at resonance. These simulation results show almost independence of the seismic mass m, demonstrating that m needs not to be known exactly in order to accurately simulate the optimum operation of the kinetic energy harvesting system shown in Figure 1. This optimum operation simulation accuracy allowed by the normalized model is of major importance for the design of an efficient MPP-harvesting interface. Hence, there is no need to know m exactly or to disassemble the harvester [5] in order to extract the weight of the seismic mass.

Figure 9 presents the measured maximum active rectifier load power P_{MPP} and the corresponding voltage V_{MPP} at resonance. Furthermore, corresponding predicted values achieved by means of (6) and (7) are given. In Figure 10 simulation results of the optimal rectifier load resistance R_{MPP} and the optimal conduction angle α_{MPP} as well as the predicted results calculated with (8) and (9) are shown. The MPP prediction precision demonstrated in Figure 9 and Figure 10 verify the utility of the energy conditioning design guidelines given in Section 3.

5. Conclusion

The proposed electromagnetic kinetic energy harvester model allows precise co-simulations with pure resistive and nonlinear loads such as active rectifiers also for variable excitations. Fully packaged harvesters can be parameterized by means of the presented transducer characterization without harvester disassembly. The proposed optimal energy conditioning design methodology allow appropriate interface circuitry in order to increase the harvested electrical energy output.

Acknowledgments

This work is supported by the German Research Foundation (DFG) under grant no. GRK1322.

References

[1] Gungor V C and Hancke G 2009 *IEEE Trans. Ind. Electron.* **56** 4258–4265
[2] Manoli Y 2010 *Proc. IEEE Eur. Solid-State Circuits Conf. (Seville)* pp 27–36
[3] Spreemann D, Hoffmann D, Folkmer B and Manoli Y 2008 *J. of Micromech. and Microeng.* **18** 104001 (10pp)
[4] Cepnik C, Yeatman E and Wallrabe U 2012 *J. Intel. Mat. Systs. Str.* **23** 1533–1541
[5] Cheng S, Wang N and Arnold D P 2007 *J. of Micromech. and Microeng.* **17** 2328–2335
[6] Maurath D, Becker P F, Spreemann D and Manoli Y 2012 *J. IEEE Solid-State Circuits* **47** 1369–1380
[7] Williams C, Shearwood C, Mellor P, Birch T and Yates R 2001 *Proc. IEE Circuits Devices Syst.* **148** 337–342
[8] Stephen N 2006 *J. Sound Vib.* **293** 409–425
[9] Cheng S, Wang N and Arnold D P 2007 *Proc. PowerMEMS (Freiburg)* pp 89–92
[10] Leicht J, Maurath D and Manoli Y 2012 *Proc. IEEE Eur. Solid-State Circuits Conf. (Bordeaux)* pp 101–104

Power Estimation for Piezoelectric Energy Harvesters in Flexure Mode with Large Displacement Amplitude

M Hara[1], L V Minh, H Oguchi and H Kuwano

Department of Nanomechanics, Tohoku University, Sendai, JAPAN

hara@nanosys.mech.tohoku.ac.jp

Abstract. We propose a new method to estimate the output power from piezoelectric energy harvesters using flexure mode of the cantilever. In the energy harvester, displacement amplitude is too large for Bernoulli-Euler hypothesis to hold true. Hence, it is not easy to derive a theoretical solution of output power. In this study, applying the correction coefficient which is a function of flexure rigidity to a conventional theoretical solution, simple equation to estimate the output power was derived. This equation was applied to design a practical $(K,Na)NbO_3$ based energy harvester. Measured output from the harvester was in good agreement with the calculated value.

1. Introduction

Recently, cantilever type micro generators using piezoelectric material are actively researched to harvest the energy from environmental vibrations [1], and spotlighted as a power source for various applications, such as tire pressure monitoring systems [2] or implantable medical devices [3]. For these applications, it is required for the harvester to simultaneously meet many specifications, which are size, height, resonant frequency and output power. However, a theoretical solution to relate these specifications is not established well. In the micro energy harvester, since amplitude of vibration is too large to apply Bernoulli-Euler hypothesis, deriving the solution from the strain in the piezoelectric film becomes complicated. Therefore we have to spend much time for numerical calculation to design such harvester.

In this study, applying the correction coefficient which is a function of flexure rigidity to a conventional theoretical solution, simple equation to estimate the output power was derived. Then, the derived equation was applied to design a practical $(K,Na)NbO_3$ based energy harvester as a verification.

2. Theoretical solution of output power

Figure 1 shows a schematic illustration of the cantilever based piezoelectric harvester. In the harvester, induced charge in the piezoelectric film can be calculated from a following equation,

$$Q = b\int_0^L e_{31}S_x + \varepsilon_{33}E_z dx = \frac{bhe_{31}}{2}\frac{\partial^2 F}{\partial x^2}\bigg|_{x=L} - bL\varepsilon_{33}\frac{V}{d} \qquad (1),$$

where, S_x is strain in x-direction, E_z is electric field in z-direction, and F is a mode function of cantilever. Assuming a fundamental mode, the mode function F is as below:

---

[1] Corresponding author: Tel & Fax +81-22-795-4771, E-mail hara@nanosys.mech.tohoku.ac.jp

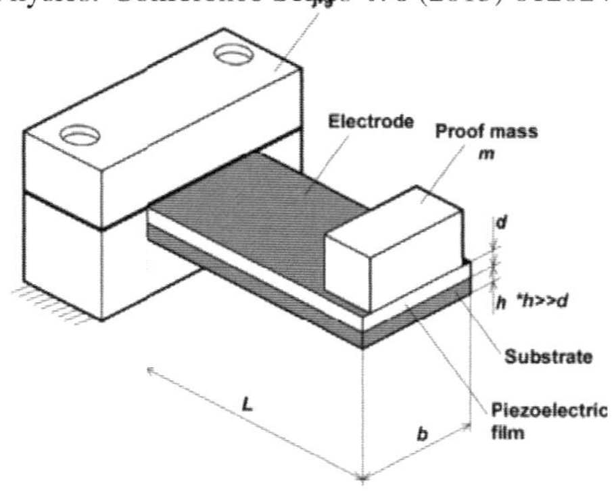

Figure 1. Schematic of cantilever based piezoelectric energy harvester

$$F(x) = C_0\left[K\left(\cosh\frac{\lambda}{L}x - \cos\frac{\lambda}{L}x\right) + \left(\sinh\frac{\lambda}{L}x - \sin\frac{\lambda}{L}x\right)\right] \quad (2),$$

where, K and λ are -1.3622 and 1.875, respectively. C_0 is also constant determined with vibration conditions. In this equation, a term of time $e^{-j\omega t}$ was omitted.

When the harvester is connected to the external load R, output power P can be obtained using following relation:

$$P = RI^2 = R\left(\frac{\partial Q}{\partial t}\right)^2 \quad (3).$$

Substituting (1) and (2) into (3) and using following relations

$$\text{Ohm's low: } V = RI, \quad \text{Matching condition: } R = \frac{1}{\omega C} = \frac{d}{\omega\varepsilon_{33} bL} \quad (4)(5),$$

P is derived as

$$P = \frac{1}{32}\frac{\omega b d h^2 e_{31}^2}{\varepsilon_{33} L} C_0^2\left[K\frac{\lambda}{L}(\sinh\lambda + \sin\lambda) + \frac{\lambda}{L}(\cosh\lambda + \cos\lambda)\right]^2 \quad (6).$$

3. Introduction of correction coefficients

Table 1 shows a performance of cantilever based energy harvester reported from other researchers [4-9]. The constant C_0 and displacement of the top of cantilever $F(L)$ were calculated from (2) and (6). In this figure, C_0 and $F(L)$ were overestimated by using Eqs. (6). This overestimation was noticeable in the smaller harvester. This is interpreted as that Bernoulli-Euler hypothesis was broken by deformation of the mode function with large displacement amplitude in the micro device. We must reconstruct the output power equations. However, it is complicated and not so easy to derive the strain distribution under the large displacement amplitude without Bernoulli-Euler hypothesis.

In this study, we try to empirically expect the output power. In Fig. 2, the C_0 was normalized with vibration condition and plotted for the flexure rigidity EI. From this figure, it was found that plotted points were ordered on one line. Using this linear relation, Eqs. (6) can be modified as below:

$$P = \frac{1}{32}\frac{\omega b d h^2 e_{31}^2}{\varepsilon_{33} L}\left\{m_{eq} a L^3 A (EI)^B\right\}^2\left[K\frac{\lambda}{L}(\sinh\lambda + \sin\lambda) + \frac{\lambda}{L}(\cosh\lambda + \cos\lambda)\right]^2 \quad (7),$$

where, A and B are approximation coefficients obtained from Fig. 2, a and m_{eq} are a vibration acceleration and an equivalent mass, respectively.

Table 1. Performance of cantilever based piezoelectric energy harvester

Author	*Material*	f_r (kHz)	a (m/s^2)	m_{eq} (mg)	P (μW)	L (mm)	EI (μNm^2)	C_0	$F(L)$ (mm)
M. Marzencki *et. al.* [4]	AlN/Si	1500	19.6	0.48	0.45	0.52	1.08×10^{-3}	1.48×10^{-5}	106
H. Fang *et. al.* [5]	PZT/Si	609	9.8	1.61	2.16	1.7	3.79×10^{-2}	1.22×10^{-5}	26.9
D. Shen *et. al.* [6]	PZT/Si	463	19.6	1.37	2.15	3.2	0.186	2.90×10^{-5}	34.0
R. Elfrink *et. al.* [7]	AlN/Si	573	9.8	35.4	30.0	3.5	4.94	1.69×10^{-4}	181
Z Cao *et. al.* [8]	AlN /SUS	74.1	9.8	1010	5.13	4.2	10.1	1.94×10^{-4}	173
I. Kanno *et. al.* [9]	KNN/Si	1036	10	6.16	1.1	16.5×10^3	295	5.45×10^{-5}	2.77
	PZT/Si	982	10	6.54	1.0	17.5×10^3	295	7.74×10^{-6}	1.66

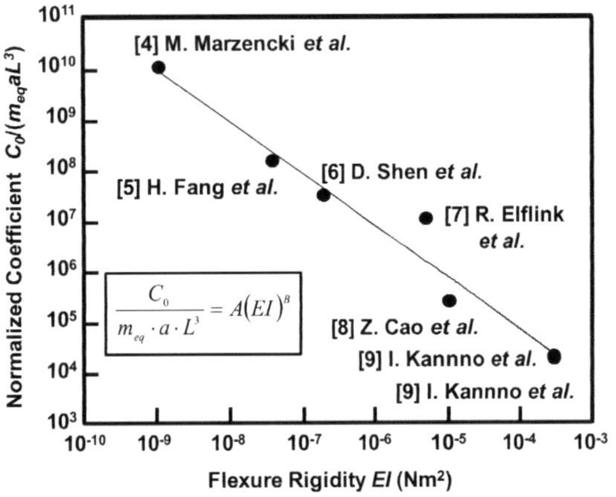

Figure 2. Relationship between the C_0 and flexure rigidity in the cantilever based piezoelectric energy harvester

4. Design and fabrication of the cantilever based energy harvester

Using derived equation, we practically design the cantilever based energy harvester to drive the communication chip based on IEEE 8002.15.4 standard. Table 2 shows a size of designed harvester. EI and C_0 were 2.11×10^{-2} μNm^2 and 2.93×10^{-6}, respectively. Output power was designed to 980 nW. According to the specifications of the communication chips [10][11], it was enough to drive them that the power was 50 mW at transmission/receiving mode and 3 μW at sleeping mode. When communication time and interval were respectively set to 5 msec. and 1 min., the power required for the harvester was about 7 μW. By connecting 10 harvesters in series, the energy to drive the

Table 2. Size of the cantilever based energy harvester

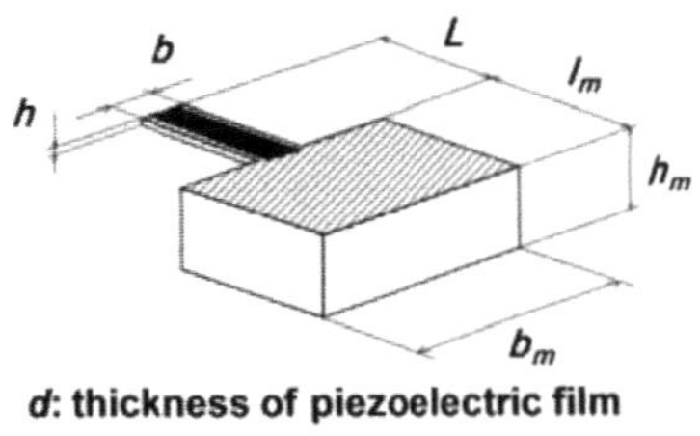

<Cantilever>	
Material	Si
b	210 μm
h	28 μm
L	1000 μm
<Proof mass>	
Material	Si
b_m	1000 μm
h_m	500 μm
l_m	600 μm
<Piezo film>	
Material	(K,Na)NbO$_3$
d	2 μm

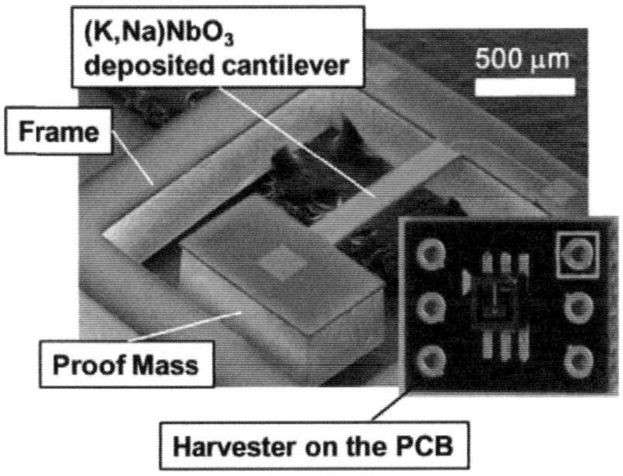

Figure 3. SEM image of the cantilever based energy harvester and photograph of the harvester mounted on the PCB for evaluation.

communication chip can be obtained sufficiently. Figure 3 shows a scanning electron micrograph of the energy harvester fabricated based on the design. This harvester was fabricated with fully Si bulk-micromachining [12].

5. Verification

Figure 4 (a) shows output power for various load resistance when vibration acceleration was 9.8 m/s^2. Using load resistance of 90 kΩ, maximum output power of 730 nW was obtained. Figure 4 (b) shows the relationship between the C_0 and *EI* including the result of this study when the optimal lord resistance was applied. It was confirmed in Fig. 4 (b) that the result was plotted on the vicinity of approximation line. From these results, though this method is suitable for power estimation in the prototype fabrication, it can be mentioned that the accuracy was insufficient to optimize the output power. The relationship shown in Fig. 2 included the data of the harvesters using various piezoelectric materials in order to obtain the enough plotting points. If you can correct the data of the harvesters with same materials, you will obtain the more accurate approximation constants and be able to optimize the design using this method.

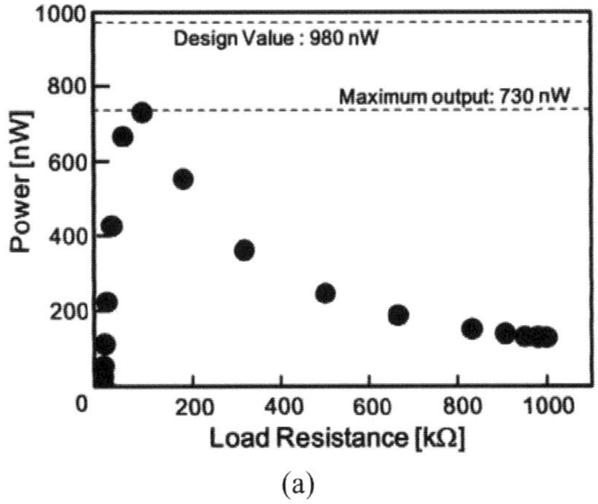

(a)

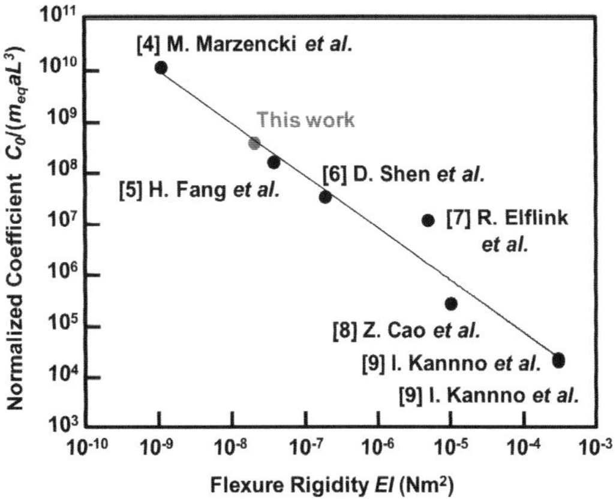

(b)

Figure 4. Evaluation results of fabricated resonator: Output power as varied load resistance (a), and relationship between the C_0 and *EI* (b)

6. Conclusion

A cantilever based piezoelectric energy harvesters is a key device for autonomous micro devices. The harvester must be design to satisfy many specifications such as size, resonant frequency, output power, and output impedance. However, it becomes complicated in the micro energy harvester to theoretically calculate the output power from the strain in the piezoelectric film since amplitude of vibration is too large to apply Bernoulli-Euler hypothesis. In this study, newly introducing the correction coefficient which is a function of flexure rigidity to the conventional theoretical solution, simple equation to estimate the output power was derived.

Then, this equation was applied to design a practical (K,Na)NbO_3 (KNN) based energy harvester. The designed harvester was fully fabricated with bulk micromachining. The harvester had the cantilever size of 210 μm × 28 μm × 1000 μm, the proof mass size of 1000 μm × 500 μm × 600 μm, and the thickness of KNN film of 2 μm. As an evaluation result, maximum output power of 730 nW was achieved when the load resistance, applied vibration frequency and the vibration acceleration were 90 Ω, 1.5 kHz and 9.8 m/s^2. And, the obtained output power was in good agreement with the design value.

Acknowledgement
We are grateful to Dr. Fumimasa Horikiri, Dr. Kenji Shibata, and Dr. Tomoyoshi Mishima at Hitachi Metal, Ltd. for their excellent advice and support. A part of this work was performed at micro/nano-machining research and education center, Tohoku University, Japan. This work was conducted under the project "Research of a nano-energy system creation" (No.18GS0203), funded by the Ministry of Education, Culture, Sports, Science and Technology (MEXT).

References

[1] S. Roundy *et al.* 2003 Computer Communications **26** 1131
[2] R. Vullers *et al.* 2013 *Proc.* the 17th *int. Conf.* on solid-state sensors, actuators and microsystems (Transducers 2013) 685
[3] L. Beker *et al.* 2013 *Proc.* the 17th *int. Conf.* on solid-state sensors, actuators and microsystems (Transducers 2013) 1663
[4] M. Marzencki *et al.* 2008 *Sens.* Actuators A **145** 363
[5] H. Fang *et al.* 2006 *Microelectron. J.* **37** 1280
[6] D. Shen *et al.* 2008 *J. Micromech. Microeng.* **18** 055017
[7] R. Elfrink *et al.* 2009 *J. Micromech. Microeng.* **19** 094005
[8] Z. Cao *et al.* 2012 *Sens.* Actuators A **179** 178
[9] I. Kanno *et al.* 2012 *Sens.* Actuators A **179** 132
[10] http://www.lapis-semi.com/en/company/news/news2012/r20120620.html
[11] http://www.jennic.com/products/protocol_stacks/selection_guide
[12] L. Minh *et al.* 2013 *J. Micromech. Microeng.* **23** 035029

Best voltage bias-flipping strategy towards maximum piezoelectric power generation

Junrui Liang[1] and Henry Shu-Hung Chung[2]

[1] School of Information Science and Technology, ShanghaiTech University, 319 Yueyang Road, Shanghai 200031, China
[2] Centre for Smart Energy Conversion and Utilization Research, City University of Hong Kong, Kowloon Tong, Hong Kong, China

E-mail: liangjr@shanghaitech.edu.cn

Abstract. In piezoelectric energy harvesting (PEH) systems, energy extracted from piezoelectric structure can be increased by making piezoelectric voltage in phase with vibration velocity and raising the voltage amplitude. Such voltage manipulations can be realized by synchronously flipping the piezoelectric voltage with respect to a bias dc source at every displacement extremum. Given that net harvested energy is obtained by deducting dissipated energy from total extracted energy, a sophisticated voltage bias-flipping scheme, which can maximize extracted energy at low dissipative cost, is required towards harvested energy optimization. This paper extends the state of the art by proposing the best bias-flip strategy, which is delivered on conceptual synchronized multiple bias-flip (SMBF) interface circuits. The proposed strategy coordinates both requirements on larger voltage change in synchronized instant for more extracted energy and smaller voltage change in each bias-flip action for less dissipated energy. It not only leads to further enhancement of harvesting capability beyond existing solutions, but also provides an unprecedented physical insight on maximum achievable harvesting capability of PEH interface circuit.

1. Introduction

In piezoelectric energy harvesting (PEH), various harvesting interface circuits have been developed one after another, refreshing the record of harvesting capability [1]. Among these solutions, the synchronized switch harvesting on inductor (SSHI) has set a milestone in circuit development of PEH systems [2]. Recent development has shown that, compared to SSHI case, more energy might be rewarded by subtly returning a part of the reclaimed energy to the vibrating structure at right moments. Such solutions were called energy injection [3], pre-biasing [4], or energy investment [5]. The control of energy flow in these solutions is achieved by successively flipping the voltage across piezoelectric element twice through two different bias dc voltages: one for harvesting energy, called *passive* bias-flip action in this paper; and the other for investing energy, called *active* bias-flip action. Given the seemingly endless development trend of PEH interface circuit, it is curious that:

a) whether PEH capability can go beyond the state-of-the-art solutions?
b) which kind of solution yields the best PEH capability, passive or hybrid (passive + active)?
c) whether PEH capability in the latest double bias-flip solutions has already been optimized?
d) is there any limit for the improvement of PEH interface circuit?

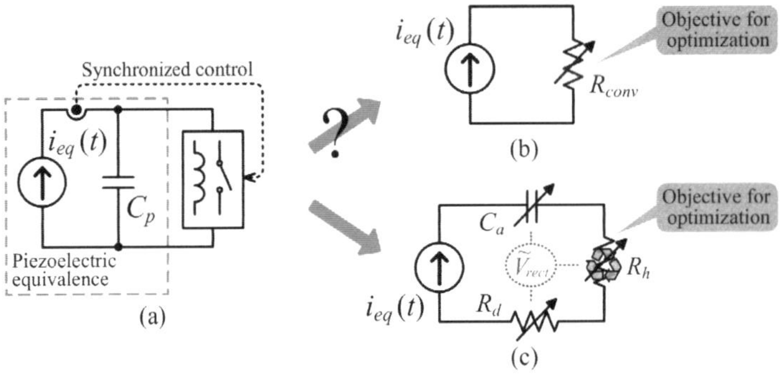

Figure 1. Bias-flip solution for PEH. (a) Principle. (b) Equivalence targeted at power conversion (extraction). (c) Equivalence targeted at energy harvesting [11].

Inspired by these curiosities, this paper generalizes the state-of-the-art bias-flip solutions with a conceptual synchronized multiple bias-flip (SMBF) circuit topology. The best voltage bias-flipping strategy towards maximum PEH capability is then developed based on this generalized SMBF topology. Through these theoretical development, new insights are gained for better understanding and further improvement of PEH interface circuit [3, 4, 5].

2. Objective for PEH optimization

The bias-flip technique was originally developed to enhance piezoelectric conversion for vibration damping purpose [6, 7]. Its principle is shown in figure 1(a). By synchronously flipping the piezoelectric voltage with respect to zero [6] or a bias dc voltage source [7] at zero-crossing instants of vibration velocity, the combination of piezoelectric capacitance C_p and connected circuit might behave like a tunable resistance R_{conv}, as shown in figure 1(b). Given the correlation between energy harvesting and vibration damping [8], the increase of electrically induced damping was once regarded as target for PEH circuit improvement in some studies [8, 9]. Yet, distinguishing the effects between harvested energy and dissipated energy (unavoidable during power conditioning) in total extracted energy, only the harvested portion should be identified as objective for PEH [10], as shown in figure 1(c). Therefore, exactly speaking, PEH circuit improvement should be specified to maximize the electrically induced damping *at low dissipative cost*. Synchronously inverting the piezoelectric voltage through an inductor provides the possibility towards this goal [2, 3, 4, 5].

3. SMBF interface circuits

From energy cycle analysis [10], given a PEH system vibrating with constant amplitude, more extracted energy requires larger voltage change before and after voltage flipping state; on the other hand, less dissipated energy requires smaller voltage change before and after each bias-flip action, because dissipated energy in a bias-flip action *parabolically rise* with such voltage change. The only way to satisfy both requirements is to make a large voltage change by combining multiple small steps, which leads to the proposal of SMBF interface circuits.

The circuit topologies of SMBF are shown in figure 2. The two configurations of series SMBF (S-SMBF) and parallel SMBF (P-SMBF) are quite similar to S-SSHI and P-SSHI, respectively, except that there are multiple bias dc sources for continuous bias-flip actions. In every zero-crossing instant of i_{eq}, the piezoelectric voltage v_p are flipped for M times with respect to the bias dc sources from $V_{b,1}$, $V_{b,2}$ to $V_{b,M}$ in sequence. These bias voltages are assumed unknown variables at start; source symbols in figure 2 are only for indicating positive directions. Half of the switch paths with switch S_{+m} are for positive current flippings, while the others with switch

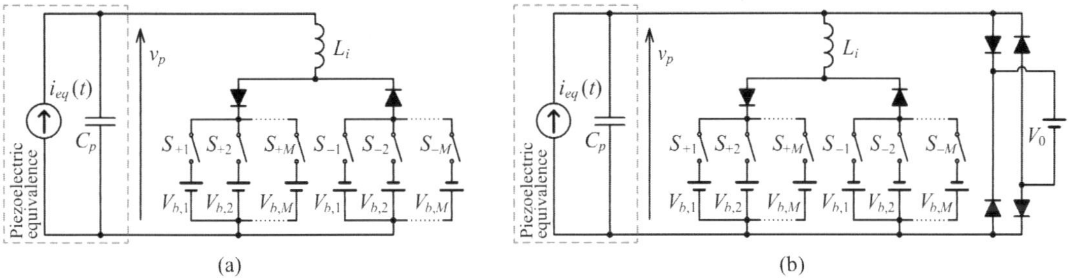

Figure 2. Conceptual SMBF interface circuits. (a) S-SMBF. (b) P-SMBF.

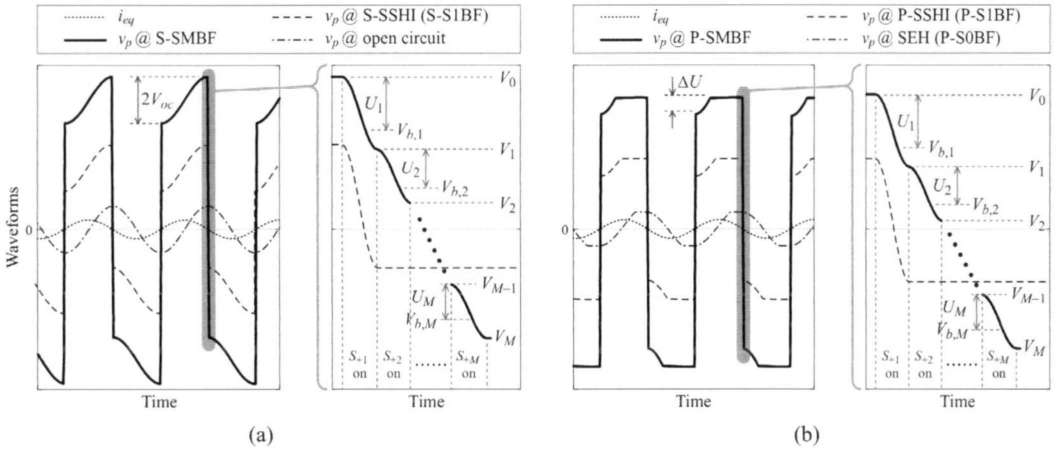

Figure 3. Waveforms in SMBF. (a) S-SMBF. (b) P-SMBF.

S_{-m} are for negative ones. Each switch-on time is half of an LC cycle, i.e., $\pi\sqrt{L_iC_p}$, for making largest voltage change with lowest dissipative cost. Therefore, flipping factor $\gamma = -e^{-\pi/(2Q)}$.

Figure 3 shows the waveforms in SMBF. Compared to conventional SSHI, SMBF can further boost the amplitude of v_p, which implies that more power can be extracted from the piezoelectric structure. In fact, S-SSHI can be regarded as S-S1BF (special case of S-SMBF with $M = 1$); P-SSHI and standard energy harvesting (SEH) interface can be regarded as P-S1BF and P-S0BF (special cases of P-SMBF with $M = 1$ and 0), respectively. The switch schedule in a synchronized instant is also zoomed in and shown in figure 3. The intermediate voltages in an SMBF process are denoted as V_m $(m = 0, 1, 2, \cdots, M)$; their differences to the bias voltages $V_{b,m+1}$ are denoted as U_{m+1}.

4. Best voltage bias-flipping strategy

The advancement of SMBF on piezoelectric power generation needs to be mathematically justified. With the denotations about intermediate voltages of v_p and bias dc sources, which were given in figure 3, voltage relations among these unknown voltages in all SMBF solutions are summarized by (1). Regarding all voltages are nondimensionalized with respect to the open circuit voltage V_{oc}; and nondimensionalizing all energy amounts with respect to $C_pV_{oc}^2$, maximum harvestable energy in one cycle by using SEH, the net total energy income of all bias dc sources in figure 2 is given by (2), where each $V_{b,m}$ can be expressed as functions of U_1, U_2, ..., U_M, and ΔU. Since U_1, U_2, ..., U_M, and ΔU are independent variables, optimal harvesting condition can be obtained by solving (3), where $m = 1, 2, ..., M$. In P-SMBF, all U_m and ΔU can be

$$\begin{bmatrix} 1 & & & & & & 1 \\ 1 & -1 & & & & & \\ & \ddots & \ddots & & & & \\ & & 1 & -1 & & & \\ & & & \ddots & \ddots & & \\ & & & & & 1 & -1 \end{bmatrix} \begin{bmatrix} V_0 \\ V_1 \\ \vdots \\ V_m \\ \vdots \\ V_M \end{bmatrix} = \begin{bmatrix} \Delta U \\ (1-\gamma)\, U_1 \\ \vdots \\ (1-\gamma)\, U_m \\ \vdots \\ (1-\gamma)\, U_M \end{bmatrix} \tag{1}$$

$$E_h = 2\left[\sum_{m=1}^{M} (1-\gamma)\, U_m V_{b,m} + (2-\Delta U)\, V_0\right] \tag{2}$$

$$\begin{cases} \dfrac{\partial E_h}{\partial U_m} = 0 \\ \dfrac{\partial E_h}{\partial \Delta U} = 0 \end{cases} \Rightarrow \begin{cases} U_{m,opt} = \dfrac{1}{1+\gamma} \\ \Delta U_{opt} = 1 \end{cases} \tag{3}$$

$$(E_{h,max})_{\text{P-SMBF}} = M\frac{1-\gamma}{1+\gamma} + 1\,; \qquad (E_{h,max})_{\text{S-SMBF}} = M\frac{1-\gamma}{1+\gamma} \tag{4}$$

$$(E_{h,max})_{\text{SSPB}} = -\frac{8\gamma}{1+\gamma^2} \tag{5}$$

tuned for approaching this optimal harvesting condition; while in S-SMBF, ΔU is fixed at 2. Under optimal harvesting condition, maximum harvested energies in both cases are derived as (4). The purpose of energy injection [3], pre-biasing [4], and energy investment [5] is to achieve the S-S2BF case. Yet, in these previous solutions, the intermediate voltage V_1 is mandatorily fixed at zero, rather regarded as free for optimization. In the single-supply pre-biasing (SSPB) solution [4], maximum harvested energy is given by (5).

Based on (4) and (5), figure 4(a) illustrates the nondimensional maximum harvested energy as function of γ in optimal P-, S-SMBF, and SSPB. It shows that, under the same bias-flip number M, optimal P-SMBF is able to harvest more energy than its series counterpart. All $E_{h,max}$ approach infinity as γ approaches -1 (ideal lossless flipping case), except the $M = 0$ cases. The larger M, i.e., more bias-flip actions for v_p in each synchronized instant, the larger $E_{h,max}$ under same γ. For SSPB, as observed from figure 4(a), it outperforms optimal SEH (P-S0BF), S-SSHI (S-S1BF), and P-SSHI (P-S1BF) when $\gamma < -0.123$, -0.172, and -0.333, respectively. Yet, it cannot catch up with optimal S-S2BF, whose circuit topology and operation are the same, except V_1 is free rather then fixed in optimization. Figure 4(b) shows the maximum harvested energy $E_{h,max}$ as function of V_1 in general S-S2BF configuration. Simulation results obtained with PSIM software show good agreement with theoretical prediction. More energy can be harvested when $V_1 = 1$ rather than 0, regardless of γ. Therefore, rather than presuming any intermediate voltages, equation (3) should be regarded as a guideline towards maximum energy harvesting capability for SMBF interface circuits.

5. Conclusion

This paper has discussed some crucial questions towards better understanding and future development of piezoelectric energy harvesting (PEH) interface circuit. Through the theory development, it is found that:

a) PEH capability might exceed state-of-the-art solutions by involving more bias-flip actions in every synchronized instant;

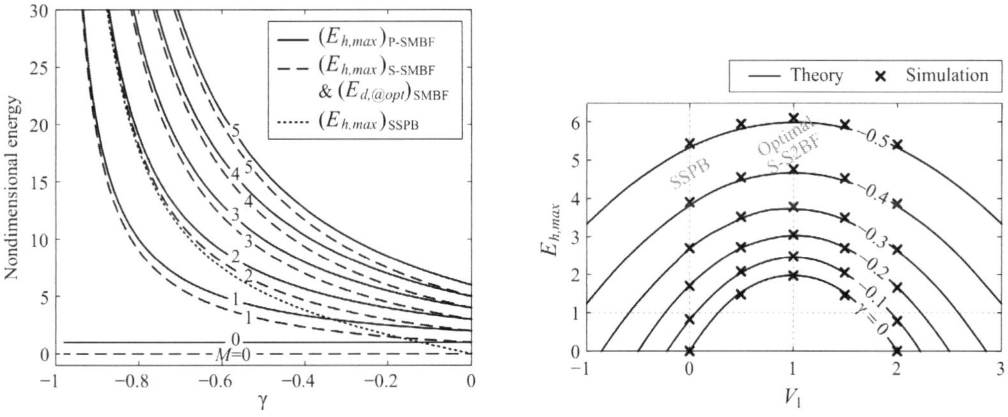

Figure 4. Maximum harvested energy in (a) different solutions, (b) S-S2BF under different intermediate voltage V_1.

b) for flipping number $M < 2$, passive bias-flip solutions give better harvesting capability, while for $M \geq 2$, hybrid solutions perform better;
c) harvested power in the latest S-S2BF solutions can be further optimized by adjusting the intermediate voltage;
d) under constant amplitude vibration, harvested power might approach infinity by either decreasing the inversion factor γ or increasing the flipping number M in SMBF; yet, under force excitation, it is confined by mechanical dynamics.

Given the rational guidance provided by the proposed strategy, more efforts should be taken on multi-source implementation, switches management, etc. towards technical realization of SMBF circuits and the best harvesting strategy delivered on them.

References

[1] Szarka G D, Stark B H and Burrow S G 2012 Review of power conditioning for kinetic energy harvesting systems *IEEE Trans. Power Electron.* **27** 803–815
[2] Guyomar D, Badel A, Lefeuvre E and Richard C 2005 Toward energy harvesting using active materials and conversion improvement by nonlinear processing *IEEE Trans. Ultrason. Ferroelectr. Freq. Control* **52** 584–595
[3] Lallart M and Guyomar D 2010 Piezoelectric conversion and energy harvesting enhancement by initial energy injection *Appl. Phys. Lett.* **97** 014104 (pages 3)
[4] Dicken J, Mitcheson P, Stoianov I and Yeatman E 2012 Power-extraction circuits for piezoelectric energy harvesters in miniature and low-power applications *IEEE Trans. Power Electron.* **27** 4514–4529
[5] Kwon D and Rincon-Mora G A 2012 Energy-investment schemes for increasing output power in piezoelectric harvesters *IEEE 55th International Midwest Symposium on Circuits and Systems* pp 1084–1087
[6] Richard C, Guyomar D, Audigier D and Ching G 1999 Semi-passive damping using continuous switching of a piezoelectric device *Smart Structures and Materials 1999: Passive Damping and Isolation* vol 3672 (SPIE) pp 104–111
[7] Ji H, Qiu J, Adrien B and Zhu K 2009 Semi-active vibration control of a composite beam using an adaptive ssdv approach *J. Intell. Mater. Syst. Struct.* 401–412
[8] Wang Y and Inman D J 2012 A survey of control strategies for simultaneous vibration suppression and energy harvesting via piezoceramics *J. Intell. Mater. Syst. Struct.* **23** 2021–2037
[9] Miller L M, Mitcheson P D, Halvorsen E and Wright P K 2012 Coulomb-damped resonant generators using piezoelectric transduction *Appl. Phys. Lett.* **100** 233901 (pages 4)
[10] Liang J R and Liao W H 2011 Energy flow in piezoelectric energy harvesting systems *Smart Mater. Struct.* **20** 015005
[11] Liang J R and Liao W H 2012 Impedance modeling and analysis for piezoelectric energy harvesting systems *IEEE/ASME Trans. Mechatron.* **17** 1145–1157

Architecture-independent power bound for vibration energy harvesters

E Halvorsen[1], C P Le[1], P D Mitcheson[2] and E M Yeatman[2]

[1] Department of Micro and Nano Systems Technology, Vestfold University College, Raveien 197, N-3184 Borre, Norway

[2] Department of Electrical and Electronic Engineering, Imperial College, London SW7 2AZ, United Kingdom

E-mail: `Einar.Halvorsen@hive.no`

Abstract. The maximum output power of energy harvesters driven by harmonic vibrations is well known for a range of specific harvester architectures. An architecture-independent bound based on the mechanical input-power also exists and gives a strict limit on achievable power with one mechanical degree of freedom, but is a least upper bound only for lossless devices. We report a new theoretical bound on the output power of vibration energy harvesters that includes parasitic, linear mechanical damping while still being architecture independent. This bound greatly improves the previous bound at moderate force amplitudes and is compared to the performance of established harvester architectures which are shown to agree with it in limiting cases. The bound is a hard limit on achievable power with one mechanical degree of freedom and can not be circumvented by transducer or power-electronic-interface design.

1. Introduction

In designing energy harvesters, an important question is how to adjust the parameters of a chosen harvester architecture in order to optimize the performance? This question has been resolved for a number of harvester architectures under various operating conditons [1, 2, 3, 4] and also resulted in guidelines on which among the known architectures to choose for given operating conditions [2]. Another important question is to what extent it is possible to significantly improve ouput power beyond that of the known architectures by inventing new device concepts, either related to the mechanics, the electromechanical conversion or the power electronic interface? In order to answer this question, it is necessary to know the ultimate limits on power without having to specify the details of the transducer and electronic interface, i.e. an architecture-independent power bound is needed. Such a bound based on input-power exists [5, 6], but does not account for parasitic losses which can amount to as much as half the input power even under optimal conditions [1]. To tighten the bound, losses should be accounted for. In this contribution we present a new theoretical bound that takes linear mechanical damping into account without making any a priori assumptions on the the details of the transducer or the electronic sub-system beyond the transducer having only one mechanical port. The optimal output powers of the established harvester architectures are compared to this bound.

2. Power bound

We consider energy harvesting systems characterized by a single mass m and a single mechanical displacement x. The mass is acted upon by an external force F which would usually be the inertial force $F = -m\ddot{y}$ due to the displacement y of the device frame attached to a vibrating body. The proof-mass motion drives an electromechanical transducer which may or may not include an elastic suspension. Unwanted parasitic loss is modelled by the linear damping force $b\dot{x}$ characterized by a constant b. An equivalent circuit for the system is shown in figure 1.

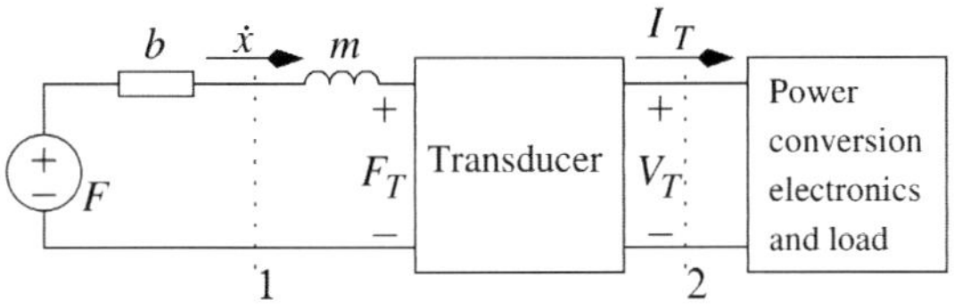

Figure 1. Energy harvesting system.

Newton's second law for the proof mass is $m\ddot{x} = -F_{\rm T} - b\dot{x} + F$ where $F_{\rm T}$ is the transducer force. As opposed to optimizing $F_{\rm T}$ for a given architecture, we here seek a least upper bound on the output power when we are not limited to already known architectures. This problem requires an approach that doesn't hinge on a priori knowledge of $F_{\rm T}$. It is made possible by the power balance which follows from observing that the time average of the mechanical input power minus the time average of power lost in parasitic damping equals the average power delivered to the rest of the system. We therefore base the analysis on this energy difference expressed as

$$E[x;t_{\rm b},t_{\rm a}] = \int_{t_{\rm a}}^{t_{\rm b}} \left[F(t)\dot{x}(t) - b(\dot{x}(t))^2\right] {\rm d}t = \int_{t_{\rm a}}^{t_{\rm b}} (F(t))^2 {\rm d}t/4b - b\int_{t_{\rm a}}^{t_{\rm b}} \left[\dot{x}(t) - F(t)/2b\right]^2 {\rm d}t. \quad (1)$$

for a time interval $[t_{\rm a}, t_{\rm b}]$ and seek the displacement waveform x that maximizes this quantity. This approach is equivalent to optimizing the average of the power $(F - b\dot{x})\dot{x}$ transferred at "1" in figure 1 and for long times it gives an upper bound on the power flow at later stages, such as the opportunity power at the output of the electromechanical transducer ("2" in the figure) or the final output power to a load or to a storage unit.

2.1. Unrestricted proof mass motion

The energy is always less than or equal to the first term on the r.h.s. of (1), so the average power is bounded by $\bar{P}(x;t_{\rm b},t_{\rm a}) = E[x;t_{\rm b},t_{\rm a}]/(t_{\rm b} - t_{\rm a}) \leq F_{\rm rms}^2/4b$ where $F_{\rm rms}$ is the rms force. The equality of the bound is reached if one can realize dynamics that give $\dot{x}(t) = F(t)/2b$. With a harmonic force $F = F_0 \cos\omega t$ of amplitude F_0 and angular frequency ω, $\bar{P} = F_0^2/8b$ when $t_{\rm b} - t_{\rm a} \to \infty$ and coincides with the optimal power point of the velocity damped generator (VDRG) [1, 2]. Therefore, no transducer or power electronic interface design can result in a better performance than the optimal VDRG when the forcing is harmonic and the displacement is unconstrained.

For a possibly irregular F, we insert $\dot{x}(t) = F(t)/2b$ into Newton's second law and solve for $F_{\rm T}$ to determine the transducer force under optimum operation. The result is $F_{\rm T} = F/2 - m\dot{F}/2b = b\dot{x} - m\ddot{x}$ which in the frequency domain amounts to complex conjugate load matching and cannot be achieved exactly over a band of frequencies with a passive load. The problem is evident for white noise which has infinite $F_{\rm rms}$. It is conceivable to closely approximate the optimal force over a limited band of frequencies with an active control strategy as proposed in [7] for another target $F_{\rm T}$. Such a system has bounded input power even with white noise input [8, 9].

2.2. Strict limits on proof mass motion and periodic forcing

We now seek an optimal proof mass trajectory $x = x(t)$ subject to the constraint that $-Z_l \leq x(t) \leq Z_l$ for a limit of displacement Z_l. We start with maximizing the energy (1) over the period T of the force $F(t)$. General solution methods exist for this type of inequality-constrained nonlinear programming problem [10, 11]. Based on the key ideas of these general methods, we present a specific approach for our problem.

The displacement constraint is turned into an equality constraint $x^2+s^2 = Z_l^2$ by introducing an auxiliary real function $s = s(t)$ and is enforced by introducing a Lagrange multiplier $\lambda = \lambda(t)$. We first look for x, s and λ that make $S = \int_0^T \left[F\dot{x} - b\dot{x}^2 + \lambda(x^2+s^2-Z_l^2)\right] dt$ stationary with $\dot{x}$ piecewise differentiable and everywhere continuous. The first order variation of S w.r.t x, s and λ is then zero if

$$2b\ddot{x} + 2\lambda x = \dot{F}, \tag{2}$$

$$\lambda s = 0, \tag{3}$$

$$x^2 + s^2 = Z_l^2. \tag{4}$$

Suppose now that x, s, λ solve (2)-(4), and consider another candidate periodic displacement waveform $\tilde{x}$ which we write $\tilde{x} = x + \Delta x$ and which also fulfils $|\tilde{x}| \leq Z_l$. We permit $\dot{\tilde{x}}$ and $\Delta\dot{x}$ that have a finite number of step discontinuities, so that cases with impulse forces on the proof mass are within the set of admissible $\tilde{x}$. The converted energy (1) can then be written $E[\tilde{x};T,0] = E[x;T,0] - 2\int_0^T \lambda x \Delta x \mathrm{d}t - b\int_0^T \Delta\dot{x}^2 \mathrm{d}t$ by help of an integration by parts, the periodic boundary condition, and (2). The rightmost integral is manifestly non-negative. When $\lambda \neq 0$, then $s = 0$ by (3), we have $|x| = Z_l$ and therefore $x\Delta x \leq 0$. This leads to the conclusion that if $\lambda(t) \leq 0$ for all t, then also $\int_0^T \lambda x \Delta x \mathrm{d}t \geq 0$ and we have $E[\tilde{x};T,0] \leq E[x;T,0]$ for all admissible $\tilde{x}$. Hence, a solution of (2-4) with $\lambda(t) \leq 0$, if it exists, maximizes the energy per cycle.

We reiterate the above arguments for a possibly aperiodic $\tilde{x}$ and find that the average power $\bar{P}$ over an arbitrary time interval $[t_a, t_b]$ can be bounded as $\bar{P}[\tilde{x}; t_b, t_a] \leq E[x; t_b, t_a]/(t_b - t_a) + (F - 2b\dot{x})\Delta x|_{t=t_a}^{t_b}/(t_b - t_a)$. It follows that

$$P[\tilde{x}] \equiv \lim_{t_b - t_a \to \infty} \bar{P}[\tilde{x}; t_b, t_a] \leq P[x] = E[x;T,0]/T. \tag{5}$$

Therefore the specified periodic solution also maximizes the long-time average power. It remains to show that such a solution actually exists. We will do so for a time-harmonic force.

2.3. Optimum operation for a time harmonic force

For a small enough force, the solution for unrestricted motion still applies, and with $F = F_0 \cos\omega t$ it is $x = (F_0/2b\omega)\sin\omega t$ up to a constant. Displacement-constrained operation will be reached at a critical force amplitude $F_c = 2b\omega Z_l$ with a critical power of $P_c = F_c^2/8b = b\omega^2 Z_l^2/2$.

For $F_0 \geq F_c$, we note that while $\lambda \neq 0$ so that $s = 0$ by (3), the proof mass is at rest at one of the displacement limits according to (4). We can then use (2) to write the requirement of nonpositive λ as $\lambda = \dot{F}/2x = -F_0\omega\sin(\omega t)/2x \leq 0$ and conclude that the proof mass may only rest at $x = Z_l$ if $\sin\omega t \geq 0$ and at $x = -Z_l$ if $\sin\omega t \leq 0$. When $\lambda = 0$, we integrate (2) twice to get

$$\dot{x} = F_0(\cos\omega t - \cos\theta_1)/2b \quad \text{and} \quad x = F_0(\sin\omega t - \omega t\cos\theta_1)/2b\omega \tag{6}$$

where the integration constants were determined by requiring zero velocity at the limits. The limit $x(t_1) = Z_l$ is reached at a time t_1 and a phase angle $\theta_1 = \omega t_1 \in [0, \pi/2]$ given by

$$\sin\theta_1 - \theta_1\cos\theta_1 = F_c/F_0. \tag{7}$$

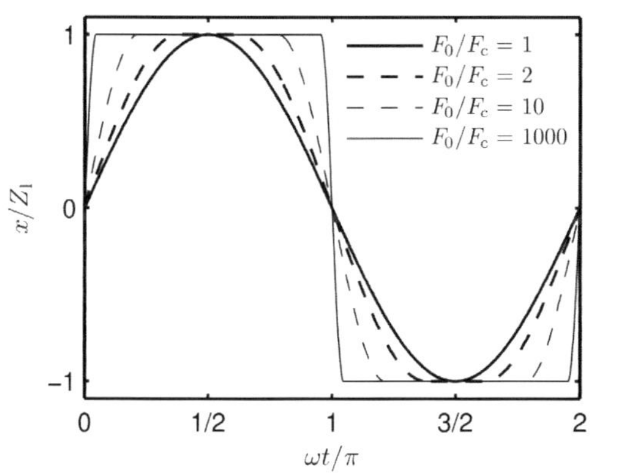

Figure 2. Displacement waveforms that maximize power delivered into the transducer.

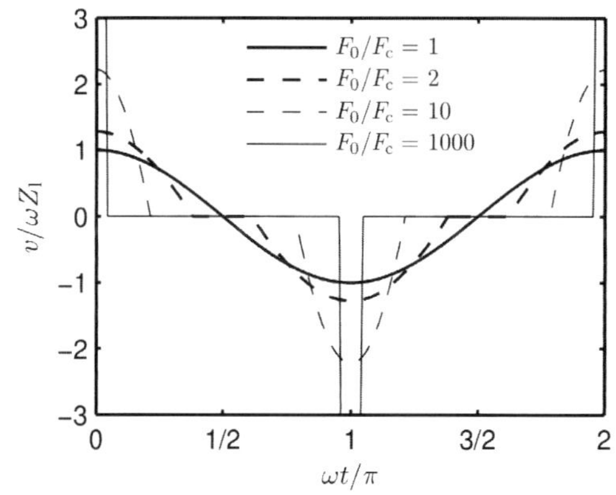

Figure 3. Velocity waveforms that maximize power delivered into the transducer.

Because $F(t-\pi/\omega) = -F(t)$, we have that $-x(t-\pi/\omega)$ is a solution of (2) if $x(t)$ is. We can therefore generate solutions for any $\omega t \in [n\pi - \theta_1, n\pi + \theta_1]$ with integer n by repeated application of this symmetry operation to (6). For the intervening intervals, the proof mass is at rest at either of the limits which is also a feature of the Coulomb-force parametric generators (CFPGs) [2]. The resulting waveforms are shown in figures 2 and 3.

To determine the output power bound for $F_0 \geq F_c$, we insert the optimal solution into (1) and divide by T. One way to write the result is

$$P/P_c = \frac{1}{\pi}\left(\frac{F_0}{F_c}\right)^2 (\sin 2\theta_1 - 2\theta_1 \cos 2\theta_1)\,. \tag{8}$$

Figure 4 shows this power bound over a range of force amplitudes. It is compared to the exact result for the optimal VDRG [2] which coincides with the bound for $F_0 \leq F_c$ as discussed above. For larger F_0, the initial asymptotics agrees to first order in $F_0/F_c - 1$, but as $F_0/F_c \to \infty$ the deviation increases and the power bound approaches asymptotically the mechanical input-power bound $P_{\text{in,bound}}/P_c = 8F_0/\pi F_c$ [5, 6] which is $4/\pi$ larger than for the VDRG. The CFPG has output power $P_{\text{CFPG}} = \beta P_{\text{in,bound}}$ with the parameter $\beta \to 1$ as the vibration amplitude goes to infinity. Hence, its performance coincides with the bound in this limit.

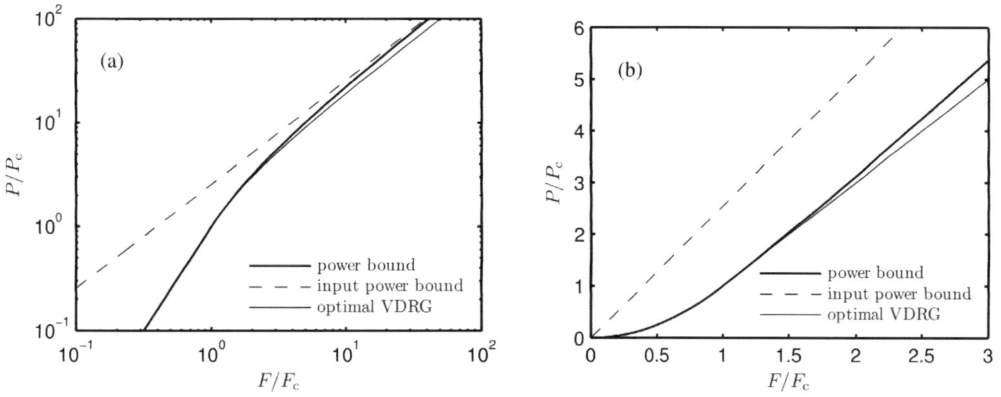

Figure 4. Output-power bound compared to optimal VDRG and input-power bound. (a) Wide force-amplitude range. (b) Moderate force amplitudes.

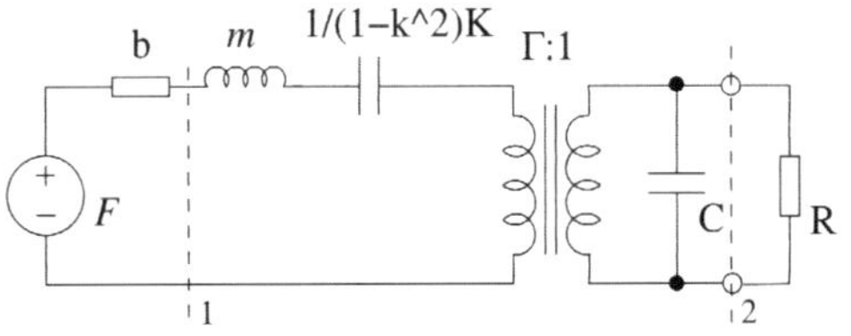

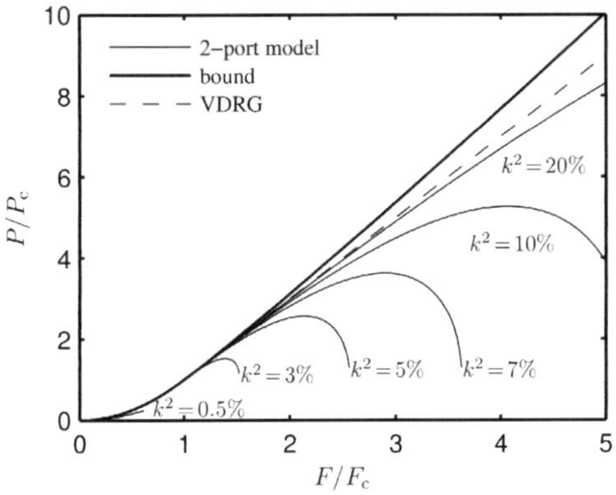

Figure 5. Linear two-port model. (a) Equivalent circuit. (a) Output power with $Q_m = 100$.

3. Linear two-port devices

The linear-two port model has the equivalent circuit in figure 5a and performs equally with the VDRG without constraints and for $k^2Q_m > 1 + \sqrt{1-k^2}$ with $k^2 = \Gamma^2/KC$, $Q_m = m\omega_0/b$ and $\omega_0 = \sqrt{K/m}$ [3]. Its output power is shown in figure 5b where the load resistance has been adjusted to avoid displacement beyond the limit. The drive frequency is held fixed at the value near the resonant frequency that is optimal for unconstrained motion [3]. The two-port model does not perform better than the VDRG and even has a power drop before sufficient damping becomes impossible to achieve. However, the higher the coupling is, the further into the high-force regime is it possible to operate and the closer the model follows the VDRG. Therefore higher coupling than the critical one is advantageous for displacement-limited operation.

4. Concluding remarks

We have derived a new architecture-independent power bound which at small to intermediate force-amplitudes greatly improves on using input-power bound as a bound for output power. VDRG performance is equal or close to the new bound for small force amplitudes while the CFPG performance is close to it when the force approaches infinity. The bound is a fundamental limit which can not be circumvented by transducer or power-electronic-interface design. Multiple degrees of freedom and explicit treatment of other parasitic damping than linear are interesting extensions to pursue in future work.

Acknowledgments

This work was supported by the Research Council of Norway under grant no. 191282.

References

[1] Williams C B and Yates R B 1996 *Sens. Actuators A* **52** 8–11
[2] Mitcheson P D, Green T C, Yeatman E M and Holmes A S 2004 *J. Microelectromech. Syst.* **13** 429–440
[3] Renno J M, Daqaq M F and Inman D J 2009 *J. Sound Vibr.* **320** 386 – 405
[4] Halvorsen E 2008 *J. Microelectromech. Syst.* **17** 1061–1071
[5] Mitcheson P D, Yeatman E M, Rao G K, Holmes A S and Green T C 2008 *Proc. IEEE* **96** 1457–1486
[6] Ramlan R, Brennan M, Mace B and Kovacic I 2010 *Nonlinear Dynamics* **59**(4) 545–558
[7] Okamoto H, Onuki T, Nagasawa S and Kuwano H 2009 *J. Microelectromech. Syst.* **18** 1420 –1431
[8] Scruggs J 2009 *J. Sound Vibr.* **320** 707 – 725
[9] Halvorsen E 2013 *Phys. Rev. E* **87**(4) 042129
[10] Léonard D and Long N V 1992 *Optimal control theory and static optimization in economics* (Cambridge University Press)
[11] Zalmai G J 1985 *Journal of mathematical analysis and applications* **109** 426–452

Journal of Physics: Conference Series **476** (2013) 012027 doi:10.1088/1742-6596/476/1/012027

Tilting pad gas bearing design for micro gas turbines

M J H W Nabuurs[1], F Al-Bender[1], and D Reynaerts[1]

[1]Dept. Mechanical Engineering, Katholieke Universiteit Leuven, Leuven Belgium

E-mail: marius.nabuurs@mech.kuleuven.be

Abstract. This paper presents the results of a dynamic stability investigation of a micro gas turbine supported by two flexible tilting pad bearings. The pad flexibility allows centrifugal and thermal shaft growth of the rotor but can also introduce undesirable rotor instabilities. An eigenvalue analysis on the linearised rotor-bearing dynamics is performed to estimate the required pad stiffness and damping for stability. Results of the eigenvalue analysis are evaluated by fully nonlinear orbit simulations.

1. Introduction

Bearings for micro gas turbines have to operate at extremely high rotational speeds while maintaining dynamic rotor stability over a wide temperature range. Rotational speeds above 200.000 rpm and temperatures of several hundred degrees are not rare for micro gas turbines with power outputs of several kWe. Conventional oil-film or ceramic ball bearings have under these conditions a limed life-span or are causing significant frictional losses.

Aerodynamic gas bearings offer in this context an interesting alternative. They operate virtually wear-free and with relative low frictional losses at high speeds. Recent research at KU Leuven showed stable operation of an experimental 6mm round shaft on aerodynamic gas bearings for speed up to 1.2 million rpm [1]. The developed bearing solution with a fixed inner diameter is however unable to deal with excessive thermal shaft growth. Current research is therefore focused on miniature tilting pad bearings with radial pad flexibility. The flexibility will allow the pads to expand outwards when shaft growth occurs, ensuring an appropriate gap geometry between the shaft and the bearing pads for aerodynamic operation. The main challenge is to obtain the correct amount of radial pad flexibility and damping to accommodate sufficient shaft growth and to ensure dynamic rotor stability over the full speed range.

This paper will show the numerical method followed to investigate the stability of a 200 gram micro gas turbine supported by two flexible tilting pad bearings. The required radial pad stiffness and damping to obtain stability are calculated by deriving first the relation between the pad position and the linear gas-film stiffness and damping coefficients with the use of a numerical perturbation method [2]. The perturbation frequency depending coefficients are then substituted in an extended Laval-Jeffcott rotor model to obtain the overall rotor stability by calculated the eigenvalues of the linearised system. Comparison of the results with fully nonlinear orbit simulations of the rotor dynamics are performed in section 4.

2. Tilting pad geometry

The geometry of the tilting pad bearing under investigation is illustrated in Figure 1. Each individual pad of the bearing and its associated pivot point is flexible mounted in radial direction with stiffness k_p, damping d_p, and a possible pre-load force F_p. When designing such flexible tilting pad bearing it is of interest to know the range for the external pad stiffness k_p and d_p which results in dynamic stability. An efficient approach for non-flexible bearings with a fixed bearing geometry is to numerically calculating the linear gas-film stiffness and damping coefficients, followed by an eigenvalue study on the linearised rotor-bearing dynamics [2, 3, 4]. The gas-film coefficients for flexible tilting pad bearings have however a mutual dependency with the initial unknown pad position. This makes the coefficient calculation for flexible bearings a computational intensive task, especially if it is repeated for each stiffness value k_p and damping value d_p of interest.

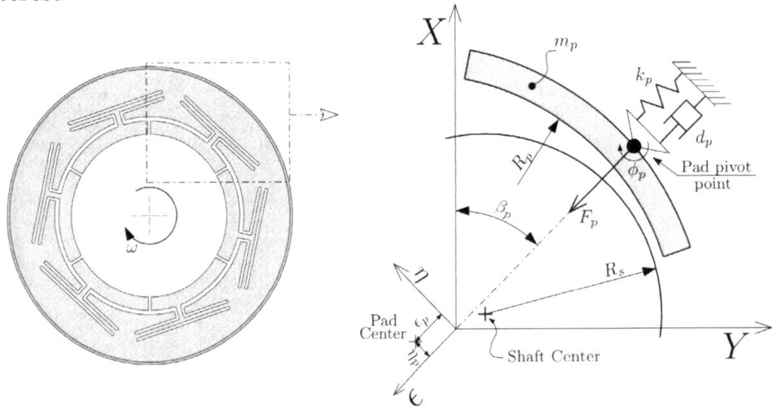

Figure 1.a 6-pad bearing example **1.b** Flexible tilting pad geometry

A less computational intensive approach is followed in this paper by first calculating the gas-film force $F_g(\epsilon)$ and the gas-film stiffness $K(\nu, \epsilon)$ and damping $D(\nu, \epsilon)$ coefficients for the full range of radial pad position. Results, for the 6-pad bearing, are shown in dimensionless form in Figure 2.

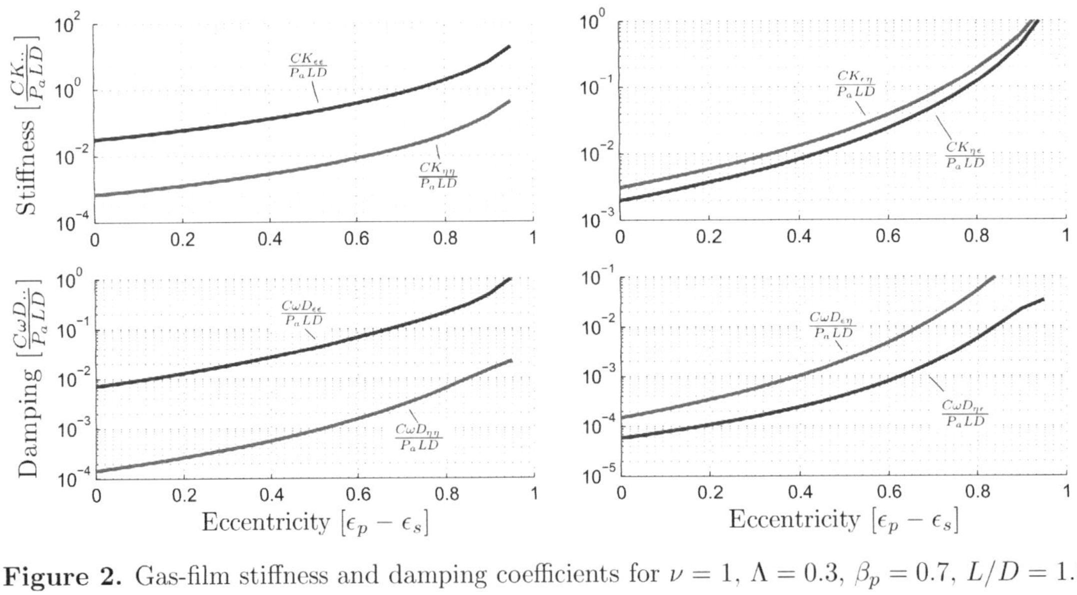

Figure 2. Gas-film stiffness and damping coefficients for $\nu = 1$, $\Lambda = 0.3$, $\beta_p = 0.7$, $L/D = 1.5$

3. Stability

Stability of the overall system dynamics can be analysed by substituting the derived gas-film coefficients in the overall rotor-bearing dynamics. The dynamic behavior of the 200 gram turbine under investigation is thereby modeled using an extended Laval-Jeffcott model where the rotor is divided into two end-masses and one center-mass connected by a flexible shaft.

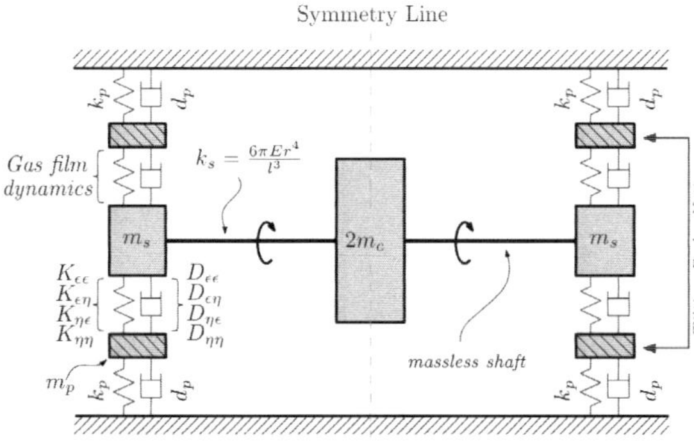

Figure 3. Rotor-Bearing model

Rotor/Bearing specifications		
Rotor speed	$\omega = 260$	krpm
Shaft diameter	$D = 10$	mm
Bearing length	$L = 15$	mm
Pad mass	$m_p = 0.02$	kg
Pad pivot stiff.	$k_t = 0$	Nm/r
Rotor mid mass	$m_c = 0.07$	kg
Rotor end mass	$m_s = 0.03$	kg
Rotor stiffness	$k_s = 5.0$	kN/m

Figure 4. Design Parameters

The extended Laval-Jeffcott model takes the pads radial degree of freedom into account while the pads pivoting motion is neglected. This effectively assumes that the pads have an infinite moment of inertia. The model will therefore under estimate the stability for pads with finite moment of inertia in return for a more simplified stability analysis method. The stability of the 6-pad bearing can for example be evaluated for a large range of k_p and d_p values by first solving equation (1) for the radial pad position ϵ_p. The gas-film stiffness and damping coefficients, corresponding with this pad position, can then be used in an eigenvalue analysis of the linearised-bearing rotor model. Here, combination of k_p, d_p, and F_p that lead to undamped eigenfrequency which collide with the perturbation frequency are defined as unstable. That is, eigenvalues that not satisfy the stability criterion (2).

$$\epsilon_p = \frac{F_p - F_g(\epsilon_p, \epsilon_s)}{k_p C} + \epsilon_s \quad (1) \qquad \mathbb{R}e\{\lambda(v)\} < 0 \ \vee \ |\mathbb{I}m\{\lambda(v)\}| \neq v \quad (\forall\ v \in (0, \omega]) \quad (2)$$

Results for the 6-pad bearing are shown in Figure 3 for a pre-tension force of $2N$ and a shaft growth of 0, 0.2, 0.4 and 0.6% of its original diameter. Notice that combination of pad stiffness k_p and damping d_p which are located below the lines will stabalise the rotor dynamics.

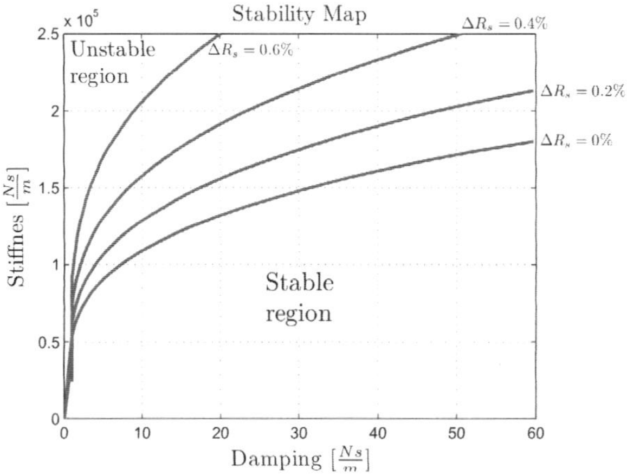

Figure 5. Stability map for a symmetric rotor (Table 4), supported on two 6-pad bearings with radial pad flexibility

4. Comparison with nonlinear orbit simulation results

The results of the eigenvalue study are validated with a fully nonlinear orbit simulation of the shaft whirl movements [5]. For the orbit simulation, the gas-film forces on the rotating shaft are calculated by solving the general Reynolds equation over time and including all the degrees of freedom for the pads and rotor. The masses and stiffness of the rotor and pads are equal to the linearised case, i.e. as specified in Table 4 with the addition of an unbalance of $0.05gmm$.

The calculated orbits trajectories of the shaft center-mass and one of the end point-masses are shown in Figure 6. Transcendent behavior, during the initial face of the simulation, is plotted with a light-colored dashed line while the steady state behavior is plotted with a solid line. The system is said to be stable if the mass center positions convert to a single point, or to a limit cycle in case of shaft unbalance.

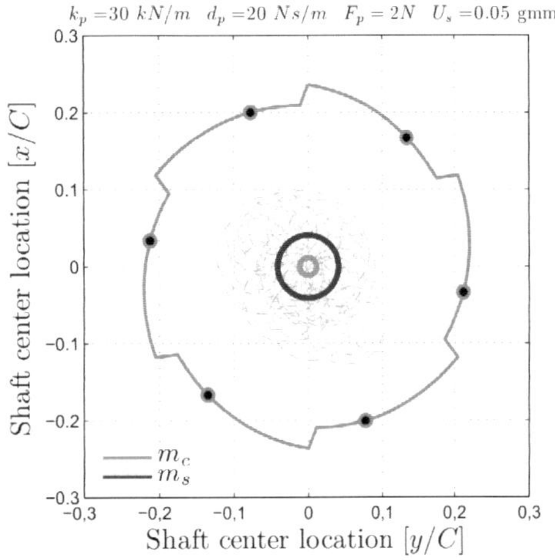

Figure 6. Shaft center orbit path

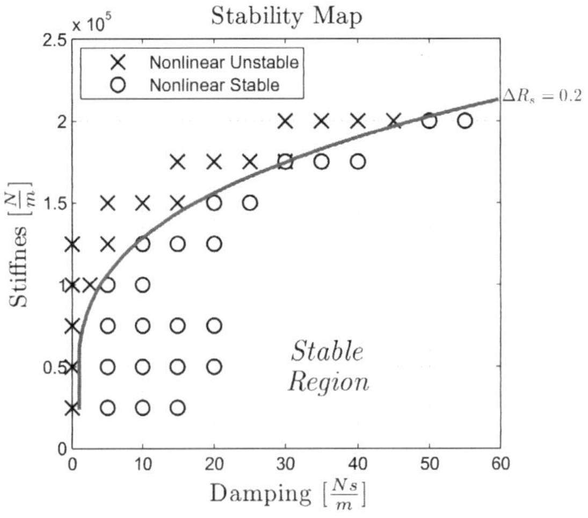

Figure 7. Orbit vs. eigenvalue analysis

The eigenvalue method of the previous section is validated by repeating the nonlinear orbit simulation for multiple points in the stability map. An '○' is placed in Figure 7 at each simulated combination of k_p and d_p where the nonlinear orbit simulation converted to a limit cycle or a single stable point. For unstable rotor behavior, an '×' is placed. As can be seen, all the simulated combination of k_p and d_p are in good agreement with the stable region calculated by the eigenvalue method. Interesting to note is that the stability map of Figure 3 was calculated within 10 seconds after the linearised bearing coefficients where obtained. The 40 orbits simulations of Figure 7 where in comparison taking over 8 hours of continues Matlab computing.

Reference

[1] Vleugels P, Waumans T, Peirs J, Al-Bender F and Reynaerts D 2011 *Journal of Micromechanics and Microengineering* **21** 104014
[2] Czołczyński K 1996 *Wear* **201** 265–275
[3] Waumans T, Peirs J, Al-Bender F and Reynaerts D 2009 *Proc. of the 9th Int. Workshop on Micro and Nanotechnology for Power Generation and Energy Conversion Applications (Washington DC)* 83–86
[4] Lund J 1968 *ASME Journal of Lubrication Technology* 793–803
[5] Kim D and Sim K 2007 *Journal of tribology* **129** 112–119

Self–Powered Sugar Indicator Using CNT–Enzyme Ensemble Film

Y Ogawa[1], S Yoshino[1], K Kato[1], T Magome[1], T Yamada[2,3], T Miyake[1,3], K Hata[2,3] and M Nishizawa[1,3]

[1] Tohoku University

[2] The National Institute of Advanced Industrial Science

[3] Japan Science and Technology Agency

E-mail: ogawa@biomems.mech.tohoku.ac.jp

Abstract. We report the stepwise modification of Os-complex mediator (polyvinylimidazole-[Os(bipyridine)$_2$Cl] (PVI-[Os(bpy)$_2$Cl]) and glucose oxidase (GOD) within the inner nano-space of a carbon nanotube forest (CNTF) film. Owing to the controlled alignment of enzyme/mediator/electrode in the ensemble, the prepared film electrode has both a high-efficiency (turnover rate of ca. 650 s^{-1}) and a large net oxidation current (ca. 15 mA cm^{-2}). The previous GOD electrodes developed by monolayer-based and polymer-based approaches have either of the performances (efficiency or net activity). In addition, the present GOD electrode is a flexible film that could be used by winding on needle devices.

1. Introduction

Controlling the electrical contact of redox enzymes with electrodes is a critical issue for enzymatic biodevices such as biofuel cells and biosensors. (1-4) The mutual positioning between enzyme molecules, mediator molecules (not always necessary), and electrode surface determines the efficiency, reproducibility, and stability of the bioelectrocatalysis systems. A conventional engineering for accelerating the electron transfer to the redox enzymes is the inclusive immobilization with mediator polymer matrices, in wallhich the successive electron exchange between the neighboring mediator groups connects the enzyme redox center and the electrode surface. For example, the Os-complex-pendant polymers are successful mediating matrices for glucose oxidase (GOD) to provide glucose oxidation current at mA cm^{-2} level. (1-3) However, because of random mutual positioning in the 3D composite, many enzyme molecules are often isolated from the molecular network for continuous bioelectrocatalysis. On the other hand, direct immobilization of enzyme monolayer on the electrode surface has improved the efficiency of enzyme utilization. A striking example is the reconstituting apo-GOD on a relay-FAD monolayer linked to electrode surfaces. (1, 3) Since all the enzyme unites are oriented in an optimal position with respect to the electrode surface, a high electron-transfer turnover rate comparable to that for bulk GOD reaction (~ 700 s^{-1} at 25 °C) has been achieved. However, the drawback of such 2D monolayer engineering is the lower bioelectrocatalytic performance due to the limited amount of immobilized enzymes.

We present herein an enzyme/mediator/electrode ordered ensemble that shows both "high turnover rate" and "large catalytic current". In order to satisfy both of these requirements, the larger amount of enzymes than monolayer should be immobilized with keeping effective contact with electrodes. We realize such ideal condition by taking advantage of a film of well-aligned carbon nanotube forest (CNTF) [(5)] consisting of single-walled CNTs arrayed with a pitch of 16 nm. The CNTF was synthesized by water-assisted chemical vapor deposition on a line-patterned Al_2O_3 / Fe catalyst on silicon wafers (see Experimental for details). [(5)] As shown in **Figure 1a**, the synthesized CNTF film (1.5 mm × 1 mm) was pulled from the substrate and pinched by inverse operating tweezers (electrical lead), to produce an exposed electrode geometric area of ca. 2 mm^2 (sum of both faces of a 1mm × 1 mm sheet). The thickness of the CNTF films (4μm, 12μm or 20μm) was determined by the width of the line-patterns of Al_2O_3 / Fe catalyst. Recently, we reported that the intraspace of the CNTF is useful for immobilization of fructose dehydrogenase and laccase, which are the direct electron transfer (DET)-type enzymes. [(4)] Although there are a few recent reports that also GOD is capable of direct communication with electrodes, repeated attempts to prepare a workable GOD/CNTF ensemble electrode without any mediators have failed. Then we have developed the stepwise process to construct molecular architecture with polyvinylimidazole-[Os(bipyridine)$_2$Cl] (PVI-[Os(bpy)$_2$Cl], MW: 15000) and GOD (EC:1.1.3.4, MW: 186 kDa), as illustrated in **Figure 1b and 1c**. The PVI-[Os(bpy)$_2$Cl] was synthesized according to a literature method, with a molar ratio of imidazole group to [Os(bpy)$_2$Cl] of 5.

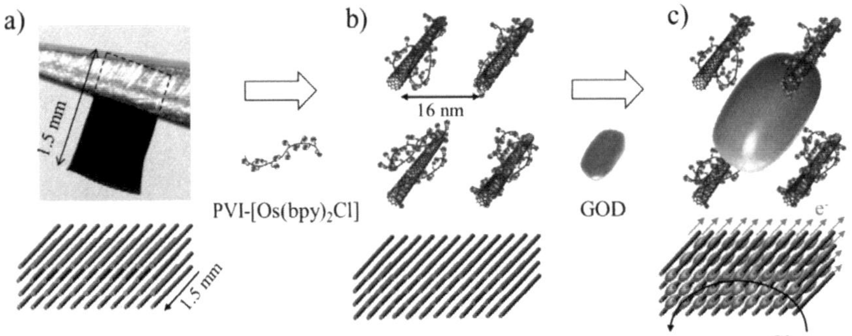

Figure 1. Schematic illustration of the stepwise process for constructing bioelectrocatalytic composite inside a CNTF film.

2. Results and discussion

2.1. Adsorption of PVI-[Os(bpy)2Cl] inseide CNTF films.

The CNTF film was first treated with 0.1 % Triton X-100 to be hydrophilic, and then soaked in a stirred phosphate buffer solution (PBS, pH7.0) containing 1 mg ml^{-1} PVI-[Os(bpy)$_2$Cl] at 4 °C. As shown in **Figure 2a**, the cyclic voltammogram (CV) of the treated CNTF showed a symmetric shape typical for adsorbed redox species. In fact, the amplitude of peak currents were proportional to the scan rates (**Figure 2b**). The amount of PVI-[Os(bpy)$_2$Cl] adsorbed within the CNTF films were estimated by integrating the CV currents and is plotted in **Figure 2c** against the soaking time in the 1 mg ml^{-1} PVI-[Os(bpy)$_2$Cl] solution. The amount of PVI-[Os(bpy)$_2$Cl] in a CNTF film increased with the soaking time and reached a maximum after 2 hours. Importantly, these values are proportional to the CNTF film thickness (7.2 × 10^{-10} mol for a 12 μm thick film and 12.8 × 10^{-10} mol for a 20 μm thick film), indicating that the PVI-[Os(bpy)$_2$Cl] molecules can entirely and uniformly adsorbed inside the CNTF films, as illustrated in Figure 1b. A part of the free imidazole groups of the mediator polymer would adsorb on CNT surfaces via π-π interaction. The adsorption density of PVI-[Os(bpy)$_2$Cl] calculated using the effective inner surface area of the CNTF films (8.2 cm^2 for 20μm thick film) [(5)] was 1.6 ± 0.1 × 10^{-10} mol cm^{-2}, which is comparable with the value for a PVI- [Os(bpy)$_2$Cl] film adsorbed on a flat Au surface (3.2 × 10^{-10}mol cm^{-2}) .

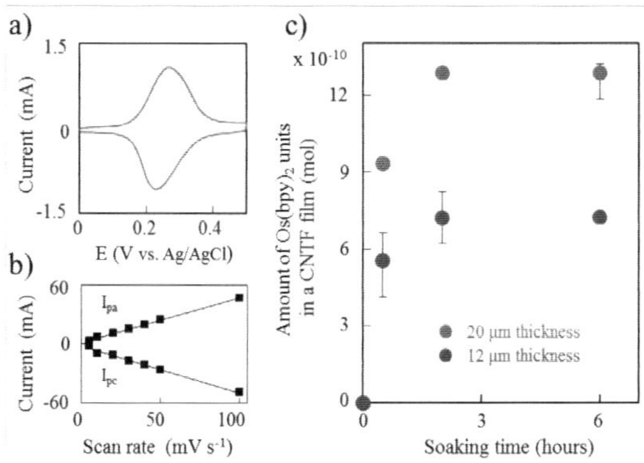

Figure 2. (a) Cyclic voltammograms at 10 mV s^{-1} of the PVI-[Os(bpy)$_2$Cl]-modified CNTF film (20 μm thickness) in phosphate buffer solution (pH7.0). A CNTF film was soaked in the 1 mg ml^{-1} PVI-[Os(bpy)$_2$Cl] PBS solution for 6 hours. (b) Redox peak currents of the CVs as a function of scan rate. (c) The amount of Os(bpy)$_2$ unit inside a CNTF film (film thickness: 12 μm and 20 μm) as a function of soaking time for 1 mg ml^{-1} PVI-[Os(bpy)$_2$Cl] PBS solution. The mean values (±standard deviation) of three independent specimens are given.

2.2. Electrocatalytic activity of GOD/PVI-[Os(bpy)$_2$Cl]/CNTF ensemble films.

Subsequent loading of the enzyme GOD was conducted by immersing the PVI-[Os(bpy)$_2$Cl]-adsorbed CNTF films in a stirred PBS (pH 7.0) solution containing 3 mg ml^{-1} GOD for 1 hour. **Figure 3a** shows the CVs of GOD/PVI-[Os(bpy)$_2$Cl]/CNTF ensemble films at 10 mV s^{-1} in a stirred 200 mM D-glucose PBS solution. The catalytic current for glucose oxidation increased in response to the thickness of CNTF films (3.7 mA cm^{-2} for 4 μm thickness and 14.7 mA cm^{-2} for 20 μm thickness), indicating that also GOD can entirely penetrate inside the PVI-[Os(bpy)$_2$Cl]-modified CNTF films. For example, the content of GOD incorporated in a 20μm-thick film was measured as ca. 0.86 μg by a C-6667 Protein Quantitation Kit, the value being a little below the case when GOD molecules (6.7 × 6.7 × 21 nm^3) align to form lines in the interspace of CNTs (1.17μg). The current density was enhanced to as high as 26.7 mA cm^{-2} by turning up the buffer temperature to 37.5 °C. Importantly, more than 90 % of the electrode activity could be maintained even after 6 days storage in an air-saturated PBS solution (**Figure 3b**), proving the stability of bioelectrocatalytic architecture with the composite of PVI-[Os(bpy)$_2$Cl] polymer and GOD. The anionic GOD molecules could be stabilized by cationic Os complex of the mediator polymer anchored on the CNT surface via π-π interaction.

The electron turnover rate for the 20 μm-thick film was calculated from the current value at 25 °C (0.29 mA), the Faraday constant (96500 C mol^{-1}), the molecular weight of GOD (186000 g mol^{-1}), and the content of GOD molecules in a piece of the ensemble film (ca. 0.86 μg). The derived turnover rate was ca. 650 s^{-1}, being comparable with that of GOD in bulk solution containing an electron acceptor of O_2 (700 s^{-1}) or ferrocene (600 s^{-1}) at 25 °C. These results indicate that most of ca. 3 × 10^{12} GOD units within the film could efficiently work to the full, presumably owing to the molecularly ordered alignment of enzyme/mediator/electrode in the ensemble. Such a high efficiency of the present GOD electrode resulted in a resistance to oxygen inhibition, as shown in **Figure 3c**. The catalytic performance was almost identical in N_2-saturated, air-saturated and even O_2-saturated solutions. In general, glucose oxidation with GOD-modified electrodes is often disturbed by dissolved O_2, which is troublesome for glucose sensing. However, the ordered Os(bby)$_2$ groups in the present ensemble electrode could effectively accept the electron from GOD in preference to O_2, resulted in excellent O_2 resistance.

2.3. Application as a windable anode of biofuel cells.

The present free-standing, bioelectrocatalytic film could be used for miniature biofuel cell devices. We demonstrate here the application of the film to a self-powered sugar indicator designed for inserting into a fruit. For indicating the glucose concentration, the net performance of the biofuel cell system should be controlled by the glucose anode. Because the oxygen in fruits is limited to a lower concentration than glucose, we employ a gas-diffusion biocathode for utilizing the abundant oxygen in air outside of the fruits (see Experimental for details). **Figure 4a** shows the biofuel cell performance measured using 200 mM glucose PBS solution with a couple consisting of a GOD/PVI-[Os(bby)$_2$Cl]

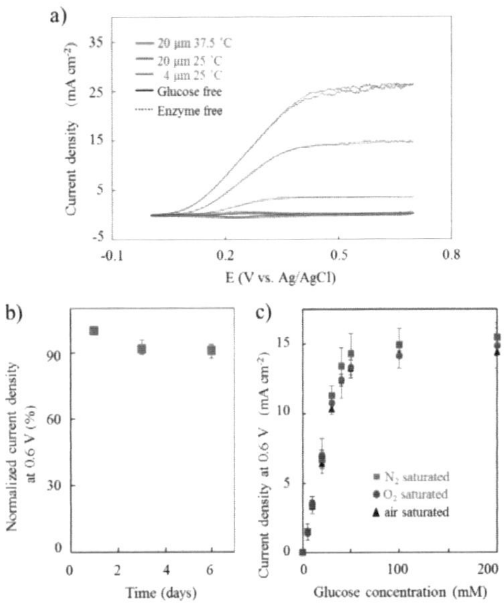

Figure 3. (a) Cyclic voltammograms of GOD/PVI-Os/CNTF ensemble films at 10 mV s^{-1} in stirred air-saturated 25°C PBS containing 200 mM D-glucose. The thickness of CNTF films were 4 μm or 20 μm. A control voltammogram without glucose and a voltammogram taken at 37.5°C are also shown. We used the total geometric area of the pinched film (2 mm^2) for the calculation of the current densities. (b) The oxidation current densities at 0.6 V vs. Ag/AgCl for the GOD/PVI-[Os(bby)$_2$Cl]/CNTF film (20μm thickness) in a stirred 200 mM D-glucose PBS solution, periodically measured during 6 days of storage in PBS solution. (c) The current densities at 0.6 V for the 20 μm GOD/PVI-[Os(bby)$_2$Cl]/CNTF film as a function of the glucose concentration, measured in O_2-, N_2-, and air-saturated stirred PBS (pH7.0) solutions. The mean values (± standard deviation) of three independent specimens are given.

/CNTF film anode (20μm thickness) and a cathode made from bilirubin oxidase (BOD)-modified carbon fabric (1cm × 1cm). The open-circuit voltage of the cell was 0.5 V in agreement with the difference between the potentials at which glucose oxidation and oxygen reduction start to occur in cyclic voltammetry. The maximum output current (0.27 mA) indicates that the system is limited by the anode even in 200 mM glucose, a concentration that markedly higher than that found in raw fruits (a few tens of mM). As shown in **Figure 4b**, a piece of GOD/PVI-[Os(bby)$_2$Cl]/CNTF film was wound on one lead of a light-emitting-diode (LED) device, whose blinking interval is inversely proportional to the power of the biofuel cell. The other lead was connected to the BOD-based gas-diffusion cathode. The blinking interval of the LED upon inserting the device to a grape was coincident with that for the extracted juice (**Figure 4c**), proving that this device could serve as a sugar indicator by simply being inserted into a grape. This principle of the self-powered sensor could be applied to more important blood sugar monitoring applications. We are planning to develop a GOD/PVI-[Os(bby)$_2$Cl]/CNTF-based device structure suitable for low-invasive insertion into a blood vessel through skins.

3. Conclusion

The larger amount of enzyme units than monolayer were successfully immobilized with keeping the effective electrical contact with the electrode (CNTs). In particular, we have succeeded in forming the entirely uniform bioelectrocatalytic architecture with PVI-[Os(bby)$_2$Cl] and GOD inside a CNTF film. The voltammograms of the PVI-[Os(bby)$_2$Cl]-modified CNTF indicated the uniform adsorption of PVI-[Os(bpy)$_2$Cl] on the CNT surface via π-π interaction with the density of ca. $1.6 \pm 0.1 \times 10^{-10}$ mol cm^{-2}. The subsequent modification of GOD seemed to form at the interspaces of PVI-[Os(bby)$_2$Cl]-modified CNTs, as judged from the protein quantitation assay. Owing to such ordered positional relationship between GOD, PVI-[Os(bby)$_2$Cl] and CNT, the composite film showed both "high activity" for glucose oxidation (ca. 15 mA cm^{-2}) and "high turnover rate" (ca. 650 s^{-1}), indicating almost every enzyme molecules within the film could work to the full.

4. Experimental section

CNTF preparation: CNTF was synthesized in a 1-in. tube furnace by water-assisted chemical vapor deposition at 750 °C with a C_2H_4 carbon source and an Al_2O_3 (10 nm)/Fe (1.0 nm) thin-film catalyst grown on silicon wafers. [5] We used He with H_2 as the carrier gas (total flow 1000 standard cubic

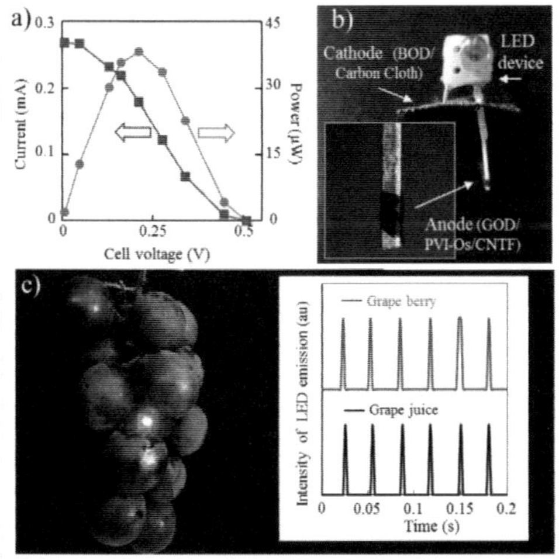

Figure 4. (a) Performance of a biofuel cell composed of an anode of GOD/PVI-[Os(bby)$_2$Cl]/CNTF film (20μm thickness) and a cathode of BOD-modified carbon cloth (1cm × 1cm) in 200 mM glucose PBS solution, measured by changing the external resistance (28Ω~46kΩ). (b) Photograph of the LED-based self-powered sugar indicator, at the tip of which the GOD/PVI-[Os(bby)$_2$Cl]/CNTF film was wound. (c) The device assembly was inserted in a grape and the LED blinking was measured (inset). The time course of LED emission, taken using an extracted juice, is also shown.

centimeters per minute (sccm) at 1 atm with a controlled amount of water vapor with ethylene (100 sccm) for 10 min. *Quantitative Analysis of the Entrapped Enzymes:* The quantitative analysis of GOD was conducted as explained in our previous paper. [4] The enzyme-incorporated CNTF film was first washed and immersed in 20 mM sodium phosphate buffer (pH 9.3) containing 0.1 M sodium borate and 1% sodium cholate and dispersed with an ultrasonic homogenizer for 15 min. The GOD in the dispersion was then analyzed using a C-6667 Protein Quantitation Kit (Molecular Probes), using 5 mM (3-(4-carboxybenzoyl)-quinoline-2-carboxaldehyde) (ATTO-TAG CBQCA) and 20 mM KCN to label the enzyme with CBQCA. After 1.5 h of incubation, the fluorescent intensity was measured by a luminescent image analyzer system (Fuji Photo Film, LAS-3000 mini), and the amount of enzyme was determined by referencing a calibration curve. *Preparation of gas-diffusion carbon fabric (CF) cathodes:* The preparation of the cathode basically followed the procedures used for our previous work. A 40 □l aliquot of a 10 mg ml^{-1} multiwalled CNT solution was put on a CF strip and dried in air, followed by thoroughly washing out the surfactant by soaking in an ethanol solution for more than 1h with stirring. The surface of the CNT-modified CF electrode was further modified with a 0.1 ml solution of 5 mg ml^{-1} bilirubin oxidase (BOD, EC 1.3.3.5, 2.5 U/mg, from Myrothecium) in vacuum oven (0.09 MPa, 35 C°). The strip was additionally coated with the CNT solution to make the surface hydrophobic. *Electrochemical Measurements:* The GOD/PVI-[Os(bpy)$_2$Cl]/CNTF ensemble films, anchored at the edge with SUS316L fine tweezers, was analyzed by a three-electrode system (BSA, 730C electrochemical analyzer) in stirred solutions using a Ag/AgCl reference and a platinum counter electrode. The gas-diffusion cathode (BOD-modified CF strip) was put on an air-saturated solution so as to contact the solution by the BOD-modified face during the cyclic voltammetry. The performance of a biofuel cell constructed from an GOD-based CNTF anode and an BOD-based CF cathode was evaluated on the basis of the cell voltage upon changing the external resistance between 1 kΩ and 2 MΩ at the time step of 60 s. Unless otherwise indicated, the electrochemical measurements were carried out at room temperature (25 °C).

References

[1] I. Willner, Y. M. Yan, B. Willner and R. Tel-Vered, *Fuel Cells* **9**: 7-24, 2009.
[2] A. Heller and B. Feldman, *Acc. Chem. Res.* **43**: 963-973, 2010.
[3] A. Zebda, C. Gondran, A. L. Goff, M. Holzinger, P. Cinquin and S. Cosnier, *Nat. Commun.* 2: 1-6, 2011.
[4] T. Miyake, S. Yoshino, T. Yamada, K. Hata and M. Nishizawa, *J. Am. Chem. Soc.* **133**: 5129-5134, 2011.
[5] D. N. Futaba, K. Hata, T. Yamada, T. Hiraoka, Y. Hayamizu, Y. Kakudate, O. Tanaike, H. Hatori, M. Yumura and S. Iijima, *Nat. Mater.* **5**: 987-994, 2006.

Enhanced Heat Rejection of Microscale Geometries in Convective Flow Boiling Evaporators

L M Safford Smith[1], W J Connacher[1,2], J C Cheng[1,3] and A P Pisano[4]

[1] Department of Mechanical Engineering, University of California, Berkeley, 5101 Etchevery Hall, Berkeley, CA , USA 94720-1774
[2] Department of Physics, University of California, Berkeley, USA
[3] Department of Electrical Engineering, University of California, Berkeley, USA
[4] Professor of ME and ECE, Dean of Engineering, University of California, San Diego, 9500 Gilman Drive, La Jolla, CA, USA 92093-0403

E-mail: saffordl@berkeley.edu

Abstract. Four surfaces have been designed, fabricated and tested under convective flow boiling (CFB) conditions in an open loop configuration. They contain features in the 10 micron range and were tested with flow velocities under 3 mm/s. To accomplish these flow rates, this work utilizes a constant pressure potential driving flow, instead of the constant flow rate imposed with a syringe pump. This limited device flooding. The evaporation surfaces were tested to the point of dry-out at three different pressure potentials: 150, 650, and 1150 Pa, across a range of powers from 25 W/cm^2 to 50 W/cm^2. Temperature data was collected from an IR Camera and showed that fluctuations in the wall temperatures exceed 5 °C in more than 50% of the tests and reached differences as high as 23 °C. The wall temperature instabilities in CFB indicate that one temperature may be inaccurate and that by including time as a variable a better understanding of the behaviour at this scale may be revealed

1. Introduction

High power densities of current and future power systems and high performance electronics have created a continuing need for novel thermal management. Historically, these problems emerged as transistor sizes decreased and the density of transistors increased. Each transistor, now closely packed, leads to an ever increasing heat flux that must be removed, or devices will overheat. Within the large arena of solutions to this problem is some very promising research utilizing phase change devices, such as loop heat pipes and vapor chambers. [1-3]. These direct phase change liquid cooling schemes have been a leading topic of research for many years, including a large body of work focused on pool boiling with different surfaces. The goal of most pool boiling work is to raise the critical heat flux (CHF) as high as possible [4]. This critical heat flux marks the point where devices can no longer remove additional power and the turning point where the textured surface overheats. Recent work in this area has pushed the CHF higher than ever thought possible [5]. With this great progress, it is time for a shift to be made to translate these surfaces into phase change devices.

One proposed solution is to start testing these surfaces in more realistic conditions by moving away from pool boiling and into convective flow boiling (CFB). Testing patterned surfaces with both evaporation and flow is more realistic to looping systems, and will hopefully help with the long-term goal of integrating these surfaces into systems with circulating fluid. Researchers have tried for many years in other CFB applications to gain insight by using correlations from large scale systems [6,7]. These correlations have been useful when trying to get a general sense of how a device will work, but there are many anomalies between large and micro scale flow and heat transfer patterns [3].

One large issue between conventional large systems and MEMS based systems has to do with the stability of micro-scale flow behavior. Studies that have explored convective flow in pin fins defer the issues associated with instabilities by taking averages or measurements over short time periods. The work presented here aims to begin showing some instabilities that arise when temperature is measured in these devices over an extended period of time.

2. Background

Loop heat pipes (LHPs) are a passive phase change thermal management device similar to heat pipes. In these devices, evaporation and the subsequent transfer of latent heat is used to efficiently remove heat from a critical area of a device [8]. The looping fluid is driven by capillary forces and thus the fluid flow rates are very low. The efficiency of the LHP is balanced between three areas: proper condensation, wicking ability, and efficient evaporation. The third challenge is addressed in this paper through the implementation of an experiment to examine thin-film evaporation.

This idea of utilizing thin-film evaporation stems from a desire to avoid macro to micro correlations and begin building surfaces solely for use on the micro scale. With this in mind, the surfaces have been based on a theory of thin film evaporation, called interline or sub region evaporation, where nucleation bubbles are avoided [9]. In this case, a film is created that allows liquid to transition straight from liquid to vapor. Promising work from Purdue showed that between 66% and 87% of all heat transfer in an evaporation system occurs in a very small region [10]. This work drove the focus on designing small features that create this thin film during convective flow boiling. To create this behavior, a model was used to influence the height to diameter ratio. The final chosen ratio was 2.5 with a height to feature spacing ratio of 2. The choice of these ratios has to do with minimizing friction and decreasing the possibility of pressure instabilities due to meniscus recession [3, 11]. This work lead to the development of several pin fin surfaces that were tested using a new method that is described in this paper.

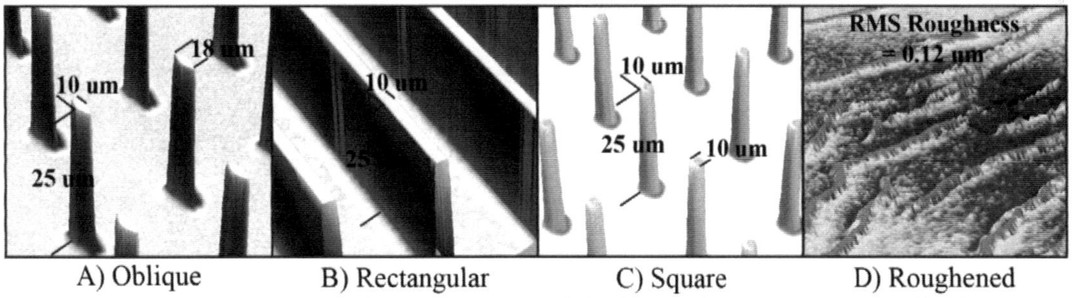

Figure 1. A 3D laser confocal microscope image of the four surface structures tested in this work.

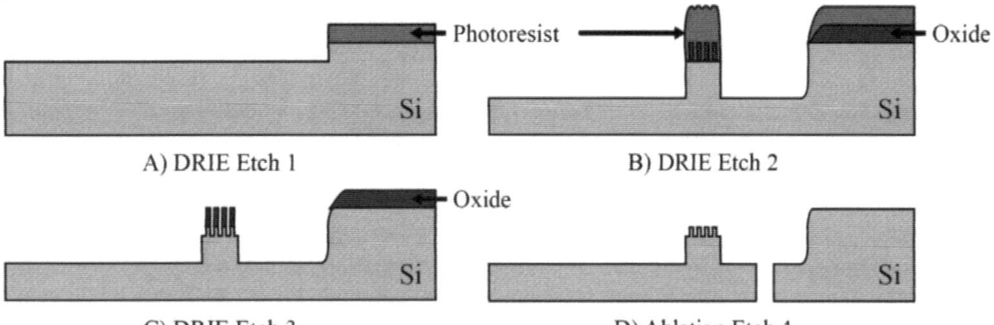

Figure 2. A simple schematic of the fabrication process used to make the tested devices.

3. Fabrication and Overview of the Pin Fins and Roughened Surfaces

The chosen pin fin structures were, oblique, rectangular, and square shaped pins, with the fourth surface blank, but roughened by deep reactive ion etching (DRIE). The size, shape and layout of the final surface are shown in figure 1. The roughened surface has an RMS roughness of 0.12 ± 0.02 μm with a maximum valley to peak distance of 1.25 ± 0.02 μm.

The pin fin surfaces were fabricated in a one sided, n-type, silicon wafer using a three mask process. The final structure has four different depths created using two masking materials, oxide and photoresist, and a laser ablation system. In order to create these different depths the wafer was processed using standard CMOS fabrication techniques including photolithography, reactive ion etching (RIE) for selective oxide removal and deep reactive ion etching (DRIE) for selective silicon removal. The general concept of masking allows patterns to be created by photolithography and then any part of the wafer that is uncovered by the masking material is removed during the etching process (see figure 2, step A).

In order to save time and avoid a long DRIE etch, all the way through the wafer, a laser ablation method was used to cut 2 mm holes in the silicon wafer. The laser system used a Universal PLS 6MW laser with a multi-wave 1.06 μm , 30 W fiber laser. This method turned out to be very effective. It allowed for ten holes to be drilled in about 20 minutes, drastically reducing the time required for DRIE etching. One downside of this process was the rough edges of the through holes. However, the holes were still effective. Once the through holes were created the wafer was cleaned and a 1.1 mm thick glass wafer was anodically bonded to the wafer to seal the device.

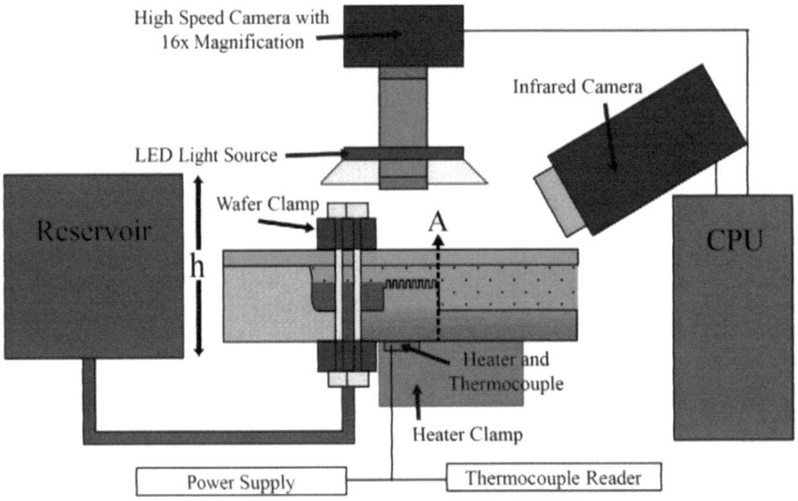

Figure 3. A schematic of the testing set up for the evaporation tests. The IR camera measurements were recorded at the point labeled A, approximately 3mm away from the hottest point.

4. Experimental Set Up and Procedure

Figure 3, shows a schematic of the experimental test set-up. The major equipment used was a high speed camera, an infrared camera, a high accuracy thermocouple reader, a power supply and a high temperature heater. For testing, the glass capped chip is placed into the small wafer clamp shown in figure 3. One of these small clamps was placed on each of the four tested chips. These clamps could then be easily moved in and out of the larger Teflon heater clamp, allowing for all of the chips to be tested without having to move any of the larger equipment. The heater clamp was anchored to a larger structure and held the heater and thermocouple. These were attached to a thermocouple reader and power supply. The fluid was supplied from a hydrostatic reservoir, where the height of the reservoir supplied a pressure that set the flow rate. The pressure was calculated using the following equation.

$$P = \rho g H \tag{1}$$

To begin testing, the flow from the reservoir was started, by opening the valve, and the chips were flooded. The IR camera began recording data and then the power supply that was attached to the heater was programmed and switched on to begin the evaporation process. The location of the temperature measurement was at the end of the surfaces indicated on figure 3 with the letter A.

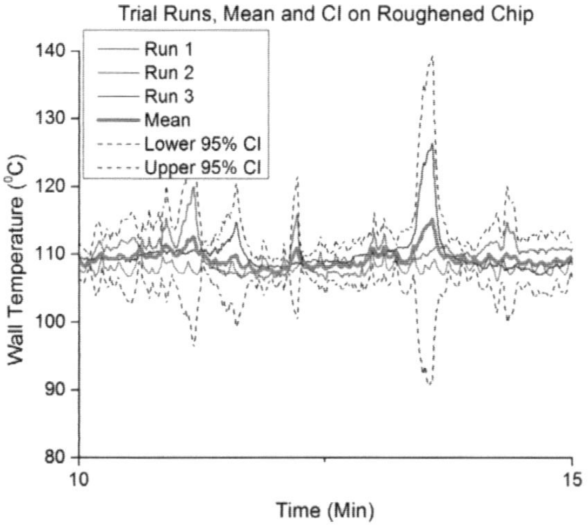

Figure 4. Raw temperature data for roughened chip at 40W/cm^2 and 1150 Pa.

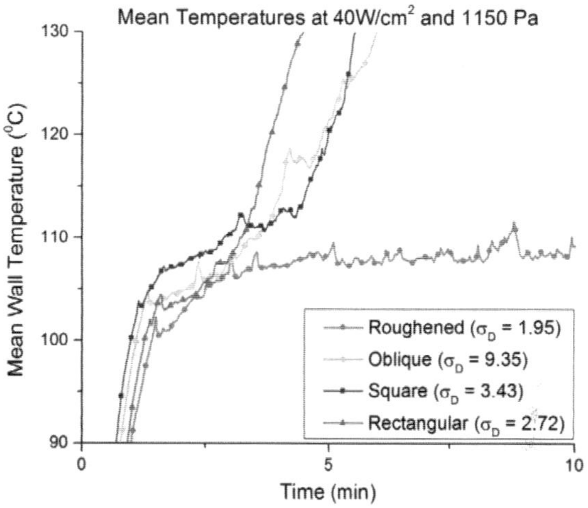

Figure 5. Mean temperature data for all four surfaces at 40W/cm^2 and 1150 Pa.

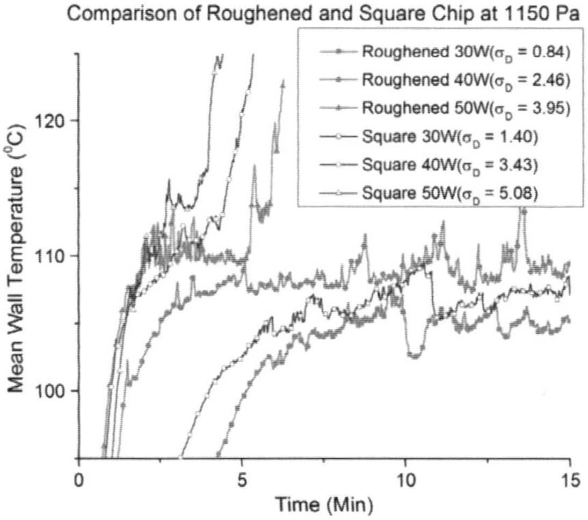

Figure 6. Mean temperature data for the square and roughened chips at three power densities.

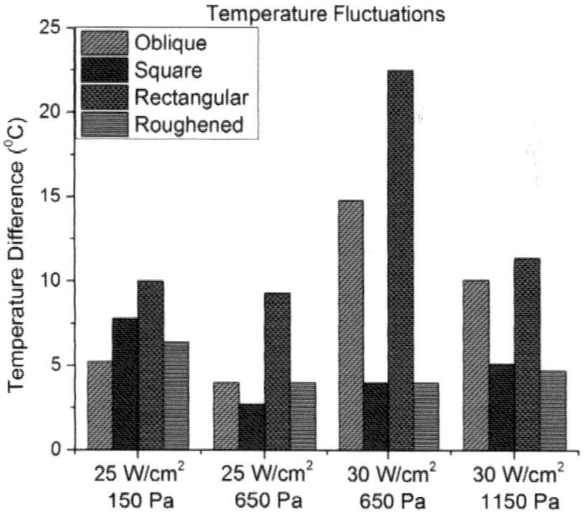

Figure 7. Temperature fluctuations above the mean wall temperature for all of the devices.

5. Results and Discussion

The results follow a path from raw data to a final comparison of the two most promising surfaces. Figure 4 shows three individual trials that were run for 60 minutes each (although the image only shows a 5 minute section). The complete data set includes the device heating up and many peaks. By zooming into a section, more important aspects become increasingly visible. Here, the mean is visible as well as the upper and lower bounds of the confidence interval. Within the confidence interval, areas with small deviations have closer confidence intervals indicating better data. Figure 5 shows the average temperatures of all four chips at 40 W/cm^2 and 1150 Pa. This is the highest power and pressure where all of the chips were tested. The oblique and rectangular chips were not tested at 50

W/cm², because they dried out quickly, and it was assumed that at a higher power they would also dry out. These two chips also had a lot of temperature fluctuations in lower power tests, indicating that these chips were not performing well. Figure 5 also shows the square chip drying out quickly, but earlier tests with small fluctuations showed that this was the second most promising chip. Figure 6 is the final analysis, where average temperatures for the square and roughened chip are compared at 1150 Pa and three different fluxes: 30, 40 and 50 W/cm². This data shows that the roughened chip survived higher power densities for longer periods of time with very good stability.

This data can further be manipulated to detail the maximum fluctuations at the measurement spot, (figure 3). Fluctuations were calculated by subtracting the maximum peak temperature from the average wall temperature over the entire test. The fluctuations are shown for the lower power and lower pressure potential tests in figure 7. These lower tests are shown because the fluctuations do not include dry out or device failure. By definition, fluctuations include the devices reaching high temperatures from which the chips recovered. Recovery means that the temperatures went back down to a temperature close to the mean. The higher power and pressure tests resulted in chip failure that makes it difficult to calculate the wall fluctuations.

An advantage of taking time dependent data is the availability of these fluctuation results. It is clear from the data that wall temperatures do not stay constant, as fluctuations of 5 or more degrees occurred in more than 50% of the tests. From this information, it seems logical that temperature data taken over short time periods may be missing profile details. It is also possible that data taken during a high fluctuation could reveal different wall temperatures than an average over a longer time frame. For this reason, a longer time test may be valuable for convective flow boiling tests on pin fin surfaces.

6. Conclusion

A testing procedure and results for convective flow boiling over roughened pin fin surfaces are presented. The results show that, of four tested pin fin structures, a roughened surface withstood higher power densities for longer time periods than the other structures under the presented conditions. Temperature fluctuations were shown as a possible indication of device and boiling behavior. However, the real conclusion of the temperature fluctuations was that valuable information on the stability of the flow over a surface is indicated when a time variable is added to the study. These results suggest that further exploration into this testing procedure and apparatus could yield better models of flow boiling in pin fins and reveal additional correlations.

References

[1] Krishnan S, Garimella S V, Chrysler G M and Mahajan R V 2007 *IEEE Transactions on Advanced Packaging* **30** 462-474
[2] Pastukhov V, Maidanik Y, Vershinin C and Korukov M 2003 *Applied Thermal Engineering* **23** 1125–1135
[3] Koşar A and Yoav P 2006 *Int. J. Heat and Mass Transfer* **49** 3142-3155
[4] Honda H and Wei J J 2004 *Experimental Thermal and Fluid Science* **28** 159-169
[5] Chu K, Enright R, and Wang E 2012 *Applied Physics Letters*, 241603-1 - 241603-4.
[6] Peles Y, Koşar A, Mishra C, Kuo C J and Schneider B 2005 *Int. J. Heat and Mass Transfer* **48** 3615-3627
[7] Arana L R, Schaevitz S B, Franz A J, Schmidt M A and Jensen K F 2003 *J. Microelectromech. Syst.* **12** 600-612
[8] Dhillion N S, Hogue C, Cheng J C and Pisano A P 2011 *Proceedings of PowerMEMS* (Seoul, Republic of Korea)
[9] Wang H, Garimella S V and Murthy J Y 2007 *Int. J. Heat and Mass Transfer* **50** 3933-3942
[10] Dhavaleswarapu H K, Garimella S and Murthy J Y 2009 *Birck and NCN Publication* Paper **384.**
[11] Nilson R H, Tchikanda S W, Griffiths S K and artinez M J 2006 *Int. J. Heat and Mass Transfer* **49** 1603-1618

Micro reactor integrated µ-PEM fuel cell system: a feed connector and flow field free approach

A. Balakrishnan*, C. Mueller, H. Reinecke
[1]Laboratory for Process Technology, *IMTEK-Department of Microsystem Engineering*, University of Freiburg, Georges-Koehler-Allee 103, 79110 Freiburg, Germany

E-mail: arvind.balakrishnan@imtek.uni-freiburg.de

Abstract. A system level microreactor concept for hydrogen generation with Sodium Borohydride ($NaBH_4$) is demonstrated. The uniqueness of the system is the transport and distribution feature of fuel (hydrogen) to the anode of the fuel cell without any external feed connectors and flow fields. The approach here is to use palladium film instead of feed connectors and the flow fields; palladium's property to adsorb and desorb the hydrogen at ambient and elevated condition. The proof of concept is demonstrated with a polymethyl methacrylate (PMMA) based complete system integration which includes microreactor, palladium transport layer and the self-breathing polymer electrolyte membrane (PEM) fuel cell. The hydrolysis of $NaBH_4$ was carried out in the presence of platinum supported by nickel (Ni-Pt). The prototype functionality is tested with $NaBH_4$ chemical hydride. The characterization of the integrated palladium layer and fuel cell is tested with constant and switching load. The presented integrated fuel cell is observed to have a maximum power output and current of 60 mW and 280 mA respectively.

1. Introduction

In the recent decades the need for energy sources with high energy density and extended operation period had been increased for various applications in the micro and millimetre scale. As a response many types of micro fuel cell have been under research, like Direct Methanol Fuel Cell (DMFC), Solid Oxide Fuel Cell (SOFC) and Polymer Electrolyte Fuel cell (PEMFC), out of which the hydrogen driven PEMFC has acquired a wide interest owing to its ambient condition performance whereas the formers have the drawback of high temperature working conditions of fuel cell hydrogen based, micro Polymer electrolyte membrane Fuel Cells (µ-PFCs) have been the same or more promising candidate than the miniaturised lithium ion accumulators and other energy systems due to its high volumetric energy density in smaller area for autonomous microsystem, eco-friendliness e.g. [1-2]. The energy carrier hydrogen for the PEMFC has various ways of storing namely compressed, liquefied and hydrides. Hydrides are further classified as metal alloy hydrides and chemical hydride. The major requirements for portable micro power sources are its functionality at ambient conditions and its practical feasibility in terms of stability. Based of storage materials, chemical and metal alloy hydrides have the advantages of its high volumetric storage density. Chemical hydrides (e.g. NH_3, NH_2-NH_2, H_3NBH_3, LiH, $NaAlH_4$, NaH, CaH_2, MgH and $NaBH_4$) have been identified as potential hydrogen carrier for micro power source. In particular $NaBH_4$ had gained more popularity among other due its high gravimetric density, easy handling, non-toxic and other storage capabilities. The exothermic hydrolysis reaction of $NaBH_4$ at ambient conditions results in pure gaseous hydrogen (equation (1)), the reaction and shelf-life is controlled the pH of the solution; under ideal ambient conditions hydrolysis of 100 mg of $NaBH_4$ will generate 237 ml of hydrogen [3-5].

$$NaBH_4 + 2H_2O \rightarrow NaBO_2 + 4H_2 + (-300\ KJ/mol) \quad (1)$$

Many researchers have been developing micro reactors using various technologies and base materials (e.g. silicon, photosensitive glass, ceramics, polymers, metals and plastics etc.). One of the major issues with the micro reactor technology is the complexity of the fully integrated system; every component introduced into the system increases the chance of arising a new issue. There is relatively less research focus, on reduction of the complexity of the system also not compromising the quality and system performance. This has given us the motivation to develop new concept which is able to fulfil the above mentioned two important factors.

1.1. New concept

In this work, a system with fully integrated microreactor and PEMFC is presented, eliminating the external hydrogen feed connectors and flow field for the fuel cells. The schematic view of the concept shown in figure 1, system in which the produced hydrogen from the microreactor (1) is transported and evenly distributed[6] with the help of a thin palladium layer (2) to the fuel cell anode (3) without any external hydrogen feed connectors and flow fields; all integrated in one system. The conceptual design proof, fabrication and functionality evaluation is carried out with PMMA substrate. The implementation of the presented concept on a silicon substrate, fabricated by MEMS technology is also discussed. Some of the major advantages of the presented system design are that we are able to eliminate components like gas liquid phase separators, filter membranes and external hydrogen feed connectors from the microreactor to the flow field of the PEMFC system.

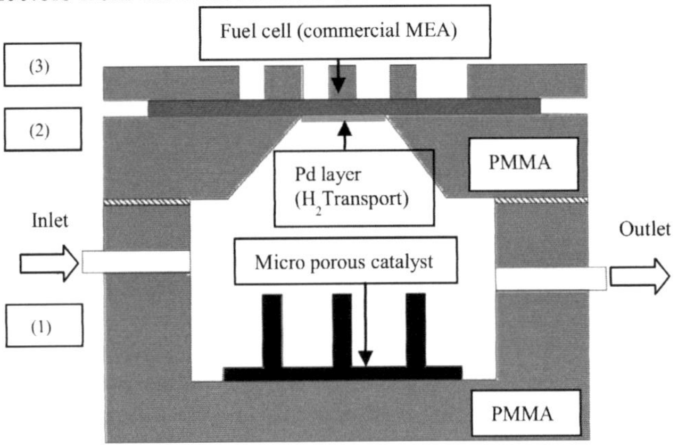

Figure 1. Schematic view of the system concept (prototype with PMMA substrate) with the 3 major components (1) base reactor (reactant inlet/outlet and micro pours catalyst layer), (2) Pd layer (H_2 transporter), (3) fuel cell component

2. Experimental

The prototype design for verifying the functionality of the presented approach is demonstrated in figure 1. The prototype system is a 3 modular system namely, base microreactor, palladium layer as the hydrogen transporter and the fuel cell. The fabrication of these components is done by CAD aided micro milling technique.

2.1 Fabrication

The base micro reactor has a reaction chamber which includes micro porous catalyst and an inlet\outlet for the reactant. The micro porous catalyst layer is fabricated with 95% porous nickel foam as a support followed by electroplating of platinum on it. The electroplating of platinum on a cleaned nickel foam is performed in a commercially available, Umicore Platina® RT electrolyte operating at a temperature of 30°C and a current density 8 mA/cm². The module with the pd layer is fabricated by first milling a through hole conical structures on a PMMA plate ures and then a palladium thin membrane is glued. Palladium thin membrane is also fabricated by electroplating process; one side protected thin copper sheet (30-45 μm thick) is used as a seed layer for electroplating the palladium on the side exposed to the electroplating electrolyte. The copper layer is selectively etched off resulting in a thin free standing palladium layer (approximately 4-5 μm thick), schematic view of the fabrication process is shown in figure 2 (a) and the fabricated palladium layer in figure 2 (b). The electroplating of palladium is performed in commercially available, DODUCO Doduform® Palladium electrolyte operating at a temperature of 45°C and current density of 4 mA/cm². The membrane thickness is optimized to about 5 μm, owing to the fact of stability and the possibility of achieving a closed membrane surface. The fuel cell component is an assembly of two nickel mesh current collectors and conventional membrane electrode assembly (MEA) of area 9 cm². A sandwich of nickel current collectors and MEA is inserted between a pair of patterned PMMA sheets. Both the PMMA sheets had patterned openings which act as a window for the fuels to come in contact with the MEA, but on the anode side of the setup there is the palladium membrane which acts as a filter, flow field and a buffer storage for hydrogen, shown in figure 3. The sealing is done by applying instant polymer sealant at the side of the two PMMA sheets. The individual modules are mechanically clamped together constituting one complete system. The inlet/outlet of the system is connected with polymer tubing in which the reactant is pumped. The complete assembly shown in figure 4 constitutes; catalyst embedded micro reactor (1), palladium membrane (2) and fuel cell stack with external contacts (3) to the measurement system.

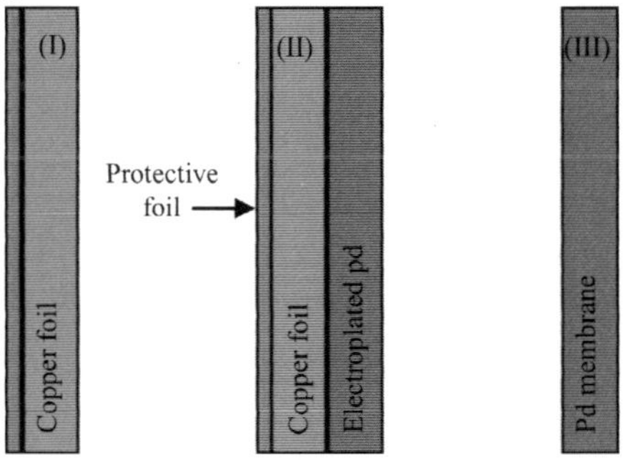

Figure 2 (a). Schematic view of 3 step fabrication of palladium membrane, (I) copper foil with one side covered with protective foil, (II) electroplating of pd on the exposed side of copper foil, (III) selective etching of the copper foil

Figure 2 (b). Picture of the palladium membrane after the selective etching of copper foil for the back side, resulting in a free standing & mechanically stable palladium membrane.

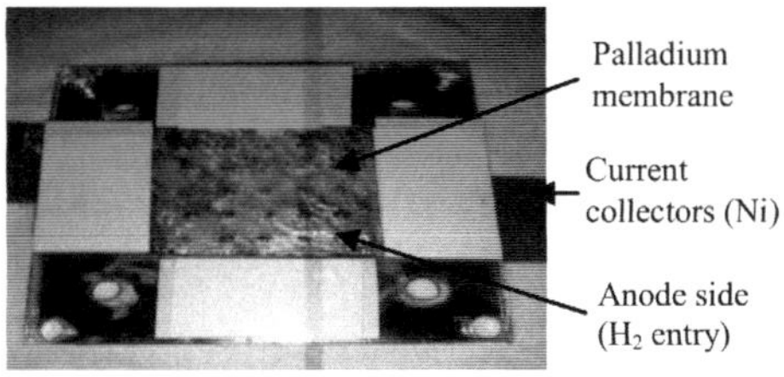

Figure 3. These two figures have been placed side-by-side to save space. Justify the caption.

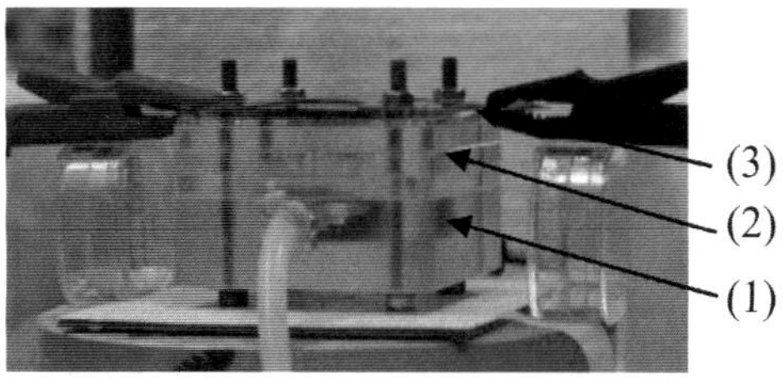

Figure 4. These two figures have been placed side-by-side to save space. Justify the caption.

2.2 Performance measurement

The assembled system is tested for its performance using an experimental set up, consisting of syringe and a computer controlled potentiostat. The system is tested with pure hydrogen source for characterization of the fuel cell module before using the chemical hydride. The chemical hydride, $NaBH_4$ solution is supplied to the reactor using the syringe. The reactant $NaBH_4$ solution is prepared by dissolving in high concentration sodium hydroxide (NaOH) and a small amount of 99% pure sodium dodecyl sulphate (surfactant). The reason for using the surfactant is to reduce the surface energy of the catalyst ultimately reducing the by-product clogging effect on the catalyst active sites. The micro reactor setup is tested with pure hydrogen flow for obtaining the performance characterization of the fuel cell module; polarization characterization. Then the system is then tested with 20 ml chemical hydride solution as a source of hydrogen; current and voltage values of the fuel cell are recorded under fixed electrical load (set based on the information obtained from polarization characteristics).

3. Result & discussion

The experiments are performed as described in the above section, the performance characteristics of the fuel cell module is tested with pure hydrogen flow to the anode side and ambient air on the cathode side. In figure 6 (a) shows the polarization characterises, on observation the maximum power output of the system is relatively small; on detailed examination of the setup the issue is identified that the structured PMMA plate (system 1) causes lack of air on the cathode side. The partnering of the PMMA plate was modified in such a way that there is enough air able to reach the cathode of the fuel cell (system 2). Replacing the PMMA plate design 1 with the PMMA plate design 2, the system 2 is again tested with pure hydrogen flow; as extected the polarization characterization of the system 2 shows a factor of 6 higher maximum power output measured and a factor of approximately 4 higher in the maximum current than system 1, the polarization characterization is shown in figure 6 (b). The system 1 & 2 is tested for its performance with the hydrolysis of $NaBH_4$ (20 ml) in the presence of Ni-Pt catalyst. A constant electrical resistance act as a load and the corresponding drop in the output voltage from the open circuit voltage is recorded. There are two variations; one with constant load and other with switching load. Figure 7 (a-b) shows the systems performance for the above mentioned types of load; figure 7 (a) shows that system 1 (PMMA plat design 1) response for constant load and figure 7 (b) shows the system 2 (PMMA plat design 2) response for switching load. The electrical load resistance are set in such a way that it is same or close to the internal resistance, obtained from the polarization characteristic of the respective systems (e.g. for system 1 it is approx. 10 ohm and for system 2 it is approx. 3 ohm). The system 1 response for constant electrical load is shown in figure 7 (a), a load sweep from 40-160 ohm is done showing a step change in the measured output voltage. The system 2 response for switching load is shown in figure 7 (b), where a 3 ohm electrical resistance is set for 2 minutes with an interval of 1 minute followed by a 5 ohm resistance was set for 1 minute with an interval of 5 minutes. The load response of the chemical hydride driven fuel cell system with an integrated palladium layer for hydrogen transport is very satisfactory, and seems to be promising for longer periods of operation.

4. Conclusion & outlook

An integrated microreactor-fuel cell system is fabricated eliminating external feed connectors, fuel cell flow fields and filter members but only using a thin palladium membrane instead. The system functionality (proof of concept) of using such an approach is demonstrated using the PMMA prototype system. The system behaviour with pure hydrogen source and chemical hydride source is clearly demonstrated. The characterization of the integrated fuel cell is demonstrated by the polarization characteristics (external supply of pure hydrogen) followed by the system performance for constant and switching electrical load resistance (microreactor with chemical hydride delivering hydrogen on hydrolysis). The repeatability of the system performance showed no degradation on long term measurments. This approach significantly reduces the complexity of the system also not compromising its performance.

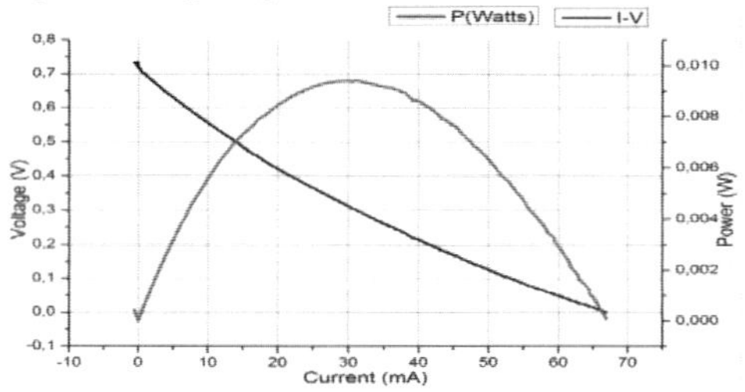

Figure 6 (a). Polarization characterization of fuel cell with pure hydrogen flow, with PMMA cathode plate design 1. Maximum power output in the range of approximately 0,009 Watts

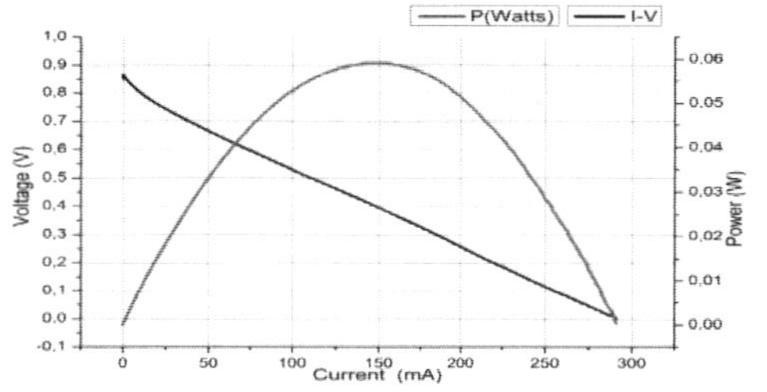

Figure 6 (b). Polarization characterization of fuel cell with pure hydrogen flow, with PMMA cathode plate design 2. Maximum power output in the rage of approximately 0,06 Watts

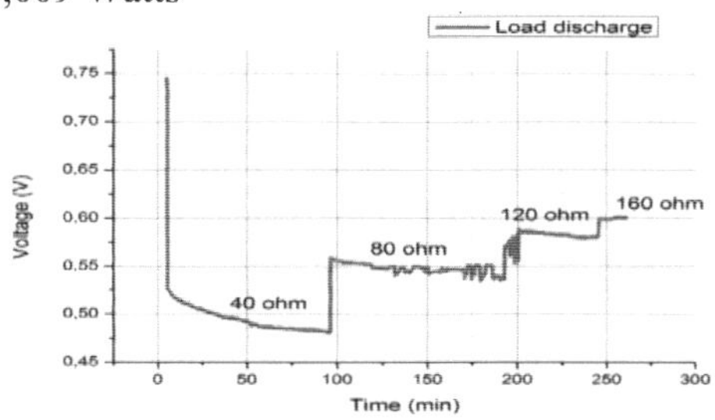

Figure 7 (a). System 1 (PMMA plat design 1) response for constant load

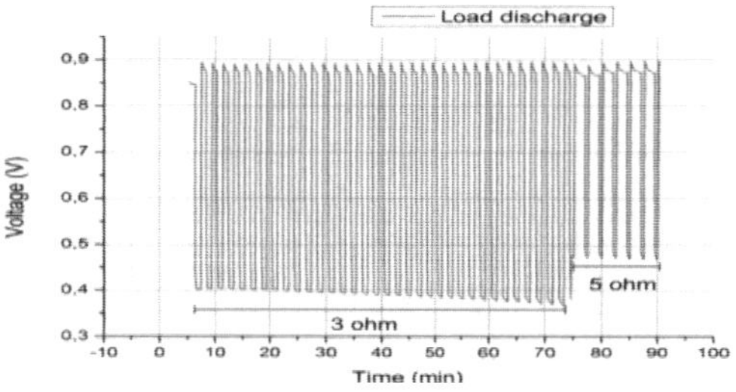

Figure 7 (b). System 2 (PMMA plat design 2) response for switching load.

This work provides an evident knowledge about the presented approach which can be scaled up or scaled down depending on the application demand. The next level is to implement the presented device concept on silicon substrate (shown in figure 8) fabricated by MEMS technology. This sets the stage for creating a new generation of Chip Integrated micro PEM Fuel Cell System (CIμ-PFCS) for realizing the future of energy sufficient autonomous microsystems which are long term stable.

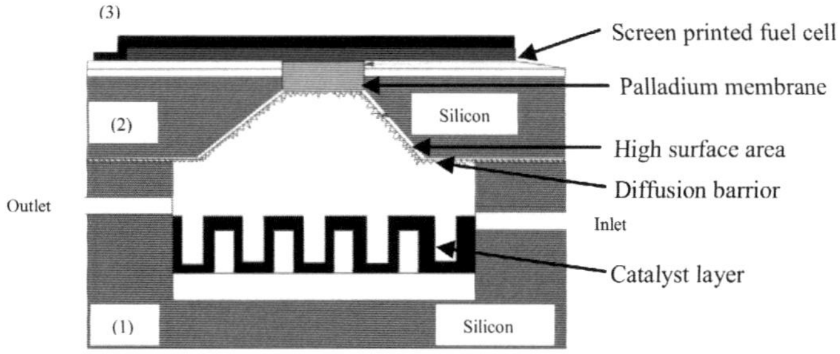

Figure 8. Schematic view of the system concept with silicon substrate, (1) base reactor (reactant inlet/outlet, micro pours catalyst), (2) high surface area palladium hydrogen transport layer, (3) screen printed fuel cell.

5. Reference

[1] Arvind Balakrishnan at.al, *Advanced Materials Research* **Vols. 608-609 (2013)** *pp 904-912*
[2] T. Kim, *International journal of hydrogen energy* **37(2012)** *2400-2446*
[3] R.Oronizio, *International journal of hydrogen energy* **34 (2009)** *4555-4560*
[4] Zhang JS, *Journal of power sources 2007:***165(2)***:844*
[5] Zhang Q, Smith G, Wu Y, Mohring, *International journal of hydrogen energy 2006;* **31**; *961.*
[6] B.D.Morreale.et.al *"Journal of Membrane Science* **212 (2003)** *87-97.*
[7] Arvind B, *In: Proc power MEMS 2012;* 2012. *pp.* ***379-382*** *Atlanta, USA,*

Screen printed flexible Bi_2Te_3-Sb_2Te_3 based thermoelectric generator

Zhuo Cao[1], E. Koukharenko, M.J. Tudor, R.N. Torah and S.P. Beeby
Electronics and Electrical Engineering Group, Electronics and Computer Science, University of Southampton, Southampton, SO17 1BJ, UK

E-mail: zc2e09@soton.ac.uk

Abstract. This paper reports the fabrication and testing of Bismuth Tellurium (Bi_2Te_3) – Antimony Tellurium (Sb_2Te_3) based thermocouples using screen printing technology. In this study, screen printable thermoelectric pastes were developed and the transport properties of cured material were measured. The dimension of each planer thermoleg is 39.3 mm × 3 mm with a thickness of 67 μm for Bi_2Te_3 leg and 62 μm for Sb_2Te_3 leg. A single thermocouple with this dimension can generate a voltage of 6 mV and a peak output power of 48 nW at a temperature difference of 20°C. The calculated Seebeck coefficient of a single thermocouple is in the range of 262 – 282 μV/K. The Seebeck coefficient at room temperature were measured to be -134 – -119 μV/K and 128 – 134 μV/K for Bi_2Te_3 and Sb_2Te_3 respectively. This work demonstrates that the low-cost screen printing technology and low-temperature materials are promising for the fabrication of flexible thermoelectric generators (TEGs).

1. Introduction

Based on Seebeck effect, the thermoelectric modules have ability to directly transfer heat into electric energy. Thermoelectric Generators (TEGs) are very attractive devices for harvesting waste thermal energy applications, for instance, powering wireless autonomous sensors [1]. The TEG devices have advantages such as silently operation, no moving parts and high reliability. It is the best approach for powering the devices in locations where there are poor levels of illumination but sufficient waste heat.

In order to evaluate the performance of a thermoelectirc material, the figure of merit ZT ($Z=\alpha^2/(\rho\cdot\lambda)$, where α is the Seebeck coefficient, ρ is the electrical resistivity, λ is the thermal conductivity and T is the temperature) is applied. In room temperature range (293 K to 303 K), Bismuth Tellurium/Antimony Tellurium (Bi_2Te_3/Sb_2Te_3) system has the highest ZT value among reported thermal active materials [1, 2]. Several papers have reported fabricating flexible Bi_2Te_3/Sb_2Te_3 based TEGs using different fabrication technologies (e.g. dispenser printing, photolithography and electrochemical deposition) [2, 3]. However, the energy transfer efficiency of conventional rigid TEGs is only around 5% [4]. Flexible TEGs are supposed to have even less transfer efficiency. Hence, in order to achieve high power output, large quality of TEG devices need to be electrically connected together. Then, an easy fabrication process is essential to low down the cost.

Screen printing is a low-cost process that is well suited for large area fabrication. It involves the deposition of synthesized thermoelectric inks that consist of thermoelectric material powders in a binder and solvent matrix. After printing only a curing process is required. The simplicity and low energy consumption is quite attractive for fabricating TEGs. The Seebeck effect and structure of a screen printed thermocouple is illustrated in figure 1. In this work, screen printable pastes filled with thermoelectric materials were formulated. A planer TEG with 4 thermocouples was fabricated by screen printing technology and the thermal performance tested.

2. Experimental

2.1. Material synthesis

The screen printable thermoelectric pastes reported here contains thermoelectric materials Bi_2Te_3 and Sb_2Te_3 alloy powders and an epoxy binder system. A solvent was also added in to the paste to adjust the viscosity to a screen printable level.

For n-type bismuth tellurium alloy, the ratio of tellurium (Te) affects the thermoelectric performance of the device. When the atomic ratio of Te is a 64%, the bismuth tellurium alloy will have the highest negative Z value [5]. The actual formulation of this n-type semiconductor material is $Bi_{1.8}Te_{3.2}$. The powders are from Testbourne Ltd, with the size of 325 mesh and a purity of 99.99%.

An EXAKT 50 triple roll mill was used to mix the paste to achieve a uniform particles distribution. The viscosity range of these screen printable pastes is from 9000 cP to 15000 cP, measured by Brookfield High Shear CAP-1000+ viscometer.

2.2. Device fabrication

The printer used was DEK-248 screen printing machine. A polyester screen with 325 mesh was used for deposition. Kapton was used as the substrate.

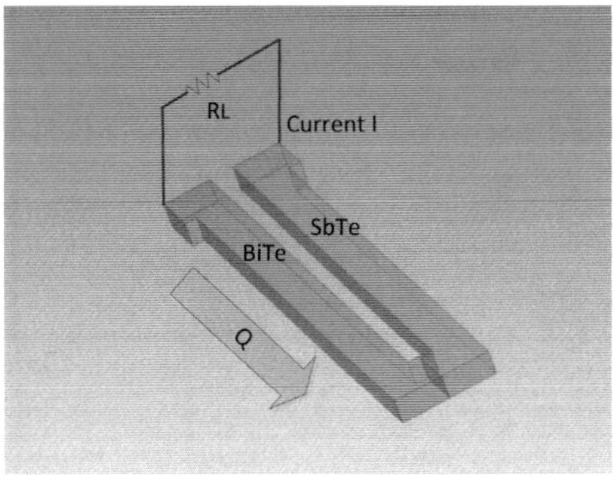

Figure 1. Operation principle of screen printed Bi_2Te_3/Sb_2Te_3 thermocouples

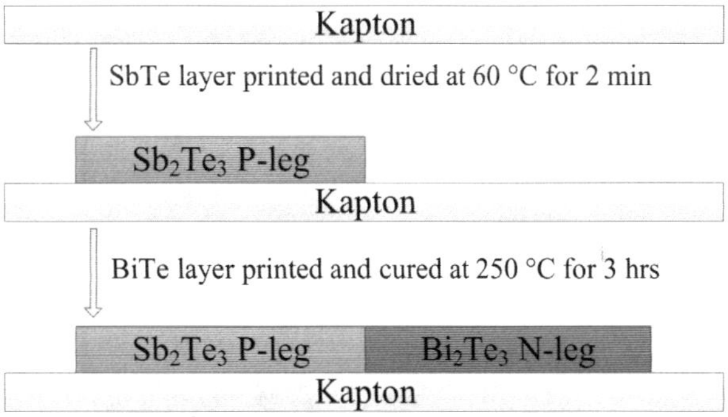

Figure 2. Fabrication process of screen printed Bi_2Te_3/Sb_2Te_3 thermocouple

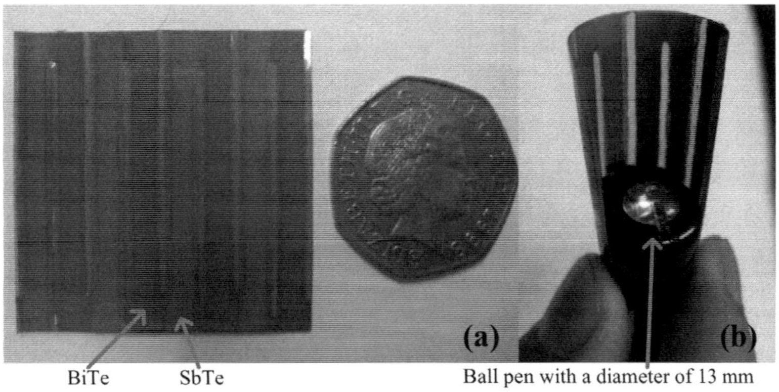

Figure 3. Printed Bi_2Te_3/Sb_2Te_3 thermocouples (a): sample size comparing with a 50 pence coin; (b): sample rolling on a 13 mm diameter ball pen

The screen printed Bi_2Te_3/Sb_2Te_3 fabrication process in shown in figure 2. It only involvs deposition of two thermal material layers and a curing process. The thickness of the active thermoelectric layers can be built up by drying each layer at 60 °C to evaporate the solvent and then repeating the printing step. After the two thermolegs have been deposited, they are cured at 250 °C for 3 hours. The screen printed sample with 4 thermocouples is shown in figure 3.

2.3. *Printed Thick-films and Device Characterization*

The surface and the cross-section of the screen printed films were observed in an EVO Scanning Electron Microscope (SEM). The SEM images in figure 4 show the homogenous mixture of the powders and binder. Also, the average particle size can be identified to be around 10 μm.

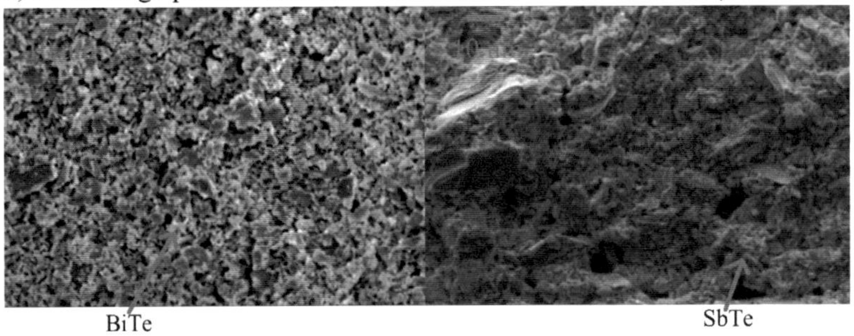

Figure 4. SEM images of printed thick-film. Left: surface of BiTe thick-film; Right: cross-section of SbTe thick-film

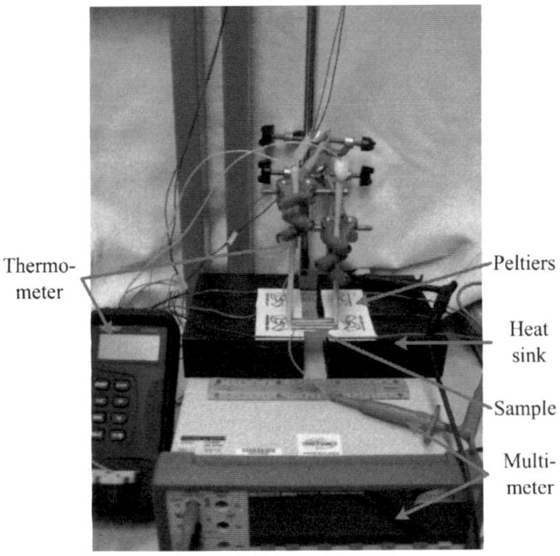

Figure 5. Seebeck Voltage measurement setup

Table 1. Seebeck coefficient of screen printed Bi_2Te_3 and Sb_2Te_3 thick-films

α (μV/K)	Bi_2Te_3	Sb_2Te_3
Day 1	-134 – -119	128 – 134
Day 100	-132 – -129	132 – 137
Bulk value [6]	-227	110

From the introduction it can be seen that the figure of merit consists there parameters: the Seebeck coefficient α, the electrical resistivity ρ and the thermal conductivity λ. In this experiment, Hall Effect measurements were performed to test the transport properties of the printed thick films, including resistivity, using a commercial Hall Effect measurement system (HMS 3000 from Ecopia). The Seebeck coefficients of all the material samples were measured using custom-made equipment. For these measurements 1 cm × 1 cm square samples were printed and tested.

The Seebeck voltage of the printed sample devices were tested by a bespoke set-up shown in figure 5. Once the device reached steady state with a constant temperature gradient, the Seebeck voltage of the device was measured using a digital multimeter. Peltier modules from European Thermodynamics provided the thermal gradient across the device. The temperatures gradient across the sample were using a Tenma digital thermometer with dual type K input. The gap between the two peltiers was fixed at 2 cm. All mechanical interfaces were sealed with thermal compound to reduce the heat loss. The electrical power output from the thermocouples was calculated from the voltage across a variable load resistance in series. All test were processed in open air at room temperature.

3. Results

3.1. Thermoelectric properties of printed thick-films

The Seebeck coefficient measurement results in comparison with the bulk value are shown in table 1. The negative sign of Seebeck coefficient indicates that the $Bi_{1.8}Te_{3.2}$ alloy powders acting as n-type material in a thermocouple, while the positive α value means that the Sb_2Te_3 alloy is p-type thermoleg. The absolute α value of $Bi_{1.8}Te_{3.2}$ thick-film is smaller than that of the bulk material, while the Seebeck coefficient of Sb_2Te_3 thick-film is higher than the bulk alloy. This is compatible with the results reported by Chen et al. [3].The variation of the average α value is with 5% within 100 days.

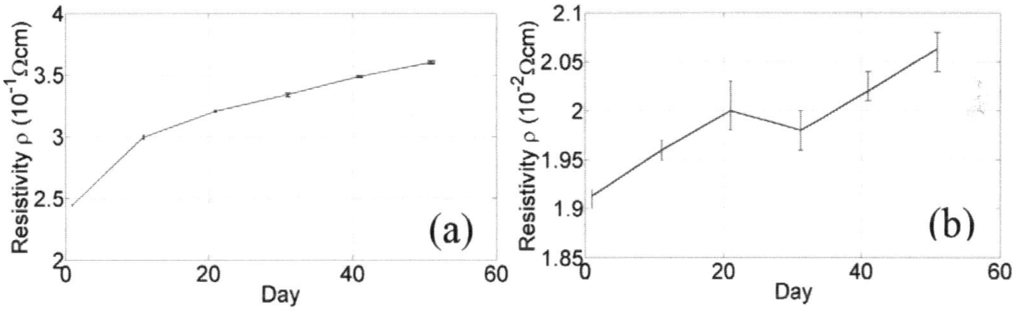

Figure 6. Resistivity of printed thick-films varying with time. (a):$Bi_{1.8}Te_{3.2}$; (b):Sb_2Te_3.

Figure 6 shows the resistivity of both thick-films and their variation with time. Both materials also show an increase in resistivity with time, while this is larger for the BiTe material. In 50 days, the $Bi_{1.8}Te_{3.2}$ thick-film shows an increase ρ of 47% compared with the original value. For Sb_2Te_3 thick-films, the variation is only 7.8% over 50 days.

3.2. Device characterization

Figure 7 shows the tested Seebeck voltage of 4 screen printed BiTe/SbTe thermocouples. A single Bi_2Te_3-Sb_2Te_3 thermocouple can generate 17 mV voltage at a temperature difference of 60 °C, which is better than the 5 mV of a single Bismuth-Tellurium thermocouple achieved by our group at the same temperature difference [6]. From the definition of Seebeck Effect (α=V/ΔT, where V is the Seebeck Voltage, ΔT is the temperature difference), for a single thermocouple of this sample, the Seebeck coefficient is in the range of 262−282 μV/K (figure 8).

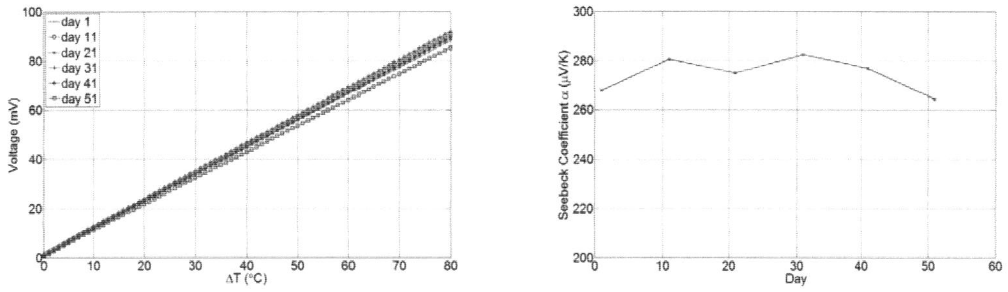

Figure 7. Generated Seebeck voltage of four printed thermocouples, varying with time

Figure 8. Calculated Seebeck coefficient for single thermocouple varying with time

The power output versus load resistance at a temperature difference of 20 °C is shown in figure 9. The voltage across the load increases with the increasing load resistance and the maximum output power occurs when the load resistance matched the TEGs resistance. The power output P is calculated by $P=V^2/R$, where V is the voltage across the load resistance and R is the load resistance. For a single thermocouple, the original power is 48 nW. This value decreased to 23 nW after 50 days.

The reason for the low power mainly comes from the high resistivity of the BiTe thick-film. From figure 4 it can be seen that there are pores existing between nearby particles, which prevent a better Ohm contact between particles.

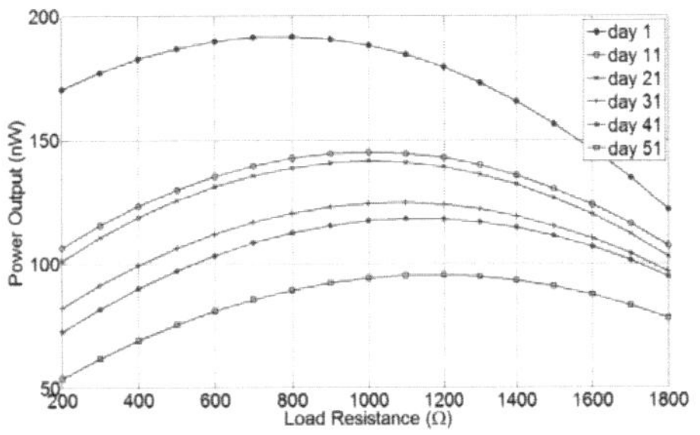

Figure 9. Power output of 4 printed BiTe/SbTe thermocouples varying with load resistance

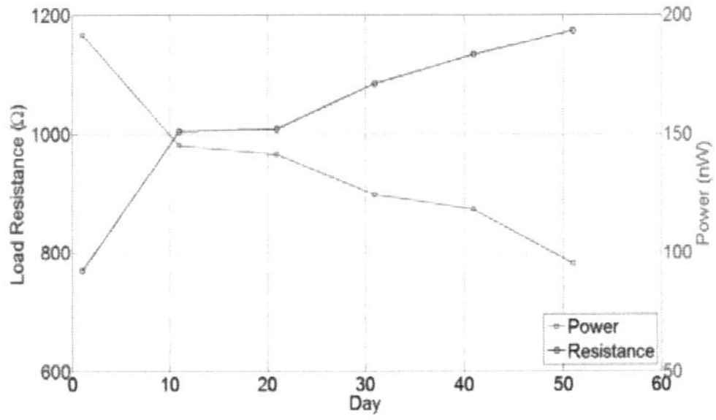

Figure 10. Maximum power output and self-resistance of 4 printed BiTe/SbTe thermocouples varying with time

4. Conclusion

This work demonstrates that the low cost screen printing technology can be used to fabricate BiTe/SbTe based TEGs for room temperature energy harvesting applications. Although the power output is low for a single thermocouple, the flexibility allows the screen printed TEGs to be connected in series and rolled in to a coil. Then, higher power output can be achieved by deploying more thermocouples in this way. Future work includes the measurement of thermal conductivity which will enable the calculation of ZT.

However, the high and increasing electrical resistivity of $Bi_{1.8}Te_{3.2}$ thick-film is an issue for reaching a higher power output for this screen printed TEG. Oxidization of bismuth tellurium powders is the likely cause. Future work will explore the optimization of design and material processing to decrease the resistivity. To reduce the resistivity variation and prevent oxidization, an encapsulation layer on top of the printed thermal materials layer will be investigated.

5. Acknowledgment

The authors thank the EPSRC for supporting this research with grant reference EP/I005323/1. We also thank Kai yang and Luyi Yang for their contributions.

6. References

[1] S. Beeby and N. White 2010 *Energy Harvesting for Autonomous Systems* (Norwood, MA: Artech House)

[2] Glatz W, Schwyter E, Durrer L, Hierold C and Member S 2009 *Microelectromechanical Syst.* **18**, 763

[3] Chen A, D. Madan, P. K. Wright, and J. W. Evans 2011 *J. Micromechanics Microengineering* **21**, 2011.

[4] Rowe D *Renew. Energy* **16**, 1251

[5] Fleurial J P, Gailliard L, Triboulet R, Scherrer H and Scherrer S 1988 *J. Phys. Chem. Solids* **49**, 1237

[6] Cao Z, Koukharenko E, Torah, R and Beeby SP 2013 *The 17th Int. Conf. on Solid-State Sensors, Actuators and Microsystems (Barcelona)* pp.478-481

Micro-fabrication of Flexible Coils with Copper Filled Through Polymer Via Structures

Q. S. Zhu[1, 2], A. Toda[3], Y. Zhang[1, 2 *], T. Itoh[1, 2], R. Maeda[1, 2]
[1]NMEMS Technology Research Organization, Japan
[2]National Institute of Advanced Industrial Science and Technology (AIST), Japan,
[3]Meltex Inc., Japan

Email: yi.zhang@aist.go.jp

Abstract. In this work, we present one flexible 3D micro-coil. This 3D micro-coil is successfully prepared in a thin polymer film with a thickness of 120μm. The flexible coil is expected to be used in current sensing and energy harvesting MEMS those require a large deformation degree to wrap target object. A typical micro-machined 3D coil is composed of bottom, vertical and top windings. We firstly adopt through polymer vias (TPVs) and metal filling technology to fabricate the vertical windings. A high-speed copper electrodeposition technology of TPVs is developed to obtain void-free vertical windings.

1. Introduction

A high performance inductor is the key component for various radio frequency (RF), sensor, and energy harvesting MEMS. Silicon-based technologies are the mainstream for integrated circuit designs catering to both mass consumer and commercial applications. For the last two decades, polymer materials have been gaining more attention in inductor fabrication technology for their merits in some factors. The application of low-loss polymer materials can be an alternative solution to achieve improved Q-factor and lower self-resonance frequency (SRF). In addition, the low stiffness and excellent flexibility can be especially attractive in the applications where the MEMS devices are directly paste or wrap the elements, such as insulated electrical wire surface for non-touch current sensing. In the fabrication of MEMS-based inductors, planar structure is usually adapted. Micro solenoid-type coils are often prepared by using wire wound coils, but this method will be limited in smaller feature size.

Recently, the 3D micro-coil fabrication has been exploring by different approaches. K. Kratt et al have successfully prepared the solenoid micro coils by using a wire bonder [1]. Zhang et al developed a micro coils on a fiber below 100 um diameter using a rotation lithograph method [2]. Recently, an air core power inductors was fabricated by metal encapsulated polymer via [3]. Using laser irradiation, T. Kikuchi et al. also successfully fabricated a platinum grid-shaped microstructure, a micro-spring, and a cylindrical network microstructure with 50-100μm line width [4]. Matsumoto et al. [5] developed a three-dimensional X-ray lithography method using X-rays from a synchrotron radiation facility as a light source for lithographic exposure and an X-ray mask for patterning cylindrical micro-coil structures with a high aspect ratio. However, these methods for a 3D micro-coils structure through lithography method are usually involved of expensive equipments or limited resolutions. Therefore, it is necessary to explore a relatively cheap and easy approach to produce the 3D micro-coils structure. In this study, we developed a 3D micro-coil in the polymer film using through polymer vias (TPVs) and metal filling method.

2. Design of 3D micro-coil structure

As shown in Fig.1, a flexible sensor is aimed to freely wrap the conductor wire. The intended functions of current sensing and self-powering are based on the Faraday law of electromagnetic induction. The embedded 3D micro-coil within a flexible polymer film is designed. This 3D coil comprises top windings, bottom windings, and vertical windings. The metal filled through polymer via (TPVs) structure is designed to use as the vertical windings. The top and bottom windings are micro-fabricated by the metal film patterning method.

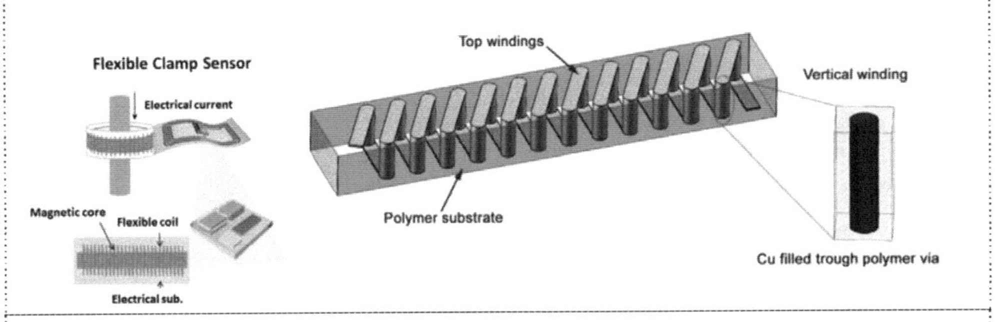

Figure 1. Schematic of a flexible clamp sensor for energy monitoring application and a 3D micro-coil using Cu filled vias as vertical windings.

As shown in Fig.1, a flexible sensor is aimed to freely wrap the conductor wire. The intended functions of current sensing and self-powering are based on the Faraday law of electromagnetic induction. The embedded 3D micro-coil within a flexible polymer film is designed. This 3D coil comprises top windings, bottom windings, and vertical windings. The metal filled through polymer via (TPVs) structure is designed to use as the vertical windings. The top and bottom windings are micro-fabricated by the metal film patterning method.

3. Experimental

The fabrication process of this 3D micro-coil is illustrated in Fig.2. The flexible polyimide sheet with a thickness of 120 μm was selected to use as the substrate. The through polymer vias (TPVs) were micro-fabricated through laser drilling process. The polymer sheet with the formed via array was shown in Fig.3. The diameter and pitch of the via were designed to be 30 μm and 60 μm respectively.
A defined "two-step" Cu electro-deposition formula was used to fill the TPVs. In this process, a Cu seed layer with a thickness of 100nm was sputtered in advanced on the bottom surface. Then the Cu electrodeposition was firstly performed on the pre-sputtered bottom side using a relatively higher current density until the vias entrances were completely blocked. Following that, the second-step electrodeposition was carried out on the top side to obtain a complete copper filling within the vias. Then the Cu overburden on the surface was removed through a chemical-mechanical polishing process to avoid the interconnect between the adjacent vias.
The top and bottom windings were micro-fabricated by lithograph process. In this process, a 20 nm Cr film and 180 nm Au film were firstly sputtered on the top and bottom surface respectively. A resist was sprayed on both sides and one side were patterned after lithography. After removing the resist and etching Cr/Au layers, the layout of parallel metal lines with a width of 20 μm was fabricated, where the line was precisely located in order to interconnect the terminals of the vertical windings. This

finished side was protected by the resist in advanced, and then the other side was metal patterned by the same way.

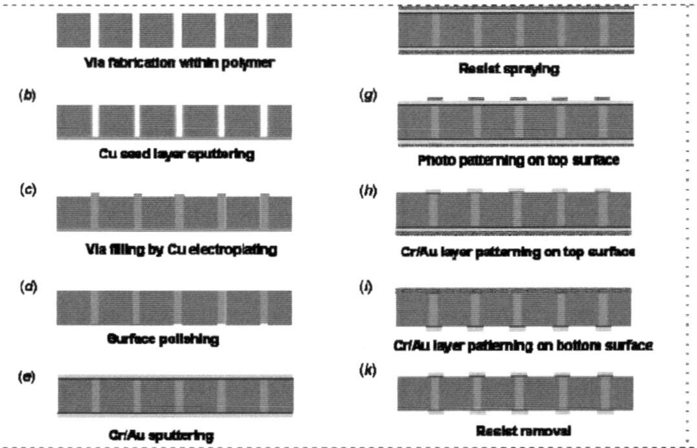

Figure 2. Fabrication process of 3D micro-coil within polymer film.

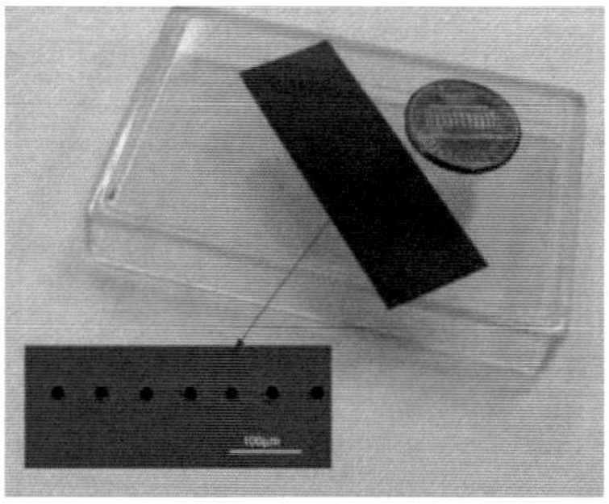

Figure 3. Polymer film with through polymer vias fabricated by laser drilling.

4. Results and discussion

During a "two-step" Cu electrodeposition process for filling the vias, a rapid deposition was firstly performed on the back side for the below considerations. Firstly, the entrances at the bottom are blocked so that a TSV-like electrodeposition filling process can start at the front entrances. Secondly, a "V" growth profile was aimed to obtain, which is needed for a void-free filling. As expected, the highest current density occurred at the entrance and decreased from back to front side during the first-step deposition. Then a "V" filling profile was formed while the back entrance had been completely sealed. In the second-step deposition, the formed "V" filling profile increased the step coverage value, which made it possible to fill completely at deeper location before the sealing of the entrance. Fig.4(a) shows the front surface after the second-step deposition. The Cu grew from the bottom to up and overflew the vias to form the interconnection between the adjacent vias. Fig.4(b) shows the cross-section of the through polymer holes after the second-step deposition. A complete and void-free filling effect was obtained using this "two-step" deposition formula. This void-free filling effect would effectively increase the electrical conductivity and reliability of such a micro-coil structure. After mechanical polishing process, the growth "overburden" on the surface was completely removed. The surface after polishing process is shown in Fig.4(c). The interconnection was removed to avoid "short circuit" between the adjacent vias.

Fig.5 shows the finished top and bottom windings after the patterning process. The lines were precisely aligned to contact the terminal of the filled vias. Of course, due to the slightly etching effect to the Cu during Au and Cr etching process, a shallow indentation could be observed at the junctions, which might bring the risk of poor contact or even disconnection. In the following improvements, the line sizes at the junctions will be enlarged to cover the via, avoiding the Cu etching during the top and bottom windings fabrication process.

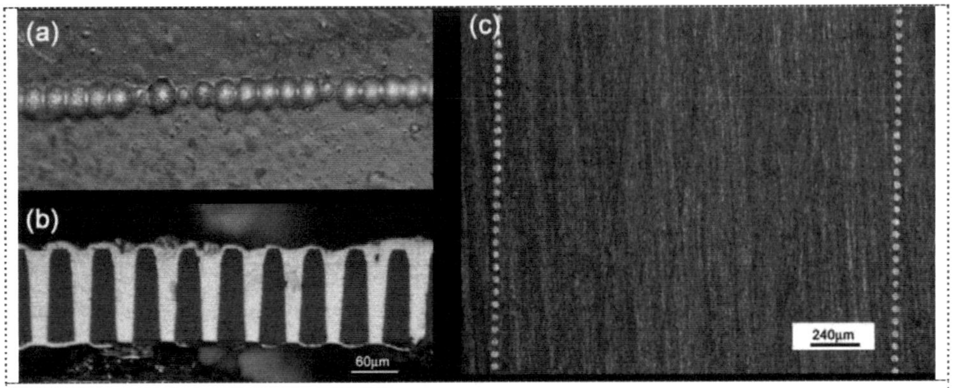

Figure 4. Optical image of vertical windings with copper filled trough polymer vias: (a) surface after Cu electrodeposition, (b) cross-section of filling TPVs, (c) surface after polishing.

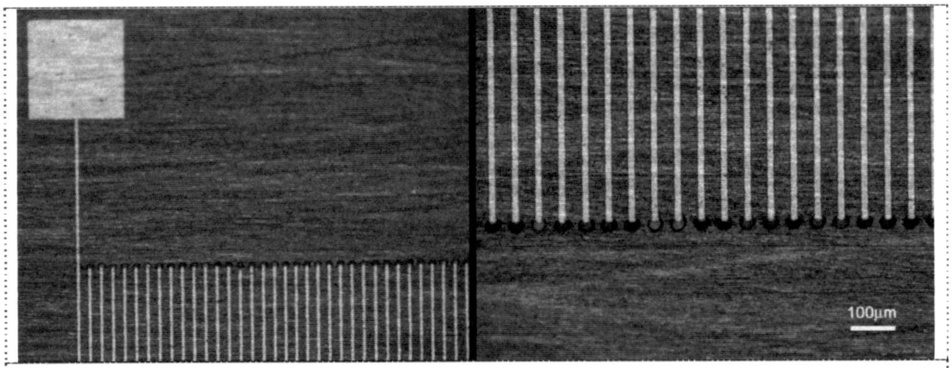

Figure 5. Surface windings fabricated using a lithography method.

5. Conclusions

In this work, we presents one fabrication method of 3D micro solenoid-type coils which is intended to use as current sensing or self-powering module. The 3D micro-coil was fabricated in a thin polymer film with a thickness of 120 µm. The vertical windings were realized through laser-drilled through polymer vias (TPVs) and the following via filling process. A complete and void-free filling effect was obtained using a "two-step" Cu electrodeposition formula. The top and bottom windings were fabricated by a metal patterning process. After a precise alignment, the surface windings connected with the terminals of the filled vias, producing a 50-turn 3D micro-coil within polymer film substrate.

Acknowledgments

This work is supported by New Energy and Industrial Technology Development Organization (NEDO). One of authors, Zhu, also acknowledges International Affairs Division, National Institute of Advanced Industrial Science and Technology (AIST) for offering a fellowship.

References

[1] Kratt K, Seidel M, Emmenegger M, Wallrabe U, 2008 Technical Digest MEMS 996.
[2] Zhang Y, Lu J, Hiroshima H, Itoh T, Maeda R, 2009 Technical Digest Power MEMS 253.
[3] Kim J, Herrault F, Yu XH, Kim M, Shafer RH, Allen MG, 2013 J. Micromech. Microeng 23 1.
[4] Kikuchi T, Takahashi H, and Maruko T 2007 Electrochim. Acta 52 2352.
[5] Matsumoto Y, Setomoto M, Noda D a As part of the production system for *Journal of Physics.*

PowerMEMS 2013 IOP Publishing
Journal of Physics: Conference Series **476** (2013) 012033 doi:10.1088/1742-6596/476/1/012033

Formic acid microfluidic fuel cell based on well-defined Pd nanocubes

A Moreno-Zuria[1], A Dector[1], N Arjona[1], M Guerra-Balcázar[2], J Ledesma-García[2], J P Esquivel[3], N Sabaté[3], L G Arrriaga[1] and A U Chávez-Ramírez[1]

[1] Centro de Investigación y Desarrollo Tecnológico en Electroquímica, 76703 Querétaro, México.
[2] División de Investigación y Posgrado, Facultad de Ingeniería, Universidad Autónoma de Querétaro, 76010 Querétaro, México.
[3] Instituto de Microelectrónica de Barcelona, IMB-CNM (CSIC), Campus UAB, 08193 Bellaterra, Barcelona, España.

E-mail:achavez@cideteq.mx

Abstract. Microfluidic fuel cells (μFFC) are emerging as a promising solution for small-scale power demands. The T-shaped architecture of the μFFC promotes a laminar flow regimen between the catholyte and anolyte streams excluding the use of a membrane, this property allows a simplest design and the use of several micromachining techniques based on a lab-on-chip technologies. This work presents a combination of new materials and low cost fabrication processes to develop a light, small, flexible and environmental friendly device able to supply the energy demand of some portable devices. Well-defined and homogeneous Pd nanocubes which exhibited the (100) preferential crystallographic plane were supported on Vulcan carbon and used as anodic electrocatalyst in a novel and compact design of a SU-8 μFFC feeded with formic acid as fuel. The SU-8 photoresist properties and the organic microelectronic technology were important factors to reduce the dimensions of the μFFC structure. The results obtained from polarization and power density curves exhibited the highest power density (8.3 mW cm^{-2}) reported in literature for direct formic acid μFFCs.

1. Introduction

A special kind of miniaturized fuel cells is emerging: the microfluidic fuel cells (μFFC); their operation mode is based in the liquid-liquid (fuel and oxidant) interface under laminar flow regimen, this particular property, is precisely the main advantage over static membrane fuel cells [1- 4], since their operation requires a smaller number of components and sporadic maintenance events[1, 5-8]. The aim of the μFFCs is and has been to supply the energy demand of portable devices such as cell phones, sensors and medical transplants [1, 9, 10]. The main problem of μFFCs is the long-times durability and their performance. In this work we attacked the performance problem using Pd nanocubes which is known that exhibit high current densities due to the presence of (100) crystallographic planes using formic acid as fuel due to its high theoretical energy density (2.086 kW L^{-1}) and low cross-over effect compared with methanol [11].

2. Methodology

2.1 Fabrication process

The use of polymeric materials like polymethyl methacrylate (PMMA) and poly-dimethylsiloxane (PDMS) has recently been proposed to produce micro-fuel cell components [3, 6, 7, 13]. However UV-sensitive resists such as SU-8 have demonstrated to be excellent candidates to build microfluidic fuel cells since they present some important advantages like flexibility and the capability to be sealed each other by a hot-pressing technique [12, 14]. The μFFC assembly used a silicone polymer microchannel sandwiched between two SU-8 current collectors. The SU-8 fabrication process was previously detailed by Esquivel et al. [12]; where three main sections (fig. 1) were assembled to build the μFFC: the silicone polymer microchannel (section 1) sandwiched between two SU-8 current collectors (sections 2 and 3). The SU-8 microfabrication process was based on photolithography. Ti-Au-Ni layers (50nm each) were used as current collectors using metal sputtering deposition. To assure homogeneity of the catalyst deposition, this was performed by a numerical control spray system.

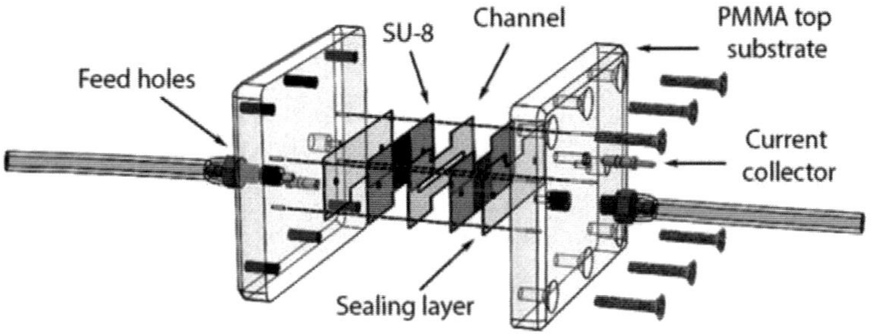

Figure 1. Assembly diagram and Photograph of the SU-8 microfluidic fuel cell with Pd nanocubes and Pt electrodes (anode and cathode) fuelled by 0.5 M formic acid and O_2 oxidant. The SU-8 microchannel is 900 μm wide, 2 cm long and 50 μm depth; the electrode active area is 0.09 cm^2.

3. Experimental

3.1. Synthesis of Pd nanocubes

The synthesis of homogeneous and well-ordered cubic-shaped Pd nanoparticles (namely Pd nanocubes for simplification purposes) was carried out by typical chemical reduction in aqueous medium [14]. In a usual synthesis two solution are made, in a first solution the surfactant (0.105 g polyvinylpyrrolidone, Sigma-Aldrich, average mol wt 44,000 g mol^{-1}) and the reductive agent (0.023 g L-ascorbic acid, Sigma-Aldrich, BioXtra, ≥99.0%) were mixed in 8 mL of water and the temperature was raised at 80°C. The second solution was prepared mixing the Pd source (0.056 g Na_2PdCl_4, Aldrich, 98%) and the additive (0.2496 g of the NaBr, Sigma-Aldrich, SigmaUltra, >99.0%) in 3 mL of water and added to the first solution when the desire temperature of the first solution was reached. The final solution was then magnetically stirred for 3 hours at 80°C. After that, Vulcan carbon was added and the solution was stirred for 30 minutes, washed with isopropyl alcohol (1:5 v/v solution and alcohol), centrifuged at 10,000 rpm and dried at 60°C.

3.2. Physicochemical characterization

Pd nanocubes were characterized by X-ray diffraction using a Bruker D8 Advanced diffractometer operated at 30 kV and 30 mA. HR-TEM images and EDX analysis were acquired through a Phillips Model CM200 transmission electron microscope.

3.3. Electrochemical characterization
The electrocatalytic activity of Pd nanocubes/C and commercial Pd/C was determined by half-cell experiments using a PGSTAT AutoLab 302 Potentiostat/Galvanostat (Metrohm®). The experiments were carried out in a typical three electrodes electrochemical cell employing 0.5 M H_2SO_4 (99.7 %, J. T. Baker) as the electrolytic solution to determine the Pd nanocubes/C and commercial Pd/C behavior in acidic media. The Hg/Hg_2SO_4 saturated in K_2SO_4 electrode was used as the reference electrode, a Pt wire as the counter electrode and the anode current collector as working electrode.

3.4. Evaluation of the SU-8 microfluidic fuel cell performance
The deposition of the catalyst was performed by spraying a catalyst ink over the current collectors. The catalytic ink was composed of 7 µL Nafion® 5% (Sigma–Aldrich), 73 µL isopropyl alcohol (J.T. Baker) and 1 mg electrocatalyst materials, mixed in ultrasonic bath for 20 min. Two micro-fuel cells were constructed in order to compare our synthesized Pd nanocubes against the Pd/C XC-72 material. The cathode electrode was maintained in both cells with a loading of 1 $mgcm^{-2}$ using Pt/C (30 wt. % E-TEK). Anode electrodes were constructed using Pd/C XC-72 (20 wt. %, E-TEK) and Pd nanocubes/C (40 wt. %) with a Pd loading of 1 mg cm^{-2} respectively.

0.5 M formic acid in 0.5 M H_2SO_4 and oxygen (4.3 U.A.P. Praxair) dissolved in 0.5 M H_2SO_4 were used as fuel and oxidant solutions, respectively. The anolyte and catholyte streams were fed to the corresponding inlets in laminar flows with a Pressure-driven fluid flow of 600 µL min^{-1} using syringe pump (NE-4000, New Era Pump Systems Inc.). All tests were performed at 25 °C. Voltage and current measurements were performed using an Epsilon potentiostat, version 1.40.67 from Bioanalytical Systems.

4. Results and discussion

4.1. Physicochemical characterization
XRD diffraction pattern for Pd nanocubes/C and TEM images are presented in Figure 2. The typical face-centered cubic Pd structure was found (Fig. 2-I). The (111), (200), (220) and (311) crystallographic planes were located at 40.04, 46.5, 68.04 and 82.19 degrees, respectively. A (200) preferential plane was found, which was expected due to the cubic-shape. The crystal size was calculated through (111) and (220) planes using the Scherrer's equation. Also, the lattice parameter using the (220) planes was calculated and showed in Table 1. HR-TEM images (Fig. 2-II) confirmed the presence of Pd nanocubes, where other geometries were not observed. Furthermore, the particle size and the interplanar distance were calculated through HR-TEM images and are also presented in Table 1. The interplanar distance 1.93 Å is in agreement with the (100) plane families.

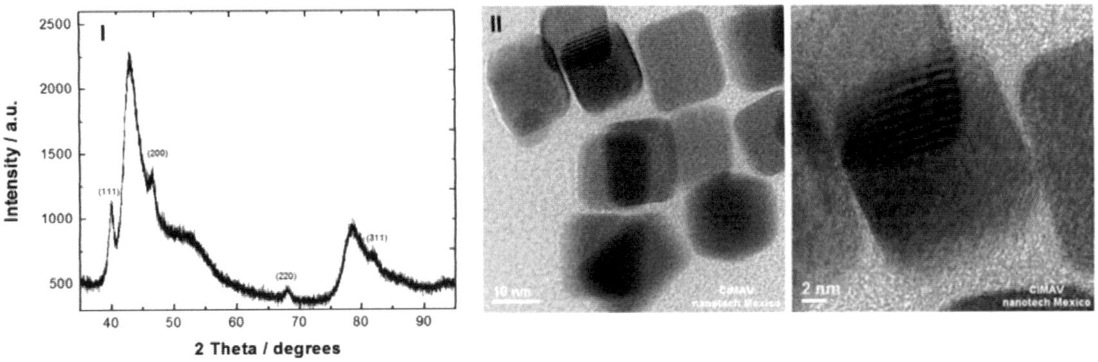

Figure 2. I) X-ray Diffraction patterns for Pd nanocubes and II) TEM images of Pd nanocubes

Table 1. Particle size, lattice parameter and interplanar distances for Pd nanocubes.

Parameter	
XRD	11.5 nm
TEM	10.6 nm
Lattice parameter (XRD)	3.8802 Å
Interplanar distance	1.93 Å

4.2. Electrochemical characterization

The electrochemical behavior in acidic media is illustrated in Fig. 3. Three well-known zones were observed, the hydrogen adsorption/desorption zone (marked in dashed rectangle), the double layer zone and the Pd oxides formation and their reduction zone. Also, the characteristic peak attributed to the (100) plane is observed at 0.28 V vs. NHE [15]. The electrochemical surface area (ECSA) was calculated through the Pd oxides reduction peak using the theoretical value of 424 μC cm^{-2} [16] resulting in an area of 10.60 cm^2. The electrocatalytic activity of Pd nanocubes was evaluated for the formic acid electrooxidation reaction at 0.5 M formic acid concentration using sulfuric acid as electrolyte (Fig. 3). The electrocatalyst showed good catalytic properties with a high current density.

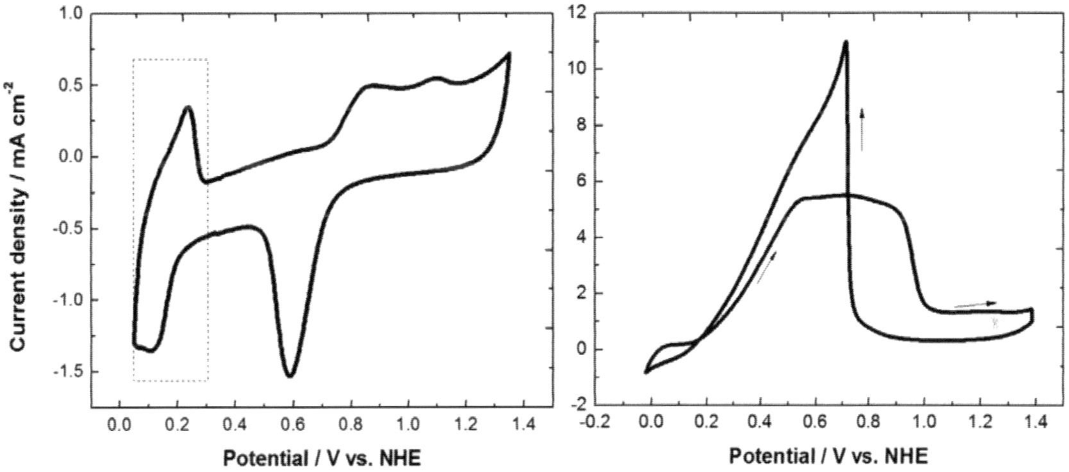

Figure 3. a) Cyclic voltammogram in acidic media (0.5 M H_2SO_4) and b) Cyclic voltammogram for the formic acid electrooxidation (0.5 M formic acid in 0.5 M H_2SO_4 as electrolyte) using Pd nanocubes as electrocatalyst.

4.3. SU-8 Microfluidic Fuel Cell Performance.

The polarization curves recorded for these SU-8 microfluidic fuel cells have the same characteristic shape of typical fuel cells, with kinetically, ohmic, and mass transport limited regions. Fig. 4 shows the polarization and power density curves for Pd nanocubes/C and Pd/C. At 0.5 M formic acid concentration is interesting to note that the open circuit potential (OCP) is similar for both microfluidic fuel cells. However, the current density values show an increase for the fuel cell using Pd nanocubes/ C than Pd/C, as a consequence of mass transport limitations (15.4 and 19.57 mW cm^{-2} respectively). Besides, the same figure shows power density values divided by Pd loading, following the same performance when is used both catalyst.

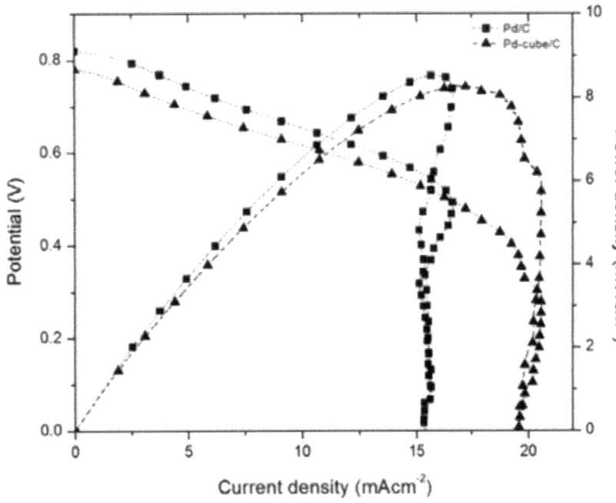

Figure 4. µFFC performance using anodes of Pd nanocubes/C at 20 wt.% (▲) and Pd/C at 20 wt.% (■); both cathodes of Pt/V XC-72 (30 wt.% from E-TEK).

5. Conclusions

A small, light and flexible direct formic acid µFFC was developed by using photosensitive SU-8 platform as current collectors and thin films of silicone where the "T" shape microchannel is delineated. Two configurations of the µFFC were compared, the first one employed commercial Pd/C as anode electrocatalyst, the second one used well-defined and homogeneous Pd nanocubes with (100) preferential crystallographic plane supported on Vulcan; both µFFC presented excellent performance (around 8.3 mW cm^{-2}); compared against similar works, this is the highest value reported so far, however the µFFC Pd-cubes based showed a slightly improvement in current density. In this way, there is an indicative of the importance of the mass activity of Pd-cubes and is expected depurate it to improve the µFFC performance.

6. Acknowledgements

The authors gratefully acknowledge the financial support from the Mexican Council of Science and Technology through FOMIX-QUERETARO (grant 193148) and ANR-CONACYT (grant 163114).

7. References

[1] Kjeang E, Djilali N, Sinton D 2009 *J Power Sources* **186** 353
[2] Morales-Acosta D, Morales-Acosta M D, Godinez L A, Alvarez-Contreras L, Duron-Torres S M, Ledesma-García J, Arriaga L G *2011 J Power Sources* **196** 9270
[3] Choban E R, Spendelow J , Gancs L, Wieckowski A, Kenis P J 2005 *Electrochim Acta* **50** 5390
[4] Gago A S, Morales-Acosta D, Arriaga L G, Alonso-Vante N 2011 *J Power Sources* **196** 1324
[5] Shaegh S A M, Nguyen N T, Chan S H 2011 *Int J Hydrogen Energy* **36** 5675
[6] Dector A, Esquivel J P, González M J, Guerra-Balcázar M, Ledesma-García J, Sabaté N, Arriaga L G 2013 *Electrochim Acta* **92** 31
[7] Ho B, Kjeang E 2011 *Cen. Eur. J. Eng* **2** 123
[8] Squires T M, Quake S R 2005 *Reviews of Modern Physics* **77** 977
[9] Dyer C K 2002 *J Power Sources* **106** (1-2) 31
[10] Kerzenmacher S, Ducrée J, Zengerle R, von Stetten F 2008 *J Power Sources* **182** 1
[11] Zhu Y, Su Y. Ha, Richard I. Masel 2004 *J Power Sources* **130** 8
[12] Esquivel J P, Senn T, Hernández-Fernández P, Santander J, Lörgen M, Rojas S, Löchel B, Cané C, Sabaté N 2010 *J Power Sources* **195** 8110
[13] Hecker K, Breitung S 2011 *Organic and PrintedElectronics, 4td Ed,* (Frankfurt: VDMA Verlag GmbH)
[14] Arjona N, Guerra-Balcázar M, Ortiz-Frade L, Osorio-Monreal G, Álvarez-Contreras L, Ledesma-García J, Arriaga LG, *J Mater Chem* **Submitted**
[15] Erikson H, Sarapuu A, Tammeveski K, Solla-Gulón J 2011 *Electrochem Commun* **13** 734
[16] Rand D A J and Woods R 1971 *J Electroanal Chem* **31** 29.

High output power AlN vibration-driven energy harvesters

Z. Cao[1,2], J. He[1,*], Q. Wang[1], M. Hara[1], H. Oguchi[1], and H. Kuwano[1]

[1] Department of Nanomechanics, Graduate School of Engineering, Tohoku University, Sendai 980-8579, Japan
[2]Institute of Communication technology, Nanjing University of Posts and Telecommunications, Nanjing 210003, China
*Presenting Author: hejj@nanosys.mech.tohoku.ac.jp

Abstract. This paper presents miniature AlN harvesters for harvesting low-frequency and two-dimensional vibration energy. A high fracture toughness and high yield strength stainless steel substrate was used to enhance output power and reduce resonate frequency of vibration energy harvesters. The thickness of 1.89 μm AlN films were deposited on 50 μm thick stainless steel (SUS) substrates for fabricating the harvesters. The Al/AlN/SUS multi-layer sheet was made into long and thin plate-like cantilevers with heavy proof masses attached at their free ends. The devices can collect vibration energy efficiently not only under perpendicular direction to the plate surface of cantilevers but also under the parallel direction. When vibration acceleration was 1.0 g, output power was 28.114 μW for perpendicular vibration and 51.735 μW for parallel vibration. When the acceleration of parallel vibration was 1.6 g, output power was 89.339 μW.

1. Introduction

To convert ambient vibration into electric energy, piezoelectric harvesters have taken great attention due to their simple configuration and high energy converting efficiency.[1-3] A piezoelectric harvesters generally employ the structure of a plate-like cantilever because resonant frequency of a cantilever is much lower than that of most of other structures such as a bridge or a diaphragm, as the cantilever structure easily enables the vibration of the device to match with vibration in environment. For common piezoelectric harvesters, researchers have been thinking that they are only sensitive to vibration perpendicular to the plate surface of the cantilever due to their plate-like cantilever structures. Consequently, it is required that the plate surface of cantilevers is aligned perpendicular to the direction of the exciting vibration to realize optimal energy harvesting. However, proper alignment is sometimes prevented by the lack of knowledge about the dominant external vibration direction.

As piezoelectric harvesters are expected to be miniaturized, a challenge occurs: how to make resonant frequency of microscale or milliscale devices as low as possible? The piezoelectric element in these miniature harvesters is generally in the form of thin films that are prepared on Si substrates. High brittleness of Si (fracture toughness: about 0.9 MPa· $m^{1/2}$) incurs that a miniature harvester with the cantilever structure cannot be designed to have low resonant frequency to match the frequency of most environmental vibration sources (below 200 Hz).[4-7] This is because low resonant frequency requires very thin thickness of the cantilever beam, or a heavy proof mass attached at the free end of the cantilever. However both approaches make the cantilever very vulnerable.

In this study, a lead-free AlN was employed to fabricate an energy harvester. Stainless steel (SUS 304) was used as the substrate for AlN thin films to fabricate small size piezoelectric harvesters with the plate-like cantilever structures. High fracture toughness (about 80 MPa· $m^{1/2}$) and high yield strength (about 205 MPa) of SUS 304 enable the cantilevers to endure heavy proof masses attached at their free end to extremely reduce the resonant frequency.[8-10] As shown in equation (1),

$$f = f_0\sqrt{0.236m/(0.236m + M)} \qquad (1)$$

where f and f_0 are resonant frequencies with and without a proof mass, m is the mass of a cantilever, M is mass of the proof mass. In addition, we investigate the power generating capability of the energy harvesters for the vibration perpendicular and parallel to their plate surface intending to realize two-dimensional vibration energy harvesters.

2. Fabrication of AlN vibration-driven energy harvesters

50 μm thick SUS304 thin sheets (length × width: 20.0 × 20.0 mm) were used as substrates. AlN thin films were deposited using an electron cyclotron resonance (ECR) sputtering system in which plasma was produced by combination of static magnetic field and microwave. The details of the deposition parameters have been described in our previous report.[11] Deposition parameters including sputtering pressure, power, gas flow rate and bias electric field applied on the substrate have been investigated to produce a high degree *c*-axis orientation AlN film. Table 1 shows the deposition condition of ECR sputtering. It is known that the ECR sputtering yields atoms and ions with high kinetic energy (up to 20 eV), which may lead to high residual compressive stress in thin films. In this study, 1.89 μm thick AlN thin films bent substrates convex upward with the bending deflection of about 4.7 mm in central of 20.0 × 20.0 mm AlN/SUS sheet. It is worth mentioning that intrinsic stress-induced bent structure of cantilever-based energy harvester is key point to collect two-dimensional vibration energy harvesters efficiently.[12] A layer of 200 nm thick Al top electrode was deposited on the surface of AlN/SUS using RF magnetron sputtering. Then Al/AlN/SUS multi-layer sheet was cut into small pieces, and each piece was fixed at one end with a clamper to form a cantilever. A copper proof mass was bonded on the free end of each cantilever. Metal wires were bonded on the top electrode and SUS substrate to form a harvester. Figure 1(a) and 1(b) are the schematic illustration and the optical photograph of an obtained AlN harvester, respectively. The cantilever beam length × width is 6.6 mm × 9.4 mm, and the Cu proof mass length × width × height is 4.0 mm × 9.4 mm × 4.8 mm.

Table 1. AlN film deposition condition of ECR sputtering

Ar flow (sccm)	50
N_2 flow (sccm)	7.5
Microwave power (W)	500
RF power (W)	500
Substrate temperature (°C)	280
Bias voltage (V)	36

3. Characterization of AlN films

Figure 2(a) shows the X-ray diffraction pattern of AlN thin films deposited on SUS substrates. The pattern showing only (0002) peak of AlN wurtzite phase is indicating that the AlN thin films have *c*-axis out-of-plane orientation. The inset shows rocking curve of the AlN (0002) peak. Small full width at half maximum (FWHM) of 5.84° implies the high degree *c*-axis orientation of the AlN thin films suitable for fabricating high output piezoelectric harvesters. Figure 2(b) demonstrates the FE-SEM cross-sectional micrograph of the AlN/SUS structure. The micrograph indicates that columnar grains with the diameter of tens of nanometer are grown densely along the direction perpendicular to the substrate.

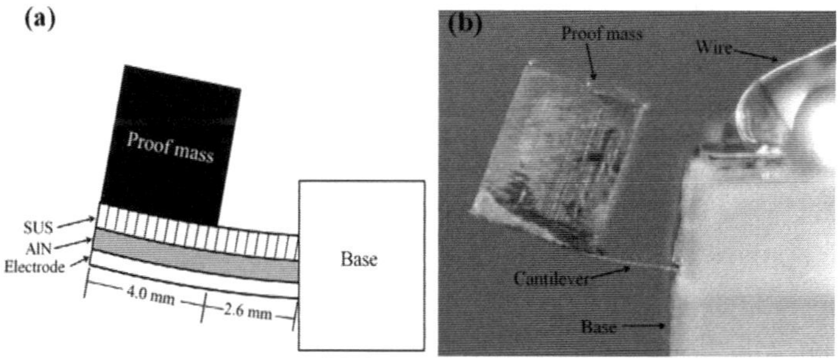

Figure 1 (a) schematic illustration, and (b) optical photo of an AlN based harvester using SUS304 as substrate with a Cu proof mass.

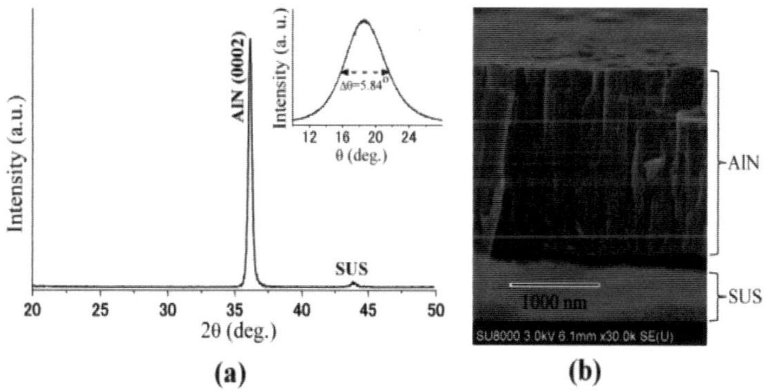

Figure 2 (a) X-ray diffraction, and (b) FE-SEM cross-sectional micrograph of AlN thin films deposited on SUS304 substrates using ECR deposition.

4. Characterization of AlN vibration-driven energy harvesters

Vibrational output characteristics of the energy harvesters were tested using in-home setup shown in Figure 3. In this study vibration was applied along two directions: perpendicular (Figure 3b) and parallel (Figure 3c) to the plate surface of the cantilevers. For impedance matching, output power was observed for various resistive loads. Then an optimal resistive load showing maximum output power was identified to be 0.588 MΩ.

Figure 4 shows output power of the AlN based harvester as a function of vibration frequency at 0.2 g acceleration. Under perpendicular vibration condition, output power reached the maximum of 2.545 μW at resonant frequency of 66.4 Hz. Under parallel vibration condition, output power increased to 7.053 μW at the same resonant frequency. It should be noted that harvesters were more sensitive to the parallel vibration, which has been considered as improper direction for cantilever type energy harvesters. Higher output power under parallel vibration condition seemingly resulted from the bent cantilever structure and heavy proof mass used in this study. Park group also reported curled cantilever structure is important reason for efficiently harvesting parallel direction vibration.[12] For a common piezoelectric harvester with the cantilever structure, the mass of the proof mass is usually only a few times that of the cantilever beam. However, in this study the mass of the proof mass is about 170 times that of the cantilever beam.

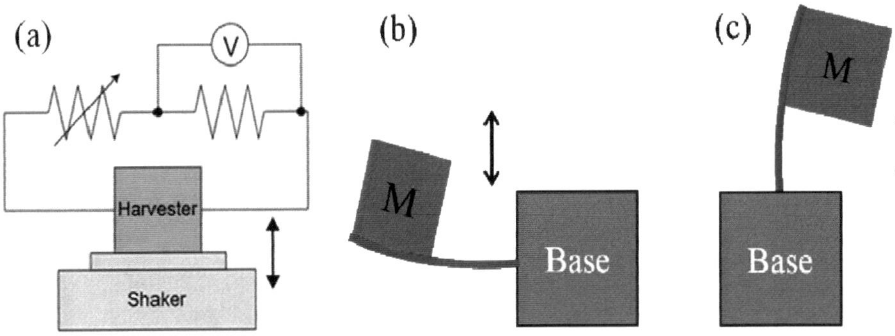

Figure 3 Schematic of (a) vibrational output characteristics measurement set up and energy harvester mount configuration of (b) vibration in perpendicular direction and (c) vibration in parallel direction.

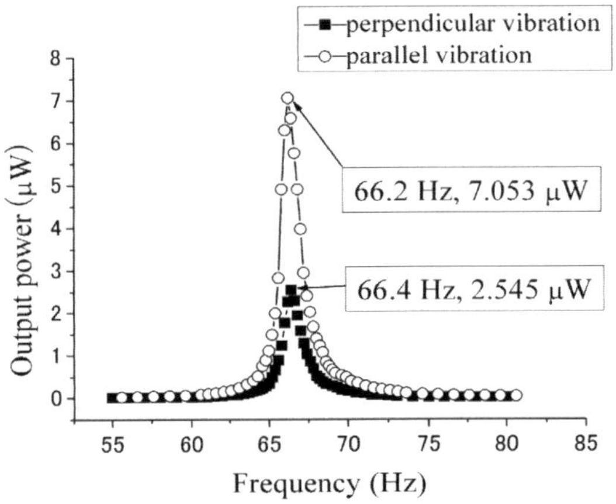

Figure 4 Output power vs. vibration frequency of an AlN harvester under perpendicular and parallel vibration at 0.2 g acceleration.

Figure 5 displays output power as a function of vibration acceleration. Under both perpendicular and parallel vibration conditions, output power increased at an exponential rate with the increase of acceleration. When the acceleration was increased to 1.0 g, the power was 28.114 μW for perpendicular vibration and 51.735 μW for the parallel vibration. When the acceleration of the parallel vibration was 1.6 g, the power was 89.339 μW, which exceeds the record of 85 μW for AlN/Si harvesters with the comparable size.[10] As shown in Figure 5, at low acceleration level (0.1~0.7 g), the slope of the double logarithmic plot for "perpendicular driven-vibration case" is equal to 1.68 (line l_1), indicating the output power increases almost quadratically with acceleration. At higher acceleration levels (0.8~1.0 g), the slope of line l_2 seems to reduce toward a value of 1, may be due to increased air damping at violent vibration and smaller piezoelectric constant under larger stress. However, the slope of line l_3 was 1.17 for "parallel driven-vibration case", indicating stronger damping occurred in this type driven-vibration whether it is in the low or high acceleration levels. The output power of 85 μW in ref. 10 is a result measured in vacuum. This literature also showed that output power in air dramatically decreased by almost one order of magnitude due to air damping.[10] Therefore, we plan to package our devices in vacuum to eliminate air damping and enhance output power in the future work.

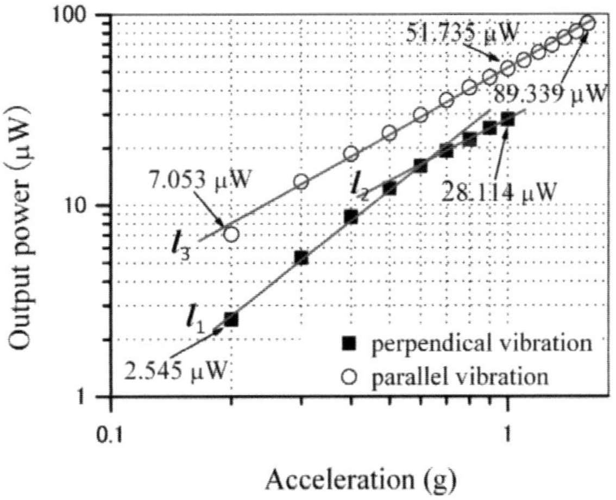

Figure 5 Output power vs. vibration acceleration of an AlN harvester under perpendicular and parallel driven-vibration.

5. Conclusions

In summary, the energy harvesting of low-frequency and two-dimensional vibration was realized in millimeter-scale AlN based harvesters with the cantilever structure. The SUS substrate ensured that a cantilever can endure a heavy proof mass at its free end to extremely reduce the resonant frequency. Moreover, the heavy proof mass and bent cantilever structure made the device sensitive to two-dimensional vibration: the perpendicular and parallel to the plate surface of the cantilever. When vibration acceleration was 1.0 g, output power was 28.114 μW for perpendicular vibration and 51.735 μW for parallel vibration. When the acceleration was 1.6 g, output power was 89.339 μW for parallel vibration. The results in this study will give valuable information to develop high output power vibrational energy harvesters.

References

[1] S. R. Anton, and H. A. Sodano, Smart Mater. Struct. **16**, R1 (2007)
[2] K. A. Cook-Chennault, N. Thambi, and A. M. Sastry, Smart Mater. Struct. **17**, 043001 (2008)
[3] N. S. Hudak, and G. G. Amatucci, J. Appl. Phys. **103**, 101301 (2008).
[4] Y. B. Jeon, R. Sood, J. h. Jeong, and S.-G. Kim, Sens. Actuators A **122**, 16 (2005).
[5] H. B. Fang, J. Q. Liu, Z. Y. Xu, L. Dong, L. Wang, D. Chen, B. C. Cai, and Y. Liu, Microelectrics Journal **37**, 1280 (2006).
[6] D. Shen, J. H. Park, J. Ajitsaria, S. Y. Choe, H. C. Wikle III, and D. J. Kim, J. Micromech. Microeng. **18**, 055017 (2008).
[7] S. L. Kok, N. M. White, and N. R. Harris, Meas. Sci. Technol. **20,** 124010 (2009).
[8] M. Marzencki, Y. Ammar, and S. Basrour, Sens. Actuators A **145-146**, 363 (2008)
[9] R. Elfrink, T. M. Kamel, M. Goedbloed, S. Matova, D. Hohlfeld, Y. V. Andel, and R. V. Schaijk, J. Micromech. Microeng. **19**, 094005 (2009).
[10] R. Elfrink, M. Renaud, T. M. Kamel, C. de Nooijer, M. Jambunathan, M. Goedbloed, D. Hohlfeld, S. Matova, V. Pop, L. Caballero, and R. V. Schaijk, J. Micromech. Microeng. **20**, 104001 (2010).
[11] J. Y. Zhang, Z. P. Cao, and H. Kuwano, Jpn. J. App. Phys. 50, 09ND18 (2011)
[12] J.C. Park, S. Khym, and J.Y. Park, Appl. Phys. Lett. 102, 043901(2013).

Piezoelectric Materials for Nonlinear Energy Harvesting Generators

S Neiss, F Goldschmidtboeing, M Kroener and P Woias
Laboratory for Design of Microsystems, IMTEK, University of Freiburg, Georges-Koehler-Allee 102, 79110 Freiburg, GERMANY
E-mail: sebastian.neiss@imtek.uni-freiburg.de

Abstract. Nonlinear piezoelectric energy harvesting generators can provide a large bandwidth combined with a good resonant power output. In an experimental study, the influence of the piezoceramic material on these two parameters is investigated. The results prove hard piezoceramics to be better suited as converting element compared to soft piezoceramics. Their improved mechanical quality compensates for their low piezo-mechanical coupling leading to both, a larger bandwidth and a higher power output of the generator.

1. Introduction

Nonlinear mass-spring systems have found growing interest in recent research for vibrational energy harvesting [1, 2, 3, 4]. In contrast to linear vibrational generators, a nonlinear stiffness is intentionally introduced in the system. By this, such systems inherently comprise a large bandwidth combined with a good resonant power output [1, 2]. This distinguishes them from linear devices which show a contrary behavior of bandwidth and conversion efficiency [3]. As a consequence, generators with a nonlinear restoring force are better suited for energy harvesting from vibrations with a varying dominant frequency compared to a device with a linear stiffness.

Current research on nonlinear generators is mainly focused on the system dynamics. However, there is no investigation of the influence of the different system components on overall performance. Only Mann [2] points out that strong mechanical damping can suppress beneficial nonlinear effects. This work investigates the influence of soft and hard piezoceramic materials on bandwidth and power output of a nonlinear energy harvesting device using the example of lead zirconium titanate (PZT). The two material categories differ strongly in the mechanical and piezoelectric properties. Soft piezoceramics are characterized by a high mobility of the ferroelectric domains while hard piezoceramics are characterized by lower mobility of the single domains. As results, soft ceramics show stronger piezoelectric coefficients but a lower mechanical quality compared to hard ceramics. In this work, an experimental approach to determine the best material choice based on frequency response and power curve measurements is proposed.

2. Theory

The dynamic displacement $z(t)$ of mechanical oscillators with nonlinear stiffness under external excitation F_{ext} is usually described by the Duffing equation:

$$m \cdot \ddot{z}(t) + c \cdot \dot{z}(t) + k_1 \cdot z(t) + k_3 \cdot z(t)^3 = F_{\text{ext}} \cdot \cos(\omega \cdot t) \tag{1}$$

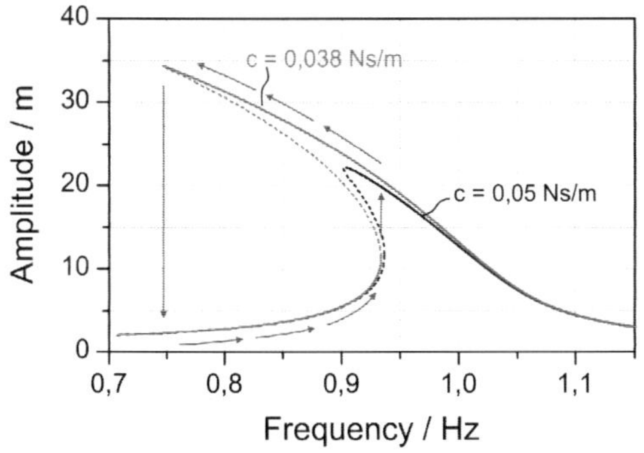

Figure 1. Frequency responses of systems with $k_1 = 1\,\mathrm{N/m}$, $m = 1\,\mathrm{kg}$, $F_{\mathrm{ext}} = 1\,\mathrm{N}$, and $k_3 = -0.0005\,\mathrm{N/m^3}$ for two different damping constants.

In this equation, m is the effective mass, c is the damping constant, and k_1 and k_3 are the linear and nonlinear spring constants, respectively. Such systems are classified depending on the type of the non-linearity as softening ($k_3 < 0$) or hardening ($k_3 > 0$).

Figure 1 shows two frequency responses of systems with softening characteristic and different damping constants. The characteristic hysteresis loop is indicated with arrows. The graph shows the strong difference in bandwidth between the two systems. A small overall damping leads to a strongly increased bandwidth and displacement amplitude.

Consequently, the overall damping of the nonlinear device has to be minimized in order to obtain a large bandwidth. For an energy harvesting device, the overall damping is the sum of the oscillator's mechanical damping and the electrical damping caused by the load. However, electrical damping depends on the piezoelectro-mechanical coupling present within the system. A strong coupling can increase the output power but will also increase the electrical damping and thus reduce the bandwidth. As a consequence, the choice of the piezoceramic material has a strong influence on the dynamic behavior of the system.

3. Nonlinear generator

The characterized harvester comprises two parts. The generator itself is shown schematically in figure 2. It is made of a laser-cut steel spring (1.4310 steel) with the ceramic glued at the base of the spring. Generators with hard PZT ceramic (PIC181, PI GmbH) and soft PZT ceramic (PIC255, PI GmbH) are fabricated. A soft-magnetic proof mass with a triangular shape made of pure iron is additionally attached to the free end of the beam.

The nonlinear stiffness is generated by the magnetic spring which is shown schematically in figure 3. The proof mass oscillates in the gap of the magnetic circuit. The total reluctance of the circuit is a function of the position d of the mass which leads to a restoring force F on

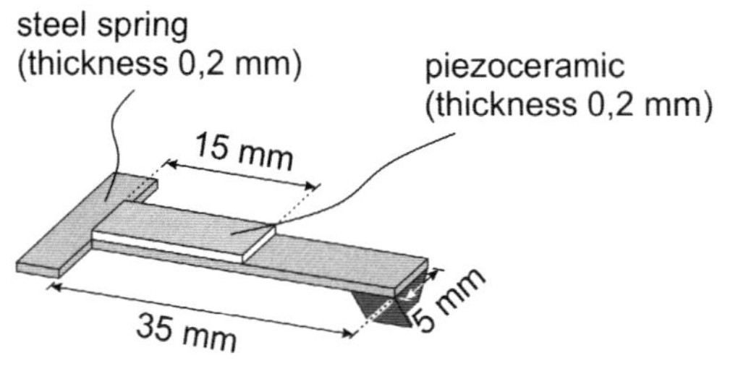

Figure 2. Schematic of the fabricated generator.

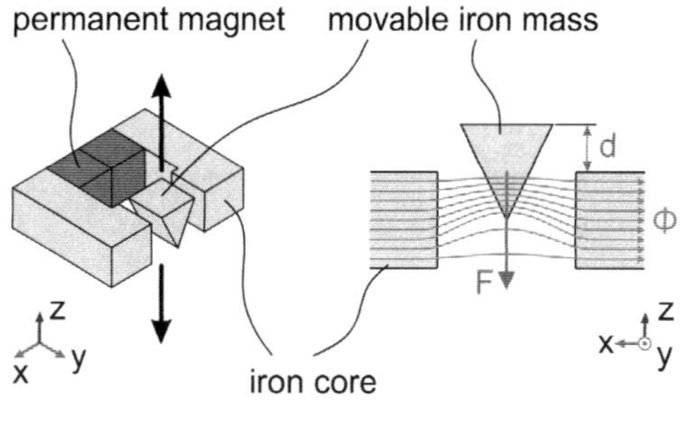

Figure 3. Schematic of the nonlinear magnetic spring.

it [5]. Depending on its geometry and idle position, a nonlinear restoring force on the proof mass is obtained [6]. All parts of the circuit are made of pure iron and have a cross section of $(5 \times 5)\,\mathrm{mm}^2$. The gap between proof mass and iron core is 0.3 mm and the magnet is a NdFeB permanent magnet (N35, $B_r = 1.17\,\mathrm{T}$).

Figure 4 shows the configuration of the two components in the measurement setup. The idle state of the mass is located about 0.5 mm beneath the top edge of the magnetic spring.

4. Dynamic characterization

Two different measurements are conducted to determine bandwidth and power output of the generators. A frequency response measurement in open-loop configuration is used to determine the influence of the ceramic's mechanical quality on the bandwidth. In addition, a power curve is recorded to measure the maximal output power. In this measurement both, piezoelectrical and mechanical properties affect the measurement results.

The test setup is shown schematically in figure 5. The generator is mounted to a shaker (TV51110, TIRA GmbH) as shown in figure 4. The sinusoidal input signal is measured with an acceleration sensor (Typ8636C5, Kistler AG). The piezoelectric voltage is measured with a PC interface (NI DAQmx9172). In addition, a resistance decade box can be used to record power curves.

Figure 4. Picture of the generator with magnetic spring.

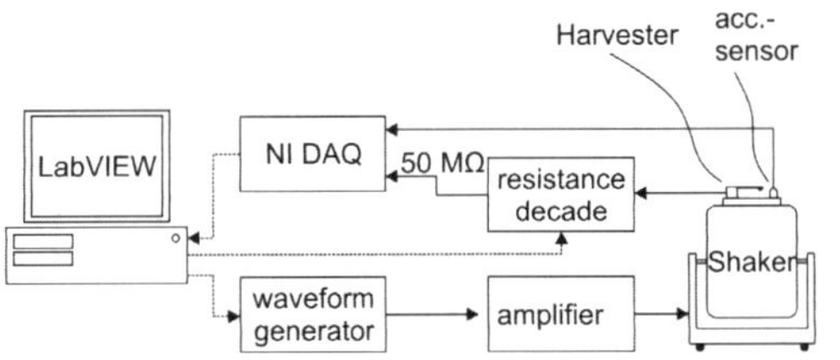

Figure 5. Schematic of the measurement setup.

5. Frequency response

Figure 6 shows the frequency response of the open-loop voltage for generators with both types of piezoceramics at a sinusoidal excitation with a peak acceleration of $5\,\mathrm{m/s^2}$. The generator with hard piezoceramic has a slightly higher resonance frequency which is caused by the higher stiffness of the ceramic. Also, the hard ceramic transducer delivers a larger bandwidth compared to the soft ceramic transducer. As a measure for the bandwidth of the system, we introduce the relative hysteresis width:

$$H = \frac{f_{\mathrm{jump,up}} - f_{\mathrm{jump,down}}}{f_{\mathrm{jump,up}}} \tag{2}$$

The generator using a hard piezoceramic transducer reaches a values of $H = 10.9\,\%$ while the generator using a soft ceramic reaches a values of only $H = 4.7\,\%$. Since there is no electrical damping involved, the increased bandwidth is a result of the better mechanical quality of the hard ceramic. In addition, the hard ceramic also produces a larger output voltage. However, since the displacement amplitude is larger as well, this is not a result of a better piezoelectro-mechanical coupling.

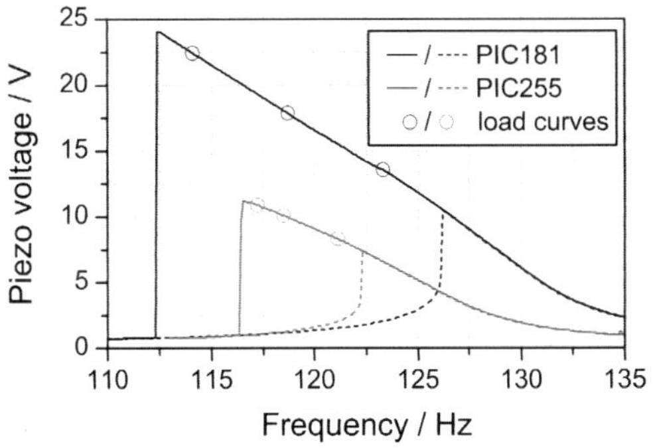

Figure 6. Measured frequency response for vibrational generator with hard piezoceramic (PIC181) and soft piezoceramic (PIC255) under sinusoidal excitation with a peak acceleration of $5\,\mathrm{m/s^2}$. Solid lines indicate a frequency down-sweep and dashed lines an frequency up-sweep, respectively.

6. Power curves

Power curve measurements at a peak acceleration of $5\,\mathrm{m/s^2}$ are shown in figure 7 for the generator with hard ceramic and in figure 8 for the generator with soft ceramic, respectively. The power curves are obtained for both, an up-sweep of the load resistance and a down-sweep of the load resistance. Prior to the measurement, the working point of the generator is set in a frequency down-sweep to be located on the upper branch of the frequency response.

For both harvester types, the power output shows an abrupt break-in at a certain load for an operation frequency close to the open-loop jump-down frequency. This is a result of the increasing electrical damping which leads to a reduction of the bandwidth. At the critical load, the system's bandwidth is reduced beneath the point of the operation frequency and the oscillation drops to the low power state. Only when increasing the operating frequency to a certain level, the optimal load resistance can be used to extract electrical energy from the system.

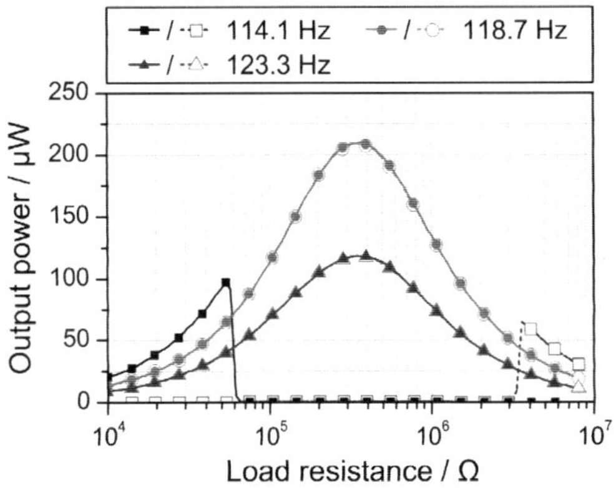

Figure 7. Power curves for a vibrational generator with hard piezoceramic (PIC181). Solid lines indicate a frequency down-sweep and dashed lines a frequency up-sweep, respectively.

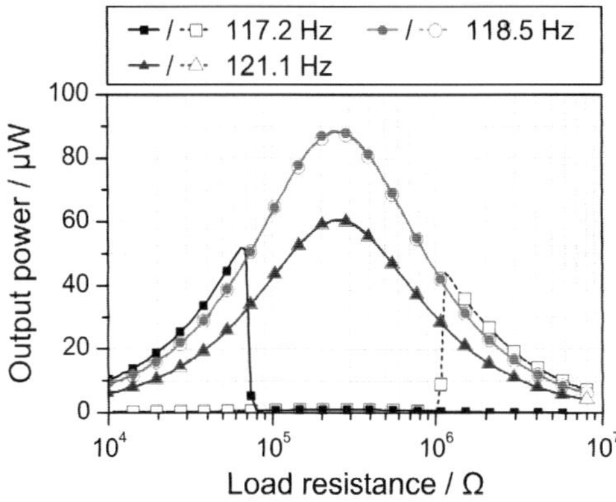

Figure 8. Power curves for a vibrational generator with soft piezoceramic (PIC255). Solid lines indicate a frequency down-sweep and dashed lines a frequency up-sweep, respectively.

The power curve measurements show that the optimal load resistance differs for an up-sweep and a down-sweep of the load at excitation frequencies close to the jump-down frequency. Thus, the generator cannot be operated in the frequency range close to the open-loop jump-down region. To obtain the maximum working range, power curves are recorded for several frequencies in the bistable region. The optimal up- and down-sweep load resistance obtained

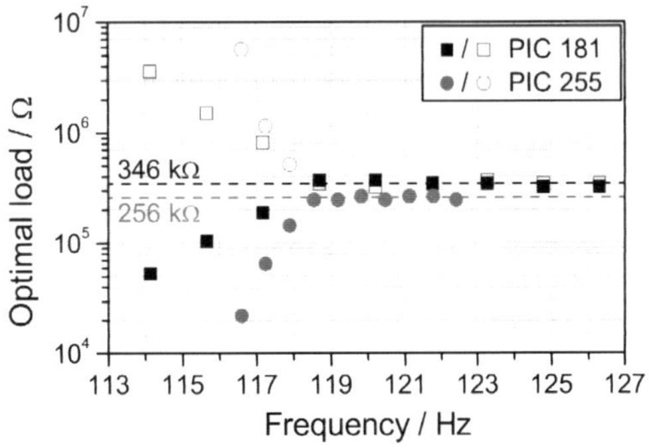

Figure 9. Optimal load resistance at different excitation frequencies for vibrational generator with hard piezoceramic (PIC181) and soft piezoceramic (PIC255) at a peak acceleration of $5\,\mathrm{m/s^2}$. Filled markers correspond to an up-sweep of the resistance and hollow markers to a down-sweep of the resistance, respectively.

in these measurements is plotted against the excitation frequency in figure 9. The maximum working range is given as the region without bifurcation in optimal load resistance. In addition, the maximum power output and optimal load resistance can be determined. The system with the hard piezoceramic transducer still reaches a relative hysteresis width of $H = 5.9\,\%$ at an optimal load resistance of $346\,\mathrm{k\Omega}$ and a maximum output power of $210\,\mu\mathrm{W}$. The generator with soft piezoceramic transducer only reaches a relative hysteresis width of $H = 3.1\,\%$ at an optimal load resistance of $256\,\mathrm{k\Omega}$ and a maximum output power of $88\,\mu\mathrm{W}$.

7. Conclusion

The results of the frequency response measurements show the superior mechanical quality of hard PZT ceramics. They lead to an approximately twofold open-loop bandwidth and voltage amplitude compared to soft PZT ceramics. Thus, the energy in the system with hard ceramic is approximately four times as much as in the soft ceramic generator. However, power curve measurements show that hard ceramics only extract about 2.4 times more electrical energy from the oscillation compared to soft ceramics which means that electro-mechanical coupling is stronger in generators with soft ceramics. Still, the lack of electro-mechanical coupling in hard ceramic generators is over-compensated by the lower mechanical damping giving to an increased power output.

Acknowledgments

We gratefully acknowledge the financial support from the German Research Association (DFG) within the Research Training Group (GRK) 1322 "Micro Energy Harvesting".

References

[1] Ramlan R 2009 *Potential benefits of a non-linear stiffness in an energy harvesting device* (University of Southampton: PhD thesis).
[2] Mann B, Barton D and Owens B 2012 Uncertainty in performance for linear and nonlinear energy harvesting strategies *J. Intell. Mater. Syst. Struct.* **23** pp 1451-60.
[3] Zhu D, Tudor M and Beeby S 2010 Strategies for increasing the operating frequency range of vibration energy harvesters: a review *Meas. Sci. Technol.* **21** pp 022001.
[4] Ferrari M, Ferrari V, Guizzetti M, And B, Baglio S and Trigona C 2010 Improved energy harvesting from wideband vibrations by nonlinear piezoelectric converters *Sensor. Actuat. A Phys.* **162** pp 422-31.
[5] Furlani E P 2001 *Permanent Manget and Electromechanical Devices* (London: Academic Press).
[6] Neiss S, Kleber J, Woias P and Kroener M 2012 Reluctance Springs for Nonlinear Energy Harvesting Generators *Proc. PowerMEMS (Atlanta, GA)* Dec 02-05 2012 pp 153-156.

Fabrication of High Temperature Thermoelectric Energy Harvesters for Wireless Sensors

J.E. Köhler[1], R. Heijl[1], L.G.H. Staaf[1], S. Zenkic[2], E. Svenman[2], A.E.C. Palmqvist[1] and P. Enoksson[1]

[1] Chalmers University of Technology, Göteborg, Sweden
[2] GKN Aerospace, Trollhättan, Sweden

E-mail: elof@student.chalmers.se

Abstract.
Implementing energy harvesters and wireless sensors in jet engines could simplify development and decrease costs. A thermoelectric energy harvester could be placed in the cooling channels where the temperature is between 500-900°C. This paper covers the synthesis of suitable materials and the design and fabrication of a thermoelectric module. The material choices and other design variables were done from an analytic model by numerical analysis. The module was optimized for 600-800°C with the materials $Ba_8Ga_{16}Ge_{30}$ and La-doped $Yb_{14}MnSb_{11}$, both having the highest measured zT value in this region. The design goal was to be able to maintain a temperature gradient of at least 200°C with high power output. The La-doped $Yb_{14}MnSb_{11}$ was synthesized and its structure confirmed by x-ray diffraction. Measurement of properties of this material was not possible due to insufficient size of the crystals. $Ba_8Ga_{16}Ge_{30}$ was synthesized and resulted in an approximated zT value of 0.83 at 700°C. Calculations based on a module with 17 couples gave a power output of 1100mW/g or 600mW/cm^2 with a temperature gradient of 200K.

1. Introduction

Measuring pressure, temperature, stress and acceleration is essential for safe operation of jet engines. For test engines in development this means thousands of sensors and kilometers of wire connecting the sensors. In the middle and the back of the jet engine it is more complicated to place sensors due to both high temperature and the need for extensive cable wiring. Cable wiring and the cost and weight of copper wire is a problem but can be reduced by replacing the wired sensors with wireless sensors coupled with energy harvesters as power sources.

In the hot parts of the jet engine the walls are cooled down with air through cooling channels. Placing thermoelectric energy harvesters inside the cooling channels would give a hot side of 800-900°C at the wall with cooling air of 500-600°C on the cold side. This means that the thermoelectric module could have a temperature gradient of 200°C if the heat transfer through the module is low enough.

This paper is a summary of an ongoing attempt to design and fabricate a small thermoelectric module optimized for the temperature range of 600-800°C with high power output rather than high efficiency, and with capability to withstand a temperature gradient of 200°C.

2. Design

To optimize the design for the temperature range and the environment an analytic model was constructed. The model included heat input, heat output, Joule heating, heat conduction [1], heat transfer in base plates, parasitic thermal conductivity [2] and contact resistance [3], the model also included load resistance matching, the resistance of cables and DC-DC conversion losses when connected to a voltage step up booster circuit. The step up booster is used to increase the voltage to an appropriate level to power the sensors, in this case 3.3V.

The choice of materials and variables such as number of couples in square devices, height of thermoelectric legs, area of legs, area ratio between n-type and p-type material, thickness of electrodes (resistance of contacts), load resistance, base plates material, electrode material and material between couples (parasitic conductance) was done by numerical analysis of the model. These materials and variables were chosen to give the highest power output possible when connected to a DC/DC converter circuit. The thermoelectric materials was chosen based on simulations, and the best results came from n-type $Ba_8Ga_{16}Ge_{30}$ [4] and p-type La-doped $Yb_{14}MnSb_{11}$ [5], both with measured figure of merit above 1.2 in the temperature range. These materials also have similar thermal expansion, with 14.2μm/K [6] and 17.5μm/K [7] respectively.

3. Synthesis

$Ba_8Ga_{16}Ge_{30}$ was synthesized by heating barium, gallium and germanium with an 2.5% excess of barium to 1050°C to ensure melting before cooled to 963°C were it was held for 38h and then cooled at -100°C/h to room temperature. This synthesis method has been shown to produce good results [8]. The result was 20g of large $Ba_8Ga_{16}Ge_{30}$ single crystals with sizes up to 1cm in length, see fig. 1. The synthesis of La-doped $Yb_{14}MnSb_{11}$ followed a recipe [5] using a tin flux with an Yb:La:Mn:Sb:Sn-ratio of 13:1:6:11:86. The crystals from this synthesis were smaller than the $Ba_8Ga_{16}Ge_{30}$ crystals and had tin impurities remaining on the crystals, see fig. 2.

Figure 1. 20g of $Ba_8Ga_{16}Ge_{30}$ single crystals. The size of the crystals was relatively consistent with some having a length of 1 cm making measurement of Seebeck coefficient and resistivity possible.

Figure 2. La-doped $Yb_{14}MnSb_{11}$ crystals covered in a thin film of Sn. No crystals were big enough to enable measurement of Seebeck coefficient, electrical resistivity or thermal conductivity.

4. Fabrication

The module was built with alumina base plates which were cut into a size of 1cm^2. Because La-doped $Yb_{14}MnSb_{11}$ reacts with most materials but is known to be stable with graphite, alumina and molybdenum [9] the choice of electrode material was molybdenum. 80μm thick electrodes were glued on to the base plates with high temperature ceramic glue. The thermoelectric legs were cut with a diamond saw and placed in a grid of alumina paste which has been seen to suppress sublimation of $Yb_{14}MnSb_{11}$ [10]. The grid also stabilized the module and kept the thermoelectric legs in place during the fabrication process, see fig 3.

Figure 3. A grid is made out of an alumina paste and placed on the base plates. The thermoelectric materials is placed in a grid which keeps the legs in place and stabilizes the module when the top base plate has been placed.

To ensure connection between the electrodes and the thermoelectric materials the electrodes were shaped as springs. Also, a thin layer of graphite was placed between the electrodes and thermoelectric legs to improve the connection further. The force needed to get a good connection through the device was quite high and a large weight or a vise was necessary for low contact resistance. Sealing the device was done with alumina paste and two different types of ceramic glues and was done to keep oxygen away from the thermoelectric materials and the electrodes.

5. Results

5.1. Simulation

The module design was, based on the analytic model, optimized for maximum power output with a temperature gradient of 200°C. Highest power output was given with 17-couples, 1mm leg height and the optimal n-type:p-type area ratio of 0.27:0.73. Simulation of power output, voltage and efficiency based on these values gave a power output of 750mW, voltage of 640mV and an efficiency of 3.3% with a 200°C temperature gradient of the environment, see fig. 4.

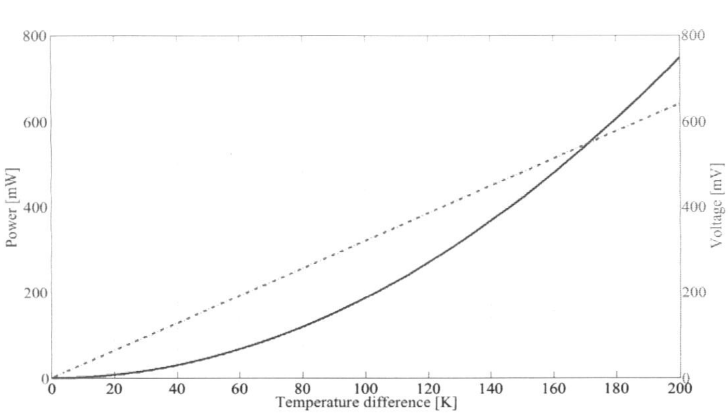

Figure 4. Estimated power output (blue line) and voltage (green, dashed line) as a function of the temperature gradient of the environment. No DC/DC converter circuit is connected in this simulation.

When including losses from cables and a low voltage DC/DC converter with a load resistance of 2.5Ω the power output is decreased to 123mW with a voltage of 1060mV. In this simulation the conversion efficiency in the step up converter was 30%.

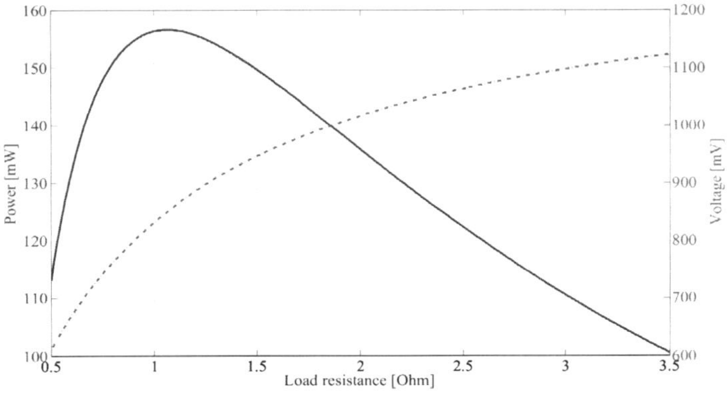

Figure 5. Estimated power output (blue line) and voltage (green, dashed line) as a function of the load resistance. The calculations are based on 2m long high temperature cables commonly used in jet engines with a wire resistance of 0.16Ω/m and a low voltage step up converter with 30% efficiency.

5.2. Synthesis

The synthesis of $Yb_{14}MnSb_{11}$ resulted in crystals that were too small to do any measurements of Seebeck coefficient, resistivity or thermal conductivity. The material was identified by x-ray diffraction measurements. The synthesis of $Ba_8Ga_{16}Ge_{30}$ gave single crystals large enough to measure Seebeck coefficient and resistivity but not thermal conductivity. The measurement was done from room temperature to 725°C and back to room temperature and showed negligible hysteresis between the heat up and the cool down measurement. The measured Seebeck coefficient and resistivity during the cool down can be seen in fig. 6 and 7.

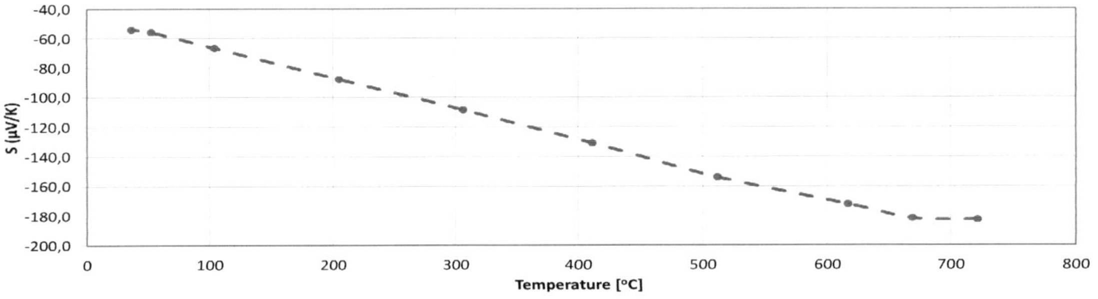

Figure 6. Seebeck coefficient measurement on $Ba_8Ga_{16}Ge_{30}$. The measurement was done from 725°C to room temperature. No hysteresis could be seen between heat up and cool down.

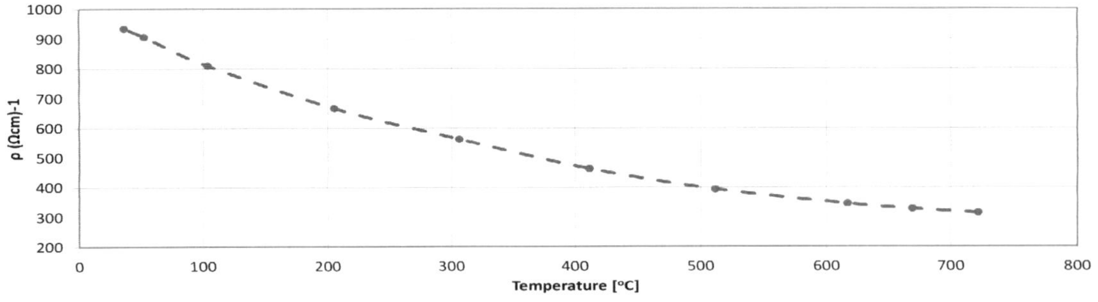

Figure 7. Resistivity measurement on $Ba_8Ga_{16}Ge_{30}$. Measured between 725°C and room temperature while cooling down.

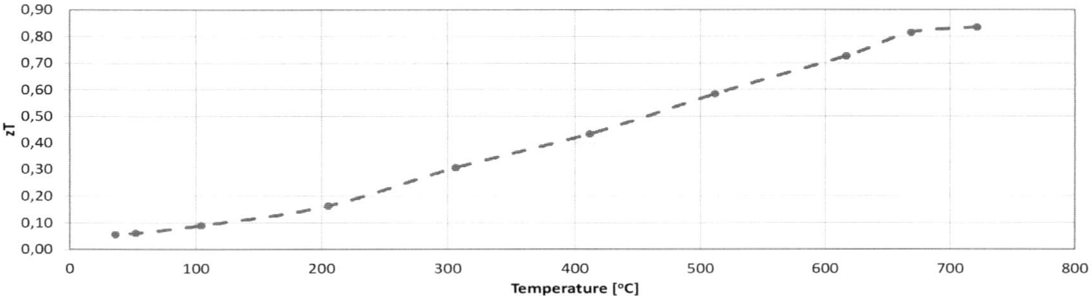

Figure 8. Approximated figure of merit of $Ba_8Ga_{16}Ge_{30}$, based on typical thermal conductivity [4] values from single crystals and the measured Seebeck coefficient and resistivity from this synthesis.

With no measurement of thermal conductivity the figure of merit of $Ba_8Ga_{16}Ge_{30}$ could not be calculated. However, the difference in thermal conductivity for single crystal $Ba_8Ga_{16}Ge_{30}$ is small between different syntheses, and an approximation based on known thermal conductivity results [4] was done. The approximation shows a figure of merit of 0.7-0.83 in the temperature range 600-725°C, see fig. 8.

6. Discussion and conclusion

To power a wireless sensor with a thermoelectric energy harvester the voltage should be stable and in the region of 2.5-3.5V. The simulations when connected to a DC/DC converter circuit shows a huge decrease in power output, mainly because of the poor conversion efficiency from the low voltage and from the poor load resistance matching. This problem can be improved by connecting several thermoelectric modules in series to increased the voltage to make it possible to use more efficient DC/DC converters. Simulations with 4 modules in series give a power output of approximately 2W with 3.3V output with 200°C temperature gradient. As most sensors require only a few mW of power, this energy harvester solution would give enough power for a large sensor network.

The approximated figure of merit of 0.83 at 700°C for $Ba_8Ga_{16}Ge_{30}$ is higher than what other common thermoelectric materials can achieve at this temperature. When coupled with La-doped $Yb_{14}MnSb_{11}$ these materials could prove to be a powerful combination in the temperature range of 600-800°C. The successful synthesis of the n-type material and the steps so far tested, shows promise for the complete assembly of a functional high temperature thermal harvester.

References

[1] Bitschi A 2009 *Modelling of thermoelectric devices for electric power generation* (Zurich: Swiss Federal Institute Of Technology Zurich) chapter 2
[2] Omer S A and Infield D G 1998 *Solar Energy Materials and Solar Cells* **53** p 67-82
[3] Rowe D M 2006 *Thermoelectrics handbook:macro to nano* (Boca Raton: Taylor & Francis Group) ch 39 p 1-15
[4] Saramat A, Svensson G, Palmqvist A E C, Stiewe C, Mueller E, Platzek D, Williams S G K, Rowe D M, Bryan J D and Stucky G D 2006 *J. Appl. Phys.* **99** 023708
[5] Brown S, Kauzlarich S, Gascoin F and Snyder J 2008 *Appl. Phys. Lett.* **93** 062110
[6] Okamoto N L, Nakano T, Tanaka K and Inui H 2008 *J. Appl. Phys.* **104** 013529
[7] Ravi V, Firdosy S, Caillat T, Brandon E, Van Der Walde K, Maricic L and Sayir A 2009 *J. Elec. Mat.* **38, No 7** p 1433-42
[8] Heijl R, Cederkrantz D, Nygren M, Palmqvist A E C 2012 *J Appl. Phys.* **112(4)** 044313
[9] Paik J, Brandon E, Caillat T, Ewell P and Feurial J Life testing of $Yb_{14}MnSb_{11}$ for high performance thermoelectric couples 2011 *Proceedings of Nuclear and Emerging Technologies for Space*
[10] Paik J and Caillat T 2010 *NASA tech briefs* **aug** p 22-23

Flexible electret energy harvesters with parylene electret on PDMS substrates

Yi Chiu[1] **and Shih-Hsien Wu**
Department of Electrical and Computer Engineering
National Chiao Tung University, Hsin Chu, Taiwan, R.O.C.

E-mail: yichiu@mail.nctu.edu.tw

Abstract. Currently, most vibrational energy harvesters have rigid and resonant structures to harvest energy from periodic motions in specific directions. However, in some situations the motion is random and aperiodic; or the targeted energy source is the strain energy in deformation, rather than the kinetic energy in vibration. Therefore we propose and demonstrate a PDMS-based flexible energy harvester with parylene-C electret that can be attached to any deformable surfaces to harvest the stain energy caused by external deformation. The proposed flexible harvester was fabricated and characterized. The measured power at 20 Hz is 0.18 μW and 82 nW in the compression and bending modes, respectively. Such a harvester has the potential for wearable and implantable electronics applications.

1. Introduction

Due to the advance in low-power VLSI technology, the power consumption of integrated circuits can be reduced to the μW or sub-μW levels. Therefore it becomes feasible to power these electronic circuits by harvesting energy in the environment. The most common energy source for harvesting is the vibration or movement of buildings, machinery, and human bodies. Currently, most vibrational energy harvesters have rigid and resonant structures to harvest energy from periodic motions in specific directions. In some situations, however, the motion is random and aperiodic; or the targeted energy source is the strain energy in deformation, rather than the kinetic energy in vibration. To harvest the energy from a deformed object such as human body, flexibility of the harvesters is essential. Flexible energy harvesters have been demonstrated by using piezoelectric thin films deposited on flexible substrates [1, 2] or micro-structured PDMS with voids for charge storage [3]. However, piezoelectric materials and their preparation are not compatible to the conventional IC processes in general. Therefore, in this paper we propose and demonstrate a flexible electret-based energy harvester that is more compatible to the IC process and thus has a lower cost of fabrication.

2. Principle and design

Electret harvesters are capacitive harvesters that are biased by the semi-permanent charge in the electret material [4-7]. They convert the mechanical energy due to displacement or deformation into electrical energy stored in a variable capacitor. The principle of a typical gap-closing electret harvester is shown in figure 1. When the gap between the upper electrode and the electret is reduced due to

[1] To whom any correspondence should be addressed.

external force or deformation, the induced charge on electrode increases and represents a current flow through the external load. The gap-closing configuration has the advantage of robust structure, simple fabrication processes and thus low cost. The flexibility is achieved by using flexible PDMS substrates for the electrodes and flexible parylene thin film deposited on the electrode as the electret material. Thus it can be compressed or bended to change the electrode gap.

The harvester has a capacitor area of 1 cm × 1 cm and gap of 1 mm. The Ansys simulation of the device upon compression (figure 2) shows that the deformation occurs only in the region close the spacer. The gap distance remains constant across most of the capacitor area. Therefore this device can be regarded as a simple variable parallel-plate capacitor upon compression. Figure 3 shows the measured and calculated load force vs. displacement. For a load of 32 N, the displacement is about 0.8 mm and thus the capacitor gap is reduced to 0.2 mm. The force-displacement relationship in figure 3 is used in a Simulink model to simulate the output of the harvester for various applied force, frequency, and loading resistance [6, 7]. Figure 4 shows the simulated output power for applied displacement of 0.8 mm and load resistance of 1000 MΩ for various excitation frequency. The output power increases quadratically with frequency, indicating that the harvester can be modelled as a constant current source in this quasi-static operation condition.

3. Device fabrication

Figure 5 shows the fabrication processes of the proposed flexible electret harvester. First, the 1-mm-thick PDMS substrates were cast by using an acrylic mold. Al/Cr electrodes were then deposited on the substrate by sputtering (Figure 5(a)). A 1-mm-thick cast PDMS spacer was then bonded to the metalized electrode (Figure 5(b)). A 5-μm-thick parylene-C as the electret layer was then coated on top of the metal layer. Electret charge was deposited on the parylene layer by corona discharge at 100 °C with V_{needle} = -4 kV and V_{grid} = -500 V (Figure 5(c)). Finally, another metalized PDMS electrode

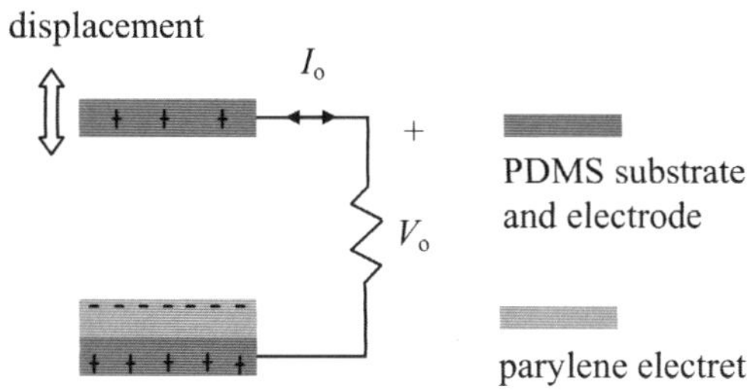

Figure 1. Schematic of the proposed flexible energy harvester.

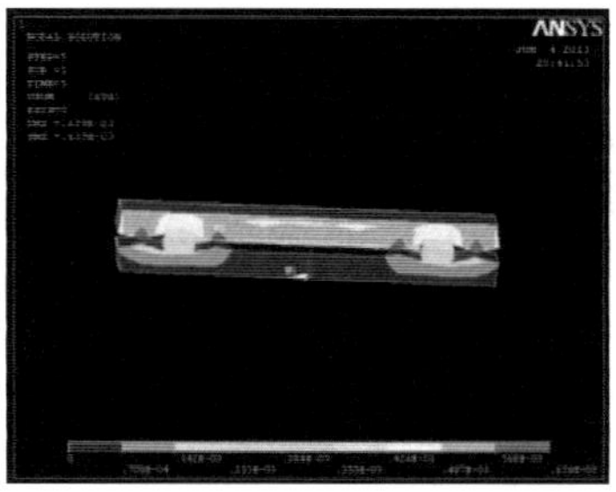

Figure 2. Ansys simulation of the flexible harvester.

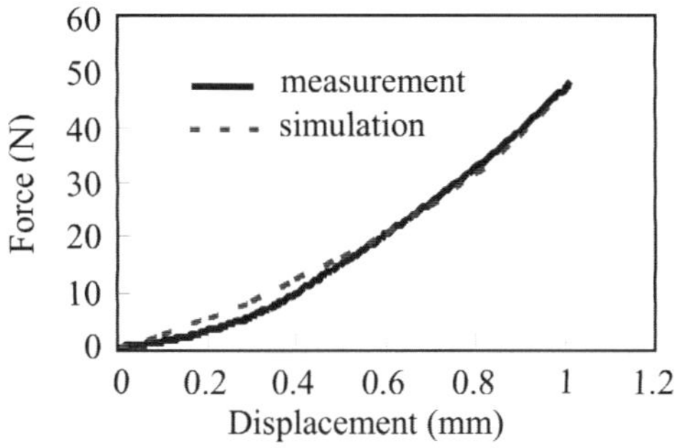

Figure 3. Measured and calculated force-displacement relationship.

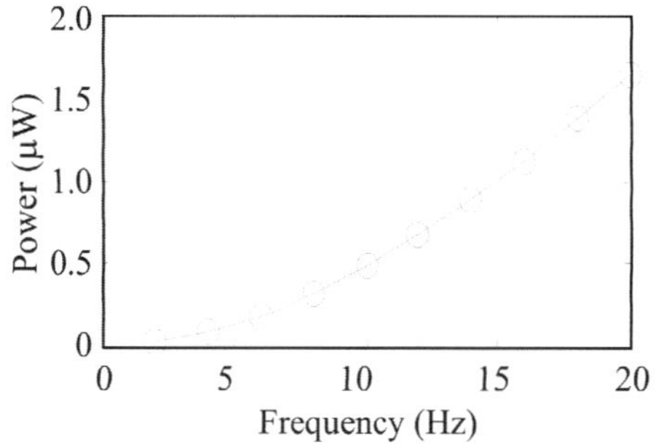

Figure 4. Simulink simulation of the harvester output.

was bonded to the bottom structure and served as the upper electrode (Figure 5(d)). Figure 6 shows the assembled and packaged harvester. Figure 7 shows the stability of the parylene-C electret. The surface potential is -227 V (64% of the initial value) after 7 months of charging, which is comparable to the data in [8]. It is noted that the surface potential was stabilized after the initial decay within one hour of charging and maintained about 10% fluctuation for the rest of the measurement period.

4. Measurement and discussion

A Bose Electroforce 3200 dynamic mechanical analyzer (DMA) (figure 8(a)) was used to apply a sinusoidal displacement on the device for mechanical and electrical characterization. A normal displacement was first applied to measure the device output in response to the compressive force (figure 8(b)). The measured open-circuit output voltage for 2-Hz displacement with 0.8-mm stroke and 1000-MΩ load is shown in figure 9. Higher excitation frequency induces more current and power. Figure 10 shows the output power vs. frequency for a 0.8-mm displacement and 1000-MΩ load resistance. The maximum power of 0.18 μW is obtained at 20 Hz.

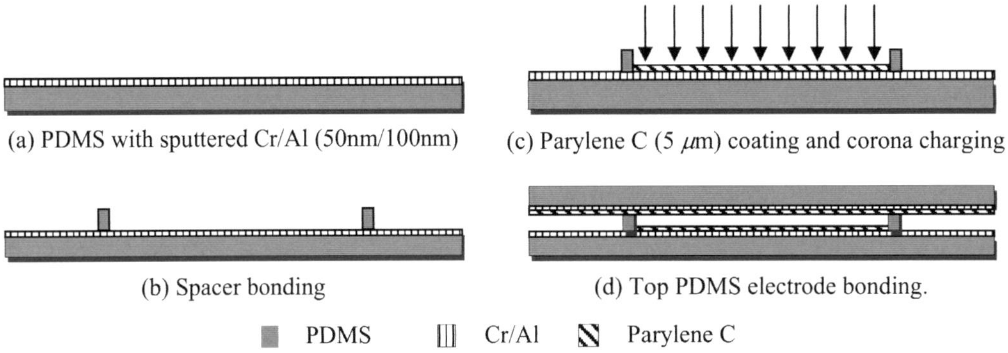

Figure 5. Fabrication processes of the proposed harvester.

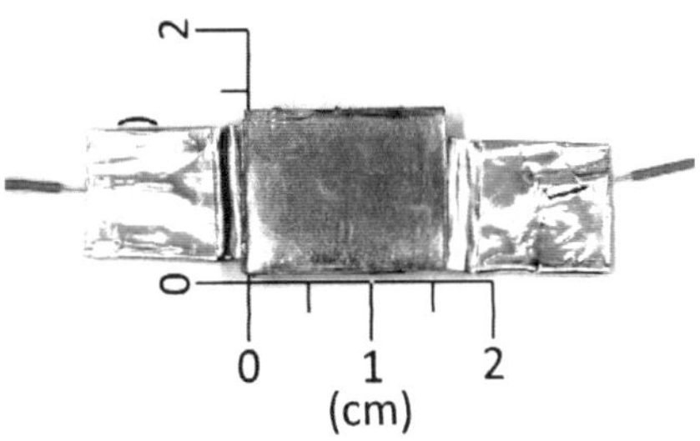

Figure 6. Photograph of assembled harvester.

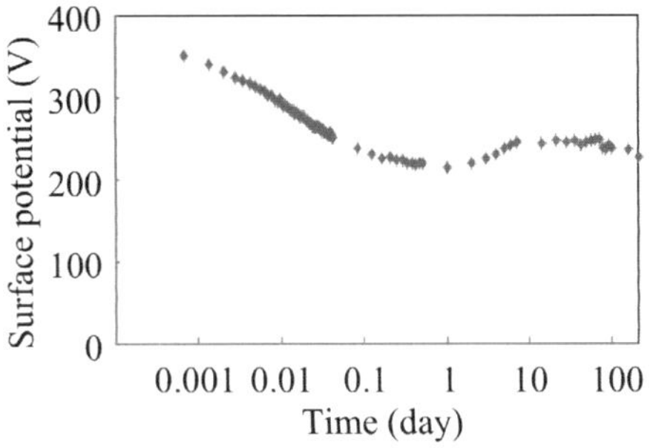

Figure 7.Charge stability of parylene-C electret.

(a) (b) (c) (d)

Figure 8. (a) Dynamic mechanical analyzer, (b) compression test, (c) and (d) bending test.

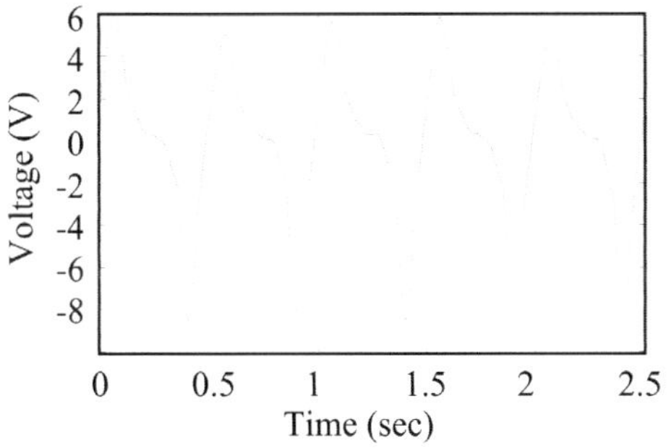

Figure 9. Output voltage for 0.8-mm compressive displacement at 2 Hz.

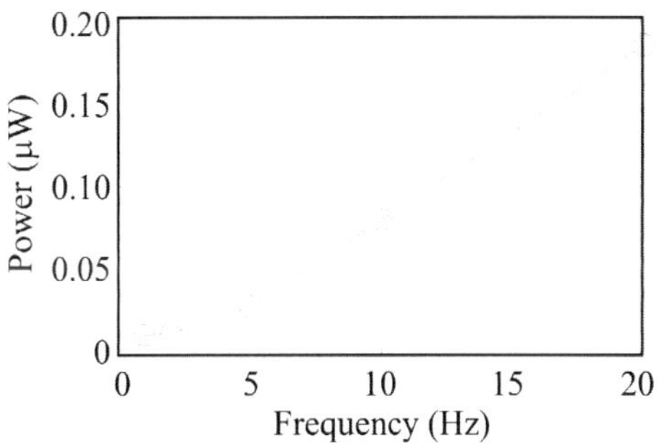

Figure 10. Output power vs. load for 0.8-mm displacement.

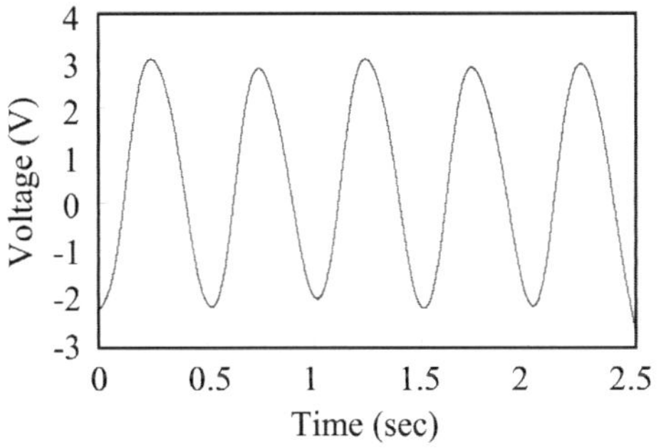

Figure 11. Output voltage in bending test (stroke = 3mm).

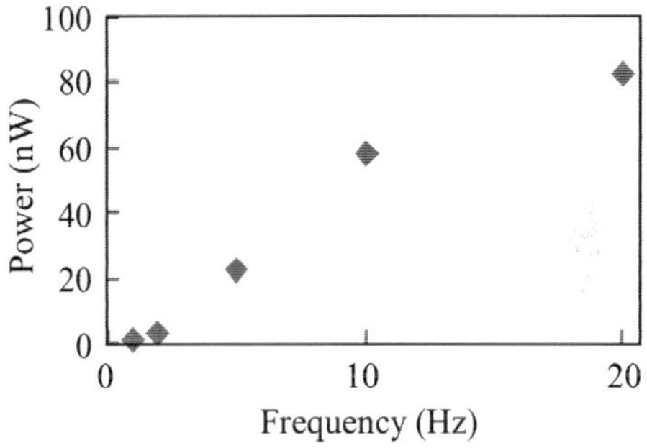

Figure 12. Output power in bending test (stroke = 3 mm).

The flexible harvester was also tested in bending situations as shown in figures 8(c)-(d) to simulate the bending of body parts such as arms or legs. The harvester was placed in a test fixture and the DMA applied a displacement at the middle of the harvester to bend it. Figure 11 shows the output voltage for a 3-mm stroke of DMA at 2 Hz. The output power for various frequency is shown in figure 12. Since the two ends of the harvester placed in the test fixture were not fixed, the required external force was relatively small and thus the power was reduced compared to that in the compressive test. The maximum power of 82 nW is obtained at 20 Hz in the bending test.

The demonstrated flexible harvester generates an output current when compressed or deformed because of the variation of the electret-induced charge on the electrodes. Therefore, it can also be regarded as a piezoelectric structure. The effective piezoelectric d_{33} coefficient can be estimated from the measured output power, assuming uniform force and charge density, to be

$$d_{33} = \frac{\Delta Q / A}{F / A} = \frac{\sqrt{2P / R\omega^2}}{F} = 12.8 \text{ pC/N},$$

where ΔQ is the charge variation, F is the applied force, A is the device area, P is the rms output power, R is the load resistance, and $\omega = 2\pi f$ is the angular frequency. The calculated d_{33} coefficient is low compared to that reported in [3, 4] where the implanted charge is stored in the micro voids in polymer substrates. Further optimization of our current design can improved the piezoelectric property.

5. Conclusion

In conclusion, a flexible PDMS electret energy harvester with stable parylene-C electret has been demonstrated. The measured power at 20 Hz is 0.18 μW and 82 nW in the compression and bending modes, respectively. Further optimization of the device design and fabrication processes can improve the output power and piezoelectric coefficient.

Acknowledgements
The project was supported in part by the National Science Council, Taiwan, R.O.C., by grant No. NSC 101-2221-E-009-171 and NSC 102-2221-E-009-191. The authors were grateful to the use of facilities at the Chip Implementation Center, the National Center for High-performance Computing, the National Nano Device Laboratory, and the National Chiao Tung University Nano Facility Center, Taiwan, R.O.C.. The authors appreciate the assistance of Mr. Sheng-Tsai Wu of Industrial Technology Research Institute, Taiwan, R.O.C., in the DMA measurement.

References
[1] Chen X, Xu S, Yao N, and Shi Y 2010 1.6 V nanogenerator for mechanical energy harvesting using PZT nanofibers *Nano Lett.* **10** 2133-2137
[2] Lee M, Chen C-Y, Wang S, Cha W S, Park Y J, Kim J M, Chou L J, and Wang Z L 2012 A hybrid piezoelectric structure for wearable nanogenerators *Adv. Mater.* **24** 1759-1764
[3] Tsai J-W, Wang J-J, Su Y-C 2013 Piezoelectric rubber films for autonomous physiological monitoring systems *Sens. Actuators A: Phys.* http://dx.doi.org/10.1016/j.sna.2013.08.044
[4] Feng Y and Suzuki Y 2013 All-polymer soft-X-ray-charged piezoelectret with embedded PEDOT electrode *Proc IEEE MEMS 2013* 865-868
[5] Yang Z, Wang J, and Zhang J 2011 A high-performance micro electret power generator based on microball bearings *J. Micromech. Microeng.* **21** 065001 (9pp)
[6] Chiu Y and Lee Y-C 2013 Flat and robust out-of-plane vibrational electret energy harvester *J. Micromech. Microeng.* **23** 015012 (8pp)
[7] Chiu Y, Bargayo R P, Hong H-C 2012 Stacked electret energy harvesting system fabricated with folded flexible printed circuit board *Proc Power MEMS 2012* 149-152
[8] Genter S and Paul O 2012 Parylene-C as an electret material for micro energy harvesting *Proc. Power MEMS 2012* 317-320

Dynamic Response of a Pendulum-Driven Energy Harvester in the Presence of Noise

M Borowiec[1], G Litak[1], A Rysak[1], P D Mitcheson[2], and T T Toh[2]

[1] Lublin University of Technology, Nadbystrzycka 36, 20-618 Lublin, Poland

[2] Department of Electrical and Electronic Engineering, Imperial College London, London SW7 2AZ, UK

E-mail: m.borowiec@pollub.pl

Abstract. This paper presents time- and frequency-domain analyses of a previously reported wave energy harvesting system in the presence of noise. The energy harvester comprises a pendulum that drives two DC generators connected electrically in series. A mechanical model of the wave energy harvester is developed and simulated in Matlab. The output voltage and output power of the energy harvester is evaluated and compared for two input signal types: a deterministic case and a stochastic case. The latter consists of a white noise random excitation source that is bounded in amplitude at a fixed standard deviation. Simulation results indicate that a stochastic input signal shows considerable influence on the output characteristics of the energy harvester. The simulation models developed in this paper can be used to complement the design of resonant frequency tuning electronics for energy harvesting systems.

1. Introduction

The presence of noise within the source excitation frequency will degrade the performance of an energy harvesting system that is designed to operate at its mechanical resonant frequency. This is evident in practical applications, whereby the source vibration frequency may have additional time varying and sometimes stochastic components [1, 2]. Tuning the resonant frequency of such energy harvesters have been demonstrated mechanically [3] or electronically [4, 5]. However, those methods were implemented under sinusoidal input excitations.

In this paper, which is an extension of the work reported in [5, 6], we will discuss the behaviour and efficiency of the electronic tuning circuit in the presence of stochastic excitations. The tuning circuit is represented by a parallel combination of a resistor, capacitor and inductor which are connected across two DC generators connected electrically in series to increase the generated voltage. A photograph of the setup is shown in figure 1(a) and this prototype was designed to be deployed in an unmanned surface vehicle (USV).

2. Mathematical model

A mathematical model of the pendulum-driven energy harvester is given by equation (1).

$$\ddot{\theta} = -d\dot{\theta} + b\Omega^2 \sin\left(\Omega t + \Gamma\left(t\right)\right) - aI_L - c\sin\left(\theta + b\sin\left(\Omega t + \Gamma\right)\right) \qquad (1)$$

where a, b, c and d are constants, θ is the relative angular displacement between the pendulum and frame, $\dot{\theta}$ is the angular velocity, $\ddot{\theta}$ is the angular acceleration and Ω is the input excitation

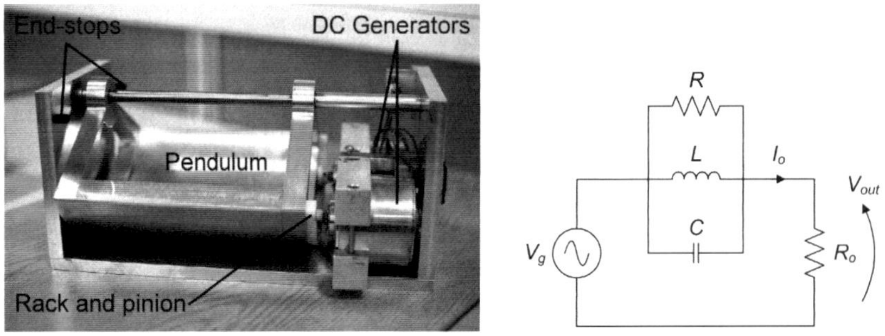

(a) (b)

Figure 1. Photograph and electrical equivalent circuit of the pendulum-driven energy harvesting system.

frequency, $2\pi f$. Finally, $\Gamma(t)$ is a time-dependent, Gaussian distributed white noise component which is introduced into the mechanical system [7, 8] through the input excitation signal Ψ which is defined in equation (2).

$$\Psi(t) = b\sin(\Omega t + \Gamma(t)) \tag{2}$$

Figure 1(b) shows the electrical equivalent circuit of the DC generators, a synthesised RLC impedance to represent an electronic tuning circuit and an output resistor R_o to represent a load application. The generated voltage V_g from the DC generators is given by equation (3).

$$V_g = K_e\dot{\theta} \tag{3}$$

The output current I_o is expressed in equation (4), where K_e is the motor constant of the DC generators and R, L and C represents the synthesised load resistance, inductance and capacitance respectively. The armature inductance of the DC generators can be neglected because the pendulum's oscillation frequency is typically less than 2 Hz for the intended application in an USV. For simplicity, the armature resistance was grouped with R_o.

$$I_o = \frac{V_g - V_{out}}{R} + C\frac{\mathrm{d}}{\mathrm{d}t}(V_g - V_{out}) + \frac{1}{L}\int_0^t (V_g - V_{out})\,\mathrm{d}t \;;\; V_{out} = I_o R_o \tag{4}$$

3. Numerical simulations

Matlab simulations of equations (1), (3) and (4) were performed using the Euler-Maruyama algorithm [9]. Two types of input excitation signals, $\Psi(t)$, were used: a harmonic signal and a stochastic signal with standard deviation $\sigma_\Gamma = 0.4\mathrm{rad}/\sqrt{s}$. This type of stochastic signal was previously investigated in [7, 8] and the time- and frequency-domain plots of the signal is shown in figure 2(c) and figure 2(d). The parameters used in the simulations are listed in table 1.

The output voltage V_{out} across a 1 kΩ output resistor, from a deterministic input signal and from a stochastic input signal is presented in figure 3. The time-domain plots of the input signals are on the left whereas the graphs on the right are their Fourier-transformed equivalents.

The phase portraits of the pendulum under harmonic and stochastic input excitations are shown in figure 4. As expected, the harmonic (deterministic) case shows a symmetrical phase portrait centred at the origin. Whereas for the stochastic excitation case, the phase portrait shows an open and wandering trajectory.

The instantaneous output power for the two input signals are shown in figure 5, with the stochastic case demonstrating large and non-harmonic fluctuations. Interestingly, the average

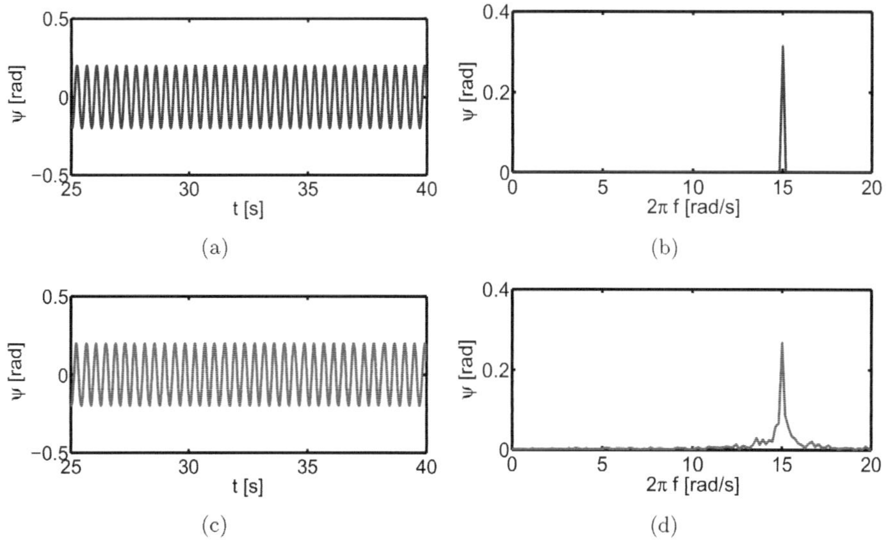

Figure 2. A harmonic signal (top) and a stochastic signal (bottom) were used as the input. Note that time-domain signals are not identical and this can be confirmed by comparing their respective frequency spectra.

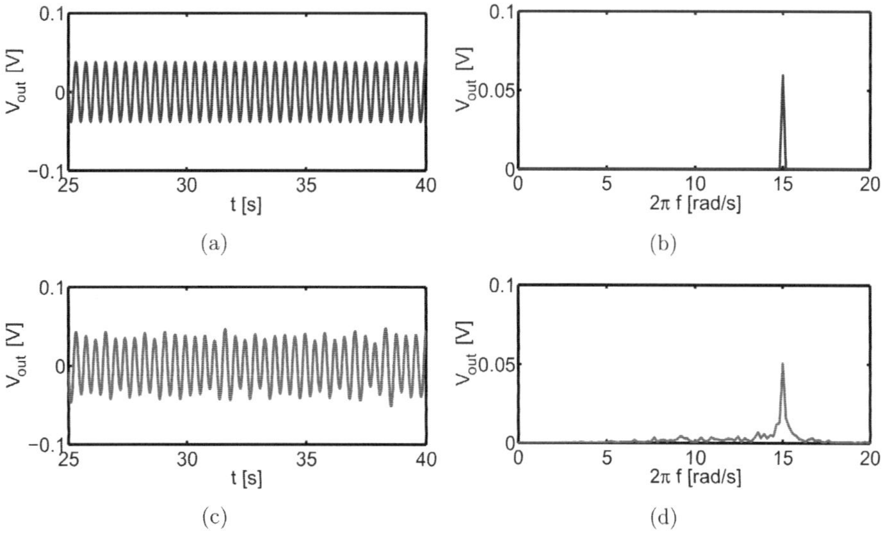

Figure 3. Time- and frequency-domain responses of the output voltage V_{out} for a harmonic (top) and stochastic (bottom) input signals. In the latter, σ_Γ is 0.4 rad/$\sqrt{s}$.

output power for the deterministic and stochastic cases are similar: 0.66 μW and 0.69 μW respectively.

Figure 6 shows the variation of average output power against output resistance at three values of σ_Γ. The results for the periodic and stochastic cases are very similar but upon further inspection, the stochastic case appears to be more efficient. This could be due to $\Gamma(t)$ having a random phase which contributes additional excitation frequencies. Consequently, this increases the available energy from the pendulum that can be harvested by the DC generators.

Table 1. List of simulation parameters and values.

Parameter	Value	Parameter	Value
a	0.6 rad·s^{-2}/A	R	50 Ω
b	0.2 rad^{-1}	L	10 mH
c	79.8 rad·s^{-2}	C	1 mF
d	2.3 s^{-1}	R_o	1 kΩ
Ω	15 rad·s^{-1}	K_e	13 mV/rad·s^{-1}

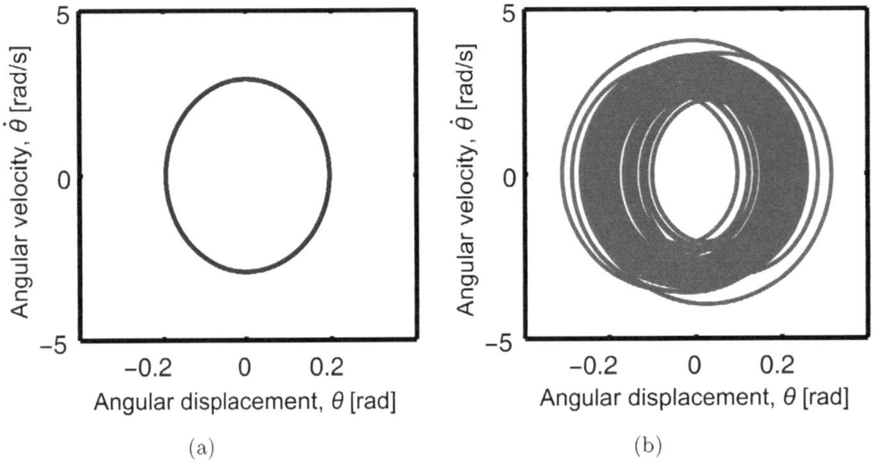

Figure 4. Phase portraits of the pendulum for: (a) deterministic and (b) stochastic input signals. The standard deviation σ_Γ is 0.4 rad/$\sqrt{s}$.

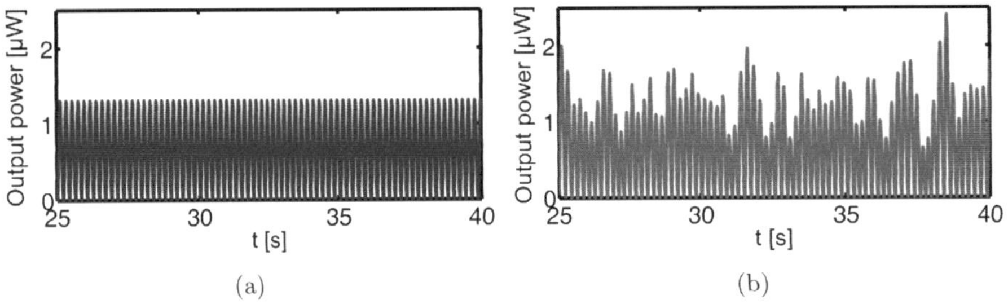

Figure 5. Instantaneous output power from a deterministic (left) and a stochastic (right) input signal.

4. Conclusions

A comparison of the output characteristics of a pendulum-driven energy harvester was performed using a harmonic and a stochastic input signal. The results show considerable influence of the noise component on the output voltage. The simulated average output power in both cases is similar and this is because of the small σ_Γ. This work demonstrated that, under stochastic excitations, the average output power is slightly higher compared to the deterministic case and

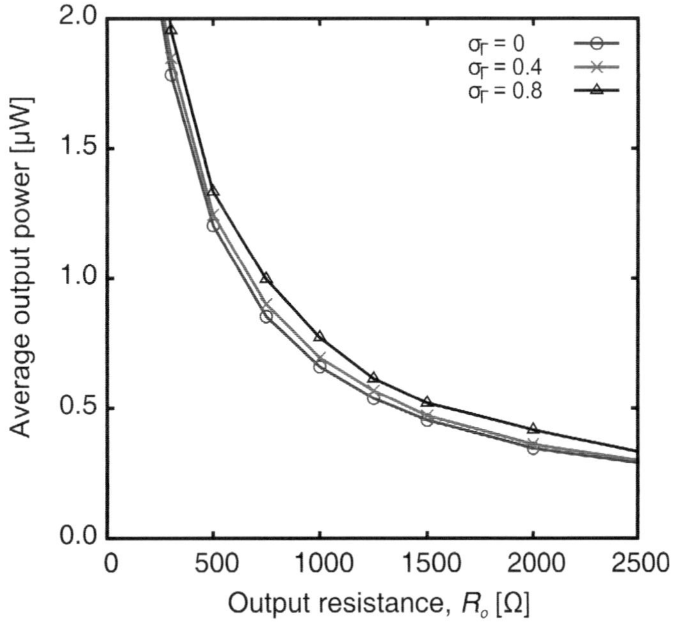

Figure 6. Average output power of the harmonic ($\sigma_\Gamma = 0.0$, circles) and stochastic ($\sigma_\Gamma = 0.4$, crosses & $\sigma_\Gamma = 0.8$, triangles) excitation cases versus the output resistance R_o.

even more so at larger σ_Γ values for the parameters used in the simulations. Further work is needed to establish the parameters of a Gaussian distributed input noise that will affect the output power generation of an energy harvester. Such an analysis will complement the design of a controller that can accurately determine the appropriate synthesised load impedance for stochastic excitation conditions.

Acknowledgements
This work is supported by The Royal Society International Exchanges Scheme, reference number: IE130292.

References

[1] Beeby S P, Wang L, Zhu D, Weddell A S, Merrett G V, Stark B, Szarka G and Al-Hashimi B M 2013 *Smart Mater. Struct.* **22** 075022
[2] Neri I, Travasso F, Mincigrucci R, Vocca H, Orfei F and Gammaitoni L 2012 *J. Intell. Mater. Syst. and Struct.* **23** 2095–2101
[3] Ayala-Garcia I N, Zhu D, Tudor M J and Beeby S P 2010 *Smart Mater. Struct.* **19** 115005
[4] Cammarano A, Burrow S G, Barton D A W, Carrella A and Clare L R 2010 *Smart Mater. Struct.* **19** 055003
[5] Mitcheson P D, Toh T T, Wong K H, Burrow S G and Holmes A S 2011 *IEEE Trans. Circuits and Systems II* **58** 792–796
[6] Toh T T, Mitcheson P D, Dussud L, Wright S W and Holmes A S *Proc. PowerMEMS 2011* pp 383–386
[7] Litak G, Borowiec M and Wiercigroch M 2008 *Dynamical Systems* **23** 259–265
[8] Kwuimy C A K, Litak G, Borowiec M and Nataraj C 2012 *Appl. Phys. Lett.* **100** 024103
[9] Naess A and Moe V 2000 *Probabilistic Engineering Mechanics* **15** 221–231

Multilayer out-of-plane overlap electrostatic energy harvesting structure actuated by blood pressure for powering intra-cardiac implants

M Deterre[1,2], S Risquez[1], B Bouthaud[2], R Dal Molin[2], M Woytasik[1] and E Lefeuvre[1,3]

[1] Univ Paris-Sud - CNRS, Institut d'Electronique Fondamentale, rue Ampère, bat. 220, 91405 Orsay Cedex, France
[2] Sorin CRM, za noveos, 4 avenue Reaumur, 92140 Clamart, France

E-mail: elie.lefeuvre@u-psud.fr

Abstract. We present an innovative multilayer out-of-plane electrostatic energy harvesting device conceived in view of scavenging energy from regular blood pressure in the heart. This concept involves the use of a deformable packaging for the implant in order to transmit the blood pressure to the electrostatic transducer. As shown in previous work, this is possible by using thin metal micro-bellows structure, providing long term hermeticity and high flexibility. The design of the electrostatic device has overcome several challenges such as the very low frequency of the mechanical excitation (1 to 2 Hz) and the small available room in the medical implant. Analytical and numerical models have been used to maximize the capacitance variation, and hence to optimize the energy conversion. We have theoretically shown that a 25-layer transducer with 6-mm diameter and 1-mm thickness could harvest at least 20 mJ per heart beat in the left ventricle under a maximum voltage of 75 V. These results show that the proposed concept is promising and could power the next generation of leadless pacemakers.

1. Introduction

Devices harvesting energy from the body [1] are drawing increasing attention as industrials such as Sorin Group are investigating their integration into in vogue miniaturized leadless pacemakers that can be placed directly in the heart. Therefore, this paper focuses on an energy harvester which size is compatible with an intravenous introduction, i.e. which diameter is below 6 or 7 mm in order to fit in a catheter (figure 1). As the major part of the energy consumed by the heart is used to expel blood to the organs, we are proposing to exploit the regular blood pressure variations during the cardiac cycle to generate energy. Compared to the inertial approach, which has previously been considered to harvest mechanical energy in the heart [2], the regular blood pressure variation presents the advantage of being a very stable and predictable power source. Moreover, contrary to inertial devices which require very good matching of their resonance frequency with the frequency of the mechanical excitation, the non-resonant principle of operation of our device guarantees that the amount of energy harvested by heartbeat cycle is unaffected by heartbeat frequency changes. The targeted overall volume for the implant is 1 cm^3, and its average power consumption is 10 μW. Given the size constraints, the power

[3] To whom any correspondence should be addressed.

needs and the very low frequency of the mechanical energy source, the design of the mechanical-to-electrical energy converter appears to be a real challenge. For the cylindrical shape volume of the considered implant, the design of a device with three dimensional features is more advantageous. As implanted devices should be compatible with MRI, miniature electromagnetic generators are discarded for this application. Our previous works showed that piezoelectric materials are promising candidates [4]. However, long term reliability of this technology must be assessed in depth before industrial development of piezoelectric energy harvesters for medical implants can be considered.

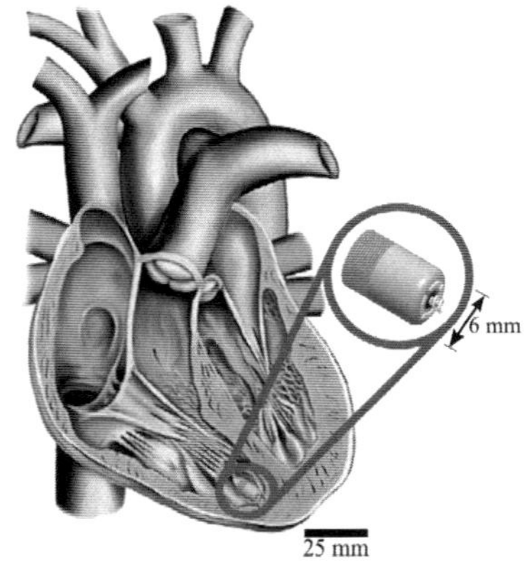

Figure 1. Schematic of a device placed in a heart cavity with deformable packaging.

Electrostatic energy conversion is also a promising way for implanted application. This technology has a high integration potential and can be easily miniaturized. Contrary to piezoelectric energy conversion which involves high mechanical stress of the active materials, electrostatic generators may be designed in order to get low mechanical stress and thus lower risks of mechanical failure. In this paper we present the study of a completely 3D electrostatic device. It takes the advantage of the advanced 2D patterning techniques while stepping in the third dimension by bringing into play an innovative and virtually unlimited stacking approach. Hence, a much higher output power can be expected compared to 2D devices.

2. Electrostatic structure design

2.1. Converted energy

An electrostatic energy harvester is a variable capacitor which capacitance is changed by relative mechanical displacement of its electrodes. To maximize the energy transduction, the capacity variation needs to be as large as possible. The two most common modes of operation are the *voltage constrained* and the *charge constrained* cases [5]. In the *voltage constrained* case, a voltage V_{max} is applied across the capacitor when its maximum capacity C_{max} is reached. Then, the capacitor is discharged when its minimum value C_{min} is reached. The energy W_{Volt} extracted from the electrostatic system is given by equation (1). In the *charge constrained* case, a voltage V_0 is applied across the capacitor when its maximum capacity C_{max} is reached. The capacitor is then left in open-circuit and the movement induces the transition to C_{min} while the voltage increases to the maximum value V_{max}. Then, the capacitor is discharged. The energy W_{Charg} extracted in this case is given by equation (2).

$$W_{Volt} = \frac{1}{2}(C_{max} - C_{min})V_{max}^2 \tag{1}$$

$$W_{Charg} = \frac{1}{2}\frac{C_{max}}{C_{min}}(C_{max} - C_{min})V_{max}^2 \tag{2}$$

Considering same values of maximum voltage, we can easily observe from (1) and (2) that the charge constrained case generates less energy in theory. However, the efficiency of the power management circuits related to both modes of operation should be taken into account to make a choice. As the mode of operation has no major influence on the design of the electromechanical transducer, we have arbitrarily chosen to consider the voltage constrained case in the following developments.

2.2. Flexible packaging

The packaging of the pressure harvesting device has to be flexible to engender a displacement upon variation of the pressure load, while transmitting the largest possible force to the transducer. The more flexible the packaging will be, the more mechanical energy will be transmitted to the electrostatic device. However, long term medical implant packaging is subjected to very stringent requirement in term of permeability. This is why the packaging of long-term implants is traditionally made of metals (for instance titanium). Flexible materials such as polymers are not hermetic enough for this application. Therefore, we have developed a thin metal packaging with micro-bellows structure in order to get very low stiffness. Experimental measurements carried out on 8-μm thick nickel bellows packaging with 7 corrugations showed that stiffness as low as 135 N/m are attainable using metal packaging [4].

2.3. Out-of-plane overlap electrostatic structure

In the goal of optimizing the available volume in the implant, we propose to have a transducer structure that fits the just the capsule and which the geometry is a disc with a 6-mm diameter and height as small as possible (up to one millimetre). Due to the design of the flexible packaging, the mechanical movement is generated along the symmetry axis of the capsule. Therefore, we have opted for an out-of-plane electrostatic structure. Also, aiming for the highest possible capacitance variation invites us to have a multilayer or a comb-like structure, which significantly increases the facing surfaces of the capacitor. In figure 2 is illustrated this structure for a single layer implementation (i) and for a multi-layer implementation (ii). The illustration of this principle adapted to the cylindrical geometry that we target is shown in figure 3 (i). The fixed part (in blue) is connected to the rigid base of the implant, and the moving part (in red) is connected to the flexible part of the implant. The gap between the fingers stays constant but their overlapping surface is dependent on the movement.

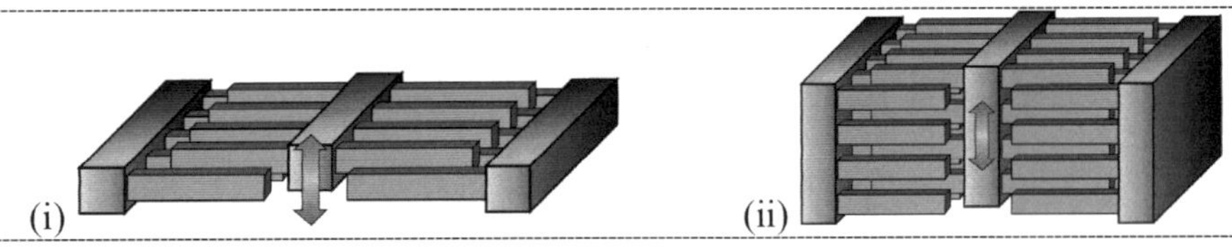

Figure 2. Principle of the proposed out-of-plane overlap electrostatic structure : single layer structure (i) and multi-layer structure (ii).

2.4. Capacitance variation

For a structure with a number n_l of layers as schematically depicted in figure 3 (ii), the total capacitance was first analytically calculated using the first order approximation of "infinite" parallel plate capacitors. As this assumption breaks down relatively fast due to electrostatic side effects, numerical simulation have been run on the electrostatic module of COMSOL software in the goal of determining the actual profile of the capacitance. As we consider that the fingers are very long compared to their width and height, the numerical model is bi-dimensional along a cross-section perpendicular to the fingers. Numerical simulations were used to determine the domain of validity and the accuracy of the analytical model, and also to carry out calculations when analytical assumptions are not accurate. An example of the linear capacitance C_l as a function of the displacement w of the mobile part is plotted in figure 4 in the case of a 6-layer electrostatic structure with w_f=2 μm h_f=20 μm, h_i=30 μm and g=2 μm. With these dimensions, the results given by the analytical model are close to those given by the numerical model, indicating that minor electrostatic side effects occur.

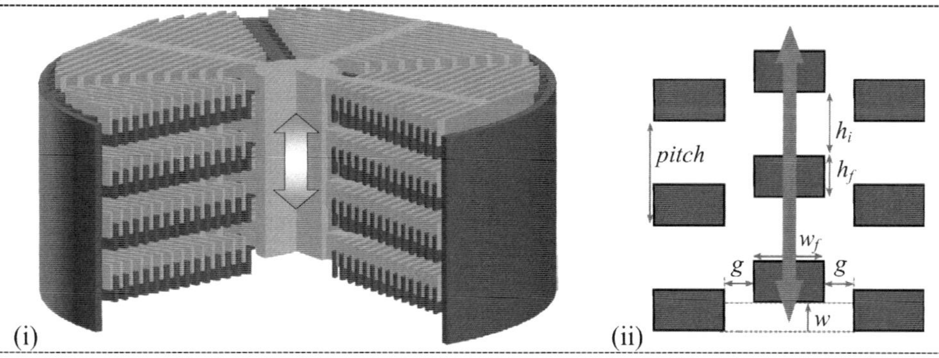

Figure 3. Multi-layer structure adapted for cylindrical geometries (i), and schematic cross-section view showing the dimensions parameters of the system (ii).

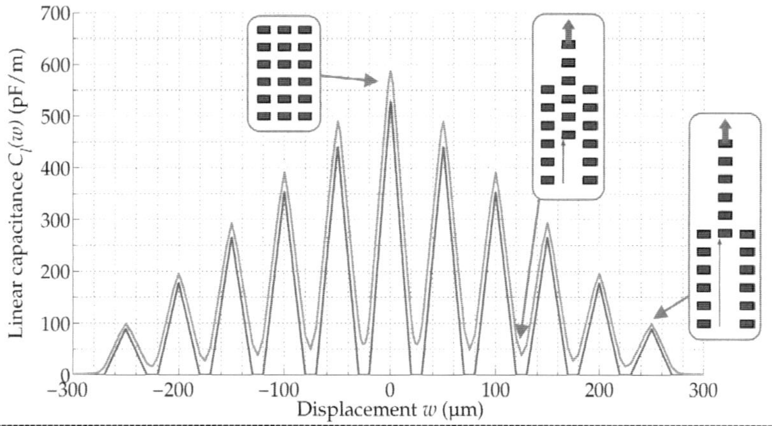

Figure 4. Linear capacitance of a 6-layer structure versus the displacement calculated using the analytical model (in blue) and the numerical model (in yellow).

2.5. Parameters optimization considerations

For the constrained voltage case, the extracted energy is proportional to the sum of (C_{max}-C_{min}) calculated for the full displacement of the mobile part. Examples of dimensions and corresponding sum of capacitance variation for full displacement of the mobile electrodes are given in Table 1. Several considerations could be drawn from simulation results (not entirely presented in this paper for the sake of compactness). First, we observed that the finger width w_f has only a small influence on the linear capacitance. As expected, the gap plays a big role and a closer investigation tends to prove that the capacitance increases roughly in a $h_f.g^{-1}$ fashion. As also anticipated, the number of layers n_l plays linear behaviour in the capacitance variation. The inter-layer height h_i is not trivial to determine. Indeed, if the layers are very close, important side effects occur and therefore the total capacitance variation is small. On the other hand, if the interlayer space is large, side effects are negligible and the capacitance variation per pitch is large, but the total capacitance variation for a given displacement will be small. Hence, there is an optimal inter-layer size for a given displacement. Furthermore, the interlayer gap has a strong effect on the total height of the system. Consequently, the optimization strategy will be different if the device limitation comes from the maximum allowable height or the maximum of comb layers (due to fabrication considerations for instance).

Table 1. Dimensional parameters and corresponding sum of the capacitance variations for maximum displacement (10 layer structure).

h_f (µm)	w_f (µm)	g (µm)	h_i (µm)	$\Sigma(C_{max} - C_{min})$ (pF/m)
20	20	10	80	910
10	5	2.5	10	650
40	20	20	120	640
60	6	30	120	440
10	5	10	40	370

As shown by (1), the energy converted is proportional to V_{max}^2. However, high voltage values may induce two undesirable effects: fingers electrostatic instability and electrical breakdown. Taking into account all the factors discussed in this subsection, we have calculated that more than 20 µJ per heart beat could be converted by our electrostatic energy harvester implanted in the left ventricle, working with a maximum voltage of 75 V. This electrostatic device is 6 mm in diameter and has 25 layers. Its other characteristic dimensions are h_f=10 µm, w_f=30 µm, g=7.5 µm and h_i=30 µm.

2.6. Fabrication considerations

Two fabrication processes were considered. The first one is based on wafer stacking, each silicon wafer being first micromachined using DRIE process. The second one consists in layer-by-layer electroplating of structural and sacrificial metals. The 3D structure is built by repeating 2D surface micromachining and polishing steps. Then, the sacrificial material is etched away, which releases the device. As the stacking of numerous wafers seems challenging, we have chosen to focus on fabrication method based on layer-by-layer electroplating. This process is currently under development at the Institut d'Electronique Fondamentale of the University of Paris-Sud.

3. Conclusion

We have presented an innovative multilayer out-of-plane gap overlap electrostatic transducer for harvesting the regular blood pressure variation. This device was thoroughly studied in considering its numerous geometrical degrees of freedom. The design for a maximum variation of capacitance, and hence for an optimal energy conversion, was performed in considering a layer-by-layer electroplating fabrication technology. Limitations such as electrostatic pull-in and voltage breakdown were taken into account. We showed that it would be possible to generate more than 20 µJ per heart beat using an electrostatic device with 6-mm diameter and 1-mm thickness. These results show that this concept is promising and applicable to industrial devices. With further development on manufacturing of the multilayer device and the associated electronics circuitry, this would provide enough power for the next generation of leadless pacemakers.

4. References

[1] Starner T and Paradiso J A 2004 *Low-power Electronics Design* (Boca Raton, FL: CRC Press)
[2] Deterre M, Dal Molin R, Boisseau S, Chaillout J-J, Lefeuvre E and Dufour-Gergam E 2011 Energy Harvesting system for cardiac implant applications *In Proc. DTIP* pp 387-391
[3] Karami A and Inman D 2012 *Appl. Phys. Lett.* **100** 042901
[4] Deterre M, Lefeuvre E, Zhu Y, Woytasik M, Bosseboeuf A 2013 Micromachined piezoelectric spirals and ultra-compliant packaging for blood pressure energy harvesters powering active implantable medical devices *In Proc IEEE MEMS 2013* vol. 1 pp 249-252
[5] Mitcheson P D, Green T C, Yeatman E M and Holmes A S 2004 *J. Microelectromech. Syst.* **13** n°3 429

Journal of Physics: Conference Series **476** (2013) 012040 doi:10.1088/1742-6596/476/1/012040

Packaging strategy for maximizing the performance of a screen printed piezoelectric energy harvester

Z Zhang, D Zhu, M J Tudor and S P Beeby
Electronics and Electrical Engineering Group, Electronics and Computer Science, University of Southampton, Southampton, SO17 1BJ, UK

E-mail: zz2e08@ecs.soton.ac.uk

Abstract: This paper reports the extended design and simulation of a screen printed piezoelectric energy harvester. The proposed design was based on a previous credit card sized smart tag sensor node, and packages the power conditioning circuit in the free space above the tungsten proof mass layer. This approach enables electronic components to be mounted onto the cantilever beam, which provides additional weight at the tip of the cantilever structure. The design structure contains a T-shape cantilever beam with size of 47 mm x 30 mm x 0.85 mm which is fabricated using screen printing. ANSYS simulation results predict the revised architecture can generate 421.9 μW approximately twice of the RMS power produced by the original design along with a higher open-circuit RMS Voltage of 8.0 V while the resonant frequency is dropped to 53.4 Hz.

1. Introduction

Piezoelectric materials are suitable for energy harvesting application and the majority of energy harvesters reported so far use this approach [1]. Research activities are focused on methods to enhance the output power and scaling down the size of harvesters. Such methods include utilizing d_{33} piezoelectric configuration instead of the d_{31} mode [2], developing more effective beam shapes, refining the power conditioning circuit designs to match the impedance between the circuits and generators and using more efficient piezoelectric materials [3].

Zhu et al. [4] reported a smart tag sensor node which was developed to harvest vibration energy to supply to a Health and Usage Monitoring Systems (HUMS) for aeronautical applications. The system contained a screen-printed bimorph piezoelectric energy generator which is the subject of this paper, and used a thick-film PZT paste, previously developed at the University of Southampton. A 70 μm thick PZT layer was printed on each side of the beam with a d_{31} coefficient -60CN^{-1} after polling. The total dimension of the harvester was 45 mm x 30 mm x 0.85 mm and this produced 243 μW RMS power at its resonant frequency of 67.4 Hz under an input acceleration of 3.9 ms^{-2}. In the system, the power conditioning circuit and super capacitor were placed separately to the side of the harvester.

In this paper, an initial investigation, simulation and design of a novel architecture that utilizes the free area of 47 mm x 10 mm on the proof mass to integrate the power management circuit and increase the seismic mass of the harvester.

This paper reports a strategy for moving the power conditioning electronics onto the energy harvesting structure which reduces system size and increases the inertial mass. Conventionally, circuit are placed separately and realized on traditional PCB boards which increases system size. The conventional approach is also a waste of material that can increase the seismic mass since circuits can be utilized as extra weight to increase the power output of generator along with lowing resonant frequency.

2. Design structure

2.1. Overview

The proposed new architecture consists of three extra layers, i.e. insulation, circuit and encapsulation layers, on top of the original bi-stable energy harvester presented in [4] (figure 1). The circuit and encapsulation layer are made of polymer materials which avoid high temperature and associated thermal stresses and are printed on one side whilst super capacitor is placed on the other side. The sequence of layer fabrication is shown in figure 1 which also indicates the number of different layers.

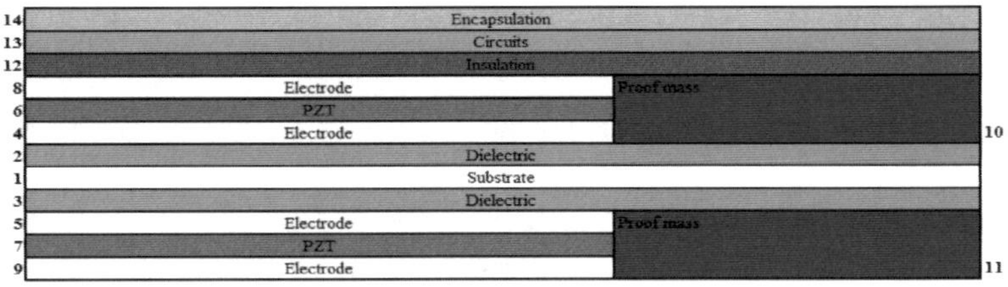

Figure 1. A cross-section of the harvester showing the material layers used (Numbers illustrate the fabrication sequence)

The power conditioning circuit was previously designed [4] which contains a bridge rectifier, a step-up voltage regulator and a cold start circuit.

An overall view of the new generator architecture designed using L-edit is shown in figure 2. From the top view, a circuit layer can be seen on top of the proof mass with two wires leading out to connect other peripheral circuits. Four electrodes (two on each side), are provided to enable polling of the PZT material (see figure 2(a)).

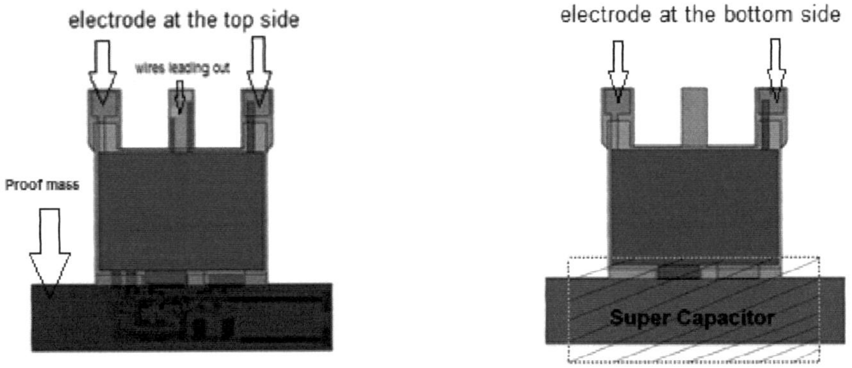

(a) Top view (b) Bottom view

Figure 2. L-edit design of the new architecture generator

2.2. Paste selections for each layer

The substrate is 110 μm thick and is fabricated from stainless steel 430S17 without excessive oxidation which can withstand the high firing temperature at 850 °C during the printing process. Stainless steel is conducting and must therefore be insulated before the bottom electrode can be deposited and fired. The dielectric paste ESL4924 from Electro Science Laboratories was used to insulate the particular stainless steel substrate. The other pastes used a high temperature gold conductor (ESL8836) for the bottom electrode and a low temperature silver polymer (ESL1901-S) for the top electrode. The PZT paste which was specific made in out lab. It was blended with different particle size based upon a piezoelectric powder type (PZT-5H). This composite enhanced the piezoelectric coefficient d_{33} up to 131pCN^{-1} and d_{31} coefficient of -60CN^{-1} [5, 6]. The proof mass layer was made of a bespoken tungsten paste with a density of 9550 kgm^{-3} [4]. DuPont 5000 silver polymer was used to print the circuit tracks in this design because it has a low sheet resistance of 15 mΩsq^{-1} and can be cured under 120 °C in box oven. Polyurethane paste is used for both the insulation and encapsulation layers which can be cured at 60 °C at UV condition.

3. Simulation

The proposed architecture was simulated using ANSYS APDL to investigate how the new design would impact on the performance of the original energy harvester of the smart tag. The influence on the center of gravity was also investigated using Autodesk INVENTOR. A comparison of results has been plotted in figure 3 showing the peak power and peak voltage output generated in the original design, with the circuit only placed on the inertial mass, with the super capacitor only placed on and with both circuit and super capacitor placed on the energy harvester.

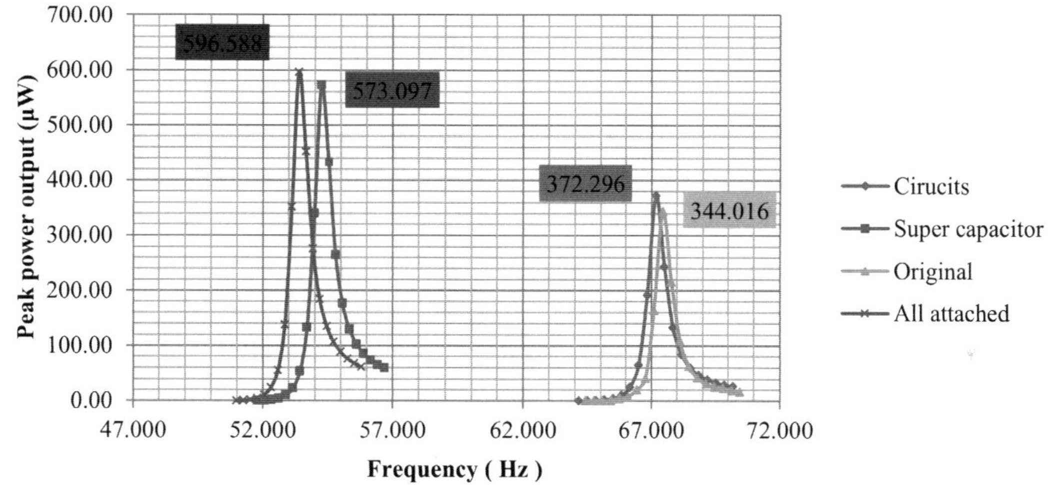

Figure 3. Comparison of optimum peak power with different components attached

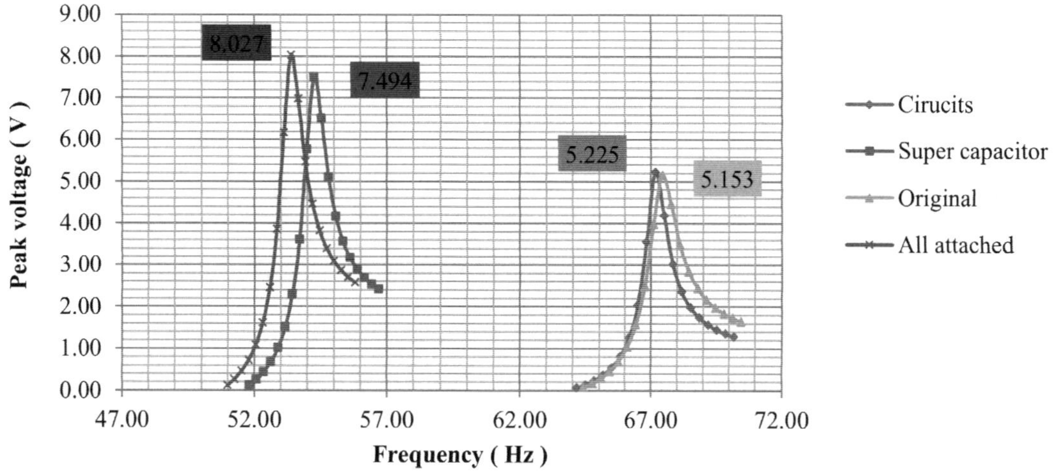

Figure 4. Comparison of open circuit peak voltage with different components attached

4. Discussion and Evaluation

A comparison of data has been provided in Table 1 and numerically illustrates the impact of the different attached components on the generator. With the super capacitor attached to the generator, the RMS power has increased by 67% from 344.02 µW to 573.10 µW and RMS voltage has increased by 45%. Resonant frequency has reduced by 17%. The circuit on its own does not contribute much to power output (less than 1%).

Components attached	Extra weight attached (g)	Optimum Load (kΩ)	Peak Voltage (V)	RMS Voltage (V)	Peak Power (µW)	RMS Power (µW)	Resonant Frequency (Hz)
Original	0.0	77	5.15	3.64	344.02	243.26	67.44
Cirucit	0.0494	77	5.23	3.69	372.30	263.30	67.32
Supcap	6.3313	97	7.49	5.28	573.10	405.31	54.25
Circuit+Supcap	6.3807	102	8.03	5.68	596.59	421.92	53.39

Table 1. Comparison of parameters of different attached object

Figure 5 shows the effect of the components and circuit has a small change on the center of gravity of the harvester. The effect of this change will be observed experimentally in the future.

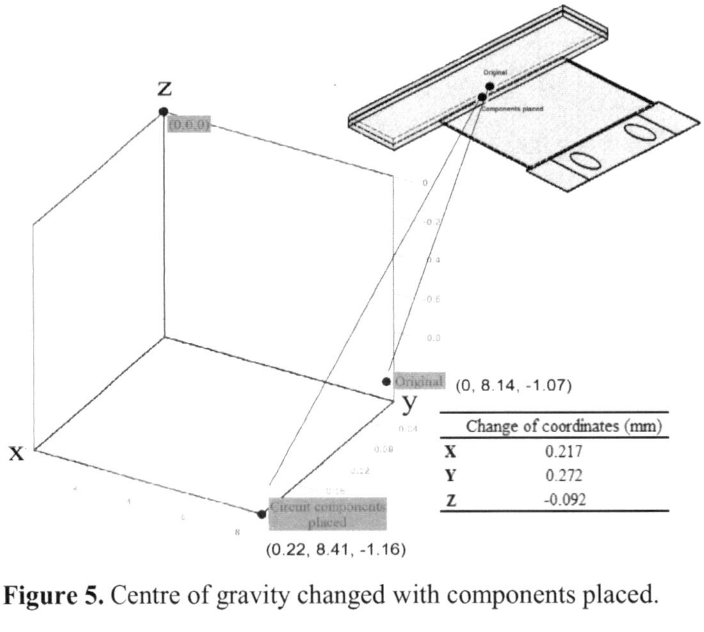

Figure 5. Centre of gravity changed with components placed.

The other improvement of this new architecture is the reduction in system area at the cost of total thickness (figure 6). An area of 345 mm^2 was used for the original power management circuit PCB and a further 520 mm^2 is saved by mounting the super capacitor. Overall, a total system area is reduced by 865 mm^2 which is nearly 20% of the smart tag area. A reduced area of original energy harvester has been shown in figure 5. The volume of original smart tag device has a total volume of 14,025 mm^3 and the new architecture takes 13,587.3 mm^3. The volume has not been reduced much due to the increase thickness of placed super capacitor.

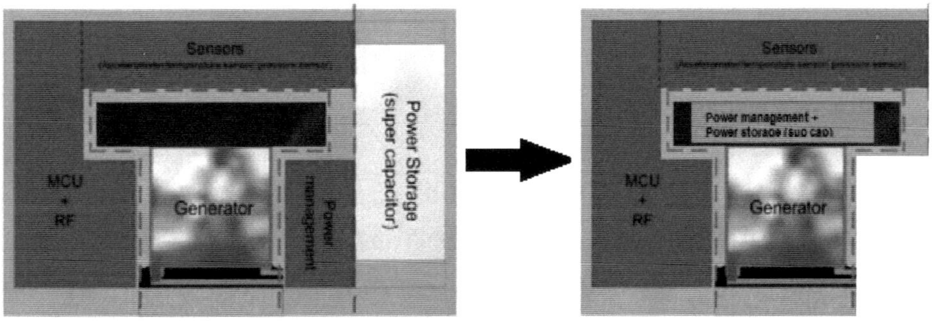

Figure 6. A reduction in size of the original energy harvester system

5. Fabrication process

The screen printing process follows a sequence of layer by layer deposition with the paste, being dried and fired before each subsequent part. The dielectric insulating layer is firstly deposited on the stainless steel substrate and fired at 850 °C through the belt furnace giving a thickness of 20 μm on both side of the beam. The gold paste was deposited as the bottom electrode 10 μm thick. The PZT layer is sandwiched by the bottom and top electrode with a thickness of 70 μm. Then, the tungsten film is deposited to form up the proof mass and this is cured at temperature of 140 °C. A layer of insulation using dielectric paste between top electrode and the circuit tracks is printed. Next, the circuit tracks are then printed and cured at 125 °C. Finally, the encapsulation is deposited which contains the same paste as insulation layer and is cured. Fabrication devices are currently being tested.

6. Conclusions and future work

The proposed architecture of integrating power management units on the generator was simulated and results were discussed. Both output power and voltage of the energy harvester are increased with a reduction of the system area. Increasing the inertial mass was for improving performance and this approach does so without increasing total volume of the system. This is much more straightforward than, for example improving the PZT material properties. This approach can be also used to tune the resonant frequency instead of adding extra tungsten material of the proof mass. In the future, devices will be tested on a shaker (Labworks ET-126) with a programmable resistance box and a PC with LabVIEW software collecting data. The results of experiment and simulation will be compared to get a better understanding of its performance.

7. Acknowledgement

The authors thank the EPSRC for supporting this research with grant reference EP/I005323/1.

8. References

[1] Beeby S P, Tudor M J and White N M 2006 *Meas. Sci Technol.* **17** R175-R195
[2] Goldschmidtboeing F and Woias P 2008 *J. Micromech. Microeng.* **18** 104013
[3] Anton S R and Sodano H A 2007*Smart Mater. Struct.* **16** R1-R21
[4] Zhu D Z, Beeby S P, Tudor M J and Harris N R 2011*Sensors and Actuators A: physical* **169(2)** 317-325
[5] Torah R N, Beeby S P, Tudor M J and White N M 2007*Journal of Electroceramics* **19(1)** 97-112
[6] Torah R N , Beeby S P and White N M 2005*IEEE transactions on ultrasonics* **52(1)** 10-16

Ferroelectric Dipole Electrets Prepared from Soft and Hard PZT Ceramics in Electrostatic Vibration Energy Harvesters

H Asanuma, H Oguchi, M Hara and H Kuwano

Department of Nano-Mechanics, Tohoku University, 6-6-01, Aoba, Aramaki, Aza-Aoba, Sendai, 980-8579, Japan

E-mail: haruhiko.asanuma@nanosys.mech.tohoku.ac.jp

Abstract. Aiming at longer stability of surface potential, we propose a ferroelectric dipole electret (FDE) prepared from hard ferroelectric material. We compared output power of electrostatic vibration energy harvester and surface potential stability between FDEs prepared from soft and hard PZT ceramics, as well as a CYTOP polymer electret. The hard FDE showed a seven-fold increase in output power over the soft FDE and nine-fold increase over the CYTOP polymer electret. The hard FDE also showed longer stability of surface potential than that of the soft FDE, whereas the stability of the hard FDE was not yet comparable to that of CYTOP polymer electret. A FDE prepared from harder PZT ceramic (with higher coercive electric field and Curie temperature) may provide more stability in surface potential.

1. Introduction

A vibration energy harvester (VEH) is considered to be viable power sources for operating sensor nodes in autonomous wireless sensor networks because of its high power density and inexhaustibility in operational environment [1–3]. In actual operation environment, VEHs are expected to generate power from low-level mechanical vibration (typical frequency and acceleration are below 200 Hz and 9.8 m/s^2, respectively) [2]. An electrostatic vibration energy harvester (E-VEH) using a charged dielectric called "electret" is being recognized as a promising solution to satisfy the aforementioned demand because of its lower operating frequency and wider frequency bandwidth, compared to piezoelectric or electromagnetic one [4–10]. However, its output power still remains lower than that obtained from other VEHs. Increase in surface charge density of an electret is one of the key factors to enhance output power in the E-VEH. Nevertheless, surface charge density of the electret so far developed is limited by the relative permittivity of its dielectric material, such as polymers [11] or silicon-based inorganics [12, 13]. Ferroelectrics could be highly promising because their relative permittivity and the resulting surface charge density exceed 1000 and 100 mC/m^2, respectively.

Recently, we developed an electret consisting of a polarized lead zirconate titanate (PZT) ceramic, and named it ferroelectric dipole electret (FDE) [14, 15]. The E-VEH using the FDE showed an eight-fold increase in output power over a conventional polymer electret; on the other hand, the FDE exhibited poor stability in surface potential [15]. In previous studies, we chose a so-called "soft" PZT to form a FDE for two reasons: high value of remnant polarization (P_r) and easy poling condition [low coercive electric field (E_c) and low Curie temperature (T_c)]. In general, soft PZT can be obtained by donor doping, for example, by substitution of La^{3+} for Pb^{2+} and/or Nb^{5+} for Ti^{4+}/Zr^{4+}. The donor doping will enhance the domain wall mobility by reducing the concentration of oxygen vacancies playing a role as pinning center for the domain wall motion by forming defect dipole [16–18]. Thus, soft PZT has high P_r, low E_c and T_c. These ferroelectric properties may result in easy depoling and thus the previously-reported poor stability in surface potential. Alternatively, "hard" PZT is generally tuned by acceptor doping, for example, by substitution of K^+ for Pb^{2+} and/or Fe^{3+} for Ti^{4+}/Zr^{4+}, resulting in intentionally increasing oxygen vacancies and thus high E_c and T_c [16, 18]. These properties of hard PZT may improve the stability in surface potential of the FDE.

In this study, we compare output power of E-VEHs and surface potential stability between FDEs prepared from soft and hard PZT ceramics, as well as a CYTOP polymer electret.

2. Fabrication

We employed unpoled 1-mm thick PZT ceramics with no electrodes [Fuji Ceramics Corp., Japan, No.C-6 (soft) and No.C-2 (hard)]. As detachable electrodes to form the FDEs, we attached the copper foil tapes with conductive adhesive (Hitachi Maxell, Japan, SLIONTEC No.8701) on both surfaces of the PZT ceramics [14]. Poling treatment was conducted in silicone oil bath at room temperature for the soft PZT ceramic and at 120 °C for the hard PZT ceramic, respectively. Applied poling electric field and treatment time are 2 kV/mm and 1 hour, respectively. The hard PZT ceramic was cooled down to room temperature for an hour while keep applying the poling electric field. After the poling treatment, we wiped off the silicone oil remained on both surfaces of the PZT ceramics using KimWipe sheets. Then, poled PZT ceramics served as the FDEs by peeling off the copper foil tapes.

For comparison, we also prepared a conventional electret consisting of CYTOP polymer (Asahi Glass, Japan, CTL-809M). An 8.2-μm thick CYTOP polymer was formed by spin-coating on a 1-mm thick copper rigid plate and then charged by using point-to-grid corona discharging method. The detailed set-up of the point-to-grid corona discharging is reported in the reference [10, 11]. Applied needle voltage and grid voltage were set to -2.3 kV and -600 V, respectively. All electrets have the same surface area (length: 20 mm, width: 20 mm).

3. Experimental

We use non-contacting electrostatic voltmeter (Trek Inc., Model 347) to measure the surface potential of electrets. Effective surface charge density σ is calculated from $\varepsilon_r\varepsilon_0 V_s/t$, where ε_r is the relative permittivity of an electret material, ε_0 the permittivity of vacuum, V_s the measured surface potential, and t the thickness of each electret. In this study, the negatively-charged surface is defined as top surface and the opposite is bottom surface.

Figure 1 shows our experimental configuration for evaluating vertical-type E-VEH [7–9, 14, 15]. An electret was put on a lower electrode attached to a shaker controlling vertical vibration, and set to face up a negatively-charged top surface against an upper electrode mounted on a translational stage equipped with a micrometer. Then initial air gap (center of the vibration) between the upper electrode and an electret was adjusted using the micrometer. A goniometer installed under the lower electrode was also used to keep the electret parallel to the upper electrode. Consequential change in the air gap distance under vertical vibration is accompanied by change in the capacitance between the upper electrode and the lower electrode, and generates alternate current. Output voltage V_{out} was calculated from $(r_1+r_0)V_{osc}/r_0$,where r_1 is the variable load resistance (3.3–99 MΩ), r_0 the fixed load resistance (0.5 MΩ), and V_{osc} the voltage measured by the oscilloscope having the internal impedance of 10 MΩ. Output power P_{out} is calculated as follows.

$$P_{out} = \frac{1}{T}\int_0^T \frac{V_{out}^{\;2}(t)}{R}dt , \tag{1}$$

where T is the measuring time period (0.5 second) and R is the total load resistance (r_1+r_0).

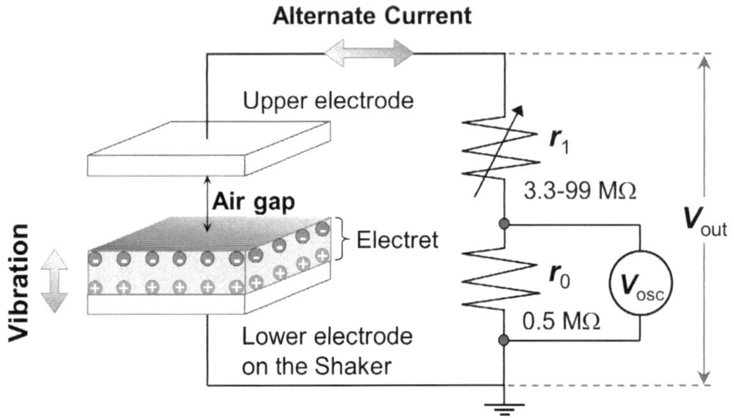

Figure 1. Schematic of the experimental configuration for output power evaluation of vertical-type electrostatic vibration energy harvester.

4. Results and discussion

Table 1 summarizes the measured surface potential and the effective surface charge density of each electret. For each FDE, top and bottom surfaces are charged with similar magnitudes but opposite signs. In contrast, only top surface is effectively charged in the CYTOP polymer electret (CPE). The difference between these electrets is that the FDE utilizes a dipole-orientation by poling, whereas the CPE does implanted charges by corona discharging. The σ of the hard FDE was twice higher than that of the soft FDE.

Table 1. Relative permittivity (ε_r), thickness (t), measured surface potential (V_s) on top and bottom surface, and effective surface charge density (σ) on the top surface of hard and soft ferroelectric dipole electret (FDE) and CYTOP polymer electret (CPE). All electrets have the same surface area (length: 20 mm, width: 20 mm).

	ε_r	t (mm)	top V_s (V)	bottom V_s (V)	σ (mC/m^2)
Hard FDE	1536	1.0	-731	+713	-10
Soft FDE	2281	1.0	-241	+231	-4.9
CPE	2.1	0.082	-580	+32	-1.3

Figure 2 (a) shows waveforms of output voltage at R = 30.2 MΩ, and Figure 2 (b) shows output power as a function of total load resistance. Applied frequency, acceleration, and initial air gap were set to 20 Hz, 4.9 m/s^2, and 0.35 mm, respectively. Maximum output powers are 59 μW for the hard FDE, 9.0 μW for the hard FDE, and 6.9 μW for the CPE. The higher output power of the hard FDE may be attributed to retaining the higher surface charge density even after removing the conductive tapes by the hardening of the domain wall motion due to the defect dipole pinning [17].

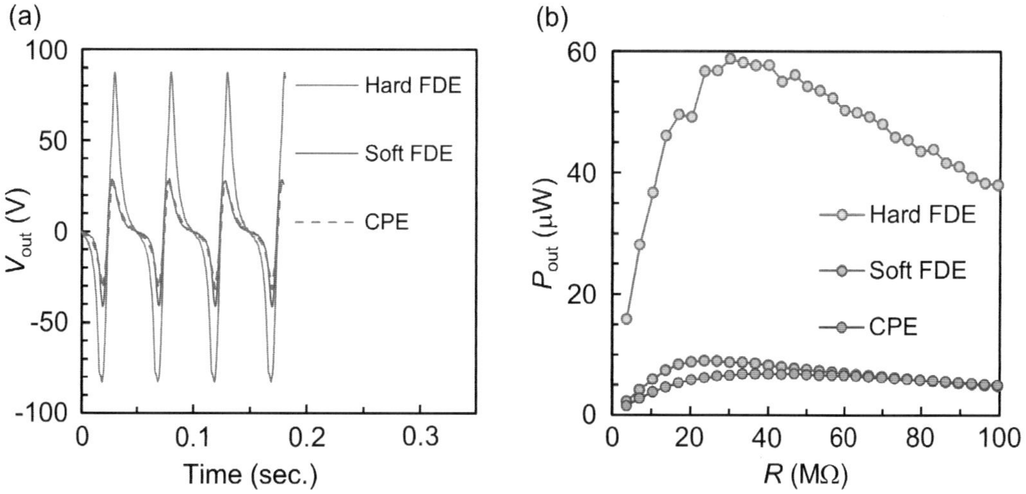

Figure 2. (a) Waveform of output voltage (V_{out}) and **(b)** output power (P_{out}) vs total load resistance (R).

Figure 3 shows surface potential stability over 800 hours. Samples were placed in an experimental room (usually kept 23-25 °C and 50-60 %RH). The surface potential at each measuring point was normalized with the initial value shown in table 1. While the surface potential of the soft FDE started decreasing from one hour later, the surface potential of the hard FDE started decreasing from 30 hours later. The result also may be attributed to the hardening of the domain wall motion due to the defect dipole pinning [17]. The hard FDE showed longer stability in surface potential than that of the soft FDE, whereas the stability of the hard FDE was not yet comparable to that of CYTOP polymer electret. A FDE prepared form harder PZT ceramic (with higher E_c and T_c) may provide further improvement in surface potential stability.

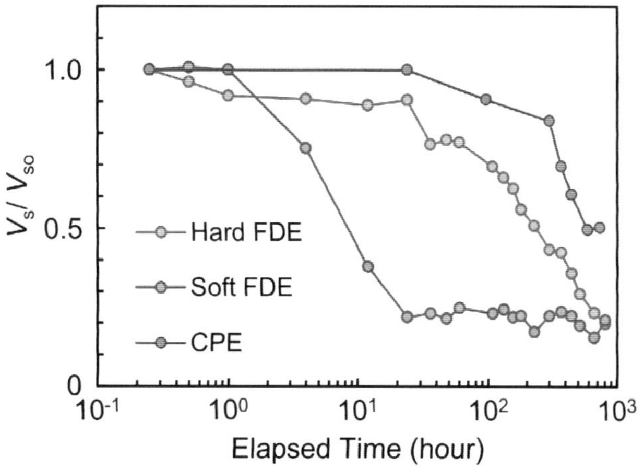

Figure 3. Surface potential stability. Surface potential (V_s) was normalized with the initial value (V_{so}) shown in table 1.

5. Conclusion

We investigated output power of E-VEHs and surface potential stability of hard and soft FDEs, as well as a CPE. The effective surface charge density of the hard FDE reached -10 mC/m^2, which was twice higher than that of the soft FDE. The maximum output power of the hard FDE normalized with surface area of electrets was 15 μW/cm^2, a seven-fold increase over the soft FDE and a nine-fold increase over the CPE. The hard FDE showed longer stability in surface potential than that of the soft FDE. The result may be attributed to the hardening of the domain wall motion due to the defect dipole pinning. However, the stability of the hard FDE was not yet comparable to that of CPE. A FDE prepared from harder PZT ceramics (with higher coercive electric field and Curie temperature) may provide more stability in surface potential.

Acknowledgements

We would like to thank the Japan Society for the Promotion of Science (JSPS). This work was supported by Grant-in-Aid for JSPS fellows (No.H3254346).

References

[1] Kuwano H 1996 *Proc. of IEEE Seventh International Symposium on Micro Machine and Human Science* (Nagoya, Japan, 2-4 October 1996) pp 21–8
[2] Roundy S, Wright P and Rabaey J 2003 *Computer Communications.* **26** 1131
[3] Boisseau S, Despesse G and Seddik B 2012 *Electrostatic Conversion for Vibration Energy Harvesting, Small-Scale Energy Harvesting, Edited by Dr. Mickaël Lallart* (Rijeka: InTech) p 91
[4] Suzuki Y 2011 *IEEJ Transactions on Electrical and Electronic Engineering* **6** p 101
[5] Masaki T, Sakurai K, Yokoyama T, Ikuta M, Sameshima H, Doi M, Seki T and Oba M 2011 *J. Micromech. Microeng.* **21** 104004
[6] Boisseau S, Despesse G and Sylvestre A 2010 *Smart Mater. Struct.* **19** 075015
[7] Boisseau S, Despesse G, Ricart T, Defay E and Sylvestre A 2011 *Smart Mater. Struct.* **20** 105013
[8] Chiu Y and Lee Y 2013 *J. Micromech. Microeng.* **23** 015012
[9] Takahashi T, Suzuki M, Nishida T, Yoshikawa Y and Aoyagi S 2013 *Proc. of The 17th International Conference on Solid-State Sensors, Actuators and Microsystems (TRANSDUCERS)* (Barcelona, Spain, 16-20 June 2013) pp 1340–3
[10] Wada Y, Oguchi H, Hara M, Asanuma H and Kuwano H 2012 *Proc. of PowerMEMS 2012 Workshop* (Atlanta, USA, 2-5 December 2012) pp 504–7
[11] Wada Y, Hamate Y, Nagasawa S and Kuwano H 2011 *Proc. of The 16th International Conference on Solid-State Sensors, Actuators and Microsystems (TRANSDUCERS)* (Beijing, China, 5-9 June 2011) pp 2626–9
[12] Leonov V, Schaijk R and Hoof C 2013 *IEEE Sensors Journal* **13** pp 3369–76
[13] Boisseau S, Chaillout J, Danel J, Legras J and Despesse G 2013 *Proc. of The 17th International Conference on Solid-State Sensors, Actuators and Microsystems (TRANSDUCERS)* (Barcelona, Spain, 16-20 June 2013) pp 1942–5
[14] Asanuma H, Oguchi H, Hara M and Kuwano H 2013 *Proc. of The 17th International Conference on Solid-State Sensors, Actuators and Microsystems (TRANSDUCERS)* (Barcelona, Spain, 16-20 June 2013) pp 1332–5
[15] Asanuma H, Oguchi H, Hara M, Yoshida R and Kuwano H 2013 *Appl.Phys.Lett.* accepted
[16] Shrout T and Zhang S 2007 *Journal of Electroceramics* **19** pp 111–24
[17] Kitanaka Y, Noguchi Y and Miyayama M 2010 *Phys. Rev. B* **81** 094114
[18] Pramanick A, Damjanovic D, Daniels J, Nino J and Jones J 2011 *J. Am. Ceram. Soc.* **94** pp 293–309

Journal of Physics: Conference Series **476** (2013) 012042 doi:10.1088/1742-6596/476/1/012042

Novel optimized design of a piezoelectric energy harvester in a package for low amplitude vibrations

G Murillo[1], H Campanella[2], J Esteve[1] and G Abadal[3]

[1] Department of Micro and Nanosystems, Instituto de Microelectrónica de Barcelona IMB-CNM (CSIC), Bellaterra, 08193, Spain.
[2] Institute of Microelectronics, A*STAR (Agency for Science Technology and Research), 11 Science Park Road, Singapore Science Park II, Singapore 129195.
[3] Departament d'Enginyeria Electrònica, Universitat Autònoma de Barcelona, Bellaterra, 08193, Spain.

gonzalo.murillo@csic.es

Abstract. This paper presents a novel piezoelectric energy harvesting device created with the flip-chip bonding of two different parts, one is a MEMS die which plays the role of inertial mass and the other is an associate CMOS chip anchored to the vibrating environment. The flip-chip bonding is performed between the MEMS die, which consists of four piezoelectric beams connected to four PADs or anchor points, and a test PCB, which is used to validate the feasibility of the whole assembled system. The resulting system in package is a proof of concept of a novel design concept that increases the extracted power from an ambient vibration. FEM simulations have been carried out to study the mechanical behaviour of the whole system. Moreover, the fabrication of the piezoelectric die and the test PCB has been successfully performed, as well as their flip-chip integration.

1. Introduction

Nowadays, the use of green energy is becoming almost a requirement for our society. Especially interesting is to scavenge mechanical energy from our environment, because it is virtually omnipresent. On the other hand, the world of electronics devices is moving towards smaller power consumptions. The convergence of these two facts is attracting the interest from researchers in energy harvesting devices at the micro- and mm-scale from ambient vibrations [1]. There are three well-kwon mechanisms to convert mechanical energy into electricity: they are the electrostatic, electromagnetic and piezoelectric transductions.

In this work, a piezoelectric energy harvester has been designed and its behaviour simulated. Also, the fabrication of a proof of concept device and a preliminary characterization are presented here.

2. Design advantages

The piezoelectric energy harvesting system presented here is created with the flip-chip bonding of two different parts. One is a MEMS die which plays the double role of inertial mass and transduction method. The other is an associate CMOS chip anchored to the vibrating environment. In this way, the resonance motion appears in the way of relative displacement between both chips.

This paper is mainly focused on the MEMS die which consists of four beams connected to four PADs. These PADs work as electrical connection and support point with the CMOS chip where the MEMS die should be mounted or vice versa. Figure 1 shows the 3D model of the whole proposed assembled system, where a CMOS chip with the power management circuitry is mounted on top of a micromachined die with four suspensions covered by a piezoelectric layer.

The use of a whole die as inertial mass has different advantages:

- decrease of the resonance frequency. It is due to the extra mass load provided by the chip substrate itself. If we assume the device to be as a spring-mass system, the resonance frequency is inversely proportional to the squared root of the inertial mass. This mass can be easily increased by changing the size or weight of the die or by adding an additional mass on top of it.
- mechanical energy transferred from the environment to the inertial mass increases. The energy captured from a sinusoidal excitation over the base of a spring-mass system is directly proportional to its inertial mass.
- the majority of the chip area can be utilized for energy transduction, increasing the maximum converted power. The active area is not utilized to create an inertial mass, but to form transduction and suspension elements.
- using different anchor points makes the bandwidth broader. It is due to the superposition of different adjacent resonance modes coming from each anchor element.

When this concept is applied thought heterogeneous integration to two different chips, the resulting system can be called Energy Harvester in a Package (EHiP) [3].

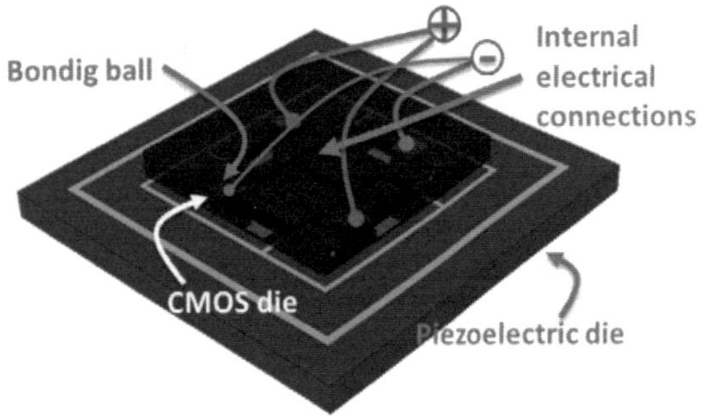

Figure 1. 3D model of the whole system (EHiP) for simulating in COVENTOR®.

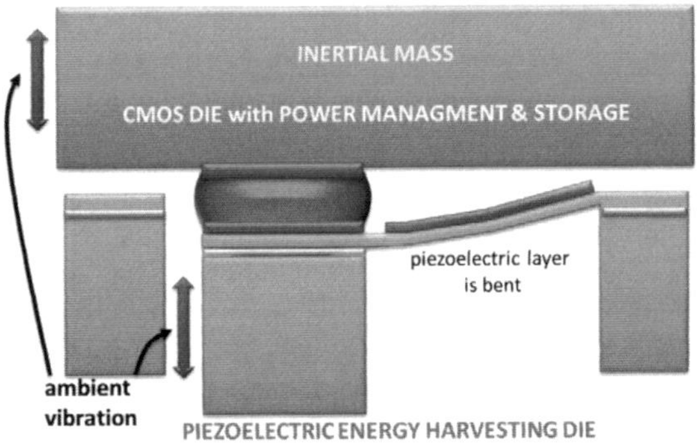

Figure 2. Cross-section of a suspension after flip-chip and deflection due to an ambient vibration.

3. MEMS fabrication process and system assembling procedure

The fabrication technology of the piezoelectric die is a variation of a process designed to fabricate Film Bulk Acoustic Resonators (FBARs) with the purpose of achieving mass sensors and RF resonators [2-3]. It is based on an SOI wafer that is both-side patterned with dry etching and consists of a bulk substrate of 500 μm, a silicon dioxide layer of 1 μm and a device silicon layer of 15 μm. Over this wafer a layer of 1 μm is made of AlN and sandwiched between two Pt-Ti layers. Therefore, the new fabrication process can monolithically integrate energy scavenging systems together with FBAR elements.

The technology used to fabricate the piezoelectric device has been recently developed in our group for the integration of FBARs with energy harvesting applications [2]. By using a similar technology, we presented in [3] an energy scavenger with a traditional architecture, i.e. a mass connected to the substrate by means of a beam. It was reported a scavenged power of 0.2 μW extracted at a resonance frequency of 515 Hz and an input acceleration of 0.64 m/s^2. That device obtained performance results comparable to values extracted from some works with high generated powers in the literature [4-5].

In figure 2 appears a detailed view of cross section of one the support beams. Figure 3 shows a picture of the fabricated piezoelectric die consisting of four support beams that are covered for a 1µm-thick layer of AlN and internally interconnected in the way that the electrical signals of the top and bottom electrodes are available through the PADs to the CMOS die. When an ambient vibration is present, the whole system resonates along the out-of-plane direction. This resonance motion forces the support beam to be deflected and the AlN layer is bent, getting a strain over this layer that is translated into a piezoelectric potential between the top and bottom electrodes.

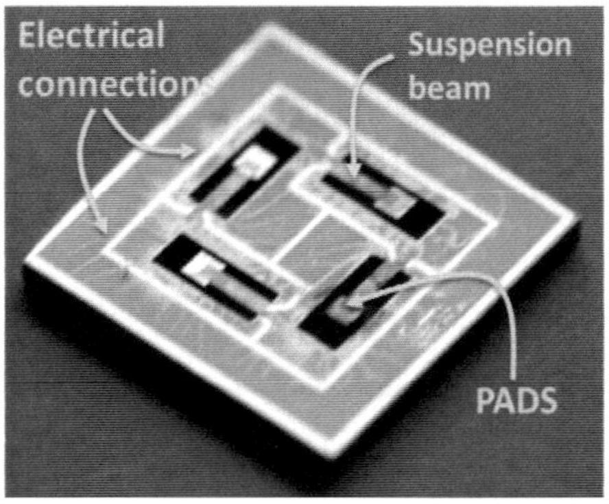

Figure 3. Detailed picture of the fabricated piezoelectric die before performing the flip-chip bonding (~5 mm x 5 mm x 515µm, 28 mg).

Figure 4. Photograph of the fabricated PCB (right) and piezoelectric die (left) on top of a 1Euro cent coin (Ø 16.25 mm).

In order to make the flip-chip step more feasible, the role of the CMOS chip was replaced by a test PCB fabricated using a circuit board milling plotter. The PCB has a size of 3 mm x 3 mm, a thickness of around 500 µm and a mass of 23 mg, which is quite similar to the piezoelectric die mass (figure 4). Moreover, in order to gain a security space between both chips, this PCB has four pillars of a height of 1 mm and a width of less than 300 µm.

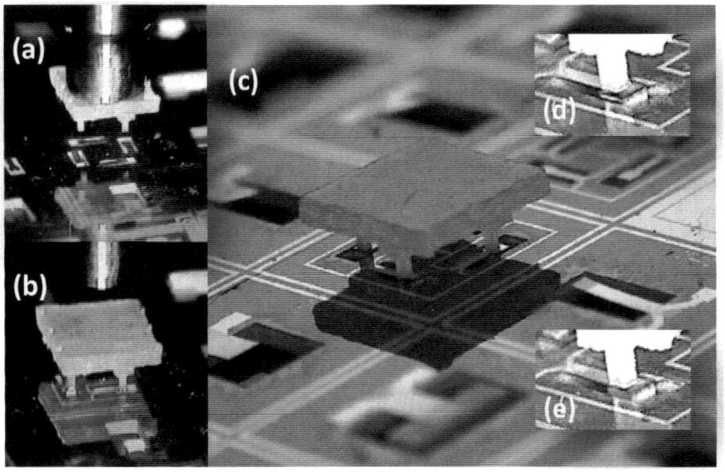

Figure 5. Assembling process of the test PCB at component alignment system: approaching (a), release of the PCB over the chip after alignment (b), final pasted PCB-chip system (c). Detail of one support beam without external vibration (d) and with a low-level vibration at 5322 Hz (e).

A component alignment system (ERSA PL 550) allowed us to place with precision the PCB over the piezoelectric die. In the case reported in this work, a conductive epoxy has been used to glue both elements. The PCB was pushed over a thin layer of this epoxy was coated on a flat surface in order to get the PCB PADs imprinted. Afterwards, the flip-chip procedure was carried out by aligning the PADs of the PCB and chip (at wafer level), approaching the PCB to the wafer (figure 5(a)), and releasing the PCB over the chip (figure 5(b)) to get the final assembled system (figure 5(c)).

With the goal of avoiding possible issues from the flip-chip step, a test PCB instead of a CMOS chip has been used but the system performance changes from the expected by using the chip. In order to reduce the new resonance frequency, a heavier PCB or an external mass can be utilized. Also an SOI wafer with an active silicon layer of 5 μm will drop off this frequency.

4. FEM simulations and preliminary characterization

COMSOL® and COVENTOR® software tools were used to simulate the mechanical behaviour of the device. Both motion strategies were explored, the single MEMS die anchored by its PADs, i.e. using the MEMS die as inertial mass, and the anchoring of this aforesaid die under the mounted PCB, i.e. the PCB works as inertial mass.

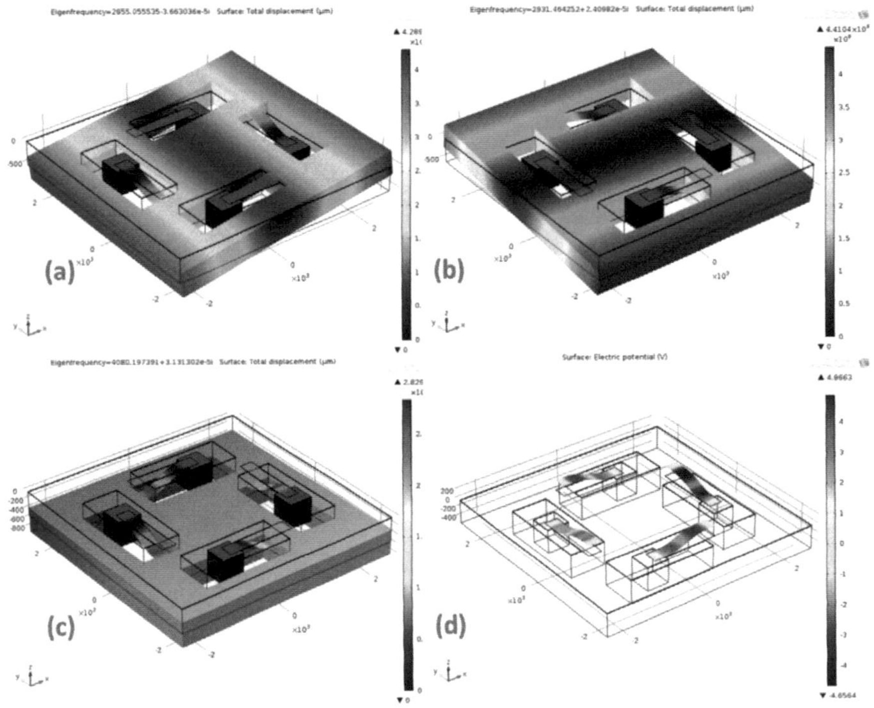

Figure 6. First (a), second (b) and third (c) resonance modes at 2855 Hz, 2931 Hz, 4080 Hz respectively (in case of using PCB as inertial mass) and electric potential of piezoelectric layers (d) simulated with COMSOL®.

The FEM-based modal simulations show that instead of having a single resonance motion, we have three resonance modes that can be used to harvest mechanical energy from an external vibration. The two first resonance modes (2855 Hz and 2931 Hz, with the PCB as inertial mass, or 2417 Hz and 2481 Hz, with the chip acting as mass) are shown in figure 6(a) and figure 6(b), where the mass is tilted and only two beams are deflected. The out-of-plane resonance motion is obtained at the third mode (4080 Hz, for PCB mass and 3454 Hz for the case of the chip) shown in figure 6(c). In this case, the four beams are bent in the typical way of a clamped-guided beam, obtaining a uniform stress gradient along the beam that is zero at the middle point and get opposite signs at every beam side. This is translated into an opposite charge density in each piezoelectric layer, and subsequently opposite electric potentials. The use of single electrodes covering the piezoelectric layers of every beam would suppose a cancelation of the majority of the generated charges. To overcome this adverse effect, each beam should have two different electrodes at every beam side. Thus, eight different piezoelectric potentials can be combined to get a global charge generator. Figure 6(d) shows this opposite potential effect and generated voltages of ±4.8V (at open circuit) for a mass vertical displacement of 10 μm.

By using a piezoelectric buzzer, a weak vibration was produced over the wafer close to the assembled PCB (figure 5(c)). A low-amplitude resonance motion was observed (figure 5(d) and figure 5(e)) when applying a sinusoidal wave at 5322 Hz. Regarding its shape, this resonant motion is supposed to be the third mode of resonance of the whole system with the PCB as inertial mass (around 4080 Hz).

5. Conclusions and future tasks

This paper presents a novel piezoelectric energy harvesting device consisting of two different parts bonded by flip-chip. One part is a MEMS die which plays the role of inertial mass and the other is a test PCB used to validate the feasibility of this heterogeneous integration. The resulting system in package is a proof of concept of a novel design concept that increases the extracted power from an ambient vibration. FEM simulations have been carried out to study the mechanical behaviour of the whole system. Also, the fabrication of the piezoelectric die and the test PCB has been successfully performed, as well as their flip-chip integration.

On the other hand, a more sophisticated double side PCB has been designed to allow the electrical connectivity with the piezoelectric chip through the PCB pillars. Moreover, a diode bridge, a capacitor and a SMD circuit for the control and management of piezoelectric energy scavenger will be added. At the same time, an ad-hoc electronic chip will be fabricated in our CNM 2.5-μm CMOS technology.

Finally, the full electromechanical characterization of the presented system is being carried out at this time. Thus, the measurement of frequency and generated power at different resonance modes is our main on-going task.

6. Acknowledgments

Thanks to Joan Puertas and Ana Moya for their support on the PCB fabrication and the creation of the characterization setup. This work has been partly funded by the GICSERV program, a JAE-DOC contract (CSIC) and OPACMEMS project (ENE2009-14340-C02-02/MIC).

References

[1] Beeby S P, Tudor M J and White N M 2006 Energy harvesting vibration sources for microsystems applications *Meas. Sci. Technol* **17** R175-95

[2] Murillo G, Abadal G, Torres F, Lopez J L, Giner J, Campanella H, Uranga A, Esteve J and Barniol N 2008 *Proc. 8th Int. Workshop on Micro and. Nanotechnology for Power Generation and Energy Conversion Applications (PowerMEMS)* (Sendai, Japan, 9-12 November 2008) 225-8

[3] Murillo G, Campanella H, Agusti J, Esteve J and Abadal G 2012 *Proc. IEEE Int. Conf. on Micro Electro Mechanical Systems (MEMS)* (Paris, France 29 January – 2 February 2012) 1253-56

[4] Roundy S, Wright P K and Rabaey J M 2004 *Energy Scavenging for Wireless Sensor Networks: With Special Focus on Vibrations* Springer, 2004.

[5] Elfrink R, Renaud M, Kamel T M, de Nooijer C, Jambunathan M, Goedbloed M, Hohlfeld D, Matova S, Pop V, Caballero L and van Schaijk R 2010 *J. Micromech. Microeng.* **20** 104001

Design of RF energy harvesting platforms for power management unit with start-up circuits

Alessandra Costanzo and Diego Masotti

DEI, University of Bologna, Italy

alessandra.costanzo@unibo.it

Abstract. In this contribution we discuss an unconventional rectifier design dedicated to RF energy harvesting from ultra-low sources, such as ambient RF sources which are typically of the order of few to few tens of μW. In such conditions unsuccessful results may occur if the rectenna is directly connected to its actual load since either the minimum power or the minimum activation voltage may not be simultaneously available. For this reason a double-branch rectifier topology is considered for the power management unit (PMU), instead of traditional single-branch one. The new PMU, interposed between the rectenna and application circuits, allows the system to operate with significantly lower input power with respect to the traditional solution, while preserving efficiency during steady-state power conversion.

1. Introduction

In those situations where available RF energy in the ambient is the only source the systems autonomy by energy scavenging is almost impractical, even for typical distributed embedded systems activities, such as sensors read out, which usually operate at very low duty-cycles, This is due to low available DC voltages or power, with respect to the minimum requirements.

For this reason a PMU is typically adopted between the rectenna and the actual user [1], [2], to provide the rectenna with the maximum power point (MPP) tracking condition. But a traditional solution with single rectifier designed and optimized for a single specific load condition is not able to ensure minimum supply voltage of PMU. Therefore we propose a new rectifier design approach to reach this goal.

The scenario we refer to is schematically depicted in Figure 1.

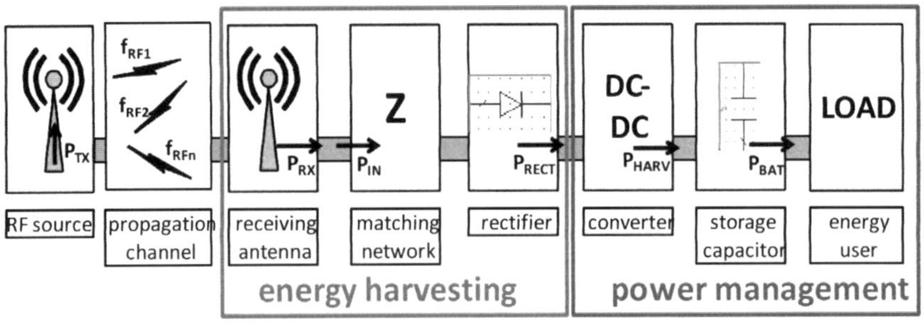

Figure 1. Block diagram of an energy autonomous system wirelessly powered by randomly distributed RF sources

For the entire system to be energetically autonomous, the PMU is equipped with a start-up stage, slow and less efficient, designed to consume few μW and to activate a more efficient boost converter stage with maximum power point tracking (MPPT) capabilities, with higher operating voltages and power consumptions requirements. These two stages provide two highly different loading conditions to the RF rectifier, namely the start-up impedance and the optimum load, whose operations require two different rectifier designs as well. In turn they represent two different source impedances for the base-band operation of the PMU, due to the nonlinear nature of the RF rectifiers. Therefore a joint RF/base-band design of the nonlinear rectifiers and of the two PMU sub-systems, is recommended to preserve the overall system efficiency maximization. The latter can be computed with reference to the schematic block diagram of Fig. 1, by considering the power quantities at the interconnecting sections:

$$\eta_{EH} = \eta_{TX-RX} \bullet \eta_{RX-DC} \bullet \eta_{DC-DC} = \frac{P_{RX}}{P_{TX}} \bullet \frac{P_{RECT}}{P_{RX}} \bullet \frac{P_{HARV}}{P_{RECT}} \quad (1)$$

The first factor, is dependent on the radio channel and is highly variable due to the unknown source location and to fading effects, especially significant in indoor environments. It can be accurately evaluated by resorting to the reciprocity theorem as in [3], but is not a concern of the present paper.

The second and third factors are strictly interdependent. In fact η_{RF-DC} is the RF-to-DC rectifier conversion efficiency with P_{RECT} being the RF power at the rectifier output port, which is strongly dependent on the rectifier load. The latter consists of the PMU input impedance, which varies whether the start-up or the boost stages are operating. Similarly the third factor in (1) is the DC-DC converter efficiency which is a function of the rectenna output impedance.

We thus design a parallel connection of two different rectifier/matching network assemblies to operate alternatively during the start-up and the boost converter operations.

2. The switched-load rectifier

The block representation of the new subsystem is shown in Fig. 2.

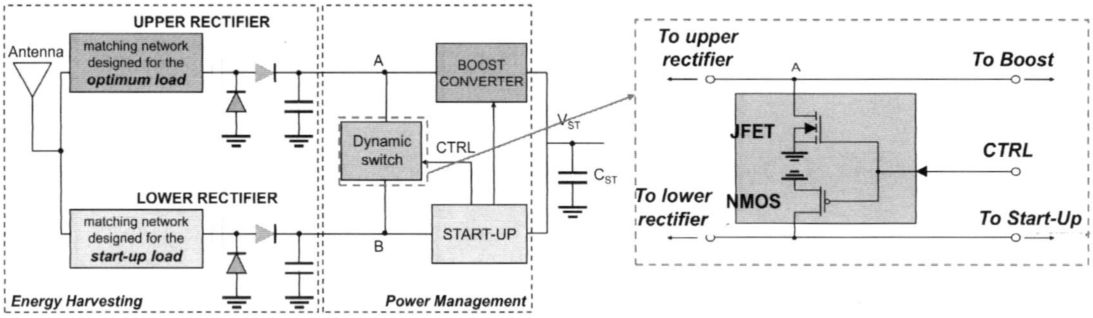

Figure 2. Block diagram of a switched-load rectenna

Two identical rectifiers, consisting of a traditional voltage doubler topology, are used in the two branches of the new proposed topology. One is designed together with the rectenna optimum load to reach the maximum RF-to-DC power conversion efficiency, thus allowing the DC-DC converter to dynamically provide the rectenna MPP condition. The other one is designed for a fixed load, namely the converter start-up circuit input impedance (≈ 200 kΩ), to provide the DC voltage and power needed for an autonomous start-up operation. The switching between the two arrangements is controlled at the load side by dynamically driving a couple of mutually exclusive field-effect transistors by acting on the CTRL signal (inset of Fig. 1). This way the selection of two separate antenna matching networks is possible: when CTRL is low, the lower branch is active and the upper one is shorted to ground; when CTRL is high the role of the branches is flipped. In this way the actual nonlinear behavior of the rectenna in the time-domain design of the PMU is rigorously accounted for. Similarly the rectenna regime in its actual loading conditions can be optimized.

In Fig. 3 the final layout of the optimized double branch matching network in microstrip technology is reported, with the two networks put into evidence: double-stub and single-stub solutions are adopted for the boost converter and the start-up branches, respectively.

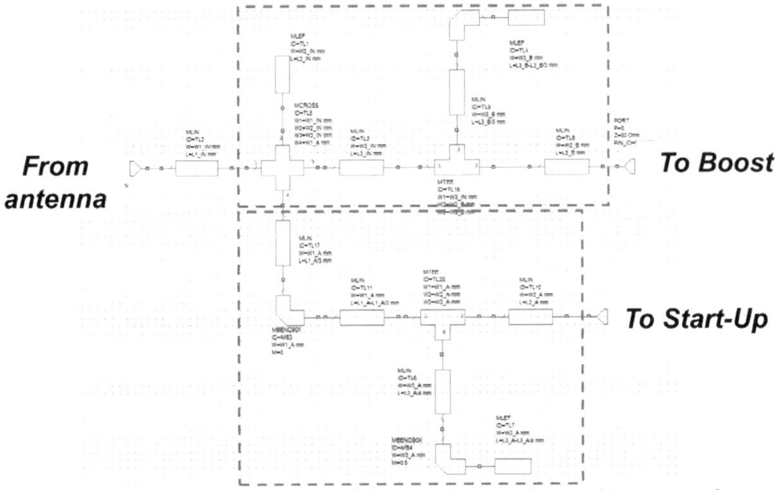

Figure 3. Distributed layout of the double-branch matching network.

2.1. Simulation results

The system operation is demonstrated by numerical simulation based on the nonlinear/electromagnetic (EM) co-design in the 900-MHz band, where RF sources are commonly available. Some significant results are plotted in Fig 4: the optimized rectenna efficiencies and output DC voltages as a function of the load current consumption, (for P_{RF} = -15 dBm) are plotted for the start-up (Fig. 4(a)) and boost (Fig. 4(b)) operating conditions: a higher DC voltage (> 1V) is reached when the start-up is loading; conversely at the MPP the best RF-to-DC efficiency is realized with the rectifier loaded by the boost, but the corresponding DC voltage is significantly lower (< 0.5 V).

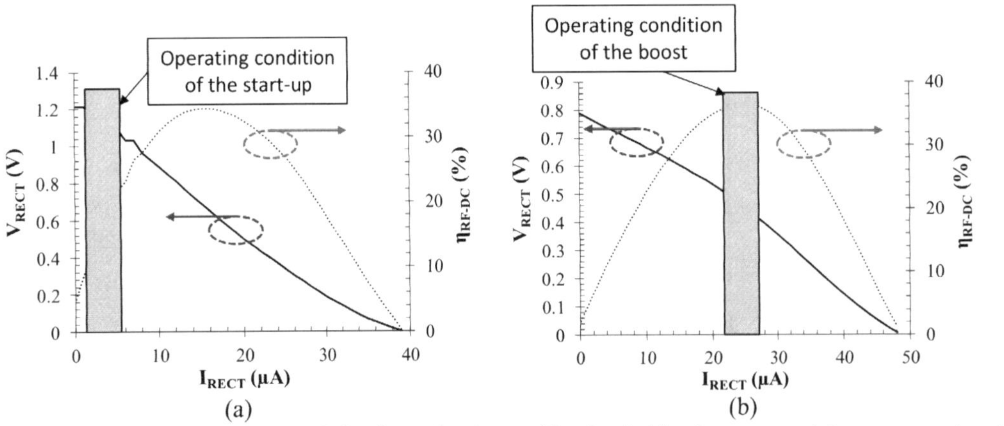

Figure 4. Efficiency and output DC voltage for the rectifier loaded by the start-up (almost open-circuit load) (a) and the boost (optimum load) (b).

This situation is of course strongly dependent on the level of the incoming RF signal (P_{RX}). In order to show this dependence results similar to the previous ones are plotted in Fig. 5, where the DC voltage (Fig. 5(a)) and the efficiency (Fig. 5(b)) are given as a function of P_{RX}, focusing on low power levels, typical of harvesting scenarios.

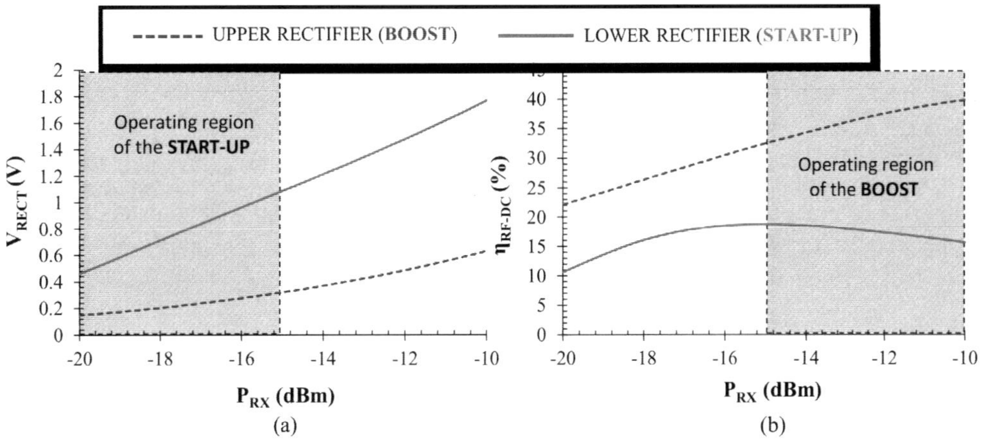

Figure 5. Output DC voltage (a) and efficiency (b) for the load-modulated rectenna, for different RF power levels.

The lower efficiency values pertaining to the start-up phase (active for low P_{RX} levels) are not a concern, since in this phase we are interested in having high DC voltage values (which are more than twice the corresponding values obtained in the boost phase again for low P_{RX} levels) (Fig. 5(a)). Once the wake-up of the system is guaranteed by the start-up, the convenience in switching to the boost phase is demonstrated by Fig. 5(b) where the higher efficiency of the boost rectifier is the desired target.

Finally, in Fig. 6, the effectiveness of the proposed approach is demonstrated by the time-domain simulation of the entire system loaded by a DC-DC converter, during the converter start-up phase. An input power P_{RX} = -15.5 Bm allows to reach an output voltage (V_{ST} of Fig. 2) equal to 2 V in a charging time of almost 20 s. This roughly corresponds to an average rectified power (lower than the corresponding instantaneous value P_{RECT}) of about 3μW, evaluated as the energy stored in the storage capacitor (C_{ST} of Fig. 2) of 33 μF in the charging period. These results show that by the present solution half the RF power needed by standard rectifier design (for optimum load) is required with the new two-branch solution.

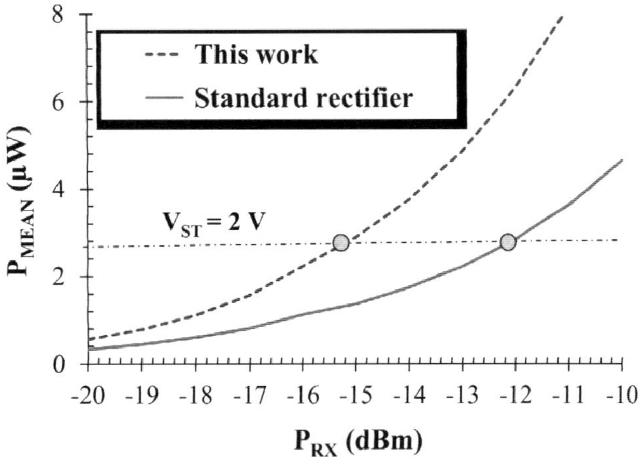

Figure 6. Converter wake-up operations

References

[1] T. Paing, E. A. Falkenstein, R. Zane, and Z. Popovic, “Custom IC for Ultralow Power RF Energy Scavenging,” IEEE Transactions on Power Electronics, vol. 26, no. 6, pp. 1620–1626, 2011.
[2] A. Costanzo, A. Romani, D. Masotti, N. Arbizzani, and V. Rizzoli, “RF/baseband co-design of switching receivers for multiband microwave energy harvesting,” Sensors and Actuators A: Physical, vol. 179, pp. 158–168, Jun. 2012.
[3] V. Rizzoli, A. Costanzo, D. Masotti, F. Donzelli, "Integration of numerical and field-theoretical techniques in the design of single- and multi-band rectennas for micro-power generation, " EuMA Int. J. of Microw. and Wirel. Tech. vol. 2, no. 3-4, pp. 293-303, July 2010.

Laccase/AuAg Hybrid Glucose Microfludic Fuel Cell

B. López-González[1], F.M. Cuevas-Muñiz[2], M. Guerra-Balcázar[+3], A. Déctor[2], N. Arjona[2], J. Ledesma-García*[3], L.G. Arriaga[2]

[1]División de Investigación y Posgrado, Facultad de Química, Universidad Autónoma de Querétaro

[2]Centro de Investigación y Desarrollo Tecnológico en Electroquímica Parque Tecnológico Querétaro Sanfandila, Pedro Escobedo 76703, Querétaro, Mexico

[3]División de Investigación y Posgrado, Facultad de Ingeniería, Universidad Autónoma de Querétaro, Cerro de las campanas s/n, 76010 Santiago de Querétaro, Mexico

[+]minbalca@yahoo.com.mx and *janet.ledesma@uaq.mx

Abstract. In this work a hybrid microfluidic fuel cell was fabricated and evaluated with a AuAg/C bimetallic material for the anode and an enzymatic cathode. The cathodic catalyst was prepared adsorbing laccase and ABTS on Vulcan carbon (Lac-ABTS/C). This material was characterized by FTIR-ATR, the results shows the presence of absorption bands corresponding to the amide bounds. The electrochemical evaluation for the materials consisted in cyclic voltammetry (CV). The glucose electrooxidation reaction in AuAg/C occurs around - 0.3 V vs. NHE. Both electrocatalytic materials were placed in a microfluidic fuel cell. The fuel cell was fed with PBS pH 5 oxygen saturated solution in the cathodic compartment and 5 mM glucose + 0.3 M KOH in the anodic side. Several polarization curves were performed and the maximum power density obtained was 0.3 $mWcm^{-2}$.

1. Introduction

The interest in micro devices has growing in the last years. Microfluidic fuel cells have emerged as an excellent option due to their high efficiencies and energy densities [1]. In this sense, microfluidic bio fuel cells are an attractive alternative, because they can operate with enzymatic catalysts. This particularity can be exploited for their use in medical devices as *ex vivo* or *in vivo* µTAS o bio MEMS [2]. In this sense, these kinds of devices can operate at high rates with high selectivity [3, 4]. Nevertheless, their main disadvantage is poor stability and low performance [5].

Hybrid biofuel cells, consisting in an enzymatic catalyst with an abiotic one which exhibit higher output power density due to a faster reaction kinetics and a long-term stability [5, 6] resulted from combining the best properties of both kind of catalysts. Gold-based inorganic materials exhibit good catalytic activity for glucose oxidation reaction (GOR). In specific, AuAg/C decrease the oxidation potential of glucose and the associated prime cost [7].

The aim of this work was to develop a hybrid microfluidic fuel cell that combines two of the best electrocatalysts, AuAg/C for the glucose electro-oxidation and Lac-ABTS/C for the oxygen reduction reaction, in order to obtain a microfluidic device with a good performance.

2. Experimental

2.1. Synthesis of catalytic materials

The cathodic material was prepared as follows: ABTS/C {2,2'-azinobis (3-ethylbenzothiazoline-6-sulfonate) ($ABTS2^-$) was prepared adsorbing 800 μL of 5 mM ABTS solution in phosphate buffer solution (PBS pH 5) and dried at room temperature for 24 hours. After that, 800 μL of laccase 1 U mL^{-1} solution was adsorbed on the ABTS/C, namely Lac/ABTS/C; this material was dried at room temperature.

On the other hand the AuAg/C was prepared according to previously reported procedure [6]. Briefly, 0.25 mM $HAuCl_4 \cdot 3H_2O$ and 0.063 mM $AgNO_3$ (J.T. Baker) aqueous solutions were added to tetra-octyl-ammonium bromide (TOAB, 98% Fluka) in toluene (99.98% Aldrich) as phase-transfer agent. Dodecanethiol (98%, Aldrich) was incorporated to this solution, followed by the addition of $NaBH_4$ in excess. The reaction was maintained at 2 °C for 2.5 h. Finally, a dark brown dispersion was obtained which was purified several times with ethanol (J.T.Baker) and then supported on XC-72 Vulcan carbon

2.2. Physicochemical characterization

The physicochemical characterization of AuAg/C material was carried out by X-ray diffraction (XRD) on an X-pert MPD Phillips diffractometer and TEM analyses which were conducted on a Jeol JSM – 35CF, transmission electron microscope (TEM) operated at 200 kV. In the case of Lac/ABTS/C, FTIR-ATR experiments were done in a Spectrum 100 (Perkin Elmer) in order to determinate the presence of the absorption signals of ABTS and laccase in the cathode. All the experiments were run at room temperature, 16 scans from 4000 to 650 cm^{-1} were performed with 4 cm^{-1} resolution for the cathodic catalyst.

2.3. Electrochemical characterization

Electrochemical experiments consisted in cyclic voltammetry (CV) done in a standard three-electrode glass cell using a BAS Epsilon Potenciostat/Galvanostat (Bioanalytical Systems). A Hg/HgO electrode (Radiometer®) and a Pt wire were used as reference and counter electrode respectively

2.4. Microfluidic hybrid fuel cell evaluation

Microfluidic fuel cell presented in this work consisted in the use of a PVC chip featuring adhesive aluminium electrode collectors and several sheets of polymeric materials. The microfluidic fuel cell geometry is identical to that reported in a previously reported work [7]. An image of the microfluidic fuel cell was showed on figure 1.

The device was assembled for characterization aligning the chip and the several layers of different laminate materials and placing them between two methacrylate pieces that are tighten by screws.

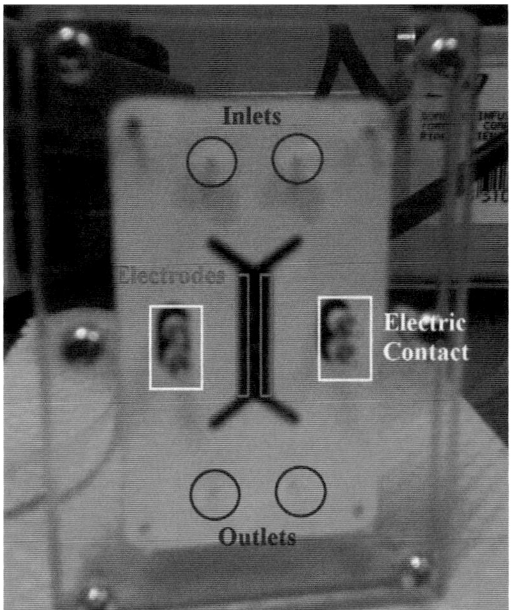

Figure 1. Image of the microfluidic fuel cell

Alkaline solution of 5mM glucose in 0.3M KOH which was previously saturated with nitrogen was passed through the anode with a flow rate of 25 μL min^{-1}. Cathodic compartment was fed with phosphate buffer pH 5 previously saturated with oxygen at the same flow rate. Polarization curves were obtained at 0.02 V s^{-1}.

3. Results and discussion

3.1. Physicochemical characterization

Infrared spectra for the cathodic material Lac/ABTS/C are shown in Figure 2. The composite Lac/ABTS/C presents characteristic peaks attributed to peptide bonds: Amide I (1600–1690cm^{-1}), Amide II (1480–1575 cm^{-1}) and Amide A (3300 cm^{-1}) which correspond to the enzyme [8]. The Lac/C material exhibited a decrease in transmittance compared to carbon due to presence of organic material adsorbed on the surface of Vulcan carbon. Peaks ascribed to shifts in ABTS was observed: SO_2 symmetrical elongation (1176 cm^{-1}), C = N (2260 – 2240 cm^{-1}) and CH aromatic (3050 – 3150 cm^{-1}). In the case of Vulcan, it is observed C-C and C-O peaks, indicating the presence of acid groups.

On the other hand, figure 3 shows the diffraction pattern and the TEM micrographs for AuAg/C electrocatalyst. The diffraction peaks located at 38.0, 44.2, 64.4, 77.3 and 82° corresponds to the characteristic FCC metals (1 1 1), (2 0 0), (2 2 0) (3 1 1) and (2 2 2) crystallographic planes. The average particle size obtained by TEM is around10 nm.

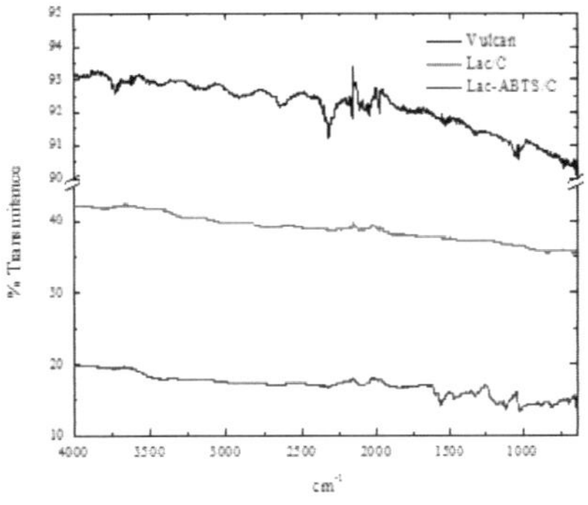

Figure 2 FTIR – ATR spectra of Vulcan, Lac/C, and Lac/ABTS/C.

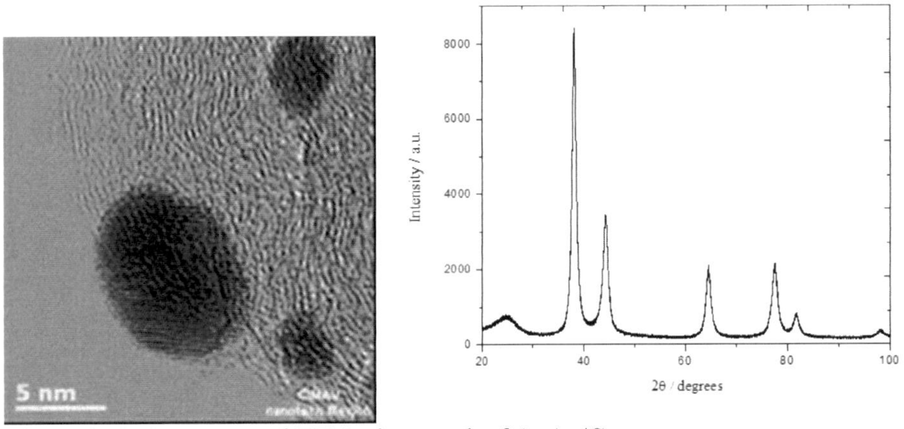

Figure 3 XRD pattern and TEM micrograph of AuAg/C

3.2. Electrochemical evaluation

The electrochemical behaviour towards glucose electrooxidation reaction is presented in Figure 4. In absence of glucose is possible to observe the typical profile of this material on basic media. The AuAg/C material shows an onset potential for the glucose electrooxidation reaction at -0.45 V, this value is according to already reported results [7].

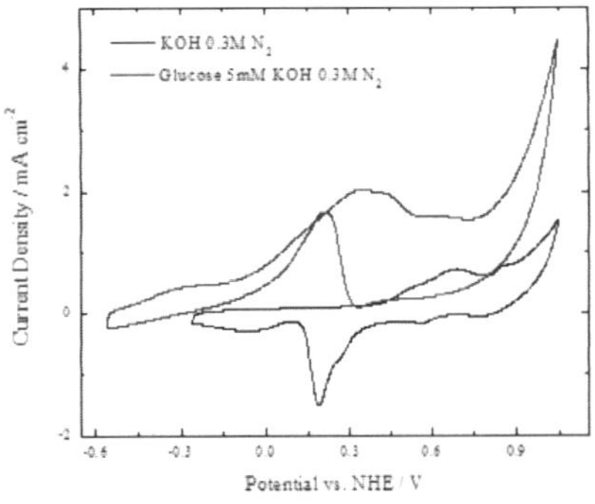

Figure 4 Cyclic voltamograms of AuAg/C in absence and presence of glucose 5 mM in basic media.

3.3. Hybrid fuel cell evaluation

Figure 5 shows the polarization and power density curve for the microfluidic hybrid fuel cell. The OCP was around 1.2 V and the maximum current density was 0.3 mWcm^{-2}.

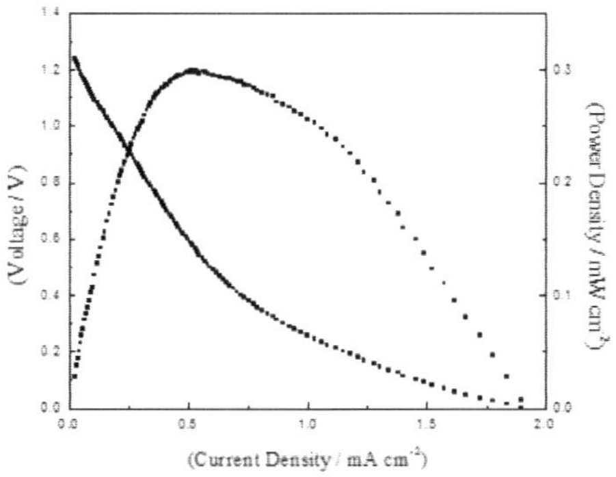

Figure 5 Polarization and power density curves for the microfluidic hybrid fuel cell.

4. Conclusions

Laccase and ABTS were absorbed in Vulcan carbon successfully. This material was evaluated in an alkaline-neutral hybrid fuel cell harvesting 0.3 mW cm^{-2}. The use of two different kinds of catalysts allows reach highest power with respect to conventional enzymatic fuel cells.

Acknowledges

The authors gratefully acknowledge the financial support of SEP-CONACYT (Grant 179921), SEP-CONACYT (Grant 133310), and ANR CONACYT 2011(Grant 163114).

References

[1]J.P. Esquivel, T. Senn, P. Hernández-Fernández, J. Santander, M. Lörgen, S. Rojas, B. Löchel, C. Cané, N. Sabaté, J. PowerSources, 195 (2010) 8110–8115.
[2]A. Zebda, L. Renaud, M. Cretin, C. Innocent, R. Ferrigno, S. Tingry, Sens. Act. B, 149 (2010) 44–50.
[3]A. Habrioux, G. Merle, K. Servat, K.B. Kokoh, C. Innocent, M. Cretin et S. Tingry, J. of Electroanal. Chem., 622 (2008) 97-102.
[4]D. Wen, X. Xuaet S. Dong, Energy Environ. Sci., 4 (2011) 1358.
[5]Y. Choi, G. Wang, M. H. Nayfeh, S. T. Yau, Biosens. Bioelectron., 24 (2009) 3103–3107.
[6]F. M. Cuevas-Muñiz, M. Guerra-Balcázar, F. Castaneda, J. Ledesma-García, L. G. Arriaga, J. Power Sources, 196 (2011) 5853-5857.
[7]M. J. González-Guerrero, J. P. Esquivel, D. Sánchez-Molas, P. Godignon, F. X. Muñoz, F. J. del Campo, F. Giroud, S. D. Minteer, N. Sabaté, Lab Chip, 13 (2013) 2972-2979.
[7]J. Kong and S. Yu, ActaBioch. Bioph.Sin., 39(2007) 549–559.

Advanced combinational microfluidic multiplexer for fuel cell reactors

D W Lee,[1,†,‡] I Doh,[2,†] Y Kim,[1] Y –H Cho[1]

[1]Korea Advanced Institute of Science and Technology, Daejeon, Republic of Korea
[2]Korea Research Institute of Standards and Science, Daejeon, Republic of Korea
[†]Both authors contributed equally to this work
[‡]Present Address: Samsung Electromechanics Co., Ltd., Suwon, Korea

E-mail: nanosys@kaist.ac.kr

Abstract. An advanced combinational microfluidic multiplexer capable to address multiple fluidic channels for fuel cell reactors is proposed. Using only 4 control lines and two different levels of control pressures, the proposed multiplexer addresses up to 19 fluidic channels, at least two times larger than the previous microfluidic multiplexers. The present multiplexer providing high control efficiency and simple structure for channel addressing would be used in the application areas of the integrated microfluidic systems such as fuel cell reactors and dynamic pressure generators.

1. Introduction

The micro fuel cell system requires simple and effective fluidic control system for the optimal amount of fuel delivery into multiple reaction chambers. Without effective fluidic control, the micro fuel cell requires numerous fluidic components in proportional to the number of the chambers, thus increasing costs for fuel cell systems. The effective microfluidic control system is essential to develop cheaper and more reliable micro fuel cell system.

Several microfluidic multiplexers for channel addressing have been proposed. The binary multiplexers analogic to electronic multiplexers are based on digital operation of pneumatic valves. They addressed up to $2^{n/2}$ channels using n control lines [1]. Previously, we had reported a high-radix microfluidic multiplexer based on the ternary or quaternary operation of pneumatic valves [2]. The valves in the high-radix multiplexers have different threshold pressures according to the membrane areas, thus providing high-radix operation. The ternary and quaternary multiplexers were capable to address $3^{n/2}$ and $4^{n/2}$ channels, using n control lines, respectively.

In this work, the proposed advanced combinational multiplexer based on the pressure combination in the control lines. By using 4 control lines, the present advanced combinational multiplexer can address up to 19 fluidic channels, at least three times larger than the previous binary and combinational multiplexers capable of only 4 and 6 channels, respectively. Even the ternary multiplexer was able to address 9 channels, a half of the proposed multiplexer using the same control lines. The present advanced combinational multiplexer having the high efficiency and simple structure can be widely used in the application areas of the integrated microfluidic systems such as fuel cell reactors and dynamic pressure generators.

2. Working Principle and Design

The present multiplexer is composed of control, membrane and channel layer (Fig. 1). The pneumatic pressure supplied from an external pressure source is delivered to membrane and the membrane deforms to the bottom channels by acting like pneumatic valves. The threshold pressure for valves is the minimum control pressure required to block the channel completely. Since a small membrane is hard to deform, it requires higher threshold pressure than larger membrane as shown in Fig. 1.

As an example for working principle of the advanced combinational multiplexer, two different pressures (50 and 100 kPa) and four control lines (C_1, C_2, C_3, C_4) are used as shown in Fig.2. Single inlet is branched to 19 channels and, in each control line, pneumatic valves having 50 and 100 kPa of threshold pressures ($V_{50\ kPa}$ and $V_{100\ kPa}$) are located. Among the total eight options (two different levels of pressures and four control lines), the selection of 4 maximizes the number of the pressure combinations and consequently the number of addressable fluidic channels [3].

Unfortunately, all selections of control pressures are not independent. When the control pressure of 100 kPa is selected in a control line, both $V_{100\ kPa}$ and $V_{50\ kPa}$ should be selected in the control line, because 100 kPa is higher pressure and covers the lower range of pressure. Therefore, selection of 100 kPa in a control line includes the selection of both $V_{50\ kPa}$ and $V_{100\ kPa}$ in the control line. All the channels having $V_{50\ kPa}$ or $V_{100\ kPa}$ in the control line are blocked. When we select two control lines for 100 kPa, all 4 options were chosen, since both $V_{50\ kPa}$ and $V_{100\ kPa}$ in two lines are selected. By

Figure 1. Cross-sectional view of pneumatic valves in the multiplexer, where $V_{100\ kPa}$ valve having smaller membrane area than $V_{50\ kPa}$ has higher threshold pressure for valve operation.

a)

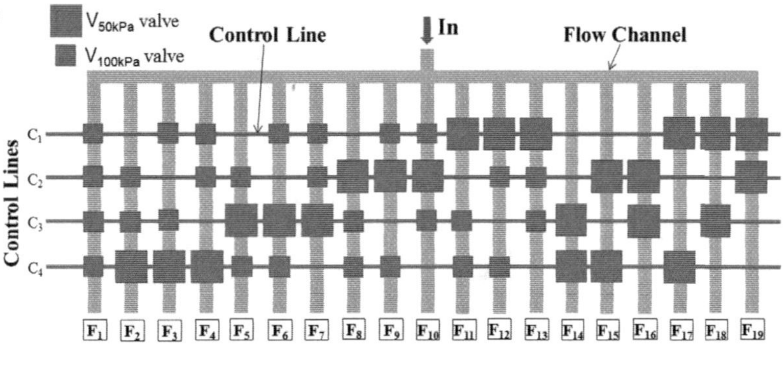

b)

C_1	50	100	50	50	100	50	50	100	50	50	-	-	-	100	100	100	-	-	-
C_2	50	50	100	50	50	100	50	-	-	-	100	50	50	100	-	-	100	100	-
C_3	50	50	50	100	-	-	-	50	100	50	50	100	50	-	100	-	100	-	100
C_4	50	-	-	-	50	50	100	50	50	100	50	50	100	-	-	100	-	100	100
	Case 1 (a_1=4, a_2=0)	Case 2 (a_1=2, a_2=1)												Case 3 (a_1=0, a_2=2)					

Figure 2. Schematic view of the present multiplexer capable to address 19 fluidic channels using 4 control lines: (a) Top view of the multiplexer; (b) Pressure combinations (kPa) in 4 control lines according to fluidic channels.

considering this dependency, the combination of four option selection needs to be divided into the following three cases,

Case 1) Four control lines are 50 kPa
All four control lines are selected as 50 kPa. The number of control pressure combination is ${}_4C_4 = 1$.

Case 2) One control line is 100 kPa, two control lines are 50 kPa
Due to the dependency of control pressure as we explained above, 100 kPa in a control line requires the selection of 50 kPa in the same control line. Once we choose one among four control lines for 100 kPa, 50 kPa is automatically selected and we lose one option. So cases are to choose two control lines among the four control lines for 50 kPa (${}_4C_2$) and choose one control lines among the remaining two (${}_2C_1$). Therefore, overall number of pressure combinations is ${}_4C_2 \times {}_2C_1 = 12$.

Case 3) Two control lines are 100 kPa
In this case, we need two control lines with 100 kPa. Each 100 kPa require 50 kPa in the same control line and no other options are left. The pressure combination is to choose two among four control lines, ${}_4C_2 = 6$.

Totally, the present advanced combinational multiplexer using 4 control lines can address 19 fluidic channels:

$$F = {}_4C_4 + ({}_4C_2 \times {}_2C_1) + {}_4C_2 = 19$$

3. Experimental Results

The device was fabricated by multi-layer soft lithography using Polydimethylsiloxane (PDMS). The SU-8 (Microchem. Co.) negative photoresist was patterned on a silicon wafer by UV-lithography process using manufacturer's recommendation and used as a mold master for soft lithography [2]. Figure 3 shows the fabricated device with an enlarged view of advanced combination multiplexer. In order to facilitate the visualization of channel flow, DI water was dyed by ink and injected to the channel layer in the multiplexer. There were no leaks of sample at the valves over time during the entire experiments.

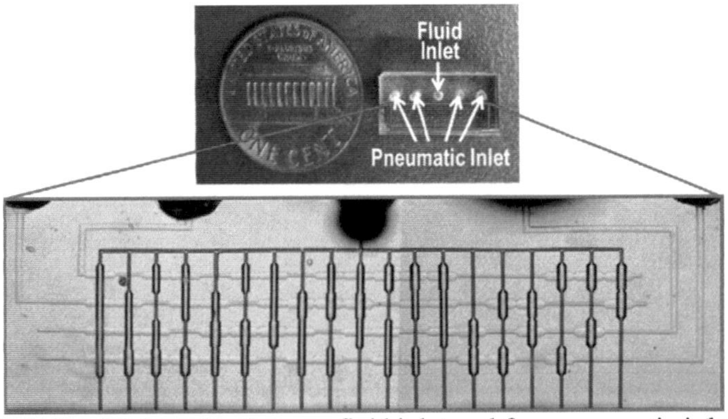

Figure 3. The fabricated multiplexer having one fluid inlet and four pneumatic inlets for flow control, with an enlarged view of the control lines and valve combination.

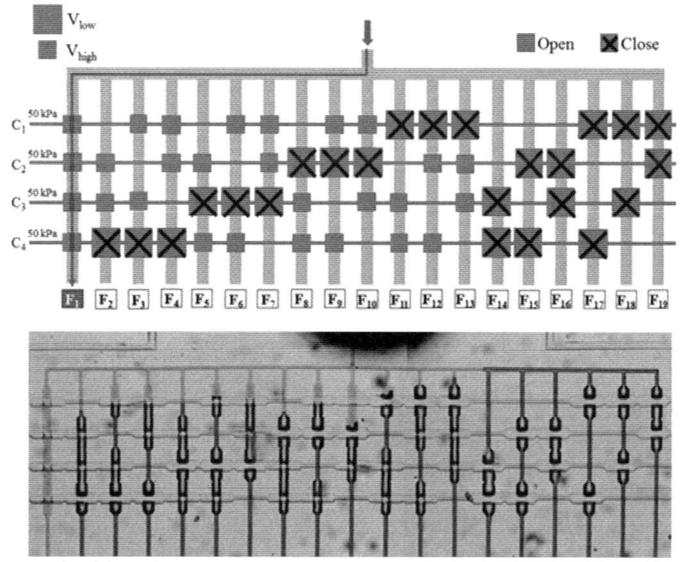

Figure 4. Fluidic channel addressing using the advanced combinational multiplexer: (a) F1 channel addressing where all control lines were pressurized to 50 kPa; (b) F18 channel addressing where only two control lines were pressurized to 100 kPa.

When we apply control pressure of 50 kPa in all control lines, all $V_{50\,kPa}$ were closed, as shown in Fig.4. For example, F_2~F_4, F_{14}, F_{15}, F_{17} channels were blocked by $V_{50\,kPa}$ valves in the control line C_4. C_3 blocked F_5~F_7, F_{16}, F_{18}. The remaining F_8~F_{13}, F_{19} were also blocked by C_1 or C_2. Since F_1 was the only channel having $V_{100\,kPa}$ in all control lines, it was the only channel which was not blocked by the control pressure of 50 kPa in all control lines. Therefore, the solution was supplied from the inlet flowed through F_1.

4. Conclusion

As a proof-of-concept, we had verified the advanced combinational multiplexer could address up to 19 fluidic channels using four control lines, which is the largest number of channels so far. The present advanced combinational multiplexer having the high efficiency and simple structure will be used in the application areas of the integrated microfluidic systems.

Acknowledgements

This research was supported by the Converging Research Center Program funded by the Ministry of Education, Science and Technology (Project No. 2013K000273), and by the Korea Research Institute of Standards and Science (Project Grant 13011045).

References

[1] Unger M A, Chou H-P, Thorsen T, Scherer A and Quake S R 2000 Monolithic Microfabricated Valves and Pumps by Multilayer Soft Lithography *Science (80-.).* **288** 113–6

[2] Lee D W and Cho Y-H 2009 High-radix microfluidic multiplexer with pressure valves of different thresholds. *Lab Chip* **9** 1681–6

[3] Hua Z, Xia Y and Srivannavit O 2006 A versatile microreactor platform featuring a chemical-resistant microvalve array for addressable multiplex syntheses and assays *J. Micromechanics Microengineering* **16** 1433–43

Energy Autonomous Wireless Water Meter with Integrated Turbine Driven Energy Harvester

P. Becker[1,3], B. Folkmer[1], R. Goepfert[1], D. Hoffmann[1], A. Willmann[1] and Y. Manoli[1,2]

[1]HSG-IMIT Institute of Micromachining and Information Technology, Villingen-Schwenningen, Germany,
[2]Fritz Huettinger Chair of Microelectronics, Department of Microsystems Engineering IMTEK, University of Freiburg, Germany

Philipp.becker@hsg-imit.de

Abstract. Accurate meter reading is the fundamental task of the home water system for the handling of payments. Meters need to be read correctly, to avoid an effect of adding events that increase unnecessary cost and create customer dissatisfaction. This paper presents a fully integrated wireless, energy autonomous water metering system based on the European Standard EN 13757 "Communication systems for meters and remote reading of meters". The system can be used in multiple water metering scenarios. No maintenance will be required and the system will provide precise and secure data transmission as well as timely and accurate recording of the consumption of water. The identification of any leakages will be improved through the analysis of the actual quantity supplied and recorded by the meters. The system is powered by an energy harvester, based on a water driven turbine wheel that is directly coupled to an electromagnetic energy transducer. The power delivered by the generator is dependent of the amount of flowing water and the pressure in the water pipes. Therefor the power is commonly non-continuous, fluctuant and unstable in the voltage amplitude. To be able to report the meter readings at all times, the system needs to be powered not only in times when the energy harvester delivers energy. Therefor an energy buffer, that stores the harvested energy, is installed to compensate the energy requirement between the actual generator output and the energy consumption of the application. Besides a complete system overview, the presentation will focus on the power management and energy aware battery charging circuitry. The design, fabrication, measuring results and the preparations for field tests in rural and urban environment will be presented and discussed.

1. Introduction

Energy efficiency is one of the major issues for wireless metering applications. A connection to the electrical grid can't be provided in all cases. Therefor the energy consumed by the gas, water and heat meters can be delivered by batteries. As a matter of course the lifetime of the batteries should be as long as the maintenance or calibration cycle of the meter, which typically is between 8 and 12 years and depends on the legal conditions for the specific medium and the specific country. This lifetime is hard to achieve for meters with regular read-out cycles. For these, energy harvesting can be a

[3] To whom any correspondence should be addressed.

significant advancement to extend the battery's lifetime. Therefore, the WiMBex project ("Remote Wireless Water Meter Reading Solution Based on the 13757 Standard, Providing High Autonomy, Interoperability and Range") [1] pursues the development of an automatic water meter reading (AWMR) technology. The presented system is based on a cost efficient integrated energy harvesting system powered by the available water flow. Eight companies entered this venture together from December 2011 to November 2013. During a two-year period, the WiMBex consortium aims to develop a remote wireless water meter reading solution, based on the EN13757 Standard, providing high autonomy, interoperability and range. To achieve this goal, the five companies (CAS, JCB, MIB, LBS and MHWS) guide the three research centres (CRIC, HSO and HSG-IMIT) [2].
This paper focuses on the system concept and power management. In addition, the authors present a separate paper which describes the rotational, radial flux energy harvester.

2. System Overview

A system overview of the complete wireless meter device, which is powered by a turbine wheel driven electromagnetic energy harvester is shown in figure 1. The power delivered by the energy harvester is dependent of the amount of flowing water and the pressure in the water pipes. Therefore the power is commonly non-continuous, fluctuant and unstable in the voltage amplitude. Because of this unpredictable nature of the generator an energy buffer is installed, which compensates the energy requirement between the actual generator output and the energy consumption of the application. In addition an efficient rectification-, voltage conversion- and battery charging interface circuit is added to the system. The complete power supply unit is shown in figure 2.

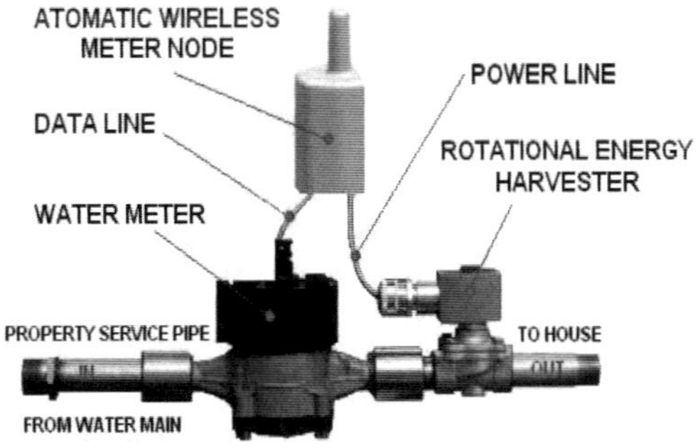

Figure 1. System overview of the wireless, energy autonomous water metering system (WiMBex).

Figure 2. Integrated energy harvesting unit with power management PCB.

2.1. Power Supply

A schematic block diagram of the power supply unit is shown in figure 3. The system is completely self-powered and able to work without any pre charging of the system.

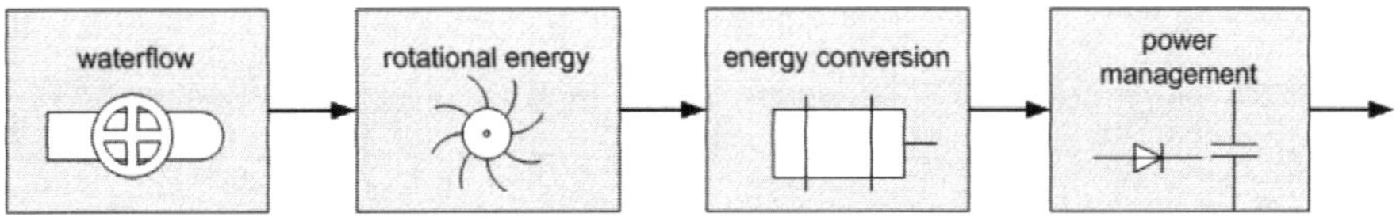

Figure 3. Block diagram of the energy harvester power unit elements.

The power supply unit can be described by three functional sections: the fluid section, the electromagnetic energy conversion section and the power management section.

2.1.1. Fluid section: The kinetic energy is transferred from the flowing water to the energy harvester by using commercial impeller device housing. Thus the manufacturing cost of the energy harvesting unit can be kept at the lowest possible level. To convert the kinetic energy of the water flow, a new impeller wheel design is implemented to guarantee a better transmission of the energy.

2.1.2. Electromagnetic energy conversion: An electromagnetic energy transducer is used to convert the kinetic energy of the impeller wheel into electrical energy. The transducer consists of three induction coils connected in a star circuit and a two-pole ring-magnet. Thus, a three-phase voltage and current will be induced in the coils. Assuming a fixed load resistance, the power output increases with the rotating speed of the induction magnet, as a function of the water flow rate.

2.1.3. Power management: The power management circuit has to meet different requirements. These requirements are efficient voltage rectification, voltage conversion and stabilization, protection against deep discharge and high voltage peaks, monitoring of the energy balance and efficient charging of the energy storage. *A lithium polymer secondary battery (LiPo) is used as main energy storage of the system.* Figure 4 shows the different parts of the power management pictured in a block diagram.

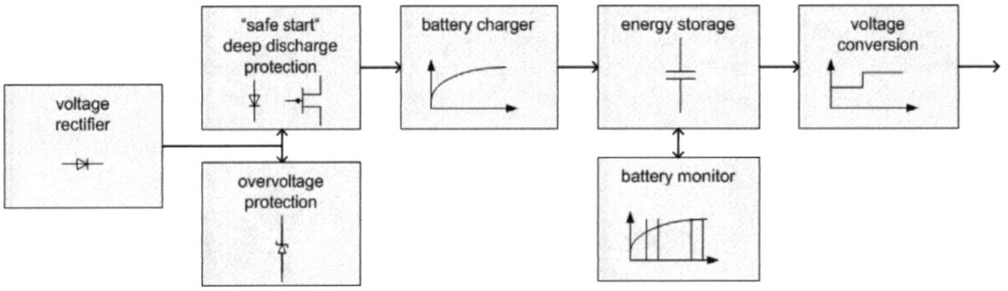

Figure 4. Power Management with its different function blocks.

The different function blocks of the power management are designed with regard to low energy consumption. Therefore all of the blocks are able to act in "active" or "sleep" modes.
A "safe start" function is implemented to the system to ensure a secure start-up of the system at first-use without any initial energy stored at the energy storage. The converted energy provided by the electromagnetic energy transducer will be stored in ceramic capacitors until a defined start voltage level is reached. After this level is reached the battery charger is connected to the ceramic capacitors and the energy transfer to the storage cell is started. Once the start voltage level is reached, the "safe start" circuit allows the voltage to decline even below the defined start level in order to transfer the complete converted energy to the storage cell.

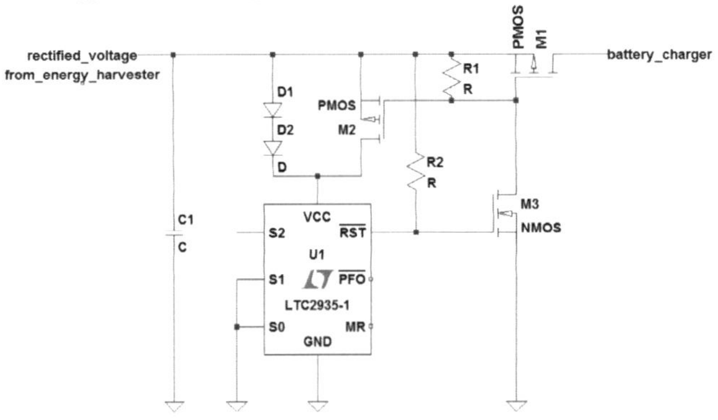

Figure 5. "safe start" circuit: A voltage supervisor is used to supervise and control the energy flow to the battery charger.

2.2. Wireless M-Bus

WiMBex exploits the powerful new features of the P- and Q-mode of the Wireless M-Bus standard and in this manner, extend the use and impact of the EN 13757 [4]. By introducing a new network Q-mode protocol which enables precise network time synchronization the high power requirements typically associated with multi hop wireless networks can be significantly reduced by a time division multiple access (TDMA) MAC protocol and an efficient energy aware routing protocol. WiMBex accommodates the new part of the standard, which specifies operation in the newly available 169 MHz band, and which will enable a radio range of 100+ meters (even in the presence of physical barriers). This is a great advantage compared to the vast majority of existing solutions, which operate at 868 MHz or 2.4 GHz. This approach will reduce the need for repeater nodes and their associated technical and commercial disadvantages. WiMBex supports the DLMS standard suite to provide an open, manufacturer independent, interoperable, device and interface specification for electricity, gas, heat and water meters.

3. Measurement results

Measurement results can be seen in figure 6 and 7. The presented graphs show the over-all system balance, since the measurement results include the complete energy consumption of the active electronic circuitries. Figure 6 shows the power measured at the input of the LiPo. The graph reveals the linear behaviour and dependency between the generated power and the applied water flow rate. Figure 6 displays the input voltage, current and power in comparison to the voltage, current and power applied to the rechargeable battery. A maximum power transfer of 310mW with an efficiency of 83% can be achieved.

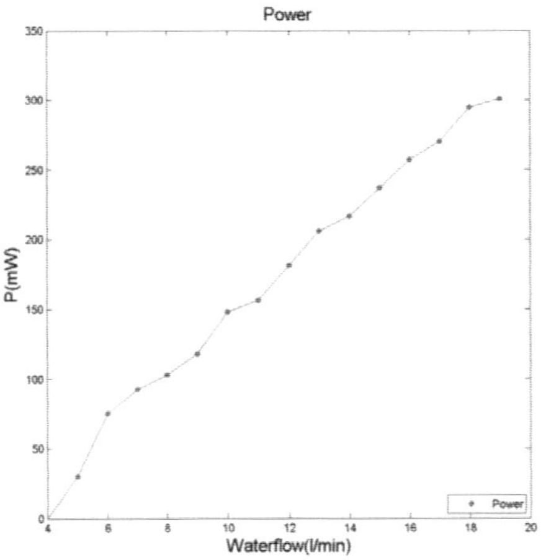

Figure 6. Power transfer to Lithium Polymer rechargeable battery in dependency of the water flow rate.

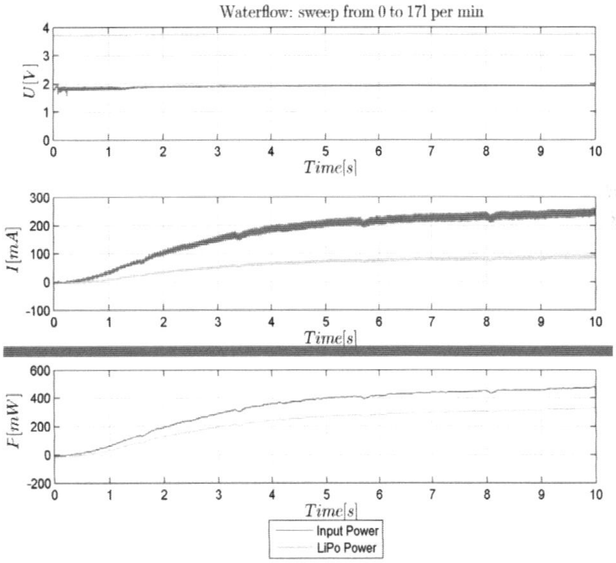

Figure 7. Charging process. A sweep of the water flow from 0 to 17l per minute is shown. a) Input voltage level and voltage level of the rechargeable LiPo battery. b) Input current und current flow into the LiPo battery. c) Input power and power transferred to the LiPo battery.

4. Conclusion and Future Work

The measurement results show that energy harvesting systems are capable to be applied also in real-life industrial applications. A maximum power transfer of 310mW with an efficiency of 83% was achieved. Real cost advantages are promised, which leads to a good interest from the side of commercial companies. It is expected that especially for the water metering application, energy harvesting approaches will become a new important feature.

At the time of submission of this paper a field test validation takes place in the water distribution network of WiMBex project partner Meath Hill Water Scheme (MHWS). MHWS is a rural water utility located on County Meath (Ireland). The water meters are installed in underground boundary boxes near the road (figure 8). The wireless motes and the 10 energy harvesters will be placed inside the boxes and will be connected to a meter, a Sensus 620, through a pulse sensor wire. First measurement results will be available in December 2013.

Figure 8. Boundary box and water meter for the field test validation.

5. Acknowledgement

The authors would like to thank all project partners for the excellent cooperation. The project is co-funded by the European Commission under reference number FP7-SME-2011-286753.

References

[1] www.wimbex.com
[2] www.wimbex.com/wimbex-team
[3] E. Bischur , S. Pobering , M. Menacher , N. Schwesinger 2010, *Active and Passive Smart Structures and Integrated Systems*, Proc. SPIE 7643
[4] *Communication systems for meters and remote reading of meters*; EN-13757: Part 1: Data exchange, Part 2: Physical and link layer, Part 3: Dedicated application layer, Part 4: Wireless meter readout, Part 5: Relaying, Part 6: Local Bus.

Design, analysis and testing of a piezoelectric flex transducer for harvesting bio-kinetic energy

A Daniels[1], M Zhu[2*] and A Tiwari[1]

[1] Department of Manufacturing and Materials, Cranfield University, Bedfordshire, MK43 0AL, UK.
[2] College of Engineering, Mathematics and physical sciences, Exeter University, Exeter, EX4 4QF, UK.

*E-mail: m.zhu@exeter.ac.uk

Abstract. The increasing prevalence of low power consuming electronics brings greater potential to mobile energy harvesting devices as a possible power source. A new piezoelectric energy harvesting device, called the piezoelectric flex transducer (PFT), is presented and developed. A Finite Element Model (FEM) was developed to design and analyse the PFT. The PFT consists of a piezoelectric element sandwiched between substrate layers and metal endcaps that are able to amplify the axial force on the piezoelectric element. Based on the concept of the Cymbal transducer, the PFT can withstand higher forces, was retrofitted into a shoe and used to power a wireless sensor module whilst the subject with a body weight of 760N was wearing the shoe and ran at 3.1mph (1.4HZ on shoe), the PFT produced an average maximum power of 2.5mW (over 2MΩ load).

1. Introduction

Based on the concept of the cymbal device [1], the piezoelectric flex transducer (PFT) is composed of top and bottom endcaps with a central piezoelectric element. Unlike the cymbal, the PFT has additional substrate layers which enable the device to be scaled up and therefore withstand forces of up to 1kN. This is because the substrate layers can increase the adhesion contact area, therefore reducing delimitation risks. The PFT has rectangular shape for an improved packing factor.

The value of mobile energy harvesting is in the benefits of autonomy, reduced monetary and environmental costs. The power output of bio-kinetic mobile energy harvesting devices is an important factor for applications so we take a system level design approach to increase the power output and decrease power consumption at every level. This paper presents the design, analysis, fabrication and testing of the PFT. The energy generated by the PFT in gait (bio-kinetic energy) when retrofitted into a shoe was demonstrated to power a custom developed wireless sensor module [2].

2. Piezoelectric Flex Transducer (PFT)

The PFT consists of a piezoelectric element sandwiched between substrate layers and two metal endcaps, shown in figure 1. The stainless steel endcaps have two main functions: 1) to amplify and redirect the downward applied mechanical force to a horizontal force so that the majority of the

* Author to whom any correspondence should be addressed

available force acts to stretch piezo-element horizontally, the converse action happens when the force is removed and the PFT tends back to a physically neutral state, and 2) to suspend the piezo-element and substrate so that they can actuate more freely, generally speaking a higher straining corresponds to a higher charge produced.

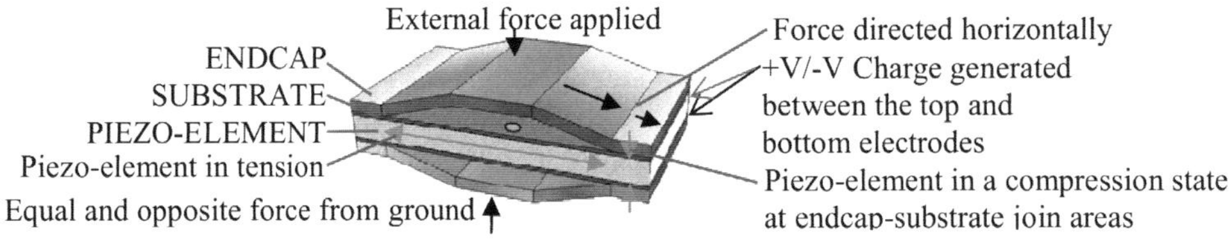

Figure 1. Function of the PFT when a downward compressive force is applied.

The endcaps and substrates were made from stainless steel by Rye-Guard Ltd (Bedfordshire, UK), whereby sheet metal bending was used to form the endcaps. The piezoelectric elements with silver fired electrodes were tailor-made from DL-35HD material by DeL Piezo Specialties, LLC (Florida, USA). Following the welding together of the substrates and endcaps by The Welding Institute (Cranfield University, UK) surfaces were textured and a 3M Scotch-Weld™ DP-460 structural adhesive was spread on the substrate-piezoelectric interfaces and baked (approx. 50 minutes at 90°C). A clamp was used to hold the device together whilst baking. Wire connections to the electrodes were made through pre-drilled 2mm diameter holes in the substrates. Two types of endcap-substrate weld methods were experimented with; laser and resistance spot welding, both performed similarly, however, fatigue analysis was not performed.

3. Design analysis

ANSYS (Version 13), a multi-physics finite element analysis (FEA) software package, was used to develop a coupled piezoelectric-circuit finite element model (CPC-FEM) of the PFT for design analyses. The CPC-FEM is comprised of the 3-dimensional structure of the PFT with a load resistor connected in parallel between the 'electrodes' of the piezo-element, shown in figure 2. Running a harmonic analysis the CPC-FEM has the ability to calculate the electrical output of the PFT whilst it was subjected to an external force excitation, such possible outputs were: the peak voltage, current and power. Static analyses were also carried out to establish the mechanical stress of the PFT.

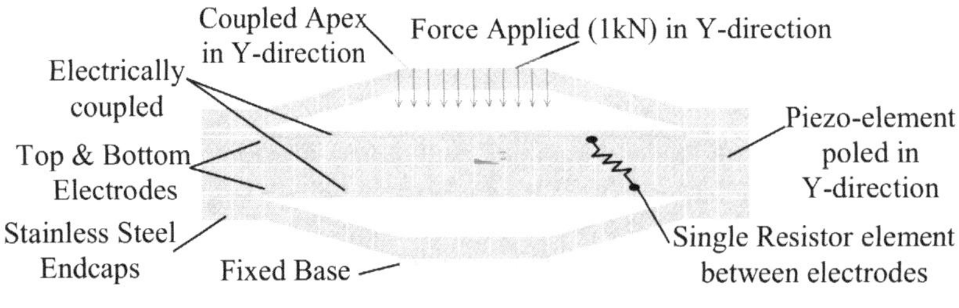

Figure 2. PFT as seen in CPC-FEM with boundary conditions and electrical resistor connected in parallel between top and bottom electrically coupled surfaces.

The element type used for the piezo-element was SOLID226. SOLID226 is a couple-field element type, capable of analysing piezoelectric structural and electrical performance. Secondly, SOLID95 was used for the endcaps. Finally, CIRCU94 was used for the resistor element connected between the piezo-element electrically coupled node surfaces. The PFT was modelled with the lower endcap fixed and a universally distributed load applied to the upper surface of the top endcap. The force applied to the top endcap was 1kN, which was an approximation of what an average male might exert on the ground whilst in gait. Because electrode and adhesion layers were very thin, they were ignored in the model. Stainless steel input properties used in the CPC-FEM were: Young's modulus 193GPa, density 8030 kg m^{-3} and Poisson's ratio 0.24.

Table 1. Dimensions of PFT in CPC-FEM and in experiment.

Total Length	D	52mm	Endcap Thickness	t_c	2mm
Total Width	D_w	30mm	Join Length	J	6mm
Piezo Thickness	t_p	4mm	Cavity Length	D_c	40mm
Apex Length	D_a	14mm	Angle of the Endcap	Θ	8.75°

Table 2. Piezoelectric material properties used in CPC-FEM.

Piezoelectric Material: DeL Piezo DL-53HD						
Elastic Compliance ($\times 10^{-12}$ m^2 N^{-1})		Piezoelectric strain constants ($\times 10^{-12}$ m V^{-1})	Relative Dielectric Constant (constant stress)	Density (Kg m^{-3})	Piezoelectric Coupling Coefficient	Mechanical quality factor
s_{11} 15.1	s_{33} 24.8	d_{15} 810	ε_{11}^{T} 3550	ρ 7900	k_{31} 0.4	Q 20
s_{12} -4.5	s_{44} 37.1	d_{31} -300	ε_{33}^{T} 3850			
s_{13} -9.4	s_{66} 39.2	d_{33} 680				

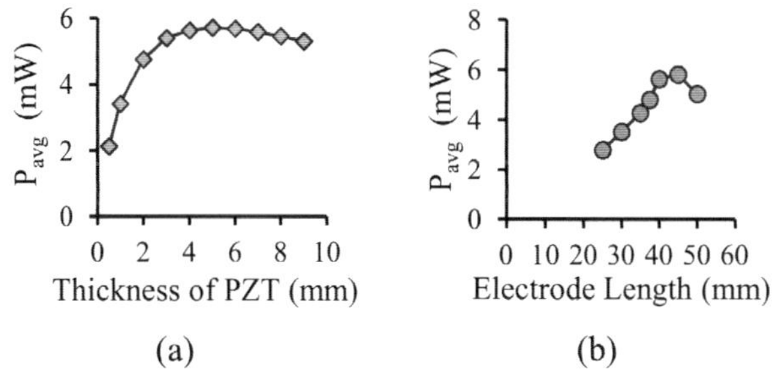

Figure 3. (a) and (b) Power output of the PFT for different piezo-element thicknesses and different electrode lengths along piezo-element, respectively. The PFT was subject to a 1kN force at 2Hz. For each point on the graphs the optimal resistance was used.

In figure 3 (a) the power output increased with an increase in piezo-thickness for a PFT under a 1kN force load at 2Hz. When the thickness of the piezo-element was increased, the capacitance decreased and the internal resistance increased. This resulted in a variation of voltage and current produced, where by an increased thickness caused an increased voltage but a decreased current. When the electrode area coverage was varied across the piezo-element top and bottom region, figure 3 (b), results showed that the most effective coverage, 40mm, was one that excluded the clamped region (join length) of the piezo-element. Piezoelectric material in the join length was effectively clamped and thus contributed negligible charge when the PFT was in use. It was thought the reason why the partial electrode fared better was that if the partial electrode PFT produced the same charge as a full electrode PFT but with a reduced capacitance—due to its smaller electrodes—it must have an increased voltage and thus also an increased power output.

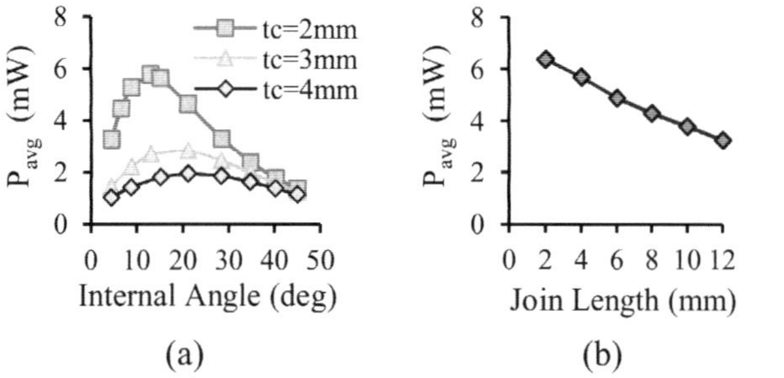

Figure 4. (a) and (b) Power output for different endcap thickness and internal angles (10MΩ), and endcap-substrate join length (8MΩ), respectively. PFT subject to a 1kN force at 2Hz. Internal angle remained constant when changing join length.

Enhanced PFT power outputs were found with the following: reduced endcap thickness led to less energy waste through unneeded bulk, figure 4 (a), though not shown here this was similar with substrate thickness. Reduced endcap internal angle gave rise to more horizontal force along the piezo-element due to leverage effects, figure 4 (a) and finally reduced join length decreased wasteful clamping of the piezoelectric element, figure 4 (b). Though not shown here the current and voltage showed a similar trend to power output when varying the endcap dimensions as in figure 4 (a) and (b). This was expected as the internal properties of the piezo-element remained constant, it was only the force applied to the piezo-element that changed. Factors that limited the choice of the join length and substrate were fabrication difficulty. The main limiting factor of choosing the piezo-element thickness was the internal resistance i.e. very thick piezo-elements had a high internal resistance, which was un-desired. Finally mechanical stress was the limiting factor when choosing the thickness and internal angle of the endcaps. Following that the stainless steel used had a Yield Strength of 125MPa (already with safety factor of 2 applied) for chosen dimensions see table 1. Though not shown here four metallic materials were considered for the endcaps; aluminium, titanium, stainless steel and copper alloy. It was found that making the endcaps thinner had a greater effect on power output than having a reduced stiffness, so possessing a high Yield Strength was important in order to get the thinnest possible endcap. Titanium performed best however, a compromise based on cost meant that stainless steel was used. Stainless steel was also used to make the substrate for similar reasons. Piezoelectric material selection was based on guidelines set out in previous works [3] with the additional compromise of availability.

4. Shoe Application Demonstration

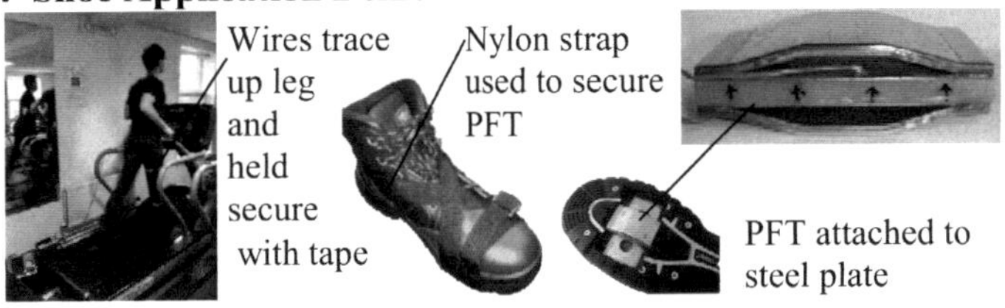

Figure 5. Testing of PFT in a shoe using a treadmill. PFT retrofitted into shoe with extra steel plate and held in place with nylon strap.

The PFT was retrofitted into a shoe and connected by wire to a wireless module [2]. The wireless module housed three sensors to measure acceleration, light and temperature. The measurement data were then wirelessly transmitted to a base-station equipped with a JN5148 wireless microcontroller (compliant with IEEE 802.15.4 protocol at 2.4GHz frequency). The transiever was liked via USB with a laptop for further analysis.

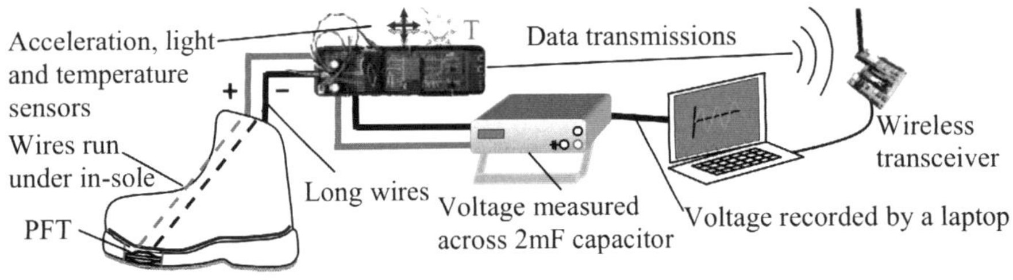

Figure 6. Testing set-up of PFT retrofitted in a shoe and the energy generated was used to power a wireless sensor module whilst the subject ran on a treadmill.

During the test the voltage across a capacitor in the wireless module was measured using a multi-meter. This capacitor was charged by electrical energy from the PFT and the energy was also used to power the module. Therfore, its voltage level gave an indication of what function the wirless module was carrying out.

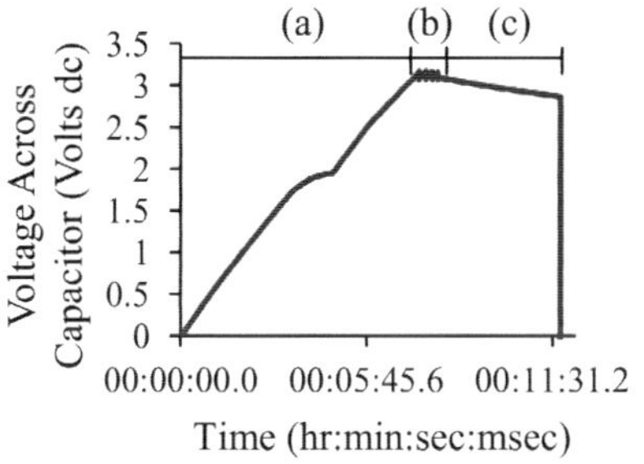

Figure 7. Voltage across 2mF capacitor connected to wireless module powered by PFT when subjected to approximately 760N at 1.4Hz whilst in shoe: (a) cold start, (b) data transmissions and (c) capacitor discharge; subject stopped running.

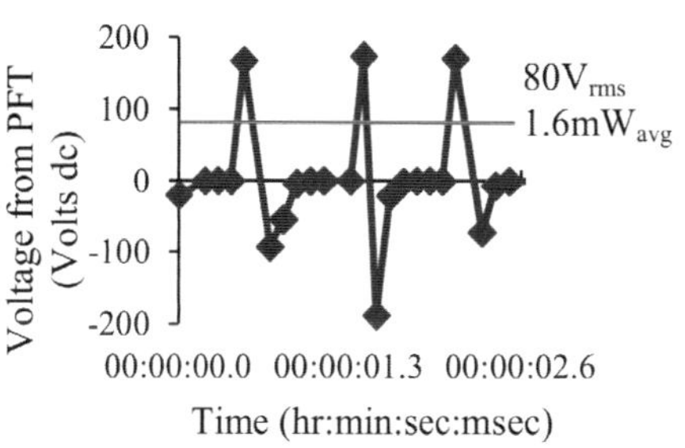

Figure 8. Voltage measured across a 4MΩ resistor connected in parallel to the PFT (4MΩ replacing the resistive load previously supplied by wireless module). When subjected to approximately 760N at 1.4Hz whilst in shoe.

Figure 7 shows the voltage across the 2mF capacitor connected to a wireless module with respect to time. The initial cold start time of 7minutes, zone (a), meant that the subject had to run for 7minuets continuously (exerting a force of 760N at 1.4Hz on the PFT) to initially charge the capacitor. As soon as the capacitor reached wireless module functioning threshold-voltage, ~3.2V, sensor reading and transmission of the data were performed, zone (b). The wireless module warm start is represented in the graph by the ascending voltage from ~3.08V to ~3.2V when the capacitor is energy recovering between module sensor readings and data transmission. The energy level provided by the subject running was enough to provide a short warm start of approximately 10s. The power consumed by the wireless module during sensor reading and data transmission is represented as a voltage drop from ~3.2V to ~3.08V.When the subject stopped running the voltage dropped to zero, zone (c). The data from each sensor was received at the laptop base station and recorded. The voltage fluctuations in Figure 8 correspond to foot strikes, where a 4MΩ resistor was connected to the PFT as an external resistor load replacing the wireless module in order to measure the power produced by the PFT directly. Though not shown here, when tested over a resistance range of 1-10MΩ (1MΩ intervals) the PFT produced a maximum power of 2.5mW_{avg} over 2MΩ (71V_{rms} 36μA_{rms}).

5. Conclusion

This paper used a coupled piezoelectric-circuit finite element model (CPC-FEM) to design and analyse the novel piezoelectric flex transducer's (PFTs) performance with focus on some key dimensional parameters. The PFT was fabrication, retrofitted into a shoe and the energy generated was used to power a wireless sensor module whilst the subject wearing the shoes ran at approximately 3.1mph.

Acknowledgments

The authors gratefully acknowledge support from SmartWater: "Smart Sensor Network with energy harvesting for real time monitoring in urban water infrastructure" (FP7-People-2-12-IRSES project, EC318985) and to Wojczech Suder at the Welding Institute (Cranfield University) for welding the PFT.

6. References

[1] Kim H, Priya S, Stephanou H, Uchino K 2007 *IEEE. T. Ultrason. Ferr* **54** (9) 1851-1858
[2] Giuliano A, Marsic V, Zhu M 2012 *IEEE. Int. Conf. iThings* 681-684
[3] Daniels A, Zhu M, Tiwari A 2013 *IEEE. T. Ultrason. Ferr* (Accepted)

Glucose Fuel Cells with a Microchannel Fabricated on Flexible Polyimide Film

Ryohei Sano[1], Yudai Fukushi[1],Tsubasa Sasaki[1],Hiroshi Mogi[1], Syohei Koide[1],Ryuta Ikoma[1], Wataru Akatsuka[2], Seiya Tsujimura[3], and Yasushiro Nishioka[1]

[1]Department of Precision Machinery Engineering, College of Science and Technology, Nihon University, 7-24-1 Narashinodai, Funabashi, Chiba 274-8501, Japan
[2]College of Engineering Sciences, University of Tsukuba, 1-1-1 Tennodai, Tsukuba, Ibaraki 305-8573, Japan
[3]Division of Materials Science, Faculty of Pure and Applied Sciences, University of Tsukuba,1-1-1 Tennodai, Tsukuba, Ibaraki 305-5358, Japan

E-mail nishioka@eme.cst.nihon-u.ac.jp

Abstract. In this work, a glucose fuel cell was fabricated using microfabrication processes assigned for microelectromechanical systems. The fuel cell was equipped with a microchannel to flow an aqueous solution of glucose. The cell was fabricated on a flexible polyimide substrate, and its porous carbon-coated aluminum (Al) electrodes of 2.8 mm in width and 11 mm in length were formed using photolithography and screen printing techniques. Porous carbon was deposited by screen printing of carbon black ink on the Al electrode surfaces in order to increase the effective electrode surface area and to absorb more enzymes on the electrode surfaces. The microchannel with a depth of 200 μm was fabricated using a hot embossing technique. A maximum power of 0.45 μW at 0.5 V that corresponds to a power density of 1.45 $\mu W/cm^2$ was realized by introducing a 200 mM concentrated glucose solution at room temperature.

1. Introduction

Microelectromechanical systems (MEMS) have been extensively studied in recent years for biomedical applications such as diagnosis, treatment via targeted drug delivery [2] and embedded systems [1]. Amongst these, enzymatic biofuel cells that use glucose in a human body to produce electricity for embedded microsystems have been of special interest [3]. This is because they work under mild conditions of room temperature, neutral pH and atmospheric pressure, which make them amenable to the human body [4-19]. MEMS technologies are advantageous for biofuel cells as they enable miniaturization and mass production. Tago et al. have reported such a biofuel cell with microchannels and enzymes-modified platinum electrodes fabricated on a glass substrate [4]. However, when this device was used in vivo and subcutaneously, the fragile and inflexible glass substrate caused problems with biocompatibility. In comparison, polyimide (PI) as a substrate material offers good biocompatibility, high thermal stability, and flexibility. However, to transfer microstructures onto PI is not very easy. Nevertheless, recent reports on thermal nanoimprint techniques have shown to accurately transfer micropatterns onto PI substrates [20,21]. In this study, we report the fabrication and characterization of glucose fuel cells with a microchannel fabricated on a PI substrate using a thermal nanoimprint technique. The fuel cell investigated here was composed of a bilirubin oxidase (BOD)-adsorbed cathode and a glucose oxidase (GOD) and osmium complex-adsorbed anode [8,9]. GOD immobilized on the anode oxidizes glucose molecules, as a result electrons and protons are produced. The electrons are transferred to the anode through an osmium complex mediator. Then they are conducted from the anode to the cathode via an external circuit generating electricity. Oxygen (O2) in the fuel solution is reduced by BOD immobilized on the cathode to water on reaction with electrons and protons.

2. 2. Experimental

2.1 Design of glucose fuel cells with a microchannel.

Figure 1 shows the layout of the cathode and anode. The anode has a width of 2.8 mm and a length of 1 mm, and the cathode has a width of 2.8 mm and length of 10 mm. These electrodes were fabricated by photolithography on a PI sheet, coated with aluminum (Al) and then modified by carbon black and respective enzymes. The gap between the anode and cathode was 0.5 mm. The microchannel was fabricated on another PI sheet, and it was designed to have a width of 3 mm, a depth of 200 μm, and a length of 11.5 mm. These two patterned PI substrates were stacked to form a fuel cell as shown in figure 2. The cathode was designed to be larger than the anode to account for the lower concentration of oxygen in comparison to that of glucose; to reduce the flux of oxygen in the upstream to the anode and consequently to improve the glucose oxidation performance [22].In order to investigate the characteristics of the anode and cathode independently using cyclic voltammetry, one of the electrode areas was fabricated to be a silver|silver chloride (Ag|AgCl) reference electrode by coating the Al surface with an ink of Ag|AgCl (BAS Inc.)

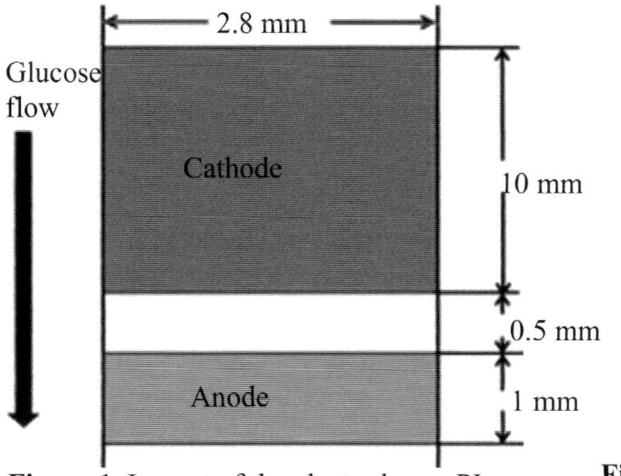

Figure 1. Layout of the electrodes on PI substrate.

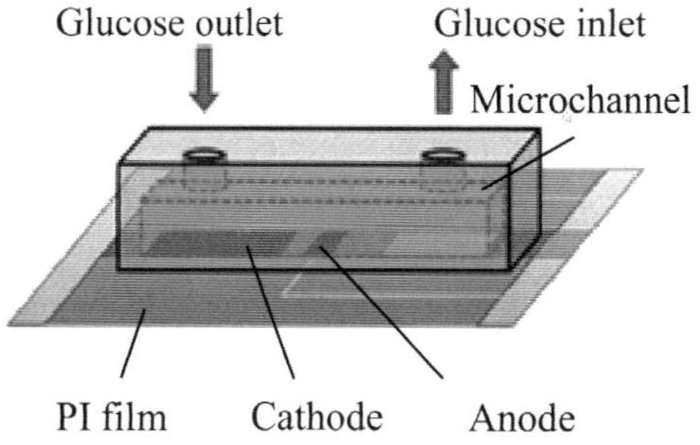

Figure 2. Design of glucose fuel cell with a microchannel

2.2 Fabrication processes

In order to increase the amount of enzymes adsorbed on the electrode surfaces, the effective surface area was increased by coating the electrode surfaces with a carbon black ink. The carbon-modified electrodes were fabricated on a PI sheet using conventional MEMS fabrication processes. Porous structures of the carbon-coated Al electrode surface were clearly observed by a scanning electron microscope (SEM), which suggest an increase in the effective electrode area.

1 mg of GOD was mixed with 25 μl of phosphate buffer solution (PBS, 50 mM, pH 7.0). Then, 6 mg of the osmium complex mediator (Polyvinylimidazole -Os(2,2'-bipyridine)2Cl) was mixed with 25 μl of PBS. Finally, 2 μl of the resulting solutions was dropped on the anode (surface area of 2.8 mm2), and left to dry for 20 min. Similarly, 1 mg of BOD was mixed with 25 μl of PBS. 2 μl of the resulting

solution was dropped on the cathode (surface area of 28 mm^2) and left to dry for 20 min. The microfluidic channel was fabricated using hot embossing techniques as previously described in our previous report [23].

2.3Measurements

Electrochemical measurements were performed in a 200 mM solution of glucose at room temperature. The electrochemical properties of the electrodes were characterized using a potentiostat (Electrochemical Analyzer, HA-1518, Hokuto Denko) having a three electrodes system, wherein the enzyme-modified electrode was the working electrode, the Ag|AgCl modified electrode was the reference and platinum was the counter electrode. The glucose solution was introduced into the reaction chamber through the microchannel by pulling a syringe pump.The power generated was evaluated by measuring the cell voltage while varying the external load resistance between 0 and 3 MΩ. The following equation was used:

$$W=V^2/R \quad (1),$$

where R is the load resistance, and V is the generated voltage measured between the terminals of the fuel cell. The power density was calculated by dividing W with the total area of the anode and the cathode, which was 30.8 mm^2.

3. Results and discussion

Figure 3 shows the cyclic voltammogram of the GOD/osmium complex-modified anode in 200 mM glucose solution. The catalytic oxidation of glucose can be clearly seen at potentials higher than 0.12 V. The maximum current density was 1457 μA/cm^2. This superior performance was due to the promotion of oxidation reactions by GOD and fast electron transfer by the mediator osmium complex [24,25]. The high current density saturated at a potential about 0.2 V and then showed a decrease from about 0.3 V. In this experiment, it is presumed that the reduction of the current density was due to the removal of the GOD/osmium complex mediator from the anode and the exhaustion of glucose near the surface of anode. Better immobilization of the enzymes to the electrode surface will help prevent their exfoliation.

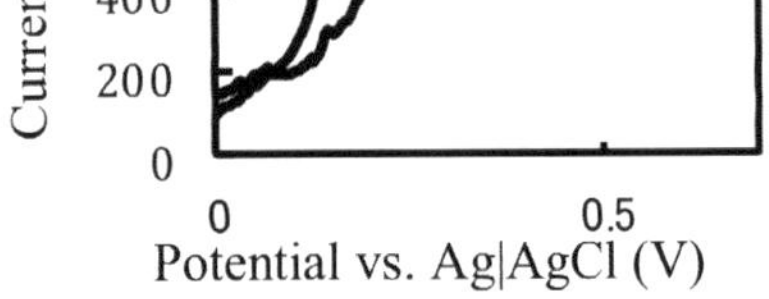

Figure 3. . Cyclic voltammogram of the GOD/ osmium complex -modified anode in 200 mM glucose solution.

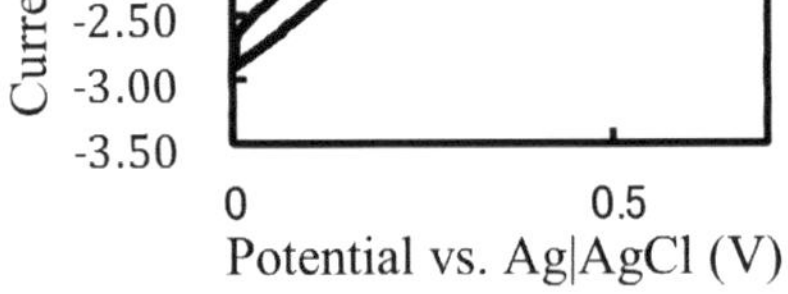

Figure 4. Cyclic voltammogram of the BOD- modified cathode in 200 mM glucose solution.

The cyclic voltammogram of the BOD-modified cathode in 200 mM glucose solution is shown in figure 4. The starting potential of the O_2 reduction current was about 0.5 V. The maximum current density was 1.45 μA/cm^2. This value was almost three orders of magnitude lower than that of the anode. The stirring of fuel and bubbling of oxygen, have resulted in a high current density as reported before [6, 22, 26-29]. In this study, a high current density on reduction was not realized because the oxygen concentration in the fuel was probably low. Alternatively, perhaps the carbon ink did not increase the effective electrode surface area of the cathode. A low resistive Al film was used as the current collector for the glucose biofuel cells in this study to achieve a low cost of fabrication. However, it is possible that during the operation of the fuel cell, the Al electrodes might have eroded on interaction with the glucose solution. Al is considered to have a high ionization tendency for electrochemical corrosion. It may be necessary to use noble metals such as platinum (Pt) or gold (Au), which have low ionization tendency to keep the fuel cell performance from degrading. In this study, since we used Al as a current collector, it was presumed that the Al electrode was completely covered with the carbon ink.

The relationship between the current and voltage of the fuel cell during operation is shown in figure 5. While the dependence of the power density on output voltage is shown in figure 6. The maximum power density was 1.45 μW/cm^2 at 0.5 V when the 200 mM glucose solution was introduced into cell by syringe pump. This maximum power density is lower by twenty factors when compared to the previously reported results [4]. This could be attributed to the extremely low current density at the cathode. In addition, the power measurements were made at room temperature, which could be another reason for the low value of the power generated in this study. Togo et al. have reported that the temperature of their fuel was maintained at 37 °C during power generation [4], which might have increased the catalytic activity of the enzymes

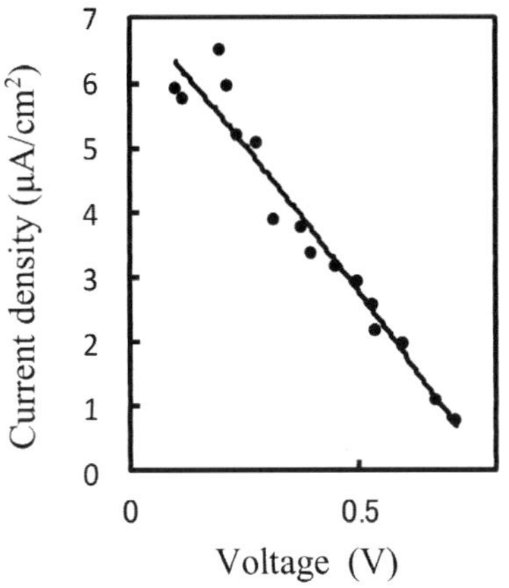

Figure 5. Current-voltage relationship during fuel cell operation.

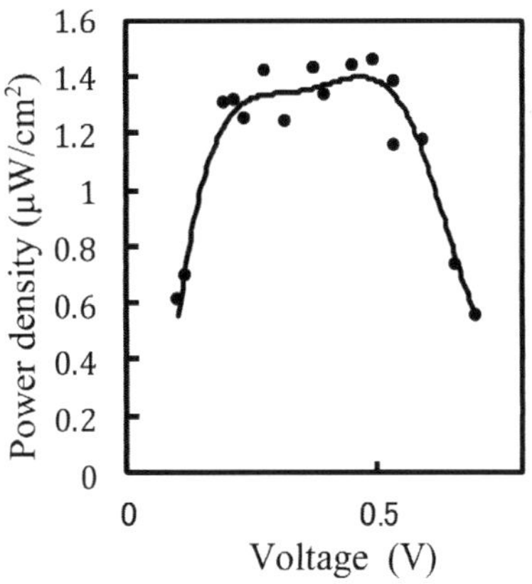

Figure 6. The dependence of power density on the output voltage.

4. Conclusion

A glucose fuel cell was fabricated on a flexible PI substrate using MEMS microfabrication processes. The fuel cell had a microchannel to introduce an aqueous glucose solution. Al electrodes of width 2.8 mm and length 11 mm were formed on the PI substrate using photolithography and screen printing. Porous carbon was deposited by screen printing of carbon black ink on the Al electrode surfaces in order to increase the effective electrode surface area and to absorb more enzymes. The microchannel with a depth of 200 μm was fabricated using a hot embossing technique. A maximum power of 0.45 μW at 0.5 V that corresponds with the power density of 1.45 μW/cm^2 was realized by introducing a 200 mM concentrated glucose solution at room temperature.

5. Acknowledgments

This research was conducted in supports of the staffs in the Micro Functional Device Research Center of the College of Science and Technology, Nihon University.

References

[1] Wei X and Liu J 2008 *Front Energy Power Eng. China,* **2(1)** 1
[2] Tsai N and Sue C 2007 *Sensors and Actuators.* **A 134** 555
[3] Tokita Y, Nakagawa T , Sakai H , Sugiyama T , Matsumoto R and Hatazawa T 2008 *ECS Trans.* **13**. Issue 21 89
[4] Togo M, Takamura A, Asai T, Kaji H and Nishizawa M 2007 *Electrochim. Acta* **52** 4669
[5] Lim K G and Palmore G T R 2007 *Biosens. Bioelectron.* **22** 941
[6] Kamitaka Y, Tsujimura S, Setoyama N, Kajino T and Kano K 2007 *PCCP Phys. Chem. Chem. Phys.* **9** 1793
[7] Sato F, Togo M, Islam M.K, Matsue T, Kosuge J, Fukasaku N, Kurosawa S and Nishizawa M 2005 *Electrochem. Commun.* **7** 643
[8] Moore C.M, Minteer S.D and Martin R.S 2005 *Lab Chip* **5** 218
[9] Heller A 2005 *AiChE* J. **51** 1054
[10] Heller A 2004 *PCCP Phys. Chem. Chem. Phys.* **6** 209
[11] Barton S.C, Gallaway J and Atanassov P 2004 *Chem. Rev.* **104** 4867
[12] Mano N, Mao F, Shin W, Chen T and Heller A 2003 *Chem. Commun. (Cambridge, United Kingdom)* **4** 518
[13] Tsujimura S, Kano K and Ikeda T 2002 *Electrochemistry* **70** 940
[14] Mano N, Mao F, Heller A and Am J 2002 *Chem. Soc.* **124** 12962
[15] Tsujimura S, Fujita M, Tatsumi H, Kano K and Ikeda T 2001 *PCCP Phys. Chem. Chem. Phys.* **3** 1331
[16] Katz E, Buckmann A.F, Willner I and Am J 200*1 Chem. Soc.* **123** 10752
[17] Chen T, Barton S.C, Binyamin G, Gao Z.Q, Zhang Y.C, Kim H.H, Heller A and Am J 2001 *Chem. Soc.* **123** 8630
[18] Palmore G.T.R, Bertschy H, Bergens S.H, Whitesides G.M and Electroanal J 1998 *Chem.* **443** 155
[19] Tamaki T, Ito T and Yamaguchi T 2007 *J. Phys. Chem.* **B111** 10312
[20] Youn S.W, Noguchi T, Takahashi M and Maeda R 2008 *Microelec. Eng.* **85** 918
[21] Becker H and Heim U 2000 *Sens, and Actuators A Physical.* **83** 130
[22] Kontani R, Tsujimura S and Kano N 2009 *Bioelectrochemistry* **76** 10
[23] Ikoma R, Komatsuzaki H, Suzuki K, Komori T, Kuroda K, Saitou H,Youn S W, Hiroshima H, Takahashi M, Maeda R and Nishioka Y 2013 *Journal of Photopolymer Science and Technology.* **25** 255
[24] Kuwahara T, Ohta H, Kondo M and Shimomura M 2008 *Bioelectrochemistry* **74** 66
[25] Ohara T.J, Rajagopalan R and Heller A 2003 *Anal. Chem.* **1003, 65** 3512
[26] Tsujimura S, Kano K and Ikeda T 2005 J. *Electroanal. Chem.* **576** 113
[27] Tsujimura S, Nakagawa T, Kano K and Ikeda T 2004 *Electrochemistry* **72** 437
[28] Kamitaka Y, Tsujimura S, Kataoka K, Sakurai T, Ikeda T and Kano K 2007 *J. Electroanal. Chem.* **601** 119
[29] Tominaga M, Otani M, Kishikawa M and Taniguchi I 2006 *Chem. Lett.* **35** 1174

Ultra-broadband electromagnetic MEMS vibration energy harvesting

Huicong Liu [1, 2], **Lokesh Dhakar** [2, 3] **and Chengkuo Lee** [2]

[1] Robotics and Microsystems Center, School of Mechanical & Electric Engineering, Soochow University, Suzhou, P. R. China
[2] Department of Electrical and Computer Engineering, National University of Singapore, Singapore
[3] NUS Graduate School for Integrative Sciences and Engineering, Singapore

E-mail: elelc@nus.edu.sg; hcliu078@suda.edu.cn

Abstract. This paper presents the design, fabrication and characterization of an electromagnetic MEMS energy harvester with ultra-broad operating bandwidth. The beam stretching induced nonlinear spring stiffness enables the resonance to extend from 65 Hz to 340 Hz and 400 Hz at accelerations of 0.5g and 1.0g, respectively. The ultra-wide bandwidth could benefit the harvester device for more efficient vibration energy harvesting.

1. Introduction

In the ambient environment, there are various vibration sources exhibiting in buildings, machines, transportations and human motions [1]. Generally, these vibration sources have random or irregular frequency peaks distributed in a certain bandwidth range. Vibration energy harvesting via electromagnetic, piezoelectric and electrostatic approaches have attracted more research attention for low-power electronics [2]. No matter which transduction mechanism was accomplished, most of the previous research work focused on the development of a linear oscillation system, where the maximum power output occurs at its resonant frequency. In the case that the ambient excitation frequency varies slightly, the output performance of the energy harvester can drop significantly. Since most of the vibration sources are present in a random or frequency varying scenario, it becomes a great challenge to broaden the operating bandwidth for more efficient vibration energy harvesting [3-4]. The use of nonlinear stiffness as a means to enlarge the bandwidth of energy harvesters has been reported by a number of works. They can be categorized into three main approaches, which are based on magnetic force [5-6], end-stopping of proof mass [7] and nonlinear beam stretching [8-9]. The first two approaches require extra assembly of bulk magnet or mechanical stoppers, thus they are usually implemented in macro-scale prototypes. Liu et al [10-11] have developed micro-scale broadband piezoelectric energy harvesters by incorporating the metal package base as the amplitude stopper of the MEMS PZT cantilever. Miki et al [12] have reported a MEMS in-plane electret energy harvester with nonlinear spring. The spring hardening effect can be induced as the silicon mass engages with the fixed-free secondary springs. For the nonlinear beam stretching method, Nguyen et al [13] have studied the spring hardening and softening behaviours of the broadband MEMS energy harvesting devices based on the design of special beam geometry. This work presents an ultra-broadband electromagnetic MEMS vibration energy harvester. The nonlinear beam stretching at three resonance modes results in an ultra-wide range of operating frequencies compared to other nonlinear energy harvesting devices.

2. Device configuration and microfabrication

The schematic illustration of the ultra-broadband harvester device is shown in Fig. 1(a). It consists of two pairs of clamped beams which are connected to a mass frame through the connecting hinges at two sides. The length, width and thickness of each clamped beam are 1.9 mm, 0.2 mm and 3 μm, respectively. The mass frame of 0.75 mm wide is patterned with metal coils of 5 μm in width and 5 μm in space. A pair of permanent magnets with two opposite poles is assembled such that a magnetic field along the *y*-direction is applied across the device. Hence, the electric current will be generated in the metal coils via the torsional vibrations of the mass frame along the connecting hinges. The center mass with reflective surface is used to observe the oscillating behaviour of the mass frame by the reflection of laser light. The wide bandwidth of the oscillation system can be explained by the Duffing stiffening induced by the clamped beam stretching during torsional and out-of-plane vibrations at resonances.

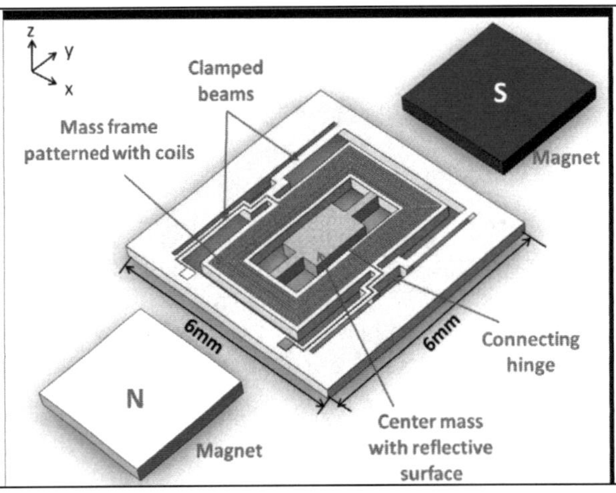

Figure 1. Schematic illustration of the ultra-broadband energy harvester device.

The energy harvesting chip of 6 mm×6 mm×0.4 mm is micro-fabricated by the micro-machining processes as shown in Fig. 2. It is first started by an 8 inch SOI wafer with 3 μm device layer, 1.1 μm buried oxide (BOX) layer and 725μm handle layer. In Fig. 2(a), a 0.2 μm thermal oxide (SiO_2) layer is initially deposited on the frontside surface by plasma-enhanced chemical vapour deposition (PECVD). Then physical vapour deposition, patterning and reactive ion etching (RIE) are conducted to form the first layer Al coils as shown in Fig. 2(b). In Fig. 2(c), 0.5 μm SiO_2 layer is deposited by PECVD and pad opening is formed by RIE of SiO_2. It is followed by the deposition and patterning of a second layer of Al coils as shown in Fig. 2(d). In Fig. 2(e), the passivating layers of SiO_2 and Si_3N_4 of each 0.5 μm thick are successively deposited by PECVD. This is followed closely by RIE to form pad opening and reflective surface. In Fig. 2(f), a series of RIE processes of Si_3N_4, SiO_2 and Si are conducted to define the frontside structural features. Later on, the SOI wafer is backside grinded and polished to be approximately 400 μm. A 2μm PECVD SiO_2 layer is deposited and patterned at the backside surface (in Fig. 2(g)) and backside Si deep reactive ion etch (DRIE) is followed to pattern and release the back structural features (in Fig. 2(h)). The DRIE process is conducted in a time-controlled manner so as to make sure the process is stopped effectively at the BOX layer. Figure 3 (a) is a photograph of the fabricated and packaged energy harvesting chip and (b) is an enlarged image of the clamped beams connected by hinge.

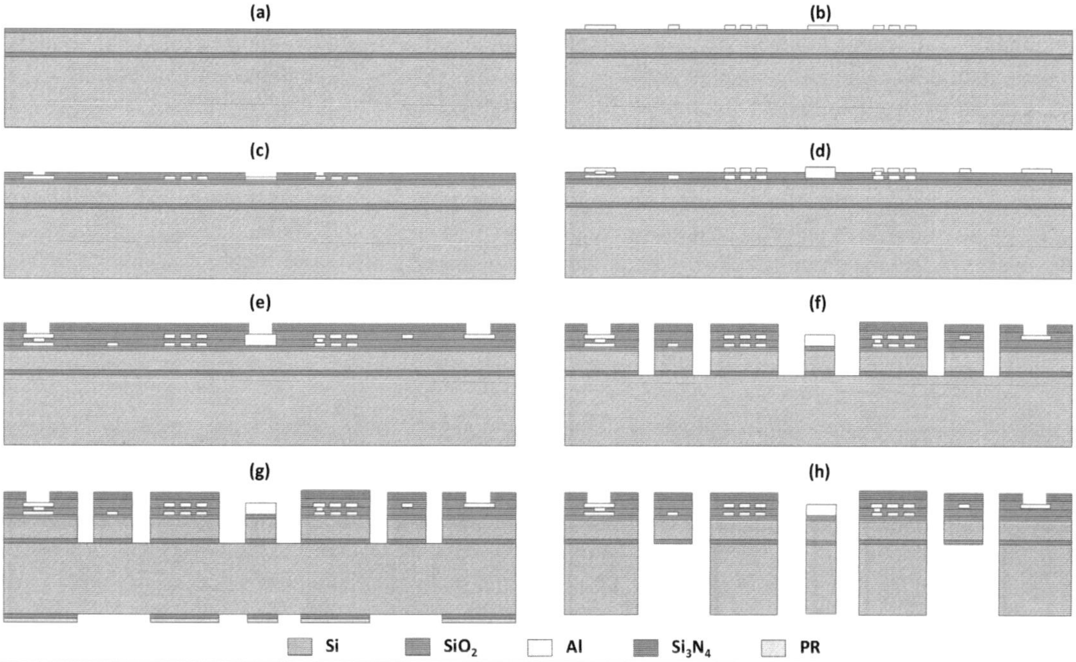

Figure 2. Microfabrication process of the energy harvesting chip.

Figure 3. Photograph (a) and enlarged image (b) of the fabricated harvesting chip

3. Experiment and results

The packaged device is tested by a closed-loop vibration control system as illustrated in Fig. 4, which is able to provide constant excitation accelerations to the harvester device. It consists of a vibration shaker driven by a controller through a power amplifier. An accelerometer is attached on the shaker to give the signal feedback to the controller. The controller is connected to a laptop installed with sweep-sine control software to perform a sweep-sine within a specific frequency range at constant vibration acceleration.

The first three resonant modes of the harvester device are around 65 Hz, 76 Hz and 220 Hz, respectively. It is seen from the experimental result in Fig. 5 that unlike a linear oscillation system, the output voltage shows a broadband response instead of three voltage peaks. It is because the beam stretching induced nonlinearity broadens the resonance of 65 Hz towards a higher bandwidth range. The broadened resonance overlaps successively with the second and third resonance modes, i.e. 76 Hz and 220 Hz. Such interaction further enables the resonance to extend towards an ultra higher resonance. At an acceleration of 0.5g, as the excitation frequency sweeps up, the resonance is pushed towards a high-energy stable region until A_1 at 340 Hz. After that, the instability overcomes the nonlinear stiffness effect of the system and thus the oscillation jumps down from the high-energy state to a low-energy state. Comparing to the resonance of the frequency down-sweep response (93 Hz at A_2), the operating bandwidth is widely broadened of 365%. Similarly, as the acceleration increases to 1.0g, the resonance of 400 Hz at B_1 for frequency up-sweep is much higher than that of 230 Hz at B_2 for frequency down-sweep. The peak voltage at points A_1 and B_1 are obtained as 0.19 mV and 0.24 mV, respectively. The corresponding peak powers are obtained as 0.18 nW at point A_1 and 0.3 nW at point B_1, with respect to the match load resistance of 195 Ω.

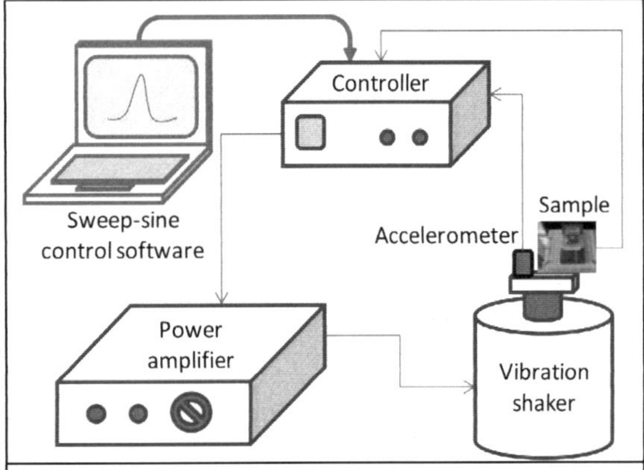

Figure 4. Closed-loop vibration control system for sweep-sine test with constant excitation accelerations.

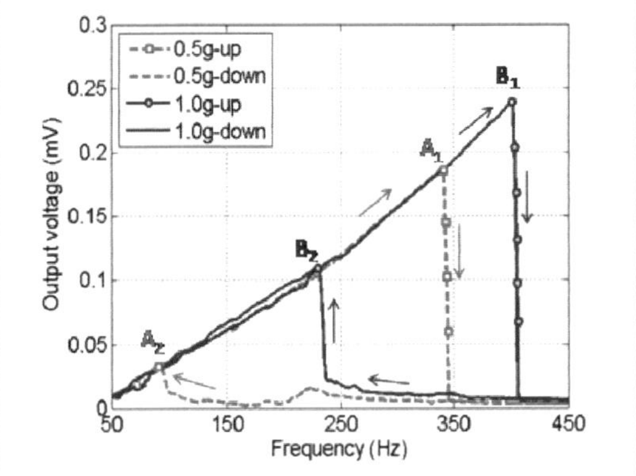

Figure 5. Output voltage against frequency sweeping up and down at accelerations of 0.5g and 1.0g.

4. Conclusion

In this paper, an ultra-broadband electromagnetic MEMS energy harvester is fabricated and characterized. Comparing to frequency down-sweeps, the bandwidth is 365% and 174% wider than frequency up-sweeps. It is due to the nonlinear beam stretching at the first three resonant modes, the operating bandwidth is extended from the original resonant frequency of 65 Hz to be as wide as 340 Hz at 0.5g and 400 Hz at 1.0g.

References

[1] J. A. Paradiso, T. Starner 2005 *IEEE Pervas. Comput.* **4** 18-27
[2] S. P. Beeby, M. J. Tudor, N. M. White 2006 *Meas. Sci. Technol.* **17** R175-95
[3] D. Zhu, M. J. Tudor, S. P. Beeby 2010 *Meas. Sci. Technol.* **21** 022001
[4] S. D. Nguyen, E. Halvorsen 2011 *J. Microelectromech. Syst.* **20** 1225-7
[5] X. Xing, J. Lou, G. M. Yang, O. Obi, C. Driscoll, N. X. Sun 2009 *Appl. Phys. Lett.* **95** 134103
[6] S. C. Stanton, C. C. McGehee, B. P. Mann 2009 *Appl. Phys. Lett.* **17** 174103
[7] M. S. M. Soliman, E. M. Abdel-Rahman, E. F. El-Saadany, R. R. Mansour 2008 *J. Micromech. Microeng.* **18** 115021
[8] L. G. W. Tvedt, D. S. Nguyen, E. Halvorsen 2010 *J. Microelectromech. Syst.* **19** 305-16
[9] A. Hajati, S.-G. Kim 2011 *Appl. Phys. Lett.* **17** 174103
[10] H. Liu, C. J. Tay, C. Quan, T. Kobayashi, C. Lee 2011 *J. Microelectromech. Syst.* **20** 1131-42
[11] H. Liu, C. Lee, T. Kobayashi, C. J. Tay, C. Quan 2012 *Microsyst. Technol.* **18** 497-506
[12] D. Miki, M. Honzumi, Y. Suzuki, N. Kasagi 2010 *IEEE MEMS 2010* 176-9
[13] D. S. Nguyen, E. Halvorsen, G. U. Jensen, A. Vogl 2010 *J. Micromech. Microeng.* **20** 125009

Nonlinear vibration energy generators with mechanical stops for low-frequency broadband applications

B Leistritz[1,2] **and H Toepfer**[1]

[1]Department of Advanced Electromagnetics, Technische Universität Ilmenau, Helmholzplatz 2, 98693 Ilmenau, GERMANY
[2]Institut für Mikroelektronik- und Mechatronik-Systeme gemeinnützige GmbH, Ehrenbergstraße 27, 98693 Ilmenau, GERMANY

E-Mail: bianca.leitritz@tu-ilmenau.de

Abstract. A variety of energy-autonomous sensor systems are excited by a spectrum of excitation frequencies. Especially in the low-frequency range, an optimal system design is important because the available energy for conversion is relatively low. Because the maximum deflection for multimodal excitations is dependent on the stochastic phase, mechanical stops must be implemented to prevent an overloading of the springs. Due to the nonlinear effect of the mechanical stops, only an expected value of the mean output power can be specified. The standard deviation increases with the occurrence of contacts between the seismic mass and the stops. This paper presents a comparison between a linear approach and nonlinear simulations.

1. Introduction

With the increase of widespread decentralised sensors in industrial processes and in daily life, benefits in the fields of monitoring, ease of use, security and conservation of energy and resources are expected. Especially at hard-to-access places alternative power supplies are required because of the finite life time of batteries. A viable option is to use vibration energy generators which convert kinetic energy available in the vicinity of the sensor system into electrical energy. Depending on the magnitude of the required output power, the mechanical-electrical conversion is carried out by micro-mechanical or mesoscopic systems.

Most of the vibration energy generators are considered as resonant systems for sinusoidal excitation with fixed excitation frequency [1, 2]. However, in practical applications this assumption is not true. In so-called broadband or multimodal excitations, different properties can be distinguished. Based on the characterisation of various application scenarios, a linear approach and a nonlinear calculation method for energy generators with mechanical stops are presented and compared. They allow the calculation of system parameters depending on the required mean output power, the excitation spectrum and the permitted deflection.

2. Characterisation of excitation spectrums

In order to characterize specific applications, 3D acceleration measurements were made. The first application scenario is the realisation of energy-independent sensors close to the wheel of a car. The excitations were measured on different vehicles, test routes and speeds. The vertical acceleration shows a continuous power spectral density (PSD) in low-frequency range up to 30 Hz excited by road unevenness. This result agrees very well with the specified frequency range between zero and about 25 Hz given in [3]. Figure 1 shows the sensor on the wheel suspension and the vertical acceleration spectrum measured in a test vehicle driving at 120 km/h on a highway.

A second application scenario is a sensor for sportspersons. The study of the vertical acceleration of human ankles while jogging on a tartan track exhibits a discrete spectrum. The maximum acceleration values can reach multiple of g (gravity of earth). Human activities are also an example for low-frequency excitations. Figure 2 shows the measurement setup and found PSD distribution.

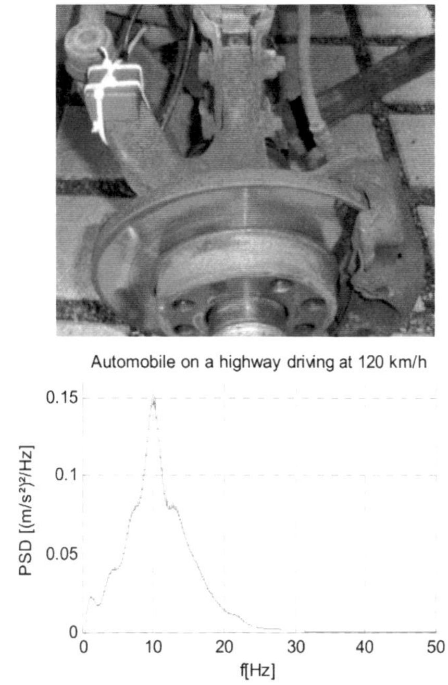

Figure 1. Accelerometer on a wheel suspension and vertical excitation spectrum while driving at 120 km/h on a highway

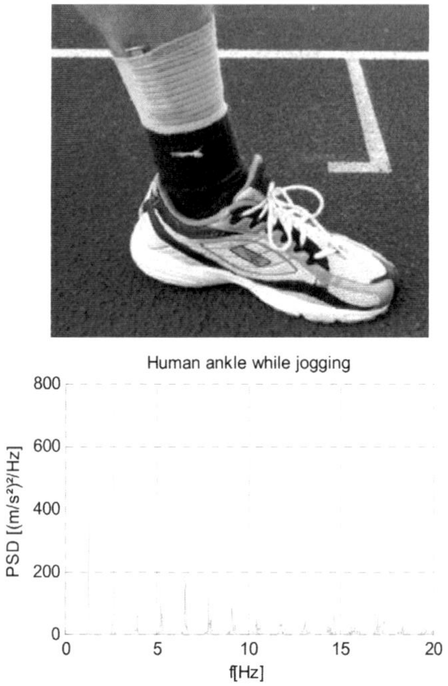

Figure 2. Accelerometer on human ankle and vertical excitation spectrum while jogging on a tartan track

Both applications show frequency spectrums with different characteristics. Broadband excitations in our study are vibrations, which results from a superposition of different frequency components. It does not refer to sinusoidal excitations with variable frequency.

A differentiation between continuous and discrete spectrums is difficult based on the time signals. The description of excitations with the help of power spectral density allows an easy comparison between stochastic signals. The PSD of acceleration (PSD_A) can be calculated as

$$PSD_A(f_i) = \frac{a_i^2}{2\Delta f} \tag{1}$$

where a_i denotes amplitude at the frequency f_i and discretisation Δf. The principal advantage of using PSD is that the power content of a bandwidth can be directly calculated by integrating the PSD over the frequency range. In contrast to the time behaviour

$$a(t) = \sum_{i=1}^{n}(a_i \sin(2\pi f_i t + \varphi_i)), \tag{2}$$

the PSD does not deliver characterisation of the maximum acceleration because of the absence of the information about the phase of the different frequency components φ_i. Therefore, the results of calculations based on PSD do not fully describe the behaviour in the time domain.

3. Design strategy for linear systems

System parameters for energy harvesters with multimodal excitation can easy calculated, when a linear spring with a constant stiffness is assumed. In an inertial mass-spring-damping system, the seismic mass *m* is connected with the housing by a spring with stiffness *k* and the mechanical (*dm*) and electrical (*de*) damping (shown in Figure 3).

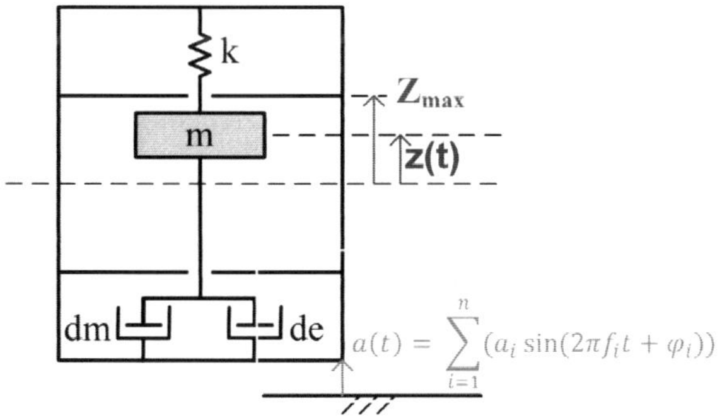

Figure 3. Mass-spring-damping system with a multimodal acceleration $a(t)$

System parameters are dimensioned using the limitation of a root mean square (RMS) of the deflection, expressed by the integral of the power spectral density of the deflection (PSD_Z):

$$RMS_Z = \sqrt{\frac{1}{T}\int_{to}^{t0+T} z^2(t)\mathrm{d}t} = \sqrt{\int_f PSD_Z \,\mathrm{d}f} \tag{3}$$

Because of the linear spring stiffness, PSD of deflection (PSD_Z) and PSD of velocity (PSD_V) can be directly expressed by transfer function and the PSD of acceleration as

$$PSD_Z = \left|\frac{m}{ms^2 + (dm + de)s + k}\right|^2 PSD_A \tag{4}$$

and

$$PSD_V = \left|\frac{ms}{ms^2 + (dm + de)s + k}\right|^2 PSD_A. \tag{6}$$

For a given excitation spectrum PSD_A, mass *m* and an assumed mechanical damping *dm* the optimal spring stiffness and the needed electrical damping by the conversion system can be calculated, so that the mean output power in steady state

$$P_e = de\sqrt{\int_f PSD_V \,\mathrm{d}f} \tag{7}$$

is maximized. With the calculated system parameters, the design parameters can be adjusted.

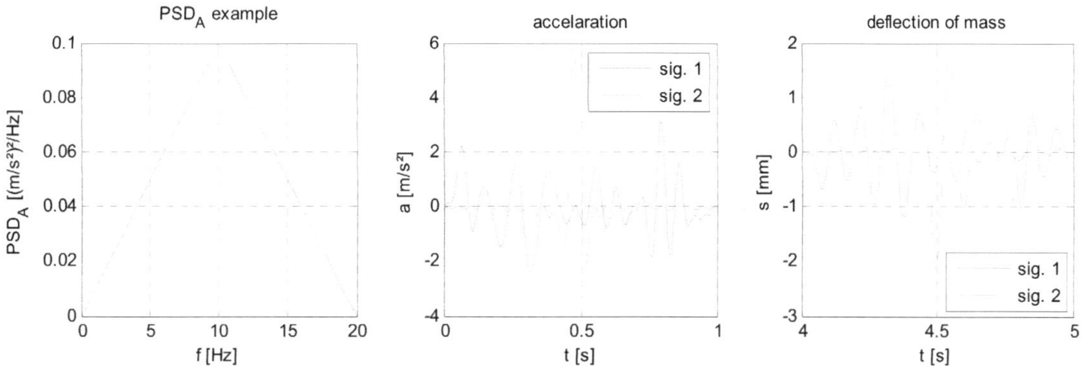

Figure 4. Demonstration example: different time signals (sig. 1 and sig. 2) for a given PSD_A. RMS_Z (= 0.5 mm) and mean output power in steady state (= 1.7 mW) of both signals are equal.

However, only the root mean square can be constrained. Figure 4 shows an example with two signals having the same PSD_A distribution. Although RMS_Z and the mean output power are equal, different maximum deflections are reached. Since the maximum deflection cannot be predicted, the springs have to be protected against the overloading by an implementation of mechanical stops.

4. Nonlinear system with mechanical stops

In [4], mechanical stoppers are examined for a realization of a wideband vibration energy generator. Besides nonlinear springs and bi-stable structures, the mechanical stoppers give also a nonlinear opportunity to increase the bandwidth of the power spectrum. Simulations of the limited model enable to find optimal system parameters. However, the mean output power cannot be calculated in the same way as in the linear case solely by the PSD characterisation of the excitation, because of the only piecewise-linear stiffness.

The excitation must be further described by a phase distribution. In our study, a uniform distribution of the excitation is assumed and stochastic excitation time signals are generated with MATLAB®. The mechanical stops are implemented as springs with a few times higher stiffness at the limit of deflectionZ_{Stop}. An increase of the mechanical damping at stop has not yet been taken into account and needs to be investigated experimentally. The mean output power in steady state for each random excitation is evaluated in the time domain by numerical simulations using Simulink®. The implemented equations can be found in Figure 5. Simulations have been started 1000 times to enable analysis of the expected value and the standard deviation of the mean output power.

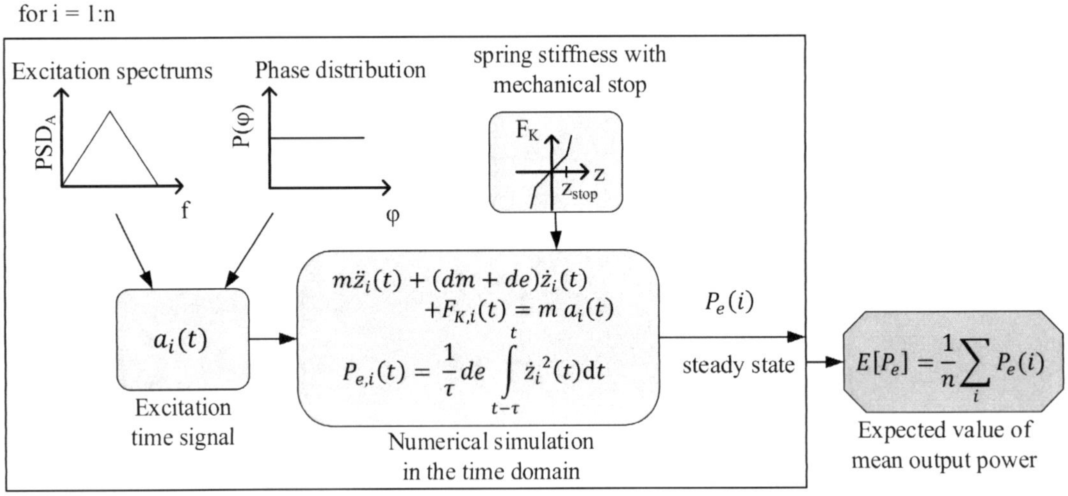

Figure 5. Principle steps of calculations of the expected value of mean output power for a nonlinear vibration energy generator with multimodal excitation and mechanical stops

5. Results for a demonstrator and comparison

For a new multimodal excited electromagnetic demonstrator with a given mass of 50 g and the limitation of deflection to 1.4 mm the optimal electric damping and spring stiffness are searched. The mechanical damping of the system is assumed to 0.1 kg/s. The excitation spectrum is approximated by a triangular shaped profile in a range of 0 to 20 Hz with a maximum value 0.12 $(m/s^2)^2/Hz$ at 10 Hz.

First, an approximate solution is determined using linear approach. In this case, a RMS value of the deflection of 0.5 mm (about one-third of the maximum deflection) is assumed. Figure 6 shows the dependence of mean output power on the spring stiffness, respectively the natural frequency. In addition, the required electrical damping values are added. For the given boundary conditions, the maximum output power is reached at the natural frequency of 10.2 Hz and the electrical damping value of 0.98 kg/s.

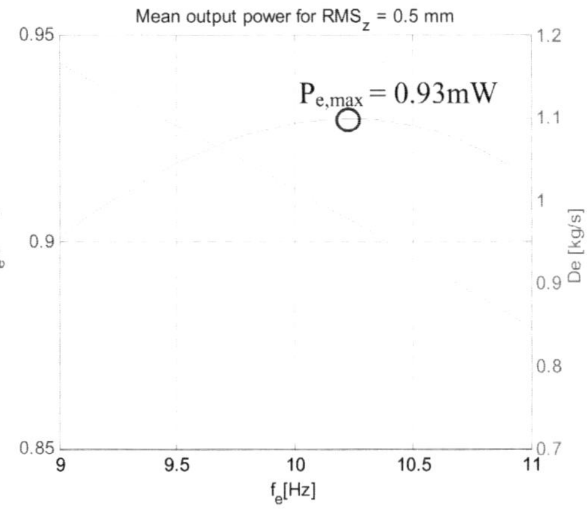

Figure 6. Maximum output power of the linear approach without mechanical stops depending on the natural frequency and the corresponding electrical damping

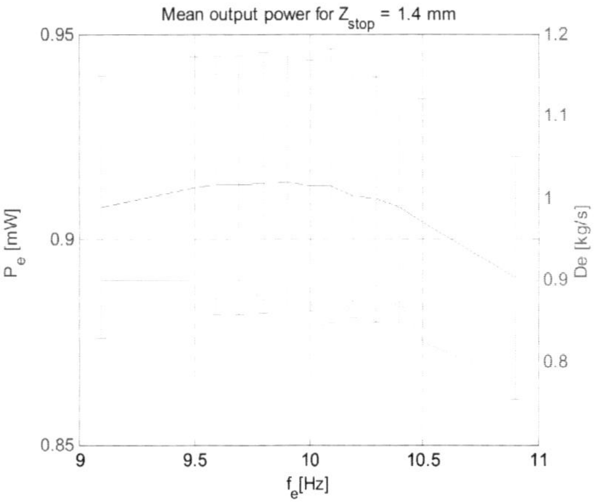

Figure 7. Expected value and standard deviation of the nonlinear calculations depending on the natural frequency and the corresponding electrical damping

Figure 7 shows results of the nonlinear study with mechanical stops at ± 1.4 mm. The maximum mean output power in steady state has a good agreement with the estimation based on the linear approach. The standard deviation is relative high compared with the differences in consequence of the variation of spring stiffness. It is a result of a frequent occurrence of contacts between the seismic mass and the stops. For the system design using linear dimensioning with RMS_Z significant less than one-third of the maximum deflection, the expected value corresponds to the linear result. However, the found power density of the system in this case is not optimal.

6. Conclusions

In the paper, the need for a proper design strategy for multimodal excited energy generators is presented. As a result of a stochastic distribution of phase shifts between the individual frequency components, only the expected value and the standard deviation can be specified for the mean output power in steady state. In our study, the system parameters could be well approximated by a linear approach using a ratio between the maximum deflection and RMS value of 2.8. However, this ratio is not fixed, but depends on the allowable deflection. In ongoing studies these effects will be studied in more details to allow an approximation of the suitable system parameters using the linear method instead of very time consuming nonlinear dimensioning.

Acknowledgement

This work has been carried out in the research groups PORT (No. 2011 FGR 0120) and GreenSense (No. 2011 FGR 0121) with financial support from the Thuringian Ministry of Economics, Labor and Technology and the European Social Fund (ESF).

References

[1] Williams C and Yates R 1995 Analysis of a micro-electric generator for microsystems *Sensors and Actuators A: Physical* **52** 8-11
[2] Mitcheson P D, Green T C, Yeatman E M and Holmes A S 2004 Architectures for vibration-driven micropower generators *J. Microelectromech. Syst.* ***13*** 429-440
[3] Mitschke M 1997 *Dynamik der Kraftfahrzeuge – Bd. B. Schwingungen* vol 3 (Berlin: Springer) p 1
[4] Tang L, Yang Y and Soh C K 2013 *Advances in Energy Harvesting Methods* ed N Elvin and A Erturk (New York: Springer Science + Business Media) chapter 2 pp 17-61

Interface circuit using SMFE technique for an inductive kinetic generator operating as a frequency-up converter

D Sanchez[1], E Jodka[1], D Hoffmann[2] and Y Manoli[1,2]

[1]Fritz Huettinger Chair of Microelectronics, Department of Microsystems Engineering - IMTEK, University of Freiburg, Germany
[2]HSG-IMIT - Institute of Micromachining and Information Technology, Villingen-Schwenningen, Germany

E-mail: `daniel.sanchez@imtek.de`

Abstract. This paper presents a novel interface circuit for an inductive kinetic energy harvester, which is designed based on the SMFE (Synchronized Magnetic Flux Extraction) technique. The generator is designed to fit into a shoe sole to power a telemetric device. It operates as a frequency-up converter and possesses a small output series resistance, allowing the SMFE scheme to be applied as an energy harvesting enhancement technique. A prototype has been implemented and simulated using discrete components in order to demonstrate the SMFE concept and compare it with typical standard interfaces used (Full Bridge and Voltage doubler rectifiers). Compared to the standard schemes, the SMFE method is able to extract a considerably higher amount of energy for a wide range of output voltages. Assuming one walking step per second, the complete system could provide 18.8 mW output power using the SMFE interface, compared to 8 mW with the full bridge and 12.3 mW using the voltage doubler.

1. Introduction

The increasing interest in recent years to develop wireless sensor networks capable to be autonomous in a reduced size, have turned the eyes to energy harvesting from ambient sources. Vibration energy harvesting is such a source and interesting new applications where miniaturization plays a key role have been widely proposed [1, 2]. Kinetic generators are mainly based on the piezoelectric, electrostatic and electromagnetic principles. The electromagnetic type has been documented to entail a high power density per cubic volume for devices in the centimeter size range [1].

Full bridge diode rectifiers and voltage doublers are considered as standard interfaces that rectify the AC output of an electromagnetic generator [3]. These schemes are efficient only at a specific output voltage at the buffer V_{out} that depends on the open circuit voltage of the generator. Therefore their behavior strongly relies on a continuous excitation or additional circuitry to fix the output voltage. Maximum power point tracking (MPPT) is an enhancement technique to overcome this, setting the output voltage to its efficient value [4]. Nevertheless, this technique has only been studied for excitations that have a continuous behavior for at least a short period of time. For the case of frequency-up conversion in which the generator voltage magnitude varies depending on the mechanical and electrical damping, MPPT schemes presented until now are not able to operate efficiently.

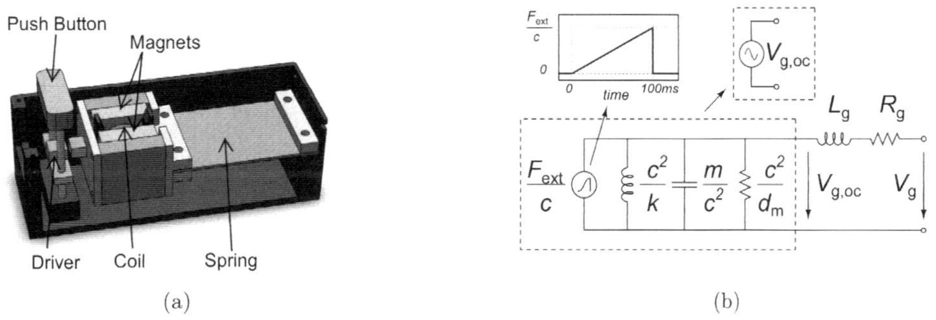

Figure 1: Proposed electromagnetic generator (a) and its equivalent circuit model (b).

The large coil series resistance R_g of many electromagnetic generators ($> 500\,\Omega$) limits the harvesting capabilities of the different electric interfaces, with MPPT being the scheme that achieves most promising results. The generator used in this work possesses a small R_g ($5\,\Omega$) which allows the SMFE scheme to be applied as an energy harvesting enhancement technique [5]. This paper presents the development of the SMFE interface for this specific generator, proposing a control circuit using discrete components, and analyzing the results obtained by comparing them with the standard interfaces. The generator is also modeled, and the special cases where the SMFE scheme is suitable are discussed.

2. Electromagnetic Generator

The generator used for this work is shown in figure 1a. It operates as a frequency-up converter which allows the extraction of energy from low mechanical frequencies by means of a conversion to a resonant higher frequency mode [6]. It is designed to fit into a shoe sole and power a telemetric device that wirelessly provides the location of the user to an external receiver. The generator is excited through a push button when the user takes a step. A beam with attached magnets then becomes deflected and is released to self-oscillate around a coil.

The kinetic generator is modeled using a low damped spring-mass-damper system [7]. Using the electrical analogy of the mechanical system [8], the generator can be modeled using electric components as shown in figure 1b. The values of the generator parameters are summarized in

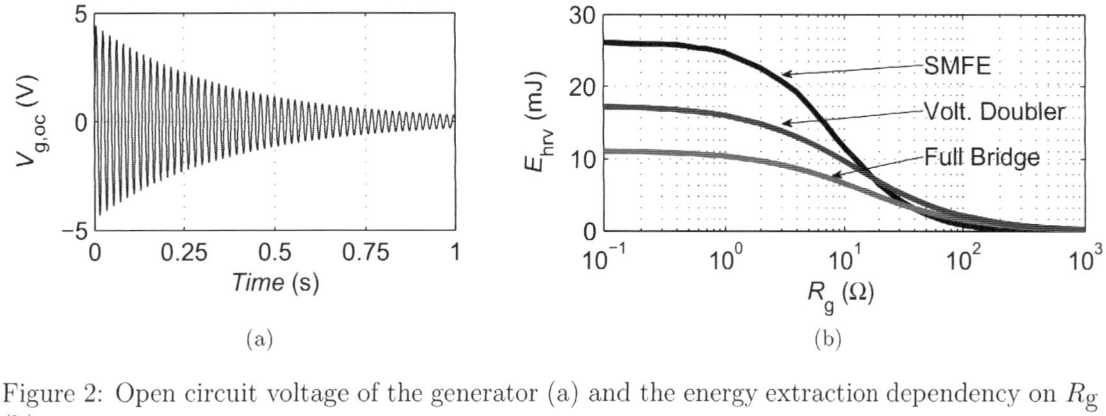

Figure 2: Open circuit voltage of the generator (a) and the energy extraction dependency on R_g (b).

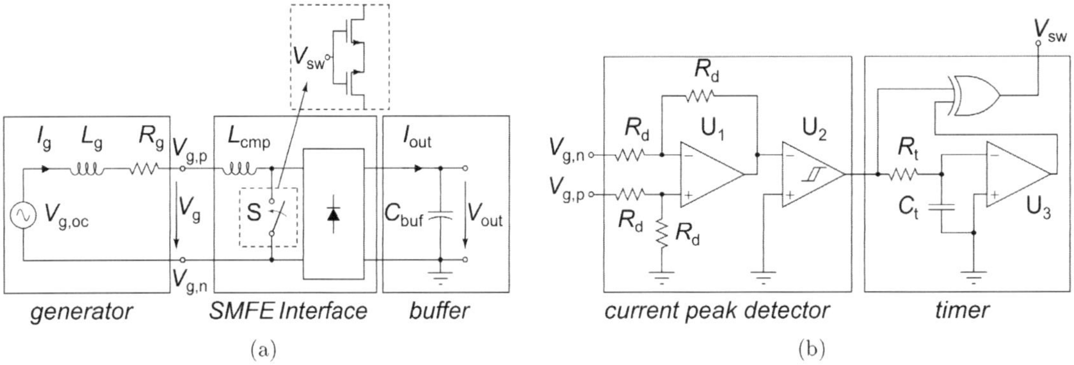

Figure 3: Schematic of the SMFE interface (a) and its control circuit (b).

table 1. F_{ext} represents the external force that the driver exerts to the tip of the cantilever. The mechanical parameters are the spring stiffness k, the mass m, the mechanical damping d_m, and the coupling coefficient c. The inductance of the coil and its series resistance are represented by L_g and R_g respectively.

For continuous resonant systems the excitation is modeled by a sinusoidal current source with a magnitude F_{ext}/c. However, in frequency-up conversion the system is excited by a pulse, and the force profile is similar to a saw-tooth signal with a relatively slow rise time, very fast fall time and with a peak magnitude F_{ext}/c. Hence the system emulates the deflection and release of the beam, which can then self-oscillate until its energy is damped. The resultant waveform of the voltage across the generator terminals in open circuit condition $V_{g,oc}$ can be observed in figure 2a.

As stated in the previous section, the series resistance of the generator limits the energy extraction capability. This behavior can be observed in figure 2b, where the energy harvesting capability of the two standard techniques and the one of the SMFE are presented with respect to a variation in the value of R_g. It can be observed that for low values the SMFE scheme is better suited as an enhancement technique, whereas for large R_g the standard interfaces perform better. Since the proposed generator possesses a $5\,\Omega$ series resistance, the SMFE interface is suitable as an efficient interface.

3. Electronic Interface

The SMFE scheme (figure 3a) consists of adding a switch S at the output terminals of the generator and before the rectification stage. After an excitation, S is closed, thus shorting the

Table 1: Generator parameters and discrete components used.

Electromagnetic generator		Discrete components used	
$F_{ext} = 25\,\mathrm{N}$	$c = 2.7\,\mathrm{Vs/m}$	Schottky diodes	BAT54
$m = 46\,\mathrm{g}$	$L_g = 97.54\,\mathrm{nH}$	Comparators U_1, U_2, U_3	LPC662A
$k = 5000\,\mathrm{kg/s^2}$	$R_g = 5\,\Omega$	n-MOS Transistor	BS170
$d_m = 0.25\,\mathrm{Ns/m}$	$f = 52.4\,\mathrm{Hz}$	p-MOS Transistor	BS250
width $= 30\,\mathrm{mm}$	lenght $= 80\,\mathrm{mm}$	R_d, $R_t = 100\,\mathrm{M\Omega}$	$1\,\mathrm{pF} \leq C_t \leq 20\,\mathrm{pF}$
height $= 26\,\mathrm{mm}$		$L_{cmp} = 61\,\mathrm{mH}$	

generator outputs and storing energy in L_g. When the generator current I_g peaks (See figure 4), S is opened and the energy stored in L_g is transferred to the buffer capacitor C_{buf}. A compensation inductor L_{cmp} is added in series to match its reactance with the impedance of the generator when S is closed, thus achieving an efficient energy transfer.

The energy harvested every half wave cycle can be calculated using equation (1), where X_{L_g} and $X_{L_{cmp}}$ are the reactances of L_g and L_{cmp} respectively. The oscillation frequency is represented by f, and V_d represents the diode forward voltage.

$$E_{hrv} = \frac{1}{2}(L_g + L_{cmp})\left(\frac{\widehat{V}_{g,oc}}{X_{L_g} + X_{L_{cmp}} + R_g}\right)^2 - \frac{I_{out}V_d - I_{out}^2 R_g}{2f} \qquad (1)$$

4. Implementation

The proposed scheme was implemented using discrete components. The switch S is implemented by means of two n-MOS transistors, with their sources connected together to cancel the effect of the bulk diodes (figure 3a). The schematic of the proposed control circuit is presented in figure 3b. It consists of a current peak detector and a precision timer. If short circuited, I_g peaks when V_g crosses zero. The current peak detector monitors V_g with comparator U_1, while U_2 detects any crossing of the ground level. U_2 is configured as a hysteresis comparator in order to avoid instability problems. R_t, C_t and U_3 are used as an externally adjusted timer to turn off S after the energy has been transferred.

The rectification stage is implemented using a full bridge diode rectifier with schottky diodes. The comparators are powered with external power supplies of 10V and −10V. The XOR gate is implemented using n-MOS and p-MOS transistors and is powered with an external 10V source.

5. Results

The proposed circuit was simulated in PSpice to demonstrate the SMFE concept and compare it with the standard interfaces. The waveforms of the generated current I_g, the current flowing to the output storage I_{out}, V_g and V_{sw} (switch control signal) are presented in figure 4.

The energy extraction profile after a single excitation is presented in figure 5a, whereas the standard interfaces and the SMFE have been configured to optimum conditions (Optimal V_{out}). The SMFE curve is well above the other two, harvesting around 135% and 53% more energy than the full bridge and voltage doubler rectifiers respectively after a single excitation.

The energy harvested by the different schemes at different output voltage levels is presented in figure 5b. It can be observed that for low output voltages both standard techniques harvest a considerably higher amount of energy. However, as the output voltage is increased the SMFE

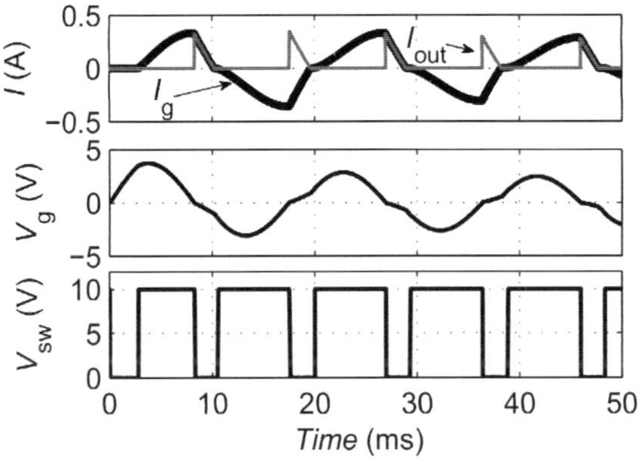

Figure 4: Waveforms of generated current I_g, output current I_{out}, generator voltage V_g and the switch control signal V_{sw}.

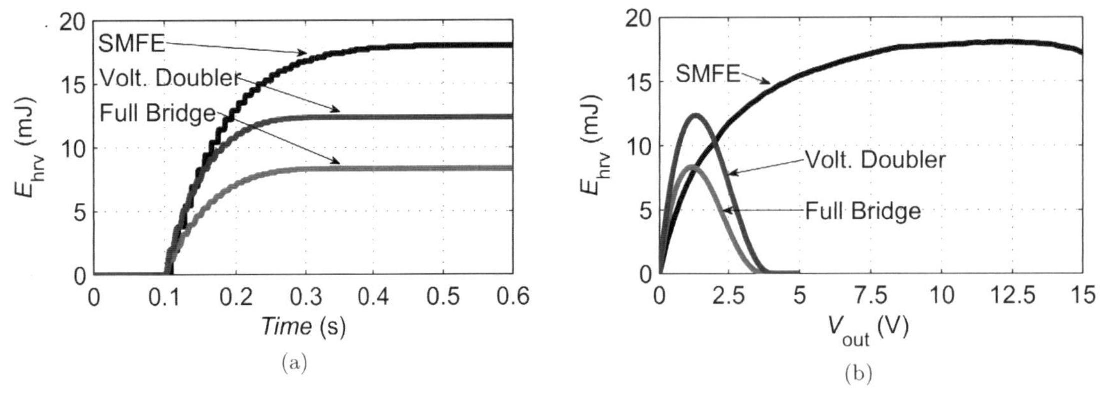

Figure 5: Energy harvested after a single excitation cycle in time domain (a) and at different output voltage levels (b).

scheme is able to outperform the standard schemes. This technique is also able to harvest a high amount of energy for a wide range of output voltages.

Assuming one walking step per second and using $P_{avg} = \triangle E_{hrv}/\triangle time$, the system could provide 18.8 mW output power using the SMFE technique in optimum conditions ($V_{out} = 10\,V$). The full bridge and the voltage doubler rectifiers are able to provide roughly 8 mW and 12.3 mW respectively. The power consumption of the control circuit including the external power supplies is around 2 mW.

6. Conclusion

A novel interface circuit for kinetic electromagnetic generators with a low series resistance is proposed. The generator has been modeled using a mass-spring-damper system, providing special emphasis on the modeling of the input force for frequency up conversion.

The electronic interface for the SMFE has been implemented together with its control circuit using discrete components. The simulation results show a significant advantage when using the SMFE technique compared to standard schemes. It also harvests a large amount of energy with low sensitivity to changes in the output voltage.

Acknowledgment

Funding by the "Bundesministerium für Bildung und Forschung" (BMBF) within the framework concept IKT 2020 is gratefully acknowledged. This work is also supported by the "Consejo Nacional de Ciencia y Tecnología" (CONACYT) and the "Deutscher Akademischer Austausch Dienst" (DAAD).

References

[1] Manoli Y 2010 *Proc. IEEE Eur. Solid-State Circuits Conf. (Seville)* pp 27–36
[2] Qian S, Patil S, Stoute S, Nian-Xiang S and Lehman B 2012 *Proc. IEEE Energy Conversion Cong. and Exposition (Raleigh)* pp 394–400
[3] Maurath D, Becker P F, Spreemann D and Manoli Y 2012 *J. IEEE Solid-State Circuits* **47** 1369–1380
[4] Leicht J, Maurath D and Manoli Y 2012 *Proc. IEEE Eur. Solid-State Circuits Conf. (Bordeaux)* pp 101–104
[5] Arroyo E and Badel A 2011 *Sensors and Actuators A: Physical* **171** 266 – 273 ISSN 0924-4247
[6] Sanchez D, Hehn T, Frey A and Manoli Y 2012 *Proc. Power MEMS 2012 (Atlanta)* pp 432–435
[7] Leicht J, Hehn T, Maurath D, Moranz C and Manoli Y 2013 *J. Phys.: Conference Series (PowerMEMS 2013, London)*
[8] Williams C, Shearwood C, Mellor P, Birch T and Yates R 2001 *Proc. IEE Circuits Devices Syst.* **148** 337–342

Sensorless Estimation and Nonlinear Control of a Rotational Energy Harvester

Kameswarie Nunna[1], **Tzern T Toh**[1], **Paul D Mitcheson**[1], **Alessandro Astolfi**[1,2]

[1] Department of Electrical and Electronic Engineering, Imperial College London, London SW7 2AZ

[2] The Dipartimento di Ingegneria Civile e Ingegneria Informatica, Università di Roma "Tor Vergata", Via del Politecnico, 1 00133 Rome, Italy

E-mail: kameswarie.nunna09@imperial.ac.uk

Abstract. It is important to perform sensorless monitoring of parameters in energy harvesting devices in order to determine the operating states of the system. However, physical measurements of these parameters is often a challenging task due to the unavailability of access points. This paper presents, as an example application, the design of a nonlinear observer and a nonlinear feedback controller for a rotational energy harvester. A dynamic model of a rotational energy harvester with its power electronic interface is derived and validated. This model is then used to design a nonlinear observer and a nonlinear feedback controller which yield a sensorless closed-loop system. The observer estimates the mechancial quantities from the measured electrical quantities while the control law sustains power generation across a range of source rotation speeds. The proposed scheme is assessed through simulations and experiments.

1. Introduction

Energy harvesting is a promising solution to extend the operational lifetime of low-power sensor networks, which over the past few decades have relied on finite energy sources such as batteries. Examples of ambient energy sources that can be converted into useable electrical power include solar, radio frequency waves and vibrations [1]. Elimination of sensors and transducers, *i.e.* sensorless operation, in these harvesters reduces the cost and increases the system ruggedness and reliability. The main approaches to sensorless operation are based on saliency, Kalman filter and model reference techniques [2]. However, the estimation of mechanical quantities from measured electrical quantities still remains challenging because the existing techniques are computationally intensive, require special construction for estimation or require proper initialisation [3].

A rotational energy harvester which is based on balancing gravitational and motor torques has been reported in [4]. The control scheme discussed in [4] focuses on the impedance matching for optimal power transfer between the load and includes a current limit to prevent the mass from flipping around. The nonlinear controller presented here is based on Lyapunov stability criterion and aims to improve the harvester's efficiency by using the estimated mechanical states of the harvester thereby providing sensorless operation.

2. Development and validation of the harvester's mathematical model

The rotational energy harvester consists of a DC generator with its stator coupled to a continuously rotating source and a semicircular mass m attached to the rotor at a distance l from the axis of rotation (see Figure 1a). When power is drawn from the generator the torque between the stator and rotor (motor torque) is counteracted by the torque generated by the gravitational force acting on the offset mass (gravitational torque). The difference between these two torques creates a difference in the angular speeds of the stator and the rotor that can be tapped off as power. The generated power is collected in a storage supercapacitor for future use [4]. To ensure optimal power transfer from the harvester to the load, the load resistance, R_L, should be closely matched to the harvester's armature resistance, R_a. The input impedance of a boost converter R_{in} can be controlled by varying its duty cycle δ as $R_{in} = R_L(1-\delta)^2$ [5]. Therefore, it is used as a power electronic interface circuit between the harvester and the load.

The experimental set-up for the harvester and the interface electronics is illustrated in Figure 1b. For a more detailed explanation of the construction and the choice of the circuit components, please refer to [4] and references therein.

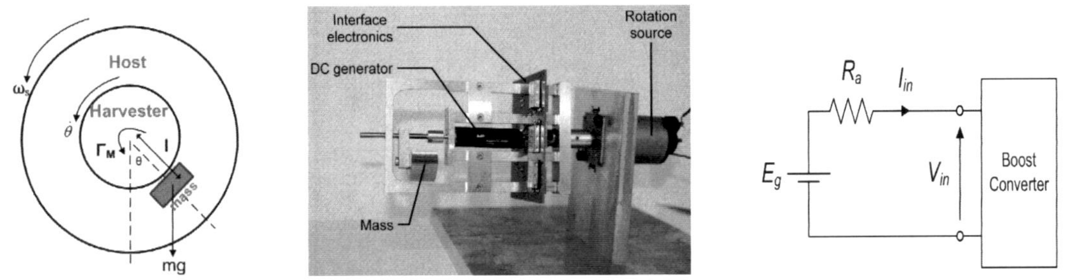

(a) A schematic diagram of the rotational energy harvester.

(b) Experimental set-up.

(c) Equivalent circuit diagram to calculate I_{in}.

Figure 1: Rotational energy harvester.

From the free body diagram of the offset mass in Figure 1a, the torque balance on the mass attached to the rotor of the harvester is given by $J\dot{\omega} = \Gamma_M - mgl\sin\theta$, where J is the moment of inertia of the mass, ω is the angular velocity, and θ is the deflection angle of the mass measured from the vertical axis. The motor torque Γ_M is calculated as $\Gamma_M = -k_T I_{in}$, where k_T is the torque constant of the motor, and I_{in} is the current drawn by the DC generator. The negative sign indicates that the current flows out of the generator and into the boost converter. The voltage generated by the harvester, E_g, when it is attached to a continuously rotating source at a speed ω_s is calculated as $k_E(\omega_s - \omega)$, where k_E is the "motor constant" of the DC generator.

Application of Kirchoff's voltage law to the circuit in Figure 1c and use of the relations $I_{in} = \dfrac{V_{in}}{R_{in}}$ and $R_{in} = (1-\delta)^2 R_L$ yields

$$V_{in} = \frac{k_E(\omega_s - \omega)(1-\delta)^2 R_L}{(R_a + (1-\delta)^2 R_L)}\,. \tag{1}$$

The standard averaged model that describes the dynamics of the boost converter composed of an inductor and a supercapacitor is (see [5] and references therein)

$$\dot{I}_L = -\frac{(1-\delta)V_C}{L} + \frac{1}{L}\underbrace{\frac{k_E(\omega_s - \omega)(1-\delta)^2 R_L}{(R_a + (1-\delta)^2 R_L)}}_{V_{in}}\,,\quad \dot{V}_C = \frac{I_L}{C} - \frac{V_C}{R_L C}\,, \tag{2}$$

where V_C is the voltage across the supercapacitor of capacitance C, I_L is the current flowing through the inductor of inductance L, and V_{in} is as in (1).

From (1) and (2), the mathematical model of the harvester with the interface circuit is given by the equations

$$\dot{x}_1 = x_2, \qquad \dot{x}_2 = -\frac{mgl\sin x_1}{J} + \frac{k_E k_T(\omega_s - x_2)}{R_a + (1-\delta)^2 R_L},$$
$$\dot{x}_3 = -\frac{(1-\delta)x_4}{L} + \frac{k_E(\omega_s - x_2)(1-\delta)^2 R_L}{L(R_a + (1-\delta)^2 R_L)}, \qquad \dot{x}_4 = \frac{(1-\delta)x_3}{C} - \frac{x_4}{R_L C}, \tag{3}$$

where x_1, x_2 describe the angular position θ and the angular velocity ω of the mass, respectively, and x_3, x_4 describe the inductor current I_L and the output capacitor voltage V_C of the boost converter, respectively.

It can be inferred from the $\dot{x}_2$ equation that the operating condition: $x_1 = \pm\pi/2$ and $x_2 = 0$ maximises the amount of gravitational torque, *i.e.* $\frac{mgl\sin x_1}{J}$, generated by the mass and ensures the largest difference in rotational speeds between the stator and rotor, *i.e.* $\omega_s - x_2$. As the source rotation increases, maintaining the angular position of the mass at $\pi/2$ prevents it from flipping over and synchronising with the source rotation.

To validate the model (3) the experimental set-up of the harvester and the simulated model are driven by the same input and the output data are collected. The relative errors of I_L and V_C for two selections of source rotations and fixed duty cycles are plotted in Figure 2. These errors, calculated as $e_{i,rel} = \frac{|x_{i,meas} - x_{i,sim}|}{x_{i,meas}}$, for $i = 3, 4$, are below 0.2% which suggests that the mathematical model gives an accurate description of the system.

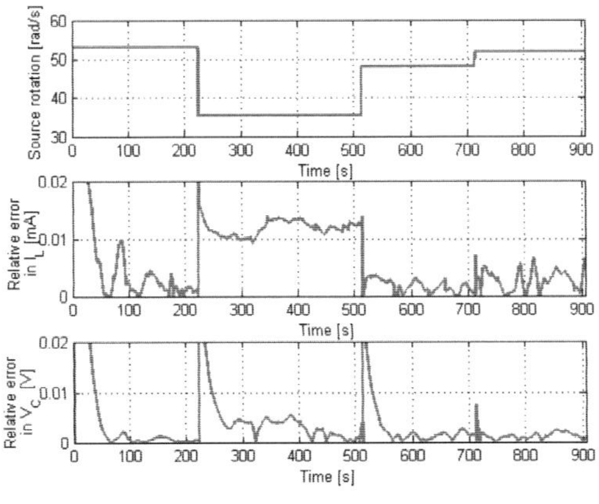

(a) Duty cycle = 70%

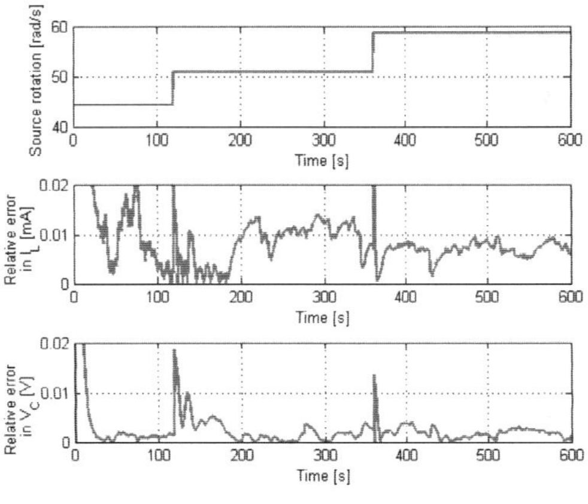

(b) Duty cycle = 80%

Figure 2: Time histories of the source rotations (top), relative errors of the inductor current (middle), and relative errors of the output capacitor voltage (bottom).

3. Sensorless estimation

The observer design is based on the method described in [6]. We aim to accurately estimate the mechanical variables *i.e.* angular position and velocity, from the measured electrical variables *i.e.* inductor current and capacitor voltage.

Proposition 1: Consider the rotational energy harvester system as described in (3). The dynamical system

$$\begin{aligned} \dot{\hat{x}}_1 &= \hat{x}_2 + \beta_2(x_3) - \frac{\partial \beta_1}{x_3} x_{3o} \,, \\ \dot{\hat{x}}_2 &= -\frac{mgl \sin(\hat{x}_1 + \beta_1(x_3))}{J} - \frac{\partial \beta_2(x_3)}{x_3} x_{3o} - \left(\frac{k_E k_T (\omega_s - \hat{x}_2 - \beta_2)}{R_a + (1-\delta)^2 R_L} \right) , \end{aligned} \tag{4}$$

with states $[\hat{x}_1(t),\ \hat{x}_2(t)]^\top \in \mathbb{R}^2$, $b_1 > 0$, $b_2 < 0$, inputs: $x_3(t) \in \mathbb{R}$, $x_4(t) \in \mathbb{R}$, $\delta(t) \in [0,\ 1]$ and $\omega_s(t) \in \mathbb{R}$,

$$\beta_1(x_3) = \frac{b_1 - 1}{\rho} x_3 \,, \qquad x_{3o} = -\frac{(1-\delta)x_4}{L} + \rho(\omega_s - \hat{x}_2 - \beta_2) \,, \tag{5}$$

$$\beta_2(x_3) = \left(\frac{b_2}{\rho} - \frac{k_E k_T}{\rho(R_a + (1-\delta)^2 R_L)} x_3 \right) , \qquad x_{4o} = \frac{(1-\delta)x_3}{C} - \frac{x_4}{R_L C} \,, \tag{6}$$

is such that $\lim_{t\to\infty} (x_1(t) - \hat{x}_1(t)) = 0$ and $\lim_{t\to\infty} (x_2(t) - \hat{x}_2(t)) = 0$, *i.e.* $\hat{x}_1$ and $\hat{x}_2$ are asymptotically converging estimates of x_1 and x_2 respectively.

Proof. Define $z = [\beta_1(x_3) + \hat{x}_1 \,, \beta_2(x_3) + \hat{x}_2]^\top - [x_1 \,, x_2]^\top$, and note that

$$\begin{bmatrix} \dot{z}_1 \\ \dot{z}_2 \end{bmatrix} = \begin{bmatrix} 0 & 1 - \rho \frac{\partial \beta_1}{x_3} \\ -\frac{mgl}{J} & \frac{k_E k_T}{R_a + (1-\delta)^2 R_L} - \rho \frac{\partial \beta_2}{x_3} \end{bmatrix} \begin{bmatrix} z_1 \\ z_2 \end{bmatrix} - \frac{mgL}{J} \begin{bmatrix} 0 \\ \Gamma(x_1, z_1) \end{bmatrix} z_1 \,. \tag{7}$$

where $\rho = \dfrac{k_E(1-\delta)^2 R_L}{L(R_a + (1-\delta)^2 R_L)}$ and $\Gamma = \dfrac{\sin(x_1 + z_1) - \sin(x_1) - z_1}{z_1}$. The variables b_1 and b_2 are chosen such that $b_2 < 0$ and $\dfrac{Jb_2}{mglb_1} \le \dfrac{1}{\| \Gamma \|_2}$ The claim then follows from standard arguments on stability of passive systems, see [7].

4. Controller design

The equations of the model in (3) suggest that the angular position of the mass x_1 can be adjusted by changing the amount of torque acting on it. This can be changed using the control variable δ.

Proposition 2: Consider the rotational energy harvester system (3). Assume ω_s is constant. The control law $\delta = 1 - \sqrt{\dfrac{k_E k_T}{\mu R_L} - \dfrac{R_a}{R_L}}$ where $\mu = -\dfrac{mgl(1 - \sin x_1)}{(\omega_s - x_2)} + \eta\,(\omega_s - x_2)\cos x_1 + \dfrac{mgl}{\omega_s}$ for $0 < \eta << 1$ globally asymptotically stabilizes the equilibrium $(\pi/2, 0)$ of the (x_1, x_2) subsystem. In addition $\delta(t) \in [0,\ 1]$ and $x_2(t) < \omega_s$ for all $t \ge 0$.

Proof. Consider the candidate Lyapunov function $V(x_1, x_2) = (1 - \sin x_1) + \dfrac{J}{\eta}\left(\ln|\omega_s - x_2| - \dfrac{\omega_s}{\omega_s - x_2} \right)$. Note that $V \ge 0$ for $0 < \eta << 1$, and $V = 0$ for $x_1 = \pi/2 + k\pi$ and $x_2 = 0$. In addition $\lim_{x_2 \to \omega_s} V = \infty$ The time derivative of V along the closed-loop trajectories is

$$\dot{V} = -\frac{mglx_2^2}{(\omega_s - x_2)^2} \le 0 \,.$$

The claim then follows from standard LaSalle arguments, see [7].

Remark 1: The physical characteristics of the harvester along with the choice of δ as the control variable restrict the operational range of the harvester. This limitation can be overcome if a saturation control scheme such as $\min\left(\max\left(\mu, \dfrac{k_E k_T}{(R_L + R_a)} - \dfrac{mgl}{\omega_s} \right), \dfrac{k_E k_T}{R_a} - \dfrac{mgl}{\omega_s} \right)$ is used.

5. Experimental results of the nonlinear observer and controller

Figure 3a demonstrates the performance of the observer described in Proposition 1 with measured electrical quantities, $\omega_s = 75$ rad/s, $\delta = 0.8$, $b_1 = 50$, and $b_2 = -70$. The observer is used to implement the control law in Proposition 2. Figure 3b shows the angular position of the mass being held at approximately $\pi/2$ rad for a source rotation $\omega_s = 75$ rad/s. The error in angular position of the mass in the experimental implementation is due to rounding errors in the microcontroller. Figure 3c illustrates the closed-loop currents and voltages from the experiments and simulations.

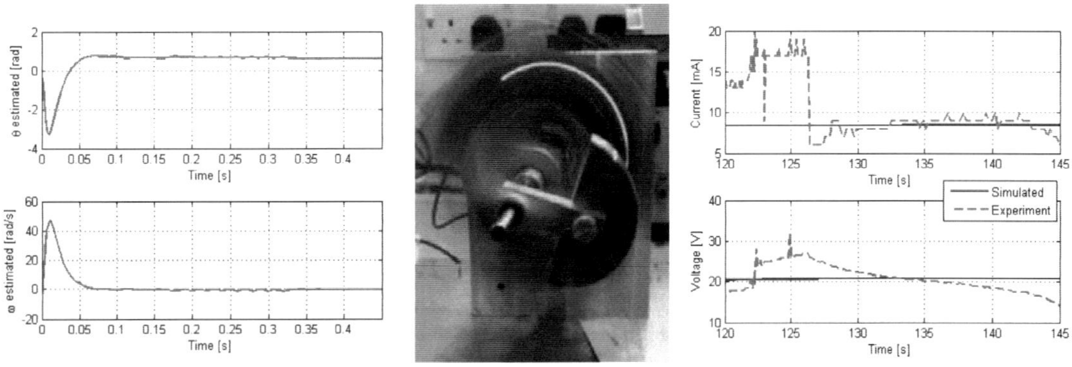

(a) Estimated angular position (top) and angular velocity (bottom).

(b) Mass being held at an angular position of approximately $\pi/2$.

(c) Simulated/measured currents (top) and voltages (bottom).

Figure 3: Performance of the proposed observer-controller scheme.

6. Conclusions

A model of a rotational energy harvester with its power electronic interface has been derived and validated. This model has been used for the design of an observer and a controller, the performances of which have been demonstrated via experiments. The proposed observer method can be adapted for other types of harvesters.

References

[1] Paradiso J A and Starner T 2005 *Pervasive Computing, IEEE* **4** 18–27
[2] Rashed M and Stronach A F 2004 *IEE Proc.-Electr. Power Appl.* **151** 685–693 ISSN 1350-2352
[3] Degner M W and Lorenz R D 1998 *IEEE Trans. Ind. Appl.* **34** 1097–1104
[4] Toh T T, Mitcheson P D, Holmes A S and Yeatman E M 2008 *J. Micromech. Microeng.* **18** 104008
[5] Rodriguez H, Ortega R, Escobar G and Barabanov N 2000 *Systems & Control Letters* **40** 1 – 8 ISSN 0167-6911
[6] Astolfi A, Karagiannis D and Ortega R 2008 *Nonlinear and adaptive control with applications* (Springer)
[7] Khalil H K 2002 *Nonlinear systems* vol 3 (Prentice hall Upper Saddle River)

Journal of Physics: Conference Series **476** (2013) 012053 doi:10.1088/1742-6596/476/1/012053

Modeling and analysis of silicon-embedded MEMS toroidal inductors

M. Araghchini[1] and J. H. Lang[1]

[1] Massachusetts Institute of Technology, Cambridge, MA, 02139, USA

E-mail: `mohamad@mit.edu`

Abstract. This paper presents the modeling and analysis of three-dimensional silicon-embedded toroidal inductors designed for power converter applications. Special attention is given to modeling phenomena associated with the presence of silicon, namely an increase in loss and parasitic capacitance. Silicon-embedded inductors can be fabricated with silicon inside the donut-shaped toroidal core and inside the donut hole, as well as with silicon above, below and outside the inductor. It is argued here that, with the exception of the losses in the core at high doping densities, the losses in the silicon can be tolerated in many power applications, making fully-integrated silicon-embedded air-core inductors viable for power applications. An equivalent circuit model is presented for such inductors which captures the stored magnetic energy, the parasitic electric energy stored between the windings and the silicon, the loss in the toroidal windings, and the electrically- and magnetically-driven losses inside the silicon. The model developed here is verified against experimental data, and the comparison shows a good match over the frequency range of interest to power electronics applications.

1. Introduction

Efforts to embed microfabricated solenoidal and toroidal inductors were previously reported in [1, 2] where fabrication limitations led to low inductances near several nH. Additionally, the inductors were widely separated from the surrounding silicon or suspended on a membrane over an open cavity to minimize losses [2]. Higher inductance and less restrictive fabrication is required for next-generation integrated power converters operating at frequencies near 10-100 MHz [3]. Correspondingly, this paper focuses on embedded inductors such as those fabricated more tightly within a deep recess in the silicon substrate as reported recently in [4]. There, a 50-nH inductor was demonstrated; its inductance was extendable to hundreds of nH using more turns. Analytic models for such inductors are essential for understanding the impact of their surrounding silicon, and for their optimized design. Models of the stored magnetic energy, the winding conduction loss, and the losses induced by adjacent windings have already been developed in [5, 6] for non-embedded inductors. These models can also be used for silicon-embedded inductors. Therefore, this paper focuses on modeling the additional phenomena that arise due to the presence of the silicon substrate. This includes losses in the silicon and increased parasitic capacitance. The original and extended models are combined to form an equivalent circuit model for silicon-embedded inductors.

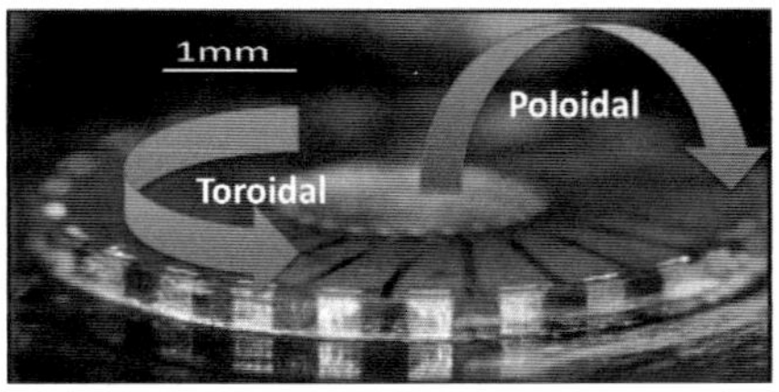

(a)

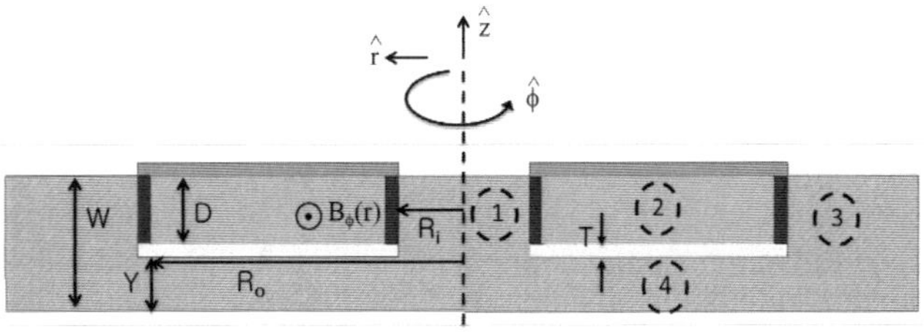

(b)

Figure 1. (a) Poloidal and toroidal directions in a microfabricated toroidal inductor; image from [5]. (b) Cross section of a silicon-embedded toroidal inductor. Not shown is a thin oxide layer having thickness Δ between the winding conductors (orange and yellow), winding vias (blue) and silicon substrate (gray).

2. Losses

As seen in Figure 1(a), the winding in a toroidal inductor spirals around the toroidal core in the poloidal direction as the spiral progresses in the toroidal direction. Thus, there exists a single-turn toroidal current which creates poloidal flux outside the toroidal core and a multi-turn poloidal current which creates toroidal flux through the toroidal core. Both fluxes drive currents and hence losses in the silicon. Additionally, the magnitude of the potential of the winding turns progresses linearly from one turn to the next in the toroidal direction. Capacitive coupling of this temporally- and spatially-varying potential also drives losses in the silicon. The geometry shown in Figure 1(b) is used to model these losses.

The magnetically-driven losses in the silicon are modeled under the assumption that the skin depth in silicon is comparable to or larger than R_o or smaller dimensions at the operating frequencies of interest; R_o is typically 5 mm. Assuming a doping density of 10^{16} cm^{-3} and an operating frequency of 30 MHz, the skin depth of silicon is approximately 7 mm, and so the large skin depth assumption is justified. At a doping density of 10^{18} cm^{-3}, for example, the skin depth drops to 1 mm, and magnetic diffusion must be considered.

The magnetically-driven losses in the toroidal core, Region 2 ranging over $R_i < r < R_o$, is modeled under the assumption that the skin depth of silicon is larger than $D/2$; D is typically 1 mm or less. Inside the core, the time-varying magnetic flux density $\bar{B}$ is assumed to be $\hat{\phi}$-directed and dependent only on r. Further, since $(R_o - R_i) \gg D$, the electric field $\bar{E}$ is assumed to be $\hat{r}$-directed and dependent only on r and z; the $\hat{z}$-directed $\bar{E}$ which is dominant near $r = R_i$ and $r = R_o$ is ignored. With these assumptions, $\bar{E}$ may be determined directly from $\bar{B}$ using Faraday's Law. The time-average Ohmic power dissipation density is then computed as $\sigma \bar{E} \cdot \bar{E}/2$ where σ is the silicon conductivity. Finally, the power dissipation density is integrated over the volume of the toroidal core to yield P_{M2}, the magnetically-driven loss in the core, according to

$$P_{M2} = \sigma\, \omega^2\, \mu_o^2\, N^2\, I^2\, D^3\, \ln(R_o/R_i) \,/\, (48\,\pi) \tag{1}$$

where NI is the peak number of Ampere-Turns around the core. As a typical example, for a toroidal inductor with $N = 25$, $R_o = 5$ mm, $R_i = 0.5$ mm, $D = 1$ mm, a silicon doping density of 10^{18} cm^{-3} which yields $\sigma = 5 \cdot 10^3$ S/m, $\omega = 2\pi \times 30$ MHz, and $I = 1$ A, (1) yields $P_{M2} = 2.7$ W, whereas a doping density of 10^{16} cm^{-3} which yields $\sigma = 200$ S/m, yields $P_{M2} = 0.1$ W. Therefore, due to the significant loss for doping densities above approximately 10^{16} cm^{-3}, the option of having a silicon core is eliminated and not considered further here.

The magnetically-driven losses in Regions 1 and 3 are modeled under the assumption that the skin depth of silicon is larger than R_i and R_o, respectively. In these regions, a nominally

$\hat{z}$-directed time-varying poloidal $\bar{B}$ drives a $\hat{\phi}$-directed $\bar{E}$. The resulting losses are analyzed here based on the inductance L_{P} [9], associated with the energy stored in the exterior poloidal magnetic fields given by

$$L_{\mathrm{P}} = 0.5\,\mu_{\mathrm{o}}\,(R_{\mathrm{o}} + R_{\mathrm{i}})\,(\ln(8(R_{\mathrm{o}} + R_{\mathrm{i}})/(R_{\mathrm{o}} - R_{\mathrm{i}})) - 2) \quad . \tag{2}$$

Assuming that the skin depth of silicon is larger than R_{i}, the peak poloidal flux Λ_{P}, and the magnitude of the approximately uniform poloidal magnetic field in Region 1, $r < R_i$, can be determined from

$$\Lambda_{\mathrm{P}} = \pi\,R_{\mathrm{i}}^2\,|\bar{B}| = L_{\mathrm{P}}\,I \quad . \tag{3}$$

Faraday's Equation in combination with (2) and (3) then yields $|\bar{E}|$ from which the magnetically-driven Ohmic loss in Region 1, P_{M1}, may be computed. This computation yields

$$P_{\mathrm{M1}} = \sigma\,W\,\omega^2\,L_{\mathrm{P}}^2\,I^2/(16\,\pi) \quad . \tag{4}$$

In Region 3, the magnitude of the magnetic flux density is assumed to fall off as $B_{\mathrm{o}}(R_{\mathrm{o}}/r)^3$. Thus it decays in r as it would away from a magnetic dipole assuming that the skin depth of silicon is larger than R_{o}. The coefficient B_{o} can then be determined from

$$\Lambda_{\mathrm{P}} = \int_{R_{\mathrm{O}}}^{\infty} B_{\mathrm{o}}(R_{\mathrm{o}}/r)^3\,2\pi r dr = 2\,\pi\,B_{\mathrm{o}}\,R_{\mathrm{o}}^2 = L_{\mathrm{P}}\,I \quad . \tag{5}$$

Faraday's Equation in combination with (2) and (5) then yields $|\bar{E}|$ from which the magnetically-driven Ohmic loss in Region 3, P_{M1}, may be computed. This computation yields

$$P_{\mathrm{M3}} = D\,\sigma\,\omega^2\,L_{\mathrm{p}}^2\,I^2/(8\,\pi) \quad . \tag{6}$$

The summation of (4) and (6) is used to model the magnetically-driven losses in silicon.

The electrically-driven losses are driven by the time- and spatially-varying winding turn potentials coupled capacitively to the silicon through the insulator that separates the turns from the silicon. Modeling the associated electric fields and Ohmic loss begins by expressing the electric potential around the inductor in the $\hat{\phi}$ direction as a Fourier sum of single harmonic functions. Following [7], this potential is then used to determine two-dimensional ($\hat{r}$,$\hat{\phi}$) potential solutions to Laplace's Equation in Regions 1, 3 and 4. The electric fields and the electrically-driven Ohmic losses can then be found directly from the potentials in the three regions. The details of this approach are given in [8]. For brevity, only the results are detailed here.

In Regions 1 and 3, the electric potential is modeled as a Fourier sum of terms taking the form $(\phi_+ r^m + \phi_- r^{-m})\sin(m\phi)$, where m is the Fourier expansion index, and ϕ_+ and ϕ_- are coefficients. A separate expression having this form is used for each silicon region and each insulating silicon-dioxide region between the silicon and the winding turns. The solutions are stitched together through boundary conditions at material interfaces, and the coefficients are determined by matching boundary conditions. The $\hat{r}$- and $\hat{\phi}$-directed electric fields in the silicon, and the associated Ohmic losses, may then be calculated directly from the potential solutions. For Region 3, $r > R_{\mathrm{o}}$, this approach yields

$$P_{\mathrm{E3}} = \frac{2\pi\,D\,\sigma\,V^2\omega^2\epsilon_{\mathrm{I}}^2}{m\,(\Gamma_{\omega\epsilon}^2 + \Gamma_{\sigma}^2)} \quad , \quad \Gamma_{\sigma} = \sigma_{\mathrm{Si}}\left(\left(\frac{R_{\mathrm{o}} + \Delta}{R_{\mathrm{o}}}\right)^m - \left(\frac{R_{\mathrm{o}}}{R_{\mathrm{o}} + \Delta}\right)^m\right) \quad ,$$
$$\Gamma_{\omega\epsilon} = \omega\left(\frac{R_{\mathrm{o}}}{R_{\mathrm{o}} + \Delta}\right)^m (\epsilon_{\mathrm{I}} - \epsilon_{\mathrm{Si}}) + \omega\left(\frac{R_{\mathrm{o}} + \Delta}{R_{\mathrm{o}}}\right)^m (\epsilon_{\mathrm{I}} + \epsilon_{\mathrm{Si}}) \tag{7}$$

for the electrically-driven loss P_{E3} where V is the terminal voltage of the inductor, ϵ_{Si} is the permittivity of silicon, and ϵ_{I} is the permittivity of the silicon-dioxide insulator. The same

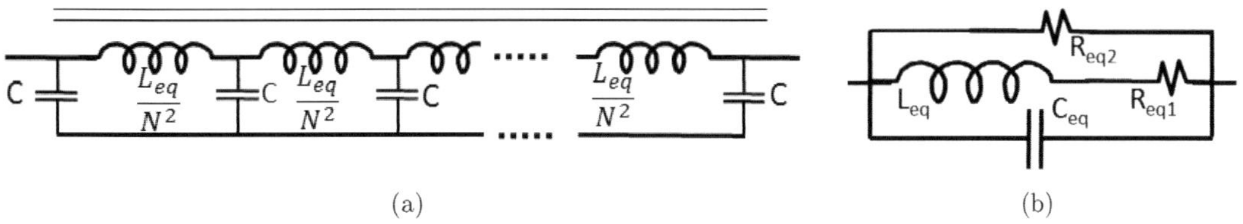

Figure 2. (a) Semi-distributed reactive circuit model for a silicon-embedded inductor. (b) Simplified equivalent circuit for a silicon-embedded inductor.

approach can be used for computing the loss P_{E1} in Region 1, $r < R_i$. Indeed, P_{E1} can be found from (7) by replacing $(R_o + \Delta)$ with $(R_i - \Delta)$ and by replacing R_o alone with R_i alone.

To determine the electrically-driven loss P_{E4} in the silicon layer underneath the inductor in Region 4, a structure involving three layers is considered. Beneath the winding turns is a thin insulating layer of silicon dioxide having a thickness of Δ above a silicon layer having a thickness Y, which is in turn above an infinitely-thick air layer. An approach similar to that used above for Regions 1 and 3 is used here as well. This approach is detailed in [8]. The analytical results for the underlying layer are not included here for brevity as they are quite long.

3. Equivalent Circuit & Capacitance

Equivalent circuits for sinusoidal-steady-state inductor operation are useful for power circuit analysis, co-optimization and design. They also capture what is naturally measured as the inductor impedance. Figure 2(a) shows a semi-distributed equivalent reactive circuit modeling a silicon-embedded toroidal inductor. This circuit treats the inductor as a leakage-free N-winding transformer with one transformer winding per inductor turn. A lumped parasitic capacitor C exists between each turn and the silicon adjacent to that turn; the silicon substrate is treated as a common node. Analysis of the inductor in Figure 1(b) and the circuit in Figure 2(a) leads to an equivalent capacitance of

$$C_{eq} = C\,(N^2 - 1)\,/\,(12\,N) \quad , \quad C = 2\,\pi\,F\,\epsilon_I\,(R_oD + R_iD + 0.5(R_o^2 - R_i^2))\,/\,(N\,\Delta) \quad . \tag{8}$$

as used in the equivalent circuit in Figure 2(b) where F is the lateral filling factor fraction of the area of the winding turns above the silicon. The inductance L_{eq} accounts for the toroidal and poloidal magnetic flux [5, 8]. Ohmic winding losses [5, 8], and P_{M1} and P_{M3} are driven by the inductor current and properly appear as parts of R_{eq1} in series with L_{eq} as shown in Figure 2(b). Their total time-average loss is divided by $I^2/2$ to determine their contributions to R_{eq1}. Losses P_{E1}, P_{E3} and P_{M4} are driven by V, and properly appear as parts of R_{eq2} in parallel with L_{eq} and R_{eq1}, along with C_{eq}. Their total time-average loss is divided by $V^2/2$ to determine their contributions to R_{eq2}. In all cases, the losses are frequency dependent, and so are evaluated at the frequency of power circuit operation or impedance measurement prior to determining R_{eq}.

4. Experiments & Model Comparisons

The measured impedance of toroidal inductors embedded in 1-10 Ω·cm standard-resistivity (SR) wafers and 2-3 kΩ·cm high-resistivity (HR) wafers, having skin depths of 1.6 cm and 46 cm, respectively, was reported in [4]. This data is shown in Figure 3(a), together with theoretical predictions based on the models presented above. A 0.65-mΩ contact resistance per vertical via joint is added to R_{eq1} to improve the experiment-to-theory match. The embedded inductors have $N = 25$, $W = 500\,\mu$m, $D = 300\,\mu$m, $R_o = 6$ mm, $R_i = 2$ mm, $\Delta = 12\,\mu$m and an average copper winding thickness of $T = 30\,\mu$m. The inductors exhibit a measured $\Im(Z)/\omega$ of 45 nH at

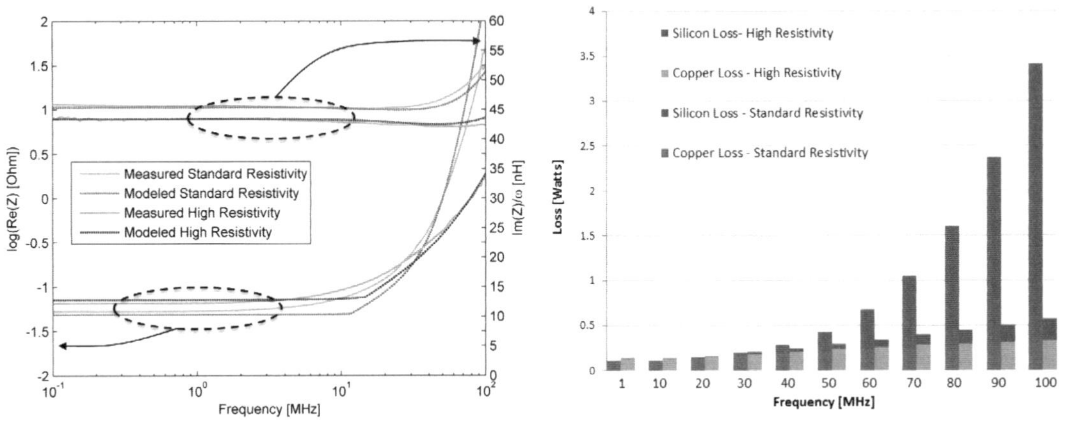

(a) (b)

Figure 3. (a) Measured and modeled impedance of embedded inductors in SR and HR wafers. (b) Decomposition of modeled losses for the cases of SR and HR wafers.

low frequency and a DC resistance of 300 mΩ. Figure 3(b) decomposes the modeled losses into winding and silicon losses. The inductors in the SR and HR wafers exhibit peak quality factors of 16 near 40 MHz and 22 near 80 MHz, respectively. Note that the silicon losses in the two inductors rise quickly near 40-50 MHz. Further discussion of these results appears in Section 5.

5. Conclusions

Several important conclusions can be drawn from the work described here. First, the equivalent circuit model matches well the measured behavior of a silicon-embedded toroidal inductor. Second, as long as it is removed from the core of the inductor at high doping densities, the close proximity of silicon need not significantly degrade inductor quality factor. Winding loss dominates the quality factors modeled and measured here for frequencies up to 40 MHz, making winding optimization more important than silicon selection; see Figure 3. For higher frequencies, quality factor is a strong function of silicon resistivity as expected, making its selection more important than winding design. At these frequencies, electrically-driven silicon losses appear to be dominant, making large Δ an important fabrication objective.

Acknowledgments

The authors gratefully thank Ms. X. Yu, Dr. F. Herrault, Dr. J. K. Kim, and Prof. M. G. Allen from the Georgia Institute of Technology for providing the measurement data shown in Figure 3(a) and the underlying image in Figure 1(a).

References

[1] Liang Y C, Zeng W, Ong P H, Gao Z, Cai J and Balasubramanian N, *IEEE Elec Dev Let*, 2002, **23**, 700-03.
[2] Gu L, Li X, *Microelectromecs Systems*, 2007, **16** 1162-72.
[3] Araghchini M, et al., *J. IEEE Transactions on Power Electronics*, 2013, **28(9)**, 4182-4201.
[4] Yu X, Araghchini M, Herrault F, Kim J, Lang J H and Allen M G , *Proc. PowerMEMS 2012*, 2012, 58 - 61
[5] Araghchini M, et al., *Proc. IEEE Energy Conversion Congress and Exposition (ECCE), 2012*, 3293-3300.
[6] Araghchini M, et al., to appear in *J. IEEE Transactions on Industry Applications, 2013*
[7] Woodson H and Melcher J R, 1968, *Electromechanical Dynamics, Vol 2*, (John Wiley).
[8] Araghchini M, 2013, *(MEMS) Toroidal Magnetics for Integrated Power Electronics*, PhD Thesis, MIT.
[9] Whinnery R and Duzer V, 1965, *Fields and Waves in Communication Electronics*,(John Wiley).

Radioisotopic battery for long-life, buried autonomous power

R. Walton[1], C. Anthony[1*], D. Chapman[2], N. Metje[2] and M. Ward[3]

[1]School of Mechanical Engineering, University of Birmingham, Edgbaston, Birmingham, B15 2TT
[2]School of Civil Engineering, University of Birmingham, Edgbaston, Birmingham, B15 2TT
[3]School of Engineering, Design and Manufacturing Systems; Faculty of Technology, Engineering & the Environment; Birmingham City University, Birmingham, B4 7XG

[*]c.j.anthony@bham.ac.uk

Abstract. Indirect conversion radioisotopic batteries (ICRBs) are investigated for use as long-life power source for autonomous buried applications. As part of this work the optimum configuration of this class of battery has been experimentally investigated. An ICRB was buried at a depth of 90cm for two months during which time its voltage was monitored, with these results presented. The ICRB successfully demonstrated the buried operation of this class of battery and suggested that the power of the device halves at twice the half-life of the radioisotope used.

1. Introduction

Radioisotopic batteries have been investigated before, however indirect conversion radioisotopic batteries (ICRBs) have received little notice. Much focus has been given to thermal cycle types of radioisotopic battery, such as those used by NASA [1], which require a large mass of radioisotope to effectively operate and are therefore unsuited to terrestrial use. The other types of radioisotopic batteries commonly reported either consist of charging a capacitor with the charged particles of radioactive decay [2] (which results in unusably high voltages, often 10kV or more) or involve generating electricity in a PV cell directly from the kinetic energy of the radioactive decay particles [3]. This latter type of radioisotopic battery produces radiation damage and therefore premature degradation of the PV cells, limiting the useful lifetime of the battery [4]. ICRBs do not suffer from these drawbacks.

ICRBs consist of a radioisotope in close proximity to a phosphor which generates photons when impacted by the decay particles of the radioisotope [5]. The photons travel to a PV cell where they are then converted into electrical energy in the conventional manner [6]. This energy may then be stored in a capacitor to be used when required. A schematic demonstrating this process is shown in figure 1.

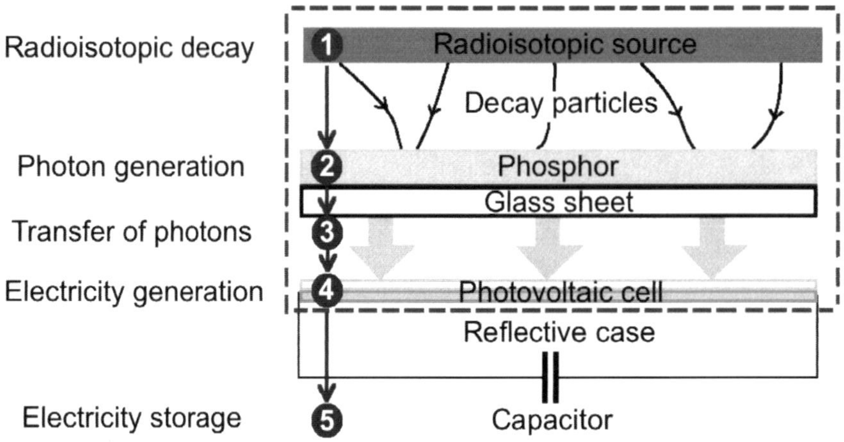

Figure 1. Operating principle of an indirect conversion radioisotopic battery.

Gaseous tritium light sources (GTLSs) consist of a glass ampule coated on the inner surface with a thin layer of phosphor and filled with pressurized tritium gas. GTLSs therefore comprise two of the processes necessary to produce an ICRB and were used during this research due to their sturdiness and availability for use under UK law [7].

2. Radioisotopic battery configuration selection

A preliminary investigation of the most effective configuration of ICRBs was undertaken using commercially available PV cells and GTLSs. The layout of the PV cells and GTLSs relative to each other was investigated in terms of the power generated by the manufactured ICRBs. The power of these ICRBs was measured indirectly by charging a capacitor and measuring the voltage in the capacitor at regular and timed intervals. The power (P) needed to alter the voltage (and hence energy stored) in the capacitor over the time interval between readings was then calculated through application of equation (1) (where C is the capacitance, V_2 is the measured voltage, V_1 is the previously measured voltage and dt is the time interval between readings). A bespoke measurement system was used to measure the voltage of these ultra-low power devices while charging a 1μF 50V aluminium electrolytic capacitor.

$$P = \frac{\frac{1}{2}C(V_2^2 - V_1^2)}{dt} \tag{1}$$

The investigated configurations of ICRBs are displayed in figure 2, where (a) used two GTLSs on a PV cell without reflective casing, (b) used two GTLSs on a PV cell with reflective casing, (c) used two GTLSs sandwiched between two PV cells surrounded by reflective casing, (d) used one GTLS surrounded by three PV cells and reflective casing, and (e) used two GTLSs surrounded by four PV cells and reflective casing. The results of testing are shown in figure 3, which plots the power generated against the voltage at which that power was produced. The PV cells used during this testing were of amorphous silicon type, chosen due to their superior performance at converting low intensity light into electricity.

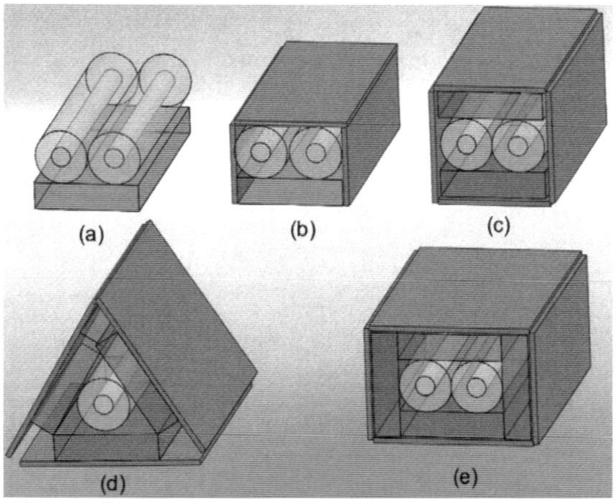

Figure 2. Investigated configurations of ICRBs showing (a) two GTLSs and a PV cell, (b) two GTLSs and a PV cell surrounded by reflective material, (c) two GTLSs and two PV cells, (d) one GTLS and three PV cells and (e) two GTLSs and four PV cells.

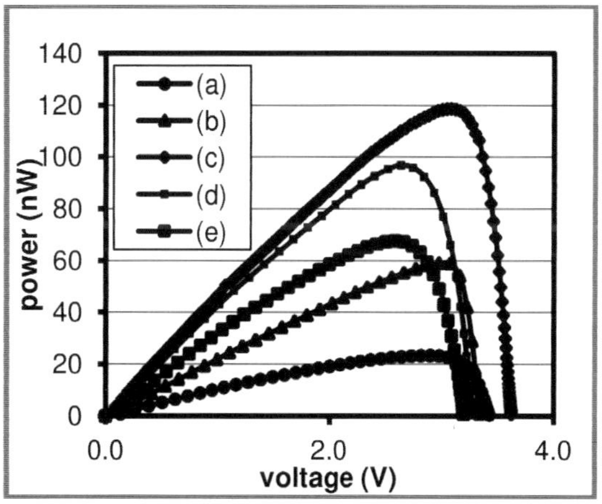

Figure 3. Power against voltage curves for the experimentally tested configurations of ICRBs as measured by charging a capacitor.

The highest power was generated by having two GTLSs sandwiched between two PV cells (configuration (b)) and this was therefore the design which was taken forward for long-life buried testing.

3. Buried, long-life testing

The ICRB was installed in a waterproof enclosure conforming to IP65 standard and was potted with a transparent polyurethance resin (as shown in figure 4) to prevent moisture intruding during the buried operation of the device. The ICRB was then buried at a depth of 90cm in soil of 19% moisture content (at time of burial) and was secured to a water pipe with plastic ties and a jubilee clip, as shown in figure 5. The hole was then backfilled and the performance of the buried ICRB was monitored over a period of two months via a bespoke sensor node, transmitter and data storage system.

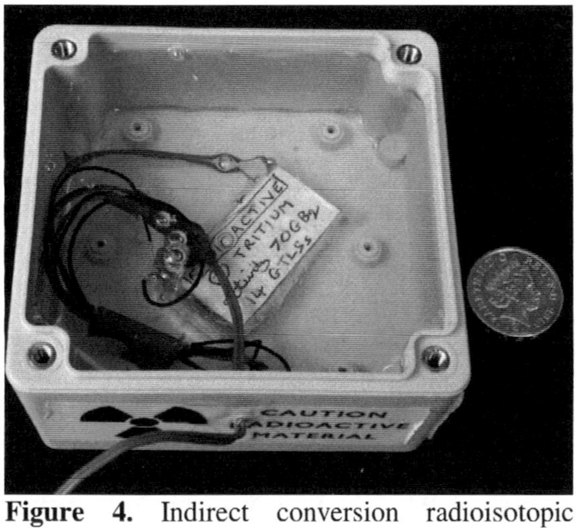

Figure 4. Indirect conversion radioisotopic battery in waterproof enclosure and potted with markings ready for burial (lid not shown).

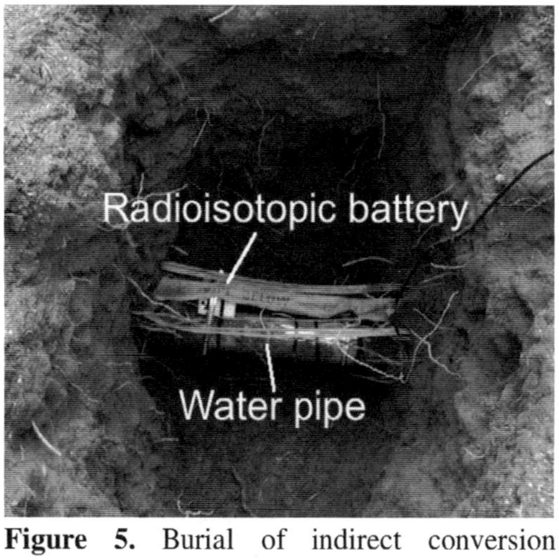

Figure 5. Burial of indirect conversion radioisotopic battery showing the device secured to a water pipe at a 90cm depth.

The complete set of measured voltage readings from the buried ICRB is presented in figure 6 along with the temperature local to the battery which was read by a temperature sensor attached to the outside of the battery enclosure.

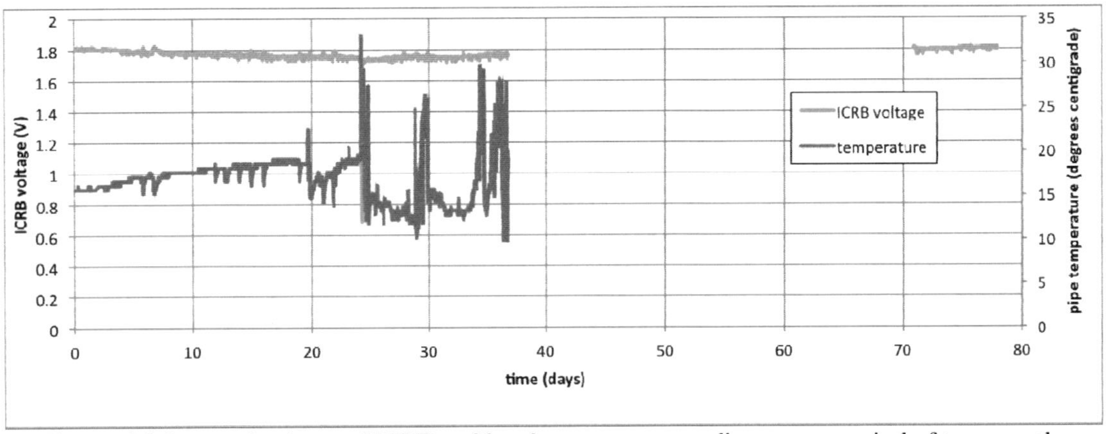

Figure 6. Measured voltage of ICRB and local temperature reading over a period of two months.

After approximately 20 days the temperature sensor degraded, most likely due to water encroaching onto the measurement mechanism, and thereafter produced unreliable data as evidence by the sharp spikes in temperature readings. The gap in readings is a result of a power loss in the logging equipment. Meanwhile the ICRB voltage shows a decline over the period of measurement of 24.4mV from a starting reading of 1.8115V. At this rate of change it would take 7.2 years for the voltage to be reduced to half of its starting value. The radioisotope used in this ICRB was tritium, which has a half-life of 12.3 years. Therefore, the voltage of the ICRB declines at a rate 1.7 times faster than the half-life of the radioisotope. This is primarily due to slow radiation damage of the phosphor layer through the creation of impurities and the ejection of phosphor atoms from the phosphor lattice [8].

A further phenomenon revealed during this testing was the close relationship between the voltage output of the ICRB and the local temperature. This relationship is shown more clearly in figure 7, where it can be seen that each drop in temperature correlates to a rise in generated voltage. In addition there is a general rise in temperature reported up to 20 days, which has also decreased the voltage output of the ICRB beyond the decline expected by battery aging. The effect of temperature on the generation of voltage by PV cells was experimentally investigated by Shaari et al. [9], which revealed similar results. For this reason it is believed that this phenomena is primarily due to the temperature dependent operation of the PV cells in the battery.

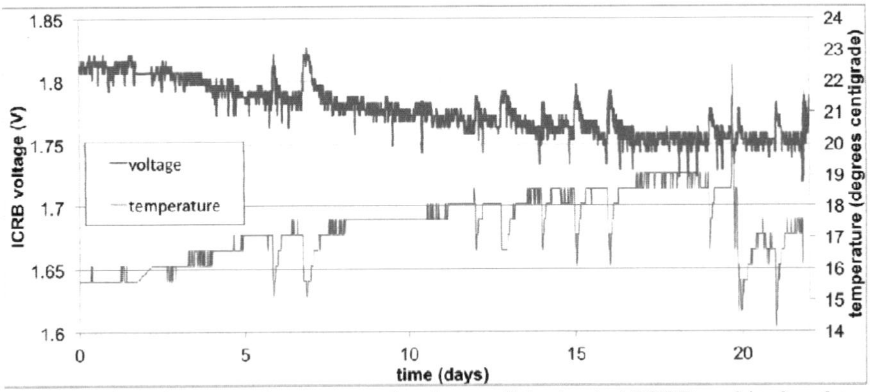

Figure 7. A comparison between the measured voltage of the buried ICRB and its local temperature.

4. Conclusions

Indirect conversion radioisotopic batteries were experimentally investigated as to their optimum configuration which was found to be a sandwich of GTLSs between two PV cells. An ICRB was successfully buried and tested over a two month period proving the durability of this class of battery to this type of environment. Buried long-life testing of the ICRB found that the device would halve in voltage output every 7.2 years. This decline in voltage was due to the decrease in activity of the radioisotope (its half-life) and slow radiation damage of the phosphor layer. The voltage generated by the ICRB is affected by the temperature at which it was operating, which is due to the temperature dependent performance of the PV cells.

Acknowledgments

United Kingdom Water Industry Research (UKWIR) and the University of Birmingham financially supported this research.

References

[1] Bennett G L, Lombardo J J, Hemler R J, Silverman G, Whitmore C W, Amos W R, Johnson E W, Schock A, Zocher R W, Keenan T K, Hagan J C and Englehart R W 2006 *Proc. 4th Int. Energy Conversion Engineering Conference* 720

[2] Kavetskiy A, Yakubova G, Lin Q, Chan D, Yousaf S M, Bower K , Robertson J D, Garnov A and Meier D 2009 *J. Applied Radiation and Isotopes* **67** 6 1057

[3] Honsberg C, Doolittle W A, Allen M and Wang C 2005 *Proc. 31st IEEE Photovoltaic specialists conf.* 102

[4] Deus S 2000 *Proc. 28th IEEE Photovoltaic specialists conf.* 1246

[5] Lambe J, Klick C C 1955 *Phys. Rev.* **98** 909

[6] Luque A and Hegedus S 2011 *Handbook of Photovoltaic Science and Engineering* (Wiley & Sons)

[7] United Kingdom government 2011 Environmental Protection, England and Wales: The Environmental Permitting (England and Wales) Regulations, No. 2043

[8] Brese N E, Rohrer C L and Rohrer G S 1999 *Solid-State Ion. Diffus. React.* **123** 19

[9] Shaari S, Sopian K, Amin N and Kassim M N 2009 *Am. J. of App. Sci.* **6** 586

Thermomechanical stability and integrability of an embedded ceramic antenna with an integrated sensor element for wireless reading in harsh environments

P. Sturesson[1, 2, 3], Z. Khaji[2], S. Knaust[2], J. Sundqvist[1, 2], L. Klintberg[2], and G. Thornell[1, 2]

[1]Ångström Space Technology Centre, Dept. of Eng. Sciences, Uppsala University, Uppsala, Sweden

[2] Division of Microsystems Technology, Dept. of Eng.Sciences, Uppsala University, Uppsala, Sweden

[3] Swedish National Defence College, Stockholm, Sweden

Corresponding author: peter.sturesson@angstrom.uu.se

This paper reports on the design, manufacturing and evaluation of a small, wirelessly powered and read resonating antenna circuit with an integrated pressure sensor. The work aims at developing miniature devices suitable for harsh environments, where high temperature prevents the use of conventional, silicon-based microdevices. Here, the device is made of alumina with platinum as conducting material. Ceramic green tapes were structured using high-precision milling, metallized using screen printing, and subsequently laminated to form stacks before they were sintered. The device's frequency shift as a function of temperature was studied up to 900°C. The contributions to the shift both from the thermomechanical deformation of the device at large, and from the integrated and, so far, self-pressurized sensor were sorted out. A total frequency shift of 3200 ppm was observed for the pressure sensor for heating over the whole range. Negligible levels of thermally induced radius of curvature were observed. With three-point bending, a frequency shift of 180 ppm was possible to induce with a curvature of radius of 220 m at a 10 N load. The results indicate that a robust pressure sensor node, which can register pressure changes of a few bars at 900°C and wirelessly transmit the signal, is viable.

1. Introduction

Miniature devices for harsh environments are facing an increasing interest. Applications are found in in, e.g., space exploration, geothermal research, and monitoring of combustion engines [1, 2]. In harsh environments high temperature, corrosive substances, high pressure and radiation may be confronted. For electronics and sensors, high temperatures can be regarded as a prime challenge.

Above about 600°C there are two main challenges on the generic level: lack of suitable materials and difficulty of interfacing. Beside thermomechanical effects that can jeopardize the operation of the component itself, signal and power transfer by leads becomes a challenge.

Silicon, which may deform slowly at these temperatures if loaded mechanically [3], can be replaced either by Low-Temperature Co-fired Ceramics (LTCC) or by thick film technologies with pre-sintered ceramics [4], or, above 800°C, by High Temperature Co-fired Ceramics (HTCC), which should withstand temperatures well beyond 1000°C [5].

Fonseca, Radosavljevi, Xiong and Tan et al. have demonstrated wireless ceramic two-membrane pressure sensors exhibiting frequency shifts in the range of 1000-10000 ppm per bar [6-9].
Here, the thermomechanical behaviour of a resonant LC antenna circuit made of alumina with printed platinum conductors for use in a sensor node is studied, as this is vital for the design of the device itself, as well as for its integration in physical system architectures. Since the resonance frequency shift should be used for reading of the sensor node's sensor, its dependency on temperature and mechanical deflection is studied.

To demonstrate the integrability with a sensor element, one of the antenna devices has been furnished with a cavity sealed by a metallised membrane, which consists of a capacitor plate to enable wireless pressure readings at high temperatures.

2. Design

The antenna is an inductively coupled LC resonance circuit consisting of a multi-turn planar square loop in series with a 3.6×3.6 mm, centred capacitor, figure 1 (left). The loop pattern including one capacitor plate is printed on a ceramic substrate, and the second capacitor plate, connected through a via, is printed on the backside, figure 1 (right). The coil conductor width is 250 μm and the substrate thickness is 250 μm.

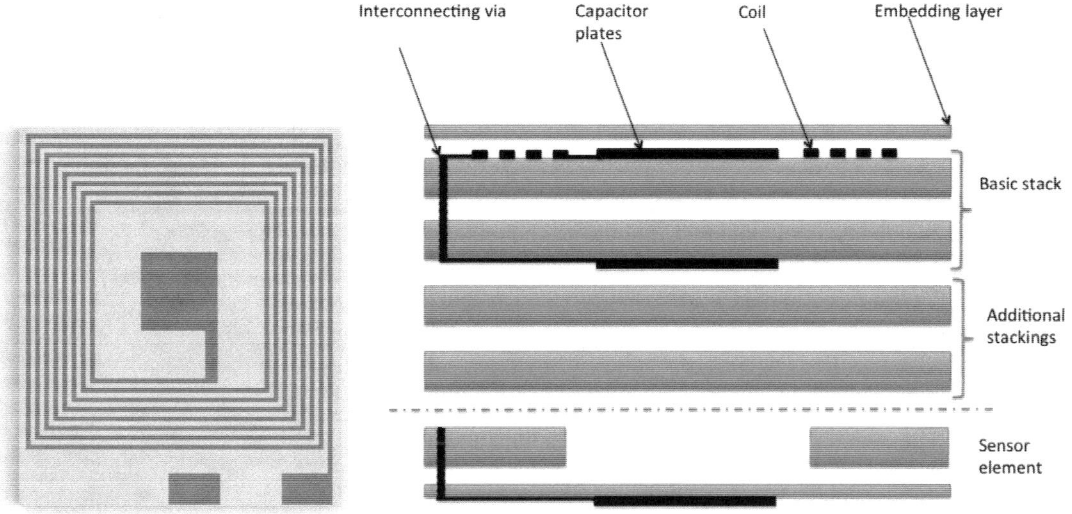

Figure 1. Top view of device, showing the antenna LC circuitry (left), and, in cross section, the stacking of the antenna part with sensor element layers and intermediate structural layers (right).

The frequency of the antenna is determined by its inductance, L, and its capacitance, C, according to:

$$f = \frac{1}{2\pi\sqrt{LC}}, \quad (1)$$

where the inductance depends on geometry only [10]. The dimensions of the loop are estimated to follow the thermal expansion of the much thicker alumina which amounts to 0.8% over 1000°C [11]. The capacitance is given by:

$$C = \varepsilon_0 \varepsilon_r \frac{A}{d}, \quad (2)$$

where A is the area of the capacitor plates, d their spacing, ε_0 the vacuum permittivity, and ε_r the relative permittivity for alumina. If the thermal expansion can be neglected, the main contribution to the change in resonance frequency will come from changes of the dielectric constant of the substrate. Chen et al. have shown that the dielectric constant of alumina is stable up to 550°C for 1 MHz, after which it starts to increase [12]. The onset temperature for the increase appears to increase with increasing frequency.

Here, with one of the capacitor plates placed on a membrane suspended across a cavity, an increased outside pressure will cause the membrane to deflect, hence change the capacitance and, thereby, the resonance frequency. Similarly, when heating the air inside the cavity, a thermally induced pressure increase will cause the membrane to deflect outwards.

3. Materials and fabrication

The devices were laminated from 99.9% 50, 130 and 150 µm thick alumina green tapes (ESL 44007-150/130/50, Electro Science Laboratories, USA). Platinum paste (ESL 5571, ESL Electro Science Laboratories, USA) and a stainless steel screen with a mesh of 325 lines per inch were used for metallization. After printing, the paste was dried at 50°C for 15 minutes.

The pressure sensor element consists of a 130 µm thick sheet with a 5-mm diameter, circular cavity milled with a PCB plotter (Protomat S100, LPKF, Germany) and sealed with a 50 µm thick tape. As a support for the membrane, a 125 µm thick fugitive material (ESL 49000, Electro Sciences Laboratories, USA) was used.

Lamination was performed at 21 MPa and 70°C for 20 minutes in a hydraulic laminator (RMP 210, Bungard, Germany). For devices without sensor elements, the printed sheets were first aligned and laminated in one step, and then laminated with a 50 µm thick top sheet and 4 or 5 150 µm thick sheets from behind. For the device with the sensor element, the sheet with the cavity was laminated with the coil pattern sheet in one step and then laminated with the 50 µm thick embedding and membrane layers (see figure 1). After lamination, the stacks were contoured in the PCB plotter into 23×21 mm devices.

Following a temperature profile with a ramping pace of 1.5°C/min to a dwell temperature of around 500°C to allow the organic binder and fugitive material to dispose, the ramping was increased to 2°C/min to 1550°C where the samples were sintered for two hours.

The resulting devices and their resonance frequencies and thicknesses are presented in table 1.

Table 1. Device IDs, total thicknesses, and resonance frequencies.

Device number	Frequency [MHz]	Thickness [ìm]
1	45	410
2	44	760
3	46	890
4 (w pressure sensor)	63	500

4. Characterization

For monitoring the general frequency dependency of the antennas, a small in-house designed and fabricated furnace was used, figure 2. The heat elements were clamped with sintered alumina nitride plates with a centered clearing for the devices to avoid magnetic interferences. An in-house fabricated tunable, 12-cm diameter loop antenna made of a 4-mm diameter copper rod was placed above the furnace, at about 5 cm distance from the devices, and connected to a vector network analyzer (Fieldfox 9923, Agilent, USA). The resonance frequency of active antenna was tuned to the respective device antenna with a variable (5.5-65 pF) capacitor, and an auxiliary 2.5-cm diameter loop antenna, milled from FR-4 PCB, was used to match the impedance to 50 Ω. The devices were passively powered by the backscattered signal from the active antenna. The network analyzer logged the S11 parameter.

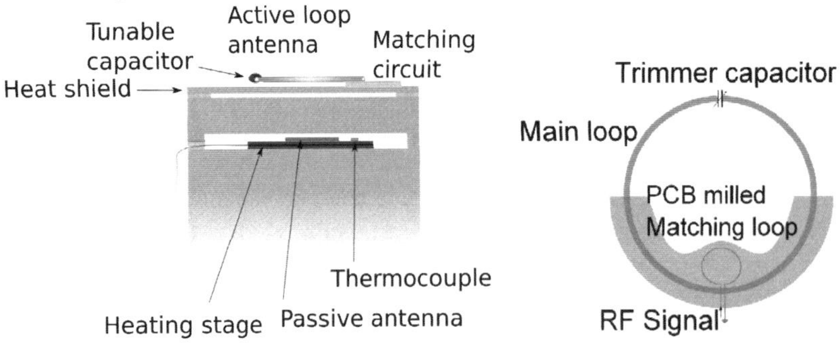

Figure 2. Measurement set up (left), with an active loop antenna (right), placed on top of a furnace containing the sample devices. Fans are used to keep the loop antenna below 50°C.

Device 3 was evaluated in a stress meter (FSM 500TC, Frontier Semiconductor, USA) using laser profilometry to measure the thermally induced deflections of the device for temperatures up to 380°C. In addition, the resonance frequency shift was measured on Devices 1, 2 and 3 during three-point bending instrument (AGS-X, Shimadzu, USA) with the load increasing from 0-10 N in steps of 0.5 N.

5. Results

The Q value of the devices, being the ratio of obtained resonance frequency to the 3-dB bandwidth, Table 1, was 6.7 in average with a standard deviation of 0.61.

Figure 3 shows the relative frequency shift of the three antenna devices over a temperature range of room temperature to 900°C. All three follow the same trend, with the frequency decreasing for up to 300°C, then slightly increasing between 400 and 600°C, and finally levelling out above 600°C. This is also the behaviour of the active antenna. The device with the pressure sensor element (No 4) exhibits an additional frequency shift of, in total, 3200 ppm, from 300°C to 900°C.

The thermally induced radius of curvature changed from 190 to 182 m for Device 3, figure 4. No frequency shift could be resolved for this sample in the bending tests, figure 5. For Devices 1 and 2, however, the frequency shows a shift of -80 and -180 ppm, respectively.

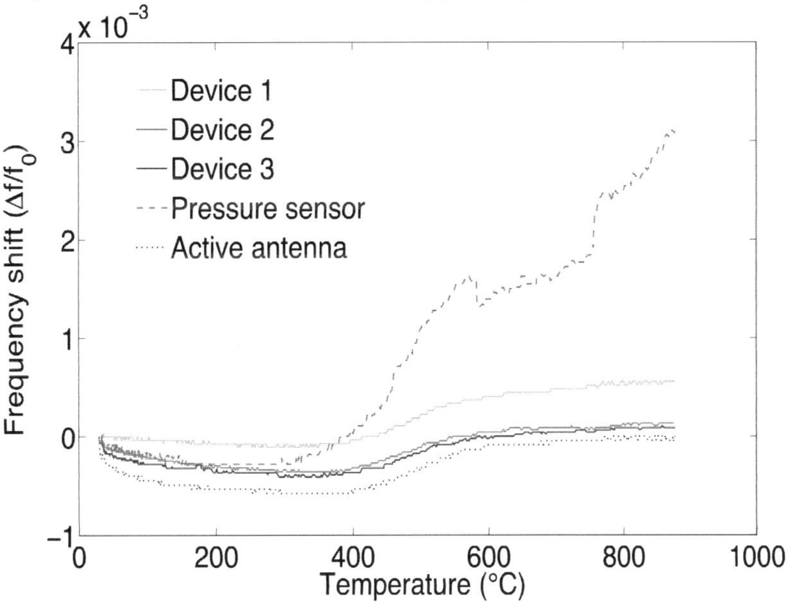

Figure 3. Frequency shift versus temperature for Devices 1-4 and the active, loop antenna.

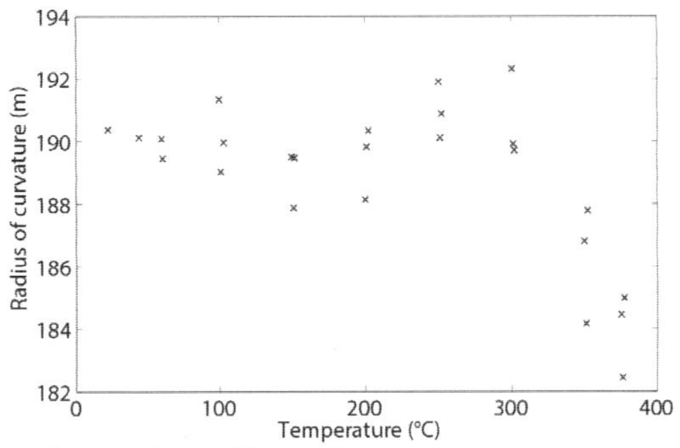

Figure 4. Radius of curvature versus temperature for Device 3.

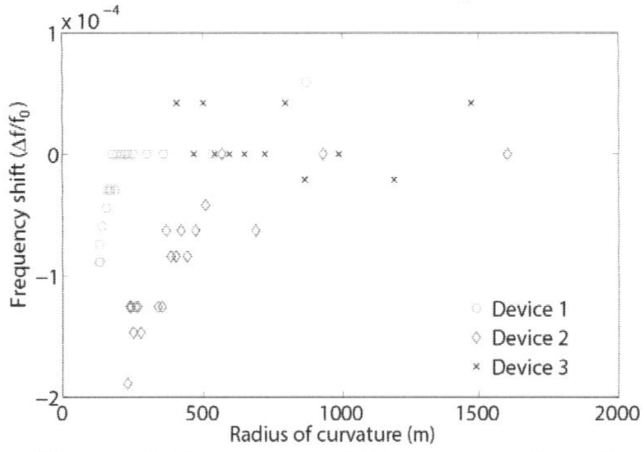

Figure 5. Frequency shift versus radius of curvature during three-point bending deformation of Devices 1-3.

6. Discussion & Conclusions

The similar behavior of Devices 1-3 and the active antenna, figure 3, suggests that the drift of the active antenna dominates. This effect could be attributed to the large bandwidth of the passive antennas, which makes the detection of small changes more difficult. In general, planar spiral loops have low Q values [13]. This can be improved by choosing a material with higher conductance than platinum, for instance silver or tungsten. They would, however, require a completely different manufacturing concept.

Throughout all these measurements, including the active antenna measurement, an increased frequency shift is observed between 400 and 600°C. This is believed to be caused by a sudden increase in conductivity of the AlN used in the furnace because of impurities, as reported by [14].

The thermally induced radius of curvature observed is very small, figure 4, even considering the relatively large change above 300°C, implying that the ceramic laminate manufacturing scheme provides thermomechanically rather robust components - at least when these are thick. Whether or not this holds also for thinner devices or if the amount of metallization or cavities and membranes is increased, remains to be investigated.

As for the three-point bending, which spans a much larger range of curvature, figure 5, the frequency shift is an order of magnitude smaller than that of the sensor-containing device, implying that these devices may be subjected to significant deformation, resulting, e.g., from mounting stresses or thermal mismatch when integrated, without severe performance losses.

The frequency shift of Device 4 is of a magnitude comparable with that of [6-9] and indicates a capacitance change from the thermally induced inner pressure deflecting the sensor's membrane. This device has a thinner membrane than those it is compared with. These, however, have two membranes and a lower Young's modulus. Hence, the result suggests a possible pressure reading at temperatures up to 900°C but the effect needs further investigation.

References

[1] Landis GA 2006 Acta Astronaut **59** 570-579
[2] Spang III H A and H Brown 1999 *Control of jet engines* **7** 1043-1059
[3] Domnich V, Aratyn Y, Waltrand M K, and Gogotsi Y 2008 *Rev. Adv. Mat. Sci.* **17** 33-41
[4] Khoong L E, Tan Y M and Lam Y C 2010 *J. Eur. Ceram. Soc.* **30** 1973-1987
[5] Lekholm V, Persson A, Palmer K, Ericson F, and Thornell G 2013, *J. Micromech. Microeng.* **23**
[6] Fonseca M A, English J M, von Arx M, and Allen M G 2002 *J. Microelectromechanical Systems,* **11** 337-343
[7] Radosavljevic G J, Zivanov L D, Smetana W, Maric A M, Unger M, and Nad L F 2009 *IEEE Sensors Journal* **9** 1956-1962
[8] Xiong J, Zheng S, Hong Y, Li J, Wang Y, Wang W, and Tan Q 2013 *J. Zhejiang University-SCIENCE C (Computers & Electronics)* **14** 258-263
[9] Tan Q, Kang H, Xiong J, Qin L, Zhang W, Li C, Ding L, Zhang X, and Yang M 2013 *Sensors* 13 9896-9908
[10] Greenhouse H M 1974 *IEEE Trans. Parts, Hybrids, Packages* **10** 101-109
[11] Werdecker W and Aldinger F 1984 *IEEE Trans. Compon., Hybrids, Manuf. Technol.* **7** 399-404
[12] Chen L Y and Hunter G W 2004 *2004 MRS Meeting* Boston MRS
[13] Sunderajan S M, del mar Hershenson M, Boyd S P, and Lee T H 1999 *IEEE J. Solid-State Circ.* **34** 1419-1424
[14] Francis R W and Worell W L 1976 *J. Electr. Chem. Soc.* **123** 430-433

Design considerations to enhance the performances of thin circular piezoelectric energy harvester diaphragms in harsh liquid environments

N Besse[1], A P Pisano[2] and N F de Rooij[1]
[1]Institute of Microengineering (IMT) - Sensors, Actuators and Microsystems Laboratory (SAMLAB), Ecole Polytechnique Fédérale de Lausanne (EPFL), Neuchâtel, Switzerland
[2]Berkeley Sensor and Actuator Center (BSAC), University of California, Berkeley, USA

E-mail: nadine.besse@epfl.ch

Abstract. Thin circular piezoelectric energy harvester diaphragms undergoing large deflection in a harsh liquid environment are investigated in this paper. A material set combining AlN as transducer, SiC as electronics, Mo as wiring and Si as holder is considered. A highly accurate analytical model, which presents less than 5% error compared to FEM simulations in COMSOL, is first developed to study thoroughly flat diaphragms. Consequently, etching the wafer and adding a corrugation are proposed to reduce both the stress concentration at the edge and the influence of residual stress on the device behavior, respectively. Both ideas are predicted to increase the power density compared to the standard flat case by at least a factor of 5 to 10.

1. Introduction

Several monitoring applications necessitating MEMS technologies are operating under hostile conditions; specifications which make it impossible to consider deploying wired and battery-powered silicon devices or accessing them once disposed. These characteristics are typically encountered in geothermal wells, in oil/gas exploration, in gas turbines or in automotive/jet engines where temperature of 600 °C, pressures of 300 atm, shocks of 50'000 G and pH between 1 and 10 are common. Hence, solutions incorporating high robustness to a wide range of operating conditions and presenting almost an infinite lifetime are required; one of the reason for energy harvesters coupled with wireless sensor networks using wide bandgap semiconductors to become recently a topic debated in the field of microsystems [1, 2]. Therefore, intensive research is conducted in Pisano's lab to create a micro-scale autonomous telemetry platform using aluminum nitride (AlN) as transducers and silicon carbide (SiC) as electronics. Inertial, pressure or temperature sensors [3, 4, 5], radio-frequency components [6], electronic circuits [7] and energy harvesters [8] based on this material set have been successfully fabricated.

In this work, a novel approach to design the most efficient thin circular piezoelectric energy harvester diaphragm vibrating in a liquid environment is proposed by combining theoretical modeling and FEM simulations in COMSOL. First, given a set of initial specifications, the dimensions of the standard flat structure are optimized thanks to an accurate analytical model. Then, two novel concepts to improve the global performances of the system are confronted: etching the wafer to reduce the stress concentration at the edge and adding a corrugation to modify the stress repartition within the diaphragm.

2. Theoretical modeling

Thin and flat circular piezoelectric energy harvester diaphragms driven by low frequency liquid pressure pulsations are studied. The design, illustrated in Figure 1, is divided in two models: a mechanical and an electrical one (both being related together thanks to the piezoelectric constitutive laws [9]). The mechanical analysis considers a bi-layer structure with an internal stress T_0, a radius a and an overall thickness h supported by uniformly distributed springs (k_r, k_θ, k_z) at the edge and driven by an harmonic pressure load of amplitude p_0 and frequency f_e. The fluid-structure interaction is represented by a thin massive layer m_f added on the surface. The equivalent electrical circuit is composed by four nodes ($i = 1, 2, 3, 4$) connected to the ground (GND) through pure capacitors C_i emulating the piezoelectric material and pure resistive loads R_i transforming the generated current I_i into a voltage V_i.

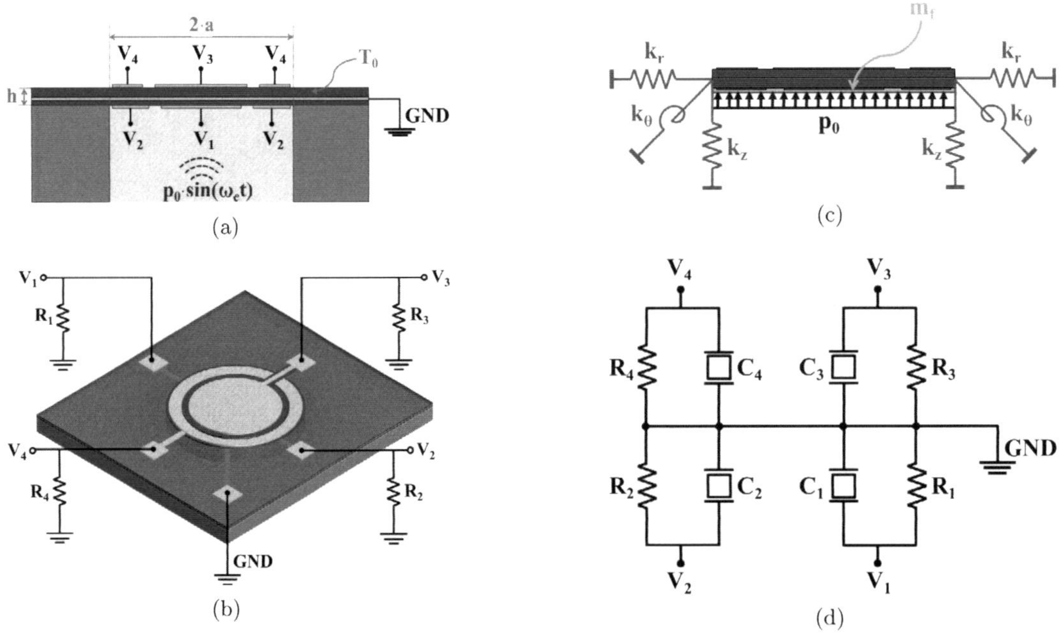

Figure 1: Schematic of the device. (a) Main parameters of the system. (b) Electrical connections. (c) Equivalent mechanical model. (d) Equivalent electrical model.

2.1. Design methodology

The design methodology, reported in Figure 2, is built on several existing theories including notably large deflections of circular diaphragms [10, 11], piezoelectric energy harvesters [12, 13] and fluid-structure interactions [14]. Firstly, given the characteristics of the studied device and the environmental conditions, the tension parameter K and the boundary conditions are set [10]. Secondly, the nondimensional spatial functions $W(\xi)$ and $S_r(\xi)$ are calculated [10, 11]. Thirdly, the nondimensional added virtual mass incremental (NAVMI) factor Γ and the corresponding added mass m_f are found [14]. Fourthly, the temporal function $\Phi(t)$ is obtained [12]. The three first steps are iterated a second time by substituting the tension parameter K by the equivalent tension parameter K_E accounting for the midplane tension to calculate the deflection w, the radial force N_r and the radial displacement u_r [10, 12]. Fifthly, the amount of stress-charges Q_i generated on each electrode is derived [12]. Sixthly, the optimal load $R_{opt,i}$ and the open-circuit voltage $V_{oc,i}$ are found [12]. Finally, the total power recovered P_{tot} is obtained by summing the maximal power $P_{opt,i}$ recovered by each electrode [13].

2.2. Model validation

In order to estimate the accuracy of the analytical model, FEM simulations have been implemented in COMSOL. An energy harvester with an overall thickness h of 5 μm, a radius a of 500 μm, a residual stress T_0 of 100 MPa and an excitation frequency f_e of 60 Hz is reported as an example. The percentage of error between the theoretical model and COMSOL is illustrated in Figure 3. Errors of less than 5% were obtained over a broad range of pressures p_0 going from 0.1 atm to 10 atm, demonstrating the interest for implementing twice the mechanical model to account simultaneously for bending and stretching within the diaphragm. Additional iterations are useless, since they demand more computations without lowering significantly the error.

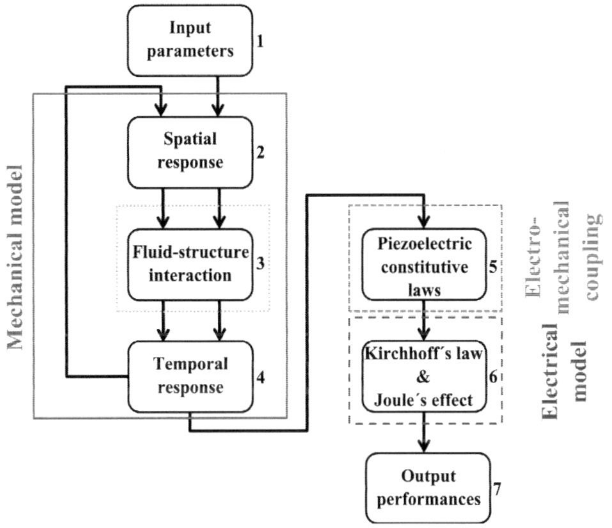

Figure 2: Process flow to model the piezoelectric energy harvester. A double iteration is implemented to improve the model accuracy.

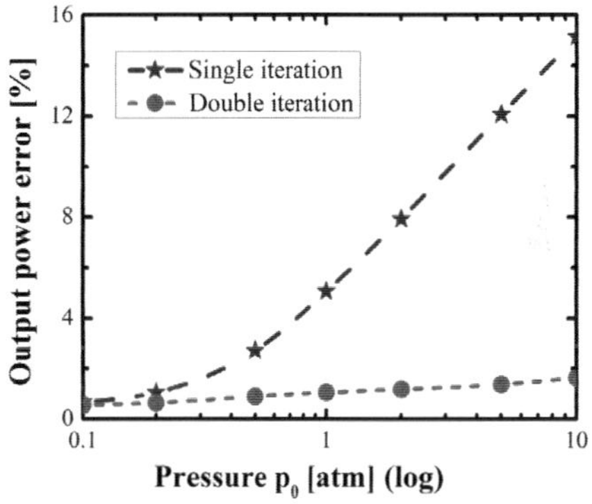

Figure 3: Percentage of error between the theoretical model and FEM simulations in COMSOL.

3. Further proposed improvements

Using our theoretical model, it was observed that the best piezoelectric energy harvester is an hinged symmetrical bimorph structure having the least amount of residual stress and two optimal electrodes patterned on either side. In order to emulate this design and thereby improve the global performances of the system, two concepts showing significant improvements compared to the standard flat diaphragm are proposed: etching the wafer and adding a corrugation. These advanced structures are illustrated in Figure 4 and were modeled using COMSOL.

3.1. Wafer etching

The goal of etching the wafer is to lower the high stress concentration at the edge by rendering it more flexible, while enhancing the output power and allowing rotation alike in a pinned edge situation. In fact, reducing the stress at the outer radius authorizes to fabricate wider diaphragms for a given maximal stress T_{max} (usually the tensile strength of the material). This approach differs from the conventional ones, since, instead of modifying the design of the active device, the shape of the holder is engineered. In order to compare several manufacturing processes, two parameters have been defined: α which is the ratio between the initial radius of the reference device and the corresponding one of the novel energy harvester (remember both are presenting the same maximal stress T_{max} at the anchor) and β which is the ratio between the lateral etching rate and the vertical one.

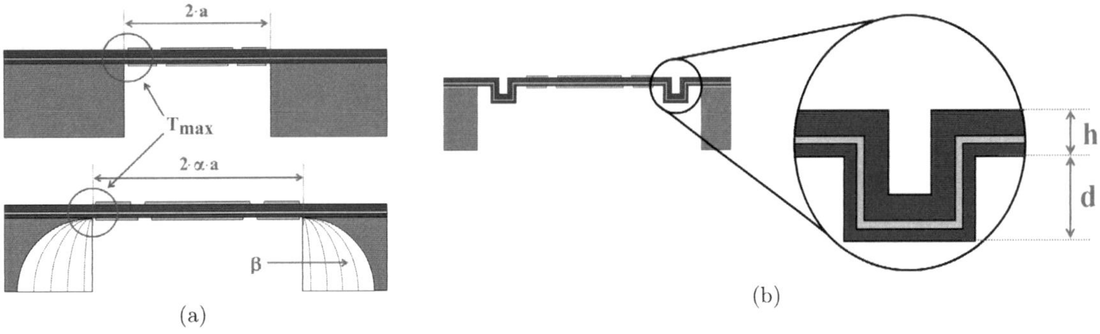

Figure 4: Novel proposed concepts. (a) Etching the wafer. (b) Adding a corrugation.

For the studied case, an enhancement factor G of almost 7 in terms of power density is expected at a pressure of 1 atm by using a simple wet isotropic etching step ($\beta = 1$). Better performances can be obtained by using highly anisotropic etching processes ($\beta > 1$), but they require a more complex fabrication. These observations are reported in Figure 5.

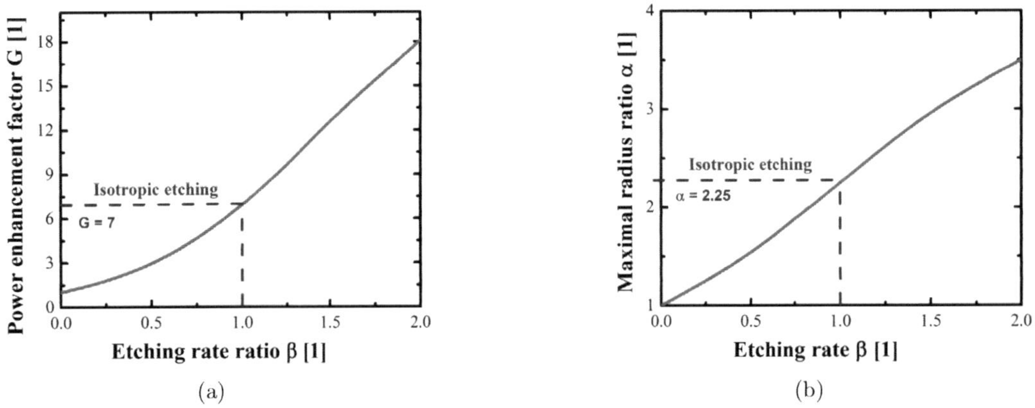

Figure 5: Improvements obtained by etching the wafer. (a) Enhancements in terms of power density compared to the flat case. (b) Corresponding increase in radius. For $\beta > 2$, the diaphragm will break first at the center, no more at the edge.

3.2. Corrugation

The behavior of thin and flat diaphragm are highly sensitive to environmental fluctuations. By adding a single corrugation of depth d at the outer edge, the stress repartition within the structure is modified [15], resulting in performances stable over a wider range of temperatures and pressures. Actually, the presence of a corrugation generates a non-uniform stress [15] which allows lateral displacement, stretches the midplane and thereby creates additional charges. Meaning that under small deflections where bending is dominating (i.e. low pressure or high tensile stress), the performances are enhanced and that under large deflections where stretching is dominating (i.e. high pressure or low tensile stress), they are slightly reduced. Hence, more power can be recovered by a structure operating at low pressure in its tensile mode; a configuration commonly encountered.

For the studied case, with a residual stress T_0 of 100 MPa, a pressure load p_0 of 1 atm and a deep corrugation ($d/h > 5$), an enhancement factor G of almost 6 in terms of power density is obtained. These results are given in Figure 6 in which several values of pressure and residual stress are implemented. Each curve compares the performances of the corrugated diaphragm with respect to the flat one.

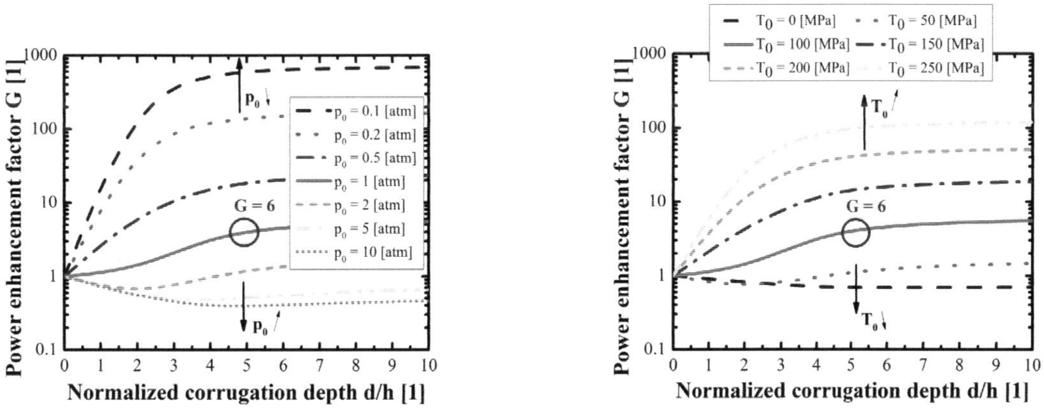

(a) (b)

Figure 6: Improvements obtained by adding a corrugation. (a) Influence of the corrugation depth and the pressure load on the performances at a residual stress T_0 of 100 MPa. (b) Influence of the corrugation depth and the residual stress on the performances at a pressure load p_0 of 1 atm. Both red continuous lines are representing the same case.

4. Conclusion

A novel theoretical model to design thin circular diaphragm piezoelectric energy harvester has been reported and compared to FEM simulations in COMSOL for validation. Consequently, two concepts have been proposed to improve the global performances of the system. By judiciously choosing between isotropically etching the wafer or adding a corrugation, an enhancement by a factor 5 to 10 in terms of power density is obtained.

In the future, the new proposed ideas should be considered for the next generations of energy harvesters and several devices should be fabricated to properly validate the design methodology. The wet isotropic etching is thought to be the most powerful solution for the short term, whereas the corrugations will probably gain importance in the future; once their fabrication will be better controlled and more reliable.

References

[1] Senesky D G 2012 *ECS Trans.*, **50**(6) 233
[2] Wijesundara M and Azevedo R 2011 *Silicon Carbide Microsystems for Harsh Environments* (New York: Springer)
[3] Goericke F T, Vigevani G and Pisano A P 2013 *Appl. Phys. Lett.* **102** 253508
[4] Goericke F T, Vigevani G, Izyumin I I, Boser B E and Pisano A P 2012 *Proc. IEEE Int. Ultrason. Symp.* (Dresden: IEEE) 1067
[5] Wodin-Schwartz S, Chan M W, Mansukhani K, Pisano A P and Senesky D G 2011 *Proc. Int. Conf. on Energy Sustainability* (Washington DC: ASME) 965
[6] Lin C-M, Chen Y-Y, Felmetsger V V, Lien C-M, Riekkinen T, Senesky D G and Pisano A P 2013 *J. Micromech. Microeng.* **23** 025019
[7] Lien W-C, D G Senesky and Pisano A P 2012 *Proc. PRiME* (Honolulu: ECS) 3032
[8] Lai Y-L, Li W-C, Lin C-M, Felmetsger V V and Pisano A P 2013 *Proc. Conf. Transducers'13* (Barcelona: IEEE) 2268
[9] IEEE Standards Board 1987 *IEEE Trans. Ultrason., Ferroelectr., Freq. Control* **Std. 176**
[10] Sheplak M and Dugundji J 1998 *J. Appl. Mech. Trans. ASME* **65** 107
[11] Chia C-Y 1980 *Nonlinear Analysis of Plates* (New-York: McGraw-Hill)
[12] Xue H and Hu H 2008 *IEEE Trans. Ultrason., Ferroelectr., Freq. Control* **55**(9) 2092
[13] Kim S, Clark W W and Wang Q-M 2005 *J. of Intell. Mater. Syst. and Struct.* **16** 847
[14] M. Amabili M and AND M. K. Kwak M K 1996 *J. of Fluids Struct.* **10** 743
[15] Zhang Y and Wise K D 1994 *J. Microelectromech. Syst.* **3**(2) 59

A miniature airflow energy harvester from piezoelectric materials

H Sun, D Zhu, N M White, and S P Beeby
Electronic and Electrical Engineering Group, Electronics and Computer Science, University of Southampton, Southampton SO17 1BJ, UK

E-mail: hs8g09 @soton.ac.uk

Abstract. This paper describes design, simulation, fabrication, and testing of a miniature wind energy harvester based on a flapping cantilevered piezoelectric beam. The wind generator is based on oscillations of a cantilever that faces the direction of the airflow. The oscillation is amplified by interactions between an aerofoil attached on the cantilever and a bluff body placed in front of the aerofoil. A piezoelectric transducer with screen printed PZT materials is used to extract electrical energy. To achieve the optimum design of the harvester, both computational simulations and experiments have been carried out to investigate the structure. A prototype of the wind harvester, with the volume of 37.5 cm^3 in total, was fabricated by thick-film screen printing technique. Wind tunnel test results are presented to determine the optimum structure and to characterize the performance of the harvester. The optimized device finally achieved a working wind speed range from 1.5 m/s to 8 m/s. The power output was ranging from 0.1 to 0.86 μW and the open-circuit output voltage was from 0.5 V to 1.32 V.

1. Introduction

Energy harvesting, also known as energy scavenging, describes the process of capturing the ambient energy and then converting it into electrical energy. Large scale energy harvesting has been applied for centuries in the form of windmills, watermills and passive solar power systems etc. With the advance of microelectronic technology such as wireless sensors, micro-energy harvesting attracts more attention [1].

For airflow energy harvesting, the most common and widely used devices are wind turbines. Although most wind turbine generators are used in large-scale applications, some researchers have designed wind turbines in smaller scales. Holmes *et al* [2] from Imperial College have reported some micro wind turbines with electromagnetic generators, of which the rotor area was down to approximately 1.5 cm^2. An output power of 4.3 mW at average flow speed of 10 m/s was achieved. Priya *et al* [3] have combined small windmills with piezoelectric material for energy harvesting and managed to achieve a 5 mW continuous power at flow speed of 4.5 m/s. For wind turbine generators, due to the increased friction losses of the bearings and the reduced surface area of the blades, the efficiency of conventional turbines reduces with size.

This concern is addressed in many novel approaches for wind energy harvesting based on vortex induced vibration, aeroelastic flutter, wake galloping and other aero-elastic instability phenomena. The windbelt which utilizes aeroelastic flutter phenomenon showed power outputs in milliwatt range, but suffered from a threshold wind speed of 3 m/s [4]. Li *et al* [5] suggested a flapping piezo-leaf-based wind generator using PVDF material, which produced an output power of 1.8 μW at a flow speed of

3.5 m/s due to the low piezoelectric strain coefficient of flexible materials. A device by Bryant and Garcia [6] used a cantilevered piezoelectric bender based on aeroelastic flutter vibrations which showed a minimum threshold wind speed lower than 2 m/s. However the bender was over 25 cm long which was too large for microelectronic systems and the efficiency at small scale was unclear. Zhu *et al* [7] proposed a promising flow-induced energy harvester using oscillatory cantilevered wing that can operate at wind speeds as low as 2.5 m /s with a corresponding electrical output power of 470 μW. When the wind speed was 5 m/s, the output power was 1.6 mW.

Most researchers reviewed above have investigated airflow induced energy harvesters with electromagnetic transducer or piezoelectric transducer made from flexible materials such as PVDF. This paper report an oscillating airflow generator utilizing screen printed PZT materials. The strategy in this work lies in building the harvester with an optimum structure for a wider air speed range for operation. To do this, both computational simulations and experiments are carried out. Final the test results of the prototype are presented and discussed.

2. Theory

The wind generator presented here is based on oscillations of a cantilever that faces the direction of the airflow. An aerofoil is attached to the free end of the cantilever spring while the other end is clamped as shown in figure 1. If there is no wind, the aerofoil will slightly bend downward due to gravity. When air is flowing towards the aerofoil, it will bend upward and the degree of bending is related to the spring constant, mass and the lift/drag force form the aerofoil caused by the flow. In this condition, however, the cantilever will remain still with a static deflection without oscillation. In order to oscillate the cantilever, a bluff body is required to be placed in front of the cantilever which reduces the flow of air behind the bluff body and thus the lift force, causing the cantilever to operate primarily under inertial effects and spring back. However when the cantilever is springing back to the original position where the aerofoil can be fully exposed to the airflow again, energy can be extracted from the wind and the cycle is repeated. Such a configuration can self-start and sustain the necessary oscillations under uniform and steady flow conditions. In addition, this device can potentially operate at a lower flow speed. By attaching the aerofoil to the free end, the deflection of the beam can be larger and its performance can be better tuned for various flows. The position of the bluff body, both its distance to the cantilever and its height affect the oscillations of the cantilever.

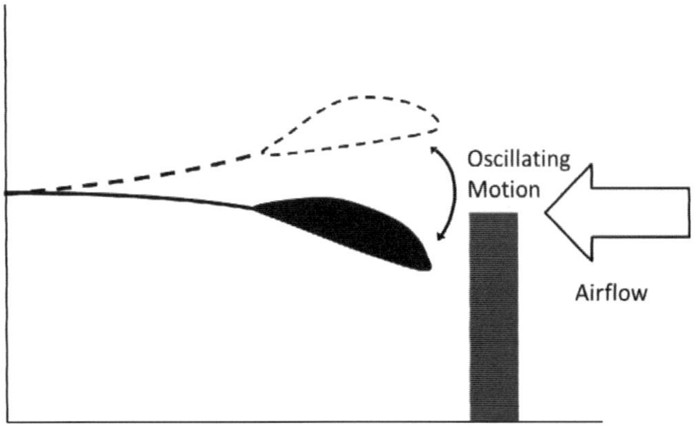

Figure 1. Operation principle of the flapping generator.

3. Design and Fabrication

3.1. Simulation Model

To assess the effect of the bluff body and the attack angle of the aerofoil, series of finite element simulations were conducted using ANSYS CFX. For simplification, a static analysis model was set up to estimate the lift force applied on the wing, which was used as a guide to determine the threshold wind speed of the device. Since the consequence that a large lift force acting on aerofoil will provide a relatively low threshold wind speed, configurations generating larger lift force should correspond with those having lower threshold wind speed. The simulation parameters to be studied included the distance from the cantilever, the height of the bluff body, and the attack angle of the aerofoil. Results from different configuration arrangements were compared and plotted using ANSYS commercial package for Design Optimization. Figure 2 and 3 show how the lift force changes with the position of the bluff body and the attacking angle respectively. As seen in these figures, the maximum values of the lifting force, which indicate lower threshold wind speed, occur at distances from 3 mm to 10 mm and heights from 18mm to 21mm, and attack angle from 5°-8°. These results will be compared with the experimental results in the later section. It should be noted that simulation in this section was purely on mechanical structure and the transducer was not taken into account.

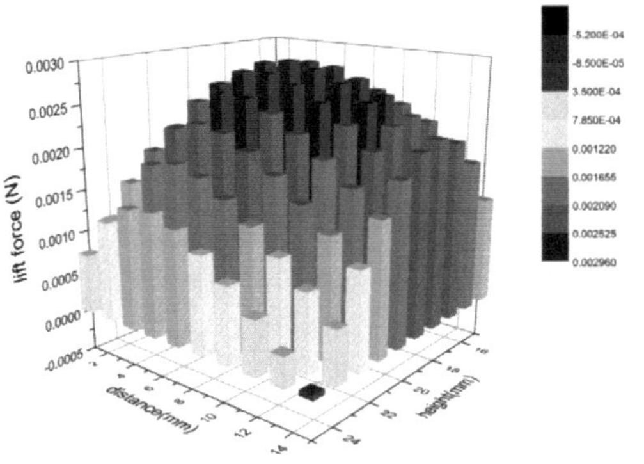

Figure 2. 3D bar chart of the lift force with different heights and distances from bluff body. (Attack angle of 5° is taken as an example.)

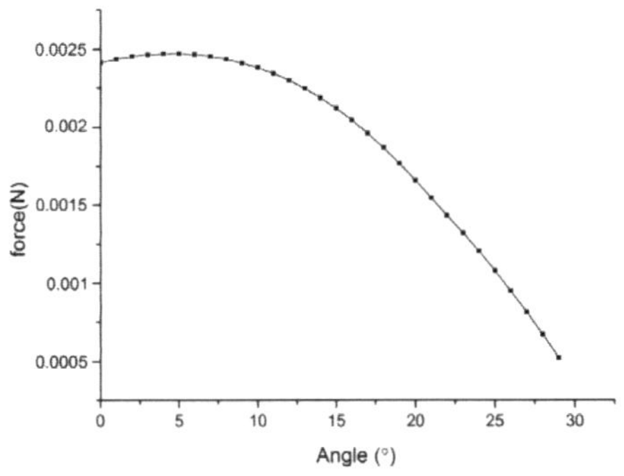

Figure 3. Lift force at different attack angles with the same height and distance. (Distance of 28 mm and height of 15 mm is taken as an example.)

3.2. Fabrication Process

A prototype of the wind harvester, with the volume of 37.5 cm^3 in total is shown in figure 4. Both the base of the generator and the bluff body were made of acrylic. The cantilever was made of type 304 stainless steel and had dimensions of 30 mm × 5 mm × 0.08 mm. The wing, made of balsa wood, was an aerofoil with shape of profile NACA 4424 whose dimensions were 30 mm × 11 mm × 5 mm. The overall mass of the wing was 0.25 grams. The piezoelectric transducer was fabricated by normal thick-film screen printing steps including paste preparation, screen-printing deposition, drying, firing and poling process [8]. After printing, samples were poled using a hot plate at 225 °C and an applied electric field of 3.5 MV/m. Finally, 6 samples that achieved maximum piezoelectric property d_{33} of 130 pC/N approximately were selected for testing.

3.3. Experimental Work

Tests were carried out in the centre of a wind tunnel, where the air flow generated by a centrifugal fan was conducted by a steel air duct of approximately 2 m in length. Another transparent dust was connected to the steel duct so that oscillation of the harvester can be observed during the test. The cross section of the opening window was 20 cm×35 cm. Some small holes in the ducting allowed for an anemometer probe to be inserted measuring the flow speed. The anemometer has a 2% precision error. Airspeed can be adjusted by the fan using a controller, which allows tests from 0 to 8 m/s. Since the speed level of air turbulence changes slightly at different position in the wind tunnel, the harvester was placed near to the end of the steel duct where the air flow is most steady. During the tests, the voltage across a load resistance was measured with an oscilloscope in order to determine the power generated by the wind energy harvester, meanwhile the operation wind speeds of the harvester were measured by the anemometer.

4. Results

The objective of the experimental analysis was to explore the influence of the bluff body and attacking angle on the power output and threshold wind speed. In this work, the threshold wind speed is defined as the airflow speed at which a peak-to-peak displacement of the wing is over 1 cm. After testing and comparison, the experimental results showed relatively good correlation with the simulation in ANSYS (shown in table 1). Both results verify that to achieve the lowest threshold wind speed, optimum positions of the bluff body are approximately between 3 mm to 10 mm in horizontal distance and 18 mm to 20 mm in vertical distance with attack angle around 6°- 8°, which validates the simulation model used in this study. For further investigation of the performance, four configurations were designed using results above. As seen in figure 5, the power outputs with various bluff body positions were tested. The maximum output power occurred in 7m/s wind speed from a configuration with distance of 11 mm and height of 16 mm, which was chosen for final tests. It can also be observed that at distance smaller than 5 mm, performance at high wind speed was not stable, thus resulting in even lower power output. After testing, the optimized device achieved a working wind speed range from 1.5 m/s to 8 m/s. The power output with optimal load (800 kΩ) was ranging from to 0.1 μW to 0.86 μW and the open-circuit output voltage was from 0.5 V to 1.32 V (shown in figure 6 and figure 7 respectively). Results verified the harvester can effectively convert wind energy into large amplitude mechanical vibration without strict frequency matching constraints. However, the power generated is not as large as expected due to the added stiffness of the beam from the presence of the printed piezoelectric layers and this will be improved in the future by increasing the length of the piezoelectric part of the beam.

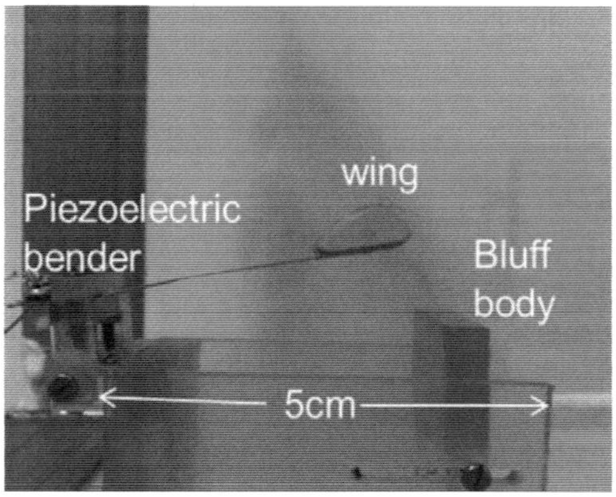

Figure 4. Prototype of the energy harvester oriented for airflow from right to left.

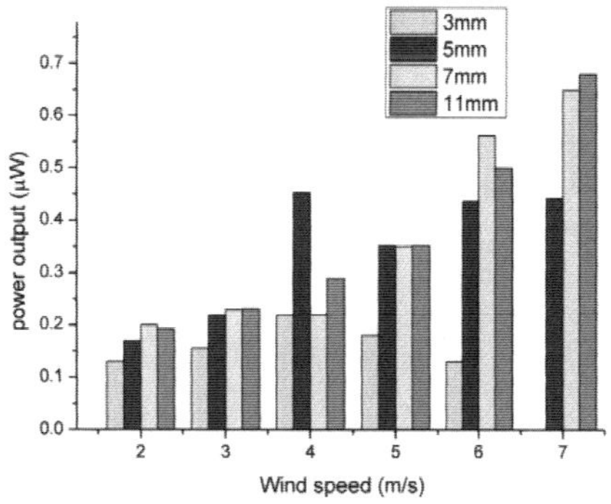

Figure 5. RMS power outputs in varying flow speed conditions of four configurations (Attack angle is 7°, distances are 3mm, 5mm, 7mm and 11mm and heights are 19 mm, 18 mm, 17 mm and 16 mm respectively)

Table 1. Comparison of optimal configurations of the harvester in simulation and test results.

	Simulation Analysis	**Wind Tunnel Tests**
Attack angle of aerofoil	5°-8°	6°-10°
Height of the bluff body	18 mm-21mm	17 mm-20 mm
Distance between bluff body and aerofoil	3 mm-10 mm	3 mm-11mm

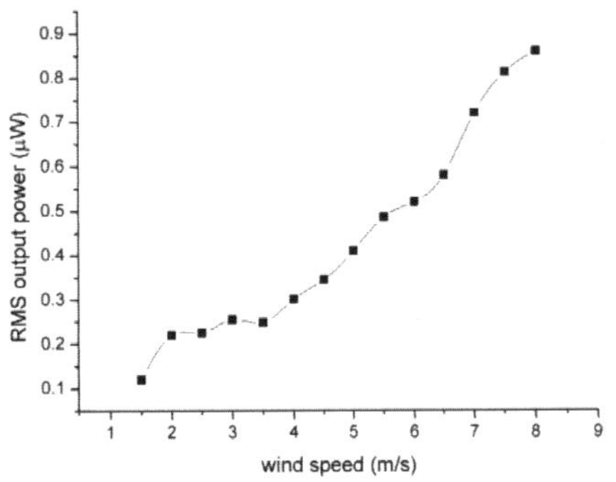

Figure 6. RMS Output power in varying flow speed conditions with optimal configuration and load (plotted line was used to indicate the trend)

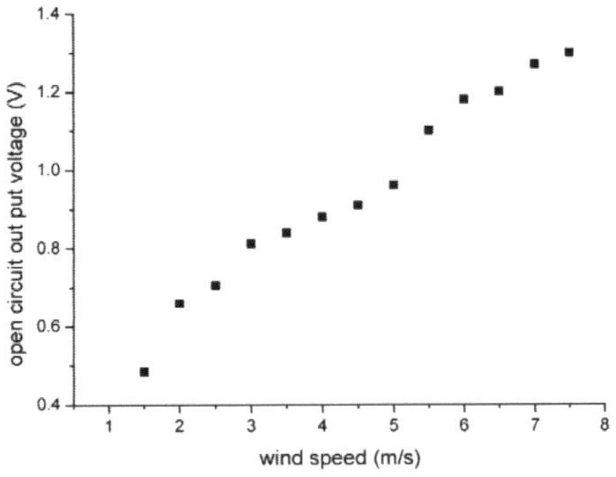

Figure 7. Open circuit output voltages in varying flow speed conditions with optimal configuration and load

5. Conclusion

In this report, a miniature, scalable piezoelectric wind generator was proposed and investigated theoretically, numerically and experimentally. The volume of the device was 37.5 cm^3 in total. The significance of this work lied in building the harvester with an optimum structure for wider wind speed range for operation. Final device achieved a working wind speed range from 1.5m/s to 8m/s with optimum structures and a maximum power output of 0.86 μW. The open-circuit output voltage was from 0.5 V to 1.32 V.

References

[1] Beeby S and White N 2010 *Energy Harvesting for Autonomous Systems* (London: Artech House)
[2] Holmes A, Howey D and Bansal A 2011 *Smart Materials and Structures* **20** 1-12
[3] Priya S, Chen C, Fye D and Zahnd J 2005 *J. Appl. Phys.* **44** 104-7
[4] μicroWindbelt Data Sheet 2008 Humdinger Wind Energy
[5] Li S and Lipson H 2009 *Proc. ASME/AIAA Conf. Smart Mater., Adapt. Struct. Intell. Syst.* vol 1 (California: Oxnard) p 611-19
[6] Bryant M, Wolff E and Garcia E 2011 *J. Vibration and Acoustics* **133** 1
[7] Zhu D, Beeby S, Tudor J, White N and Harris N 2010 *Proc. IEEE Sensors* (Waikoloa: Hawaii) p 1415-8
[8] Zhu D, Beeby S, Tudor J and Harris N 2011 *Sensors and Actuators A Physical* **169** 317-25

Analysis of Pulsed vs. Continuous Power Delivery from an Electromagnetic Generator

A Cimpian[1], G Ó Laighin[1] and M Duffy[2]

[1] Bioelectronics Research Cluster, Electrical & Electronic Engineering, National University of Ireland Galway, Ireland

[2]Power Electronics Research Centre,Electrical & Electronic Engineering National University of Ireland Galway, Ireland

ana.cimpian@gmail.com

Abstract. The purpose of this work is to present alternative power conversion techniques for an energy harvester optimized to the power requirements of an activity monitoring device. Many energy harvesters designed to use human motion provide a pulsed type of output waveform, as the signal will be strongly related to the pattern of motion used for harvesting energy. Due to this type of discontinuous signal it is considered that wearable sources have the potential to provide higher energy values by pulses rather than continuous form. For this work an electromagnetic generator system was optimised to power a monitoring device located in the shoe. Rectification techniques as well as coil parameters design have been employed and powering conditions have been analysed. The generator system can provide pulses of power high enough to sustain a low power consumption device.

1. Introduction

The concept of energy harvesting has known an increase in popularity with the development of portable electronics. Although there are infinite sources of energy that have been considered for energy harvesting applications, in the context of portable electronics the human body represents the most obvious source. According to Starner [1] power can be obtained by unobtrusively converting the energy the human body consumes while performing every day activites. Based on literature [1] one of the most deliberate human actions that can offer a good potential for energy harvesting is walking. During walking power can be taken from rotation of joints, swinging motion of arms and legs or pressure applied at different points. It is calculated that by performing the swinging motion of the legs a human needs between 1.07 and 34.9 J/step. This is the equivalent of 2.1 to 69.8 W for each joint during a gait cycle [2]. In an investigation with accelerometers Romero et al. proved that the power values from walking at the ankle area are approximately 10 mW/cm^3, which is twice as much as for most of other body locations [3]. This work is centred on developing and optimizing an energy harvesting system that is capable of providing the highest amount of energy from walking and that is designed for integration in the shoes.

Energy can be harvested from human motion by using thermoelectric, piezoelectric, electromagnetic or electrostatic energy conversion [4], [5]. Previous work [6] has concluded that for an energy scavenger designed to convert the energy consumed during walking, an electromagnetic system is the most suitable. Although electromagnetic energy harvesting for walking is not new [7], the scope of this work is to demonstrate how a shoe-integrated electromagnetic generator system can be designed to provide useful levels of pulsed power for wearable sensing applications.

The industry of wearable sensors has known a rapid development over the past decade, with individual components decreasing in size and becoming more powerful. Nowadays, wearable technology has reached the point where some of these sensors can be employed in clinical applications. Of particular interest in this work are activity monitoring devices which aim to provide both diagnostic and monitoring functions for recovering patients or elderldy people who want to continue living independently [8], [9]. As monitoring devices need to be able to work for extended periods of time, the provision of a self-sustained system is desirable. Therefore, this work investigates the performance of a shoe-integrated electromagnetic generator in powering such a system. Due to the discontinuous nature of generated voltage pulses on one hand, and the intermittent power demand profiles of wearable sensors, it is proposed to deliver the generator power in large bursts (as it's generated) rather than to filter the output to provide an averaged constant power. The remainder of the paper deals with optimizing the generator parameters and with analysing rectification techniques employed to obtain the maximum power under continuous and pulsed power conditions.

2. Power requirements of activity monitoring devices

Wearable activity sensors can have diverse functions, covering the care and monitoring of elderly people [10, 11], sports activities and performance monitoring [12-14], disease monitoring during daily activities [15], or gait analysis [16]. A typical wearable wireless sensor will contain a data control and processing block, a sensing block and a transmission unit. Although the power profile of recently developed microcontrollers and sensing modules are suggesting that energy harvesting solutions are applicable, the transmission modules are still representing a challenge in the development of self-sustained systems. Other research groups [17-19] have recently developed systems with particular consideration for the power consumption. Overall current consumption typically ranges between 3 mA for idle state and 20 mA for listening and data transmission modes. Similarly, transmission protocols specially adapted for wearable sensor nodes that minimize the power dissipation and compress data are emerging [20-21], [17].

In this context, the newest generation of microcontrollers require current values ranging from 0.19 mA for idle state to 6.5 mA for the active state [22]. An accelerometer needs 0.002 mA for stand by state and 0.2 mA for active mode [23]. Typical power consumption for transmission modules are 17mA for RF [24] or 14.6 for Bluetooth 4.0 [25]. However, the latter are peak values needed during listening and receiving the data and are usually needed for durations shorter than 1ms. A typical sensing cycle in the case of a wearable generator the MCU might be activated by the input voltage generated by the user walking. The movement would be recorded with the aid of an accelerometer, followed by data processing and transmission to indicate that the user has taken a step. The highest amount of power is required during transmission even if the active mode time interval is considerably smaller than those of the other components (1ms for active mode in the case of Bluetooth). Taking the supply voltage ranging between 1.8V and 3.6V the instantaneous power levels vary from 0.34mW for the microcontroller and sensing unit to a peak of 60mW for data transmission. Taking a typical wireless sensor power profile, with a microcontroller operating time of 100ms, sensor operation over 80ms and data transmission lasting for 10ms, the overall energy per sensing cycle is estimated as 1.04mJ.

3. Electromagnetic generator design

The generator prototype proposed in this work is based on the sliding magnet principle described in [26]. Based on the theory that as long as the mass of the magnet is considerably smaller than the mass of the foot, continuous movement of the magnet can be sustained by external forces due the movement of the foot with no additional or deliberate effort from the user.The prototype generator was assigned a volume of $50x15x15mm^3$ and 10mm diameter disk shaped NdFeB magnets were determined to be optimum for the generator's cross sector. Applying optimization rules from [26], it was determined that the most suitable generator prototype consists of 1 magnet of length 15mm and one 17.3mm long coil, centred along the 50mm generator length. In this way, the magnet has sufficient scope to travel into and out of the coil at either end, while the coil is just long enough to capture the majority of magnetic flux gradient. Therefore, the generator is optimised to provide the maximum output voltage by power product. A 300 turn version of this generator shown in Figure 3(a) has been tested with the generator attached to the outer heel of a shoe for walking speeds ranging from 2 to 12km/h [6].

It was observed previously that the maximum peak power was achieved at values approximately equal with half the value of the open circuit voltage. However, maximum loaded voltage pulses of 0.7V peak are outside of the 1.8V to 4.1V operating voltage range required by most sensor system components. In order to counteract this issue the number of generator winding turns is increased by a factor of 5 in this work. However, with the space constraints remaining the same, the diameter of the wire needs to be reduced to 0.1mm, resulting in a coil of 10 layers with 150 turns per layer and an estimated resistance of 160Ω. Simulations have concluded that the new design will produce output voltage pulses of approximately 8V peak under open circuit conditions; this corresponds to a maximum load power voltage of ~4V as required for sensor operation. Predicted open circuit waveforms are included in the results of Figure 2 later. These are based on measurements for the 300 turn generator at a walking speed of 8km/h.

4. Rectifier design for pulsed and continuous power delivery conditions

Most energy harvesting systems apply AC/DC conversion stages to provide a steady DC output voltage. However as described in Section 2 the power requirements of most wireless sensor loads are more pulsed in nature. Given that the generator is producing a pulsed output, the design of a simple half-wave rectifier is investigated with a view to matching the pulsed load demand every time a step is taken. The average power under pulsed and continuous conditions is compared in simulations in Figure 1(a), with corresponding peak voltage levels included in Figure 1(b). These were predicted for a range of constant RC values of a capacitive filter, where for operation under continuous conditions, the choice of RC is based on the repetition period of a walking step (1 step per second), while a lower RC constant is chosen in the case of pulsed power with the aim of eliminating ripple within an individual generator voltage pulse. Two values of the ripple are included for each case. It is clear that the maximum average power settles at around 1mW for all values of RC. However this is achieved for R= 1925Ω and C=0.65mF (RC = 1.25) for the continuous case and for R= 769Ω and C=65μF (RC=0.05) under pulsed conditions. The corresponding peak voltages for the maximum power points are 3.6V for the pulsed case and 2.1V for continuous conditions.

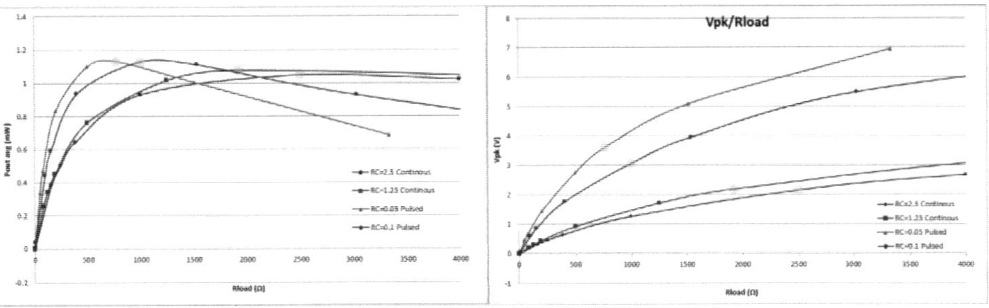

Figure 1: (a) Output average power for a half-wave rectifier under pulsed and continuous conditions (b) Output peak voltages for a half-wave rectifier under pulsed and continuous conditions

The high values of R and C needed to achieve the maximum average power point under continuous conditions combined with the lower values of output voltage are suggesting that the pulsed power delivery method is more practical. Simulated output voltages are compared for pulsed and continuous operation in Figure 2. The voltages produced under pulsed conditions provide instantaneous peak power levels of 17 mW with a corresponding current level of 3.3 mA compared to 8.9 mW and a current level of 1.1 mA for the continuous case. In terms of time, it is found that the pulsed power output lasts ~150 ms per walking step, giving an overall energy per step of approximately 0.7 mJ.

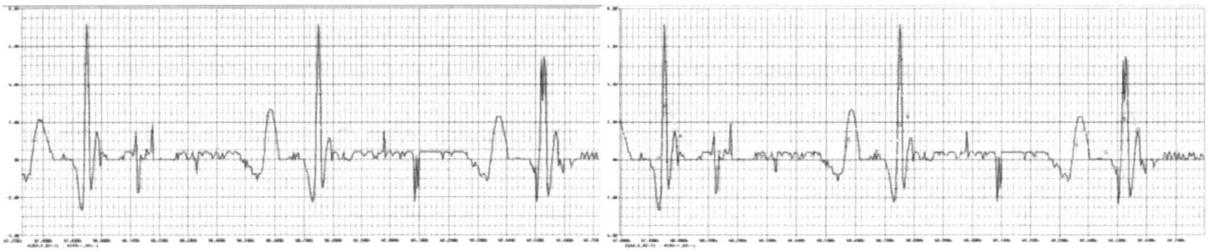

Figure 2: Input and rectified waveforms for the maximum power points for a half-wave rectifier under (a) pulsed and (b) continuous conditions

As discussed in Section 2, typical energy levels needed per wireless sensing & transmission operation are of the order of 1 mJ and therefore the application of the generator for supplying this energy is plausible. The predicted performance of the system is summarised in Figure 3(b) where available current and voltage levels for the generator are compared with those required by various sensing components. It is clear that the continuous case supports sleep mode operation while pulsed operation is more suited to active power consumption.

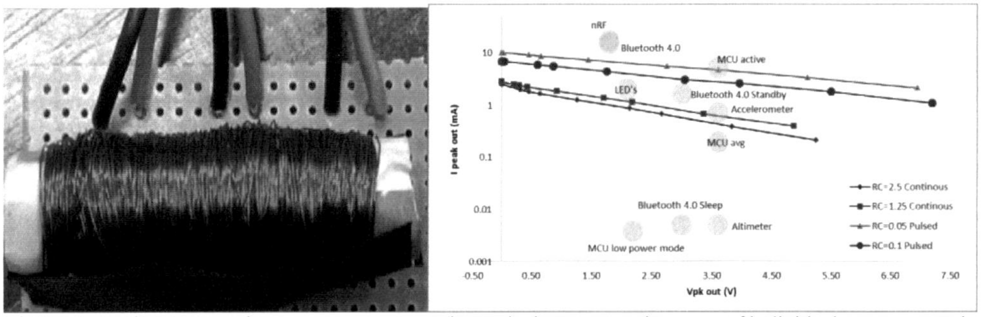

Figure 3: (a) Electromagnetic generator prototype (b) Typical power requirements of individual components during operation compared with simulated current levels produced by the generator system

5. Measurements results

Measurements have been conducted with a prototype consisting of a 300 turn coil that occupies the pre allocated volume of 50x15x15 mm^3. Results are presented in Figure 4 for a speed of 10 km/h with the RC product designed for maximum power under pulsed and continuous conditions. It is clear that the trends predicted by simulations are observed. The peak voltage under the pulsed conditions is of 0.5V and is higher than the 0.3V peak achieved for continuous conditions resulting in instantaneous peak power values of 5.5mW and 0.37mW respectively. The maximum average power also confirms the trends shown in simulations with similar values of ≈220μW for both cases. Lower power levels than predicted in Section 4 are explained by a higher percentage source voltage drop across the rectifier diodes.

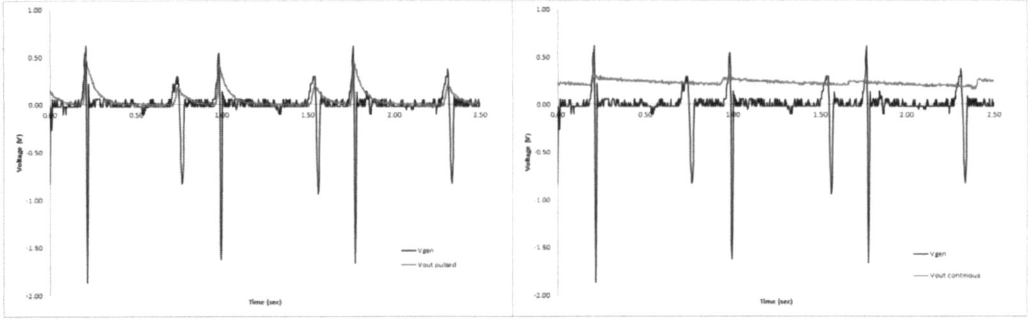

Figure 4: Measured results for (a) pulsed and (b) continuous conditions

6. Conclusions and future work

An electromagnetic generator system has been designed and optimised for a typical wearable sensor load. Alternative power delivery methods have been analysed and simulations show that using a rectifier designed to deliver pulses of power each time the generator is activated can satisfy typical power requirements of a wearable sensor.

Work is on-going to investigate full-wave rectifiers which are expected to provide a higher output power. Future work will also be focused on verifying the generator capabilities by measurements for a demonstrator wireless sensor.

References:

[1] T. Starner and J.A. Paradiso, "Human-Generated Power for Mobile Electronics," *Low Power Electronics Design*, C. Piguet,ed., CRC Press, 2004, chapter 45, pp. 1–35

[2] P. Niu, P. Chapman, R. Reimer, and X. Zhang, "Evaluation of motions and actuation methods for biomechanical energy harvesting", in Proc. 35th Annu. IEEE Power Electron. Specialists Conf., Aachen, Germany, Jun. 2004, pp. 2100–2106.

[3] E. Romero, R. O. Warrington, and M. R. Neuman, "Powering Biomedical Devices with Body Motion", 32nd Annual International Conference of the IEEE EMBS Buenos Aires, Argentina, August 31 - September 4, 2010

[4] Harb A., "Energy harvesting: State-of-the-art, Renewable Energy (2010), doi:10.1016/j.renene.2010.06.014

[5] K. Reimann, "Energy considerations for small electromechanical generators that use human body motion", Koninklijke Philips Electronics N.V. 2005

[6] A. Cimpian, B. Granier, G. Ó. Laighin, and M. Duffy, "Comparison of Energy Harvesting Techniques for Wearable Activity Monitoring Devices," 6th International Joint Conference BIOSTEC 2013, February 11-14, 2013, Revised Selected Papers.

[7] S. W. Angrist, "Direct energy conversion", Allyn and Bacon, Inc. Boston, 1971

[8] S. Patel, H. Park, P. Bonato, L. Chan, and M. Rodgers, "A review of wearable sensors and systems with application in rehabilitation," *Journal of NeuroEngineering and Rehabilitation*, vol. 9, no. 1, p. 21, 2012.

[9] P. Bonato, "Wearable sensors and systems. From enabling technology to clinical applications.," *IEEE Engineering in Medicine and Biology Magazine*, vol. 29, no. 3, pp. 25–36, 2010.

[10] C. Doukas and I. Maglogiannis, Advanced patient or elder fall detection based on movement and sound data. IEEE, 2008, pp. 103–107.

[11] A. Wood, J. Stankovic, G. Virone, L. Selavo, Z. H. Z. He, Q. C. Q. Cao, T. D. T. Doan, Y. W. Y. Wu, L. F. L. Fang, and R. Stoleru, Context-aware wireless sensor networks for assisted living and residential monitoring, vol. 22, no. 4. IEEE, 2008, pp. 26–33.

[12] http://www.fitbit.com/uk

[13] A. Ahmadi, D. D. Rowlands, and D. A. James, "Investigating the translational and rotational motion of the swing using accelerometers for athlete skill assessment," 2006 5th IEEE Conf. Sensors, pp. 980–983, 2007.

[14] F. Michahelles and B. Schiele, Sensing and monitoring professional skiers, vol. 4, no. 3. IEEE Computer Society, 2005, pp. 40–46.

[15] http://www.shimmersensing.com/

[16] A. Salarian, H. Russmann, F. J. G. Vingerhoets, C. Dehollain, Y. Blanc, P. R. Burkhard, and K. Aminian, Gait assessment in Parkinson's disease: toward an ambulatory system for long-term monitoring., vol. 51, no. 8. 2004, pp. 1434–1443.

[17] N. Stohs, A. Raij, and M. Absi, "AutoSense : Unobtrusively Wearable Sensor Suite for Inferring the Onset , Causality , and Consequences of Stress in the Field," Changes, pp. 5–6, 2011.

[18] V. Shnayder, M. Hempstead, B. Chen, G. W. Allen, and M. Welsh, "Simulating the power consumption of large-scale sensor network applications," Proc. 2nd Int. Conf. Embed. networked Sens. Syst. SenSys 04, p. 188, 2004.

[19] A. Prayati, C. Antonopoulos, T. Stoyanova, C. Koulamas, and G. Papadopoulos, "A modeling approach on the TelosB WSN platform power consumption," J. Syst. Softw., vol. 83, no. 8, pp. 1355–1363, 2010.

[20] B. P. D. Mitcheson, M. Ieee, E. M. Yeatman, S. M. Ieee, G. K. Rao, S. M. Ieee, A. S. Holmes, and T. C. Green, "Human and Machine Motion for Wireless Electronic Devices," Proc. IEEE, vol. 96, no. 9, pp. 1457–1486, 2008.

[21] http://www.nordicsemi.com/eng/Products/2.4GHz-RF/nRF2401A

[22] http://www.ti.com/lsds/ti/microcontroller/16bit_msp430

[23] http://www.freescale.com/webapp/sps/site/prod_summary.jsp?code=MMA7660FC

[24] http://www.nordicsemi.com/eng/Products/2.4GHz-RF/nRF2401A

[25] http://www.nordicsemi.com/eng/Products/Bluetooth-R-low-energy/nRF8001

[26] Carroll. D, Duffy. M, Modelling, design, and testing of an electromagnetic power generator optimized for integration into shoes, Journal Article 2011, 8 September 9, 2011, Proceedings of the Institution of Mechanical Engineers, Part I: Journal of Systems and Control Engineering, Volume 226 issue 2, pages 256-270

A piezoelectric pulse generator for low frequency non-harmonic vibration

Hao Jiang and Eric M Yeatman

Department of Electrical and Electronic Engineering, Imperial College London, South Kensington Campus, London SW7 2AZ, UK

E-mail: hao.jiang10@imperial.ac.uk

Abstract. This paper reports a new piezoelectric prototype for pulse generation by energy harvesting from low frequency non-harmonic vibration. The pulse generator presented here consists of two parts: the electromechanical part and the load circuit. A metal rolling rod is used as the proof mass, moving along the substrate to achieve both actuating of the piezoelectric cantilever by magnetic coupling and self-synchronous switching of the circuit. By using this new approach, the energy from the piezoelectric transduction mechanism is regulated simultaneously when it is extracted. This allows a series of tuneable pulses to be generated, which can be applied to self-powered RF wireless sensor network (WSN) nodes.

1. Introduction

With increasing demand of highly mobile and portable electronic devices, it has become a popular topic to seek for new power approaches such as motion energy harvesting instead of using batteries as the external power of traditional WSN nodes as in figure 1(a) [1]. Previous work of our group has realised a self-powered WSN node platform using electrostatic transduction without any external power sources [2][3]. It is the first ever device that uses a rolling rod as the proof mass to allow operation with non-harmonic vibration. As shown in figure 1(b), signals from the sensor are supplied as the priming voltage to the electrostatic energy harvester and can be amplified and transmitted wirelessly through the device. However, the power density of this prototype is significantly restricted by the amplitude of the priming voltage from the sensor, and weakness of the electrostatic transduction due to the low variable capacitances achievable, and only 2 nJ per pulse can be extracted from it. Taking advantage of the rolling rod actuation mechanism, another energy harvester from our group using piezoelectric transduction has managed to increase the power density dramatically, compared to the electrostatic transduction [4]. This device can only function as an energy harvester due to its non-regulated output, since the waveform of the piezoelectric energy directly depends on the mechanical behaviour of the rolling rod and the piezoelectric cantilever. A piezoelectric pulse generator in [5] can provide a well regulated output for wireless sensing, but uses a mass-spring-damping construction, which cannot operate at a random low frequency. The piezoelectric pulse generator proposed here combines features of the two prototypes in [3] and [4], and is designed to generate a series of reliable and tuneable pulses which is aimed at power for WSN nodes as shown in figure 1(c). This proposed architecture connects the sensor and the pulse generator in parallel instead of in series, which eliminates the priming voltage restriction of the electrostatic prototype. A simple load circuit is designed to work together with the piezoelectric transduction mechanism, in order to achieve self-regulation of the piezoelectric energy.

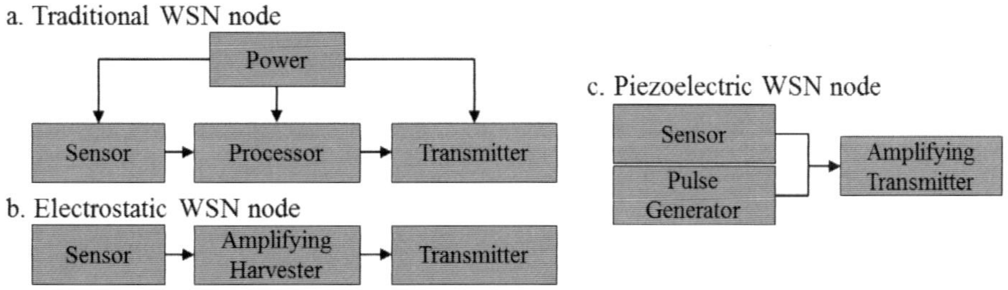

Figure 1. Comparison of different WSN node architectures.

This paper presents the operating mechanism of the proposed pulse generator, and experimental data measured from the scalable prototype are illustrated. Simulation results are provided for further application to the WSN node.

2. Operating mechanism of the pulse generator

The pulse generator consists of two parts. The electromechanical part is used to actuate the piezoelectric cantilever for energy generation, and the electrical part, which is the load circuit, is connected to the piezoelectric cantilever for instant energy storage and regulation.

Figure 2 shows the architecture of the electromechanical part. A metal rolling rod is used as the proof mass, and aligned on the side wall of the substrate to travel back and forth. A block magnet is attached to the tip of the piezoelectric beam for magnetic coupling together with the rolling rod. When the rod is excited by the motion source and rolls from one end to the other (one charge and discharge cycle), it attracts the tip magnet to deflect the piezoelectric cantilever and afterwards release the cantilever to vibrate. A fixed amount of piezoelectric energy is generated both from the deflection and vibration, since the surface charge is proportional to the tip deflection of the cantilever [6].

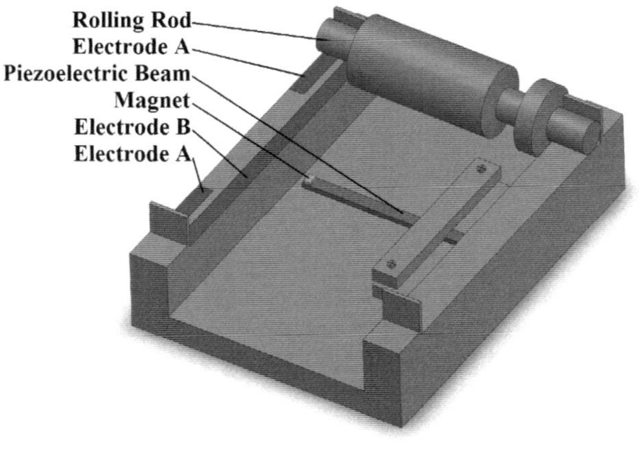

Figure 2. Electromechanical part of the pulse generator.

The electrical part shown in figure 3 is connected to the piezoelectric beam as a storage load. When the piezoelectric beam is deflected, diode D_1 is forward-biased and the piezoelectric charge is shared with the load capacitor, C_{load}, instantaneously for each of the charge and discharge cycles. As the rod moves away to let the beam vibrate, the charge is reversed, and D_1 is reverse-biased, disconnecting C_{load} from the beam, while D_2 is forward-biased allowing the beam to reset to its neutral axis.

The electrode pair *A* and *B* illustrated both in figure 2 and 3 are made of copper tape for synchronous switching of the circuit together with the metal rod. After the energy is shared and stored in C_{load}, the rod approaches one of the electrode pair as shown in figure 2, connects the two electrodes, and forms a switch as shown in figure 3. Then the stored energy can be discharged to the load of the pulse generator.

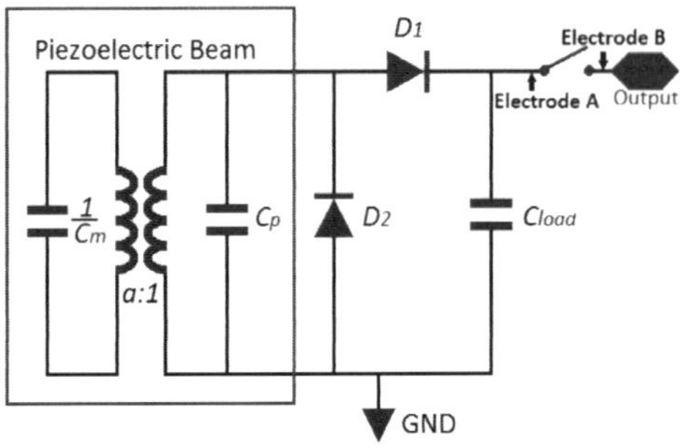

Figure 3. Schematic of the electrical part of the pulse generator.

3. Experimental set-up and measurement results

A large scale prototype is used to demonstrate the proposed concept of the pulse generator as shown in figure 4. The testing circuit is fixed on the side wall of the electromechanical stand, and is used as the track of the rolling rod. The piezoelectric beam is clamped by a small platform, which is mounted on the substrate and can be removed to adjust for piezoelectric beams of different dimensions. The cantilever under test is a piezoelectric bimorph from Johnson Matthey. The free length of the beam is 38 mm with ~45nF capacitance for each piezoelectric layer. The dimension of the tip magnet (N45 neodymium) is 5×4×1.5 mm. The load capacitor used in the set-up is 47 nF to match the piezoelectric capacitance, and the mass of the rod is 0.39 kg.

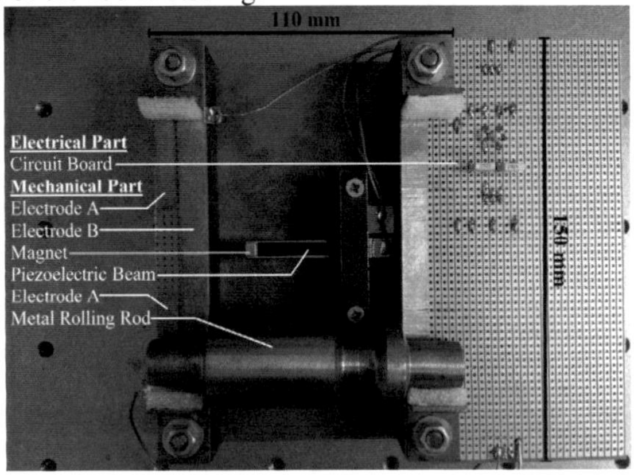

Figure 4. Large scale prototype of the pulse generator.

Figure 5 illustrates an output of the present prototype but actuated in the manner presented in [4]. The first pulse with two positive peaks is generated during the deflection of the piezoelectric beam. After the beam is free to vibrate a series of resonating waves are generated. According to figure 5, the width of the deflection pulse depends on the speed of the external rolling rod, whilst the frequency of the resonating waves is the natural frequency of the beam. Although this approach provides an excellent performance for energy harvesting, it leaves the output result not directly suitable for WSN nodes, due to its unclean shape with many resonating cycles, and its width cannot be tuned since it is dependent on either the deflection speed caused by the magnet-rod coupling or the natural frequency of the beam.

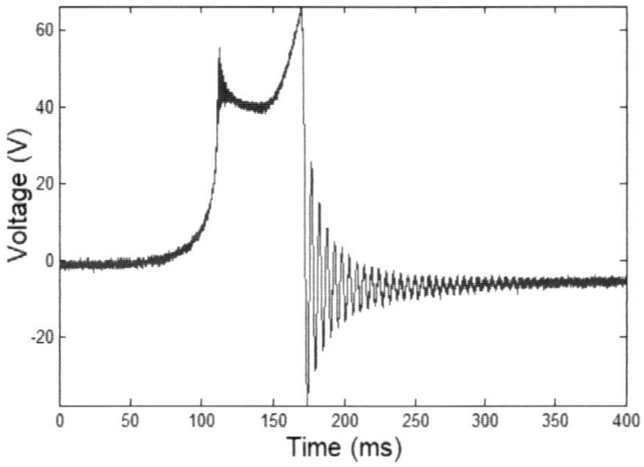

Figure 5. Output voltage without the load circuit.

By contrast, the piezoelectric energy extracted from the proposed prototype in figure 4 is stored simultaneously as the generator is operating, and is regulated and discharged to the load of the device synchronously from the self-switching of the rod-electrode pair mechanism. Figure 6 illustrates the output pulses with different load impedances.

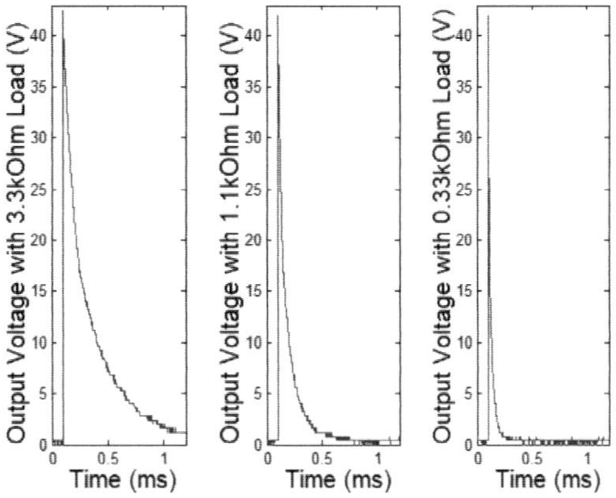

Figure 6. Output pulses from the generator connected with different load resistors.

During the storage and self-regulation of the energy, a charge penalty is inevitable. As can be seen from figure 6, the amplitude of the pulses decreases to 42 V compared to the highest peak in figure 5, which is 62 V, due to charge sharing between two capacitors. The energy extracted in each charge and discharge cycle is the energy shared and stored to the load capacitor, and up to 41.5 μJ per pulse can be achieved with the 47 nF load capacitor, which is still high enough to power a WSN node. In addition, since the energy is stored in a load capacitor before discharge, the pulse width is dependent on the electrical components rather than the physical behaviour of the beam, making it tuneable with different load impedances as shown in figure 6.

4. Signal amplification and simulation

In order to realise signal processing, a single MOSFET amplifier as illustrated in figure 7 is used to simulate the application. To simplify the simulation, a load capacitor charged by a DC voltage is used to represent the pulse generator, which is connected to the amplifier as a discrete power source. Capacitors C1 and C2 are used to provide a bias voltage for the signal from the sensor and the synchronous switching by the rod-electrode pair mechanism is represented by voltage controlled switches.

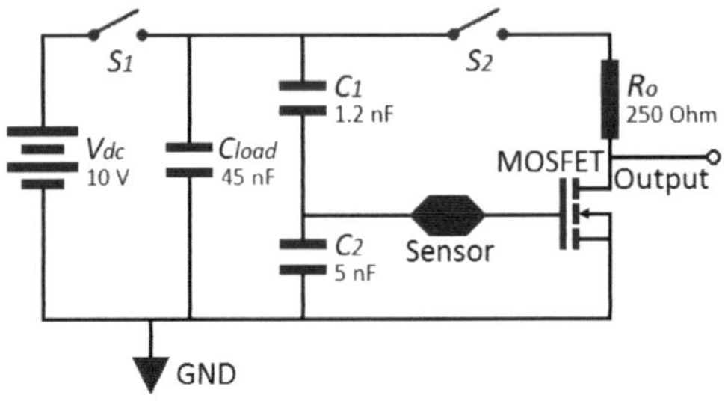

Figure 7. Schematic of the amplifier.

Instead of powering a MOSFET amplifier with a DC voltage source, the pulse generator can inject a series of pulses to the amplifier to achieve signal amplification and sampling at the same time. Figure 8 illustrates the simulation results of the schematic. As it is shown, the blue sine wave represents the unbiased sensor signal, whose peak-to-peak amplitude is 100 mV, and the red pulsed signal is the output of the circuit. It can be observed that the original signal from the sensor is amplified and sampled through the circuit, and the peak-to-peak amplitude of the pulsed sine wave is 400 mV, providing an amplification of 4. By using the circuit presented, the original signal can be processed for wireless transmission with a resonant radio frequency oscillator as in [3].

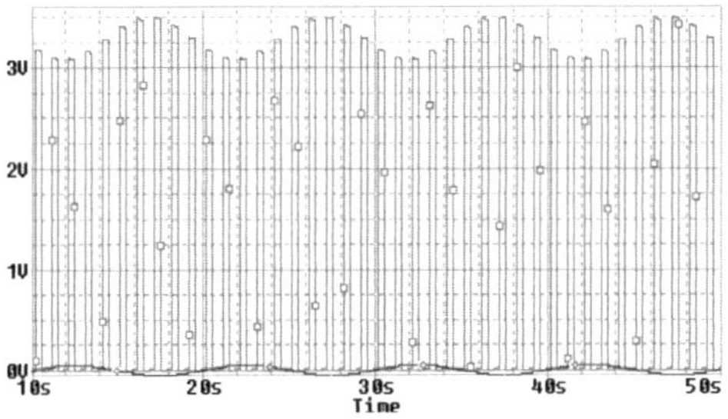

Figure 8. Simulation result of the amplifier.

5. Conclusion

A new pulse generator is proposed to enhance the performance for self-powered WSN nodes by taking advantage of the piezoelectric transduction and the external rolling proof mass. A testing prototype is built to demonstrate that the pulse generator can achieve the function as expected in theory, and simulation results are illustrated to introduce the application of this pulse generator. Self-synchronous switching is successfully implemented using an external proof mass. A reliable pulse with high energy (41.5 μJ) can be generated repeatedly from the model and is capable of being tuned to satisfy the requirement of wireless transmission by load impedance matching.

The geometry of the device is scalable by choosing different dimensions of piezoelectric beams for further miniaturization. In addition, a charge penalty is induced by the charge sharing mechanism, and alternative load circuits may be required to potentially enhance the efficiency of the energy harvesting.

References

[1] Mitcheson P D, Yeatman E M, Rao G K, Holmes A S and Green T C 2008 *Proc. IEEE.* **96** 1457–86
[2] Kiziroglou M E, He C and Yeatman E M 2009 *IEEE Trans. Ind. Electron.* **56** 1101–8
[3] He C, Kiziroglou M E, Yates D C and Yeatman E M 2011 *IEEE Sensors J.***11** 3437–45
[4] Pillatsch P, Yeatman E M and Holmes A S 2012 *IEEE 25th Int. Conf. on MEMS (Paris)* pp 1205–1208
[5] Pinkston C S and Engel T G 2005 *IEEE 15th Int. Pulsed Power Conf. (Monterey)* pp 767–770
[6] Erturk A and Inman D J *Piezoelectric Energy Harvesting* (New York: John Wiley & Sons)

Start-up and Self-sustain Test of 500 W Ultra-Micro Gas Turbine Generator

JeongMin Seo, Jun Young Park and Bum Seog Choi

Korea Institute of Machinery & Materials, Daejeon, Republic of Korea

E-mail: jmseo@kimm.re.kr

Abstract. This paper provides the performance test for start-up and self-sustaining of 500W ultra-micro gas turbine (UMGT) generator. Each component of UMGT, a centrifugal compressor, a radial turbine, an annular combustor and a shaft is already designed, manufactured and tested to meet design requirements in previous researches. However, they are not tested to work in an integrate system. Currently, integrated test unit with a compressor, a combustor and a turbine, is developed to find the proper condition of start-up and self-sustain. Ignition sequence depending on rotating speed is designed. Performance test for start-up and self-sustain is designed based on the ignition possible condition. An air impingement starter and a hot bulb inginer are applied. LPG is used as main fuel.

1. Introduction

The increase of power consumption in mobile devices with motion parts, such as UGV (Unmanned Ground Vehicle) and humanoid robot, requires new mobile power sources with high power density and high energy density. Lithium ion batteries are currently the most convenient energy sources, but their energy densities are usually under 250 Wh/kg. Fuel cell, another energy source for mobile devices, has high energy density but has poor power density. MTG (Micro Turbine Generator) is suggested as a new energy source by many research groups [1-9].

500W class UMGT (Ultra-Micro Gas Turbine) generator is under development by KIMM (Korea Institute of Machinery & Materials). A centrifugal compressor, a radial turbine, an annular combustor and a shaft are developed [10-12]. However, each component is designed and developed based on design point. To increase speed of the integrated gas turbine from stop to design points, start-up procedure and additional auxiliary equipment is required and should be tested. This study presents the test of start-up and self-sustain of integrate gas turbine system for 500 W UMGT generator.

2. Specifications of UMGT

Figure 1 shows the schematic of 500 W class ultar-micro gas turbine generator of KIMM. The target volume of MTG (Micro Turbine Generator) is 1,000 cm^3 for mobility. It is composed of a centrifugal compressor, a turbine, a combustor, six recuperators, a motor/generator and a control unit. The compressor, the turbine and the motor/generator are all directly connected in a single shaft. The design speed of the MTG is 400,000 rpm. The pressure ratio of the compressor with 7+7 blades and vaneless diffuser is 3.0. The radial turbine consists of nozzle and rotor. The blade number of the nozzle and rotor is 20 and 8 respectively. The tip clearance of the compressor and the turbine is 0.1mm. The blade of the compressor and the turbine is made by Inconel 718. The combustor is designed in annular type with 12 nozzles. TIT (turbine inlet temperature) is 1200K. Kerosene is used as main fuel and ignition fuel. The mass flow rate of the main fuel is 0.001945 kg/s at design point. A graphite hot bulb is used as an igniter. Air foil bearings are scheduled to be used for radial and thrust bearings and are under development. Generator is designed to work as a motor at start-up.

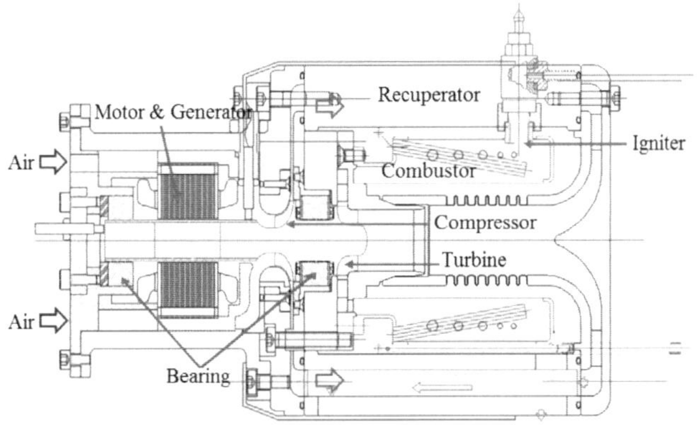

Figure 1 Schematics of 500 W MTG.

3. Test equipment

3.1. Test unit

A test unit is developed to test start-up and self-sustain evaluation as shown in figure 2. It consists of the compressor, the turbine and the combustor developed for 500 W UMGT. Some other parts are designed for this test equipment because they are under development. A radial and thrust all-in-one static air bearing is applied. Compressed Nitrogen gas is supplied for static pressure.

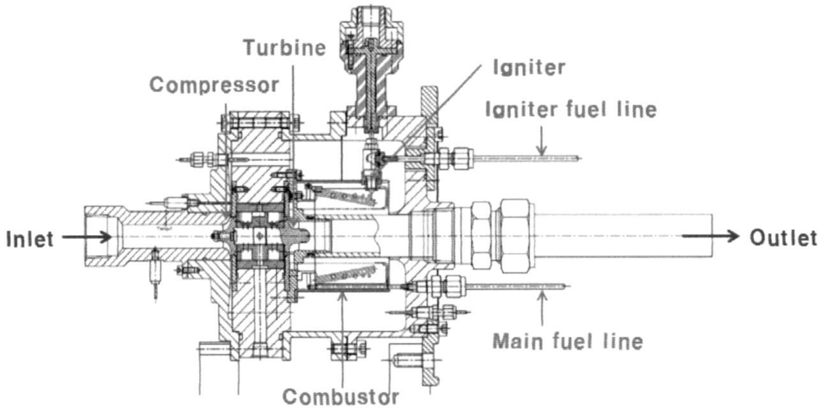

Figure 2 Integrate test unit for start-up and self-sustaining.

3.2. Starter

In start-up, the generator is used as a starter motor in 500 W MTG and is under development. Therefore, a substitute of a starter is needed. An air impingement starter is used to the test unit. Compressed air is supplied and air flow rate is controlled by a throttle value. A pipe from compressed air is connected directly to the inlet of the test unit. By the compressed air, the compressor can be driven from zero to 240,000 rpm, 60% of design speed. In the test, if the rotor reaches to a certain target speed for ignition by the starter, the pipe can be disconnected from the inlet of the test unit and an ambient air can be inhaled through the inlet of the test unit.

3.3. Igniter
Electric sparks are generally used in gas turbines as igniters, but graphite hot bulb igniter is applied in the test unit. It can reduce the weight for mobility because of absence of a voltage step-up transformer. Another reason is flash temperature. Total pressure ratio of the compressor at design point is 3.0 and the ratio at the start of ignition is about 1.3. The increase of temperature by air compression is about 10~20 °C and the flash point of kerosene is between 37 ~ 65 °C. So, the compressor outlet temperature may be insufficient for ignition depending on ambient temperature conditions. Therefore, a hot bulb type igniter is used. Kerosene is used for igniter fuel.

3.4. Combustor
An annular type combustor with 12 nozzles is developed. Mass flow rate of main fuel and igniter fuel is controlled by MFC (Mass Flow Controller). 500 W MTG is designed to use kerosene for igniter fuel and main fuel, but in the test unit, LPG is used for main fuel. The tilt angle and length of flame holder are optimized and the diameter and type of air admission hole are modified to increase the stability of flame by enhancement of air flow and swirl effect. The performance test of combustor is conducted to fine the ignition possible condition. Depending on the speed, air mass flow rate and total pressure ratio are generally determined and a proper amount of fuel mass flow rate is obtained.

3.5. Test bench
Test bench is established as shown in figure 3. Temperatures at compressor outlet, turbine inlet and exhaust out are measured. Compressor outlet pressure is measured. Rotating speed and mass flow rate of main fuel and igniter fuel are also monitored. Moving pipe line for air impingement is installed at inlet of test unit.

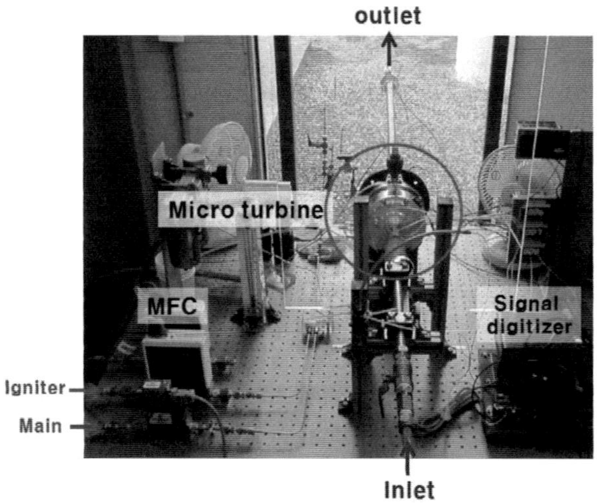

Figure 3 Test bench for start-up and self-sustaining test.

4. Experimental results

4.1. Starting sequence
Figure 4 shows the starting sequence of the MTG. Rotor speed is accelerated to target speed 100,000 rpm, 25% of design speed, for ignition by air impinging starter and the speed is held by throttle valve. The graphite hot bulb igniter is heated and then, igniter fuel is injected. After some time passes, main fuel is injected and compressed air supply is disconnected simultaneously. Rotating speed decreases by free run down, but the speed is recovered by combustion. If the main fuel is sure to be ignited, the igniter is turned off.

In case of electric motor type start, the starter can provide sufficient acceleration to spin the rotor up to a light up speed, and then continue to provide useful torque to enable the engine to accelerate to self-sustain speed smoothly by controller. But in case of air impingement type starter, it is hard to control the speed smoothly. Therefore, free run down is needed. In 500 W UMTG, an electric motor and generator will be used instead of air impingement starter and the starting sequence should be modified and retested.

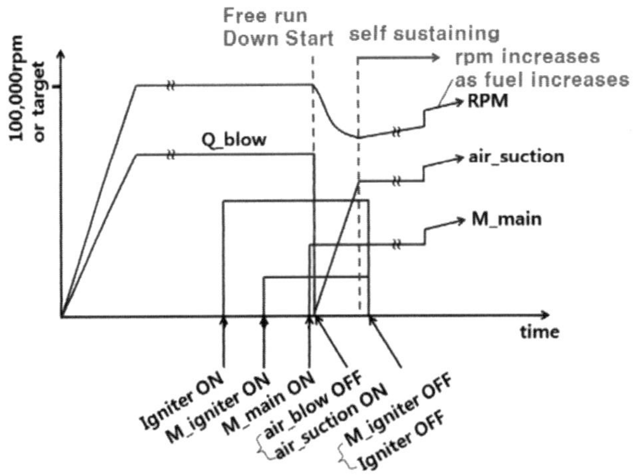

Figure 4 Starting sequence logic.

4.2. Self-sustaining test

Many self-sustaining tests are conducted. One of the test results is shown in figure 5. Free run down and ignition starts at 100,000 rpm and dramatic drop of speed by free run down does not happen. But sudden increase of speed happens by combustion. Self-sustaining successfully continues for 160 second around 200,000 rpm.

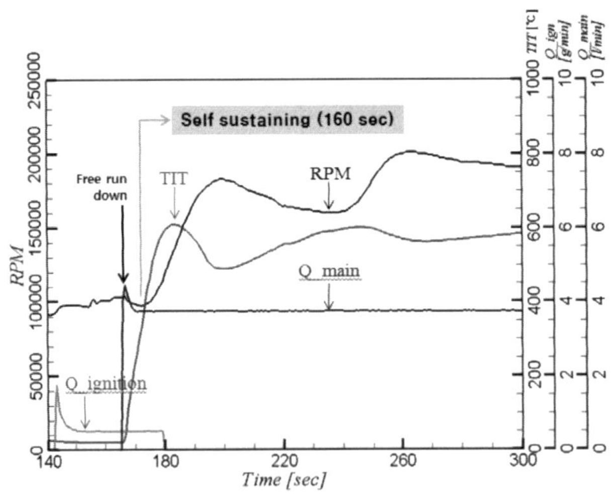

Figure 5 Duration graph for self-sustaining test.

4.3. Boosting test

Boosting test is conducted to examine the smooth speed acceleration from self-sustaining speed to the design speed by increase of main fuel. Figure 6 shows the results of boosting test. Slight increase of main fuel produces slight increase of rotating speed. But, the boosting test is conducted till 230,000 rpm. The test unit uses a radial and thrust all-in-one static air bearing and its durable temperature is about 800 K and its maximum rotating speed is about 250,000 rpm. Therefore the boosting test cannot be carried out in higher speed range. However, the result shows the successful increase of speed by increase of main fuel.

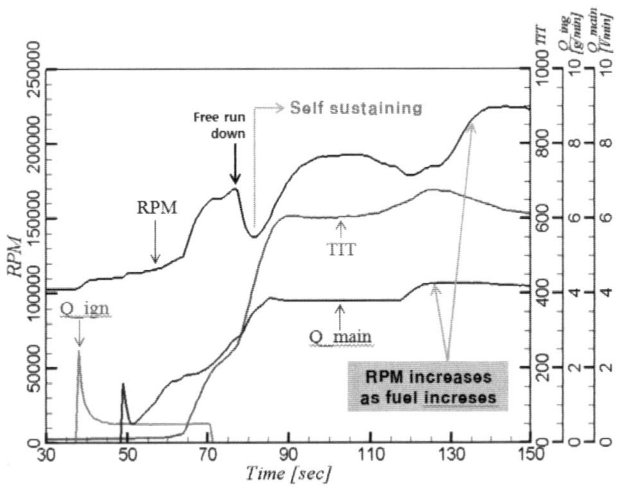

Figure 6 Duration graph for boosting test.

5. Conclusion and perspectives

An integrate test equipment with a centrifugal compressor, a radial turbine and an annular combustor is developed to conduct a performance test for start-up and self-sustain of 500W ultra-micro gas turbine (UMGT) generator. Air impingement starting is used as a starting method and a hot bulb igniter is employed. LPG is used as fuel. Starting sequence is designed based on ignition possible condition and the test results show that the start-up and self-sustain procedure works properly. New integrate test unit with a motor/generator and a new bearing system will be developed and tested in the near future.

ACKNOWLEDGEMENT

This work was supported by the Next Generation Military Battery Research Center program of Defense Acquisition Program Administration and Agency for Defense Development.

References

[1] Kang P, Tanaka S and Esashi M 2005 *J. Micromechanics Microengineering* **15** 1076–87
[2] Monroe M A, Epstein A H, Kumakura H and Isomura K 2005 *Proc. ASME Turbo Expo* **1** 981–8
[3] Onishi T, Burguburu S, Dessornes O and Ribaud Y 2005 *Proc. ASME Turbo Expo* **1** 847–55
[4] Shan X C, Wang Z F, Jin Y F, Wu M, Hua J, Wong C K and Maeda R 2005 *J. Micromechanics Microengineering* **15** S215–S221
[5] Isomura K, Murayama M, Teramoto S, Hikichi K, Endo Y, Togo S and Tanaka S 2006 *J. Micromechanics Microengineering* **16** S254–S261
[6] Shan X C, Zhang Q, Sun Y and Wang Z 2006 *J. Phys. Conf. Ser.* **34** 316–21
[7] Zwyssig C and Kolar J W 2006 *J. Micromechanics Microengineering* **16** S297–S302
[8] Gong Y, Sirakov B T, Epstein A H and Tan C S 2013 *Proc. ASME Turbo Expo* **6** 95–102
[9] Peirs J, Waumans T, Liu K and Reynaerts D 2008 *PowerMEMS* 413-6
[10] Han S, Seo J, Park J, Choi B and Do K 2010 *PowerMEMS* 247-50
[11] Park C, Choi S and Ham S 2011 *PowerMEMS* 443-6
[12] Seo J, Park J, Choi B, Park M and Han S 2011 *PowerMEMS* 447-9

An improved method for piezoelectric characterization of polymers for energy harvesting applications

E Gusarova[3], B Gusarov[1,2], D Zakharov[1,2], M Bousquet[3], B Viala[3], O Cugat[1,2], J Delamare[1,2] and L Gimeno[1,2]

[1] G2Elab, Univ. Grenoble Alpes, F38000, Grenoble, France
[2] CNRS, Univ. Grenoble Alpes, F38000, Grenoble, France
[3] CEA, LETI, MINATEC Campus, F38000, Grenoble, France

E-mail: boris.gusarov@g2elab.grenoble-inp.fr

Abstract. This work presents an improved method for measuring the direct piezoelectric voltage and energy of flexible polymers. Well-controlled stress is applied with a four-point bending system and voltage is measured in real open-circuit conditions. The presented method separates the piezoelectric part from the measurement part by introducing a mechanical switch, allowing instantaneous post-deformation discharge measurements. Oscilloscope and contact-less electrostatic voltmeter are compared. Direct piezoelectric measurements under open-circuit conditions have been performed on commercial PVDF (polyvinylidene fluoride) and its copolymers. Significant differences to data sheet values (close-circuit conditions) are reported and commented.

1. Introduction

Conventional direct piezoelectric characterization can be hampered by large electrical losses and uncontrolled charge leakage into the circuit during measurements. This may result in lower detected voltage and possible underestimation of the material properties. Taking into account the low permittivity or capacitance of polymers compared to piezoelectric ceramics, this problem becomes even more important to solve for flexible organic piezoelectric films.

Usually, direct piezoelectric response measurements for polymers are made with a standard oscilloscope with closed-circuit connection [1, 2], or by using sophisticated charge-amplifier circuits [3, 4]. Here we propose a simple and accurate method for measuring the voltage after mechanical deformation of flexible organic films with real open-circuit conditions, virtually with no charge leakage. For this purpose, we introduce a mechanical switch into the measurement circuit to separate the piezoelectric part (generator) from the measuring part (collector). Two technics using oscilloscope and contact-less electrostatic voltmeter are compared. Results for commercial PVDF and copolymer of P(VDF-TrFE) are reported and compared to data sheet values (measured in closed-circuit conditions).

2. Experimental

2.1. Experimental set-up

The proposed method for measuring the voltage output introduces into the circuit a mechanical switch, which separates charges generation from charges collection for voltage measurement. The switch is kept open during the mechanical deformation of the polymer while space charges are produced. When the deformation is completed, the switch is closed and space charges are converted to free charges at the electrodes, and transferred at once into the measurement circuit.

The output voltage is measured with two technics: first using a standard oscilloscope (for this work Agilent Technologies DSO1014A was used) and second with a non-contact electrostatic voltmeter (in this work a TREK 370 voltmeter). The latter measures the electrostatic sample surface potential without physical contact, making the impedance of such system virtually infinite, thus eliminating potential residual charge leakage that could take place at the oscilloscope load input.

A schematic representation of the circuit is shown on figure 1.

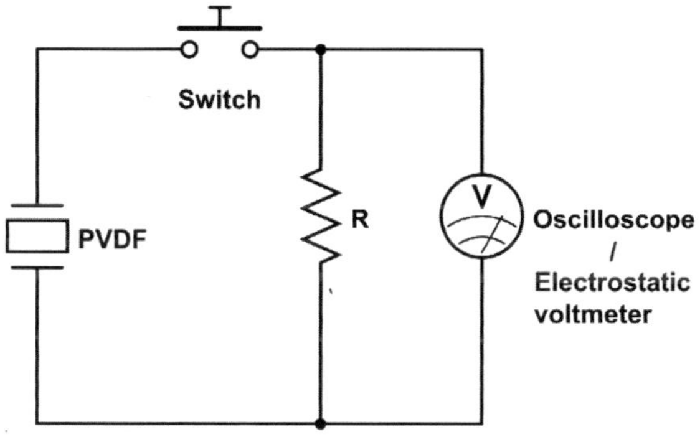

Figure 1. Schematic presentation of the measurement circuit with a switch, an oscilloscope and a contact-less electrostatic voltmeter.

This real open-circuit set-up allows ideal space charges generation when applying stress to the polymer material. For the purpose of applying stress, we use a controlled high precision four-points bending system. Once the charges are generated and the switch is closed (A), the discharge takes place over few tens milliseconds (B) as shown in figure 2. The corresponding electrical energy can be calculated from time dependent voltage by equation (3) with integration limits from A (start of discharge) to B (discharge completed).

$$E = 1/R \int_A^B V^2 dt \tag{1}$$

As can be seen from figure 2, both methods lead to identical remarkable high voltage output range of the order of hundred volts (here with commercial PVDF at 0.5% strain). This is considerably higher than that of the conventional oscilloscope measurement which reduces to the order of ten volts (see dotted line in figure 2).

When comparing real open-circuit measurement results with the oscilloscope and the electrostatic voltmeter, one see a difference in the time-dependence of charges decay, as it greatly depends on the load resistance (10 MΩ was used for oscilloscope and 1 MΩ for electrostatic voltmeter). However, the calculated energy remains resistance-independent. During this work most of the piezoelectric measurements were performed with the switch and oscilloscope set-up with a load resistance of 10 MΩ.

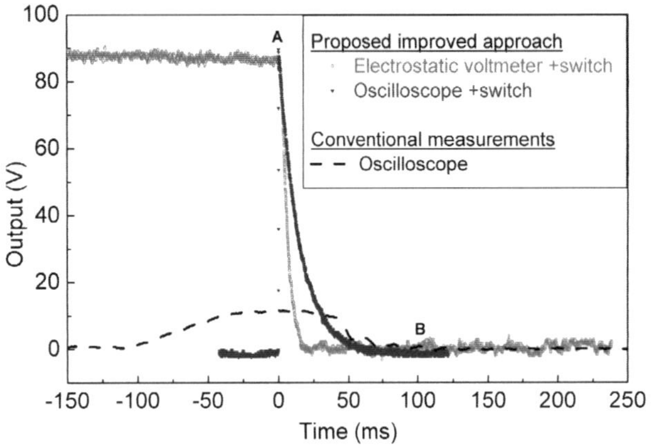

Figure 2. Typical discharge curves of PVDF after 0.5% deformation measured with oscilloscope and electrostatic voltmeter. For both cases the switch is closed at time zero. The curve with oscilloscope was made with 10 MΩ, and the curve with electrostatic voltmeter was made with 1 MΩ resistance to demonstrate the impact of resistance on discharge behavior. A conventional closed-circuit measurement with an oscilloscope (R=10 $M\Omega$) is shown in dash for comparison.

Table 1. Properties of a commercial PVDF sample used in present work [5, 6, 7].

Name	Length(mm)	Width(mm)	Thick(mm)	d_{31} (pm/V)	g_{31} (Vm/N)	k_{31} (%)	YM (GPa)
LDT1	30	12.19	0.028	23	0.216	12	2-4
PVDF	20	15	0.040	6 ±10%	0.06 ±20%	10-15	2.5 ±20%
P(PVDF-TrFE)	20	15	0.050	6 ±20%	0.09 ±20%	10-15	1 ±20%

2.2. Experimental results

Different commercial PVDF and P(VDF+TrFE) samples (summarized in table 1) were measured with the proposed method in 3-1 mode.

To apply deformation samples were glued to a 1 mm thick Plexiglas® substrate and bent with the four-point bending machine. As the curvature radius is constant, the linear deformation ε of the sample can be calculated as

$$\varepsilon = \frac{h}{2R}$$

where h is the substrate thickness, R is the curvature radius.

Assuming free deformation (stress T is constant) and open electrical circuit (electrical displacement D is constant), the piezoelectric coupling matrix can be writen in the following simplified form:

$$S = s^D T + g_{ij}^t D$$

$$E = -g_{ij} T + \beta^T D$$

where D is electric displacement field, E is electric field, T is mechanical stress, S is mechanical strain, s is flexibility, β is permeability constant, and g_{ij} is piezoelectric coefficient.

From the second term under our experimental conditions where the switch is kept open during deformation, the electric displacement D is zero, and the voltage can be extracted as:

$$V = -g_{31} \cdot S \cdot YM \cdot t \qquad (2)$$

where t is the sample thickness, YM is the Young's Modulus.

Table 2. Comparison of piezoelectric coefficients from supplier data-sheets and experimentally obtained for different samples.

Name	Data-sheet values		Experimental values	
	g_{31}, Vm/N	k_{31}, %	g_{31}, Vm/N	k_{31}, %
LDT1 (28 μm) MeasSpec	0.216	12	0.22	13
PVDF (40 μm) Piezotech	0.06 $\pm$20%	10-15	0.09 (0.16)	5.5 (10)
P(PVDF-TrFE) 70-30 (50 μm) Piezotech	0.09 $\pm$20%	10-15	0.4	12

The electrical energy produced by the piezoelectric material can be estimated by multiplying the elastic energy by the electromechanical coupling coefficient k^2, and is given by equation 3.

$$U_{electrical} = \frac{1}{2} \cdot YM \cdot S^2 \cdot k^2 \tag{3}$$

3. Discussion

Output voltage and the corresponding generated electric energy density as function of applied strain are presented on figures 3 and 4 respectively. The classical linear voltage strain-dependence (as from equation (2)) is confirmed for all samples, but with unusual high voltage values of up to 120V. The slopes of curves reflect at first order the differences in films thickness and also g_{31} coefficients.

Considering the electrical energy density, experimental results also agree with the expected parabolic strain dependence as predicted by equation (3). Consistently with unusual high voltages, one notes remarkable high values of the electrical energy density with about 0.5 to 1 mJ/cm^3 for strain of 1%.

Finally, g_{31} voltage coefficient and k_{31} mechanical coupling factor were calculated from results of figures 3 and 4, respectively. The results of calculation are shown in table 2 and compared with those of data sheets.

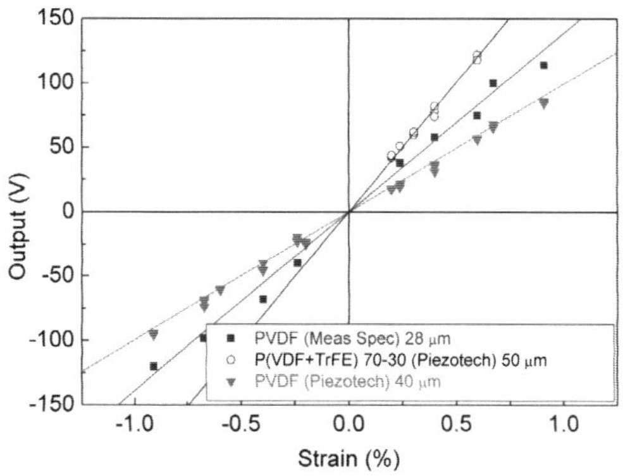

Figure 3. Voltage output as a function of applied strain in 3-1 mode.

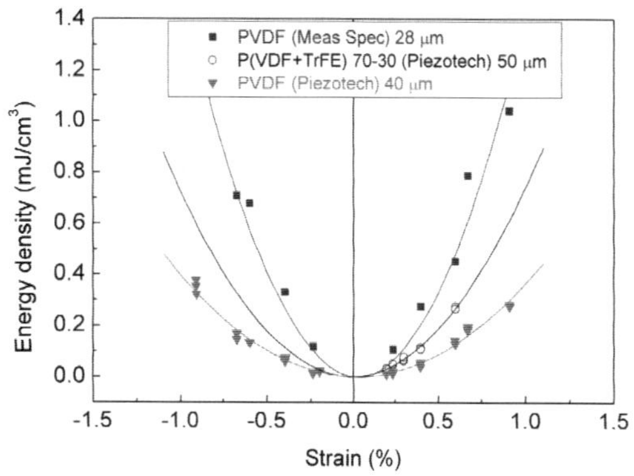

Figure 4. Generated electric energy density as a function of applied strain in 3-1 mode.

The proposed measurement method successfully allows reproducible measurements in quasi-static conditions of the piezoelectric coefficients of the polymer films in real open-circuit conditions. When comparing, one notes g_{31} coefficients being significantly larger than on data sheet. It is a clear feature of the P(VDF-TrFE) sample, for example. Considering k_{31} coupling factor, experimental values are in the range of the data sheet, except for the "PVDF (40 μm) Piezotech" sample. The too low value may indicate improper YM value in this case. When using 1.4 GPa instead of 2.5 GPa, k_{31} agrees with expectation (equation (3)). This indicates that g_{31} value is possibly even much larger (shown in brackets in table 2). It is therefore desirable, in order to improve the accuracy of the measurements, to assess with higher precision the YM of the samples prior to piezoelectric measurements. Also, the analytic estimations made for the strain and energy calculation may alter the precision of calculated coefficients.

Coupling factors have also been measured by classical frequency-impedance measurements. However, as resonance and anti-resonance with polymers are usually broaden, it does not give adequate precision. Therefore only orders of magnitude can be confirmed.

In terms of measurement accuracy, the main limit of the proposed method is set mostly by the measurement equipment. For instance, in the present work a TREK 370 voltmeter was used which is capable of measuring in 0-3 kV range with 1 V resolution.

4. Conclusion

An improved method for measuring the direct piezoelectric voltage and energy of flexible polymers is reported. Stress is applied with a high precision four-points bending system and voltage is measured with real open-circuit conditions using a mechanical switch. This technic allows exploring direct piezoelectric effect in quasi-static conditions, without use of high-precision equipment nor sophisticated measurement circuits. As a clear feature, we report unusual large strain-induced voltage and electric energy density for PVDF and copolymers. Remarkably high voltage and energy density values of up to 120 V and 1.2 mJ/cm^3 were obtained for 40 μm PVDF sample with 0.7 % of strain in 3-1 mode.

Finally, we conclude that the proposed method allows direct piezoelectric coefficients evaluation, which is adequate for generator or harvester considerations. The experimental results of this work suggest that usual voltage coefficient of PVDF and P(VDF-TrFE) copolymers may be underestimated.

References

[1] Guigon R, Chaillout J J, Jager T and Despesse G 2008 *Smart Mater. Struct.* **17** 015039
[2] Vatansever D, Hadimani R L, Shah T and Siores E 2011 *Smart Mater. Struct.* **20** 055019
[3] Canavese G, Stassi S, Cauda V, Verna A, Motto P, Chiodoni A, Marasso S and Demarchi D 2012 *Sensors, IEEE* pp 1–4
[4] Jiang Y, Shiono S, Hamada H, Fujita T, Higuchi K and Maenaka K 2010 *PowerMEMS*
[5] Piezo film sensors technical manual URL `http://www.meas-spec.com/downloads/Piezo_Technical_Manual.pdf`
[6] Piezo sensor - ldt series URL `http://www.meas-spec.com/product/t_product.aspx?id=2484`
[7] Piezoelectric film technical information URL `http://www.piezotech.fr/image/documents/22-31-32-33-piezotech-piezoelectric-films-leaflet.pdf`

Piezoelectric and electrostatic bimetal-based thermal energy harvesters

A. Arnaud[1,3], S. Boisseau[2], S. Monfray[1], O. Puscasu[1], G. Despesse[2], J. Boughaleb[1], Y. Sanchez[1], F. Battegay[1], M. Fourel[1], S. Audran[1], F. Boeuf[1], J. Delamare[3], G. Delepierre[4], G. Pitone[4], T. Skotnicki[1]

[1] ST Microelectronics, 850 rue J. Monnet - 38926 Crolles, France
[2] CEA, Leti, Minatec Campus, 17 rue des Martyrs - 38054 Grenoble Cedex 9, France
[3] G2ELab Grenoble, 25 av. des Martyrs - 38042 Grenoble Cedex 9, France
[4] Delta Concept, 6 rue de Chamechaude, ZI de l'argentière – 38360 Sassenage, France

E-mail: arthur.arnaud@st.com

Abstract. This paper reports on innovative thermal energy harvesters (TEH) turning heat fluxes into electricity in a two-step conversion, involving (i) a curved bimetallic strip converting thermal gradients into mechanical oscillations, which are then (ii) converted into electricity by a piezoelectric or an electret-based electrostatic transducer. This work mainly focuses on (i) the optimizations of the piezoelectric devices, (ii) a first demonstration of a Wireless Sensor Node powered by our electrostatic transducers, validating the viability of bimetal-based thermal energy harvesters, and (iii) the possibility of future scaled scavengers by a micrometric silicon approach to improve efficiencies and power densities.

1. Introduction

Thermal energy harvesting is a field of growing interest aimed at turning ambient thermal gradients into electricity to develop fully-autonomous electronic devices, such as wireless sensor networks (WSN). This topic is today dominated by thermoelectric energy harvesters based on bimetallic junctions that generate thermoelectric voltages (Micropelt, Thermolife…).

Bimetal-based thermal energy harvesters (b-TEH), presented in this paper, are an alternative to these devices, employing standard and fully available materials, and compatible with mass production over large areas. The novelty of b-TEH lies in the use of curved bimetallic strips to perform a first thermal-to-mechanical power conversion, turning thermal gradients into mechanical movements. Bimetals are the combination of two metals characterized by different coefficients of thermal expansion (CTE) that are bonded together. This CTE difference makes bimetallic strips thermal actuators, deforming with temperature, and acting as thermo-mechanical converters.

In fact, curved bimetallic strips have long been known as thermal actuators able to quickly switch between two states (snapping and snapping back) according to their temperature and with a hysteresis cycle. Inserted in a cavity with a hot lower plate ($T=T_h$) and a cold upper plate ($T=T_c$), curved bimetallic strips become the working substance of a heat engine, kept in an unstable thermodynamic state when exchanging thermal energy either with the hot or the cold source and thus switching constantly. The cycle starts when the bimetallic strip is in its lower state (figure 1a) and in contact with the lower plate; its temperature increases till reaching its snapping temperature (T_s): it suddenly snaps to its upper state, entering in contact with the upper plate (figure 1b) where it is cooled down until reaching its snapping-back temperature (T_{sb}) and returning to its lower state. Then, a new cycle restarts, and the thermal gradient is turned into mechanical oscillations. The hysteresis cycle overviewing states and temperatures is presented in figure 1c.

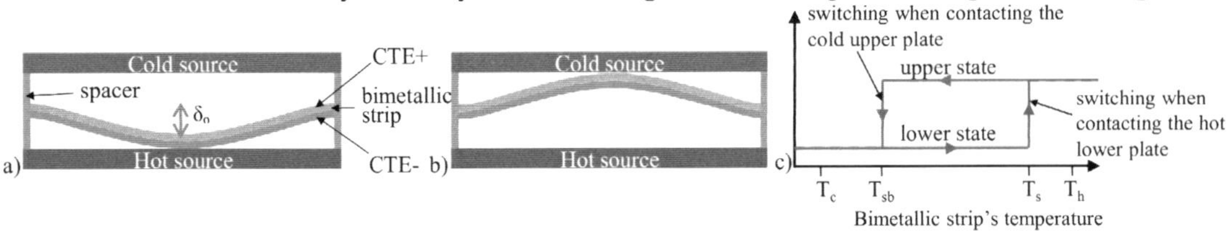

Figure 1. Curved bimetallic strip in a cavity subjected to a thermal gradient (a) lower state and (b) upper state; and (c) hysteresis cycle of the bimetallic strip enabling the oscillations

These oscillations can then be converted into electricity by a standard electromechanical converter such as a piezoelectric material, an electrostatic converter or a coil-magnet architecture. We have chosen to focus on the two first ones, converting the shock of the bimetallic strip on the upper plate with a piezoelectric material (section 2) or its movement by an electret-based electrostatic converter (section 3).

2. Piezoelectric devices

2.1. Principle

The bimetallic strip switching from an unstable equilibrium position to a stable one is followed by a conversion of strain energy into kinetic energy. Our first approach [1] has consisted in harvesting this kinetic energy release into electric energy by making the bimetallic strip shock a piezoelectric oscillator when snapping [2], as shown in figure 2a. In such a structure, the piezoelectric oscillator performs both functions of electro-mechanical transducer and thermal dissipator by enabling the bimetallic strip to cool down and so to snap back. This double function offers the device the advantage of not requiring massive heat sink to maintain its internal thermal gradient (figure 2b) and then to obtain thin prototypes (4mm-thick prototype for a 25mm-long bimetal).

2.2. Piezoelectric transducer mechanical optimization

The transducer global efficiency is mainly linked to the quality of the kinetic energy transfer between the bimetallic strip and the transducer, and to the conversion efficiency of the transducer strain energy into electric energy. During this conversion, the energy is either harvested or lost in the clamps. We have focused here on optimizing the piezoelectric transducer to limit the energy losses in the structure.

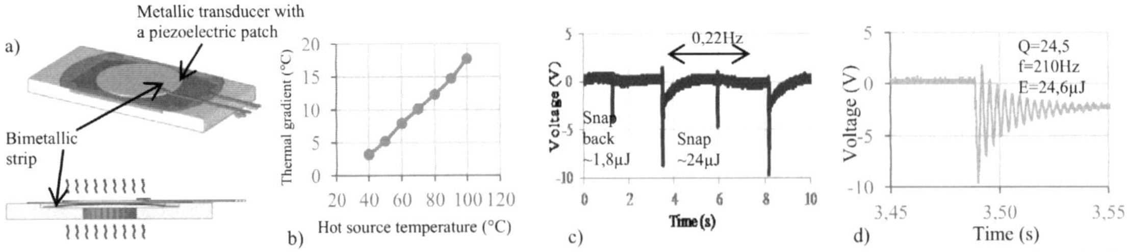

Figure 2. (a) Piezoelectric device principle. (b) Thermal optimization of the device. (c) Output signal of the transducer. (d) Detail of the signal after a bimetallic strip snap

This optimization has led to the design of a new transducer that minimizes the stress exercised on the oscillator clamp and so to the diffusion of energy into the Teflon cavity, and increases the efficiency of the piezoelectric conversion. This geometry has enabled to raise the harvested energy to 24.6 µJ per snap (vs.

1.8μJ with the conventional devices). This approach is currently under investigation to power a Wireless Sensor Node.

3. Electrostatic devices

Electret-based electrostatic converters are built on a capacitive architecture made of two plates separated by an air gap and polarized by an electret (figure 3a) [3]. The electret is a permanently charged dielectric that induces charges on the two plates, with $Q_i=Q_1+Q_2$, where Q_i, Q_1 and Q_2 are respectively, the charges on the electret, the lower electrode and the upper electrode; its equivalent electric model is presented in figure 3b. A relative movement between the two plates induces a new repartition of charges between the two electrodes, converting the mechanical movement into electric charges that are collected by a load (R).

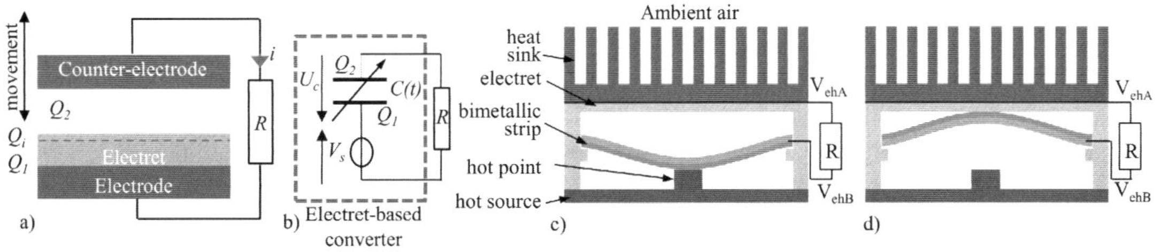

Figure 3. (a) Electret-based converter, (b) equivalent electric model of the electret-based converter, be-TEH in (c) its lower state and (d) its upper plate.

The above-mentioned bimetal-based heat engine can be easily combined with an electret-based converter (thereby becoming a bimetal-and-electret-based TEH [be-TEH]) by adding an electret on the upper plate and by connecting a load between the upper plate and the bimetallic strip, turning the bimetallic strip's oscillations into electricity. First proofs of concept have already been presented in [4], but the devices required forced convection to work. Therefore, in order to increase the thermal gradient between the upper and the lower plates, so as to ensure the bimetallic strip's oscillations without forced convection, a heat sink and a localized hot point have been added (figures 3c and 3d).

Prototypes have been manufactured from B72M (CTE=26.4×10^{-6})/Invar(CTE=2×10^{-6}) curved bimetallic strips measuring 1cm×2cm×115μm, characterized by (T_s, T_{sb})=(47°C, 43°C). The electret is made of a 25μm-thick FEP Teflon layer charged by a corona discharge at 500V and directly glued on the heat sink. The device (figure 4a) is set on a hot source at 70°C and, as expected, the bimetallic strip oscillates without forced convection, turning the thermal gradient into electricity. The output power is in the 5μW range and the snapping frequency is about 2Hz (figure 4b).

As the energy harvester's output voltage is not compatible with standard electronic circuits, a power management circuit implementing SECE (Synchronous Electric Charge Extraction [5]) and withstanding devices in parallel has been manufactured (figure 4c). Ten bimetallic strips have been connected to this power management circuit to power a wireless sensor node consuming 80μJ per emission every 40 seconds (figure 4d), definitely validating the suitability of be-TEH to power WSN from ambient thermal energy [6].

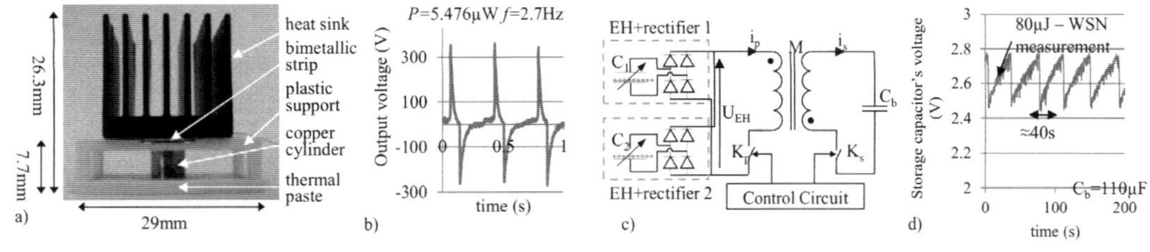

Figure 4. (a) prototype, (b) output voltage (c) power management circuit implementing SECE and (d) WSN supplied by 10 devices in parallel

Nevertheless, at this scale, the capacitance variation between the two equilibrium states is around 50%, limiting the efficiency of the power management circuit due to a large impact of parasitic capacitances: the scaling of the structure down to micrometer scale should enable to increase the capacitance variation and then to reduce the impact of these parasitic capacitances. Moreover, according to Timoshenko's bimetallic strips modeling [7], and as demonstrated in [8], the bimetallic strips scaling helps achieving higher thermal energy transfer rates, switching speeds, snapping frequencies, and electrets-bimetallic strips capacitance variations, and will result in a global increase of our devices efficiency and power densities.

4. Miniaturization of electrostatic devices

4.1. Bimetallic strips modeling

As demonstrated by Timoshenko in [9], if its initial curvature δ_o (see figure 1a) exceeds a critical ratio such that $\delta_o > t/\sqrt{3}$ (t: total beam thickness), the bimetallic strip will have a snap action. Otherwise, if it is lower, the snap action disappears and the bimetallic strip only buckles slowly. Thanks to this model, the evolution of snapping and snapping-back temperatures for all the different parameters of the bimetallic strips can be plotted in order to evaluate the best materials couple and the bimetallic strips dimensions. The influence of the initial curvature δ_o on the stability of a bimetallic strip is shown in figure 5a while figure 5b depicts the sensibility of the snapping temperature to the CTE difference and to the bimetallic strip's length.

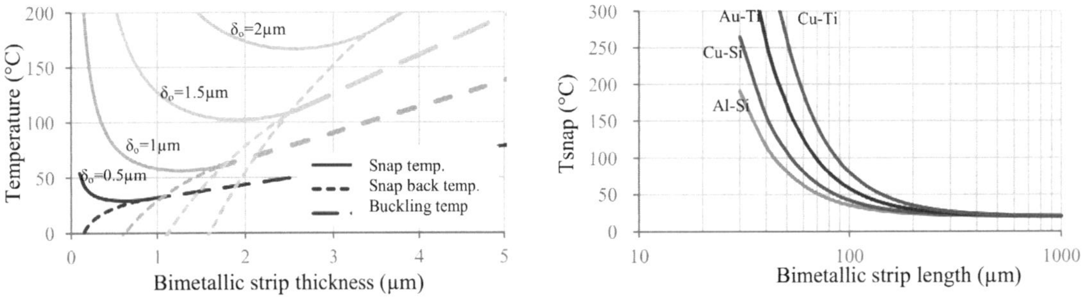

Figure 5. (a) Stability of Ti-Au bimetallic strips (L=200µm) as a function of the dimensions (thickness and middle deflection). (b) Snapping temperature (for t= $\delta_o\sqrt{3}$=0.5µm) as a function of the bimetallic strip length.

4.2. Micro-bimetallic strips process

As previously explained, the bimetallic strip mechanical instability is mainly linked to the ratio between the thickness of each metal layer and its curvature. The development of a micro-bimetallic strip process based on the use of a resist [10] enables to control this last parameter with a good precision. Figure 6a summarizes the different steps of the bimetallic strips fabrication. The coated resist is first patterned into arrays of rectangles, forming parallelepipeds of resist. Baking the resist at 140°C makes it melt and turns parallelepipeds of resist into rounded shapes for the bimetallic strips. Both metal layers are then deposited on the resist by PVD at 210°C to avoid the metal layers suffering from excessive residual stress and the resist from collapsing. The bimetallic strips are then shaped thanks to lithography and finally released after resist stripping.

a) 1. Resist coating | 2. Resist patterning | 3. Resist baking 140°C | 4. Au layer PVD 210°C | 5. Ti layer PVD 210°C | 6. Bimetallic strip patterning and etching | 7. Resist stripping and bimetallic strips release

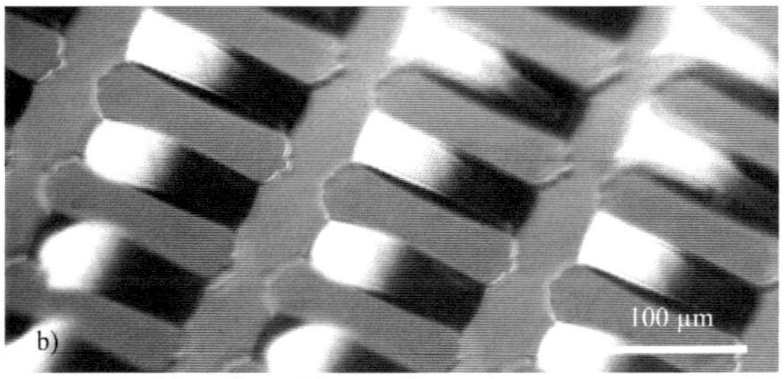

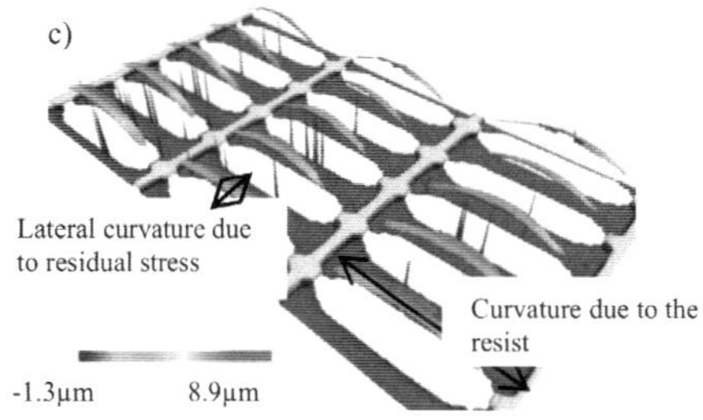

Figure 6. (a) Micro-bimetallic strips process steps (b) Images of Ti-Au bimetallic strips obtained thanks to the new process. (c) Interferometry measurement of the 210μm-long Ti-Au bimetallic strips curvature at 70°C.

We have already developed the first bimetallic strips with a middle curvature (δ_0) of 8 μm to validate the process feasibility. First characterizations realized on Ti-Cu bimetallic strips have pointed out a problem of copper oxidation, modifying irreversibly the mechanical properties of the bimetallic strips and so preventing them from snapping. Choosing Ti and Au, which have a CTE difference close to the one of Ti-Cu, has enabled to tackle this issue (figures 6 b and c). Next steps will consist in (i) reducing the height of the micro-lenses to lower the snapping and snapping-back temperatures of the bimetallic strips and (ii) in creating the electret needed to harvest thermal energy.

5. Conclusion

Last optimizations and developments realized on piezoelectric and electrostatic prototypes have shown the relevance and the effectiveness of bimetal-based devices as thermal energy harvesters. The modifications performed on the piezoelectric devices have enabled to broaden their operating temperature ranges and their output powers (about 5μW at 0.22Hz). Similar output powers (5μW at 2Hz) have been reached with electret-based prototypes. The power level of ten parallelized electrostatic devices at ambient temperature has enabled to supply a wireless sensor node thanks to our own power management circuits, making be-TEH suitable for wireless sensors applications. Finally, because of their manufacturing easiness and their theoretical efficiency, electret-based energy harvesters at micro-scale are our next step. A new microelectronic process based on Timoshenko's bimetallic strip modeling has been developed in this perspective.

Acknowledgements
The project was partially funded by the French Inter-Ministerial Fund (FUI) through HEATec project.

References

[1] Puscasu O *et al* 2012 *Proc. Electron Devices Meeting (IEDM), 2012 IEEE* (10-13 Dec. 2012) 12.5.1-12.5.4
[2] Gu L, Livermore C 2011 *Smart Mater. Struct.* **20** 045004
[3] Boisseau S *et al* 2012 *Small-Scale Energy Harvesting,* ed M. Lallart, Intech, doi:10.5772/51360.
[4] Boisseau S *et al* 2013 *Smart Mater. Struct.* **22** 025021
[5] Lefeuvre E *et al* 2005 *J. Intell. Mater. Syst. Struct.* **16** 865
[6] Boisseau S *et al* Bimetal-and-electret-based thermal energy harvesters – Application to a battery-free Wireless Sensor Node, Submitted to Smart Materials and Structures
[7] Timoshenko S 1925 *JOSA* **11** 233-55
[8] Puscasu O *et al* 2012 *Silicon Nanoelectronics Workshop (SNW), 2012 IEEE* (10-11 June 2012) 1
[9] Timoshenko S and Gere J 1961 *Theory of elastic stability* (New York: McGraw-Hill) 312
[10] Audran S *et al* 2010 *J. Micromech. Microeng.* **20** 095008

Implantable Glucose BioFuel Cells for Medical Devices

P. Cinquin[1], S. Cosnier[2], N. Belgacem [3], M. L. Cosnier [4], R. Dal Molin [5], D.K. Martin[1]
[1] TIMC-IMAG, UJF/CNRS
[2] DCM, UJF/CNRS
[3] LGP2, UJF/G-INP/CNRS,
[4] CEA-LETI,
[5] SORIN-CRM

E-mail: philippe.cinquin@imag.fr

Abstract. An Implantable BioFuel Cell (IBFC) is a device that produces power only from the chemicals that are naturally occurring inside the body. We have been working on two approaches to creating an IBFC. The first approach is to use chemicals such as glucose and oxygen to provide the fuel for an enzymatic IBFC. The second approach is to use electrolytes such as sodium to provide the fuel for a biomimetic IBFC.

1. Introduction

We have demonstrated the feasibility of implanting in rats the enzymatic biofuel cells that are capable of utilising glucose and oxygen from the physiological fluids to create power for other implanted medical devices. We were able to obtain a power to volume ratio of the order of magnitude of 1 μW.μL-1 and to demonstrate the stability of the device during 3 months. We are presently optimizing the device, notably by using printing techniques for the fabrication of the electrodes and by exploring the potential of porous silicon for the packaging.

Another type of IBFC involves a biomimetic approach. This biofuel cell generates electricity by imitating salt transport that occurs naturally in the cells of the body: electrolytes such as sodium are used to provide the fuel for power generation. At the heart of the device is a lipid bilayer containing transport proteins similar to the ones naturally occurring in a cellular membrane. Gradients of sodium are then transformed into voltage and power. Preliminary results show that a biomimetic membrane of 0.3 cm^2 can produce an output of 59 μW, which is encouraging since such membranes are able to be easily stacked in series and in parallel.

The very interesting power to volume ratio opens interesting perspectives of applications of IBFCs, for instance to power implantable medical devices such as lead-less pacemakers or robotic artificial urinary sphincters.

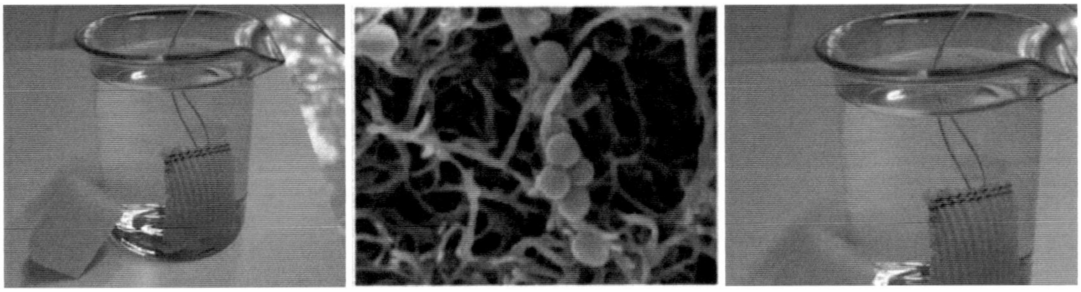

(a) (b) (c)

Figure 1. (a) **Enzymatic Glucose BioFuel Cell ready for implantation**
(b) **Enzymes (spheres) in the nanostructured glucose bioelectrode**
(c) **Prototype biofuel cell containing nanostructured biomimetic membranes**

2. About The Speaker

Prof Philippe Cinquin, 57, is Professor of Medical Informatics at Grenoble (France). He heads TIMC-IMAG, UMR5525, a Research Unit of CNRS and of Université Joseph Fourier, and co-heads CIC-IT 803 (Centre of Clinical Investigation - Technological Innovation) of INSERM and Grenoble's University Hospital. He holds a PhD in Applied Mathematics and is a Medical Doctor. In 1984, he launched a research team on Computer-Assisted Medical Interventions (CAMI), which led to innovative surgical practice, benefiting to more than 100 000 patients, thanks to the creation of several startup companies. He recently turned on intra-body energy scavenging in order to power implanted medical devices. Dr. Cinquin was the recipient of the 1999 Maurice E. Muller Award for excellence in computer-assisted orthopedic surgery, of the 2003 CNRS Silver Award and of the 2013 CNRS Innovation Award.

High Performance Nanocatalysts Supported on Micro/Nano Carbon Structures Using Ethanol Immersion Pretreatment for Micro DMFCs

Liang-You Lin[1], Yi-Shiuan Wu[1], Chaun Chang[1] and Fan-Gang Tseng[1,2,3]
[1]Department of Engineering and System Science, National Tsing Hua University, Hsinchu, Taiwan
[2]Institute of NanoEngineering and MicroSystems, National Tsing Hua University, Hsinchu, Taiwan
[3]Division of Mechanics, Research Center for Applied Sciences, Academia Sinica, Taipei, Taiwan

E-mail: r20008@hotmail.com

Abstract. In this paper, highly dense platinum (Pt) nanocatalysts were successfully deposited on the hydrophilically-treated nano/micro carbon supports with an ethanol (EtOH) immersion pretreatment and an acidic treatment for the performance improvement of methanol oxidation reaction (MOR). In order to thoroughly immerse the three-dimensional, interwoven structures of the carbon cloth fibers with a 6 M sulfuric acid surface modification, which increasing more oxygen-containing functional groups on the surfaces of the carbon supports, the EtOH immersion pretreatment of the carbon supports was utilized prior to the sulfuric acid treatment. Subsequently, Pt catalysts were reduced on the modified carbon supports by a homemade open-loop reduction system (OLRS) [1] For comparisons, carbon cloth (CC) and carbon nanotube on CC (CNT/CC) supports were employed with and without EtOH immersion pretreatments before Pt catalyst reduction. In the cyclic voltammetry (CV) curves, the electrosorption charges of hydrogen ion (Q_H) and the peak current density (I_P) of the fabricated Pt/CC and Pt/CNT/CC electrodes with the EtOH immersion pretreatments can efficiently be enhanced due to more active Pt sites for electrocatalytic reactions.

1. Introduction

With the rising price of the fossil, the alternative energy sources have become a significant issue for scientific and technological developments. Direct methanol fuel cells (DMFCs) utilize methanol as the anode fuel, mainly because it offers a high energy density, low pollution, fast refueling, and a low operating temperature [1-3]. Compared to proton exchange membrane fuel cell (PEMFC) using hydrogen gas as the anode fuel, DMFC has some advantages such as the safety of fuel storage, operation at room temperature, and so on. Therefore, DMFCs have potential applications for portable electronics. To obtain smaller particles and uniform dispersion of noble metal catalysts for the achievement of better utilization and activity of electrocatalysts, carbon materials become candidates for catalyst supports such as carbon black, carbon fiber, and carbon nanotube (CNT). In recent years, CNT providing a novel nano-structure with large specific surface area and good chemical resistance has been widely used in many applications. Most of CNTs are directly grown on the carbon cloth [2, 3, 7], on the silicon (Si) [1, 4, 10], or on the graphite substrates [5] as the catalyst supports. However, the

CNT and CC must adopt the hydrophilic treatment due to its instinct hydrophobic property prior to depositing the catalysts. The commonly used oxidants are sulfuric acid, nitric acid, or mixed solutions of H_2SO_4 and HNO_3, and hydrogen peroxide, etc. However, it is still difficult to thoroughly immerse the three-dimensional, interwoven structures of carbon cloth fibers in the acidic solution, resulting in an obvious gradient distribution of the Pt nanoparticles between the inner and outer surfaces of the carbon fibers [2,3]. This result led to low catalyst utilization of Pt and limited electrocatalytic activity for MOR. In order to eliminate the gradient distribution, the EtOH immersion pretreatment of the carbon supports was utilized prior to the above-mentioned hydrophilic surface modification. This can efficiently remove surface impurities of CNT, reducing environmental changes of reduction, and making more catalyst adhere to the surface of CNT due to the oxygen-rich functional groups [6]. To enhance Pt catalyst efficiency, small size (1-5 nm) and uniformly distributed Pt nanoparticles are highly desired. Conventionally, there are two common methods applied for Pt nanoparticle preparation, including chemical reduction [3,7] and electro-deposition [8]. Chemical reduction provides the advantages of easier preparation, direct reaction, and uniform size distribution of the nanocatalysts. However, it usually takes much longer time (24 h) for catalyst preparation due to slow nucleation and growth process at low temperature (60 °C). As a result, recent developed reflux technology [9] can react at higher reaction temperature (160 °C), substantially shorten the time required for catalyst preparation (~3-4 h). The corresponding fabrication process of Pt electrocatalysts is shown in *Fig. 1*.

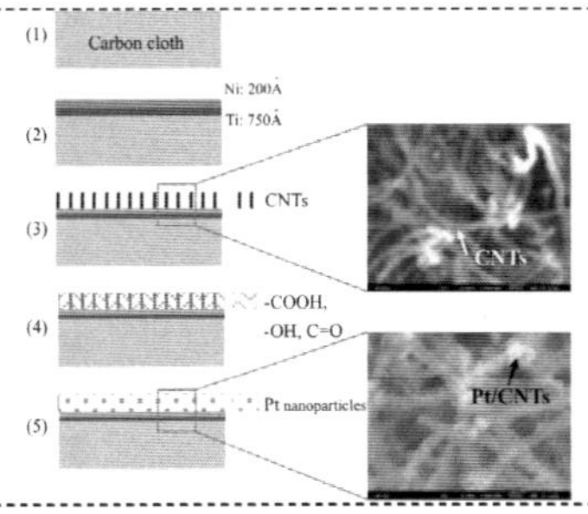

Figure 1. The preparation process and the corresponding SEM images of Pt/CNTs/CC electrodes: (1) CC substrate, (2) Sputter Ni/Ti metal films, (3) CNTs growth, (4) Hydrophilic treatment of CNTs, and (5) Deposit nano-scale Pt catalysts [1, 3].

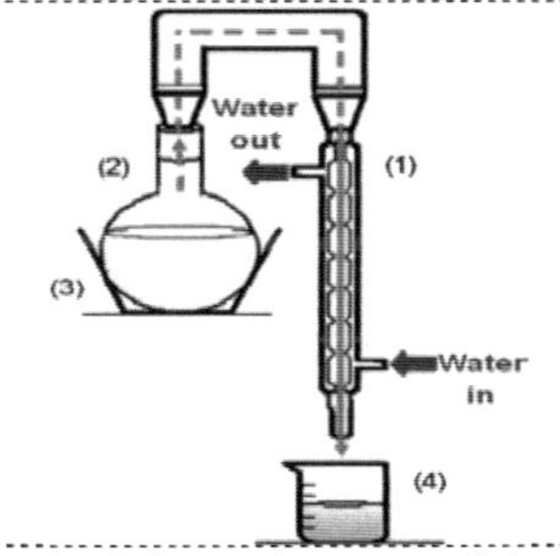

Figure 2. Schematic setup of the open-loop reduction system (OLRS) for Pt catalyst preparation. The red arrows in the system represent the directions of vapor (dash line) and condensed liquid (solid line). (1)Condenser, (2)Round-bottom flask containing reactants, (3)Heating Mantle, and (4)Beaker containing condensate.

2. Experimental

2.1. Growth of Carbon Nanotube

We used a 330 μm-thick carbon cloth (CC) as the substrate followed by respectively coating with Ni/Ti (film thickness: 15/75 nm) for carbon nanotube (CNT) growth by thermal chemical vapor deposition (TCVD) [7, 10]: First, we purged argon (Ar) and heated the quartz tube from room temperature to 800 °C, then maintained the growth temperature at 800 °C while leading to Ar and ammonia (NH_3). The first two steps are called thermal treatment. Ni was deposited as the catalyst for the growth of CNT with ethylene (C_2H_4). The growth time of CNT was 30 min, and this technique is known as the catalytically-growth CVD.

2.2. Ethanol Immersion Pretreatment & Hydrophilic Treatment of Carbon Supports

We used a 6 M sulfuric acid (H_2SO_4) solution as the oxidant for hydrophilic treatment of the CNT/CC supports. In order to remove air bubbles from the surface of the carbon supports to effectively immerse the whole surfaces of the supports in a H_2SO_4 solution for better Pt catalyst adhesion, vacuum means was added in the hydrophilic treatment process [1, 10]. However, it is still difficult to thoroughly immerse the three-dimensional, interwoven structures of carbon cloth fibers with CNTs in the acidic solution, resulting in an obvious gradient distribution of the Pt nanoparticles between the inner and outer surfaces of the carbon fibers [2, 3]. This result led to low utilization of carbon supports and Pt catalyst so as to limit electrocatalytic activity for MOR. In the present work, the EtOH immersion pretreatment for the CC and CNT/CC carbon supports was utilized prior to the above-mentioned hydrophilic modification of the carbon supports. The wetting properties of the CC surfaces with, without the hydrophilic treatments and with the EtOH immersion pretreatment and hydrophilic treatment were measured by contact angle goniometry (Model Firsttenangstroms, FTA32).

2.3. Open-Loop Reduction System for Platinum Catalyst

According to a homemade open-loop reduction system (OLRS) in our previous study [1], as shown in *Fig. 2*, we used 0.04392 g $H_2PtCl_6•6H_2O$ and 0.1 mM ethylene glycol as a Pt catalyst precursor and a reducing agent, respectively. Compared with a traditional reflux system that maintains the ratio of water to ethylene glycol (EG) at ~160°C for ~4 h, the gradual concentration increase of EG in the precursor solution can be accomplished by distilling off water in the OLRS while increasing the temperature to 130°C. This process with simultaneous increases in precursor concentration and in reaction temperature rendered high-quality Pt nanoparticles to precipitate with a high-density dispersion on the pretreated CNTs. The OLRS is not only able to only shorten the reduction time (<1.5 h) but is also able to enhance the electrocatalytic activity of the electrodes by creating a preferential orientation of Pt (1 1 1) facets for the methanol oxidation reaction (MOR).

2.4. Electrochemical Measurements

Cyclic voltammetry (CV) was carried out to identify the electrochemical properties of the electrocatalysts. The working principle is to impose a variable voltage at the working electrode (WE) and analyze the current signal received with time to determine the oxidation-reduction (redox) status in the electrochemical reaction. In this case, we imposed time-varying triangular wave of potential in the WE, and then observed the relationship between potential (E, V_{SCE}) and current density (I, mA/cm^2) to realize the potential for the redox reaction, electrochemical mass activity (MA, A/g_{Pt}) and reaction rate [1, 4]. In the CV curves, we can also obtain the active electrochemical surface area (ESA) of Pt catalysts from the hydrogen ion electrosorption reactions (charge transfer density of H^+, Q_H, mC/cm^2) in a H_2SO_4 aqueous solution. The three-electrode cell was setup for all of the electrochemical tests. Saturated calomel electrode (SCE, 0.241 V vs. standard hydrogen electrode, SHE) and Pt-coated Ti mesh are used as the reference electrode (RE) and the counter electrode (CE), respectively. All of the electrochemical measurements were purged with nitrogen (N_2) gas to remove oxygen for extra oxidation reaction.

3. Results and discussion

The contact angle (CA) of the CC electrodes was measured with a droplet of deionized (DI) water.The FTA images and their corresponding CAs are shown in *Table 1*.Compared to cases 1 and 2 (CAs: 136.74°, 48.72°), the CA of the CC electrode w/ the EtOH immersion pretreatment is smaller (18.05°), suggesting that the wetting surface of carbon fibers is probably prone to absorb many hydroxyl (-OH) and carbonyl (>C=O) oxygen-containing groups for more Pt catalyst nucleation.

Table 1. Different hydrophilic properties of the CC electrodes treated at different conditions.

Electrodes	(1)CC w/o H_2SO_4 hydrophilic treatment	(2)CC w/ H_2SO_4 hydrophilic treatment	(3)CC w/ EtOH immersion pretreatment & H_2SO_4 hydrophilic treatment
FTA image	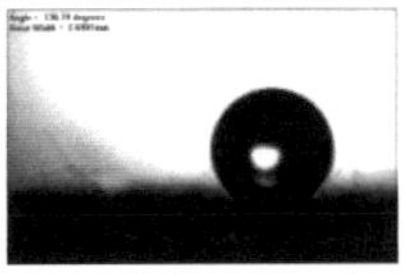	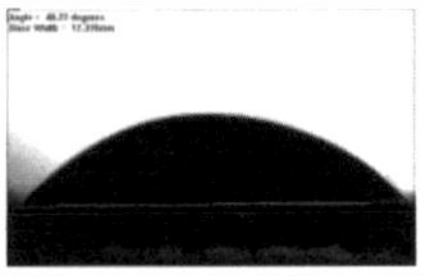	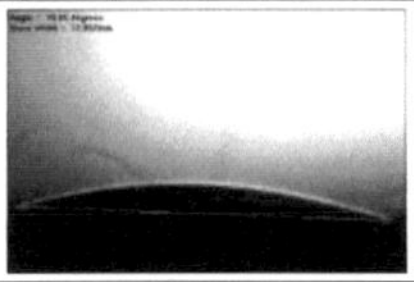
Contact angle	136.74°	48.72°	18.05°

Both the Q_H and the peak current density (I_P) of the fabricated Pt/CC and Pt/CNT/CC electrodes with the EtOH immersion pretreatments obtained in the CV curves, as shown in *Fig. 3* and *Fig. 4*, respectively, can be significantly improved compared to those without the EtOH immersion pretreatments. *Table 2* lists the electrocatalytic properties of the fabricated electrodes at different hydrophilic treatments, depicting that the values of Q_H and I_P for the best performed electrodes of Pt/CNT/CC with the EtOH immersion pretreatment are about 4 and 5 times higher than those of Pt/CC without the EtOH immersion pretreatment, respectively. This is attributed to both nano and micro structured supports together with the enhanced EtOH immersion technique, which provide more active Pt sites for electrocatalytic reactions based on the Q_H. This can also be proved from the increasing Pt loading (mass density of Pt, M_{Pt}) by 7.6 times. However, the values of ESA and MA for the Pt/CC and Pt/CNT/CC electrodes with the EtOH immersion are less than those without the EtOH immersion. This may result from different agglomeration degrees of the Pt clusters based on higher Pt loading.

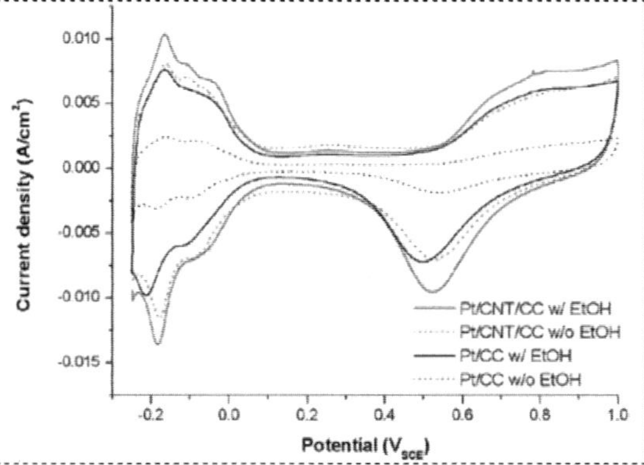

Figure 3. Cyclic voltammograms of the four different electrodes. Electrolyte: 0.5 M H_2SO_4 aqueous solution. Scan rate: 50 mV/s. Temperature: 25°C.

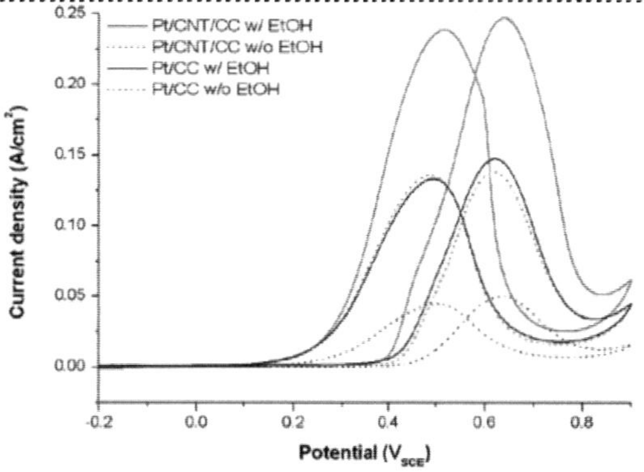

Figure 4. Cyclic voltammograms of the four different electrodes. Electrolytes: 1 M CH_3OH + 0.5 M H_2SO_4 aqueous solutions. Scan rate: 20 mV/s. Temperature: 25°C.

Table 2. Electrocatalytic properties of the electrodes pretreated at different conditions.

Electrodes	*Treatments	Q_H mC/cm^2	E_P V_{SCE}	I_P mA/cm^2	M_{Pt} mg/cm^2	ESA m^2/g$_{Pt}$	MA A/g$_{Pt}$
Pt/CC	H	6.61	0.64	50	0.095	33.1	526.3
	E+H	17.08	0.62	147	0.431	18.9	341.1
Pt/CNT/CC	H	19.67	0.62	138	0.373	25.1	370.0
	E+H	24.93	0.64	247	0.723	16.4	341.6

*E: ethanol solution (C_2H_6O), H: sulfuric acid solution (H_2SO_4).

As shown in *Table 3*, low-magnification scanning electron microscopic (SEM) images illustrate that denser Pt nanoparticles were apparently deposited on both the inner and outer surfaces of the hydrophilically-treated CC and CNT/CC supports with the EtOH immersion pretreatments. More uniform dispersion of the Pt nanoparticles on a larger surface area results from both micro and nano structured CNT/CC supports. In addition, the corresponding high-magnification images show that the aggregation size and degree of the Pt clusters on the CNT/CC supports was less serious than those on the CC supports. To further solve this problem, a proper sonication technique would be utilized to agitate the Pt reduction process and increase its nano-scale dispersion on the CNT/CC supports.

Table 3. Low- and high-magnification (1K &100K) SEM images of the Pt catalysts on the different carbon supports w/ and w/o the EtOH immersion pretreatments.

Electrodes	Pt/CC (1K)	Pt/CNT/CC (1K)	Pt/CC (100K)	Pt/CNT/CC (100K)
w/o EtOH	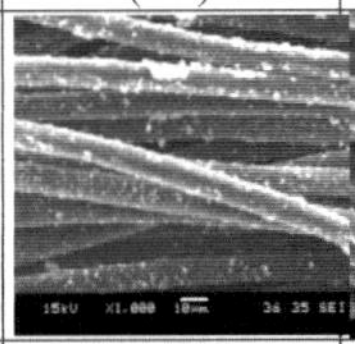	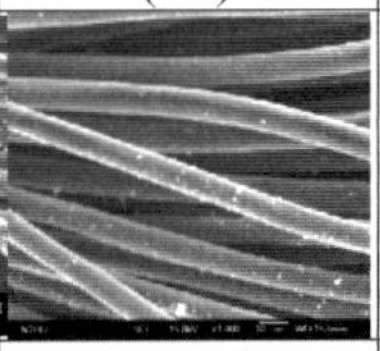	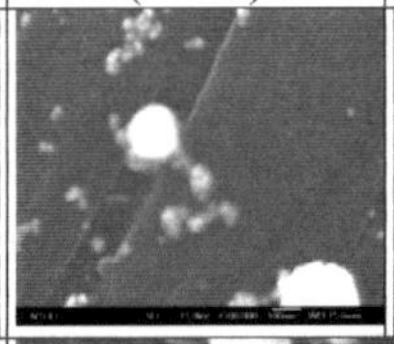	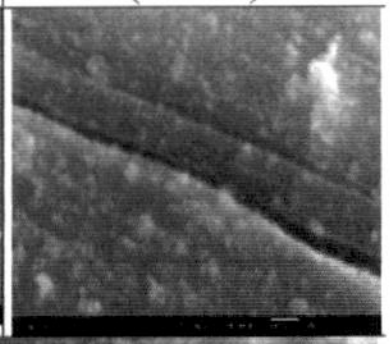
w/ EtOH	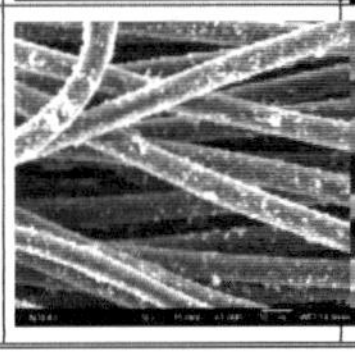	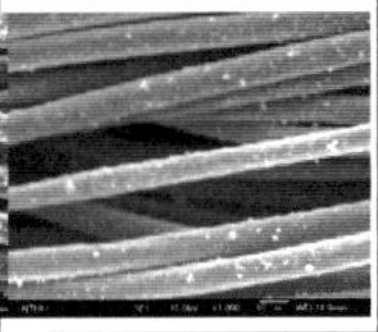	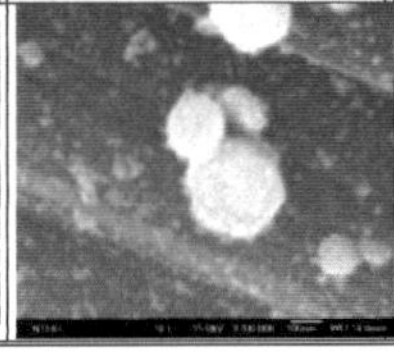	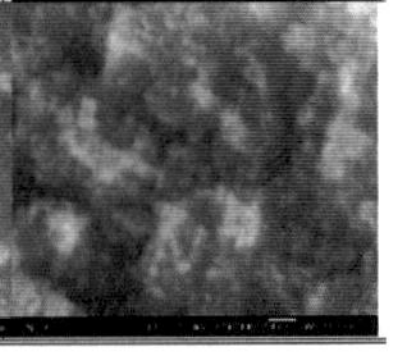

4. Conclusion

The EtOH immersion pretreatment on nano/micro carbon supports prior to the hydrophilic treatment in an acidic solution was thoroughly to immerse the three-dimensional, interwoven micro structures of CC fibers with CNTs for the improved Pt deposition. The pretreatment helps more Pt catalysts adhere to the nucleation sites of the immersed CC and CNT/CC supports, which were created with the oxygen-rich functional groups by the hydrophilic treatment. For the electrocatalytic properties of the fabricated electrodes, the values of the Q_H and the I_P for the Pt/CNT/CC with the EtOH immersion pretreatment are higher than those of the Pt/CC without the EtOH immersion pretreatment (Q_H: 24.93 vs. 6.61 mC/cm^2, I_P: 247 vs. 50 mA/cm^2). This is mainly attributed to higher mass loading of the Pt catalysts (M_{Pt}: 0.723 vs. 0.095 mg/cm^2). Results from the SEM measurements show that smaller and highly denser distribution of the Pt nanoparticles were clearly observed on the hydrophilically-treated CNT/CC supports with the EtOH immersion pretreatment compared to the CC supports without the EtOH immersion pretreatment.

Acknowledgment

This research was financially supported by the National Science Council (NSC), Taiwan under grant NSC 101-3113-E-007-002. The authors also thank the Instrument Center at National Tsing Hua University for FE-SEM and ICP-MS examinations.

References

[1] Y. S. Wu et al., "Highly efficient platinum nanocatalysts synthesized by an open-loop reduction system with a controlled temperature loop," *Electrochimica Acta* 64 (2012) 162–170.
[2] C. H. Wang et al., "High performance of low electrocatalysts loading on CNT directly grown on carbon cloth for DMFC," *Journal of Power Sources*, vol. 171, pp. 55-62, 2007.
[3] M. C. Tsai et al., "A catalytic gas diffusion layer for improving the efficiency of a direct methanol fuel cell," *Electrochemistry Communications*, vol. 9, pp. 2299-2303, 2007.
[4] S. K. Wang et al., "Electrocatalytic properties improvement on carbon-nanotubes coated reaction surface for micro-DMFC," *Journal of Power Sources*, vol. 167, pp. 413-419, 2007.
[5] Z. B. He et al., "Deposition and electrocatalytic properties of platinum nanoparticles on carbon nanotubes for methanol electrooxidation," *Materials Chemistry and Physics*, vol. 85, pp. 396-401, 2004.
[6] N. Jha et al., "Pt–Ru/multi-walled carbon nanotubes as electrocatalysts for direct methanol fuel cell," *International Journal of Hydrogen Energy*, vol. 33, pp. 427-433, 2008.
[7] M. C. Tsai et al., "An improved electrodeposition technique for preparing platinum and platinum–ruthenium nanoparticles on carbon nanotubes directly grown on carbon cloth for methanol oxidation," *Electrochemistry Communications,* vol. 8, pp. 1445–1452, 2006.
[8] M. Watanabe et al., "Preparation of highly dispersed Pt-Ru alloy clusters and the activity for the electrooxidation of methanol," *Journal of Electroanalytic Chemistry*, vol. 229, pp. 395-406, 1987.
[9] C. Bock et al., "Size-selected synthesis of PtRu nano-catalysts: reaction and size control mechanism," *Journal of the American Chemical Society,* vol. 126, pp. 8028-8037, 2004.
[10] Y. S. Wu et al., "Characteristics of controlled Nafion® coating inside nano structured anode for three-phase zone arrangement in micro DMFC", *Proceedings of the 2010 5th IEEE International Conference on Nano/Micro Engineered and Molecular Systems*, Xiamen, China, pp. 698-701, January 20-23, 2010.
[11] X. Wang et al., "Surfactant stabilized Pt and Pt alloy electrocatalyst for polymer electrolyte fuel cells," *Electrochimica Acta*, vol. 47, pp. 2981-2987, 2002.
[12] Z. L. Liu et al., "Physical and electrochemical characterizations of microwave-assisted polyol preparation of carbon-supported PtRu nanoparticles," *Langmuir,* vol. 20, pp. 181-187, 2004.

A Flexible Ascorbic Acid Fuel Cell with a Microchannel Fabricated using MEMS Techniques

Hiroshi Mogi, Yudai Fukushi, Syohei Koide, Ryohei Sano, Tsubasa Sasaki and Yasushiro Nishioka

Department of Precision Machinery Engineering, College of Science and Technology, Nihon University, 7-24-1 Narashinodai, Funabashi City, Chiba 274-8501, Japan

E-mail : nishioka@eme.cst.nihon-u.ac.jp

Abstract. We fabricated a miniature ascorbic acid fuel cells equipped with a microchannel for the circulation of ascorbic acid (AA) solution using micro electronic mechanical system techniques. The fuel cell was fabricated on a flexible polyimide substrate, and its porous carbon-coated aluminium (Al) electrodes of 2.8 mm in width and 11 mm in length were formed using photolithography and screen-printing techniques. The porous carbon was deposited by screen-printing of carbon-black ink on the Al electrode surfaces in order to increase the effective electrode surface area and to absorb more enzymes on the cathode surface. The microchannel with a depth of 200 μm was fabricated using a hot-embossing technique. A maximum power of 0.60 μW at 0.58 V that corresponds to a power density of 1.83 $\mu W/cm^2$ was realized by introducing a 200 mM concentrated AA solution at room temperature.

1. Introduction

Recently, great attention has been paid to the appearance of power sources, especially for human wearable or portable electronic equipment. These power sources have to correspond to ubiquitousness for the next generation, and it is necessary to develop highly convenient and miniaturized devices using safe materials for a human body. An ascorbic acid (AA), also known as vitamin C, is an environmentally and biologically friendly compound. Furthermore, it has been well known that the AA oxidation advanced on the carbon electrodes without metal catalysts [1,2]. Fujiwara et al. have reported a fuel cell adopting AA and without catalysts on the anode [3-6]. This device does not require a precious metal catalyst on the anode, and does not release any toxic product. Moreover, Fukushi et al. have reports a miniature glucose fuel cell with a microchannel fabricated on polyimide (PI) substrates using micro electronic mechanical system (MEMS) technologies [7]. This paper reported on relatively low cost and miniaturized fuel cells adopting enzymes instead of precious catalysts and MEMS fabrication processes. The fuel cells were superior in flexibility and disposability. We also report the fabrication and characterization of miniaturized ascorbic acid fuel cells (AAFCs) with microchannels fabricated on PI substrates using thermal nanoimprinting techniques [8-10] and MEMS technologies.

2. Experimental

2.1. The power generation mechanism and the schematic of the AAFCs

Figure 1 describes the possible power generation mechanism of the AAFCs. The AAFCs was composed of an anode with no catalysts and a bilirubin oxidase (BOD)-adsorbed cathode [11-14]. On the anode, the AA was oxidized on the electrodes surface, and dehydro ascorbic acid (DHAA) was generated. As a result, the electrons and protons were produced on the anode. The electrons were transferred from the anode to the cathode via an external circuit to extract electricity. Oxygen (O_2) in the fuel solution was reduced by BOD immobilized on the cathode to water (H_2O) on reaction with electrons and protons. Figure 2 shows the layout of the electrodes and the schematic of the AAFCs with the microchannel and the photography of the fabricated AAFCs. The cathode has a width of 2.8 mm (W) and length of 10 mm (L_1), and the anode has a width of 2.8 mm (W) and a length of 1 mm (L_3). These electrodes were fabricated by photolithography on a PI film (Kapton, DuPont), coated with aluminium (Al) and then modified by carbon black. The gap between the cathode and anode was 0.5 mm (L_2). The microchannel was fabricated on another PI film (Kapton, DuPont), and it was designed to have a width of 3 mm, a depth of 200 μm, and a length of 11.5 mm. These two patterned PI substrates were adhered to form a fuel cell by double-sided Kapton tapes (Sumitomo 3M). The cathode was designed to be larger than the anode to account for the lower concentration of oxygen in comparison to that of AA; to reduce the flux of oxygen in the upstream to the anode and consequently to improve the AA oxidation performance [15]. In order to investigate the characteristics of the anode and cathode independently using cyclic voltammetry, one of the electrode areas was fabricated to be a silver|silver chloride (Ag|AgCl) reference electrode by coating the Al surface with an ink of Ag|AgCl (BAS Inc.).

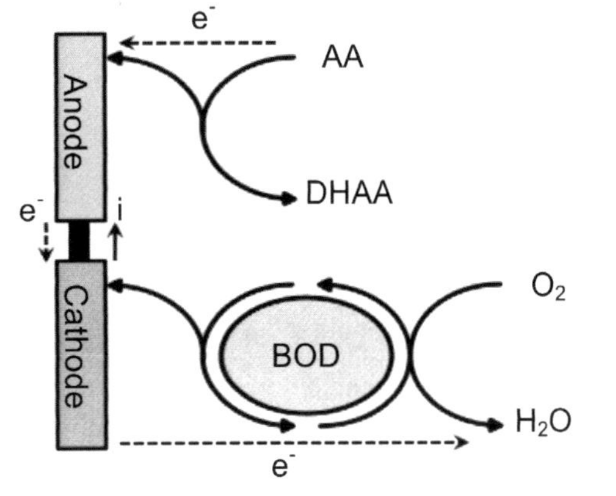

Figure 1. The possible power generation Mechanism.

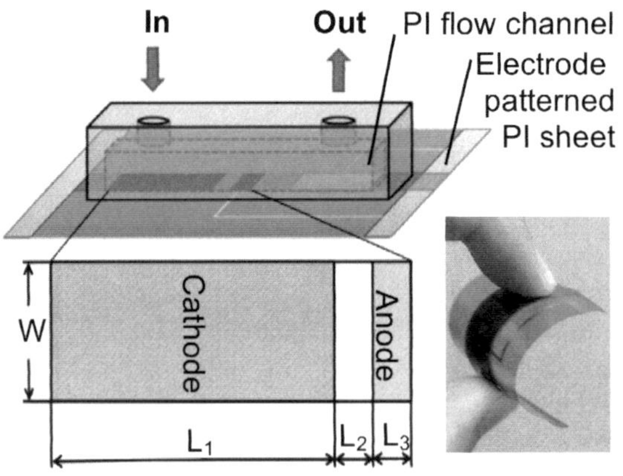

Figure 2. The layout of the electrodes and the schematic of the AAFCs with the microchannel and the photography of the fabricated AAFCs.

2.2. Fabrication processes of the carbon-modified electrodes and microchannel

The carbon-modified electrodes were fabricated on a PI sheet using conventional MEMS fabrication processes [7]. In order to increase the effective electrode surface area and the amount of enzymes on the electrode surfaces, we coated the fabricated Al electrode surfaces with the carbon-black ink [15-18]. Porous structures were clearly observed by a scanning electron microscopy, which suggested an increase in the electrode surface area. The microfluidic channel was also fabricated on the PI substrate using photolithography and hot-embossing as previously reported [7].

2.3. Fabrication processes of the enzyme-modified cathode

0.03 mg of BOD was mixed with 20 µl of phosphate buffer solution (50 mM, pH 7.0). 20 µl of the BOD solution was dropped on the cathode (surface area of 28 mm^2). Then the cathode was subjected to UV-ozone treatment for 15 min and left to dry for 40 min. According to a previous report, BOD was stable in circumneutral pH and didn't generate reaction intermediates during oxygen reduction process [19].

2.4. Measurements

Electrochemical measurements were performed in 200 mM ascorbic acid solution (AAS) and 50 mM pH 7.0 phosphate buffer solution (PBS) at room temperature. The electrochemical properties of the electrodes were characterized using a potentiostat (Electrochemical Analyser, HA-1518, HokutoDenko) having a three electrodes system, wherein the enzyme-modified electrode was the working electrode, the Ag|AgCl modified electrode was the reference and platinum was the counter electrode. The power measurement systems were performed introducing the AAS into the reaction chamber through the microchannel by pulling a syringe pump. The power generated was evaluated by measuring the cell voltage while varying the external load resistance between 0 and 3 MΩ. The following equation was used:

$$W=V^2/R \tag{1}$$

where R is the load resistance, and V is the generated voltage measured between the terminals of the fuel cell. The power density was calculated by dividing W with the total area of the anode and cathode, which was 30.8 mm^2.

3. Results and discussion

Figure 3 shows the cyclic voltammograms of the anode. The potential was scanned toward positive potential direction from -0.6 V to 0.8 V. The electrochemical measurements were performed on (a) porous carbon modified electrodes (PCE) in AAS, (b) glassy carbon electrodes (GCE) in AAS, and (c) PCE in PBS, respectively. The maximum current density on the PCE was 24 mA/cm^2 at 0.80 V, and it was approximately 2.5 times higher than that of the GCE. The potential at the AA oxidation reaction on the PCE was 0.13 V, and that on the GCE was 0.31 V. The potential at the oxidation reaction of the AA on the PCE was 0.18 V lower than that on the GCE. It was considered that these results were due to the increase of the active surface areas of the porous carbon. The current density in the PBS was very small compared to others. In contrast, the increasing of the current density associated with the oxidation reaction appeared clearly in the AAS. The current was assumed to be generated by oxidation of AA, because phosphoric acid did not contribute to the oxidation reaction.

Figure 4 shows the cyclic voltammograms of the cathode. The potential was scanned toward negative potential direction from 0.8 V to -0.6 V. The electrochemical measurements were performed on PCE in (A) AAS and (B) PBS, respectively. The maximum current density was -1.2 mA/cm^2 at -0.59 V. The current density of the cathode was 20 times lower than that of the anode. According to the previous reports, the higher current density of the cathode was obtained by stirring or bubbling the fuel [2,19-22]. Based on these results, it was assumed that the oxygen supply to the cathode was not enough because the cathode of this device in the microchannel was not exposed to the air. The lack of the oxygen supply caused the decrease of the current density. In order to solve this problem, it may be possible to adopt hydrofuge processes to the cathode or supply oxygen near the cathode. The reduction potential measured in the AAS was 0.36 V lower than that of the PBS. The possible reason for this may be due to the fact that AA oxidation occurred instead of the reduction of O_2 on the cathode. Fujiwara et al. reported that this unfavorable oxidation reaction at the cathode can be avoided by the use of an ion-exchange film (Nafion, DuPont) that repels AA by the electrostatic repulsion [3].

Therefore, we assume that the current density on the cathode of the cell investigated here can be further increased by the use of the electrolyte films by suppressing AA oxidation on the cathode.

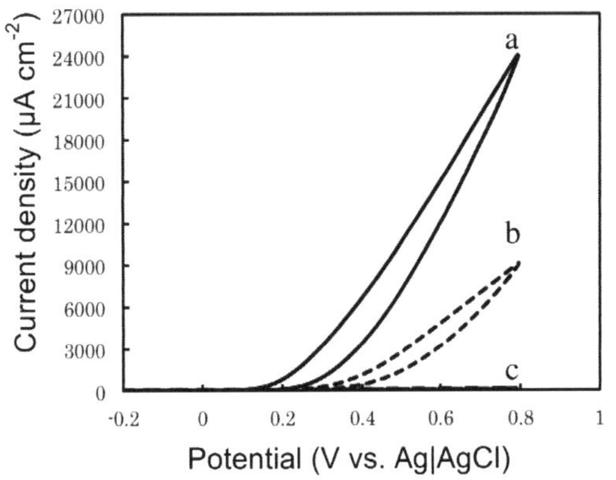

Figure 3. Cyclic voltammograms of (a) the anode for PCE, (b) GCE in a 200 mM AAS, and (c) PCE in a 50 mM pH 7.0 PBS.

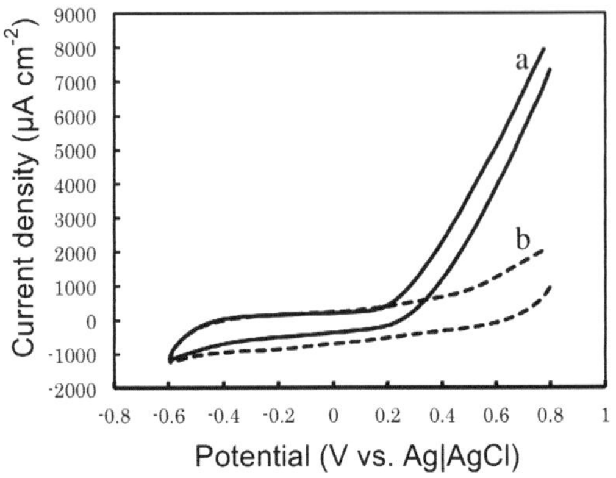

Figure 4. Cyclic voltammograms of (a) the cathode for PCE in a 200 mM AAS, and (b) PCE in a 50 mM pH 7.0 PBS.

Figure 5 shows the relationship between the current and voltage of the fuel cell during operation and the dependence of the power density on the output voltage. The maximum current density and power (power density) were 1.04 μA and 0.60 μW (1.83 μW/cm^2) at 0.58 V in the AAS at room temperature. The current density of the cathode was lower than that of the anode. The electromotive force of the fuel cell is defined by the potential difference between the oxidation potential, 0.13 V of the anode and the reduction potential, 0.17 V of the cathode, and it was estimated to be 0.04 V. This may be the reason for the relatively low power generation of the cell investigated here. On the other hand, it was reported that the crossover current density of AA did not cause the reduction of the power generation performance since it was smaller (~1/100) compared to that of methanol [3]. However, it was presumed that AA was oxidized on the cathode and the crossover phenomenon should have occurred causing reduction of the electric power density based on the results of the cyclic voltammogram measurements of the electrodes.

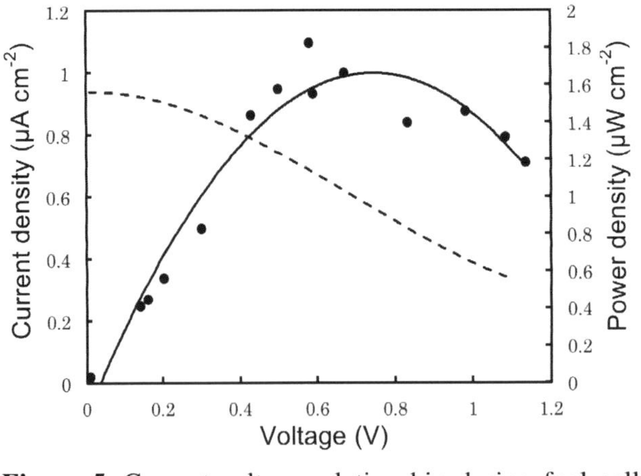

Figure 5. Current-voltage relationship during fuel cell operation and the dependence of power density on the output voltage.

4. Conclusion

The AAFCs was fabricated on a flexible PI substrate using MEMS microfabrication processes. The fuel cell had a microchannel to introduce an aqueous of AAS. Al electrodes were formed on the PI substrate using photolithography and screen-printing. The porous carbon was deposited by screen-printing of carbon-black ink on the Al electrode surfaces in order to increase the effective electrode surface area. The microchannel with a depth of 200 μm was fabricated using a hot-embossing technique. The maximum power of 0.60 μW at 0.58 V that corresponds with the power density of 1.83 $\mu W/cm^2$ was realized by introducing AAS at room temperature.

References

[1] Deakin M R, Kovach P M, Stutts K J and Wightman R M 1986 *Anal. Chem.* **58** 1474
[2] Hu I F and Kuwana T 1986 *Anal. Chem.* **58** 3235
[3] Fujiwara N, Yamazaki S, Siroma Z, Ioroi T and Yasuda K 2007 *J. Power Sources* **167** 32
[4] Fujiwara N, Siroma Z, Ioroi T and Yasuda K 2007 *J. Power Sources* **164** 457
[5] Fujiwara N, Yasuda K, Ioroi T, Siroma Z, Miyazaki Y and Kobayashi T 2003 *Electrochem. Solid-State Lett.* **6** A257
[6] Fujiwara N, Yamazaki S, Siroma Z, Ioroi T and Yasuda K 2006 *Electrochem. Commun.* **8** 720
[7] Fukushi Y, Koide S, Ikoma R, Akatsuka W, Tsujimura S and Nishioka Y 2013 *J. Photopolym. Sci. Technol.* **26** 303
[8] Youn S W, Noguchi T, Takahashi M and Maeda R 2008 *Microelec. Eng.* **85** 918
[9] Becker H and Heim U 2000 *Sensors and Actuators A Physical.* **83** 130
[10] Ikoma R, Komatsuzaki H, Suzuki K, Komori T, Kuroda K, Saitou H, Youn Sung-won, Hiroshima H, Takahashi M, Maeda R and Nishioka Y 2012 *J. Photopolym. Sci. Technol.* **25** 255
[11] Moore C M, Minteer S D and Martin R S 2005 *Lab Chip* **5** 218
[12] Heller A 2005 *AiChE Journal* **51** 1054
[13] Bullen R A, Arnot T C, Lakeman J B and Walsh F C 2006 *Biosensors and Bioelectronics* **21** 2015
[14] Kontani R, Tsujimura S and Kano N 2009 *Bioelectrochemistry* **76** 10
[15] Tsujimura S, Tatsumi H, Ogawa J, Shimizu S, Kano K and Ikeda T 2001 *J. Electronal. Chem.* **496** 69
[16] Mano N, Mao F, Shin W, Chen T and Heller A 2003 *Chem. Commun.* **4** 518
[17] Tsujimura S, Kano K and Ikeda T 2002 *Electrochemistry* **70** 940
[18] Katz E, Buckmann A F and Willner I 2001 *J. Am. Chem. Soc.* **123** 10752
[19] Tsujimura S, Kano K and Ikeda T 2005 *J. Electroanal. Chem.* **576** 113
[20] Tsujimura S, Nakagawa T, Kano K and Ikeda T 2004 *Electrochemistry* **72** 437
[21] Kamitaka Y, Tsujimura S, Kataoka Z, Sakurai T, Ikeda T and Kano K 2007 *J. Electroanal. Chem.* **601** 119
[22] Tominaga M, Otani M, Kishikawa M and Taniguchi I 2006 *Chem. Lett.* **35** 1174

Miniature fuel cell with monolithically fabricated Si electrodes - Alloy catalyst formation -

Daiki Ogura[1], Takahiro Suzuki[1], Noboru Katayama[2], Kiyoshi Dowaki[3], and Masanori Hayase[1]

[1]Department of Mechanical Engineering
[2]Department of Electrical Engineering
[3]Department of Industrial Administration
Faculty of Science and Technology
Tokyo University of Science
2641 Yamazaki, Noda, Chiba 278-8510, Japan

E-mail: mhayase@rs.noda.tus.ac.jp

Abstract. A novel Pd-Pt catalyst formation process was proposed for reduction of Pt usage. In our miniature fuel cells, porous Pt was used as the catalyst, and the Pt usage was quite high. To reduce the Pt usage, we have attempted to deposit Pt on porous Pd by galvanic replacement, and relatively large output was demonstrated. In this study, in order to reduce more Pt usage and explore the alloy catalyst formation process, atomic layer deposition by UPD-SLRR (Under Potential Deposition – Surface Limited Redox Replacement) was applied to the Pd-Pt catalyst formation. The new process was verified at each process steps by EDS elemental analysis, and the expected spectra were obtained. Prototype cells were constructed by the new process, and cell output was raised to 420mW/cm^2 by the Pd-Pt catalyst from 125mW/cm^2 with Pd catalyst.

1. Introduction

Miniature fuel cells get a lot of attention for mobile devices, and we proposed a Si based fuel cell that has both fuel channel and catalyst layer on a Si substrate as shown in figure 1. Recent reports at Power MEMSs[1,2] showed that good performance was available by our miniature fuel cells. However, in our prototypes, porous Pt was used as a catalyst and the Pt usage was quite high. We have discussed Pd-Pt alloy catalyst for reduction of the Pt usage in our miniature fuel cells[3]. In the study, we developed the porous Pd layer formation process. Then, Pt was deposited on the porous Pd layer by galvanic replacement reaction, and estimated Pt usage was reduced to 0.16mg/cm^2. Some prototype fuel cells were constructed, and more than 400mW/cm^2 peak output was obtained.

In the previous study, the reduction of Pt was successfully demonstrated. However, probably, the Pt was deposited disorderly with the galvanic replacement reaction. In order to achieve further efficient Pt use, controllable atomic layer deposition of Pt on the porous Pd is desired. Besides, alloy catalysts are useful for not only Pt reduction, but application of various fuels, such as Pt-Ru alloy catalyst for methanol. Such alloy catalysts are also useful for biomass-derived hydrogen, because it tends to contain high carbon monoxide.

In order to reduce more Pt usage and explore the alloy catalyst formation process, atomic layer deposition by UPD-SLRR (Under Potential Deposition – Surface Limited Redox Replacement) [4,5]was applied to the Pd-Pt catalyst formation as shown in figure 2.

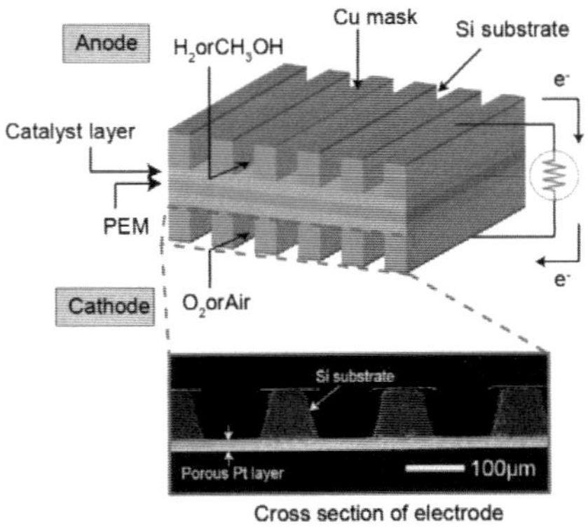

Figure 1. Schematic view of the miniature fuel cell and the cross section of monolithic fuel cell Si electrodes.

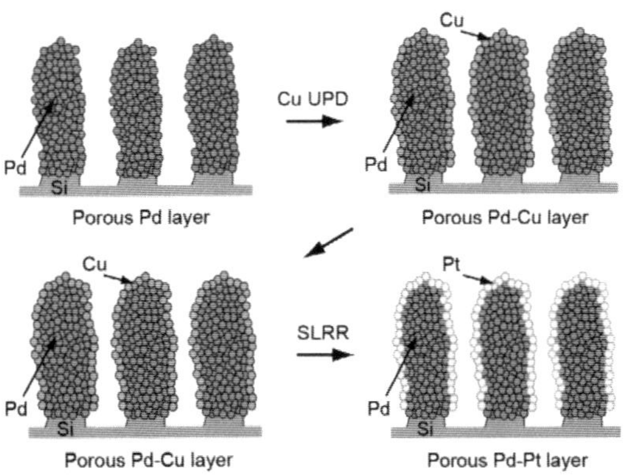

Figure 2. Strategy for atomic layer Pt deposition on porous Pd by UPD-SLRR.

2. Fabrication process

2.1. Si electrode with porous Pd

Figure 3 illustrates the fabrication process of the Si electrode. The electrode chip has quite simple structure, in which a catalyst layer and fuel channels are monolithically fabricated on a Si wafer. The silicon wafers used in this work were N-type (Resistivity 0.004-0.006 Ohm-cm), (100) oriented, double side mirror polished 100um thick ones. After removing oxide by HF, Cu thin film is deposited on the Si wafer by sputtering. Using usual photolithographic patterning with photoresist and wet etching, fuel channel mask for plasma etching is made with the Cu thin film. Porous Si layer is formed on the opposite side of the Si wafer by anodization in an electrolyte containing HF. The porous Si layer is subsequently immersed into a Pd plating bath, and porous Pd layer is obtained. Process condition for the porous Pd layer formation was reported elsewhere [3]. Fuel channels are opened by applying plasma etching on the backside of the porous Pd layer with the cupper thin film mask. The reaction area was 1mm^2. Conventional parallel plate reactive ion etching system (Samco RIE-10N, Japan) is used and 18sccm of SF_6 and 4sccm of O_2 gases are supplied for the etching. In this plasma etching process, porous Pd layer is supposed to work as a stopping layer of the etching, because the etching rate is low at the porous metal layer, and through-chip porous Pd layer can be relatively easily fabricated. In this way, monolithic Si based electrodes were prepared. Some text.

2.2. UPD-SLRR for Pt deposition on Pd

Process conditions for the UPD-SLRR Pd-Pt catalyst layer formation is shown in Table 1. Dimension of the porous Pd layer formed on a Si chip was about 1 cm in diameter and 10um in thickness. In the Cu UPD process, as a reference electrode, Cu wire was employed, and Pt wire was used as a counter electrode. An electrolyte solution of 0.1M H_2SO_4+0.1M $CuSO_4$ was prepared for the Cu UPD, and the solution was bubbled with nitrogen gas for 30 min prior to the experiment to eliminate dissolved oxygen and carbon dioxide. Before the UPD process, the chip with the porous Pd layer was cleaned by a 1M H_2SO_4 solution. Then, the chip was immersed into the electrolyte solution in a vessel filled with nitrogen gas. The electrode potential was varied to determine the appropriate electrode potential for the UPD process.

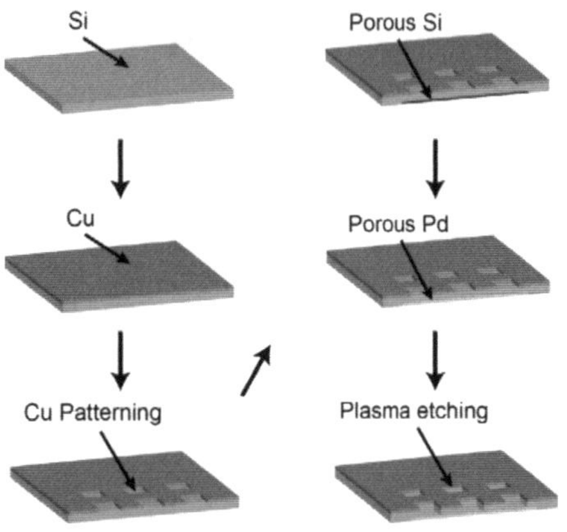

Figure 3. Fabrication process of the Si electrode.

Table 1. Condition for atomic layer Pt deposition on porous Pd layer by UPD-SLRR.

Copper UPD for the porous Pd layer	
Working Electrode	Chip formed porous Pd layer
Reference Electrode	Cu wire
Counter Electrode	Pt wire
Composition of solution	0.1M H_2SO_4+ 0.1M $CuSO_4 \cdot 5H_2O$
Sampling frequency [Hz]	40
Scan rate [mV/s]	5
Scan range [mV]	From 400mV to 0 mV
SLRR for the after UPD processed porous Pd layer	
Working Electrode	Copper UPD processed chip
Reference Electrode	Pt wire
Counter Electrode	Pt wire
Composition of solution	2mM H_2PtCl_6 + 0.1M H_2SO_4
Potential [mV]	-273
Plating time [min]	1

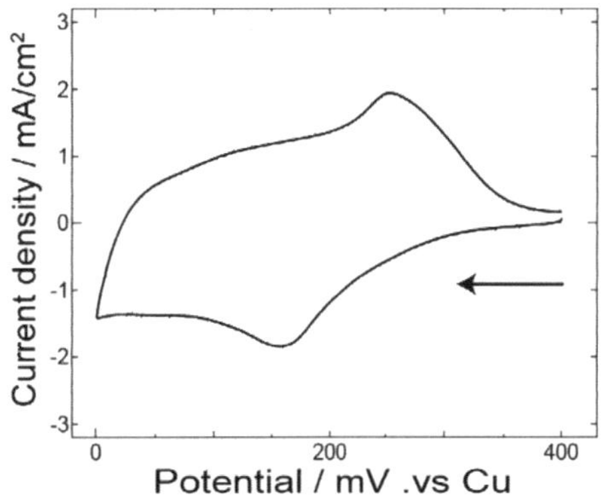

Figure 4. Cyclic voltammogram for Cu UPD on the porous Pd layer.

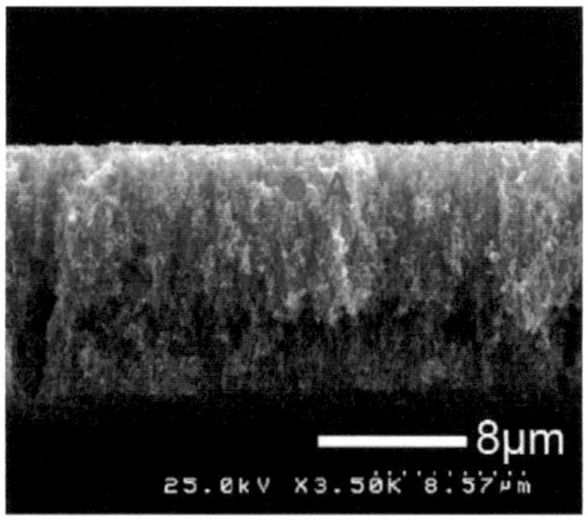

Figure 5. Typical cross sectional SEM image of the porous Pd-Pt catalyst layer. Point A shows the location for the EDS analyses.

After the Cu UPD process, the porous Pd chip was moved to the SLRR process. Since a replacement reaction of Pd to Pt occurs automatically in a Pt plating bath, the electrode potential of porous Pd layer was shifted to cathodic side, and the chip was immersed into the 2mM Pt plating bath for one minute. All electrochemical experiments were controlled by a potentiostat (Hokuto Denko, HA150G) with in house software.

2.3. Prototype assembly
A thin gold film was deposited by sputtering on the edge of the catalyst layer to have electrical contact to the low resistive Si substrate, because direct electrical contact between the porous catalyst layer and Si substrate is poor. Then, two Si electrodes are hot-pressed onto either side of PEM sheets. In order to obtain good contact between the catalyst layer and PEM, the double PEM treatment [6] was employed.

3. Results and discussion

3.1. Cu UPD
Figure 4 shows the cyclic voltammogram of the porous Pd in the Cu UPD bath. Current peak was observed at +160 mV vs. Cu. The electrode potential at the peak was positive to the equilibrium

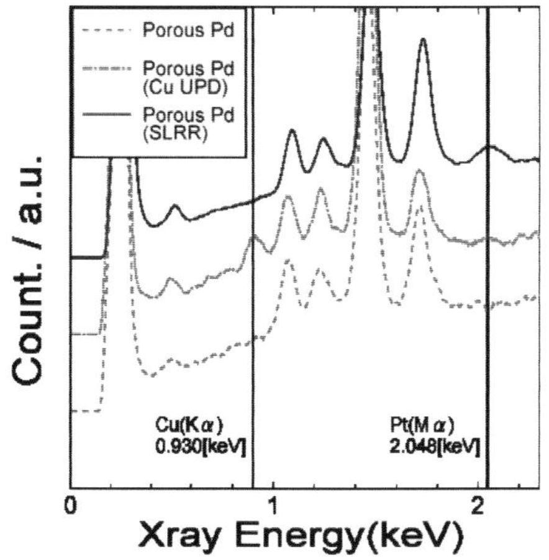

Figure 6. EDS spectra before and after the Cu UPD and after SLRR.

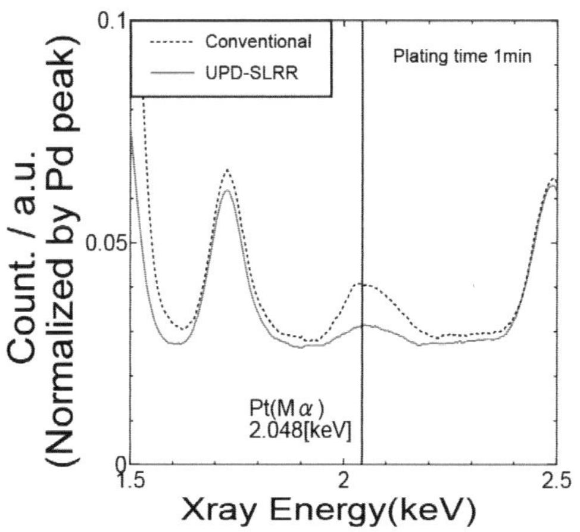

Figure 7. Comparison of the Pt usage between conventional and novel catalyst.

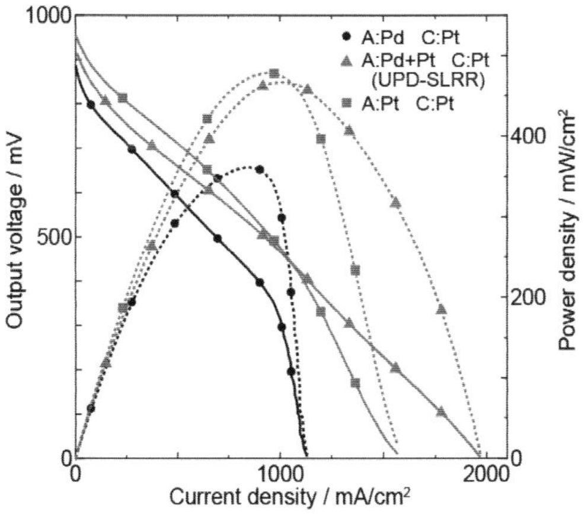

Figure 8. Polarization curve of prototype cells used Pd-Pt as anode at 313K.

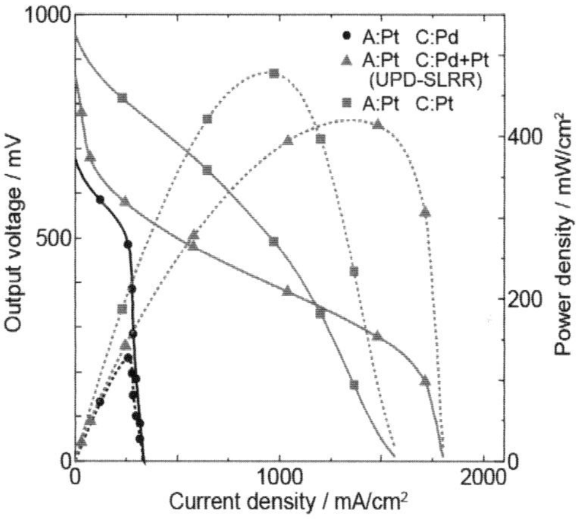

Figure 9. Polarization curve of prototype cells used Pd-Pt as cathode at 313K

potential, and we assumed that the current peak was caused by the UPD. We decided to perform the UPD process by following potential modulation. The electrode potential was swept from 400mV to 50mV, and was kept constant for 1min until the electric current became enough small.

3.2. EDS elemental analyses

Figure 5 shows the typical cross sectional image of the porous Pd-Pt layer, and EDS elemental analysis was performed around the point A. Figure 6 shows the EDS spectra. Green line shows the spectrum of porous Pd before the Cu UPD process. Red line shows the spectrum after the Cu UPD. Blue line shows the spectrum after the SLRR. As we expected, small Cu peak appeared after the UPD process and disappeared after the SLRR reaction. After the SLRR process, small Pt peak was observed. These results supports that the expected UPD-SLRR reaction occurred. Figure 7 shows the comparison between present Pd-Pt(UPD-SLRR) catalyst and previously proposed Pd-Pt catalyst, in which Pt was deposited by just galvanic replacement reaction for 1 min. Dashed line shows the spectrum of Pd-Pt prepared by previously proposed process, while red line shows the spectrum of Pd-Pt(UPD-SLRR). It was found that the Pt signal was smaller in the novel UPD-SLRR process than the one observed in previous process. It can be said that Pt deposition was successfully limited by UPD-SLRR.

3.3. Power Generation Test

Finally, prototype cells were produced with the novel UPD-SLRR process, and power generation was tested. Prototype cells were set into the aluminium casing at 313K in a temperature controllable chamber. Pure hydrogen and oxygen were supplied to anode and cathode respectively. Three catalysts, Pt, Pd and Pd-Pt(UPD-SLRR), were prepared. Figures 8 and 9 show the polarization curves. Figure 8 shows the impact of anode catalysts, while figure 9 shows the cathode impact. In both figures, though Pt catalyst showed the best results, Pd-Pt(UPD-SLRR) showed comparable results to pure Pt catalyst. Generally, cathode (oxygen side) needs higher catalytic activity, and the catalyst performance can be demonstrated. In figure 9, Pd catalyst showed only 125mW/cm^2, while Pd-Pt(UPD-SLRR) showed 420mW/cm^2. Obvious power increase was observed with Pd-Pt(UPD-SLRR) from simple Pd catalyst, and it was found that Pt deposition by UPD-SLRR was effective in catalyst improvement and Pt usage reduction.

4. Conclusion

In this study, we attempted to deposit Pt atomic layer on porous Pd by UPD-SLRR. From the results of CV and EDS elemental analyses, it was assumed that Pt atomic layer deposition was successfully realized. It was also suggested that the novel process showed more Pt usage reduction compared to previously proposed simple galvanice replacement reaction process. Power generation was examined, and prototype cell with UPD-SLRR Pt showed comparable performance to the simple Pt catalysts, though the Pt usage was significantly reduced. The new process showed promising results, and further study to improve catalyst performance will be performed.

References

[1] M. Kobayashi, R. Teranishi and M. Hayase, 2012 *Intl. workshop on PowerMEMS 2012*, pp. 395-398.
[2] Ryohei Ogawa, Shin Aoki, Masanori Hayase, 2012 *Intl. workshop on PowerMEMS 2012,* pp. 391-394.
[3] T. Honjo, T. Matsuzaka, M. Hayase, 2010 *Intl. workshop on PowerMEMS 2010*, 231-234.
[4] S.R. Brankovic, J.X.Wang, R.R.Adžić, 2001 *Surface Science*, vol.**474**, L179-179 (2001)
[5] D. Gokcen, S.E. Bae, S.R.Brankovic, 2010 *J. Electrochem. Soc.*, **157**(11), D582-587 (2010)
[6] M. Hayase, K.Horikoshi, T. Fujii and Y.Tajima, 2009 *9th intl. Workshop on PowerMEMS 2009*, pp. 13-16.

A long-term stable power supply μDMFC stack for wireless sensor node applications

Z. L. Wu[1], X. H. Wang[12]*, F. Teng[1], X. Z. Li[1], X. M. Wu[1] and L. T. Liu[1]

[1]Institute of Microelectronics, Tsinghua University, China
[2]Tsinghua National Laboratory for Information Science and Technology, China

Email: wxh-ime@mail.tsinghua.edu.cn

Abstract. A passive, air-breathing 4-cell micro direct methanol fuel cell (μDMFC) stack is presented featured by a fuel delivery structure for a long-term & stable power supply. The fuel is reserved in a T shape tank and diffuses through the porous diffusion layer to the catalyst at anode. The stack has a maximum power output of 110mW with 3M methanol at room temperature and output a stable power even thought 5% fuel is the remained in reservoir. Its performance decreases less than 3% for 100 hours continuous work. As such, it is believed to be more applicable for powering the wireless sensor nodes.

1. Introduction

Direct methanol fuel cell (DMFC) is one type of electrochemical energy conversion device which directly converts the chemical energy to electricity. The μDMFC has been the promising power source for wireless sensor nodes work in extreme environment, because of its high energy density, instantaneous recharging, the simple system design and environmentally friendly[1-2]. In the last decades, many researchers have been attracted to develop the active μDMFCs and their components, such as catalysts, membrane electrode assembly (MEA) membranes, anode and cathode plates, etc [2-4]. More recently, the passive air-breathing μDMFCs have been more attractive than active ones because they don't need extra pump or air blower acting as ancillary devices and consequently have much simpler structures, lower cost and higher fuel utilization [3, 4]. In fact, a single unit μDMFC can only output a limited potential and power. Therefore, the passive air-breathing μDMFC stacks are required to meet the practical application.

Several studies on the development of passive μDMFC stacks have been reported. Baglio et.al reported a monopolar three-cell stack connected in series with different catalyst loading and methanol concentration, it can provide a peak power density more than 20mW/cm^2 at ambient conditions [5]. Zhu et.al proposed a novel assembly method for μDMFC stack, which is beneficial to decreasing the internal resistance by using multi-layer bonding technique [6]. N. Hashim et.al developed a six-cell stack with laminated structures connected in series, and the mount of single cells in the stack can be easily extensible for realizing a higher output [7]. However, there remain some common challenges in passive μDMFC stacks, i.e. fuel and oxygen self-supply, enough power and long-term & stable output, which are the most important and most difficult also [8]. Many reported stacks employed small reservoirs and operated only several minutes or hours [6, 9-10].

Therefore, we design and fabricate two kinds of passive, air-breathing 4-cell μDMFC stack with different fuel delivery structure for a long-term stable power supply. The stacks were operated completely at room conditions. The constant current discharge, the constant voltage output and long-term operation ability of the two prototypes were tested. In addition, one of the prototypes can output a stable power even thought the fuel in the reservoir is nearly used up. The results have potential contribution for the development of passive DMFCs stack.

2. Design

We design a passive μDMFC stack, consisting of 4 single cells with back-to-back distribution and a shared fuel reservoir sandwiched in the middle, as shown structure A in Fig. 1. The fuel diffuses through the diffusion layer to the catalyst at anode, and the oxygen is self air-breathed by the cathode, on which the hydrophilic capillary channels are arrayed along the ribs of the windows to remove excessive water out.

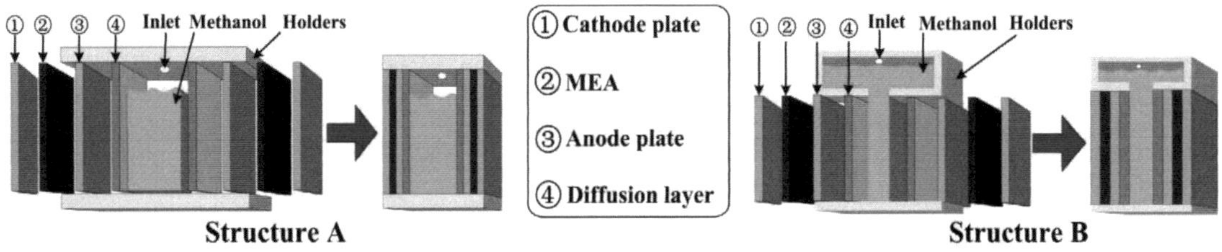

Fig.1. A passive μDMFC stack consists of 4 single cells with back-to-back distribution and a shared reservoir sandwiched in the middle. A T shape fuel reservoir above the stack is designed improvably to supply fuel to the stack continuously, as shown in structure B.

The structure A has many characteristics which is benefit for a passive stack for practical application; for example, the back-to-back cells distribution is help for maximizing the air-breathing area of cathode as well as feeding the fuel by a simple and shared reservoir, furthermore, the stack works in vertical direction can make sure each cell is supplied with uniform methanol, which can make the stack output a relative big overall power.

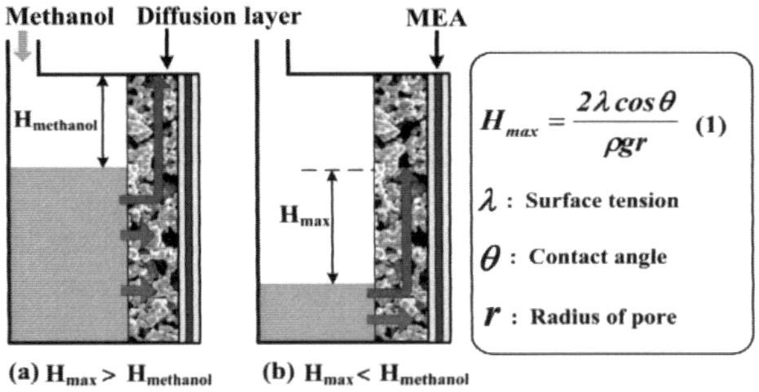

Fig.2. Schematic of a vertical passive μDMFC with a porous diffusion layer in the case of the solution does not cover the whole surface of the diffusion layer after long-term work, the μDMFC may be supplied with insufficient fuel.

However, an actual question has exposed after a very long time operation for this passive stack: the liquid level of the methanol solution in the reservoir will change with operation time, and the μDMFC may be supplied with insufficient fuel in the case of the methanol solution is not completely full in the reservoir. As described in Fig. 2(a), when the liquid level of methanol are not very low, methanol solution can be sucked to the top by the capillary force, making sure that the whole catalyst layer can be supplied with fuel. However, the maximum sucking height of the solution, H_{max}, is usually not big enough to suck the fuel to the top while the level of the methanol solution is low, as shown in the Fig. 2(b). This is because the radius of diffusion layer pores is often a little big, avoiding the shortage of methanol diffused to the catalyst layer, as the equation (1) in Fig. 2(b) expressed. Therefore, a new structure, a T shape fuel reservoir with very narrow lower part and fat upper part (volume ratio,1: 9), is designed improvably to feed the passive stack continuously, as shown structure B in Fig 1. In this way, the liquid level will not decrease until more than 90% fuel in the reservoir has been used. Furthermore, it is also easy for adding an extra reservoir in this structure for realizing a longer work time if necessary, so the stack with structure B is believed to have a long-term and stable output.

3. Experimental

3.1. Fabrication

The structure and fabrication of the silicon based plates of the μDMFC is very similar to our previous work [8]. Both the anode and cathode plates are fabricated by using MEMS techniques, detailed in Fig 3, while the fabrication process of the silicon plates consist of four steps.

(a) 20μm thick thermal oxide and 100μm thick Si_3N_4 layers are deposited on both sides of a 500 μm thick 4 inches double-polished (100) silicon wafer, acting as the mask layers for KOH etching.
(b) Double side lithography technology is introduced to form the flow structure of the anode and cathode plates. In the cathode plate, the patterns of windows and capillary channels in both sides are formed.
(c) KOH-timed etching is employed to anisotropically etch the silicon wafer until the windows are through and capillary channels are formed.
(d) Then 20nm thick Cr and 200nm thick Au are deposited on back side of wafer acting as the current collecting layer.

The fabricated anode and cathode plates are shown in Fig 4(a) and 4(b) respectively. Fig 4(c) shows the SEM photograph of the capillaries channel along the rib of the cathode window.

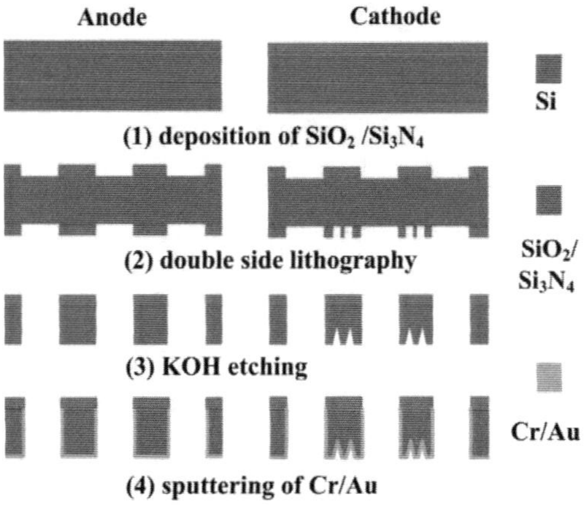

Fig.3. The fabrication processes of the silicon anode and cathode plates.

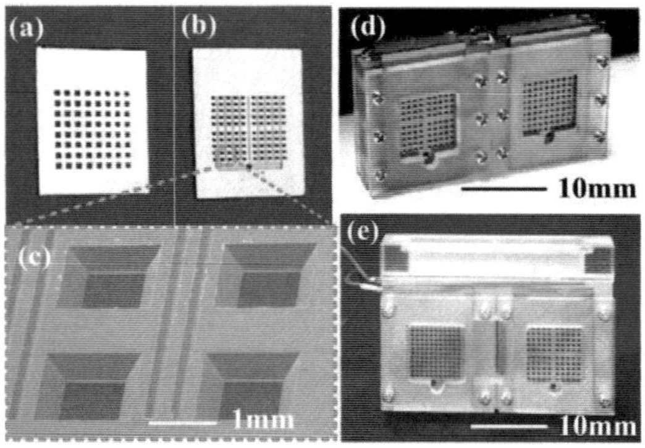

Fig.4. The photographs of fabricated anode plate (a), cathode plate (b) and the SEM image of the capillary channels on the cathode plate; two prototypes, stack A (d)with structure A and stack B(e) with structure B are assembled.

3.2. MEA preparation

The Nafion 117 membrane (DuPont) is used as the proton exchange membrane (PEM), with catalyst loading of 5.0 mg cm^2 Pt/Ru black at the anode and 4.0 mg cm^2 Pt black at the cathode. The PEM with catalyst is sandwiched between two carbon papers (TGP-H-090, Toray), which served as gas diffusion layers (GDLs), and then hot-pressed at 140◦C and 30 atm for 2 min to form the MEA. The carbon papers on cathode side are waterproof while those on anode sides are modified to form a hydrophilic surface. The whole area of the MEA is 18mm × 24mm, with an active area of 14mm × 15mm.

3.3. Assembly

The fabricated methanol and water plates are bonded together using PDMS bonding technique to form anode structure. Then the anode structure is assembled with the cathode plate using PMMA holders and PDMS material, sandwiched with a MEA in between. Two prototypes are assembled, i.e. stack A with structure A (Fig 4(d)) and stack B (Fig 4(e)) with structure B.

4. Test and discussion

The performances of two prototypes with different fuel delivery structure are tested in completely passive mode, i.e. the anode reservoir is filled with pure methanol and the cathode is air-breathing, at room temperature and 40% relative humidity.

Firstly, in order to test the output performance of prototype A, the reservoir is full of 3M methanol solution, the constant current is applied and the output voltage is monitored for a period of 1 minute until the final steady-state value is recorded. The I-V curve and power curve are shown in Fig 5. The prototype A achieves an open circuit voltage of 1.1V, maximum current of more than 800mA and maximum power of more than 110mW. Then we test the output stability of prototype A as fewer and fewer methanol solution is remained in reservoir. The corresponding I-V curves and power curves are recorded when remained methanols are about 50% and 5%, as shown in Fig 5. The maximum output power has a 20% drop when 50% methanol is remained in the reservoir, while that is only about 20mW as 5% methanol is remained in the reservoir. This obvious decrease of power output is attribute to the amount of methanol diffused to the anode catalyst is not enough, with the existence of gravity.

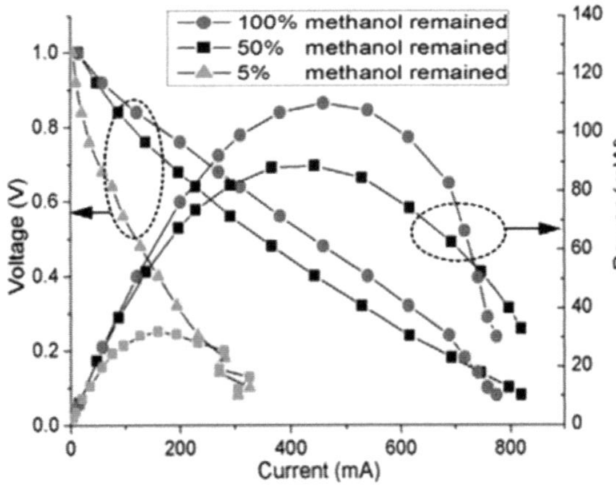

Fig.5. The prototype with structure A has a maximum power output of 110mW, and it decreases seriously as the amount of fuel decreases.

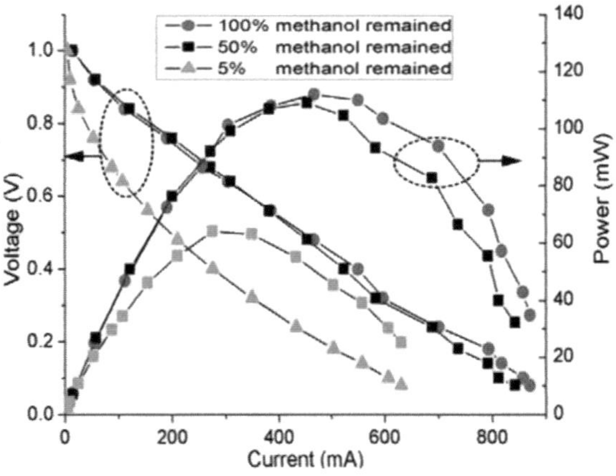

Fig.6. The stack with structure B can employ a stable power output even thought the remained fuel is about 5%.

Accordingly, the I-V curve, power curve and the output stability of prototype B are tested with the same work condition as prototype A. When the methanol is 100% full in reservoir, it has the same maximum power about 110mW, and there is almost no decrease for power when remained methanol is about 80% (Fig 6). Furthermore, it still has an output about 60mW when remained methanol is only 5%.It indicate that the prototype B has a much better output stability than prototype A.

Finally, the long time performances of prototypes are tested. The prototypes worke under a constant current condition in completely passive mode, at room temperature and 40% relative humidity, and the voltage was monitored and recorded till it fell down to 0V.The tested μDMFCs worked under a constant current of 250mA, with a certain amount of 3M methanol 4mL. As shown in Fig 7, the prototype A can only work about 180 minutes, while the prototype B can work as long as 300 minutes. These measurements are done for several times and the average value is chosen. In addition, the performance decreases of prototype B less than 3% for 100 hours continuous work.

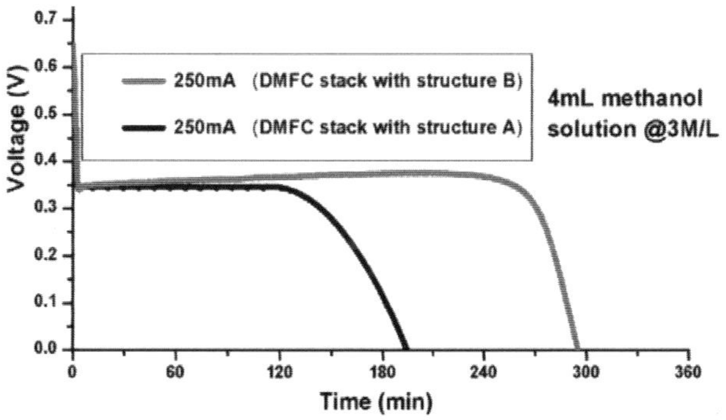

Fig.7. Both stacks are worked in a constant current, filled with 4mL methanol solution, stack B can work about 300 minutes stably, while stack A worked only 180 minutes.

5. Conclusions

We successfully demonstrate a passive, air-breathing 4-cell micro direct methanol fuel cell (μDMFC) stack featured by a fuel delivery structure for a long-term stable power supply. The fuel is reserved in a T shape tank and diffuses through the porous diffusion layer to the catalyst at anode. The stack has a maximum power output of 110mW with 3M methanol at room temperature and output a stable power even thought 5% fuel is the remained in reservoir. Its performance decreases less than 3% for 100 hours continuous work. As such, it is believed to be more applicable for powering the wireless sensor nodes.

Acknowledgment

This project is supported by the 973 program (No.2009CB320304), the National Natural Science Foundation of China (No. 60936003), and the 863 program (No.2009AA04Z319).

References

[1] Qian Zhang Design, optimization and microfabrication of a micro-direct methanol fuel cell with microblocks in anode structure, *Sensors and Actuators A* 154 (2009) 247–254
[2] D.D. Meng, A degassing plate with hydrophobic bubble capture and distributed venting for micro fluidic devices, *J. Micromech. Microeng.* 16(2006) 419–424.
[3] Y. Jiang et.al, Design, fabrication and testing of a siliconbased air-breathing micro direct methanol fuel cell, *J. Micromech. Microeng.* 16 (2006) 233–239.
[4] Yan'an Zhou et.al, A water collecting and recycling structure for silicon-based micro direct methanol fuel cells, *International journal of hydrogen energy* 37(2012) 967-976
[5] Baglio Optimization of properties and operating parameters of a passive DMFC mini-stack at ambient temperature , *Journal of Power Sources* 180 (2008) 797–802
[6] Zhu et.al, A novel assembly method using multi-layer bonding technique for micro direct methanol fuel cells and their stack, *Sensors and Actuators A* 188 (2012) 246– 254
[7] N. Hashim et.al, Design, fabrication and testing of a PMMA-based passive single-cell and a multi-cell stack micro-DMFC, *International journal of hydrogen energy* 34 (2009) 8263 – 8269
[8] Xiaohong Wang et.al, An air-breathing micro direct methanol fuel cell stack employing a single shared anode using silicon microfabrication technologies, *Journal of Micromechanics and Microengineering*, 19 (2009), 094012
[9] Zhu et.al, Development of a passive direct methanol fuel cell (DMFC) twin-stack for long-term operation, *Journal of Power Sources* 193 (2009) 649–655
[10] T.S.zhao et.al, A small mono-polar direct methanol fuel cell stack with passive operation, *Journal of Power Sources* 178 (2008) 118–124

PowerMEMS 2013 IOP Publishing
Journal of Physics: Conference Series **476** (2013) 012068 doi:10.1088/1742-6596/476/1/012068

Self-tuning behavior of a clamped-clamped beam with sliding proof mass for broadband energy harvesting

P Pillatsch[1], L M Miller[2], E Halvorsen[3], P K Wright[2], E M Yeatman[1] and A S Holmes[1]

[1]Imperial College London
[2]University of California, Berkeley
[3]Vestfold University College

E-mail: p.pillatsch10@imperial.ac.uk

Abstract.
Real world systems rarely vibrate at a single resonance frequency and the frequencies drift over time. Tunable devices exist, but generally need additional energy to achieve frequency adaptation. This means that the benefits in power output from this tuning need to be large enough to power the mechanism itself. Passively self-tuning systems go into resonance without requiring active control. This paper focuses on a passively self-tuning system with a proof mass that can slide freely along a clamped-clamped beam. Under external vibration, the slider moves along the beam until the system goes into resonance. A proof-of-concept design is introduced using either a copper or a steel beam and a 3D-printed ABS thermoplastic proof mass. Successful self-tuning is demonstrated in both cases. The frequencies range from 80 – 140 Hz at accelerations as low as 0.007 g rms. Results show the resonance of the beam and the position of the slider along the beam with time. Furthermore, the dynamic magnification and the proof mass position at resonance are discussed, together with the inherent non-linearities of double-clamped beam resonators. The findings support the hypothesis that the effect of the ratio between proof mass and beam mass outweighs the Duffing spring stiffening effects.

1. Introduction

One of the problems faced by resonant vibration energy harvesting is that real world systems exhibit changes in frequency over time. This causes a mismatch between the excitation frequency and the narrow optimal operation frequency of many of these devices [1]. Various tunable designs offer potential solutions [2], but most of the proposed tuning mechanisms require some form of energy to operate. This means that in order for tuning to be viable, the system needs to improve the power output sufficiently to accommodate for the energy that goes into the tuning itself. Another approach is frequency up-conversion, where a random input gets converted into an ideal operation frequency for the transducer [3]. However, this strategy is best for situations where resonant devices are not suitable in the first place. For instance, an advantage of resonant harvesters is the dynamic magnification of small vibration amplitudes, which is beneficial when the internal proof mass displacement limit is large compared to the input excitation.

One method for passive self-tuning, which means that the system naturally goes into resonance without the need to supply additional energy, uses spring-stiffening effects on a microelectromechanical device [4]. In contrast, our work expands on a system, first presented in [5], that allows passive self-tuning of the device by incorporating a proof mass that can slide along a double-clamped beam. The principle is similar to work presented in [6], where the resonance of a set-up with a freely sliding bead on a string was investigated. In [7], a spring-suspended slider was used with a simple cantilevered beam and [8] investigated a number of systems, one based on a sphere moving inside a cylinder. The latter two attempts in particular proved the importance of carefully selecting the system parameters. Both showed dissipation of energy, thus achieving self-damping of the system instead of self-tuning. This paper builds on experimental results from [5], in which passive self-tuning is reproducibly shown at various frequencies in three prototype systems, to present new data on the potential role that non-linear spring stiffness plays in the system.

2. Experimental Setup

A schematic of the system setup is shown in figure 1 and a photograph of the prototype is given in figure 2. A clamped-clamped beam carries a freely sliding proof mass.The position of this slider affects the resonance frequency of the beam. Under external vibration the slider will progressively migrate into a position such that the beam matches the driving frequency and goes into resonance. This paper is based around two different types of beams, steel and beryllium copper ($60 \times 3 \times 0.1$ mm and $60 \times 3 \times 0.2$ mm respectively); a third configuration is discussed in [5]. It was found that the steel beam can sag through under its own weight during clamping to the U-shaped holder with two bars screwed in on either side. The resulting compressive load causes the beam to buckle and needs to be avoided for the system to work properly. This was much easier to achieve with the thicker copper beams. Furthermore, due to the lower Young's modulus of copper and the higher specific weight, the operation frequencies stay in a similar range.

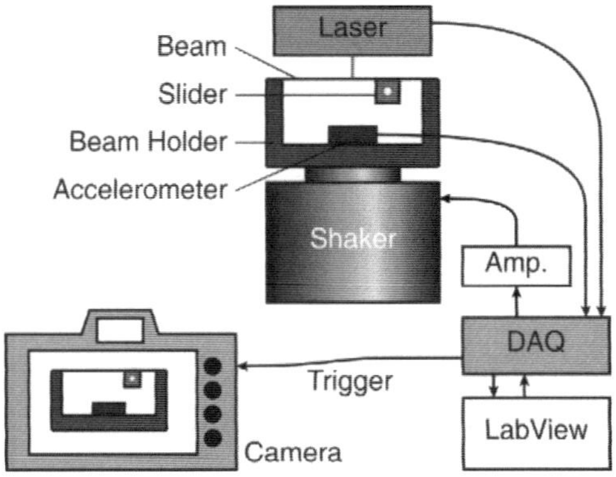

Figure 1: Schematic of the measurement set-up

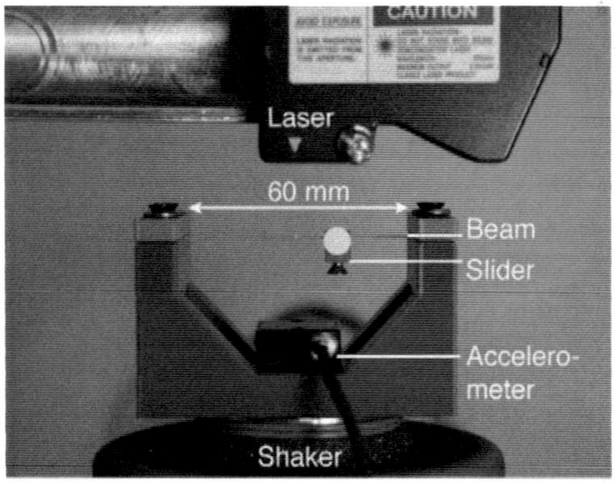

Figure 2: Photograph of the measurement set-up

The 3D-printed slider carries a screw at the bottom that acts as an additional mass and allows the adjustment of the clearance with the beam. It was marked with a white dot on the side for image processing in Matlab. The position of this white dot can be read from photographs taken at one second intervals, triggered by a Labview program. The results were used to plot the relative position of the slider on the beam over time together with the vibration amplitude at the beam center recorded with a Keyence LK-H087 laser displacement sensor (LDS). The vibration shaker (Labworks ET-126) and accelerometer (Kistler 8315A) were also operated through the same Labview program.

3. Measurement Results

Passive self-tuning was successfully achieved for both steel and copper beams. Figures 3 and 4 show the measurements for the steel beam at the upper and lower frequency limit respectively. The corresponding graphs for the copper beam are shown in figure 5 and figure 6. The mid-point displacement is shown along with the proof mass position, normalized over the length of the beam. It can be observed that the mass slides away from the center until the system hits resonance. The difference between the copper and the steel beam is in the frequency range over which tuning was successfully achieved. In both cases the same sliding mass was used, which means that the proof mass to beam weight ratio is twice as high for the steel beam as for the copper one. A relatively heavier proof mass results in a wider range of frequencies over which tuning will occur. A similar effect of the proof mass weight was observed in [6].

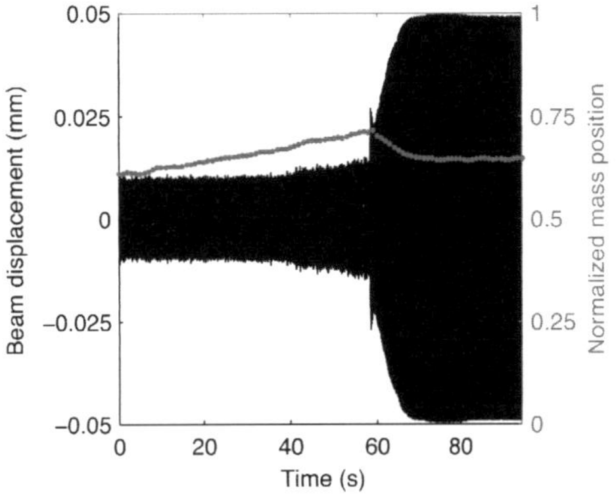

Figure 3: Self-tuning at 140 Hz on a steel beam at 0.0084 g rms acceleration

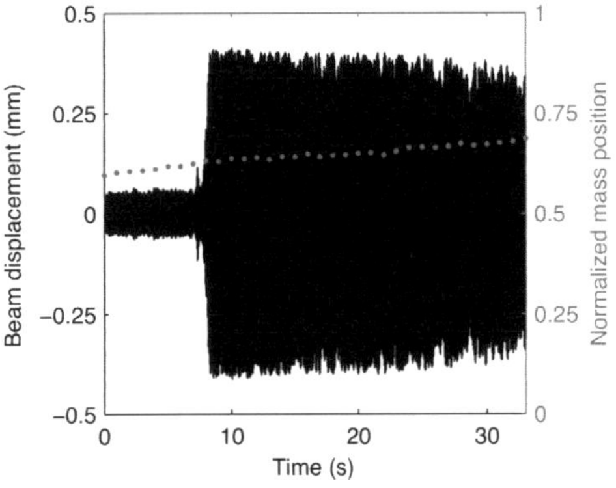

Figure 4: Self-tuning at 80 Hz on a steel beam at 0.012 g rms acceleration

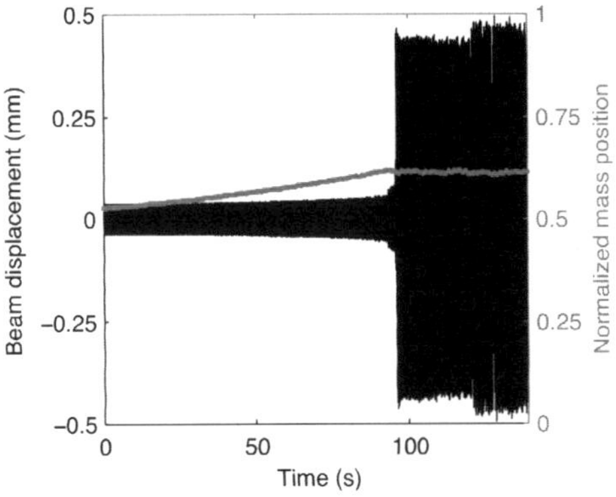

Figure 5: Self-tuning at 135 Hz on a copper beam at 0.007 g rms acceleration

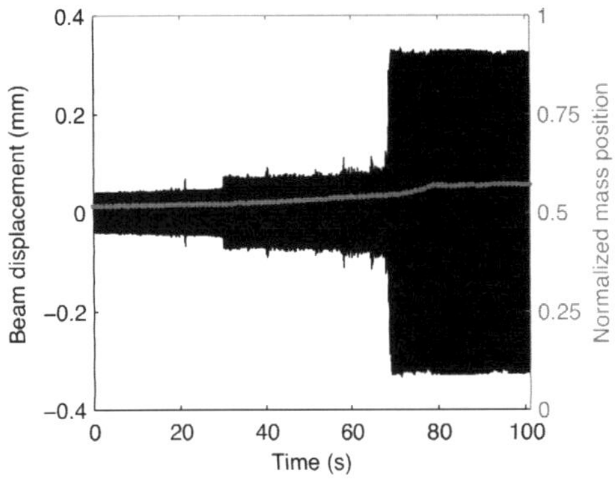

Figure 6: Self-tuning at 100 Hz on a copper beam at 0.011 g rms acceleration

The normalized position of the slider when the system tunes in to resonance for the tested driving frequencies is shown in figure 7. The graph is restricted to the right hand side of the beam. In theory, the proof mass direction of travel should be symmetric, not favouring either side, when the initial position is in the middle of the beam. During the experiments, the slider had a tendency to move to the right. This might be caused by the fact that the initial position was not exactly in the center because the laser beam of the displacement sensor needed a clear path. However, as expected, there is a trend of the slider to move towards the end of the beam with an increase of frequency. The explanation is simple; if the mass were fixed on the beam, a position in the middle would yield the lowest possible resonance frequency, whereas a position at the end of the beam would yield the highest one. The same trend was observed for a larger aluminum beam prototype in [5].

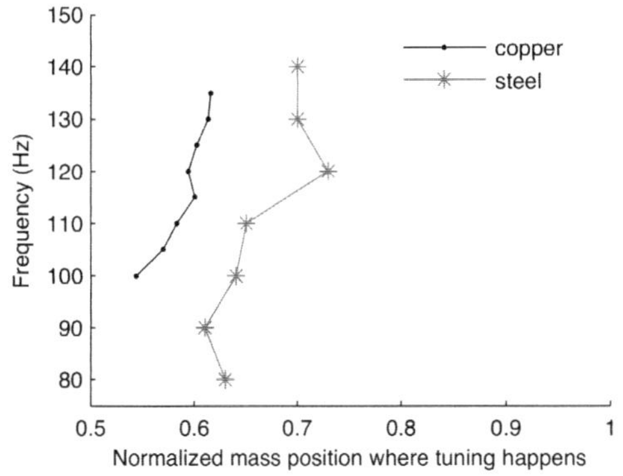

Figure 7: Position of the slider on the beam when tuning occurs

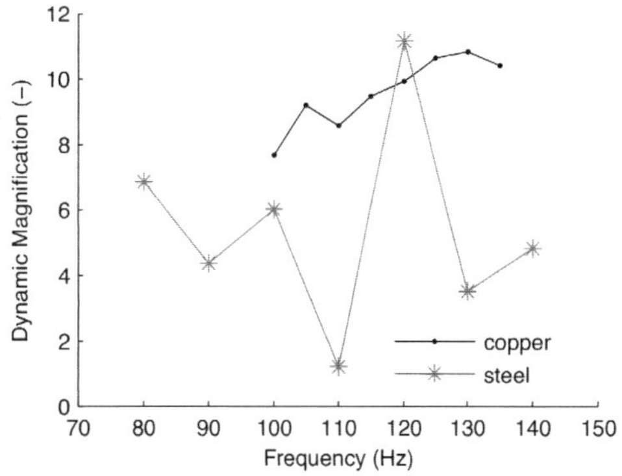

Figure 8: Magnification of the displacement amplitudes when the system goes into resonance

Figure 8 depicts the magnification of the beam center displacement in resonance for all the frequencies where tuning was achieved. The results achieved with the copper beam are more consistent and show a slight increase of magnification with an increase of frequency. The reasons behind this are not entirely clear and need further investigation. The scattering of the results for the steel beam could be due to the previously mentioned difficulties in clamping the beams, whereas the increase in magnification for the copper beams is likely due to the higher energy present in the system at higher frequencies.

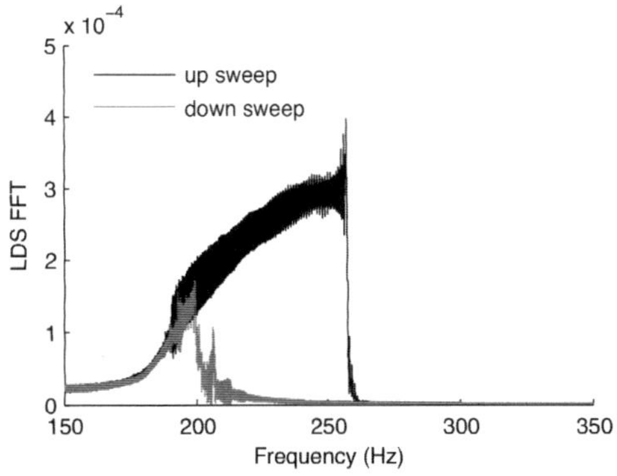

Figure 9: Frequency sweep on a steel beam without the proof mass, FFT of the linear displacement sensor output

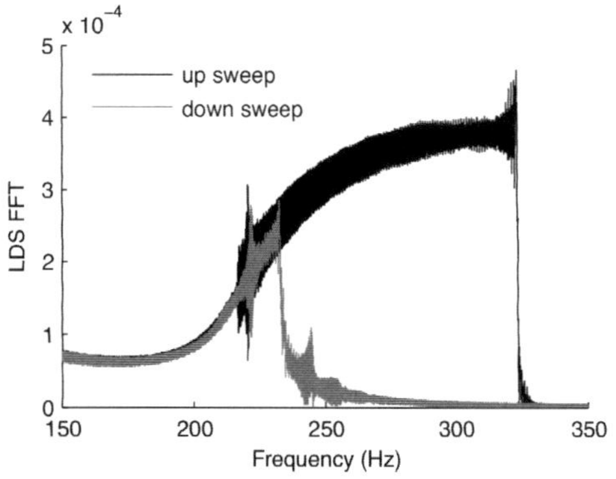

Figure 10: Frequency sweep on a copper beam without the proof mass, FFT of the linear displacement sensor output

Figures 9 and 10 were obtained by performing a frequency up-sweep and a down-sweep on beams without a proof mass and exhibit the inherent non-linearity of the frequency response of double-clamped beams. The first finding from these graphs is a good consistency with figure 8; the FFTs show a larger amplitude for the copper beam than for the steel beam. It is assumed that these effects originate from different damping values between the copper and steel set-ups. Those might simply be due to different material damping coefficients or caused by the difficulties in clamping the steel beams. Furthermore, in the FFTs without attached mass, the steel beam exhibits a significantly smaller frequency bandwidth than the copper beam. However, in the case of the self-tuning results previously shown, the opposite was found. What this suggests is that the effect of proof mass to beam mass ratio far outweighs the effects of the non-linearity and the bandwidth of the frequency response of a double clamped beam.

4. Conclusions

This paper presents an experimentally verified self-tuning mechanism based on a sliding proof mass on a beam. It was found that the crucial points are the clamping of the beam to avoid any compression or tensioning of the beam, and the clearance between the slider and the beam. The latter could easily be adjusted during the experiments by tightening or loosening a screw in the base of the slider. However, this process required a fair amount of fine tuning before the self-resonance was successful. If the clearance is too tight, the slider will not move, if it is too large, the slider will start rattling and dissipates energy. The possibility of lubrication could be investigated in a next step.

Further work is required to determine the effects of spring-stiffening and to fully characterize the behaviour of the system. That being said, the results are encouraging, showing self-tuning over a range between 80 – 140 Hz for a steel beam and between 100 – 135 Hz for a copper beam at low acceleration levels. In addition, the comparison between measurements without the proof mass and the results when self-tuning occurred consistently shows the differences in magnification between copper and steel and suggests that the effects of non-linearity are only secondary to the effects of the proof mass to beam mass ratio. The next step for energy harvesting could be the inclusion of piezoelectric patches on the beam to generate power.

Acknowledgments

This research was supported by the California Energy Commission Contract 500-01-43, the Demand Response project DR-03-01, the Research Council of Norway Grant 191282 and SIU Contract No. NNA-2008/10004, the Engineering and Physical Sciences Research Council (EPSRC) and the Institution of Engineering and Technology.

References

[1] Miller L M, Halvorsen E, Dong T and Wright P K 2011 *Journal of Micromechanics and Microengineering* **21** 045029 ISSN 0960-1317
[2] Zhu D, Tudor M J and Beeby S P 2010 *Journal of Measurement Science and Technology* **21** 022001 ISSN 0957-0233
[3] Pillatsch P, Yeatman E M and Holmes A S 2012 *Journal of Smart Materials and Structures* **21** 115018 ISSN 0964-1726
[4] Marzencki M, Defosseux M and Basrour S 2009 *Journal of Microelectromechanical Systems* **18** 1444–1453
[5] Miller L M, Pillatsch P, Halvorsen E, Wright P K, Yeatman E M and Holmes A S 2013 *Journal of Sound and Vibration*
[6] Boudaoud A, Couder Y and Ben Amar M 1999 *The European Physical Journal B* **9** 159–165 ISSN 1434-6028
[7] Miranda E C and Thomsen J O N J 1998 *Journal of Nonlinear Dynamics* 167–186
[8] Kozinsky I 2009 *PowerMEMS* pp 388–391 ISBN 0974361151

Self-powered resonant frequency tuning for Piezoelectric Vibration Energy Harvesters

B. Ahmed-Seddik, G. Despesse, S. Boisseau, E. Defay

CEA-LETI, Minatec Campus, Grenoble, France

Ghislain.despesse@cea.fr

Abstract. This paper reports on the design, fabrication and testing of an innovative 33-mode piezoelectric vibration energy harvester (VEH). This system is able to change its resonant frequency in real time to follow the main frequency of a vibration source. The system proposed in this paper enables to adapt VEH characteristics (resonant frequency, electrical damping) to vibration parameters variations (frequency and amplitude) in order to optimize the extraction of energy and then the output power at any time. This solution allows up to 40% of resonant frequency tuning ratio; moreover, the adaptation is made in real time and the consumption of the regulation electronic is less than 10% of the VEH output power (480µW@0.1g-276Hz).

1. Introduction

During the last decades, a large attention has been given to micro-energy harvesters. This is mainly due to the increasing number of implemented low power sensors in different fields such as health, buildings, transportation…. These sensors are in most cases powered by batteries. The main drawback of batteries is the need for a continuous control of their state of charge, their recharge and their replacement that may become tedious and expensive. Thus, in order to overcome these limitations, one of the most promising solutions is to harvest the surrounding energy. In our environment, we can find many types of harvestable energy, for example: mechanical energy, thermal energy and radiant energy (solar, infrared, radiofrequency). This paper is dedicated to the harvesting of mechanical vibrations produced by cars, trains, variable speed engines and so on.

Theoretical models show two important points to improve VEH efficiencies [1]: *(i)* equality between vibration frequency and resonant frequency, and *(ii)* equality between mechanical damping and electrical damping.

For the first point, many works have been carried out (most of them are summarized in [2]):

- Increasing the bandwidth by increasing the mechanical damping. This solution induces an output power always lower than a less damped system.
- Increasing the bandwidth by increasing the electrical damping (part of mechanical energy stored in the mechanical spring turned into useful electrical energy) [3]. This solution induces a higher output power outside of the resonant frequency and less power on the resonant frequency unless adjusting in real time the electrical damping (the optimal electrical load from a highly coupled system).
- Increasing the bandwidth by adding a non-linear effect [4-5]. This solution induces in many cases a relative displacement saturation effect, reducing the maximum power peak.

Yet, it does not really increase the output power outside of the resonant frequency unless in some specific cases [6].

- Increasing the bandwidth by dynamically adjusting the resonant frequency to the input frequency [7-10]. This solution enables to get the maximum output power even if the input vibration frequency changes over time. Nevertheless it requires a feedback loop that consumes one part of the output power and is not appropriate for random vibrations.

From measurements we made in different environments, it appears that the vibration source has its energy concentrated on one frequency that may change over time. The main vibration frequency can change for example if the system has a mobile part moving at variable speeds (i.e. car engines, variable speed processing machines…). Finally, at a given time, the energy is mainly present at only one frequency; it is then interested to make the harvester able to resonate on that frequency to get the maximum output power. A suitable solution is a system able to adapt its resonant frequency in real time, without compromising neither the operating frequency bandwidth nor the power density. The aim of the work presented in this paper is to show that it is possible to make a system able to follow in real time the main vibration frequency of the vibration source on a quite large frequency band (up to 40% of frequency variation), with a power consumption that does not exceed 10% of the harvested power.

For the second point, that is to say, adapting in real time the electrical damping forces in order to maintain an optimum point of electrical energy extraction, only one study has been carried out [11].

The system we propose in this paper is based on an active method for tuning the resonant frequency and the electrical damping at the same time in order to keep, in real time, the best working conditions for the VEH, maximizing its efficiency and its output power.

2. Resonant frequency change mechanism

Piezoelectric materials are characterized by an elastic compliance *s* which is dependent on the applied electrical conditions. The piezoelectric bulk stiffness is lower when the material is placed in short circuit (s^E) than in open circuit (s^D). s^D and s^E are linked by equation (1) where *d* is the piezoelectric coefficient, ε the dielectric permittivity and *k* the electromechanical coupling coefficient.

$$s^D = s^E - d^2/\varepsilon^T \Leftrightarrow s^E = \frac{s^D}{1-k^2} \qquad (1)$$

Then, the piezoelectric material's stiffness can be adjusted between $1/s^E$ and $1/s^D$ by connecting it to an adjustable electrical load.

Actually, the choice of the electrical load type is primordial: as choosing a dissipative electrical load (e.g. resistor) could damp the mechanical system and dissipate its energy. It is then appropriate to connect a non dissipative load, for example an inductor or a capacitor. In order to easily reach an impedance value close to the piezoelectric one (the optimal load effect is around the piezoelectric impedance), we have chosen to connect a variable capacitor C_s named *shunt capacitance* in parallel with the piezoelectric material. This capacitor is adjustable in order to reach the required resonant frequency between the short circuit and open circuit resonant frequencies. It can be modeled as an increase of the effective dielectric permittivity of the piezoelectric material and then it modifies the piezoelectric material stiffness as shown by the following equation (2).

$$\left.\begin{array}{l} s^D = s^T - d^2/\varepsilon^T \\ C = C_p + C_s \\ C_p = \varepsilon^T \dfrac{A}{t_p} \end{array}\right\} \Rightarrow s^D = s^E - \frac{d^2 A}{t_p(C_p + C_s)} \Rightarrow E_p = \left(s^E - \frac{d^2 A}{t_p(C_p + C_s)}\right)^{-1} \qquad (2)$$

Parameter	**Definition**
C_p	*The piezoelectric capacitance of a piezoelectric layer*
C_s	*The shunt capacitance*
t_p	*The piezoelectric thickness*
A	*The piezoelectric electrode surface*
E_p	*Piezoelectric Young's modulus*

To maximize the rate of the resonant frequency change, the most part of the mechanical energy has to be stored in the piezoelectric material. To achieve this purpose, a new design has been proposed and has already been presented in [9, 10].

3. Electrical damping change mechanism

As mentioned above, in addition to track the resonant frequency, the electronic circuit must adapt the electrical damping to the mechanical damping. For this purpose, we propose to adjust the VEH output current by using a second reconfigurable capacitor (C_2) in series with the load (Figure 1). Hence, we suppose that the VEH open circuit voltage is significantly higher than the output voltage on the storage element (e.g. V_{bat}@3V). Finally, the electrical damping can be adjusted in real time in order to maintain the electrical damping equal to the mechanical damping and then to maximize the VEH efficiency.

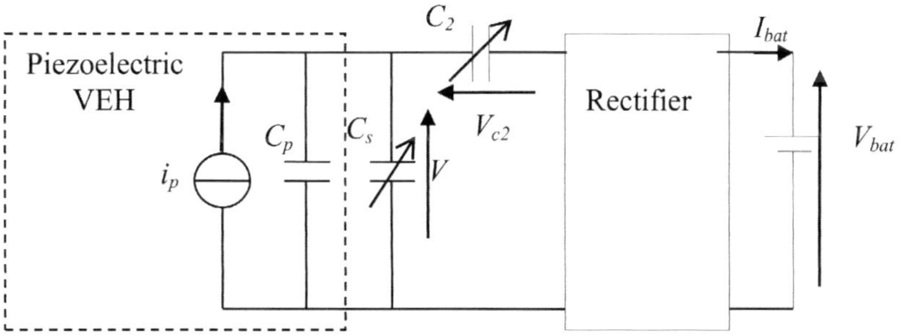

Figure 1. Combined electrical damping / resonant frequency adjustment mechanism

C_2 has also an effect on the resonant frequency, like C_s. Yet, its effect is automatically compensated by the feedback loop on C_s.

4. Feedback loop mechanism

C_s and C_2 are composed of a reconfigurable C, $2C$, $4C$.. $2^N C$ network and are adjusted by two feedback loops working as a Maximum Power Point Tracker (MPPT) with a significantly different feedback time constants (Figure 2).

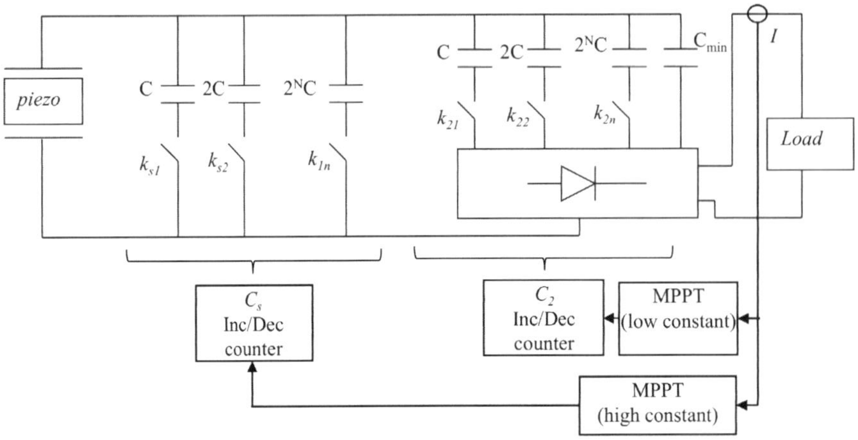

Figure 2. Feedback loops mechanism

The Feedback loops controllers are digital with a clock generated by the vibration itself and with a digital input of '0' if the output power decrease and '1' if it increase. The feedback loop for C_2 is actuated each 8 vibrations periods while the feedback loop for C_s is actuated each 64 periods. The operation principle is as follow: if the last capacitance change induces an increase of the output power, then the capacitance is changed in the same direction; otherwise the capacitance is changed in the opposite direction. C_s and C_2 are directly controlled by the output bits of 4 bits digitals counters.

The current flowing through the storage is first converted into a voltage signal by using a current to voltage converter (CVC). The resulted voltage is then filtered in order to remove all the variations related to the periodic oscillation of the VEH and to keep only the variations due to the resonant frequency and the electrical damping changes.

5. Experimental setup

As already mentioned in section 2, in order to maximize the interaction between the electrical conditions and the mechanical behavior, a triangular shape structure has been chosen to homogenize and maximize the energy stored in the piezoelectric material compared to the substrate material. Moreover, to further increase the electromechanical coupling of the device the longitudinal piezoelectric mode (33-mode) has been exploited (Figure 3a). The prototype is presented in figure 4b.

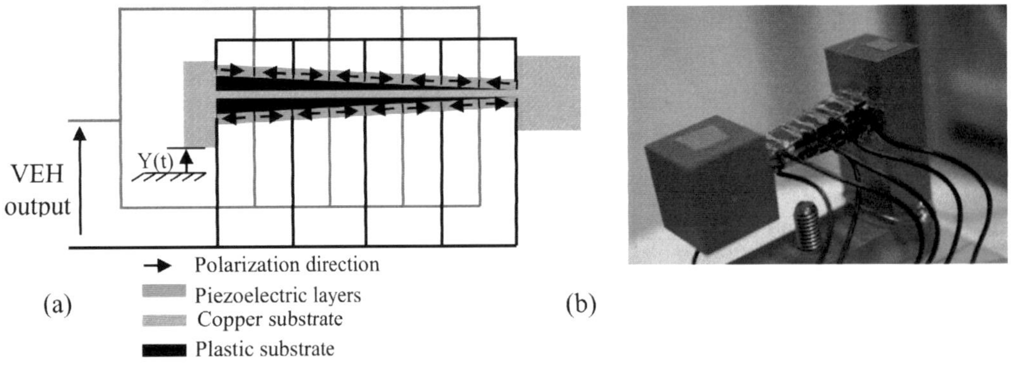

Figure 3. VEH with a triangular shape structure (a) principle and (b) prototype

6. Experimental results

Figure 4 shows the harvested power between 200Hz and 300Hz in closed and open loops. The difference between the harvested power and the net output power is the power consumed by the control circuit to adapt in real time the resonant frequency to the vibration source frequency and to provide at the same time the output power on a 3V-storage with an optimized current load. In closed loop, the output power is nearly always higher than in open loop, even by taking into account the full electronic circuits consumption (50µW); moreover, the frequency bandwidth is multiplied by 3 and, outside this frequency range, the output power is clearly higher in closed loop (Figure 4). These results prove that it is possible to electrically adjust in real time the resonant frequency of a piezoelectric harvester system with less than 10% of the harvested power (480µW@0.1g-276Hz). The resonant frequency adjustment is performed every 64 mechanical periods (i.e. around every 350 ms; every 22s was reached by [8] in 2011).

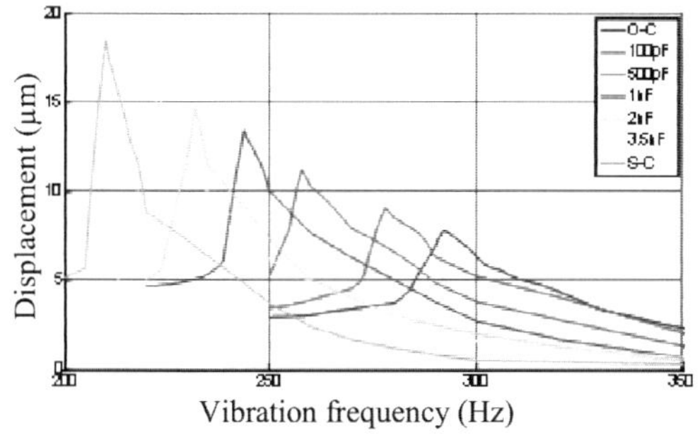

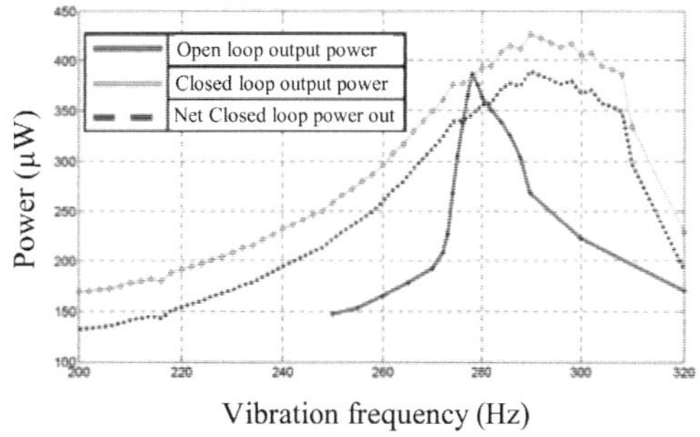

Figure 4. (a) Frequency response as a function of the electrical conditions @ 0.1g and (b) Output power as a function of the input vibration frequency in open and closed loops, with and without the power management consumption

7. Conclusions

The approach presented in this paper allows to efficiently scavenge a vibration which has a main frequency changing over time, and is particularly suitable for transportation systems which work at different speeds. The resonant frequency adjustment is carried out by switching 'on' or 'off' some transistors every few periods of the mechanical vibration. The implementation of this strategy shows that it is feasible to make a complete automatic feedback loop circuit that consumes less than 10% of the VEH output power and able to operate over up to 40% of frequency change and maintaining at the same time the electrical load optimal. The system produces more than 150µW of net power from 220Hz to 320Hz @ 0.1g.

References

[1] Williams C B *et al.* 1996 *Sens. Actuators A: Phys*, **52** 8
[2] Zhu D *et al.* 2010 Strategies for increasing the operating frequency range of vibration energy harvesters: a review *Meas. Sci. Technol.* **21**
[3] Despesse G *et al.* 2008 Mechanical vibrations energy harvesting and power management *Proc. IEEE Sensors* 29.
[4] Blystad L *et al.* 2010 *IEEE Trans on Ultrasonic, Ferroelectric and Frequency Control* **57** 908
[5] Shahruz S 2006 *J. Sound. Vib.* **292** 987
[6] Boisseau S *et al.* 2013 Nonlinear H-Shaped Springs to Improve Efficiency of Vibration Energy Harvesters *J. Appl. Mech* **80** (6) 061013
[7] Morgan B and Ghodssi R 2008 *J. Micrelectromech. Syst.* **17** (1), 85-92
[8] Eichhorn C *et al.* 2011 A smart and self-sufficient frequency tunable vibration energy harvester, *J. Micromech. Microeng*, **21**
[9] Ahmed Seddik B *et al.* 2011 Increased bandwidth of mechanical energy harvester, *Journal of Sensors & Transducers* **13**, Special Issue 62-72
[10] Ahmed Seddik B *et al.* 2012 Wideband mechanical energy harvester based on piezoelectric longitudinal mode *New Circuits and Systems Conference (NEWCAS), IEEE 10th International* **17-20** 453-456
[11] Mitcheson P D *et al.* 2011 *Circuits and Systems II: Express Briefs, IEEE Transactions on* **58**, no 12 792-796

A broadband electromagnetic energy harvester with a coupled bistable structure

D Zhu[1] and S P Beeby
Electronics and Electrical Engineering Group, Electronics and Computer Science, University of Southampton, SO17 1BJ, UK

E-mail: dz@ecs.soton.ac.uk

Abstract. This paper investigates a broadband electromagnetic energy harvester with a coupled bistable structure. Both analytical model and experimental results showed that the coupled bistable structure requires lower excitation force to trigger bistable operation than conventional bistable structures. A compact electromagnetic vibration energy harvester with a coupled bistable structure was implemented and tested. It was excited under white noise vibrations. Experimental results showed that the coupled bistable energy harvester can achieve bistable operation with lower excitation amplitude and generate more output power than both conventional bistable and linear energy harvesters under white noise excitation.

1. Introduction

Mechanical energy from ambient vibrations can be converted into electrical energy and this is often an attractive approach for powering wireless sensors. A linear energy harvester generates maximum power when its resonant frequency matches the ambient vibration frequency and they are typically designed to have a high Q-factor [1]. Its output power drops dramatically if the ambient and device frequencies do not match. However, real world vibrations do not normally consist of only one peak at a certain frequency [2]. Performance of linear energy harvesters will be affected significantly when excited at broadband vibrations or vibrations with multiple peaks [3]. This drawback severely limits the practical applications of linear vibration energy harvesters.

Many methods have been developed to increase operational frequency range of vibration energy harvesting [4], one of which is to use bistable structures. A bistable structure is a type of nonlinear structure that has two potential wells. These two potential wells allow the inertial mass in the system to travel between the two equilibrium positions under external mechanical vibrations regardless of their frequencies. It can be realised with a pre-stressed structure or by applying an external nonlinear force to a linear structure. Conventional bistable structures have been found to have broader operating frequency ranges than linear structures and harvest more energy than linear structures under white noise excitation [5].

This paper presents a vibration energy harvester with a coupled bistable structure. A prototype device was implemented and tested under white noise excitation of various amplitudes. Its output power was compared to those of a linear energy harvester and a conventional bistable energy harvester.

2. Coupled bistable structure

2.1. Overview

A coupled bistable structure proposed in [6] consists of two cantilevers, the main cantilever and the assisting cantilever as shown in Figure 1(a). Two magnet sets are placed at free ends of the two cantilevers to produce a repelling force in-between and thus create two potential wells. The cantilevers jump between the two equilibrium positions to give bistable operation. In conventional bistable structures one magnet is fixed as shown in Figure 1(b), whereas in the device presented here both sets of magnet sets are fee to move in the direction of the vibrations

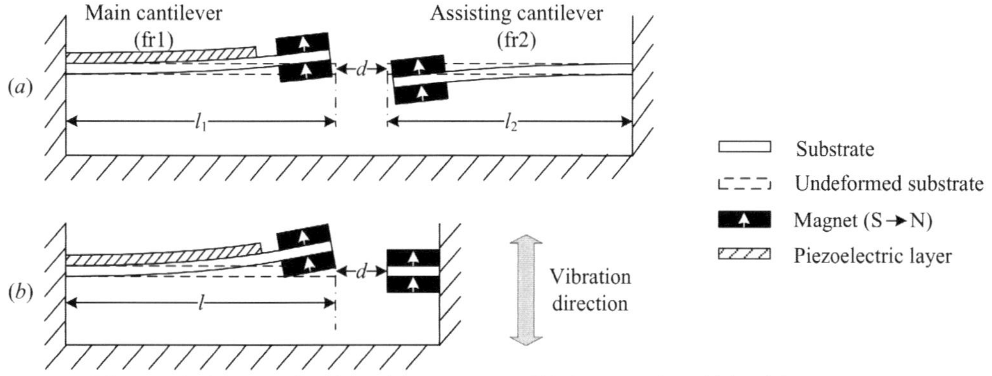

Figure 1. (a) Coupled bistable structure (b) Conventional bistable structure.

2.2. Theory

Figure 2 shows the dynamic model of the coupled bistable structure. As shown in Figure 2(a), the main cantilever and the assisting cantilever have initial displacements of x_m and x_a, respectively. The interaction between the two sets of magnets can be regarded in such way that the two magnets are connected via a magnetic spring, k_m, whose initial length is l_0. When an external force, F, is applied to the base, m_1 has a displacement of Δx_m and the length of the spring k_1 becomes x_m. m_2 has a displacement of Δx_a, the length of the spring k_2 becomes x_a and the length of the magnetic spring becomes l'. Due to limited space in this conference paper, only final equations are presented here. Detailed derivation will be presented in a future journal paper.

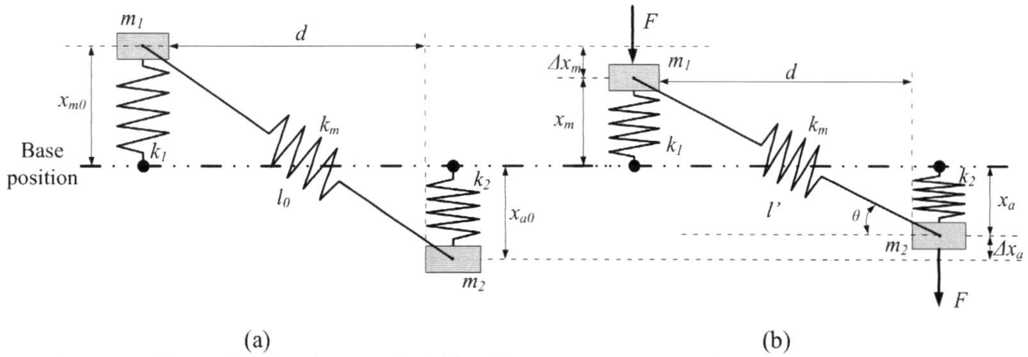

Figure 2. Dynamic model for the coupled bistable structure (a) equilibrium position (b) force acting on the spring.

The potential energy of the two linear springs, E_{ls1} and E_{ls2}, are:

$$E_{ls1} = \frac{1}{2}k_1(x_{m0} - x_m)^2 \quad (1)$$

$$E_{ls2} = \frac{1}{2}k_2(x_{a0} - x_a)^2 \quad (2)$$

The potential energy of the virtual magnetic spring, E_{ms}, is:

$$E_{ms} = -C\left[\frac{1}{\sqrt{d^2+(x_m+x_a)^2}} - \frac{1}{\sqrt{d^2+(x_{m0}+x_{a0})^2}}\right] \tag{3}$$

where C is a constant related to the magnetic force.

The overall potential energy of the system, E_p, is

$$E_p = E_{ls1} + E_{ls2} + E_{ms} \tag{4}$$

$$E_p = \frac{1}{2}k_1(x_{m0} - x_m)^2 + \frac{1}{2}k_2(x_{a0} - x_a)^2 - C\left[\frac{1}{\sqrt{d^2+(x_m+x_a)^2}} - \frac{1}{\sqrt{d^2+(x_{m0}+x_{a0})^2}}\right] \tag{5}$$

Figure 3 shows an example of the potential energy of a coupled bistable structure with different ratios of k_2 to k_1. It is obtained using Eq. (5). Total potential energy in each case is normalised by dividing the potential energy of the conventional bistable structure and then compared.

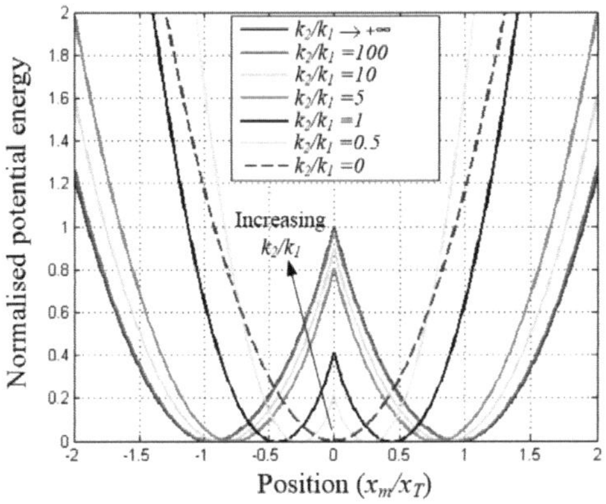

Figure 3. Normalised potential energy with variation of ratios of spring constants of the two cantilevers.

It is found that the lower the ratio of k_2 to k_1 is, the closer the equilibrium position is to the base position and the lower the potential energy barrier is. The potential energy barrier is defined as the energy required to trigger movement of the inertial mass between the two equilibrium positions. In other words, in a coupled bistable structure, the more flexible the assistant cantilever is compared to the main cantilever, the less energy is required to trigger the inertial mass to jump from one equilibrium position to the other.

3. Design of a broadband energy harvester with a coupled bistable structure

3.1. Overview

The proposed coupled bistable energy harvester has three cantilevers shown in Figure 4(a), i.e. two main cantilevers for energy harvesting and a T-shape assisting cantilever in the middle. All cantilevers are made of BeCu. Resonant frequencies of main cantilevers and the assisting cantilever are 28.9 Hz and 16 Hz, respectively. Two 1 mm thick cylinder NdFeB magnets were fixed to the free end of each main cantilever. Same magnets were placed on the both wings of the T-shape assisting cantilever and were axially in line with those magnets on main cantilevers to produce repelling force between them. The distance between magnets on main and assisting cantilevers was 5 mm. Figure 4(b) shows the electromagnetic transduction mechanism. Each main cantilever has a 10 mm thick cylinder NdFeB magnet with a diameter of 10 mm. One coil was placed underneath each main cantilever. Both coils are 5 mm thick and have the outer and inner diameters of 28 mm and 16 mm, respectively. Each coil was wounded with 60 μm thick copper wire and has around 5400 turns. The two coils have resistances of 2200 Ω and 2340 Ω, respectively. Magnets travel through coils, which induces a current in coils.

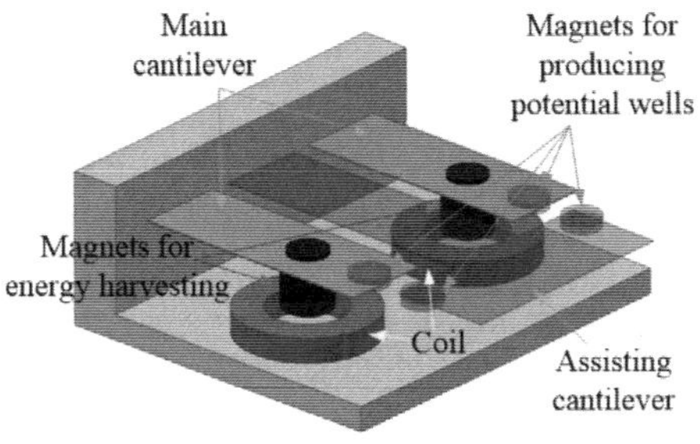

(a)

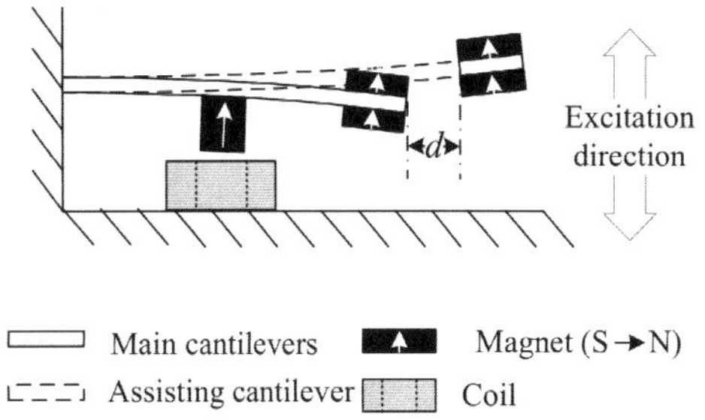

(b)

Figure 4. Proposed vibration energy harvester with a coupled bistable structure (a) perspective view (b) side view.

The energy harvester was excited under white noise excitation with three different amplitudes of 6.9 $m \cdot s^{-2}_{rms}$ (signal 1), 1.23 $m \cdot s^{-2}_{rms}$ (signal 2) and 1.87 $m \cdot s^{-2}_{rms}$ (signal 3). Figure 5 shows test setup and one example of the excitation signal spectrum and waveform. Output of the two coils were connected in series and rectified to charge a 1mF electrolytic capacitor. The charging rate of the capacitor was recorded and compared with that of a linear harvester (by removing the magnets on the assisting cantilever) and a conventional bistable harvester (when the tip of the assisting cantilever was clamped).

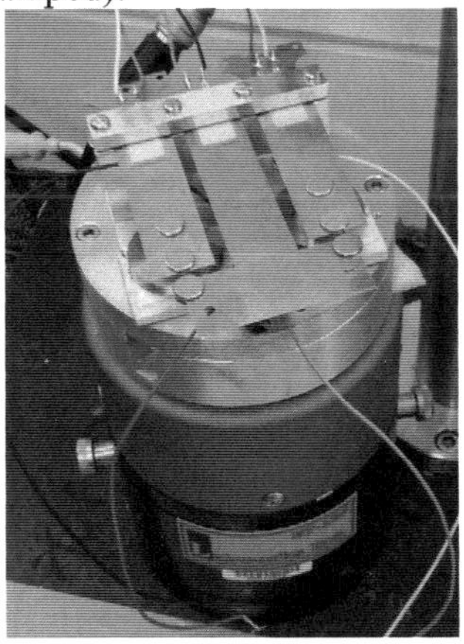

(a) test setup

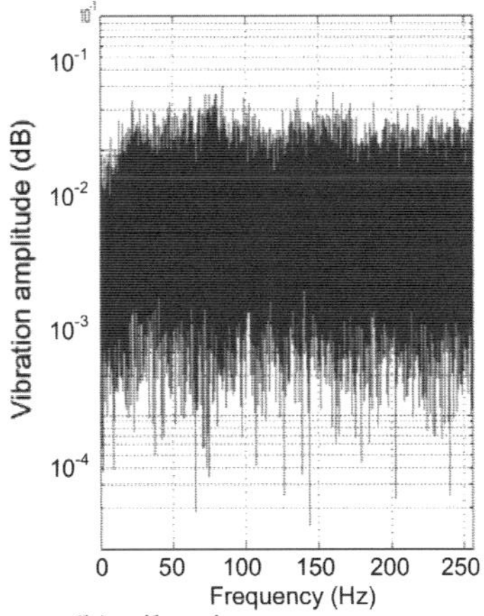

(b) vibration spectrum

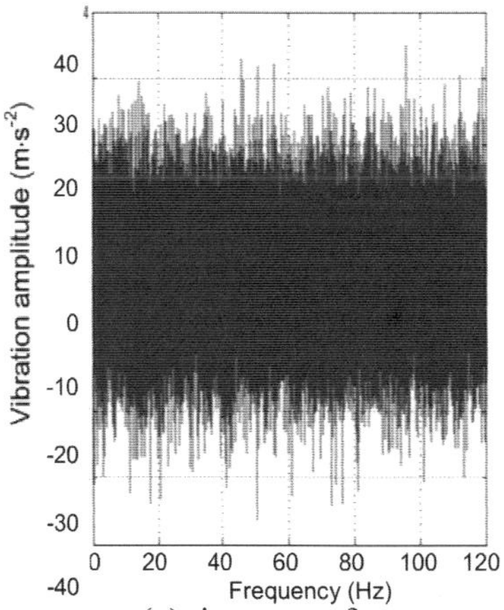

(c) time waveform

Figure 5. Test setup and excitation signal.

Figures 6 compares the capacitor charging rates when charged by a coupled bistable energy harvester, a linear energy harvester and a conventional bistable energy harvester under the three excitation signals. When vibration level was low, no bistable operation was triggered in any case. Therefore, the charging rate was the same for the linear and the coupled bistable harvester. However, since cantilevers in the conventional bistable harvester were stiffer than those in the coupled bistable harvester, its charging rate was 56% lower by comparison (Figure 6a). For excitation signal 2, bistable operation occurred occasionally in the coupled bistable harvester. Its charging rate was 2.5% higher than the linear harvester. Meanwhile, the cantilevers in the conventional bistable harvester were still trapped in one potential well, resulting in a 56.7% lower charging rate than the coupled bistable harvester (Figure 6b). When the vibration level reached 18.3 $m \cdot s^{-2}_{rms}$, bistable operation occurred more frequently in the coupled bistable harvester and its charging rate is 46.6% higher than that of the linear harvester. Bistable operation occurred occasionally in the conventional bistable harvester under this excitation but its charging rate was still 45% and 18.8% lower than the coupled bistable harvester and the linear energy harvester, respectively (Figure 6c).

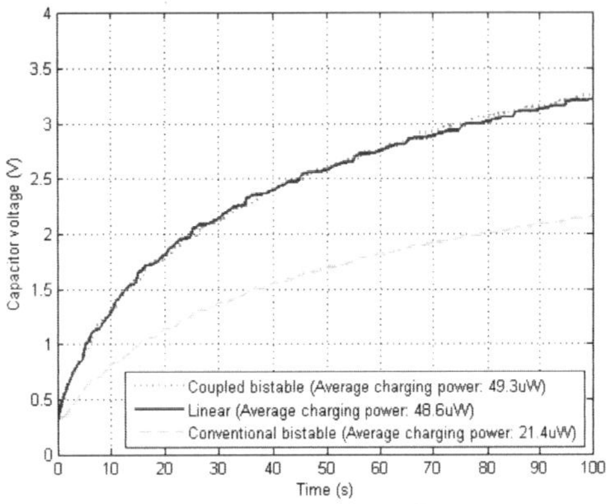

(a) signal 1: 6.9 $m \cdot s^{-2}_{rms}$

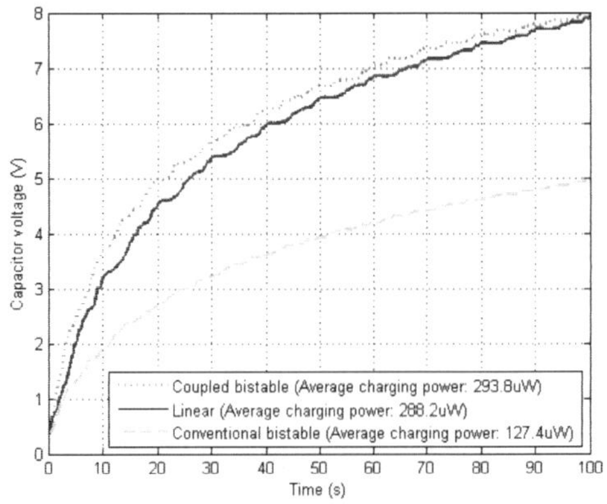

(b) signal 2: 12 $m \cdot s^{-2}_{rms}$

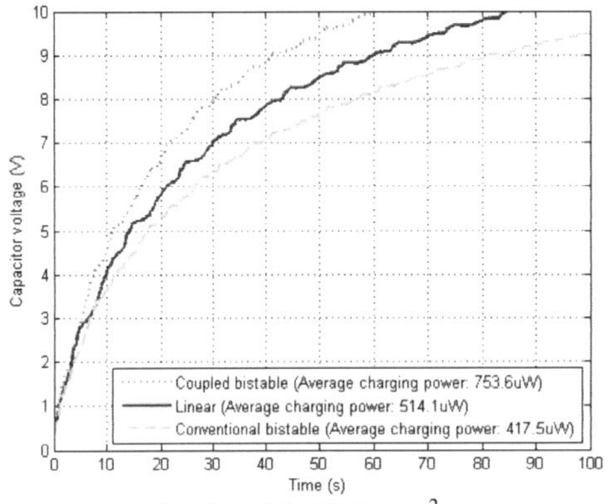

(c) signal 3: 18.3 $m \cdot s^{-2}_{rms}$

Figure 6. Comparison of charging rates under different excitation signals.

4. Conclusions

This paper presented a compact electromagnetic energy harvester with a coupled bistable structure. Experimental results showed that the coupled bistable energy harvester can achieve bistable operation under lower excitation levels and generate more output power than both conventional bistable and linear energy harvesters under white noise excitation. The next step will be miniaturisation of the couple bistable energy harvester. However, there is a trade-off between miniaturising the harvester and improving its performance, which will be investigated in future work.

References

[1] Beeby S P, Tudor M J, White N M 2006 Energy harvesting vibration sources for microsystems applications *Meas. Sci. Technol.* **17** 175-195

[2] Energy Harvesting Network Data Repository [online], http://eh-network.org/data (Accessed: 1 October 2013)

[3] Zhu D, Harris N and Beeby S 2012 Performance of Linear Vibration Energy Harvesters under Broadband Vibrations with Multiple Frequency Peaks *Eurosensors XXVI, Krakow, Poland, 09 - 12 September 2012*

[4] Zhu D, Tudor M J and Beeby S P 2010 Strategies for increasing the operating frequency range of vibration energy harvesters: a review *Meas. Sci. Technol.* **21(2)** 022001

[5] Ferrari M, Ferrari V, Guizzetti M, Andò B, Baglio S and Trigona C 2010 Improved Energy Harvesting from Wideband Vibrations by Nonlinear Piezoelectric Converters *Sens. Actua. A* **162** 425-431

[6] Zhu D and Beeby S P 2013 A coupled bistable structure for broadband vibration energy harvesting *Transducers 2013 & Eurosensors XXVII, Barcelona, Spain,, 16 - 20 Jun 2013*

Bandwidth of a Nonlinear Harvester with Optimized Electrical Load

A. Cammarano*, **A. Gonzalez-Buelga, S.A. Neild, S.G. Burrow**
University of Bristol, University Walk, BS8 1TR, Bristol, UK
E-mail: *andrea.cammarano@bristol.ac.uk

D.J. Inman
University of Michigan, Ann Arbor, MI 48109-2140, United States

Abstract. Many researchers have investigated the possibility of amplifying ambient vibrations and converting the associated kinetic energy into usable electric energy. The vast majority of vibration harvesting devices use mechanical oscillators to boost the amplitude of vibration; however, this can result in a rather narrow band of excitation over which the harvesting device is effective. One approach proposed to overcome this limitation is to substitute the conventional linear oscillator with an oscillator featuring a non-linear compliance characteristic: these mechanisms produce broader frequency responses. The design and optimization of non-linear energy harvesting devices is however not trivial and there is no consensus among the publish works that the benefits of non-linear oscillators can be realized in the energy harvesting context. This work attempts to further develop understanding of nonlinear energy harvesters by investigating the optimum resistive load. The definition of an optimal load for the nonlinear device is first considered, given due consideration to bandwidth and stability of the operating point, and comparisons with linear devices is shown. Finally, the issue of multiple solutions in the frequency response is addressed.

1. INTRODUCTION

Energy harvesting from vibrations has become a viable option for powering small devices or low power electronics. Harvesters based on resonant oscillators, in particular, are widely used to obtain high efficiency devices that deliver maximum power to the electrical load. A key strength of this technique is that it can be adapted to different transduction mechanisms. However there are also disadvantages: for the oscillator to amplify efficiently the input excitation, its damping coefficient has to be small. This results in the device having a narrow bandwidth of operation.

Methods to improve the bandwidth involve changing the geometrical characteristics of the oscillator [1], using external mechanism like magnets [2] or piezoelectric materials [3] to change the stiffness of the oscillator, and designing the system so that multiple frequency are amplified (multiple degree of freedom systems) [4]. Another possible solution to the problem is to exploit the electrical force reflected on the oscillator by the transducer [5, 6]. Finally, exploiting the characteristics of nonlinear oscillator to widen the frequency response of energy harvesting, has been studied in several works [7, 8]. Compliance characteristics allowing changes from hardening to softening [3, 9] and bistable configurations [10] have been proposed.

This work focuses on energy harvesters with hardening compliance - one of the most common forms of nonlinearity. In the literature it is suggested that the presence of hardening nonlinearity results in an increase of bandwidth. We first discuss the optimal electrical resistive load for harvesting devices with a hardening nonlinearities, following[11]. We then provide information on how to select both the stiffness and the resistive load of the nonlinear harvester to achieve maximum power. A comparison of the bandwidth of an optimal linear and nonlinear device is then presented. One major drawback with the nonlinear device is the presence of a low power stable solution. A potential method of forcing the nonlinear harvester response onto the desirable high power solution is presented.

2. Optimal load for a nonlinear harvester

For energy harvesters the efficiency of the device is heavily influenced by characteristics of the load it is powering, as this governs the current and hence the force that is reflected back on the mechanical oscillator.

2.1. Optimal resistive load

Considering a purely resistive load, for a linear harvester featuring an electromagnetic transducer, it has been shown that maximum power is delivered to the load when the energy absorbed in the electrical side is equal to that dissipated mechanically. This is not necessarily true for a nonlinear device, where the response often depends on the input excitation and initial conditions [12, 13]. In addition, the frequency response is highly influenced by the forcing and dissipation levels. Here we consider a hardening nonlinearity with the compliance governed by $F_e = kx + k_{nl}x^3$, where x is the displacement of the oscillator. Specifically we set $k = 300N/m$, k_{nl}=1.02×10^8N/m^3, the mechanical damping to be 4.8Ns/m and θ=8.9Vs/m. A frequency response of the harvester in terms of the relative displacement between the coil and stator is shown in figure 1(a) for the case where the sinusoidal input excitation has amplitude 1.75 μm. When the harvester is excited open circuit, such that only mechanical damping losses occur, the displacement of the oscillator reaches its maximum. With a 30Ω resistive load, the additional electrical energy losses result in a reduced amplitude of the response. Note that the maximum amplitude is reached at different frequencies for different electrical loads. This implies that the optimal load resistance varies with the characteristics of the harvester a also with the amplitude and frequency of excitation. This has been shown in [11], where an approximated expression for the optimal load was found to be

$$R_L = \frac{\theta^2}{\dfrac{m\, Y_0\, \Omega_t\, \sqrt{3\,\alpha}}{2\,\sqrt{\Omega_t^2 - \omega_n^2}} - c_m}, \tag{1}$$

where θ is the electromechanical coupling coefficient, c_m is the mechanical damping coefficient, m is the moving mass of the oscillator, ω_n is the underlying linear natural frequency and $\alpha = k_{nl}/m$. Y_0 is the amplitude of the sinusoidal base input. The frequency Ω_t is the tuning frequency (the frequency at which maximum power is harvested). This frequency varies with resistive load as shown in figure 1(b). Note that at the two limits, open and short circuit, the power delivered to the load is zero.

2.2. Optimal linear natural frequency

To widen the range in which the harvester can be tuned to the frequency of excitation, without decrease the power harvested, one possible solution is to change the underlying natural resonant frequency. Using the optimal resistance, equation 1, the optimal natural frequency and the

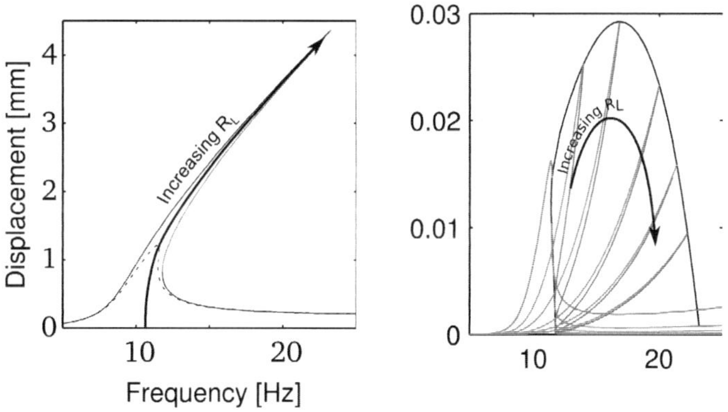

Figure 1. Frequency response of the nonlinear energy harvester in terms of relative displacement between the coil and the stator (a) and of the power delivered to the load (b). In panel (a) the displacement occurring open circuit (solid line) and powering a resistance of 30 Ω (dashed line) is shown. The red lines are representative of unstable solutions. Panel (b) shows the power vs frequency response of the harvester powering different load resistances.

optimal power can be found to be

$$\omega_n^2 = \Omega_t^2 - \frac{3}{16}\frac{Y_0^2\alpha}{c_m^2}; \qquad P_{opt} = \frac{1}{8}\frac{\Omega_t^2 Y_o^2 m}{c_m}. \tag{2}$$

Note that the maximum power does not depend on the nonlinear coefficient whereas the optimal linear natural frequency is a function of both the nonlinear coefficient and the amplitude of excitation. Interestingly, the maximum power delivered by a tunable nonlinear harvester to the optimized load is the same produced by a tunable linear harvester. The power envelope

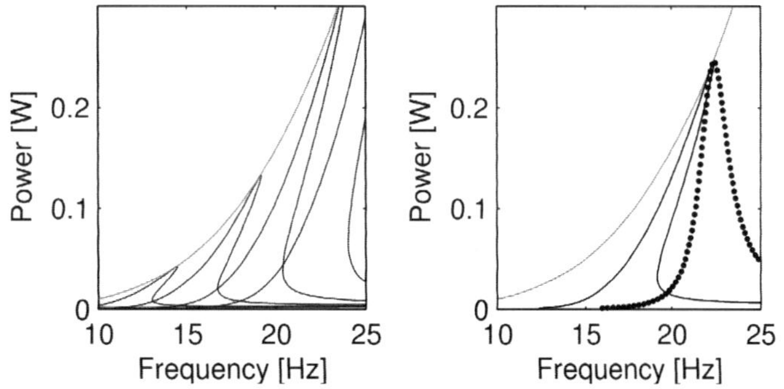

Figure 2. Maximum power with tunable harvesters: envelope of the power vs frequency curves tuned at different frequencies (a) and comparison between the response of a linear (bullets) and nonlinear (solid line) device tuned at the same frequency (b). In both panels the red curve represent the envelope of the maximum achievable power.

described by equation (2) is plotted in red in figure 2, where the amount of power delivered to the load are the same for the linear and nonlinear devices, however for the nonlinear device the underlying linear frequency is significantly smaller than ω_t.

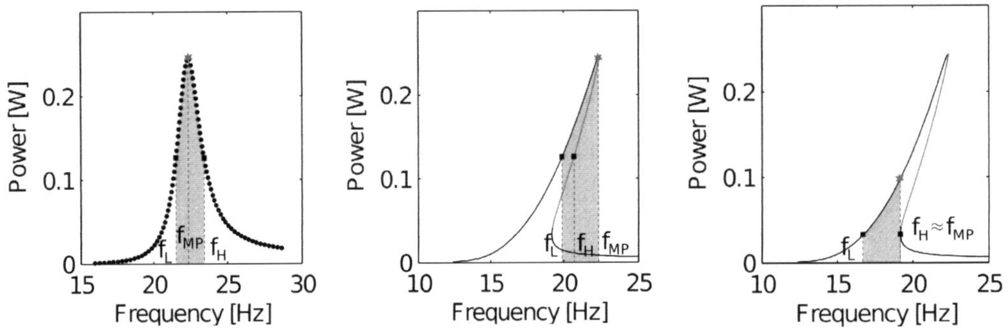

Figure 3. 3dB-bandwidth for the linear (a) and nonlinear (b) device. The linear power vs frequency curve is drawn with black bullets. The nonlinear stable oscillations are drawn with a black solid line whereas the unstable branches are drawn in red. Panel (c) gives a graphical representation of the MPUS definition. The maximum power is marked with a red star and the half power with black squares.

3. Bandwidth: definition and comparison

In the previous section we demonstrated that, when optimally tuned for maximum power extraction, the linear and nonlinear harvester generate the same amount of power. We now consider the bandwidth of the optimized linear and nonlinear harvesters. Firstly, the classical definition of 3dB-bandwidth is extended to nonlinear harvester, where the frequency response is strongly asymmetric about the maximum power.

3.1. Bandwidth definition for nonlinear harvesters

The definition of 3dB-bandwidth requires that the frequencies at which half the maximum power is delivered to the load are found. Due to the near-symmetry of the resonant peak of the linear device, see figure 3(a), these frequencies, Ω_L (or f_L) and Ω_H, span the resonant frequency giving the bandwidth $[\Omega_L : \Omega_H]$.

In contrast, for the nonlinear device, figure 3(b), $\Omega_L < \Omega_H < \Omega_{MP}$ hence the maximum power occurs at one extreme of the bandwidth $[\Omega_L : \Omega_{MP}]$. Note that over a portion of the bandwidth of the nonlinear energy harvester more than one stable solution exists. This definition of bandwidth does not take into account that the response could be on the lower branch. An alternative definition of bandwidth is to limit it to the region where only one solution exists is possible [14], which we term MPUS-bandwidth (Maximum Power Unique Solution - bandwidth), see figure 3(c). Note that as for the previous definition, the maximum power is delivered to the load at one edge of the bandwidth but in addition the maximum power achieve with this bandwidth is lower than the maximum achievable power and hence lower than the maximum power harvested with the equivalent linear device.

3.2. Comparison with linear devices

Figure 4(a) compares the bandwidth of the optimal linear and nonlinear devices; however; this does not tell the full story due to the presence of lower amplitude solutions within the bandwidth of the nonlinear device. To assess this we consider the MPUS-bandwidth definition. Here a comparison with the linear device bandwidth in figure 4(a) would be unfair, since the linear harvester produces more power than the nonlinear device. For this reason the MPUS-bandwidth is compared with the bandwidth of a linear device where the mechanical and the electrical damping have been increased so that its maximum power is the same as the maximum

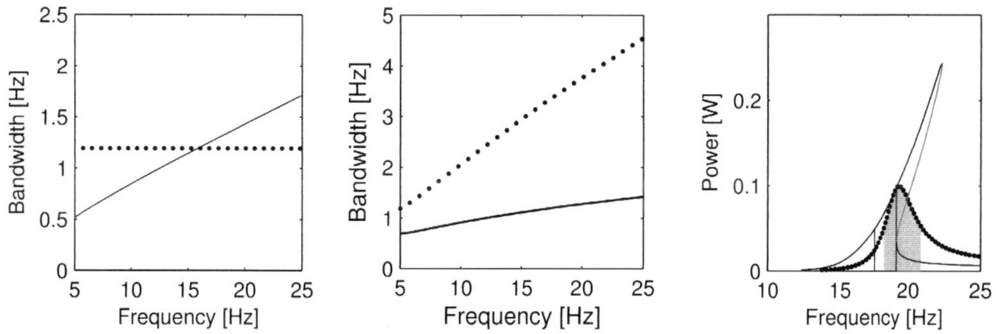

Figure 4. Comparison between the bandwidth of the linear (dots) and the nonlinear (solid line) device. The bandwidth of the response in of the linear and nonlinear devices are shown in (a) for the case where the nonlinear bandwidth in figure 3(b) is used. A comparison using the MPUS-bandwidth, figure 3(c), is shown in (b) for the case where the linear harvester has been tuned to achieve the same maximum power as the nonlinear harvester at the same frequency, as shown in (c).

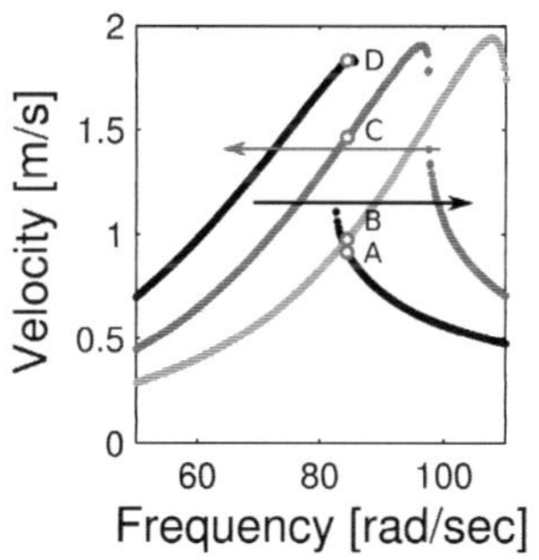

Figure 5. The effect on the frequency response of a nonlinear device when subjected to modification of the underlying linear natural frequency ω_n

P_{MPUS} (see figure 4(b)). In this case the linear device out-performs the nonlinear device in terms of bandwidth over the whole range of tuning frequencies. Hence, to make use of a wider bandwidth via nonlinearity, it is necessary to operate in the region where multiple solutions exist.

4. Ensuring a higher solution response

As shown in the previous section, a nonlinear harvester offers real advantages over the linear device only if it is possible to exploit the region in which multiple solutions exist. In order to operate in this region, a controller that induces the device to respond with high amplitude oscillations is required. Here, a possible method to force the response of the nonlinear harvester on the high energy branch is suggested. The method is based on the observation that at low frequency only one solution exists and that if the frequency is slowly increased from this point, the solution climbs the high energy solution. As the excitation frequency cannot be controlled, the suggestion here is to shift the underlying natural frequency instead.

This is shown graphically in figure 5. Consider the black line as the normal operation response of the device, which is tuned to operate close to 82rad/s. If the response drops, due to a perturbation, onto the lower branch then underlying natural frequency is adjusted upwards such that the gray response is achieved. Here there is only one solution at the excitation frequency. Now the natural frequency is slowly reduced to its normal value, hence the curve moving back towards and onto the black curve. During this reduction in the natural frequency the device response 'climbs' the response curve back to the desired operation point.

5. CONCLUSIONS

In this work a comparison between optimized linear and nonlinear energy harvester is provided. It is discussed that in the case of tunable harvesters, a linear and nonlinear device deliver the same maximum power to the optimal resistive load. Using this result, the bandwidth of optimally tuned linear and nonlinear harvesters are compared. The results showed that there exists a specific frequency range over which the nonlinear harvester has a wider bandwidth than the linear one. However it requires that the device operating is a regime where a lower response is also possible. Removing this possibility by restricting operation to regions where only one solution exists results in a narrower bandwidth for the nonlinear device.

Considering the importance of operating in the multiple solution region, for the nonlinear device, a possible strategy to force the harvester to response with high amplitude oscillations is suggested. Although this is presented as a preliminary study, the results achieved are encouraging. In future work we will discuss a mechanism for implementing the linear frequency shifting and a complementary control strategy to ensure the harvester stays on the upper branch.

References

[1] J. Baker, S. Roundy, and P. Wright. Alternative geometries for increasing power density in vibration energy scavenging for wireless sensor networks. In *3rd International Energy Conversion Engineering Conference. San Francisco, CA, USA*, 2005.

[2] V.R. Challa, MG Prasad, Y. Shi, and F.T. Fisher. A vibration energy harvesting device with bidirectional resonance frequency tunability. *Smart Materials and Structures*, 17(1), 2008.

[3] S. Roundy and Y. Zhang. Toward self-tuning adaptive vibration-based microgenerators. In *Proceedings of SPIE*, volume 5649. SPIE, 2005.

[4] H. Wu, L. Tang, Y. Yang, and C.K. Soh. A novel two-degrees-of-freedom piezoelectric energy harvester. *Journal of Intelligent Material Systems and Structures*, 24(3), 2013.

[5] A. Cammarano, S.G. Burrow, D.A.W. Barton, A. Carrella, and L.R. Clare. Tuning a resonant energy harvester using a generalized electrical load. *Smart Materials and Structures*, 19, 2010.

[6] P.D. Mitcheson, T.T. Toh, K.H. Wong, S.G. Burrow, and A.S. Holmes. Tuning the resonant frequency and damping of an electromagnetic energy harvester using power electronics. *Circuits and Systems II: Express Briefs, IEEE Transactions on*, 58(12), 2011.

[7] B.P. Mann and N.D. Sims. Energy harvesting from the nonlinear oscillations of magnetic levitation. *Journal of Sound and Vibration*, 319(1), 2008.

[8] S.G. Burrow and L.R. Clare. A resonant generator with non-linear compliance for energy harvesting in high vibrational environments. In *Electric Machines & Drives Conference, 2007. IEMDC'07. IEEE International*, volume 1. IEEE, 2007.

[9] S.C. Stanton, C.C. McGehee, and B.P. Mann. Reversible hysteresis for broadband magnetopiezoelastic energy harvesting. *Applied Physics Letters*, 95(17), 2009.

[10] A. Cammarano, S.G. Burrow, and D.A.W. Barton. Modelling and experimental characterization of an energy harvester with bi-stable compliance characteristics. *Proceedings of the Institution of Mechanical Engineers, Part I: Journal of Systems and Control Engineering*, 225(4), 2011.

[11] A. Cammarano, S.A. Neild, S.G. Burrow, D.J. Wagg, and D.J. Inman. Optimum resistive load for vibration-based electromagnetic energy harvesters with a stiffening nonlinearity. *Journal of Intelligent Material Systems and Structures*, 2013, in press.

[12] Francis C Moon. *Chaotic and fractal dynamics*. Wiley. com, 2008.

[13] John Guckenheimer and Philip Holmes. Nonlinear oscillations, dynamical systems, and bifurcations of vector fields. 1983.

[14] A. Cammarano, S.A. Neild, D.J. Inman, and D.J. Burrow. The bandwidth of optimized nonlinear vibration-based energy harvesters. *Smart Material and Structures*, 2013, in review.

Planar SiC MEMS flame ionization sensor for in-engine monitoring

D A Rolfe[1] S Wodin-Schwartz[1], R Alonso[1], and A P Pisano[2]

[1]Department of Mechanical Engineering, University of California, Berkeley, 5101 Etcheverry Hall, Berkeley, California 94720, USA
[2]Professor of ME and ECE, Dean of Engineering, University of California, San Diego, 9500 Gilman Drive, La Jolla, CA, USA 92093-0403

Email: rolfe@berkeley.edu

Abstract. A novel planar silicon carbide (SiC) MEMS flame ionization sensor was developed, fabricated and tested to measure the presence of a flame from the surface of an engine or other cooled surface while withstanding the high temperature and soot of a combustion environment. Silicon carbide, a ceramic semiconductor, was chosen as the sensor material because it has low surface energy and excellent mechanical and electrical properties at high temperatures. The sensor measures the conductivity of scattered charge carriers in the flame's quenching layer. This allows for flame detection, even when the sensor is situated several millimetres from the flame region. The sensor has been shown to detect the ionization of premixed methane and butane flames in a wide temperature range starting from room temperature. The sensors can measure both the flame chemi-ionization and the deposition of water vapour on the sensor surface. The width and speed of a premixed methane laminar flame front were measured with a series of two sensors fabricated on a single die. This research points to the feasibility of using either single sensors or arrays in internal combustion engine cylinders to optimize engine performance, or for using sensors to monitor flame stability in gas turbine applications.

1. Introduction

Hydrocarbon combustion is still used nearly-ubiquitously in transportation, heating and power generation. Though it is inexpensive and convenient, hydrocarbon combustion is the primary generator of CO_2, NO_x, and SO_x: leading contributors to global warming, smog, and acid rain. Though an ultimate solution may come from phasing out such fuels, a more immediate method to mitigate environmental harm may be to improve the efficiency of combustion reactions in internal combustion engines.

A flame is an ionic reaction in which fuel and oxygen are converted into carbon dioxide and water. Flame dynamics are heavily governed by the kinetics of the reaction, which have a highly sensitive dependence on temperature and the initial concentration of the hydrocarbon fuel and oxidizer. Because of this dependence, flame speed can be used to determine engine conditions such as pressure, temperature and air-fuel ratio.

Flame ionization sensors were first developed in 1920, when MacKenze and Honaman installed two ionization probes on an engine's cylinder head [1]. With the advent of computerized data logging, flame ionization sensors became capable of monitoring combustion in engines on a cycle-by-cycle basis. In the 1970s, spark plug-based sensors, which used a non-firing spark plug with a bias voltage on the order of 100V to detect flame conductivity, began being used in engines [2]. In the 80s and 90s, Douaud (1983), Witze (1993) and Meyers (1993) among others began experimenting with multiple ionization sensors, placed on the gasket head of the engine [1]. Though these sensors provided spatial data including swirl and flame speed, they could not survive for more than a few hours in the harsh environment of an internal combustion engine.

Silicon carbide (SiC) offers an approach for a sensor capable of surviving the heat and soot of an engine cylinder. SiC does not melt, but has a sublimation point of 2830°C [3], and is a functional semiconductor at temperatures up to 800°C [4]. When kept at 300°C on the wall of an internal combustion engine, SiC has been shown to develop an electrically insulating oily residue layer at roughly half the rate of silicon due to its low surface energy [5]. In the past, SiC devices such as pressure sensors [6] have been tested successfully in internal combustion engines.

2. Flame conductivity

A flame, like any plasma, is conductive. When a bias voltage is introduced, current is carried by the charged species of the flame reaction. Electrons are particularly mobile because of their low mass and high mobility. The current density of a flame is given by the equation [7]:

$$\boldsymbol{j} = -e \sum_k Z_k n_k \boldsymbol{v}_{Dk}$$

where n_k, Z_k, and $\boldsymbol{v}_{Dk}$ are the number density, charge, and drift velocity, of an ion species, k. For a flame where the relative concentrations of all constituent ions are invariant to the rate and size of the combustion reaction, the current across a flame ionization sensor is proportional to the concentration of ions in the flame.

Chemi-ionization (as discussed above) does not occur in the quenching layer that exists between a flame and a cold surface (such as an engine wall), but electrons and other species scattered out of the flame front can still conduct current some distance away from the flame. Franke proposed a two-zone model for flame ionization conduction shown in Figure 1 [7].

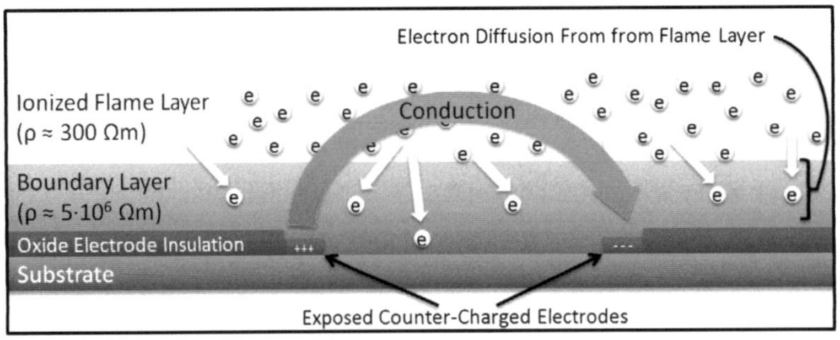

Figure 1. A schematic of the conduction mechanism of a flame ionization sensor. The flame travels from one electrode, through the flame and boundary layer, toward the other electrode. The boundary layer is roughly four orders of magnitude more resistive than the flame layer, though this varies based on boundary layer thickness.

Because the quenching layer is much less conductive than the flame itself, the flame ionization sensor response can be highly sensitive to the quenching distance. When the quenching distance is much larger than the gap of the flame ionization sensor, conduction occurs primarily though the quenching layer. When the quenching distance is much smaller, current will travel through the flame (chemi-ionization) layer and conductivity improves significantly. This is shown in Figures 2 and 3.

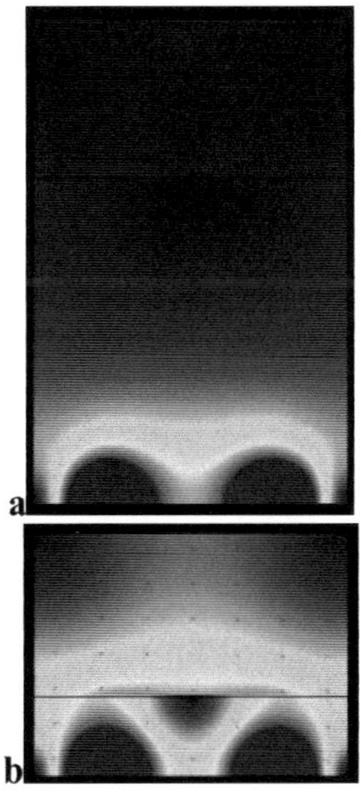

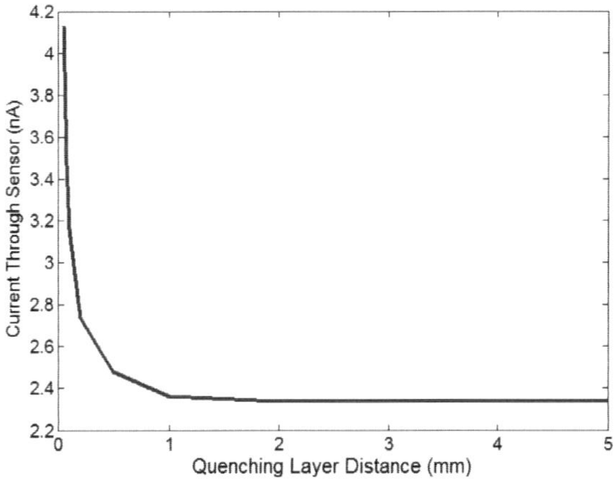

Figure 2. FEM current flux map for a quenching layer (a) twice the size of the electrode gap, with most current in the quenching layer and (b) half the size of the electrode gap, with most current in the flame.

Figure 3. Plot of current from finite element models like those shown in Figure 2. The plot shows a sharp increase in conductivity as the quenching layer becomes smaller than the sensor gap size. This is caused by the increased current though the flame layer, like the upper conduction layer in Figure 2(b).

In an internal combustion engine, quenching distances vary from 100μm to 300μm [8], and for a wall at room temperature, a quenching layer of 5mm can be expected [9] Since all the electrode gap sizes were smaller than the quenching distance, the flame sensor response was not very dependent on the electrode gap size. This has been corroborated by both experimental response (as seen in Section 4) and the experimental results of Witze [1]. The quenching layer could be shortened potentially, if the sensor were heated, as discussed in Section 5.

In the future, the sensor could be constructed with a vacuum insulation layer directly below the sensor's surface. An insulation layer as thin as 500μm could raise the temperature of the sensor 300°C, increasing the sensor's sensitivity and further limiting soot formation on the sensor's surface.

3. Design, fabrication and testing

Flame ionization sensor arrays, consisting of six sensors of varying electrode gap sizes were fabricated on a silicon substrate using a lithography process. A 3μm oxide layer for was deposited using LPCVD on a 6in silicon wafer for electrical isolation. 3μm of SiC was then deposited in a 2-step (to reduce stress) LPCVD process. The SiC was etched using an oxide mask, and a layer of oxide was deposited and patterned to insulate the SiC leads leading from the wire-bonding pads to the electrode gaps. The pads themselves were made of platinum with a chromium adhesion layer. Electrode gaps ranging from 20μm to 4mm were evaluated as well as electrode sizes between 5000μm^2 and 52000μm^2.

As electron absorption from flames has been shown to cause signal interference in flame ionization sensors [10], great care was taken to keep any electrical connection isolated from the flame. The sensors were attached to a data acquisition system with a set of electrical probes covered in dielectric epoxy, as shown in Figure 4.The sensor response was measured as the drop in voltage across a 1 MΩ resistor in series with the flame ionization sensor. This voltage divider configuration was used so that the output voltage would intuitively correspond to flame conductivity.

An acrylic "boom tube" was used to expose the sensor to a laminar premixed methane flame. The flame was lit roughly 0.5m from the sensor, and the flame front traveled through the tube perpendicular to the sensor's surface.

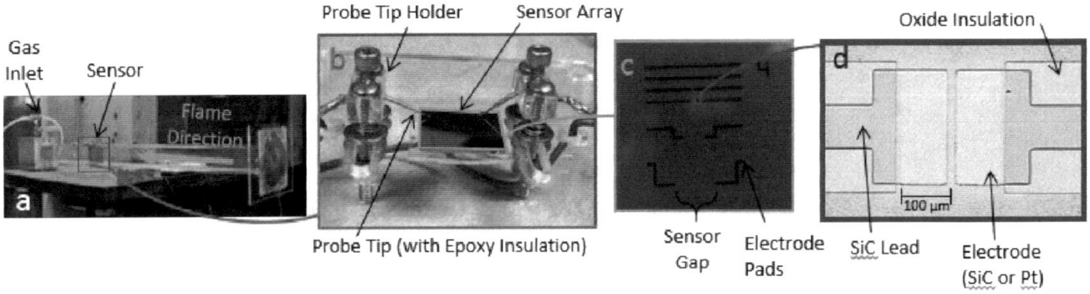

Figure 4. a) The sensor is mounted to the surface of an acrylic tube though which a laminar premixed flame propagates. It is connected to a data acquisition system via epoxy-insulated probe tips (b). Each die has six sets of sensors (c) that each consist of a pair of electrode gaps (d) connected to contact pads by SiC electrodes insulated with a layer of SiO_2.

4. Results and discussion

Flame ionization sensors with varying geometries were tested in the setup shown in Figure 4. Flame resistivity was on the order of 1MΩ. The sensor response was not sensitive to electrode size or gap size. The insensitivity to electrode size shows that at the quenching distances measured, the resistance of the bulk of the quenching layer is greater than the resistance across the electrode-quenching layer interface. The insensitivity to the electrode gap size is a result of large quenching distance (5mm) relative to the electrode gap of the sensor (as shown in Figures 2 and 3).

Water is a byproduct of any hydrocarbon combustion reaction, and at room temperature the sensor conductivity increases more from water deposition on the surface of the sensor after the flame passes than from the flame's chemi-ionization. This can be remedied by raising the temperature of the sensor, as shown in Figure 5. At 25°C, the flame ionization conductivity (highlighted in red on Figure 5) is almost imperceptible compared to the moisture conductivity. At 60°C, less water deposits on the sensor, and the chemi-ionization conductivity and moisture conductivity are at roughly the same magnitude. In an engine, the wall on which the sensor would be mounted would be roughly 300°C, and the conductivity of deposited moisture would be significantly limited and the quenching layer would be an order of magnitude smaller, increasing measured ionic conductivity significantly.

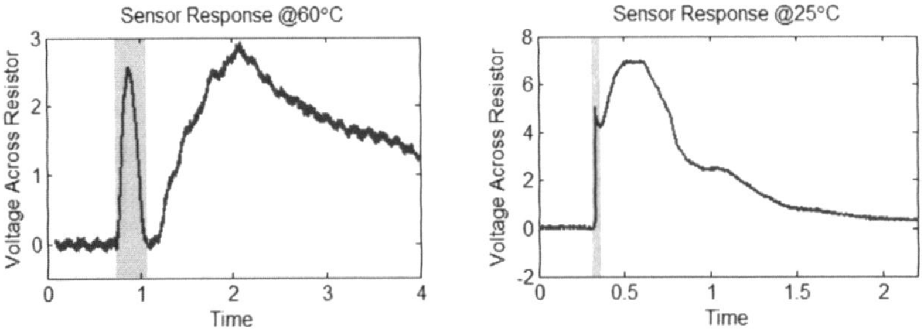

Figure 5. A comparison of the sensor response to a flame at two different temperatures, 60°C (left) and 25°C (right). At temperatures approach 100°C, the flame ionization peak (highlighted in red) becomes increasingly pronounced compared to the vapour deposition peak that follows it.

Flame speed, which is an important indicator of a variety of engine conditions (as described in Section 1) may be measured two different ways using this sensor. Global flame speed can be measured by determining the time it takes for a flame to propagate from the spark plug of an engine to a sensor at a known location. Local flame speed can be measured by two sensors mounted to a single die. Figure 6 shows an example of such a measurement with two sensors located 11.1 mm away from each other. Further measurements comparing sensor response to flame images are needed to determine the accuracy of this technique, but preliminary results show that the flame speed measured by the two-sensor arrangement is in agreement with the range of possible local flame speeds.

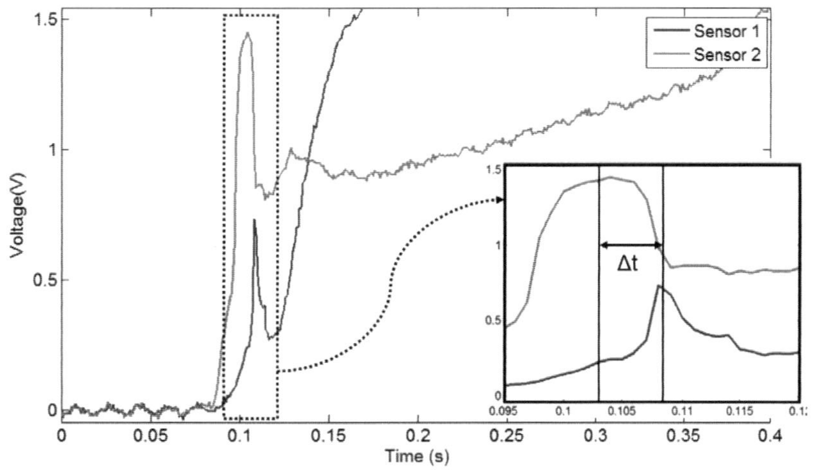

Figure 6. A preliminary demonstration of the response of two sensors (the top and bottom electrode pairs from Figure 4c) on the same die, measured simultaneously. A close up (right) shows how the time difference between the apexes of the ionization peaks may be used to measure the speed of the flame. The broader peak of Sensor 2 owes to the fact that Sensor 2 has an electrode gap 200 times the size of the electrode gap in Sensor 1.

5. Conclusions

Room temperature testing in a constant pressure laminar flame shows that microscale planar flame ionization sensors present a viable approach for measuring flame speeds, even when the electrode gap of the sensor is many times smaller than the quenching layer of the flame. In an engine environment, where quenching distances become two to five times smaller due to the increased temperature and pressure of the flame, which should increase the effectiveness of such a sensor. Though in-engine measurements would require new packaging and wiring, the sensors discussed in this paper should survive an engine environment longer than previous designs due to their SiC construction.

References

[1] Witze P O 1996 *In cylinder diagnositcs for production spark ignition engines Unsteady Combustion* ed Culick F et al (New York: Springer) chapter 15 pp 333-368
[2] Eriksson L and Nielsen L 1997 *Control Eng. Practice* **5** pp 1077-1113
[3] Aylward G H et al *SI chemical data 4th ed* (New York: Wiley)
[4] Neudeck P G 2006 *The VLSI handbook* ed Chen W (Boca Raton: CRC Press) chapter 5
[5] Wodin-Schwartz S et al *Proc. PowerMEMS 2008 (Sendai)* (London: Imperial College)
[6] Chen L and Mehregany M 2008 *Sensors and Actuators A: Physical* **145-146** pp 2-8
[7] Franke A 2002 Characterization of an electrical sensor for combustion diagnostics *Lund Reports Combustion Physics* **80**
[8] Daniel W A 1957 *Proc. Int. Symposium on Combustion* **6** pp 886-894
[9] McAllister S et al 2011 *Fundamentals of Combustion Processes* (New York: Springer)
[10] Xie R et al 2009 *Proc. US Nat. Combustion Meeting* **6** pp10-13

Patternable Temperature Sensitive Paint using $Eu(TTA)_3$ for the Micro Thermal Imaging

T Tsukamoto[1] and S Tanaka[2]

[1]Micro System Integration Center, Tohoku University, 6-6-01 Aoba Aza Aramaki Aoba-ku Sendai, Miyagi, Japan
[2]Department of Bioengineering and Robotics, Tohoku University, 6-6-01 Aoba Aza Aramaki Aoba-ku Sendai, Miyagi, Japan

E-mail: t_tsuka@mems.mech.tohoku.ac.jp

Abstract. This paper reports a novel patternable temperature sensitive paint (PTSP) which can be photolithographically patterned. The proposed PTSP consists of europium (III) thenoyltrifluoroacetone ($Eu(TTA)_3$) as a luminescent material and SU-8 as a photosensitive matrix. The optimal patterning condition was investigated and the PTSP microstructures as small as 2 μm were successfully patterned. The luminescence from the micro-patterned PTSP was observed and the temperature sensitivity was measured as high as $-0.8\%/^{\circ}$C.

1. Introduction

Micro thermal imaging is important for the evaluation of a microdevice such as power MEMS and LSI [1]. Because the heat capacity of such microdevices is usually quite small, a non-contact thermal sensing method is required. A far infrared thermography is widely used, but emissivity correction is needed for accurate temperature measurement. In many cases, however, both temperature and emissivity are unknown parameters, and thus precise temperature measurement by thermography is difficult. High speed measurement is often required due to the small thermal time constant. However, thermography using a bolometer is too slow to measure such a high speed phenomenon. In addition, high cost of far infrared optics usually made of Ge is limiting the application field.

A temperature sensitive paint is one of the candidates for high-speed and high-resolution thermal imaging [2–4]. The intensity of luminescence from the TSP is modulated by the temperature, therefore, the temperature distribution can be converted to an optical image. The wavelength of the luminescence is usually in a visible range, and thus a CCD or CMOS imager and a microscope for the visible wavelength can be used for imaging, which can drastically reduce the cost of the system. The lifetime of luminescence is of the order of sub-millisecond [5, 6], which promises a high temporal resolution [3].

Recently, we developed a flash exciting method using a slow-scan CCD camera for high-speed, high resolution and low-cost thermal imaging [4]. The TSP used in the system was manually patterned to open electric contact pads, and the patterning resolution was about a few millimeters. For the temperature measurement of microdevices, however, the micropatterning of the TSP is required. In addition, the micropatterning is essential for a novel low cost thermal imaging device made of TSP, in which the self-suspended TSP microstructures are heated by incident infrared light and their temperature is read out by an optical method [7].

The conventional TSP can be patterned by reactive ion etching (RIE) as shown in Fig. 1. First, the TSP is spin-deposited (Fig. 1 (a)), and then a protective material such as a parylene is deposited on it to protect the TSP from the organic solvent in a photoresist (Fig. 1 (b)). A metal thin film is deposited (Fig. 1 (c)), and the photoresist is patterned on it (Fig. 1 (d)). The metal layer is wet-etched by chemicals (Fig. 1 (e)), and the photoresist is removed. The parylene and TSP are etched by oxygen plasma RIE using the metal layer as an etching mask (Fig. 1 (f)). Finally, the metal layer is removed (Fig. 1 (g)). However, the fabrication process has some disadvantages: The fabrication steps to prepare and remove the etching metal mask are required. The TSP should be protected for the photolithography process, and the plasma process sometimes induces a damage in the TSP. Therefore, we proposed a novel patternable TSP (PTSP), which can be directly patterned by the photolighography technique.

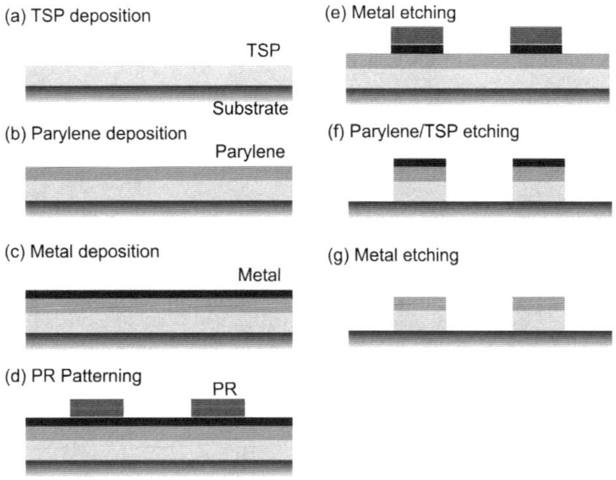

Figure 1. Patterning process of conventional TSP.

2. Principle of the patterning

The TSP consists of a luminescent material and a matrix. We chose europium (III) thenoyltrifluoroacetone ($Eu(TTA)_3$) as a luminescent material because of its high temperature sensitivity [5]. $Eu(TTA)_3$ is dispersed in the matrix (i.e. polymer) to make a film structure. The matrix should be transparent for both excitation UV light (350 nm) and luminescence (610 nm). Polymethylmethacrylate (PMMA) or polyvinylbutyral (PVB) are widely used as the matrix due to their high transparency in a region between near UV and visible wavelength [8, 9]. However, these materials cannot be applicable as the matrix of the PTSP, because they have no photosensitivity. Therefore, we chose SU-8 as the matrix, of which the transparency in near UV and visible wavelength region is excellent [10]. In addition, the protective layer as mentioned in the previous section is not needed due to the high chemical resistivity of SU-8.

Figure 2 illustrates the patterning principle of the PTSP. First, 0.03 mg of $Eu(TTA)_3$ (Thermo Fisher Scientific Inc., USA) was mixed with 1.0 mL of SU-8 (SU-8 2000.5, MicroChem Corp., USA) to make a PTSP solution. The prepared PTSP solution was spun-deposited on a substrate, baked at 95°C for 1 minute to vaporize the solvent (Fig. 2 (a)), and then exposed to UV light through a photomask (Fig. 2 (b)). Subsequent post exposure baking (PEB) at 95°C activated a cross-linking reaction in the exposed parts (Fig. 2 (c)). Finally, unexposed SU-8 was removed by the developer (Fig. 2 (d)). $Eu(TTA)_3$ in the unexposed SU-8 was also removed by the developper, while that in the exposed area remained in the SU-8 film.

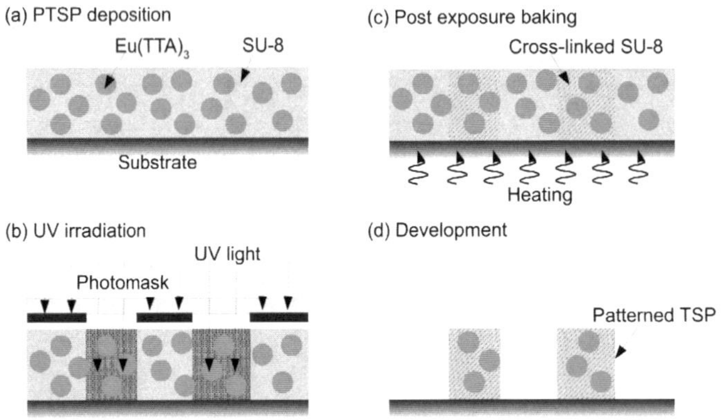

Figure 2. Photolithographic patterning principle of TSP.

3. Patterning conditions

The wavelength used for photolithography was 365 nm, which is close to that of $Eu(TTA)_3$ excitation (350 nm). Therefore, the exposure dose should be higher than that of the normal SU-8 process. In addition, the PEB condition should be also modified, because the cross-linking reaction is inhibited by $Eu(TTA)_3$. These two parameters are critical for successful patterning and need to be optimized. Table 1 shows the experimental conditions for the photolithography of the PTSP. The exposure dose and the PEB time were chosen as variables, while the other parameters were fixed. Figure 3 shows the patterning results on a Si substrate, where the open circles and crosses show successful and failure conditions, respectively. When the PEB time was short, all patterns disappeared during development (Region (A) in Fig. 3). On the other hand, PEB time longer than about 15 minutes made the unexposed part insoluble in the developer (Region (B) in Fig. 3). Only when the exposure dose was higher than 2400 mJ/cm^2, there was a process window within PEB time between 5 and 15 minutes (Region (C) in Fig. 3).

Using the obtained process parameters, PTSP micro structures as small as 2 μm were successfully patterned as shown in Fig. 4. The film thickness of the PTSP was about 300 nm.

Table 1. Conditions for the photolighography of PTSP.

Procedure	Condition		(Normal SU-8 condition		)
Spincoating	3000 rpm	30 s	(3000 rpm	30 s	)
Soft baking	95°C	1 min	(95°C	1 min	)
Exposure Dose	800 $\sim$ 4000 mJ/cm^2		(60 mJ/cm^2		)
Post exposure baking	95°C	5 $\sim$ 32 min	(95°C	1 min	)

4. Luminescence property measurement

Figure 5 shows an experimental setup for measuring luminescence from the PTSP at different temperatures. The PTSP was patterned on a Si substrate, and placed on a hot plate with a device holder. The temperature of the substrate was measured by a thermocouple clamped together with the substrate. The PTSP was excited by a UV-LED (SSL-LXTO46355C, Lumex Inc., USA), and luminescence from the PTSP was captured by a cooled CCD camera (BU-50LN, Bitran Corp., Japan) through an optical filter to eliminate the excitation UV light. Figure 6 shows the captured luminescent image, where only the patterned area are bright due to the luminescence. Figure 7 shows the measured temperature dependency of the luminescence from the PTSP. The intensity is normalized to 1.0 at a room temperature. The intensity reduced with increasing the temperature, which is known as a thermal quenching [11]. The temperature coefficient of intensity (TCI) was measured as high as $-0.8\%/^{\circ}C$, which was smaller than the previous reported values for the normal TSP ($-1.2 \sim -2.0\%/^{\circ}C$) [4, 5].

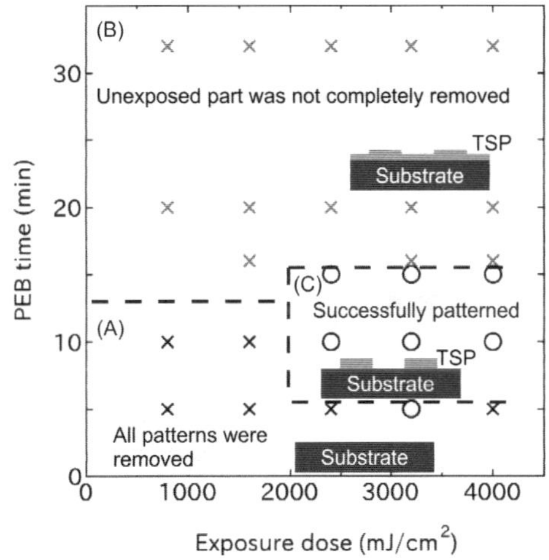

Figure 3. Process window of PTSP patterning. Open circles and crosses show successful and failure conditions, respectively.

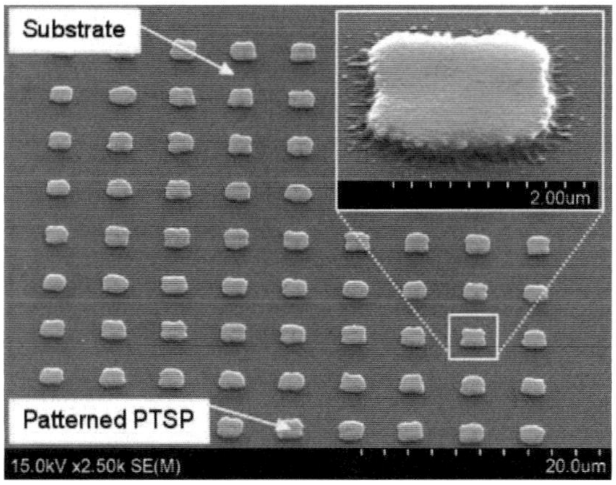

Figure 4. SEM of the patterned PTSP. Exposure dose and PEB time were 2400 mJ/cm^2 and 15 minutes, respectively

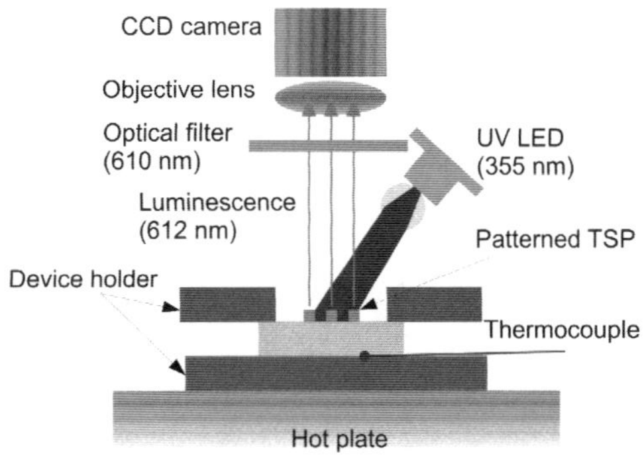

Figure 5. Experimental setup for characterizing the PTSP.

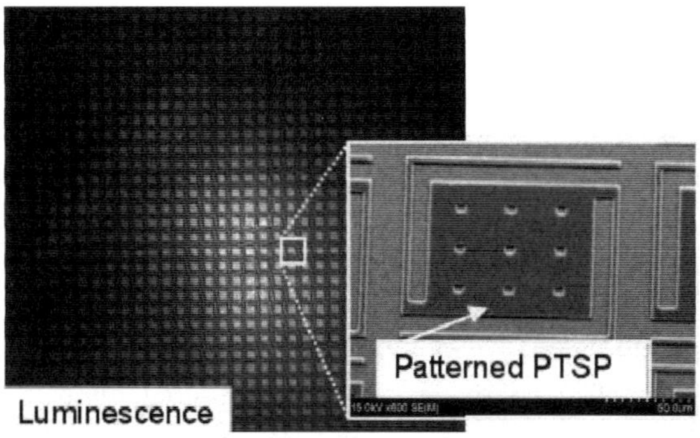

Figure 6. Captured luminescent image of the array of patterned PTSP shown in the inset SEM.

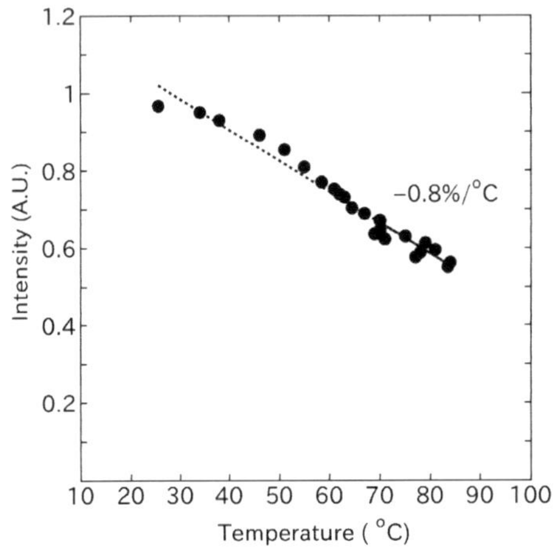

Figure 7. Temperature dependency of the luminescence.

5. Conclusion

We developed a patternable temperature sensitive paint (PTSP) using $Eu(TTA)_3$ and SU-8 as a luminescent material and a photosensitive matrix, respectively. The optimal patterning conditions were investigated, and the maximum resolution as small as 2 μm was demonstrated. The temperature sensitivity was measured as high as $-0.8\%/°C$. Using this technique, various microstrucutres of the TSP can be easily fabricated without plasma damage. Thus, the proposed PTSP can be applicable for a wide variety of micro thermal imaging applications.

References

[1] Kato F, Kikuchi K, Nakagawa H and Aoyagi M 2012 Hot spots suppression by high thermal conductivity film in thin-sub strate CMOS ICs for 3D integration *2011 IEEE International 3D Systems Integration Conference* (Osaka, Japan) pp 1–4
[2] Kolodner P and Tyson J A 1982 *Appl. Phys. Lett.* **40** 782–4
[3] Nagai H, Ohmi S, Asai K and Nakakita K 2008 *J. Thermophys. Heat Transfer* **22** 373–81
[4] Tsukamoto T, Esashi M and Tanaka S 2012 High spatial, temporal and temperature resolution thermal imaging method using $Eu(TTA)_3$ temperature sensitive paint *PowerMEMS 2012* (Atlanta, USA) pp 84–7
[5] Basu B B J and Vasantharajan N 2008 *J. Lumin.* **128** 1701–8
[6] Hadjichristov G B, Stanimirov S S, Stefanov I L and Petkov I K 2008 *Spectrochimica Acta Part A: Molecular and Biomolecular Spectroscopy* **69** 443–8
[7] Tsukamoto T, Esashi M and Tanaka S 2013 Infrared-to-visible transducer using temperature sensitive $Eu(TTA)_3$ on self-suspended thin film for inexpensive thermal imaging device *MEMS 2013* (Taipei, Taiwan) pp 421–4
[8] Zidan H M and Abu-Elnader M 2005 *Physica B: Condensed Matter* **355** 308–17
[9] French R H, Rodríguez-Parada J M, Yang M K, Derryberry R A and Pfeiffenberger N T 2011 *Solar Energy Materials and Solar Cells* **95** 2077–86
[10] Parida O P and Bhat N 2009 Characterization of optical properties of SU-8 and fabrication of optical components *International Conference on Optics and Photonics* (Chandigarh, India) PS3.E.8
[11] Khalil G E, Lau K, Phelan G D, Carlson B, Gouterman M, Callis J B and Dalton L R 2004 *Rev. Sci. Instrum.* **75** 192–206

Acknowledgement

This research was partly supported by Creation of Innovation Centers for Advanced Interdisciplinary Research Areas Program and MEXT Regional Innovation Strategy Support Program.

A Method for Microscale Combustion of Near Stoichiometric Energy Dense Liquid Fuel Mixtures

E D Tolmachoff, W R Allmon and C M Waits

United States Army Research Lab Sensors and Electronic Devices Directorate, 2800 Powder Mill Rd Adelphi, MD 20912, USA

E-mail: erik.d.tolmachoff.ctr@mail.mil

This paper reports on the potential of a heterogeneous/homogeneous (HH) reactor for use as a fuel-flexible heat source, meeting the needs of the next generation of high temperature thermal-to-electric (TEC) portable power converters. In this class of reactor, low activation energy catalytic reactions provide a means to stabilize high activation energy homogeneous reactions. Diffusion limited surface reactions play a critical role in HH reactor operation. Surface conversion must be sufficiently fast to generate the high temperatures (~1000 K) necessary to initiate gas phase reactions. Therefore, fuel diffusivity and the reactor dimension are important parameters in governing HH reactor operation. We examine the performance of an HH reactor fuelled by propane and n-dodecane, representing two extremes of liquid hydrocarbon diffusivity, as a function of confining reactor dimension. Unburned fuel/air mixtures are close to stoichiometric, which is an important factor in minimizing the amount of excess air and, therefore, balance of plant energy costs. At moderate levels of confinement, the reactor is capable producing high, uniform temperatures for both fuels.

1. Introduction

To date, most microcombustion based energy conversion devices have focused on available thermoelectric (TE) modules limiting, them to ~200-500 °C. As both TE and thermophotovoltaic (TPV) improve, efficient TEC energy conversion will require higher operating temperatures (800 to >1000 °C) [1]. Reaching these high temperatures requires a high mass flow rate through a small reactor. High flow rates limit residence time of the reacting flow and diminishing reactor size results in reaction quenching heat losses as the ratio of surface area to volume increases; both factors adversely affect fuel conversion, reactor stability and temperature. A HH flow reactor with catalyst-lined walls separated by small distance in order to tune reactor operation may provide a platform to fully oxidize fuel in very small reactors with large heat losses. Low activation energy catalytic reactions at the reactor's inner surface provide stability against heat loss while heating up the unburned gas to a point where significant homogeneous reactions occur. Homogeneous reactions are less energetic since the fuel content has been partially depleted by surface reaction; they consume fuel and CO that may otherwise escape a small residence time heterogeneous reactor, completing the conversion of chemical energy to heat while avoiding the extreme gas phase temperatures characteristic of ordinary flames ~2000 K. The net result is that HH reactors are capable improved fuel conversion and heat release over the length of the reactor [2-5].

It is well established in the field of gas phase combustion that while hydrocarbon reaction pathways are complex and fuel specific, overall fuel conversion rates are largely determined by reaction rates of propane and smaller species [6]. It follows that it may be possible to engineer a reactor so that a variety of fuels decompose in such a way that they yield similar distributions of these rate determining smaller species, and therefore, that the reactor behaves in similar ways with different fuels and is fuel-flexible. For fuel flexibility to be achieved this way in a HH reactor, the confining dimension between the catalytic walls would likely play a critical role. This distance, g, determines the characteristic transverse diffusion time scale ($\tau \propto g^2/D$ where D is the diffusivity). Transport to the catalytic surfaces must be fast enough to generate the high temperatures required to initiate gas phase reactions. Since diffusive transport plays such a large role in determining the stability limits of HH reactors [4], it is not immediately clear if there exists a gap such that a HH reactor can operate in a fuel flexible manner over a significant range of practical fuels. Here, we investigate the role of gap size on the performance of a HH reactor. Propane and dodecane (a jet fuel surrogate) were examined to assess how the reactor operates under conditions of light and heavy fuels.

2. Experimental

The reactor used in this study was designed after the reactor in [7]. It consists of two thin stainless steel plates (0.9 mm thick) separated by an interchangeable stainless steel gasket. Welded to the top plate are stainless steel tubes, one for reactant inlet and one for exhaust outlet. Inside the reactor chamber, stainless steel screens (250 μm openings) are placed immediately before and after the reactor section in order to straighten flow and prevent "jetting". In the middle, the reaction zone consists of catalyst coated alumina inserts separated by thin strips of fibrous alumina insulation. The length of the reaction zone is 5 cm, the width is 1cm and the gap between catalyst inserts is governed by the thickness of the gasket chosen. The assembly is held together by a series of screws around the perimeter. Inside, compressed alumina insulation maintains a snug fit between the catalytic inserts and the stainless steel walls. A photograph and schematic of the reactor are seen in figures 1 and 2.

The catalytic inserts are developed on solid thin amorphous alumina sheets (15x50x0.65 mm). A thin porous layer of γ-alumina nanoparticles are screen printed and annealed onto the solid alumina sheets. The inserts are masked and 0.2 ml of chloroplatinic acid (Aldrich, 8 wt. % in H_2O) is delivered on the porous layer. After drying, the loaded inserts are baked for 5 hours at 923 K to reduce the chloroplatinic acid into platinum. Each insert is loaded with 5 mg of platinum.

The unburned fuel/air stream is preheated to 423 K in order to vaporize the dodecane; the stream is also preheated when running propane for consistency. A butane torch is used to heat the reactor to a point where autothermal operation is possible. Six K type thermocouples welded along the centre of the bottom plate of the reactor are used to monitor reactor temperature. The thermocouples are placed in 1 cm intervals coinciding with the location of the catalytic insert. The reactor assembly is sandwiched between two 5 cm thick needled silica mats (AVSil AB100S) wrapped in aluminium tape. Here we examine equivalence ratio (φ=fuel/oxidizer)/(fuel/oxidizer)$_{stoichiometric}$) φ=0.9 and three gap heights (0.5, 1.0, 2.0 mm) with volumetric flow rates varying from 1.5 to 3 SLPM. The conditions chosen are power input rates, flow rates, and gap distances that are representative of values that would be present in a small scale portable power converter (Table 1).

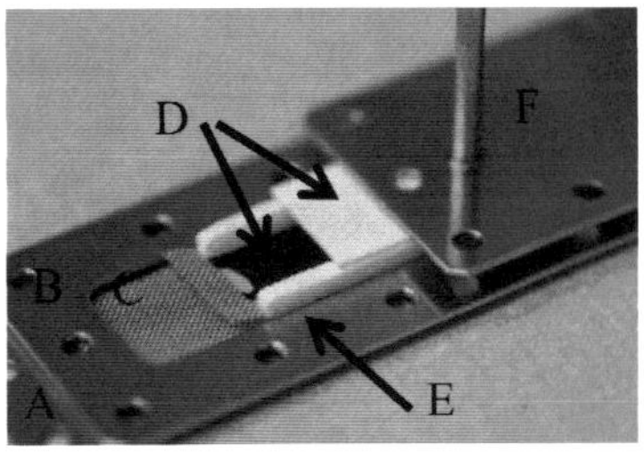

Figure 1. Disassembled HH reactor. A. Reactor bottom. B. Gasket controlling the gap size between reactive walls. C. Flow straightening mesh. D. Catalytic inserts. E. Alumina foam separating catalytic inserts. F. Reactor top.

$\dot{V}$ (SLPM)	$\dot{H}$ (W) dodecane	$\dot{H}$ (W) propane
3	156	169
2.5	130	141
2	104	113
1.5	78	85

Table 1. Summary of reactor conditions. The parameters chosen are representative of values expected to be present in a TPV based portable power conversion device.

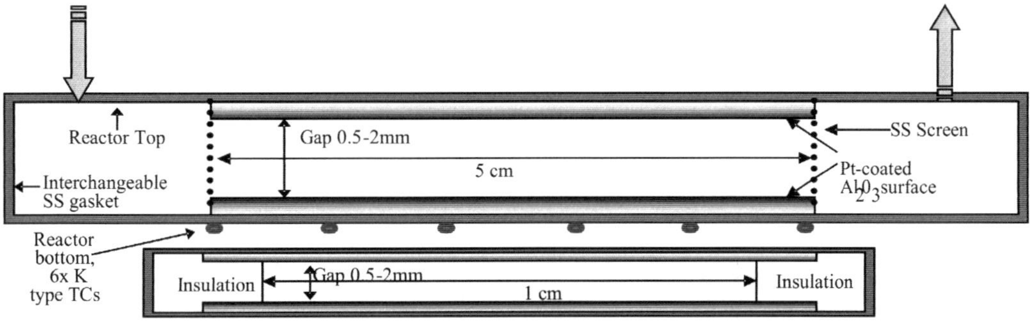

Figure 2. (Top, lengthwise cross section) fuel/air enters from the left, passing through a SS flow-straightening mesh, through the reaction region, finally passing through a SS mesh and out the reactor. Thermocouples record location-dependent temperature. The gap between the reactor walls is controlled by an interchangeable gasket. (Bottom, cross section of width, not to scale).

3. Results and Discussion

The results examined here are based primarily on the steady state location-dependent temperature of the reactor, though some transient behaviour is worth mentioning. For both fuels, transients approaching high steady state temperatures for the reactors with g = 1mm and 2 mm showed temperature discontinuities near the middle of the reactor, suggesting that flames were established nearby. Similarly, the transients from high to low steady states temperatures show downward temperature discontinuities, suggesting a flame inside the reactor was blown out (extinguished).

Each condition was run multiple times. The data is the average of 2-4 sets of experimental values with the error bars representing two standard deviations. At high temperatures uncertainty is rather large, however, it is comparable with similar systems [7] and has been attributed to deviations of both flow conditions and catalyst uniformity [8]. It should be noted that steady operation could not be achieved for any conditions without catalyst.

In spite of being well insulated, heat losses are significant and unavoidable for small scale reactors. Norton *et al.* [7] calculated a Biot number of ~100 for a thin walled reactor similar to the ones examined here. The same analysis yields Biot numbers >10 for the reactors in this study. Large Biot numbers are characteristic of systems in which the effects of conduction can be largely ignored. A CFD analysis of the reactor (not shown) confirmed this fact. Since the effects of heat conduction are minimized, wall temperatures are highly representative of the phenomena occurring locally. Since microcombustors cannot accommodate traditional combustion diagnostics and numerical modelling is computationally intractable for complex reacting flows, analysis of wall temperatures in a high Biot number HH reactor can provide valuable information about its operation. For example, it can be shown that, under conditions in which the heat releasing reactions are limited by mass diffusion of reactants to the reactor walls, temperature is flow rate independent. Similarly, a strong flow rate dependence on reactor temperature is evidence that the reactor is not transport limited and that a significant amount of homogeneous reactions occur.

Figure 3 shows the data for g=2 mm. Under the least confined conditions in this study, surface reactions play a smaller role and favour lighter species. For dodecane, reactor temperatures are essentially independent of flow rate, a characteristic of transport limited surface reactions. Fuel decomposition, which readily occurs at high temperatures, is not expected to be appreciable at the reactor temperatures observed for dodecane in Figure 3 [9]. With transport limited operation and minimal homogeneous reactions, reactor temperatures are low. Under propane operation, however, there is a strong dependence of reactor temperature on flow rate, with temperatures rising from 1.5 SLPM to 2.5 SLPM. The residence time dependent temperature behaviour suggests that gas phase reactions play an important role for propane. This is consistent with the understanding that homogeneous reactions are highly sensitive to fuel diffusivity as reactors become less confined [4]. The drop in reactor temperature seen for the flow rate of 3 SLPM, however, suggests that these homogeneous reactions are susceptible to blowout for short residence times. In the case of intermediate confinement g=1 mm (figure 4), reactor temperatures are in good agreement for both dodecane and propane. This is the desired behaviour for fuel flexible operation. While the experimental results are promising, modelling work is underway to try to understand the underlying mechanisms allowing this fuel flexible operation.

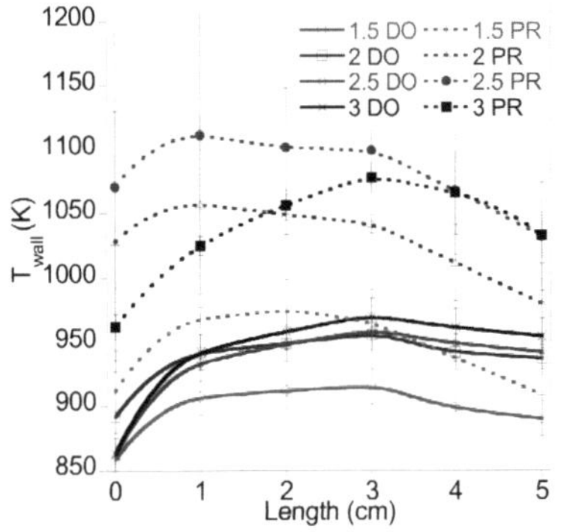

Figure 3. Results for g=2 mm. The reactor operating on dodecane (DO) is limited by transport of fuel species to the reactor wall, evidenced by the flow rate independent temperature profiles. Gas phase reactions play a significant role in the reactor operating on propane (PR). However, at flow rates above 2.5 SLPM, the gas does not have sufficient residence time to react and homogeneous reactions are blown out, leading lower temperatures.

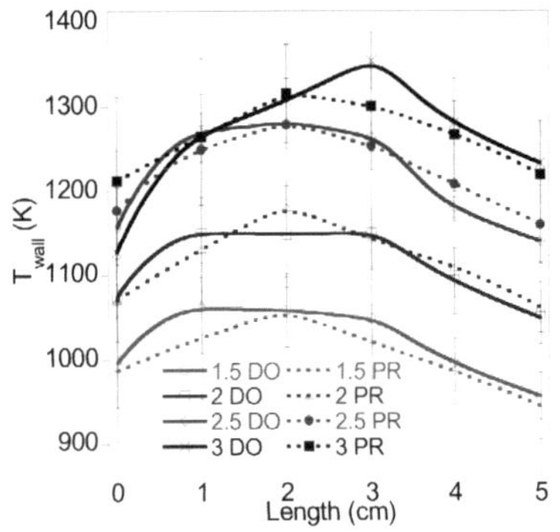

Figure 4. Results for g = 1 mm (DO = dodecane, PR = propane). At this gap size, temperature jumps in transients, occurring near the middle of the reactor, suggest flames can be sustained under certain conditions. The temperature characteristics are quite similar for the two fuels, though the underlying reason for this is not well understood at this time.

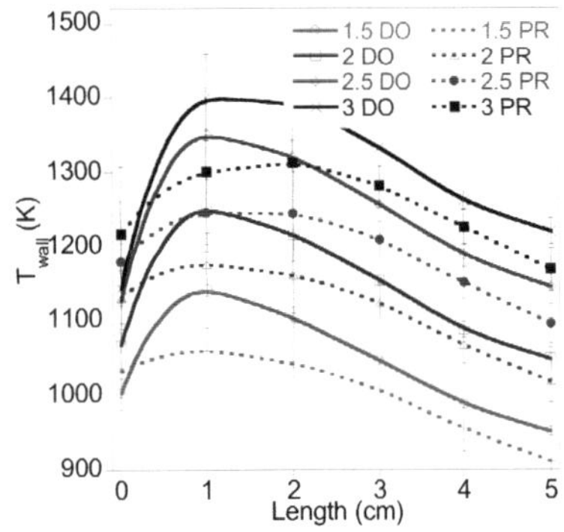

Figure 5. Results for g = 0.5 mm reactor. It is remarkable that propane (PR), the lighter fuel, results in lower peak and average temperatures. This suggests that, dodecane (DO) readily decomposed into smaller highly diffusive species near the reactor inlet, resulting in a rapid temperature rise.

The results for the g=0.5 mm reactor are shown in figure 5. It is remarkable that when the reactor is fuelled by dodecane, both the peak and mean temperatures are greater than when the reactor is fuelled by propane. Since propane is a smaller molecule with a greater diffusivity than dodecane, it would be expected that it is more readily transported to the reactor walls and, therefore, that it would lead to higher reactor temperatures. Modelling work on this reactor shows that a transport limited C_2H_4 (the most abundant intermediate of dodecane decomposition [9]) reaction can account for this rapid temperature rise near the reactor inlet. Although a heat source in a power conversion device would experience greater heat losses than those encountered in this work, the rapid coupling between surface and gas phase reaction and temperature gradients seen in both modelling and experiments can be detrimental to the performance and longevity of a heat source. Therefore, there may be a lower limit to level of confinement for a HH heat source to act in a way that is both fuel flexible and optimal.

4. Conclusions

HH reactors show fuel flexible characteristics, capable of maintaining similar reactor temperatures with dissimilar alkane fuels. The underlying reason for the observed fuel flexibility cannot be conclusively determined without detailed modeling or advanced diagnostics capable of seeing into the microcombustor, however, it is clear that the confining dimension in these reactors plays an important role. There is a fuel dependent upper limit on the confining dimension in HH reactor operation. Above this limit, fuel cannot diffuse quickly enough to the reactor walls to react and generate the temperatures necessary to sustain gas phase reactions. There may also be lower limit to this dimension, below which the heat generated by surface reactions causes large and slowly diffusing but unstable fuel species to decompose fast enough to generate undesirably high temperatures and temperature gradients.

Acknowledgements
This research was supported in part by an appointment to the U.S. Army Research Laboratory Postdoctoral Fellowship Program administered by the Oak Ridge Associated Universities through a cooperative agreement with the U.S. Army Research Laboratory.

References

[1] C. M. Waits, *Thermophotovoltaic Energy Conversion for Personal Power Sources*, Army Research Lab Adelphi Md Sensors and Electron Devices Directorate, 2012.
[2] G. D. Stefanidis, D. G. Vlachos, *Ind. Eng. Chem. Res.* 48 (2009) 5962-5968.
[3] S. Karagiannidis, J. Mantzaras, G. Jackson, K. Boulouchos, *Proc. Combust. Inst.* 31 (2007) 3309-3317.
[4] S. Karagiannidis, J. Mantzaras, K. Boulouchos, *Proc. Combust. Inst.* 33 (2011) 3241-3249.
[5] A. D. Benedetto, V. D. Sarli, G. Russo, *Catal. Today* 147 (2009) 156-161.
[6] C. K. Law, *Combustion Physics*, Cambridge University Press, 2006.
[7] D. G. Norton, E. D. Wetzel, D. G. Vlachos, *Ind. Eng. Chem. Res.*, 45 (2006) 76-84.
[8] D. G. Norton, E. D. Wetzel, D. G. Vlachos, *Ind. Eng. Chem. Res.*, 43 (2004) 4833-484.
[9] X. You, F. N. Egolfopoulos, H. Wang, *Proc. Combust. Inst.* 32 (2009) 403-410.

Experimental demonstration of a tailored-width microchannel heat exchanger configuration for uniform wall temperature

S Riera[1], J Barrau[1], J I Rosell[1], M Omri[2] and L G Fréchette[2]

1 University of Lleida, Solar Energy Group, Edifici CREA, C/Pere de Cabrera, s/n, 25001 Lleida, Spain

2 Université de Sherbrooke, Dépt génie mécanique, 2500 boul. Université, QC J1K 2R1, Canada

E-mail: sriera@macs.udl.cat

Abstract. In this work, an experimental study of a novel microfabricated heat sink configuration that tends to uniform the wall temperature, even with increasing flow temperature, is presented. The design consists of a series of microchannel sections with stepwise varying width. This scheme counteracts the flow temperature increase by reducing the local thermal resistance along the flow path. A test apparatus with uniform heat flux and distributed wall temperature measurements was developed for microchannel heat exchanger characterisation. The energy balance is checked and the temperature distribution is analysed for each test. The results show that the wall temperature decreases slightly along the flow path while the fluid temperature increases, highlighting the strong impact of this approach. For a flow rate of 16 ml/s, the mean thermal resistance of the heat sink is $2{,}35\cdot10^{-5}$ $m^2\cdot K/W$ which enhances the results compared to the millimeter scale channels nearly three-fold. For the same flow rate and a heat flux of 50 W/cm^2, the temperature uniformity, expressed as the standard deviation of the wall temperature, is around 6 ºC.

1. Introduction

Solar cell arrays need to maintain acceptable operating temperatures and temperature uniformity to maintain the reliability, which implies dissipating high heat fluxes from small areas. That demands efficient and innovative cooling systems with high capacities for heat removal. Within the cooling technologies that have high heat flux removal capacities, microchannels are one of the most reliable and commonly used.

Microchannel heat sinks were first suggested by Tuckerman and Pease [1]. The improvement of the cooling performance is based on the fact that the convective heat transfer coefficient scales inversely with channel width. The major drawbacks of this cooling scheme are the temperature increase in the direction of the fluid flow and the relatively large pressure drop that they generate. In order to reduce the temperature non uniformity of the cooled object, Ryu et al. [2] proposed to alternate inlets and outlets along the microchannels, adding some manifolds dividers. Missagia and Walpole [3] reduce temperature non-uniformity by alternating flow directions through the channels in a single layer, while Vafai and Zhu [4] proposed a two-layered microchannel structure with opposite directions flows. Lee et al. [5] enhances the heat transfer performance by implementing oblique fins into a microchannel.

Along these lines, Barrau et al. [6] developed a new hybrid jet impingement/channel cooling scheme. In this design the fluid enters through the slot located in the symmetry plane of the heat sink and travels along the channels to leaves the heat sink at its extremities. The fluid flows through a series of microchannels with variable width along the flow path. As the fluid heats up, the smaller channel diameter decreases to consequently decrease the convective thermal resistance. The authors demonstrated the capacity to achieve temperature profiles of the target object that remain constant or even decrease in the fluid flow direction of the coolant, maintaining the global high heat removal capability. The numerical study [7] showed that the temperature distribution provided by the heat sink can be tailored according to the needs of the system by varying the local heat removal capability.

Some subsequent studies, such as that conducted by Karathanassis et al. [8], studied experimentally and numerically a hybrid system based on the design of Barrau et al. [6], but with a radial disposition of the channels. They confirmed that the introduction of stepwise varying channels increases the thermal performance of this heat-sink configuration without introducing a severe pressure drop penalty. The ability of this design to provide a quite uniform temperature profile of the cooled object has been confirmed by Ji et al [9]. Those last investigations display the enhancement of implementing variable channel width.

Those previous studies were done at the millimeter scale. But more and more, heat flux densities tend to increase, suggesting the use of smaller scale channels to increase the convective heat transfer rate. The current design is therefore implemented at the micrometer scale in order to evaluate the impact of the channel scale on the performance of the device, looking for the same effect of thermal resistance reduction at small scale. This work presents an experimental analysis of such a novel microfabricated heat exchanger configuration that tends to make the wall temperature uniform even under lateral flow with increasing flow temperature. This is the first demonstration of this approach in a Si microchannel heat exchanger.

2. Test module description

The schematic design of the experimental setup used for this study is shown in figure 1.

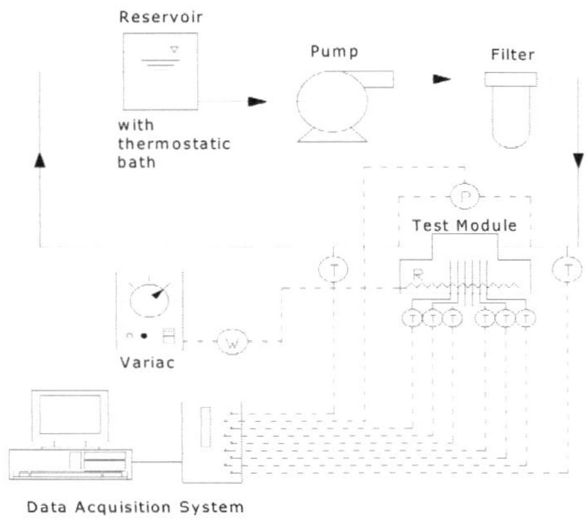

Figure 1. Test module setup.

The test module is cooled with water which is stored in a reservoir with a thermostatic bath to maintain a constant temperature. The water circulates in the loop with the aid of a variable speed peristaltic pump (JP Selecta PERCOM N-M), which can give a flow of 1 l/min. Then, the liquid passes through a 3-μm filter before entering the test module.

In the module, the heat flux needed to simulate the solar irradiation on a cell is generated by an advanced ceramic heater (Watlow Ultramic 600), which can give more than 100 W/cm^2.

This cooling scheme entails an optimized array of variable width microchannels (Fig. 2). The main objective of this optimized cooling scheme is to improve the temperature uniformity of the cooled object, while maintaining thermal resistance levels achieved in the previous works [6, 7] or less. The cooling scheme consists of a series of microchannel sections with stepwise varying width. The range of width of the microchannel sections was designed from 1.53 mm to 140 μm to increase the convective heat transfer coefficient and surface to compensate the increase in flow temperature. The microchannel pattern is etched in silicon by lithography and DRIE, creating a constant depth (300 μm) microchannel array. In figure 2 is shown the microchannel heat exchanger scheme.

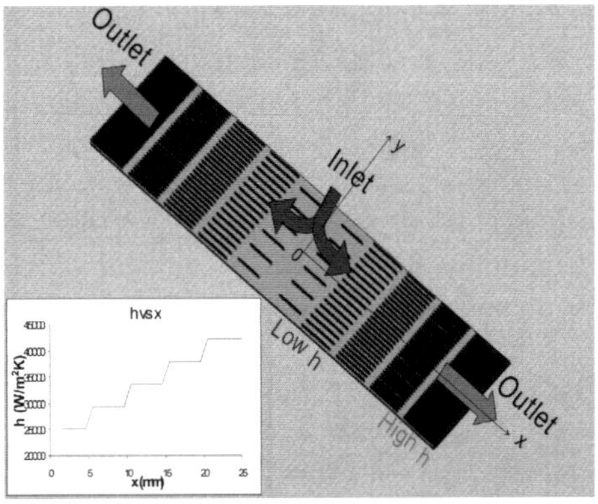

Figure 2. Microchannel heat exchanger scheme.

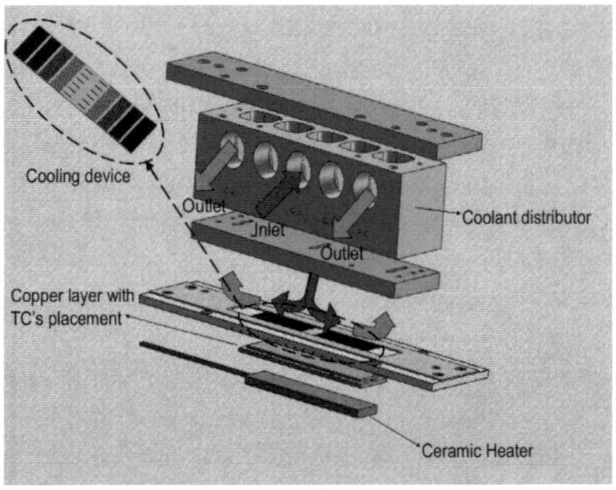

Figure 3. Disassembled test module.

The microchannel heat exchanger is sealed on the copper layer of the test module through a thin layer of Thermal Interface Material (TIM). The thickness of the TIM layer has been measured with a microscope (50 μm, ±5 μm) and its thermal conductivity has been verified (λ_{TIM}=0,82 W/m·K).

The module has one cooling liquid inlet. Water is guided directly to the microchannel heat exchanger center through a slot in its bottom. The flow is then divided in two, following the microchannels path to both ends of the module. Two outlets collect the liquid leaving the extremities of the device after passing through the microchannels.

Type-K thermocouples are used to measure the water temperature at the inlet and outlet of the microchannel heat exchanger and the temperature distribution of the copper layer along the flow path, at different positions (x) of the centerline (y=0 mm) of the microchannel heat exchanger. All this information is acquired by a datalogger Campbell CR23X and sent to a computer to be stored and analyzed. Finally the flow is channeled again to the reservoir with the thermostatic bath, closing the loop. Figure 3 shows a global scheme of the test module sited in the laboratory, disassembled.

3. Results and discussion

To study the temperature distribution and thermal resistance of the heat sink described previously, three experimental conditions are considered by varying the flow rate. Multiple tests were done in similar conditions to assure repeatability of the results, but only the conditions for three validated tests are listed in Table 1.

Table 1. Test conditions.

	TEST 1	TEST 2	TEST 3
Q (m³/s)	8,80E-06	1,12E-05	1,60E-05
q''_{eff} (W/cm²)	50,06	50,05	49,96
T_{in} (°C)	21,7	21,9	22,6

3.1. Temperature distribution and temperature uniformity

For each experiment, the energy balance is checked to validate the tests, and the temperature distribution is analyzed.

The temperature maps for the three tests in Table 1 are presented in Figure 4. As the distribution is symmetric with respect to the symmetry plane, the temperatures are displayed only for one side of the heat sink. It is clearly seen that even though the water temperature increases along the heat exchanger, the wall temperature is maintained nearly constant.

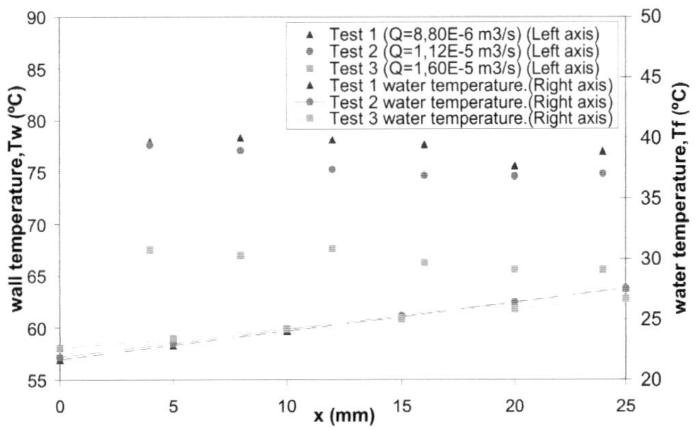

Figure 4. Temperature of the experimental microchannel heat exchanger for a heat flux of 50 W/cm^2, and several mass flow rates. The graph shows that near-uniform wall temperature (Tw) was achieved even though the water temperature (Tf) increases along the heat exchanger.

Table 2 presents the results of temperature uniformity at the height of the copper layer in the heat sink. To express the temperature uniformity, the standard deviation σ_T is calculated with respect to the average temperature and its difference to each data point. Included in the table, $\Delta T_{in\text{-}out}$ shows the difference between coolant temperature from inlet to outlet, $T_{i,m}$ the mean wall temperature at copper layer height and $\Delta T_{min\text{-}max}$ the maximum difference in the wall temperature.

Table 2. Temperature uniformity at copper layer height.

		TEST 1	TEST 2	TEST 3
$\mathbf{\Delta T_{in\text{-}out}}$	**(ºC)**	5,8	5,6	4,1
$\mathbf{T_{i,m}}$	**(ºC)**	76,3	83,4	74,3
$\mathbf{\Delta T_{min\text{-}max}}$	**(ºC)**	14,9	15,2	14,1
$\mathbf{\sigma_T}$	**(ºC)**	5,8	5,9	5,4

The fact that there's a decreasing temperature distribution shows that it would be possible to shape a uniform one, or tailor the temperature distribution to the user's need depending of different applications. That shows the effectiveness of the proposed variable microchannel width scheme design procedure.

3.2. *Thermal resistance coefficient*

The thermal resistance is a way of showing and comparing the performance of the module in cooling the desired surface. Calculation of the global average thermal resistance coefficient of the system is carried out by applying the following equation:

$$R_{t\,m} = \frac{(T_{w,m} - T_{in})}{q''_{eff}} \tag{1}$$

where q''_{eff} is the heat flux, T_{in} is the inlet water temperature and $T_{w,m}$ is the average temperature of the wall on which the heat flux is applied.

This global thermal resistance is ascribed to the heat sink, but also to the thermal interface material and, with a minor impact, to copper. This result must be corrected by removing the effect of the TIM and the copper layer that exists between the base of the microchannel heat exchanger and the thermocouples location.

The average thermal resistance coefficients for the four different tests conditions listed in Table 1 are calculated, subtracting the effect of the copper layer and TIM. The value obtained for a flow rate of $1{,}6 \cdot 10^{-5}$ m^3/s is $2{,}35 \cdot 10^{-5}$ $m^2 \cdot K/W$. These values are in concordance with the existing reports and literature on microchannel cooling, enhancing most of the millimeter-scale results found. In effect, for the same flow rate, the thermal resistance coefficient was $7{,}2 \cdot 10^{-5}$ $m^2.K/W$ [6,7].

4. Conclusions

Following the previous investigations done by Barrau et al. [6-7], this paper explores the viability of a variable-diameter microchannel heat exchanger in micrometer scale to achieve high heat fluxes and uniform wall temperature with only one fluid inlet. Experimental tests done with the test module in different conditions of flow rate are studied. The mean thermal resistance of the heat sink for a flow rate of 16 ml/s and a heat flux of 50 W/cm^2 is $2{,}35 \cdot 10^{-5}$ $m^2 \cdot K/W$, which enhances the results compared to the millimeter scale channels nearly three-fold. Another parameter evaluated is the temperature uniformity along the heat sink. The standard deviation σ_T is calculated and for the same flow rate and a heat flux of 50 W/cm^2, the temperature uniformity, expressed as the standard deviation of the wall temperature, is around 6 ºC. It can even be noticed that the wall temperature decreases slightly along the flow path, highlighting the effectiveness of the proposed variable microchannel width scheme. This shows that design optimization can lead to more uniform wall temperatures. Also, a specific profile of the microchannel width distribution could be fitted to the requirements of specific applications, such as adapting it to a non-uniform heat flux along the heated surface.

References

[1] Tuckerman D B and Pease F W 1981 High-Performance Heat Sinking for VLSI *Electron. Dev. Lett.* EDL-2 pp 126–9.

[2] Ryu J H, Choi D H and Kim S J 2003 Three-Dimensional Numerical Optimization of a Manifold Microchannel Heat Sink *International Journal of Heat and Mass Transfer* 46 pp 1553–62.

[3] Missaggia L J and Walpole J N 1991 A Microchannel Heat Sink with Alternating Directions of Water Flow in Adjacent Channels *Integrated Optoelectronics for Communication and Processing* 1582 pp 106-11.

[4] Vafai K and Zhu L 1999 Analysis of Two-Layered Micro-Channel Heat Sink Concept in Electronic Cooling *International Journal of Heat and Mass Transfer* 42 pp 2287-97.

[5] Lee Y, Lee P and Chou S 2009 Enhanced Microchannel Heat Sinks Using Oblique Fins *Proceedings of ASME IPACK2009–89059,* San Francisco, CA(2009)

[6] Barrau J, Chemisana D, Rosell J, Tadrist L and Ibañez M 2010 An Experimental Study of a New Hybrid Jet Impingement/micro-channel Cooling Scheme. *Applied Thermal Engineering* 30 pp 2058-66.

[7] Barrau J, Omri M, Chemisana D, Rosell J, Ibañez M and Tadrist L 2012 Numerical Study of a Hybrid Jet Impingement/micro-channel Cooling Scheme. *Applied Thermal Engineering* 33-34 pp 237-45.

[8] Karathanassis I K, Papanicolaou E, Belessiotis E V and Bergeles G C 2013 Three-Dimensional Flow Effects on Forced Convection Heat Transfer in a Channel with Stepwise-varying Width. *International Journal of Thermal Sciences* 67 pp 177-91.

[9] Ji J, Wang Y, Chow T, Chen H and Pei G 2012 A Jet Impingement/channel Receiver for Cooling Densely Packed Photovoltaic Cells under a Paraboloidal Dish Solar Concentrator. *Heat Transfer Research* 43(8) pp 767-78.

PowerMEMS 2013 IOP Publishing
Journal of Physics: Conference Series **476** (2013) 012076 doi:10.1088/1742-6596/476/1/012076

Tools for Analytical and Numerical Analysis of Electrostatic Vibration Energy Harvesters: Application to a Continuous Mode Conditioning Circuit

D. Galayko[1], E. Blokhina[2], P. Basset[3], F. Cottone[3], A. Dudka[1], E. O'Riordan[2] and Orla Feely[2]

[1] LIP6-UPMC — Sorbonne Universités, France
[2] University College Dublin, Ireland
[3] Université Paris-Est, ESYCOM, ESIEE Paris, France

E-mail: `elena.blokhina@ucd.ie`

Abstract. This paper reports the application of different analytical tools to a basic continuous conditioning (CC) circuit for electrostatic vibration energy harvesters (e-VEHs). We address the fundamental issues of this conditioning circuit and give design advice that enhances the performance of e-VEHs employing this circuit. This circuit is widely used for harvesters with or without an electret layer. Despite its wide use, its fundamental problems have been weakly addressed even for simple configurations of e-VEHs since it is impossible to solve the corresponding equations in closed form. As a consequence, appropriate semi-analytical methods that provide an insight into the physics of the system are required.

1. Introduction

Vibration energy harvesting is a promising technique for the generation of electricity from ambient vibrations for the supply of autonomous microsystems. Capacitive (electrostatic) vibration energy harvesters appear to be the best candidates for miniatuarization. They are not as bulky as electromagnetic harvesters, and they do not suffer from fatigue or depolarization like piezoelectric devices. A large number of studies have investigated electrostatic vibration energy harvesters, and many have focused on the design of a transducer device including a mechanical resonator and a capacitive transducer. However, the main challenge of e-VEH implementation is related to the design of conditioning and interface electronics. Currently, there are several families of circuits for the conditioning of capacitive energy converters. We can group them into three categories. (i) Constant-charge or constant-voltage operation mode [1], (ii) Circuits employing a charge pump [2, 3] and (iii) "Basic" conditioning circuit obtained by connecting a capacitive transducer to a source of voltage in series with a load resistor [4].

The first two families of circuits require complex control electronics in order to correctly sequence different phases of energy conversion. They are considered to be an encouraging prospect for the implementation of smart adaptive e-VEH systems, able to react to a variation in the input vibrations. On the other hand, the operation of a basic conditioning circuit does not require any smart electronics. Its cost is low, however its efficiency is also lower in comparison with the other families. The basic conditioning circuit shown in fig. 1a has been extensively used in two contexts. Firstly, when the main research or design work is focused on a transducer or a resonator, this circuit is used for the testing of the device [4]. Secondly, this circuit is often used in e-VEH employing an electret layer for transducer biasing [5]. In this case, the voltage source is incorporated into the model of a variable capacitor with a charged electret layer.

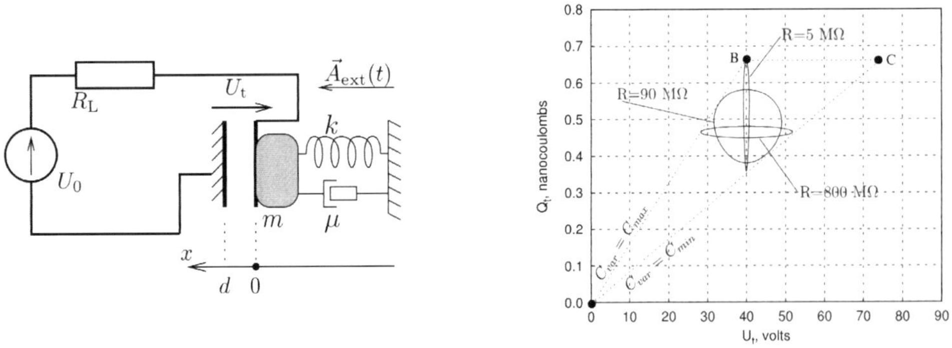

Figure 1. Schematic view of the continuous conditioning circuit (left) and charge-voltage diagrams for different values of load resistance at U_0=40 V (right). X=12 μm, f = 150 Hz.

In spite of its extreme simplicity and wide use, there are no established or formalised methods that will allow a designer to choose optimal parameters of the circuit (the optimal resistance and bias voltage) or analytically predict the circuit performance (the converted power). This paper presents a study of the basic conditioning circuit and its operation in conjunction with a mechanical resonator. Three methods are employed for the analysis: charge-voltage diagram method, mechanical impedance method and multiple-scale method.

2. Mathematical Model

The system is described by two equations. One models the displacement of a resonator, and the other models the circuit itself:

$$\begin{cases} \dot{v} = A_{\text{ext}}(t) - k/m \cdot x - \mu/m \cdot v + 1/m \cdot F_{\text{t}}(x) \\ \dot{x} = v \\ \dot{q} = (U_0 - q/C_{\text{var}}(x))/\,R_L \end{cases} \qquad \text{where } F_t(x) = \frac{1}{2}\frac{q^2}{C_{\text{var}}^2}\frac{\partial C_{\text{var}}}{\partial x} \qquad (1)$$

In these equations, q is the instantaneous charge of the transducer capacitance, x and v are the instantaneous displacement and velocity of the mobile mass respectively, A_{ext} is the instantaneous acceleration of the external vibrations, m, k and μ are the mass, stiffness and damping factor of the resonator. The function $C_{\text{var}}(x)$ gives the capacitance of the transducer as a function of the displacement x: $C_{\text{var}} = \varepsilon_0 S/(d - x)$, where d is the initial gap between the plates, S is the area of overlap between the plates. The parameters of the device are as follows: $m = 66 \cdot 10^{-6}$ kg, $\mu = 7.9 \cdot 10^{-3}$ Nsm^{-1}, $k = 68$ Nm^{-1}, $d = 43.5 \cdot 10^{-6}$ m, the natural frequency is $f_0 = \omega_0/(2\pi) = \sqrt{k/m}/(2\pi) = 161.5$ Hz, the rest capacitance is $C_0 = 1.08 \cdot 10^{-11}$ F.

3. Analysis in the Electrical domain: Method of Charge-Voltage Diagrams

Even in the absence of mechanical coupling, i.e., when the mobile mass vibration amplitude does not depend on the electrical processes, the analysis of the "basic" conditioning circuit is very difficult. The latter is described by a linear ordinary differential equation with time-variable coefficients, which cannot be integrated in closed form even for a simple $C_{\text{var}}(x)$ function. Charge-voltage diagrams are used to give a graphical representation of the state evolution of a variable capacitor in the (U, Q) plane. For the study of an energy harvester, charge-voltage diagrams are plotted in the steady-state mode when the mobile electrode of a transducer oscillates with a given sinusoidal law: $x(t) = X\cos(\omega t)$ and $C_{\text{var}}(t) = \epsilon_0 S/(d - X\cos(\omega t))$. In this way, the transducer capacity varies between the two known values, C_{max} and C_{min}. The charge-voltage diagram for different values of the load resistance R_L is shown in fig. 1. A typical QV diagram is a closed cycle. The area of the cycle is equal to the energy converted by the transducer during the cycle – this energy is dissipated on the R_L. It can be seen that for small R_L, the transducer voltage is equal to U_0, and the QV cycle is degenerated to a vertical line. For very large R_L,

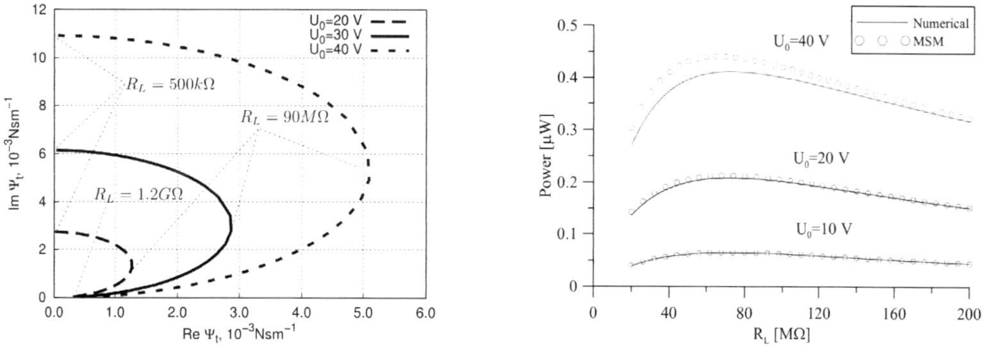

Figure 2. Impedance of gap-closing transducer plotted for R_L in the range (0.5 MΩ, 1.2 GΩ), for different U_0 (left). The converted power as a function of R_L for three different U_0 values (right). The solid lines show the converted power obtained from numerical simulations of the system (1) while the circles show the the converted power calculated using the multiple scales method.

the charge on the transducer is virtually constant, and the cycle is degenerated to a horizontal line. There is an optimal R_L for which the converted power is maximal. It can be seen, that the energy converted in the optimal cycle is much lower than the energy converted with a constant charge conditioning circuit (the triangle OBC) [1], having the same bias voltage and the same amplitude of the transducer capacitance variation.

4. Mechanical Impedance Method

The mechanical impedance method for analysis of a vibration energy harvester was proposed in [6]. The method assumes that a high-Q resonator responds only to the first harmonic of the external actuation. In the mechanical impedance method, the mobile mass is supposed to move following the sinusoidal law (see above). All forces applied to the mass are represented by mechanical impedances Ψ equal to the ratio between minus the first (fundamental) harmonic of the force expressed by its complex amplitude ($\dot{F}^{\omega}$) and the complex velocity of the mass motion ($\dot{V}$). Therefore, the impedance and the power of mechanical energy conversion, related to a force with impedance Ψ, is given by:

$$\Psi = -\dot{F}^{\omega}/\dot{V}, \quad P = |\dot{V}|\text{Re }\Psi. \tag{2}$$

The power P is positive if the energy leaves the mechanical domain. The imaginary part of the impedance provides important information about the resonance frequency shift due to the nonlinearity of the resonator/transducer.

More generally, the mechanical impedance of the transducer Ψ_t is calculated by analysing the system composed of a transducer whose mobile electrode moves following the sinusoidal law. As explained in [6], Ψ_t depends on the motion amplitude X, since the considered system is usually nonlinear. Knowing the $\Psi_t(X)$ function, it is possible to find the amplitude of mobile mass vibration in the coupled electromechanical mode. It is important to understand that although the transducer impedance relates two mechanical quantities, it is obtained by analyzing the electrical side of the system.

Figure 2 presents an example of loci of the transducer impedance for three values of bias voltages. The load resistance R_L varies from 0.5 $M\Omega$ to 1.2 $G\Omega$. As expected, the real part of the impedance is positive (the transducer removes energy from the mechanical domain), and it increases with the bias voltage. There is an optimal value of the resistance R_L at which *Re* Ψ_t is maximal. This value corresponds to the maximal power which can be converted by the conditioning circuit at the given mobile mass vibration amplitude. It can be seen that the imaginary part of the impedance varies with the load resistance as well, from a maximal value towards a value close to zero. The imaginary part of the impedance is positive: it highlights the shift of the resonance frequency toward the low values. This highlights a very important phenomena which has never been studied: the value of the load resistance has an impact on the resonance frequency shift of the vibration energy harvester.

5. Perturbation Technique: Multiple Scales Methods

The method of multiple scales (MSM) is a type of perturbation technique that is often applied for the analysis of weakly nonlinear oscillators [7]. Its approach is based on presenting oscillations in a quasi-harmonic form and find adjustments to oscillation characteristics, such as amplitude and phase, that result from the nonlinearity. In order to apply MSM, we introduce different time scales $T_k = \varepsilon^k t$ that reflect the shift in the frequency and present the solution in the form of a series $x(t) = \sum_{k=0}^{N} \varepsilon^k x_k(t)$ that accounts for the change in the amplitude of oscillations due to nonlinearity. This approach allows one to analyse the system simultaneously in the mechanical and electrical domains. As a result, we are able to find both the displacement of the resonator and the instantaneous charge on the transducer as functions of time. In addition, MSM allows one to take into account various additional nonlinearities such as, for instance, mechanical nonlinearity.

The method is applied to dimensionless equations and we rewrite system (1) using normalised parameters and variables [8]:

$$y'' + 2\beta y' + y = \alpha \cos \Omega\tau + \nu Q^2(y), \quad Q' = 1 - \rho(1-y)Q \tag{3}$$

where we introduced the following normalised variables: time $\tau = \omega_0 t$, dissipation $\beta = \mu/(2m\omega_0)$, normalised external vibration frequency $\Omega = \omega_{\text{ext}}/\omega_0$, normalised external acceleration amplitude is $\alpha = A_{\text{ext}}/(m\omega_0^2 d)$, the coefficients $\rho = (C_0\omega_0 R_L)^{-1}$ and $\nu = U_0^2/(2mC_0\omega_0^4 d^2 R_L^2)$, dimensionless displacement $y = x/d$ and dimensionless charge $Q = q/(U_0\omega_0 R_L)$. The solution of equations (3) yields the displacement of the resonator and the charge:

$$y(\tau) = Y_0 + Y\cos(\Omega\tau + \psi_0) \tag{4}$$

where the amplitude Y of oscillations is found from the equation

$$\frac{\alpha^2}{4} = \left(\beta f_0(Y) + \frac{\nu h_1(Y)}{2}\right)^2 + \left(Y\sigma + \frac{\nu g_1(Y)}{2}\right)^2 \tag{5}$$

and f_0, g_1 and h_1 are the coefficients of the Fourier series for the transducer force $F_t = \nu Q^2$. The charge Q is found from the equation:

$$Q(\tau) = \left[\frac{a_0}{\rho} + \sum_{n=1}^{\infty} a_n \cos n\tau - \sum_{n=1}^{\infty} b_n \sin n\tau\right] \times \left[\frac{a_0}{\rho} + \sum_{n=1}^{\infty} \frac{a_n(\rho\cos n\tau + n\sin n\tau)}{n^2+\rho^2} + \right. \\ \left. + \sum_{n=1}^{\infty} \frac{b_n(\rho\sin n\tau - n\cos n\tau)}{n^2+\rho^2}\right] \tag{6}$$

Here, a_0, a_i and b_i are the coefficients of the Fourier series of the function $\exp[-\beta a \sin t]$, and they are the functions of the amplitude of oscillations Y. The first few coefficient are given below:

$$a_0 = I_0(\rho Y), \quad a_1 = 0, \quad b_1 = -2I_1(\rho Y) \quad a_2 = -2I_2(\rho Y), \quad b_2 = 0 \tag{7}$$

where $I_n(z)$ is the modified Bessel function of the first kind. The coefficients a_i and b_i decay as the index i increases. The series (6) must be truncated when the corresponding coefficients are small enough. The dynamics of the system strongly depend on the product $\rho \cdot Y$ where Y is the amplitude of oscillations in the resonator. For instance, when $\rho \cdot Y \approx 1$, it is enough to account only few first terms in (6), and in this case $Q(\tau)$ represents almost harmonic oscillations. In addition, we note that system (1) can display stiffness, and it is difficult to solve this system numerically. The stiffness of the system is also determined by this product (it is stiff if $\rho \cdot Y \gg 1$).

The power dissipated at the resistor is given by the formula

$$P = \frac{1}{T} \cdot \frac{U_0^2}{R_L} \int_{T_0}^{T_0+T} (\dot{Q})^2 d\tau \tag{8}$$

where $T = 2\pi/\Omega$ is the dimensionless period of oscillations and T_0 is the time required to reach a steady state. The comparison of the converted power obtained by numerical solution of system (1) and the converted power yielding from the application of MSM is given in fig. 2. A very good correspondence is seen, in particular, for the plot obtained at $U_0 = 10$ and $U_0 = 20$ V. In all cases, there is an optimal value of the load resistance R_L at which the converted energy is maximal. This value is $R_L = 73$ MΩ and it does not depend on the applied constant voltage U_0. It is interesting to mention, that at the peak of the converted power in fig. 2, $\rho \cdot Y \approx 1$ for all values of A_{ext} and U_0.

6. Conclusions

We have presented three different techniques for the analysis of a basic continuous conditioning circuit for electrostatic vibration energy harvesters. This circuit is widely used for harvesters with or without an electret layer. Despite its wide use, its fundamental problems have been weakly addressed even for simple configurations of e-VEHs as it is impossible to solve the corresponding equations in closed form.

The method of charge-voltage diagrams is used for the graphic representation of the state evolution of a variable capacitor in the (U, Q) plane. Though it is very simple, it provides the qualitative understanding of the operation of the system. However, there is no closed form expression providing the value of the energy converted in one cycle or the optimal value of the resistance. The mechanical impedance method starts from an electrical analysis of the system, and provides the information about how the conditioned transducer impacts on the mechanical dynamics of the resonator (power conversion, resonance frequency shift). The multiple scale method, although advanced, allows one to solve the governing equation simultaneously in the mechanical and electrical domain and obtain the displacement of the resonator and the charge on the transducer as function of time. Based on the obtained solutions, we can calculate the converted power and find the optimal value of the load resistance R_L at which the converted power is maximised.

References

[1] S. Meninger, J. Mur-Miranda, R. Amirtharajah, A. Chandrakasan, and J. Lang, "Vibration-to-electric energy conversion," *IEEE Trans. Very Large Scale Integr. Syst. (VLSI)*, Vol. 9, Pp 64 – 76, 2001.

[2] S. Roundy, P. Wright, and K. Pister, "Micro-electrostatic vibra- tion-to-electricity converters," *in Proc. 2002 ASME Int. Mechanical Engineering Congr.*, Nov. 2002, pp. 487 – 496.

[3] B.C. Yen and J.H. Lang, "Avariable-capacitancevibration-to-electric energy harvester." *IEEE Trans. on Circ. and Syst. I*, Vol. 53, Pp. 288 – 295, 2006.

[4] L. G. W. Tvedt and D. S. Nguyen and E. Halvorsen, "Nonlinear Behavior of an Electrostatic Energy Harvester Under Wide- and Narrowband Excitation", *IEEE/ASME Journal of MEMS*, vol. 19, Pp. 305 – 316, 2010.

[5] M. Renaud, G. Altena, M. Goedbloed, C. de Nooijer, S. Matova, Y. Naito, T. Yamakawa, H. Takeuchi, K. Onishi and R. van Schaijk "A High Performance Electrostatic MEMS Vibration Energy Harvester With Corrugated Inorganic SiO2-Si3N4 Electret " *In proc. of Transducers/Eurosensors 2013 conference*, June 2013, Barcelona, Spain, Pp. 693 – 696.

[6] D. Galayko and P. Basset, "A general analytical tool for the design of vibration energy harvesters (VEHs) based on the mechanical impedance concept," *IEEE Trans. in Circ. and Syst. I* Vol 58, Pp. 299 – 311, 2011.

[7] A. Nayfeh, *Introduction to Perturbation Techniques.* New York: Wiley, 1993.

[8] E. Blokhina, D. Galayko, P. Basset, and O. Feely "Steady-State Oscillations in Resonant Electrostatic Vibration Energy Harvesters", *IEEE Trans. on Circ. and Sys. I*, Vol. 60, P. 875 - 884, 2013.

Low Electrostatic Dumping Force Energy Harvester Using Bipolar Charged Electret

K Sonoda, T Onishi, K Minami, T Fujita, K Kanda and K Maenaka
Department of EECS, Graduate School of Engineering, University of Hyogo
2167 Shosha, Himeji, Hyogo 671-2280, Japan

E-mail: eu10x003@steng.u-hyogo.ac.jp

Abstract In this paper, a bipolar charging method for electrostatic vibratory energy harvester and its analysis result of electrostatic force are described. To improve the harvesting power, higher voltage charged electret with large capacitance change structures are extensively studied. However, the higher voltage also shows a higher electrostatic force, which prevents the device movement for harvesting and causes pull-in behavior. From the result of FEM analysis on bipolar charged electret, the electrostatic force for vertical and horizontal directions were reduced by 15% and 10% from negative only charged device, respectively.

1. Introduction

Wireless sensor networks (WSNs) are attracted to many fields such as healthcare, infrastructure, and transportation to remotely monitor various parameters [1]. A power supply for WSNs is an issue in order to extend the lifetime and to reduce the maintenance cost. Utilization of WSNs are restricted by lifetime of the battery and installation place where the battery is hard to replace. Ambient energy harvesting technology is possible to overcome for these problems. There are several sources for energy harvesting such as mechanical vibration, thermal gradients, electromagnetic fields, and fluid flow. Mechanical vibration sources are ubiquitous and easy to use for conversion to electrical energy. In order to obtain large energy from the vibration by electret vibratory energy harvester (EVEH), higher charging voltage of electret with large capacitance change structures are typically used [2]. However, the higher electret charge causes some failures on EVEH's operation by electrostatic force increasing or pull-in behavior especially on the MEMS EVEH with small and light weight mass structure [3].

In this paper, we present a novel EVEH with low electrostatic dumping force by using bipolar charged electret [4]. From a preliminary experiment and an FEM analysis, advantages of bipolar charged electret for reducing the electrostatic force are demonstrated.

2. Bipolar Charging Method

The bipolar charging method is carried out with two electrodes that buried under an electret material. Two buried grid electrodes (BGEs) for negative and positive charging are abbreviated as the BGE-N and the BGE-P, respectively. At first, the BGE-N is grounded, the BGE-P and grid electrode for corona discharging are set to –400 V, and a needle voltage of –8 kV is applied. The electret material on the BGE-N is charged with –400 V after the first negative charging step. Then the BGE-N is set to +800 V and the BGE-P is grounded to prevent a neutralization of the negatively charged area. After that the grid is set to +400 V and the needle voltage of +8 kV is applied. The surface potential of the negatively charged electret is the subtraction of the charged electret potential with –400 V from the BGE-N biased with +800 V that is +400 V. Finally, the electrets on the BGE-N and the BGE-P are charged with –400 V and +400 V, respectively (figure 1).

Journal of Physics: Conference Series **476** (2013) 012077 doi:10.1088/1742-6596/476/1/012077

Figure 2 shows the CYTOP (CTL-809M; Asahi glass Co., Ltd. Japan) test specimen with a thickness of 2.5 μm, which was deposited on the patterned 0.5 μm thickness aluminum BGEs on a 0.5 mm thickness glass substrate. The BGE-N and BGE-P have 1 mm line width and 100 μm spacing dimensions. Each charging process was performed for 3 minutes on a hotplate with 80 °C. Figure 3 shows the surface potential profiles for (a) only the negative charged sample of –400 V and for (b) the bipolar charged sample of ±400 V. By using a counter electrode (CE) with 1 mm line width and 1.2 mm spacing and 50 μm air gap, a preliminary harvesting test was performed. Figure 4 shows the harvesting result with a load resistance of 1 MΩ for (a) the negative charged sample and (b) the bipolar charged sample with 30 Hz, 1 mm_{p-p} sinusoidal vibration. The output voltages for negative and bipolar charged electrets were 248 mV_{p-p} and 528 mV_{p-p}, respectively. This result shows the harvested voltage is obviously proportional to the voltage difference of adjusting electrets.

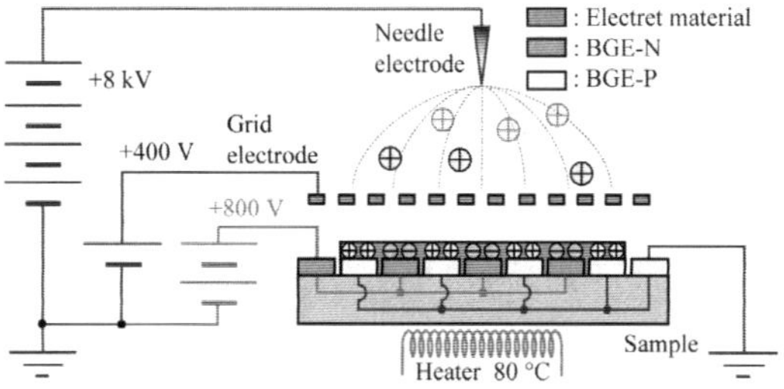

Figure 1. Bipolar charging diagram. +400 V positive charging with +800 V biased BGE-N after -400 V negative charging step with -400 V biased BGE-N and grounded BGE-P.

Figure 2. 2.5 μm thickness CYTOP test specimen for bipolar charging method.

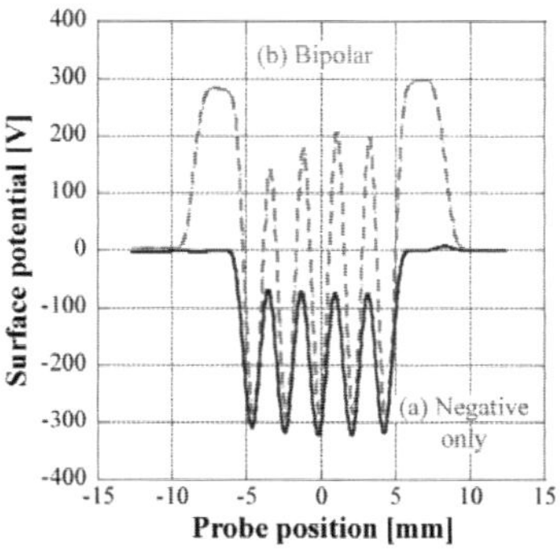

Figure 3. Surface potential measurement of the electret charged by (a) negative only and (b) bipolar.

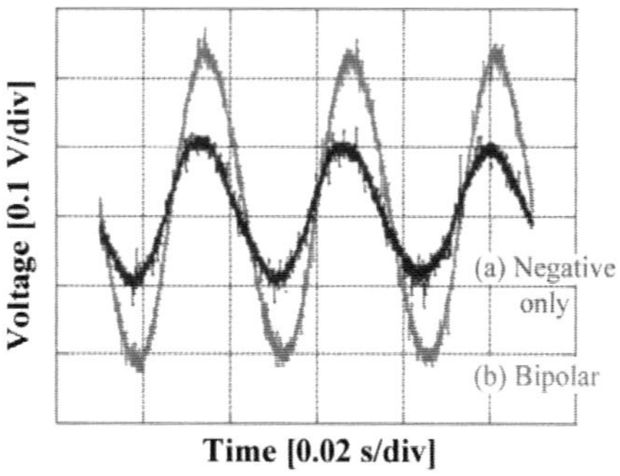

Figure 4. Harvesting result of test specimen for (a) only negative and (b) bipolar charged electret at 30 Hz, 1 mm_{p-p} excitation with gap of 50 μm.

3. Electrostatic Force Analysis

Electrostatic force is calculated by FEM analysis (ANSYS 14.0®). Figure 5 shows the analysis model consists of 50 pieces of CE with 10 mm in depth. The CEs are periodically located and move for X-axis direction. The BGE-P and BGE-N are set to the GND (0 V) and –600 V for negative only device, +300 V and –300 V for bipolar device, respectively. Figure 6 shows the vector plots of the electric field directions. The vector arrow direction depicts a electric field direction, length and color shows intensity. From the results in figure 5(a) and (d), (b) and (e), (c) and (f), electric fields between the CEs and the BGE-N are weak for bipolar device.

Figure 7 shows the electrostatic force between the BGEs and the CEs for (a) Y-axis and (b) X-axis. The 0 μm position on the horizontal axis depicts that the CE and the BGE-N are completely overlapped in the center. The electrostatic force for Y and X-axes are reduced by approximately 15% and 10%, respectively at maximum force point of the negative only result.

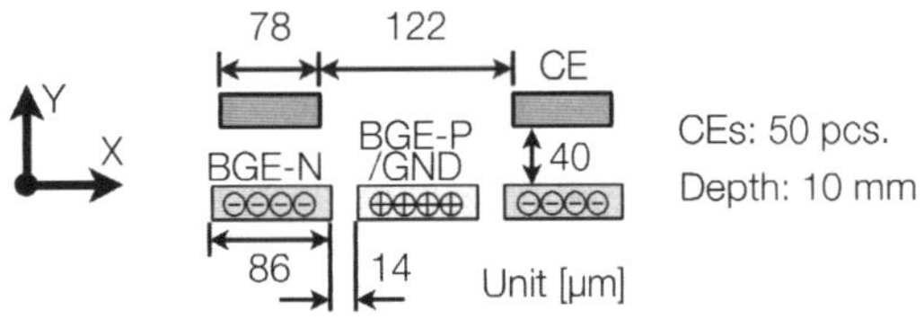

Figure 5. Dimension of the analysed device for the bipolar/negative only charged electrostatic force analysis. CEs move at on X-axis.

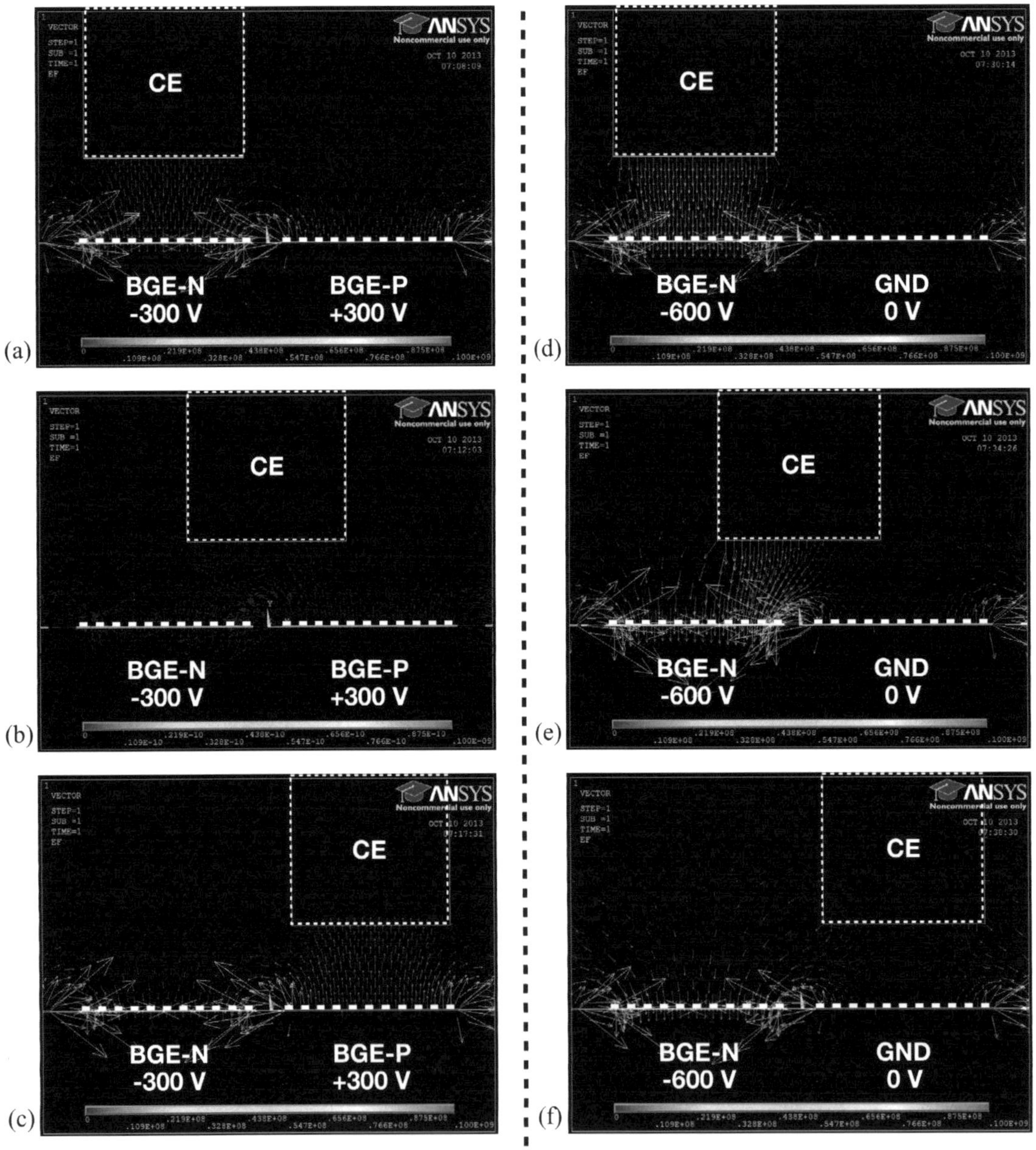

Figure 6. Vector graphics of the CE displacement of the bipolar and the negative only charged device (a), (b), and (c) are the bipolar of the displacement of 0, 50, and 100 μm, respectively. (d), (e), and (f) are the negative only of the displacement of 0, 50, and 100 μm, respectively.

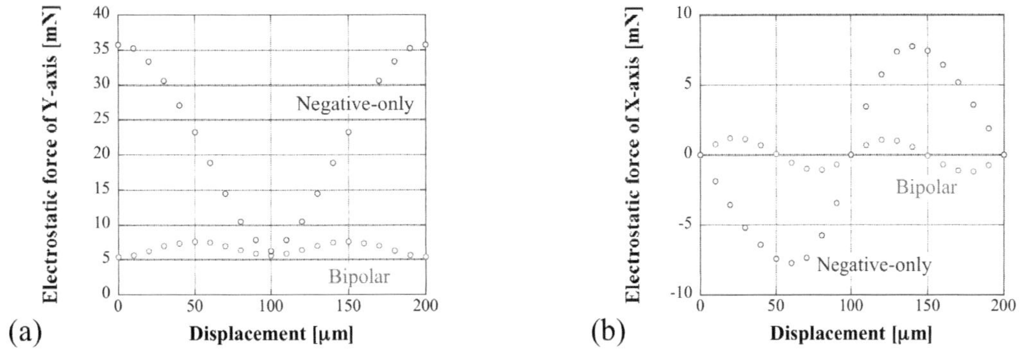

Figure 7. Electrostatic force analysis results for (a) Y-axis and (b) X-axis between the BGEs and the CEs with an air gap of 40 μm. The bias voltages for negative only and bipolar electret are −600 V and ±300 V, respectively.

In previous work, the CE is designed for mono-polar (negative only) electret as shown in figure 8(a). In order to improve a space efficiency, we designed novel CEs for bipolar electret (figure 8(b)), which has separated electrodes of CE-P and CE-N for BGE-P and BGE-P, respectably. FIgure 8(c) shows the 3-D diagram of the CE-N. The comb electrode like slits and two-step etching for air gap will be fabricated by DRIE. The CE-P and CE-N are isolated by dicing process after the wafer level bonding process.

Figure 9(a) shows conceptual diagram of wiring for the bipolar electret device. The harvested voltage from the CE-P and the CE-N are individually supplied to load resistances. Figure 9(b) shows the estimated output voltage with opposite phase waveforms for CE-P and CE-N.

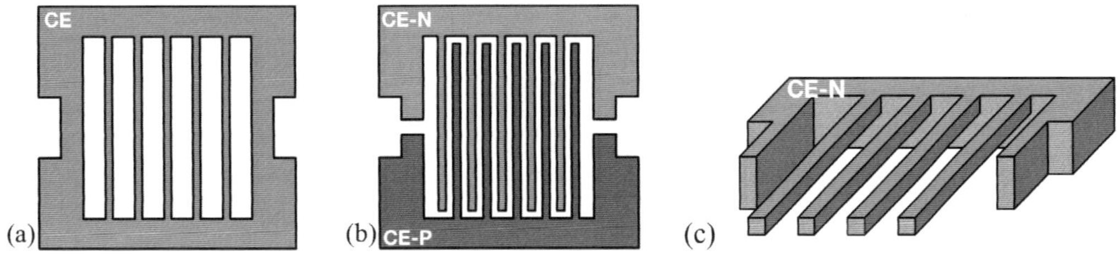

Figure 8. CE structures (a) the conventional CE top surface, (b) the novel CE top surface, and (c) the novel CE-N structure. The slit between the CE-P and the CE-N is formed by dry etching. The both electrode are completely divided by dicing process.

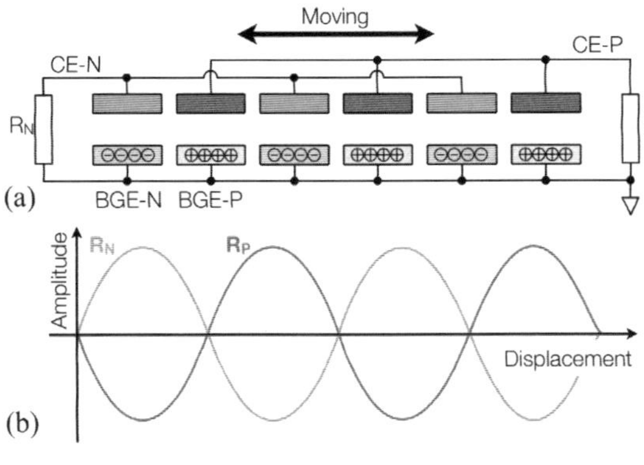

Figure 9. (a) Connection diagram of CE-P and CE-N for bipolar charged device and (b) estimated output voltage waveforms on the load resistance R_P and R_N.

4. Conclusion

We present a bipolar charging method for electret energy harvester with low electrostatic force. From the result of FEM analysis, we also demonstrate the electrostatic dumping force could be drastically reduced by using the bipolar charged device than the negative only charged device. The electrostatic forces for Y and the X-axes are reduced by approximately 15% and 10%, respectively.

References

[1] Beeby S P, Tudor S J and White N M 2006 *Meas. Sci. Technol.* **17** 175-95
[2] Suzuki Y, Miki D, Edamoto M and Honzumi M 2010 *J. Micromech. Microeng.* **20** 104002 (8pp)
[3] Fujii K, Onishi T, Sonoda K, Fujita T, Kanda K, Higuchi K and Maenaka K 2012 *Proc. ICETET'2012* 48-51
[4] Fujita T, Onishi T, Fujii K, Sonoda K, Katsuma H, Kanda K, Higuchi K and Maenaka K 2012 *Proc. PowerMEMS 2012* 436-9

Design improvements for an electret-based MEMS vibrational electrostatic energy harvester

G Altena, M Renaud, R Elfrink, M H Goedbloed, C de Nooijer and R van Schaijk

Holst Centre / imec, P.O. Box 8550, 5605KN, Eindhoven, The Netherlands

geert.altena@imec-nl.nl

Abstract. This paper presents several improvements to the design of an electret-based MEMS vibrational electrostatic energy harvester that have led to a two orders of magnitude increase in power compared to a previously presented device. The device in this paper has a footprint of approximately 1 cm^2 and generated 175 μW. The following two improvements to the design are discussed: the electrical connection principle of the harvester and the electrode geometrical configuration. The measured performance of the device is compared with simulations. When exited by sinusoidal vibration, a device employing the two design improvements but with a higher resonance frequency and higher electret potential generated 495 μW AC power, which is the highest reported value for electret-based MEMS vibrational electrostatic energy harvesters with a similar footprint. This makes this device a promising candidate for the targeted application of wireless tire pressure monitoring systems (TPMS).

1. Introduction

Tire pressure monitoring systems (TPMS) is a very promising application area for vibration energy harvesters. Presently, TPMS is already mandatory in the United States for all new cars and upcoming legislation in Europe and parts of Asia makes that these regions will follow soon. This will open up a multi-million units per year market with a demand for low-power miniaturized sensor systems. The currently used, valve-mounted systems need a large-sized battery which needs to be replaced every 3-7 years [1]. Integrating battery-operated TPMS in the inner liner of the tire means that only smaller, low-capacity batteries could be used, thereby increasing the replacement frequency and therefore the costs. An energy harvester can replace the batteries to avoid the high replacement costs and also provide the autonomy for the TPMS during the life time of the tire or car.

Mechanical vibrations are abundant in the (internal) environment of the tire and therefore, vibrational energy harvesters are prime candidates to convert the energy present in these vibrations into useful electrical power for the TPMS. There are three main methods to convert mechanical energy into electrical energy: based on electromagnetic [2]; piezo-electric [3] and electrostatic [4] transduction. We have already shown the feasibility of using a piezo-electric harvester as a power source for TPMS in [5] where several 10's of microwatts were generated from harvesting the mechanical shocks in the tire that occur during driving. This paper will show the potential of a device using the electrostatic conversion method as a power source for TPMS. The electrostatic energy harvester has the advantage over the piezo-electric harvester that the mechanical parts of the device can be independently designed and optimized from the electrical parts whereas in a piezo-electric device, the design and optimization of these parts are interconnected.

Being able to measure the pressure in a tire using a harvester-powered sensor system is the stepping stone towards a truly intelligent tire, one which is not only able to measure its pressure but also can sense the forces which act upon it and detect changes in the contact area with the road surface, thereby providing input for the various control systems (e.g. ABS and ESP) in the automobile thus improving the overall safety and comfort when driving.

2. Device description

The device, previously published in [6], is schematically shown in figure 1. It consists of a mass-spring system acting as one of the plates of a variable capacitor, itself polarized by an electret. The harvester is essentially a 3-terminal transducer (figure 2). It is fabricated from a stack of three wafers with two metal electrodes on the glass bottom wafer and one connection to the bulk silicon of the middle wafer which contains the electret; the device is capped by a glass top wafer. Glass wafers were chosen to reduce the effect of parasitic capacitances in the device. For the electret, we use a stack of SiO_2-Si_3N_4 for its compatibility with CMOS/MEMS processing [7,8]. This stack is deposited on a corrugated Si wafer. Both the corrugation for the electret and the suspension springs in wafer W2 are fabricated by deep reactive ion etching (DRIE). The corrugations are 100 μm deep; the suspension springs are etched through the entire Si wafer. A more detailed description of the device fabrication is given in [6].

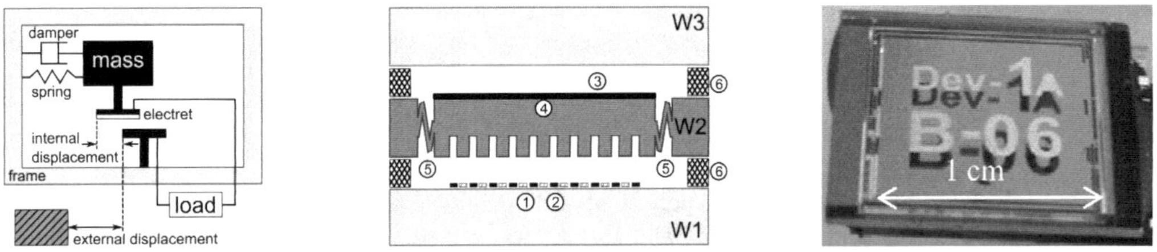

Figure 1. Left: schematic operation of the harvester; middle: drawing of the device showing glass wafer W1 containing electrodes (1) and (2); Si wafer W2 containing backside electrode (3), proof mass (4), springs (5) and corrugated electret; and glass capping wafer W3. The three wafers are bonded together using polymer bond frames (6); right: photograph of a fabricated device (top view).

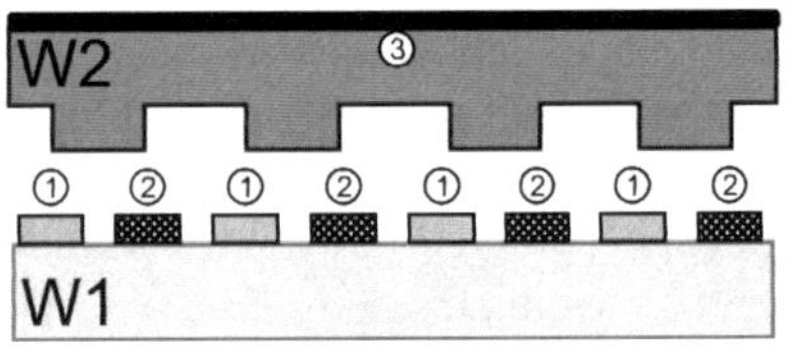

Figure 2. Schematic overview of the device showing electrodes (1) and (2) on the glass bottom wafer W1 and electrode (3) on the backside of the corrugated Si electret wafer W2.

3. Design improvements

In the following sections the two improvements to the design of the previous device presented in [9], the electrical connection principle and the geometric configuration of the electrodes, will be discussed in detail. In addition to these two improvements, compared to the previous device, also the electret was changed to a corrugated electret fabricated by deep-etching of the Si substrate. This change of the electret is also discussed extensively in [6].

3.1. Electrical connection principle

Two connection schemes for the load circuit are investigated: the 'slot-effect' scheme [10], where the external load is connected between the two electrodes on wafer 1 (figure 3a and b); and the 'cross-wafer' scheme, where the load is connected between one of the bottom electrodes and the connection to the Si bulk of wafer 2 (figure 3c and d). In the slot-effect scheme, the remaining terminal of the device is the backside of the electret wafer; in the cross-wafer scheme, the remaining terminal is one of the electrodes on wafer 1 (which is then generally referred to as the 'guard' electrode). In both schemes, this remaining terminal can be left floating (figure 3a and c) or be connected to ground (figure 3b and d).

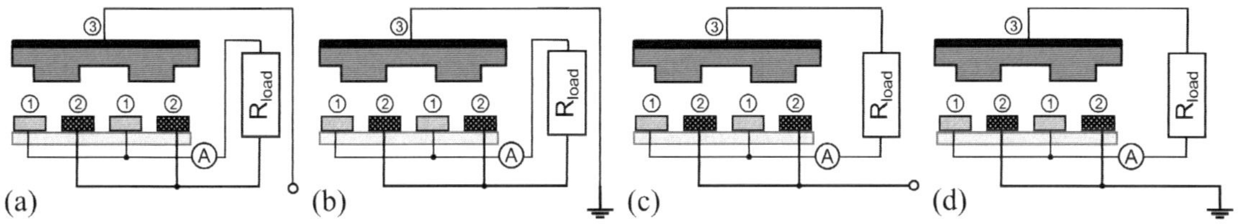

Figure 3. Slot-effect and cross-wafer connection schemes. a: slot-effect, electret floating; b: slot-effect, electret grounded; c: cross-wafer, guard electrode floating; d: cross-wafer, guard electrode grounded.

Figure 4 shows the experimentally measured output power using the different connection schemes as a function of the applied frequency (sinusoidal vibration with an acceleration of 2 G at the device's optimal load). Clearly, the slot-effect connection scheme (with floating electret backside connection), used in our present devices, is superior in output compared to the cross-wafer scheme as used in our previous devices. Additionally, grounding the guard electrode in the cross-wafer scheme does improve the performance of this scheme by approximately 40 % but even then the slot-effect scheme delivers almost a factor of two more power.

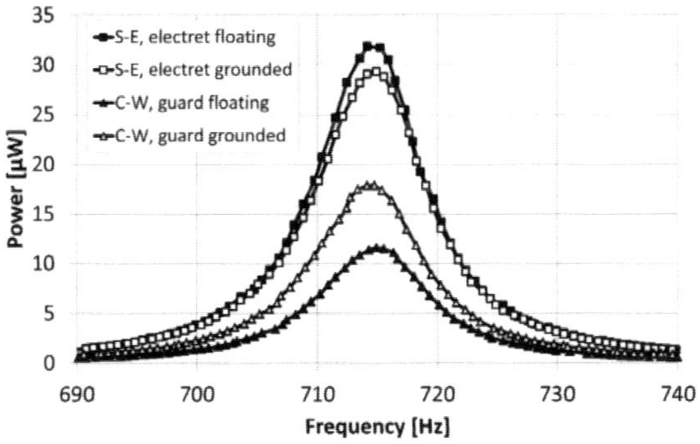

Figure 4. Power output of slot-effect scheme (S-E) and cross-wafer (C-W) scheme. The device was excited at an acceleration of 2 G. The optimal load resistances were 5 MΩ for S-E and 4 MΩ for C-W. The power is computed by measuring the electrical current through the load resistor as shown in figure 3.

3.2. Electrode geometrical configuration

The design of the electrodes/electret geometrical arrangement has a significant influence on the device's output power. The dimensions of the electrode (relative to the maximum peak-to-peak mass displacement d_{max}) need to be optimized to ensure maximum power output. Figure 5 schematically shows the different possibilities that were investigated. The lateral dimensions of the electrodes can be equal to d_{max} (Type 1) or smaller than d_{max} (Type 2). These two types have been simulated using a combination of analytical expressions and FEM calculations.

Figure 5. Schematic for two different electrode configurations: Left: Type 1, electrode width=d_{max}; right: Type 2, electrode width=$0.5d_{max}$. The displacement is represented by the dashed outline.

The modelled output power as function of the mass displacement and the voltage waveforms are shown in figure 6 using a resonance frequency of 1 kHz and d_{max}=100 µm. The waveform for the Type 1 device (middle graph) has a frequency of 1 kHz; in the waveform for Type 2 (right graph) higher frequency components are visible because of multiple crossings of the electrodes during one period. The optimal resistive load from the theoretical model was 8 MΩ for Type 1 and 5 MΩ for Type 2; experimentally, slightly lower values were obtained.

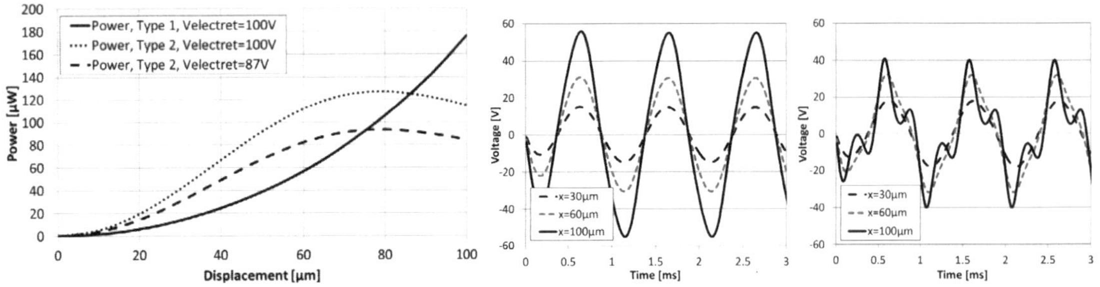

Figure 6. Modelled output of the electrostatic harvester for a resonance frequency of 1 kHz, d_{max}=100 µm and optimal resistive load of 8 MΩ for Type 1 and 5 MΩ for Type 2. Left: power output versus mass displacement of the two device types for different electret potentials; middle, right: voltage waveforms of type 1 devices (middle) and type 2 devices (right).

Figure 7 shows the measured output power for the two configurations (for optimal resistive load and slot-effect connection scheme). The output of the Type 1 device (squares) continues to rise with increasing displacement while the Type 2 output (triangles) starts to level off when the displacement equals the electrode dimension, as predicted by the model. The lower measured power output of Type 2 is partly attributed to a lower electret potential which is also shown in the modelled output (figure 6).

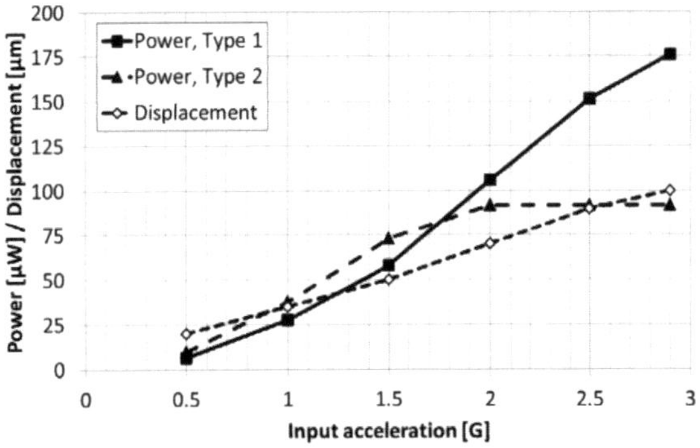

Figure 7. Measured output power of the two different electrode configurations. The resonance frequency and optimal load for the Type 1 device were 730 Hz and 5 MΩ, respectively. The resonance frequency and optimal load for the Type 2 device were 710 Hz and 3 MΩ, respectively.

For a Type 1 device from a next generation of devices employing the same discussed design improvements but with a higher resonance frequency and a higher value of the electret potential, a record output power of 495 µW was obtained using an input acceleration of 1187 Hz with 2.5 G amplitude and an optimal resistive load of 3.2 MΩ. The electret potential is estimated to be approximately 200 V, using the measurement method described in [11].

4. Discussion

In table 1 the generated power output of the device in this work is compared to the power output of devices listed in a recently published overview of state-of-the-art vibrational-type electret-based harvesters (data taken from [12], sorted by power output in µW). Note that the top-most harvester that generates 700 µW (Sakane [13]) is not an integrated device but a macroscopic structure where the electret and the counter electrode are mounted on a shaker and a 6-DOF positioner, respectively, while being placed in a SF_6-gas-filled box to prevent discharge between the electret and the counter electrode.

It is clear that the device presented in this paper performs very well, in particular when the footprint (active surface) is taken into account. For TPMS, the total device size, and therefore the active area, of the device should be small as not to interfere with the alignment and balance of the tire. The main challenge will be the reliability of the harvester and the sensor system when mounted in the tire. It needs to withstand very high accelerations and for automotive applications the highest acceleration specification is 2900 G [14].

Table 1. Comparison of this work with state-of-the-art electret-based vibrational energy harvesters (data taken from [12])

Author	Acceleration [G]	Frequency [Hz]	Active surface [cm²]	Electret potential [V]	Output power [μW]
Sakane	0.94	20	4	640	700
This work	**2.5**	**1187**	**1**	**200**	**495**
Boisseau	0.1	50	4.16	1400	50
Naruse	0.4	2	9	600	40
Tsutsumino	1.58	20	4	1100	38
Lo	4.9	50	6	1500	18
Edamoto	0.87	21	3	600	12
Boland	7.1	60	0.12	850	6
Kloub	0.96	1740	0.42	25	5
Lo	14.2	60	4.84	300	2.26

5. Conclusions

An electret-based MEMS vibrational energy harvester has been modelled and characterized with sinusoidal excitation. The harvester generated a power output of 175 μW. Compared to a previously published device, a two orders of magnitude increase in power output was obtained. This increase in output power is the result of the combination of two improvements with respect to the previous device: a different electrical connection principle of the harvester and an optimized geometrical configuration of the electrodes. Another device from a next generation, employing the same improvements but with a higher resonance frequency and electret potential, produced an output power of 495 μW. This makes this device a promising candidate for the targeted application of wireless tire pressure monitoring systems (TPMS).

References

[1] Schaijk R van, Elfrink R, Oudenhoven J, Pop V, Wang Z, and Renaud M In: Schmid U, Sánchez de Rojas Aldavero JL, Leester-Schaedel M, editors. *Proc. SPIE 8763, Smart Sensors, Actuators, MEMS VI, 876305 (May 17, 2013)*, vol. 8763, 2013, p. 876305–10

[2] Arnold DP *IEEE Trans Magn* 2007;**43**:3940–51

[3] Anton SR and Sodano HA *Smart Mater Struct* 2007;**16**:R1–R21

[4] Suzuki Y *IEEJ Trans Electr Electron Eng* 2011;**6**:101–11

[5] Elfrink R, Matova S, Nooijer C de, Jambunathan M, Goedbloed MH, van de Molengraft J, Pop V, Vullers RJM, Renaud M, and Schaijk R van *2011 Int. Electron Devices Meet.*, IEEE; 2011, p. 29.5.1–29.5.4

[6] Renaud M, Altena G, Goedbloed MH, Nooijer C de, Matova S, Naito Y, Yamakawa T, Takeuchi H, Onishi K, and Schaijk R van *Solid-State Sensors, Actuators Microsystems Conf. (TRANSDUCERS), 2013 17th Int.*, Barcelona, Spain: IEEE; 2013, p. 693–6

[7] Leonov V and Schaijk R van *IEEE Trans Dielectr Electr Insul* 2010;**17**:994–1000

[8] Leonov V, Fiorini P, and Hoof C Van *IEEE Trans Dielectr Electr Insul* 2006;**13**

[9] Altena G, Hohlfeld D, Elfrink R, Goedbloed MH, and Schaijk R van *Solid-State Sensors, Actuators Microsystems Conf. (TRANSDUCERS), 2011 16th Int.*, Beijing, China: IEEE; 2011, p. 739–42

[10] Jefimenko O *Proc West Virginia Acad Sci* 1968;**40**:345–8

[11] Shchelko NS, Sokolova IM, and Zakrzhevskiy VI *Proc. 8th Int. Symp. Electrets (ISE 8)*, IEEE; 1994, p. 1028–33

[12] Boisseau S, Despesse G, and Seddik BA In: Lallart M, editor. *Small-scale Energy Harvest.*, InTech; 2012, p. 1–39

[13] Sakane Y, Suzuki Y, and Kasagi N *J Micromechanics Microengineering* 2008;**18**:104011

[14] Automotive Electronics Council AEC-Q200 Rev D. 2010

Electrostatic Thermal Energy Harvester Using Unsteady Temperature Change

Junya Yoshida, Kenichi Morimoto, and Yuji Suzuki[1]
Department of Mechanical Engineering, The University of Tokyo, Tokyo, Japan
E-Mail: ysuzuki@mesl.t.u-tokyo.ac.jp

Abstract: A model electrostatic thermal generator using unsteady temperature change is proposed. The device consists of a capacitor based on high-permittivity ceramics, and an electret layer serving as a permanent voltage source. Connecting them in series, permittivity change by temporal temperature change alters the amount of induced charges on the electrode thereby produces electric current in the external circuit. Optimum design parameters of the system have been obtained using a simplified circuit model. An early prototype using $BaTiO_3$ ceramic as the dielectric and SiO_2 as the electret is microfabricated, and its response is compared with the model prediction.

Keywords: Thermal energy harvesting, Electret, Permittivity change, $BaTiO_3$

1. INTRODUCTION

Thermal energy harvesting from waste heat or body heat is useful for powering low-power electronics such as wireless sensor nodes [1]. The Seebeck effect is usually employed for getting electric power from steady temperature gradients, where thermal impedance matching is an important design strategy [2]. MEMS thermoelectric devices with thermal insulation structures are developed for that purpose [3, 4]. On the other hand, transient temperature gradient is also relevant as energy sources. Moser et al. [5] propose thermoelectric harvesters from temperature difference between ambient air and large backside thermal capacitance such as road tunnel wall, and reported that the thermal cut-off frequency plays an important role in their design. Temporal temperature change during aircraft flights is also an interesting application, because the air temperature is decreased to -50 °C at the cruising altitude [6-8]. Samson et al. [7] developed a thermoelectric generator prototype using phase change material (PCM) as the backside thermal capacitance, and made a series of in-flight tests by installing the device on fuselage.

Pyroelectric is an alternative method to extract electricity from transient temperature change, where temporal temperature change can be directly used for energy conversion. Sebald et al. [9] made detailed comparison between thermoelectric and pyroelectric energy conversion, and reported that the conversion efficiency with pyroelectric can be much higher than that with thermoelectric. Thermal engines using pyroelectric device such as PZT or P(VDF-TrFE) are also proposed [10, 11].

In the present study, we propose electret-based electrostatic induction for thermal energy harvesting from unsteady temperature change. If compared with the pyroelectric principle, the present power generation principle should be more robust against excessive temperature variation.

[1] Correspondence to: Yuji Suzuki E-mail: ysuzuki@mesl.t.u-tokyo.ac.jp
Department of Mechanical Engineering, The University of Tokyo
7-3-1, Hongo, Bunkyo-ku, Tokyo 113-8656, Japan

2. PRINCIPLE AND OPTIMIZATION

Figure 1 shows configuration of the present device with a capacitor (capacitance: C) using high-permittivity material and metalized electret (capacitance: C_e, surface charge density: σ_e) serving as a permanent voltage source. They are connecting in series with an external load R. The temperature change of the capacitor leads to the permittivity change, and the amount of induced charges on the electrode is changed. Thus, the external current I is produced when temperature of the dielectric is changed. A circuit model of the generator is developed based on the Kirchhoff's law, the Gauss's law, and the conservation law of charges:

$$V = IR = d_1E_1 - d_2E_2,\quad I = -S_1\frac{d\sigma_1}{dt} = S_2\frac{d\sigma_2}{dt}, \quad (1)$$

$$\sigma_1 = \varepsilon_0\varepsilon_1E_1,\quad \sigma_2 = \varepsilon_0\varepsilon_2(t)E_2, \quad (2)$$

$$S_1\sigma_e = -\left(S_1\sigma_1 + S_2\sigma_2\right), \quad (3)$$

where d_1, S_1, E_1, σ_1, and ε_1 are the thickness, the area, the electrical field, the induced charge, and the permittivity of the electret, while d_2, S_2, E_2, σ_2, and $\varepsilon_2(t)$ are those of the dielectric. Eqs. (1-3) can be integrated into a first-order differential equation, i.e.,

$$\frac{d\sigma_1}{dt} = -\frac{1}{R}\left\{\left(\frac{1}{C_e} + \frac{1}{C(t)}\right)\sigma_1 + \frac{1}{C(t)}\sigma_e\right\} \quad (4)$$

Eq. (4) can be non-dimensionalized with $t = RC_e t^*, \sigma_1 = \sigma_e\sigma_1^*, C(t) = C_e C^*(t^*)$, and we get

$$I^* = V^* = \frac{d\sigma_1^*}{dt^*} = -\left(1 + \frac{1}{C^*(t^*)}\right)\sigma_1^* - \frac{1}{C^*(t^*)} \quad (5)$$

For simplicity, capacitance of the dielectric is assumed to change linearly in time, i.e.,

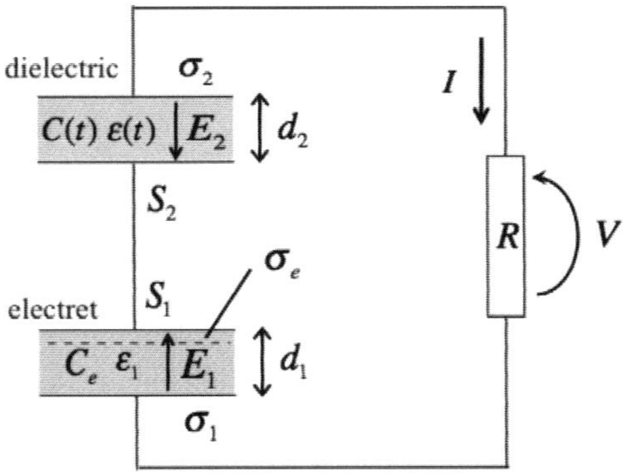

Figure 1. Circuit model of thermal electret generator.

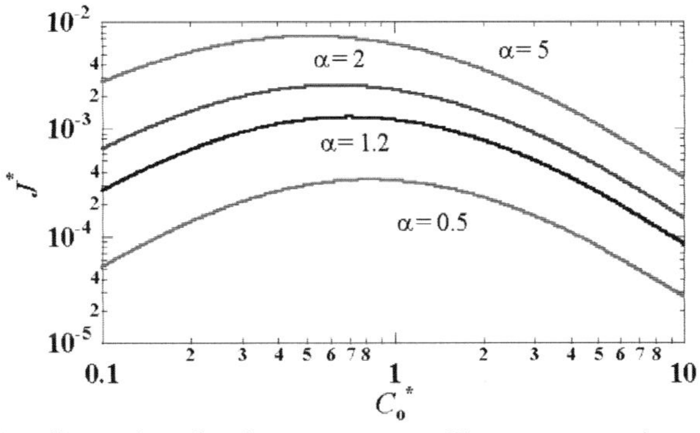

Figure 2. Nondimensionalized output power J^* versus capacitance ratio C_o^*.

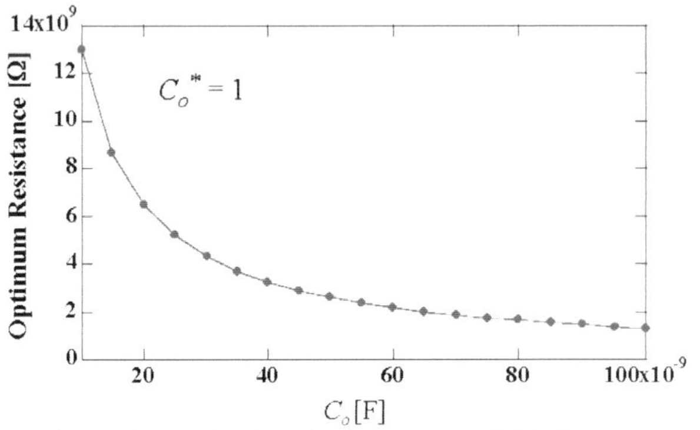

Figure 3. Optimum load resistance versus initial capacitance.

Figure 4. $BaTiO_3$ capacitor (15 mm dia., 0.1 mm in thickness).

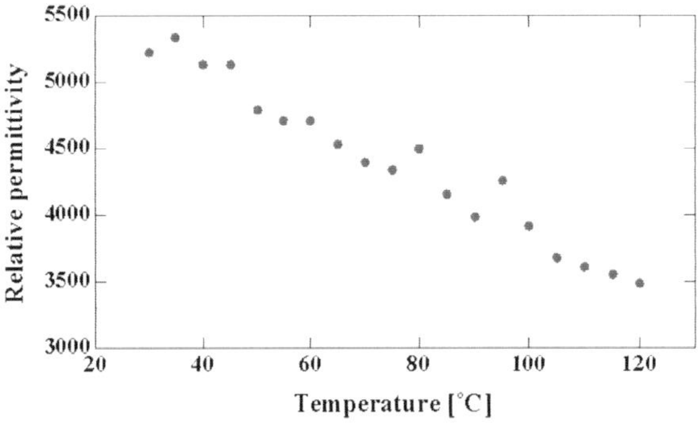

Figure 5. Relative permittivity of the $BaTiO_3$ capacitor versus temperature.

$$C^*(t^*) = \left(1+\alpha\frac{t^*}{\tau^*}\right)C_O^* \qquad (6)$$

where α, τ*, and C_0* are the capacitance increase rate, time scale of the temperature change non-dimensionalized with the time constant RC_e, and the initial capacitance ratio of the dielectric to C_e (C_0*=C_0/C_e).

Figure 2 shows simulation results for the non-dimensionalized output power J^* $(=\varepsilon_1\varepsilon_0 / d_1 S_1 \sigma_e^2 \cdot J)$ versus C_0*. It is found that the optimum C_0* exits, and it is around unity for α=0.5, and decreased with increasing α. In addition, when α increases, the output power is dramatically increased. As shown in Fig. 3, the optimum resistance is decreased with C_0, indicating that high-permittivity material is required for this type of generator.

3. PROTOTYPE DESIGN AND EXPERIMENTS

Figures 4 shows the $BaTiO_3$ capacitor (Chiba Ceramic MFG, 15 mm dia., 0.1 mm in thickness) used in our preliminary experiment. Pure $BaTiO_3$ has its Curie temperature at around 130 °C, where

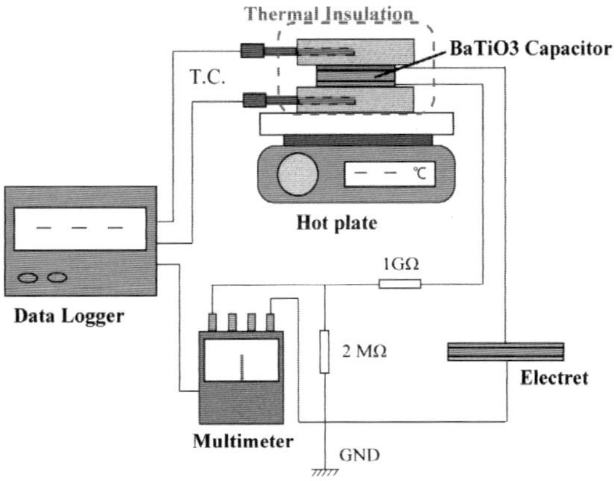

Figure 6. Schematic of the experimental setup.

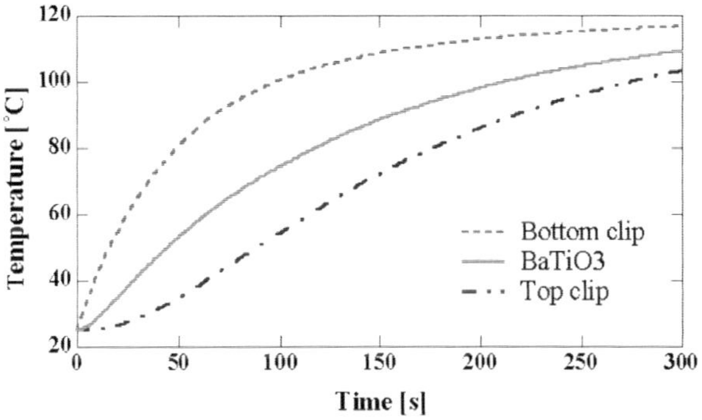

Figure 7. Estimated temperature of $BaTiO_3$ capacitor versus elapsed time.

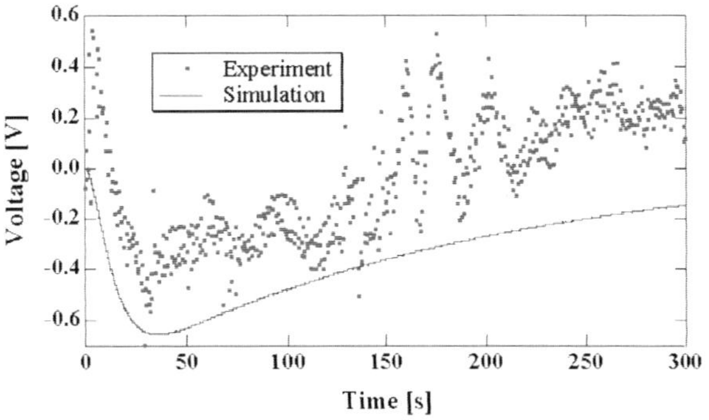

Figure 8. Output voltage versus elapsed time.

the permittivity is drastically changed. Figure 5 shows the DC permittivity of the $BaTiO_3$ capacitor versus the temperature. Due to additives into the present $BaTiO_3$, α is not so high and about 0.5.

An electret plate is fabricated using a 4 cm square Si substrate with 1.5 μm-thick thermally-grown SiO_2. After corona charging, the surface potential is 420 V, which corresponds to the surface charge density σ_e of 9.5 mC/m^2. The metalized electret is formed by stacking another Si substrate with 1.5 μm-thick SiO_2 onto the electret plate. In the preliminary experiment with the present $BaTiO_3$ and electret plates, C_0*=0.2, which is far from the optimum design.

Figure 6 shows the experimental setup. The $BaTiO_3$ capacitor is sandwiched with copper plates and heated on a hot plate. Temperature of the copper plates is measured with thermocouples, and the $BaTiO_3$ temperature is estimated using one-dimensional heat conduction equation with contact resistance. Figure 7 shows the measured temperature of the top/bottom copper plates and the estimated temperature of the $BaTiO_3$ capacitor, which is increased from 20 to 100 °C in about 300 s.

Figure 8 shows the output voltage across a 1 GΩ resistive load. Up to 0.6 V is obtained, corresponding to about 30 nJ in 300 s. Although amount of the generated power in this early prototype is limited, the results are in accordance with prediction of the circuit model. Thus, the output power can be much increased by using the optimum C_0*, much higher α, and higher surface potential with CYTOP electrets [13, 14]. It is also noted that the Currie temperature of Zr-added $BaTiO_3$ could be as low as 60 °C, and α can be as large as 5.

4. CONCLUSION

A new electret thermal generator using temporal temperature change has been proposed. A simplified circuit model is developed for parameter optimization. It is found that the optimum capacitance ratio is around unity and the output power is significantly increased with the capacitance increase rate. A preliminary experiment using BaTiO3 ceramic as the dielectric layer is carried out. It is demonstrated that the experimental data are well predicted with the present model.

This work is partially supported through JSPS NEXT Program. Photo masks are made using F5112+VD01 EB writer at VDEC of the University of Tokyo.

References

[1] V. Leonov, and R. J. M. Vullers, 2009, Wearable electronics self-powered by using human body heat: The state of the art and the perspective, J. Renew. Sust. Energ., Vol. 1, 062701.
[2] M. Lossec et al., 2010, Thermoelectric generator placed on the human body: System modelling and energy conversion improvements, Eur. Phys. J. Appl. Phys., Vol. 52, 11103.
[3] Z. Wang et al., 2009, Characterization and optimization of polycrystalline $Si_{70\%}Ge_{30\%}$ for surface micromacined thermopiles in human body applications, J. Micromech. Microeng., Vol. 19, 094011.
[4] J. Xie et al., 2010, Design, fabrication, and characterization of CMOS MEMS-based thermoelectric power generators, J. Microelectromech. Syste., Vol. 19, 317-324.
[5] A. Moser et al., 2012, Thermoelectric energy harvesting from transient ambient temperature gradients, J. Electr. Mater., Vol. 41, 1653-1661.
[6] N. Bailly et al., 2008, Energy scavenging based on transient thermal gradients: Application to structual health monitoring of aircraft, PowerMEMS2008, Sendai, 205-208
[7] D. Samson et al., 2012, Flight test results of a thermoelectric energy harvester for aircraft, J. Electr. Mater., Vol. 41, 1134-1137.
[8] M. E. Kiziroglou et al., 2014, Design and Fabrication of heat storage thermoelectric harvesting device, IEEE Trans. IE-61, 302-309.
[9] G. Sebald et al., 2009, On thermoelectric and pyroelectric energy harvesting, Smart Mater. Struct. Vol. 18, 125006.
[10] S. K. T. Ravindran et al., 2011, A self-sustaining micro thermomechanic-pyroelectric generator, Appl. Phys. Lett., Vol. 99, 104102.
[11] H. Nguyen et al., 2010, Pyroelectric energy converter using co-polymer P(VDF-TrFE) and Olsen cycle for waste heat energy harvesting, Appl. Therm. Eng., Vol. 30, 2127-2137.
[12] J. Xie et al., 2010, Performance of thin piezoelectric materials for pyroelectric energy harvesting, J. Intell. Mater. Syst. Struct., Vol. 21, 243-249.
[13] Y. Sakane et al., 2008 The development of a high-performance perfluorinated polymer electret and its application to micro power generation, J. Micromech. Microeng., Vol. 18, 104011.
[14] K. Kashiwagi et al., 2011, Nano-cluster-enhanced high-performance perfluoro-polymer electrets for energy harvesting, J. Micromech. Microeng., Vol. 21, 125016.

Self-starting power management circuits for piezoelectric and electret-based electrostatic mechanical energy harvesters

S. Boisseau, P. Gasnier, M. Gallardo, G. Despesse
CEA, Leti, Minatec Campus, 17 av. des martyrs, 38054 Grenoble, France

E-mail: sebastien.boisseau@cea.fr

Abstract. This paper reports on an innovative power management circuit for piezoelectric and electret-based mechanical energy harvesters able to self-start and to power battery-free Wireless Sensor Nodes (WSN) from scratch without any initial energy. The key elements of this circuit are a depletion-mode MOSFET combined with self-powered Schmitt triggers that enable to switch between (i) a non-optimized passive diode-bridge-capacitor configuration to start the system and (ii) an active power conversion path to maximize the energy extraction from mechanical energy harvesters. A discrete circuit implementing this architecture is presented and its operation is validated on simple piezoelectric and electret-based devices. An ASIC, based on the same architecture, has finally been designed, fabricated and validated.

1. Introduction

Mechanical Energy Harvesting is an increasing field of research since the 2000s whose main objective is to develop autonomous wireless sensors powered by ambient vibrations, stress, strains or shocks… The output power of piezoelectric and electret-based mechanical energy harvesters (EH) is generally in the 10µW-100µW range and their high AC output voltages are not compatible with standard electronic circuits requirements (3V-DC supply source needed). As a consequence, a power management circuit (PMC) is essential to turn the EH raw output into a viable supply source for wireless sensors.

In fact, two main types of PMCs can be employed to achieve this electrical conversion:

- Passive PMCs (e.g. diode bridge rectifier + storage capacitor) that do not require power to operate, but which suffer from a non optimal energy extraction resulting in a low usable output power (figure 1a).
- Active PMCs that have high conversion efficiencies, and which enable to implement nonlinear techniques such as SSHI (Synchronized Switch Harvesting on Inductor), SECE (Synchronous Electric Charge Extraction) [1] to increase the extraction of energy from the energy harvester [2]. Figure 1b shows an example of such a circuit: a flyback converter driven by a control circuit. Unfortunately, this type of circuit must be powered to operate, for example by a rechargeable battery which stores the harvested energy while supplying the electronic functions: the power balance (harvested power – PMC power consumption) is generally much larger than the output power obtained with a passive PMC.

Yet, energy harvesting has the great benefit of enabling the deployment of battery-free devices compatible with harsh environments (e.g. high temperatures) and with a theoretical unlimited lifetime. But, a problem arises when starting an energy harvesting system from scratch, without battery and any initial energy: the active PMC cannot be powered and the power converter (e.g. flyback converter) cannot be controlled. The solution we propose consists in switching between the passive and the active PMCs as soon as enough energy has been stored in a startup capacitor C_s to power the control circuit.

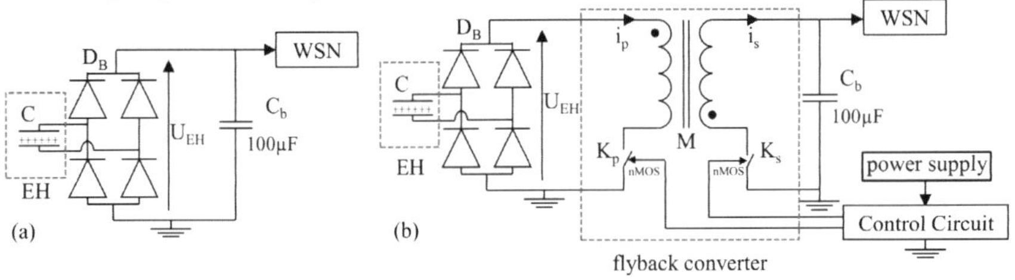

Figure 1. (a) passive PMC and (b) active PMC (flyback converter)

2. Self-starting power management circuit architecture

The whole architecture of the self-starting PMC we propose is depicted in figure 2a. A flyback converter implementing SECE [1] has been chosen as the active PMC: all the energy stored into the EH capacitance is transferred to C_b through the coupled inductors when U_{EH} reaches its maximum. This operation mode has proven its benefits for piezoelectric [3-4] and electrostatic [5] devices.

In our case, two capacitors are used: (i) C_s which powers the control circuit and (ii) C_b which supplies the WSN; they are connected by a diode (D_C). A depletion-mode MOSFET (dMOS) K_{bp} is employed to bypass the flyback converter during the startup, in order to charge C_s. Then, this circuit is able to switch between the passive and the active PMC according to U_{Cs} (the voltage across C_s); this switching is controlled by the 'Startup Control' which is a Schmitt trigger. The operation mode of this circuit can be summarized in 5 steps and the typical voltages on C_b and C_s during startup are presented in figure 2b.

- *State 0 – no-energy state.* At the beginning, there is no energy, K_{bp} is closed, K_{cc} is open, K_{app} is open, U_{Cb}=0 and U_{Cs}=0.
- *State 1 – non-optimized diode-bridge-capacitor conversion.* The energy scavenger starts to harvest energy. As K_{bp} is closed and K_{cc} is open, the flyback converter is bypassed by K_{bp}. The PMC behaves like a diode-bridge-capacitor circuit; the energy goes directly from the EH to C_s and U_{Cs} increases.
- *State 2 – optimized flyback conversion.* When U_{Cs} reaches U_{Cs}^+, 'Startup Control' opens K_{bp} and closes K_{cc}. The Control circuit is supplied by C_s and the flyback conversion starts. U_{Cb} increases. As C_s is not supplied anymore while powering the control circuit, its voltage decreases until falling below U_{Cs}^-. The Startup control switches to state 1 to recharge C_s. This commutation between state 1 and state 2 continues until U_{Cb} reaches $U_{Cs}^-+V_{thDc}$ (threshold voltage of D_C).
- *State 3 – end of startup.* The PMC remains in the flyback conversion configuration, optimizing power extraction from the energy harvester. K_{cc} stays closed, K_{bp} open.
- *WSN measurement.* When U_{Cb} reaches U_{Cb}^+, 'WSN Control' (also a Schmitt trigger) closes K_{app} and makes the WSN perform its measurement and transmission cycle.

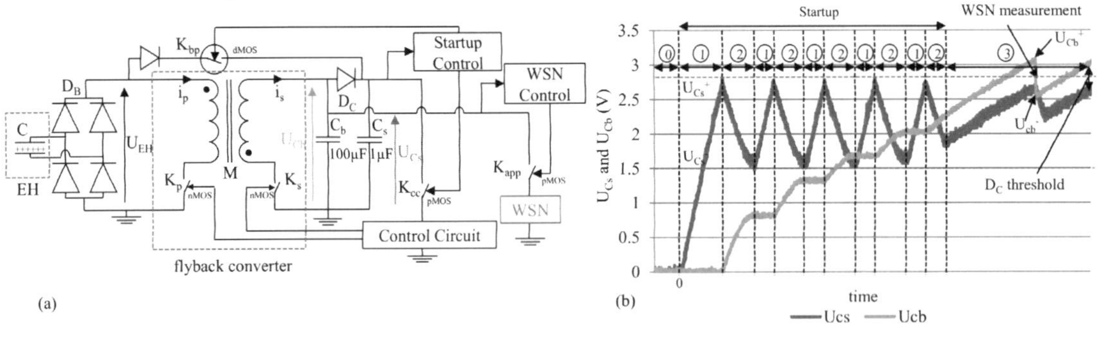

Figure 2. (a) self-starting PMC architecture and (b) mode of operation (extracted from experimental results)

3. Electronic circuit made with Discrete Commercial Off-The-Shelf components

The challenge of this circuit lies in the design of the 'startup control' and of the 'control circuit' that must be low power, autonomous and able to work from a zero-energy state.

3.1. Self-powered Schmitt trigger - 'Startup Control' and 'WSN Control'

As previously explained, *'Startup Control'* and *'WSN Control'* are Schmitt triggers. It is quite easy to make Schmitt triggers with low-consumption comparators with internal reference (e.g. MAX917) and by adding resistors to induce a hysteretic behavior. Yet, the power consumptions of these components are in the 700-800nA range, which is quite elevated for a startup circuit whose interest is limited to the startup phase.

Our Schmitt triggers have been made from discrete commercial off-the-shelf components. Their core concept is a dMOS-based self-powered voltage comparator (figure 3a). The dMOS is here used as a voltage limiter: its gate is connected to the ground and a high-impedance load (R_0>100MΩ) is placed at its source. In this configuration, U_{dmos} will be limited by the dMOS threshold voltage (V_{thdmos}): $U_{dmos}=\min(U_{in}, V_{thdmos})$. Two CMOS inverters, powered by U_{in} and connected to the dMOS source, act as a buffer switching at $U_{sw}=U_{in}/2$.

When $U_{in}<V_{thdmos}$, $U_{out}=U_{dmos}=U_{in}$. Then, for $U_{in}>V_{thdmos}$, U_{dmos} saturates at V_{thdmos}; U_{out} remains equal to U_{in}. When U_{in} reaches $2V_{thdmos}$, U_{dmos} is still equal to V_{thdmos} and U_{out} switches to 0. A hysteretic behavior is achieved by adding resistors in this circuit; diodes can also be added to shift the switching levels. The complete self-powered Schmitt trigger is presented in figure 3c. Its power consumption is in the 50-100nA range depending on the switching levels (10 times lower than Schmitt triggers made from comparators).

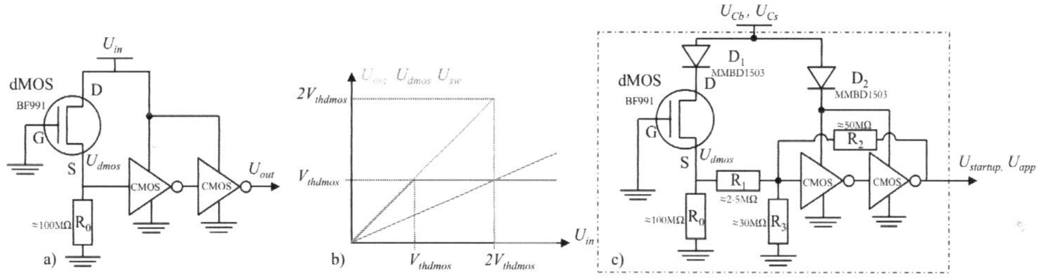

Figure 3. (a) Self-powered voltage comparator and (b) mode of operation. (c) Schmitt trigger

3.2. Control Circuit

The Control Circuit is the other key element of this architecture. In SECE, the control circuit is aimed at detecting the output voltage's maximums and to control at the right time K_p and K_s. The control circuit is made of a RC differentiator (R_D-C_D) and a comparator (figure 4a) to detect the zero crossings of U_{EH}'s derivative, i.e. its maximums. Three delay cells (figure 4b) generate the control times T_1 and T_3 for K_p and K_s (figure 4c). The circuit has been made from standard electronic components and its power consumption is comprised between 500nA@3V and 1μA@3V.

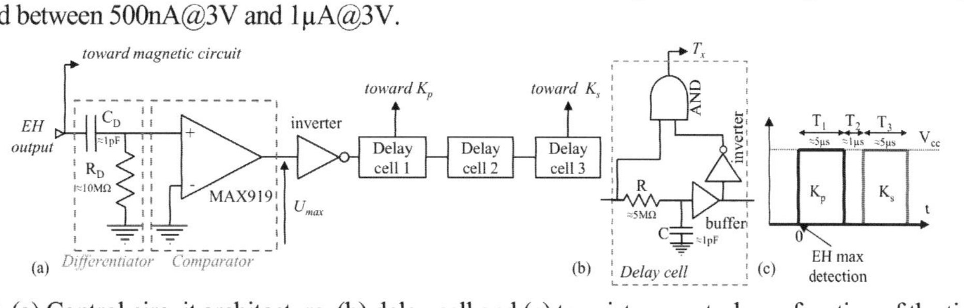

Figure 4. (a) Control circuit architecture, (b) delay cell and (c) transistors control as a function of the time

4. Experimentations on simple electret-based and piezoelectric mechanical energy harvesters

4.1. Electret-based vibration energy harvester

A simple electret-based vibration EH [6] (figure 5a) has been connected to the self-starting PMC. It oscillates at 35Hz and harvests 23μW of raw output power on a 50MΩ-load (figure 5b). $U_{Cs}(t)$ and $U_{Cb}(t)$ are presented in figure 5c validating the good operation of the self-starting PMC. And then, it proves that even with this simple device, it is possible to power a Wireless Sensor consuming 112μJ after 200s (figure 5c), and every 5s after the startup phase.

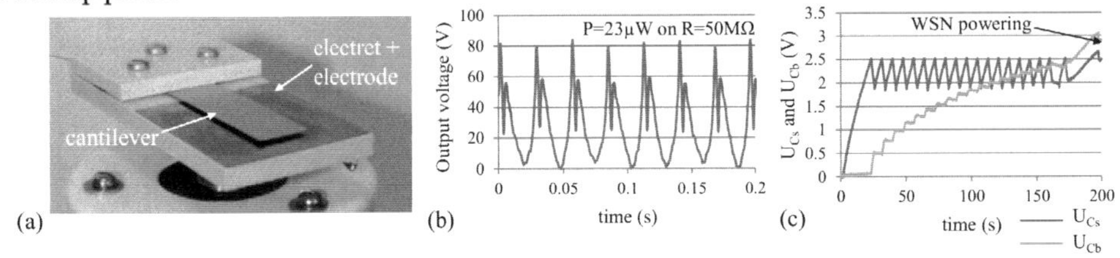

Figure 5. (a) Simple electret-based vibration energy harvester, (b) output voltage on a 50MΩ-load after the diode bridge and (c) U_{Cs} and U_{Cb} as a function of the time and WSN measurement

4.2. Piezoelectric strain energy harvester

The same experiment has been performed on a simple piezoelectric energy harvester (figure 6a) deformed by hand at 7Hz and harvesting about 100μW. Our self-starting PMC is again validated (figure 6c), and logically, as the output power is much higher than in the previous case, the startup duration is highly reduced (about 12s) and a measure can be performed about every second after the startup phase.

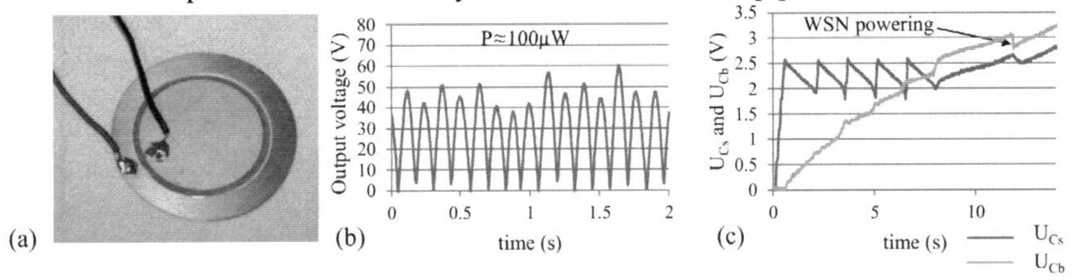

Figure 6. (a) Simple piezoelectric energy harvester (buzzer from Murata), (b) output voltage on a 10MΩ-load after the diode bridge and (c) U_{Cs} and U_{Cb} as a function of the time and WSN measurement

5. ASIC implementing the self-starting PMC

5.1. Architecture

An integrated PMC [7] has been fabricated in AMS 0.35μm CMOS process (figure 7a). The autonomous integrated circuit, whose architecture is depicted in figure 7b, is mainly composed of the control circuit, the Startup control and the WSN control functions. All its blocks are powered by C_S, except the WSN control which is dependent on the voltage across the WSN storage capacitor (C_b). The Flyback circuit and its diode bridge, the two buffer capacitors, the derivation capacitor C_D and the dMOS K_{bp} are off-chip, and K_{cc} is removed. Thanks to these external components, the architecture is not limited in terms of tolerated harvester voltages, enabling the use of a low cost and easily available CMOS technology.

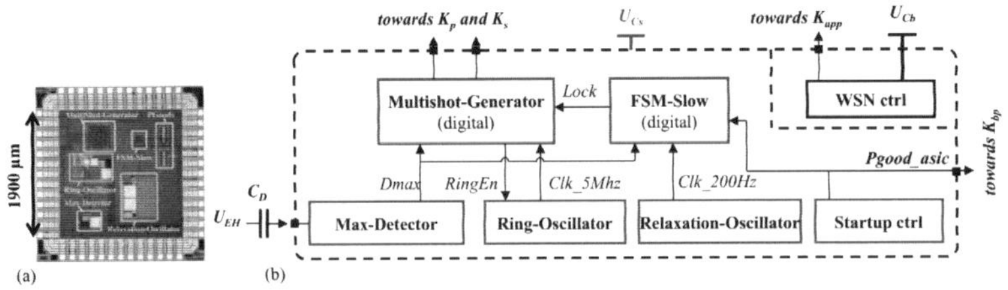

Figure 7. (a) Die of the ASIC implementing the self-starting PMC and (b) its internal architecture

5.2. Flyback control

The chip detects when the harvester has reached a maximum voltage thanks to the block "*Max-Detector*". This block, described in [8], shows a power consumption of 150nW @ 3V with a low detection delay (50 μs). It uses a transconductance amplifier biased by a 20nA current source. Instead of using delay cells (as in 3.2), the Flyback control is performed by a digital state machine clocked by a 5 MHz ring-oscillator. The ring is powered at a low stabilized voltage (≈ 1.2V) enabling a low power operation (200 μW) and a clock frequency independent of U_{Cs} level. Two input vectors of 8-bits set the number of clock cycles corresponding to the closing duration of K_p and K_s. This block has the particularity to switch the ring oscillator only during the Flyback control. It enables a very low power operation of the Flyback control since the harvester is discharged at a very low duty cycle (2% max) compared with the EH mechanical period.

5.3. Start-up control and WSN control

The Startup and WSN control functions are performed with two very low power POR (Power On Reset) which consume 50 nA@3V, and detect when U_{Cs} and U_{Cb} exceed 3V and fall below 2V to control K_{bp} and K_{app}. A power cut separates U_{Cs} and U_{Cb} in two power supply domains avoiding the pads' ESD diodes to conduct if a voltage (U_{Cs} or U_{Cb}) becomes greater than the other one. This enables to use a low threshold voltage diode (D_C) between C_b and C_s.

5.4. Harvesting energy from very low frequency deformations

Figure 8 shows an example of a complete operation of the proposed system harvesting energy from a piezoelectric harvester (figure 6a) excited with a mechanical frequency of 1.6 Hz and providing an input power of 30 µW. It validates the good operation and the compatibility of the PMC we propose with ASICs.

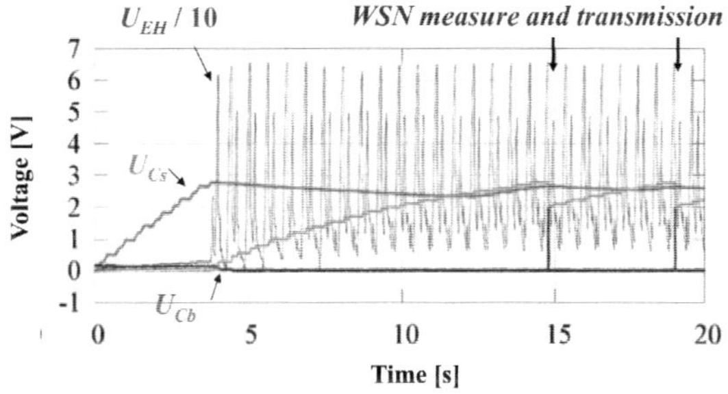

Figure 8. Experimental time-domain waveforms for a piezoelectric buzzer excited at 1.6 Hz

6. Conclusions

We have proposed an autonomous and self-starting PMC architecture employing a dMOS as a bypass of its power converter. A Schmitt trigger using a dMOS as a voltage limiter has been developed to control the startup with an ultra-low power consumption not exceeding 100nA. A PMC implementing a SECE on a flyback converter has also been designed; its power consumption is in the 500nA-1µA range. The operation has been validated with an electret-based vibration EH at 35Hz and a piezoelectric EH at 7Hz. This architecture has finally been implemented in an ASIC and validated on a low-frequency piezoelectric energy harvester. This new PMC is completely autonomous and can be made from standard off-the-shelf components or in ASICs. The proposed architecture is able to turn most of piezoelectric and electret-based mechanical EH into viable supply sources for battery-free WSN.

References

[1] Lefeuvre E *et al* 2005 Piezoelectric energy harvesting device optimization by synchronous electric charge extraction *J. Intell. Mater. Syst. Struct.* **16** 865.
[2] Guyomar D *et al* 2005 *Transactions on Ultrasonics, Ferroelectrics, and Frequency Control* **52** 584.
[3] Gasnier P et al 2012 Power conversion and integrated circuit architecture for high voltage piezoelectric energy harvesting *Proc. NEWCAS* 377.
[4] Hehn T *et al* 2011 A fully autonomous pulsed synchronous charge extractor for high voltage piezoelectric harvesters *Proc. ESSCIRC* 371.
[5] Boisseau S *et al* 2012 Electrostatic conversion for vibration energy harvesting *Small-Scale Energy Harvesting, ed M. Lallart, Intech*, doi:10.5772/51360.
[6] Boisseau S *et al* 2011 Cantilever-based electret energy harvesters *Smart Mater. Struct.* **20** 105013.
[7] Gasnier P *et al* 2013 An Autonomous Piezoelectric Energy Harvesting IC based on a Synchronous Multi-Shots Technique *Proc. ESSCIRC*
[8] Gasnier P *et al* 2013 An ultra low power maximum voltage detector for piezoelectric and electrostatic energy harvesters *Proc. NEWCAS*

A low-voltage boost converter using a forward converter with integrated Meissner oscillator

P Woias, M Islam, S Heller, R Roth

University of Freiburg, Dept. of Microsystems Engineering – IMTEK, Laboratory for Design of Microsystems, Georges-Koehler-Allee 102, 79110 Freiburg, Germany

E-mail: woias@imtek.de

Abstract. This paper describes a novel boost converter to be used with energy harvesters that provide only low output voltages. The device is self-supplied from electric power delivered to its input. With peak power conversion efficiencies above 30% at start-up voltages down to 10 mV this circuit sets best values in comparison with the state-of-the-art. This is achieved by the novel combination of a Meissner oscillator, used as stand-alone in most low-voltage step-up converters today, with a forward converter usually applied in high power systems.

1. Introduction

Depending on the application scenario, generators used for energy harvesting may deliver relatively low output voltages that are variable under a variable ambient power input. This is the case for thermoelectric energy harvesting, as shown e.g. in [1] where a low temperature difference of appr. 1 K generates output voltages of only a few 10 mV at a mid-size thermoelectric generator. Similar conditions exist for photovoltaic cells under low-light conditions or biofuel cells [2,8]. For these generators it is mandatory to use an electronic boost converter to raise the low delivered voltage up to a higher voltage for the supply of an embedded system. The requirements for such a step-up converter are very different from the state-of-the-art defined by various - even low voltage - off-the-shelf components: A step-up converter used in conjunction with energy harvesting should be self-starting from input voltages as low as possible without using any auxiliary power, provide a suitable step-up ratio and exhibit an acceptable efficiency of power conversion. Also, the input and output impedance may have to be adjusted for an optimal load matching or for an optimal voltage step-up ratio.

2. A brief review on low-voltage step-up converters

The parameters start-up voltage and self-supply together with the used circuit concept are important criteria for comparing different step-up converters. Exploring low-voltage step-up converters is still a topic of active ongoing research with only few devices being commercially available. These converters have in parts impressive start-up voltages as low as 20 mV [3]. An early circuit from the mid 1970s uses a Meissner oscillator with junction field effect transistors (JFETs) instead of bipolar junction transistors (see Fig. 1) [4]. This combination is favourable with respect to three important criteria: (1) JFETs provide a conducting channel with zero gate voltage thus enabling the onset of oscillations at low voltages. (2) The self-resonance of a Meissner oscillator does not require any auxiliary power for start-up and (3) a favourable transition of the oscillation from a linear into a non-linear regime can be accomplished by a suitable feedback circuitry, as shown especially in [5].

The latter effect helps to reduce internal losses and therefore to increase power conversion efficiency. Consequently, the Meissner oscillator, is used today in the majority of low-voltage step-up converters, sometimes with only minimal modifications of the type and number of transistors or the design of the feedback circuitry [5-8]. Start-up voltages as low as 6 mV have been achieved with a purely JFET-based Meissner oscillator [6], however with the drawback of large and bulky transformers. An early modular combination of a Meissner oscillator with a standard boost converter is presented in [5]: There, the Meissner oscillator serves as a start-up circuit providing the supply voltage for a separate, inductor-based step-up converter from an input voltage of 300 mV. A fusion of both converters is presented in [7], with the Meissner oscillator and the boost converter sharing the same transformer. The start-up voltage of this circuit is 70 mV.

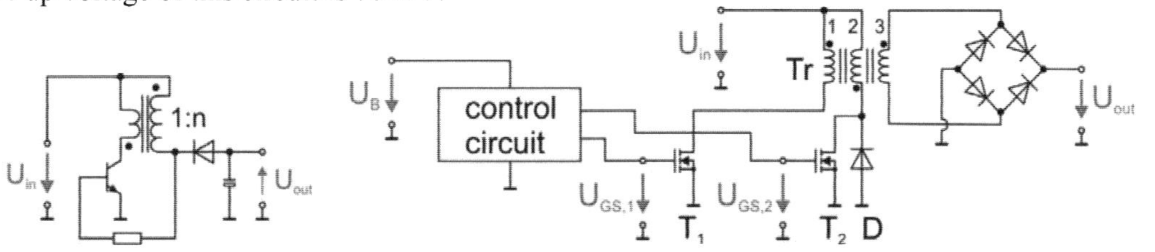

Figure 1. Basic circuit of a Meissner oscillator (left) and a forward converter (right)

3. Design concept

The low-voltage step-up converter developed within this study uses, as far as the authors know for the first time, the combination of a self-resonant Meissner oscillator and a self-resonant forward converter. This concept was chosen for several reasons. (1) A stand-alone Meissner oscillator will only draw an intermittent input current in its nonlinear operation regime, as the feedback circuit is continuously disrupting and re-establishing the current flow through its transistor. Therefore, the magnetic core of the transformer is biased in one direction with the danger of non-optimal use or even magnetic saturation. Also, the connected generator will only be loaded in an intermittent fashion, which may be detrimental and non-efficient. Both effects are not present in a forward converter (see Fig. 1) that is using a transformer with two input windings that are driven in an alternating way. (2) A Meissner oscillator starts at low input voltages but may operate inefficiently at low voltage levels. The Meissner oscillator , however, is able to work as a kick-off starter circuit for a more efficient forward converter, which would not start operation from such low input voltages. (3) An appropriate design of the feedback circuitry will allow to draw power not only from the output winding of the transformer but also from the feedback winding without compromising the self-oscillation of all circuits. As shown within this work such a behavior will increase the power conversion efficiency of the whole converter.
Two versions of the designed step-up converter, “design A” and “design B”, are presented in Fig. 2. In both circuits a Meissner oscillator is formed by connecting an array of five p-channel JFETs T_1 to T_5 to input winding 2 and feedback winding 1 of a transformer. Following the design concept presented in [5] the low end of the feedback winding is connected to ground by an RC tank in design A. As an improvement to this concept the low end of the feedback winding is connected to the output capacitor C_{out} via a parallel circuit of a capacitor C_F and a Schottky diode D_F in design B. The forward converter is realized by adding one p-channel MOSFET T_6 and one n-channel MOSFET T_7. While the drain and source of the PMOS transistor T_6 are connected in parallel to the JFET array, the NMOS transistor T_7 is connected to an additional input winding 3 (see also Fig. 1). Output power is drawn from the transformer through winding 4, using a simple one-way rectifier (Schottky diode D_R). With the connection scheme shown in Fig. 2 the feedback winding 1 is capable to deliver a common drive signal for all transistors, without the need of a separately powered control circuit (see Fig. 2). The gates of the JFET array are directly connected to the feedback winding, whereas the MOSFET gates are connected to the feedback winding by a common high-pass filter constituted by R_G and C_G.
The operation principle for design A is described in the following: At low input voltages the Meissner oscillator starts operation first. If the input voltage is high enough it undergoes a transition from its lin-

ear into a non-linear operation regime. The details of this start-up phase are described in [5] for a similar circuit and will not be repeated here. During start-up the gate-source diode of the JFET array acts as a rectifier for the AC signal present at the feedback winding 1. In consequence, a positive DC voltage U_F eventually builds up at C_F. This growing DC voltage shifts the operational set-point at the gates of the JFET array into the blocking region and eventually enforces the nonlinear operation of the Meissner oscillator. The bleeding resistor R_F prevents the build-up of a DC voltage at C_F that is large enough to keep the AC signal at the JFET gates in the blocking regime. In this case the JFETs would no longer be turned on and oscillation would stop.

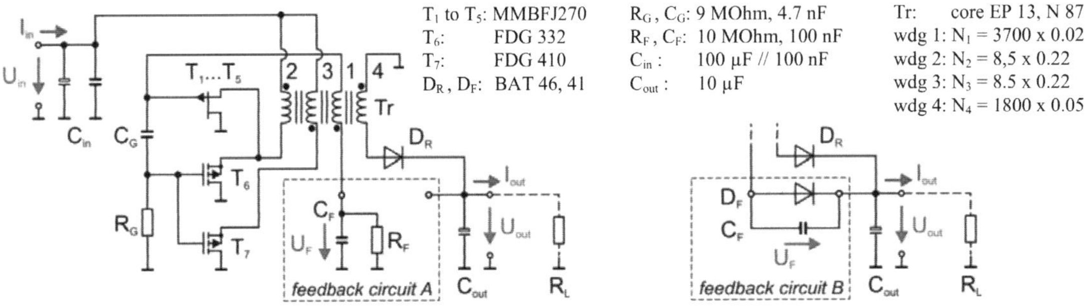

Figure 2. Meissner-forward converter: design A (left) and modifications for design B (right)

During the start-up phase the amplitude of the feedback signal at winding 1 is gradually increasing. As soon as the signal amplitude exceeds the threshold voltages of the MOSFET transistors, they start to turn on and off in an alternating way, such setting the forward conversion into operation. These MOSFETs exhibit a much lower on-resistance than the JFET array. Hence, the input currents and AC voltages at the input, feedback and output windings of the transformer are abruptly increased. The rectified output voltage will also increase up to an operational set-point defined by input voltage U_{in} and load resistance R_L.

Contrary to the circuit presented in [5] p-channel JFETs are used within this work. A drawback of such transistors is their much higher channel resistance compared to equivalent n-channel JFETs. This is compensated by connecting several transistors in parallel. The advantage of using p-channel JFETs is, however, that the DC voltage at C_F is positive with respect to ground. Therefore, the charge accumulated at C_F can be easily forwarded to the output capacitor C_{out}, e.g. by connecting a diode between C_F and C_{out}. Also, as done in design B, the low end of C_F can be directly connected to C_{out}. Additionally a Schottky diode D_F is connected in parallel to C_F to allow for a fast build-up and a limit of U_F to its threshold voltage. The current through D_F and C_F is directly fed into the output capacitor with a net increase of the available output power. In design A, as in [5], power is dissipated in the bleeding resistor R_F instead. If the load resistance is too low, the start-up phase of the converter is prolonged or even completely prohibited at low input voltages. It is therefore advisable to monitor the state of the converter and to connect the load after full start-up only. This can be accomplished by monitoring either the voltage at the output or the voltage at the feedback capacitor.

4. Experimental

A printed circuit board was designed and manufactured to characterize the step-up converters shown in Fig 2. Access to several test points was provided together with various jumpers to change device settings, and pin sockets to exchange MOSFET transistors, transformer and feedback circuitry (Fig. 3). Transformers were wire-wound on a commercial coil winder. Different core sizes, types, and materials, as well as different winding ratios and numbers of windings were used to find a sufficient set of parameters. For all transformers the input windings 2 and 3 were fabricated first in a bifilar winding pattern using enamelled copper wire. On top of this layer, first the feedback winding and then the output winding were wire-wound. Layers of 25 µm thick adhesive tape were applied on top of all windings as flat surface for the next winding. The required low input voltages were provided by a custom-made power supply. A large electrolyte capacitor (4700 μF) was connected to the output of this power supply to average the delivered current for increasing the accuracy of current measurements. Multimeters were used to measure input vol-tage, input current and output voltage. The load resistance was changed with a switchable resistor box 10 equidistant values between 10 kOhm and 100 kOhm. For the output voltage range of the step-up converters (0.5....3 V) these resistance values generate the typical power requirements of an energy autonomous system (10...200 μW).

Figure 3. Test board (design A, B)

5. Results and Discussion

For both designs A and B the start-up of the Meissner oscillation did set in, under no-load conditions at the output, at an input voltage of appr. 6.2 mV. For input voltages up to appr. 10 mV the Meissner oscillator was not capable of initiating the start-up of the forward converter. For input voltages of 10 mV and above a gradual increase of the AC voltage at the feedback winding and of the DC voltage at C_F was observed, with a rapid on-set of forward conversion as soon as a certain DC threshold was reached. During this stepwise transition the AC feedback voltage changed shape from a sinusoidal over a distorted sinusoidal into a square-like waveform. For an input voltage of 10 mV and with the connected rectifier and a discharged output capacitor C_{out} the start-up procedure took appr. 25 seconds under no-load conditions. The output voltage was app. 360 mV at this point, i.e. together with the threshold voltage of the used diode D_F, the DC feedback voltage $U_F + U_{out}$ was appr. 500 mV.

Fig. 4 shows the most important characteristics of design B. Depending on the delivered output power and the available input voltage the step-up ratio varied between 50 and 136. For the respective maximal output power P_{out}, whose value varied between 15.6 μW for U_{in}=10 mV and 189.7 μW for U_{in}=30 mV, a fairly constant step-up ratio of 80 was reached. Under no-load condition much higher step-up ratios ranging between 618 and 792 could be achieved. However, at input voltages exceeding 19 mV clipping did set in, i.e. intermittent oscil-lations occurred under no load, due to a periodically too high DC part of the feedback voltage $U_F + U_{out}$. Such high step-up ratios under no-load can be used favourably to charge the output capacitor C_{out} with a high voltage that may then be used to kick-start a self-supplied boost converter requiring higher start-up voltages. This concept is realized in [5] and [9] with a much higher start-up voltage of 300 mV.

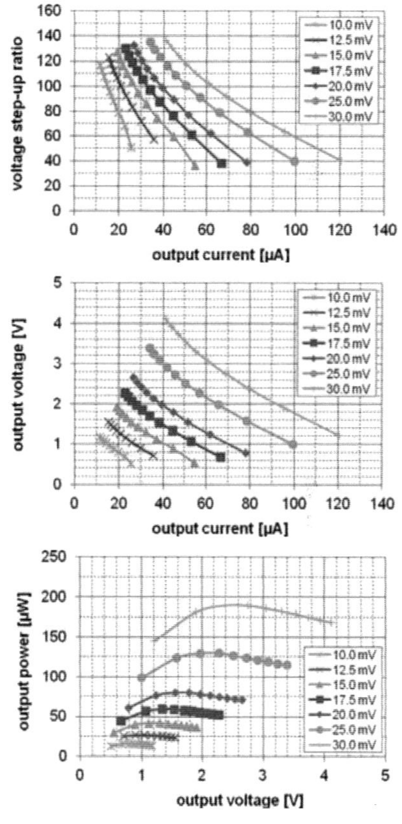

Figure 4. Measured data of design B: step-up ratio U_{out}/U_{in} and output voltage U_{out}, both vs. output current I_{out} and input voltage U_{in}, output power $P_{out}=U_{out}\cdot I_{out}$ vs. U_{in} and U_{out}.

As expected, output voltage and output power are highly depending on the input voltage as well as on each other. For the lowest input voltage (10 mV) the minimal values of U_{out} and P_{out} occurred at the same set point, with 0.509 V at 12.95 μW. For the highest input voltage (30 mV) the maximum output voltage and output power were 4.109 V at 168.9 μW and 189.8 μW at 2.386 V, respectively.

A comparison of the power conversion efficiencies P_{out}/P_{in} for design A and B (Fig. 5) shows the improvement obtained with design B. The maximal power conversion efficiency of design A does significantly depend on the input voltage and falls drastically from its highest values of 32% at 10 mV to only 14% at 20 mV. This is, for instance, also found in [6]. At input voltages slightly above 20 mV design A showed clipping under load, with a further, drastic reduction of power conversion efficiency and, overall, an unstable operation. The main reason for clipping is a too high feedback voltage at C_F. Design B avoids this problem by the measures discussed before. Power conversion efficiencies for design B were consistently higher than for design A. Also the efficiency values were stable in a much larger range of load currents and input voltages. Clipping was not observed with design B, for the tested ranges of output current and input voltage.

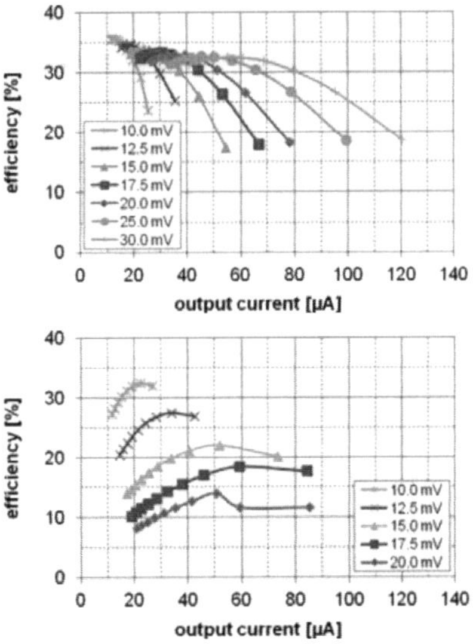

Figure 5. Power conversion efficiency P_{out}/P_{in} for design B (top) and design A (bottom) versus output current I_{out}.

6. Conclusion

A novel design of a low-voltage step-up converter was presented, using a combination of a Meissner oscillator and a forward converter. The realized circuit starts self-resonant operation as Meissner oscillator at an input voltage of 6.2 mV and forward conversion at 10 mV. In difference to the state-of-the-art that frequently uses Meissner oscillators alone the maximum power efficiency of an improved design presented within this work was almost constant. It exceeded 30% for a large range of input voltages. An input voltage between 20 and 25 mV is sufficient to supply a typical energy-autonomous embedded system with a power consumption of approximately 100 µW at a voltage of 2.2 V [10]. Future research will concentrate on optimization and miniaturization, and on an application in energy-autonomous systems.

7. Acknowledgement

The DFG Cluster of Excellence “BrainLinks-BrainTools” (EXC 1086) has partly funded this research.

8. References

[1] Moser A, Erd M, Cobry K, Kroener M, Woias P 2012 Thermoelectric Energy Harvesting from Transient Ambient Temperature Gradients *Journal of Electronic Materials* **41 (6)** 1653-1661
[2] Kerzenmacher S, Zehnle S, Volk T, Jansen, D, von Stetten, F, Zengerle R 2008 An efficient low-power DC-DC converter enables operation of a cardiac pacemaker by an integrated glucose fuel cell *Proc. PowerMEMS 2008 (Sendai, Japan, 9-12 Nov. 2008)* pp 189-192
[3] LTC3108 2010 *datasheet* Linear Technology Corp. Milpitas CA USA
[4] Parker B, Furman S 1975 Pacemaker electronic circuits *Advances in Pacemaker Technology* ed M Schaldach and S Furman (Berlin: Springer-Verlag) pp 55-72
[5] Damaschke J M 1997 Design of a Low-Input-Voltage Converter for Thermoelectric Generator *IEEE Transactions on Industry Applications* **33 (5)** 1203-1207
[6] Grgić D, Ungan T, Kostić M, Reindl L M 2009 Ultra-low input voltage DC-DC converter *Proc. PowerMEMS 2009 (Washington DC, USA, 1-4 December 2009)* pp 265-268
[7] Pollak M, Mateu L, Spies P 2008 Step-up DC-DC-Converter with Coupled Inductors for low Input Voltages *Proc. PowerMEMS 2008 (Sendai, Japan, 9-12 Nov. 2008)* pp 145-148
[8] Degrenne N, Buret F, Morel F, Adami S-E, Labrousse D 2011 Self-Starting DC:DC Boost Converter for Low-Power and Low-Voltage Microbial Electric Generators *Proc. IEEE Energy Conv. Congress and Exposition (ECCE)* pp 889-896
[9] S-882Z 2005-2010 *datasheet* Seiko Instruments Inc. Chiba Japan
[10] Woias P, Wischke M, Eichhorn C, Fuchs B 2009 An energy-autonomous wireless temperature monitoring system powered by piezoelectric energy harvesting *Proc. PowerMEMS 2009 (Washington, DC, USA, 1-4 December 2009)* pp 209-212

Switched-mode impedance synthesis for electrical tuning of a vibration energy harvester

J A Bowden, S G Burrow, and L R Clare
Department of Aerospace, Faculty of Engineering, University of Bristol, UK.

E-mail: james.bowden@bristol.ac.uk

Abstract. Switched-mode power circuits are able to efficiently synthesise a variable complex load impedance that can tune a vibration energy harvester, whilst also providing rectification and feeding the harvested energy into a DC store. The electrical tuning system presented in this paper is based upon a boost rectifier configured as a variable power factor converter. Its performance is benchmarked against a more basic resistance emulator where it is demonstrated that electrical tuning provides an increase in power bandwidth of over three times. The paper describes the experimental results of electrical tuning in some detail and elucidates the design challenges for these systems.

1.1. Introduction

Tuning of vibration energy harvesters via the input impedance of the electrical load is one method of broadening the frequency response of a harvesting system. This approach has been most widely reported for vibration absorbers [1] were the goal is to suppress structural vibrations and since in this application power autonomy is often not required, there is considerably more freedom in how the impedance synthesis is implemented. However, for an energy harvesting application it is necessary to harvester all the available power and hence energy-efficient switched-mode power circuits must be used. In this paper we report on the latest experimental work toward the development of a self-contained electrical tuning system.

1.2. Theoretical and simulated behavior

The concept of an electrically tuned energy harvester is illustrated by the diagram of figure 1. The mechanical components comprises of a base excited mass/spring system with parasitic damping, and coupling into the electrical domain. In this case electromagnetic coupling is considered. The transducer will introduce parasitic resistance (R_{coil}), and inductive reactance (L_{coil}). The variable impedance load is shown represented by resistance, capacitance and inductance. The electrical load can be expressed as: $Z_l = R_l + jX_l$, (noting for inductors and capacitors, $X_L = \omega L$ and $X_C = -1/\omega C$, respectively), with values ranging $0 < R_L < \infty$ and $-\infty < X_L < \infty$. Detailed analysis of this system has been undertaken in [2], where it has been shown that the optimum value of load impedance, to maximise power generation is given by (1).

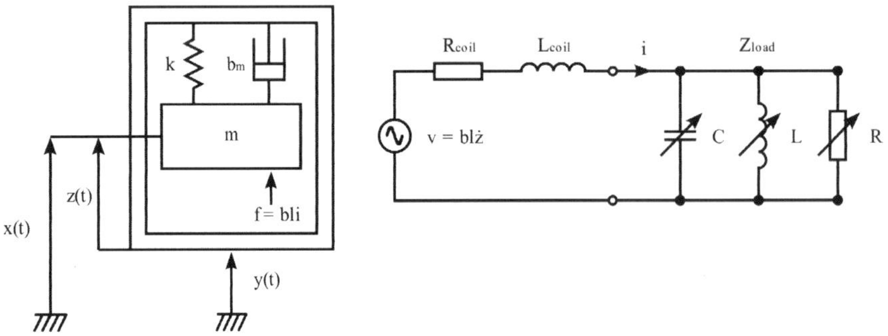

Figure 1. Representation of an electromagnetic energy harvester with variable impedance electrical load. y = input excitation, k = spring compliance, c = mechanical damping, m = mass, bl = electro-magnetic coupling coefficient, L_{coil} = coil impedance, R_{coil} = coil resistance, R_l = load resistance, L_l = load inductance, C_l = load capacitance.

$$Z_l = R_l + jX_l = R_{coil} + \frac{bl^2 c\omega^2}{\left(k - m\omega^2\right)^2 + \left(c\omega\right)^2} + j\left(\frac{bl^2\left(m\omega^2 - k\right)\omega}{\left(k - m\omega^2\right)^2 + \left(c\omega\right)^2} - \omega L_{coil}\right) \tag{1}$$

Where, ω is the frequency of excitation.

Figure 2 plots the apparent, reactive and true input power into the optimum electrical load impedance. These results are derived from the numerical simulation of the system in figure 1 using the parameters given in table 1, measured on a prototype energy harvester shown in figure 5. The prototype harvester develops a nominal 20mW of power into an optimal resistive load with harmonic base excitation of 70μm. Referring to figure 2 it is evident that when tuning away from the harvester's original resonant frequency significantly more reactive power flows between the harvester and its electrical load as the optimum load impedance becomes more reactive, where at its peak it is almost four times higher than true power delivered to the electrical load resistance. This results in a very low power factor, P.F., as given by equation (2) and plotted in figure 3 where |S| is the apparent input power and P the true input power.

$$P.F. = \frac{P}{|S|} = \frac{R_{opt}}{|Z_{opt}|} \tag{2}$$

Thereby the design requirements of the converter are driven by the apparent power that it must handle rather than the true power that will be delivered to the load. Figure 4 plots the magnitude of the input current the converter must handle which shows additionally that the converter must also handle very high input currents with very little loss to achieve highly efficient operation. This is particularly true above the harvester's original resonant frequency the input current increases to over five times higher than that at of a unity power factor, this results in the requirement of the converter times higher than that at of a unity power factor, this results in the requirement of the converter to have extremely low conduction losses in order to tune to higher frequencies above mechanical resonance.

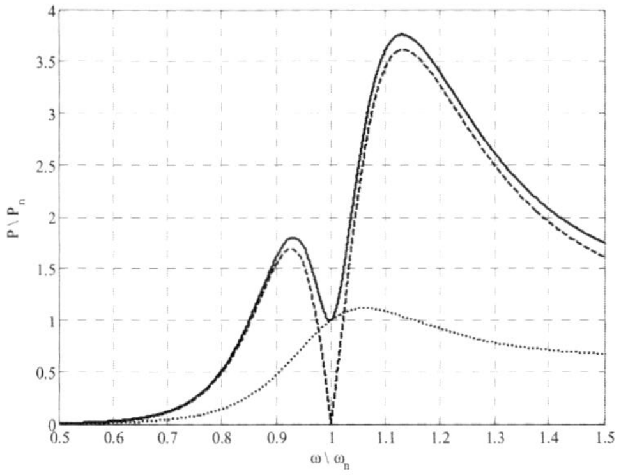

Figure 2. Normalised input power to the bridge with the energy harvester optimally loaded versus frequency, apparent power (VA) (|), reactive power (VAR) (- -), true power (W) (...).

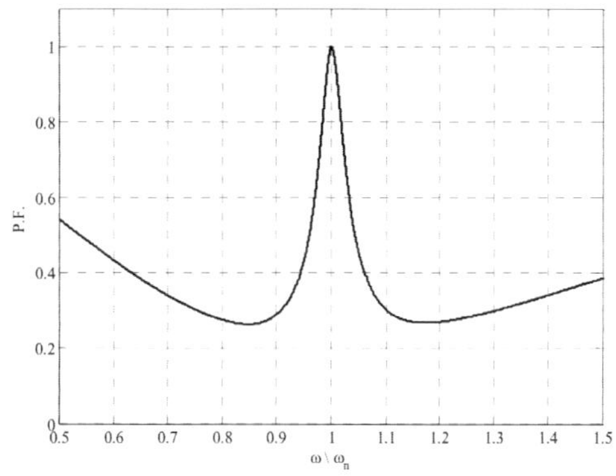

Figure 3. Input power factor versus normalised frequency. Power factor falls rapidly from unity away from the harvester's mechanical resonant frequency.

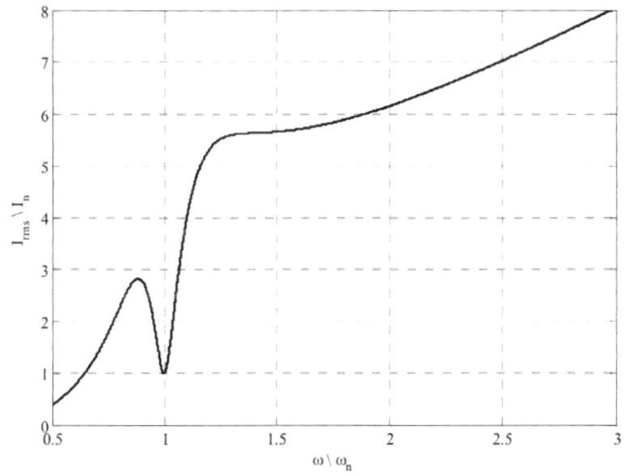

Figure 4. Normalised rms input current versus frequency.

Figure 5. The physical energy harvester used in the experimental validation..

Table 1. Harvester parameters.

Property	Symbol	Value (unit)
Inertial mass	m	22g
Spring compliance	k	3371Nm^{-1}
Mechanical damping	c	0.345Nsm^{-1}
Coil inductance	L_{coil}	250μH
Coil resistance	R_{coil}	1.6Ω
Coupling coefficient	bl	5.64Vsm^{-1}

1.3. Experimental setup

A four quadrant boost rectifier configured in average current mode control is used to form the power system, figure 6 and 7 respectively. Current demand is derived from the output voltage of the harvester, modified by the reciprocal of the desired impedance to be synthesised. Careful choice of power devices, micro-power current sensing and gate driver circuits are required to achieve operation at this very low power level. A compensated feedback control, implemented using micro power analogue circuitry, forces the converter's input current to the required value. Referring to figure 7 a dSPACE system is used to derive the current demand from the measured harvester voltage (admittance block in figure 7), this allows for ease of modifying the parametric values of load impedance in real-time during experimentation.

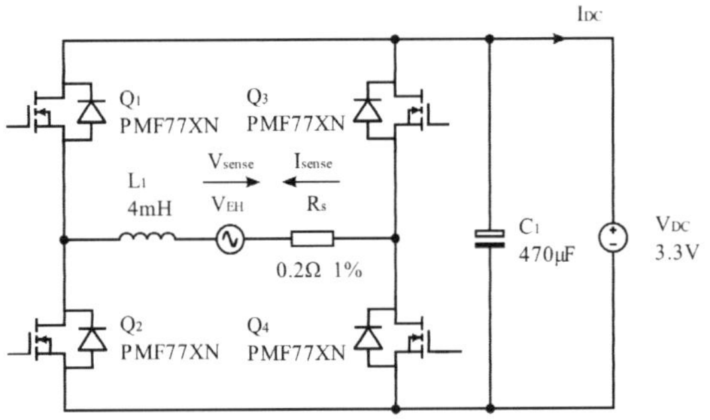

Figure 6. Four quadrant boost rectifier power stage. The 3.3V DC link is maintained by a power supply that is both capable of sinking and sourcing power. The MOSFET's drive electronics is not shown.

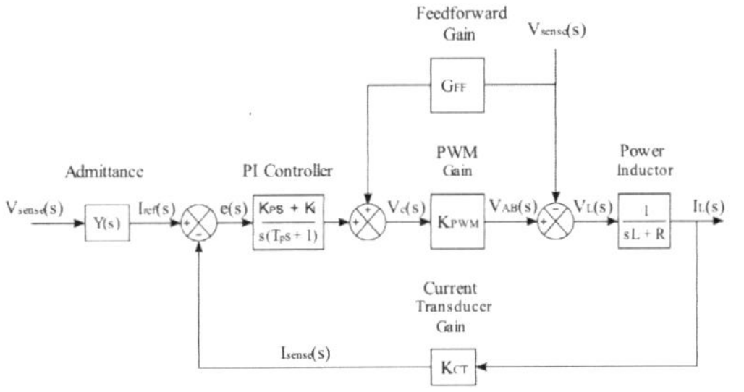

Figure 7. The controller structure for the impedance synthesiser. Blocks are implemented in dSPACE, except for error amplifier which is constructed from discrete components..

1.4. Results

To contrast the performance of the electrical tuning technique at providing useable DC power with conventional power conditioning circuits, a resistance emulator, based on the flyback topology, similar to that reported in [3]. Experimental results are given in figure 8. With reference to figure 8, there are two groups of three curves representing the theoretical power output, measured power output of the harvester (AC side) and the useful DC side power, for the resistance emulator circuit and the electrical tuning circuit. It can be seen that electrical tuning extends the frequency bandwidth of usable power over the resistance emulator.

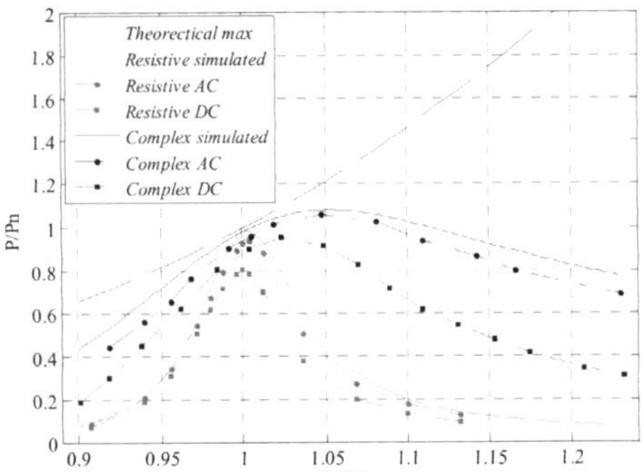

Figure 8. Comparison of power bandwidth of tunign converter and resistance emulator

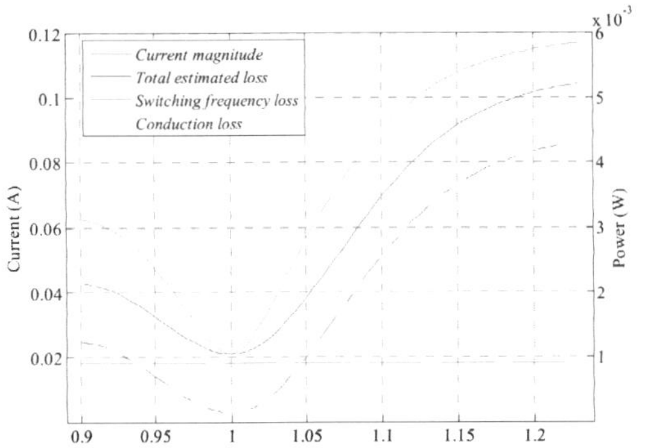

Figure 9. Losses of the tuning converter

1.5 Losses in the switched-mode converter

Close to the resonant frequency of the harvester the converter is able to provide most the input power at its DC output: at its peak an efficiency of 91.6% was achieved. However, moving away from resonance, the percentage of useful power on the DC side drops. This is due to the fact that as the harvester is excited further away from its resonant frequency current magnitude increase and thus conduction losses. Figure 9 shows the AC current magnitude, showing the large increase in magnitude away from resonance, and made up of: 1) conduction losses from fundamental current component; 2a) conduction and magnetic losses from switching frequency components; 2b) switching loss in the MOSFETs, measured experimentally. Details of the parameters and values used in the loss analysis are given in table 2. It can be seen that around resonance, the quiescent losses dominate, but as the current magnitude increases the conduction loss at the fundamental frequency starts to dominate. The values calculated for loss by this method show good agreement with the experimental data in figure 8 around resonance; however for higher current levels found at higher frequencies they predict only 75% of the losses seen in practice.

Table 2. Parameters and measured values used in loss analysis.

Property	Symbol	Value (unit)
Resistance of R_{sense}	R_{sense}	0.2Ω
DC resistance of L_1	R_{LDC}	0.278 Ω
Voice coil resistance	R_{VC}	1.6 Ω
MOSFET R_{DSON}	R_{DSON}	77mΩ
Switching loss	P_{Lsw}	470μW
Switching ripple current losses	P_{Lrip}	440μW

1.5. Conclusion

Implementation of electrical tuning with switching converters offers the ability to extend the bandwidth of an energy harvester. The system presented in this paper demonstrate a switched-mode converter operating as a variable impedance synthesiser can increase the half-power bandwidth of the harvester to over three times that of a unity power factor converter operating as a fixed load resistance. However, the constraint is that the design of the converter must be based upon the apparent power rather than the real power that is delivered to the DC link to achieve efficient operation whilst operating with very low power factors. In the case described in this paper the converter needed to handle currents in excess of five times higher than a converter designed just to operate with the harvester just at mechanical resonance. This is made worse when tuning harvesters with high quality factors. This is in direct contrast to the rather more relaxed power handling capabilities of the unity power factor converter. To achieve tuning over wide bandwidths the design of converter topologies with very high efficiencies is required.

[1] D Niederberger, S Behrens, A J. Fleming, S. O. Reza Moheimani, and M Morari 'Adaptive Electromagnetic Shunt Damping'. IEEE/ASME Transactions on Mechatronics, vol.11, No.1,Feburary 2006 pp 103-104.
[2] Cammarano A, Burrow S G, Barton D A W, Carrella A, Clare L R. "Tuning a resonant energy harvester using a generalized electrical load" Journal of smart materials and structures, vol. 19, 055003, March 2010.
[3] S Burrow, L Clare 'Open-loop power conditioning for vibration energy harvesting' Electronic Letters, vol.45, Issue 19, pp999-1000. September 2009.

Impedance matching for broadband piezoelectric energy harvesting

F Hagedorn[1], J Leicht[1], D Sanchez[1], T Hehn[2] and Y Manoli[1,2]

[1] Fritz Huettinger Chair of Microelectronics, Department of Microsystems Engineering – IMTEK, University of Freiburg, Germany

[2] HSG-IMIT – Institute of Micromachining and Information Technology, Villingen-Schwenningen, Germany

E-mail: hagedorn@imtek.de

Abstract. This paper presents a system design for broadband piezoelectric energy harvesting by means of impedance matching. An inductive load impedance is emulated by controlling the output current of the piezoelectric harvester with a bipolar boost converter. The reference current is derived from the low pass filtered voltage measured at the harvester terminals. In order to maximize the harvested power especially for nonresonant frequencies the filter parameters are adjusted by a simple optimization algorithm. However the amount of harvested power is limited by the efficiency of the bipolar boost converter. Therefore an additional switch in the bipolar boost converter is proposed to reduce the capacitive switching losses. The proposed system is simulated using numerical parameters of available discrete components. Using the additional switch, the harvested power is increased by 20 %. The proposed system constantly harvests 80 % of the theoretically available power over frequency. The usable frequency range of ±4 Hz around the resonance frequency of the piezoelectric harvester is mainly limited due to the boost converter topology. This comparison does not include the power dissipation of the control circuit.

1. Introduction

A piezoelectric harvester (PEH) transforms the kinetic energy stored in mechanical vibrations into electrical energy using the piezoelectric effect. Usually a PEH consists of a piezo material mounted to a mechanical resonator which can be a cantilever beam for instance. In case a PEH is excited near its resonant frequency, a power in the range between 10 µW and 10 mW can be harvested [1, 2]. This power is sufficient to supply wireless sensors nodes in a vibrating environment [3, 4, 5]. But due to the mechanical resonance behavior, the harvested power decreases considerably for nonresonant excitation frequencies. Since many vibrational environments do not have an a priori fixed frequency but rather a shifting frequency or even a random behavior, the PEH bandwidth must be increased to conform to the practical conditions [6].

The mechanical design approaches for broadening the bandwidth are summarized in [7, 8] and can be grouped as follows: resonance tuning, multi-modal and multi-frequency harvesting, frequency up-conversion and bistable oscillations. However the power spectrum of the PEH is not only afected by the mechanical resonator but also by the electrical load. Even at resonance the power harvested using a state of the art interface circuit (diode rectifier, SECE or SSHI [9, 10]) varies and depends mainly on the mechanical characteristics of the PEH. The theoretical power limit is achieved when the electrical load behaves like the conjugate complex electrical impedance of the PEH [11, 12, 13]. Furthermore with a conjugate load impedance the harvested power at nonresonant frequencies could be theoretically the same compared to the excitation at resonance. Previously reported designs use the amplitude ratio and phase shift between the excitation acceleration and the PEH voltage to optimize the harvested power [11, 12]. Therefore an acceleration sensor is needed which is unnecessary in the proposed system.

This paper is organized as follows. First, the conjugate complex impedance matching of the PEH is described. Then, the control of a bipolar boost converter emulating a inductive load impedance for a constant frequency is proposed and simulated. Afterwards the optimization algorithm adjusting the emulated load impedance to maximize the harvested power also for changing frequencies is presented. Finally, the simulation results are compared with a similar system design presented in [13].

2. Emulating an inductive load impedance

The electromechanical model of the PEH is shown in figure 1(a), where m, k and d are the mass, spring and damping constants of the mechanical oscillator and $\dot{x}$ its velocity. The electromechanical coupling and the capacitance of the piezoelectric material is modeled by γ and C_p. The voltage and current on the PEH terminal are denoted by v and i.

The excitation acceleration a is assumed to be sinusoidal with the amplitude $\hat{a}$ and the frequency f. Then, the maximum power limit P_lim is harvested when the PEH is connected to a load impedance Z_0 which is equal to the conjugate complex PEH impedance Z_PEH:

$$P_\mathrm{lim} = \max \overline{vi} = \frac{(m\hat{a})^2}{8d} \qquad \Longleftrightarrow \qquad Z_0 = Z^*_\mathrm{PEH}(f) \ . \tag{1}$$

As the PEH impedance is frequency dependent the load impedance must be adjusted whenever frequency changes. Since the power limit P_lim is frequency independent, the same power for resonant as well as nonresonant excitation could be harvested. The applicable frequency range is limited due to the reactive power flowing alternately between the energy reservoirs of the load and the PEH. It can be shown that the reactive power needed for impedance matching increases for nonresonant frequencies. Since the power transfer between the PEH and the energy storage in the load is associated with losses the usable harvested power decreases.

In order to realize a complex load impedance, the bipolar boost converter in figure 1(b) is able to transfer power from the PEH into the buffer storage V_buf and vice versa. That means, for positive and negative PEH voltages v, the current i through the inductor L can rise and fall as long as $|v| < V_\mathrm{buf}$. The resistive and capacitive parasitics of the inductor L and the switch S are summarized by R_par and C_par modeling the power losses of the converter.

According to equation (1) the load impedance Z_0 must be inductive since the PEH impedance Z_PEH is capacitive. A possible realization of an inductive load impedance would be $Z_0 = R_0 + sL_0$ where R_0 and L_0 are adjustable to satisfy equation (1). If the PEH voltage v is measured then

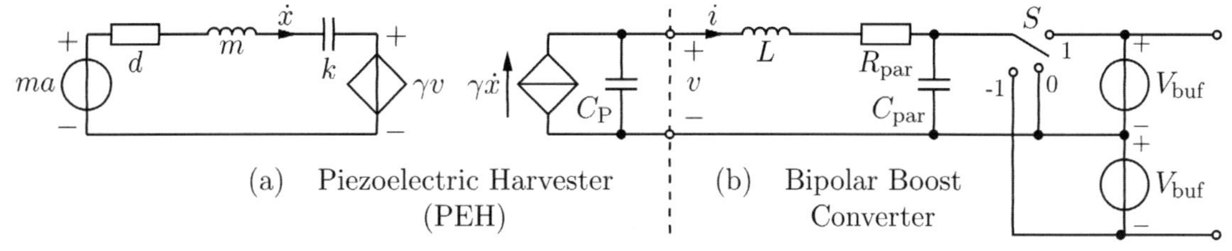

Figure 1. Model of the piezoelectric harvester and the bipolar boost converter.

the current i_0 through Z_0 is calculated by

$$I_0(s) = \frac{V(s)}{R_0 + sL_0} = A_0 \frac{V(s)}{1 + s/\omega_0} \quad \text{with} \quad A_0 = \frac{1}{R_0} \quad \text{and} \quad \omega_0 = \frac{R_0}{L_0} .$$

Thus, the current i_0 through the emulated load impedance is determined by low pass filtering the PEH voltage v. Since the gain A_0 and the cutoff frequency ω_0 must be adjustable, a digital filter realization with the sampling time $t^k = kT_s$ is preferred. Before the measured PEH voltage v^k is low pass filtered it is extrapolated a half time step into the future compensating for the time delay of the sampled output i_0^k. The extrapolated PEH voltage $\tilde{v}^k \approx v^k(t^k + T_s/2)$ is estimated by a cubic polynomial fitted to the last eight measurements.

After determining the sampled current i_0^k flowing through the emulated load impedance Z_0, the bipolar boost converter in figure 1(b) is used to pull this current out of the PEH in average. This is achieved by a hysteresis control which sets the switch position S. Using only the two switch positions $S \in \{-1, 1\}$, the current i can be easily controlled in a hysteresis window ΔI around i_0^k as follows. In the switch position $S = -1$ the current i rises as long as $|v| < V_{\text{buf}}$. If i reaches the upper hysteresis boundary $i_0^k + \Delta I$, the switch position $S = 1$ forces the current i to fall towards to the lower boundary $i_0^k - \Delta I$. With that two-switch hysteresis control the capacitive energy loss in each switching cycle is equal to $4V_{\text{buf}}^2 C_{\text{par}}/2$. This energy loss can be reduced by a factor of four if the additional switch $S = 0$ is used. The finite state machine of the three-switch hysteresis control is shown in figure 2(a). If the PEH voltage v is positive then the current i is controlled by the switch positions $S \in \{0, 1\}$, otherwise the switch positions $S \in \{-1, 0\}$ are used. A simulation of the proposed system design is shown in figure 2(b). The applied parameters are given in Table 1 in which R_{par}, C_{par} and L are extracted from available

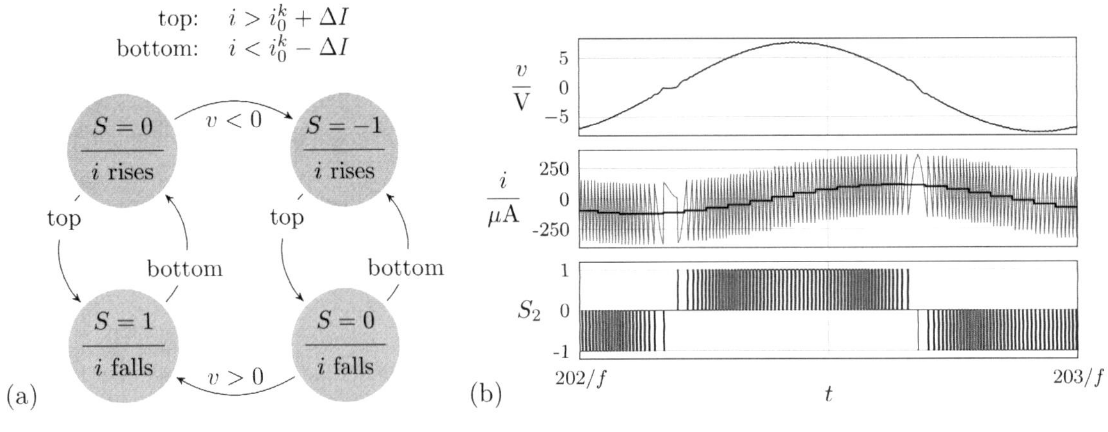

Figure 2. (a) Finite state machine of the three-switch hysteresis control. (b) Quasi steady state of the proposed system. The thick line in the middle plot shows the sampled current i_0^k.

Table 1. Applied numerical parameters for available discrete components.

Piezoelectric Harvester (Mide V22B)		Bipolar Boost Converter		Load Impedance
$m = 1.8 \cdot 10^{-3}\,\text{kg}$	$\hat{a} = 4\,\text{m/s}^2$	$V_{\text{buf}} = 10\,\text{V}$	$L = 180\,\text{mH}$	$f = 175\,\text{Hz}$
$d = 0.0426\,\text{kg/s}$	$C_{\text{P}} = 23\,\text{nF}$	$R_{\text{par}} = 15\,\Omega$	$\Delta I = 250\,\mu\text{A}$	$R_0 = 19.7\,\text{k}\Omega$
$k = 2076.5\,\text{N/m}$	$\gamma = 1.07 \cdot 10^{-3}\,\text{C/m}$	$C_{\text{par}} = 15\,\text{pF}$		$L_0 = 55\,\text{H}$
$f_{\text{res}} = 170.94\,\text{Hz}$	$P_{\text{lim}} = 152.11\,\mu\text{W}$			$T_{\text{s}} = 250\,\mu\text{s}$

discrete components. The mean power flowing out of the harvester is $0.99\,P_{\mathrm{lim}}$. The gap to P_{lim} results from a small time shift between the emulated load current i_0 and the mean of the modulated current i. This is caused by the irregular switching at the zero crossings of v.

3. Broadband impedance matching

According to equation (1), the load impedance $Z_0 = R_0 + sL_0$ must be adjusted whenever frequency changes since the PEH impedance $Z_{\mathrm{PEH}}(f)$ is frequency dependent. If the PEH model parameters are known in advance the frequency f can be measured and R_0, L_0 are determined by a simple look-up table. However, in practice the model parameters are not sufficiently known and slightly depend on the bias point of the PEH. For this reason the averaged output power $\overline{p} = \overline{vi}$ is maximized by adjusting the load parameters R_0 and L_0 with a two-dimensional optimization method. The utilized algorithm is known as coordinate-search method [14]. Alternately, one of the two coordinates R_0, L_0 is changed with the constant step size ΔR or ΔL as long as the averaged power $\overline{p}_0$ increases. If a local maximum along one coordinate is found further optimization is done along the other coordinate. The time delay between two steps T_{d} must correspond to the settling time of the total system which is shorter for smaller step sizes. For this reason the coordinate-search method is chosen over the simplex algorithm where both coordinates are changed at the same time resulting in higher transient overshoots.

The averaged output power $\overline{p} = \overline{vi}$ is approximated by averaging the instantaneous power through the emulated load impedance $p_0 = R_0(i_0^k)^2$. Using the approximation $\overline{p}_0 \approx \overline{p}$ which is caused by the phase shift between i_0 and i, an extra measurement of the fast changing current i is avoided. The averaging of p_0 is done by a third order Bessel low pass filter with a cutoff frequency $f_p \ll f$. Figure 3 shows the averaged output power $\overline{p}$ while optimizing $\overline{p}_0$ after a frequency step from f_1 to f_2 at $t = 0$. The steady state power ripple is necessary to detect the next frequency change, but can be further reduced with smaller step sizes.

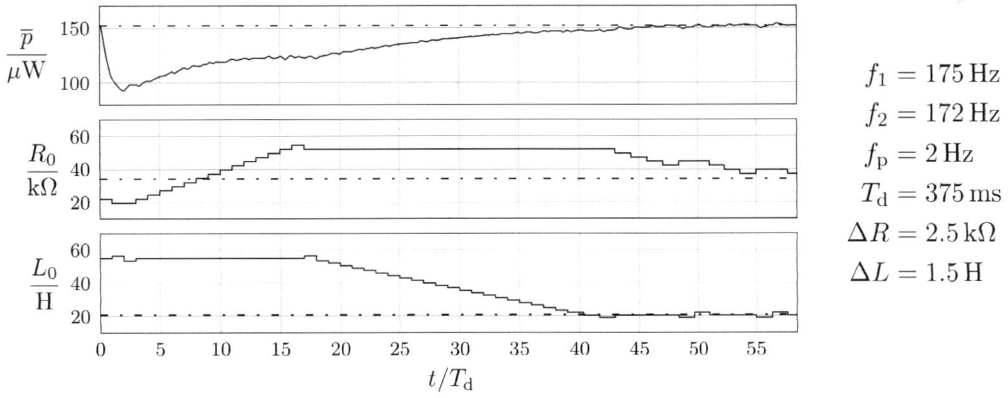

Figure 3. Illustration of the coordinate search algorithm after a frequency step from f_1 to f_2 at $t = 0$, maximizing the averaged PEH output power $\overline{p} = \overline{vi}$. The dash dotted lines mark P_{lim}, $R_0 = \mathrm{Re}\,Z_{\mathrm{PEH}}(f_2)$, $L_0 = -\mathrm{Im}\,Z_{\mathrm{PEH}}(f_2)/(2\pi f_2)$, from top to down.

By means of this optimization the emulated load impedance is matched to the conjugate complex impedance of the PEH for arbitrary frequencies around the PEH resonant frequency. Figure 4 compares the harvested power flowing into the two buffer sources V_{buf} for three different operating conditions. This comparison does not consider the power consumption of the required control circuitry. The dotted line shows the power which is harvested with the system concept presented in [13]. By using only the two-switch hysteresis control this concept emulates a *constant* load impedance which is equal to the conjugate complex impedance of the harvester at resonance.

The next two cases show the improvements of the proposed system. First, the dash dotted line illustrates the improvement by using a optimized load impedance. Finally, the latter case is further improved (solid line) by using the three-switch hysteresis control which reduces the capacitive power losses. In contrast to the description of equation (1) the harvested power spectrum is approximately flat and limited to the shown frequency range. Since the reactive power needed for impedance matching increases for nonresonant frequencies the amplitude $\hat{v}$ of the PEH voltage also increases. The bipolar boost converter stops working properly if $\hat{v} \geq V_{\text{buf}}$. However, if $\hat{v} \lesssim V_{\text{buf}}$ then the efficiency increases due to the lower switching frequency.

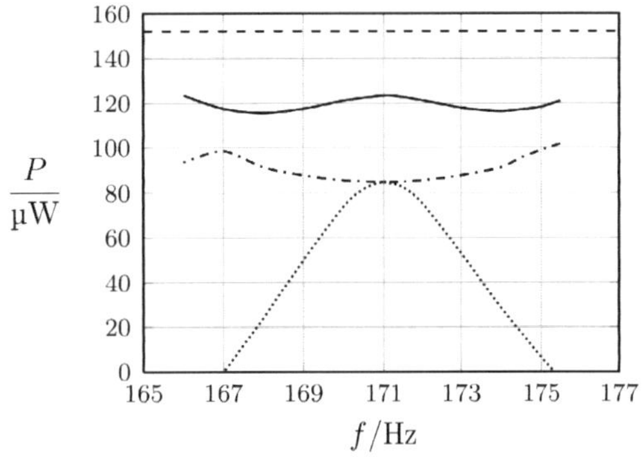

Figure 4. Theoretical power limit P_{lim} of the PEH (dashed), harvested power with the three- (solid) and two- (dash dotted) switch hysteresis control, harvested power with a similar system concept presented in [13] (dotted) using a hysteresis window of $\Delta I = 1.5\,\text{mA}$.

4. Conclusion

A system design has been presented optimizing the power harvested from a nonresonant, sinusoidally and continuously excited PEH. The proposed concept harvests a power of $0.8\,P_{\text{lim}}$ which is approximately constant over frequency. This power is 40 % larger than the peak power of the concept presented in [13]. An experimental validation of the presented system design has to be performed. The power consumption of the required control circuitry is expected to be much higher compared to implementation of the SECE scheme which needs less then 5 µW [9]. Hence the switching frequency must be decreased with a more sophisticated hysteresis control.

Acknowledgement

This work is part of the graduate program GRK 1322 Micro Energy Harvesting which is funded by the German research Foundation (DFG).

References

[1] Becker P, Hymon E, Folkmer B and Manoli Y 2013 *Sens. Act.* A (*Preprint* `doi:10.1016/j.sna.2013.04.030`)
[2] Roundy S and Wright P K 2004 *Smart Mater. Struct.* **13** 1131–42
[3] Leicht J, Heilmann P, Hehn T, Li X, Maurath D, Moranz C, Thewes M, Scholl G and Manoli Y 2013 *J. Phys.: Conf. Series (PowerMEMS 2013, London, UK)*
[4] Stojcev M, Kosanovic M and Golubovic L 2009 *Proc. Int. Conf. on TELSIKS, 2009 (Nis, Serbia)* 65–72
[5] Arms S W, Townsend C P, Churchill D L, Galbreath J H and Mundell S W 2005 *Proc. SPIE* **5763** 267–275
[6] Manoli Y 2010 *Proc. IEEE Eur. Solid-State Circuits Conf. (Seville, Spain)* pp 27 – 36
[7] Tang L, Yang Y and Soh C K 2010 *J. of Intell. Mater. Syst. and Struct.* **21** 1867–97
[8] Zhu D, Tudor M J and Beeby S P 2010 *Meas. Sci. Technol.* **21** 22001–29
[9] Hehn T, Hagedorn F, Maurath D, Marinkovic D, Kuehne I, Frey A and Manoli Y 2012 *IEEE J. Solid-State Circ.* **47** 2185 – 2198
[10] E Lefeuvre, A Badel, A Benayad, L Lebrun, C Richard and D Guyomar 2005 *J. Phys. IV France* **128** 177–186
[11] Luo C and Hofmann H 2011 *IEEE Trans. Ultrason. Ferroelectr. Freq. Control* **58** 1294–1301
[12] Zhang L, Zheng G and Li J 2012 *Proc. IEEE ICAL 2012 (Zhengzhou, China)* 131–135
[13] Chang S C, Ocalan M, Pabon J and Lang J H 2012 *Proc. PowerMEMS 2012 (Atlanta, USA)* 72–75
[14] Luenberger D and Ye Y 2008 *Linear and Nonlinear Programming* 3rd ed (New York: Springer)

Influence of a SiN Drift-Barrier on the Charge Decay in Parylene-C Electret Layers

S Genter, O Paul

Department of Microsystems Engineering - IMTEK, Microsystem Materials Laboratory, University of Freiburg, Germany

E-mail: simon.genter@imtek.uni-freiburg.de, paul@imtek.de

Abstract. This paper reports on the attempt to improve the charge carrier life time in Parylene-C electret layers by implementing a drift barrier layer close to the surface. The barrier layer is intended to avoid the rapid discharge by ionic drift through the electret layer. For this purpose, we fabricated test chips comprising silicon nitride drift barriers with thicknesses of 50 nm, 100 nm and 200 nm sandwiched between a 7.1-µm-thick Parylene-C base layer and capping Parylene-C electret layers of three thicknesses, charged using the corona discharge method. Experimentally, it turns out that there is no pronounced influence on the charging process due to the drift barrier. Surface potential measurements were repeated over a period of 20 days. Up to $t = 10$ days a weak improvement of the remaining surface potential can be interpreted into the data. However, thereafter the surface potentials decrease rapidly. Surface potential line scans over the samples show that the decreased width of the charge carrier distribution, likely due to an enhanced horizontal drift or diffusion, is the reason. This effect is not observed on charged Parylene-C layers without drift barrier.

1. Introduction

Recently, the development of micro energy harvesting devices transducing ambient energy into the electrical domain has received much attention [1, 2, 3, 4, 5]. These devices need no costly maintenance and thus are suitable for deployment in wireless sensor nodes and in medical implants such as cardiac pacemakers. They can contribute to extend the battery lifetime and may even avoid the use of dedicated batteries at all.

Vibrational energy can be harvested by electrostatic transduction relying on charged electret materials [6]. An electret material is able to store charge carriers quasi-permanently [7]. The stability of implanted ionic charge carriers is mandatory to guarantee the long-term functionality of the device. Among other materials such as silicon oxide and nitride and polymers like Cytop and Teflon, Parylene-C has been considered as a promising electret material [8, 9, 10]. Previously, the charge decay in simple Parylene-C layers with thicknesses up to 23 µm has been investigated [11]. The highest charge carrier stability was achieved using 11-µm-thick samples charged at 100°C with a decay of 20% after 140 days. However, the charge decay with time has to be further reduced for this material to be suitable for micro energy harvesters. An attempt to use a drift barrier layer close to the electret surface is described in this work.

2. Theory

2.1. Electret Charging

Various techniques to charge electret materials are available [7]. One of them is the corona discharge method, whereby an inhomogeneous electrical field between a sharp needle tip and a planar electrode generates a discharge in air at atmospheric pressure. A negative potential between the needle tip and the planar electrode accelerates the dominantly generated CO_3^- ions towards the electret layer covering the planar electrode where they are implanted close to the surface [7, 12]. A grid-shaped electrode between the needle tip and the planar electrode serves to control the implanted charge carrier density. As the surface potential of the electret layer rises, the potential difference against the applied grid potential and thus the accelerating force on the ions decreases until the charging process finally stops.

2.2. Discharging

The distribution of the stored charges decays due to several mechanisms [7]. Electret materials are non-ideal insulators and thus resistive losses occur depending on the presence of positive and negative intrinsic charge carriers available in the valence or conduction bands, respectively. Second, the implanted electrical charges build up an electrical field within the electret layer which lets the stored charges drift towards the planar electrode where they are neutralized. A third phenomenon is diffusion. The initially pronounced concentration gradient of the charge carriers implanted close to surface progressively smoothes out with time. Finally, compensation charges are attracted from the environment and cause a further charge decay. The discharging effect in Parylene-C predominant over longer times has been identified to be ionic drift [7].
The implanted ionic charge carriers drift through the electret layer with the drift velocity $v_D = \mu_I E(x,t)$ with the mobility μ_I under the self-induced electrical field $E(x,t)$ depending on the momentary charge carrier distribution $\rho_I(x,t)$. Thus, implementing a drift barrier layer close to the surface is conjectured to be a suitable approach to impede the charge carrier movement through the complete layer. Materials with low ionic mobilities and high dielectric constant such as silicon nitride or other ceramics are promising candidates. The lower electrical field within the barrier layer is expected to result in even lower drift velocities.

3. Test Chip Fabrication

To test this hypothesis, test chips implementing a silicon nitride (SiN) barrier layer were prepared. The test chips, schematically shown in Figure 1, were fabricated using four-inch Pyrex substrates with a thickness of 500 µm covered by a bottom electrode consisting of 20 nm of Cr, 300 nm of Au and 40 nm of Cr. Next, a Parylene-C layer with the thickness $d = 7.1 \pm 0.5$ µm is

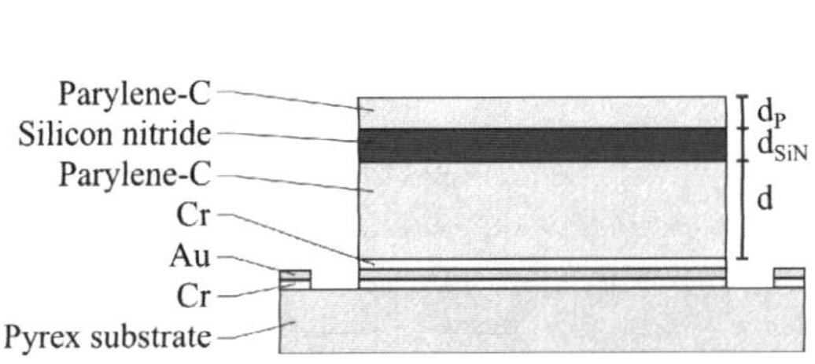

Figure 1. Schematic drawing of a cross-section of the test chip showing the composition of the stacked materials.

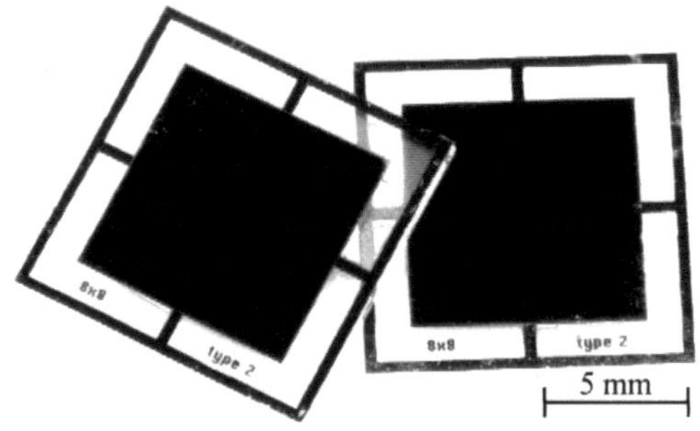

Figure 2. Photograph of two fabricated test chips for the characterization of the charge stability.

deposited in a *Labcoter 2* from *SCS Coatings* at a pressure of 22 mTorr. Then, SiN playing the role of a drift barrier is deposited by PECVD with thicknesses d_{SiN} of 50 nm, 100 nm and 200 nm. In view of the wafer-to-wafer Parylene-C layer thickness variation, SiN is etched back on one half of each wafer using reactive ion etching (RIE) to obtain test chips with and without barrier layer on each wafer with matching Parylene-C thickness. Subsequently, capping Parylene-C layers with thicknesses $d_P = 0.3\,\mu m$, $0.6\,\mu m$ and $1.2\,\mu m$ are deposited. The layer stack is structured using RIE to $8 \times 8\,mm^2$ squares. The exposed Cr layer is removed completely and the remaining Au and Cr layers are removed by wet etching except for a contact frame along the chip edges. Finally, the wafer is diced into $11.5 \times 11.5\,mm^2$ large chips as shown in Figure 2.

4. Experimental

4.1. Test Chip Charging

The fabricated test chips were charged using the custom-made corona discharge setup shown in Figure 3. The high potentials are generated by the two high-voltage sources *LNC-10000-2 neg* and *LNC 1200-20 neg* from *Heinzinger*, Germany. The potential $U_{Needle} = -8\,kV$ is applied between the sharp needle tip and the grounded electrode at the bottom of the electret layer on the test chip. At the same time the potential U_{Grid} is applied to the grid-shaped electrode. For the experiments U_{Grid} values ranging from $-200\,V$ to $-1000\,V$ in steps of 200 V and a charging duration of 3 min at room temperature (RT) were chosen.

4.2. Surface Potential Measurements

The measurement of the surface potential U_S was first carried out directly after the charging process and repeated after increasing time spans. For this purpose the electrostatic voltmeter *Isoprobe 279* from *Monroe Electronics* is used. The circular measurement spot of the probe has a diameter of 500 µm. Thus, the measured value corresponds to an average within this area. As shown in Figure 4, the voltmeter probe is mounted above a chip carrier able to handle up to 16 chips at once. Using two LabView-controlled linear stages the voltmeter probe scans the electret surface with a horizontal step size of 200 µm to map the surface potential.

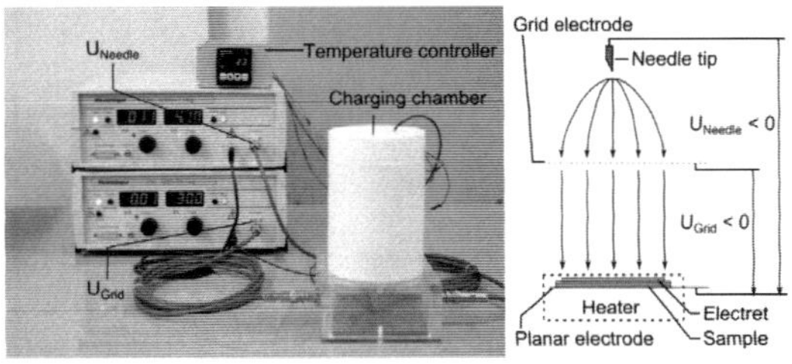

Figure 3. Photograph of the corona discharge setup (left) and a schematic view of the charging chamber (right).

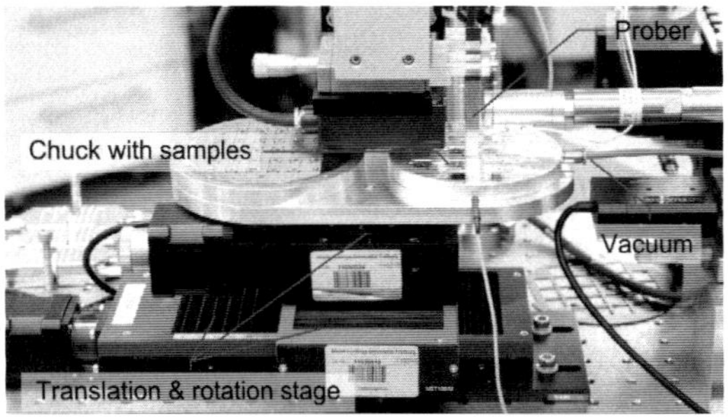

Figure 4. Photograph of the measurement setup of the surface potential of charged electret layers with drift barrier.

5. Results

5.1. Dependency of U_S on U_{Grid}

To observe the influence of the drift barrier layer on the charging and discharging process the measured surface potential U_S is repeatedly compared to identically charged test chips without drift barrier. In Figure 5 the dependency of the initial surface potential value, $U_{S,initial}$, on the grid potential U_{Grid} is shown for test chips with a $d_{\mathrm{P}} = 0.3\,\mu\mathrm{m}$ thick capping Parylene-C layer and for the different d_{SiN} values. No significant differences among three drift barrier thicknesses and the sample without drift barrier can be observed. Calculating the mean $U_{\mathrm{S,inital}}$ for each U_{Grid} value among all charged test chips and linearly fitting the dependency of $U_{\mathrm{S,initial}}$ versus U_{Grid} results in a slope of 683 V/kV and a standard deviation among the samples of 23.8 V. These results are in agreement with the charging behavior of test chips with pure Parylene-C electret layers [11]. Charging of the test chips with different d_{P} values resulted in similar dependencies.

5.2. Influence of the Drift Barrier on the Charge Decay

The measurement of the surface potential U_{S} was repeated over a time span of 20 days. In Figure 6, the discharging behavior for test chips charged at $U_{\mathrm{Grid}} = -1000\,\mathrm{V}$ at RT with a capping Parylene-C thickness $d_{\mathrm{P}} = 0.3\,\mu\mathrm{m}$ and drift barrier thicknesses of 50 nm, 100 nm and 200 nm are shown exemplarily. Except for one measurement point, all chips seem to discharge at roughly the same rate for times below $t = 0.03$ days. This similar behavior is likely due to the thickness of the capping Parylene-C layer which has to be traversed and has the same thickness for all shown test chips. Later, the discharging curves spread. At $t = 10$ days, the test chips with SiN drift barriers show remaining surface potentials U_{S} increased slightly by

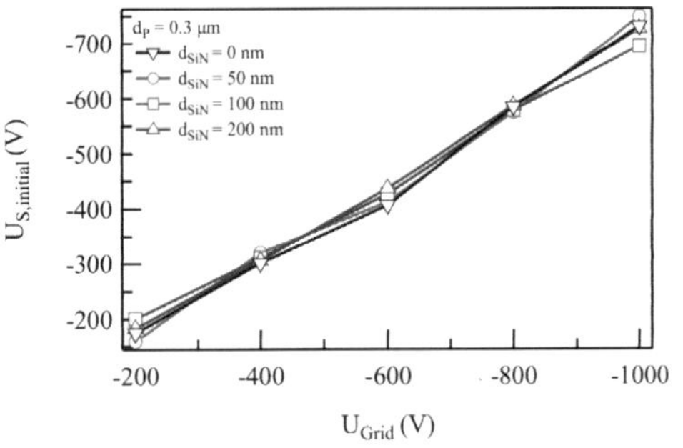

Figure 5. Measured initial surface potential $U_{\mathrm{S,initial}}$ as a function of the grid potential U_{Grid} for 7.1-µm-thick Parylene-C with different barrier thicknesses charged at RT.

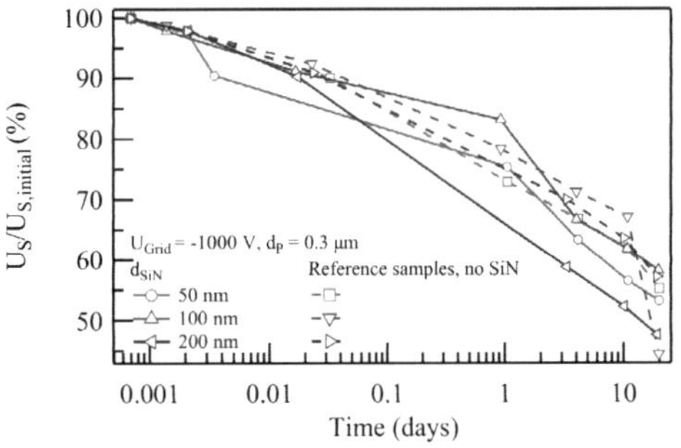

Figure 6. Measured maximum surface potential U_{S} for samples without and with SiN thicknesses of 50 nm, 100 nm and 200 nm charged at RT at different time spans.

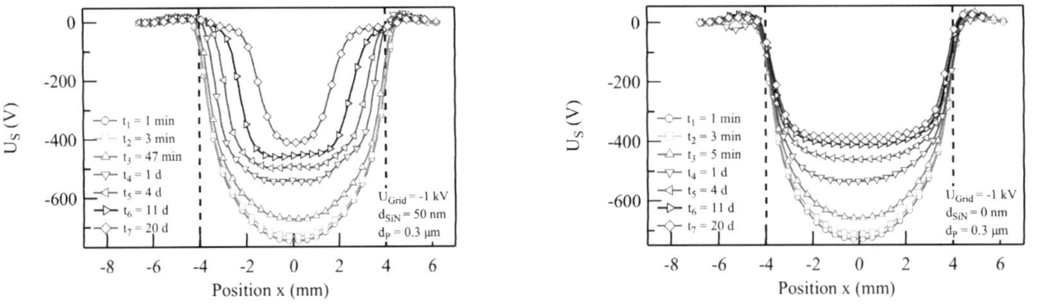

(a) (b)

Figure 7. Measured surface potential U_{S} scanned over the electret layer on the test chips (a) with and (b) without SiN barrier at seven increasingly spaced times. The lateral extent of the charged electret is indicated by the dashed vertical lines.

5.2% (d_{SiN} = 50 nm), 5.3% (d_{SiN} = 100 nm) and 11.3% (d_{SiN} = 200 nm) over those of the corresponding reference test chips without barriers. In the period between 11 days and 20 days, the surface potential of the sample with the 100 nm barrier begins to drop rapidly, reducing the remaining U_S below that of the reference sample. This behavior was observed for other thicknesses d_P and d_{SiN}, too. Among the reference test chips differences in discharging rates are observed as well. These are compatible with wafer-to-wafer thickness variations known to lead to different discharging rates [11]. Nevertheless, test chips with barrier and the corresponding references have the same Parylene-C thickness and can thus be directly compared.
Figure 7 (a) shows scans of the surface potential U_S of a chip with a 50 nm barrier. Results at different times are shown. The potential profile is observed to progressively narrow with time. After t = 20 days the effective width of the profile has decreased to significantly below 4 mm. This behavior was observed for all test chips with a SiN drift barrier. In contrast, the width of the charge carrier profile of barrier-free reference test chips, as shown in Figure 7 (b), remains constant over time. We conjecture that the drift barrier offers a new, horizontal discharging path parallel to the electret.

6. Conclusion

In this work an approach to reduce the charge carrier decay in Parylene-C electret layers using SiN drift barriers was investigated. Test chips with different drift barrier thicknesses and different thicknesses of the capping Parylene-C layer were fabricated, charged and the surface potential U_S was measured repeatedly. The drift barrier does not affect the charging process and $U_{S,initial}$ values are in agreement with those achieved for pure Parylene-C layers [11]. Comparing the discharging behavior of test chips with and without drift barrier, SiN layers as barrier provide a slightly higher stability of the charge carriers for short times. However, a horizontal discharge mechanism ruins this effect and decreases the long term stability dramatically.

Acknowledgments

We gratefully acknowledge the financial support by the German Research Foundation (DFG) through grant no. GRK 1322.

References

[1] Matsumoto K, Saruwatari K and Suzuki Y Vibration-Powered Batteryless Sensor Node using MEMS Electret Generator 2011 *Tech. Dig. PowerMEMS* (Seoul, South Korea, 15-18 Nov 2011) 134-7
[2] Woias P Micro Energy Harvesting as Core Technology for Energy-Autonomous Embedded Systems 2011 *Tech. Dig. PowerMEMS* (Seoul, South Korea, 15-18 Nov 2011) 98
[3] Bartsch U, Gaspar J and Paul O Low-Frequency Two-Dimensional Resonators for Vibrational Micro Energy Harvesting 2010 *J. Micromech. Microeng.* **20** (3) 035016
[4] Beeby S P, Tudor M J and White N M Energy Harvesting Vibration Sources for Microsystems Applications 2006 *Meas. Sci. Technol.* **17** R175-R195
[5] Crovetto A, Wang F, Triches M and Hansen O MEMS Fabricated Energy Harvesting Device With 2D Resonant Structure 2012 *Tech. Dig. PowerMEMS* (Atlanta, USA, 2-5 Dec 2012) 125-8
[6] Bartsch U Electret-Based Resonant Micro Energy Harvesting in Two Dimensions 2010 (Der Andere Verlag, Germany) *MEMS Technology and Engineering Series*, **15** 66
[7] Sessler G M 1998 *Electrets - Third Edition* (Laplacian Press)
[8] Wada Y et al Aging Characteristics of Electret Used in a Vibration-based Electrostatic Induction Energy Harvester 2011 *Dig. Tech. Papers Transducers* (Bejing, China, 5-9 June 2011) 2626-9
[9] Yamashita K, Honzumi M, Hagiwara K, Iguchi Y and Suzuki Y Vibration-driven MEMS Energy Harvester with Vertical Electrets 2010 *Tech. Dig. PowerMEMS* (Leuven, Belgium, 30 Nov - 3 Dec 2010) 165-8
[10] Lo H and Tai Y Parylene-based electret power generators 2008 *J. Micromech. Microeng.* **18** (10), 104006
[11] Genter S, Paul O Parylene-C As An Electret Material For Micro Energy Harvesting 2012 *Tech. Dig. PowerMEMS* (Atlanta, USA, 2-5 Dec 2012) 129-132
[12] Skalny J, Hortvth G and Mason N J Spectra of Ions Produced by Corona Discharge 2006 *AIP Conf. Proc.* **876** 284-93

Development and manufacture of printable next-generation gel polymer ionic liquid electrolyte for Zn/MnO_2 batteries

R Winslow[1], C H Wu[2], Z Wang[1], B Kim[3], M Keif[4], J Evans[3] and P Wright[1]

[1] Department of Mechanical Engineering, University of California, Berkeley, California, 94720, USA
[2] Green Energy and Environmental Laboratories, Industrial Technology Research Institute, Hsinchu, Taiwan
[3] Department of Material Science and Engineering, University of California, Berkeley, California, 94720, USA
[4] Graphic Communication Department, Cal Poly State University, 1 Grand Avenue, San Luis Obispo, CA 93407

E-mail: rwinslow4@berkeley.edu

Abstract. While much energy storage research focuses on the performance of individual components, such as the electrolyte or a single electrode, few investigate the electrochemical system as a whole. This research reports on the design, composition, and performance of a Zn/MnO_2 battery as affected by the manufacturing method and next-generation gel polymer electrolyte composed of the ionic liquid [BMIM][Otf], ZnOtf salt, and PVDF-HFP polymer binder. Materials and manufacturing tests are discussed with a focus on water concentration, surface features as produced by printing processes, and the effect of including a gel polymer phase. Cells produced for this research generated open circuit voltages from 1.0 to 1.3 V. A dry [BMIM][Otf] electrolyte was found to have 87.3 ppm of H_2O, while an electrolyte produced in ambient conditions contained 12400 ppm of H_2O. Cells produced in a dry, Ar environment had an average discharge capacity of 0.0137 mAh/cm^2, while one produced in an ambient environment exhibited a discharge capacity at 0.05 mAh/cm^2. Surface features varied significantly by printing method, where a doctor blade produced the most consistent features. The preliminary results herein suggest that water, surface roughness, and the gel polymer play important roles in affecting the performance of printed energy storage.

1. Introduction

With the strain on global electrical grids increasing year after year, there is an imperative need to develop and install reliable, long-lasting energy storage solutions. The utilization of new materials such as ionic liquid electrolytes [1] and gel polymers [2] with additive manufacturing techniques [3] exhibit the potential for creating flexible, rechargeable energy storage solutions can fit many geometries and scales. In order to succeed at that goal, it is imperative that the parameters that influence system performance, from electrochemical to production, be well-modeled and understood.

Energy storage research is often conducted on a single component. These experiments test performance separate from full cells, thus leaving a knowledge gap in understanding how one part can affect the system as a whole. For instance, much research surrounding ionic liquid-based electrolytes explores proof of concept functionality among various material options instead of determining robustness with respect to many parameters. Such qualities are important to know for manufacturing control to attain consistent device performance.

Printed electrochemical energy storage composed of a manganese dioxide (MnO_2) slurry cathode, zinc (Zn) foil anode, and an ionic liquid gel polymer electrolyte were produced with several different compositions, environments, and manufacturing methods. This work details the variability between processes and the identification of the influential parameters as determined through cell discharge capacity.

2. Experimental

Cathode ink composed of 9 g of MnO_2 powder, 0.5 g of carbon black, 6.6 g of modified styrene-butadiene copolymer (PSBR), and 4 g of DI water was deposited via flexographic printing, doctor blade casting, and dispenser printing to create cathode films. The ink was mixed in a ball milling machine for 120 min at 30 Hz to evenly disperse the components. Zinc foil with a purity of 99.95%+ and thickness of 50 μm was purchased from Goodfellow Corp. and used as the anode.

The electrolyte was composed of 0.2 g of zinc trifluoromethanesulfonate (ZnOtf) (purchased from Sigma-Aldrich) dissolved in 3 g of 1-butyl-3-methylimidazolium trifluoromethanesulfonate ([BMIM][Otf]) ionic liquid (IL) (purchased from EMD Chemical). A Mettler Toldeo DL39 Karl Fischer coulometer was used to determine the amount of water present in each electrolyte composition. Electrolytes were produced in a dry, Ar-filled glovebox and in ambient laboratory conditions.

The gel polymer was produced by mixing 1 g of poly(vinylidene fluoride-co-hexafluoropropylene) (PVDF-HFP) polymer to 5 g of n-methylpyrollidone (NMP) organic solvent. 1 g of the electrolyte was dispensed into the gel polymer mixture to yield the gel polymer electrolyte with a ratio of 1 g of electrolyte to 1 g of PVDF-HFP polymer.

Coin cells (CR2330) with a 1.27 cm diameter punched cathode disc and a 1.43 cm diameter punched anode disc were assembled in a dry, Ar-filled glovebox. A 26 μm thick Celgard separator insulated the electrodes from each other while still allowing for ions to conduct between interfaces. Electrolyte was dispensed onto both sides of the separator to provide a good interface with the electrodes. These cells were charged and discharged, from 1.0 V to 1.8 V, 100 times on a commercial Maccor tester.

Sandwich cells were produced in ambient conditions by casting gel polymer electrolyte with a doctor blade on cathode and anode strips that were cut to have a width of 1 cm. The electrodes with gel polymer electrolyte were dried at 80 °C for 6 h to form a solid electrolyte film. After drying, the electrodes were stacked orthogonally to produce a cell with an active area of 1 cm^2. These cells were cycled, from 1.0 V to 1.8 V, for 48 h on a custom galvanostat/potentiostat [4].

An Olympus LEXT OLS4000 laser confocal microscope was used to retrieve 643 μm^2 area primary surface profiles of cathodes produced via flexographic printing, dispenser printing, and doctor blade casting. Matlab was used to run a Fast Fourier Transform (FFT) with a cutoff wavelength of 80 μm to produce the waviness profile. Thus the roughness profile was calculated by subtracting the waviness profile from the primary profile.

3. Results and Discussion

3.1. Water Saturation

[BMIM][Otf] is known to be hygroscopic, but it is currently unknown how environmentally sensitive [BMIM][Otf]-based electrolytes are [5]. Water absorbed from the environment has the

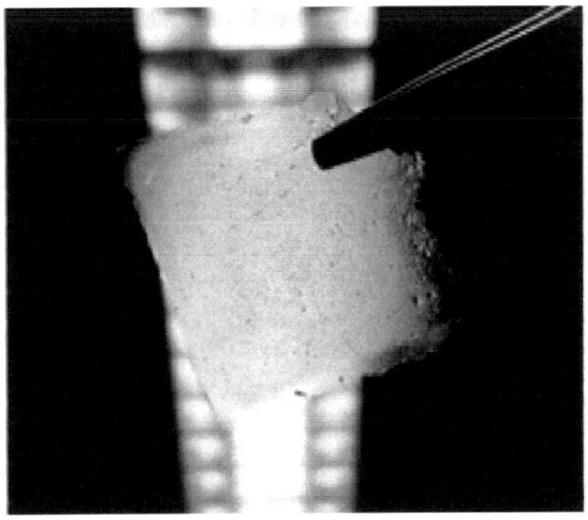

Figure 1. Gel polymer ionic liquid electrolyte.

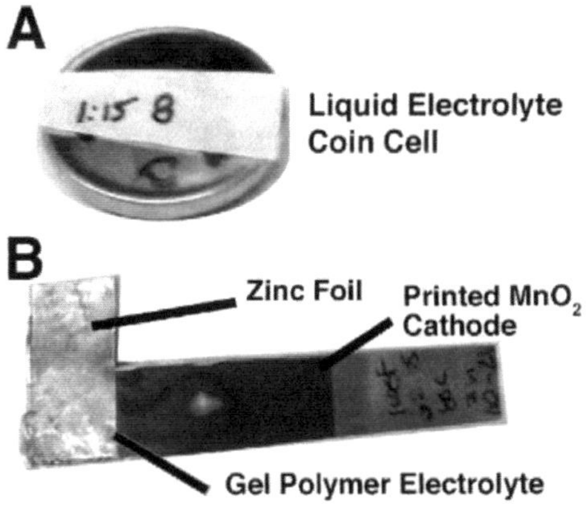

Figure 2. Test cell form factors. A) Coin cell containing liquid electrolyte (no PVDF-HFP gel component); B) Sandwich cell.

potential to take part in the reactions of the battery and may affect the discharge performance of the battery cells, thus it is important to determine how cells respond to the presence of water at various concentrations.

The compositions produced and their respective water contents are shown in Table 1. The saturated sample produced in the ambient laboratory environment was exposed to air with an average relative humidity of 50%. All cells produced had open circuit voltages between 1.0 V and 1.3 V. Cells of each electrolyte composition were produced and cycled 100 times to determine their average discharge capacity normalized per unit area, as shown in Figure 3.

As expected, the addition of the ZnOtf salt to the ionic liquid increased the water content from 37.5 ppm of H_2O to 87.3 ppm of H_2O. Interestingly, the electrolyte was able to hold up to 12,400 ppm of H_2O and, as indicated in Figure 3, the cell produced with that electrolyte had the highest discharge capacity. These results indicate that exposure to a humid environment can have a strong effect on the composition of battery cells produced. While the trend over this subset of data indicates that water has a beneficial effect on cell discharge capacity, it is unknown whether a humid manufacturing environment will affect other aspects of the cell such as cycle life or thermal stability. Further experiments are required to answer these questions and to determine if any optimum composition exists.

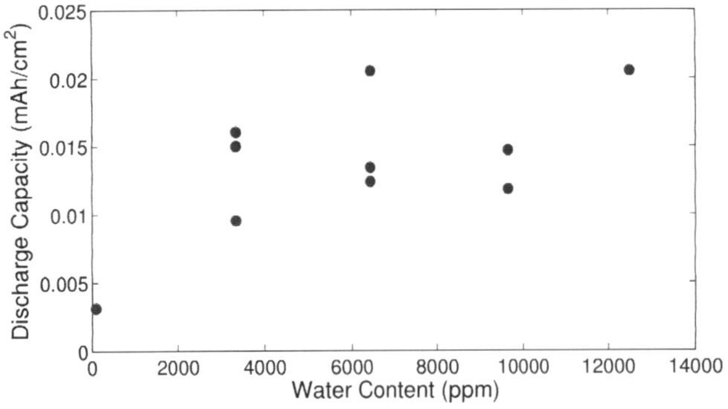

Figure 3. Discharge capacity of coin cells without a gel polymer. Each cell was cycled 100 times.

Table 1. Amount of water present in neat [BMIM][Otf] and several electrolyte samples.

Sample Label	Composition	Water Content (ppm)
Neat IL	Neat [BMIM][Otf] in dry Ar only	37.5
Dry	Electrolyte with all components dried in Ar	87.3
Saturated	Electrolyte produced in ambient environment	12400
1:3 Mixture	1 g of Dry mixed with 3 g of Saturated	9640
2:2 Mixture	2 g of Dry mixed with 2 g of Saturated	6430
3:1 Mixture	3 g of Dry mixed with 1 g of Saturated	3340

3.2. Surface Roughness

With many printing options available, it is important to determine how surface characteristics are influenced manufacturing method. Roughness and waviness are common metrics used to quantify such features. The Figures 4 and 5 show the average line roughness and average line waviness for a 643 μm^2 sample for each manufacturing method. The dispenser printer has both the least uniform roughness and waviness across the sample area, with variation on the order of 2.5 μm for average line roughness and 50 μm from highest to lowest average line waviness. While flexographc printing and doctor blade casting appear approximately equivalent in terms of surface roughness, the flexographic printer exhibits a greater degree of non-uniformity than the doctor blade cast with two distinct troughs and peaks ranging from approximately -5.99 μm to 6.88 μm. This may be the result of viscous fingering, a phenomenon where surface features are formed on the printer as air is pressed between the rollers with the ink.

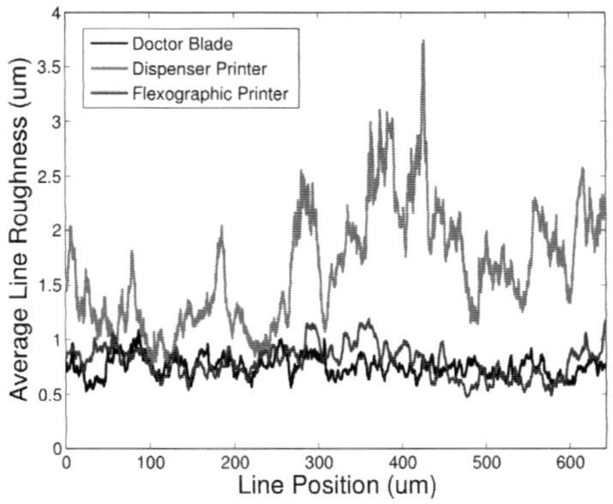

Figure 4. Comparison of resulting cathode average line roughness (R_a) as produced by doctor blade, dispenser printer, and flexographic printer.

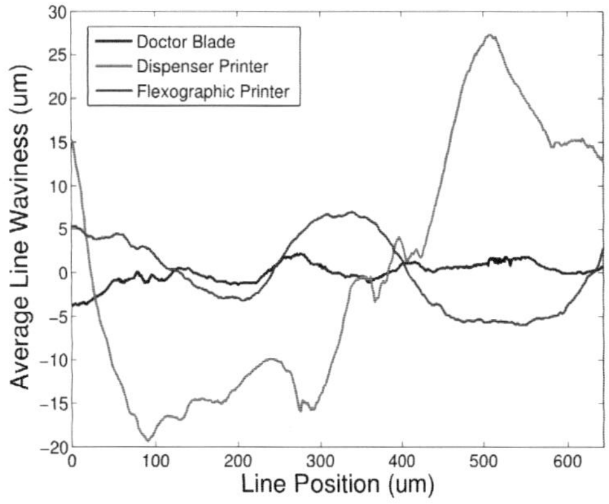

Figure 5. Comparison of resulting cathode average line waviness (W_a) with a zero average as produced by doctor blade, dispenser printer, and flexographic printer.

3.3. Gel Polymer Performance

Each of the samples in Figure 6 utilized a doctor blade cast cathode. The sandwich and ambient coin cells were cycled with the custom galvanostat/potentiostat, while the coin cell produced in

Ar was cycled on a Maccor. All cells used the same cycling procedure with a charge to 1.8 V with constant 0.1 mA current, hold at a constant 1.8 V for 3 minutes, 20 seconds, discharge to 1.0 V with constant 0.1 mA, then rest for 5 minutes. The coin cells, neither of which contained any gel polymer component, have an expected drop in performance after the first cycle. Interestingly, the sandwich cell improves in performance as the cycles continue, although not smoothly, from 0.024 mAh/cm^2 to 0.077 mAh/cm^2. This phenomenon has been seen in other work with similar materials and composition [3].

The effect seems to be connected to the inclusion of the gel polymer component, although the specific mechanism related to the materials has cannot yet be determined from this initial investigation. Further experiments will be conducted with a focus on interfacial interactions.

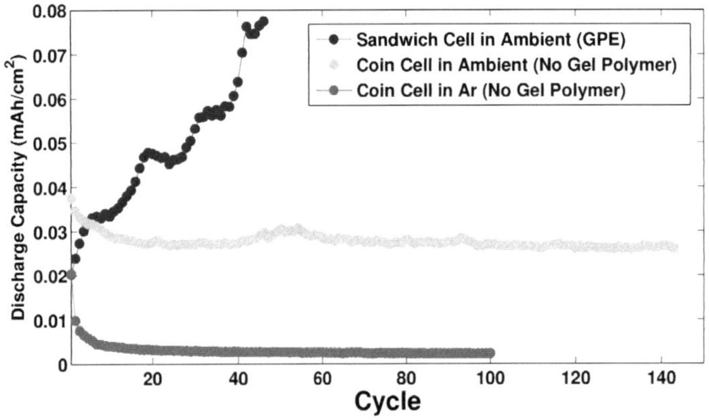

Figure 6. Comparison of performance of a sandwich cell produced with a gel polymer electrolyte in ambient laboratory conditions, a coin cell produced with a liquid electrolyte (no gel polymer) in ambient laboratory conditions, and a coin cell produced with a liquid electrolyte in an Ar atmosphere.

4. Conclusion

The printed Zn/MnO_2 battery with an ionic liquid-based gel polymer electrolyte was successfully fabricated and tested to determine parameters that most influence performance. A greater amount of water dissolved into the electrolyte seems to improve cell performance, although the current subset of data does not yet indicate an optimum composition. Doctor blade casting exhibited the most consistent roughness and waviness, while dispenser printing exhibited the most variability with the highest range. Both doctor blade casting and flexographic printing may be improved with process controls. The gel polymer component seems to provide a boon to cell performance, causing the discharge capacity to increase with number of cycles. Future experiments investigating these phenomena and the underlying mechanisms are underway.

5. Acknowledgements

The authors thank the California Energy Commission for supporting this research under award 500-01-43. We would also like to thank Dr. Vince Battaglia, Prof. Xiaoying Rong, and Dr. Jay Keist for their contributions.

References

[1] Kuboki T, Okuyama T, Ohsaki T and Takami N 2005 *Journal of power sources* **146** 766–769
[2] Kumar G G and Sampath S 2003 *Journal of The Electrochemical Society* **150** A608–A615
[3] Ho C, Evans J and Wright P 2010 *Journal of Micromechanics and Microengineering* **20** 104009
[4] Steingart D A, Redfern A, Ho C, Wright P and Evans J 2006 *ECS Transactions* **1** 17–22
[5] Cao Y, Chen Y, Sun X, Zhang Z and Mu T 2012 *Physical Chemistry Chemical Physics* **14** 12252

Design, Fabrication and Testing of Silicon-integrated Li-ion Secondary Micro Batteries with Side-by-Side Electrodes

K Hoeppner[1], M Ferch[1], M. Eisenreich[2], K Marquardt[2], R Hahn[2], P Mackowiak[1], B Mukhopadhyay[2], H-D Ngo[2], R Gernhardt[2], M Toepper[2] and K-D Lang[1, 2]

[1]TU Berlin, Forschungsschwerpunkt Technologien der Mikroperipherik, Gustav-Meyer-Allee 25, 13355 Berlin, Germany
[2]Fraunhofer IZM, Gustav-Meyer-Allee 25, 13355 Berlin, Germany

katrin.hoeppner@izm.fraunhofer.de

Abstract. This paper reports the design, fabrication, and testing of silicon-integrated lithium ion secondary micro batteries with a side-by-side electrode setup. Two cavities separated by a narrow silicon spacer served as a containment for the electrodes and were etched into <110>-Si by wet chemical etching using aqueous KOH solution. The etched silicon battery containment was passivated by a stress-compensated layer of SiO_2/Si_3N_4. Ti/Pt-current collectors were applied by E-beam evaporation and lift-off structuring. A volumetric dispenser served to fill the cavities with slurries of the active materials - lithium cobalt oxide ($LiCoO_2$) as the anode and graphite as the cathode material. Encapsulation, electrolyte filling, and electrochemical characterisation of the finished cells took place in an Ar-filled glove box. The fabricated batteries have shown a rate capability of up to 5C and a linear capacity loss rate of <1 % per cycle over 30 full-cycles. Battery containments with different cavity and spacer widths have been fabricated.

1. Introduction

The fabrication of silicon integrated batteries appears to be of interest for applications where a small volume, high capacity energy supply is needed. Examples are small energy autarkic sensors, e.g. implantable sensors for medical applications [1], or miniaturized sensor networks. Such applications cannot be supplied by state of the art solid state batteries due to their limited capacity per area [1].

Previously we have investigated the fabrication of Li-ion batteries with stacked electrodes, separated by a porous polymer separator [2]. In this study, a design with anode and cathode lying side by side, separated by a silicon spacer has been implemented. This design facilitates assembly and encapsulation considerably since no polymer separator is needed and both electrodes can be contacted from downside. The fabricated structures could also be used in materials research, e.g. for high-throughput screening and in-operando electrode surface studies by applying different lid materials.

2. Design

The investigated battery design with a side-by-side electrode setup implies the lithium ions moving laterally from one electrode to the other, crossing a silicon spacer of reduced surface height with respect to the wafer surface. This is expected to result in a different dependency between geometry and impedance of the cell as compared to a conventional setup with electrodes stacked on top of each other and the lithium ions moving back and forth between the electrodes perpendicularly to the electrode surface (Fig.1).

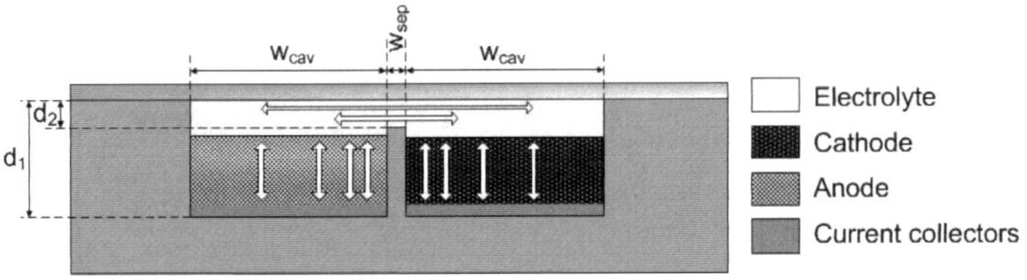

Fig.1: Schematic profile of a battery with side-by-side electrode setup (arrows indicate paths of Li-ion movement)

We chose to fabricate structures as shown in Fig. 1 from a <110>-silicon substrate by applying two steps of wet chemical etching. The <110>-orientation of the wafer allows for the fabrication of long and narrow cavities bordered by {111} sidewalls that are vertical at the long sides and offer a slope of 35° at the front sides. Such geometry ensures a good utilization of area (i.e. a high specific capacity) as well as an easy patterning of the current collectors leads via the front side slopes. In the first etch step, two cavities separated by a Si-spacer of full height should be formed, whereas the second etch step should give rise to a clearance above the Si-spacer. The unmasked exposure of convex edges to KOH etch solution is known to result in a rounding of these edges. This effect was expected to be of benefit considering cell impedance because sloped edges diminish the dimension of the very thin electrolyte layer above the Si-spacer.

To confirm the feasibility of our concept, we simulated the etching process with the aid of the software ACES (Anisotropic Crystalline Etching Simulation) [3]. ACES works on an algorithm based on the Continuous Cellular Automata method and is available for free download in the internet [4]. The anisotropic ratio between {100} and {110} crystal plane etch rates was adapted to comply with our experimental conditions (33wt% KOH in pure water, 80°C) using the rate law proposed by Seidel et al. [5]. The etch rate of the {311}-planes was set to equal that of the {110} plane as proposed by ACES and the data given in literature [6]. This resulted in etch rate ratios of 1:1.535:1.535:0 for {100}:{110}:{311}:{111}, respectively.

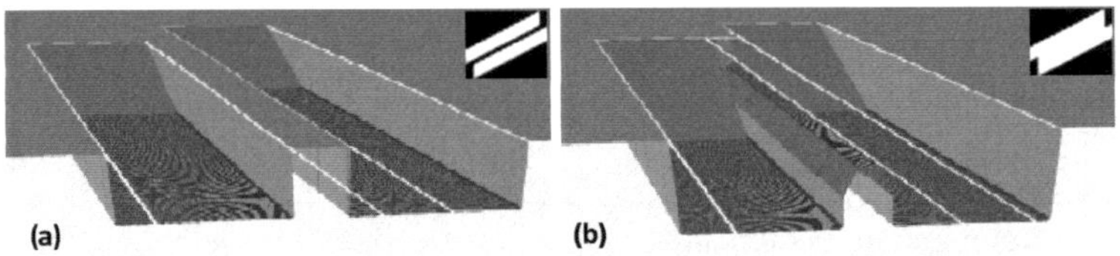

Fig. 2: ACES simulation results for battery containments with w_{cav} = 300 µm and w_{sep} = 100 µm a) after 1st etch step b) after 2nd etch step (yellow lines: mask outline; insets: mask shape)

The simulation results confirm the build-up of slanted edges during unmasked exposure of the Si-spacer to the etch solution. The slants are forming an angle of 29.5° with the next {111} plane and correspond to the uncovering of {311} crystal planes.

3. Fabrication

Fig. 2 gives a short survey on the fabrication process flow. The starting substrate is a p(B)-doped CZ-grown <110>-Si wafer with a resistivity of 1-30 Ohm cm. In a first step, the exact crystal orientation of the wafer is determined and alignment marks are applied by deep reactive ion etching (DRIE, not shown in Fig. 2). A 1 µm thick oxide layer is grown by wet thermal oxidation and structured using buffered HF solution (**B**, Fig. 2). Then a thin layer of silicon nitride (Si_3N_4) is deposited by low pressure chemical vapour deposition (LPCVD) and structured selectively by DRIE (step C, Fig. 2). The Si_3N_4 bar for structuring the Si-spacer is deposited directly on the silicon surface. Doing so allows for a two-step wet etch process without deposition or growth of another masking layer for the second step. This helps to prevent contamination of deposition equipment by metal ions.

During the first KOH etch step, two 200 µm deep side-by-side cavities separated by a narrow spacer are formed. After this, the Si_3N_4 masking layer is removed by DRIE (**D**, Fig. 2). A second KOH etch step provides for a further 100 µm deepening of both, the cavities and the spacer (**E**, Fig. 2).

The remaining oxide mask is removed prior to passivation of the substrate by the growth of 100 nm of thermal SiO_2 and the deposition of 180 nm of Si_3N_4 by LPCVD (**F**, Fig. 3). The thickness of the individual layers is adapted to compensate for the CTE mismatch between the different materials. Deposition and patterning of Ti/Pt-current collectors by E-beam evaporation and lift-off conclude the fabrication process of the bottom substrate of the battery housing (**G**, Fig.3).

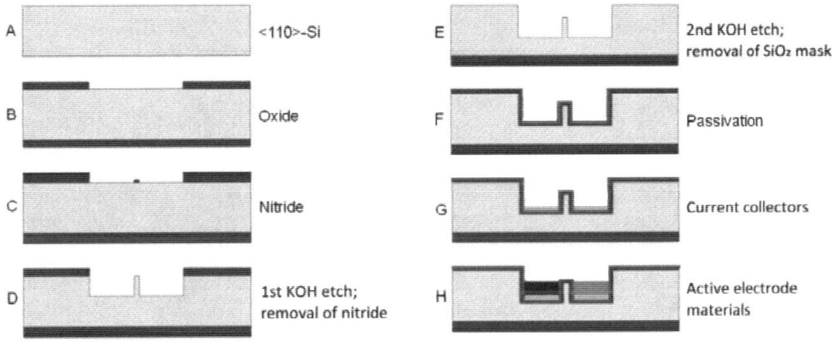

Fig. 3: Main processing steps for the fabrication of the battery containment

Fig. 4 shows cross-section polishes of the etched battery containments. Featuring the simulated effects, the etched substrates basically confirm simulation. There are, however, differences in geometry at the separator region. Firstly, the separator is slightly narrower than simulated due to an undercut at the nitride-on-silicon masking bar. Secondly, the slopes at the separator edges form angles different from 29.5° with the vertical {111} walls. In fig. 4b, also a rounding of the separator edges can be observed. The authors believe these effects to be caused by a step flow etch mechanism as proposed by Horn et al [7] rather than by the etching of high-index crystal planes. Fig. 4d shows the surface quality of the sloped edges of a separator.

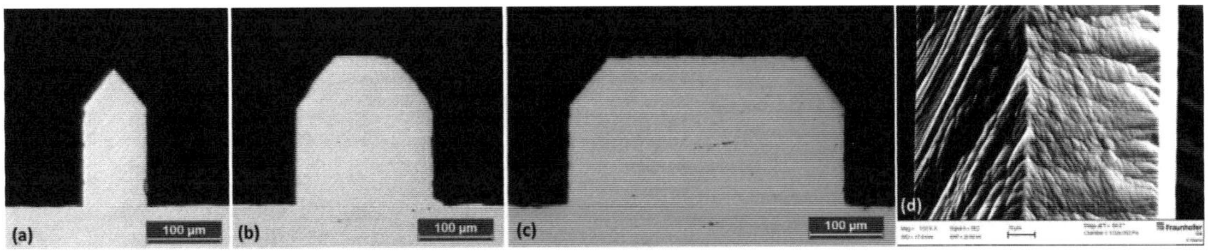

Fig. 4: Cross-section polishes of battery containments with different spacer widths
a) 100 µm b) 200 µm c) 400 µm d) surface of the sloped edges of a 100 µm spacer

3.1. Deposition of active materials and assembly

Prior to the deposition of the active materials, a slurry of conductive carbon, water, and polyvinylpyrrolidone (PVP) as a binder was dispensed on top of the current collectors. Having a final thickness of some 30 µm after solvent removal, this coating provides for a better contact between current collectors and active electrode materials. The active materials for cathode and anode were prepared form slurries of commercially available powders (lithium cobalt oxide, LCO, for the cathode and graphite for the anode, both from MTI corp. [8]), polyvinylidene difluoride (PVDF) as a binder, and N-methyl-2-pyrrolidone (NMP) as a solvent [9]. The slurries were applied to the cavities by volumetric micro dispensing (Ultimus IV, Nordson) and the solvent removed by appropriate heat treatment. We observed a good separation between the resulting electrodes by measurement of isolation resistance and SEM investigation (Fig. 5a). The filling process provides for an excess capacity on the anode side ranging from 10 to 45% of the nominal capacity of the cathode (Fig. 5b).

The chips with the active electrodes were introduced into an Ar-filled glove box. Hereafter, a heating step of 12 h at 80°C and under vacuum was applied to allow for all materials to dry completely. A glass lid with holes for the deposition of electrolyte was provided with a filled, UV-curable epoxy resin by dispensing. After alignment, lid and bottom substrate were attached to each other and the resin cured by UV-exposure at room temperature (Fig. 5c). The cavities were evacuated and liquid electrolyte was filled in using a home-made holder with a standard GC-septum and a syringe. We applied 1M $LiPF_6$ dissolved in a mixture of ethylene carbonate and dimethylcarbonate (EC:DMC, 1:1wt%) as the electrolyte. To allow for the removal of gas generated during formation, the holes for electrolyte filling were finally sealed not until after formation of the batteries.

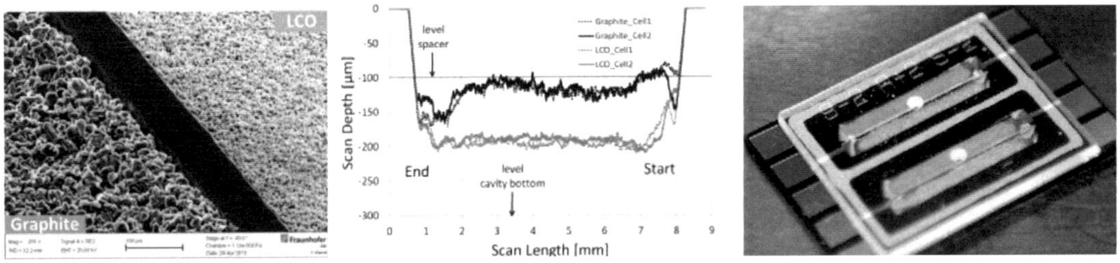

Fig.5 a) Separator region after application of electrodes (w_{sep}=200µm; final electrode distance 75 µm)
b) Height profiles of different cavities filled with active material (black: graphite, red: LCO)
c) Photography of a chip with two sealed cells (chip size: 15x10 mm)

4. Testing

4.1. Cycling behaviour

The fabricated batteries were tested using a multichannel battery test system (CTS-LAB, Basytech GmbH). To investigate the cycling behavior, we applied CC-CV charge and CC discharge steps between 3 and 4.1 V at a rate of 0.1C, respectively. Linear capacity loss rates ranging from 0.5 to 1% per cycle were observed over 30 cycles (fig. 6a).

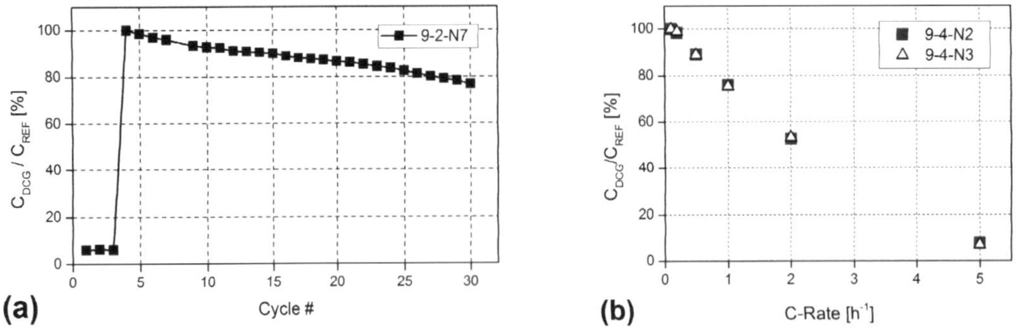

Fig. 6a) Discharge capacity loss over 30 cycles (cycles 1 to 3: formation)
b) Rate capability of two different cells (both w_{sep}=400 µm, w_{cav}=900 µm)

4.2. Rate capability
Different batteries have been tested to determine their rate capability. To do so, we cycled the cells between 3 and 4.1 V with CC/CV-charge steps at a rate of 0.2C and increasing discharge rates ranging from 0.1 to 5 C. The reference capacity for calculating the C-rates was determined during a first cycle at a discharge rate of 0.1C. At this rate the value basically equaled the nominal capacity of the material (145 mAh/g) as given by the supplier. For each discharge rate two complete cycles were run through. Due to the impedance of the cells, only part of the nominal capacity is extractable at higher C-rates. The capacity was found to drop linearly with the C-rate (fig. 6b). We observed a reproducible rate capability behavior for cells of the same geometry and on the same chip.

5. Conclusion
Lithium ion secondary batteries with a side-by-side electrode setup have been fabricated by the anisotropic wet etching of <110>-silicon, the automated dispensing of electrodes from slurries, and room temperature adhesive bonding. First cycling tests have shown that the concept is feasible and results were found to be reproducible for different cells with the same set-up.

The influence of the mask geometry on the final shape of the Si-spacer was investigated. For realizing designs with spacers narrower than 175 μm, the duration of each of the two KOH-etch steps must be adapted in compliance with the geometry-dependent etch behavior at the separator tip.

For cells with a side-by-side electrode setup the capacity drop with discharge rate is more pronounced than for conventional cells with stacked electrodes due to the higher impedance. However, the extractable discharge capacity during pulse load tests is expected to be higher than observed during continuous CC-discharge due to relaxation effects. As the supply of sensors coupled to radio modules is considered an important application for Si-integrated micro batteries, the characterization of the discharge performance of the cells during pulse load tests will be subject of upcoming investigations. To gain insight into the relation between cell set-up and impedance, a more comprehensive testing of cells with different geometry or differently applied electrodes is necessary. Here, the measurement and analysis of impedance spectra at different operation points is considered to be of value to optimize cell performance and intended in the future.

References
[1] F. Albano et al., A fully integrated microbattery for an implantable microelectromechanical system, Journal of Power Sources, 185 (2008), pp. 1524–1532.
[2] K. Marquardt et al., Development of near hermetic silicon/glass cavities for packaging of integrated lithium micro batteries. Microsystem Technologies, 2010, 16:1119{1129.10.1007/s00542-009-0954-7.
[3] Z. Zhu et al., An efficient PC-based preferential-etch simulator using dynamic cellular automata method, Proc. 5th Solid-State and Integrated Circuit Technology Conference 1998, pp. 943-946.
[4] Homepage of the Brigham Young University, Department of Electrical and Computer Engineering; http://www.cleanroom.byu.edu/KOH.phtml (as at Oct. 04, 2013)
[5] Seidel et al., Anisotropic Etching of Crystalline Silicon in Alkaline Solutions, I. Orientation Dependence and Behavior of Passivation Layers, J. Electrochem. Soc., Vol. 137, No. 11, November 1990.
[6] E. Herr et al., KOH etch rates of high-index planes from mechanically prepared silicon crystals", Transducers (1991) 807-810.
[7] A. Horn et al., Simulation of Orientation-Dependent Etching of Silicon Using a New Step Flow Model of 3D Structuring, Proceedings of the International Conference on Modeling and Simulation of Microsystems (MSM), Chapter 2: Process Modeling, 2000, pp. 63-66.
[8] Homepage of MTI Corporation: http://www.mtixtl.com/
[9] K. Hoeppner et al., Development and Investigation of Electrode Slurries for the Batch Fabrication of Li-ion Secondary Micro Batteries, EnMat Conference, 2013, Abstr. No 3276.

High power primary lithium ion microbatteries

James H Pikul[1], **Paul V Braun**[1,2] **and William P King**[1,2]

[1] Department of Mechanical Science and Engineering, University of Illinois at Urbana – Champaign, 1206 W. Green st., Urbana, IL, USA 61801
[2] Department of Materials Science and Engineering, University of Illinois at Urbana – Champaign, 1304 W. Green st., Urbana, IL, USA 61801

E-mail: jpikul2@illinois.edu

Abstract. Microbatteries are a critical component for powering autonomous and remote microdevices, but current microbatteries lack the power density and capacity required for high performance use. This paper reports primary microbatteries with high power density and capacity. The microbatteries are based on interdigitated three-dimensional nanoporous bicontinuous electrodes, which enable short ion diffusion lengths and high discharge power. The microbatteries utilize a high capacity manganese oxide cathode and lithium anode to achieve a capacity of up to 29.5 $\mu Ah/cm^2\mu m$, which is 10X the average capacity of the highest capacity rechargeable microbatteries reported.

1. Introduction

The capacity and power density of microbatteries directly impact the operation time and performance of the devices they power. The capacity of a battery is dependent on the total volume and capacity of the anode and cathode electrolytic materials, materials that undergo an electrochemical reaction. Thin film microbatteries made by layering films, about 1 – 10 um, of cathode, electrolyte, and anode on metal current collectors have high volumetric capacity (16 – 21 $\mu Ah/cm^2\mu m$) [1, 2]. The 2-D nature of thin film batteries gives them poor areal capacity, often making them much larger than the devices they power [3]. Recent research on microbatteries has focused on creating 3D electrodes, so that the areal capacity is high (up to 3,500 $\mu Wh/cm^2$) and the battery is of similar shape and size to the device it powers [4-7]. The focus on high areal capacity has led to microbatteries with volumetric capacities 5 – 5,000 X lower than conventional lithium ion batteries [8]. The low capacity of microbatteries is due to the fabrication difficulty of packing high volumes of anode and cathode material within 10's of μm spacing and the use of low capacity materials.

Microbatteries also suffer from low power density, typically below 3.5 $\mu W/cm^2\mu m$ [4-7]. Specifically engineered structures with short ion transport lengths and highly conductive electrodes are required to achieve high power density [9]. Only recent microbattery work with an interdigitated 3D bicontinuous nanoporous architecture has demonstrated power densities up to 7.4 $mW/cm^2\mu m$, which was 2,000 X the power density of other 3D microbatteries and similar to supercapacitors [8].

Here we report primary, single discharge, microbatteries that use high capacity manganese oxide conversion compounds to achieve high power and high energy density. The microbatteries are based on interdigitated 3D nanoporous bicontinuous electrodes, which enable short ion diffusion lengths and high discharge power with a large volume of electrolytic material. The microbatteries can be pulsed at 20 mA/mm^3 current densities for over 100 times to deliver high on-demand power. The microbatteries utilize primary battery chemistry to achieve a capacity of up to 29.5 μAh/cm^{2}μm, 10X the average capacity of the highest capacity rechargeable microbatteries previously reported by us [8].

Figure 1 shows the microbattery design. The porous electrodes are interdigitated with alternating anode and cathode. The electrodes are composites composed of a highly porous bicontinuous nickel current collector conformally coated with lithium or manganese oxide, for the anode and cathode. The microbattery architecture offers short ion diffusion lengths in the electrolytic material and electrolyte and highly conductive continuous electron pathways in the nickel scaffold to reduce transport resistances and achieve high power density [8-10]. The high capacity primary chemistry, low cut-off voltage, and large volume fraction of electrolytic material enables high volumetric capacity.

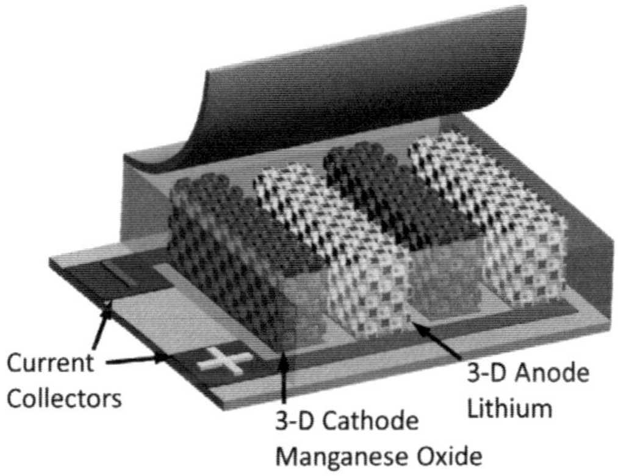

Figure 1. Microbattery design. The nanoporous microbattery electrodes consists of an electrolytically active layer (red and yellow) coated on an electrically conductive bicontinuous nickel scaffold (blue). The nickel scaffold acts the current collector attached to an outside circuit. A manganese oxide is used as the cathode (red), and lithium as the anode (yellow).

2. Methods

The microbatteries were fabricated by electrodepositing electrolytic material onto porous metal scaffolds. Electrodeposition allowed for precise, conformal deposition of the active materials so that a large volume of active material could be deposited without shorting the microbattery. Figure 2 shows the microbattery fabrication, which followed a similar procedure developed by us in [8]. An opal structure was first self-assembled onto a glass substrate with interdigitated gold templates. The interdigitated templates had 5 mm long and 5 μm wide parallel rectangles, called fingers, connected to two perpendicular 4 mm wide contacts. The glass slide with interdigitated gold templates was then cut into multiple samples, piranha cleaned for ten minutes, immersed in Millipore water with 2.2 % by weight 3-mercapto-1-propanesulfonic acid, sodium salt for 3 hours and rinsed. The PS opal was self-assembled onto the substrates by placing them vertically in a colloid solution of 500 nm in diameter PS spheres set on a hot plat at 55 °C, covered, and left for 24 hours. The substrates were then sintered at 96 °C for 10 hours to increase the interconnect size between the PS spheres. A larger interconnect size allowed for a thicker active material coating. The porous, bicontinuous nickel current collectors were fabricated by electrodepositing nickel through the PS opal at a constant -2.0 volts versus a nickel reference electrode for 18 minutes in commercial plating solution. The PS was removed by immersing the substrates in two toluene baths for 3 hours each followed by a toluene and tetrahydrofuran rinse.

Voltage controlled, pulsed electrodeposition was used to deposit the lithium and manganese oxide active material to ensure conformal coating throughout the 3D structure. The nickel scaffold was first primed for manganese deposition by a applying a 1.2 and -0.5 pulsed voltage to the nickel electrode for 0.1 and 3.0 seconds and 50 cycles with a Ag/AgCl reference and steel counter electrode in a 0.1 M manganese acetate and 0.1 M sodium sulfate solution. Manganese oxide was deposited on to the cathode nickel scaffold using a three electrode deposition, where a sheet of platinum 3 cm^2 was the working electrode, Ag/AgCl was the reference electrode, and the microbattery substrate attached to a 25 cm^2 steel sheet was the counter electrode. A 1.3, 1.1, and 0 volts pulsed voltage signal was applied to the platinum for 0.1, 0.5, and 10 seconds for 275 cycles. Manganese oxide was plated onto the substrate during the application of 0 volts to the platinum electrode. The pulsed and exponentially decreasing current applied to the nickel current collector allowed for a highly conformal and thick deposition, which is critical for achieving a large volume fraction. A 1 mm thick polyethylene mask with a 0.5 to 4 mm^2 through hole and acrylic monomer adhesive was pressed on to a 110 °C pre-heated microbattery substrate and then cooled to room temperature. Figure 3 shows scanning electron microscopy images of the microbattery. Figure 3a is a top-down view of the interdigitated electrodes with a representation of the area covered by the mask. Figure 3b shows a cross-section of the electrodes, before lithium deposition. The electrodes are 15 um tall and 30 μm wide with a 40 μm period. The nickel current collector has 500 nm pores.

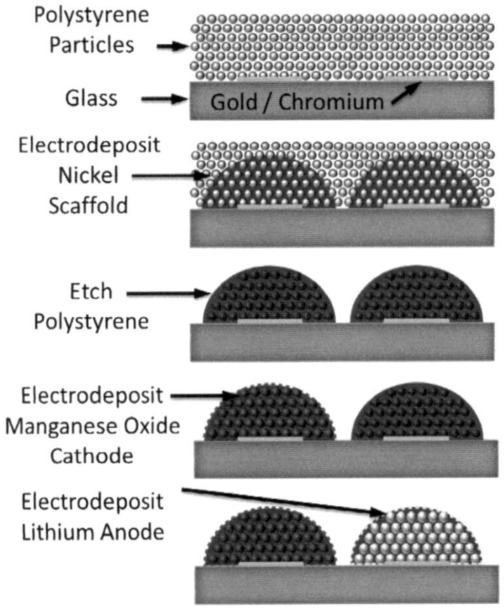

Figure 2. Schematic of the fabrication process where the nickel scaffold defines the battery architecture and the active materials are electrodeposited onto the nickel scaffold for precise integration of the electrodes on a single substrate. The nickel scaffold is formed by electrodepositing nickel through a polystyrene scaffold on glass with gold interdigitated electrodes.

The microbattery substrate was moved to an Argon glovebox and lithium was deposited on to the anode nickel scaffold using a pulsed voltage of -0.5 for 4 seconds and 0 amp off current for 6 seconds versus lithium metal in 1:1 ethylene carbonate (EC) and dimethyl carbonate (DMC) with 1 M LiClO4 for 40 cycles. 1:1 EC:DMC with 1 M LiClO4 liquid electrolyte was then dropped in to the mask opening and the microbattery cell was tested by galvanostatically discharging the cells to 0.5 volts at various current densities and pulsed discharges.

The energy and power density of the microbattery cells were obtained by measuring the voltage and current during the cell discharge, calculating the total energy and power, and normalizing the energy and power to the volume of the microbattery cell, which includes the total volume occupied by the electrodes (anode, cathode, Ni current collector, and electrolyte in the porous 3D electrode) and the electrolyte in the separation between the electrodes. The volume was calculated by multiplying the height of the electrodes by the total area of the cell not covered by the mask.

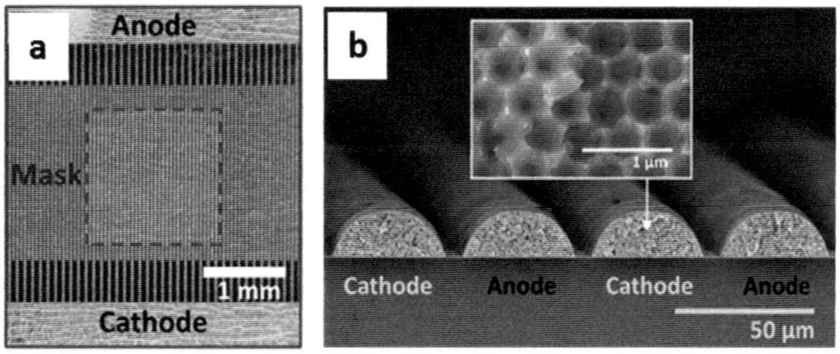

Figure 3. (a) A top-down SEM image of the interdigitated electrodes with the anode electrodes connecting at the top, cathode electrodes connecting at the bottom, and the interdigitated anode and cathode in the middle. A polyethylene mask, illustrated in blue, is placed on the electrodes to control the active discharge area. (b) Cross-section SEM of the electrodes with a close up of the cathode.

3. Results and Discussion

Figure 4a shows the discharge curve for a primary microbattery. The total capacity stored was 29.5 µAh/cm^{2}µm. The discharge voltage range is between 3.2 and 0.5 volts, with most of the capacity stored between 1.5 and 0.5 volts. The 0.5 to 1.5 voltage plateau is similar to the plateau seen in the first 400 mAh/g capacity of other manganese oxide conversion anodes [11-14]. Manganese oxides that undergo conversion reactions have been studied as anodes for lithium ion batteries due to their high capacity and low voltage. Here we use the manganese oxide as a cathode because of its high capacity relative to conventional cathodes, typically 100 – 200 mAh/g [15]. Conventional cathodes have an advantage due to their high voltage versus lithium, ~2 – 4 volts, despite their low capacity. The high voltage output improves energy density and also interfaces well with electronics since most electronics require greater than 1.8 volts to operate. However, recent work in low power and energy harvester electronics has seen the development of boost converters that operate with >90% efficiency and can up convert from 0.3 to 5.0 volts input to a constant 1.8 to 5.5 volts output, like the Texas Instruments tps61200. These technologies allow batteries to be efficiently discharged to lower voltages and opens the possibility of using high capacity anode materials as cathodes in primary batteries. The capacity of anode materials can be up to 4,200 mAh/g for silicon and 1,300 mAh/g during the first discharge of conversion manganese oxides [11-13, 16]. Lithium is used as the anode because of its high capacity and low, constant voltage relative to the cathode.

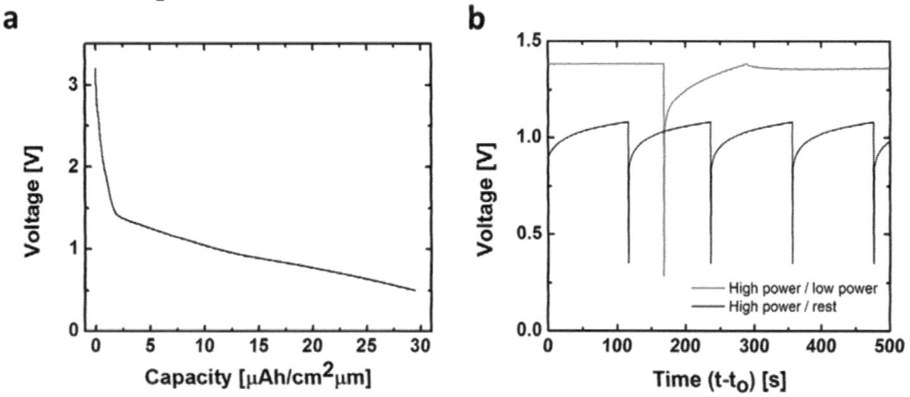

Figure 4. (a) A microbattery discharge curve. Most of the energy is stored between 1.5 and 0.5 volts. (b) Sections of pulsed discharge curves for a microbattery. The red curve shows the battery voltage after 50% discharge during a high current 1950 µA/cm^{2}µm pulse for 50 ms, followed by a 2 minute rest

and 15 minute low power discharge. The black curve shows the battery voltage after 85% discharge during a 970 μA/cm^{2}μm pulse for 50 ms, followed by a 2 minute rest.

The microbattery cells can deliver a high power density of 5,500 μW/cm^{2}μm, which is close to the 7,400 μW/cm^{2}μm highest power reported for microbatteries and superior to most supercapacitors [8]. The microbatteries can alternate between a high power pulsed discharge and a low power sleep discharge, which is critical for applications requiring wireless communication and device actuation. Figure 4b shows discharge profiles of a microbattery during various types of pulsed discharges. The microbattery was discharged at 1950 μA/cm^{2}μm current for 120 pulses and 970 μA/cm^{2}μm for 180 pulses. The microbattery was discharged to 0 volts during the high power pulses, similar to capacitors, to maximize the power performance. The microbattery voltage returned to ~75% of its voltage before the high rate discharge within 1 second of rest. The microbatteries can be repeatedly discharged at high power with no noticeable performance degradation.

4. Conclusion

High power and high capacity primary microbatteries have been fabricated by electrodepositing high capacity electrolytic materials on an interdigitated 3D nanoporous bicontinuous architecture. The high capacity manganese oxide cathode demonstrated a new way to use what are typically considered anode materials to dramatically improve the capacity of micro energy storage technologies. The high power pulsed operation of the microbatteries can improve the performance and lifetime of devices that rely heavily on actuation or wireless communication. High capacity and high power density primary microbatteries could increase the capacity of micro power sources by 20X and improve the power performance by 1,000X by replacing the rechargeable battery and energy harvester typically used to power microsystems, enabling smaller power sources, faster computation, stronger actuation, and longer data transmission distances for short to medium lifetime devices.

References

[1] Drews J, Fehrmann G, Staub R and Wolf R, *J. Power Sources,* **97–98**, 747-9
[2] Lee Y-G, Kim J, Kim S and Kim K M, *J. Power Sources,* **195**, 3715-9
[3] Arthur T S, Bates D J, Cirigliano N, Johnson D C, Malati P, Mosby J M, Perre E, Rawls M T, Prieto A L and Dunn B, *MRS Bulletin,* **36**, 523-31
[4] Chamran F, Yeh Y, Min H S, Dunn B and Kim C J, *J. Microelectromech. Syst.,* **16**, 844-52
[5] Min H S, Park B Y, Taherabadi L, Wang C L, Yeh Y, Zaouk R, Madou M J and Dunn B, *J. Power Sources,* **178**, 795-800
[6] Nathan M, Golodnitsky D, Yufit V, Strauss E, Ripenbein T, Shechtman I, Menkin S and Peled E, *J. Microelectromech. Syst.,* **14**, 879-85
[7] Kotobuki M, Suzuki Y, Munakata H, Kanamura K, Sato Y, Yamamoto K and Yoshida T, *Electrochim. Acta,* **56**, 1023-9
[8] Pikul J H, Gang Zhang H, Cho J, Braun P V and King W P, *Nat. Commun.,* **4**, 1732
[9] Braun P V, Cho J, Pikul J H, King W P and Zhang H, *Curr. Opin. Solid State Mater. Sci.,* **16**, 186-98
[10] Zhang H, Yu X and Braun P V, *Nat. Nanotechnol.,* **6**, 277-81
[11] Deng Y, Li Z, Shi Z, Xu H, Peng F and Chen G, *RSC Advances,* **2**, 4645-7
[12] Gao J, Lowe M A and Abruña H D, *Chem. Mater.,* **23**, 3223-7
[13] Chang L, Mai L, Xu X, An Q, Zhao Y, Wang D and Feng X, *RSC Advances,* **3**, 1947-52
[14] Wang H, Cui L-F, Yang Y, Sanchez Casalongue H, Robinson J T, Liang Y, Cui Y and Dai H, *J. Am. Chem. Soc.,* **132**, 13978-80
[15] Goodenough J B and Kim Y, *Chem. Mater.,* **22**, 587-603
[16] Boukamp B A, Lesh G C and Huggins R A, *J. Electrochem. Soc.,* **128**, 725-9

Analysis of Nonlinear Spring Arm for Improved Performance of Vibrational Energy Harvesting Devices

D Mallick[1], A Amann[2], S Roy[1]

[1]Micropower-Nanomagnetics Group, Tyndall National Institute, University College Cork, Cork, Ireland
[2]School of Mathematical Sciences, University College Cork, Cork, Ireland

Email: saibal.roy@tyndall.ie

Abstract. Recently, a number of attempts have been made to increase the operational bandwidth of the energy harvesting devices. Nonlinear mechanisms are one of them. In this paper, we report design and analytical formulation of stretching strain of an electromagnetic energy harvester on FR4 material under large deformation of the spring arms. It is found that nonlinearity has an inverse square dependence on thickness of the arms. Numerical solution of a monostable Duffing oscillator that governs the dynamics of such a large deformed nonlinear energy harvester showed that with decrease of load resistance, the average power output increases, where the output response depends strongly on the input force. For small input acceleration, the desired large amplitude vibration does not come into play and the response becomes linear. However, for higher input acceleration nonlinearity appears and the operational bandwidth increases, at the same time, output power level also increases.

1. Introduction

The field of Vibration-based Energy Harvesting (VEH) has been recognized as an area of intensive research for the past few years [1-4]. Most of the researches in the field of VEH involved linear resonating structures. The problem with linear systems is that the response drops significantly as the external excitation frequency shifts from the natural frequency of the structure. The mismatch of frequency is inevitable, as in the majority of practical scenarios the environmental vibration is either frequency varying or random in nature. Thus over the last few years a significant amount of research is focused on obtaining broadband response of VEH devices while increasing the output power.

2. Overview of Conventional Methods to widen the bandwidth

In this section, a brief overview of some of the commonly used methods for obtaining wideband response has been outlined. Among the state-of-the-art techniques, resonant frequency tuning [5-8], multimodal energy harvesting [9-10], and more recently nonlinear energy harvesting [11-17] have emerged as the popular solution.

Nonlinearity is generally introduced in VEH through nonlinear stiffness [11-17]. The dynamics of a general nonlinear oscillator can be described as:

$$\ddot{x} = -\frac{dU(x)}{dx} - \gamma\dot{x} + f(t) \tag{1}$$

Where x is the position of the oscillator; γ is the viscous damping coefficient; f(t) is the force input by the ambient vibration and U(x) is the potential function. For an electromagnetic (EM) generator, γ

also includes the damping caused by electromagnetic coupling. A common approach is to model the potential energy function U(x) in a quadratic form [16] as $U(x) = -\frac{1}{2}\alpha x^2 + \frac{1}{4}\beta x^4$.

The potential function U(x) is symmetric and bistable for α>0, and monostable for α≤0. For the bistable case, two minima at $x_m = \pm\sqrt{\alpha/\beta}$ are separated by a barrier at $x = 0$. For the monostable case, β>0 indicates a hardening response whereas β<0 is a softening response. Bistable nonlinear oscillator has been implemented by snap through mechanism [15], or on a piezomagnetoelastic beam through a magnetic attractive force [11] or repulsive force [16]. The advantage of using bistability is that it can be easily driven to the high amplitude branch with excitation from either large amplitude periodic force or low amplitude stochastic force. On the other hand monostable nonlinear oscillators have been realized through nonlinear magnetic levitation [12], on a piezoelectric beam with magnetic end masses [14] etc. The monostable energy harvester can only work in the condition with slow and stepped frequency sweep [21]. However monostable oscillation appears naturally in a system, whereas to obtain bistable oscillation one needs to accommodate a special configuration.

3. Design and Static Analysis of Nonlinear Stretching of the Spring Arm

In this work, we report an EM Energy Harvester that utilizes nonlinear stretching strain to obtain wideband response. The designed harvester is shown in Figure 1(a). The oppositely polarized magnets are placed on the top of the central platform (10 mm^2 square) of the vibrating structure. Under the influence of external vibrations, the magnets move up and down and flux lines are cut by a conducting coil assembled above it. As a result, voltage is induced into the coil. We have chosen FR4 (standard PCB material) as the structural material of the vibrating structure. Nonlinear springs on FR4 could be a novel approach for wideband and low frequency applications due to its low Young's Modulus (Y=22 GPa) and particularly for EM harvesters as conducting coils can be easily routed on FR4 laminates [18].The length (L), width (W), thickness (d) of the cantilever arms are 9.2 mm, 0.8 mm, 0.2 mm.

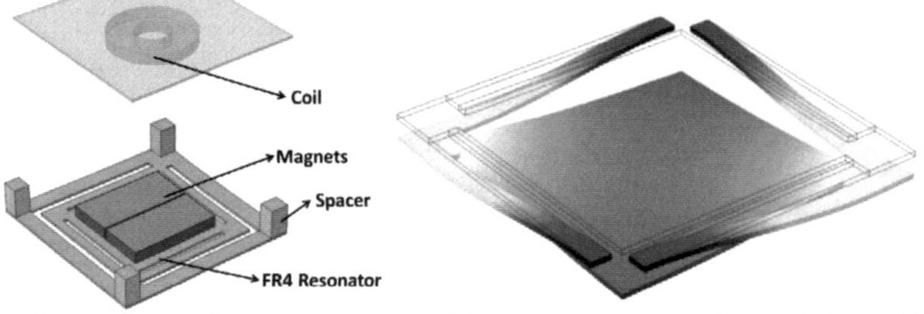

Figure 1: (a) Components of the proposed nonlinear energy harvester prototype. (b) Deformed structure in the first vibrating mode.

Large deflection appears when the proof mass deflection is comparable to, or larger than, the thickness of the spring arms. This results into stretching in addition to bending. The first vibrating mode of the device is shown in Figure 1(b).

The components of the spring force due to bending and stretching are obtained separately. For analytical modeling of the nonlinear spring force, a simplified model of fixed-fixed cantilever beam is used as shown in Figure 2(a). The contribution due to bending of the beam can be approximated by using the Euler Bernoulli's beam equation for the vertical deflection of the neutral axis u(x) along the beam is given by

$$\frac{\partial^4 u}{\partial x^4} = 0 \qquad (2)$$

For a simple bending of the beam, u(x) can be defined along the length of the beam. As seen from Figure 2(a), the boundary conditions for the problem are

$$u(0) = 0, \qquad u'(0) = 0, \qquad u(L) = d_f, \qquad u'(L) = 0$$

Solving equation (2) for given boundary conditions, we obtain an expression for the deflection along the length (x) of the beam as

$$u(x) = \frac{d_f}{L^2}\left[3x^2 - \frac{2x^3}{L}\right] \tag{3}$$

Now, the strain energy due to the bending component of the deformation,

$$E_b(d_f) = \frac{Y}{24} d^3 W \int_0^L \left(\frac{\partial^2 u}{\partial x^2}\right)^2 dx\,) = \frac{1}{2} Y \frac{d^3 W}{L^3} {d_f}^2 \tag{4}$$

Hence the component F_b of spring reaction force arising due to bending is given by

$$F_b = -\frac{\partial E_b}{\partial d_f} = -Y\frac{d^3}{L^3} W d_f \tag{5}$$

The contribution due stretching of the beam can be modeled using the same Figure 2(a).
Stretching component of the strain can be approximated by Pythagoras theorem:

$$Strain(v) = \frac{\Delta L}{L} \approx \frac{\sqrt{L^2 + {d_f}^2} - L}{L} = \sqrt{1 + \left(\frac{d_f}{L}\right)^2} - 1 \approx \frac{1}{2}\left(\frac{d_f}{L}\right)^2 \tag{6}$$

Then the energy stored due to this stretching component of strain can be written as:

$$E_S = \int \frac{1}{2} Y d W \sigma^2 dL = \frac{1}{2} Y \left(\frac{1}{2}\frac{{d_f}^2}{L^2}\right)^2 dWL = \frac{1}{8} Y \frac{dW}{L^3} {d_f}^4 \tag{7}$$

In a similar way, the component F_S of spring reaction force arising due to stretching is given by

$$F_s = -\frac{\partial E_s}{\partial d_f} = -\frac{1}{2} Y \frac{d}{L^3} W {d_f}^3 \tag{8}$$

So, combining both the components, total spring reaction force is obtained analytically.

$$F = F_b + F_s = -\frac{Y}{L^3} W \left[d^3 d_f + \frac{1}{2} d {d_f}^3\right] = -\frac{Y}{L^3} W d^4 \left[\frac{d_f}{d} + \frac{1}{2}\left(\frac{d_f}{d}\right)^3\right] \tag{9}$$

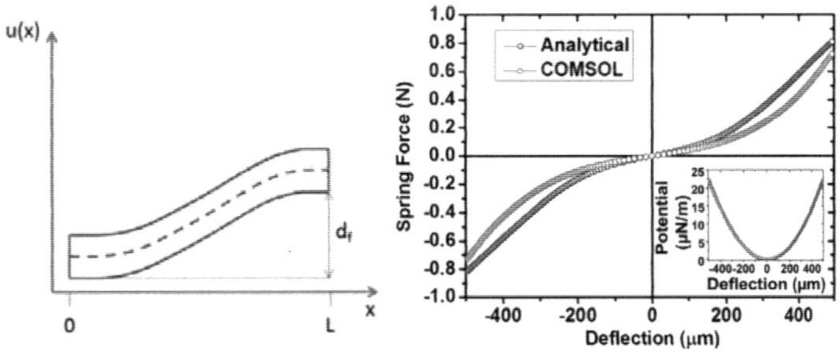

Figure 2: (a) Approximate model for analytical formulation of stretching strain. (b) Variation of spring force with deflection.

The above equation is similar to the well-known nonlinear spring force ($F = kx + k_n x^3$). When the ratio $\frac{d_f}{d}$ is less than one, the cubic nonlinearity is insignificant. When the tip deflection (d_f) is only greater than the thickness of the arm (d), or in other words, $\frac{d_f}{d}$ is greater than one, then only effect of nonlinearity comes into play. Our proposed device exploits this cubic nonlinearity due to large deformation and results in wider bandwidth. The variation of the spring force with deflection is shown on Figure 2(b). The same result is validated by finite element analysis using COMSOL. Up to almost 200 μm of deflection, the stiffness is in the linear region. Beyond that, stiffness becomes nonlinear. The Duffing Potential ($U = \frac{1}{2}kx^2 + \frac{1}{4}k_n x^4$) has monostable configuration [Figure 2(b) inset], which implies the hardening effect due to large deformations. From equation (9), the relation between linear stiffness constant ($k = \frac{YWd^3}{L^3}$) and nolinear stiffness constant ($k_n = \frac{YWd}{2L^3}$) is

$$\frac{k_n}{k} = \frac{1}{2d^2} \tag{10}$$

For our design, the values of k and k_n are found to be 180.81 N/m and $2.25 \cdot 10^9$ N/m^3 and resonant frequency is ($f_0 = \frac{\omega_0}{2\pi}$) 169 Hz.

4. Numerical Investigation of the Monostable Duffing Oscillator for EM Energy Harvesting

The governing equation for an electromechanical oscillator with restoring force $\frac{dU}{dx}$, base excitation $\ddot{Z}$, and electromagnetic conversion mechanism [19] is given by

$$m\ddot{x} + c\dot{x} + \frac{dU}{dx} + \gamma I = -m\ddot{z} \tag{11}$$

As discussed in the previous section, the restoring force is of the form $F = \frac{dU}{dx} = kx + k_n x^3$. So, substituting this in equation (11), we get the monostable Duffing oscillator equation

$$m\ddot{x} + 2m\rho\omega_0\dot{x} + kx + k_n x^3 + \gamma I = -m\omega^2 Z sin\omega t \tag{12}$$

Where, m is the moving mass, ρ is the damping co-efficient, z=Zsinωt is the sinusoidal external perturbation, ω is the external frequency. Since, we are considering EM transduction the external conversion circuit can be modeled using the following equation

$$L\dot{I} + RI - \gamma\dot{x} = 0 \tag{13}$$

Where γ is the electromagnetic coupling co-efficient R is the total resistance which includes coil resistance R_C and load resistance R_L, L is the electromagnetic inductance. We have solved equation (12) and (13) numerically using 4th order Runge Kutta method. The power dissipated in the electrical domain for the electromechanical oscillator is given by $P(t) = \frac{\gamma^2}{R}\dot{x}^2$ and consequently we have calculated the average power as $P_{avg} = \frac{1}{T}\int_t^{t+T} P(t')dt'$. Spectra of average power for different load resistances are plotted in Figure 3(a). In our design we have chosen a copper wire wound coil having wire diameter 32 μm and 450 turns. The coil resistance is R_C=60Ω. We have also considered large input acceleration of $\omega^2 Z$=20g (g=10m/s^2). The value of m, γ and ρ are found to be $16\cdot10^{-5}$ kg, $2.25\cdot10^{-3}$ Wb/m and 0.1.

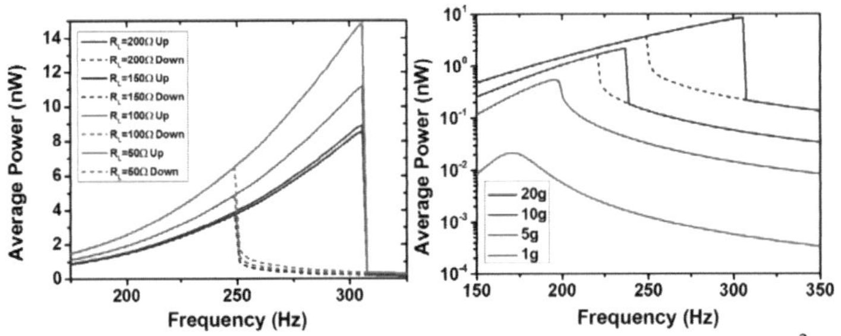

Figure 3: (a) Power spectra for different values of load resistances and fixed $\omega^2 Z$=20g. (b) Power spectra for different input accelerations and fixed R_L=200Ω.

It is seen that as we lower the load resistance, total average power increases. This is because lowering the load increases the electrical damping and as a result more power is dissipated. It can be seen that there are two steady states available in that hysteresis width of the spectrum. So, in energy harvesting application the main concern is to keep the output in the high amplitude branch for longer time. The effect of external force can be seen from Figure 3(b). In this case we have kept load resistance fixed at R_L=200Ω. For smaller input force, the nonlinearity disappears and also the average power becomes low. The effect of nonlinear stretching becomes significant only with large force and it was seen in Figure 2(b) that the large deformation occurred for such high input force. Large amplitude deformation is responsible for increased average power. However, if we increase the input force too much then in addition to the main resonance several secondary resonances also appear increasing the overall loss within the system. Similar phenomenon was addressed by Parlitz et al [20].

Figure 4 shows the variation of jump frequency on up-sweep with load resistance and input acceleration. For small accelerations, there is no jump phenomena indicated by the blue surface of the curve. The jump frequency appears as the driving force increases and the surface become more reddish. Also for the red region, jump frequency increases with increasing load resistance. This is because increase of load reduces the total damping. As a result jump frequency increases. However very low damping region is avoided in our analysis as such region includes more complex phenomenon such as sub harmonic resonances.

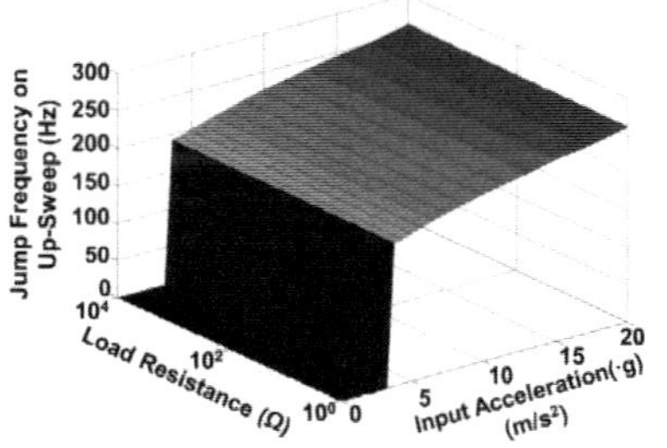

Figure 4: Variation of Jump Frequency on Up-Sweep with Load resistance and input acceleration. The blue portion implies there is no jump. As the curve become red, nonlinear phenomenon becomes prominent and jump frequency appears with high input force.

5. Conclusions

In this work, we have discussed the design and stretching strain based nonlinearity of an EM Energy Harvester. FR4 is chosen as the device material due to its suitable mechanical properties and ease of integratability with copper coils. An analytical formulation of large deformation of the spring arms has been done by calculating separately the contributions from bending and additional stretching, which results is cubic nonlinearity. The results are verified with the Finite Element Analysis. It is found that nonlinearity has an inverse square dependence on thickness of the arms. Also, numerical investigation of a monostable Duffing oscillator has been done. Results showed that with decrease of load resistance, the average power output increases, where the output response depends strongly on the input force. For small input acceleration, the desired large amplitude vibration does not come into play and the response becomes linear. However, for higher input acceleration nonlinearity appears and the operational bandwidth increases, at the same time, output power level also rises

Acknowledgement

This work is financially supported by Science Foundation Ireland (SFI) Principal Investigator (PI) project on 'Vibration Energy Harvesting' grant no SFI-11/PI/1201.

References

[1] Kulkarni S, Roy S, O'Donnell T, Beeby S, Tudor J 2006 Journ. Appl. Phys. **99**:08P511.
[2] Kulkarni S et al. 2008 Sensors and Actuators A **145–146**:336–342.
[3] Beeby S P et al. 2007 J. Micromech. Microeng. **17**:1257–1265.
[4] Koukharenko E et al. 2006 Microsystem technologies **12**;1071-1077.
[5] Roundy S and Zhang Y 2005 Proc. SPIE, 5649:373-84.
[6] Leland E S and Wright P K 2006 Smart Mater. Struct., **15**:1413-20.
[7] Challa V R, Prasad M G, Shi Y and Fisher F T 2008 Smart Mater. Struct., **17**:015035.
[8] Peters C et al. 2009 J Micromech. Microeng, **9**:094004.
[9] Sari I, Balkan T and Kulah H 2008 Sens. Actuators. A, 145-146:405-13.
[10] B Yang et al. 2009 J. Micromech. Microeng., **19**:035001.
[11] Erturk A, Hoffmann J and Inman D J 2009 Appl. Phys. Lett., **94**:254102.
[12] Mann B P and Sims N D 2009 J. Sound Vib., **319**:515-30.
[13] Marinkovic B and Koser H 2009 Appl. Phys. Lett., **94**:103505.
[14] Stanton S C, McGehee C C and Mann B P 2010 Physica D, **239**:640-53.
[15] Ramlan R, Brennan M J, Mace B R and Kovacic I 2010 Nonlinear Dyn., **59**:545-58.
[16] Gammaitoni L, Neri I and Vocca H 2009 Appl. Phys. Lett., **94**:164102.
[17] Cottone F, Vocca H and Gammaitoni L 2009 Phys. Rev. Lett., **102**:080601.
[18] Hatipoglu G and Urey H Smart Mater. Struct. 2010 **19**:015022.
[19] R L Harne and K W Wang 2013 Smart Mater. Struct. **22**:023001.
[20] Parlitz U, Lauterborn W 1985 Physics Letters **107A**:8.
[21] Tang L et al. 2010 Journal of Intelligent Material Systems and Structures **21**: 1867.

Frequency Up-Converted Low Frequency Vibration Energy Harvester Using Trampoline Effect

S Ju[1], S H Chae[1], Y Choi[1], S Jun[1], S M Park[1], S Lee[1], H W Lee[2], and C-H Ji[1]

[1] Department of Electronics Engineering, Ewha Womans University, Seoul, Korea
[2] Department of Neurology, Ewha Womans University School of Medicine, Seoul, Korea

E-mail: cji@ewha.ac.kr

Abstract. This paper presents a non-resonant vibration energy harvester based on magnetoelectric transduction mechanism and mechanical frequency up-conversion using trampoline effect. The harvester utilizes a freely movable spherical permanent magnet which bounces off the aluminum springs integrated at both ends of the cavity, achieving frequency up-conversion from low frequency input vibration. Moreover, bonding method of magnetoelectric laminate composite has been optimized to provide higher strain to piezoelectric material and thus obtain a higher output voltage. A proof-of-concept energy harvesting device has been fabricated and tested. Maximum open-circuit voltage of 11.2V has been obtained and output power of 0.57μW has been achieved for a 50kΩ load, when the fabricated energy harvester was hand-shaken.

1. Introduction

Harvesting energy from environmental vibration is being widely researched to replace the conventional energy sources for emerging sensor applications [1]. Vibration energy harvester converts ambient mechanical vibration energy into electrical energy using piezoelectric, electromagnetic, and electrostatic transduction mechanisms. Magnetoelectric laminate composites based on magnetostrictive and piezoelectric materials are also investigated as a candidate for vibration energy harvesting [2]. Magnetoelectric harvesters can produce relatively higher output voltage without mechanical impact or frequency matching due to high strain generating capability of some of the magnetostrictive materials.

Generating power from low frequency vibrations such as human-body-induced motion poses a significant challenge due to its random nature, very low input frequency, and high acceleration. In general, conventional vibration energy harvester produces power using spring-mass-damper-based resonant system, which is optimized to operate at relatively narrow frequency band near the resonance. To overcome the well-known issues of conventional systems, wide variety of alternative approaches have been reported including the implementation of multi-frequency generator, resonant frequency tuning, utilization of impact-based piezoelectric generators and frequency up-converted electromagnetic transducers [3-5].

Despite the successful deployment in magnetic and piezoelectric energy harvesters, frequency up-conversion of magnetoelectric energy harvester has not been demonstrated. Previously, we have presented a non-resonant low frequency vibration energy harvesting device based on magnetoelectric laminate composite to generate power from random vibration of the ball magnet inside a channel [6].

Although a high peak-to-peak voltage in response to various input vibrations has been observed, output power has been limited due to fast decay in output voltage and low input frequency. In this research, we present a magnetoelectric energy harvester based on mechanical frequency up-conversion using trampoline effect of aluminum springs integrated on both ends of the channel.

2. Harvester design

A schematic diagram of the proposed vibration energy harvester is shown in figure 1. The harvester consists of a spherical permanent magnet, a magnetoelectric laminate composite fabricated by stacking two MSMA (Magnetic Shape Memory Alloy) bars on top and bottom of an MFC (Macro Fiber Composite) sheet, and aluminum housing with monolithically integrated aluminum springs. The springless spherical permanent magnet inside a 25mm-long rectangular channel provides a time-varying vertical magnetic field to the magnetoelectric composite for power generation. When the magnet reaches the end of the channel, it bounces off the springs formed monolithically on the channel end. Spring constant of the S-shaped spring is 37.7kN/m. Teflon sheet covers the bottom of the channel to reduce the friction. Optimized gap between the ball magnet and MSMA is 1.5mm.

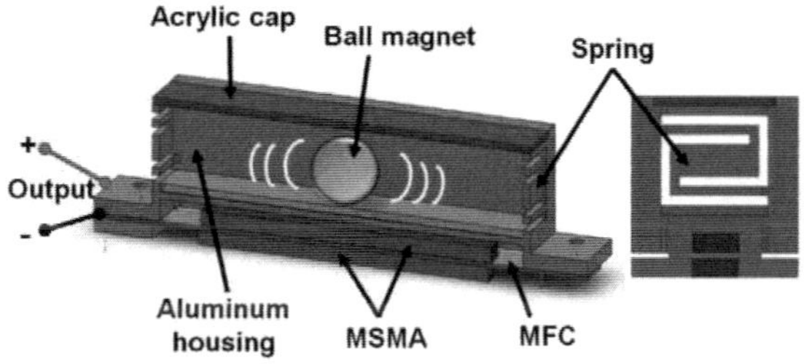

Figure 1. Schematic of the proposed energy harvester using spherical permanent magnet, integrated spring on channel end, and MSMA/MFC/MSMA laminate structure.

2.1. Bonding method of magnetoelectric laminate composite

To maximize the strain in magnetoelectric laminate composed of MSMA and MFC at given magnetic field variation, we have tested various boundary conditions and bonding configurations of constituent components, as shown in figure 2. As summarized in Table 1, harvester with cantilever-type transducer which had two MSMA bars on top and bottom of the MFC sheet and bonding regions limited to both ends of the MSMA bar produced the highest output.

Figure 2. Schematic diagram of bonding methods of magnetoelectric laminate composite.

Table 1. Comparison of open circuit voltage for various bonding methods and configurations (harvester is hand-shaken).

Bonding region		Entire MSMA bar	Both ends of the MSMA bar	Both ends and center of the MSMA bar
MSMA on top	Bridge [%]	36.5	12.5	38.5
	Cantilever [%]	65.4	92.3	107.7
MSMA on top and bottom	Bridge [%]	52.9	28.8	50
	Cantilever [%]	100	156.9	113.5

2.2. Trampoline effect
The trampoline effect refers to definite elasticity in the impacting object such as baseball bat, tennis racquet, and golf club which act like a trampoline. This phenomenon means that the stored energy of impacting object is efficiently returned to the moving mass due to reduced energy dissipation during collisions. In this research, trampoline effect has been utilized to effectively return the stored compressional energy in the housing during the impact between the ball magnet and channel ends. In place of the flat channel ends used in the previous work [6], S-shaped aluminum spring has been formed on both ends of the channel for improved coefficient of restitution.

3. Fabrication
Figure 3 shows the fabricated energy harvesting device. A proof-of-concept harvester has been fabricated by assembling a 5mm-diameter NdFeB spherical magnet, a magnetoelectric component using MSMA/MFC/MSMA laminate structure, and aluminum housing with Teflon floor. Two MSMA bars ($20x2.5x1mm^3$) have been attached to an MFC sheet ($36x5.6x0.3mm^3$) with limited bonding region on both ends of the MSMA elements. Magnetoelectric laminate has been fixed to the housing in cantilever configuration using miniature screws. Effective volume of the assembled device measures $26x9x8.5mm^3$.

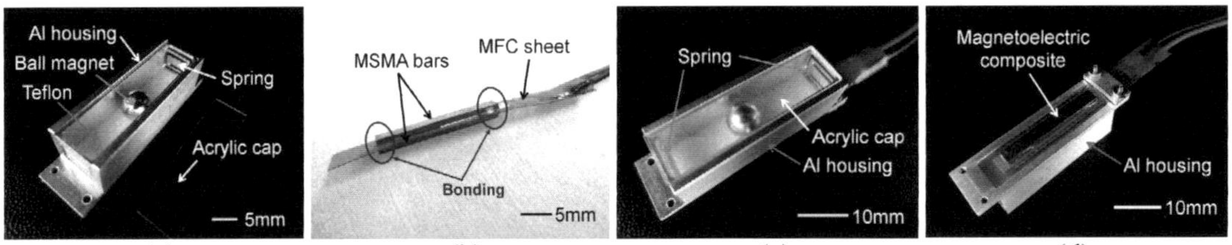

(a) (b) (c) (d)

Figure 3. Assembled device: (a) aluminum housing with spherical permanent magnet and Teflon plate at the bottom, (b) MSMA/MFC/MSMA laminate composite, (c) top-side of the energy harvester, (d) bottom-side of the energy harvester.

4. Results and discussion
Fabricated device has been tested using vibration exciter. Output from the harvester has been measured at 6-20Hz sinusoidal excitation with 1Hz step. Input acceleration at each frequency has been varied from 1g to 3g with 0.5g step. Input vibrations at frequencies under 6Hz and accelerations larger than 3g could not be applied due to the limitation of the exciter. In addition, manual vibration test has been carried out where the fabricated device has been attached to the backside of a smartphone and output has been measured when the phone was hand-shaken.

4.1. Vibration exciter test
Device with single magnet and with three magnets connected in series have been tested and compared. In addition to the change in magnetic field distribution and total mass, increased number of magnets effectively increases the length of the proof mass, thereby reducing the time required for the magnets to impact the channel end.

As shown in figure 4, peak-to-peak open-circuit voltage showed an increasing trend as the acceleration was increased in both cases, with very weak frequency dependence. In addition, increase in output voltage was pronounced for device with three magnets at higher input accelerations, although responses to frequencies higher than 16Hz require further investigation (figure 4(b)). Maximum open-circuit voltage of 1.43V has been obtained at 19Hz for device with single magnet, and 2.2V has been obtained at 20Hz for device with three magnets.

As shown in figure 5(a), decay of the output waveform has been reduced substantially compared to previous results [6], presumably due to change in resonance mode affected by changed channel end geometry. Also, improved output voltage and increased number of visible peaks per cycle have been observed with the three-magnet device, which confirms the frequency-up-conversion effect (figure 5(b)). Frequency of impact between the magnet and channel end per each cycle of input vibration has been increased by a factor of 2 due to addition of springs on channel ends and increased length of the proof mass.

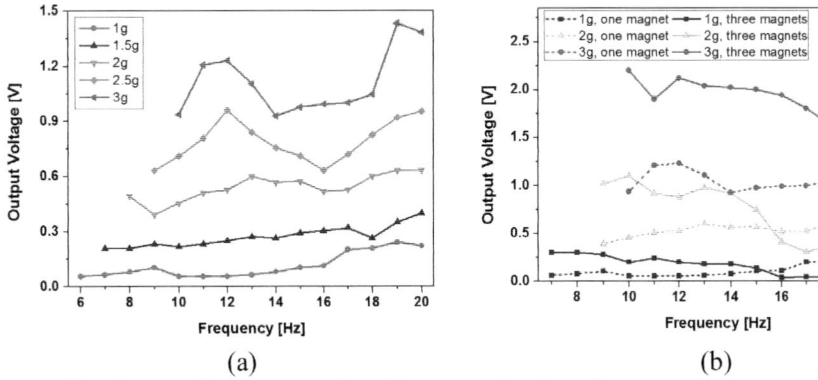

Figure 4. Vibration exciter test results: (a) open-circuit voltage vs. frequency at various input accelerations (device with single magnet), (b) open-circuit voltage vs. frequency at various input accelerations at different number of magnets (device with single magnet and three magnets).

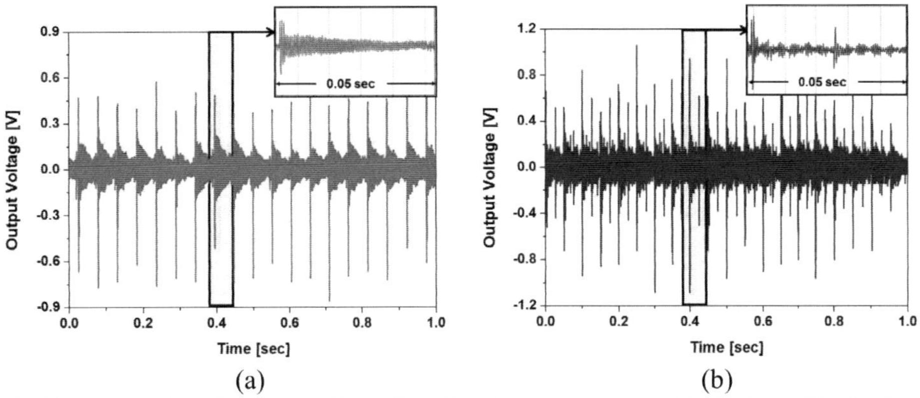

Figure 5. Open-circuit voltage waveform for vibration exciter test: (a) device with single magnet (3g acceleration at 19Hz), (b) device with three magnets (3g acceleration at 20Hz).

The RMS (root-mean-square) voltage and average power delivered to the external resistive load for device with three magnets are shown in figure 6. The maximum output power reached 0.045μW across a 50kΩ resistor when 3g acceleration at 20Hz has been applied. Average power of the device with three magnets is about 1.5 times higher than the power generated with device having single magnet.

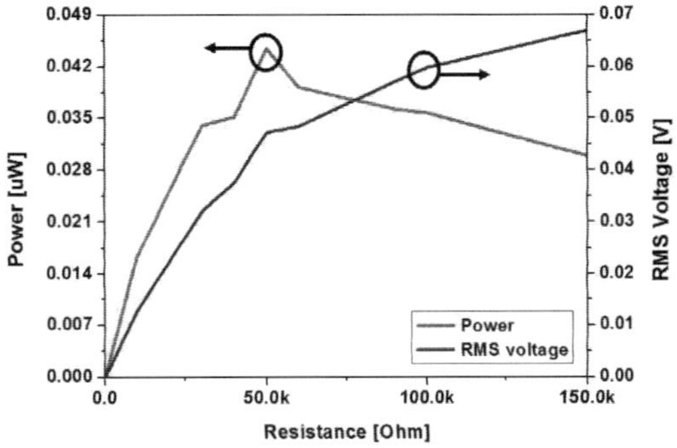

Figure 6. Vibration exciter test results: RMS voltage and average power vs. load resistance for device with three magnets (3g acceleration at 20Hz has been applied).

4.2. Manual vibration test

For the manual vibration of device with three magnets, typical output voltage and average power were 5V and 0.35μW, respectively. To further improve the output by extending the total length of the magnet while maintaining the perpendicular magnetic field, modified device with block magnet (9.5x4.8x4.8mm^3) and two hemispherical pure irons attached on both ends has been tested. Maximum open-circuit voltage of 11.2V and output power of 0.57μW have been achieved (figure 7).

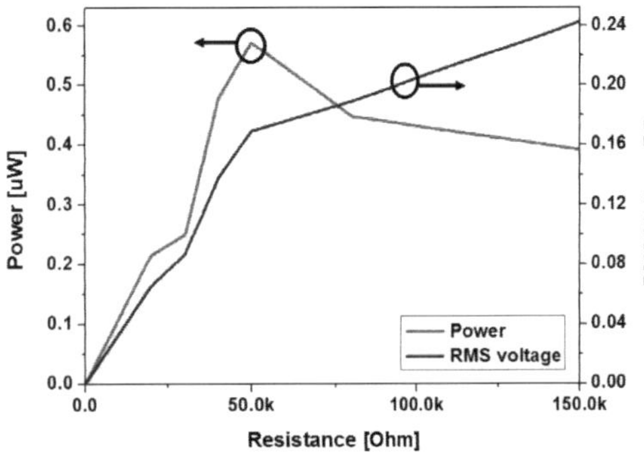

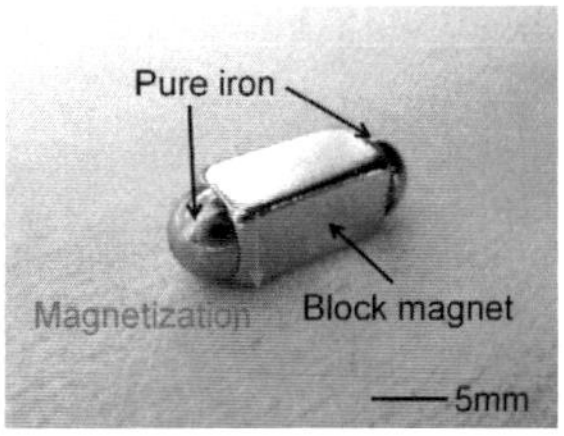

Figure 7. (left) Manual vibration test results: RMS voltage and average power vs. load resistance (device with magnet shown on the right), (right) block magnet assembly.

For further improvement of the generated power, coefficient of restitution at both ends of the cavity, geometry of the proof mass and magnetizing direction of the magnet can be optimized. Moreover, electrode configuration of the MFC sheet requires further optimization.

5. Conclusion

We have demonstrated a non-resonant magnetoelectric vibration energy harvester based on mechanical frequency up-conversion of input vibration using trampoline effect of aluminum springs integrated on both ends of the channel. A proof-of-concept vibration energy harvester has been designed, fabricated, and tested successfully. For device with three magnets, maximum open-circuit voltage of 2.2V has been obtained at 3g acceleration at 20Hz. Moreover, maximum open-circuit voltage of 11.2V has been obtained and output power of 0.57μW has been achieved for a 50kΩ load, when the fabricated device with block magnet was mounted on a smartphone and shaken manually. Although the device requires further optimization to achieve higher output voltage, we have successfully demonstrated the feasibility of frequency up-conversion using trampoline effect.

6. Acknowledgement

This work was supported by the Ewha Global Top 5 Grant 2011 of Ewha Womans University and by the Pioneer Research Center Program through the National Research Foundation of Korea funded by the Ministry of Science, ICT & Future Planning (2010-0019453).

References

[1] Najafi K *et al.* 2011 Microsystems for energy harvesting *Proc. Transducers 2011 (Beijing, 5-9 June 2011)* 1845-1850

[2] Beeby S P *et al.* 2006 Energy harvesting vibration sources for microsystems applications *Meas. Sci. Technol.* **17** 175-195

[3] Ashraf K *et al.* 2013 Improved energy harvesting from low frequency vibrations by resonance amplification at multiple frequencies *Sens. and Act. A* **195** 123-132

[4] Gu L *et al.* 2011 Impact-driven, frequency up-converting coupled vibration energy harvesting device for low frequency operation *Smart Mater. Struct.* **20** 045004

[5] Galchev T *et al.* 2011 Micro power generator for harvesting low-frequency and nonperiodic vibrations *J. Microelectromech. Syst.* **20** 852-866

[6] Ju S *et al.* 2013 Harvesting energy from low frequency vibration using MSMA/MFC laminate composite *Proc. Transducers 2013 (Barcelona, 16-20 June 2013)* 1348-1351

Multiphysics finite element model of a frequency-amplifying piezoelectric energy harvester with impact coupling for low-frequency vibrations

R Dauksevicius[1,2,*], D Briand[1], A Vásquez Quintero[1], R A Lockhart[1], P Janphuang[1], N F de Rooij[1], V Ostasevicius[2]

[1] École Polytechnique Fédérale de Lausanne (EPFL), Institute of Microengineering (IMT), Sensors, Actuators and Microsystems Laboratory (SAMLAB), Rue Jaquet-Droz 1, Neuchâtel, 2002, Switzerland.
[2] Kaunas University of Technology, Faculty of Mechanical Engineering and Mechatronics, Institute for Hi-Tech Development, Kestucio str. 27, Kaunas, 44312, Lithuania

E-mail[*]: rolanas.dauksevicius@epfl.ch

Abstract. This paper presents experimentally-verified multiphysics finite element model of a wideband vibration energy harvester with impact coupling, which operates on the principle of frequency up-conversion: under low-frequency harmonic base excitation a cantilever-type resonator (with resonant frequency of 18.8 Hz) impacts a high-frequency piezoelectric cantilever, which starts freely vibrate at its resonant frequency of 374 Hz. Such input frequency amplification enables efficient power generation under low-frequency ambient excitations. The model was implemented in COMSOL and the contact between the cantilevers was formulated by using a nonlinear viscoelastic model. Reported results of dynamical and electrical testing of the fabricated vibration energy harvester confirm the accuracy of the model as well as reveal some operational characteristics of the device under varying impact and excitation conditions.

1. Introduction

Deriving electrical energy from ambient vibrations through piezoelectric conversion represents one of the most promising approaches to power MEMS-based sensors since kinetic energy is abundant in the environment, while piezoelectric transducers exhibit the highest energy density and are amongst the most cost-effective solutions in terms of fabrication at micro and nano scales [1]. Vibration energy harvesters (VEHs) are mostly based on linear oscillators, generating maximum power only when driven at their resonant frequency. Slight deviation from the resonance leads to a significant drop in power output. Tackling the issue of narrow bandwidth currently represents one of the major challenges in the field [2]. Three main categories of broadband VEHs may be distinguished [3]: i) linear generators (e.g. generator arrays, coupled resonators), ii) nonlinear generators (e.g. mono/bi/multi-stable structures, frequency up-converting systems, generators with resonance tuning), iii) generators exploiting advanced power conditioning circuits. This study is concerned with the frequency up-conversion, whereby low-frequency (LF) resonator impulsively triggers high-frequency (HF) self-oscillations of the piezoelectric generator(s) using either magnetic [4] or impact coupling [5]. The former avoids the issue of contact wear that is inherent to the latter but the usage of ferromagnetic components increases device volume leading to reduced power density. On the other hand, impact coupling is a more straightforward approach for implementing frequency up-conversion, but care must be taken in selecting materials of contacting surfaces to avoid premature failure due to contact fatigue.

This study considers an impact-coupled vibration energy harvester (IC-VEH), where rigid end mass of the LF resonator impacts a soft polymeric substrate of the piezoelectric generator (instead of brittle piezoceramics), providing favorable conditions to prolong contact fatigue lifetime [6]. The paper is devoted to multi-physics finite element (FE) modeling of the IC-VEH. Device characterization results are presented in order to show accuracy of the developed model and provide some insight into the operational characteristics of the harvester. To the best of the authors' knowledge this model is the first attempt to simulate with FE a piezoelectric IC-VEH as a vibro-impact system consisting of two deformable structures as only analytical models are reported in the literature [7,8].

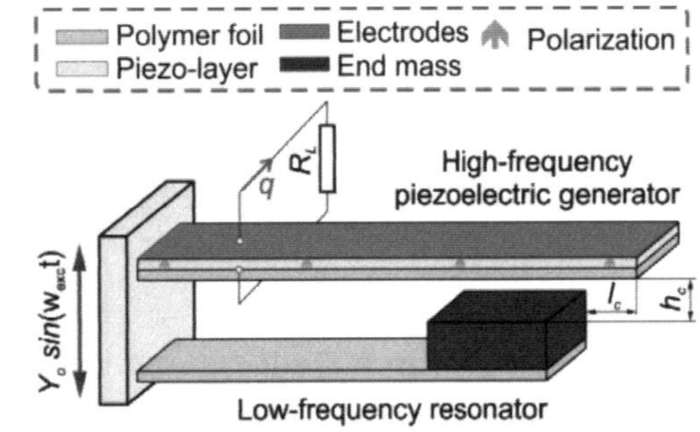

Figure 1. Schematics of the modeled and tested impact-coupled vibration energy harvester.

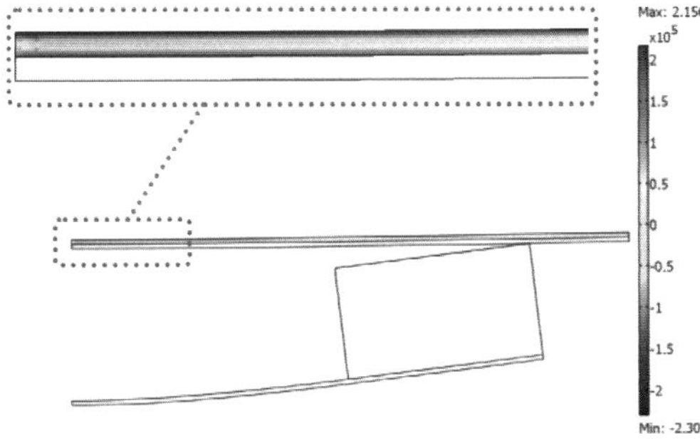

Figure 2. FE model of the IC-VEH: close-up shows electric field distribution (a = 0.5g).

2. Finite element modeling and testing of impact-coupled vibration energy harvester

2.1. Device description

The considered IC-VEH consists of polymeric LF resonator with steel end mass mounted below the HF piezoelectric generator connected to the load R_L (Figure 1). LF resonator impacts the generator at a contact point that is located at a distance l_c from the free end of the piezoelectric cantilever. Contact gap h_c is defined as a distance between the generator and resonator tip when it is displaced downwards due to gravity. Devices design (Table 1) is based on polymeric cantilevers that are patterned by laser cutting. It was realized by a variation of the process flow reported in [6]. LF resonator is fabricated from PET foil and has steel end mass, while piezoelectric generator comprises bulk PZT-5H sheet laminated on PET substrate. Dry adhesive is used for bonding of structural layers.

2.2. Model formulation

The considered IC-VEH constitutes a nonlinear vibro-impact system comprised of two elastic deformable structures undergoing dynamic contacting under harmonic base excitation. A 2D finite element (FE) model of the harvester (Figure 2) was developed in COMSOL. External resistive load was introduced into the model, enabling prediction of generated electrical power. Mechanical contact between the LF resonator and piezoelectric generator was implemented by using a nonlinear viscoelastic model of Hunt and Crossley. The model assumes small contact surface and is valid for direct central and frictionless contact. The normal contact force in Hunt-Crossley model is represented by the pure elastic Hertz's law combined with a nonlinear damping term [9]:

$$F_C = k_C z(t)^{\alpha} + c_C z(t)^{\alpha} v(t) \qquad for\ z(t) < 0\ and\ v(t) < 0 \tag{1}$$

where k_c – contact stiffness coefficient, c_c – contact damping coefficient, $z(t)$ – indentation, $v(t)$ – indentation velocity, α – force exponent depending on contact surface geometry (α=2 is assumed).

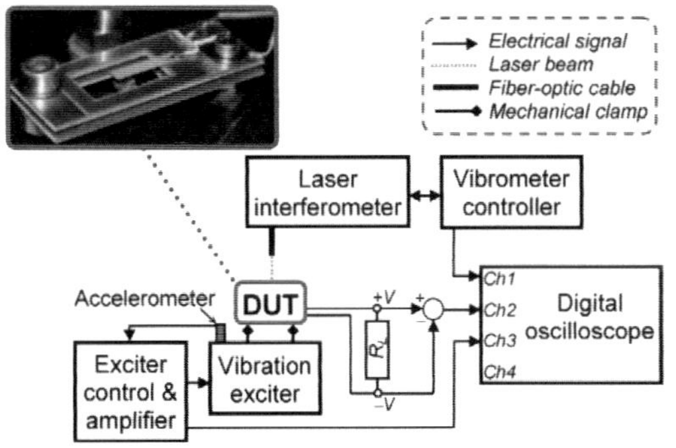

Figure 3. Experimental setup for IC-VEH testing.

Table 1. Dimensions of the simulated and tested impact-coupled vibration energy harvester.

Component	LF resonator	Piezoelectric cantilever
	l×w×t (mm)	
Polymeric substrate (PET)	13×4×0.125	15×4×0.125
Piezoelectric layer (PZT-5H)	-	15×4×0.130
Steel end mass	4.8×4×3.2	-

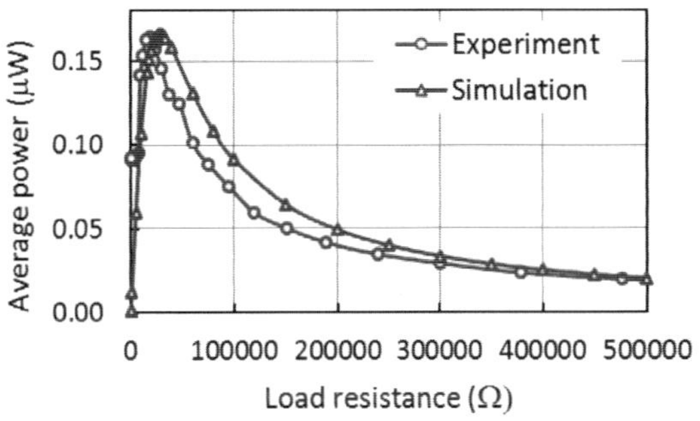

Figure 4. Experimental and simulation plots of average power vs. load resistance (a = 0.5g).

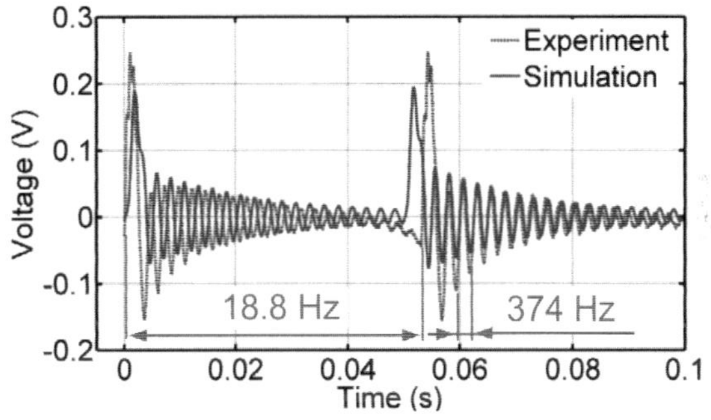

Figure 5. Measured and simulated voltage signals (h_c = 1.2 mm, a = 0.05g, R_L = 1 MΩ).

Eq. 1 indicates that damping term depends on the indentation, which is reasonable from physical point of view since contact area increases with indentation and a plastic region is more likely to develop for larger indentations. Another advantage of the model is that the contact force evolves continuously, i.e. it builds up from zero upon impact and returns to zero upon separation [9]. The model includes a single contact pair located on the right edge point of the end mass.

2.3. Model verification

Verification of the FE model was conducted in two steps. Firstly, the accuracy of piezoelectric generator model without impacting was tested by comparing results of experimental and numerical frequency response analyses performed by varying resistive load. The responses of the fabricated IC-VEH (Figure 3) were measured by means of automated vibration test bench [10]. Figure 4 indicates good agreement between experimental and simulation plots of average power as a function of load resistance. The relative error between the measured and the computed values of average power at optimal load and voltage at open circuit is less than 2%. Therefore the voltage output at open circuit is further used as a parameter for verifying and adjusting the contact model of the IC-VEH.

In the second step, values of k_c and c_c were determined by trial-and-error through multiple simulation tests. Transient voltage responses measured with the setup shown in Figure 3 were compared with the numerical responses computed under the equivalent excitation conditions. The values of $k_c \approx 10^6$ and $c_c \approx 10^{-6}$ were found to provide a good qualitative and acceptable quantitative agreement between the model predictions and measurements (Figure 5). The figure demonstrates the mechanism of frequency up-conversion in the IC-VEH. When the device is subjected to base excitation at 18.8 Hz (resonant frequency of the resonator), high-frequency decaying oscillations (374 Hz) are induced in the generator through impacting. Results of transient simulations performed with gap sizes of 1.2 mm and 1.7 mm indicate that subtle changes in RMS voltage magnitude are well reproduced for different gaps (Figures 6-7).

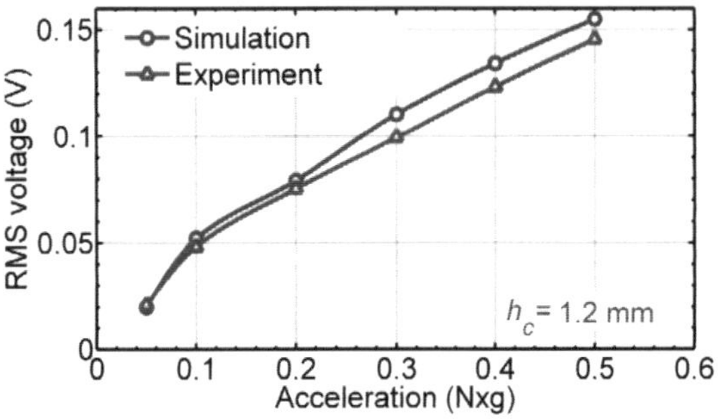

Figure 6. Measured and simulated RMS voltage values as a function of applied acceleration (h_c = 1.2 mm, l_c = 2 mm, f_{exc} = 18.8 Hz, R_L = 1 MΩ).

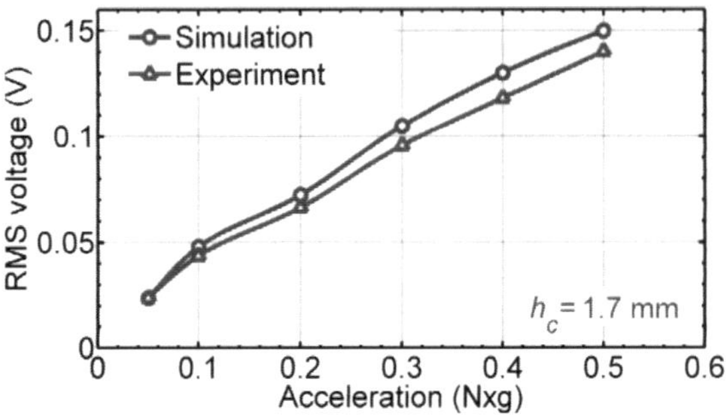

Figure 7. Measured and simulated RMS voltage values as a function of applied acceleration (h_c = 1.7 mm, l_c = 2 mm, f_{exc} = 18.8 Hz, R_L = 1 MΩ).

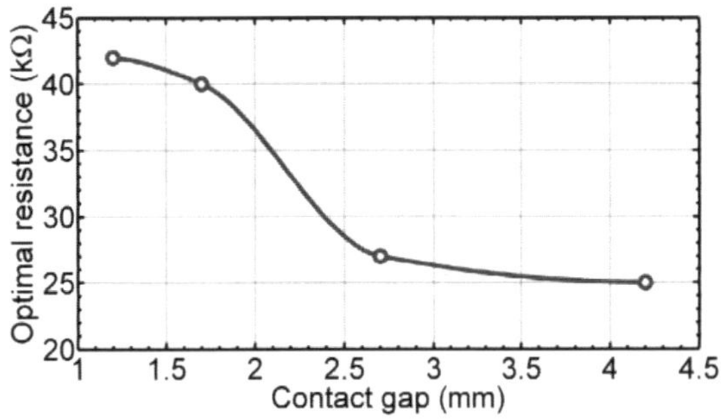

Figure 8. Variation of optimal load resistance as a function of contact gap h_c (a = 0.2g).

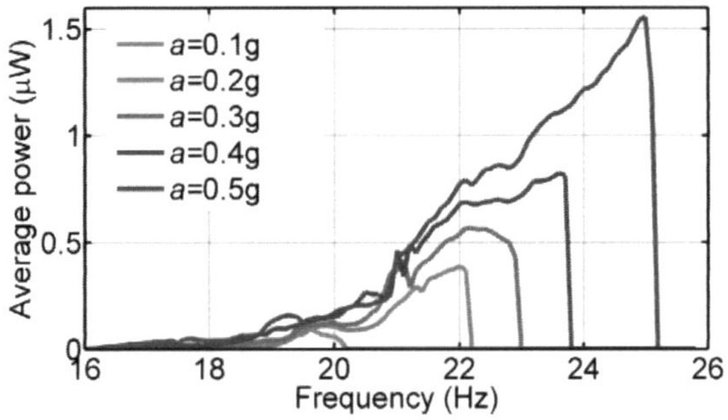

Figure 9. Measured power frequency responses during harmonic up-sweep at various acceleration magnitudes (h_c = 1.2 mm, R_{opt} = 42 kΩ).

It can be also noted from the figures that variation of discrepancy between experimental and numerical values follows the same trend for both gap sizes, i.e. slightly increases with applied acceleration. Performed verification indicates that an average relative error constitutes less than 10%.

2.4. Characterization results

Contact gap is an important design parameter in IC-VEH therefore experimental study was carried out to examine the influence of gap size on optimal resistance, power output and bandwidth. Measurements show that optimal resistance of the generator under normal resonant operation is about 22 kΩ. However results derived from frequency responses recorded using different contact gaps in 1.2–4.2 mm range reveal increasing optimal resistance with smaller gaps (Figure 7). It was reported in [7] that characteristic frequency of the coupled vibration phase (when resonator and generator move together after contact) largely determines the power output in IC-VEHs. Therefore it could be reasoned that in our case the variation in contact gap evokes changes in the coupled vibration frequency and its contribution to the resonant frequency of the overall vibro-impact system, which in turn could change optimal load resistance (R_{opt}=1/$\omega_r C$). The performance of the IC-VEH in terms of power output and bandwidth was examined both for up-sweep and down-sweep excitation cases. Figures 9-10 indicate that up-sweep is considerably more beneficial. Measurements at different contact gaps and acceleration levels demonstrate that for up-sweep excitation the peak power is on average 400% larger than for the down-sweep, while in the case of bandwidth – 300%. Figure 11 illustrates the trend of diminishing bandwidth with larger contact gaps for various acceleration magnitudes during harmonic up-sweep. The results also suggest that the most advantageous contact gap size in terms of both power and bandwidth may be derived for the considered case.

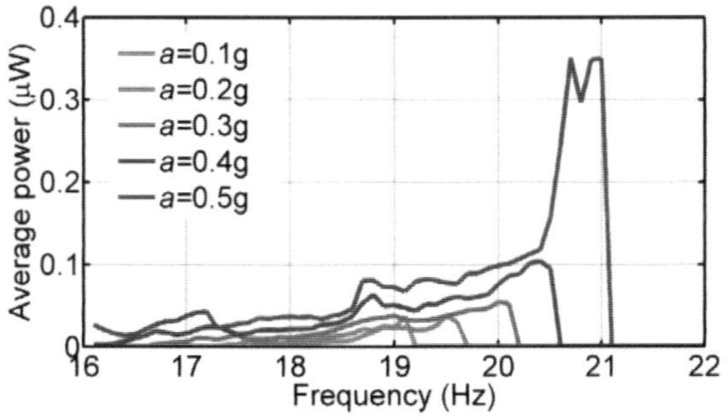

Figure 10. Measured power frequency responses during down-sweep at various acceleration magnitudes (h_c = 1.2 mm, R_{opt} = 42 kΩ).

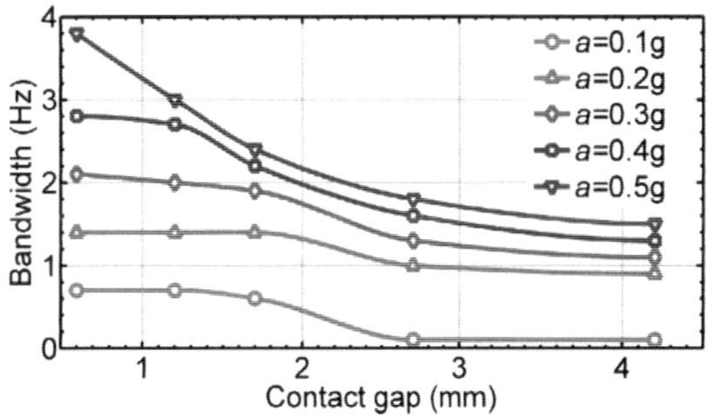

Figure 11. Variation of bandwidth as a function of contact gap h_c for different acceleration magnitudes (up-sweep).

3. Conclusions

The paper presented multiphysics FE model of the IC-VEH that constitutes an electromechanical vibro-impact system composed of two cantilevers – a low-frequency resonator and a high-frequency piezoelectric generator that undergo impacting when subjected to harmonic base excitation. Such configuration realizes frequency up-conversion mechanism, which enables large input frequency amplification leading to efficient power generation under low-frequency input excitation. Vibro-impact interaction between the resonator and generator is accurately reproduced by implementing into COMSOL a nonlinear viscoelastic (Hunt-Crossley) contact model, yielding a small relative error between the experimental and numerical values of RMS voltage produced by the generator under open circuit condition. Results of performed characterization of the fabricated IC-VEH device in the automated vibration test bench indicate strong influence of the contact gap size on optimal load, average power and bandwidth of the harvester.

The presented FE model constitutes an effective and reliable design tool enabling optimization of mechanical and piezoelectric structures of IC-VEHs, which may have arbitrary geometry and configuration (e.g. multiple resonators and generators, segmented electrodes, etc.).

Acknowledgments

This research work was supported by the Swiss Sciex-NMS^{CH} fellowship project No 12.042.

References

[1] Kim S-G, Priya S and Kanno I 2012 *MRS Bull.* **37** 1039-1050
[2] Tang L, Yang Y and Soh C K 2010 *J. Intell. Mater. Syst. Struct.* **21** 1867-1897
[3] Twiefel J and Westermann H 2013 *J. Intell. Mater. Syst. Struct.* **24** 1291-1302
[4] Galchev T, Aktakka E E and Najafi K 2012 J. Microelectromech. S. **21** 1311-1320
[5] Liu H, Lee C, Kobayashi T, Tay C J and Quan C 2012 *Sens. Actuators* A **186** 242-248
[6] Vásquez Quintero A, Briand D, Janphuang P, Ruan J J, Lockhart R and de Rooij N F 2012 *Proc. IEEE MEMS* Paris, France, 29 January – 2 February 1289-1292
[7] Gu L and Livermore C 2011 *Smart Mater. Struct.* **20** 045004
[8] Renaud M, Fiorini P, van Schaijk R and van Hoof C 2009 *Smart Mater. Struct.* **18** 035001
[9] Machado M, Moreira P, Flores P and Lankarani H M 2012 *Mech. Mach. Theory.* **53** 99-121
[10] Ruan J J, Lockhart R A, Janphuang P, Vásquez Quintero A, Briand D and de Rooij N. 2013 *IEEE Trans. Instrum. Meas.* **99** 1-8

Variable reluctance harvester for applications in railroad monitoring

M. Kroener[1], S. K. T. Ravindran and P. Woias

Group Manager, *Micro Energy Harvesting Technologies*
University of Freiburg, Department of Microsystems Engineering – IMTEK, Laboratory for Design of Microsystems, Freiburg, Germany

E-mail: michael.kroener@imtek.uni-freiburg.de

Abstract. We present the design, realization, and testing of a variable reluctance energy harvester for the detection of moving ferromagnetic parts, e.g. the wheels of a train while passing over a train passage detector. Measurements were done to determine the output voltage, the energy output per event and the power output with the frequency of the moving ferromagnetic body. Results are compared with finite element analysis (FEA) to estimate the change in magnetic flux and the output voltages. A maximum energy output of 131 µJ per pulse was measured for a simulated condition of a train wheel passing with a speed of 81.5 km/h, which results in a mean output power of 5.9 mW, with a spacing of 10 mm between wheel and the reluctance circuit. This shows that the variable reluctance principle, a well-known method used for numerous sensor applications, is also a comfortable, energy-autonomous and reliable method to detect passing train wheels or other moving ferromagnetic parts, with a simple setup and fairly high useable output power.

1. Introduction

In a previous study we have proven that vibrations of a passing train provide enough power to supply a wireless sensor node from a piezoelectric energy harvester mounted at the sleeper of a track [1]. Although initially these wireless sensor nodes were intended to be used for structural health monitoring in tunnel buildings, alternative applications, e.g. for train passage detection were proposed [1]. In continuation of this research we now present a contactless energy harvester tailored for energy autonomous train passage detection. The concept uses a modulated reluctance in a magnetic circuit: A wheel passing by the tips of a U-shaped magnetic core will redirect the flux through the magnetic circuit into wire-wound coils, thus inducing an electric voltage (Figure 1). Variable reluctances have been shown in literature for tuning the resonance frequency of vibrational harvesters [2] and in automotive applications as autonomous sensors [3]. The system design presented in [3] is similar to our design but has a clearance between the harvester and the moving parts in the order of 100 µm. In our setup gaps ranging from 6 to 14 mm have been realized, which are more suited for train surveillance applications.

[1] To whom any correspondence should be addressed.

2. Design concept and experimental setup

The concept of the realized variable reluctance circuit is shown in Figure 1: A stationary magnetic circuit, made of ferromagnetic materials, permanent magnets and one or two wire-wound coils is positioned aside from the rail (Fig. 1 left and middle). A passing train wheel will redirect the flux in the magnetic circle, and a significant part of the flux generated by the permanent magnets is now guided through the arms. After the wheel passage the magnetic flux returns into its stationary pattern, where most of the flux is running through the – shorter – air gap between the poles of the permanent magnet. Train passage will therefore produce two dynamic variations of the flux pattern inducing electric voltage in the coils. Energy can be harvested from this event and used to power up a wireless train passage detector.

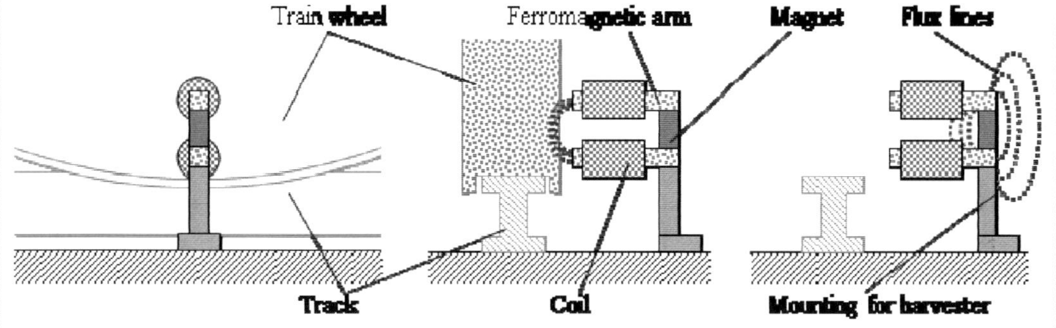

Figure 1. Schematic of the harvester mounted on a railway track. (a) shows the front view whereas (b) and (c) show the scenarios with and without the wheel, respectively.

The test set-up used in this study consists of a small piece of ferromagnetic material, called a "moving arm" that is mounted to a rotating plate, resembling the passing wheel of a train. Under rotation, the moving arm is intermittently closing the gap in a U-shaped magnetic circuit ("stationary arm"), mounted close to the trajectory of the passing wheel. The magnetic circuit consists of a U-shaped 10×10 mm^2 cross section low-grade steel reluctance core with 10×10 mm^2 N42 NdFeB (*supplier: Supermagnete.com*) magnets providing the magnetic flux.

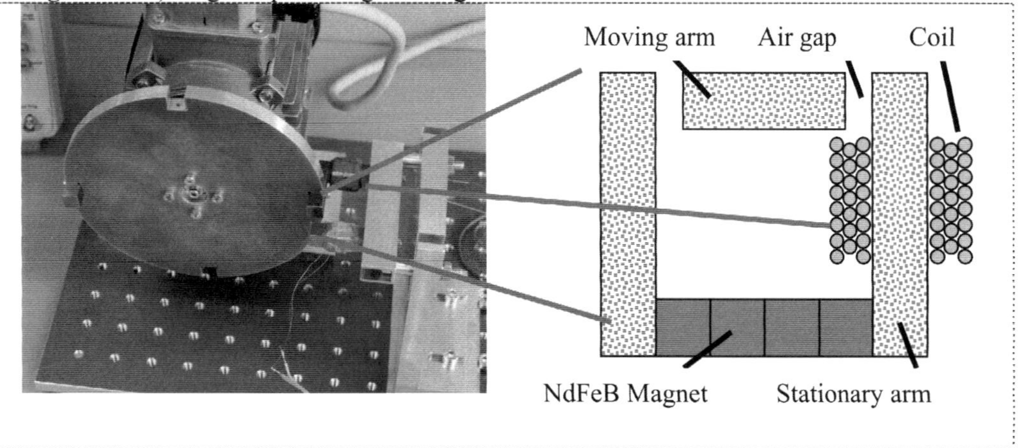

Figure 2. Test set-up for the harvester (left) and schematic top view of the reluctance circuit (right).

Different magnets with thicknesses of 10 and 2 mm were used to evaluate different magnetic fluxes and to enable different clearances between the stationary moving arms of the magnetic circuit. The coil wound around one of the stationary arms consists of 2000 turns of 150 µm diameter enameled copper wire. The coil series resistance and inductance were measured with an *Agilent 4263B* LCR meter to be R_{coil}=110.8 Ω, and L_{coil}=148.1 mH at a measurement frequency of 100 Hz. The rotating plate carrying the 30×10×10 mm^3 moving arm has a diameter of 160 mm (Figure 2 left and right) and is fabricated from aluminium. It carries two moving ferromagnetic arms to compensate the mass imbalance otherwise occurring. The air gap between the core and the shaft is set to 6, 10 and 14 mm, representing a realistic scenario for the application as a train passage detector. Smaller gap distances are assumed to be unsafe for operation in traffic applications. The induced coil voltages are measured with a *Tektronix TDS 2002* oscilloscope and also peak diagrams are recorded for later calculations of the per peak energy generation. The rotating plate is driven by a *Pophof* asynchronous motor in combination with a *Pophof i55* frequency converter control. The rotational frequencies *(f)* were varied between 5 to 45 Hz, corresponding to linear velocities of 9 to 82 km/h.

3. Simulation of the reluctance circuit

To evaluate the magnetic flux density change inside the ferromagnetic arms a magnetic simulation in ANSYS 13.0 was set up. The model comprises the parts shown in Figure 2 on the right, without the coil. The surrounding air is modelled by a sphere set around the reluctance circuit including infinity elements on the exterior. The magnetic flux density is calculated inside the static ferromagnetic part of the system by averaging the magnetic field in longitudinal direction over the whole cross-section at three different slices. By this the location of the coil is taken into account, and the induced voltage can be calculated from the change of the magnetic field with respect to the changing position of the moving ferromagnetic arm in the reluctance circuit. Two different simulations were set up:

1. A simulation that correspond to the rotation as occurring in the experiment to compare and validate simulation and experiment
2. A simulation using a linearly passing metal plate sized three times as high as the moving rotor, instead of the rotating arm moving between the stationary arms, to model a passing wheel train more precisely.

Figure 3 shows the waveforms for changes of the magnetic field dB/dθ (rotation) and dB/dx (linear), both for clearances between moving and stationary arm of 10 mm. Additionally the calculated coil voltage is shown in the graph for the linear motion.

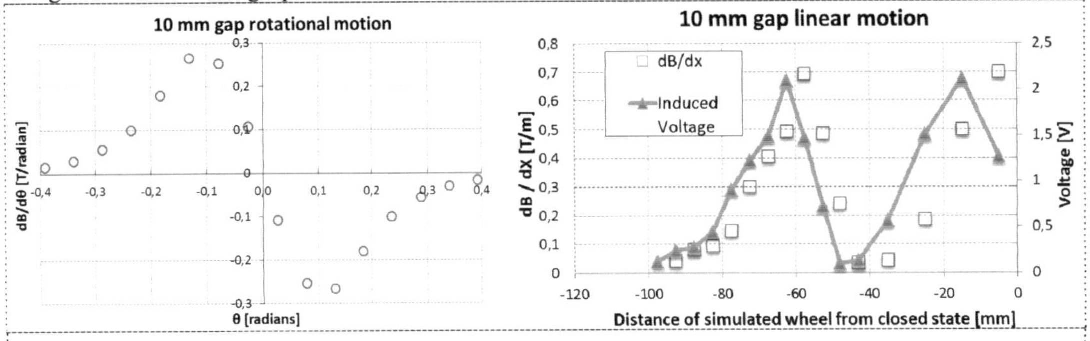

Figure 3. Simulated change of reluctance for a rotational motion (left) and linear motion (right) with 10 mm clearance widths.

From these simulations the voltage waveforms and values can be calculated according to eq. (1) (rotation) and eq. (2) (linear motion), respectively, where *N* is the number of turns and *A* the cross-sectional area of the ferromagnetic arms:

$$U_{rot} = -N \cdot A \cdot 2\pi \cdot f \cdot (\partial B / d\theta) \tag{1}$$

$$U_{lin} = -N \cdot A \cdot (\partial B / dx) \cdot (\partial x / dt) \tag{2}$$

Figure 4 shows a comparison of the measured and simulated peak-to-peak voltages for the rotational motion. Clearly the simulation shows a steeper rise in voltage.

The slower increase in the experiment might be attributed to the significant eddy currents appearing at higher rotational speeds.

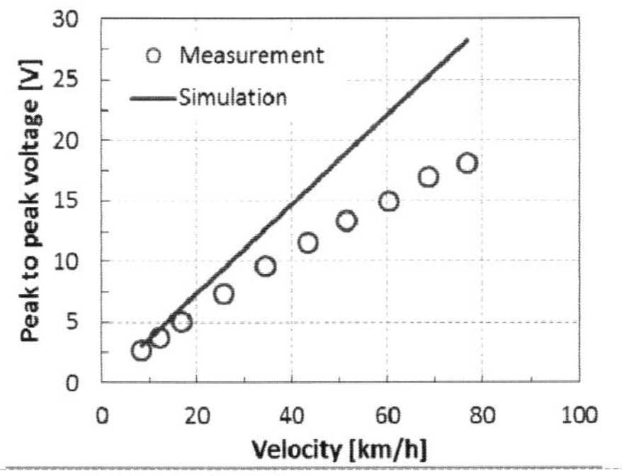

Figure 4. Comparison of simulated and measured peak to peak voltages for the rotational system. The clearance for this was 10 mm. The data used was taken from another experiment, than those for calculating the energy and power output. The alignment of stationary and moving arms has a strong influence on the coil voltages.

4. Experimental results and discussion

4.1. Results

The coil voltage was measured for a rotational frequency f between 5 to 45 Hz, corresponding to a passing wheel velocity of 9 to 82 km/h for three different clearance distances of 6, 10 and 14 mm. Figure 5 shows the measured peak-to-peak voltages (left) under no-load conditions, and, and the maximum extractable energy generated per pulse (right), with respect to the wheel speed. The pulse energy was calculated numerically from the recorded oscilloscope signals according to eq. (3), including the optimum load-matched conditions ($R_{load} = R_{coil}$).

$$E_{pulse} = \sum U^2 /(4R_{coil}) \cdot \Delta t \tag{3}$$

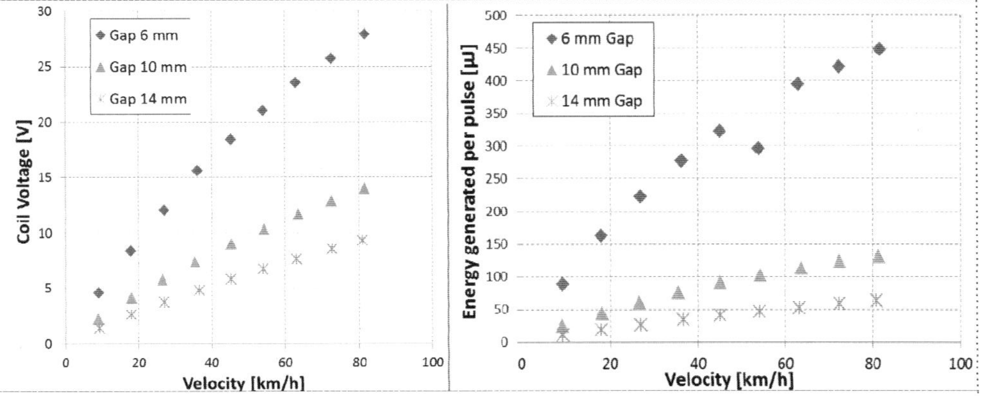

Figure 5. Measured open-circuit coil voltage (left), and energy generated per pulse (right), with respect to the train velocity for three different clearance widths between the moving and stationary ferromagnetic parts of the reluctance circuit.

The mean power output with respect to velocity is calculated straightforward according to $P_{mean} = E_{pulse} * f$, and is shown for three different measured clearances in Figure 6. Smaller clearances enhance the power output strongly.

4.2. Discussion

From previous research [1, 4] we have found that a realistic minimum voltage and power consumption for a wireless sensor node can be 2.0 V and 60 µW [4]. As traffic safety requirements may require higher RF transmit power or several sequential transmissions, a power of 500 µW at slightly higher system voltages seems to be more realistic [1]. With these data, the presented reluctance harvester is capable to detect train passages from speeds above 20 km/h at 10 mm clearance.

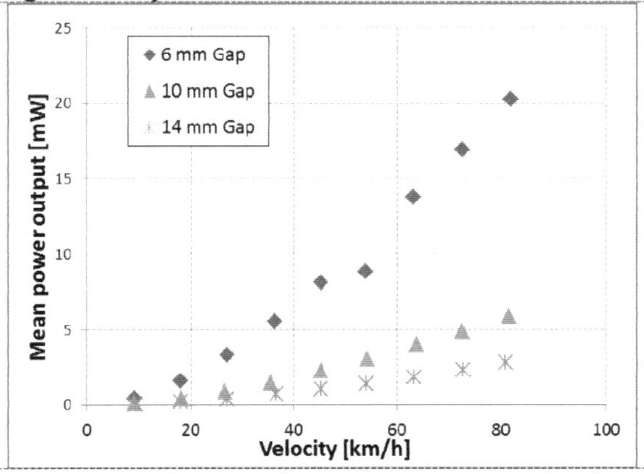

Figure 6. Measured mean power output with respect to the velocity for three different clearance widths between the moving and the static parts of the reluctance circuit.

For a proof-of-concept the temperature measurement system with RF module presented in [4] was powered from the reluctance harvester. The system could be operated with speeds corresponding to calculated optimal-matched output powers above 1 mW in Figure 6. However, one has to notice that this combination happened under non-ideal conditions, i.e. without a matched load and, hence, without an optimal energy transfer. During the experiments it was found that this low-power system, with an average power consumption of approximately 60μW, sometimes started operation without any rotation at all. As a source for this unexpected behaviour, the coil pick-up of magnetic stray field fluctuations was found, especially from the control of the asynchronous motor used. A more detailed investigation showed that these additional pick-ups did not have a remarkable influence during operation of the electric motor, but occurred more dominant in short periods of its stopped condition, when the electronic control employed an electronic holding operation for the motor. On a greater perspective this is a disadvantage for an accurate characterization of such a system under laboratory conditions, but may even be an advantage for its practical application. Using reluctance circuits obviously provides an opportunity to also harvest from close alternating magnetic fields, which may be present in a train system anyway.

5. Conclusions

The results found within this study clearly show the applicability of variable reluctance harvesters as a train passage detector, or in any other environment with rotating or periodically moving ferromagnetic parts. With power outputs beyond the mW regime, energy-autonomous systems can be easily powered. The simulations of the output voltages have a higher slope than actually measured values, which has to be investigated in more detail in future work to figure out the relatively large deviation.

[1] Wischke M, Masur M, Kroener M, Woias P, *Smart Mater. Struct.* **20** (2011), 085014.

[2] Ayala-Gacia Ivo, Mitcheson Paul, Yeatman Eric, Zhu Dibin, Tudor M.J. and Beeby S.P., Magnetic tuning of a kinetic energy harvester using variable reluctance. *Sensors and Actuators: A Physical* **189** (2013), 266-275.

[3] Parthasarathy D., Energy harvesting wheel speed sensor, *Master thesis*, Chalmers university of technology, Goeteborg, Sweden, 2012.

[4] Woias P, Wischke M, Eichhorn C, Fuchs B, An energy-autonomous wireless temperature monitoring system powered by piezoelectric energy harvesting, *Proc. PowerMEMS* (2009), 209-212.

A multi degree of freedom vibration magnetic energy harvester for transport application

M. Cueff, S. Basrour

Univ. Grenoble Alpes, TIMA, F-38031 Grenoble
CNRS, TIMA, F-38031 Grenoble

E-mail : Matthieu.cueff@imag.fr

Abstract. This paper reports a parametric study of a multi degree of freedom vibrations magnetic energy harvester for transport application. Two aspects were studied. First we focused on downsizing capability of our harvester. We showed that the output power is proportional to the scavenger volume. Typical output power of 100 $\mu W/cm^3$ was demonstrated with 0.5g harmonic excitation and 5 $\mu W/cm^3$ when harvesters were excited by the movement of a trolley. Then we studied the effect of the coil winding to optimize both output power and output voltage. Raising the number of turns and the wire radius can increase output power. Output voltage is only related to the number of turns.

1. Introduction

While resonant energy scavenger like piezoelectric [1], [2] or magnetic magnet-coils [3] cantilevers are able to harvest a large amount of energy on a specific resonant frequency and a specific direction of excitation identified in the environment, resonant harvesters are not suitable with broadband frequency signals and a chaotic displacement. Few studies deal with non-resonant harvesting method [4], [5], [6] and [7]. The most interesting study for non-resonant and multiple direction of excitation energy of vibration harvesting was presented by Bowers and Arnold [8]. In this paper, we report the design, fabrication and test of such prototypes with harmonic and real-time excitation to determine the downsizing capability of this kind of electromagnetic transducers. We finally focus on output voltage, which is a well-known drawback of magnetic scavengers.

2. Device design and fabrication

A drawing of the device is presented on Figure 1. A 1.4T neodymium (NdFeB) magnetic ball from Supermagnete company is inserted in a composite spherical cavity fabricated with a 3D printer ZBuilder. Two 600-turns coils are positioned around each piece of the support as presented on Figure 2. Both parts of the cage where then fixed with plastic screws. This geometry allows the magnet to move in the six degrees of freedom.

The coil is separated in two parts to maximize the amount of energy scavenged as proposed by Bowers [8]. 4 prototypes were fabricated with different cavity size. The ratio magnet diameter / cavity diameter was fixed to 85% as proposed by Bowers and Arnolds. Magnets and cavities diameters were summarized in table 1.

Table 1. Dimensions of the 4 prototypes

Prototype	#1	#2	#3	#4
Magnet diameter (mm)	26	19	13	10
Cavity diameter (mm)	30	22	15	11.5

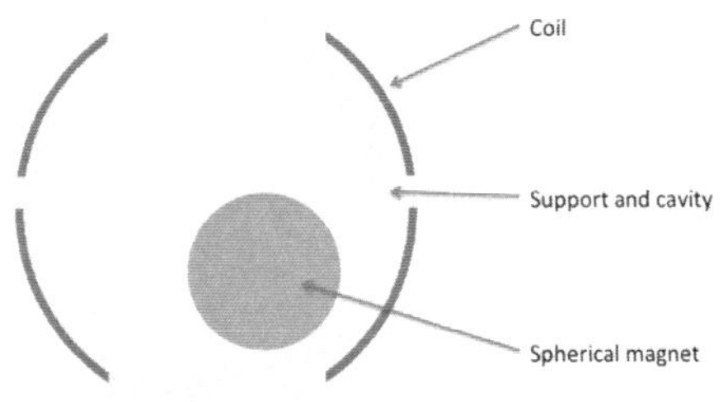

Figure 1. Schematic view of the multi-degree of freedom electromagnetic energy harvester.

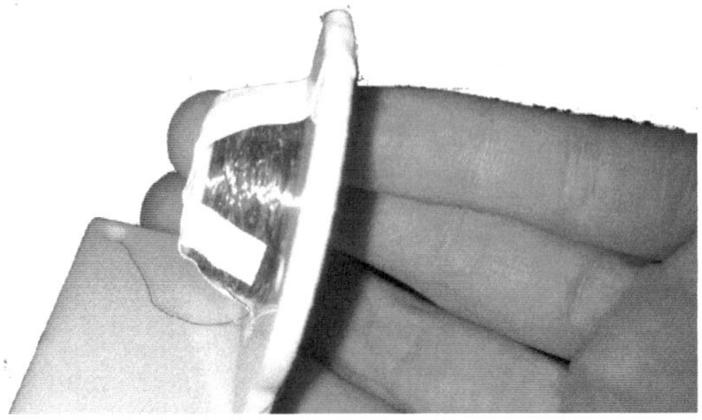

Figure 2. View of one part of the harvester #2 with the coil.

3. Harmonic analysis

Prototypes were tested at with harmonic acceleration at 15 Hz and 0.5g to determine output power on a shaker as shown on Figure 3. Below 0.5g, magnet movement is too small to harvest energy. Below 15 Hz, shaker movement is noised due to its large displacement. Output powers versus load are presented on Figure 4. Optimum load were 200 ohm for both 30 mm and 22 mm diameter prototypes and 150 ohm for both 15 et 11.5 mm diameter prototypes. They fit with coils resistances. Results are summed up on Table 2. We obtained power of 1.2 mW, 500 μW, 350 μW and 80 μW with the 30, 22, 15 and 11.5 mm cavities prototypes respectively. Power densities are closed to 100 $\mu W/cm^3$ for all the prototypes excited by an harmonic signal.

Figure 3. Prototype #1 mounted on the skaker.

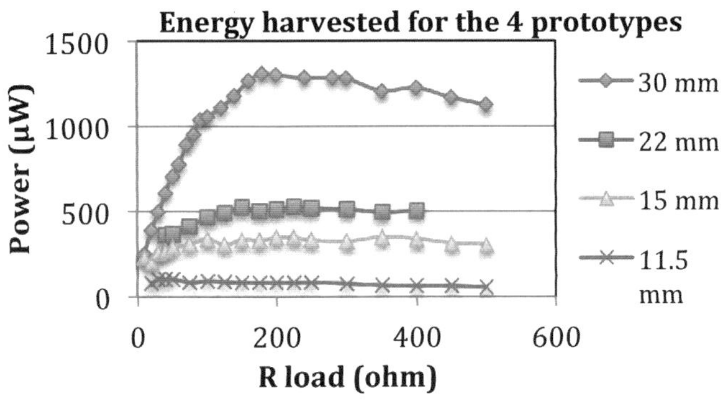

Figure 4. Output power in function of resistance load with an harmonic excitation signal of 0.5g at 15 Hz. Blue diamonds : 30 mm cavity. Red squares : 22 mm cavity. Green triangles : 15 mm cavity. Violet crosses : 11.5 mm cavity.

Table 2. Optimal load and output power measured on each prototype.

Cavity diameter (mm)	30	22	15	11.5
Optimal load (ohm)	200	200	150	150
Output power (μW)	1200	500	350	80

4. Real-time excitation

Prototypes were then mounted on a trolley, which was displaced on a metallic footbridge during 140s to simulate a non-resonant application. Harvesters were plugged to their optimum load and output voltages were measured with a DAQ interface. Acceleration was sensed with an inertial measurement unit. A specific bunch set-up has been developed on Labview to sense both harvesters output and IMU signal.

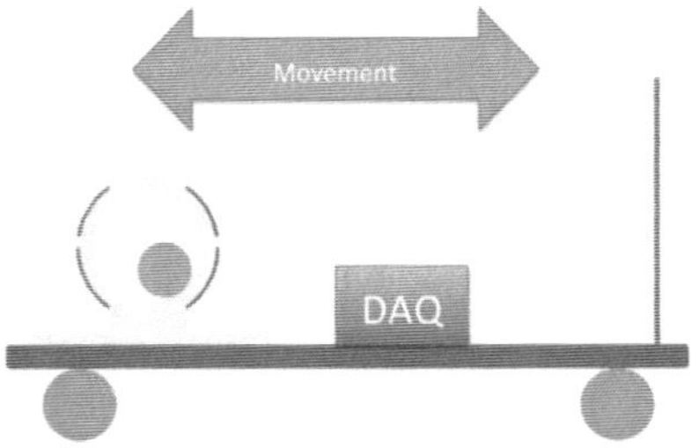

Figure 5. Schematic view of one harvester mounted on the trolley.

Acceleration along z-axis time signal and FFT are presented on Figure 6 and 7 respectively. Prototype 1 output voltage time signal and FFT are presented on Figure 8 and 9 respectively. FFT analysis of acceleration signal shows that a frequency band is excited between 10 and 20 Hz. The variation is probably due to trolley speed change. FFT analysis of both acceleration and output voltage doesn't show a matching between the two spectra.

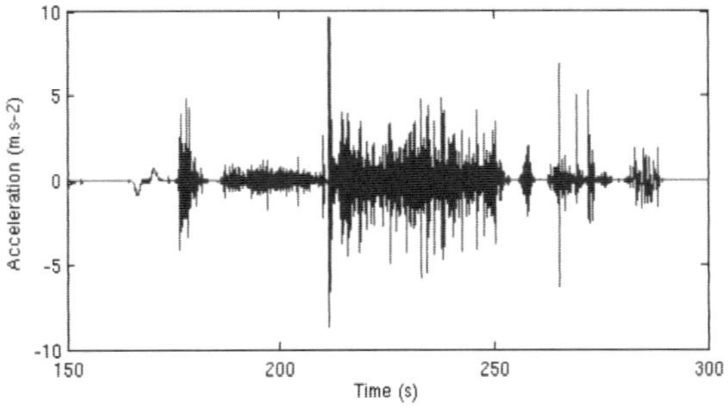

Figure 6. Acceleration signal along vertical axis during the move on the trolley.

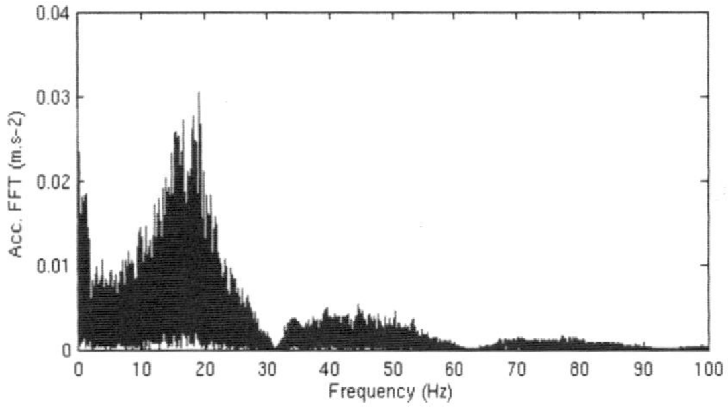

Figure 7. FFT analysis of acceleration signal along z-axis.

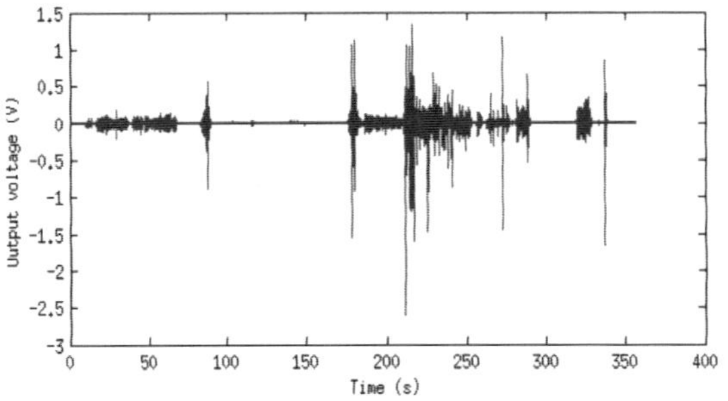

Figure 8. Prototype 1 output voltage.

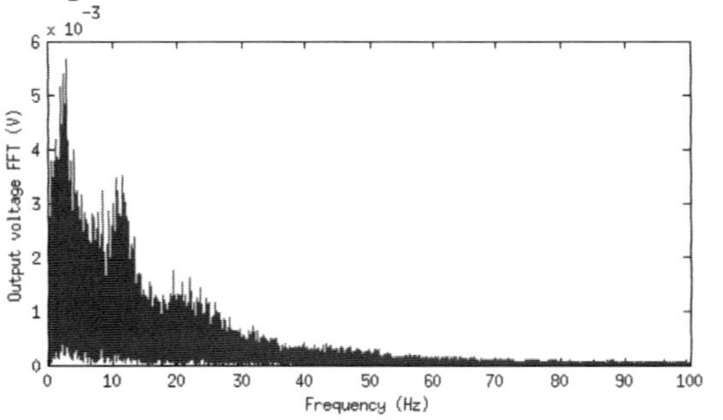

Figure 9. FFT analysis of prototype 1 output voltage.

Results were summed up on Figure 10. There is a strong decrease of the amount of energy when the cavity diameter decreases. For each cavity size, the power per unit of volume is quite constant and around 5 $\mu W/cm^3$. We showed that output power of this kind of electromagnetic transducers is proportional to cavity volume. Therefore, downsizing will not give benefits in terms of power density. Size of the prototype will only be determined by the output power required by application.

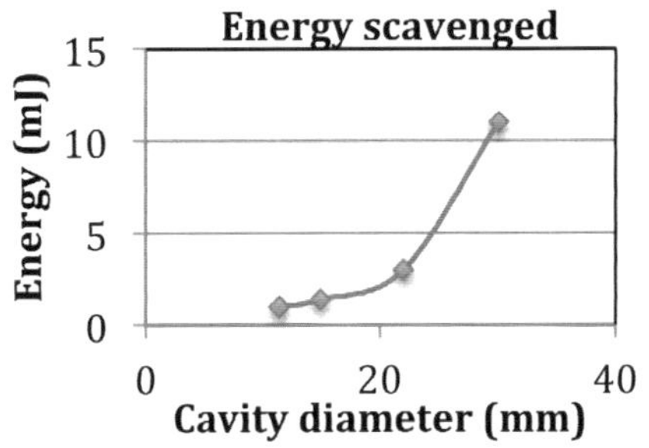

Figure 10. Energy scavenged during 140s on the trolley in function of the diameter of the cavity.

5. Modeling

In order to provide ways of improvement, we made a rough simulation of the classical damped resonant energy harvester because flux linkage gradient is hard to estimate for non-resonant and 3D displacement of our magnet. At low level of excitation, harmonic output voltage could be expected. For higher amplitude of excitation, chaotic behavior is expected. We used the Williams & Yates model [9] adapted by Saha to magnetic energy harvesters [10]. We focus on the voltage output power and how we can improve it by changing the coil winding. Output voltage is a critical issue for magnetic energy harvester because large amount of energy can be extracted at the optimal load but output voltage levels are too weak to be extracted by an electronic circuit.

To be close to the non-resonant harvester, we simulate harvester resonating at 15 Hz with a Q of 2. Electromagnetic damping is negligible compared to structural damping. Furthermore, due to the large air gap between the magnet and coils, we assume that the electromagnetic damping is weak. With these damping conditions and due to the low resonance frequency, optimal load is then equal to the coil resistance.

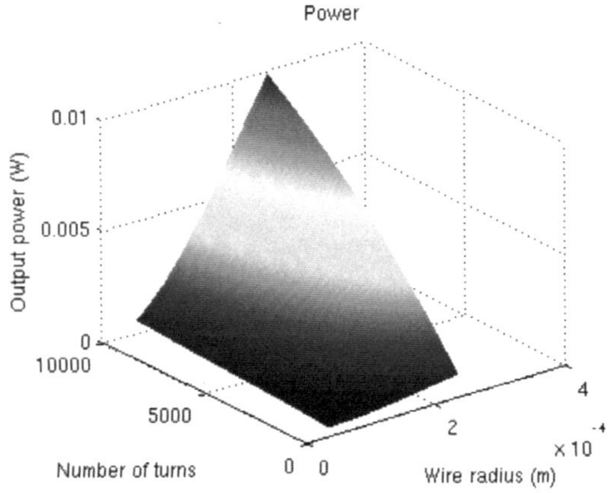

Figure 11. Output power in function of the number of turns and wire radius.

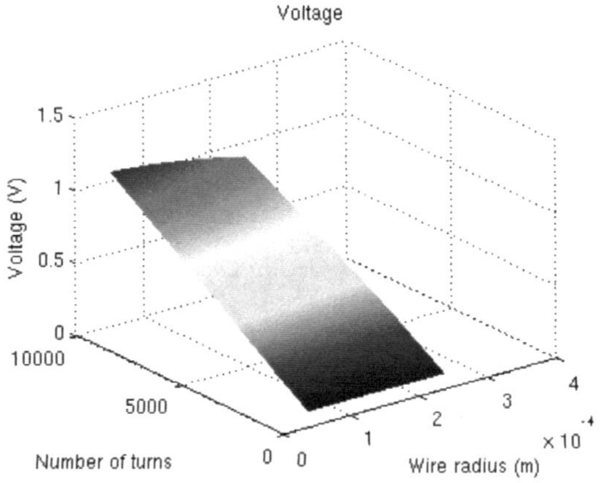

Figure 12. Output voltage in function of the number of turns and wire radius.

We changed the number of turns and the diameter of the coil wire. Output power results are presented in Figure 11. Output power increases when number of coil turns increases and wire radius increase, due to the raise of the electromagnetic damping. At the same time, as shown in

Figure 12, the output voltage only depends on the number of turns. This result is interesting to increase output power and output voltage to power conversion electronics. As shown on Figure 13, the decrease of harvester impedance is compensated by the increase of the electromagnetic coupling.

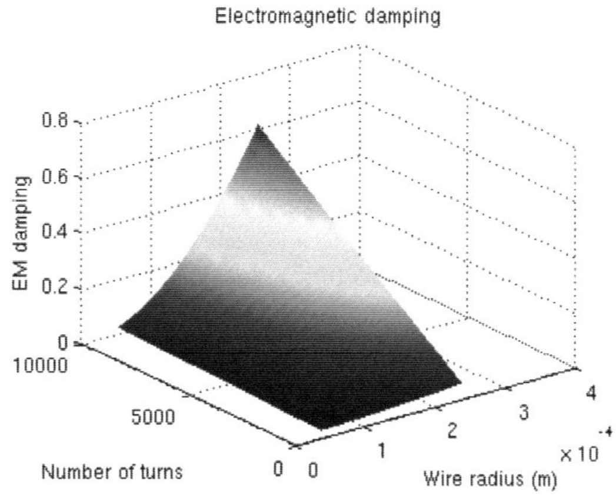

Figure 13. Electromagnetic damping in function of the number of turns and wire radius.

6. Conclusion

We showed that output power of this kind of electromagnetic transducers is proportional to cavity volume. Typical output power of 100 $\mu W/cm^3$ was demonstrated with 0.5g harmonic excitation and 5 $\mu W/cm^3$ when harvesters were excited by the movement of a trolley. Therefore, downsizing will not give benefits in terms of power density. However, the total amount of energy we can obtained with a 30 mm cavity diameter is important and can be useful to power Zigbee applications for instance [11].

We also demonstrated that the output power can be increased by increasing the number of turns and the wire radius with a rough harmonic model. The most interesting thing is the fact that the output voltage only depends on the number of turns, even if coil resistance and optimal resistance become very small when we increase the wire radius.

Acknowledgment

The authors wish to acknowledge the OSEO funding to give us the opportunity to work on this subject.

References

[1] Roundy S and Wright P K 2004 *Smart Mater. Struct.* **13** 1131
[2] Marzencki M, Defosseux M and Basrour S 2009 *J. Microelectromechanical Syst.* **18** 1444
[3] Glynne-Jones P, Tudor M J, Beeby S P and White N M 2004 *Sensors Actuators Phys.* **110** 344
[4] Halvoresen E 2008 *J. Microelectromechanical Syst.* **17** 1061
[5] Galchev T, Kim H and Najafi K 2009 *Proc. Transducers* (Denver) 632
[6] Ju S, Chae S, Choi Y, Jun S, Park S M, Lee S, Lee H and Ji C H 2013 *Proc. Transducers* (Barcelona) 1348
[7] Arrieta A F, Hagedorn P, Erturk A, and Inman D J 2010 *Appl. Phys. Lett.* **97** 104102
[8] Bowers J B and Arnold D P 2009 *J. Micromech. Microeng.* **19** 094008
[9] Williams C B and Yates R B 1996 *Sensors Actuators Phys.* **52** 8
[10] Saha C R, O'Donnell T, Loder H, Beeby S and Tudor J 2006 *IEEE Trans. Magn.* **42** 3509
[11] Lee J S, Su Y W and Shen C C 2007 *33rd An. Conf. IEEE Ind. Elec. Soc.* 46

Efficient Direct-Matching Rectenna Design for RF Power Transfer Applications

Shady Keyrouz[1,2] **and Huib Visser**[2,1]

[1]Technical University of Eindhoven, Den Dolech 2, 5612 AZ, Eindhoven, The Netherlands.
[2]Holst Centre/imec-nl, High Tech Campus 31, 5656 AE, Eindhoven, The Netherlands.

E-mail: s.keyrouz@tue.nl

Abstract. This paper presents the design, simulation, fabrication and measurements of a 50 ohm rectenna system. The paper investigates each part (in terms of input impedance) of the rectenna system starting from the antenna, followed by the matching network, to the rectifier. The system consists of an antenna, which captures the transmitted RF signal, connected to a rectifier which converts the AC captured signal into a DC power signal. For maximum power transfer, a matching network is designed between the rectifier and the antenna. At an input power level of -10 dBm, the system is able to achieve an RF/DC power conversion efficiency of 49.7%.

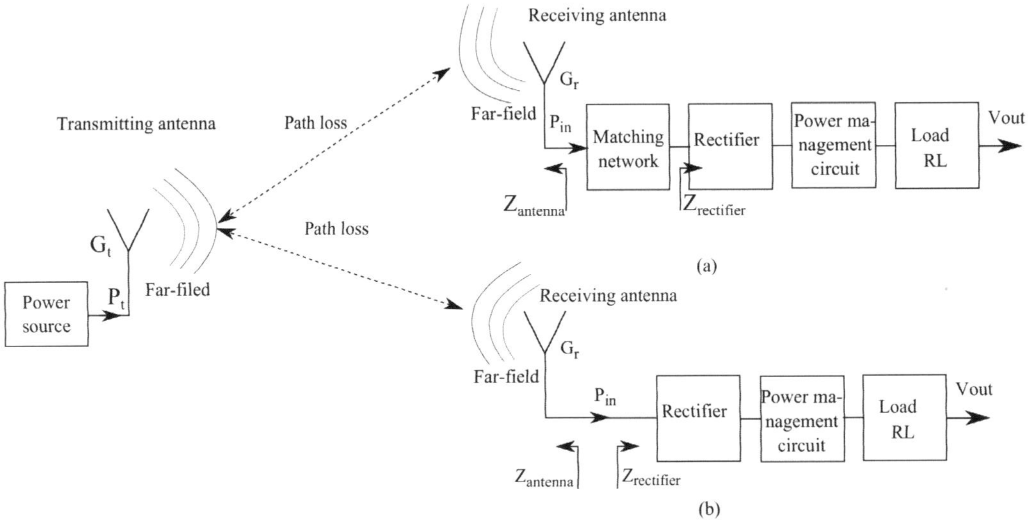

Figure 1. Conventional wireless power transmission systems.

1. Introduction

A conventional far-field RF power harvesting system is shown in Fig.1. The system consists of a transmitter that transmits RF power and a receiver that collects the transmitted power.

When the transmitter is not dedicated for wireless power transmission, e.g. TV and GSM broadcasting, the power collection and conversion is defined as wireless power harvesting [1]. When the power is intentionally transmitted using a dedicated source, this is defined as wireless power transport [2]. For both, the receiving side consists of a rectenna (rectifying antenna) i.e. an antenna connected to a rectifier and a load. A power management circuit between the rectifier and the load is needed to transfer and store the DC power. For a maximum power transfer, a matching network is needed between the antenna and the rectifier as shown in Fig. 1-a. In order to achieve a compact rectenna system, the matching network can be removed as shown in Fig.1-b, and the antenna is conjugately matched to the rectifier ($Z_{antenna} = Z^*_{rectifier}$). In this paper, a standard 50 ohm antenna is matched to the rectifier using a lumped element matching network see Fig1-a.

2. Voltage doubler and matching network

As it is clearly indicated in Fig. 1-a, a rectifier is needed to transform the RF input power into DC power. For maximum power transfer between the antenna and the rectifier, a matching network is designed, simulated, manufactured and measured. Commercially available, discrete schottky diodes HSMS 2852 [3] have been used. ADS harmonic balance and momentum have been used for simulations. The input impedance of the rectifier is simulated at an input power level of -10 dBm and an operating frequency of 868 MHz.

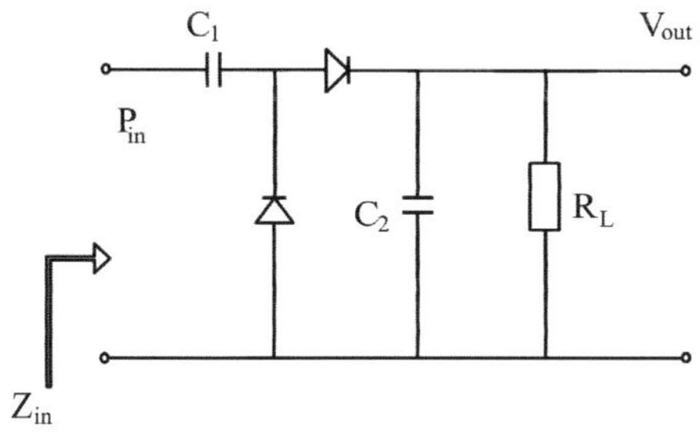

(a) - Voltage doubler configuration

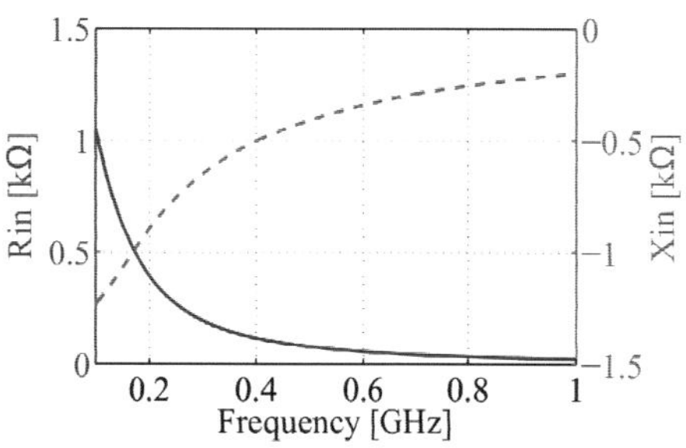

(b) - Voltage doubler impedance

Figure 2. Voltage multiplier configuration (a) and its input impedance (b) as a function of frequency. Pin = -10 dBm.

Fig. 2-a shows the suggested voltage doubler configuration. C1 and C2 are set to 100 pF. RL is set to 10 kΩ. Zin is the input impedance of the voltage doubler. Since the rectifier is a non linear device and its impedance changes as a function of frequency and as a function of input power level, an input power level of -10 dBm is chosen. Fig. 2-b shows the simulated real (solid curve) and imaginary (dashed curve) parts of the input impedance as a function of frequency. At an operating frequency of 868 MHz, the impedance of the voltage doubler is found to be 27- j222 Ω. A matching network is designed to transform this impedance to 50 ohms. Fig. 3-a shows the suggested lumped elements matching network to match the impedance of the voltage doubler to 50 ohms. The matching network consists of a 32 nH inductor and a 1 pF capacitor. The capacitor is placed in parallel with the antenna ports and the inductor is placed in series with the voltage doubler. Fig. 3-b shows the simulated and the measured reflection coefficients of the fabricated voltage doubler and the matching network as a function of frequency at an input power level of -10 dBm. It is clear from the measurement results that the designed matching network will match the impedance of the voltage doubler to that of the antenna at 868 MHz.

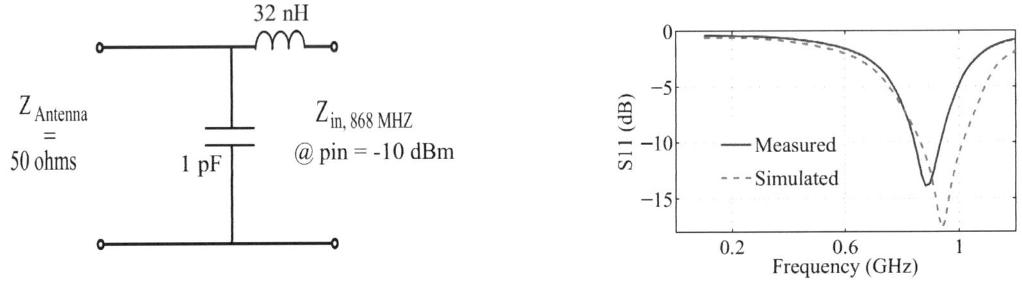

(a) - Designed matching network

(b) - Simulated and measured S11

Figure 3. Designed and manufactured (a) lumped element matching network and its reflection coefficients (b) vs. frequency.

3. Standard 50 ohm antenna design

This section presents the optimization and the design of a modified 50 ohm Yagi-Uda antenna that will be used to harvest RF power. The antenna was first introduced in [4] and used to harvest RF power from DTV signals. The main advantage of such an antenna, is that by tuning its parameters it can behave like a broadband, dual-band or a multi-band antenna [5]. The antenna is designed to resonate within the frequency band 850 MHz to 1 GHz. Fig. 4-a shows the suggested antenna configuration to cover the required frequency band. The optimized parameters are shown in Fig. 4-b.

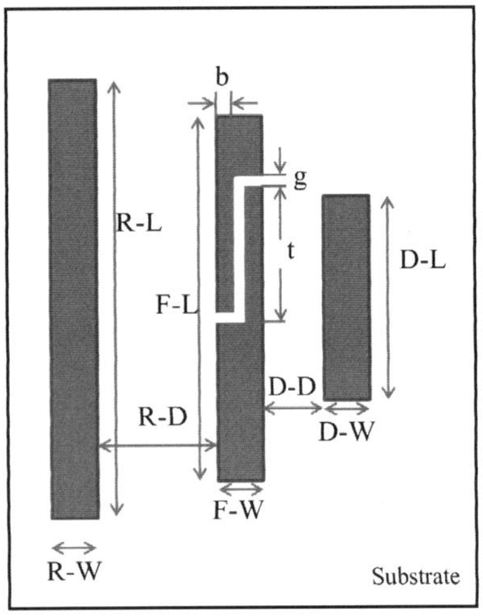

(a) - Suggested antenna configuration

Parameter	(mm)
D-L	106
D-W	8
D-D	8
R-L	167
R-W	5
R-D	20
F-L	158
F-W	10
t	39
b	2
g	1

(b) - Optimized parameters

Figure 4. (a) Suggested broad-band modified Yagi-Uda antenna and (b) its optimized parameters.

The antenna is designed, simulated and manufactured on a 1.6 mm FR4 substrate. The fabricated antenna is shown in Fig.5-a. The reflection coefficient of the fabricated antenna is measured between 600 MHz and 1.2 GHz using the Agilent PNA-X Network Analyser. The simulated (solid curve) and measured (dashed curve) reflection coefficient as a function of frequency are shown in Fig.5-b. It is clear form the figure that the measured reflection coefficients agree well with the simulated ones, which validates the simulated results. In addition, Fig. 5-b shows the simulated antenna gain (dashed-dotted curve), the simulated antenna gain is 4.23 dBi at an operating frequency of 868 MHz. This antenna gain will be used to calculate the Power Conversion Efficiency (PCE) of the rectenna.

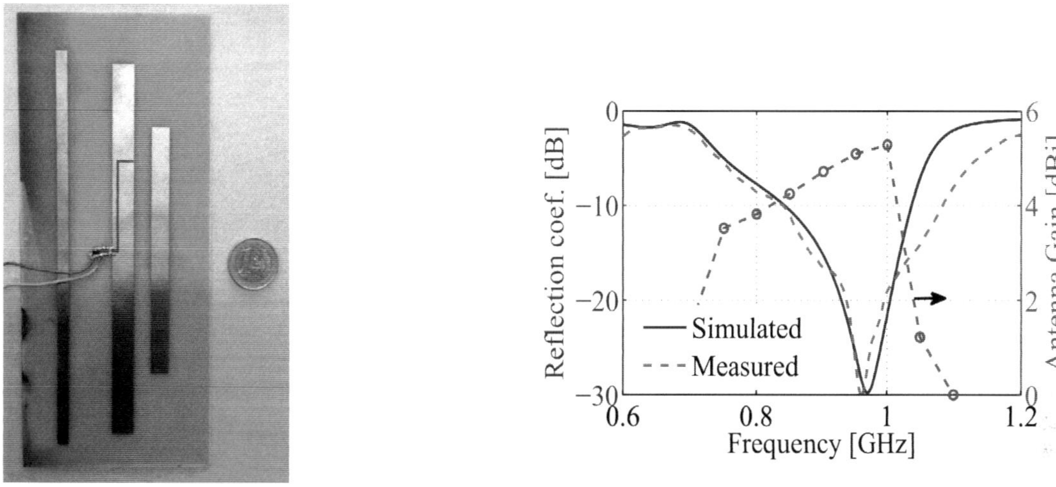

(a) - Fabricated rectenna system (b) - S11 and antenna gain

Figure 5. (a) Manufactured rectenna (b) its reflection coefficients and antenna gain.

4. RF power transport

The setup is calibrated using two identical antennas with a known gain at 868 MHz separated by 1 meter ensuring far-field conditions. A power meter and a spectrum analyzer were used to measure the available RF power. The measured power is within within ± 0.5 dB agreement with the theoretically calculated available power. The receiving rectenna system shown in Fig. 5-a consists of the antenna introduced in the previous section connected to a rectifier and to a variable load resistor RL. The DC output voltage is measured for different load resistances at an input power level of - 10 dBm. The available power $P_{available}$ on the receiving antenna terminal is expressed by:

$$P_{available} = P_T - L_P + G_T = EIRP - L_P, \tag{1}$$

P_T is the transmitted power, G_T is the maximum gain of the transmitting antenna. The product $G_T P_T$ is called the Effective Isotropic Radiated Power (EIRP). L_P is the free space path loss,

$$L_P(dB) = 32.44 + 20 \times \log(f) + 20 \times \log(R), \tag{2}$$

where f is in MHz and R is in km. The input power level P_{in} on the rectifier terminals is :

$$P_{in}(dBm) = P_{available}(dBm) + G_R, \tag{3}$$

where G_R is the gain of the receiving antenna.

Fig.6-a shows the measured output voltage vs. load resistance at an input power level of -10 dBm at an operating frequency of 868 MHz. The system output voltage reaches 0.705 V over 10 kΩ load resistance. The PCE is calculated using eq. 4 and is shown in Fig.6-b. The system PCE reaches 49.7 % over 10 kΩ load resistance. Compared to the state of the art, [6], [7] and [8]

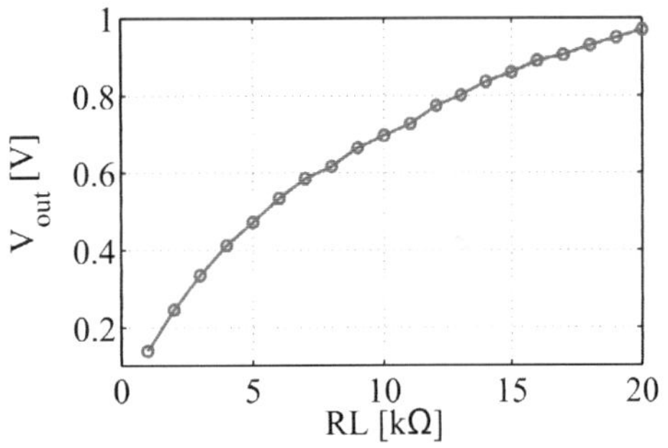

(a) - Measured output voltage

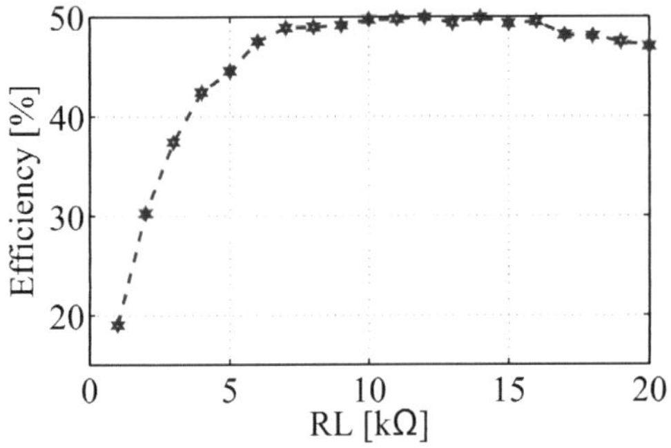

(b) - Mesured PCE

Figure 6. (a) Rectenna output voltage (b) its measured Power Conversion Efficiency (PCE).

the proposed rectenna in this paper shows higher efficiency performance. Compared to [9] our suggested rectenna achieves almost similar PCE performance at the same input power level.

$$PCE\,(\%) = \frac{P_{load}}{P_{in}} = \frac{V_{out}^2}{R_L}\frac{1}{P_{in}} \tag{4}$$

5. Conclusion

In this paper a standard 50 ohm rectenna system has been presented. Each sub-part of the rectenna system (antenna, matching network and rectifier) has been investigated independently. For maximum power transfer, a matching network is designed between the rectifier and the antenna. The system has been manufactured and tested. At an input power level of -10 dBm, and an operating frequency of 868 MHz, over a 10 kΩ load resistance,the system output voltage reaches 0.705 V and the RF/DC power conversion efficiency reaches 49.7%.

References

[1] M. Pinuela, P. D. Mitcheson, and S. Lucyszyn, Ambient RF energy harvesting in urban and semi-urban environments, IEEE Trans. Microw. Theory Tech., vol. 61, no. 7, pp. 2715-2726, 2013.

[2] G. Monti, L. Corchia, and L. Tarricone, UHF wearable rectenna on textile materials, IEEE transactions on Antennas and Propagations, vol. 61, no. 7, pp. 3869-3873, 2013.

[3] Hsms-285x, 282x surface mount zero bias schottky diodes data sheet.

[4] S. Keyrouz, G. Perotto, H. J. Visser, and A. G. Tijhuis, Novel broadband yagi-uda antenna for ambient energy harvesting, in IEEE 42nd European Microwave Conf. (EuMC), pp. 518 - 521, 2012.

[5] S. Keyrouz, H. J. Visser, and A. G. Tijhuis, Multi-band simultaneous radio frequency energy harvesting, in 7th European Conference on Antennas and Propagation (EuCAP), pp. 3058-3061, 2013.

[6] H. Kanaya, S. Tsukamoto, T. Hirabaru, and D. Kanemoto, Energy harvesting circuit on a one-sided directional flexible antenna, IEEE Micro. and Wireless Components Letters, vol. 23, no. 3, pp. 164-166,2013.

[7] C. Mikeke, H. Arai, A. Georgiadis, and A. Collado, DTV band micropower RF energy-harvesting circuit architecture and performance analysis, in IEEE International Conference on RFID-Technologies and Applications (RFID-TA),, pp. 561-567, 2011.

[8] V. Marian, B. Allard, C. Vollaire, and J. Verdier, Strategy for microwave energy harvesting from ambient field or a feeding source, IEEE Trans. on Power Electronics, vol. 27, no. 11, pp. 4481-4491, 2012.

[9] T. Le, K. Mayaram, and T. Fiez, Efficient far-field radio frequency energy harvesting for passively powered sensor networks, IEEE Journal of Solid-State Circuits, vol. 43, no. 5, pp. 1287-1302, 2008.

Capacitance variation in electrostatic energy harvester with conductive droplet moving on electret film

Z Yang, E Halvorsen[1], T Dong
Department of Micro and Nano Systems Technology, Vestfold University College, Norway

E-mail: Einar.Halvorsen@hive.no

Abstract. This work addresses numerical finite element calculations on a droplet-based electrostatic energy harvester to reveal additional characteristics that supplement previous test results. Assumptions of 2D electrode and static droplet profile have been applied to make the simulation achievable based on the real prototype. We investigate the consequences of a uniform space charge distribution in the film. Capacitance variation and open-circuit voltage of the simulation model have been determined and display respectively maximum and minimum magnitudes when the droplet is in the middle of finger gap. The sharp variation of capacitance, during which the droplet moves from the gap centre to the finger centre, can explain the narrow peaks of output voltage seen in experiments. Additionally, the influence of droplet size on the capacitance variation is also investigated.

1. Introduction

Harvesting energy from the environment and converting to electric energy has been a hot spot in both scientific and engineering fields since last two decades [1]. Various energy conversion principles, such as electromagnetic, thermoelectric, piezoelectric, and electrostatic, etc., have been explored [2]. Electrostatic energy harvesters, which mainly operate on the capacitance variation, and mostly employ a mass-spring configuration, have attracted considerable attention due to their compatibility with microfabrication process. Distinguished from the conventional mass-spring configuration, the electrostatic energy harvester utilizing droplet or liquid flow has advantages for low frequency non-resonant operation [3-7].

A fluidic electrostatic energy harvester utilizing interdigital electrodes (IDEs) and a thin, charged dielectric film was demonstrated in our previous work [8]. It operates on a rolling conductive droplet causing variations in the capacitance and open-circuit potential of the device's electrical port, and thereby converting mechanical energy into electrical energy when a load is connected. The detailed experimental characterization supported by a phenomenological lumped model was performed with the prototype and is presented in [9]. However, due to lack of available experimental data on the distribution of fixed charges in the film and on the possible deformation of the droplet in the inhomogeneous electric field, as well as the serious obstacles to analytical treatment of the electrostatics in the complicated structure, essential details that would be useful for further developments are not known.

[1] To whom any correspondence should be addressed.

To provide better understanding of this type of device, we investigate the droplet-based energy harvester concept by finite element analysis (FEA), so as to extract more features of this type of energy harvester that can't be obtained from the experimental results directly.

2. Modelling and simulation method

2.1. Model description

The droplet based electrostatic energy harvester (as shown in Figure 1a) can be modelled by a simple equivalent circuit (Figure 1b). The embedded charges inside the thin insulation film give rise to the voltage source V_e, while the resultant capacitance from the two terminals of the interdigital electrodes involving the conductive droplet can be simplified as a variable capacitance. Thus the relation between the embedded potential and the terminal potential can be expressed by,

$$V_{out}(x) = V_e(x) + Q(x)/C(x) \tag{1}$$

As the droplet moves across the thin film, the resultant capacitance varies in pace with the droplet displacement with respect to IDE fingers.

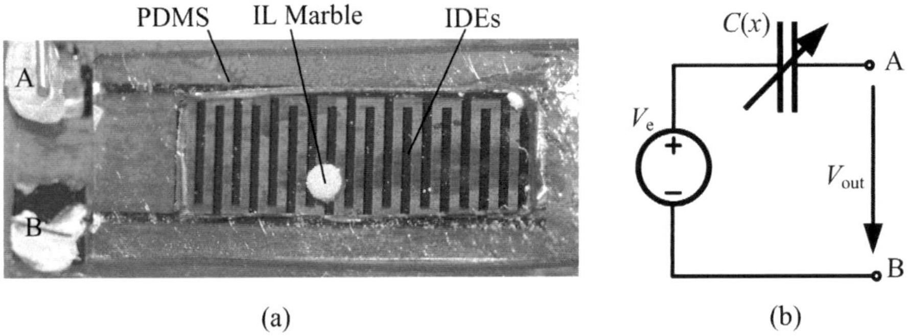

Figure 1. Droplet-based electrostatic energy harvester. (a) Optical photo of prototype; (b) Equivalent circuit characterizing the energy harvester.

2.2. Simplifying assumptions

The device has greatly varying length scales within it, posing serious challenges for FEA meshing, i.e. 100nm metal film thickness, micrometre scale film thickness, millimetre scale of droplet/electrode pattern and centimetre scale device dimensions. With the technical difficulties of meshing considered, the following assumptions were applied in the model to make FEA achievable:

- the IDE finger is treated as 2D structure with zero thickness;
- the variation of droplet profile due to the droplet dynamics or electrostatic forces has been neglected;
- and the droplet is simplified as a spherical cap with a circle contact area on the dielectric film.

As the thickness of IDE finger is rather small compared with the thickness of insulation film, the 2D model can reduce the meshing difficulty but without losing the transducing feature of IDE. And the simplifications on the droplet profile actually characterize a static droplet, without addressing the dynamic behaviour. Therefore, these assumptions are reasonable. In addition, the charge distribution in the thin film is not known. As a first investigation we therefore consider a uniform space charge density and investigate its consequences.

2.3. Simulation domain

The simulation domain is depicted in Figure 2. The droplet was treated as a conductor and PTFE was used as the insulation film. The detailed geometrical parameters are listed in Table 1.

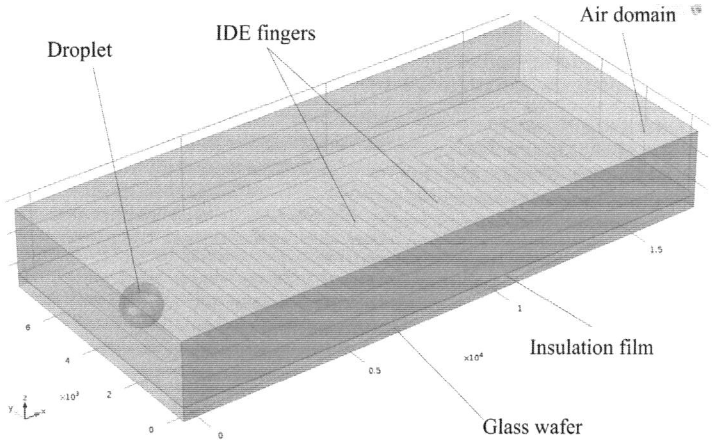

Figure 2. Schematic drawing of simulation domain.

Table 1. Parameters of the simulation domain.

Parameters	Unit	Value
Finger width of IDE	μm	400
Finger gap of IDE	μm	400
Number of finger pair	–	10
Droplet size	μm	1200
Insulation film thickness	μm	2
Glass substrate thickness	μm	400
Thickness of air domain	μm	2000

The calculation was conducted by the MEMS module of the commercial software COMSOL Ver. 4.3. Charge conservation was set for the whole simulation domain with 0-V boundary on the outer surface of the domain and also on the bottom surface of the glass substrate so as to match with practical testing where the device was placed on a conducting surface.

In order to obtain the capacitance variation with respect to the droplet displacement, a particular fine mesh was generated for IDE fingers and the thin film, as illustrated in Figure 3. As an aid in the difficult meshing of the thin film structure, virtual surfaces were used to control the mesh generation.

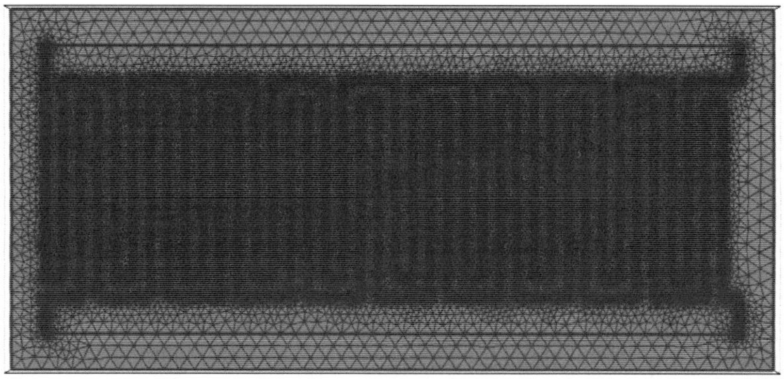

Figure 3. Meshing of the IDE fingers and thin insulation film (top view).

While the droplet moves along the central axis (*x*-direction) of the simulation domain, the capacitance was determined by applying 1.0V and 0V on the two terminals (A and B), respectively. The variation of the open-circuit potential of the prototype was checked by setting terminal A to zero charge while keeping terminal B grounded. Finally, the short-circuit charges on each electrode were checked by setting both to ground. A uniform space charge density was arbitrarily set at -1.0 C/m^3 for the insulation film. This is meaningful as the electrostatics is linear and only a scaling is necessary to correct the result to another value of embedded charge.

3. Results and discussion

3.1. Capacitance variation and open-circuit voltage

Figure 4 and Figure 5 plot the variations of the capacitance and open-circuit voltage with respect to the droplet position, respectively. It is noted that the maximum values occur at the positions where the droplet is right in the middle of the two adjacent fingers (gap center) while the minimums are corresponding to the positions where the droplet are sitting right on the top of one finger (finger center). Moreover, the capacitance varies sharply as the droplet moves from the gap center towards the finger centre, which can explain why the narrowly peaked output voltage appears in the test results [9]. According to Figure 4, the range of capacitance variation in the energy harvester is not so impressive for a droplet having a diameter of 1.20 mm, less than 0.2 pF. For the real prototype, the thickness of insulation layer was smaller, down towards 1.0 micron, thus the capacitance variation is larger.

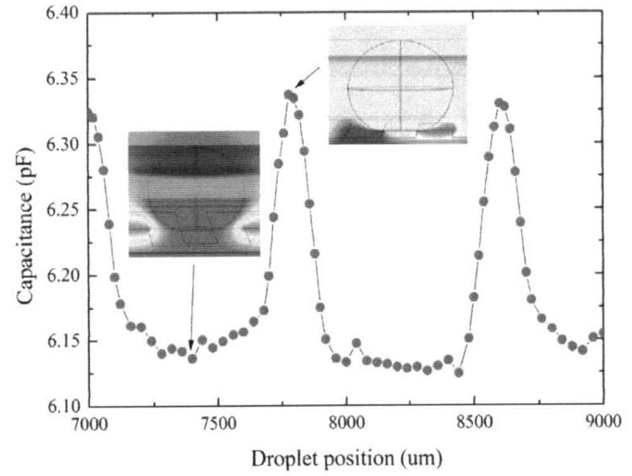

Figure 4. Capacitance variation with droplet position.

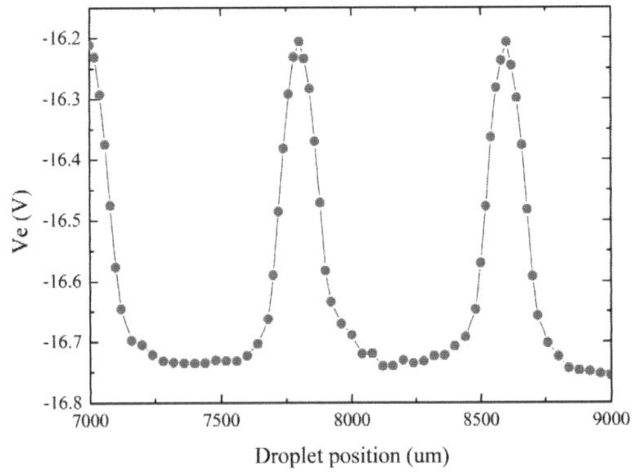

Figure 5. Open-circuit potential variation with droplet position

3.2. Charge variation on the grounded electrodes

Figure 6 plots the charge distribution on the two terminals under grounded condition with the droplet moving within one period (four times of finger width) of IDE configuration. Due to computational errors, there are considerable fluctuations in the result. The reason for the small variations, in the third decimal, is cancellation which can be verified by calculating short-circuit charge as the product of the results in figures 4 and 5 and gives a variation that drowns in numerical error (not shown). The peak in the open circuit potential of figure 4 is a notch in its magnitude that nearly cancels the corresponding peak in the capacitance. However, figure 6 still indicates some tendency of the charge variation with respect to the droplet position: the charge variations on the two terminals alternate and reach nearly the same value at the positions when the droplet is at the gap centre. This is also where the change is largest which would give peaks in the derivative of the short-circuit charge that alternate with the period shown in the experiments. Therefore, a uniform charge distribution in the film seems to be able to qualitatively account for the observed behaviour, but numerical accuracy needs improvement to make a quantitative test.

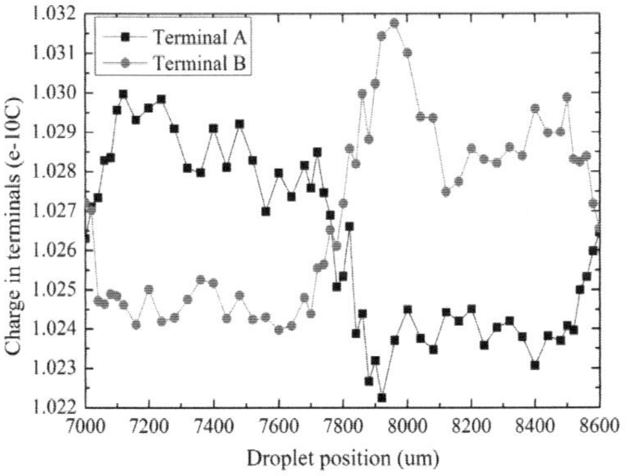

Figure 6. Charges on the grounded electrodes.

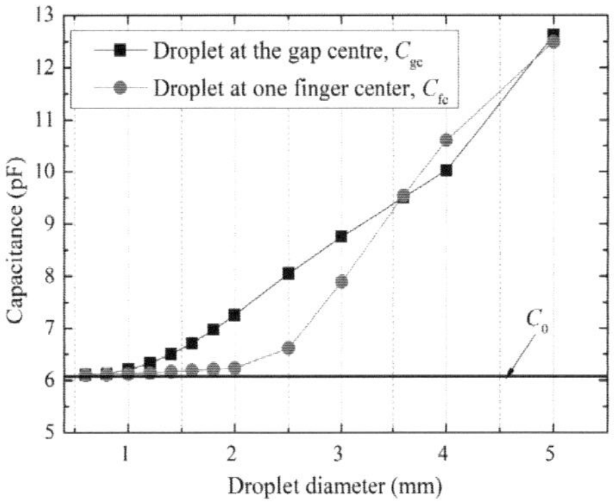

Figure 7. Influence of droplet size on the capacitance variation.

3.3. Influence of droplet size on the capacitance variation

The influence of droplet size upon the capacitance is plotted in Figure 7, in which the droplets hold a constant contact angle of 60° with both positions at gap centre (gc) and finger centre (fc). It is clear that for the IDE with finger/gap width of 400μm (C_0=6.09pF), the total capacitance variation ($\Delta C=C_{gc}-C_{fc}$) increases rapidly with the droplet size, and has a maximum of 1.4pF for a droplet size of 2.5mm. The variation drops as the droplet size keep increasing, even turns to negative when the droplet covers more than one pair of fingers. This indicates that the droplet size should be matched well to the IDE pitch and could favourably have been considerable larger than what was used in the previous experiments. The finding explains why a bigger droplet gave larger output in test [9].

4. Conclusions

Based on a previous prototype of a droplet-based energy harvester, a FEA model was built by adopting several simplifying assumptions. Numerical calculations were used to obtain variations of capacitance, open-circuit voltage and short-circuit charge by setting different boundary conditions. The findings are qualitatively consistent with the voltage peaks related to the short-circuit charge in previous tests. The numerical accuracy of the short-circuit charge needs improvement for performance estimates. It is of interest to further study the effect of charge variations in the film. It is also found that droplet size should match with IDE pitch to have a larger capacitance variation.

Acknowledgement

This work was supported by Research Council of Norway under Contract No.191282.

References

[1] Bogue R 2009 *Sensor Rev.* **29** 194
[2] Mitcheson P D, Yeatman E M, Rao G K, Holmes A S and Green T C 2008 *Proc. IEEE* **96** 1457
[3] Boland J S, Messenger J D M, Lo K W, and Tai Y C 2005 *Technical Digest: 18th IEEE International Conference on Micro Electro Mechanical Systems* 618
[4] Krupenkin T and Taylor J A 2011 *Nat. Commun.* **2** 448
[5] Löhndorf M, Kvisteroery T, Theuss H, and Blixhavn B 2009 *US patent application* US 2009/0080138A1
[6] Choi D-H, Han C-H, Kim H-D and Yoon J-B 2011 *Smart Mater. Struct.* **20** 125012
[7] Bu L, Wu X, Wang X and Liu L 2013 *J. Intel. Mat. Sys. Str.* **24** 61
[8] Yang Z, Halvorsen E and Dong T 2012 *Appl. Phys. Lett.* **100** 213905
[9] Yang Z, Halvorsen E and Dong T 2013 *J. Microelectromech. S.* DOI: 10.1109/JMEMS.2013.2273933

Electromagnetic Energy Harvester by Using NdFeB Sputtered on High Aspect Ratio Si Structure

Y Tanaka, T Fujita, T Kotoge, K Yamaguchi, K Sonoda, K Kanda, and K Maenaka
Department of Electrical Engineering and Computer Sciences, Graduate School of Engineering, University of Hyogo, 2167 Shosha, Himeji, Hyogo 671-2280, Japan

E-mail: er13l034@steng.u-hyogo.ac.jp

Abstract. This study addresses the design optimization of the electromagnetic energy harvester consisting of the sputtered NdFeB film on a high aspect ratio corrugated Si structure and Au electroplated serpentine coil. The high-aspect-ratio Si structure has advantages that the magnetic flux density change is caused by distance change between the coil and magnet film on the fine-patterned corrugated Si, and it is easier to fabricate with high yield than previous study. We also optimized design parameters such as width and depth of the trench, coil size by using FEM analysis and theoretical calculations. Assuming the mass size of 10×10 mm^2 , the trench depth of the 400 μm, the vibration amplitude of 40 $\mu m_{p\text{-}p}$ and the vibration frequency of 100 Hz, the maximum output power of 12 nW and the maximum electromotive force of 4 mV are obtained for 60 μm and 80 μm magnet widths, respectively.

1. Introduction

Recently, wireless sensor network systems are used for various applications such as a human monitoring system, a tire pressure monitoring system, and a building central monitoring system. Some applications are often installed in a place where it is difficult to maintain. Additionally, some applications are required the continuous monitoring operation. For the reason, a wireless autonomous power generator is strongly required in this field. We have developed the vibratory MEMS electromagnetic energy harvesters. The harvester has advantages for its simple structure with coil and magnet structures, and has low output impedance from the coil resistance. In the previous work, the energy harvester consisted of a mass with buried NdFeB magnet array and an Au spiral coil. This buried magnet array has been fabricated by the NdFeB sputtering on the trenched Si structure and the mechanical polishing method [1]. For the 30 μm gap between the 400 μm line and space magnet array on the 10×10 mm^2 mass and the Au coil, the output power from the harvester has measured to be 760 pW at 95 Hz and 400 $\mu m_{p\text{-}p}$ sinusoidal mass movement.

The output power of the electromagnetic energy harvester depends on vibrating frequency, the change amplitude of magnetic flux density attributed to the mass movement. In order to improve the harvesting energy, we focused on a miniaturization of the magnet array. By the miniaturization, frequency of the magnetic flux density changing becomes large. The serpentine-shaped coil instead of spiral coil has been proposed in previous work to fit the magnetic array because the serpentine coil are easy to be fabricated and has low electrical resistance without multilayer connection [3]. In this work, the high aspect ratio Si corrugated structure to realize finer pattern magnetic array. By using FEM analysis of magnetic field and theoretical calculations, the size of magnet array and coil are optimized for the maximum harvesting energy.

2. Fine fabrication process of magnetic material

2.1. Fabrication limitation of the previous work

The fine magnet array is required for higher gathering energy. Nevertheless, previously reported harvester had limitations for micro fabrication of magnetic material. Figure 1 shows fabrication process flow of magnetic array in previous work [1]. The NdFeB/Ta film sputtered on bottom of Si trench is used as buried magnet as shown in figure 1(c) and SEM image of figure 2. However, the thickness of the magnet film becomes non-uniform and hard to cover on narrow trench. Moreover, a difficulty of the control of the polishing process (figure 1(e)), which caused high yield loss.

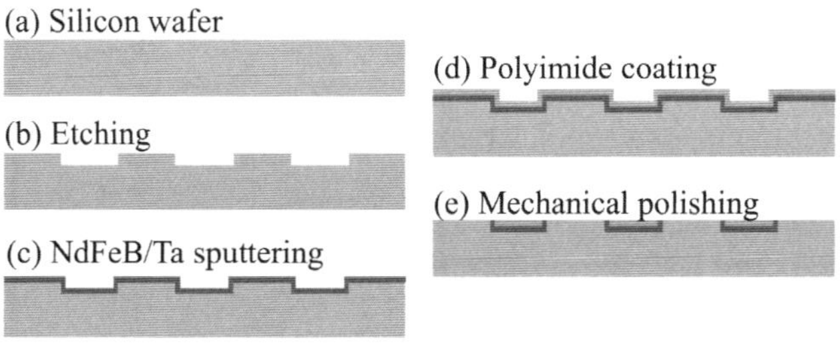

Figure 1. Fabrication process flow of the previously-reported magnet structure.

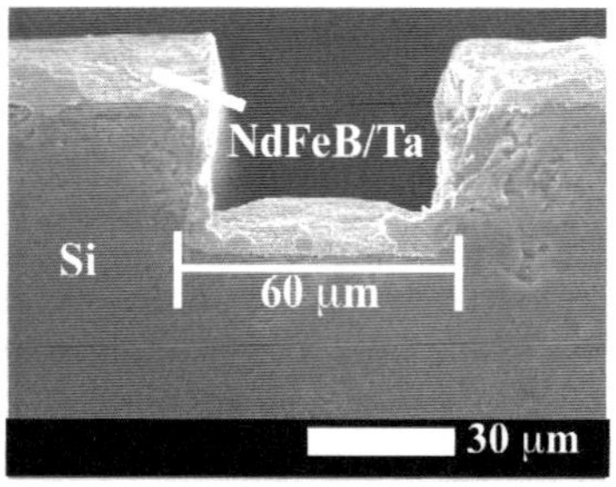

Figure 2. SEM image of NdFeB/Ta multilayer film sputtered on 60 μm width of Si trench and dike with 30 μm depth.

2.2. Magnet material on high aspect ratio corrugated Si structure

We proposed the novel fabrication process for high-aspect-ratio corrugated magnet structure as shown in figure 3. At first, a Si trench is etched to be 300 μm depth by DRIE (Deep Reactive Ion Etching). Then, NdFeB/Ta multilayer film is sputtered on the high aspect ratio corrugated Si substrate. Finally, the fine magnet array is obtained.

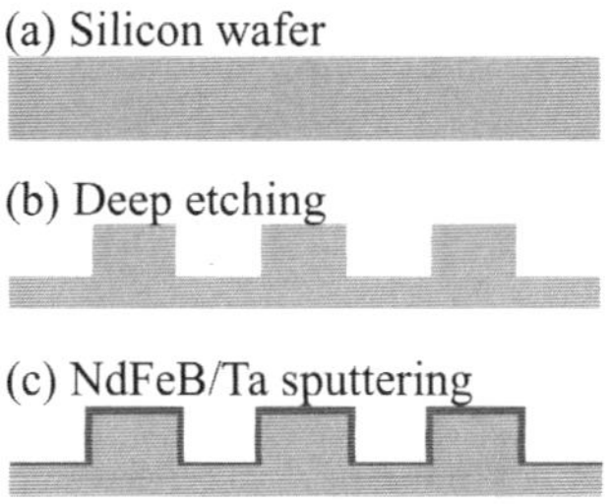

Figure 3. Fabrication process flow of the proposed magnet structure.

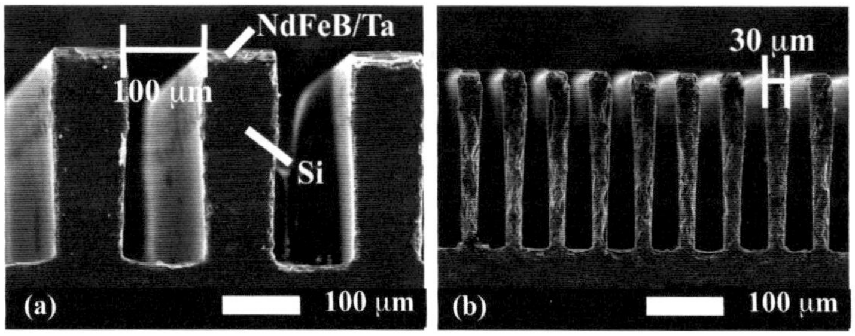

Figure 4. SEM image of NdFeB/Ta multilayer film sputtered on various dimensions of Si trenches. The pitches of Si trenches are 100 μm (a) and 30 μm (b), respectively.

Figure 4 shows the cross sectional SEM image of magnetic array with pitch of (a) 100 μm and (b) 30 μm. The thicknesses of the NdFeB/Ta deposited on the top and the bottom of Si trench are 13 μm and 3 μm, respectively. Obviously the burying the magnet film on bottom of the high aspect ratio Si trench was difficult.

3. Optimization of serpentine coil and magnetic array on corrugated Si

3.1. Schematic diagram of the coil and the magnet

The schematic diagram of the proposed harvester is shown in figure 5(a). The harvester consists of two structures, the Au serpentine coil structure on the substrate and the magnet array on corrugated Si trench on the moving mass. These structures are stacked and magnet array and coils are located with relative distance each other. Then the magnetic flux density across the coil changes according to the magnet movement. As a result the electromotive force (EMF) is generated in accordance with Faraday's law of induction. In order to obtain the maximum harvesting energy, we investigated the optimum size of the device.

Considering to improve the harvesting power, there are three trade-off parameters to optimize. (1) Width of the magnetic bump for magnetic flux density change frequency, (2) width of the magnetic trench to maximize the amplitude of magnetic flux density gradient, (3) line and space of the serpentine coil unit and its multistage connection number (figure 5(b)).

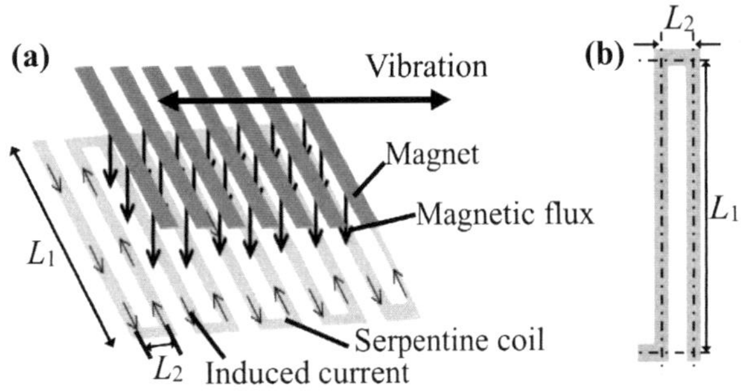

Figure 5. Schematic diagram of the power generation. Overall view (a) and definition of coil unit (b).

3.2. Design optimization of the harvester

For comparison with the previous work, we assume the volume and the area of the magnet and coil structure 14×14×0.4 mm^3 and 9×9 mm^2, respectively. In addition, the thickness and the width of the electroplated coil, air gap between magnet bump and coil are defined as 4 μm, 20 μm, and 30 μm, respectively. The FEM analysis model is shown in figure 6. The magnet thicknesses on top, bottom and sidewall are set as 13 μm, 3 μm and 5 μm, respectively from the preliminary experimental results as shown in figure 4. And then, the magnetic field intensity is defined as 1.0 MA/m from a bulk NdFeB value. Figure 7 shows some results of magnetic flux density with 30 μm air gap for 100, 60 and 40 μm line and space of corrugated array.

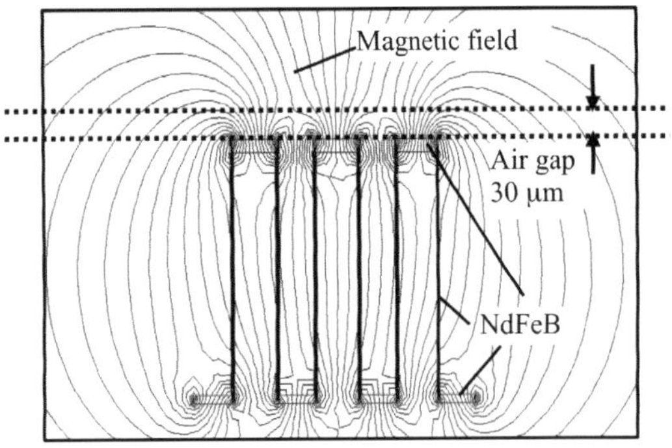

Figure 6. Resulting contour plot of the magnetic flux line calculated by FEM.

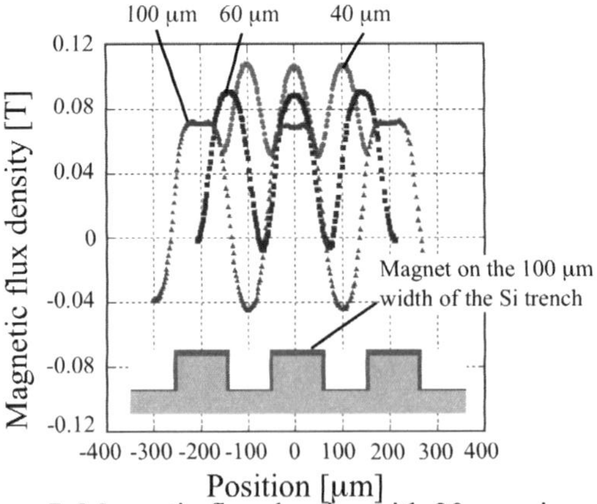

Figure 7. Magnetic flux density with 30 μm air-gap for 40 μm, 60 μm and 100 μm magnet width.

Since the magnetic flux density on position x, $B(x)$, from the FEM analysis was cyclically changed, it is represented from Fourier transform by

$$B(x)=\sum_{n=1}^{k}(a_n\cos[2\pi f_n x+\theta_n]+b_n\sin[2\pi f_n x+\theta_n]), \qquad (1)$$

where a_n is the real number component of the amplitude, b_n is the imaginary number component of the amplitude, f_n is the frequency and θ_n is the initial phase. The EMF can be calculated from the magnetic gradient and the applied vibration velocity. The serpentine coil unit can simply be divided to two wires which are outward and inward wires as shown in figure 5(b). The EMF of single unit coil, $V_1(t)$, is represented by

$$V_1(t)=v(t)B(x)L_1-v(t)B(x+L_2)L_1\ , \qquad (2)$$

where $v(t)$ is the vibration velocity of the mass and $B(x)$ is the arbitrary magnetic flux density on the position of x. Thus the EMF of the multistage of the serpentine coil, $V_n(t)$, is represented by

$$V_n(t)=n\{v(t)B(x)L_1-v(t)B(x+L_2)L_1\}, \qquad (3)$$

where n is the connection number of the unit coil. According to equation (3), the output voltage is the difference of the Lorentz force generated on one coil to adjacent coil. Finally, the effective power of the harvester, $W(t)$, is represented by

$$W(t)=\frac{V_n(t)^2}{4R_c}\ , \qquad (4)$$

where R_c is an resistance of the serpentine coil as large as optimum resistance. When connection number of n is increased, the EMF and the R_c are also increased because these are proportional to n. Therefore its trade-off. There will be optimum values for the magnet array and coil sizing. Figures 8 and 9 show the harvesting voltage and power results for applying the sinusoidal vibration of 100 Hz, the amplitude of 400 $\mu m_{p\text{-}p}$ to the harvester with 200, 300 and 400 μm trench depth, respectively.
These results are not considered the width of Au coil. Therefore, the actual result requires additional calculation for moving average of the magnetic flux density with the coil width of 20 μm. The harvesting energy is increasing as the decreasing of the magnet line and space. The maximum harvesting voltage 4 mV is obtained that the width and the depth of the corrugated Si structure are 60 μm and 400 μm, respectively. Then, the maximum harvesting output power is 12 nW when the width and the depth of the corrugated Si structure are 80 μm and 400 μm, respectively.

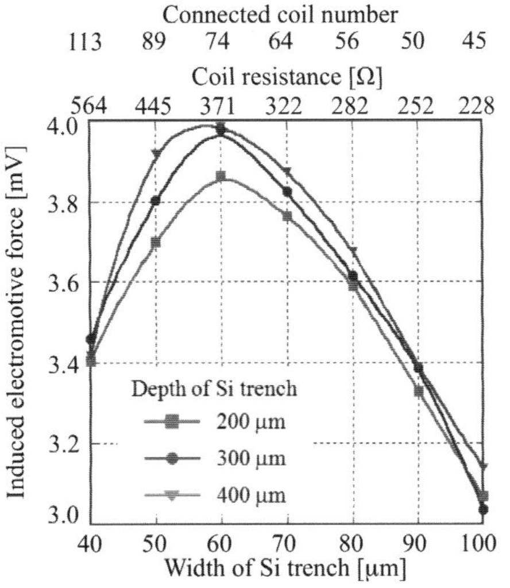

Figure 8. Induced electromotive force estimated from various width and depth of the high-aspect-ratio magnet structures.

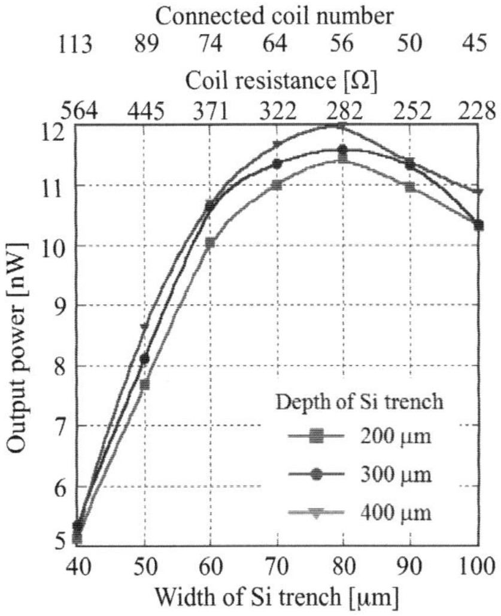

Figure 9. Output power estimated from various width and depth of the high-aspect-ratio magnet structures.

4. Conclusion

We described the energy harvester with the NdFeB array on high aspect ratio corrugated Si structure. By the optimization of the harvester design by FEM analysis and theoretical calculations, the estimated harvesting voltage of 4 mV was obtained. The output power of 12 nW with 282 Ω of optimum load was also obtained. These estimated values were approximately 16 and 15 times as large as previous device respectively.

References

[1] Miki S, Fujita T, Kotoge T, Jiang Y G, Uehara M, Kanda K, Higuchi K, and maenaka K 2012 *Tech. Dig. Int. Conf. on MEMS2012 (Paris)* pp 1221-24

[2] Tanaka Y, Fujita T, Kotoge T, Yamaguchi K, Sonoda K, Kanda K, and Maenaka K 2013 *Proc. The 2nd Int. Workshop on e-WiSe (Beijing)* pp 1656-58

[3] Challa V R and Arnold D P 2012 *Tech. Dig. Int. Conf. on PowerMEMS2012 (Atlanta)* pp 6-9

Non-resonant electromagnetic energy harvester for car-key applications

X. Li[1, 2], T. Hehn[3], M. Thewes[4], I. Kuehne[5], A. Frey[6], G. Scholl[4], and Y. Manoli[2, 3]

[1]Research & Development, Daimler AG, Sindelfingen, Germany
[2]Fritz Huettinger Chair of Microelectronics, IMTEK, University of Freiburg, Germany
[3]HSG-IMIT, Villingen-Schwenningen, Germany
[4]Electrical Measurement Engineering, Helmut-Schmidt-University, Germany
[5]Faculty of Engineering and Business Administration, Heilbronn University, Germany
[6]Faculty of Electrical Engineering, Augsburg University, Germany

E-mail: xiaoming.li@daimler.com

Abstract. This paper presents a novel non-resonant electromagnetic energy harvester for application in a remote car-key, to extend the lifetime of the battery or even to realize a fully energy autonomous, maintenance-free car-key product. Characteristic for a car-key are low frequency and large amplitude motions during normal daily operation. The basic idea of this non-resonant generator is to use a round flat permanent magnet moving freely in a round flat cavity, which is packaged on both sides by printed circuit boards embedded with multi-layer copper coils. The primary goal of this structure is to easily integrate the energy harvester with the existing electrical circuit module into available commercial car-key designs. The whole size of the energy harvester is comparable to a CR2032 coin battery. To find out the best power-efficient and optimal design, several magnets with different dimensions and magnetizations, and various layouts of copper coils were analysed and built up for prototype testing. Experimental results show that with an axially magnetized NdFeB magnet and copper coils of design variant B a maximum open circuit voltage of 1.1V can be observed.

1. Introduction

In recent years our daily routine has fundamentally changed with the use of many consumer electronic products, like smart phones, wearable electronics, and wireless sensor networks. The frequency of battery replacement or the operation time on one battery charge of such mobile devices are important factors influencing the decision in the choice of consumer products. To match these market trends the power consumption of electronic circuits and systems can be decreased due to the achievements in microelectronics on one hand but on the other hand, the battery capacity and lifetime is not keeping up in pace with the increase in performance demand. Energy harvesting which can convert ambient energy into electrical energy to recharge the battery can lead to a prolonged lifetime, or can even be used to directly supply the electronic devices and replace the battery altogether [1].

In the automobile branch there are also many potential application possibilities of energy harvesting technologies [2] with a clear target to reduce component maintenance cycles and costs, and to serve increasing requirements of monitoring and diagnostic functionality. In this paper the research on an electromagnetic energy harvester for application in a remote car-key is presented. The motivation is to extend the lifetime of the battery of the car-key or even to replace it in order to realize a fully energy autonomous, maintenance-free car-key product.

Electromagnetic vibration energy harvesters have been investigated in many research facilities over the last few years. The prototypes have varying sizes, employ different electromagnetic coupling architectures, and work under a wide range of conditions thus resulting in very diverse output voltages and powers. To obtain the maximum harvesting performance, most of these energy harvesting mechanisms are based on the resonant vibration approach, using a mass-spring-damper system to convert kinetic mechanical energy into electrical via a magnet-coil coupling interface based on Faraday's principle [3]. However, a resonant energy generator can only achieve this best performance over a very small frequency range. Because of various manufacturing limitations in small device dimensions it is rather difficult to implement a relatively big inertial mass and a soft spring for a traditional mass-spring-damper system. Thus most of the reported prototypes work efficiently only at a resonance frequency of 50Hz and above while some MEMS based micro structures have resonance frequencies of a few hundred Hz or even in the kHz range. In contrast, the available ambient vibration sources in many scenarios are normally random and have a rich frequency spectral content with a typical bandwidth of 1Hz to 20Hz. Consequently, to be targeted at practical applications in wearable electronics, some previous research has concentrated on energy harvesters for human motion. Von Büren and Tröster in 2007 proposed a linear inertial electromagnetic generator for body-worn applications with a coupling architecture based on opposite polarized magnets which provide the oscillating mass, and a flexible parallel-spring bearing. During normal walking 2-25μW of generated power has been obtained depending on the mounting position [4]. Bowers and Arnold in 2008 reported a spherical magnetic generator intended for energy harvesting from human motion and observed a maximum RMS voltage of ca. 0.7V and a power of 1.44mW [5]. Another non-resonant energy harvester by Yang and Lee presented in 2010 uses a free-standing cylindrical magnet moving inside a sealed cavity of five pieces of printed circuit boards embedded with multi-layer copper coils. This structure generates the maximum output power 0.4μW under matched load conditions [6].

2. Design and fabrication of the energy harvester

It is well-known that the development of energy harvesters is very application specific which means that the following key points of the general requirements of a mass product must be taken into account from the very beginning of the design phase for an energy harvester in a car-key.

- Production and assembly compatibility to existing product – The packaging size of the generator should be compatible to a coin battery (CR2032) and its bracket (shown in figure 1)
- Target product use case – An energy generator in a car-key works daily mostly under car-body vibrations and human motion conditions with dominant spectral content between 1 to 10Hz almost in all three moving directions (figure 2 shows the results of vibration testing of a car-key in a trousers pocket while walking at 6km/h)
- Environmental requirements – fulfilling automobile product standards (temperature, vibration durability, EMC, and so on)
- Cost-effectiveness – standard printed circuit board (PCB) technology, reliable and low cost assembly process in mass production, less special and expensive materials

2.1. Non-resonant energy harvester design

The conventional mass-spring-damper coupling architecture has always a natural resonant frequency, which can be simply written as $\omega_n = \sqrt{k/m}$. In order to have the natural resonant frequency match the spectral content of human motion for a most efficient performance in such resonant systems, one may reduce the spring constant *k* and increase the mass *m* of the oscillator. The design of the low frequency micro-power generator in [4] is based on this principle. But, if the spring in the system is removed, the system changes to a non-resonant one, which is the idea behind the generator principles in [5] and [6].

Fig. 1 Commercial series car-key (Source: Daimler AG)

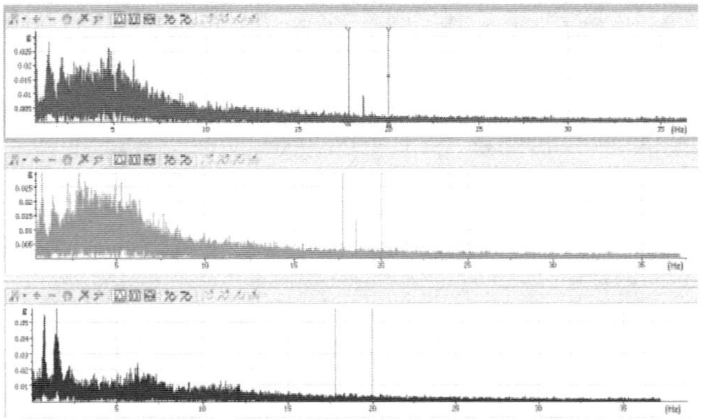

Fig. 2 Spectrum diagram of x/y/z direction in vibrational testing of car-key in trousers pocket by walking at 6km/h

A novel non-resonant electromagnetic vibration energy harvester structure intended for remote car-key application is proposed in this study. The basic idea of this non-resonant generator is to use a round flat permanent magnet which moves freely in a round flat cavity covered on the two flat sides by a PCB embedded with multi-layer copper coils and on the round side by displacement plates made by the same material of PCB substrates (figure 3). The relative motion between magnet and PCB-housing causes a time-varying magnetic flux through the copper coils and thus induces a voltage.

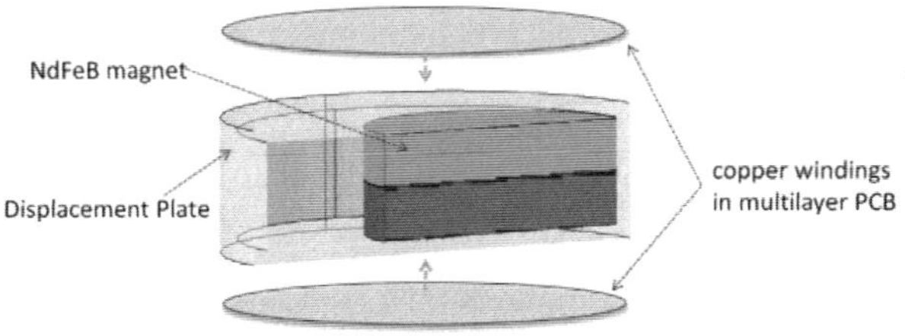

Fig. 3 Car-key power generator structure concept

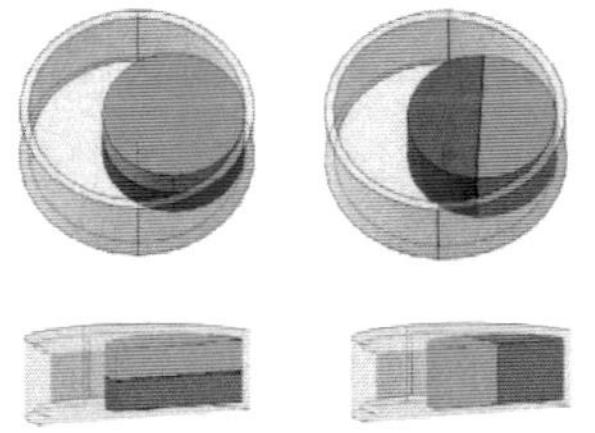

Fig. 4 Magnetic polarization in Z-Axis (left) and in X-Y plane (right)

The primary goal of this structure is to easily integrate the energy harvester within the existing electrical circuit module into the commercial car-key design feature. In other words the whole size of the energy harvester is comparable to a CR2032 coin battery plus its bracket. Consequently, this simple structure would cause no big cost impact for the series product integration and assembly.

2.2. *Experimental prototype fabrication*

For the experimental prototypes, the diameter of the inner cavity is fixed to 20mm, while the height (distance between the two multilayer PCBs) is fixed at 3.1mm. The flat and round shaped magnets, which have different diameters in each prototype, have a common height of 3mm, so that they can freely and smoothly move within the cavity. The displacement plates (shown in figure 5, type S) and the screws make the prototype assembly quite easy and the distance precisely adjustable.

In this particular non-resonant generator design approach, under random ambient excitation, the magnet will be moving in a random direction and rotating at the same time. In fact, the relative motion between magnetic field and the coils can occur only after the magnet overcomes the friction with the housing and gets an initial velocity. Based on Faraday's electromagnetic theory, two main factors play an important role for the optimization of the induction:

- The magnetic flux density and the gradient through the coils
- The form of the coils on the multilayer PCB

Journal of Physics: Conference Series **476** (2013) 012096 doi:10.1088/1742-6596/476/1/012096

Permanent magnets manufactured by the company HKCM (http://www.hkcm.de) with different diameters and polarization direction in Z-Axis or in X-Y plane (shown in figure 4) are applied for performance testing. For finding the most efficient coupling architecture four different layouts of the copper coils were analyzed and built up for testing (shown in figure 5). On the outer circle surrounding of the 4-layer PCB there are bonding points (green points in figure 5), which are used for connecting the copper winding on each layer of the circuit board. Each PCB has 8 through-holes (brown points in figure 5) for screw fixing points. The 4-layer PCB in the prototypes were made by the standard FR4 PCB substrates, and all the PCB copper windings are 100μm wide and 35 μm thick. An energy harvester prototype assembly is shown in figure 6.

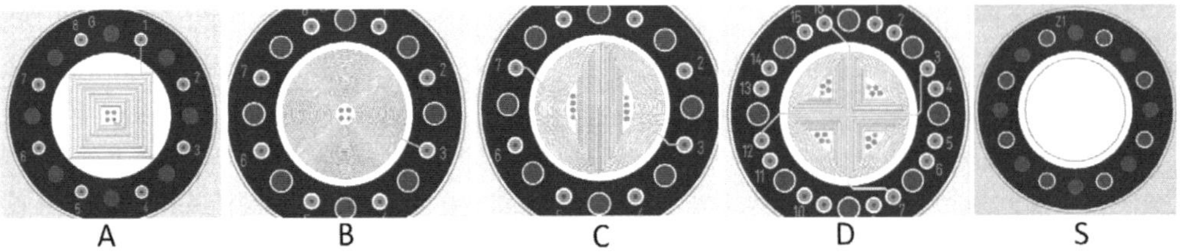

Fig. 5 type A, B, C and D: four layouts of copper coils; type S: layout of displacement plate

Fig. 6 Car-key energy harvester prototype assembly (layout type A as example)

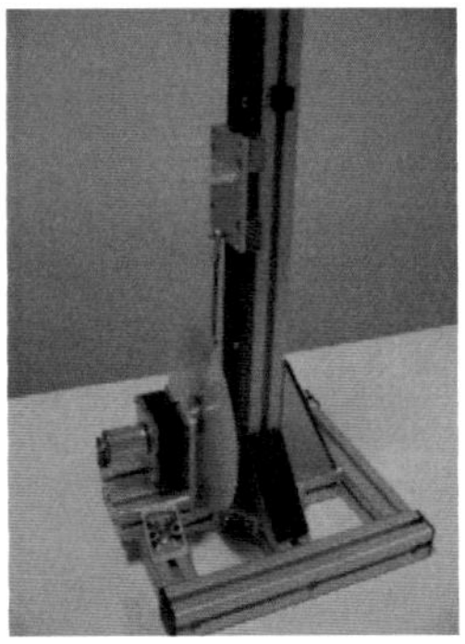

Fig. 7 Big Amplitude Shaker

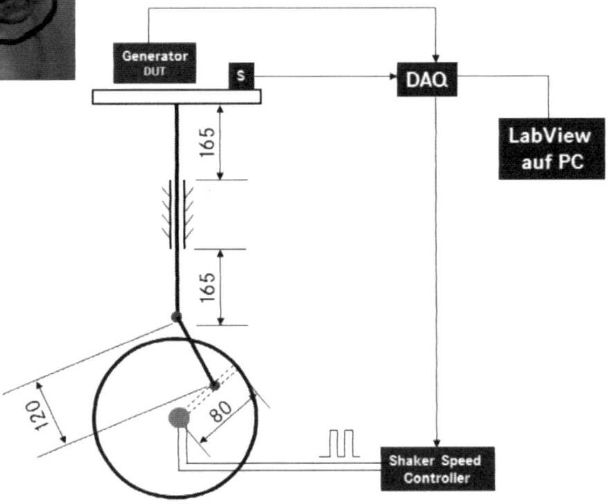

Fig. 8 Measurement setup diagram

3. Measurements and results

In the application the ambient excitation to a car-key is always a random vibration at low frequency and big amplitude. In order to simulate practical application scenarios, a "Big Amplitude Shaker" was developed (shown in Figure 7). The frequency and the amplitude of the reciprocating motion are controlled by the speed of the driving motor and adjusted by the length of the con-rod. Figure 8 shows the measurement setup diagram: the packaged prototype (DUT) is mounted vertically on the holder of the shaker, while the acceleration of the holder is detected by an accelerometer (Kistler® Type 8315A).

A LabVIEW® program running on the PC is able to control the speed of the driving motor of the shaker and to acquire the real-time data of the open circuit induced voltage of the generator as well as the accelerometer values.

Consequently, a total of 28 experimental prototypes of 7 types of magnets (various diameters and polarizations) and 4 types of winding layouts were assembled and measured. Table 1 shows the summary of the prototype matrix and the testing results of the maximum induced voltages, under the same input conditions with the experimental shaker calibrated at 3Hz vibration motion and 100mm amplitude.

The experimental results show that a maximum open circuit voltage of 1.1V was observed with a Z-Axis polarized magnet and copper coils of type B. The inductive voltage increases with the diameter of the flat magnet: comparing the results of the magnet with a diameter of 15mm to that of 3mm, the voltage is around 2~8 times higher, the reason being the larger magnetic flux. Comparing the coil layouts, the type D shows the best performance with the small size magnet, but this advantage is less with bigger size magnets. The polarization direction of the magnetic flux plays a less dominant role in the experiments.

						Coil Layout Type			
						A	B	C	D
					Coil resistance (Ohm)	55	115	92	80
Magnet Variant	Diameter (mm)	Material	B(T)	weight (g)	magnetic polarization	Inductive Voltage [V]			
	3	NI-N35	1,17	0,16	Z-Axis	0,06	0,22	0,22	0,4
	5	NI-N38SH	1,22	0,44	Z-Axis	0,15	0,4	0,25	0,8
		ND-N50	1,4	0,44	X-Y Plane	0,08	0,4	0,25	0,7
	7	ZN-N35	1,17	0,86	Z-Axis	0,35	0,7	0,4	0,64
	10	EP-N52	1,43	1,76	Z-Axis	0,53	1	0,6	0,8
		AD-N35	1,17	1,76	X-Y Plane	0,1	0,8	0,48	0,8
	15	NI-N45	1,32	3,96	Z-Axis	0,5	1,1	0,8	0,97

Tab. 1 Car-key energy harvester prototypes testing result

4. Conclusion and outlook

In summary, a novel electromagnetic energy harvester concept based on a non-resonant mechanism designed for a car-key application is demonstrated. Various magnet and coil forms were taken into account with a total of 28 experimental prototypes built up for testing. A maximum open circuit inductive voltage of 1.1V was obtained by a coil layout type B and a magnet of 15mm diameter.

Based on these results, further investigation into the theoretical method of parameter optimization is needed, e.g. the simulations model of the coupling between the electromagnetic flux and the copper coils in the random relative motion. Further investigations should include the influence of the magnet size and the configuration of the connection of copper-windings on each PCB layer.

References

[1] Manoli Y 2010 Proc. IEEE Eur. Solid-State Circuits Conf. (Seville) pp 27–36
[2] I. Kuehne et al.: *Energy autonomous sensors in the automobile,* 6. GMM-Workshop - Energieautarke Sensorik, pp.91-95, 2010
[3] D. Spreemann and Y. Manoli, *Electromagnetic vibration energy harvesting devices*, Springer Verlag 2012.
[4] T. von Büren and G. Tröster, *Design and optimiyation of a linear vibration-driven electromagnetic micro-power generator*, Sens. Acturators A135, 765-775 (2007)
[5] B. J. Bowers and D.P. Arnold, *Spherical magnetic generators for bio-motional energy harvsting*, Proc. PowerMEMS 2008, pp. 281-284.
[6] B. Yang and C. Lee, Non-resonant electromagentic wideband energy harvesting mechanism for low frequency vibrations, J. Microsyst Technol., Springer Verlag 2010.

Comparison between piezoelectric and magnetic strategies for wearable energy harvesting

G De Pasquale, A Somà, F Fraccarollo

Department of Mechanical and Aerospace Engineering, Politecnico di Torino
Corso Duca degli Abruzzi 24, 10129 Torino, Italy

E-mail: giorgio.depasquale@polito.it, aurelio.soma@polito.it, federico.fraccarollo@polito.it

Abstract. This paper introduces the design and fabrication of energy harvesters for the power generation from human body motion. Two alternative strategies are compared: piezoelectric and magnetic inductive. The generated energy is used to supply body sensors including accelerometers and temperature sensors and RF module. Two prototypes of the magnetic based generator and of the piezoelectric generator are built and tested with shaker at resonance condition and by dedicated bench reproducing joints rotation during walking. The experimental results show that the magnetic prototype can generate 0.7mW from human body motion, while the piezo harvester generates 0.22 and 0.33μW respectively for flexion and extension at angular velocity lower than 1rad/s and 45° amplitude.

1. Introduction

Many industrial and academic researches have been recently addressed to low power techniques for energy harvesting from the environment [1-3]. The available energy sources are light, wind, temperature gradients, radiofrequency waves and kinetic energy of sea waves, but in the last years other sources started to be exploited as mechanical vibrations of machines and human body motion. The small power amount obtained by harvesters is suitable for supplying electronics and portable devices with low consumption specifications, however these strategies allow reducing the number and size of batteries and eliminating cables. The first aspect has positive impacts on environment and economy due to pollution and costs related to batteries disposal; the second aspect is highly appreciated in the field of sensor networks where complicated cablings are often needed. Furthermore, energy harvesters allow supplying sensing devices in critical positions where accessibility is limited, for example in telemedicine applications and biological parameters monitoring. In the field of human body monitoring, the continuous monitoring of patients is feasible through miniaturized and wearable harvesters combined with miniaturized sensors integrated in the body without heavy and uncomfortable batteries for the power supply.

The contribution of this paper is in the field of energy harvesting from human body, which is characterized by additional design complications compared to other fields of kinetic energy harvesting, already discussed in the literature. The authors, in previous works, analyzed the performances of different energy harvesters: piezoelectric [4], magnetic inductive [5], electrostatic capacitive, and magnetically levitated [6]. Also they patented some dedicated devices [7-9]. The critical aspects of harvesting power from the motion of human body are discussed and some design strategies are proposed, with particular focus on two transduction strategies: magnetic inductive and piezoelectric. The design, prototyping and testing of two wearable energy harvesters for human body are introduced. The magnetic generator consists of a movable magnet able to oscillate in a polymeric guide between two fixed magnets situated at the ends of this guide that work as magnetic springs. This harvester has 25mm diameter and 60mm length. The magnetic field generated is used to produce electricity into a wire coil with 400 turns. The conditioning circuit for the electrical power management includes diode bridge, capacitor and battery. The piezoelectric generator is based on a composite transducer including piezoelectric fibers in polymeric matrix. This solution allows higher bending ratios of the transducer compared to traditional piezoelectric sheets and improve the device reliability. The two harvesters are tested on dedicated benches reproducing the body motion.

Finally, a small wireless sensing unit is connected to the generators and supplied with the power produced by the harvesters during the walking activity; the duty cycle of the system is then estimated.

2. Wearable harvesters design approach

The harvester design activity should be preceded by the analysis of vibrating excitation source. In general applications, as machines, vehicles or mechanical systems with variable operating regimes and velocities, the input force has random shape. In case of human body during walking or running, the excitation, although strongly irregular, is periodic because it is related to the alternate motion of limbs. For specific activities (e.g. walking at normal velocity), the frequency of the periodic excitation is almost constant and can be measured and predicted with statistical analyses (Fig. 1).

The next step is the evaluation of the average energy associated to the input signal; this calculation can be performed, for instance, by measuring the acceleration (or the overall acceleration range) in the position where the energy harvester should be attached. The acceleration allows estimating, at least roughly, the size and the mass of the generator and of its oscillating parts. In this preliminary design activity the goal is to provide the order of magnitude of masses and suspensions stiffness with reference to the response/excitation amplitude ratio.

Then, the analysis of frequency spectrum of excitation signal is needed to properly tune the dynamic response of the generator and to maximize its mechanical response (and the output power). The power spectral density (PSD) curve describes the energy distribution in the frequency domain and should be used for the fine dimensioning of the harvester and for its tuning through the modification of the mass and stiffness parameters previously defined only roughly. The dynamic parameters are finalized in detail at this stage and remain the same for the overall life of the wearable harvester.

The last important specification of the energy harvester is the bandwidth. Working as a mechanical filter, kinetic harvesters are excited dynamically only within specific frequency ranges. It is important that the harvester is able to capture a wide range of frequencies from the randomly shaped excitation source to improve its conversion efficiency. For this reason, wide band generators are more performing than narrow band generators: the firsts can catch the energy associated to large frequency ranges, the seconds can only catch the outer energy closely to their resonances. Some strategies to widen the bandwidth of harvesters have been discussed in the literature, such as modal coupling of multiple transducers, series coupling of multiple generators, and the use of bi-stable structures.

The harvester design and dimensioning must be supported by detailed information of the overall autonomous system, including the supplied devices, performances, sensors sampling rate, data transmission frequency, etc. For this reason, the definition of 'self-powered system design' is definitely better than 'energy harvester design'. Although the typology of components included in the autonomous system may vary among the applications, usually the following functional blocks can be identified (Fig. 2): current rectifier, charge reservoir (storage battery), one or more sensing devices (sensors), and transceiver device. Very frequently the energy generated by the harvester is stored in the battery before to be used; this operation requires the preliminary conversion of the alternate current produced by the harvester in continuous current. The battery is necessary to provide continuous supply even in case of irregular power generation and to reach the prescribed energy threshold needed to supply the utilizer. The energy generated and stored in the battery is suitable for the supplying only over the minimum charge threshold; this lower level of the battery charge is imposed by the energetic demand of the utilizers. Similarly, the time of energy transfer from the battery to the utilizer is function of the energy consumed in the unit time. All the mentioned constraints are to be considered when the self-power system duty cycle is designed. Power management strategies for the reduction of energy consumption are welcome for the improved working of the system: activating the system only above given battery charge thresholds (triggering), designated components switch-off during given operations (e.g. the antenna during data measurement and processing) are some examples.

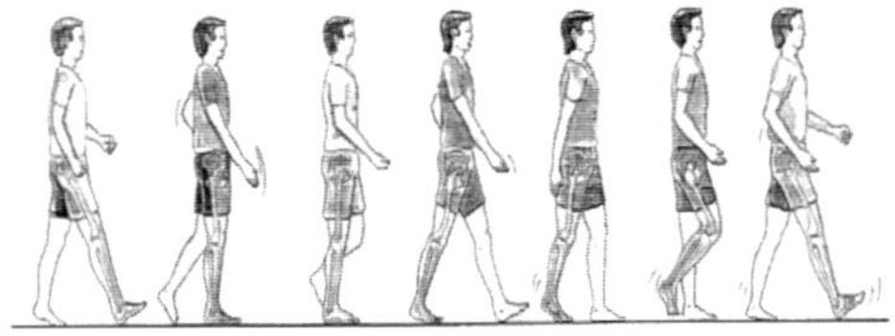

Figure 1. Kinematics of elbow and knee joints in walking activity [10].

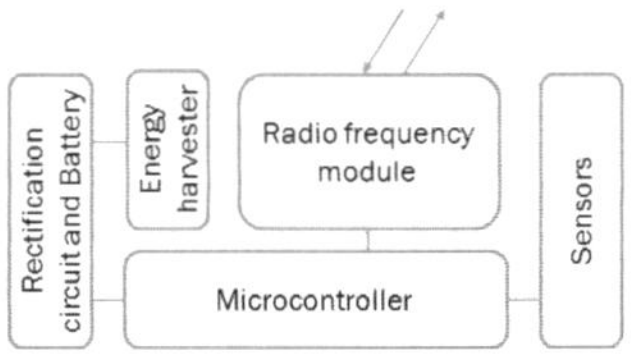

Figure 2. Typical architecture of self-powered sensing system.

3. Prototypes

3.1. Magnetic inductive generator

The magnetic inductive generator is based on the scheme of Fig. 3a, already described by the authors in [5]. The magnetic suspensions provide the mechanical stiffness, which is determined by the magnetic field associated to the magnets with opposite polarity; similarly, the electric current induced in the coil is proportional to the size of the oscillating magnets and to their relative velocity. The dimensioning of this harvester is significantly complicated by the strong electro-mechanical coupling among the components. It includes two fixed magnets (M1), the oscillating mass (M2) composed by three permanent magnets spaced with two teflon rings that moves in the polymeric guide, and the 400 turns coils (C1). The harvester in its final configuration is represented in Fig. 3b.

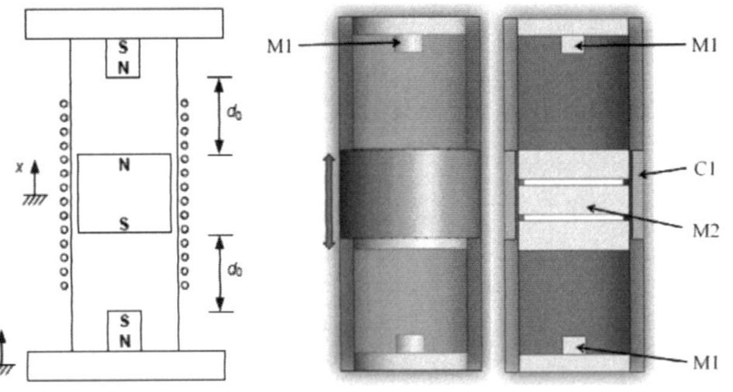

Figure 3a. Magnetic-inductive harvester scheme.

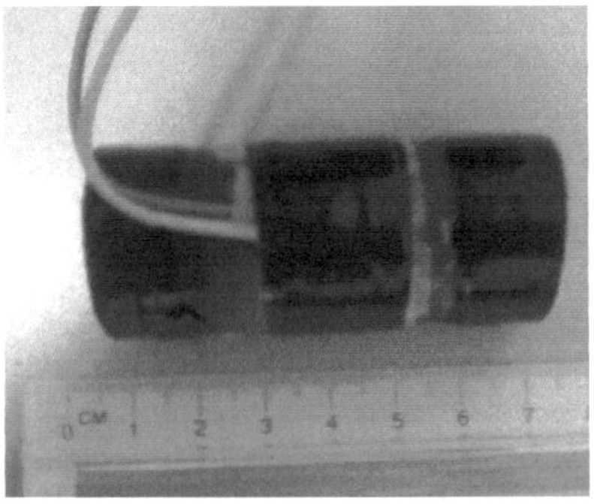

Figure 3b. Magnetic-inductive harvester prototype.

3.2. Piezoelectric generator

Differently from the first one, wearable piezoelectric harvesters are designed to be excited by the motion of joints and not necessary by vibrations [11]. The piezo transducer is embedded in clothes or other wearable fabrics and bended or extended by the motion of limbs. For the design, the detailed modeling of the joints is not needed because the transducer can be considered as adherent to the skin, similarly to clothes, and is not influenced by the motion of internal parts of the joints. Then, the only interesting parameters are the bending angles and velocity of joints. Traditional electro-mechanical transducers based on piezoelectric materials are usually rigid, heavy and brittle. These characteristics do not fit with the requirements of energy harvesting from human body motion where angular travels of joints are very large. Alternative materials with high conversion efficiency, good reliability and high flexibility are needed. This suggests moving to composite materials with polymeric matrix and piezoelectric fibers embedded. The technology called 'viscose suspension spinning process' (VSSP) allows producing fibers with diameters ranging from 10 to 250μm that possess all desirable properties of ceramic piezo materials (electrical, thermal, chemical), but that are not affected by their typical detrimental characteristics (brittleness, weight, low reliability). The samples used in this work (Advanced Cerametrics Inc.) have substrate in kapton and PZT-5A type fibers. Figure 4 reports the sample and one phase of testing on the dedicated test bench already described in [12].

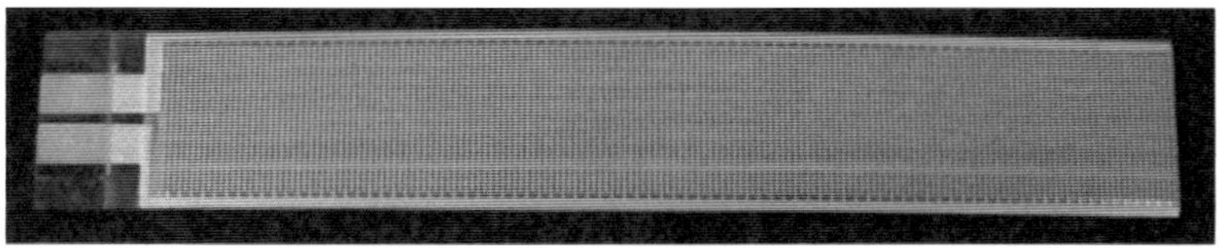

Figure 4a. Piezoelectric fiber-based transducer.

Figure 4b. Phase of testing.

4. Experimental results

The magnetic harvester is tested on an electromechanical shaker reproducing the dynamic excitation of the human body in walking activity (0.2g at 4Hz). The electric output power and current measured on the prototype are reported in Fig. 5. The piezoelectric harvester is tested on a dedicated test bench, designed to reproduce the motion of the human joints [12]; different bending angles and velocities are considered and some of the experimental results are reported in Fig. 6.

After the characterization, the harvesters performances are compared with the energetic requirements of simple MEMS sensors platform for human motion detection applied to garments (Atmel ATmega328 on Lilypad 328 Arduino) with three-axis accelerometer, temperature sensor and a RF module. For the magnetic generator, the output power is enough to supply a 1s sensing every 4.3 walking cycles and a 0.1s transmission every 20.0 cycles. This corresponds to 3.4 and 16.0m of walking distance respectively. For the piezoelectric harvester applied to four limbs of the body, the output power is enough to supply a 1s sensing every 234.4 walking cycles and a 0.1s transmission every 1092.5 cycles. This corresponds to 187.5 and 874.0m of walking distance respectively.

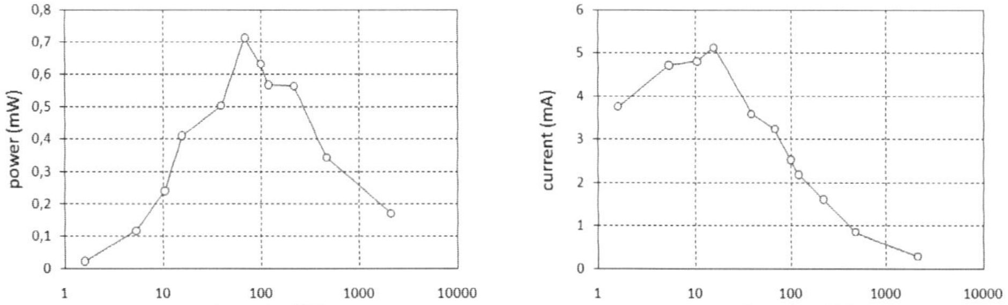

Figure 5. Experimental output power and current under 0.2g acceleration at resonance (4.1Hz) at variable electric loads for the magnetic-inductive generator.

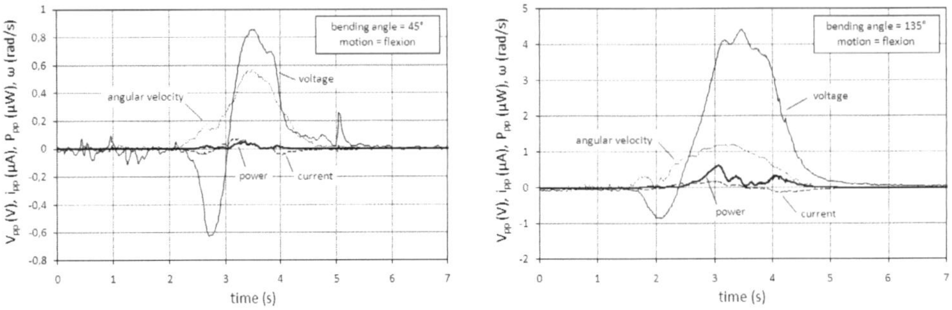

Figure 6. Experimental electric output of piezo harvester at 45° and 135° flexion angles.

5. Conclusions

The study revealed that magnetic inductive and piezoelectric transducers are suitable for harvesting energy from human body motion and supplying wearable sensors with reasonable duty cycles. The higher efficiency of magnetic generators is accompanied by higher masses and a movable part that may be felt as uncomfortable in some applications. Piezoelectric generators are completely embedded in clothes and do not impede motion of limbs, but their efficiency is quite lower as well as their power density compared to magnetic harvesters.

References

[1] S. Roundy, "On the effectiveness of vibration-based energy harvesting," J. Intel. Mat. Syst. Str. 16, 809-823 (2005).

[2] S. Roundy, P.K. Wright, J. Rabaje, "A study of low level vibrations as a power source for wireless sensor nodes," Comput. Commun. 26, 1131-1144 (2003).

[3] S. Priya, D.J. Inman, "Energy harvesting technologies," Springer, New York (2008).

[4] G. De Pasquale, A. Somà, F. Fraccarollo, "Piezoelectric energy harvesting for autonomous sensors network on safety-improved railway vehicles," P. I. Mech. Eng. C-J. Mec. 226(4), 1107-1117 (2011).

[5] G. De Pasquale, A. Somà, N. Zampieri, "Design, simulation, and testing of energy harvesters with magnetic suspensions for the generation of electricity from freight train vibrations," J. Comput. Nonlin. Dyn. 7(4), 041011 (2012).

[6] G. De Pasquale, S. Iamoni, A. Somà, "3D numerical modeling and experimental validation of diamagnetic levitated suspension in the static field," Int. J. Mech. Sci., 68, pp. 56-66, 2013.

[7] A. Somà, G. De Pasquale, "Device for diagnosing railway bogies by applying an energy-autonomous measuring and transmitting bolt, and corresponding control method," PCT Patent n. WO2011/117718 (2011).

[8] A. Somà, G. De Pasquale, F. Fraccarollo, "Ski- or trekking-rachet equipped with magnetic-inductive harvester device for electric generation," PCT Patent n. IT2012/000280 (2012).

[9] A. Somà, G. De Pasquale, F. Fraccarollo, "Self-supplied info-mobility and/or diagnostic system and harvester device for supplying such system," PCT Patent n. IT/2012/000183 (2012).

[10] V. Pirola, "Cinesiologia," Edi.Ermes Eds. Milano, 2003.

[11] P.D. Mitcheson, E.Y. Yeatman, G.K. Rao, A.S. Holmes, T.C. Green, "Energy harvesting from human and machine motion from wireless electronic devices," Proc. of IEEE, vol. 96, pp. 1457-1486, 2008.

[12] G. De Pasquale, A. Somà, "Energy harvesting from human motion with piezo fibers for the body monitoring by MEMS sensors", proc. of DTIP, Barcelona, Spain, 2013.

Two self-powered energy harvesting interfaces based on the optimized synchronous electric charge extraction technique

Yipeng Wu[1], Adrien Badel[1], Fabien Formosa[1], Weiqun Liu[1], Amen Agbossou[2]

[1] Laboratoire SYMME, Université de Savoie, 74944 Annecy-le-Vieux, France
[2] Laboratoire LOCIE, Université de Savoie, 73376 Le Bourget du Lac, France

E-mail: yipeng.wu@univ-savoie.fr

Abstract. Optimized synchronous electric charge extraction (OSECE) interface is a load weakly-dependant circuit, which is a favorable characteristic for piezoelectric wideband vibration energy harvesting. However, it introduces synchronous switches that need to be self-powered in a stand-alone system. This paper presents the design and experimental testing of two self-powered approaches for the OSECE technique. One is made of electronic switches driven by analog peak detector circuits; the other uses mechanical switches directly controlled by the ambient vibrations. Finally, advantages and drawbacks of the two approaches will be compared and discussed.

1. Introduction

During the last decade, piezoelectric vibration energy harvesting devices (VEHD) have attracted more and more research attention [1]. They are based on a piezoelectric electrical generator (PEG) that transforms mechanical energy into electrical energy, and an energy extraction circuit (EEC) that extracts and stores the generated energy. Because environmental vibrations have wide frequency spectrums usually with a strong random component, how to broaden the operating bandwidth of the VEHDs is one of the most challenging issues before their practical deployment. Many works aim at developing wide band mechanical oscillator to enlarge PEGs bandwidth [2], while the optimization of the EECs which makes them appropriate to wideband vibrations is still behind.

PEG has its output impedance, to maximize the extracted power, the impedance matching strategy is usually used. However, the piezoelectric output impedance is a function of the vibration frequency due to its capacitive behavior, while most EECs cannot tune their input impedance sensitively. So this strategy is inappropriate to wideband vibrations. The synchronous electric charge extraction (SECE) EEC successfully addresses this impedance matching issue [3], but the switching closing time in this approach needs to be accurately controlled, which makes it hard to operate autonomously in stand-alone VEHDs. For this reason, Wu et al proposed an optimized synchronous electric charge extraction (OSECE) technique that simplifies the switch control strategy [4]. The interface circuit is then easily self-powered. Moreover, the energy conversion of the PEG is also enhanced compared to the SECE technique.

This paper presents two self-powered approaches to drive the synchronous switches in the OSECE circuit. If the switches are realized using metal-oxide-semiconductor field-effect transistors (MOSFETs), two identical peak detector (PKD) circuits are designed to detect the vibration displacement extreme and drive the switches synchronously. The approach does not require any additional piezoelectric elements to generate the switch control signal. However, it introduces a phase lag in the control signal and an extra energy consumption to power the PKDs, which reduces the harvested power compared to the ideal case. These effects are particularly studied in the next section. Due to the simple switch control strategy, mechanical switches can also be used to replace the electronic switches in the OSECE EEC. The switches take advantage of the two stoppers and the moving part of the oscillator, so they can be passively controlled by the vibration itself. As a result, the added stoppers in the structure introduce a piecewise stiffness in the oscillating system. This nonlinear stiffness significantly increases the operating bandwidth of the PEG [5]. Finally, both of the approaches are experimentally realized.

2. Self-powered OSECE circuits

2.1. Ideal OSECE circuit

The OSECE EEC is shown in figure 1. A flyback transformer with two primary and one secondary windings splits this interface into two parts: a left part which is used to extracted electric charges; a right part which is a load circuit (a storage capacitor C_s and an equivalent resistance R_L represents the input impedance of the following electronic modules).

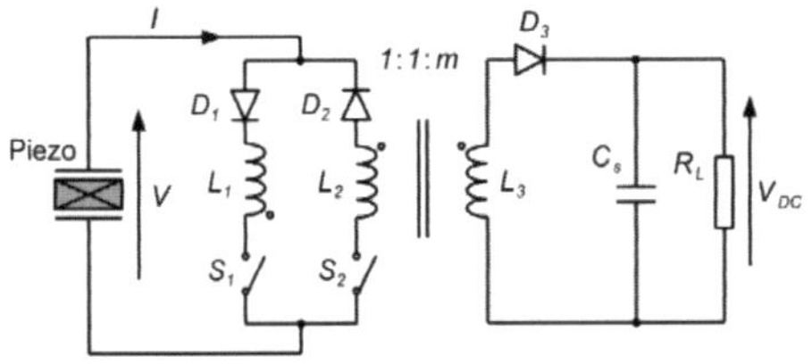

Figure 1. Schematic of the OSECE circuit

When the piezoelectric voltage reaches a maximum (minimum), the switch S_1 (S_2) is trigged to closed, the charges accumulated on the piezoelectric element are then transferred to the primary inductor L_1 (L_2). At the moment that all the electric charges have been extracted, as the switch has not been trigged to open, a fraction of the energy stored in the inductor flows back to the piezoelectric element similar to the initial energy injection technique [6]. The injection phase will be finished as soon as the primary diode in series with the closed switch is reverse-biased, the piezoelectric element is open-circuited again. So in the OSECE approach, an accurate switch closing time is not required, it just has to be shorter than half of the vibration period. From reference [4], the harvested power can finally be expressed by equation (1) in the case of a constant vibration amplitude u_M, where ω_I and Q_I are the natural pulsation and the quality factor of the primary $\{L_1, C_0\}$ oscillating circuit, m is the turns ratio of the transformer and t_m is the duration of the energy extraction phase. Although the equation still shows that the harvested power depends on the load resistance, increasing the parameter m reduces the dependence. For values of m between 1 and 10, it was shown in [4] that the dependence to the load is strongly reduced compared to the classical standard approach.

$$\begin{cases} P_{\text{ideal}} = \dfrac{2\alpha^2\omega}{\pi C_0} \dfrac{\sin^2(\omega_I t_m) e^{-\frac{\omega_I t_m}{Q_I}}}{\left[1+\cos(\omega_I t_m) e^{-\frac{\omega_I t_m}{Q_I}}\right]^2} u_M^2 \\ \omega_I t_m = \arctan\left(-m\sqrt{\dfrac{2\pi}{R_L C_0 \omega}}\right) + \pi S \end{cases} \quad (1)$$

2.2. Electronic self-powered OSECE circuit

If the synchronous switches S_1 and S_2 in the OSECE circuit are made up of the electronic switches, due to the simple switch control strategy, two identical analog PKDs whose inputs share the common piezoelectric element are designed to drive them. The EEC is then self-powered, as shown in figure 2.

Two N-channel MOSFETs are selected for S_1 and S_2. Their low on-state voltage drops do not drastically hinder the quality factor Q_I of the circuit. In addition, MOSFETs are voltage controlled that requires a smaller storage capacitor C_p in the PKD. This characteristic reduces the energy consumption.

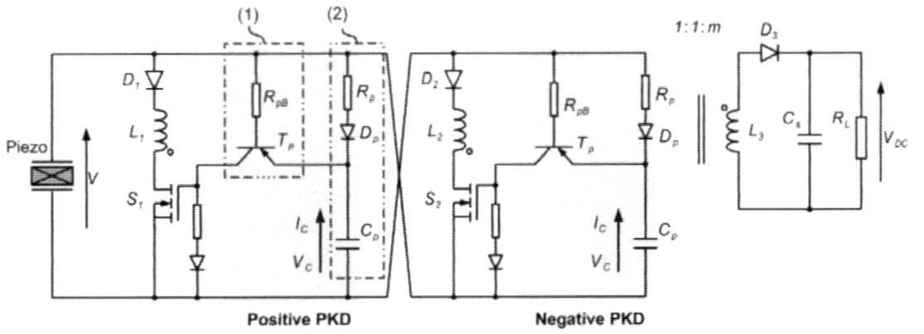

Figure 2. Electronic self-powered OSECE circuit (1) Comparator (2) Envelope detector

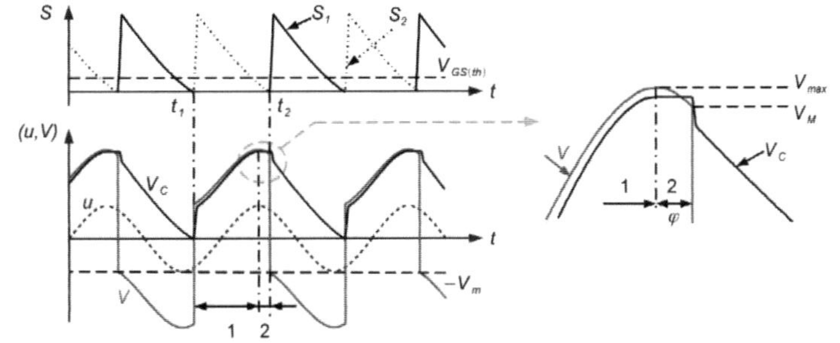

Figure 3. self-powered OSECE waveforms of the structure displacement (u), piezoelectric voltage (V), voltage across the capacitor C_p (V_c), and switch control signal (S)

$$\varphi = \cos^{-1}\left(1 - \frac{(V_{\max} - V_M)C_0}{\alpha u_M}\right) \tag{2}$$

$$E_p = \left(2 + \frac{C_p}{C_0}\right)\frac{1}{2}C_p V_{\max}^2 \tag{3}$$

$$\begin{cases} P_{\text{self}} = \dfrac{\alpha^2\omega}{2\pi C_0}\dfrac{\sin^2(\omega_I t_m)e^{-\frac{\omega_I t_m}{Q_I}}}{\left[1 + \dfrac{C_p}{C_0} + \cos(\omega_I t_m)e^{-\frac{\omega_I t_m}{2Q_I}}\right]^2} u_M^2\left[2\cos\varphi + \dfrac{C_p}{C_0}(\cos\varphi - 1)\right]^2 \\ \omega_I t_m = \arctan\left(-m\sqrt{\dfrac{2\pi}{R_L C_0 \omega}}\right) + \pi \end{cases} \tag{4}$$

Figure 3 presents the typical waveforms in the self-powered circuit. The switch control signal S_1 (S_2) is the output of the positive (negative) PKD. The MOSFET switches will be driven to close as soon as their control signal value is larger than the gate-source threshold voltage $V_{GS(th)}$. Because there exist the threshold voltages of the diode D_p and the BJT T_p in the PKD circuit, the voltage V_M just before the energy extraction phase is always lower than the maximal piezoelectric voltage V_{max}, which introduces a switching phase lag compared with the ideal case. Assuming this value is φ, and the difference between V_{max} and V_M is roughly equal to 1.4V according to the datasheets, φ can be then expressed by equation (2).

In addition, during the stage 1 shown in figure 3, the small capacitor C_p is charged by the piezoelectric element, the stored energy in C_p will finally dissipated in the PKD circuit. The piezoelectric element in this stage is not in open-circuit condition, the outgoing charge can be expressed by the approximated value $C_p V_{max}$. Considering the difference between the energy stored in C_0 if the piezoelectric element was in a true open-circuit condition and in the case where the PKDs are connected, the energy consumption in the PKD circuit can be finally evaluated by equation (3).

In the self-powered circuit, the theoretical expression of the energy extraction phase is the same as that in the ideal approach. However, taking into account the phase lag and the additional energy lost, the harvested power is finally given by equation (4). According to the above equations, figure 4 shows the theoretical harvested powers as well as the experimental values. Compared with the harvested power using standard EEC, the load dependency using OSECE techniques is much lower. The discrepancy between the self-powered and the ideal OSECE approaches is due to the above two effects. However, experimental results still confirm the potential of the self-powered OSECE approach.

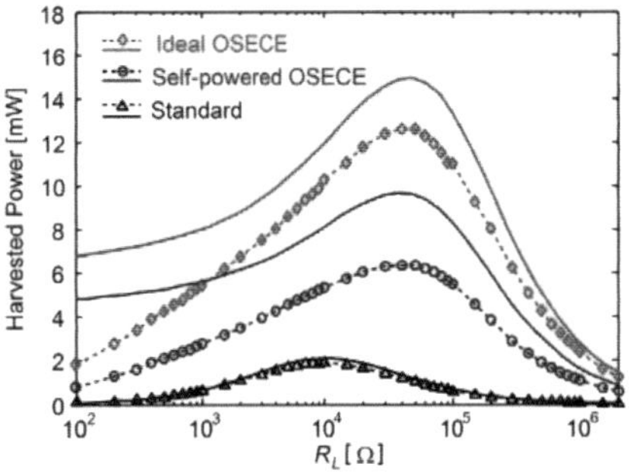

Figure 4. Harvested power as a function of the load resistance (plain lines: theoretical results; lines with dots: experimental results. m = 0.95, u_M = 1.2mm, Q_I = 5.5, C_p/C_0 = 0.03)

2.3. Mechanical self-powered OSECE circuit

The mechanical self-powered OSECE circuit is shown in figure 5. Most of the time, the piezoelectric element in the structure is in open-circuit condition, when the displacement reaches the bounded stroke, the electric charges accumulated on the PEG reach a maximum. As the tip of the beam hits one of the stoppers, the corresponding mechanical switch is closed. The OSECE EEC begins to extract the energy. After this extraction phase, even if the stopper and the beam are still engaged, the reversed diode will cut of the primary loop circuit. The operation principle is just the same as the ideal OSECE approach. As the mechanical stoppers limit the vibration displacement of the oscillator, the harvested power in this VEHD can also be expressed by equation (1), where u_M roughly equals half of the distance between the two stoppers.

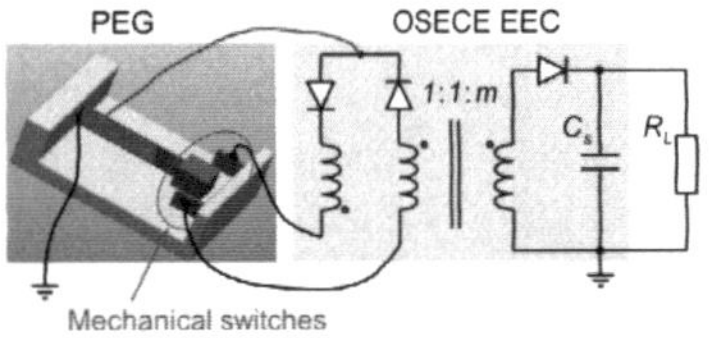

Figure 5. Mechanical self-powered OSECE circuit

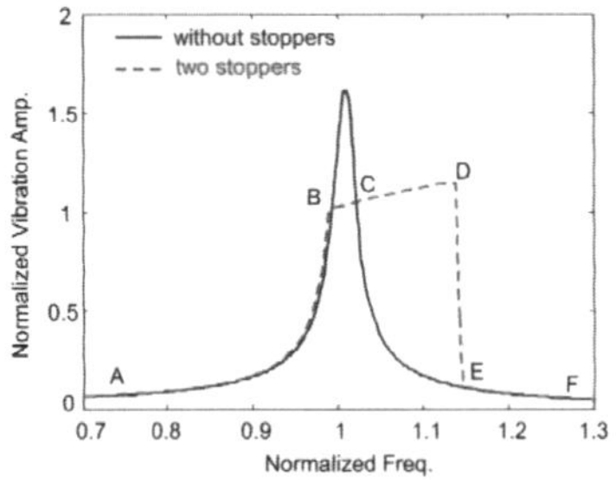

Figure 6. The frequency response of the oscillating structure

Moreover, from reference [5], the mechanical stopper in the PEG introduce a piecewise stiffness in the oscillating system, which increases the operating bandwidth of the PEG, especially when considering a forward sweep acceleration, as shown in figure 6, where the excited acceleration amplitude is 1g, mechanical damping factors of the beam and the stoppers are both 0.01, their resonance frequencies are 87Hz and 200Hz respectively. First, the beam vibration follows a linear cantilever beam behavior (without stoppers) and increases monotonically from A to B. Then the vibration amplitude reaches the maximal distance and starts to deform the stoppers. The operating bandwidth is significantly extended beyond the original frequency bandwidth. When the excitation frequency reaches to point D, the motion drops down to point E. Subsequently, the free end of the beam cannot reach the stoppers and follows the frequency response of the linear beam again. Because the mechanical stoppers are the synchronous switches of the OSECE EEC, the theoretical widest energy harvesting bandwidth is from point B to D. When the excitation frequency is not in this bandwidth, the circuit is unable to harvest any energy.

3. Experimentation and discussion

Both of the self-powered circuits are tested on the classical PEG represented in figure 5. The base of the PEG is fixed on a shaker, and the mechanical stoppers can be tuned to adjust the maximal stroke or can be removed. The external acceleration is a forward sweep signal whose frequency interval is [80, 100]Hz and amplitude is 1.2g. Three kinds of the VEHDs are compared in the experiment: linear PEG plus standard EEC, linear PEG plus electronic self-powered OSECE EEC, and nonlinear PEG plus mechanical self-powered OSECE EEC.

Figure 7 gives the experimental results of the three VEHDs. It is clearly shown that using electronic self-powered EEC, the harvested power is always higher than when using the standard EEC. Using the nonlinear PEG, due to the mechanical stoppers, the vibration amplitude around the resonance frequency is limited at the benefit of a wider bandwidth. Moreover, due to the advanced OSECE approach, even if the vibration amplitude is limited, high power can be obtained, especially when the load is larger than 50kΩ.

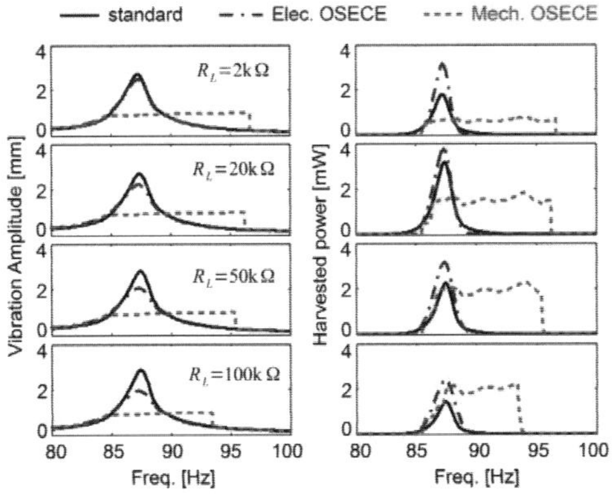

Figure 7. The comparison of the vibration amplitudes and harvested powers for the different VEHDs

4. Conclusion

Two self-powered OSECE approaches are proposed in this paper: An electronic self-powered approach as well as a mechanical one. The difference between these self-powered approaches and the ideal case are theoretically analyzed and compared. Finally, advantages and drawbacks of the two self-powered approaches are also given: compared to the electronic self-powered approach, the mechanical switches allow to increase the bandwidth at the detriment of the maximal harvested power.

References

[1] Anton S R and Sodano H A 2007 *Smart Mater. Struct.* **16** 1–21
[2] Andò B, Baglio S, Trigona C, Dumas N, Latorre L and Nouet P 2010 *J. Micromech. Microeng.* **20** 125020
[3] Lefeuvre E, Badel A, Richard C and Guyomar D 2005 *J. of Intell. Mater. Syst. and Struct.* **16** 865–76
[4] Wu Y, Badel A, Formosa F, Liu W and Agbossou A 2012 *J. of Intell. Mater. Syst. and Struct.* **24** 1445–58
[5] Liu H, Lee C, Kobayashi T, Tay C J and Quan C 2012 *Smart Mater. Struct.* **21** 035005
[6] Lallart M and Guyomar D 2010 *Appl. Phys. Lett.* **97** 014104

Mechanical Design and Material Characterization of the Piezoelectric Energy Harvestor

M Ichiki[1], E Komine[2], K Sueshige[2], K Iimura[2], T Kobayashi[1], T Kitahara[3] and S Fujimoto[4]

[1] National Institute of Advanced Industrial Science and Technology(AIST), Tsukuba, 305-8564,Japan
[2] the University of Tokyo, Tokyo, 113-8656,Japan
[3] Shonan Institute of Technology, Fujisawa,251-8511, Japan
[4] Tokyo City University, Tokyo, 158-8557 Japan

E-mail: ichiki-m@aist.go.jp

Abstract. This paper is on the proposal of the methodology for the design and fabrication of the piezoelectric energy harvesting devices in viewpoint of the justification both structural and material properties. There has been developing various kinds of energy harvesting techniques using piezoelectric films. Most of them are for the justification of the design or the circuit. The output of the piezoelectric harvester is the product of the material and structural parts. It is, therefore, necessary to justify both of them simultaneously. In this paper cantilever structure was fabricated for the improvement of the power generation in the low frequency region. The resonant frequency and output voltage and charge from the following theory formula were calculated in this study.

1.Introduction

There has been increasing technological demand to develop the energy harvestor using ambient vibration for the application for ubiquitous micro devices and systems[1][2]. Piezoelectric energy harvestor is one of the techniques of the applicable for transducer from the mechanical to electrical energy. The advantage of piezoelectric type is in its simple structure of smart materials and also the high energy conversion efficiency. MEMS piezoelectric harvestor has been developing with use of microfabrication technology and thin films[3]-[8]. Micro vibrational harvestor devices, like cantilever type have its resonant frequency over 1kHz in its earlier stages[3]. On the other hand, ambient environmental vibration has normally its resonant frequency in lower frequency region of 1-100Hz. The vibrational devices have its maximum output at its resonant frequency. It is, therefore, necessary to develop such device structure as has low resonant frequency that is useful for ambient environmental vibration. In addition to frequency requirements, it is also necessary to establish the methodology to justify the device structure and the choice of piezoelectric materials for the expected applications.

The output of the harvestor devices are shown to be as equation (1). The right term is divided into the multiple of the 2 parts ; one is on the material characterization and the other one is on the size of the device. It is, therefore, necessary to justify both the material and also the size of the device for the realization of the maximum performances. In this article two approaches will be shown for the improvement of the piezoelectric energy harvestors. One is the methodology of the justification of the size of cantilever devices. Computational solution will be obtained for the high output with low frequency. The other one is comparison of the resonant frequency obtained by simulation and experiment. The thickness of the device Si layer was shown to be effective to the output of the devices.

2. Method and Process

Output charge will be equation (1) using a simplified model of the device shown as Fig.1. Capacitance becomes equation (2) assuming that piezoelectric is a capacitor, therefore the output voltage can be calculated equation (3). It is considered by formula (3) that characteristics of the device will be determined by the structural shape and material parameters. Equation (4) is the resonance frequency equation and is derived from a spring mass model of Si only. Q is the generated charge in this system. d, E, F, K are piezoelectric constant, Young' modulus, induced force and coupling constant. Those are based on the material properties of the piezoelectric films

The dimension of the cantilever structure was examined using the finite element analysis for the justification. In this study, these theoretical calculations and finite element analysis carried out for the confirmation of the effectiveness of structural design and material composition. This structure is typical in case of piezoelectric MEMS energy harvestor using SOI wafer[ ] . The fabrication process is typical and was shown in previous work[ ]. The measurement of resonant frequency was carried out using laser Doppler vibrometer.

The justification of the size of device was made for the specification the length of the beam that is kept as total length. The proportion of the beam with mass and without mass was justified for the realization of the high output at low frequency region. The comparison of the simulated and experimental results of resonant frequency will be shown with different thickness of device Si layer of SOI wafer. Simulation was made with use of FEM method.

$$Q = \frac{d_{31} E_{PZT} F}{K_f} w(y_n - h_1)(\frac{L^2}{2} + \alpha L) \tag{1}$$

$$C = \varepsilon_{PZT} \frac{wL}{h_{PZT}} \tag{2}$$

$$V = \frac{Q}{C} = \frac{\frac{d_{31} E_{PZT} F}{K_f} w(y_n - h_1)\left(\frac{L^2}{2} + \alpha L\right)}{\varepsilon_{PZT} \frac{wL}{h_{PZT}}} \tag{3}$$

$$f_0 = \frac{1}{2\pi}\sqrt{\frac{k}{m}} = \frac{1}{2\pi}\sqrt{\frac{E_{Si} h_{Si}^3}{8\rho_{Si} \alpha h_m (L+\alpha)^3}} \tag{4}$$

3.Experimental Results

It was found that L = 2.5mm, α=1.25mm, h_{si} = 9μm, h_{PZT} = 2μm cantilever was optimal for the low resonant frequency and high output voltage by changing the dimensions in the equation (3) and (4) . Fig.2 show the output charge and voltage using the dimension value and the value of material properties. Charge is dependent on the Young's modulus and piezoelectric constant. Nb doped PZT has the largest piezoelectric constant. On the other hand, voltage is dependent on the dielectric constant, ZnO has the lowest dielectric constant among the materials in Table. According to these results, it is expected that Nb doped PZT which have high piezoelectric constant shows higher output. And similar tendency is also expected to KNN.

Output voltage was calculated by using finite element analysis software: MemsONE. Simulation procedure for the output voltage is examined and established as follows. 1) Determine the displacement by using the harmonic response analysis. 2) Voltage was determined from the displacement by piezoelectric analysis. The thickness of the device Si Layer is shown to be effective to the resonant frequency. The comparison of the resonant frequency of the simulated ans experimental is in good coincidence. Fig.3 shows the results of calculation of the resonance frequency. Among them type 2 has the highest output and low frequency. But this type is difficult because of the heavy mass. In the following sections type1, 3 and 4 will be compassioned in FEM and also the measurement of the devices. Fig.4 shows the results of the output voltage at 100Hz. Based on the results of theoretical calculation and finite element analysis, high output of the actual device is expected by changing the material composition.

Fig.5 shows the compassion of the resonant frequency and the thickness of device-Si layer of type 1 device. The resonant frequency has linear relationships with the thickness of device Si.

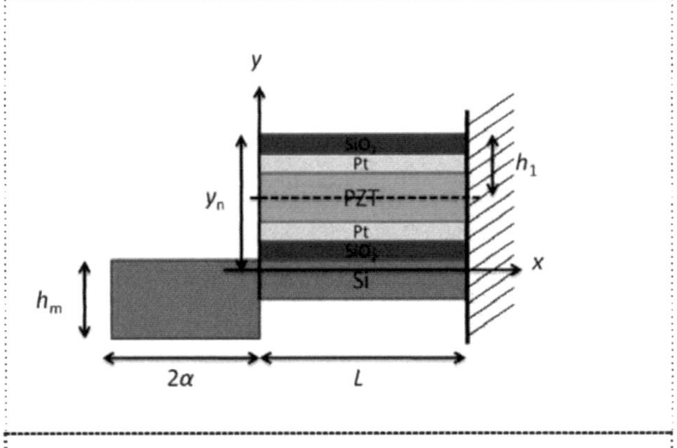

Figure 1. Simplified model of the device.

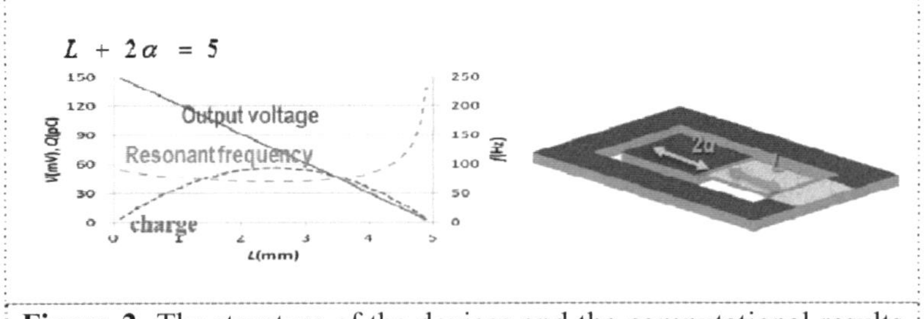

Figure 2. The structure of the devices and the computational results of the resonant frequency and output

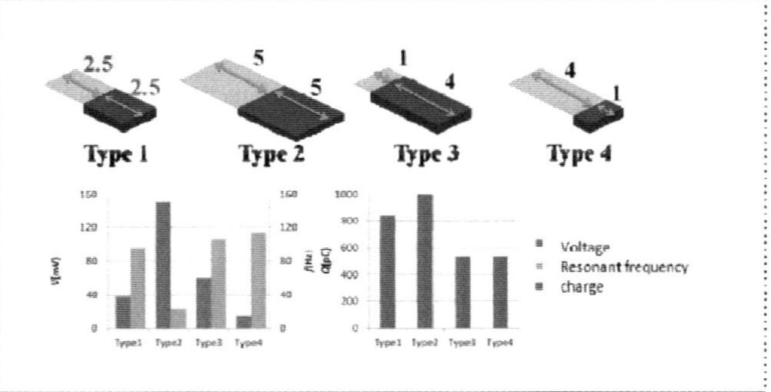

Figure 3. Simulated results of voltage, resonant frequency and charge of the 4 types of cantilever devices

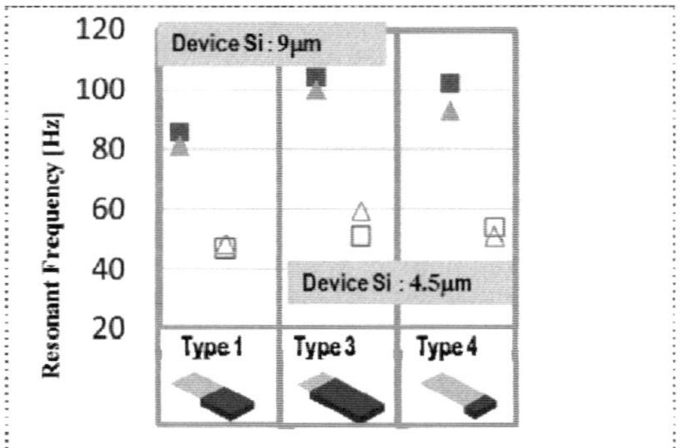

Figure 4. Comparison of the simulated and experimental resonant frequency of different size with their schematic figure

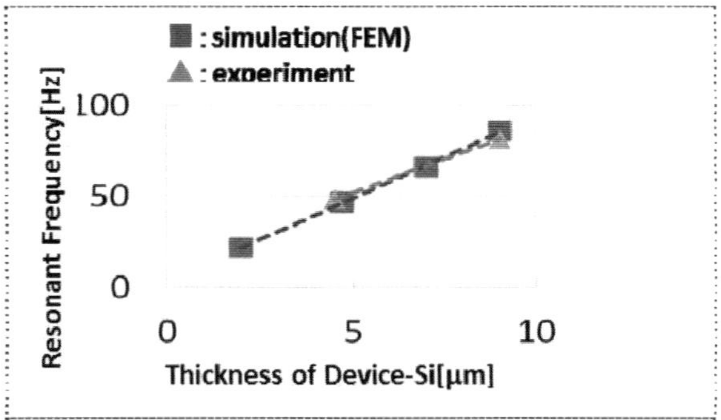

Figure 5. The relationships between the resonant frequency and the thickness of Device Si layer

4.Discussions and Summary

In this article design and simulation of the piezoelectric cantilever structure was carried out. Justification of the size was made and to be shown the most suitable structure of the energy harvestor. The resonant frequency of the same devices was simulated to obtain the resonant frequency. The resonant frequency was shown to be strongly depend on the thickness of the device Si layer. The relationship was linear. The following results was mainly shown. 1) One of the optimization methods was shown. MEMS cantilever structure was examined and shown to be its optimised size. This is estimated using maximum output voltage and low resonant frequency. 2)The justification of the simulation and experimental value of the resonant frequency was carried out. The thickness of the device Si Layer is shown to be effective to the resonant frequency. In the optimization of the device structure, the 2.5mm-mass and 2.5mm-cantilever structure was most suitable for low resonant frequency and high output. As a result, the method of the justification of the size of device was established in the above procedure. The coincidence of the simulated and experimental is in good. The thickness of the device Si layer is affective to the output of the harvestor.

References

[1] Feng Y, Hagiwara K, Iguchi Y and Suzuki Y, 2012, Appl.Phys.Lett, 100, p.262901.

[2] Liu H, Zhang S, Kathairesan, Kobayashi T and Lee C, 2012, Appl.Phys.Lett, 100, p.223905.

[3] Kobayashi T, Okada H, Masuda T and Itoh T 2009 Tech. Digest of MEMS 2009 Conference, Sorento, p.801

[4] Renaud M, Karakaya K, Sterena T, Fiorini P, Van Hoof C, and Puers R 2008 Sensors and Actuators A **145-146**, p.380.

[5] Elfrink R, Kamel T M, Goedbloed M, Motova S, Hohlfeld D, van Andel Y and van Schaijk, R, 2009, J. Micromech. And Microengineering,**19**, p.094005-14

[6] Shen D, Park J H, Noh J H, Coe s Y, Kim S H, Wikle III H C and Kim d J, 2009, Sensors and Actuators A, **154**, p.103.

[7] Miller L M, Halvorsen E, Dong T and Wright P K, 2011, J. Micromech. And Microengineering, **21**, p.045029-1

[8] Cook-Chennault K A, Thambi N, and Sastry A M, 2008, Smart Materials and Structures, **17**, p.43001-1.

Acknowledgments

This work was partially supported by the Presto JST. The authors express their thanks to K.Tanaka and M.Obata(AIST) for their preparation of the sample treatment. We will show our grateful to Dr.R.Maeda and Dr.T.Itoh(AIST) for their encouragement during this study. This work was partially supported by Grant-in-Aid for Scientific Research, 25420196(KAKENHI) program.

Modelling of a bridge-shaped nonlinear piezoelectric energy harvester

G Gafforelli[1,3], **R Xu**[2], **A Corigliano**[1] **and S G Kim**[2]
[1]MEMS modeling and design group, Department of Civil and Environmental Engineering, Politecnico di Milano, Piazza Leonardo da Vinci 32, Milano, Italy
[2]Micro and Nano Systems Laboratory, Department of Mechanical Engineering, MIT, 77 Massachusetts Avenue, Cambridge, 02139, MA, USA

E-mail: giacomo.gafforelli@polimi.it

Abstract. Piezoelectric MicroElectroMechanical Systems (MEMS) energy harvesting is an attractive technology for harvesting small magnitudes of energy from ambient vibrations. Increasing the operating frequency bandwidth of such devices is one of the major issues for real world applications. A MEMS-scale doubly clamped nonlinear beam resonator is designed and developed to demonstrate very wide bandwidth and high power density. In this paper a first complete theoretical discussion of nonlinear resonating piezoelectric energy harvesting is provided. The sectional behaviour of the beam is studied through the Classical Lamination Theory (CLT) specifically modified to introduce the piezoelectric coupling and nonlinear Green-Lagrange strain tensor. A lumped parameter model is built through Rayleigh-Ritz Method and the resulting nonlinear coupled equations are solved in the frequency domain through the Harmonic Balance Method (HBM). Finally, the influence of external load resistance on the dynamic behaviour is studied. The theoretical model shows that nonlinear resonant harvesters have much wider power bandwidth than that of linear resonators but their maximum power is still bounded by the mechanical damping as is the case for linear resonating harvesters.

1. Introduction

Piezoelectric MEMS is an attractive technology for harvesting small magnitudes of energy from ambient vibrations. This technology promises to eliminate the need for replacing chemical batteries or complex wiring in microsensors/microsystems, moving us closer toward battery-less autonomous sensors and networks by harvesting the environmental energy on-site to fulfil their tasks. New developments in electronics have considerably reduced the power consumption of devices which can be now powered by MEMS harvesters if they can robustly generate about 100 μW of continuous power from ambient vibrations [1]. In addition, harvesters should be small enough to be integrated with the electronics and the manufacturing cost should be sufficiently low for mass scale deployment. For MEMS-scale (smaller than 0.5 cm^3) energy harvesters, piezoelectric transduction is the most appropriate scenario since standard MEMS thin-film processes are available for many piezoelectric materials assuring high efficiency, high energy density and scalability. Operating frequency, frequency bandwidth, excitation level, power density and size are the key design requirements. At the present time, most of the devices reported in the literature do not meet the desired requirements. Cantilever

[3] To whom any correspondence should be addressed.

laminated beams show reasonable power generation but the frequency bandwidth is impractically small and big device size is required to have low operating frequencies. Moreover, the full strain capability of piezoelectric materials is only exploited at high (and impractical) proof mass displacements. A MEMS-scale nonlinear resonator is developed for ultra wide-bandwidth (UWB) operation [3]. By utilizing a doubly clamped beam resonator, the stretching strain triggered at large deflection stiffens the beam and transforms the dynamics to nonlinear regime and increases the bandwidth. Nonlinear resonant harvesters can also easily achieve the full strain use of PZT since higher strains can be obtained at lower displacements than is needed by a linear resonator. In this paper a first comprehensive theoretical discussion on nonlinear resonating piezoelectric energy harvesting is provided. Firstly, the electrical damping is considered as a linear dashpot added to the classical mechanical damper. In the second part, a more accurate model is built to consider the influence of an external load resistance to the dynamic behaviour of the nonlinear oscillator. The modelling shows that the power generation of the nonlinear resonant harvesters is bounded by the mechanical damping of the system and the maximum power generation can be obtained when the electrical damping inject to the system equals the mechanical one as was reported for linear resonant harvesters [5].

2. Simple model: linear electrical damping

A general nonlinear piezoelectric resonant energy harvester is modelled as a classical Duffing oscillator with an additional linear dashpot (c_E) to include the dissipation due to energy harvesting. By classical methods [4] the nonlinear dimensionless Frequency Response Function (FRF) is:

$$Y=\left\{\left[-\Omega_M^2+1+\frac{3}{4}\alpha Y^2\right]+\mathrm{i}\left[2\left(\zeta_E+\zeta_M\right)\Omega_M\right]\right\}^{-1} \tag{1}$$

where $\alpha=k_N/k_L\cdot W_0^2$ is a measure of the nonlinearity of the system, k_L and k_N are the linear and nonlinear stiffness, $W_0=F/k_L|_{kN=0,\omega=0}$ is the linear static displacement, F *is* the external inertial force, $Y=W/W_0$ is the dimensionless amplitude, $\Omega_M=\omega/\omega_n$ is the dimensionless excitation frequency, ω the excitation frequency, ω_n the linear natural frequency and $\zeta_M=c_M/2\omega_n m_{TIP}$ and $\zeta_E=c_E/2\omega_n m_{TIP}$ are the electrical and mechanical damping ratio, c_M and c_E are the mechanical and electrical damping coefficient and m_{TIP} the tip mass.

The amplitude of the oscillation is obtained by solving equation (1) and plotted in figure 1 ($\zeta_M=0.0056$, $\alpha=0.048$, $\omega=0.0016$) where it is shown the influence of the damping on the amplitude response. The higher is the damping the lower the jump-down frequency and the relative amplitude. The linear electric dashpot represents the amount of damping injected in the system by the piezoelectric material, thus the generated power is dissipated by the dashpot. The power generation increases until the system jumps down, where the power reaches the maximum. When the electrical damping is zero, no power is harvested, on the other hand when the damping is too high the oscillator does not move and no power is then scavenged. An optimum lies in the middle. Providing the power generation for all the values of electrical damping, the envelope of all peaks is obtained and shown in Figure 2. It is shown that nonlinear harvesters have a much larger bandwidth than linear harvesters which are efficient only near the resonance frequency of the system. A close form expression of the maximum power is obtained computing the energy dissipated by the electrical dashpot:

$$P_d=\frac{\zeta_E m\left|\ddot{y}\right|_{ext}^2}{4\omega_0\left(\zeta_M+\zeta_E\right)^2} \tag{2}$$

Equation (2) is the same expression that is obtained for linear harvesters [5] and the peak power occurs when $\zeta_E=\zeta_M$, as it can be seen in figure 2. In conclusion, the previous expression can be considered as a theoretical upper bound to the power generation for all kind of resonant energy harvesters. However, this does not mean that $\zeta_E=\zeta_M$ is always the optimal condition for harvesting power. Indeed, this is true only for $\Omega=\Omega_d|_{\zeta E=\zeta M}$ (for linear systems it is true only at resonance) while for other frequencies the optimal c_E can be higher ($\Omega<\Omega_d|_{\zeta E=\zeta M}$) or lower ($\Omega>\Omega_d|_{\zeta E=\zeta M}$) than c_M. This means that the optimal c_E depends on the excitation frequency. Linear systems harvest high power at resonance because inertial and restoring forces compensate each other and the solicitation is in phase with the velocity (so with the voltage); in that condition the optimum is obtained for $c_E=c_M$. Out of resonance a phase shift between the force and the velocity (and the voltage) is introduced. Thus, the electrical damping must be calibrated to compensate the shifting and assure the optimal power generation. In nonlinear harvesters the same thing happens, the electrical damping must be calibrated at each frequency to assure the maximum power generation even if the peak power is still obtained for $c_E=c_M$.

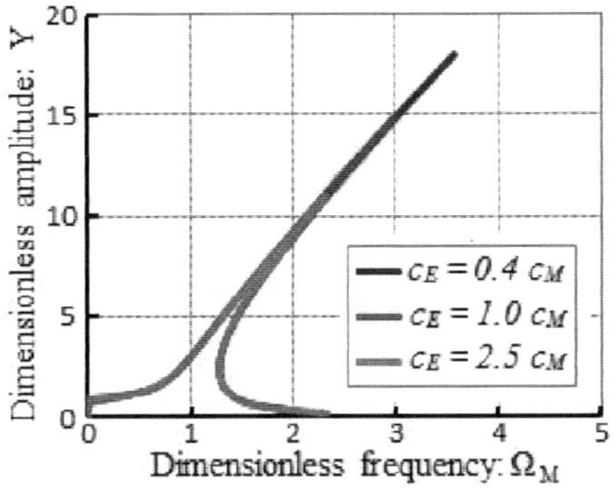

Figure 1. Amplitude response of the Duffing Oscillator for different values of electrical damping

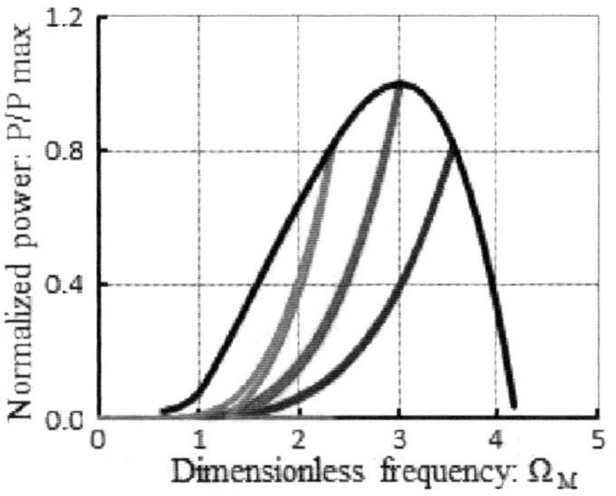

Figure 2. Power response of the Duffing Oscillator for different values of electrical damping

3. Accurate model

A more accurate model needs to be discussed since a linear dashpot is not suitable to describe the completely coupled piezoelectric behaviour [5]. This second model allows studying the influence of an external load resistance (R) on the electromechanical behaviour of the harvester. A doubly clamped beam resonator is presented in figure 3. The x_1-coordinate originates in the neutral axis and is directed downwards while x_3-coordinate lies along the beam axis. A PZT layer is placed on top of the beam substrate and is activated in d33-mode when the beam deflects.

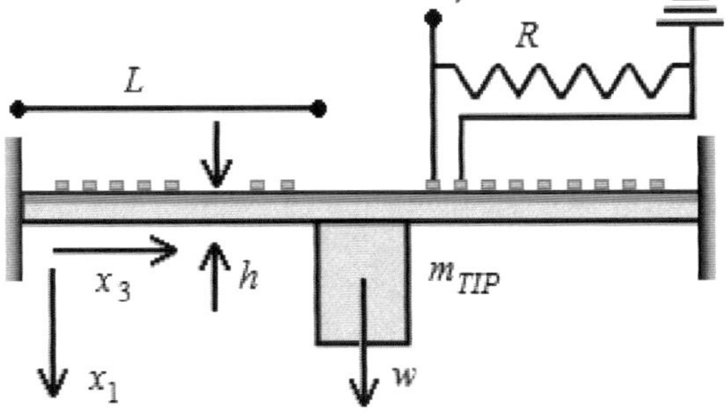

Figure 3. Doubly clamped piezolaminated beam with tip mass, m_{TIP}. L is the half-length and h the total thickness. IDT electrodes span one surface of PZT, and polarization, directed along x_3-axis, inverts from one electrode to the next. The electric field arises parallel to the x_3-axis and the generated charge is collected by the electrodes alternately connected; one series is grounded while the other is attached to an external load resistance (R).

3.1. Dynamic equilibrium equations

The beam's final stack is not homogeneous since different deposited layers are employed. The mechanical response of the layered beam can be obtained by means of a number of theories, examined in [6]. A very thin beam must be designed to activate nonlinear stretching. Therefore Bernoulli hypothesis and CLT is adopted herein, resulting significant improvement by introducing the piezoelectric coupling in PZT layer and the Green-Lagrange nonlinear strain tensor. The lumped dynamic equilibrium equations are then obtained through Rayleigh Ritz method by supposing a cubic displacement and an alternate (between IDT electrodes) electric field. Additional coupling terms proportional to the amplitude arise because of the nonlinear strain as shown below:

$$\begin{aligned} m\ddot{w}+c_M\dot{w}+\left(k_l+k_r\right)w+k_N w^3-\Theta_{\chi v}v+\Theta_{\eta v}wv&=-m\ddot{y}_{ext} \\ k_E\dot{v}+\Theta_{\chi v}\dot{w}-\Theta_{\eta v}w\dot{w}+R^{-1}v&=0 \end{aligned} \tag{3}$$

where m is the total mass; k_l, k_r and k_N are the linear elastic, the residual stress and the nonlinear stiffness; k_E is the internal capacitance of PZT while $\Theta_{\chi v}$ and $\Theta_{\eta v}$ are the linear and nonlinear coupling coefficients. All these coefficients take into account both the integration in the thickness (achieved by the CLT) and along the length (achieved by the Rayleigh Ritz method) [7].

3.2. Solution: Harmonic Balance Method (HBM)

The frequency response of the oscillator to harmonic excitations is studied through HBM. Nonlinear systems do not respond to a monoharmonic signal with a monoharmonic at the same frequency but all harmonics are involved in the response [8]. In this case the amplitude is well described by a single harmonic while two harmonics are required to get the good voltage response since stretching mode has twice the frequency of bending one. Moreover, at large amplitude the linear response (described by the first harmonic) can be neglected and only the second harmonic survives. Neglecting linear coupling terms and substituting in equation (3) the suitable trial solutions the FRF is computed. An implicit expression of the equivalent electrical damping and of the power generation is computed starting from equation (4) and using the definition of power:

$$Y=\left\{\left[1-\Omega_M^2+\left(\frac{3}{4}\alpha+\frac{\kappa_{\eta v}^2\Omega_M^2}{2\left(4\Omega_M^2+\Omega_E^2\right)}\right)Y^2\right]+\mathrm{i}\left[\left(2\zeta_M+\frac{\kappa_{\eta v}^2\Omega_E}{4\left(4\Omega_M^2+\Omega_E^2\right)}Y^2\right)\Omega_M\right]\right\}^{-1} \tag{4}$$

$$c_E=\left(\frac{\kappa_{\eta v}^2\Omega_E\Omega_M}{4\left(4\Omega_M^2+\Omega_E^2\right)}\right)Y^2 \qquad P=\frac{1}{2}c_E\left|\dot{w}\right|^2=m\omega_n^3\Omega_M^2\left(\frac{\kappa_{\eta v}^2\Omega_E}{8\left(4\Omega_M^2+\Omega_E^2\right)}\right)W^4 \tag{5a,b}$$

where $\Omega_E=1/Rk_E\omega_n$ is the dimensionless cut-off frequency of the circuit and $\kappa_{\eta v}=W_0\Theta_{\eta v}/(k_Ek_L)^{1/2}$ the effective piezoelectric nonlinear coupling coefficient.

Figure 6 and Figure 7 plot the solution of Equations (4) and (5b) while Figure 8 shows the normalized electrical damping and the normalized power generation. In short circuit (S.C., $\Omega_E\rightarrow\infty$) and open circuit (O.C., Ω_E=0) conditions, no electrical damping is introduced in the system, and the amplitude is higher while no power is generated.

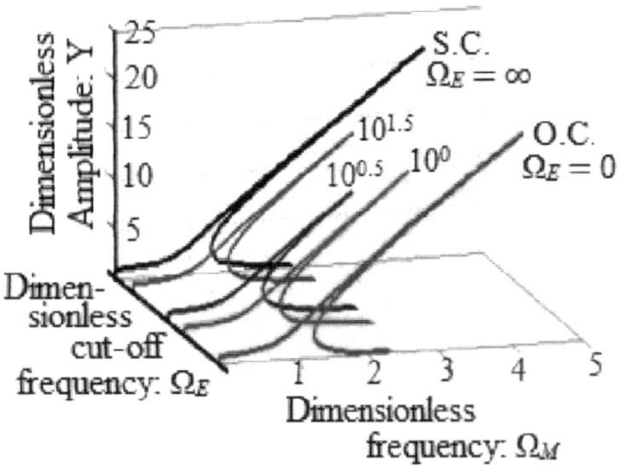

Figure 6. Amplitude response for different values of cut-off frequency

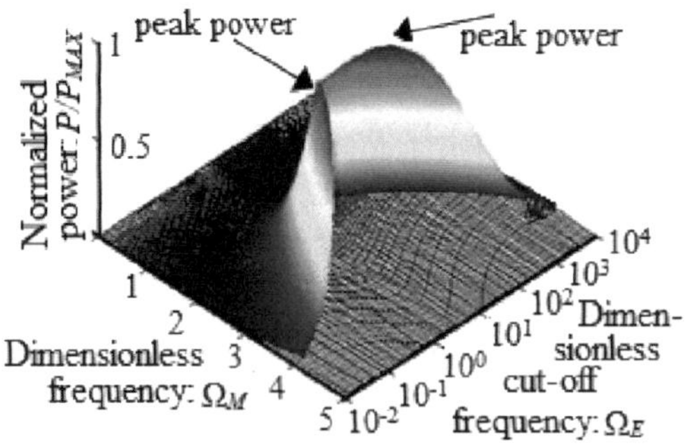

Figure 7. Normalized power generation vs. dimensionless cut-off frequency (Ω_E)

Starting from O.C. condition the damping introduced in the system increases as Ω_E increases, until a maximum is reached (and the maximum amplitude is the lowest). After this point, increasing Ω_E reduces the electrical damping until the S.C. conditions is reached. The peak power is not obtained when the electrical damping is the maximum because in that condition the high damping excessively restrains the movement of the oscillator. A trade-off between high damping (required to produce power) and low damping (required to not stop the motion of the oscillator) has to be found. As the simpler model suggests and figure 8 confirms the optimum is obtained for c_E=c_M.

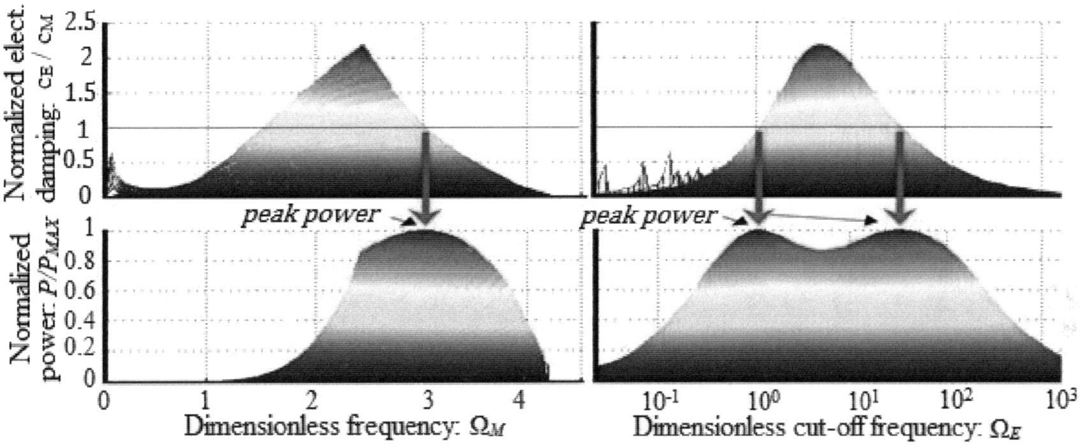

Figure 8. Normalized electrical damping (c_E/c_M: upper) and Normalized power generation (P/P_{MAX}: bottom) vs. dimensionless frequency (Ω_M: left) and dimensionless cut-off frequency (Ω_M: right)

4. Conclusion

A theoretical bounding of the power generation for nonlinear resonating harvesters has been studied through two different models. A simpler model shows that the electrical damping due to piezoelectric coupling should be equal to the mechanical damping for the system to produce the maximum power. This result is the same obtained for linear resonating harvesters. However, the accurate model shows that the near maximum production of power is spread out over a wider bandwidth in the nonlinear harvesters.

5. References

[1] Kim S G, Priya S and Kanno I 2012 Piezoelectric MEMS for energy harvesting *MRS Bull.* **37** pp 1039-50.

[2] Jeon YB, Sood R, Jeong JH and Kim SG 2005 MEMS power generator with transverse mode thin film PZT *Sens. and Act.* A **122** pp 16-22.

[3] Hajati A, and Kim S G 2011 Ultra Wide Bandwidth Piezoelectric Energy Harvesting *App. Phys. Let.*, **99**, P. 083105.

[4] Brennan M J, Kovacic I, Carrella A and Waters TP 2008 On the jump-up and jump-down frequencies of the Duffing oscillator *J. of Sound and Vibration* **318** *(4-5)* pp 1250-61.

[5] du Toit NE, Wardle B, Kim SG, 2005 Design Considerations for MEMS-scale Piezoelectric Mechanical Vibration Energy Harvesters Integrated Ferroelectric **71** pp 121

[6] Ballhause D, D'Ottavio M, Kröplin B, and Carrera E 2005 A unified formulation to assess multilayered theories for piezoelectric plates *Comp. and Struc.* **83** pp 1217-35.

[7] Ardito R, Bertarelli E, Corigliano A, and Gafforelli G 2013 On the application of piezolaminated composites to diaphragm micropumps *Comp. Struc* **99** pp 231-40.

[8] Worden K and Tomlinson G R 2000 *Nonlinearity in structural dynamics: Detection, Identification and Modelling.* (London, UK: Institute of Physics Publishing)

Characterization of dielectric properties of polycrystalline aluminum nitride for high temperature wireless sensor nodes

S Knaust[1], Z Khaji[1], P Sturesson[2], and L Klintberg[1]

[1] Division of Microsystems Technology, Department of Engineering Sciences, Uppsala University, Uppsala, Sweden

[2] National Swedish Defence College, Stockholm, Sweden

E-mail: stefan.knaust@angstrom.uu.se

Abstract. An aluminium nitride (AlN) passive resonance circuit intended for thermally matched high temperature wireless sensor nodes (WSN) was manufactured using thick-film technology. Characterization was done for temperatures up to 900°C in both a hot-chuck for frequencies below 5 MHz, and using wireless readings of resonating circuits at 15 MHz, 59 MHz, and 116 MHz. The substrate for the circuits was sintered polycrystalline AlN. Using a simplified model for the resonators where the main contribution of the frequency-shift was considered to come from a shift of the dielectric constant for these frequencies, the temperature dependency of the dielectric constant for AlN was found to decrease with increasing frequency up to 15 MHz. With an observed frequency shift of 0.04% at 15 MHz, and up to 0.56% at 59 MHz over a temperature range of 900°C, AlN looks as a promising material for integration of resonance circuits directly on the substrate.

1. Introduction

To realize high-temperature wireless sensor nodes (WSN), e. g. for monitoring of components operating at temperatures above 300°C, polymer-based materials fall short because of their low temperature tolerance and large thermal expansion. Hence ceramic alternatives are worth considering. Single-crystalline aluminium nitride (AlN), being a wide band-gap semiconductor, has attracted recent attention as surface acoustic wave resonators used for sensors and energy harvesters [1, 2, 3], currently capable of generating energies up to 2 μW [4]. Although often designed to work at room temperature, AlN-based components could theoretically sustain its piezoelectric properties up to 1150°C [5]. However, for these wide temperature ranges, integration of components into sensor nodes, packaging and communication is still a challenge. Large temperature variations requires minimizing the thermal stresses from coefficient of thermal expansion (CTE) mismatch, and for most materials, the electrical properties will vary extensively with temperature.

To ensure a good CTE matching, a homogeneous system is preferred and to enable multicomponent systems, sintered AlN is a promising substrate material for high temperature applications with a close to perfect CTE match to crystalline AlN ($5.36{*}10^{-6}/K$, 20°C - 1000°C), and also to silicon carbide ($5.18{*}10^{-6}/K$, 20°C - 1000°C) [6]. Although showing rapidly increasing conductivity above 600°C , AlN remains a reasonably good insulator [7], and together with a thermal conductivity (150 W/mK - 180 W/mK) [8] comparable to many metals, high corrosion resistance and good mechanical properties (Young´s modulus of 340 GPa, and flexural strength of 320 MPa) [9], AlN is considered as a suitable substrate material for harsh environments, both at high ambient temperatures and where the temperature is generated from attached components. Compared to the commonly used construction ceramic alumina, AlN has a higher resistance to thermal shock [10].

Although crystalline AlN is well characterized [11, 12], there are gaps to fill when it comes to understanding the properties of the sintered ceramic. In the 1960s, Taylor and Lenie characterized the drift of the dielectric constant for AlN up to 500°C [13], manufactured by hot isostatic pressure (HIP). Their AlN was reported to have 4% of impurities among which O, Si and Fe were present, giving characteristics comparable to alumina. Today, sintered AlN from major manufacturers like Maruwa and Sienna Technologies, the impurities, still about 5%, mainly consists of Y_2O_3. It is therefor of interest to verify the properties of the material.

2. Theory

For resonant circuits, the corresponding resonance frequency is

$$f = \frac{1}{2\pi\sqrt{LC}}, \tag{1}$$

where L is the inductance and C is the total capacitance in the circuit. The inductances of the loops depends primarily on the geometry [14]. Physical dimensions of the loop (the turning width, turning spacing and side size of the loop) are estimated to be linearly dependent on the CTE for AlN [6], giving rise to geometrical variations by approximately 0.5% over 1000°C. The influence of temperature on the inductance is small in comparison to the drift in capacitance, or more specific, the change in dielectric constant of the materials contributing to the capacitance, here dominated by the AlN parallel plate capacitor at the center of the passive antennas. Hence, the frequency will depend on temperature and dielectric constant according to

$$f \propto \frac{1}{\sqrt{\epsilon_{AlN}(T)}}. \tag{2}$$

3. Experimental

To evaluate the influence of temperature to the dielectric constant, three different resonance circuits were manufactured using 1 mm thick polycrystalline AlN substrates, ST-170, Sienna Technologies. Using silver thick-film paste, CN33-145, Ferro, single layers were printed and fired on both sides of the AlN substrate, and manually connected over the edges by a third deposition of silver paste, Figure 1. On the top side, the pattern consisted of a coil with the capacitor patch in the middle, and on the bottom side, the pattern consisted of only the patch for the capacitor.

The AlN in between the two printed parallel plates was used as a dielectric material for the resonating circuit, and the corresponding resonance frequencies were simulated to be close to 14 MHz, 60 MHz and 120 MHz, using Advanced Design System 2011.10. The simulated inductances of the antennas were 6 μH, 2 μH, and 1 μH, respectively.

To evaluate the devices, they were placed on a heating stage made out of an etched Kanthal foil acting as a heater having AlN plates as electrical insulators. The heating stage was placed in an alumina muffle that provided thermal insulation. To minimize the interference from the metal in the heater, the Kanthal foil was patterned in shape of a 2-dimensional torus, having a free area of only AlN below where the passive antennas were placed. The devices were monitored with respect to frequency shifts using a loop antenna outside of the muffle, actively powered through a vector network analyzer FieldFox N9923A, Agilent Technologies. To be able to match the large frequency span of the passive antennas, two different active loop antennas were used.

Figure 1. Fabricated silver-patterned, AlN-based inductively coupled antennas with the sizes of 3 x 3 cm^2 for 15 MHz, 2.5 x 2.5 cm^2 for 59 MHz, and 1.5 x 1.5 cm^2 for 116 MHz. Opposite to the patch in the middle of the chip, a patch is printed on the backside with the same dimensions, serving as a parallel plate capacitor for the circuit.

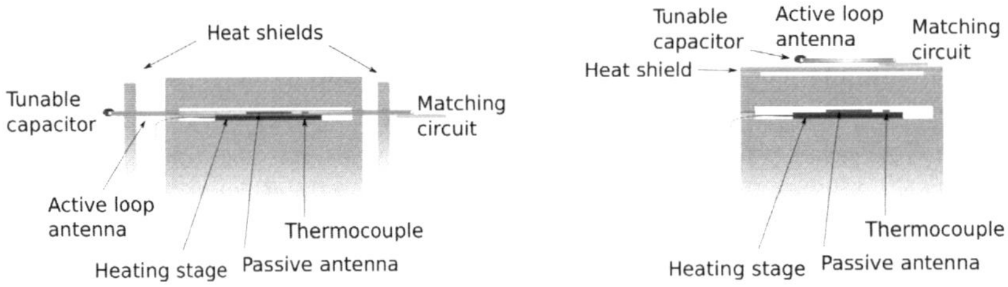

Figure 2. A cross sectional view of the setup. *(left)* The passive antenna is placed on a heating stage next to a thermocouple inside an alumina fixture. The passive antenna is read by an loop antenna consisting of a copper tube in series with a tunable capacitor, protected from thermal radiation by an alumina heat shield. *(right)* The smaller loop antenna was placed on top of the alumina fixture, centered above the passive antenna. An additional alumina sheet was placed in between the two antennas to provide protection from thermal radiation.

The large loop for readings below 20 MHz was placed around the alumina muffle, Figure 2 (left), and the smaller loop for higher frequencies was placed on top of the muffle, with an additional heat shield in between, Figure 2 (right), in order to keep the loops as close to room temperature (RT) as possible.

To account for the different resonance frequencies of the passive antennas, a variable capacitor was used to tune the active antenna. The stage temperature was varied between RT and 900°C using a constant increase of power, and the scattering parameter S11 was logged continuously. Having logged the frequency shifts and simulated the inductances, equation (2) can be used to calculate the changes of the dielectric constant.

As a complementary method for determining the dielectric constant of AlN for frequencies below 5 MHz, a 1 cm^2 plate capacitor manufactured with same material and process, was characterized on a hot- chuck connected to a B1500A, Aglient Technologies, parameter analyzer. The maximum applicable frequency was limited to 5 MHz and the maximum temperature was 600°C. Passive antennas for these low frequencies would have needed to be very large. Hence, no overlapping measurements could be conducted.

4. Results

The resonance frequencies of the passive antennas were measured to 15 MHz, 59 MHz, and 116 MHz, respectively. Figure 3 illustrates the relative frequency shift of the antennas over a

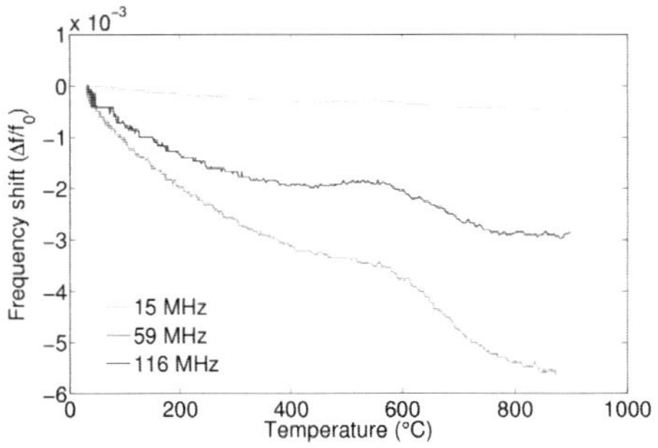

Figure 3. The frequency shift of the S11 signal versus temperature at readings for 15 MHz, 59 MHz, and 116 MHz.

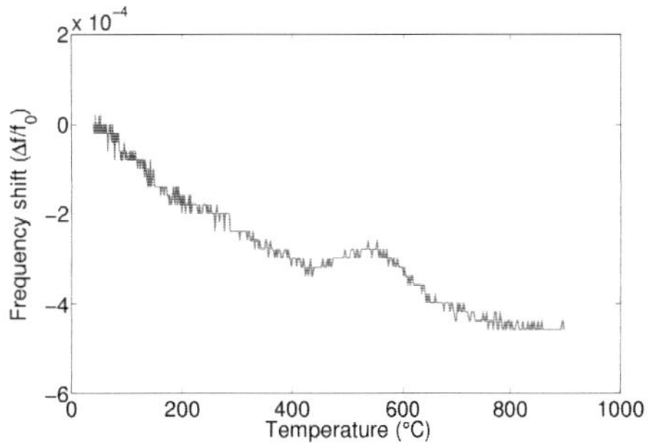

Figure 4. The frequency shift of the 15 MHz antenna versus temperature.

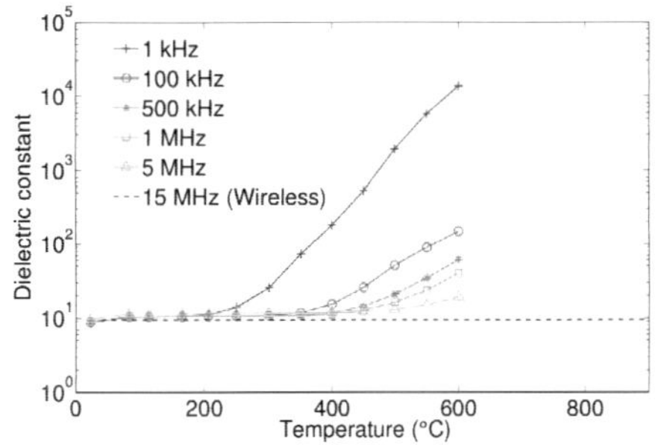

Figure 5. Dielectric constant for sintered AlN, measured in a hot-chuck probe station up to 600°C for frequencies in the range of 1 kHz to 5 MHz, and using wireless reading for 15 MHz. The temperature stability of the dielectric constant is seen to increase with frequency.

temperature range of 900°C. The same pattern for frequency shift was observed for all the three antennas, with decreasing values up to 400°C. Between 400°C and 550°C there is a plateau and above 550°C the frequency shift decreases again with slightly different rate. The 15 MHz reading was found to be most stable with a shift of only 7 kHz (0.04%) from RT to 900°C, Figure 4. This can be compared with an overall shift of 330 kHz (0.56%) and 352 kHz (0.28%) for 59 MHz and 116 MHz, respectively.

The dielectric constants extracted from the hot chuck measurement were found to be relatively constant up to 200°C where they started to increase rapidly, Figure 5. With increasing frequencies, the dielectric constant was observed to be more stable at higher temperatures. For the dielectric constant extracted for 15 MHz, only a small shift from 9.46 to 9.47 was measured, accounting for the thermal expansion of AlN over 900°C.

5. Discussion

Through out all measurements, a plateu is observed between 400°C and 600°C. This is believed to be caused by a sudden increase in conductivity for the AlN because of impurities, as reported by Francis and Worrell [7].

The approximation that the parallel plate capacitor dominates for the circuit is a simplification. For lower frequencies with large capacitance patches in the middle, other capacitive sources will not be significant. But for high frequencies with smaller patches, the total capacitance is much more complex, including the addition of e. g. overlapping tracks and neighboring lines, giving stray capacitances of comparable magnitude, and the simplified model will be invalid. Hence, for the higher frequencies, the dielectric constant was not extracted.

For the higher frequencies, the active loop antenna had to be placed on top of the muffle, close to the heat source because of the lower reading ranges. At the highest temperatures, the temperature sensitive variable capacitor used for tuning was heated up to 50°C , resulting in a large drift in the measurements as seen in Figure 3 for 59 MHz and 116 MHz. For 15 MHz, a more stable variable capacitor was used which also was more successfully cooled. Hence, a more stable measurement was acquired.

6. Conclusion

Having small shifts over a temperature range of 900°C, the approach of integrating resonance circuits directly in an AlN substrate for use in high temperature wireless sensor nodes looks promising, with increasing stability for higher frequencies.

Acknowledgments

The authors would like to acknowledge the Knut and Alice Wallenberg foundation for financing the equipment.

References

[1] Yen T T, Hirasawa T, Wright P K, Pisano A P and Lin L 2011 *J. Micromech. Microeng.* **21**
[2] Tadigadapa S and Mateti K 2009 *Meas. Sci. Technol.* **20**
[3] Elfrink R, Kamel T M, Goedbloed M, Matova S, Hohlfeld D, van Andel Y and van Schaijk R 2009 *J. Micromech. Microeng.* **19**
[4] Saadon S and Sidek O 2011 *Energy Convers. Manage.* **52** 500–504
[5] Damjanovic D 1998 *Curr. Opin. Solid State Mater. Sci* **3** 469–473
[6] Werdecker W and Aldinger F 1984 *IEEE Trans. Compon., Hybrids, Manuf. Technol.* **7** 399 – 404
[7] Francis R W and Worrell W L 1976 *J. Electrochem. Soc.* **123** 430 – 433
[8] Coppola L, Huff D, Wang F, Burgos R and Boroyevich D 2007 Power Electronics Specialists Conference (IEEE) pp 2234 – 2240
[9] Boch P, Glandus J C, Jarrige J, Lecompte J P and Mexmain J 1982 *Ceram. Int.* **8** 34 – 40
[10] Lu T J and Fleck N A 1998 *Acta mater.* **46** 4755 – 4768
[11] Cimalla V, Pezoldt J and Ambacher O 2007 *J. Phys. D: Appl. Phys.* **40** 1–50
[12] Willander M, Friesel M, ul Wahab Q and Straumal B 2006 *J. Mater. Sci. - Mater. Electron.* **17** 1–25
[13] Taylor K M and Lenie C 1960 *J. Electrochem. Soc.* **107** 308–314
[14] Greenhouse H M 1974 *IEEE Trans. Parts, Hybrids, Packag.* **10** 101 – 109

Duty-cycle passive time characterisation for input power and energy storage variation of an energy harvesting tailored wireless sensing system

V Marsic[1], **M Zhu**[2]
[1] Manufacturing and Materials Department, Cranfield University, Bedfordshire, MK43 0AL, UK.
[2] College of Engineering, Mathematics and Physical Sciences, Exeter University, Devon, EX4 4SB, UK.

E-mail: m.zhu@exeter.ac.uk

Abstract. This experimental study investigates the duty-cycle passive times modifications for a low power wireless sensing system (WSS) designed for energy harvesting technology when its input power level and energy storage size are varying. The different low power WSSs presented in the literature feature specific designs aimed at solving particular problems, and due to their specificity their performance indicators are not directly comparable. As a result of this incompatibility, one cannot identify a correlation between the input power, energy storage element size, passive and active time variations to evaluate the potential usability of the system for static or dynamic testing. The present work covers this result comparison gap induced by the incompatibility factor, providing the experimental data obtained as a result of input power level and energy storage size variation for the same low power WSS, thus generating a reference point for the advanced designer and also for the inexperienced user. The experimental results illustrate that, by varying the storage capacity of a low power WSS, its input power range can be enlarged by up to 20 times.

1. Introduction

This study focuses on a low power WSS designed to be autonomous using energy harvesting technology and provides an experimental investigation of its performance when modifying its input power level and energy storage capacity. As the diagram from figure 1 illustrates, the general structure for all applications designed to be powered by an energy harvester (EH) is similar at the block level, the differentiation being made at the component level, [1]:

- EH size, shape, material, energy conversion efficiency and excitation (e.g., light, wind speed, vibration, temperature, etc.) determine the electricity input power level provided to the system.
- Switch A can be a diode rectifying bridge, an alternative current (AC) to direct current (DC) or DC-DC converter, a mechanical or electrical switch designed to convert the EH input power to DC and/or buffer it by boosting, lowering and impedance matching the internal structure of the EH.

[2] To whom any correspondence should be addressed.

- The energy storage element, which can be a capacitor, super-capacitor or a rechargeable battery, can vary in size according to its intended functionality: immediate energy transfer to the WSS when there is enough energy available for one functioning duty-cycle, or long term energy storage in order to facilitate a timed complex operation.
- Switch B can serve the same purposes and functionality as Switch A, plus the capability of keeping the WSS disconnected while the energy storage element is charging, and reconnecting it when enough energy is stored to perform the sensor reading, local data processing and transmitting. The usual circuit used for Switch B is an operational amplifier comparator with hysteresis or a voltage supervisor.
- The wireless sensing system can also vary from the low power radio frequency identification (RFID) architecture transmitting on industrial, scientific and medical (ISM) radio bands, to the ones designed for standard communication protocols like IEEE 802.15.4, ZigBee, Bluetooth, etc.

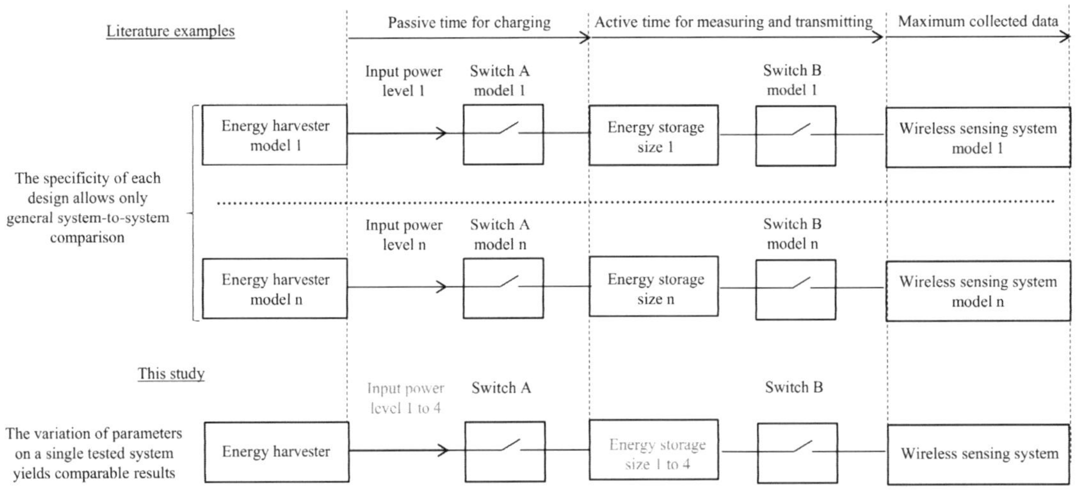

Figure 1. Diagram of the general architecture for a low power WSS powered by an energy harvester showing the advantage of a study that characterises the duty-cycle passive time when parameters like input power and energy storage vary

This study, by applying the already acknowledged methodology from prior investigations [2-4] for measuring the passive times for low power wireless communication modules, determines the trend of WSS characteristics resulted from input power level and energy storage size modification. This trend is hard to be spotted because of the large technology variation used in autonomous WSS implementations. Therefore, the findings of this work give an instant lead to future designers and nevertheless to undecided potential users.

2. WSS description

The WSS supporting this research has been implemented as previously described in [5], its general internal diagram in connection with the piezoelectric vibration EH being showed in figure 2. The vibration EH uses a thin piezoelectric macro fibre composite (MFC) patented by NASA which is bonded to an aeronautic aluminium alloy plate used as substrate. The MFC is an M8528-P2 rectangular model, commercialised by Smart Material Corporation. Its overall dimensions are 60mm x 103mm, active area being 28mm x 85mm. The active area is made from rectangular piezo-ceramic rods embedded in an epoxy matrix resin along with electrodes and polyimide film. By bonding the MFC on an aeronautical grade substrate, the EH successfully demonstrates in laboratory testing its potential usage to harvest the strain energy resulted at low frequencies during the flight of an aircraft. This harvested energy could power the WSS to perform structural health monitoring (SHM), [6].

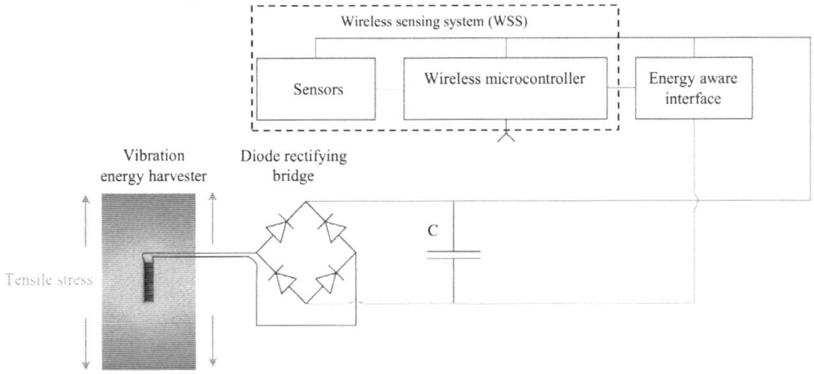

Figure 2. General architectural design for the implemented WSS powered by an EH based on a low frequency vibration MFC attached to an aluminium alloy substrate.

The Switch A, presented in figure 1, is implemented as a full wave diode rectifying bridge, therefore reducing the complexity of the circuit and transforming the AC produced by the vibration EH into DC in order to directly charge the energy storage element.

The energy storage element is implemented as an electrolytic capacitor, whose capacitance will be modified to take the values of 1, 2, 5 and 10mF. The minimum value for the electrolytic capacitor is chosen by design to provide enough energy for the WSS to perform its functions.

The Switch B is represented by the energy aware interface (EAI). The EAI monitors the voltage from the storage capacitor and acts as a voltage supervisor with two voltage thresholds. One of the thresholds, when reached during the charging of the storage capacitor, is used to connect the WSS to its ground terminal by the mediation of a 2N7000 N-MOS transistor. The other threshold value attained during the discharge of the storage capacitor due to WSS power consumption, is used to disconnect the WSS ground terminal by commanding the same N-MOS transistor. The EAI, by disconnecting the WSS, allows the energy to be regenerated in the capacitor, and, by connecting the WSS, allows the energy to be used for sensor readings, local data processing and wireless transmission of information in IEEE 802.15.4 data frame format. The voltage threshold chosen by circuit design for the EAI to connect the WSS is ~3.2V and the second threshold, for disconnecting the WSS, is ~3.08V.

The wireless sensing system is composed of two blocks. One is the wireless microcontroller, which manages the information between the external sensors and the internal wireless 2.4GHz transceiver. The other block is represented by the sensors. The sensors implemented in the WSS were chosen to have low power consumption and a fast initialisation time. In order to acquire a broad range of data useful for future aeronautical SHM implementation, a 3-axis accelerometer, a temperature and a light intensity sensor were included. Figure 3 shows the resulting prototype system. The full-wave rectifying diode bridge built from four fast switching 1N4148 diodes provides DC to the capacitor

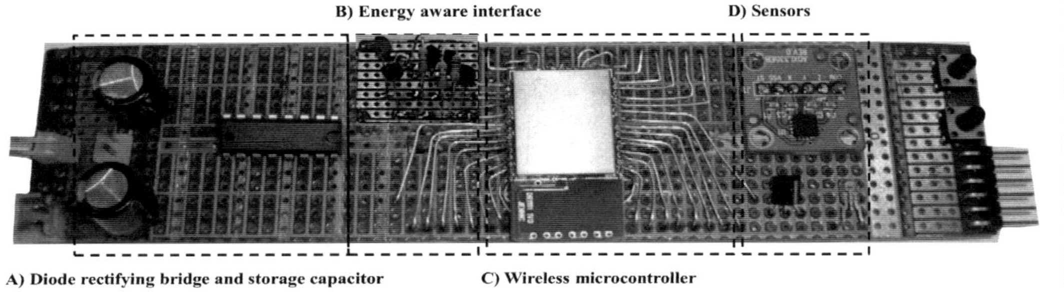

Figure 3. Tested wireless sensing system divided into its four distinct areas: (A) and (B) form the power management block, (C) JN5148 wireless microcontroller linking wirelessly the data collected from the sensor block (D) to the beneficiary user base station.

bank composed of two 1mF electrolytic capacitors. The EAI circuit, present on the back of the illustrated module, connects and disconnects the wireless microcontroller from its ground terminal by constantly monitoring the voltage on the capacitor bank. The ultra-low power consumption wireless microcontroller implemented with JN5148 from Jennic-NXP is IEEE 802.15.4/ZigBee-compliant. The wireless microcontroller manages the three external low power analogic sensors: a temperature sensor MCP9700, a 3-axis acceleration sensor ADXL335 and a light intensity sensor GA1A2S100LY.

3. WSS testing set-up

The data resulted from measuring the WSS passive times, cold start, t_c, and warm start, t_w, are experimentally examined. The diagram set-up illustrated in figure 4 was used for measuring the WSS passive time periods, t_c, and t_w. The Agilent 34410A multimeter monitors the voltage on the storage capacitor C, saving the time and voltage measurement data in a comma separated values (CSV) text file format on the computer it is connected to via USB.

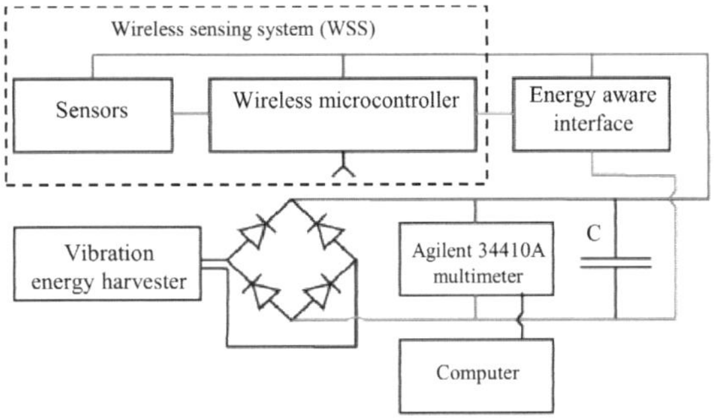

Figure 4. Testing set-up for measuring the passive times, cold and warm start of the wireless sensing system.

4. Results and discussion

The WSS cold start, t_c, is produced once at the beginning of the WSS duty-cycle. With the WSS disconnected by the EAI from the storage capacitor C and neglecting the small amount of current required by the EAI to function, ~1.19μA, t_c represents a classic case of the time required by the capacitor to charge from level zero to a predefined value (in our case ~3.2V). The WSS cold start depends only on the input power value, P_{IN} and the storage capacitor size. Table 1 presents the values obtained for the WSS cold start by varying the storage capacitor size for a 10mW P_{IN}, using the testing set-up from figure 4. The EH variation of P_{IN} is derived from the controlled frequency shift and strain deformation on the MFC based EH aluminium substrate plate produced by the user input data in accordance with real flight vibration scenarios [7]. The corresponding warm start values t_w are shown in the same table because they also vary in accordance with the power input applied to different size capacitors.

The WSS warm start t_w has a double significance:

- The time gap between two consecutive wireless data transmissions.
- The charging time for the storage capacitor C, until reaching the pre-set energy threshold validated by the EAI.

In order to analyse the effect of P_{IN} variation on the WSS warm start, t_w, the MFC based EH was subjected to the same procedure as the one described above for the cold start t_c measurements. The DC power source was used to provide the following input powers: 0.5, 1, 5 and 10mW for the WSS 2mF capacitor. The results are plotted in figure 5.

Table 1. The WSS cold and warm start measured at 10mW input power when transmitting 20 bytes (10 bytes of data payload and 10 bytes for network addressing, error check, etc.) on 2.4GHz using IEEE 802.15.4 and a current consumption of 1.19μA for all passive time duration.

C (mF)	t_c (s)	t_w (s)
1	5	0.3
2	9.6	0.4
5	24.3	0.9
10	43.2	1.8

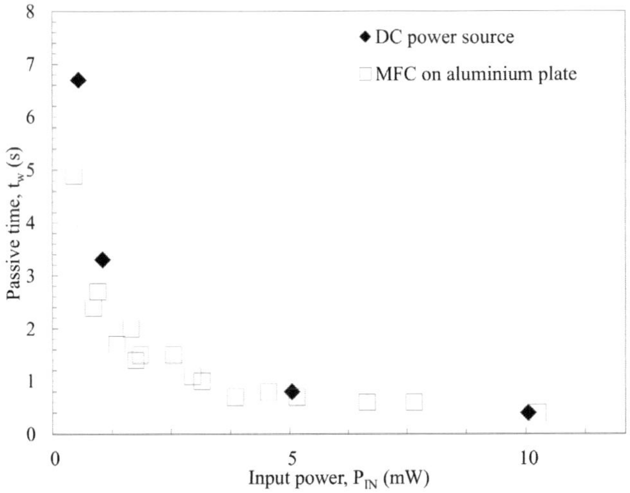

Figure 5. The time gap between two consecutive data transmissions (passive time warm start, t_w) is plotted against the variable input power provided by an MFC based vibration energy harvester (EH), and by a DC power source being used as a plotting reference for 0.5, 1, 5 and 10mW due to the wide power range generated by the EH due to the frequency and stress inputted by the tensile testing machine.

The impedance mismatch between the EH and the storage capacitor is reflected also in the WSS warm start: the vibration harvester with a maximum produced input power P_{IN} of 12.16mW measured with a resistive matching load of 66kΩ, [8], generates a t_w value similar to the one produced by the DC power source with a P_{IN} at 10mW. The values presented in table 1 and figure 5 recorded for the same WSS indicate that a low power WSS is functionally adaptive for a broad range of input powers directly correlated with its storage capacity. Considering that the basic functionality of the WSS involvs an input power of 0.5mW generated by the EH, by varying only the storage capacity the input power can reach 10mW which represents an enlargement by up to 20 times in the WSS functioning range (implying a similar increase in the amount of data that can be transmitted). By increasing the input power and preserving a fixed storage capacity, the passive times t_c and t_w decrease.

5. Conclusion

This study has experimentally investigated the duty-cycle passive times modifications for a low power WSS designed for EH technology, and demonstrated that by varying the storage capacity, the WSS input power range can be enlarged by up to 20 times. Also, the speed of consecutive data transmissions gets higher as the input power level is increased and the storage capacity remains fixed.

Further work will be done to synthesise and present by experimental results the implications of input power and storage capacity variation on the active time, energy consumed during a WSS duty-cycle and maximum data amount transmitted given different testing conditions.

Acknowledgments

The authors acknowledge the support from EPSRC trough project SMARTER EP/K017950/1.

References

[1] CATRENE Working Group on Energy Autonomous Systems 2009 *Energy Autonomous Systems: Future Trends in Devices, Technology, and Systems* 28 (Belgium: IMEC) pp 1-84
[2] Casilari E, Jose M, Cano G and Gonzalo C 2010 *J. Sensors* **10** 5443-5468
[3] Zarrabi H, Al-Khalili A and Savaria Y 2011 *18th IEEE Int. Conf. on El. Circ. and Sys.*727-731
[4] Gomez C, Oller J and Paradells J 2012 *J. Sensors* 11734-11753
[5] Marsic V, Zhu M and Williams S 2012 *J. Telfor* **4** 89 – 94
[6] Marsic V, Giuliano A, Zhu M 2013 *J. Sensors & Transducers* **18** 156-165
[7] Pozzi M, Canziani A, Durazo C and Zhu M 2012 *Proc. of SPIE* **8348** 834832–1–10
[8] Giuliano A, Marsic V, Zhu M 2012 *Proc. IEEE. Int. Conf. iThings* 681-684

Human Motion Energy Harvester for Biometric Data Monitoring

D Hoffmann[1,3], B Folkmer[1] and Y Manoli[1,2]
[1] HSG-IMIT – Institute of Micromachining and Information Technology, Wilhelm-Schickard-Str.10, 78052 Villingen-Schwenningen, Germany
[2] Fritz Huettinger Chair of Microelectronics, Department of Microsystems Engineering – IMTEK, Georges-Koehler-Allee 102, 79110 Freiburg, Germany

E-mail: daniel.hoffmann@hsg-imit.de

Abstract. In this paper we present an energy autonomous sensor system fully integrated into the heel of a shoe for biometric data monitoring. For powering the wireless sensor system a pulse-driven energy harvester was developed, which uses the acceleration-impulses from heel-strike during walking. In preparation of the device development acceleration measurements were carried out. The pulse-driven energy harvester is based on the electromagnetic conversion principle and incorporates a 4x4 coil matrix. A beam fixed at both ends is used for suspending the magnetic circuit. The geometric parameters of coil and magnetic circuit were optimized for maximum power output. For an idealized acceleration pulse with a width of 5 ms and a height of 200 m/s² an average power output of 0.7 mW was generated using a step frequency of 1 Hz. The functionality of the self-sustained sensor system is demonstrated by measuring the temperature and step-frequency of a walking person and transmitting the data to a base station. We also found that the implementation of the suspension can have a significant impact on the harvester performance reducing the power output.

1. Introduction

The idea of energy harvesting from human motion, in particular from the human gait for powering mobile electronic devices goes back to 1996 [1]. Meanwhile, a lot of work has been done around the world in this area and several demonstrators showing the feasibility of power conversion from human walking were presented [2]. However, the demonstration of useful applications is still exceptional.

The energy available in the human gait comes in different forms: acceleration pulses from heel-strike, acceleration during the swing phase of the leg and weight on the ground and shoe during downward movement. In either case, the available acceleration is non-harmonic and occurs periodically at a very low frequency around 1 Hz. Thus, operating conventional vibration based energy converters in resonance will not be feasible. However, acceleration pulses can be used to deflect the seismic mass of a vibration based mechanism out of the equilibrium position resulting in an oscillation with the Eigen-frequency of the system. In this manner, a simple and flat design is achievable, necessary for integration into the sole of a shoe. Moro et al presented such a device, which was a shoe-

[3] To whom any correspondence should be addressed.

a)

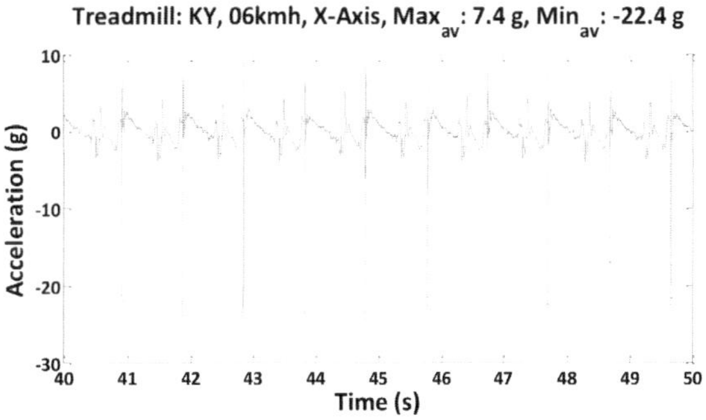

b)

Figure 1. a) Fotograph of a running shoe with mounted acceleration sensor showing the axes-orientation. b) Example of an acceleration record (x-axis) from a treadmill run at a speed of 6 km/h.

mounted piezoelectric cantilever for converting kinetic shock energy from heel-strike during walking [4]. In this work a pulse-driven energy harvester was developed, which is based on the electro-magnetic conversion principle. For the first time a self-sustaining sensor system including the energy harvester device and a wireless sensor system was successfully integrated into the sole of a shoe. The device is excited by acceleration pulses which occur from heel-strike during human locomotion.

2. Design and Modelling

2.1. Excitation conditions

Previous to the device development acceleration measurements were carried out with two different persons on a treadmill at four different motion speeds. Figure 1a shows the used running shoe with the acceleration sensor attached at the outer side of the shoe sole. A triaxial accelerometer was used with the x-axis orthogonal to the ground (figure 1a) for recording the acceleration impulses from heel-strike during motion. An example of an acceleration record is shown in figure 1b. The motion speed was 6 km/h (fast walking). The weight of the person KY was 105 kg. The average of the positive and negative acceleration peaks was 7.4 g and -22.4 g, respectively. In table 1 the peak average is summarized for all measurements carried out.

An acceleration profile of a single heel-strike is shown in figure 2a. The average width of the acceleration pulse is approximately 5 ms. Based on the measurement data idealized pulse profiles with different acceleration peaks were synthesized. An example is shown in figure 2b. Each pulse profile is designed from a half-sine with a pulse width of 5 ms. The pulse height was varied between 100 m/s² and 300 m/s². The pulse profiles were then used in subsequent numerical system simulations and for characterization of the device on a lab-shaker. For optimizing the design parameters of the energy harvester an idealized profile was used as well with a pulse height of 200 m/s² (figure 2b). The reason

Table 1. Average acceleration peaks in x-direction during 120 s treadmill runs. Weight of the two persons: KY: 105 kg, SB: 78 kg

Motion speed		Peak average positive (g)		Peak average negative (g)	
		Person KY	Person SB	Person KY	Person SB
4 km/h	(slow walking)	6.5	4.2	-9.9	-11.5
6 km/h	(fast walking)	9.2	5.6	-21.8	-17.6
8 km/h	(jogging)	8.7	4.5	-27.4	-20.2
10 km/h	(running)	10	7.3	-39.5	-24.2

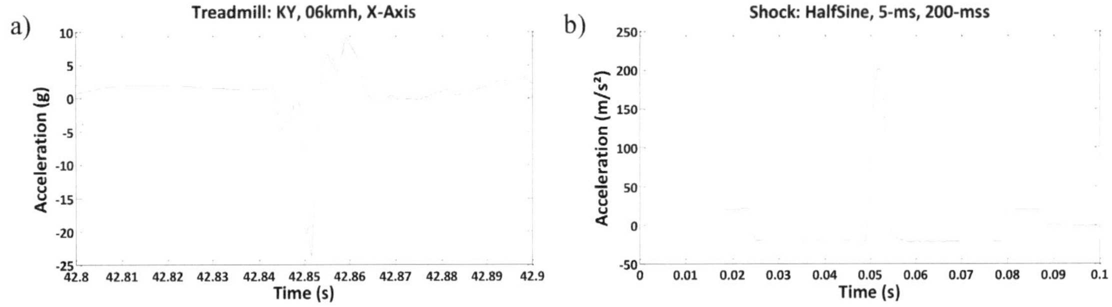

Figure 2. a) Acceleration profile of a single heel-strike. b) Idealized acceleration profile of a single heel-strike: a half-sine pulse with 5 ms pulse-width and 200 m/s² pulse-height.

for using a synthesized pulse profile for the parameter optimization process is the fact that our lab-shaker was not able to replicate the measurement data.

2.2. Design and parametric model

The energy harvester incorporates a 2x2 coil array and two identical magnetic circuits. Each magnetic circuit contains two smaller air gaps (instead of a single, larger one) for placement of the induction coils. In this manner a larger coil volume is achievable while a reduction of magnetic flux density is avoided. A schematic diagram of the energy harvester structure (cross-section of one of the two magnetic circuits) including relevant design parameters is shown in figure 3a. The total width W_{total} is limited by the size of the shoe and must not exceed 38 mm since the total device width is restricted to 40 mm. The gap between coil and magnet was chosen to be 0.5 mm. The width of the coil W_{coil} follows from W_{sheet}, W_{mag} and the gap as shown in table 2. The width of the magnet and the back iron sheet was optimized in order to achieve maximum power output for this particular design. The height of the back iron sheet is limited by the maximum height of the device, which is designed to be 20 mm. The wall thickness of the housing and the inner displacement amplitude are designed to be 2 mm and 3 mm, respectively. As a result, the maximum possible value for H_{sheet} is 10 mm. Consequently, H_{mag} must be 5 mm. The value for L_{mag} and L_{sheet} is equal to H_{sheet}. The outer and inner diameter of the coil is 6 mm and 2 mm, respectively. All relevant parameters are summarized in table 2.

In figure 3b a 2D-simulation of the magnetic field distribution is shown. The homogeneous field in the two gaps is clearly visible and the mean value of the magnetic flux density is about 0.4 T.

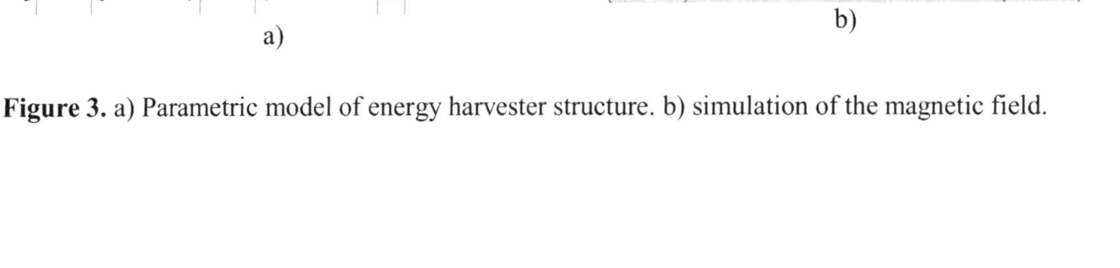

Figure 3. a) Parametric model of energy harvester structure. b) simulation of the magnetic field.

Table 2. Parameters of the energy harvester structure

Parameter	Value (mm)	Remark
W_{total}	38	limited by the available space of the shoe
W_{sheet}	4	optimized for maximum power output
W_{mag}	4	optimized for maximum power output
gap	0.5	fixed value
W_{coil}	3	$= (W_{total} - 3W_{sheet} - 4W_{mag} - 4_{gap}) / 2$
H_{sheet}	10	limited by the available space in the shoe
H_{mag}	5	$= H_{sheet} / 2$
z_{max}	3	internal displacement amplitude

3. Experimental

3.1. Prototype Development

A first prototype implementation of the heel-strike energy harvester is shown in figure 4a. Both magnetic circuits are firmly connected resulting in a single seismic mass, which is suspended using a fixed-end beam structure. The power-management circuit and the wireless transmission electronics are accommodated within the harvester housing. A temperature sensor connected by flexible wires is placed in the shoe close to the foot. The integration of the self-sustaining wireless sensor system is shown in figure 4b.

3.2. Measurements

Measurements of the transient electromechanical response of the pulse-driven energy harvester are shown in figure 5. Voltage and power output were measured using the optimum load resistance of 500 Ohm (see figure 6a). For each heel-strike the output voltage shows the response of a damped system. The instantaneous power output shows peaks up to 50 mW. For an idealized pulse profile with a width of 5 ms and a height of 200 m/s² an average power output of 0.7 mW was generated using a step frequency of 1 Hz.

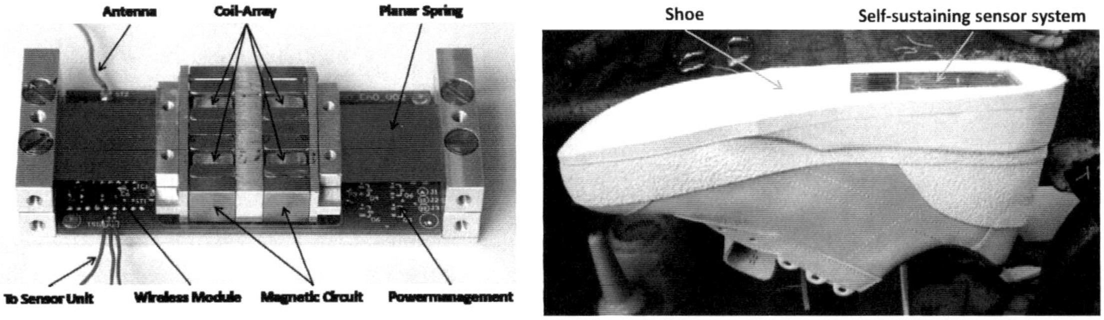

a) b)

Figure 4. a) Prototype of the energy autonomous sensor system including the heel-strike energy harvester, power-management circuit and wireless data communication. b) Energy autonomous shoe incorporating the self-sustaining sensor system.

a)
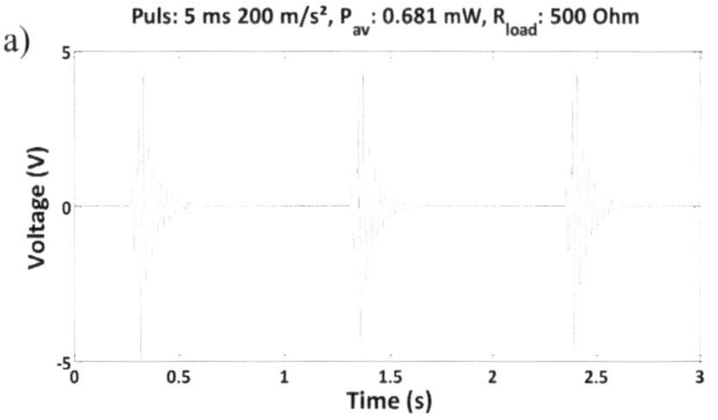

b)
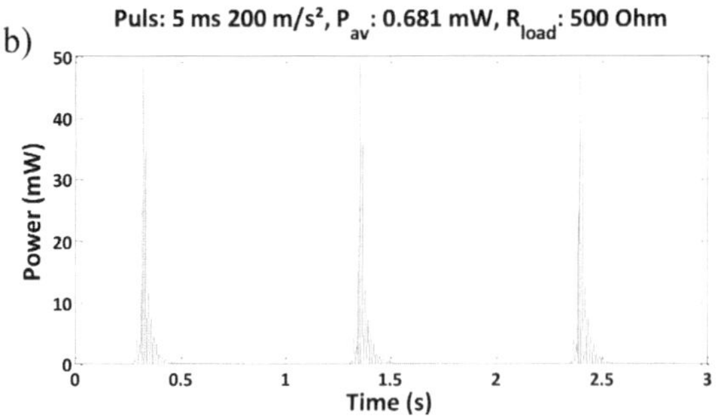

Figure 5. Time-response of the harvester using an idealized pulse profile with an acceleration peak of 200 m/s². a) Output voltage. b) Output power

According to figure 6a the optimum load resistance of the pulse-driven energy harvester is equal to the internal resistance of the induction coil (this measurement was performed using a pulse profile with a peak acceleration of 100 m/s²). Figure 6b shows the power output for pulse profiles with different peak accelerations. The power increases with increasing acceleration. However, due to the fact that the internal harvester structure reaches the limit stops at some point, the power levels off and does not increase further. We also demonstrated that the sensor system was able to transmit one data packet (sensor data) per heel-strike for motion speeds as low as 4 km/h.

a)
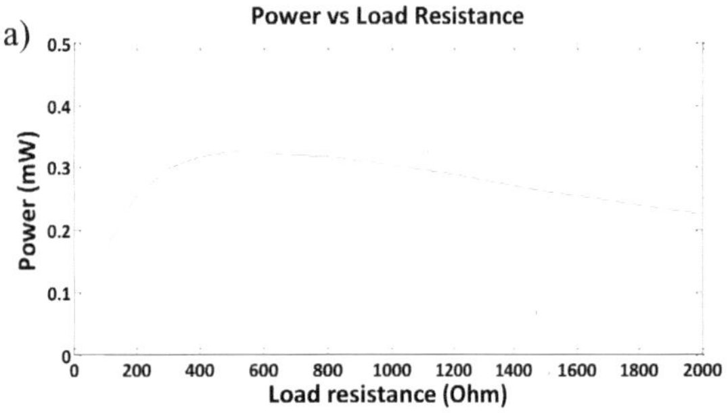

b)
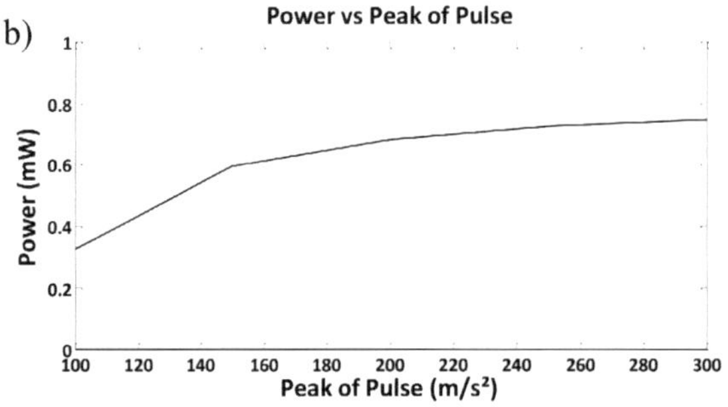

Figure 6. a) Average power versus load resistance. b) Average power versus pulse-height

4. Conclusions

We have demonstrated an energy-autonomous sensor system fully integrated into the sole of a shoe including a pulse-driven electromagnetic energy harvesting device, power-management circuit, sensor unit and wireless data transmission. The energy harvester was able to provide electrical power up to 0.7 mW on average using a step frequency of 1 Hz and a pulse profile with 200 m/s² peak acceleration. We found that the implementation of the suspension had a significant impact on the oscillation characteristic of the harvester structure reducing the power output. A fixed-end beam structure is not a practical solution for suspending the magnetic circuit. Therefore a re-design of the mechanical structure is necessary for improving the power output. Based on first simulation results we expect the power output to increase by a factor of 5 when employing a magnetic levitation principle as a suspension for the magnetic circuit.

References

[1] Starner T 1996 *IBM Systems Journal* **35** 618
[2] Riemer R and Shapiro A 2011 *J. NeuroEngineering and Rehabilitation* **8**
[3] Paradiso J A and Starner T 2005 *IEEE Pervasive Computing* **4** no. 1 18–27
[4] Moro L and Benasciutti D 2010 *Smart Mater. Struct.* **19** 115011

Energy Harvesting from Fluid Flow in Water Pipelines for Smart Metering Applications

D Hoffmann[1,3], A Willmann[1], R Göpfert[1], P Becker[1], B Folkmer[1] and Y Manoli[1,2]
[1] HSG-IMIT – Institute of Micromachining and Information Technology, Wilhelm-Schickard-Str.10, 78052 Villingen-Schwenningen, Germany
[2] Fritz Huettinger Chair of Microelectronics, Department of Microsystems Engineering – IMTEK, Georges-Koehler-Allee 102, 79110 Freiburg, Germany

E-mail: daniel.hoffmann@hsg-imit.de

Abstract. In this paper a rotational, radial-flux energy harvester incorporating a three-phase generation principle is presented for converting energy from water flow in domestic water pipelines. The energy harvester together with a power management circuit and energy storage is used to power a smart metering system installed underground making it independent from external power supplies or depleting batteries. The design of the radial-flux energy harvester is adapted to the housing of a conventional mechanical water flow meter enabling the use of standard components such as housing and impeller. The energy harvester is able to generate up to 720 mW when using a flow rate of 20 l/min (fully opened water tab). A minimum flow rate of 3 l/min is required to get the harvester started. In this case a power output of 2 mW is achievable. By further design optimization of the mechanical structure including the impeller and magnetic circuit the threshold flow rate can be further reduced.

1. Introduction

The application of an automatic remote water meter reading technology requires low-cost, secure and reliable communication solutions being independent from the electricity infrastructure. The wireless M-Bus is one of the possible solutions for water utility companies to address the special requirements of home utilities by providing a technology, which is independent from the electricity infrastructure, in terms both data communication and power supply. In this context a cost-efficient integrated energy harvesting system was developed converting energy from the available water flow in water pipelines. This enables operation of the water meter reading system independently of the main grid and eliminates the need for battery replacement.

Power generation from fluid flow confined in pipelines is not a new concept [1]. In the past, a lot of work has been done in particular in the area of pressurized gas flow. In this case mesoscale turbines with rotational speeds up to 600000 rpm were developed for power generation [2]. However, power generation systems, which can be integrated in conventional water meter devices fulfilling the power requirements of wireless water meter reading solutions based on the wireless M-Bus, are rarely to find. An automatic water meter reading system requires both permanent data acquisition and periodic data

[3] To whom any correspondence should be addressed.

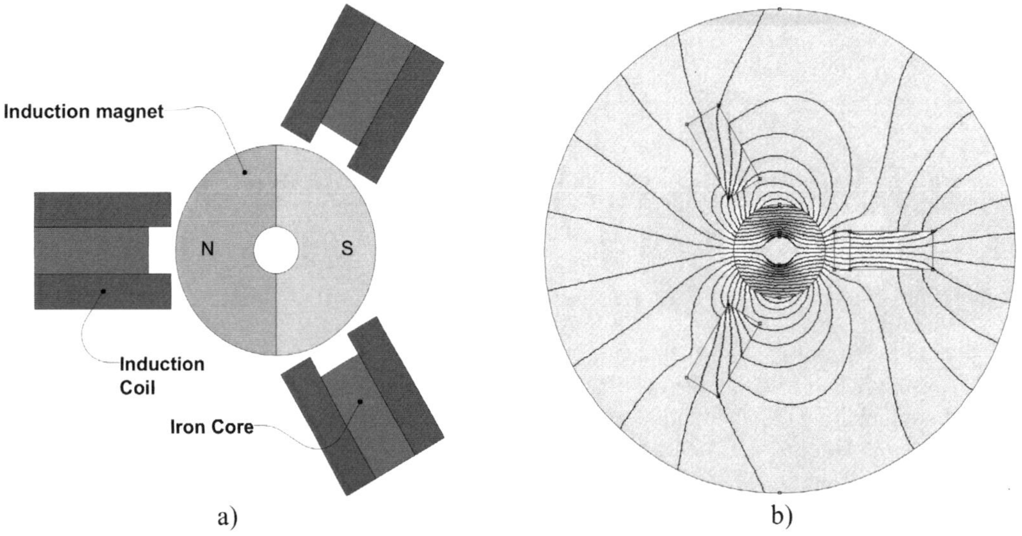

Figure 1. a) Structure of rotational radial-flux energy harvester incorporating a two-pole ring magnet. b) Magnetic field distribution of the energy harvester structure

communication. This should work independently from the energy available at particular times. As a consequence, an energy harvesting system is required, which generates and stores sufficient energy at times of water flow. Therefore, the energy harvester proposed in this work is based on an impeller wheel, which is directly coupled to an electromagnetic energy transducer. When water flow occurs, the impeller starts to rotate driving the transducer.

2. Design and Implementation

The presented rotational, radial-flux energy transducer incorporates a two-pole ring magnet and three induction coils connected in a star circuit (figure 1a). Iron core elements are used for concentrating the magnetic flux in the coils. The two-pole ring magnet is directly mounted on the impeller wheel. This particular three-phase structure was chosen for the following reasons: low-cost design, increasing the active volume with respect to the given design space, reducing cogging torque and ease of integration into conventional water meter housings including the impeller.

Table 1. Parameters of the radial flux energy harvester.

Parameter	Value
Outer diameter of ring magnet	20 mm
Inner diameter of ring magnet	6 mm
Height of ring magnet	8 mm
Outer coil diameter	17 mm
Inner coil diameter	9 mm
Coil length	20 mm
Length of iron core	17 mm
Wire diameter of coil	0.2 mm
Internal resistance of coil	37 Ohm

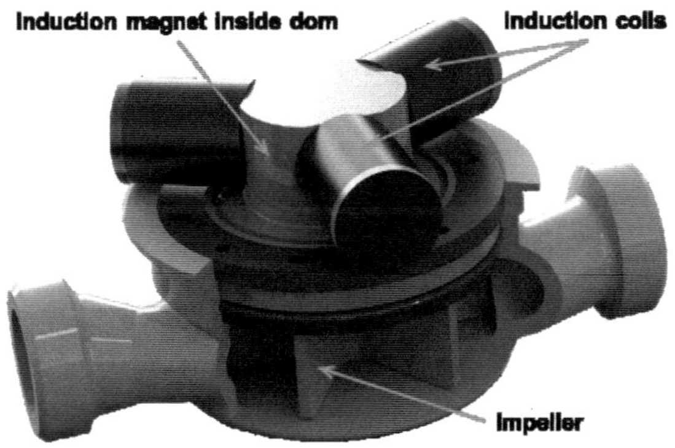

a)

b)

Figure 2. a) Design of rotational energy harvester incorporating a three-phase generation principle. b) Implementation of the energy harvester: power management and battery storage are placed in the top cap.

The complete design of the radial-flux energy harvester is shown in figure 2. The electromagnetic transducer is mounted directly on top of a water meter housing. When the induction magnet starts to rotate, a magnetic flux inversion in the iron cores are generated. This leads to the induction of a three-phase voltage and current in the coils. The induced voltage is then proportional to the rate of change of the magnetic flux density. With respect to a fixed load resistance connected to the coil, the power output increases with the rotating speed of the induction magnet.

The implementation of the flow energy harvester is shown in figure 2b. The power management and the battery storage are accommodated in the cap, which is mounted on top. The complete system was designed with respect to the given requirements (electrical parameters, robustness, integratability, costs).

For optimizing the coil parameters a parametric system model was developed in Matlab based on analytical equations. The magnetic flux density in the coil was numerically calculated using the FEMM tool (figure 1b). When the magnetic axis of the ring magnet and the axis of one of the coils get aligned, the magnetic flux density becomes maximal. In this case the maximum average magnetic flux density in the coil is 0.38 T. The minimum average magnetic flux density is about 8 μT and occurs when the magnetic axis is orthogonal to one of the coils. For the electrical part of the model a circuit including a single coil and a load resistor was considered. Voltage and power was then calculated with

Table 2. Experimental Measurements: the voltage was measured over the optimum load resistance of 70 Ohm

Flow rate (l/min)	Pressure drop (bar)	Voltage (V)	Power (mW)
3	-	0.4	2
5	0.1	1	15
8	0.4	2.3	76
10	0.6	3.2	150
12	0.9	4	228
15	1.4	5.3	400
17	1.8	6.2	546
20	2.2	7.1	720

respect to the load resistor connected to the coil in series. The development of a model configuration in which all three coils are connected by means of a full-wave three-phase rectifier will be part of the future work. The final coil parameters are summarized in table 1.

3. Experimental

For experimental characterisation, the axial-flux energy harvester was connected to a conventional domestic water pipe line (4 bar static pressure). In this set-up the flow rate can be varied between 0 l/min and 20 l/min. During the power measurements the pressure drop over the energy harvester was also measured. The voltage and power output were measured over a load resistor, which is connected to a three-phase rectifier. The optimum load resistance was 70 Ohm (see figure 3a). The measurement data is summarized in table 2. For a flow rate as low as 3 l/min a power output of 2 mW was measured. The corresponding pressure drop was not detectable. With increasing flow rate the voltage output and thus the power output increases accordingly. A maximum power of 720 mW was generated at a flow rate of 20 l/min. At this flow rate a pressure drop of 2.2 bar was measured over the energy harvester.

In figure 3b the power output (simulation and experiment) is shown with respect to a load resistance connected to a single coil. The predictions by the simulation model are in close agreement with the experimental results. The power output increases with the flow rate in a non-linear manner.

4. Conclusions

In this contribution we have presented a low-cost rotational radial-flux energy harvester based on a flow-driven impeller wheel and an electromagnetic energy converter incorporating a three-phase coil circuit. The energy converter was designed for easy integration with commercial water flow meters to harvest energy from water flow in domestic water pipe lines. The harvested energy is used for powering smart water meter systems. The power output ranges from 15 mW at 5 l/min to 720 mW at 20 l/min. At flow rates smaller than 5 l/min the power output becomes very low. In this respect, the design of the power characteristic of the energy harvester becomes an important step in line with further device optimization. Here we aim to achieve a more constant power output, which is supposed

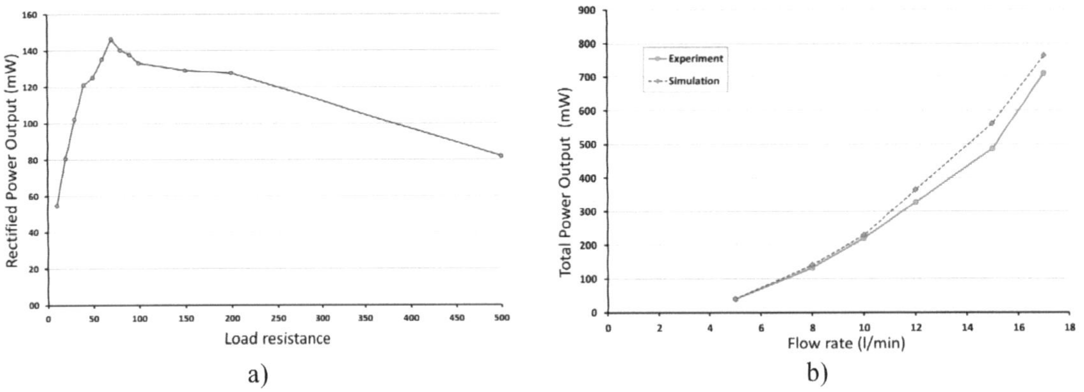

Figure 3. a) Voltage output over optimum load resistance as a function of flow rate (the three coils are connected by means of a full-wave three-phase rectifier; the load resistor is connected to the output of the rectifier. b) Power output over optimum load resistance (40 Ohm) as a function of flow rate (load resistor is connected to a single coil).

to be less dependent on the flow rate. This can be achieved, for example by a fluidic bypass. To accomplish higher power output at lower flow rates the frequency of the system requires to be increased. This can be realized, for example, by utilizing a multi-pole magnet.

The power management including a battery charging circuit as well as the overall concept of an automatic water meter system will be presented in a separate contribution.

Acknowledgments

The authors would like to thank all project partners for the excellent cooperation. The project is co-funded by the European Commission under reference number FP7-SME-2011-286753.

References

[1] Pobering S and Schwesinger N 2008 *Proc. IEEE Sensors* **35** 6858
[2] Krähenbühl D, Zwyssig C, Weser H and Kolar J W 2009 *J. Micromech. Microeng.* **19**

Super-capacitor and Thin Film Battery Hybrid Energy Storage for Energy Harvesting Applications

Wensi Wang, Ningning Wang, Alessandro Vinco, Rashid Siddique, Mike Hayes, Brendan O'Flynn, Cian O'Mathuna

Tyndall National Institute, Dyke Parade, Cork, Ireland

E-mail: `wensi.wang@tydndall.ie`

Abstract. This paper presents the design of hybrid energy storage unit (HESU) for energy harvesting applications using super-capacitor and thin film battery (TFB). The power management circuits of this hybrid energy storage unit are proposed to perform "smart" charge/discharge control in order to optimize the HESU from the perspectives of energy loss due to leakage current and equivalent series resistance (ESR). This paper shows the characterizations of ESUs for energy harvesting powered wireless sensor networks (WSN) applications. A new design of power management circuits is proposed in order to utilize the low ESR characteristics of super-capacitor and the low leakage current characteristics of the TFB in the hybrid energy storage. The average power loss due to leakage current is measured at 38μW in the proposed system. When Compared to the super-capacitor energy storage with the similar capacity, the proposed hybrid energy storage unit reduces the leakage power by approximately 45% whilst maintains a similar (<100 mΩ) ESR.

1. Introduction

In recent years, energy harvesting (EH) techniques have been frequently used for wireless sensor networks (WSN) applications in order to provide a long-term "self-powered" power supply. Within the energy harvesting system, energy storage unit (ESU) is a key component. ESU provides an energy "buffer" to temporarily store the harvested energy for WSN.

The characteristics of ESU have significant impact on the performance of the EH system. Two types of characteristics of ESU are considered in this work. The first type is the characteristics associated to system long-term lifetime, which includes energy density (and its ageing effects), operation lifetime, limit of rechargeable cycles. These parameters determine the "long-term operation lifetime" of ESU in terms of years. In addition to these parameters, the second type of characteristics determine the "efficiency" of ESU. Parameters such as leakage current and equivalent series resistance (ESR) lead to power losses when ESU is charged/discharged. These parameters contribute to the "short-term lifetime" of EH powered WSN, i.e. how long the EH powered WSN will remain active when only utilizing the stored ESU energy.

Super-capacitor and thin film solid state battery (TFB) have been proposed for the applications of energy harvesting [1, 2]. This is mainly due to their "long-term lifetime" characteristics. Super-capacitors have demonstrated lifetime more than 20 years and over one million rechargeable cycles [3]. TFB also have lifetime more than 10 years and over 10 thousands rechargeable cycles [4]. From this perspective, super-capacitor and TFB are more suitable for

EH application than state of the art rechargeable battery, which has <1000 charge cycles and less than 10 years lifetime [3].

However, from the aspect of ESU efficiency, issues exist in both super-capacitor and TFB.

1) Super-capacitor with 1-10F capacitance has a typical leakage current of 0.01-0.1 mA. For example, for Maxwell PC10 2F 5.0V super-capacitor, the average leakage current is measured at 40μA [5] in the first 12 hours of discharge. For applications charge/discharge with 0.1-10A current, e.g. camera flash light, the μA level leakage current is negligible [6]. However, in the application of EH powered WSN, the charge/discharge current is several orders of magnitude lower at 0.1mA level. Table 1. shows the power consumption of a temperature/light monitoring WSN system. In this case of low power WSN module, the leakage current of super-capacitor, 40μA, is more than 2 times higher than WSN module average current consumption, 16μA.

Table 1. The power Consumption of a Tyndall mote at 3.3V (Sleep mode is 300 seconds, i.e. 5 minute measurement interval)

Mode	Symbol.	Power (mW)	Current (mA)	Time (mSec)	Energy (mJ)
Active Mode	P_{Act}	57.1	17.1	87	4.97
Sleep Mode	P_{slp}	0.033	0.01	300,000	9.9
Average	P_{avg}	0.050	0.016	300,087	14.87

2) TFB has a low leakage current (less than 1μA). However, the ESR of the TFB is 50-75Ω in Infinite Power Solutions MEC series TFB [7]. When the active mode current of 17.1mA is discharged from TFB, 14.6mW of conduction loss ($I^2 \times R$, assuming R= 50Ω) can be attributed to the internal resistance. In this low duty cycle WSN example, 5μW power loss can be attributed to ESR. For WSN mote with higher active mode power consumption and longer active mode time, power loss due to TFB ESR is more severe.

Both the super-capacitor leakage power loss, 132μW (40μA $\times$3.3V), and the TFB ESR conduction loss, 5μW (1.6μA$\times$3.3V) contribute to the total power consumption of the WSN system. When the ESU power losses 137μW are compared with the WSN average power consumption, 50μW, the power losses of ESU are higher than the WSN power consumption.

In this work, super-capacitor and TFB hybrid energy storage is proposed with charge/discharge control circuits to reduce the power losses due to high leakage in super-capacitor and high ESR in TFB. The concepts of the hybrid ESU are: the optimized ESU with control circuits operate super-capacitor as the main storage unit and the TFB as the secondary storage unit. The super-capacitor is continuously connected to the load whilst its voltage is limited to within pre-set thresholds. Thin film battery is charged when the super-capacitor voltage excesses the voltage threshold. During high power active mode, the current consumption is mainly dissipated from super-capacitor instead of TFB.

2. Super-capacitor and Thin Film Battery Characterizations

Super-capacitor equivalent circuit model [8] has been used to simulate the leakage current of super-capacitor. In this model shown in Figure 1(a), three RC circuits are used to simulate a super-capacitor. The first RC circuit R_1, C_1 and C_V models immediate behaviour. The second RC circuit R_2, C_2 simulates the delayed behaviour with time constants on the level of 10s of minutes. The leakage current is simulated by the variable resistor R_3.

When the DC source is disconnected, charge re-distribution occurs between the first and second RC circuits. Self discharge occurs on variable leakage resistor R_3. The leakage current is determined by the terminal voltage V_O and the value of R_3. When the terminal voltage of super-capacitor is close to its voltage rating, the leakage current is significant higher than the

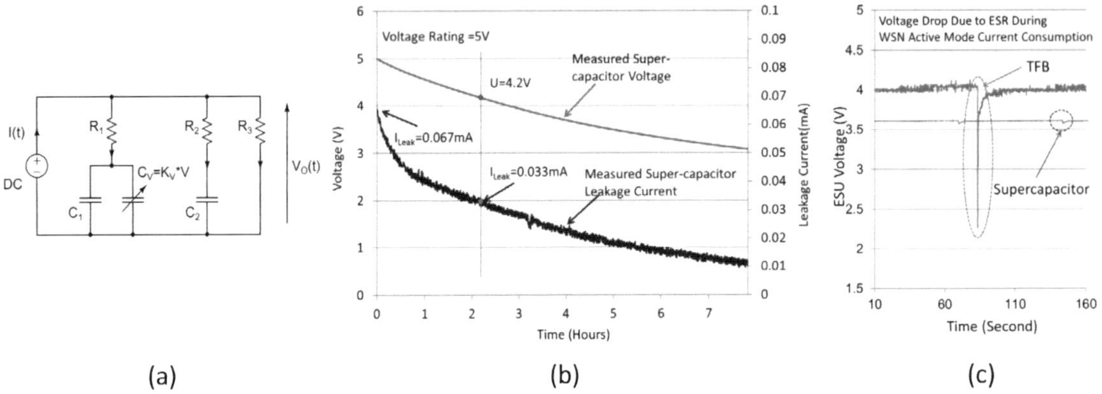

Figure 1. (a) Super-capacitor Equivalent Circuit with Leakage Resistor (b) Maxwell Super-capacitor PC10 self-discharge characterizations (c) Voltage Drop Due to ESR of Thin Film Battery

leakage current when the terminal voltage is 85% of voltage rating. Self discharge experiments have been conducted to investigate the leakage currents of the aforementioned Maxwell PC10 super-capacitor.

Figure 1 (b) shows the results of self discharge experiments conducted on Maxwell PC10 super-capacitor. The leakage current is 0.033mA when the super-capacitor voltage is 4.2V (85% of voltage rating). The leakage current at 100% of the voltage rating 5V is 0.067mA. Thus, by reducing the voltage of the super-capacitor by 15%, the leakage current can be reduced by 50%.

In order to calculate the power loss due to ESR in super-capacitor and TFB, both ESUs are tested with Tyndall WSN module. Both ESUs are connected to Texas Instruments TPS63036 buck-boost switching regulator to generate regulated 3.3V for the WSN module. The test results are shown in Figure 1 (c). Based on the measured voltage drop of 1.75V due to ESR of TFB and the peak current of WSN mote 25.1mA, the ESR of TFB is 70Ω. The power loss due to the ESR of TFB is calculated at 9.3μW,. The super-capacitor ESR is measured at 1.5Ω. The power loss due to ESR of super-capacitor is only 0.18μW, which is 47 times lower than the power loss due to TFB ESR and less than 1% of the WSN power consumption.

Based on these characteristics, if the super-capacitor is connected to the load as the main ESU and supply the active mode current, the power loss due to ESR can be reduced. In practice, this can be achieved by connecting super-capacitor and TFB in parallel. However, the "hard-wired" parallel connection is not suitable, since the TFB has operation voltage range of 3.8V to 4.2V whilst the super-capacitor operation voltage range is 0V to 5V. Over-voltage and under-voltage protection circuits must be included to limit TFB to its operation voltage range.

3. Hybrid Energy Storage Unit and Power Management Circuits

Based on the characterizations of the TFB and super-capacitor ESUs, the hybrid energy storage unit should be designed to 1) reduce the average super-capacitor voltage; 2) discharge the WSN active mode current from super-capacitor; 3) perform over-voltage and under-voltage protection for TFB.

The schematics designed for the hybrid ESU are shown in Figure 2. The super-capacitor is directly connected to the output voltage regulator. Three switches are used to connect/disconnect the TFB with the super-capacitor. These three switches are controlled by three out of four channels of a voltage comparator Maxim Max934. Max934 includes an internal reference which is set to 1.25V in this work. The total current consumption of the control logic circuits is less than $2\mu A$.

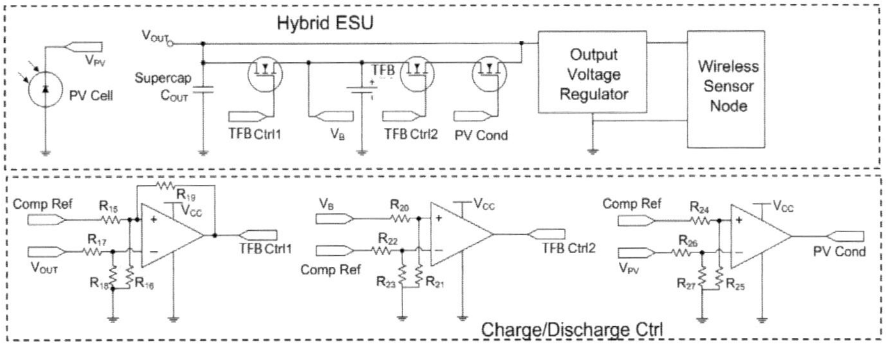

Figure 2. Schematics of hybrid ESU power management circuits

The control circuit limits the hybrid ESU voltage to a pre-set threshold of 4.2V in order to reduce the leakage current. Once the super-capacitor voltage is above the 4.2V threshold, "TFB Ctrl1" switch is turned on, the energy harvester will charge the additional energy into TFB for its low leakage current characteristics.

During the discharging, however, the TFB is always discharged first, and is connected with super-capacitor in parallel in order to minimize the ESR. This discharge only occurs when there is no input solar power. This is facilitated by "PV Cond" switch. The solar cell voltage, V_{PV} is compared with reference voltage. "PV Cond" switch is only switched on when V_{PV} is less than 1V, which indicates low light intensity condition. "TFB Ctrl2" switch is used to perform under-voltage protection of TFB. Once the TFB voltage V_B is lower than 3.8V cut-off voltage, the discharge of TFB is switched off by "TFB Ctrl2" switch.

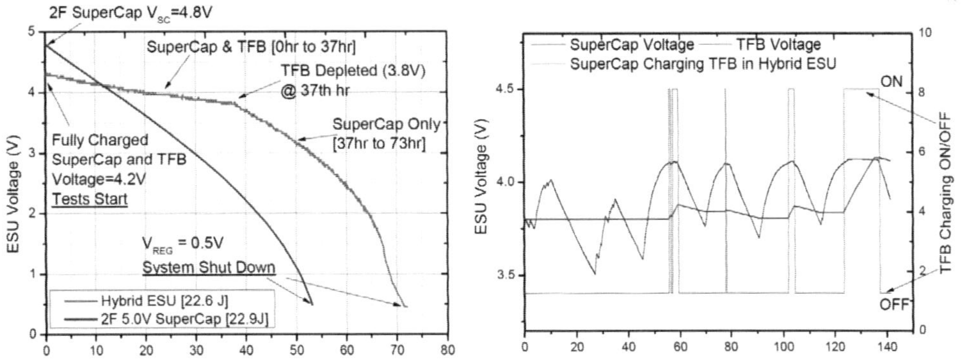

Figure 3. (a) Hybrid energy storage unit discharge experiment with a Tyndall WSN mote. (b) 1 week long charge/discharge test with 0.5mW PV cell and a Tyndall WSN mote. Black: super-capacitor Voltage; Blue: TFB Voltage; Red: super-capacitor charges TFB function ON/OFF

The prototype circuits are connected to a Maxwell 0.47F 5.0V super-capacitor and an Infinite Power Solutions TFB with 1mAh 4.2V capacity. The hybrid ESU prototype has been tested in two experiments shown in Figure 3. The first experiment shown in Figure 3(a) demonstrates the discharge process of the hybrid ESU when the load is the aforementioned Tyndall WSN module. The total operation time is 73 hours. The fully charged hybrid ESU has 22.6J capacity in total, with the super-capacitor and TFB (at 4.2V) store 8.2J and 14.4J energy, respectively. The power consumption of Tyndall WSN module and power loss in the proposed ESU is calculated at 0.088mW based on the 73 hours operation time. The average power consumption of Tyndall WSN module is 0.05mW when it operates with 5 minutes measurement interval (shown in Table 1). Thus, average power loss due to ESU is calculated at 38μW.

A Maxwell 2F 5.0V super-capacitor stores 22.9J energy when it is charged to 4.8V, which is 97% of its voltage rating. Its stored energy is 1.3% more than the 22.6J energy capacity of the proposed hybrid energy storage. The discharge process of this super-capacitor is also shown in Figure 3(a). The operation lifetime is 52 hours. The average leakage power of the stand-alone super-capacitor ESU is calculated at 69μW. The proposed hybrid ESU extends the operation lifetime from 52 hours to 73 hours in this experiment.

One week long charge/discharge experiments have been conducted to test the functionality of the control circuits. The results are shown in Figure 3(b). The super-capacitor charges the TFB at Day 3/4/5 of the experiment when it is charged to the pre-set 4.2V threshold and fully charges the TFB in Day 6. TFB is partially discharged during the night of Day 3/4/5 where no solar energy is available and the TFB voltage is higher than the 3.8V under-voltage protection threshold.

4. Conclusions

This paper presents a super-capacitor and thin film battery hybrid energy storage solution. The characterization of super-capacitor shows that the leakage current can be reduced by more than 50% by decreasing the operation voltage from 100% to 85% of its voltage rating. By connecting the thin film battery and super-capacitor in parallel in order to reduce the equivalent series resistance of the thin film battery, the power loss due to ESR can be reduced from 9.3μW to 0.2μW. By utilizing these two characteristics with the charge/discharge control circuits proposed in this work, the average power loss of ESU is only 38μW. When compared to the stand-alone super-capacitor ESU with same capacity, the power loss is 45% lower. The 22.6J capacity of the hybrid ESU powers a wireless sensor module to perform light and temperature monitoring every 5 minute for 73 hours.

References

[1] Yang H and Zhang Y 2013 *Power Electronics, IEEE Transactions on* **28** 5391–5403 ISSN 0885-8993
[2] Popovic Z, Falkenstein E and Zane R 2013 *Radio and Wireless Symposium (RWS), 2013 IEEE* pp 31–33 ISSN 2164-2958
[3] Ongaro F, Saggini S and Mattavelli P 2012 *Power Electronics, IEEE Transactions on* **27** 3944–3952 ISSN 0885-8993
[4] Abdelkefi A, Alothman A and Hajj M R 2013 *Smart Materials and Structures* **22** 095014
[5] Becker T, Kluge M, Schalk J, Tiplady K, Paget C, Hilleringmann U and Otterpohl T 2009 *Sensors Journal, IEEE* **9** 1589–1595 ISSN 1530-437X
[6] Kehlstadt F, Karcher Y, O'sullivan T and Pedrotta F 2012 Rapidly charging super cap based rechargeable device uS Patent 8,148,954
[7] Anton S, Erturk A and Inman D 2010 *Smart Materials and Structures* **19** 115021
[8] Yang H and Zhang Y 2013 *Power Electronics, IEEE Transactions on* **28** 5391–5403 ISSN 0885-8993
[9] Cisneros A 1971 *Astrophys. Space Sci.* **10** 87
[10] Morse M 1996 *Atomic Molecular and Optical Physics* (*Experimental Methods in the Physical Sciences* vol 29) ed Dunning F B and Hulet R (San Diego: Academic)

High resolution electrochemical micro-capacitors based on oxidized multi-walled carbon nanotubes

T M Dinh, D Pech*, M Brunet, A Achour

Laboratoire d'Analyse et d'Architecture des Systèmes, LAAS-CNRS
7 av du Colonel Roche, 31077 Toulouse Cedex 4, France.

*dpech@laas.fr

Abstract. This study reports the preparation of all-solid-state micro-supercapacitors in planar interdigitated configuration based on MWCNTs via the electrophoretic deposition technique. The carbon nanotubes were functionalized with carboxylic groups via a HNO_3 concentric solution in order to prevent their agglomeration and obtain a stable aqueous suspension, and to add a pseudo-capacitance contribution to the predominant double-layer capacitance. The electrode materials were characterized in a 3 electrode configuration, and the micro-devices in a two-electrode configuration using electrochemical impedance spectroscopy and cyclic voltammetry in 0.5 M sulfuric acid and a gel of PVA-H_3PO_4-H_2O doped by SiWA electrolyte. Spatial resolution down to 10 μm was obtained for the device with specific capacitance up to 1.8 mF.cm^{-2} in electrolyte based PVA and a high power density of 1.28 W.cm^{-2} in 0.5 M H_2SO_4 electrolyte.

1. Introduction

Electrochemical capacitors (also called supercapacitors) store electrical energy using the electrochemical double layer formed on electrode/electrolyte interface (electrochemical double layer capacitance EDLC) or the very fast and reversible electrochemical reaction on the surface of electrode (pseudo-capacitance) [1]. The miniaturization and the development of portable electronic devices such as wireless sensor networks and implantable medical devices, has led to an increasing demand for micro-power sources of appropriate size fulfilling specific requirements such as long lifetime, capability to charge and discharge at high rate and resistance to severe environments (low temperature in particular). On-chip micro-supercapacitors have shown therefore very interesting characteristics when it comes to these applications, because of their extended lifetime and high power density. In the last decade, the realizations of planar interdigitated micro-sized supercapacitors with resolution going from 500 μm to 5 μm were reported using the micro-fabrication process [2-13]. The best values of 10 and 5 μm have been obtained for our recent works with RuO_2 as active material using electrolytic deposition [12, 13]. One other way to obtain a high resolution for micro-supercapacitors is to use the selective electrophoretic deposition (EPD) of micrometric multi-walled carbon nanotubes (MWCNTs) particles onto patterned current collectors.

MWCNTs are well-known for their high specific surface area (can achieve 1315 m^2.g^{-1} for one individual single-walled carbon nanotube [14]), a good electronic conductivity (3300-5000 S.cm^{-2} [15]) which made them attractive electrode materials for powerMEMS devices [16]. The performance of supercapacitors based on MWCNTs is affected by physical properties of carbon electrodes as well as the pore size distribution and the surface area. Apart of the physical structure, the chemical characteristics, such as the chemical stability, the contribution of heteroatoms on carbon surface can also influent the capacitor performance. In this work, oxygen-containing functional groups have been introduced onto the surface of carbon by concentric HNO_3 treatment to obtain a stable aqueous suspension, improve wettability, chemical reactivity and electrical properties. Then, the functional MWCNTs thin film has been produced onto gold current collector by EPD which has the advantages of short formation time, simple apparatus and suitability for mass production. The micro-supercapacitor built from EPD thin film electrodes and different electrolytes such as 0.5 M sulfuric acid and a gel proton conductor composing polyvinyl alcohol (PVA), phosphoric acid doped silicon tungstic acid ($H_4SiW_{12}O_{40}$, SiWA) in distilled water (PVA-H_3PO_4-SiWA-H_2O) offered a high power density and small equivalent series resistance (ESR) [17]. In addition, the gel polymer electrolyte is a potential candidate for encapsulation of the final device.

2. Experimental

Commercial MWCNTs (Nanostructured & Amorphous Materials, Inc., Houston, USA, > 95% purity) were oxidized in 15 $mol.l^{-1}$ HNO_3stirringsolution for 8h at 120°C. The powder was then successively washed with distilled water, rinsed with ethanol and dried at 80°C. The obtained powder was named O-MWCNTs. A stable suspension of 50 mg O-MWCNTs in 100 ml distilled water were prepared by ultrasonication dispersion for 1h. The zeta potential and size distribution of the suspension were measured using Zetasizer Nano ZS (Malvern Instruments, Ltd, UK). X-ray photoelectron spectroscopy (XPS) was used to determine the surface groups on the MWCNTs before and after acid treatment.

Micro-devices were prepared as follow. SiO_2 layer was first thermally oxidized on four inch silicon wafers to provide a good electrical insolation. Different interdigitated electrode series were then realized. 100 nm Ti and 400 nm Au were evaporated onto the substrate and patterned via a lift-off process. After that, the current collectors were protected by an insulating SU8 masking layer. To improve the resolution of the micro-device, a N-LOF photoresist was deposited in the gap between the electrodes. Afterwards, 50 nm of Au layer was deposited on the whole wafer, followed by the electrophoretic deposition of the active material. Both Au and O-MWCNTs layers on the unwanted areas were then lift-off to obtain the final devices.

The electrophoretic deposition was performed between a 2 cm^2 gold plate (negative electrode) and the positive target electrode of 0.25 cm^2electrode and the micro-pattern before the second lift-off step. The DC voltage applied was 50 V and the distance between two electrodes was 0.5 cm. The EPD duration was set to 60s and 30s for 0.25 cm^2 electrodes and micro-device, respectively. The obtained deposit was annealing at 100°C for 1h to remove the remainder water.

The morphology and the thickness of O-MWCNTs thin films were observed using a field emission scanning electron microscopy Hitachi S-4800, Japan. The electrochemical behaviors of the O-MWCNTs electrodes as well as the micro-supercapacitors were analyzed using cyclic voltammetry (CV) and electrochemical impedance spectroscopy (EIS) on a potentiostat/galvanostat (SP 240 from BioLogic). All three-electrode configuration experiments were carried out using Ag/AgCl (KClsat'd) and Pt wire as reference and counter electrodes, respectively. EIS measurements of the cell were made at chosen potential values with 10 mV amplitude and frequencies ranging from 100 kHz to 10 mHz.

3. Results and Discussion

3.1. Oxidation of CNTs

The preparation of the stable dispersion of MWCNTs in a suitable solvent is a necessary prerequisite for successful EPD. The MWCNTs treated in this way were negatively polarized that electrostatically stabilizes the aqueous suspension for long periods (zeta potential of -35.7 mV). The nanoparticle average size obtained is 65.7 nm that is in favour of the EPD process and the realization of supercapacitors with the high resolution.

The oxygen functional groups can be detected by XPS shown in figure 1. The C 1s spectrum is composed of C=C or C-H (~284.55 eV) , C-O (285.90 eV), O=C-O (288.78 eV) functional groups. The results revealed that the MWCNTs have been successfully decorated by hydroxyl (-OH) and carboxyl (-COOH) groups during the concentric HNO_3 treatment at 120°C.

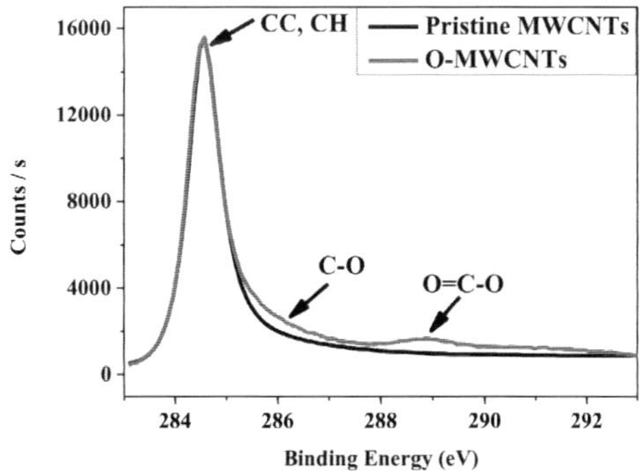

Figure 1.C 1s spectrum for MWCNT and O-MWCNT deposits.

3.2. Electrochemical characterizations of the electrodes.

The CV curve of O-MCNTs (figure 2a) displays a rectangular shape with good symmetry showings a typical electric double layer capacitive behaviour. The reversible peaks at 0.3 and 0.4V *vs.* Ag/AgCl can be attributed to the pseudo-capacitive behaviour of the carboxyl-groups incorporated on the O-MWCNT structure as the results of the nitric acid treatment [18].In an acid electrolyte, the hydrogen adsorption/desorption increases the apparent capacitance:

$$CNTs - COOH + H^+ \leftrightarrow CNTs - COOH//H^+$$

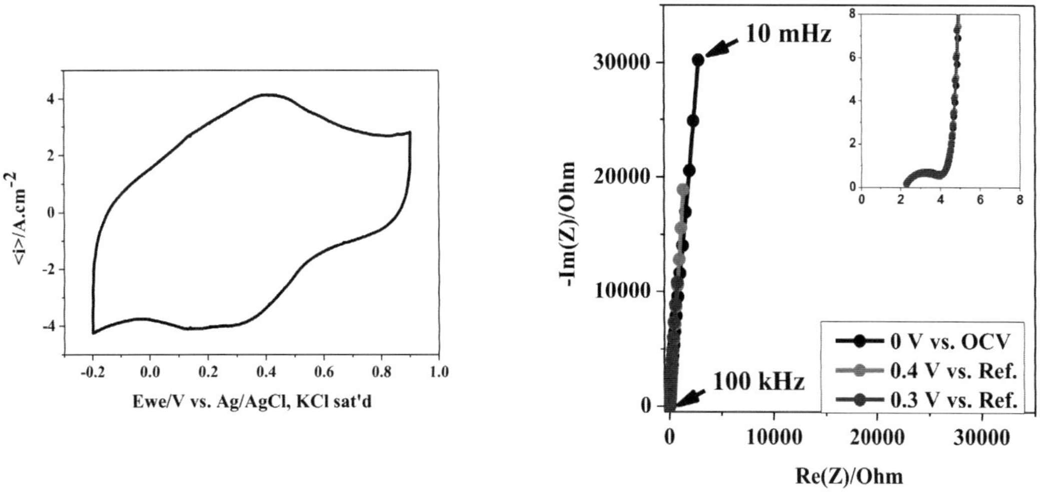

Figure 2. Electrochemical characterizations of O-MWCNT deposits in deaerated 0.5 M H_2SO_4 : a) CV curve of O-MWCNTs coated 0.25 cm^2 gold electrode at the scan rate of 500 $mV.s^{-1}$, b) Nyquist plots of O-MWCNT deposits at 0.3 and 0.4 V *vs.* Ag/AgCl and 0 V *vs.* open circuit voltage (OCV).

The capacitance characteristics of the material were shown in figure 2b with the near straight lines in the low-frequency region which also revealed the absence of the leakage current. The straight line at 0 V *vs.* OCV where no electrochemical reaction occurs is typical of the double layer capacitance behaviour. This capacitance is lower than the pseudo-capacitance observed at 0.3 and 0.4 V *vs.* Ag/AgCl.

3.3. Realization of all-solid-state high resolution micro-supercapacitors.

Figure 3 shows the image of a micro-supercapacitor based on O-MCNTs after the second lift-off with the following dimensions: 100 μm wide, 900 μm long and 10 μm interspace. A homogeneous deposition of O-MWCNTs is observed on the micro-supercapacitor, with a well-defined patterned and no short circuit between the 10 μm interspaced electrodes. The electrophoretic bath does not damage the lift-off photoresist during the EPD process, and high resolution micro-EDLCs were therefore successfully realized. The electrochemical performances of the micro-device was characterized in a conventional deaerated 0.5 M H_2SO_4 aqueous electrolyte and using an all-solid state PVA-H_3PO_4-SiWA polymer electrolyte. The gel electrolyte was obtained by immobilizing the SiWA proton conductor in a polyvinyl alcohol matrix (PVA - H_3PO_4) [17].

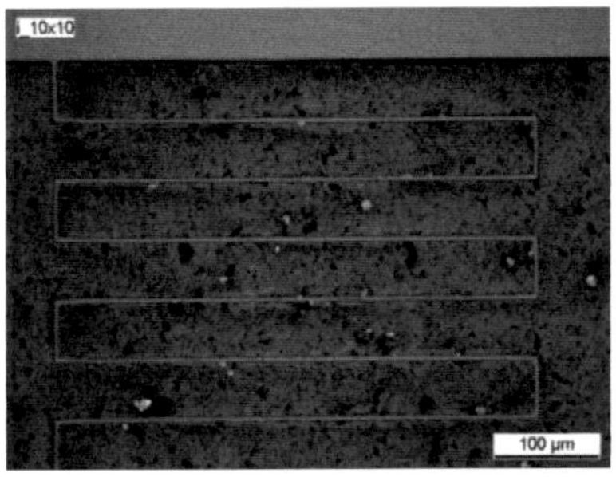

Figure 3. 10 μm interspace O-MWCNT supercapacitor.

The figure 4a presents the Nyquist plots of a 10 μm interspace micro-supercapacitor in these different electrolytes. When the frequencies decrease to 10 mHz, the imaginary part of the impedance increases according to the capacitive behavior of a capacitor. The ESR of the micro-device in sulfuric acid is smaller than polymer electrolyte that related to the ionic conductivity of these electrolytes [12]. These both small values of the ESR show the very high electronic conductivity of the electrode material. The CV curves of 10 μm interspace supercapacitor show a rectangular and symmetric shape for the capacitive behavior. For the same device, the polymer electrolyte provides a higher capacitance. This result can be explained by the oxidation/reduction of adsorbed SiWA electrolyte species next to the oxidation/reduction of –COOH groups [19]:

$$SiW_{12}O_{40}^{4-} + e + H^{+} \leftrightarrow HSiW_{12}O_{40}^{4-}$$

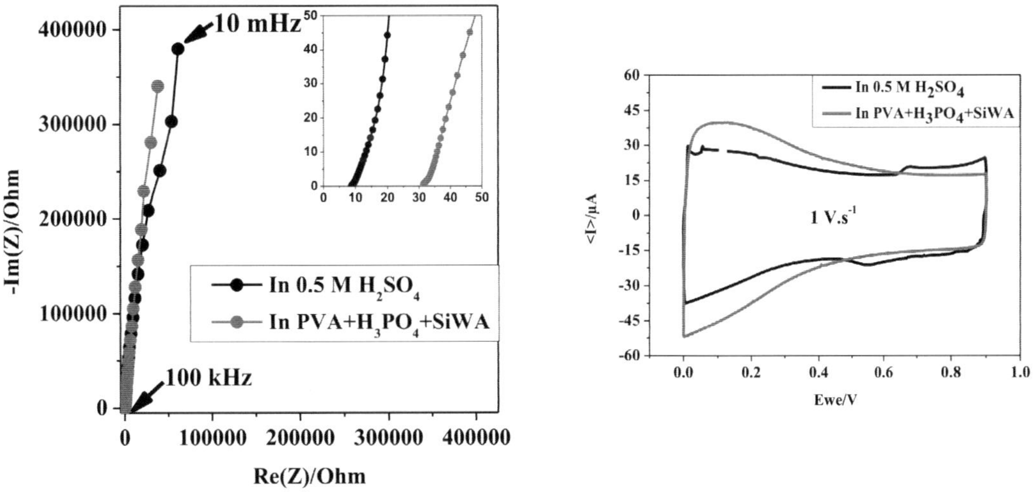

Figure 4. Electrochemical characteristics of a 10 µm interspace supercapacitor.

4. Conclusions

In this work, we have successfully developed an all-solid-state electrochemical micro-capacitor with 10 µm resolution. An energy density of 0.7 mJ.cm^{-2} and a power density of 1.28 W.cm^{-2} were obtained for the micro-cell with the PVA-H_3PO_4-SiWA electrolyte. The polymer electrolyte offered suitability for an encapsulation final device.

References

[1] Simon P and Gogotsi Y 2008 *Nature materials* **7** 845-854

[2] Wang K, Zou W, Quan B, Yu A, Wu H, Jiang P and Wei Z 2011 *Adv. Energy Mater.* **1** 1068-1072

[3] Ajayan P M and Yakobson B I 2011 *Nature Materials* **10(6)** 415-417.

[4] Lin J, Zhang C, Yan Z, Zhu Y, Peng Z, Hauge R H, Natelson D and Tour J M 2013 *Nano Lett.* **13** 72-78

[5] Huang P, Heon M, Pech D, Brunet M, Taberna P L, Gogotsi Y, Lofland S, Hettinger J D and Simon P 2013 *J. Power Sources* **225** 240-244

[6] Pech D, Brunet M, Durou H, Huang P, Mochalin V, Gogotsi Y, Taberna P L and Simon P 2010 *Nature Nanotech.* 5651-654

[7] Durou H, Pech D, Colin D, Simon P, Taberna P L and Brunet M 2012 *Microsyst. Technol.* **18** 467-473

[8] Pech D, Brunet M, Taberna P L, Simon P, Fabre N, Mesnilgrente F, Conédéra V and Durou H 2010 *J. Power Sources* **195** 1266-1269

[9] Shen C, Wang X, Li S, Wang J, Zhang W and Kang F 2013 *J. Power Sources* **234** 302-309

[10] Chen C H, Tsai D S, Chung W H, Lee K Y, Chen Y M and Huang Y S 2012 *J. Power Sources* **205** 510-515

[11] Shen C, Wang X, Zhang W and Kang F 2011 *J. Power Sources* **196** 10465-10471

[12] Dinh T M, Armstrong K, Pech D, Brunet M, Gaudet J and Guay D 2013 *Book of Abstract ISEECap* (Taormina) p 84

[13] Pech D, Brunet M, Dinh T M, Armstrong K, Gaudet J and Guay D 2013 *J. Power Sources* **230** 230-235

[14] Peigney A, Laurent Ch, Flahaut E, Bacsa R R and Rousset A 2001 *Carbon***39** 507-514
[15] Ma R Z 1999 *Materials Research Bulletin***34** 741–747
[16] Inagaki M, Konno H and Tanaike O 2010 *J. Power Sources* **195** 7880-7903
[17] Gao H, Ting Y J, Kherani N P and Lian K 2013 *J. Power Sources***222** 301-304
[18] Hu C G, Wang W L, Wang S X, Zhu W and Li Y 2003 *Diamond and Related Materials***12** 1295-1299
[19] Tian Q and Lian K 2010 *Electrochemical and Solid-State Letters***13** A4-A6

Integrated bistable generator for wideband energy harvesting with optimized synchronous electric charge extraction circuit

Weiqun Liu[1], Adrien Badel[1], Fabien Formosa[1], Yipeng Wu[1], Amen Agbossou[2]

[1]Laboratoire SYMME, Université de Savoie, France
[2]Laboratoire LOCIE, Université de Savoie, France

E-mail: weiqun.liu@univ-savoie.fr

Abstract. Bistable generators have been proposed as potential solutions to the challenge of variable vibration frequencies. In the authors' previous works, a specific BSM (Buckled-Spring-Mass) harvester architecture has been suggested. It presents some properties of interests: simplicity, compactness and wide bandwidth. Using a normalized model of the BSM generator for design and optimization at different scales, this paper presents a new integrated BSM bistable generator design with the OSECE (Optimized Synchronous Electric Charge Extraction) technique which is used for broadband energy harvesting. The experimental results obtained from an initial prototype device show that the BSM generator with the OSECE circuit exhibits better performance for low coupling cases or reverse sweep excitations. This is also confirmed by simulations for the proposed integrated generator. Good applications prospective is expected for the bistable generator with the nonlinear OSECE circuit.

1. Introduction

Autonomous wireless sensors are widely used for structures, environment, earthquake and health monitoring. They bring great facilities and conveniences to a lot of technology fields. To get rid of the battery limitation on the devices' life span and achieve less maintenance cost, vibration, light, heat and so on are now considered as the possible power sources to supply clean and renewable energy. Harvesting the vibration energy with the piezoelectric materials is one of the favorable ways because of the feasibilities of integrating the piezoelectric materials into devices [1].

To make the energy harvester robust, efficient and adaptable for different vibrations environments, one important and desired property is the wide bandwidth. Taking advantage of nonlinearities into the mechanical design is an effective way to increase the generators' operating bandwidth [2-3]. Among the developed nonlinear generators, bistable oscillators show some better performances. Cottone et al. [4] investigated a cantilever beam using a tip magnet which is arranged in face of another fixed magnet. Compared to a linear cantilever generator, a power increase of 400%-600% is obtained. The performance improvements and wide bandwidth property are also observed for chirp excitations [5] and noise excitations [6]. Buckled beams [7][8] or plates [9] are considered as another way to realize bistable oscillators. More recently, a BSM (Buckled-Spring-Mass) harvester is also proposed by the authors [10]. Experimental investigations show that this BSM bistable generator has the desired wideband property and it is simple and compact. Moreover, a developped model is able to well describe the harvester dynamics which simplify the optimization procedure of the design.

Apart from the mechanical design optimization, extraction interface circuits are another critical subject to improve the harvesting efficiency. To enhance the harvesting efficiency and decrease the load dependence, many nonlinear switching techniques are developed [11]. An OSECE (Optimized

Synchronous Electric Charge Extraction) technique has been proposed by Wu et al. [12] recently. In this new circuit, the switching strategy is much simpler while providing higher performance and maintaining the low impedance dependence property for a wide load range.

In this paper, a new integrated BSM generator design is proposed to be used with the OSECE technique. The combination of the bistable generator and the nonlinear harvesting circuit is believed to bring better performance.

2. Design and Analysis

2.1. *Integrated BSM generator*

The miniaturized BSM generator and its assembling platform have been developed and is presented in figure 1. Its unique mono-block design allows the compactness and reliability to be optimized. Four beams are used as flexible hinges which will be demonstrated as critical parts of the device hereafter. An inertial mass is located at the maximum displacement position to improve its effect whereas the piezoelectric elements strain is maximized using an elliptical shape geometry to amplify the force. Two piezoelectric stack elements are inserted under a pre-stress condition. The interdigital electrodes are separately connected to the positive and negative poles.

Before operations, the deformation of the frame is tuned by two dedicated screws on the assembling platform. The device is statically set to a buckled position and the mass then moves from the center position to a new equilibrium position as indicated by the dashed line in figure 1.

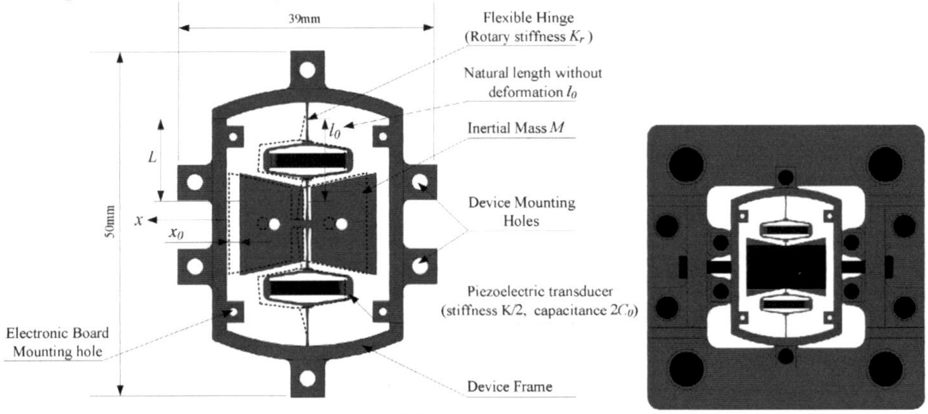

Figure 1. Integrated BSM generator and its assembling platform

According to the analysis in [2], the governing equations of the integrated BSM generator are expressed as:

$$\begin{cases} M\gamma = M\ddot{x} - \dfrac{Kx_0^2}{2L^2}x + \dfrac{K}{2L^2}x^3 + d\dot{x} + \alpha V_p \dfrac{x}{L} \\ I = \dfrac{\alpha}{L}x\dot{x} - C_0\dot{V} \end{cases} \tag{1}$$

where α is the piezoelectric coefficient; d is the damping coefficient while the other parameters are shown in figure 1. The initial buckled position is:

$$x_0 = \sqrt{l_0^2 - L^2 - \frac{8K_r}{K}} = \sqrt{\Delta l \times L - \frac{8K_r}{K}} \tag{2}$$

$\Delta l = l_0 - L$ represents the initial deformation required to be less than allowed maximum deformation of the piezoelectric transducer which limit the available displacement of the generator. It is clear that the flexible hinges play a critical role in the design. If K_r is too high, x_0 becomes negative, meaning that a bistable behaviour cannot be obtained. To satisfy the requirement, four long thin beams in Fig.1 are optimized to ensure low K_r as well as acceptable maximal stress. Adjustable PTFE stoppers have been added for safety purpose. Additional mass can be added to optimize the performances for a given vibration environment.

For a more generic scope of the model, normalization is performed using the following definitions:

$$\varepsilon=\frac{x_0}{L},\ \bar{x}=\frac{x}{L},\ \omega_0=\varepsilon\sqrt{\frac{K}{M}},\ \xi=\frac{d}{2\varepsilon\sqrt{KM}},\ \bar{I}=\frac{I}{\alpha\varepsilon L}=\frac{I}{\alpha x_0},\ \bar{V}=\frac{V_pC_0}{\alpha\varepsilon L}=\frac{V_pC_0}{\alpha x_0},\ k^2=\frac{\alpha^2}{\alpha^2+KC_0},\ \bar{\gamma}=\frac{\gamma}{L} \quad (3)$$

Then equation (1) can be written as:

$$\begin{cases}\bar{\gamma}=\ddot{\bar{x}}-\frac{1}{2}\omega_0^2\bar{x}+\frac{1}{2}\frac{\omega_0^2}{\varepsilon^2}\bar{x}^3+2\xi\omega_0\dot{\bar{x}}+\frac{k^2}{1-k^2}\frac{\omega_0^2\bar{x}}{\varepsilon}\bar{V} \\ \bar{I}=\frac{\bar{x}\dot{\bar{x}}}{\varepsilon}-\dot{\bar{V}}\end{cases} \quad (4)$$

For the BSM bistable harvester architecture, there are four critical dimensionless parameters: buckled level (ε), the electromechanical coupling coefficient of the piezoelectric components (k^2), the damping coefficient (ξ) and the characteristic frequency (ω_0).

2.2. *OSECE technique*

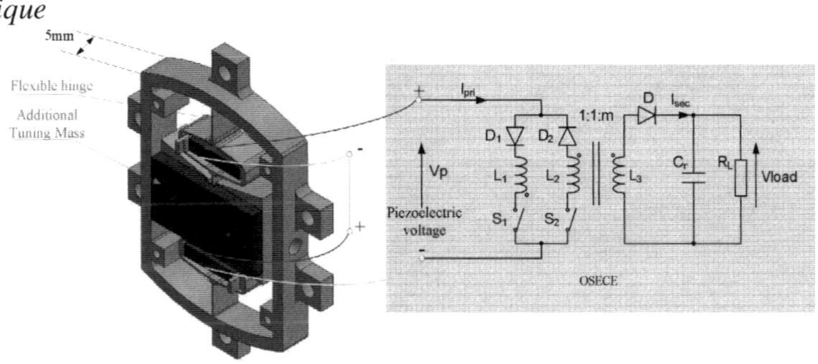

Figure 2. OSECE circuit for the integrated BSM generator

To achieve high harvesting efficiency, the nonlinear OSECE technique is selected to be used with the proposed generator as shown in figure 2. It is based on the SECE circuit with a simpler switching operation strategy. The switch S_1 is closed at the positive maximum of V_p while S_2 is kept open. As V_p arrives at the negative maximum, the status of the switches are inverted (S_1 open, S_1 closed). In this way, one part of the electric energy of the piezoelectric elements is extracted to the load and the rest is preserved at the same time for future extraction. As a result, the energy harvesting efficiency can be enhanced. Compared with the bistable generator with standard circuit, the generator's performance is expected to be improved with OSECE circuit.

3. Experimental and numerical investigations

The principle of the BSM generator was first validated using an initial large scale prototype by experiment and simulation. With the confirmed model, numerical investigations are performed for the proposed integrated miniaturized BSM generator.

3.1. *Principle validation*

Figure 3 shows the initial prototype for the model validation. It is composed of two piezoelectric transducers (APA120S, CEDRAT Technologies©), four PTFE flexible hinges and an inertial mass (45.8g). This device is subject to a chirp excitation of 4m/s^2 (4Hz-44Hz). The average harvested power values over the whole forward sweep and reverse sweep are separately calculated for the two cases of OSECE circuit and standard circuit. To investigate the performances for different coupling levels, an additional serially connected capacitance is used with the piezoelectric components to change the k^2 value. Tests are repeated for several k^2 cases and the average harvested power is plotted in figure 4.

When k^2 is low, the OSECE technique performs better than the standard circuit for both forward and backward sweeps. As k^2 increases, the energy extracted by the OSECE circuit during each switching process is enlarged. The OSECE's damping effect on the generator's motion becomes stronger for the forward sweeps and the harvested power is reduced because of the suppression on the motion. The damping effect exceeds the enhancing effect. Therefore, the standard circuit harvests more energy in the cases of forward sweep and high k^2. The situation is slightly different for the reverse sweeps because the damping effect has less influence on the motion responses. More power is obtained for OSECE circuit unless k^2 is much higher to exert more significant damping. The model is validated considering the good agreement between experiments and simulations.

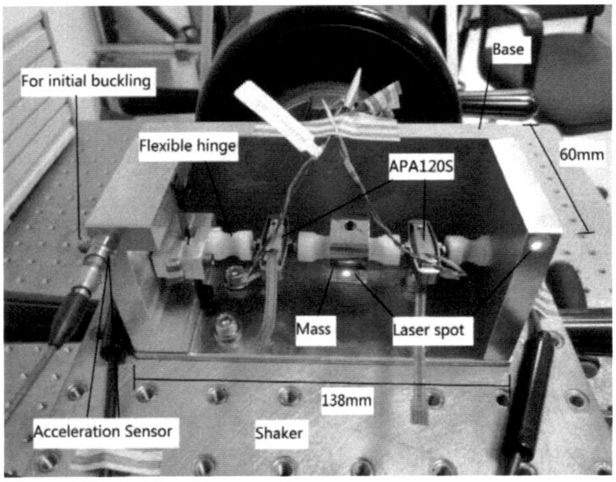

Figure 3. Initial prototype for the model validation.

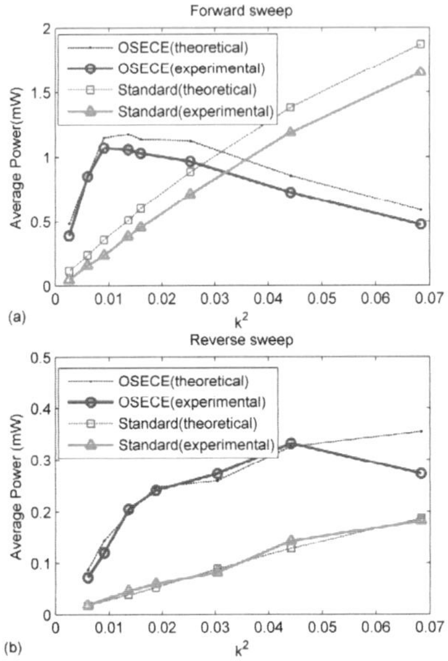

Figure 4. Average harvested power versus k^2 for chirp signal.

3.2. *Numerical investigation for the proposed integrated generator*

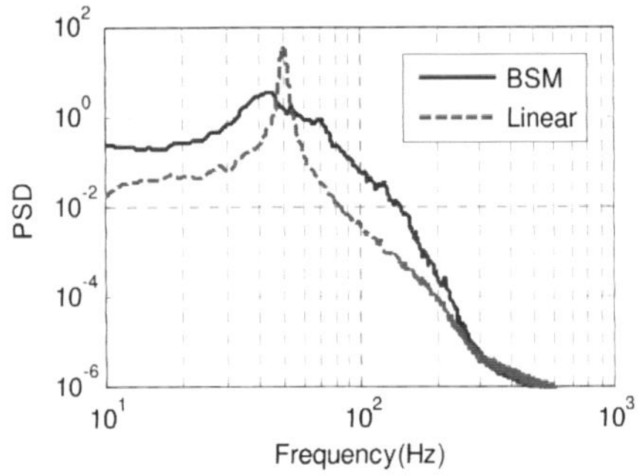

Figure 5. PSD of the piezoelectric voltage for the bistable and linear generators.

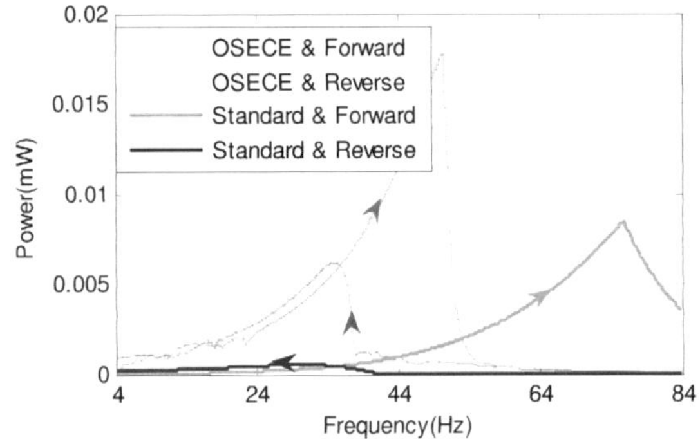

Figure 6. Power response of the integrated BSM generator for OSECE and standard circuit. The arrows represent the frequency sweeping direction.

The performance of the proposed integrated generator can be evaluated by the model. Simulations are done using the parameters listed in Table 1.

Table 1. Simulation Parameters for the integrated BSM generator

M(g)	*L(mm)*	*K(N/m)*	x_0*(mm)*	C_0*(μF)*	*α(N/V)*	*d(N•s/m)*
15	11.17	4e5	0.061	0.125	0.11	0.14

A bandlimited noise excitation (white noise passing by a 10Hz-200Hz Butterworth filter, RMS 5m/s^2) is first used to confirm the wideband property (the piezoelectric component is kept open with no load). The PSD (Power Spectrum Density) of the piezoelectric voltage for the BSM integrated generator and the linear equivalent generator is shown in figure 5. It is clearly demonstrated that the bistable generator has a much wider bandwidth than the linear equivalent one.

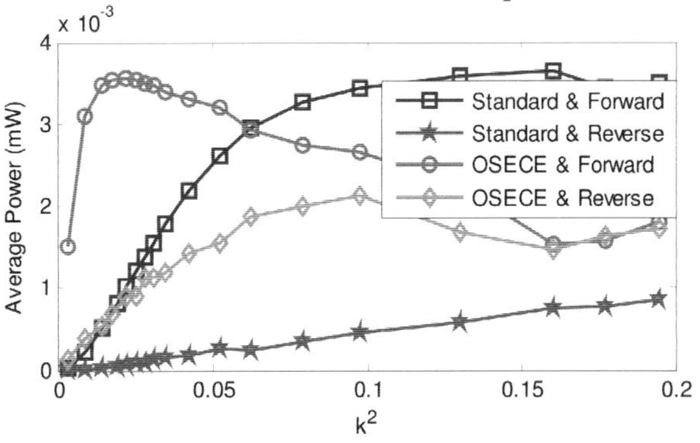

Figure 7. Average harvested power vs k^2 for OSECE and Standard circuit.

The performance of the combination of the integrated generator and the OSECE circuit is then investigated for chirp excitations (10m/s^2, 4Hz-84Hz). As is done for the initial prototype, a serial capacitance is supposed to connect to the piezoelectric element. The power response for $k^2=0.042$ is plotted and compared with the power response for the standard circuit in figure 6. More power is harvested for OSECE over most excitation frequencies. By changing k^2 in the similar way, the harvested power for different k^2 values are simulated and calculated in figure 7. Similar to the results in figure 4, better performance is achieved for low coupling and reverse sweep cases.

4. Conclusion

In this paper, an integrated BSM generator is proposed. It is used with the OSECE circuit to provide efficient and wideband energy harvesting. The principle is first validated with an initial prototype by experiments and simulations then simulations are performed for the integrated BSM generator. The results show that the combination of the BSM generator and the OSECE circuits enhances the harvested power for low coupling cases as well as reverse sweep excitations. The wideband property of the proposed BSM generator is also presented. This research is believed to be useful for the developing of the wideband generator.

References

[1] Beeby S P, Tudor M J and White N M 2006 *Meas. Sci. Technol.* 17(12) R175-R195.
[2] Tang L, Yang Y and Soh C K 2010 *J. Intel. Mat. Syst. Str.* 21 1867-1897.
[3] Harne R L and Wang K W 2013 *Smart Mater. Struct.* 22 023001.
[4] Cottone F, Vocca H and Gammaitoni L 2009 *Physical review letters* 102(8) 080601.
[5] Ferrari M, Ferrari V, Guizzetti M, Andò B, Baglio S and Trigona C 2010 *Sensors and Actuators A: Physical* 162(2) 425–31.
[6] Stanton S C, Mcgehee C C and Mann B P 2010 *Physica D* 239(10) 640-53.
[7] Masana R and Daqaq M F 2011 *J. Sou. Vib.* 330(24) 6036-52.
[8] Cottone F, Gammaitoni L, Vocca H, Ferrari M and Ferrari V 2012 *Smart Mater. Struct.* 21, 035021.
[9] Arrieta A and Hagedorn P 2010 *Applied Physics letter* 97 104102.
[10] Liu W Q, Badel A, Formosa F, Wu Y and Agbossou A 2013 *Smart Mater. Struct.* 22(3), 035013.
[11] Lefeuvre E, Badel A, Richard C, Petit L and Guyomar D 2006 *Sensors and Actuators A: Physical* 126(2) 405-16.
[12] Wu Y, Badel A, Formosa F, Liu W and Agbossou A E 2013 *J. Intel. Mat. Syst. Str.* 24(12) 1445-58

Screen-printed piezoelectric shoe-insole energy harvester using an improved flexible PZT-polymer composites

A Almusallam, R N Torah, D Zhu, M J Tudor and S P Beeby
Electronics and Electrical Engineering Group, University of Southampton, SO17 1BJ, UK

E-mail: asa1g09@ecs.soton.ac.uk

Abstract. This paper reports improved screen-printed piezoelectric composites that can be printed on fabrics or flexible substrates. The materials are flexible and are processed at lower temperature (130°C). One main PZT particle size (2μm) was mixed separately with smaller piezoelectric particles (0.1, 0.3 and 0.8μm) with different weight ratios to investigate the piezoelectric property d_{33}. The blended PZT powder was then mixed with 40% polymer binder and printed on Alumina substrates. The applied poling field, temperature and time were 8MV/m, 160°C and 10min, respectively. The optimum material gives a d_{33} of 36pC/N with particle sizes of 2μm and 0.8μm and mixed percentages of 82% and 18%, respectively. A screen-printed piezoelectric shoe-insoles (PSI) has been developed as a self-powered force mapping sensor. The PSI was simulated, fabricated and tested. ANSYS results show that one element of PSI sole can produce an open- circuit voltage of 3V when a human of average weight of 70kg makes a gait strike. Experimental results show that one element produced 2V which is less than the simulated results because of the reduction of poling field for the practical device.

1. Introduction

Exploiting ambient resources to power electronic systems using energy converters can save money (i.e. changing the battery) and time. Converting kinetic energy (i.e. vibration or moving body) into electricity is one of the techniques that can power such systems. Piezoelectric materials are one of the methods used for kinetic energy harvesters.

Screen-printing piezoelectric materials is well-suited to the mass production of the proposed devices since it is a straightforward fabrication process. High temperature screen-printed piezoelectric materials have been used in kinetic energy harvesting [1]. This paper presents the development of low temperature printable materials printed on flexible substrates and fabrics. Other work on low temperature materials have concentrated on improving the piezoelectric coefficient d_{33} of the material but have not considered the mechanical flexibility [2-4]. Our previous work [5] showed a study of both piezoelectric properties and flexibility and the optimum material, ECS-PolyPZT produced a d_{33} of 27 pC/N. Improving the piezoelectric properties can be achieved by densifying the film using smaller particles from the same piezoelectric material (i.e. PZT). Finding the optimum particle size and weight percentages is necessary for a maximum film density.

Many papers have reported [6-8] shoe-based energy harvesting but they often require complex fabrication process. The motivation of this work is to develop an improved version of ECS-PolyPZT that can be screen-printed onto flexible substrates (especially fabric-based ones) and exploit its promising piezoelectric and mechanical properties in piezoelectric shoe-insoles energy harvesters. The screen-printed materials have to be flexible, processed at low temperature (100-200°C) and exhibit a maximum d_{33}.

2. Improved Piezoelectric Composites

2.1. Formulations and Testing Devices

The proposed materials are mixtures of PZT ceramic powder (Pz29, Ferroperm Piezoceramics) mixed with polymer materials. The powder was supplied with only one particle size 2μm. Then, it was milled (i.e. using attritor mill) into three sizes (0.15, 0.3 and 0.8μm) particles which were used as the filler particles. In the experiment, the 2μm particle powder was mixed with single filler separately (e.g. 2μm PZT powder was just mixed with 0.8μm PZT powder). The blended PZT powder was then mixed with the polymer binder with a 60% overall powder by weight. Table 1 shows the formulations investigated of ECS-PolyPZT inks. The materials were printed in a capacitive structure (CS) on Alumina substrates to test the piezoelectric coefficient d_{33} of the material.

Table 1. ECS-PolyPZT (with 60% PZT) composites investigating formulations.

Particle Size (μm)	Formulation Ref. #	Large Particle Percentage (%)
0. 8	08-1	90
	08-2	82
	08-3	75
	08-4	62
0.3	03-1	98.6
None	02-1	100
0.15	015-1	90
	015-2	82
	015-3	75
	015-3	62

2.2. Poling the devices and d_{33} Measurements

These CS devices were poled with optimum poling conditions used in [5] (Electric field ~10MV/M, poling temperature = 160 °C and poling time = 10 min). Five devices were poled for each formulation. The d_{33} measurements was taken using a PM35 piezometer from PiezoTest. The piezoelectric activity was investigated taking 5 measurements of the d_{33} for each of the 5 samples of each formulation. This gives 25 measurements for each formulation. Figure 1 shows formulation 08-2 provided a d_{33} of 36 pC/N. This is a 25% increase in activity compared to our previous material and is 10% better than PVDF which is a commonly used flexible polymer piezoelectric material. The average d_{33} value for a formulation without fillers is 29 pC/N. The formulations with 0.3 and 0.15μm filler particles gave lower d_{33} values.

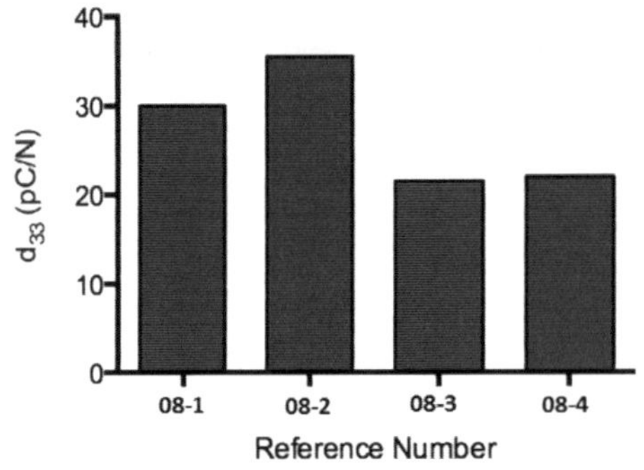

Figure 1. d_{33} values for materials with 0.8μm filler

3. Piezoelectric Shoe-Insole PSI

3.1. Design of PSI

The PSI was designed as shown in figure 2. The PSI consists of two parts, the sole and heel. These parts are considered to be the stress-active parts of the insoles [9]. Therefore, the CS was printed at these locations which reduces the materials used and therefore the overall cost of the PSI device. During walking or running, the heel and sole parts of the insoles are subjected to two types of forces, compressive and bending forces.

Only compressive force is applied to the heel part. However, both compressive and bending forces are subjected to the sole part. At the sole, the device can be used as force-mapping sensor. The sensor can measure the contact force and pressure between the foot and the shoe insole. The sensor is also self-powered by the PSI itself. The sole was divided into 8 elements that sense the force at different locations at the sole part. The number of elements was limited by the number of tracks which increases the non-exploited piezoelectric areas by the top electrode.

Four layers were printed onto these two locations of the PSI as shown in figure 2. The first one is the interface layer which makes the surface of the insole smooth and protects the CS. Then, the CS layers were printed after the interface layer, which are the bottom electrode, piezoelectric material (ECS-PolyPZT) and the top electrode.

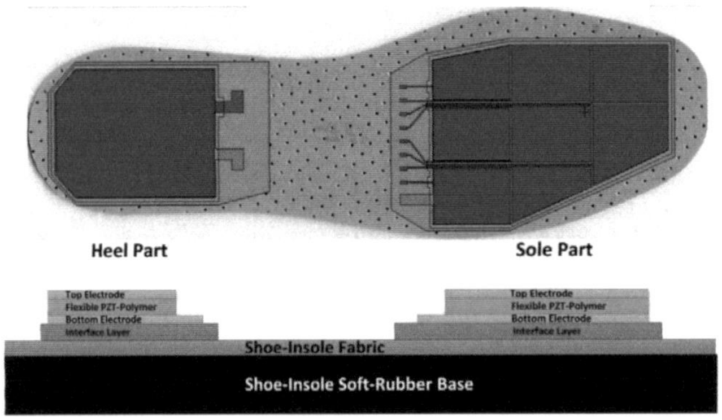

Figure 2. Schematic of top (top) and cross sectional (bottom) views of PSI

3.2. Fabrication of the Device

Screen-printing technique was used to fabricate the PSI device shown in figure 2. The complete printed PSI is shown in figures 3 and 4. Two types of interface layers were used lamination sheets and UV cured, screen-printed Polyurethane UoS-IF#4 supplied by Smart Fabric Inks Ltd [10]. The curing conditions depended upon the polymer. The thermally cured materials were placed in a box oven. The bottom and top electrodes (ELX30) were cured at 125°C for 10min. However, the piezoelectric material ECS-PolyPZT was cured at 130°C for 10min. The average thickness of the printed piezoelectric material is 170μm.

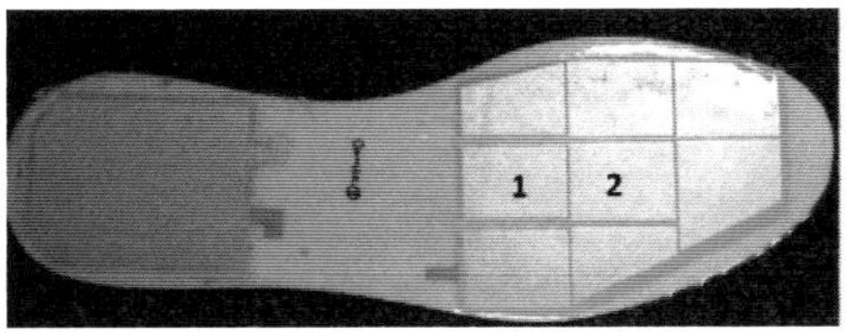

Figure 3. Top view of the printed PSI showing elements 1 and 2

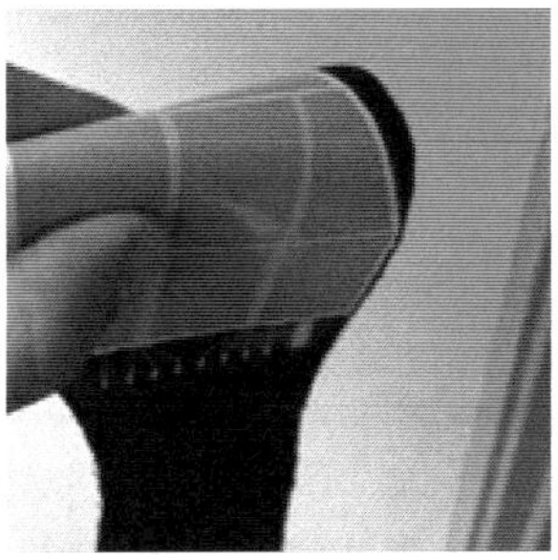

Figure 4. Showing the flexibility of the device

3.3. ANSYS APDL Simulation

The ANSYS results in figure 5 show the output of the sole part of the insole which demonstrates that the PSI can give an open-circuit voltage of 3V when a force strike of average weight of a man 70kg was applied just on element number 1 shown in figure 3 during walking.

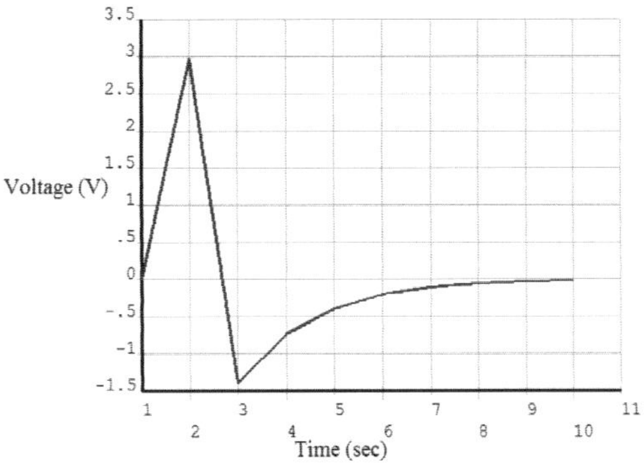

Figure 5. The open-circuit output voltage for one strike for one element of the sole part

3.4. Testing Practical PSI

Testing the practical PSI involved measuring the open-circuit voltage at the sole part of the insole. Typically this material is poled with an applied field of 8MV/m. However, in this case the maximum field was around 4 MV/m as short-circuiting occurred beyond this voltage. The results in figure 6 show that an open-circuit-voltage for one element of the sole part of around 2V. This is less than the simulated voltage because the reduced poling field lowers the piezoelectric properties of the film. This can be solved by increasing the thickness of the piezoelectric material to avoid-short circuiting for later devices. Also, it was found that connecting the sole part elements in parallel can cause the output voltage to change depending upon the magnitude of the applied pressure on each element.

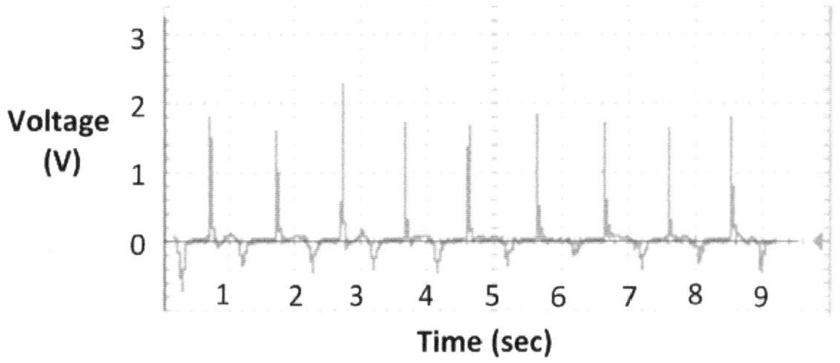

Figure 6. The open-circuit output voltage of one element of normal gait 1strik/sec

4. Conclusion

It was found by varying the filler particle size and particle weight ratios, the d_{33} can be increased. The piezoelectric property d_{33} of a screen-printed flexible piezoelectric composite ECS-PolyPZT were improved to reach 36 pC/N. Also, a new piezoelectric shoe-insole (PSI) was investigated. The sole part of the insole was divided into 8 elements; each element can provide a 2V open-circuit voltage. It was found connecting these elements in parallel will lead to a change in the output voltage due to the different pressure applied to each element.

5. Acknowledgment

The authors thank EPSRC for supporting this research with grant reference EP/IO05323/1.

6. References

[1] Zhu D, Almusallam A, Beeby S P, Tudor J and Harris N R 2010 A Bimorph multi-layer piezoelectric vibration energy harvester *PowerMEMS 2010, December 1-3, Leuven, Belgium*

[2] Papakostas T and white N 2000 Screen printable polymer piezoelectrics *Sensor Review,* vol. 20, pp. 135 - 138

[3] Son Y H, Kweon S Y, Kim S J, Kim Y M, Hong T W and Lee Y G 2007 Fabrication and electrical properties of PZT-PVDF 0–3 type composite film *Integrated Ferroelectrics,* vol. 88, pp. 44–50

[4] Dietze M and Es-Souni M 2008 Structural and functional properties of screen-printed PZT-PVDF-TrFE composites *Sensors and Actuators,* vol. 143, pp. 329-334

[5] Almusallam A, Torah R N, Yang K, Tudor J and Beeby S P 2012 Flexible Low Temperature Piezoelectric Films for Harvesting from Textiles presented at the PowerMEMS 2012, December 3-6, Atalanta, USA, Atalanta, USA

[6] Starner T 1996 Human-powered Wearable Computing *IBM SYSTEMS,* vol. 35

[7] Shenck S N and Paradiso J A 2001 Energy scavenging with shoe-mounted piezoelectrics *IEEE Mic,* vol. 21, pp. 30-42

[8] Yildiz F 2011 Energy harvesting from passive human power *Journal of Applied Science and Engineering Technology,* pp. 5-16

[9] Abdul Razak A, Zayegh A, Begg R K and Wahab Y 2012 Foot Plantar Pressure Measurement System: A Review *Sensor,* vol. 12, pp. 9885-9912.

[10] http://www.fabinks.com/.

Microfabrication of a Polymer Based Bi-Conductive Membrane for a Polymer Electrolyte Membrane Fuel Cell

S Hamel[1], T Tsukamoto[2], S Tanaka[2] and L G Fréchette[1]
[1] Université de Sherbrooke, Sherbrooke J1K 2R1, Québec, Canada
[2] Tohoku University, Sendai 980-8579, Japan

E-mail: Simon.Hamel@usherbrooke.ca

Abstract. This paper reports a novel fabrication process of a high active area ratio bi-conductive membrane for PEMFCs. The fabricated device is a 50μm thick flexible polyimide based membrane that integrates for the first time lateral electrical conductive layers on both sides with a through ionic conductive path. With the use of thermo-conductive rubber as a bonding agent allowing a quick-flip process, five configurations of double-sided multilayer metal sputtering on polyimide were tested. An approach for filling through pores in the membrane with the ionic conductor (Nafion) with a temporary reservoir was also developed. The development of these new processes allowed to fabricate a membrane with 50μm wide holes filled with ionic conductor with double-sided electrical conductive layers.

1. Introduction

Polymer electrolyte fuel cells (PEMFCs) benefit of potential advantages compared to lithium batteries to power small portable devices. In addition to their ability to deliver simultaneously high energy and power densities, PEMFCs provide power as long as they are supplied by reactants. At small scales however, packaging and peripheral components tend to dominate the size, being significantly larger than the fuel cell stack itself.

Recently, some efforts have been made to replace the bulky bipolar plates of the fuel cell system using microfabrication approaches to pattern silicon or polymer substrates [1-5]. While this approach reduces substantially the volume of the system, swelling of the membrane induces stresses on surrounding materials or generate delamination of the catalyst membrane. To bypass these issues, Esquivel et al. integrated directly ionic conductive material in the microfabrication process [6], but hot pressing assembly steps were required. To remove assembly steps, current collectors can be integrated directly to the ionic conductive membrane with a support material. Photosensitive glass and polymer [7-9] have been successfully used as a substrate but the active area in these designs covered only very small portion of total surface of the device. Large patterns etched in SOI wafers filled with Nafion and with peripheral sputtered gold used as current collectors were used by Zhu et al. [10] to obtained reasonable power density. In this case, reducing the pattern size could reduce electrical and ionic losses significantly.

To contribute to the effort of minimizing low power PEMFCs, a novel fabrication process of a flexible polyimide based bi-conductive membrane that integrates for the first time a lateral electrical conductive layers on both sides with a through ionic conductive path is presented (Figure 1). The device aims to integrate, on the same polymer sheet ; mechanical support, ionic conduction and current collection while keeping low ionic and electrical resistance and high active area ratio. The proposed membrane consists of a square 1cm^2 and 50µm thick patterned polyimide sheet (Kapton 200HN, Dupont®) between two electrically conductive 300nm thick gold layers. Holes in the polymer layer are plasma etched and filled with an ionic conductive polymer, Nafion (Dupont®). Dimensions have been selected in order to reduce ionic and electrical resistivity while easing fabrication. This design brings down the total thickness of the cell from millimetres to tens of microns. On the other hand, this important gain in volume brings up important fabrication challenges concerning metallic deposition on polyimide and Nafion filling of the holes.

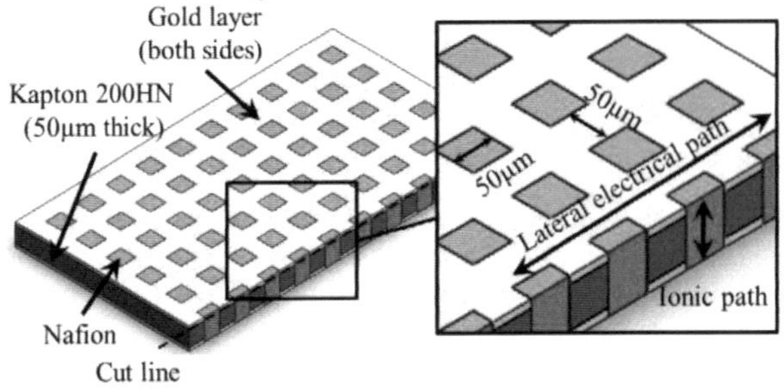

Figure 1 - Isometric cut view of the polymer-based bi-conductive membrane

In this work, we report the fabrication techniques that have been used to achieve the desired configuration. Double-sided multilayer metal sputtering on polyimide and Nafion filling are discussed in more depth.

2. Fabrication process & challenges

The fabrication process of the bi-conductive membrane is shown in figure 2. It consists of three critical steps ; metal sputtering, metal and polyimide etching, and Nafion filling.

2.1. Metal double-sided multilayer sputtering on polyimide

The process starts with a 400µm thick silicon wafer that is use as a mechanical support for the subsequent steps. A thermo-conductive adhesive rubber (λgel COH-4000 by Taica®) is laid on the silicon substrate (Figure 2a) and a Kapton 200HN 50µm thick sheet is fixed on top of the gel with a bearing roller to eliminate trapped air bubbles (Figure 2b). Gold (300 nm) and then titanium (100 nm) are sputtered on the Kapton top surface (Figure 2c). The gold layer acts as a current collector in the final device while the Ti layer is used as an etching mask for the plasma etching process. After the first sputtering step, the metalized Kapton sheet is peeled off of the λgel with tweezers (Figure 2d). The sheet is turned and bonded again on the same λgel layer (Figure 2e). The sputtering process is then repeated on the second side with the same conditions as before (Figure 2f).

A variety of techniques has been developed for a single side metallic layers deposition on polyimide [11, 12], however due to the low thermal conductivity and flexibility of the polyimide film, thermal management and fixing issues appear during the double-sided deposition process. Also, Au/Ti combination has been chosen in the final device, but other configurations have been tested as conductive/protective layers.

2.1.1. Fixing polyimide on a solid substrate and thermal management

The developed quick-flip approach using λgel lowers thermal resistance between the sample and the substrate compared to tape or photoresist bonding. The risk of sample overheating leading to damaged polyimide or alloy formation between the metal layers is thereby considerably reduced. Indeed, all samples that have been set with tape or photoresist were damage by heat generated by the sputtering process. Moreover, the bonding force between λgel and polyimide is strong enough to withstand a sputtering or evaporation process but the layers can be easily peeled and bond again on the other side for another deposition without cleaning since the λgel doesn't leave any residue on the Kapton sheet. This quick-flip process reduces greatly the handling time between two depositions.

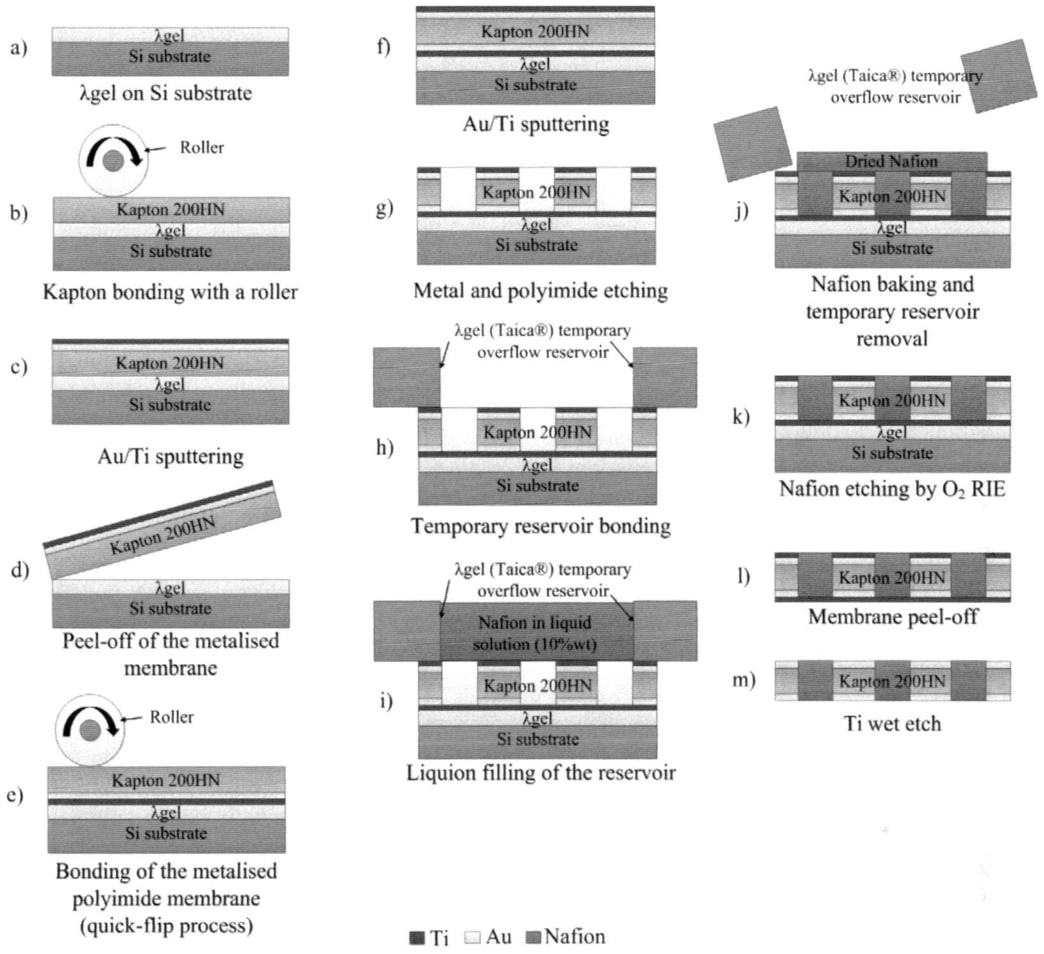

Figure 2 - Fabrication process of the membrane

2.1.2. Multilayer metal sputtering configuration

The λgel quick-flip approach was used to test five metal layer configurations (Table 1). These metal layers were sputtered on 2cm x 2cm polyimide sheets in argon plasma at 0.5Pa with no temperature control. The resulting metalized polyimide samples were evaluated under three criteria.

Table 1 – Result of double-sided multilayer deposition by sputtering with λgel as bonding layer

	Metal integrity	Polyimide integrity	Adhesion between metal and polyimide
Cr/Au/Cr/Au	Good	Good	Good
Cr/Au/Al	Alloy formation	N.D.	N.D.
Au/Al	Alloy formation	N.D.	N.D.
Cr/Au/Ti	Good	Good	Good
Au/Ti	Good	Good	Passable

Thickness (μm) : Cr : 60, Au : 300, Al : 400, Ti : 100
Deposition rate (nm/min) : Cr : 20, Au : 60, Al : 30, Ti : 15

First, metal layers integrity was verified by patterning each layer with its corresponding etchant. Unlike other samples, those containing Al formed an alloy, and were not possible to etch. Secondly, etching rate was measured to confirm the integrity of the polyimide, since heat damaged polyimide etches much faster than normal in O_2 plasma. All the samples were etched normally, which means the polyimide was not damaged by the sputtering process. Finally, adhesion between the metal layers and the polyimide was verified. An adhesion layer (Cr) increases the bond strength between metal and polyimide, but is not essential. In rare cases only, for Au/Ti samples, part of the metal layers peeled off during the etching and cleaning steps. Nevertheless Au/Ti combination was considered as the best choice since Cr etchant would damage Ti layer during etching.

2.2. Metal and Polyimide etching

After the sputtering steps, metal layers and polyimide must be etched (Figure 2g). The top metallic layers are patterned using a 4μm thick layer of photoresist (OFPR-800, Tokyo Ohka Kogyo Co., Ltd.) as an etching mask, giving 50μm squares holes spaced by 50μm large of metalized polyimide. Titanium is first etched with a diluted HF solution (49%) for approximately 20s and gold with Aurum-302 (Kanto Chemical Co., Inc.) for 10 minutes. The titanium layer is then used as masking layers for the 50μm deep polyimide etch by RIE (O_2 plasma, 120W RF power, 80sccm, 60min) giving etch angle near 90° . This good etch profile suggests that the walls could be reduced between holes, thus increasing the active surface of the membrane. Once the polyimide layer is etched through, the bottom gold layer is exposed and must be etched. Therefore, it is essential that the top Ti layer resists well to the RIE step to protect the top gold layer. The bottom Ti layer will remain to avoid bonding between Nafion and λgel, and will only be etched at the end of the process.

2.3. Nafion filling and etching

After etching, the 50μm wide holes are filled with Nafion. Esquivel et al. [6] reported filling of 250μm wide holes in a polymer substrate using a liquid solution of Nafion with a permanent reservoir to maintain the solution. In this work, the reservoir had to be removed after the filling process to expose the metallic conductive layer. A new technique has therefore been developed using a temporary reservoir.

A 3mm thick reservoir patterned in λgel is bonded to the sample (Figure 2h). The reservoir is then filled with Nafion in liquid solution (Liquion 10%wt, Ion Power) and dried at 60°C for 30min and 15min at 110°C (Figure 2i). The λgel reservoir is then peeled off (Figure 2j). Another step of RIE is required to etch Nafion that overflowed to expose the metallic layer underneath (Figure 2k). The RIE conditions are the same as the polyimide etch, but the time varies between 1 and 10 minutes depending on the thickness of the Nafion film to be removed. With this technique, holes are completely filled with Nafion, while the gold layers are exposed allowing electrical and ionic connection with the future carbon electrodes. The membrane is then peeled off from the λgel (Figure 2l) and is quickly dipped in a HF diluted solution (49%) to etch the remaining Ti (Figure 2m). A SEM image of the resulting membrane is shown at figure 3. An easy to handle bi-conductive membrane is then obtained and is ready to be integrated in a PEMFC.

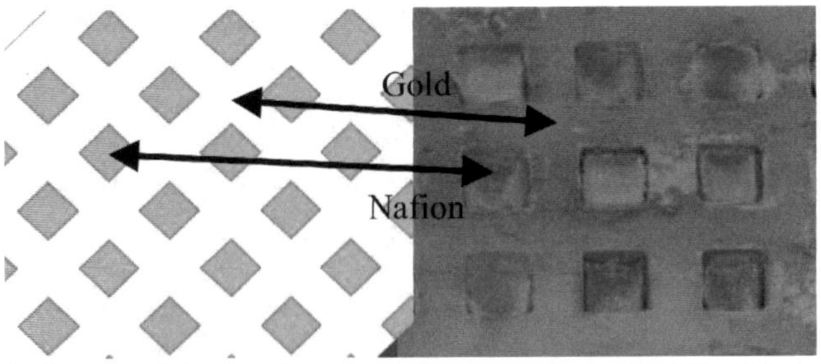

Figure 3 - SEM picture of the resulting membrane at the end of the fabrication process.

3. Conclusion

Double-sided multilayer metal deposition was achieved on polyimide sheet to provide lateral electrical conduction directly on an ionic conductive membrane for a PEMFC. With the use of λgel as a bonding agent between polyimide and a solid substrate, thermal management and fixing issues have been bypassed and three configurations of double-sided multilayer metal sputtering were achieved successfully. A novel approach using a temporary λgel reservoir to completely fill the holes with Nafion has been developed. These processes allowed the microfabrication of high active area ratio bi-conductive membrane for PEMFCs. Techniques developed here can also enable the fabrication of a variety of polymer-MEMS with through substrate vias and composite polymer structures.

Acknowledgement

This work was funded by the Natural Science and Engineering Research Council (NSERC) of Canada and the Canada Research Chair program.

References

[1] J. Yeom, G. Z. Mozsgai, B. R. Flachsbart, E. R. Choban, A. Asthana, M. A. Shannon, and P. J. A. Kenis, "Microfabrication and characterization of a silicon-based millimeter scale, PEM fuel cell operating with hydrogen, methanol, or formic acid," *Sensors and Actuators B (Chemical),* vol. 107, pp. 882-91, 2005.

[2] M. Kyong-Bok, S. Tanaka, and M. Esashi, "Fabrication of novel MEMS-based polymer electrolyte fuel cell architectures with catalytic electrodes supported on porous SiO2," *Journal of Micromechanics and Microengineering,* vol. 16, pp. 505-11, 2006.

[3] C. Weinmueller, G. Tautschnig, N. Hotz, and D. Poulikakos, "A flexible direct methanol micro-fuel cell based on a metalized, photosensitive polymer film," *Journal of Power Sources,* vol. 195, pp. 3849-57, 2010.

[4] D. B. St-Pierre, M. Paquin, and L. G. Frechette, "Design and microfabrication of a PEM fuel cell with silicon-supported electrodes in a column matrix configuration," in *PowerMEMS 2006. Sixth International Workshop on Micro and Nanotechnology for Power Generation and Energy Conversion Applications, 29 Nov.-1 Dec. 2006*, Piscataway, NJ, USA, 2006, pp. 93-6.

[5] D. Modroukas, V. Modi, and L. G. Frechette, "Micromachined silicon structures for free-convection PEM fuel cells," *Journal of Micromechanics and Microengineering,* vol. 15, pp. S193-S201, 2005.

[6] J. P. Esquivel, T. Senn, P. Hernandez-Fernandez, J. Santander, M. Lorgen, S. Rojas, B. Lochel, C. Cane, and N. Sabate, "Towards a compact SU-8 micro-direct methanol fuel cell," *Journal of Power Sources,* vol. 195, pp. 8110-15, 2010.

[7] T. Ito, K. Kimura, and M. Kunimatsu, "Characteristics of micro DMFCs array fabricated on flexible polymeric substrate," *Electrochemistry Communications,* vol. 8, pp. 973-976, 2006.

[8] T. Ito and M. Kunimatsu, "Fabrication of a micro DMFCs array made of photosensitive glass," *Electrochemistry Communications,* vol. 8, pp. 91-94, 2006.

[9] S. Tominaka, H. Nishizeko, J. Mizuno, and T. Osaka, "Bendable fuel cells: On-chip fuel cell on a flexible polymer substrate," *Energy and Environmental Science,* vol. 2, pp. 1074-1077, 2009.

[10] L. Zhu, K. Y. Lin, R. D. Morgan, V. V. Swaminathan, H. Kim, B. Gurau, D. Kim, B. Bae, R. I. Masel, and M. A. Shannon, "Integrated micro-power source based on a micro-silicon fuel cell and a micro electromechanical system hydrogen generator," *Journal of Power Sources,* vol. 185, pp. 1305-10, 2008.

[11] S. S. Yoon, D. O. Kim, S. C. Park, Y. K. Lee, H. Y. Chae, S. B. Jung, and J. D. Nam, "Direct metallization of gold patterns on polyimide substrate by microcontact printing and selective surface modification," *Microelectronic Engineering,* vol. 85, pp. 136-42, 2008.

[12] L. Supriya and R. O. Claus, "Solution-based assembly of conductive gold film on flexible polymer substrates," *Langmuir,* vol. 20, pp. 8870-8876, 2004.

Analysis of packaging effects on wire bonder made microtransformer chips

A Moazenzadeh[1], S H Moosavi[2], N Spengler[1], R Zeiser[2] and U Wallrabe[1]

[1] Laboratory for Microactuators, Department of Microsystems Engineering (IMTEK), University of Freiburg, Georges-Koehler-Allee 102, 79110 Freiburg, Germany.
[2] Laboratory for Assembly and Packaging Technology, Department of Microsystems Engineering (IMTEK), University of Freiburg, Georges-Koehler-Allee 103, 79110 Freiburg, Germany.

wallrabe@imtek.uni-freiburg.de

Abstract. We present the fabrication and characterization of microtransformer chips, based on wire bonded microcoils which are encapsulated in a Polydimethylsiloxane (PDMS) package. Our investigations focused on the degradation in electrical performance due to the higher dielectric constant of PDMS compared to air. We further studied the thermo-mechanical induced strain on the system due to thermal expansion of the package at temperatures up to 250°C. According to the measurements, encapsulating the chips decreased the resonance frequency and the quality factor near the resonance. However, both the maximum quality factor and the power efficiency were not affected by the encapsulation. Furthermore, induced thermo-mechanical strains due to the mismatched coefficients of thermal expansion of the chip and the packaging materials did not decrease the mechanical integrity of the SU-8 post and the coil itself.

1. Introduction

The trend towards further miniaturization of electronic devices makes the integration and miniaturization of reliable power convertors indispensable. Hence much effort has been spent to increase the performance of the on chip inductors and transformers. The effects of using low-κ polymers for the encapsulation of the radio frequency (RF) inductors have been investigated before in several publications [1,2]. However, increase in the capacitance of the inductor chip as a result of the packaging could compromise the performance in very high frequency regime (VHF) (30-300 MHz) as well and needs further investigation.

The fabrication of wire bonder made microcoils [3] and microtransformers [4–6] has been reported before. Apart from the increase in processing speed and the additional flexibility, manufactured devices are among the best microinductors due to the high performance in terms of inductance density, coupling factor and power efficiency [5]. In this paper, we report a method for the encapsulation of the wire bonder made microtransformer chips in PDMS for protection of the delicate wires and bonds. Meanwhile we investigate two important aspects related to the added packaging to the chips: the influence on the electrical performance and also the reliability of the chips at high temperatures.

The encapsulation of microtransformer chips provides a practical solution in order to protect them from external mechanical shocks. It also makes the chip handling easier, especially for further processes like the assembly into the target circuit. However, replacing air (ε_r=1) with a dielectric material (ε_r>1) in an inductor-based system, results in an increase of the capacitance and therefore parasitic losses of the chip. Also, induced strain due to the differences in coefficients of thermal expansion (CTE) of the chip and the packaging materials results in additional thermo-mechanical stress on the chip. Therefore, further investigations of the reliability at increased temperatures are needed.

2. Fabrication

For the fabrication, a 4 inch borosilicate wafer was gold-metallized via electroplating to implement 12 µm thick supply tracks and bond pads (figure 1a). Using a thick SU-8 process defined an array of 650 µm high cylindrical posts as well as a frame around the chip (figure 1b). After a short O_2-plasma cleaning step a modified, automated ball-wedge wire bonder (WB 3100plus, ESEC, Switzerland) was employed to wind two solenoids, each with 20 windings, in a helical trajectory around the posts (figure 1c). The coils themselves were stable, however, in order to provide a physical protection for the coils and their bond connections, Sylgard® 184 PDMS (Dow Corning Corp., USA) was used as the packaging material. Besides the low volume shrinkage, its dielectric constant of 2.65 is lower than of most epoxies and other silicone materials [7]. The PDMS was statically dispensed into the SU-8 frame at a temperature of 80 °C, and subsequently cured at 70 °C for 2 hours in an oven to form the package (figure 1d). Details for the fabrication have been reported in [5].

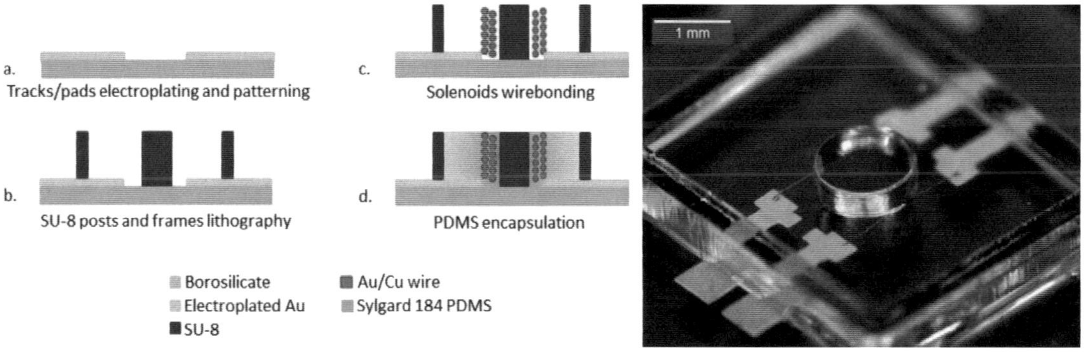

Figure 1. Left: wire bonded microtransformer manufacturing process flow (a-d). Right: a complete copper wire bonded transformer chip, encapsulated in PDMS.

3. Measurement and discussion

3.1. Packaging effects on the electrical performance

To investigate the effect of the encapsulation on the electrical performance, seven identical chips were electrically characterized before and after the encapsulation. The high frequency characterization of the transformers was performed using a 2-port on-wafer measurement method.

Encapsulating the chips led to an increase in capacitance (C), while the inductance (L) of the chip stayed constant. As a result, based on the formula (1) the resonance frequency ($f_\circ$) decreased (fig. 2).

$$f_\circ = \frac{1}{2\pi\sqrt{LC}} \qquad (1)$$

The quality factor also decreased near the resonance (fig. 3) as a result of the increased internal capacitance of the inductors, respectively. However, below 100 MHz, the quality factor was not influenced, and it's maximum of approx. 64% appeared at a constant frequency of f=*85 MHz*. Finally, the power efficiency was hardly affected by the encapsulation (fig. 3). Details of the measured values for the seven identical chips, before and after the packaging are summarized in the table 1.

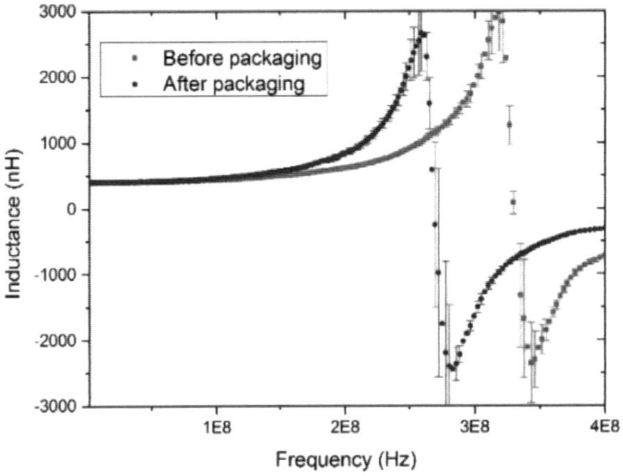

Figure 2. Measured inductance of the primary solenoid of seven identical transformer chips before and after the encapsulation. The resonance frequency was decreased by 21% after the encapsulation.

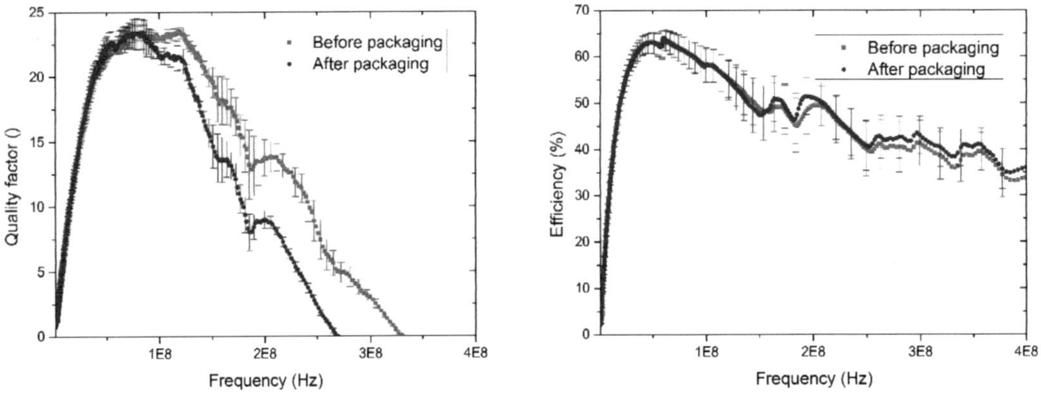

Figure 3. Measured Q-factor (left) and power efficiency (right) of the primary solenoid of seven identical transformer chips before and after the encapsulation.

Table 1. Average values for the measurements of the chips, before and after the PDMS encapsulation.

	L (nH)	Q_{max}	η_{max} (%)	Resonance
Before packaging	404±4	23.40±0.5	63.59±2.39	-
@ f (MHz)	1	85	61	312±3
After packaging	406±5	23.33±1.21	64.03±2.31	-
@ f (MHz)	1	84	60	258±2

3.2. Verification of the thermo-mechanical induced strain

To check the thermo-mechanical reliability of the package, the induced strain was measured at different temperature stages, using the Digital Image Correlation technique (DIC) where the 3D expansion of the chip was recorded and analyzed. Absolute behavior in each thermal stage can be achieved with precision of 1 μm.

During the measurement procedure, the sample was kept in a vacuum oven with a transparent lid, where the temperature was increased from room temperature up to 250°C in steps of 25°C. The final results revealed that the expansion of the PDMS package did not decrease the mechanical integrity of the SU-8 post and the coil (figures 4 and 5). Figure 4 (a-c) show that the package expanded around 40 μm in the Z direction, whereas it was around zero for the SU-8 post on which the solenoids were wire bonded. Fig. 4 (d-f) show the strain in X, and fig. 4 (g-i) in Y direction. They represent a compression between the two points on the SU-8 post and on the package in each temperature stage.

Figure 5 reveals, that the induced shear strain in the XY plane mainly appeared at the package frame, instead of the solenoids or the bonds at 250°C.

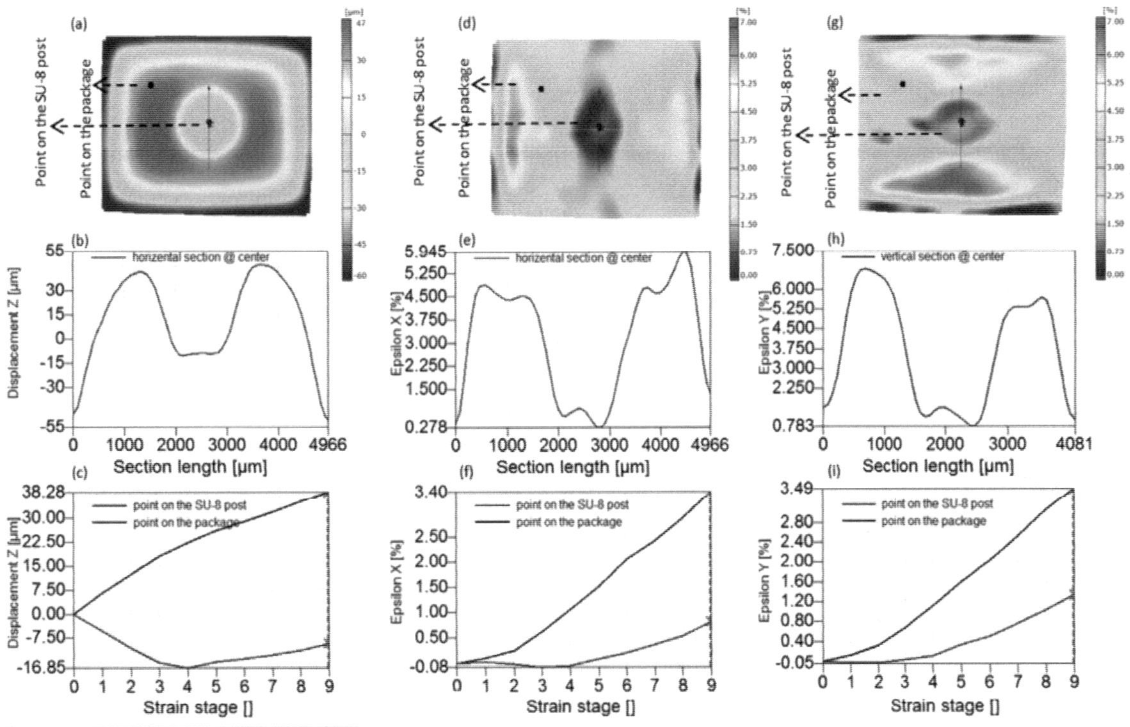

Figure 4. The package behaviour at temperature ramps. (a-b) the displacement of the chip at 250°C with respect to the room temperature. (c) The displacement of two points on the SU-8 post and on the package going through different temperature stages. (d), (e), (g) and (h): The strain imposed on the chip in the X and Y direction at 250°C. (f-i): The strain exposed on the two points on the SU-8 post and on the package going through different temperature stages in the X and Y direction respectively

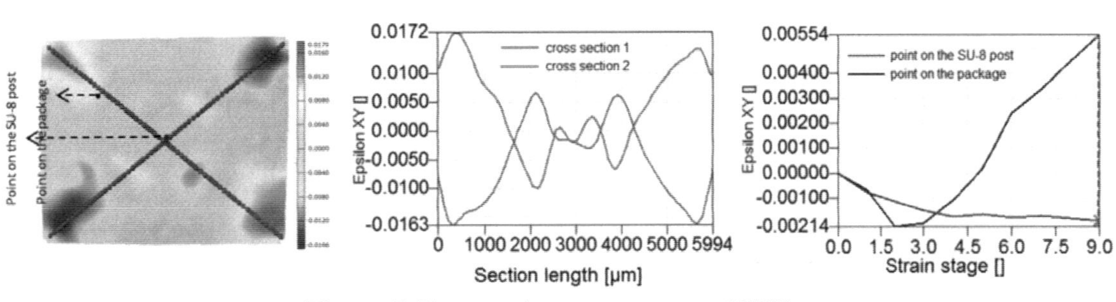

Figure 5. Shear strain measurement at 250°C

4. Conclusion

We present a successful packaging approach for the wire bonded miniaturized microtransformer chip by using PDMS as an encapsulation material. According to the measurements, encapsulating the chips decreased the resonance frequency and also the quality factor near the resonance. However, both the maximum quality factor and the power efficiency were not affected by the encapsulation. Studying the induced thermo-mechanical strain due to the thermal expansion of the PDMS up to 250°C showed that the package expanded around 40 μm in the Z direction. This value was around zero for the SU-8 post on which solenoids were wire bonded. Furthermore, induced thermo-mechanical strains due to mismatched coefficients of thermal expansion of the chip and the packaging materials did neither affect the mechanical integrity of the SU-8 posts, nor electrical connectivity of the wire bonds.

Acknowledgments

This work was supported by DFG graduate school *Embedded Microsystems* under grant Number 1103.

References

[1] Choi Y, Yoon E and Yoon J 2003 Encapsulation of the micromachined air-suspended inductors *Microwave Symposium Digest* pp 1637–40
[2] Carchon G, Raedt W and Beyne E 2004 Wafer-level packaging technology for high-Q on-chip inductors and transmission lines *Microw. Theory Tech.* **52** 1244–51
[3] Kratt K, Badilita V, Burger T, Korvink J G and Wallrabe U 2010 A fully MEMS-compatible process for 3D high aspect ratio micro coils obtained with an automatic wire bonder *J. Micromechanics Microengineering* **20** 015021
[4] Moazenzadeh A, Spengler N and Wallrabe U 2013 High-performance, 3D-microtransformers on multilayered magnetic cores *Micro Electro Mechanical Systems (MEMS), 2013 IEEE 26th International Conference on* (Taipei) pp 287–90
[5] Moazenzadeh A, Spengler N, Lausecker R, Rezvani A, Mayer M and Korvink J G 2013 Wire bonded 3D coils render air core microtransformers competitive *J. Micromechanics Microengineering* **23**
[6] Raimann M, Peter A, Mager D, Wallrabe U and Korvink J G 2012 Microtransformer-Based Isolated Signal and Power Transmission *IEEE Trans. Power Electron.* **27** 3996–4004
[7] Schneider F, Fellner T, Wilde J and Wallrabe U 2008 Mechanical properties of silicones for MEMS *J. Micromechanics Microengineering* **18** 065008

Journal of Physics: Conference Series **476** (2013) 012111 doi:10.1088/1742-6596/476/1/012111

Analytical Method for Selecting a Rectification Technique for a Piezoelectric Generator based on Admittance Measurement

Loreto Mateu, Henrik Zessin and Peter Spies

Fraunhofer Institut for Integrated Circuits IIS, Nordostpark 93, 90411 Nuremberg, Germany

E-mail: loreto.mateu@iis.fraunhofer.de, henrik.zessin@iis.fraunhofer.de, peter.spies@iis.fraunhofer.de

Abstract. AC-DC converters employed for harvesting power from piezoelectric transducers can be divided into linear (i.e. diode bridge) and non-linear (i.e. synchronized switch harvesting on inductor, SSHI). This paper presents an analytical technique based on the measurement of the impedance circle of the piezoelectric element to determine whether either diode bridge or SSHI converter harvests more of the available power at the piezoelectric element.

1. Introduction

AC-DC converters that perform the rectification of the power harvested by piezoelectric transducers are divided in two different groups: non-linear (i.e. parallel synchronized switch harvesting on inductor, SSHI) and linear (i.e. diode bridge). The paper presents a method to estimate the electrical impedance of AC-DC converters in order to predict the harvested power to be obtained with a certain piezoelectric transducer from the measurement of its impedance circle. Thus, it is possible to estimate which AC-DC converter harvests more power.

The paper is divided as follows: Section 2 describes how to calculate the elements of the piezoelectric equivalent circuit. In section 3, the impedance of the mechanical part of the piezoelectric transducer as well as the electrical part are computed when a resistive load is connected. These values are employed to calculate the harvested power ratio. Section 4 presents an analytical method to compute the electrical impedance of a diode bridge. In Section 5, the expression for the inversion voltage factor of the parallel SSHI converter is given. Finally, Section 6 summarizes the conclusions.

2. Piezoelectric Equivalent Electrical Impedance in Open Circuit

Measurements of the admittance of different piezoelectric transducers have been done with an AUTOLAB PGSTAT302N potentiostat/galvanostat. The piezoelectric equivalent circuit at the resonance frequency on the secondary side has the electrical components L'_n, C'_n, R'_n in series and C_p connected in parallel to them. The admittance of a piezoelectric element is expressed as:

$$Y(s) = \frac{s^2 + s\frac{R'_n}{L'_n} + \frac{C'_n + C_p}{C_p C'_n L'_n}}{s^2 + s\frac{R'_n}{L'_n} + \frac{1}{L'_n + C'_n}} \tag{1}$$

A fitting curve process using the measured data and equation (1) is employed to estimate the value of L'_n, C'_n, R'_n and C_p.

The ratio of the modulus of the maximum conductance to the modulus of the susceptance at the series resonance frequency corresponds to the figure of merit M [1]:

$$M = Q_m K = Q_m k^2 = \frac{|Y_{m0}|}{|Y_d|} = \frac{1/R'_n}{\omega_s C_p} \tag{2}$$

where ω_s is the series resonance frequency.

This figure of merit is employed by Guyomar et al.[2] to decide when a non-linear rectifier will harvest more power than a diode bridge since if $M \leq \pi$, the non-linear technique harvests more power than the linear technique. However, there is no prediction about the percentage of the harvested power that will be gained employing one or the other technique. Two piezoelectric cantilevers based on the QP40W and V21BL piezoelectric elements from Mide are employed to present the results obtained with a value of M equal to 0.42 and 1.13, respectively.

3. Piezoelectric Equivalent Electrical Impedance with a Resistive Load

The analysis done by Brufau-Penella et al. [3] allows calculating the power ratio that is defined as the ratio of the maximum power harvested with resistive load R_L to the maximum power harvested with a complex load employing the elements of the piezoelectric equivalent circuit. However, the argumentation related to the Thevenin equivalent model is not valid when a load is connected to the piezoelectric element since in this case the load is placed in parallel to the piezoelectric electrical capacitor C_p as it is exposed in [4].

The electrical equivalent impedance of the mechanical part of the piezoelectric transducer is the same for all the circuits under consideration and is given by the series connection of L'_n, C'_n, R'_n:

$$Z_{mech}(j\omega) = R'_n + j\omega L'_n + \frac{1}{j\omega C'_n} = R_{mech} + jX_{mech} \tag{3}$$

If a resistive load is connected to the piezoelectric transducer, the electrical equivalent impedance corresponds to the impedance of capacitor C_p in parallel with R_L:

$$Z_{elecr}(j\omega) = \frac{R_L}{1+\omega^2 C_p R_L^2}\left(1 - j\omega C_p R_L\right) = R_{elecr} + jX_{elecr} \tag{4}$$

The harvested power ratio of a resistive load to a complex load is defined as the ratio of the harvested power employing resistor R_L as a load to the harvested power employing the complex conjugate of electrical impedance Z_{mech}, which corresponds to the maximum harvested power:

$$P_{ratior} = \frac{4R_{elecr}R_{mech}}{\left(R_{mech} + R_{elecr}\right)^2 + \left(X_{mech} + X_{elecr}\right)^2} \tag{5}$$

Thus, there will be piezoelectric elements for which the power ratio is close to one and others for which is far from one. The peak value of the power ratio for two different piezoelectric cantilevers based on the QP40W and the V21BL piezoelectric elements is 0.58 and 0.91, respectively.

4. Piezoelectric Equivalent Electrical Impedance with a Diode Bridge

The connection of a diode bridge circuit with a resistive load and a filtering capacitor to a piezoelectric generator, see figure 1, has two equivalent circuits, as shown in figure 2, depending if the diodes are not conducting (circuit A) or are conducting (circuit B).

The electrical impedance for circuits A and B is:

$$Z_{eelcdbA}(j\omega) = \frac{1}{j\omega C_p} \tag{6}$$

$$Z_{elecdbB} = \frac{R_L}{1 + j\omega R_L C_L} \tag{7}$$

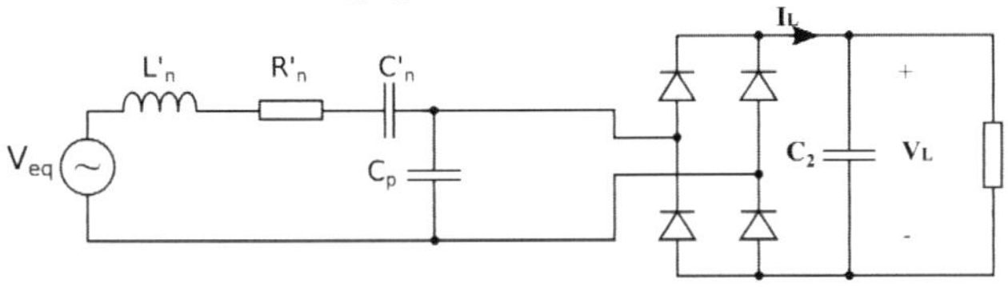

Figure 1. AC-DC converter circuit composed by a piezoelectric element connected to a diode bridge.

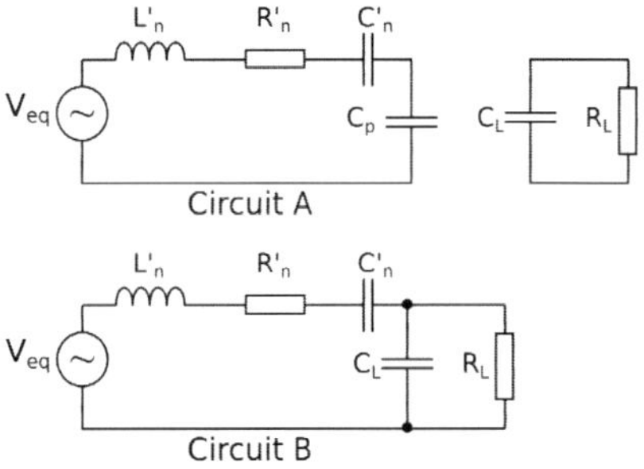

Figure 2. Two equivalent circuits of figure 1 depending on the fact that the current flows through the diodes (circuit B) or not (circuit A).

Therefore, the transfer function of the electrical impedance of a diode bridge connected to resistor R_L and capacitor C_L is given by:

$$Z_{elecdb}(j\omega) = \frac{Z_{elecdbA}(j\omega)\left(1 - e^{-j\theta}\right) + Z_{elecdbB}(j\omega)\left(e^{-j\theta} - e^{-j\omega T/2}\right)}{1 - e^{-j\omega T/2}} \tag{8}$$

where between 0 and θ only C_p is connected to Z_{mech} and between θ and T/2 the current flows through the diodes and C_L and R_L are also connected to the piezoelectric element. The transfer function of the electrical impedance has a period T/2. The value of θ is calculated combining the following two equations:

$$V_L = I_L R_L = \frac{1}{T/2} \int_{\theta/\omega}^{T/2} \frac{V_{eq}\omega}{L'_n} \frac{1}{\left(\left(\frac{C_p + C'_n}{L'_n C_p C'_n} - \omega^2\right)^2 + \left(\frac{\omega R'_n}{L'_n}\right)^2\right)^{1/2}} \sin \omega t \, dt \tag{9}$$

$$\cos\theta = 1 - 2V_L / V_{eq} \tag{10}$$

where V_{eq} corresponds to the open circuit voltage of the piezoelectric element.

The harvested power ratio for this case is defined as the ratio of the harvested power connecting a diode bridge to a piezoelectric element to the maximum harvested power:

$$P_{ratiodb} = \frac{4R_{elecdb}R_{mech}}{\left(R_{mech} + R_{elecdb}\right)^2 + \left(X_{mech} + X_{elecdb}\right)^2} \tag{11}$$

If the power ratio is close to one, then a linear technique like the diode bridge rectifier would give a result as good as a non-linear technique without the complexity at the circuit level. The harvested power is the power of R_{elecdb} that is not equal to the power of R_L. The power of the load is a fraction of the harvested power. Figures 3 and 4 show the harvested power ratio for piezoelectric elements QP40W and V21BL respectively as a function of frequency and resistor R_L when a diode bridge is employed as AC-DC converter. The peak value of the power ratio is 0.57 and 0.92 for figures 3 and 4, respectively. The results obtained from an electrical simulation of the piezoelectric transducers connected to a diode bridge have been compared to the results obtained from the equations of this section and match.

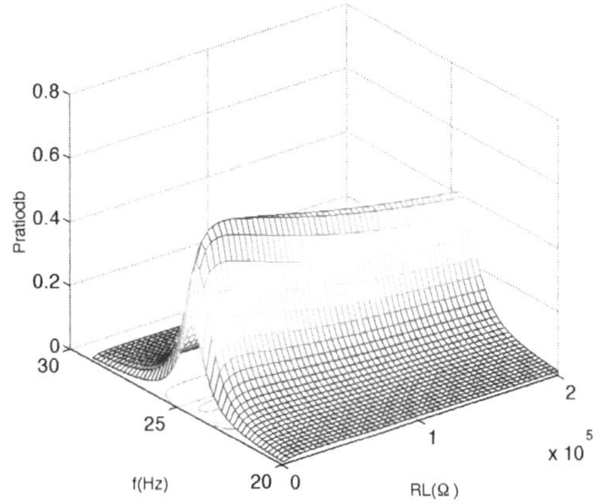

Figure 3. Power ratio of the diode bridge as a function of RL and frequency for the cantilever based on the piezoelectric element QP40W.

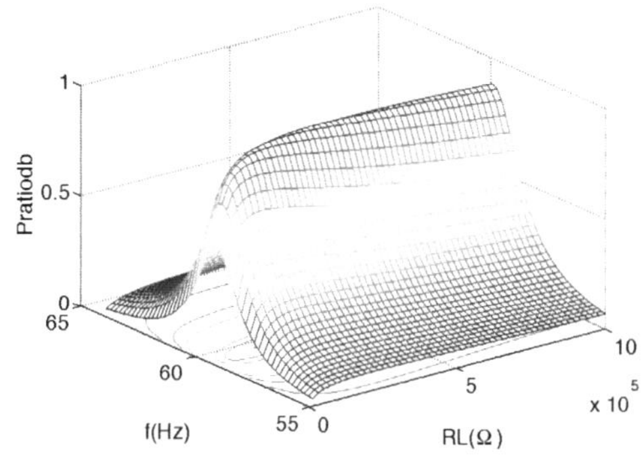

Figure 4. Power ratio of the diode bridge as a function of RL and frequency for the cantilever based on the piezoelectric element V21BL.

5. Piezoelectric Equivalent Electrical Impedance with a SSHI Converter

The connection of a SSHI circuit with a resistive load and a filtering capacitor to a piezoelectric generator, as shown in figure 5, has three equivalent circuits depending if the diodes are not conducting (circuit A), are conducting (circuit B) or the current is flowing through inductor L (circuit C), see figure 6.

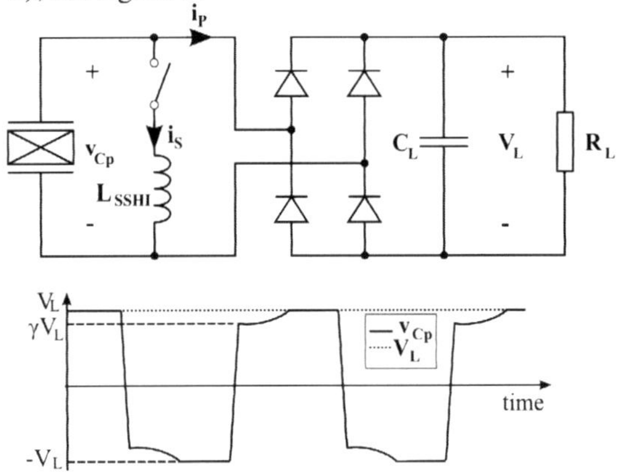

Figure 5. AC-DC converter circuit composed by a piezoelectric element connected to parallel SSHI circuit.

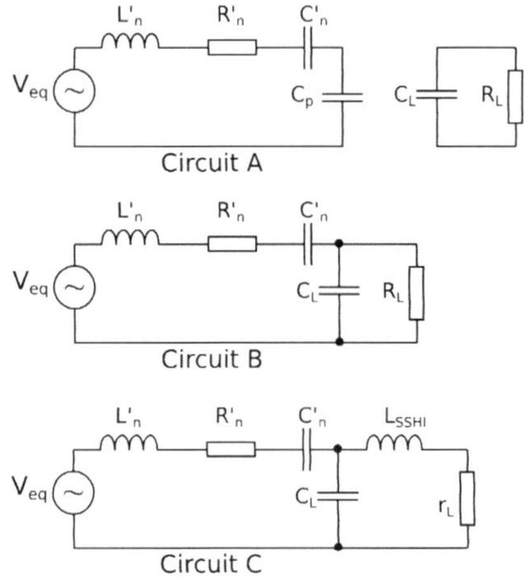

Figure 6. Three equivalent circuits of figure 1 depending on the fact that the current does not flow trough diodes (circuit A), flows through diodes (circuit B) or through inductor L_{SSHI} (circuit C).

A non-linear technique implies the inversion of the piezoelectric voltage through inductor L_{SSHI} when the maximum displacement of the piezoelectric element is reached. The voltage inversion factor γ is calculated considering that in circuit C of figure 6, capacitor C_p has an initial voltage V_L that corresponds to the rectified voltage on R_L. The value of γ is given by the following expression:

$$\gamma = -C_p L_{SSHI} \omega_n e^{-\zeta\omega T_d/2} \left(\frac{\alpha^2 - 2\alpha\zeta\omega_n + \omega_n^2}{1-\zeta^2} \right)^{1/2} \sin\left(\omega_n \left(1-\zeta^2\right)^{1/2} T_d/2 + \theta \right) \quad (12)$$

where r_L is the effective associated series resistance of inductor L_{SSHI}, $\alpha = r_L / L$, $\zeta = r_L / 2\left(C_p / L\right)^{1/2}$, $\omega_n = \left(1/\left(LC_p\right)\right)^{1/2}$, $T_d/2 = \pi/\left(\omega_n\zeta\right)$ and $\theta = \tan^{-1}\left(\omega_n\left(1-\zeta^2\right)/\left(\alpha-\zeta\omega_n\right)\right)$.

Figures 7 and 8 show the voltage on capacitor C_p as a function of time for cantilevers based on piezoelectric elements QP40W and V21BL, respectively, for different values of inductor L_{SSHI} and r_L. Piezoelectric element V21BL obtains a better voltage inversion factor γ than piezoelectric element QP40W.

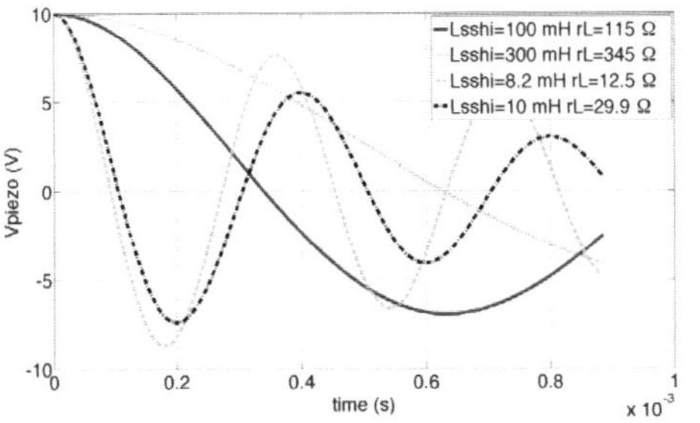

Figure 7. Inversion voltage on C_p as a function of time for cantilever based on piezoelectric element QP40W.

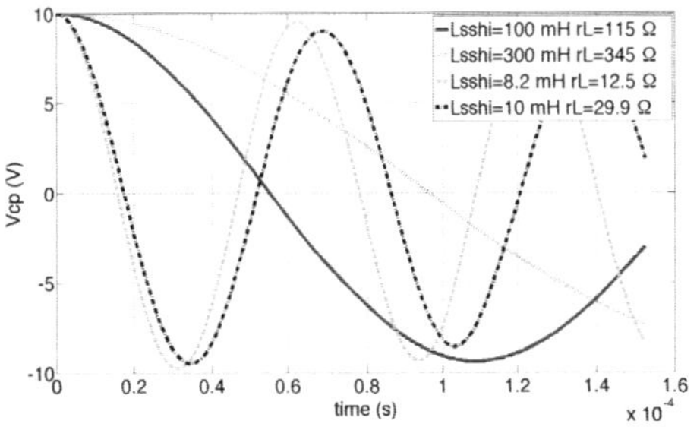

Figure 8. Inversion voltage on C_p as a function of time after closing MOSFET T_1 for cantilever based on piezoelectric element V21BL.

6. Conclusions and Future Work

The electrical equivalent impedance of a resistor and a diode bridge has been calculated. Then, the power ratio defined as the harvested power by each one of the two circuits to the maximum harvested power has been calculated and compared to the results obtained from electrical simulations. One of the piezoelectric elements harvests only a 0.57 of the available harvested power while the other one has a power ratio close to 1. If the harvested power of the diode bridge is close to the harvested power using the complex conjugate of the mechanical impedance as a load, a non-linear technique will not provide a benefit in terms of harvested power.

Further work will include the calculation of the electrical impedance of the parallel SSHI converter to determine the power ratio of this rectifying technique.

References

[1] Ikeda T 1990 *Fundamentals of Piezoelectricity* (Oxford: Oxford University Press)

[2] Guyomar D, Sebald G, Pruvost S, Lallart M, Khodayari A and.Richard C 2009 *Energy Harvesting from Ambient Vibrations and Heat* Journal of Intelligent Material Systems and Structures , 20 (5), pp 609-624

[3] Brufau-Penella J, Puig-Vidal M, *Piezoelectric Energy Harvesting Improvement with Complex Conjugate Impedance Matching* 2009 Journal of Intelligent Material Systems and Structures, 20 (5), pp 597-608

[4] Liang J, Liao W-H, *Impedance modeling and analysis for piezoelectric energy harvesting systems* 2012 Mechatronics, IEEE/ASME Transactions on 17.6, pp 1145-1157

Large-dynamic-range MEMS Electret Energy Harvester with Combined Gap-closing/Overlapping-area-change Electrodes

Qianyan Fu[1], **Yuji Suzuki**
Department of Mechanical Engineering, The University of Tokyo, Tokyo, Japan
E-Mail: qfu@mesl.t.u-tokyo.ac.jp

Abstract: A novel in-plane MEMS electret energy harvester with combined electrodes of overlapping-area-change and gap-closing types is proposed for large output power both at low and high vibration accelerations. An early prototype has been successfully micro-fabricated with the single layer silicon-on-insulator process. Soft-X-ray charging is employed to establish uniform surface potential around 60 V on vertical electrets on the sidewall of the comb fingers. Up to 1.19 μW output power has been obtained at 552 Hz and 2.15 g acceleration oscillations, which corresponds to the effectiveness as high as 27.2%.

Keywords: Combined gap-closing/overlapping-area change electrodes, vertical electrets

1. INTRODUCTION

Vibration energy harvesting is a method to convert vibration energy in the environment to electricity for powering low-power electronics such as wireless sensor nodes. Among various energy conversion techniques, electrostatic/electret principle has attracted much attention for its high output power in small dimensions [1]. In our previous studies, we employ CYTOP as the electret material, and develop an in-plane MEMS energy harvester, in which overlapping area between the electret and the counter electrode is changed. Up to 6 μW output power has been obtained under 40 Hz and 1.4 g oscillation [1, 2]. Another configuration of in-plane electrostatic/electret energy harvesters is based on comb drives, which is micro-fabricated through one-mask SOI process without any assembling. Hoffmann et al. [3] and Nguyen et al. [4] separately developed MEMS electrostatic generators with overlapping-area-change comb drives and obtained output power up to 5 μW RMS and 3.4 μW, respectively. Guillemet et al. [5] developed an in-plane gap-closing electrostatic generator and obtained up to 2.3 μW at 250 Hz. However, external voltage source for the bias voltage is needed for those generators. Recently, Suzuki et al. [6] developed an electret generator with SiO_2 electret induced by highly-doped potassium ions. But, the maximum output power was limited to 50 nW at 58 Hz.

In the present study, an in-plane MEMS electret energy harvester with combined electrodes with overlap-area-change and gap-closing types is proposed for large dynamic range response. A novel charging method using soft X-ray irradiation [7] is employed to realize vertical electrets with high surface potential on the sidewall of comb drives.

2. PRINCIPLE AND DEVICE DESIGN

In electrostatic/electret energy harvesters, capacitance change per unit displacement should be max-

[1] Correspondence to: Qianyan Fu. E-mail: qfu@mesl.t.u-tokyo.ac.jp
Department of Mechanical Engineering, The University of Tokyo, 7-3-1, Hongo, Bunkyo-ku, Tokyo 113-8656, Japan

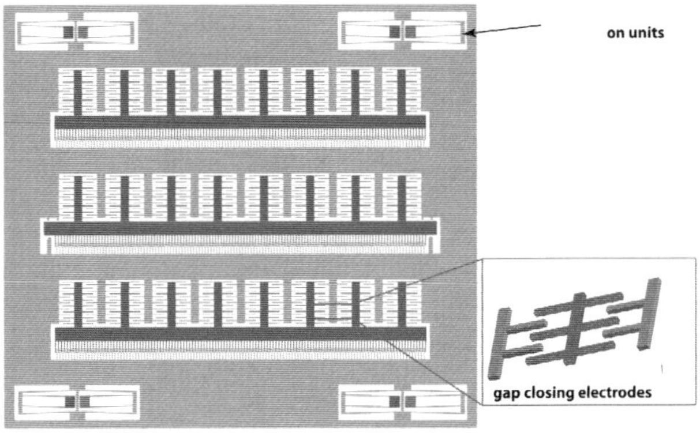

Figure 1. Schematic view of the in-plane electret energy harvester with combined electrodes.

imized for large output power [1]. Capacitance of the gap-closing converters ((2) in inset of Fig. 1) is inversely proportional to the displacement, and thus the gap-closing converter is suitable for large vibration acceleration. However, at small acceleration, the capacitance change diminishes. On the other hand, capacitance of the overlapping-area-change converters with a constant gap ((1) in inset of Fig 1) is linearly changed with the displacement, and thus its change is higher than that of the gap-closing type at low acceleration (or small amplitude). Therefore, we propose a novel electrode configuration combined with in-plane overlapping-area-change and gap-closing converters that provide relatively-high output power both at low and high vibration accelerations.

Figure 1 shows a schematic of the proposed structure. The mechanical resonator has a seismic mass of 4.25 mg, which is suspended by four suspensions at the corners. Based on the VDRG (velocity-damped resonance generator) model [8], the output power of vibration energy harvester is proportional to the seismic mass amplitude. Thus, the spring structure should be designed in such a way that large amplitude is obtained without lateral instability. For that purpose, we employ tilted folded-beam springs having a width of 10 μm and a length of 970 μm, with a tilt angle of 2 degree. The capacitive electrodes comprise two types of interdigitated comb structures: gap-closing and overlapping-area-change. For the overlapping-area-change type, fixed and movable comb fingers are designed with a gap g of 10 μm and an initial overlap L of 40 μm. For the gap-closing type, g and L are 31 μm and 75 μm, respectively. Thickness of the device layer is 70 μm. The maximum variation of capacitance ΔC is 10.2 pF in the harvester with 453 finger pairs for the overlapping-area-change type and 336 finger pairs for the gap-closing one. The oscillation amplitude X_{max} of the proof mass is limited to 25 μm by mechanical stoppers. The chip size is 1 cm by 1 cm.

3. MICROFABRICATION AND CHARACTERIZATION OF THE ELECTRET ENERGY HARVESTER

The device was fabricated through the silicon-on-insulator (SOI) MEMS technology. The process starts with a standard lithography on 4-inch SOI wafer using photoresist (Fig. 2a). 70 μm-thick device layer is etched with DRIE (Fig. 2b) to form springs, electrodes and etched holes. Then, the buried oxide layer is etched with vapor HF for releasing the structure through the etched holes on the seismic mass (Fig. 2c). This is followed by a 1.5-μm-thick parylene-C deposition as the electret material (Fig. 2d). Finally, soft X-ray charging using 9.5 keV acceleration voltages is applied with the bias voltage of 130V for 150 seconds (Fig. 2e).

Figure 3 shows the prototype device, which is then glued on a printed circuit board. After that wire bonding is employed for electrical connection. Figure 4 shows the mechanical response of the device, which indicates slight softening behavior. The resonant frequency is decreased from 668 Hz to 648 Hz after the parylene deposition, while the maximum amplitude is unchanged, which is as large as 23.6 μm. The quality factor is around 60.7.

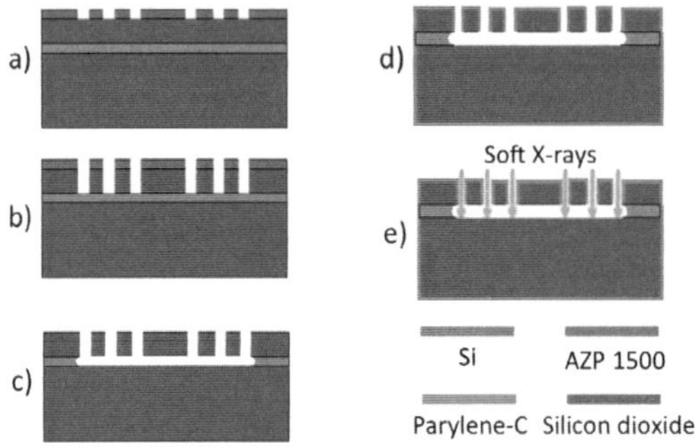

Figure 2. MEMS process flow.

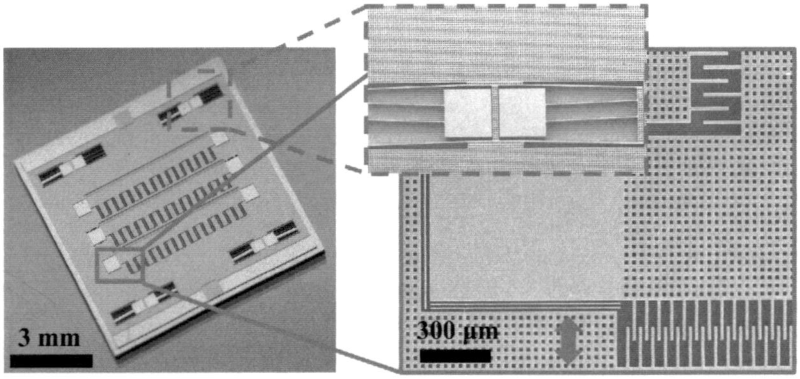

Figure 3. Overview of the present electret energy harvester prototype.

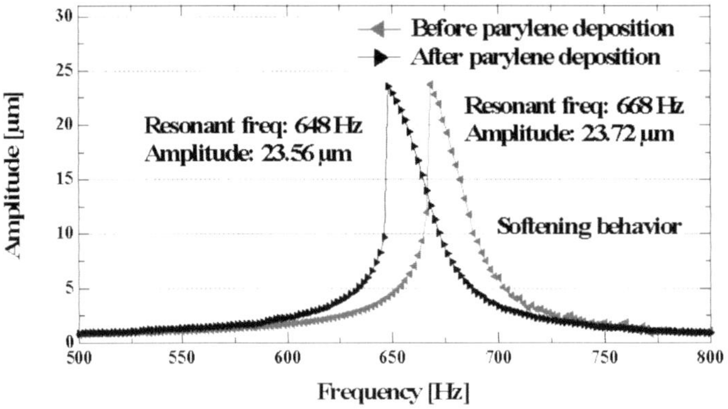

Figure 4. Frequency response before/after 1.5-μm-thick parylene deposition.

We employ soft X-ray irradiation for electret charging as shown in Fig. 5. When the soft X-ray is irradiated onto comb drives, positive and negative ions are equally generated in the gap, which are dragged toward the electrets under an imposed electric field across the gap, so that the charges are transferred to the electrets. Since the soft X-ray can penetrate into narrow gaps, even the electrets on the sidewall of high-aspect-ratio structures can be charged [7]. The bias voltage and the charging time are 130 V and 150 s, respectively.

The surface potential of the vertical electrets is measured with Kelvin force microscope (KFM). After charging, the comb fingers are intentionally broken and one comb finger is fixed onto the KFM stage. Figure 6 shows the surface potential distribution in a 1 μm-square area of the vertical electrets. High surface potential of around 60 V is uniformly distributed in the depth direction down to 65 μm.

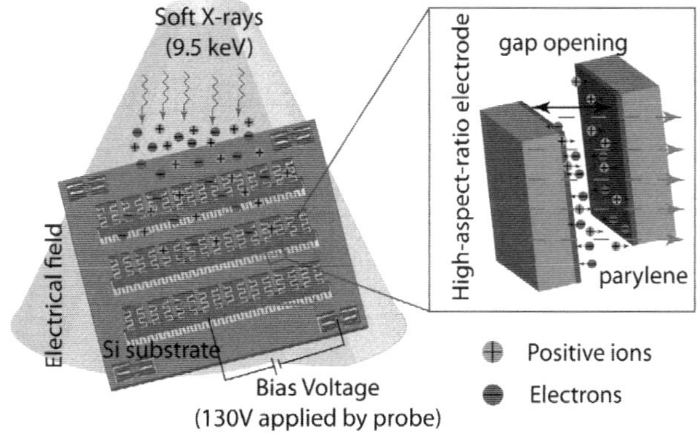

Figure 5. In-situ charging method for 'vertical' electrets using soft X-ray irradiation. Bias voltage of 130 V is applied during the charging process.

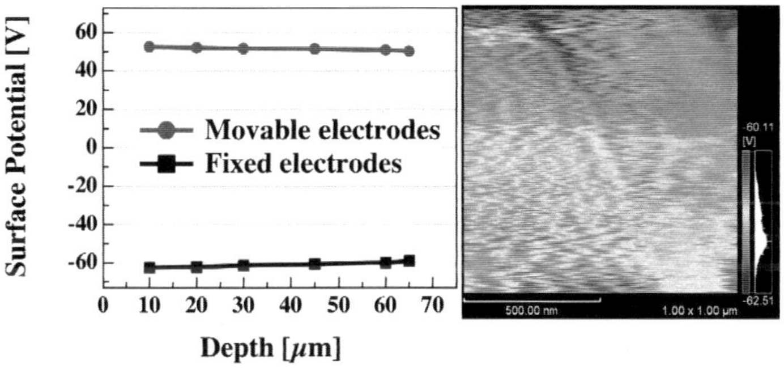

Figure 6. Surface potential of the vertical electrets measured with Kelvin force microscopy (KFM).

Figure 7a shows the output voltage waveform at the resonant frequency of 552 Hz and 2.15 g acceleration. Optimal load is 6.5 MΩ. Peak-to-peak voltage around 12 V is obtained without external bias voltage. Figure 7b shows the output power versus external acceleration at 552 Hz. Even at low acceleration of 0.6 g, 0.1 μW output power is obtained. Thanks to the gap-closing electrodes, the output power is rapidly increased with the external acceleration, and 1 μW is obtained at 2 g. Bi-stable behavior is found over 2 g acceleration, which is attributed to the mechanical response of the tilted folded-beam suspensions for large displacements, and the output power jumps from 1 μW to 1.17 μW. The maximum output power of 1.19 μW is achieved at 552 Hz and 2.15g, which corresponds to the effectiveness defined by the output power divided by the VDRG limit [8, 9] as high as 27.2%.

The output power at different vibration frequency is shown in Fig. 8. If compared with the frequency response shown in Fig. 4, the hysteresis becomes more significant, and broadened with increasing external acceleration. This is probably due to the electrical damping force, which increases for large oscillation amplitudes. By the combined effects of the electrical damping and the softening nonlinear spring of the tilted folded suspensions, the bandwidth is increased to 75 Hz at 1 g.

4. CONCLUSION

A novel in-plane electret energy harvester with combined electrodes of overlapping-area-change and gap-closing converters has been proposed in order to realize large dynamic range of the harvester. The proposed prototype is successfully microfabricated, and up to 1.19 μW output power has been obtained, which corresponds to the effectiveness as high as 27.2%.

This work is partially supported through JSPS NEXT Program. Photo-mask is made using F5112+VD01 EB writer at VDEC of the University of Tokyo.

References

[1] Y. Suzuki, 2011, Recent Progress in MEMS Electret Generator for Energy Harvesting IEEJ Trans. Electr. Electr. Eng., Vol. 6, 101-111.
[2] Y. Suzuki et al., 2010, A MEMS Electret Generator with Electrostatic Levitation for Vibration-Driven Energy Harvesting Applications, J. Micromech. Microeng., Vol. 20, 104002.
[3] D. Hoffmann et al., 2009, Fabrication, Characterization and Modelling of Electrostatic Micro-generators, J. Micromech. Microeng., Vol. 19, 094001.
[4] S. Nguyen et al., 2013, Bistable Springs for Wideband Microelectromechanical Energy harvesters, Appl. Phys. Lett., 023904.
[5] Guillemet et al., 2012, Efficient In-plane Gap-closing MEMS Electrostatic Vibration Energy Harvester, PowerMEMS 2012, Atlanta, 137-140.
[6] M. Suzuki et al., 2012 Electrostatic Micro Power Generator Using Potassium Ion Electret Forming on a Comb-drive Actuator, PowerMEMS2012, Atlanta, 247-250.
[7] K. Hagiwara et al., 2012, Electret Charging Method Based on Soft X-ray Photoionization for MEMS Applications, Trans. IEEE, Dielectr. Electr. Insul., Vol. 19, 1291-1298.
[8] P. D. Mitcheson et al., 2004 Architectures for Vibration-Driven Micropower Generators, J. Microelectromech. Syst., Vol. 13, 429-440.
[9] P. D. Mitcheson et al., 2008 Energy Harvesting From Human and Machine Motion for Wireless Electronic Devices, Proc. IEEE, Vol. 96, 1457-1486.

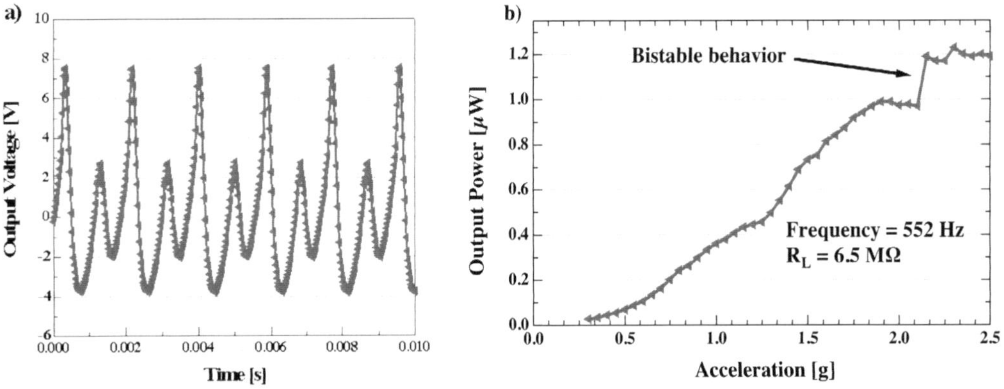

Figure 7. Power generation experiment results at 552 Hz. a) Output voltage waveform at 2.15 g external acceleration, b) Output power versus external acceleration.

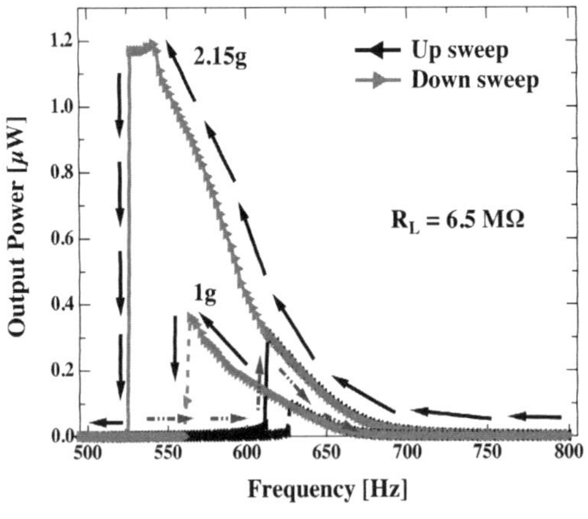

Figure 8. Output power versus the vibration frequency for both up- and down-sweeps at the external acceleration of 2.15g.

Flexible and Robust Multilayer Micro-Vibrational Harvesters for High Acceleration Environments

R. Lockhart, R. Dauksevicius, A. Vasquez Quintero, P. Janphuang, D. Briand and N. F. de Rooij

École Polytechnique Fédérale de Lausanne (EPFL), Institute of Microengineering (IMT), The Sensors, Actuators and Microsystems Laboratory (SAMLAB), Neuchâtel, Switzerland

robert.lockhart@epfl.ch

Abstract. This paper presents the fabrication and characterization of multilayer PVDF resonant micro-vibrational energy harvesters designed to withstand environments in which high levels of acceleration are present. The multilayer cantilevers are fabricated by combining two folded PVDF stacks into a multilayered, bimorph structure. This acts to increase the overall capacitance of the harvester, a problem that plaques PVDF cantilevers as a result of its low dielectric constant. Moderate powers (7 μW) are produced from the cantilevers even at high acceleration levels (20 g) due to the limited piezoelectric coefficient of PVDF; however, as a result of the high tensile strength and low elastic modulus of PVDF, the cantilevers are able to survive extremely high accelerations (> 4000 g) without breakage – a critical problem for harvesters based on brittle piezoelectric materials and substrates.

1. Introduction

Our increasing desire to measure the world around us has pushed the development of autonomous wireless sensor nodes establishing a need for self-sufficient power sources. At the same time, advances in ultra-low power communication strategies, such as the ultra-wideband radio technology (UWB), continue to reduce the energy requirements of such nodes [1]. Thus, in many environments, piezoelectric harvesters are more than capable of fulfilling the power requirements of sensor nodes; however, designing an inertial harvester to survive conditions in which low-frequency, high amplitude vibrations (or shocks) are present (i.e. transportation applications, industrial machinery, etc.) remains a challenge [2, 3]. Inertial energy harvesters based on rigid, brittle piezoelectric materials such as PZT, PMN-PT, AlN, etc. and relatively stiff, fragile substrates such as silicon require large structures, and carefully designed stoppers to limit cantilever displacement, achieving only minor gains in terms of reliability [4].

Polyvinylidene fluoride (PVDF), a flexible polymeric piezoelectric material, is often overlooked in the design of inertial harvesters due to its relatively low piezoelectric coefficient (23 pC/N), and its low dielectric constant (ε_r=10) causing it to generate low currents, and comparatively high voltages. However, PVDF offers a compliant solution enabling small, low frequency structures which are able to withstand high deformations without failure.

Thus, here we present fully packaged multilayer cantilevers fabricated by combining two folded PVDF stacks into a bimorph structure and encapsulated in a PMMA encasing. The harvesters are characterized at continuously varying moderate accelerations (20 g) and subjected to high-amplitude impacts in order to evaluate their performance and durability.

2. Design and Fabrication

The low dielectric constant of PVDF presents a concern, especially for low-frequency inertial harvesters, since the optimum load is inversely proportional to ωC, where ω is the angular frequency and C is the capacitance of the harvester. The optimum load of PVDF harvesters can often be tens of megaohms – signifying high voltages but very little current passing into the power management circuitry [5, 6]. The capacitance of the structure is proportional to the number of layers which make up the device; therefore, combining several layers in parallel lowers the optimum load of the harvester, allowing increased current to flow [7, 8]. Thus, the design of the PVDF harvesters incorporates multiple thin layers of piezoelectric material connected in parallel (as opposed to a single thick layer) in order to augment the capacitance of the device. 4-layer harvesters achieve a capacitance of 850 pF, while 2-layer devices have a capacitance of approximately 420 pF.

Piezoelectric harvesters occupying a volume of 64 mm^3 (4 x 4 x 4 mm^3 - including displacement) were fabricated by combining two folded PVDF stacks into a multilayered, bimorph structure as illustrated in figure 1. The PVDF used in these harvesters was supplied by Measurement Specialties. Their thinnest piezoelectric films were selected - 28 μm thick with Cu/Ni (70/10 nm) pre-sputtered electrodes on either side of the film. A compliant and easily applicable dry adhesive layer (ARClear) with a thickness of 50 μm was added between successive layers of PVDF to bind the structure together. Both materials, the PVDF and the dry adhesive, were patterned by laser ablation (Trotec Speedy 300).

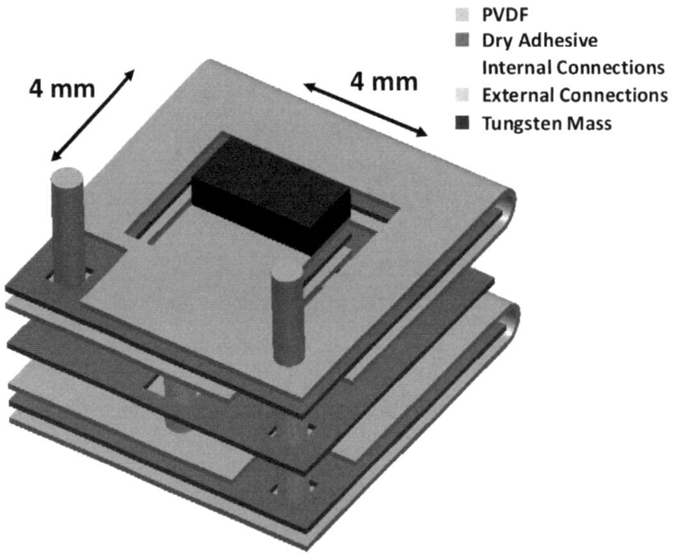

Figure 1: 3D schematic of a folded multistack PVDF harvester.

As demonstrated in figure 1, vias patterned into successive layers of PVDF and dry adhesive allow electrical connections to be made between the layers. The layers are connected in parallel; therefore, the PVDF is polarized in the same direction on either side of the neutral plane of the cantilever. For a simple 2-layer bimorph structure, the layers are electrically connected in the center plane. For multistack structures, the PVDF layers are again electrically connected in the center plane, but vias are also patterned to connect the inner electrodes of the folded PVDF stacks. An additional via is also required to make the external connections as shown in figure 1.

Tungsten proof masses (2 x 4 x 0.5 mm^3) , extending to half the length of the cantilever, were also patterned by laser machining and fixed to the end of the cantilevers with a layer of dry adhesive. A single proof mass was added in the case of the 2-layer structure in order to tune the resonance frequency to 100 Hz, while two masses were used in the case of the 4-layer structures to reduce the frequency to 350 Hz.

The harvesters were then placed in a PMMA package (1 x 1 x 0.5 cm) with a 4 mm gap between the ceiling and the floor of the cavity to limit the displacement of the mass at high accelerations (figure 2)

Figure 2: A packaged multilayer (4 layer) PVDF harvester on a 1SFr. coin (1 x 1 x 0.5 cm).

3. Characterization

Electromechanical testing of the cantilevers using a continuous sinusoidal acceleration of 20 g with and without the PMMA packaging reveals a slight degradation in the harvester performance as a result of increased damping within the package. The maximum power output is achieved by the simple bimorph structure which generates 7 μW at 100 Hz for an input acceleration of 20 g (figure 3). The quality factor of the structure (Q = 4) is very low in comparison to most resonant energy harvesters evident by the large observable bandwidth (FWHM = 25 Hz). A portion of the damping can be attributed to the packaging; however, as stated previously damping of the structure is comparable even when unpackaged. The main contributor is the PVDF which has a lower Q than ceramic piezoelectric materials. In addition, the dry adhesive also contributes to the increased damping of the structure.

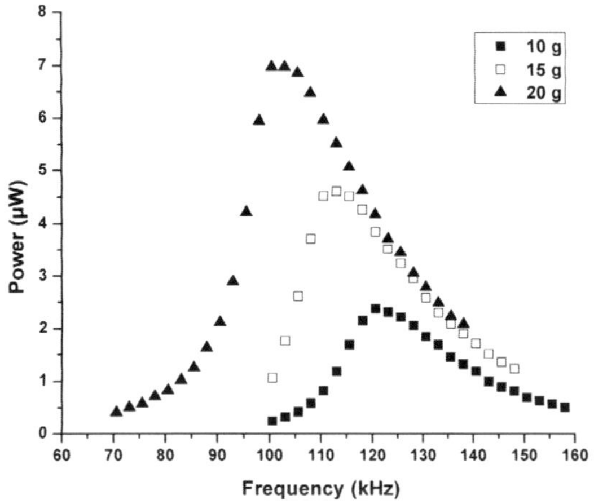

Figure 3: Generated power as a function of frequency and acceleration from a packaged bimorph PVDF cantilever excited at 20 g.

4. High Acceleration Testing

Shock tests performed by inflicting impact accelerations up to 4000 g show that the cantilevers are able to survive high acceleration environments. The harvesters were placed on a large mass (as illustrated in figure 4a) equipped with an accelerometer and dropped from 40 cm onto a cement floor. The voltage generated by the oscillating harvester was recorded and is presented along with the acceleration profile in figure 4b. After repeated high acceleration shock tests, the harvesters were retested with a continuous sinusoidal acceleration of 20 g. These tests which show equivalent performance before and after the shock tests demonstrate that harvesters are not damaged by the high acceleration impacts.

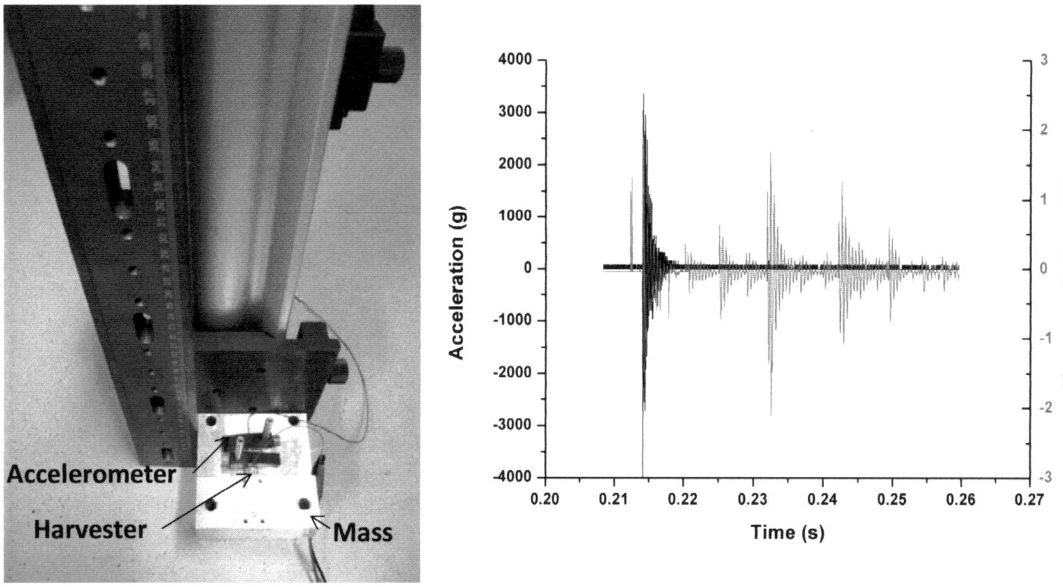

a) b)

Figure 4: a) Shock test setup to measure the effects of high acceleration impacts on vibrational energy harvesters. b) Shock impulse and the voltage signal output from the harvester.

Lifetime tests were also performed on the harvesters by providing a sinusoidal input excitation at the resonance frequency of the cantilevers over an extended period (18 hrs. at 20 g). In these tests, the harvesters undergo 6.5 million cycles. The RMS voltage generated by the harvesters was recorded periodically throughout these tests. Figure 3 shows a gradual reduction of the measured power by 50% after several hours of operation; however, upon successive testing after a period of rest the original power was recovered. The reduction observed in the output power the harvester over time is mainly due to a shift in its resonance frequency as evidenced by the increase in output power observed when the frequency of the input vibrations is adjusted after 18 hours of operation as shown in figure 5.

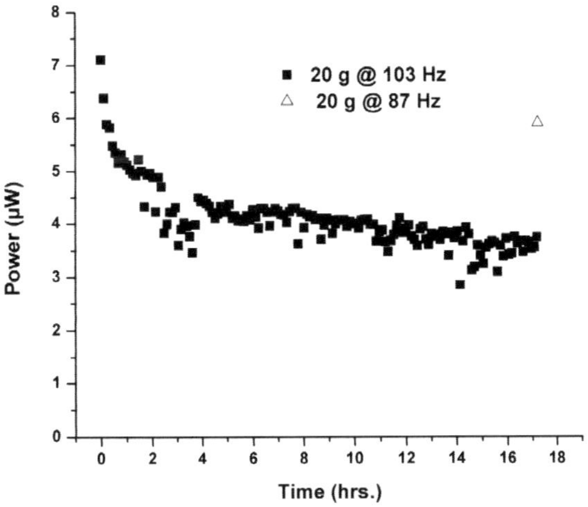

Figure 5: Accelerated lifetime tests driving the harvesters through 6.5 million cycles at an acceleration of 20 g. The power decay over time is mainly due to a shift in the resonance frequency.

5. Conclusions

The cantilevers produced in this study are small, yet flexible allowing low vibrational frequencies to be targeted while remaining robust enough to survive high acceleration environments. Optimization to improve the modest output power is currently ongoing.

References

[1] Janphuang P, Mishra B, Briand D, Lockhart R, Tasseli G, Robert C, Botteron C, de Rooij N F and Farine P A 2012 *Proc. of PowerMEMS 2012* (Atlanta, GA, USA, 2-5 December) 145-8
[2] Xu C-N, Akiyama M, Nonaka K and Watanabe T 1998 *IEEE Trans. UFFC* **45** 1068–70
[3] Roundy S, Wright P K and Rabaey J M 2004 *Energy scavenging for wireless sensor networks: with special focus on vibrations* (London: Springer)
[4] Schröder C, Senger F, Stoppel F, Wagner B and Benecke W 2012 *Proc. of PowerMEMS 2012* (Atlanta, GA, USA, 2-5 December) 343-6
[5] Jiang Y, Shiono S, Hamada H, Fujita T, Higuchi K and Maenaka K 2010 *Proc. of PowerMEMS 2010* (Leuven, Belgium, 30 Nov - 3 Dec, 2010) 375-8
[6] Tang Q C and Li X 2013 *Proc. of Transducers 2013* (Barcelona, Spain, 16-20 June) 697-700
[7] Song H-C, Kim H-C, Kang C-Y, Kim H-J, Yoon S-J and Jeong D-Y 2009 *J. Electroceram.* **23** 301-4
[8] Zhu D, Almusallam A, Beeby S, Tudor J and Harris N 2010 *Proc. of PowerMEMS 2010* (Leuven, Belgium, 30 Nov - 3 Dec) 335-8

Fabric based supercapacitor

S Yong[1], J R Owen[2], M J Tudor[1], S P Beeby[1]

[1]Electronic &Electrical Engineering Group, Electronics and Computer Science, University of Southampton, Southampton, SO17 1BJ, UK
[2]Electrochemistry, Chemistry, University of Southampton, Southampton, SO17 1BJ, UK

E-mail: sy2g11@soton.ac.uk

Abstract: Flexible supercapacitors with electrodes coated on inexpensive fabrics by the dipping technique. This paper present details of the design, fabrication and characterisation of fabric supercapacitor. The sandwich structured supercapacitors can achieve specific capacitances of 11.1F/g, area capacitance 105 mF.cm^{-2} and maintain 95% of the initial capacitance after cycling the device for more than 15000 times.

1. Introduction

Wearable electronics require the implementation of electronic functions on a flexible, wearable substrate. The electronic functions require specific energy storage devices with minimal weight and size that can be integrated on to such a flexible substrate. Research and development of a flexible energy storage system is essential for the powering of wearable electronics devices. The supercapacitor is a commonly used storage devices for smart sensor node [1] and energy harvesting systems [2]. This paper present a detailed investigation in to supercapacitor fabricated on standard polyester-cotton fabrics.

Conventionally, carbon nanotube (CNT) and/or pseudo-capacitive materials are the most common materials used in the fabrication of flexible supercapacitor electrodes. Supercapacitor electrodes made with CNT benefit from the unique properties of CNT such as porous structure, excellent electrical conductivity, high specific surface area, and good thermal and mechanical stability. Previous reports on flexible supercapacitors using CNT and/or pseudo-capacitive materials achieve high specific capacitance, but it is essential to note that these materials are expansive. The use of CNT can also reduce the device stability, as found by K Wang et al [3] in flexible electrodes of PANI-CNT-Cotton whose specific capacitance of 410 F.g^{-1} fell to 61% of its original value after 3000 cycles. Activated carbon, also known as engineered carbon, is another suitable material for flexible supercapacitor electrodes due to its large specific surface area, low electric resistivity and low cost in comparison with some specific treated materials like carbon nanotubes.

In comparison with a conventional rigid supercapacitor, a fabric based supercapacitor needs to be conducting, lightweight, and wearable and must not react with the electrolyte. Previously, researchers have focused on using substrates that are either lack of mechanical strength witch is not directly appropriate for device integration, such as non-woven fabric[3] or paper [4] to support a high mass loading of active conducting material, or have adopted complex procedures to make flexible conductive substrates, such as woven graphene fibres [5] or conductive polymers [6]. These methods permit the fabrication of flexible supercapacitors with high storage capacities and power/energy efficiencies; however they also involve high materials cost and fabrication complexity which limits their commercialisation.

2. Theory

Supercapacitors store energy with two types of capacitive mechanisms: the electrical double layer capacitance (EDLC) and redox pseudocapacitance. The EDLC is based on the accumulation and release of electrostatic charge that occurs at the interfaces between the electrode and electrolyte solution. A supercapacitor is constructed using two electrical double layer interfaces, each having opposite polarity with respect to the charge separator that contains electrolyte solution.

The exchange of electrostatic charge occurs via the balance of ions and electrons at the interface between electrode and electrolyte and introduces an internal potential difference across the interface, which acts as a dielectric. Charge on the inner surface of electrodes will be balanced by the accumulation of counter-ions in the electrolyte. The ratio of charge to the voltage difference gives the capacitance between the electrodes [7]. The maximum potential difference between the top and bottom electrodes is restricted by the thermodynamic stability of the electrolyte material [7].

This work demonstrates energy storage in wearable devices by implementing a wearable electric double layer supercapacitor into a fabric substrate. The proposed fabric electrode is made via the dipping process using various inexpensive commercial carbon powders and poly-cotton fabric substrates that are used in the clothing industry. The supercapacitors will be fully tested in order to study their operation and stability as functions of cycle rate and cycle number.

3. Device fabrication

Firstly the liquid coating mixture containing the active material carbon is prepared. In order to create a permanently conductive carbon-coated fabric, a polymer binder is required to provide adhesion between carbon powders and the yarns of the fabric. In this work the carbon powder is mixed with a polystyrene, ethyl acetate solvent and a surfactant. The carbon powders used in this work are activated carbon powder SX ULTRA and carbon black powder Shawinigan Black. SX ULTRA, processed from peat by Cabot Norit, has high purity, a mean particle size of 100μm in diameter, a neutral pH balance and an effective surface area of 1200 $m^2.g^{-1}$. Shawinigan Black has a mean particle size of 42 nm in diameter and an effective surface area of 75 $m^2.g^{-1}$.

The next step of this process is to load the carbon powders into the fabric using a dipping process. The fabric substrate in this work is poly-cotton material that combines both the advantages of polyester and cotton material. During the dipping process, the fabric material absorbs the liquid and is quickly filled with the carbon mixture as the solvent evaporates at room temperature. After the sample is fully cured, carbon powder bonds to the substrate through the polymer binder to introduce a large, conductive surface area that forms an electrical double layer on contact with the electrolyte. The photograph of original fabric and carbon fabric electrode can be seen in figure 1 (a)

Figure 1. (a) Original blue fabric substrate and black carbon fabric electrode. (b) SEM photograph of carbon fabric electrode (b).

Figure 1(b) is the SEM micrograph of the carbon fabric electrode which shows that the carbon particles adhere to the fibre to make the fabric conductive. The assembled supercapacitor device shown in figure 3 is compressed from the top to bottom electrodes using spring loaded stainless steel current collectors housed within a Swagelok® PFA tube fitting shows in figure 2(b). In figure 2(a) the white circle is the filter paper layer that acts as the charge separator to prevent short circuit of device. In our design the electrolyte is 1M lithium sulphate solution is chosen as a chemically safe and inexpensive electrolyte compared with an organic electrolyte.

Figure 2. (a) Photograph of fabric supercapacitor before assembly Swagelok® PFA tube fitting. (b) Swagelok® PFA tube fitting test rig

4. Results and discussion

The performance of the textile based supercapacitor is assessed by several methods. First, the encapsulated device is using electrochemical impedance spectroscopy (EIS), and then the stability of the device is examined by cyclic voltammetry (CV) at different scan rates. These can be seen in figure 3 and 4.

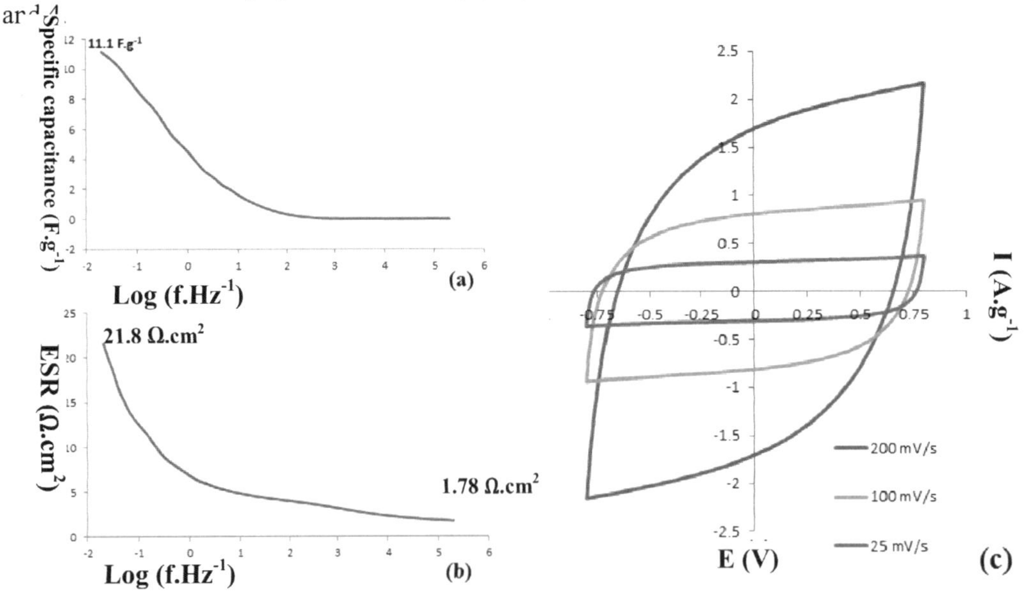

Figure 3. (a)(b) Bode plot of device from 10 kHz to 20 mHz, extract from EIS test with equivalent series resistance (ESR) marked. (c) CV test of the device between +/- 0.8 V at the scan rate of 200, 100, 25 $mV.s^{-1}$

In each tested fabric supercapacitor, carbon materials account for 17% of the electrodes weight; each electrode has the area of 0.785 cm^2. As shown in figure 3 (a) and (b) the impedance result is expressed as a series combination of the frequency dependent capacitance and a series resistance. The high frequency resistance indicates a total resistance of 1.78 $\Omega.cm^2$ due to the electrolyte and electrode materials. At the lowest frequency, the specific capacitance is found to be 11.1 $F.g^{-1}$, the area specific capacitance 105 $mF.cm^{-2}$ and a normalized ESR of 21.8 $\Omega.cm^2$ that includes the diffusion resistance. As shown in figure 3, the CV curves indicate that the device is electrochemically stable at the scan rates from 25 to 200 mV, and capacitance values of about 10 $F.g^{-1}$ according to the ratio of the maximum current to the scan rate.

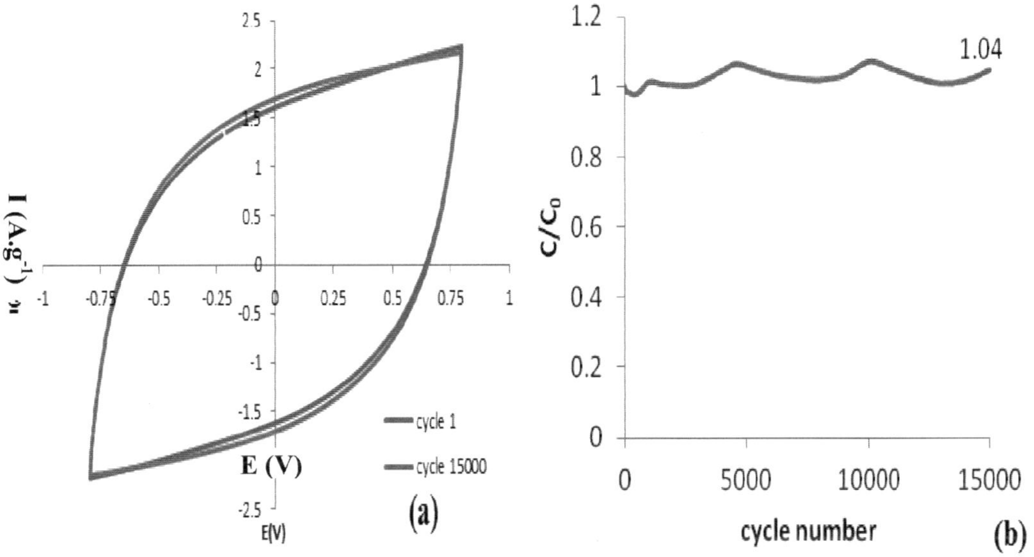

Figure 4. (a) CV test of the device for 15000 cycles between +/- 0.8V at the scan rate of 200 $mV.s^{-1}$, (b)stability of device over 15000 cycles. Co is the capacitance of device calculated at the beginning (cycle 1) of the CV test.

Figure 5(a) (b) shows the device stability with a CV response that is unchanged from its initial value after 15000 cycles. The overall variation of less than 5% in the low frequency capacitance is correlated with small temperature changes causing variation of the diffusion coefficients. The high device stability shows excellent adhesion of the carbon powders forming a continuous conducting network, which is due to an effective and stable polymer binder with a good chemical resistance both the electrolyte and carbon material.

5. Conclusion

This report presents a three layer supercapacitor on a fabric substrate. The supercapacitor fabricated in this work achieves a mass specific capacitance of 11.1 $F.g^{-1}$, area specific capacitance of 103 $mF.cm^{-2}$, a low normalized ESR of 1.78 $\Omega.cm^2$ and achieves good cycling stability of less than 5% capacitance variation over 15000 cycles. In comparison with other supercapacitors, the electrode and separator materials of the device are potentially wearable, fully scalable, and inexpensive. Future work will include developing a encapsulation structure and materials that seal the device, retain the electrolyte and protecting it against damage. The final device could see use applications in a wide range of wearable electronic systems like energy harvesters, medical sensors and energy sources for all kinds of personal wearable electronics.

6. Acknowledgment

The authors thank the EPSRC for supporting this research with grant reference EP/1005323/1.

References

[1] Zhu D, Beeby, S, Tudor, J and Harris N 2011 *Sensors and Actuators A Physical* **169(2)** 317-325

[2] Beeby, S and Zhu, D 2011, *Materials Science Foundations*, 197-218.

[3] Wang K, Zhao P, Zhou X, Wu H, Wei Z 2011 *J. Mater. Chem*, **21(41)** 16373–16378

[4] Zhang Z, Zhai T, Lu X, Yu M, Tong Y, Mai K, 2013 *J. Mater. Chem*,**1**, 505-509

[5] Cheng H, Dong Z, Hu C, Zhao Y, Hu Y, Qu L 2013 *Nanoscale*, **5**, 3428–3434

[6] Hu L, Pasta M, Mantia F, Cui L, Jeong S, Deshazer H, Choi J, Han, Cui Y 2010. *Nano Lett.* **10 (2)**, 708–714

[7] Li Z, Zhao X 2009 *Chem. Soc. Rev*, **38**, 2520–2531

Electrostatic vibration energy harvester with increased charging current

V Dragunov and V Dorzhiev

Department of Semiconductor Devices and Microelectronics, Novosibirsk State Technical University, Novosibirsk, Russia

E-mail: b.dorzhiev@gmail.com

Abstract. The analysis of the operation of the electrostatic vibration energy harvester to charge self-contained power supply is carried out. An analytical expression to estimate the average charging current taking into account diode's reverse current is obtained. The ways to increase the charging current were found. The harvester with increased charging current containing no switches and inductive elements is suggested.

1. Introduction

Recent achievements in microelectronics and microelectromechanical systems technology led to the appearance of mobile, compact and absolutely self-contained devices with very low energy consumption. Nowadays, batteries requiring periodic replacement, or recharging that is not always possible are used as self-contained power source for such devices.

The most suitable alternatives to traditional batteries are electrostatic vibration energy harvesters that allow harvesting the energy from the environment directly at the place of operation. However, in many cases, the application of traditional batteries cannot be completely excluded. Therefore, the most optimal solution of the problem of a self-contained power source is the using of electrostatic vibration energy harvesters to charge compact rechargeable battery.

The main limitations for the most such harvesters are an inductive element and the necessity to use switches. Moreover for operation of these switches the control circuits consuming energy are needed [1, 2].

One of the harvesters without these limitations is the harvester based on Bennet's doubler [3, 4], however the charging current in the harvester is limited by battery voltage. In [3, 4], a modification of such harvester with the increased charging current is proposed, but it includes an inductive element and requires a switch. Furthermore, in this harvester the charging of the battery is implemented through periodical connecting the capacitor charged to high voltage to the battery.

The paper proposes the electrostatic vibration energy harvester with increased charging current without inducting elements and switches. Its theoretical and experimental investigation is described.

2. Basic harvester based on Bennet's doubler

The electrical circuit of the basic harvester based on Bennet's doubler is shown in figure 1. This harvester contains a battery (V_0), three diodes (D_1, D_2 and D_3) and two variable capacitors (C_1 and C_2) operating in antiphase. The load is connected in parallel to a battery.

An analysis of the harvester operation is carried out in [3, 4]. However, during the analysis the diode model without reverse current was used. At the same time, our study shows that the charging current at this circuit is comparable in many cases with the magnitude of the diode's reverse current.

In this paper, the harvester operation was analyzed using the Shockley diode equation as a diode's model.

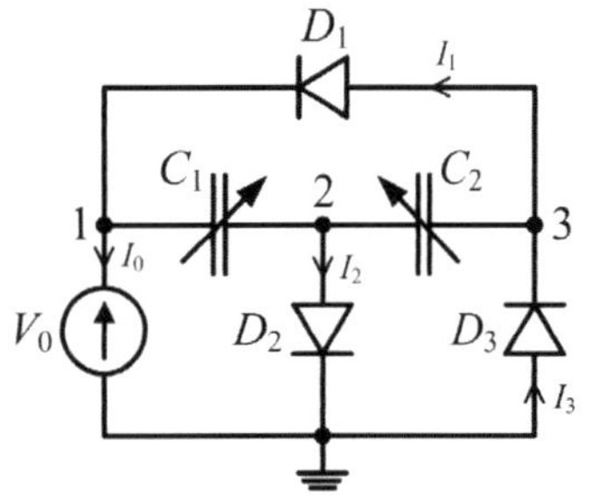

Figure 1. Basic harvester based on Bennet's doubler.

2.1. Numerical model and analysis

The differential equation system to describe the operation of the harvester taking into account diode's reverse current is:

$$\begin{cases} \dfrac{dq_1(t)}{dt} = I_S\left[\exp\left(\dfrac{\varphi_2}{m\varphi_T}\right)-1\right] + I_S\left[\exp\left(\dfrac{\varphi_3-\varphi_1}{m\varphi_T}\right)-1\right] - I_S\left[\exp\left(\dfrac{-\varphi_3}{m\varphi_T}\right)-1\right] \\ \dfrac{dq_2(t)}{dt} = I_S\left[\exp\left(\dfrac{-\varphi_3}{m\varphi_T}\right)-1\right] - I_S\left[\exp\left(\dfrac{\varphi_3-\varphi_1}{m\varphi_T}\right)-1\right] \end{cases} \quad (1)$$

where $q_1(t)$ and $q_2(t)$ are charges on C_1 and C_2 capacitors plates respectively, I_S is reverse bias saturation current, φ_i is potential of node i, m is the ideality factor, φ_T is the thermal voltage.

Simulation results using equation (1) are presented in figure 2. During the calculations the interelectrode gap varied in a sinusoidal manner, and the harvester's parameters were as follows: $q_1(0) = q_2(0) = 0$, $V_0 = 3$ V, $I_S = 3$ nA, $m = 1$, $\varphi_T = 25$ mV, maximum capacitance is $C_{max} = 100$ pF, minimum capacitance is $C_{min} = 20$ pF and frequency of mechanical vibrations is $f = 40$ Hz ($T = 1/f = 0.025$ s). Figure 2 shows that when connected battery to the circuit, the diode D_2 opens and capacitor C_1 is charged to a voltage of battery. Further operation of the harvester can be divided into two phases. The first phase corresponds to the decrease in capacitance C_1, meanwhile C_1 is discharged through the chain $C_1 \Rightarrow V_0 \Rightarrow D_3 \Rightarrow C_2$ charging the battery and the capacitor C_2. The second phase corresponds to the increase in capacity C_1, meanwhile C_1 is charged and C_2 is discharged through the chain $C_2 \Rightarrow D_1 \Rightarrow C_1$.

The analysis shows that the efficiency of the harvester decreases because of the diode's reverse current. Note that the diode D_2 theoretically required for the initial charging capacitor C_1. Thus, the disabling the circuit diode D_2 after the initial charging of the capacitor C_1 could enhance the effectiveness of the harvester. But in practice due to uncontrolled self-discharge capacitors the need for their recharge remains and the diode D_2 must not be interrupted.

From the analysis of equation (1) we have obtained an expression that allows to estimate the average charging current taking into account diode's reverse current

$$I_0 = V_0 C_{max} f \frac{\eta - 1/\eta - 1}{\eta + 1} - I_S \frac{1.5\eta + 1}{\eta + 1} \quad (2)$$

where $\eta = C_{\max}/C_{\min}$.

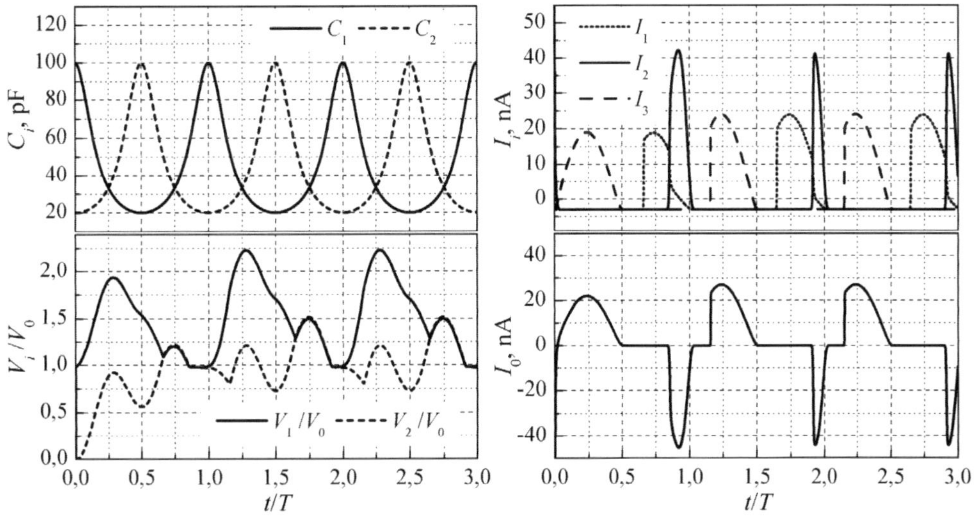

Figure 2. Dependences of the capacitances, capacitors voltages, diode currents and the charging current on the normalized time.

The second term in equation (2) is the discharge current of the battery caused by the diode's reverse current. According to equation (2) at $I_S > 2V_0C_{\max}f\left(\eta^2-\eta-1\right)/\left(3\eta^2+2\eta\right)$ charging battery becomes impossible. For example, at harvester parameters used previously, the use of diodes with reverse current value of 6 nA will lead to a negative average charging current, i.e. instead of battery charging there will be battery discharge.

In [5] the obtained expressions allow to estimate the critical values of the main characteristics of the harvester components, such as the minimum ratio η, the minimum frequency of mechanical vibration and minimum battery voltage.

2.2. Ways to increase the average charging current

According to equation (1) the average charging current can be increased through higher values of the following characteristics: battery voltage, mechanical vibrations frequency, maximum values of the capacitors and ratio $C_{\max}/C_{\min}$. However, in reality, the battery voltage and the frequency of mechanical vibrations are determined by the operation conditions of the harvester. The maximum value of the capacitors are dependent on technological capabilities and mass-dimensional characteristics of the harvester. The analysis of equation (1) also shows that an increase in $C_{\max}/C_{\min}$ more than 10 does not lead to a significant increase in the charging current.

Thus, while developing converters with higher charging current it is necessary to consider the restrictions mentioned above.

3. Electrostatic vibration energy harvester with increased charging current

The proposed harvester circuit with increased charging current is shown in figure 3. In this harvester the increase in charging current is done by enhancing the potential of node 1 by means of setting the storage capacitor C_S accumulating the charge and Zener diode D_0. As a result, the potential of the node 1 is increased, that is equivalent to the use of a battery with high voltage in harvester (figure 1).

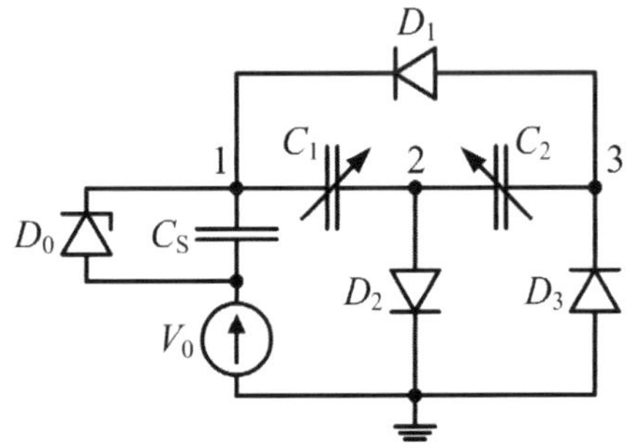

Figure 3. Electrostatic vibration energy harvester with increased charging current.

3.1. Simulation

Figure 4 and figure 5 shows the dependence of the voltage at the storage capacitor C_S and the battery charging current on the normalized time. This dependence is obtained by solving a system of differential equations describing the operation of the circuit shown in figure 2. The calculations assumed zener voltage of 90 V, capacitance, C_0 = 0.1 nF, and other parameters the same as when calculating the dependences given in figure 2.

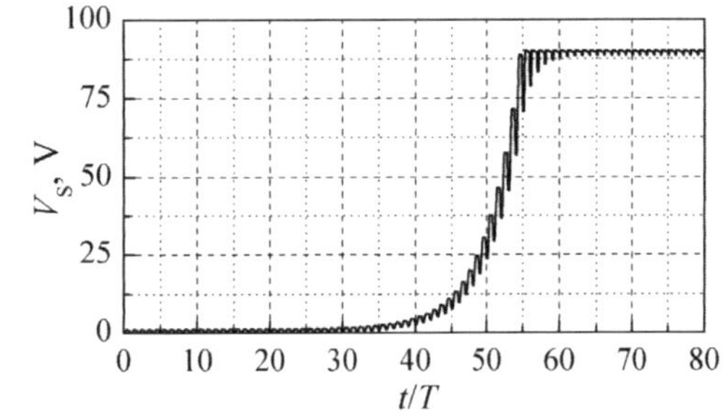

Figure 4. The dependence of the storage capacitor voltage on the normalized time.

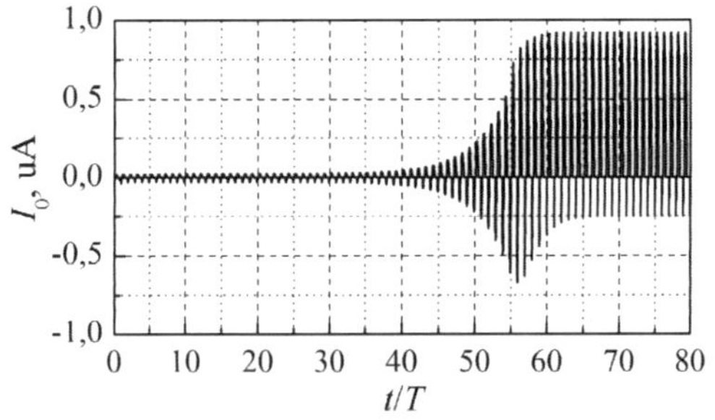

Figure 5. The dependence of the charging current on the normalized time.

Figure 4 shows that in the proposed scheme, the storage capacitor voltage in steady state is equal to the zener voltage, with an average charging current (figure 5) ≈ 227 nA, whereas without this storage capacitor, the average charging current is ≈ 3.4 nA (figure 2).

Theoretically, by increasing the zener voltage it is possible to increase the charging current of battery indefinitely. However, in reality, this current will be limited by obstinate parameters in particular. Moreover experimental studies are required to determine the attainable parameters.

3.2. Experiment

In experimental studies, as two variable capacitors with a variable capacitance in antiphase, there was used the structure of three electrodes area of 49 cm^2, two of which are rigidly interconnected with the electrode gap of 8 mm, and the third electrode is mounted there between. Electrode gap varied harmonically with frequency of 40 Hz. A series connection of five diodes 1N4004 was used as diodes and a series connection of two diodes 1N4004 was used as the Zener diode. Capacitance of the storage capacitor was C_0 = 3.3 nF, the battery voltage V_0 = 30 V.

Figures 6 and 7 show the dependence of voltage at the storage capacitor C_S and the charging current on the charging time obtained experimentally. Time difference is caused by the fact that when measuring the voltage, 1 GOm is in parallel connected to C_S. Upon that the steady state time is increasing.

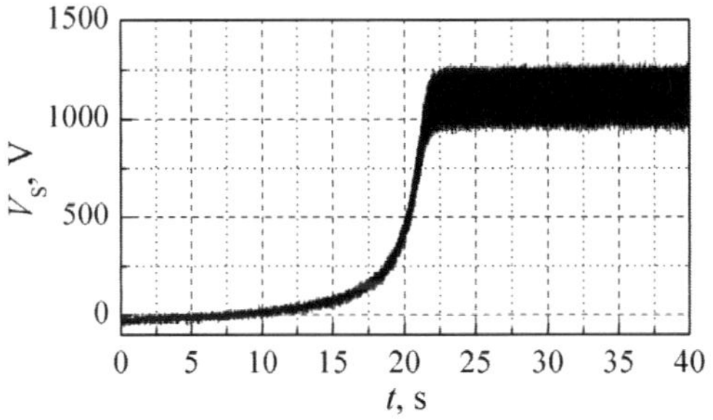

Figure 6. The dependence of the voltage at the storage capacitor on the time, the experimental.

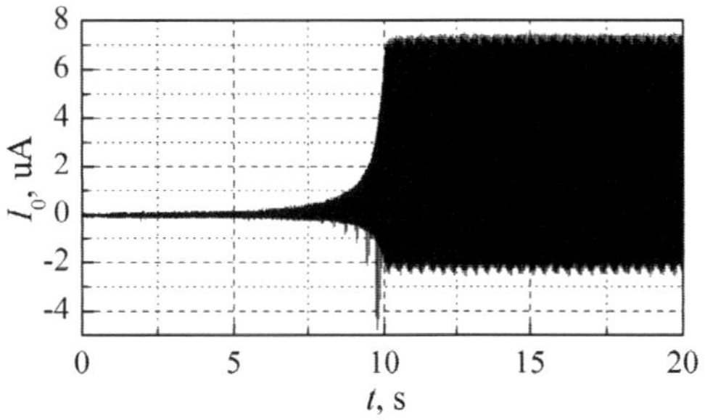

Figure 7. The dependence of the charging current on time, the experimental.

It is seen that the average voltage in the steady state is equal to 1150 V and the average charging current in the steady state is 1.8 uA. At the same time, the experiment without storage capacitor showed charging current about 1.6 nA.

Thus, in the proposed harvester, through increasing the zener voltage, and, accordingly, the voltage at the storage capacitor, charging current can be significantly increased without higher voltage of battery and without inductive elements and switches.

4. Conclusion

In this paper the simulation of the basic harvester for charging a rechargeable self-contained power supply is carried out. An analytical expression that allows to estimate the average charging current taking into account diode's reverse current was obtained. It is shown that at high diode's reverse current average charging current can be negative, i.e. instead of battery charging there will be battery discharge.

The ways to increase the average charging current were found. It is shown that the average charging current can be increased through higher values of the following characteristics: battery voltage, mechanical vibrations frequency, maximum values of the capacitors and ratio C_{max}/C_{min}. Determined that an increase in C_{max}/C_{min} more than 10 does not lead to a significant increase in the charging current.

An electrostatic vibration energy harvester, theoretically has unlimited charging current, and without inductive elements and switches was proposed. The simulation and experimental study of this harvester were carried out. A significant increase in the charging current in comparison with the basic harvester was proved. An experimental study in which the average charging current was 1.8 uA at a voltage on a storage capacitor kV.

References

[1] Dragunov V and Ostertak D 2012 Microelectromechanical Converters *Russian Microelectronics* **41** 107–21

[2] Dudka A, Galayko D and Basset P 2012 Design of controller IC for asynchronous conditioning circuit of an electrostatic vibration energy harvester *IEEE Int. Conf. on Green Computing and Communications (Besancon, France, 20–23 November 2012)* pp 660–3

[3] de Queiroz A C M and Domingues M 2011 The doubler of electricity used as battery charger *IEEE Trans. Circuits Syst. II* **58** 797–801

[4] de Queiroz A C M and Domingues M 2011 Electrostatic energy harvesting using doublers of electricity *54th IEEE Midwest Symp. on Circuits and Systems (Seoul, Korea, 7–11 August 2011)* pp 1–4

[5] Dragunov V and Dorzhiev V 2012 Microelectromechanical generator based on Bennet's doubler *Journal of nano- and microsystem technique* **11** 39–42

Limit behavior of the magnetic coupling coefficient for mid-range, near-field applications

Emily Guthrie, Anton Frolenkov and Jose Oscar Mur-Miranda
Olin College of Engineering, Needham, MA, USA
E-mail: `emily.guthrie@students.olin.edu`

Abstract. We have derived compact algebraic bounds for the limit behavior of the magnetic coupling coefficient, κ, for axially aligned multilayer coils. These bounds are validated experimentally. We have also found an expression based on the coil geometry that captures its long-range magnetic behavior. In particular, the limit behavior of the magnetic coupling coefficient is the same for any pair of axially aligned multilayer coils when the separation distance is normalized by this expression.

1. Introduction

The magnetic coupling coefficient κ is a key parameter that characterizes the performance of inductive systems. Since the efficiency of inductive power transfer is proportional to the magnetic coupling coefficient squared for large enough separation distances [1, 2], these expressions bound the performance of these systems in this range. This also suggests optimal geometric coil designs that maximize the limit behavior of the magnetic coupling coefficient.

Since the mutual inductance between two filament loops is a transcendental expression that requires elliptic integrals [3], there are no general closed-form expressions for the self and mutual inductance of coils. Furthermore, even though several approximations for both self and mutual inductances of various geometries have been found [4, 5, 8], the magnetic coupling coefficient is still calculated empirically [2], derived [6], or approximated within a specific context [7, 9]. We present a simple expression that serves as a widely applicable upper bound on the magnetic coupling coefficient for a broad class of inductive systems. Furthermore, they provide a limit test for similar design equations in other configurations, such as unaligned coils, or coils with other cross sectional geometries.

2. Derivation

Figure 1 shows the specific case under consideration, consisting of axially aligned, multi-layer coils $i = 1, 2$ with separation s and characterized by their inductance L_i, turns density ρ_{Ni}, outer radius r_i, length ℓ_i, and fill factor f_i, where $0 < f_i < 1$. In the limit as the separation between sender and receiver increases, the magnetic coupling coefficient starts to decrease inversely proportional to the distance cubed, which determines the performance of mid-range inductive systems [1, 2, 10]. At this long range, while in the near field, the magnetic coupling coefficient can be calculated by combining the mutual inductance, given by

$$L_M = \frac{\mu_0 \rho_{N1} \rho_{N2} \pi}{2} \int_0^{\ell_2} \int_{r_2(1-f_2)}^{r_2} r_2'^2 \int_0^{\ell_1} \int_{r_1(1-f_1)}^{r_1} \frac{r_1'^2}{[r_1'^2 + (s + \ell_1' + \ell_2')^2]^{3/2}} dr_1' d\ell_1' dr_2' d\ell_2',$$

with Wheeler's approximation for the self inductance [5],

$$L_i = \frac{\mu_0 (\rho_{Ni} \ell_i r_i f_i)^2 \pi r_i^2 (2 - f_i)}{3r_i + 3.51 r_i f_i + 4.51 \ell_i}, \qquad i = 1, 2.$$

The mutual inductance calculation is the integral for the cross section of each coil of the field along the center axis from a single loop multiplied by the area of a single receiving loop. Since this expression relies on Wheeler's approximation for the self inductance of a multi-layer coil, it shares the same limitations, namely, the aspect ratio (a.r.) of the coils, $\ell_i/(r_i f_i)$ must be close to 1, or $1/2 < \ell_i/(r_i f_i) < 2$. However, there are widely known approximations for other aspect ratios that would result in similar expressions [5].

The resulting expression for the magnetic coupling coefficient κ is

$$\kappa = \frac{L_M}{\sqrt{L_1 L_2}} = \frac{\sqrt{3r_1 + 3.51 r_1 f_1 + 4.51 \ell_1}}{2 - f_1} \frac{\sqrt{3r_2 + 3.51 r_2 f_2 + 4.51 \ell_2}}{2 - f_2} \frac{r_1 r_2}{4 f_1} \left(\frac{f_2^2}{3} - f_2 + 1 \right) \times \ldots$$

$$\sum_{i=1..4} (-1)^i \left[\sqrt{1 + \left(\frac{s + \ell_i}{r_1} \right)^2}^2 - (1 - f_1) \sqrt{(1 - f_1)^2 + \left(\frac{s + \ell_i}{r_1} \right)^2}^2 + \ldots \right.$$

$$\left. \left(\frac{s + \ell_i}{r_1} \right)^2 \arcsin \frac{r_1}{s + \ell_i} - \left(\frac{s + \ell_i}{r_1} \right)^2 \arcsin \frac{r_1 (1 - f_1)}{s + \ell_i} \right], \quad \ell_i = \{0, \ell_1, \ell_1 + \ell_2, \ell_2\}.$$

The first term of the Taylor series expansion of this expression provides an upper bound, κ_u, for the magnetic coupling coefficient as the separation s tends to infinity, where

$$\kappa_u = \frac{1}{2} r_1 \sqrt{3r_1 + 3.51 r_1 f_1 + 4.51 \ell_1} \left(\frac{f_1^2/3 - f_1 + 1}{2 - f_1} \right) \times \ldots$$

$$r_2 \sqrt{3r_2 + 3.51 r_2 f_2 + 4.51 \ell_2} \left(\frac{f_2^2/3 - f_2 + 1}{2 - f_2} \right) \frac{1}{s^3},$$

as shown in Figure 2. Adding the second term of the Taylor series expansion results in a lower bound for the magnetic coupling coefficient, κ_l, where

$$\kappa_l = \kappa_u \left(1 - \frac{3}{2} \frac{\ell_1 + \ell_2}{s} \right),$$

also shown in Figure 2. Even though the value of the magnetic coupling coefficient in this range is on the order of 10^{-2} to 10^{-4}, large enough quality factors in resonant systems may allow the transfer of power at high enough magnitude and efficiency for some applications [11].

The upper and lower bounds for the coupling coefficient also let us determine when the upper bound approximation for κ is an appropriate substitute. The normalized difference between the bounds is

$$\frac{\kappa_u - \kappa_l}{\kappa_u} = \frac{3}{2} \frac{\ell_1 + \ell_2}{s}.$$

This expression can be solved for s to indicate the range of separations where the magnetic coupling coefficient is arbitrarily close to its limit κ_u.

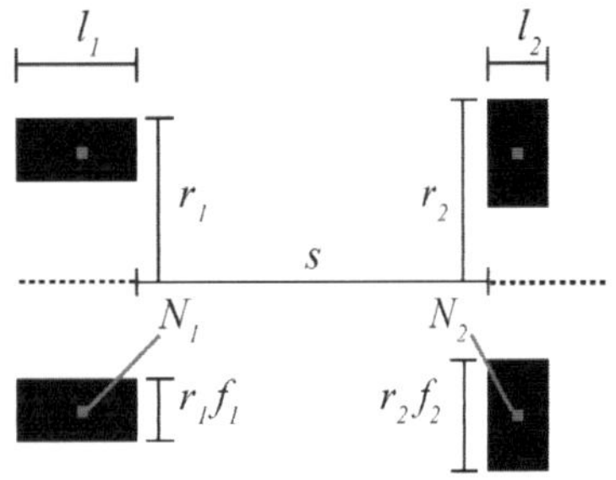

Figure 1. Cross-section diagram of the coil geometry. Sender and receiver coils are multilayer, axially aligned with length ℓ_i, outer radius r_i, turns N_i, and thickness $r_i f_i$, where $0 < f_i < 1$ is the fill. The separation between coils is s. The turns density ρ_{Ni} is defined as $N_i/(\ell_i r_i f_i)$.

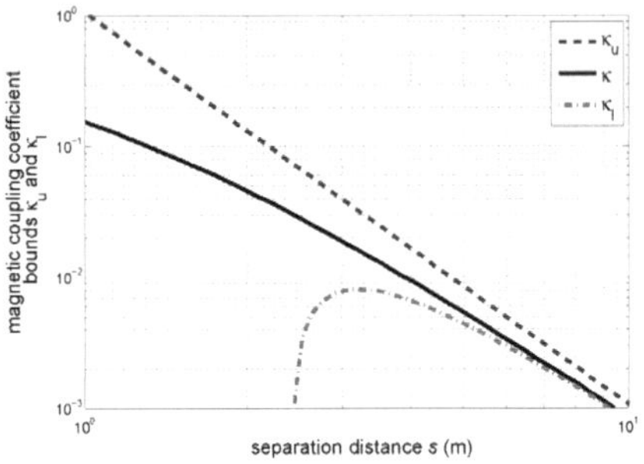

Figure 2. Long-range approximation for the magnetic coupling coefficient κ and bounds κ_u and κ_l as a function of separation distance s for identical sender and receiver coils of radii $r_i = 1$ m, aspect ratio 2, and fill $f = 0.4$. As expected, both the approximation for κ and κ_l converge to κ_u as s approaches infinity.

3. Normalization

The upper bound for the magnetic coupling coefficient, κ_u, suggests a normalized separation distance δ, where

$$\delta = s \Big/ \sqrt{(V_{\text{eff},1})^{1/3} \cdot (V_{\text{eff},2})^{1/3}}$$

and

$$V_{\text{eff,i}} = \left[3 + 3.51 f_i + 4.51 \ell_i / r_i\right] \left[\frac{f_i^2/3 - f_i + 1}{2 - f_i}\right]^2 r_i^3, \quad i = 1, 2.$$

Since the quantity V_{eff} has units of volume, we call it the *effective magnetic volume*. As shown in Figure 3, this normalization simplifies the limit behavior of any pair of coils to

$$\kappa_u = 1/2\delta^3.$$

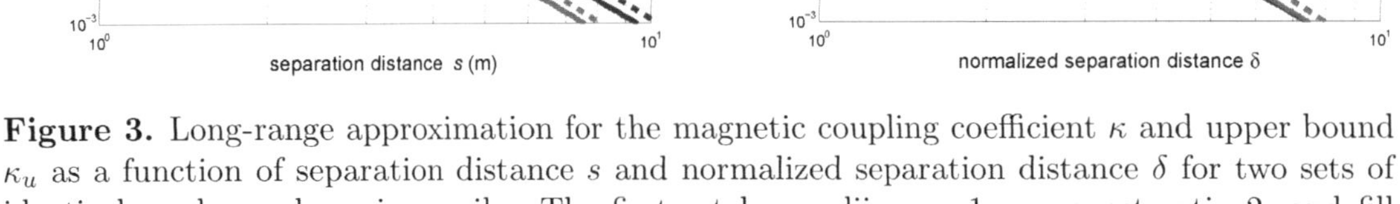

Figure 3. Long-range approximation for the magnetic coupling coefficient κ and upper bound κ_u as a function of separation distance s and normalized separation distance δ for two sets of identical sender and receiver coils. The first set has radii $r_i = 1$ m, aspect ratio 2, and fill $f = 0.4$. The second set has radii $r_i = 1$ m, aspect ratio 1/2, and fill $f = 0.9$. As expected, the limit κ_u is identical for both sets when plotted against the normalized separation distance δ.

Journal of Physics: Conference Series **476** (2013) 012116 doi:10.1088/1742-6596/476/1/012116

4. Validation

The expressions derived for the magnetic coupling coefficient and its bounds were validated experimentally by measuring the voltage induced by a sender coil on an axially aligned receiver coil as shown in Figure 4. The experiment was repeated using three different coil pairs. The circuit model of the experimental system ignores the load presented by the measuring probes and assumes no current in the receiver coil. The sender coil was driven with a sinusoidal voltage v_S of 1.64 Vrms. The frequency was adjusted until the phase difference between the voltages across the sender v_S and receiver v_2 was 45 degrees, such that

$$\left|\frac{v_2}{v_S}\right| = \frac{\kappa}{\sqrt{2}}\sqrt{\frac{L_2}{L_1}},$$

where v_2 is the voltage across the receiver coil, L_2 is the inductance of the receiver coil, L_1 is the inductance of the sender coil, and v_S is the voltage across the sender coil.

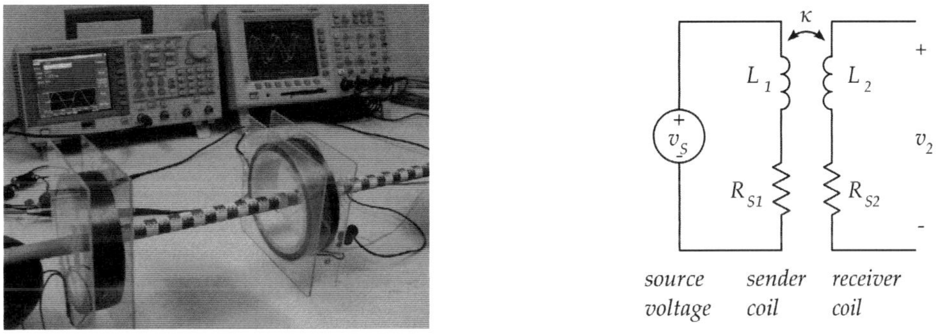

Figure 4. Experimental system for measuring the magnetic coupling coefficient between to axially aligned coils and circuit model. A sender coil of inductance L_1 is driven using a voltage source v_S. The induced voltage v_2 is measured across an axially aligned receiver coil of inductance L_2. The source coil is driven at the frequency $R_{S1}/(2\pi L_1)$ such that the measurement is independent of R_{S1}.

In two experiments, shown in red and blue in Figure 5, the sender coil had an outer radius $r_1 = 5$ cm, inner radius of 4.5 cm (for a resulting fill factor $f_1 = 0.1$), length $\ell_1 = 2.5$ cm and a measured inductance of $L_1 = 0.74$ H. The first experiment, shown in red in Figure 5, had a receiving coil with dimensions and inductance identical to the transmitting coil. The second experiment, shown in blue in Figure 5, had a receiver coil with an outer radius $r_2 = 7$ cm, inner radius of 6.5 cm (for a resulting fill factor $f_2 = 0.07$), length $\ell_2 = 2.5$ cm and a measured inductance of $L_2 = 2.76$ H.

In the third experiment, shown in purple in Figure 5, the sender coil was the receiver coil from the second experiment with a measured inductance of $L_1 = 2.76$ H. The receiver coil had an outer radius $r_2 = 6.5$ cm, inner radius of 4.5 cm (for a resulting fill factor $f_2 = 0.31$), length $\ell_2 = 8.9$ cm and a measured inductance of $L_2 = 2.5$ H.

5. Conclusion

The universal behavior resulting from the normalized separation δ can be used to maintain the same limit behavior of the magnetic coupling coefficient while changing the dimensions of either coil. In particular, the limit behavior will be the same as long as the geometric mean of effective magnetic volumes of the coils remains constant. Alternatively, the effective magnetic volume can be used to maximize the limit behavior of the magnetic coupling coefficient by optimizing the geometry of coils given any set of constraints.

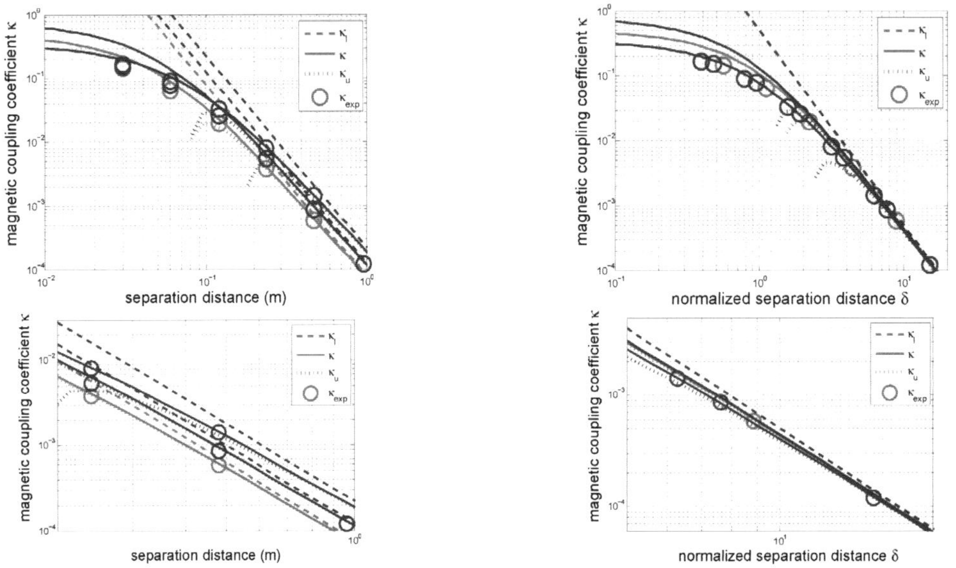

Figure 5. Magnetic coupling coefficient κ as a function of separation s and normalized separation δ for three different coil pairs. The solid lines are the long-range approximation for the magnetic coupling coefficient κ, the dashed lines are the upper bounds κ_u, the dotted lines are the lower bounds κ_l, and the circles are empirical results derived from measurements of the voltage induced in the receiver coil. The bottom graphs shown the largest measured separation to show the convergence when the separation distance is normalized.

Acknowledgements
The authors wish to acknowledge the help of Evan Dorksy in the data collection and analysis for the experimental portion of the paper.

References
[1] L. Rindorf, L. Lading and O. Breinbjerg, "Resonantly coupled antennas for passive sensors," *IEEE Sensors Conferences*, pp. 1611-1614, 2008
[2] A. Kumar, S. Mirabbasi and M. Chiao, "Resonance-based Wireless Power Delivery for Implantable Devices," pp. 25-28, 2009.
[3] J. C. Maxwell and J. J. Thompson, "Potential of two parallel circles expressed by elliptic integrals," in *A treatise on electricity and magnetism*, Oxford, 1873, pp. 338-350.
[4] E. B. Rosa, "The Self and Mutual Inductances of Linear Conductors," *Bulletin of the Bureau of Standards*, vol. 4, no. 2, pp. 302-344, 1908.
[5] H. A. Wheeler, "Simple Inductance Formulas for Radio Coils," vol. 16, no. 10, pp. 1398-1400, 1928.
[6] A. P. Sample, M. A. David and J. R. Smith, "Analysis, Experimental Results, and Range Adaptation of Magnetically Coupled Resonators for Wireless Power Transfer," *IEEE Transactions on Industrial Electronics*, vol. 58, no. 2, pp. 544-554, 2001.
[7] J. T. Conway, "Exact Solutions for the Magnetic Fields of Axisymmetric Solenoids and Current Distributions," *IEEE Transactions On Magnetics*, vol. 37, no. 4, pp. 2977-2988, 2001
[8] H. Nagaoka, "The Inductance Coefficients of Solenoids," *Journal Of the College of Science, Imperial University*, vol. 27, no. 6, pp. 1-33, 1909.
[9] S. Hui, C. K. Lee and W. X. Zhong, "Recent Progress in Mid-Range Wireless Power Transfer," pp. 3819-3824, 2012.
[10] J. O. Mur-Miranda, G. Fanti, Y. Feng, K. Omanakuttan, R. Ongie, A. Setjoadi, and N. Sharpe, "Wireless power transfer using weakly coupled magnetostatic resonators," in *IEEE Energy Conversion Congress and Expo (ECCE 2010)*, September 12-16, 2010.
[11] J. O. Mur-Miranda, and G. Fanti, "Peak wireless power transfer using magnetically coupled series resonators," in *IEEE Energy Conference and Exhibition (EnergyCon 2010*, December 18-22, 2010.

Journal of Physics: Conference Series **476** (2013) 012117 doi:10.1088/1742-6596/476/1/012117

An electroactive polymer energy harvester for wireless sensor networks

T.G. McKay[1,2] S. Rosset[1] I.A. Anderson[2] and H. Shea[1]
[1] LMTS, École polytechnique fédérale de Lausanne, [2]The Biomimetics Lab, Auckland Bioengineering Institute, The University of Auckland

thomas.mckay@auckland.ac.nz

Abstract. This paper reports the design, fabrication, and testing of a soft electroactive polymer power generator that has a volume of 1cm^3. The generator provides an opportunity to harvest energy from environmental sources to power wireless sensor networks because it can harvest from low frequency motions, is compact, and lightweight.

Electroactive polymers are highly stretchable variable capacitors. Electrical energy is produced when the deformation of a stretched, charged electroactive polymer is relaxed; like-charges are compressed together and opposite-charges are pushed apart, resulting in an increased voltage.

Although electroactive polymers have impressively displayed energy densities as high as 550 mJ/g, they have been based on films with thicknesses of tens to hundreds of micrometers, thus a generator covering a large area would be required to provide useful power. Energy harvesters covering large areas are inconvenient to deploy in a wireless sensor network with a large number of nodes, so a generator that is compact in all three dimensions is required. In this work we fabricated a generator that can fit within a 11x11x9 mm envelope by stacking 42, 11mm diameter generator films on top of each other.

When compressed cyclically at a rate of 0.5 Hz our generator produced 300 uW of power which is a sufficient amount of power for a low power wireless sensor node. The combination of our generator's small form factor and ability to harvest useful energy from low frequency motions provides an opportunity to deploy large numbers of wireless sensor nodes without the need for periodic, costly battery replacement.

With applications such as sensors monitoring the health of a building or transducers to monitor wildlife populations, wireless sensor networks provide an opportunity to obtain an enhanced awareness of both our local and remote environments. The major advantage of a wireless sensor network is that the cost and inconvenience of wiring between nodes is eliminated [1, 2].

The advantages of wireless sensor networks can be offset if frequent battery replacement is required. One solution is to integrate a compact energy harvester into the wireless sensor nodes. If hundreds or thousands of sensor nodes are to be deployed, energy harvesting devices become impractical if they are too large, thus there is a drive for energy harvesters that fit within a 1 cm^3 envelope [3].

Dielectric elastomers (DE), a class of electroactive polymers, are highly deformable variable capacitors that show great promise for sub-cm^3 scale energy harvesters. Electrical energy is produced when the deformation of a stretched, charged DE is relaxed; like-charges are compressed together and opposite-charges are pushed apart, resulting in an increased voltage (see Figure 1). DE lend themselves well to low complexity, small-scale energy harvesters because the energy harvesting mechanism is theoretically scale invariant, they can be directly coupled to rectilinear motions [4], they can produce high energy densities [5, 6], and can harvest energy efficiently from a wide range of frequencies.

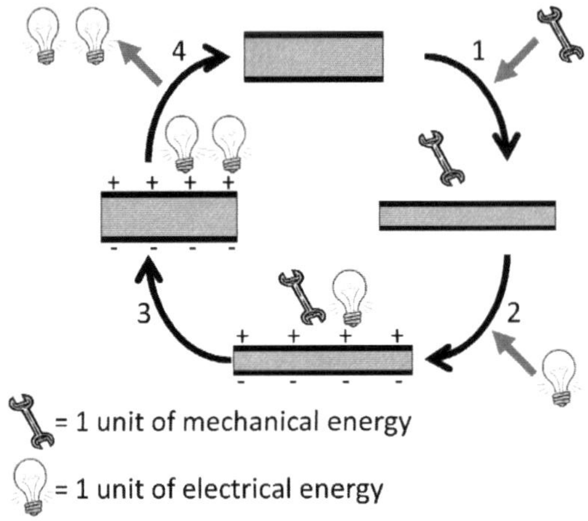

Figure 1: The DE energy harvesting cycle. 1) Mechanical energy is used to stretch the generator; 2) The generator is electrically charged; 3) the generator is mechanically relaxed which pushes opposite charges apart and compresses like charges together, this action boosts the electrical energy of the charges; 4) the charges are extracted in this higher energy state.

Although DE have displayed energy densities as high as 550 mJ/g [5], they have been based on films with thicknesses of tens to hundreds of micrometers [4-6], thus a generator covering a large area would be required to provide useful power. Furthermore, the thin film structures often require additional rigid frames to maintain their desired shape. The framing reduces the generator's energy density, deformability, simplicity, and introduces potentially catastrophic stress concentrations.

In this paper we present a stacked membrane generator configuration which allows self-supporting generators to be fabricated that do not require rigid frames. The stacked configuration also allows a large quantity of generator material to be fabricated onto a much smaller footprint than that achievable using a single thin membrane.

Our generator consisted of 42, 40 um thick, 11mm diameter DE layers stacked on top of each other and electrically connected in parallel. The generator stack was then sandwiched between 3mm thick silicone end caps (see Figure 2). The end caps were included because the most convenient mechanism for coupling a stack to a load is to adhere the ends of the stack between two structures. The constrained ends do not deform as much as the center of the stack. And as described in Figure 1, the generator harvests energy by displacing electrical charges using mechanical deformation. Thus, the constrained end-cap regions would not provide much additional energy. In-fact, if the end-caps were substituted for DE layers, they would reduce the overall proportional change of the generator's capacitance, which could reduce energy production [7].

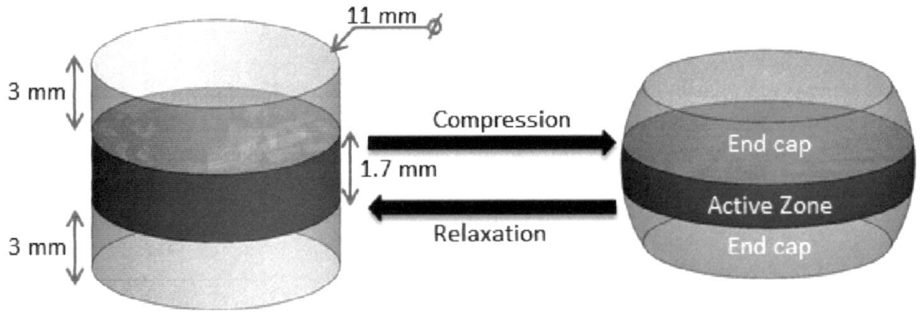

Figure 2: The stack DE generator is coupled to a mechanical energy source by adhering its ends between two bodies that move relative to each other. The active generator zone is deformed when the stack is compressed. The end caps do not deform as much as the the active zone because of the fixed boundary condition between the generator stack and the mechanical energy source.

An overview of our fabrication process is given in Figure 3. Our generator's silicone membranes (Sylgard 186, Dow Corning) were cast on to a polyimide backing using a Zehntner ZAA2300 film applicator. Electrodes consisting of 90% by weight Sylgard 186, Ketchenblack EC-300J carbon black (Akzo-Nobel) (5% by weight), and Cabot Black Pearl 2000 carbon black (5% by weight) were applied to the membranes using a TPM-101 pad printer from Teca-Print. A Diener Zepto plasma system was then used to bond the layers together with a short exposure to an oxygen plasma, until a stack of 42 active membranes was formed. A Trotec Speedy 300 laser cutter was used to cut the stack to shape before the polyimide backing film was removed. LSR-20 silicone (factor 2) endcaps were then bonded to the stack. The layers were then interconnected using the silicone/ketchenblack/cobalt carbon black electrode mix.

The stack generator was tested using the energy harvesting circuit illustrated in Figure 4. The generator transfers charge from a priming source up to a higher voltage where the energy was supplied to a load. Thus our energy harvester provides a gain in electrical energy by boosting the voltage of the charge before it is delivered to a load. We replicated the approach of Huang et al. who used a zener diode array as the load. When the voltage reaches the breakdown voltage of the zener array, charge is transferred from the generator to the zener load. We used an array of 9 200 Volt zeners (1N5388BG). The generator was primed to 1400 Volts using an EAP controller power supply (www.biomimeticslab.com). The voltage across the DE generator was measured using a 5 GΩ: 5MΩ resistor ladder. The charge on the generator was measured using a 1 μF current integration capacitor.

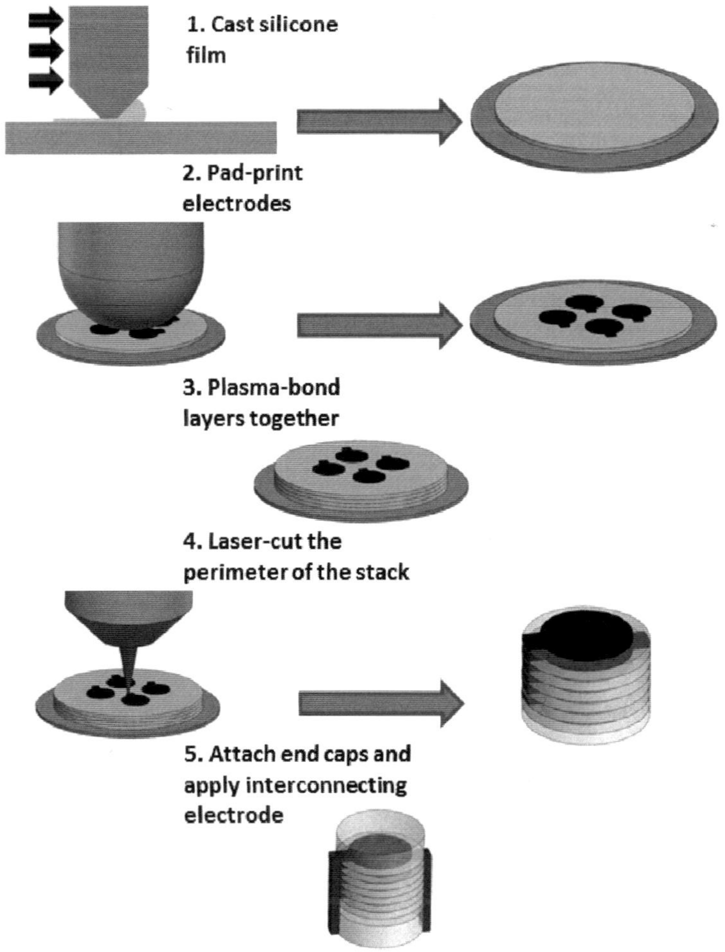

Figure 3: An overview of the process which was developed for fabrication of stacked DE generators.

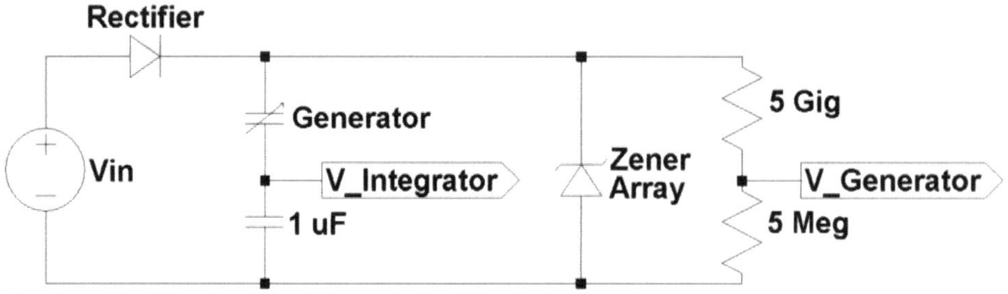

Figure 4: A schematic of the circuit used for analysis of the stack generator.

The generator was cyclically compressed by 3 mm at 0.5 Hz using an Instron 3342 dynamic test rig. The measured voltage versus charge relationship from one cycle is given in figure 4. As described by Huang et al. the energy produced by the generator in a single cycle was found by computing the area within the Voltage-charge envelope. When primed at 1400 Volts and delivering charge to the load at 1800 Volts, our generator produced 600 μJoules in a single cycle. Since the Instron cyclically deformed the DE generator at 0.5 Hertz, this energy production corresponds to a power output of 300 μWatts.

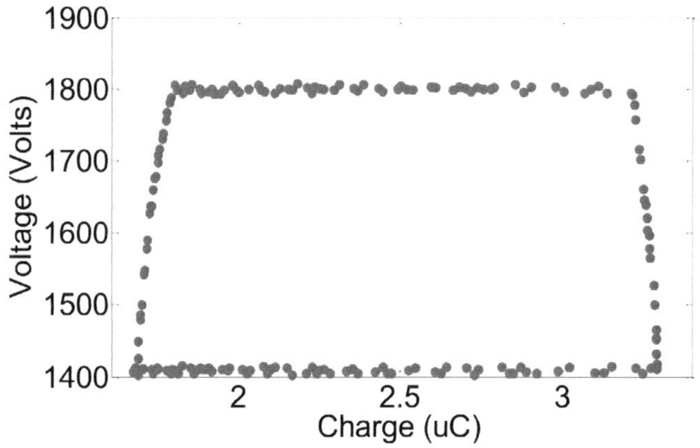

Figure 5: Voltage versus charge measured during a single compression cycle of our generator. The area within the voltage charge cycle is equal to the energy that is harvested.

The production of 300 uW is a sufficient for powering a low power wireless sensor node. For instance, Rhee et al. developed a wireless sensor network with transmitting nodes that consumed 300 uW to send 1 packet of sensor data per second [2]. Furthermore the power output of electroactive polymer energy harvesters scales linearly with frequency, so power hungry wireless sensor nodes could be powered using our generator if a higher frequency energy source was available.

The combination of our generator's small form factor and ability to harvest useful energy from low frequency motions provides an opportunity to deploy large numbers of wireless sensor nodes without the need for periodic, costly battery replacement. The ability to harvest from low frequencies opens the possibility to harvest from a wide range of energy sources. For instance, we previously used electroactive polymers to harvest energy from a tree branch swaying in the breeze [8]. Our generator coupled to swaying branches could, for example, power wireless sensor nodes that provide an alarm if a forest fire develops.

References

1. Chen, Y. and Q. Zhao, *On the lifetime of wireless sensor networks.* IEEE Communications Letters, 2005. **9**(11): p. 976-978.
2. Rhee, S., D. Seetharam, and S. Liu. *Techniques for minimizing power consumption in low data-rate wireless sensor networks.* 2004. Atlanta, GA.
3. Roundy, S., P.K. Wright, and J.M. Rabaey, *Energy Scavenging For Wireless Sensor Networks.* 2004, Boston, MA: Kluwer Academic Publishers. 212.
4. McKay, T.G., B. O'Brien, E. Calius, and I. Anderson, *Self-priming dielectric elastomer generators.* Smart Mater. Struct., 2010. **19**(055025): p. 055025.
5. Huang, J., S. Shian, Z. Suo, and D.R. Clarke, *Dielectric elastomer generator with equi-biaxial mechanical loading for energy harvesting.* 2013: p. 86870Q-86870Q.
6. Kaltseis, R., C. Keplinger, R. Baumgartner, M. Kaltenbrunner, T. Li, P. Machler, R. Schwdiauer, Z. Suo, and S. Bauer, *Method for measuring energy generation and efficiency of dielectric elastomer generators.* Appl. Phys. Lett., 2011. **99**(16): p. 162904.
7. McKay, T.G., *Soft, Low Complexity Dielectric Elastomer Generators*, in *Auckland Bioengineering Institute*2011, The University of Auckland: Auckland.
8. Anderson, I.A., I.A. Ieropoulos, T. McKay, B. O'Brien, and C. Melhuish, *Power for robotic artificial muscles.* IEEE Trans. Mech., 2011. **16**(1): p. 107-111.

Journal of Physics: Conference Series **476** (2013) 012118 doi:10.1088/1742-6596/476/1/012118

Analysis, Design and Realization of a Near-Field Focused RF Power Transfer System

R. van der Linden[1] **and H.J. Visser**[1, 2]

[1] Eindhoven University of Technology, Eindhoven, The Netherlands
[2] Holst Centre / imec, Eindhoven, The Netherlands

E-mail: huib.visser@imec-nl.nl

Abstract. This paper reports the analysis, design, prototyping and first measurement results of a linear array antenna consisting of Yagi-Uda elements, rotated and phased for focusing RF power in the near-field of the antenna. By employing Near Field Focusing more power may be transmitted before the EIRP limit is reached due to the inherent far-field difusing. A design based on four Yagi-Uda array elements is able to produce an average power density of 21.04dBm/m^2 at 2.45GHz in a box of 10x10x10cm^3 located 1m from the linear array face. The design has been verified through comparison with full-wave electromagnetic simulations. First measurements on a realized prototype confirm the functioning of the array.

1. Introduction

This paper describes the analysis, design and realization of a wireless energy transfer system that can be used for, e.g. remotely charging batteries at distances bridging about one to two meters (desk size). Since inductive [1] or non-radiative, resonant magnetic coupling [2] would require very large transmitting and receiving coils (diameters in the order of the distance to be bridged), we choose for radiative coupling. The system will be employed in the license-free 2.4 GHz Industry, Science and Medical (ISM) frequency band. Since in the ISM bands the allowed Effective Isotropic Radiated Power (EIRP) - i.e. the transmit power times the transmit antenna gain – is limited, the collection in the far-field of a few milliwatts DC-power will limit the bridging range to far below one meter. From the Friis transmission equation [3] it follows, for example, that a power density of 20dBm/m^2 at 2.45GHz, using an EIRP of 0.5W is obtained at a distance of 0.63m from the transmit antenna.

To overcome this distance-restriction we make use of the fact that EIRP is a parameter defined in the far-field of an antenna. By focusing the RF power in the near-field of a transmit antenna, the far-field of that antenna will diffuse and lead to a lower EIRP then when focused for the far-field. Thus, more power may be transmitted before reaching the EIRP limit. For focusing the RF power we use an array antenna and apply phase-shifts between the radiating elements to compensate for the different path lengths from the elements to the focal point.

In the following we will describe the basic theory of Near Field Focusing (NFF), the obtainable results and limitations and ways to overcome some of these limitations. Then we will discuss a linear NFF array design and will discuss measurement results.

2. NFF theory

For reasons of space reduction, we will work towards the design of a linear array. Figure 1 shows part a linear array antenna, positioned in the *xy*-plane of a rectangular coordinate system.

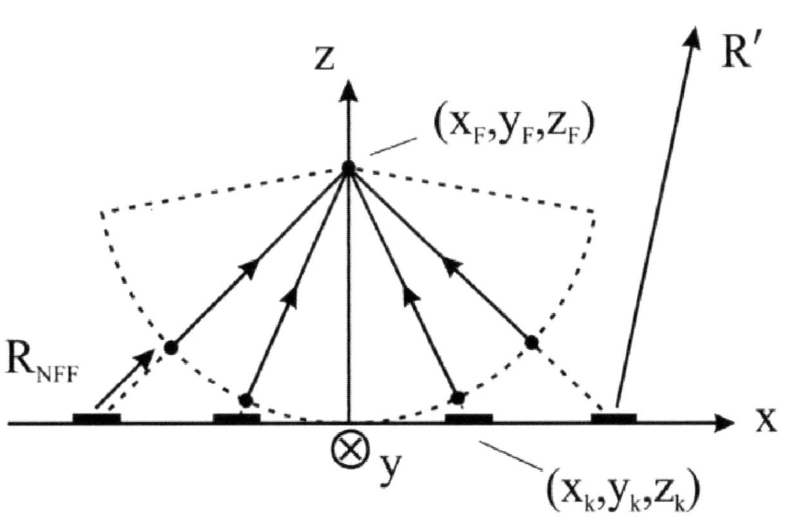

Fig. 1 – *Part of a linear array antenna to be focused at* (x_F,y_F,z_F).

The required phase shift for an element k for focusing the power (adding the signals in-phase) in the focal point (x_F,y_F,z_F) is given by [4]

$$\varphi_k = kR_{NFF_k} = \frac{2\pi}{\lambda}\left[\sqrt{(x_k - x_F)^2 + (y_k - y_F)^2 + (z_k - z_F)^2} - \sqrt{x_F^2 + y_F^2 + z_F^2}\right]. \quad (1)$$

The electric field at any position (x,y,z) in the far-field region of the individual (N) array elements is given by

$$\vec{E}(x,y,z) = \sum_{k=1}^{N} \frac{\vec{E}_k(\vartheta_k,\varphi_k)}{4\pi R'} e^{-jk(R_{NFF_k} - R')}, \quad (2)$$

where (θ_k,φ_k) describes the local orientation of the element and where

$$R' = \sqrt{(x - x_k)^2 + (y - y_k)^2 + (z - z_k)^2}\ . \quad (3)$$

A similar expression may be derived for the magnetic field and the time averaged power density is then obtained from the vector (cross) product of the electric field and the complex conjugate of the magnetic field.

3. NFF analysis results

An example of the possibilities of NFF for a (0.5W) EIRP limited system consisting of a linear array of eight short dipoles, spaced apart one wavelength, is shown in Figure 2 for a frequency of 2.45GHz. Figure 2(a) shows that the power density around the focal spot at z=1m, produced by the NFF array is 5.5dB higher than that of a single element. The Figure also shows that that in this case the array antenna focused at infinity underperforms by -11dB compared to a single element. Figure 2(b) Shows the power density at a distance z=1m parallel to the face of the array. Figure 2(c) shows that the normalized far-field power density of a near-field-focused array is more diffuse than that of an array focused at infinity, leading to a lower directivity and thus allowing to transmit more power before the EIRP limit is reached.

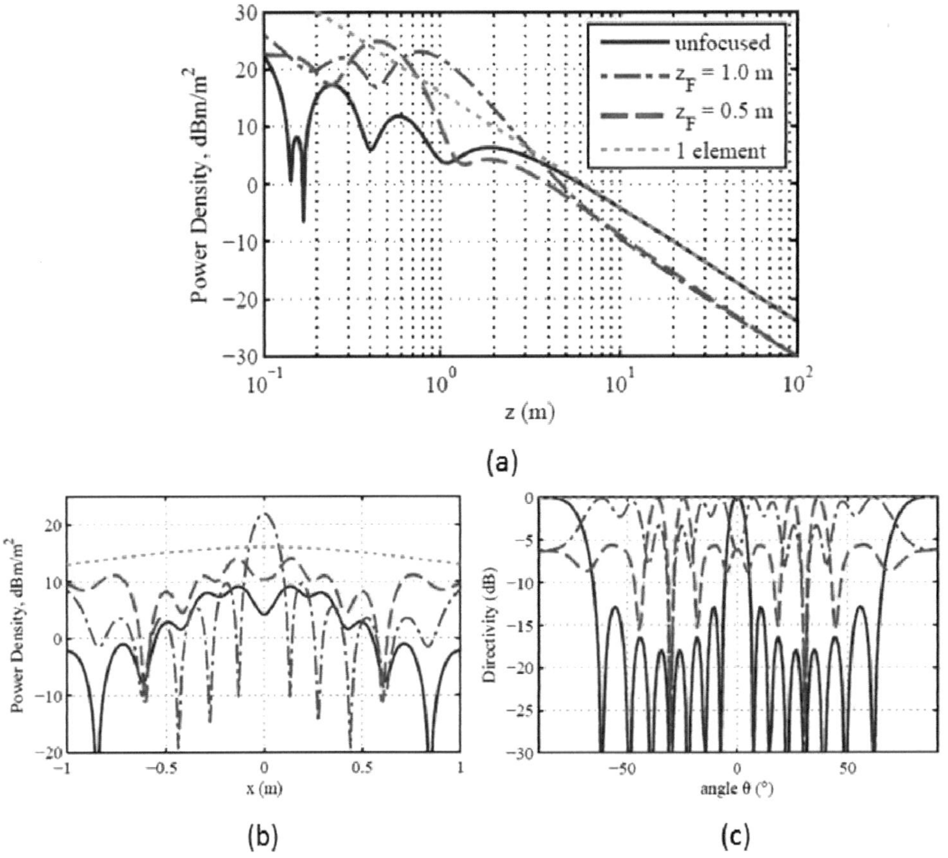

Fig. 2 – *Power densities along the axis perpendicular to the array face (a), parallel to the array face at a distance z=1m (b) and the normalized far-field in the xz-plane (c).*

For practical applications we do not need a sharp focal point but rather a focal area or volume in which a desired power density level can be guaranteed. Figure 3 shows the variation in power density averaged over a volume of 10x10x10cm^3 centered around z=1m for an eight-elements linear short-dipole array having different uniform element spacings.

4. Improvements

Although the results obtained with a 'standard' NFF array (employing realistic radiating elements) are satisfactory, we have been looking for ways to increase the focused power level. Several of the investigated methods, like amplitude tapering, the creation of multiple focal points, employment of orthogonal polarizations, employment of non-planar arrays resulted in the proposal to employ *rotated, highly directive array elements*.

The principle of this technique is shown in Figure 4. The antenna-elements are mechanically rotated towards a distance z_{rot}. The focal area is concentrated around z_F (z_F>z_{rot}). For a four-elements array consisting of 11.8dBi-gain Yagi-Uda elements, spaced 1.8 wavelengths apart, z_{rot} being 0.4m and z_F being 0.8m, the average power density in a 10x10x10cm^3 box is found to be 21.04dBm/m^2, see Figure 5. For the array focused at infinity this value is 10.8dB lower. For a single (Yagi-Uda) element the value is 5.04dB lower. The transmitted power for the focused array antenna is 46mW.

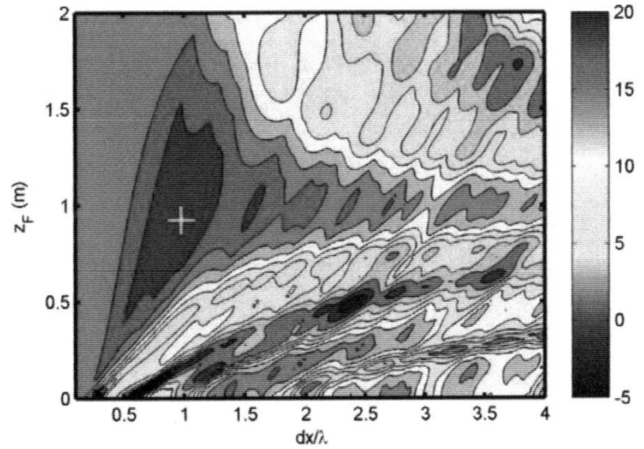

Fig. 3 – *Contourplot of the average power density in dBm/m² in a box of 10x10x10cm³ centered around z=1m for varying inter-element distance and focus distance. The array consists of eight short dipole elements. The green cross designates the maximum power density..*

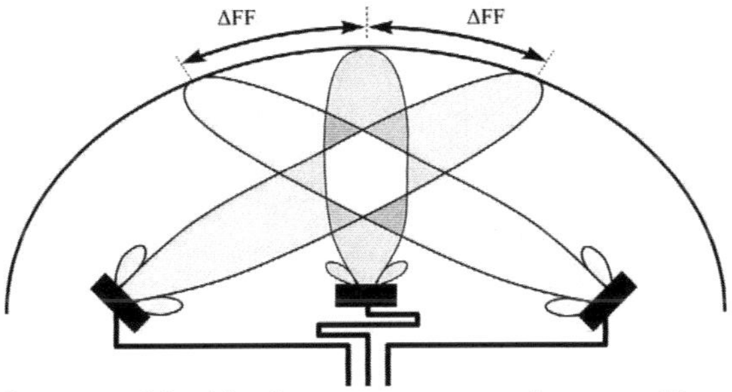

Fig. 4 – *Schematic of the rotated highly directive antenna elements. The power density increases in the target region, while simultaneously the far-field is difused.*

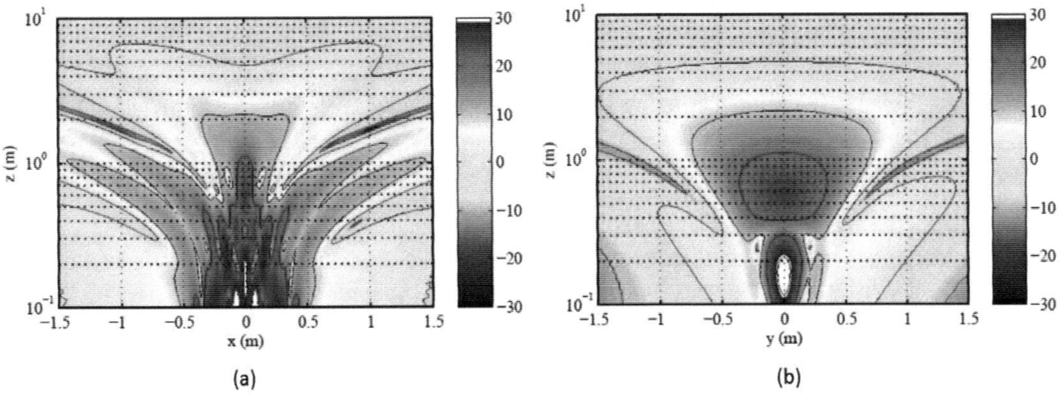

Fig. 5 – *Power densities in dBm/m² along the plane through the normal and long side of the array, i.e. the E-plane (a) and the plane through the normal and short side of the array, i.e. the H-plane (b).*

5. First measurements

The results for the array as shown in Figure 5 have been verified for several cuts through these figures with full-wave simulation results (CST Microwave Studio®). Finally, a prototype has been hand-made as shown in Figure 6. The Yagi-Uda radiating elements have been made out of copper strips on a High Impact Polystyrene (HIPS) substrate. The same substrate material has been used to create microstrip phase shifting elements for exciting the array elements.

Fig. 6 – *Prototype NFF array with phase shifter elements. Feeding network not shown.*

Figure 7 shows the simulation results for the power density in the horizontal plane (corrected for the measured phase errors in the feeding) (a) and the *relative* power density measurements sampled in a number of regions in front of the antenna (b).

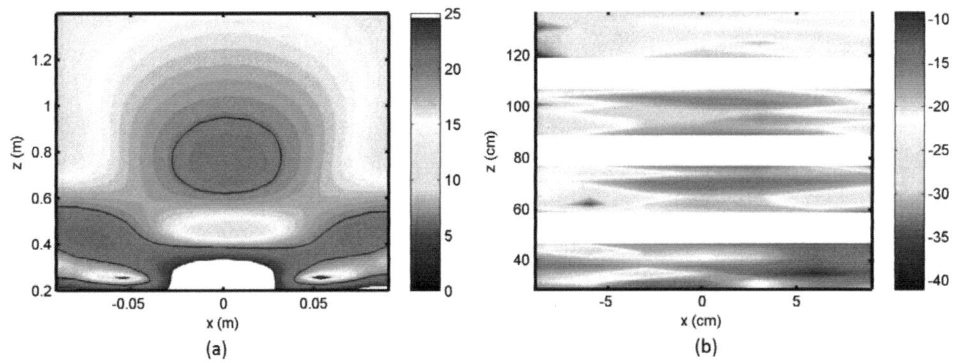

Fig. 7 – *Simulated (a) and measured power density (b).*

Given the limitations in the measurement setup, the measured results show an acceptable agreement with simulations.

6. Conclusion

Through near-field focusing, employing a (linear) array antenna, more power can be transferred over intermediate distances than will be possible by RF far-field power transfer when obeying the limitations in allowed EIRP. NFF array antenna optimization is currently going on.

References

[1] J. Hirai, T.-W. Kim and A. Kawamura, 'Wireless Transmission of Power and Information for Cableless Linear Motor Drive', *IEEE Transactions on Power Electronics*, Vol. 15, No. 1, pp. 21-27, 2000.

[2] A. Kurs, A. Karalis, R. Moffatt, J.D. Joannopoulos, P. Fisher and M. Soljačić, 'Wireless Power Transfer Via Strongly Coupled Magnetic Resonances', *Science*, Vol. 317, No. 5834, pp. 83-86, 2007.

[3] C.A. Balanis, *Antenna Theory, Analysis and Design*, 2nd edition, John Wiley & Sons, New York, 1997.

[4] A. Buffi, P. Nepa and G. Manara, 'Design Criteria for Near-Field-Focused Planar Arrays', *IEEE Antennas and Propagation Magazine*, Vol. 54, No. 1, pp. 40-50, 2012.

A frequency up-converted electromagnetic energy harvester using human hand-shaking

M A Halim and J Y Park[1]
Micro/Nano Devices and Packaging Lab., Dept. of Electronic Engineering, Kwangwoon University, 447-1 Wolgye-dong, Nowon-gu, Seoul, 139-701 Korea

E-mail: jaepark@kw.ac.kr

Abstract: We present a frequency up-converted electromagnetic (EM) energy harvester that is capable of powering various portable devices and systems by hand-shaking. It consists of a freely movable ball to impact periodically (at low frequency) on the parabolic top surface of a mass of a cantilever beam allowing it to vibrate at higher (resonant) frequency. Relative motion between a magnet attached to the cantilever and a coil induces voltage. A prototype of the energy harvester has been fabricated and characterized by applying vibration from hand-shaking. The frequency and acceleration of the applied hand-shaking vibration has been experimentally found to be 4.6 Hz and 2g, respectively. With an optimum distance between magnet and coil, a maximum 672 mV peak-peak open circuit voltage of 370 Hz frequency and a maximum 413μW peak power delivered to an 85Ω matched load resistance have been obtained, respectively.

1. Introduction

With the rapid development in the fields of electronics, portable smart devices such as mobile phones, tablet computers, wireless sensor nodes, wrist watches, audio devices, hearing aids, implanted biomedical devices etc are becoming very popular due to their smaller size, low power consuming and multi-functional features. Each and every device requires power sources to be operated. Generally, electro-chemical batteries have been using as power source to supply power to these portable and smart electronic devices. Beside the advantages (e.g., high energy density, smaller size, low cost) of the electro-chemical batteries, they have some disadvantages including limited lifetime, inconvenient in recharging or sometimes, replacing. This is why, many researchers all over the world have been investigating alternatives to the electrochemical batteries. Light, wind, heat, radio wave, vibration etc. are the energy sources readily available around our environment. Extracting energy from environmental vibration has drawn much attraction over the last few decades due to its abundance in nature and unlimited lifetime [1-4]. Different vibration sources generate vibrations of different frequencies and amplitudes, but most ambient vibrations are of low frequencies and large amplitudes with various cyclic movement in different directions. Basic human activities such as walking, running, shaking limbs, jumping, breathing etc. also generate vibration which are recently being interesting to the researchers to be used for energy harvesting [5-7]. Widely used techniques for harvesting energy

[1] To whom any correspondence should be addressed.

from ambient vibration, as well as human and machine motion are piezoelectric, electromagnetic and electrostatic mechanisms[8-10].

Vibration energy harvesters typically employ spring-mass-damper system that makes the harvesters a resonant device. To-date most of the vibration energy harvesting devices have been tested by electrodynamic shaker in the laboratory without considering the real applications e.g., harvesting energy from human-body-induced vibration. Human-body-induced vibrations are of low frequency with large amplitude which do not allow the conventional resonant harvesting devices to employ [5]. Moreover, the maximum voltage and generated electrical power of a resonant generator is strongly dependent on the externally applied vibration frequency and drops dramatically at low frequencies [11]. In this research, our proposed device overcomes both the limitations by employing the conventional spring-mass-damper system, but in a different way that is able to meet the real application. It utilizes the electromagnetic (EM) energy harvesting method with frequency up-conversion mechanism [12]. Upon excitation applied through hand-shaking, a freely moveable non-magnetic ball impacts on a cantilevered mass at low frequency on its way of movement, allowing the cantilever to vibrate with its higher resonant frequency. The cantilever vibrates in a direction perpendicular to that of the ball movement. A magnet attached to the cantilever also vibrates along with it. A pick up coil placed at the bottom of the magnet induces voltage due to the relative motion between magnet and coil according to Faraday's law of electromagnetic induction. This frequency up-converted feature of the device makes it to supply meaningful power to the portable smart devices from human-body-induced vibration such as hand-shaking.

2. Structure of the proposed EM harvester

The schematic structure of the proposed frequency up-converted EM energy harvester is shown in figure 1. It consists of a freely moveable non-magnetic ball within a rectangular channel having an opening at the centre of its bottom surface. A cubic mass with parabolic extension at the top edge is attached to (at the centre of one side) a fixed-fixed cantilever beam that works as the spring-mass-damper system. The beam is placed at such a position that the parabolic top surface of the mass can be positioned within the opening of the rectangular channel so that it would be impacted by the freely movable ball. A cylinder magnet is attached to the cantilever, opposite to the mass. A pick-up coil is also placed below the magnet, on the bottom cover of the device structure. When the device is shaken, the ball, inside the channel hits on the parabolic top of the mass allowing the cantilever to vibrate that makes the magnet to vibrate relative to the coil, inducing electromotive force within the coil. The vibration frequency (resonant frequency) of the cantilever beam is much higher than the applied vibration that can be determined by material and structural parameters of the cantilever beam.

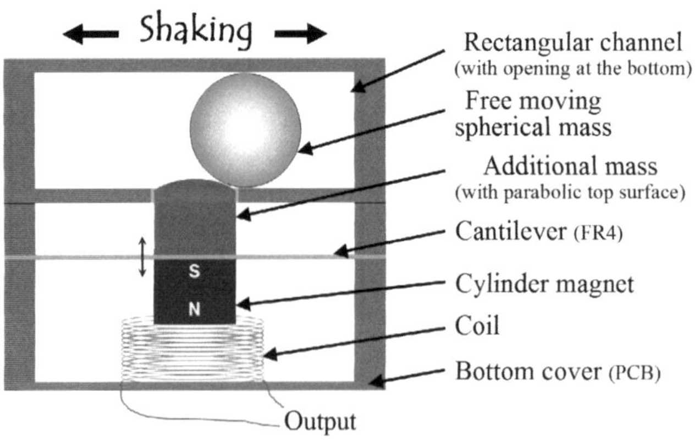

Figure 1. Schematic drawing of the proposed frequency up-converted electromagnetic energy harvester.

3. Characteristics of hand-shaking vibration

In order to characterize the behaviour of hand-shaking vibration, we measured and analyzed the vibration generated by hand-shaking using an accelerometer (LSM330DLC 3-axis accelerometer; ST Microelectronics) embedded in a smart phone (Galaxy SIII; Samsung Electronics). In fact, it is obvious to characterize the vibration behaviour generated by hand-shaking because we are intended to operate the proposed EM energy harvester under hand-shaking. We have collected the data from four subjects at 50 Hz sampling rate for one minute. Figure 1(a) shows peak acceleration values were ranged from 15 ms^{-2} (~1.5g) to 20 ms^{-2} (~2g). Frequency components of the measured acceleration values were analyzed by FFT (Fast Fourier Transform) of the measured data. Figure 1(b) shows that the frequency of hand-shaking vibrations falls between 2.5 Hz and 6 Hz range for different subjects.

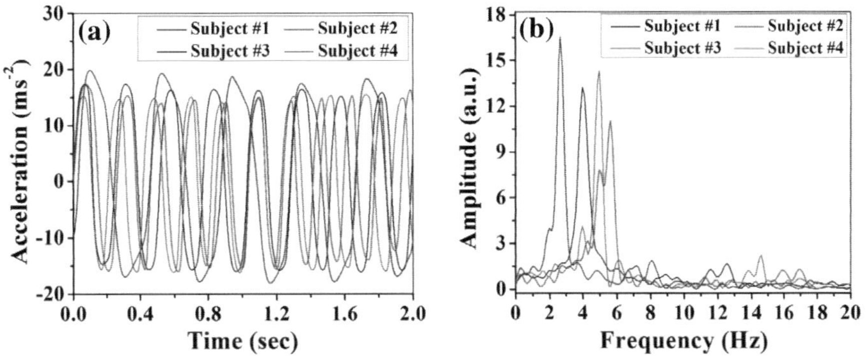

Figure 2. (a) Measured acceleration of vibration generated by hand-shaking a smart phone and (b) their frequency components obtained by FFT.

4. Prototype fabrication

A prototype was fabricated by using a Ø6x5 mm^2 NdFeB cylinder magnet and a cubic mass (Fe) of length, 6 mm with a parabolic extension of 1 mm height at the top attached on either side of a FR4 cantilever beam (40×6×0.5 mm^3), a 1000-turn coil (0.1 mm copper wire) with 8 mm bobbin diameter, and a Ø10.3mm steel (type 316) ball moving in a rectangular channel (inner area 10.5×10.5 mm^2) assembled within an aluminium structure (40×30×16mm^3) as shown in Figure 3. The prototype has been tested by manual shaking. The cantilever was assembled in such a way that the parabolic top of the cubic mass is occupied inside by the opening (7×7 mm^2) of the channel making an overlap of 0.5 mm with the ball so that the ball can impact on the top while shaking the prototype.

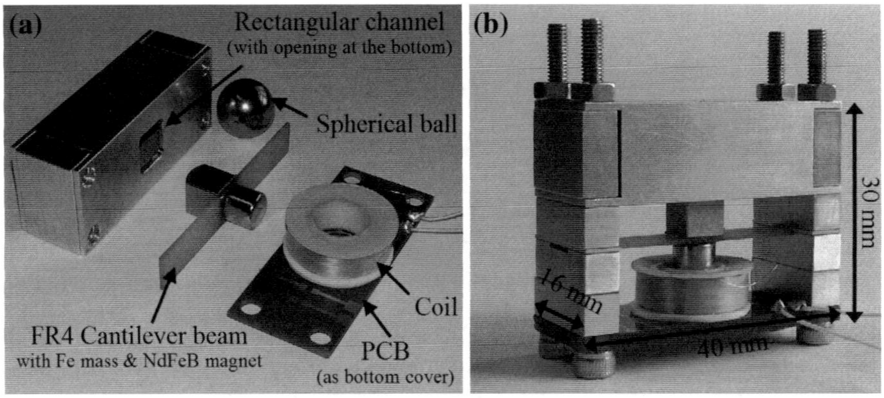

Figure 3. Photomicrographs of (a) components before assembling and (b) fabricated macroscale prototype of the proposed energy harvester.

5. Experimental results and discussion

The output terminals of the fabricated prototype were connected to a digital storage oscilloscope (Tektronix TDS5052B) to measure its output response. Vibrations could not be applied to the prototype by the electrodynamic shaker (TMS 2004E) because of its limitation in producing vibration of low frequency (<6 Hz) and higher acceleration (>10 ms^{-2}). Figure 4 shows the peak-peak open circuit voltage vs. distance to determine the optimum distance between magnet and coil. A maximum voltage of 672 mV was obtained at -1 mm optimum distance.

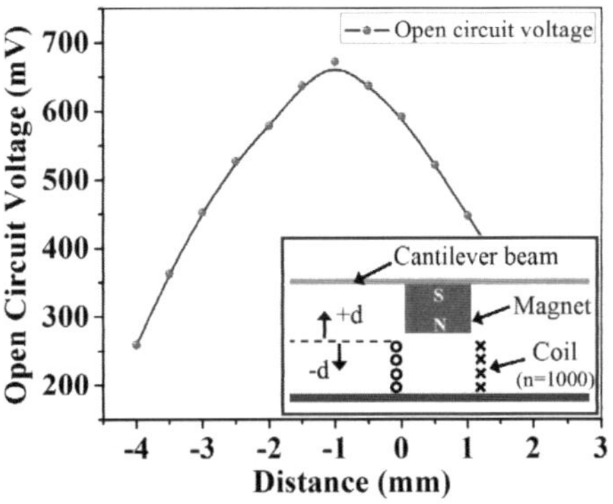

Figure 4. The measured peak-peak open circuit voltages at various distances between magnet and coil.

Figure 5 illustrates the peak-peak load voltage and peak power delivered to the load while the device was hand-shaken. The output terminals of the prototype were connected with a continually adjustable load resistor and the resistance values were swept from 50 Ω to 200 Ω range. A maximum 413 μW peak power was obtained from 85 Ω matched load resistance. Figure 6 shows the instantaneously generated voltage waveform generated across the matched load. The maximum peak-peak voltage was found to be 340 mV which was attenuated exponentially with time due to the damping of the cantilever beam vibration. As we found, the attenuation is not perfectly exponential due to the process variation in assembling the components. Two consecutive maximum peaks were generated in one cycle of the applied vibration because the ball impacted on the parabolic top of the mass twice in one cycle of its movement. By applying FFT, frequency of the generated voltage waveform was found to be 370 Hz

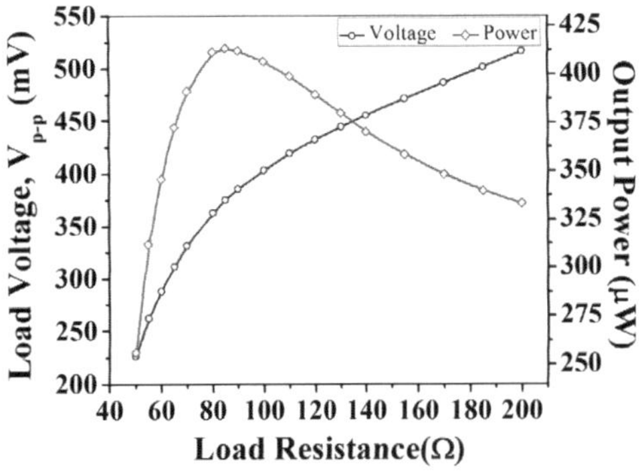

Figure 5. The measured peak-peak output load voltages and peak output powers of the fabricated device across various load resistances.

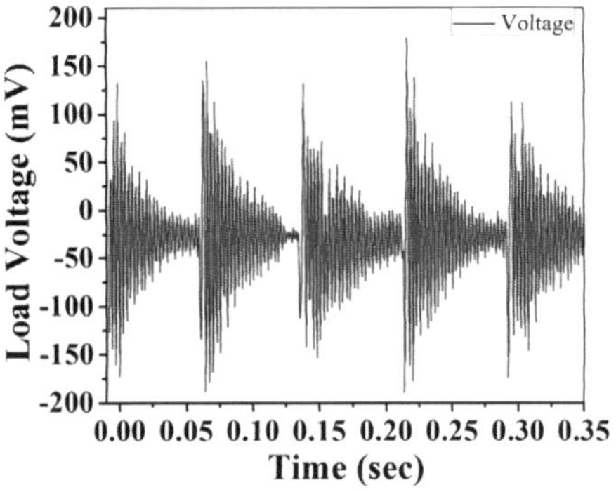

Figure 6. Waveform of the instantaneously generated output load voltage across 85Ω optimum load resistance.

whereas the frequency of the applied hand-shaking vibration was 4.6 Hz. The applied acceleration was measured as 20 ms^{-2} (~2g). It clearly indicates the frequency up-conversion behaviour of the proposed energy harvester which can be used to generate significant amount of power from hand-shaking. In order to improve its performance, the damping of the cantilever beam vibration needs to be reduced which can be done by changing the cantilever material and geometry. Beside this, the shape of the mass's top surface and the overlap area of the mass top and the ball needs to be optimized.

6. Conclusions

We have proposed and demonstrated a frequency up-converted electromagnetic energy harvester that is capable of generating significant power from hand-shaking vibration. The vibration behaviour of the hand-shaking has been characterized. A macroscale prototype has been fabricated and tested as a proof of concept. It generated maximum 672 mV peak-peak open circuit voltage of 370 Hz frequency and 413 μW peak power delivered to 85 Ω load resistance at a hand-shaking vibration with the frequency 4.6 Hz and acceleration 20 ms^{-2} (~2g). The frequency up-conversion feature increased the power flow of the device. Although the generated voltage (and power) level was not sufficient to power up an electronic circuit, the test results showed its ability in powering portable smart devices from hand-shaking. With further optimization in design parameters (e.g., stiffness, damping, overlapped contact area), it is possible to improve the performance of the proposed frequency up-converted electromagnetic energy harvesting device.

Acknowledgements

The authors are grateful to acknowledge the Basic Science Research Program (2010-0024618) and the Pioneer Research Centre Program (2010-0019313) through the National Research Foundation of Korea (NRF) funded by the Ministry of Science, ICT and Future Planning, Korea.

References

[1] Amirtharajali R, Meninger S, Mur-Miranda J, Chandrakasan A and Lang J 1998 *IEEE J. Solid-State Circuits* **33** 687-95
[2] Roundy S, Wright P K and Rabaey J 2003 *Computer Communications* **26** 1131–44
[3] Beeby S P, Tudor M J and White N M 2006 *Meas. Sci. Technol.* **17** R175-95
[4] Harne R L and Wang K W 2013 *Smart Mater. Struct.* **22** 023001
[5] Saha C R, O'Donnell T, Wang N and McCloskey P 2008 *Sens. Actuators A* **147** 248-53
[6] Bowers B J and Arnold D P 2008 *J. Micromech. Microeng.* **19** 094008
[7] von Büren T, Mitcheson P D, Green T C, Yeatman E M, Holmes A S and Tröster G 2006 *IEEE Sens. J.* **6** 28-38
[8] Renaud M, Fiorini P, van Schaijk R and van Hoof C 2009 *Smart Mater. Struct.* **18** 035001
[9] Munaz A, Lee B-C and Chung G-S 2013 *Sens. Actuators A* **201** 134-40
[10] Tashiro R, Kabei N, Katayama K, Ishizuka Y, Tsuboi F and Tsuchiya K 2000 *JSME Int. J. C.* **43** 916-22
[11] Kulah H and Najafi K 2008 *IEEE Sens. J.* **8** 261-68
[12] Halim M A, Khym S and Park J Y 2013 *J. Appl. Phys.* **114** 044902

Fabrication and characterization of the electrets material for electrostatic energy harvester

M R Ahmad[1], M H Md Khir[1], J O Dennis[1] and A Mohd Zain[2]
[1] Universiti Teknologi PETRONAS, Bandar Seri Iskandar, Tronoh Perak, Malaysia
[2] MIMOS Berhad, Technology Park Malaysia, Kuala Lumpur, Malaysia

E-mail: mohamad.radzi.ahmad@gmail.com

Abstract. Electrostatic energy harvesting devices have been found feasible to convert ambient vibrations into electrical energy. Fabrications of such devices are carried out using standard CMOS technology. This paper explains the fabrications processes of the energy harvesting components and the characterizations of the electret material for the electrostatic energy harvester device. The electrostatic energy harvester components are made up of the trapezoidal electrodes and electrets structures. The aluminum-silicon-copper (AlSiCu) electrodes reside on the Micro Electro-Mechanical System (MEMS) structure of 25 mm^2 and comprises the seismic mass and serpentine beams. On the other hand, the electrets which made up of Silicon Dioxide (SiO_2) and Silicon Nitride (Si_3N_4) material, are fabricated separately. A 0.35 µm CMOS processes technology is utilized to fabricate the electrodes and electrets structures onto 200 mm silicon wafers. The performance of the energy harvester is evaluated through the characterizations of the electrets material. The electrets are charged with the Corona charging method and the surface potential is measured. From the experiment, it is proven that the CVD oxides is chargeable at 4.7 kV Corona voltage, 200 V grid voltage, 10 minutes charging time and 70°C substrate temperature, hence it is a recommended material for electrostatic energy harvester device.

1. Introduction

Energy harvesting system is defined as a process to convert ambient energy sources such as vibrations, solar, thermal, wind, wave and RF energies into electrical energy. Energy harvesting topic has recently gained extensive research interest mainly driven by the needs of alternative power sources for ubiquitous sensor networks and low power application devices. Existing energy harvester devices are used in wireless sensor nodes (WSN) for monitoring structural health, home/building energy management system (HEMS/BEMS), automotive tire pressure monitoring systems (TPMS), and biomedical devices such as pacemakers and hearing aid [1].

Among the available ambient energy sources mentioned, the mechanical vibration energy source is of specific interest due to its availability anytime anywhere. The vibration to electrical energy conversion method known as the transduction mechanisms, are the feasible techniques used currently on the energy harvesting devices and can be categorized into electrostatic, electromagnetic and piezoelectric [2]. In this paper, the electrostatic conversion method is the target subject of study due to ease of integration into monolithic structure and fabrication into micro systems. Hence, the small scale electrostatic energy harvester is designed and called the micro energy harvester [3]. However, the conventional electrostatic conversion method has its own drawbacks, whereby a separate voltage source is required to provide initial charge to the electrodes [4]. Therefore, to overcome such drawback, *Electret* which is a form of dielectric material, is deposited onto the electrode and then charged with a Corona charging method [5].

The Corona charging process provides a surface potential with constant electrostatic field, hence the electret transforms into a quasi-permanent charged material. In this paper, the process to fabricate the components of the electrostatic energy harvester device is explained. The energy harvester electrodes and electrets structures are designed, fabricated and then the electret material is characterized with certain parameters in order to gain insights into its performance.

2. Energy harvester design

The electrostatic energy harvester described in this paper uses the *Trapezoidal* electrode design. As the name implies, the electrode is designed adopting a trapezoidal shape whereas the electrets design remains a rectangular shape as depicted in figure 1. In the electrodes structure, two trapezoidal electrodes are placed adjacent to one another. The electrodes are embedded onto a seismic mass with beams which vibrates in response to input vibrations. The electrets structure however, is made up of silicon dioxide (SiO_2) and placed on a stationary structure. The electrets are patterned and aligned to the trapezoidal electrodes structure as in figure 2.

The electrode and electrets designs are laid out with Cadence Virtuoso® software leveraging MIMOS 0.35 μm process technology. The completed layouts of the electrodes and electrets are depicted in figure 3 and figure 4 respectively. The final layouts are later sent to the mask shop where the photo masks or reticles are developed.

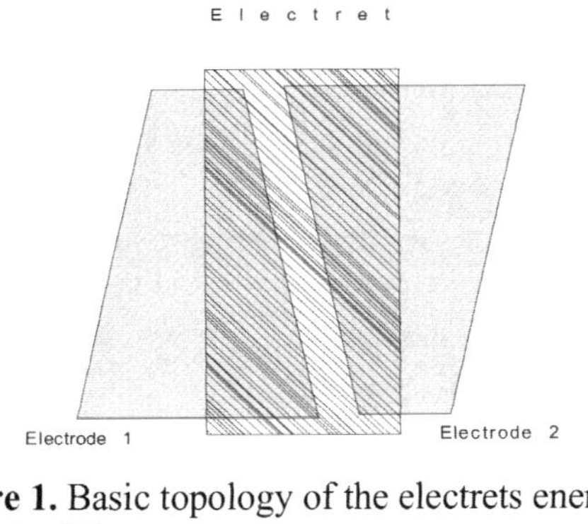

Figure 1. Basic topology of the electrets energy harvester [6].

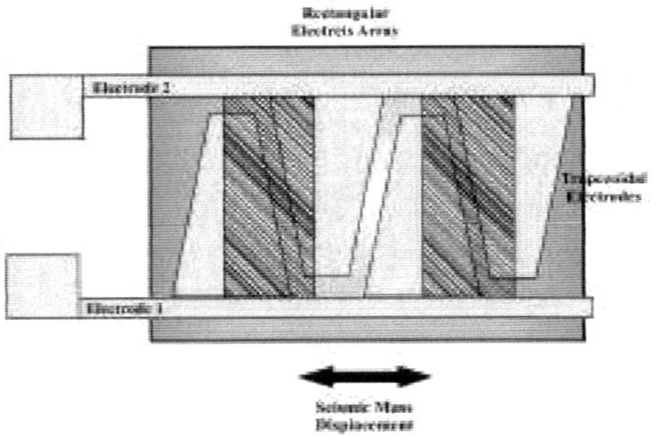

Figure 2. A preliminary design of the electrets energy harvester [6].

3. Device fabrications

Fabrications of the electrostatic energy harvester components, i.e. the trapezoidal electrodes and electrets, into silicon wafers are carried out at *MIMOS Berhad* (Malaysian Institute of Microelectronic Systems) adopting the Double-Poly-Triple-Metal CMOS process technology.

The fabrication process of the electrodes and electrets structures begins with the selection of silicon substrate. In this paper, a 200 mm Single Crystal Silicon (SCS) substrate with P-type <100> crystal orientation is selected for the fabrication. The detail fabrication processes of the energy harvester components are explained next.

3.1 Trapezoidal electrodes fabrication

Fabrication of the trapezoidal electrodes structure requires seven masking steps. Each mask is subjected to deposition, pattern transfer, develop and etch processes. The processes are repeated for each and subsequent masks. Each masking process creates either a layer or trench on the wafer, specifically on the dice leaving behind the intended pattern of the designs where positive photoresist materials are used throughout the process. The masking processes of the trapezoidal electrodes are described in table 1.

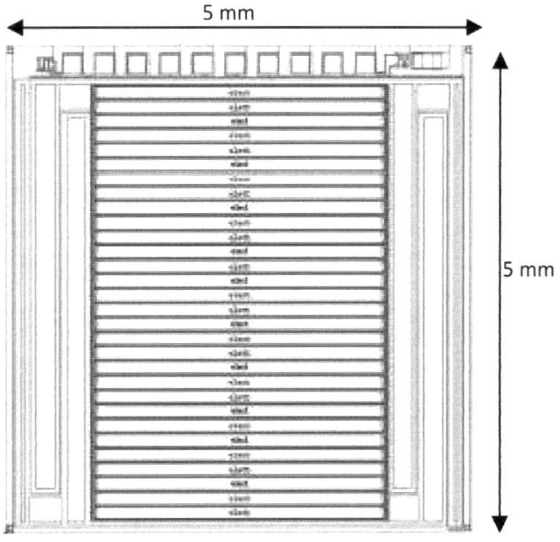

Figure 3. Layouts of the trapezoidal electrodes.

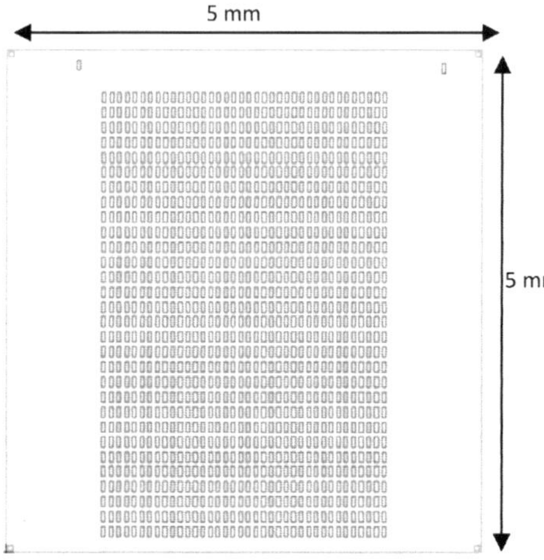

Figure 4. Layouts of the patterned electrets.

Table 1. Trapezoidal electrodes process.

Masks	Process name	Functions
Four corner guide	Developer	Pattern the four corner guide trenches
Metal 1 pattern (Photo-Lithography)	Developer	Transfer M1 interconnect pattern from the mask/reticle onto the wafer.
Via1 pattern	Developer	Pattern via hole for interconnection between M1 and M2
Metal 2 pattern	Developer	Transfer M2 interconnect pattern from the mask/reticle onto the wafer
Via 2 pattern	Developer	Pattern via hole for interconnection between M2 and M3
Metal 3 pattern	Developer	Transfer M3 interconnect pattern from the mask/reticle onto the wafer
Pad pattern	Developer	Define bond pad opening

3.2 Electrets fabrication

Fabrication of the electret structure begins with the Tetra Orthosilicate (TEOS) glass deposition by Chemical Vapor Deposition (CVD) method at 400°C, to achieve overall thickness of 20,000 Å (2 µm). Next, a thin layer of Silicon Nitride (Si_3N_4) of thickness 1500 Å is deposited onto the oxide by a CVD method using Silane (SiH_4) and Ammonia (NH_3) gases at 400°C. Then, a photoresist film of 4 µm thickness is spin coated onto the nitride layer. To pattern the rectangular shape electrets, the photoresist is first exposed with UV light through an electret mask or reticle, which is later developed and UV baked. Later, Reactive Ion Etching (RIE) also called the plasma etch process is utilized to remove the unwanted nitride layer. The RIE process condition is set as follows, 80 sccm of CF_4, 20 sccm of CHF_3, 200 sccm of Ar, 300 mT of pressure and 1100 W of RF power. The remaining photoresist left on the patterned nitride layer is further removed with oxygen plasma and acid resist strip. The patterned electrets goes through a post annealing process at 450°C in N2 atmospheric furnace for 1 hour. The annealing process re-arranges the oxide molecules which are damaged due to the etching process. Last but not least, the Aluminum of 0.5 µm thickness is deposited at the backside of the wafer using the e-beam deposition method. The aluminum layer provides a conductive surface required for the charging process. The detail process steps of the electrets fabrication are described in table 2.

Table 2. Electrets fabrication process.

Process steps	Process name	Functions
Four corner patterning	Developer	Pattern the four corner guide trenches
Four corner etching	DRIE	Create notches / trenches for the electrode / electrets alignment at each corner of the device
TEOS Oxide deposition	PECVD	Deposit thin oxide layer to form the electrets layer
Nitride deposition	PECVD	Deposit thin Silicon Nitride layer to passivate the electrets layer
Photoresist coating and Nitride patterning	Developer	Transfer the rectangular pattern of the electrets from the mask / reticle to the wafer.
Nitride layer etching	RIE	Etch away the unwanted oxide / nitride material and leave behind the electrets pattern.
Oxide Annealing	Post Anneal	To cure the oxide at 450°C in N_2 for 1 hour
Aluminum backside deposition	E-beam	To deposit thin layer of Aluminum

4. Electrets characterization

Electrets characterization is a process to charge the fabricated oxide samples by Corona charging method. During the charging process, negatively charged ions ($CO3^-$) are deposited onto the oxides surface which carries the electrostatic potential. To achieve this, several equipments are required e.g. Corona chamber with heater base, high voltage DC power supply (-5 kV), grid voltage DC power supply (-600 V), electrostatic voltmeter (Monroe Electronics Inc. USA, model 344 with probe) and probe holder. The oxide samples are first cut into 10 x 10 mm dimension for ease of handling during the experiment.

There are six parameters identified for the characterizations, i.e. Corona voltage, charging time, grid voltage, substrate temperature, grid height and needle height, as shown in table 3. The Corona voltage also called the needle voltage is first characterized. This highly negative DC voltage potential ionizes the surrounding air molecules which later broken down into positive and negative ions. The negative ions are attracted to the oxide sample due to positively polarize conductive base. The Corona voltages are set to sweep from -4 kV to -4.9 kV (limitations of the power supply), while the rest of the parameters remain constant. Once completed, the Corona voltage is set to the optimum value while characterizing the next parameters. The process continues until all the intended parameters have been characterized.

Each sample is charged for 15 minutes. When the charging time is completed, the sample is measured with the Monroe electrostatic voltmeter at five different locations. Measurements of the same sample are repeated three times and the average values are recorded and plotted into trend charts. The charging process is repeated for three different oxide samples in order to achieve consistency. The characterization results in figure 5, shows that the oxide deposited through CVD process is responded well to the Corona voltages where the optimum surface charge obtained occurs at 4.7 kV.

Table 3. Characterization parameters.

Parameters	Constant	Variables	Units
Corona Voltage	4.7	4 – 4.9	kV
Grid Voltage	200	50 - 450	V
Substrate Temp.	80	25 - 85	°C
Charging Time	15	5 - 40	mins
Needle Height	10	10 - 15	mm
Grid Height	2	1 - 5	mm

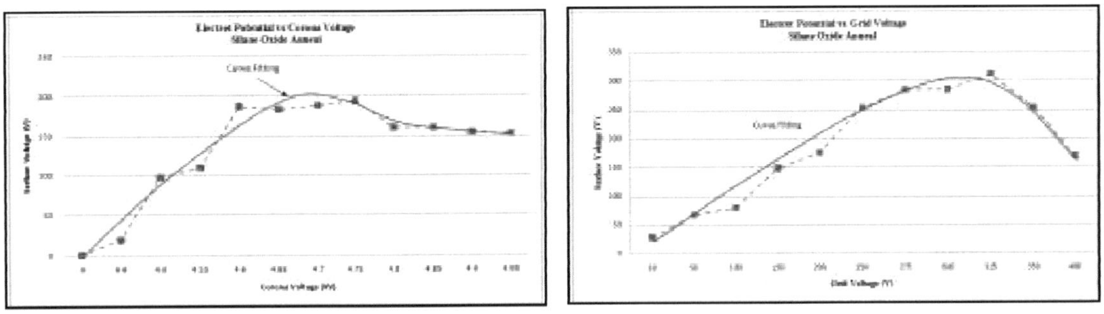

(a) (b)

Figure 5. Electrets potential with respect to the parameters. (a) Corona voltage and (b) grid voltage.

5. Conclusion

This paper described the fabrication processes of the vibration-based MEMS electrostatic energy harvester components, i.e. the trapezoidal electrodes and electrets, leveraging MIMOS 0.35 μm CMOS technology. The electrodes and electrets structures are fabricated on separate processes. Once the fabrications are completed, the performance of the electrets component is evaluated using the Corona charging method. Measurement results of the electrostatic potential show that the CVD oxides is easily chargeable and responded well to all parameters. From the experiment, it is concluded that the CVD oxide is a suitable electrets material for use on the electrostatic energy harvester device.

6. References

[1] Suzuki Y 2010 *10th Int. Workshop on Micro and Nanotechnology for Power Generation and Energy Conversion Applications (PowerMEMS 2010)* (Leuven Belgium) 47-52

[2] Mitcheson P D, Yeatman E M, Rao G K, Holmes A S and Green T C 2008 *Proc. of the IEEE* **96(9)** 1457-1486

[3] Beeby S P, Tudor M J and White N M 2006 *Meas. Sci. Technol.* vol **17** 175-195

[4] Mialtu R and Sararoiu M 2009 *CAS 2009 International* vol **1** 219-222

[5] Arakawa Y, Suzuki Y and Kasagi N 2004 *PowerMEMS 2004* (Kyoto Japan) 187-190

[6] Ahmad M R and Md Khir M H 2012 *4th International Conference on Intelligent and Advanced Systems* (Kuala Lumpur Malaysia) vol **1** 402-406

Acknowledgments

This research is partly funded by the Ministry of Higher Education (MOHE) Malaysia under grant name, *Fundamental Research Grant Scheme (FRGS)* with grant no. FRGS/1/2011/TK/UTP/18.

The authors would like to extend an appreciation to the MIMOS wafer fab personnels for their supports on the fabrication processes.

Under Voltage Lock-Out Design Rules for Proper Start-Up of Energy Autonomous Systems Powered by Supercapacitors

V.Boitier[1,3], P.Durand Estèbe[1,4], R. Monthéard[1,4], M.Bafleur[1,2] and J.M.Dilhac[1,4]
[1] CNRS, LAAS, 7 Avenue du colonel Roche, F-31400 Toulouse, France
[2] Univ de Toulouse, LAAS, F-31400 TOULOUSE, France [3] Univ de Toulouse, UPS, LAAS, F-31400 TOULOUSE, France [4] Univ de Toulouse, INSA, LAAS, F-31400 TOULOUSE, France
E-mail: vboitier@laas.fr, dilhac@laas.fr

Abstract. This paper deals with the issue of the initial start-up of an autonomous and battery-free system powered by an energy harvester associated with a storage subsystem based on supercapacitors initially discharged. A review of different low power Under Voltage Lock-Out (UVLO) solutions used to delay the load start-up and to avoid a useless discharge of supercapacitors is presented and discussed.

1. Introduction

For battery-powered systems, the supply voltage remains quite constant, whatever the state of charge of the batteries (within extreme limits) and start-up of powered systems (load) is not an issue. Conversely, in a battery-free system powered by an energy harvester, a storage unit is required as a buffer stage to adapt for current surge from the load and as a charge reservoir when the energy source is temporarily absent. This storage unit is generally based on a supercapacitor whose voltage (V_{SC}) at the terminals is charge dependent. Considering the classical topology of two supercapacitors in series, the voltage level varies typically between 0 and 5.0 V. On the other hand, the load usually works properly within a limited range [V_{MIN} - V_{MAX}], raising the following issues.

First, generally speaking, the device consumption is null or negligible for a voltage below a threshold V_0 (with $V_0 < V_{MIN}$). When the supply voltage reaches V_0, the device consumption increases to a certain level. If this level is greater or equal to the harvested power, supply voltage will never reach V_{MIN}, a necessary condition for a proper start-up of the system. Therefore, in this situation the storage device and the load must be disconnected until $V_{SC} > V_{MIN}$.

Moreover, a power peak is often required for load start-up inducing a voltage drop across the supercapacitors. As it takes place when the voltage value is close to V_{MIN}, the system is likely to stop immediately. To avoid such a problem, it is necessary to delay the start-up.

Finally, when harvested power vanishes, the system relies on stored energy only. When consequently V_{SC} falls below V_{MIN}, the load stops properly working, but its consumption will remain stable or will even increase. It will then uselessly empty out the supercapacitors, and will accordingly delay the next start-up. In such a case, it is of interest to disconnect the load as soon as the decreasing voltage reaches V_{MIN}.

2. Start-up issue in an energy autonomous system

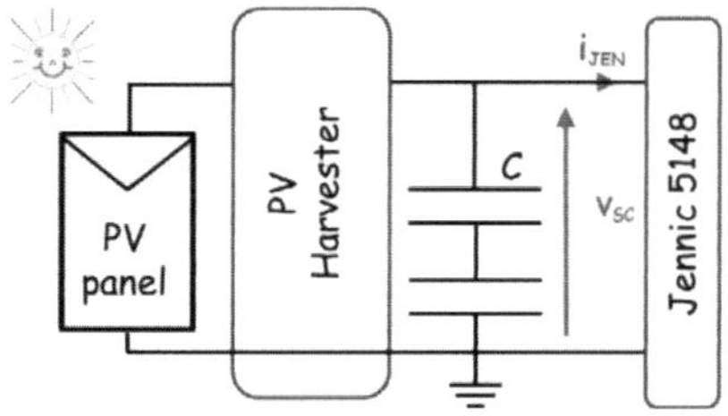

Figure 1. Schematic of the structure considered.

To illustrate the above phenomenon, we built the system described in figure 1. It is made of the following devices:

- energy storage is achieved by two Maxwell PC10 supercapacitors in series,
- the harvested energy is supplied by Solarex MSX-01 photovoltaic panels (V_{OC} = 10V, I_{SC} = 150 mA @STC,)
- an energy management device incorporating a DC/DC buck converter and a microcontroller sits between the above panels and supercapacitors. [**1**].
- the load is an JN5148 Wireless Microcontroller Module (WMM) targeted at low-power wireless networking applications, programmed for a measurement / transmission each 8 s, and switching to sleep mode in between. The discussed [V_{MIN} - V_{MAX}] supply voltage nominal interval is [2.3V - 3.6V].

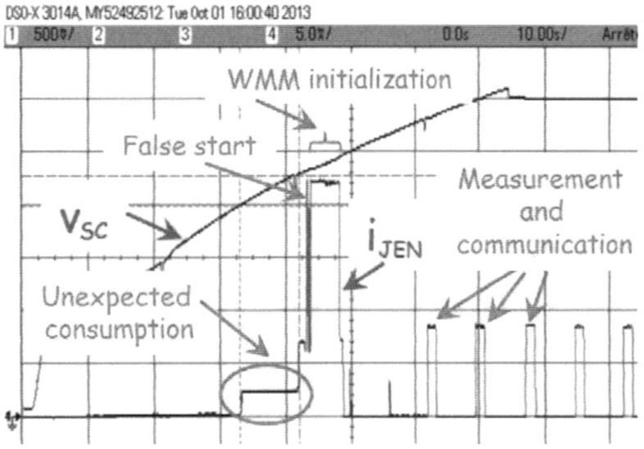

Figure 2. *Cold start* in the case of a high harvested power. V_{SC} (500 mV/div), current provided to the load, i_{JEN} (5 mA/div) vs. time (10s/div).

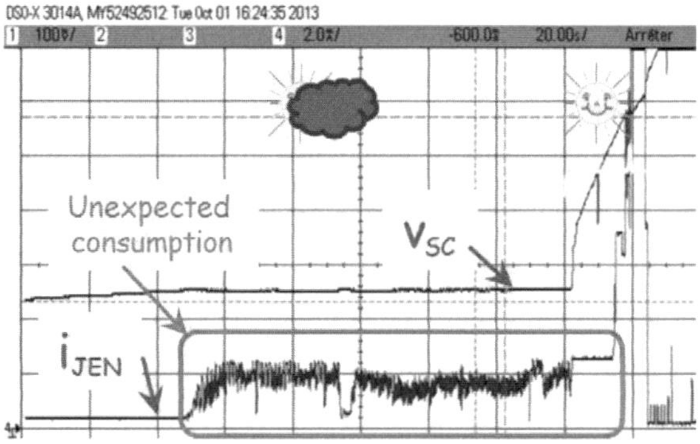

Figure 3. *Cold start* in the case of an initial low harvested power; initial V_{SC}=2.0V. V_{SC} (100 mV/div), i_{JEN} (2 mA/div) vs. time (20s/div).

In figure 2, V_{SC} voltage and current i_{JEN} supplied to the Wireless Microcontroller Module are plotted vs. time, when supercapacitors are initially empty and panels are benefiting from a good illumination (~1000 W/m^2). Initially V_{SC} regularly increases and the module draws no current. When V_{SC} reaches V_0 = 2.0 V, the supplied current establishes itself at 2mA (label "*unexpected consumption*" in Figure 3) until V_{SC} reaches the lower threshold voltage of the nominal bias voltage interval (i.e V_{MIN} = 2.3V). During this phase a few false starts corresponding to higher and abnormal current values can be observed. Then, while V_{SC} is still increasing, initialization of the Wireless Microcontroller Module takes place, requiring a high current. Later on, current peaks correspond to measurement and data transmission. It appears that a lot of energy is required during system start-up. If this energy cannot be supplied (reduced illumination), start-up is not possible (see figure 3), and because of the current drawn by successive attempts, this energy will never be stored in the supercapacitors.
The purpose of an Under Voltage Lock-Out (UVLO) strategy is to deal with this problem.

3. UVLO solutions

A UVLO system is made of two devices:

- a blocking device (switch) controlling the load biasing,
- a control logic. [**2**]

Below we present various implementations of these two functions.

3.1. Blocking system

3.1.1. Simple switch located on the ground rail.

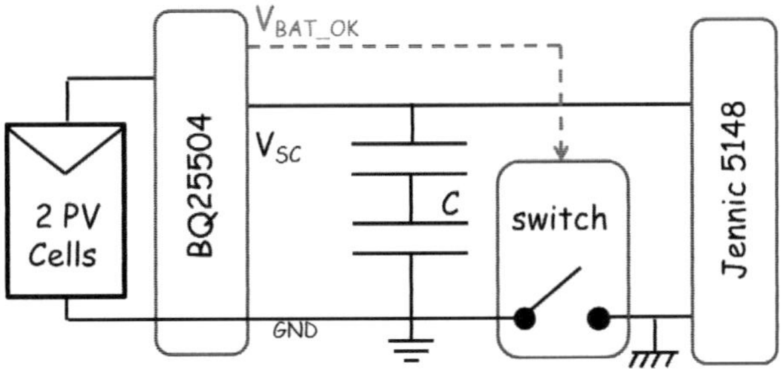

Figure 4. *Structure with the switch on the ground line.*

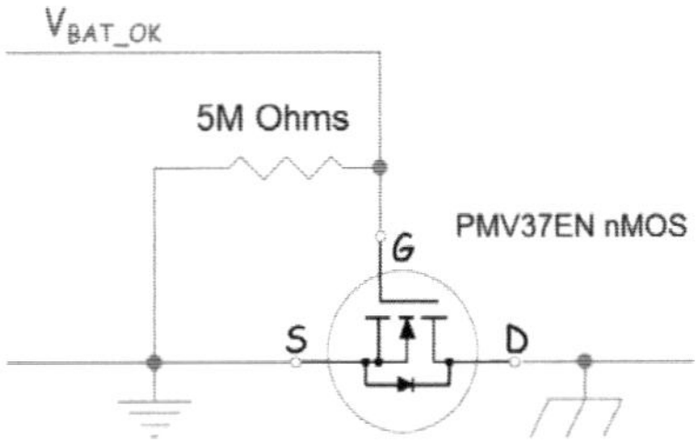

Figure 5. *Switch on the ground line.*

This is a straightforward implementation (see figures 4 and 5) using an NMOS transistor as a switch whose control voltage V_{GS} comes from the control logic. An NMOS is chosen because its body diode will never be forward biased. The 5 MOhms resistor connected to the gate guarantees an OFF state when V_{BAT_OK} is low during a *cold start*; its high value limits the current between control signal V_{BAT_OK} and ground when V_{BAT_OK} is high.
The following characteristics are required for the NMOS transistor:

- a gate threshold voltage compatible with the control logic output,
- a R_{DSON} value as low as possible to limit energy losses and voltage drop between supercapacitors and wireless module,
- a nominal current compatible with the maximum current drawn by the load,
- a low *slew rate* to limit inrush current which can be adjusted with an RC circuit [**3**].

The main drawback associated with this topology is the introduction of two ground references, likely to complicate measurement in a prototyping phase. An alternative is presented below.

3.1.2. Switch located on the supply rail.

Figures 6 and 7 present this alternative where the switch is inserted in the supply rail, preserving a single ground reference. Details about the control logic are given in figure 8.

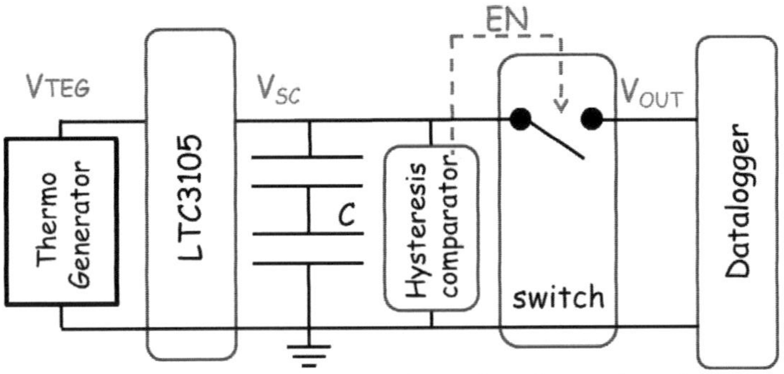

Figure 6. *Structure with the switch located on the supply rail.*

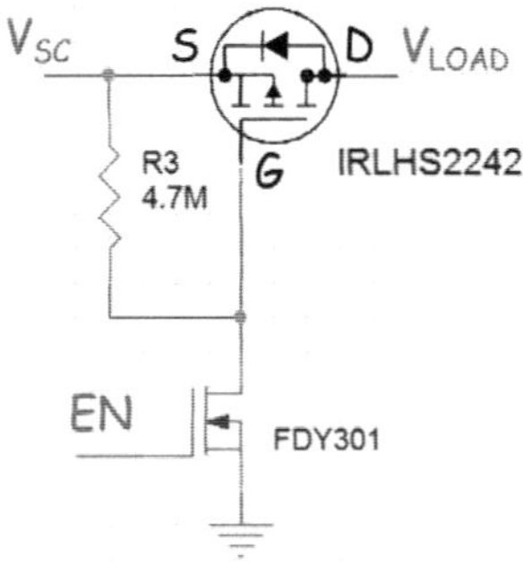

Figure 7. *Switch on the supply rail.*

The switch here is a PMOS transistor whose gate is controlled by an NMOS. The intrinsic diode of the MOS asks here for a PMOS to get an initial OFF state during system start-up. When *high*, the logic output turns ON the PMOS by pulling its gate to ground through the NMOS, and conversely turns it OFF when *low* by turning OFF the NMOS, allowing the PMOS gate to be pulled up to V_{SC} through the pull up resistor.
This structure requires the same characteristics (low R_{DSON}, NMOS threshold voltage, etc.) as in the NMOS switch presented in the previous paragraph, for the same reasons.
Such a structure can easily be implemented using discrete devices, or by choosing a dedicated integrated circuit as explained below.

3.1.3. Load switch
An integrated load switch contains a P-channel MOSFET that operates over an input voltage range (for example, 1.0 V to 3.6 V for a TPS22901). The switch is controlled by an on/off input, which is capable of interfacing directly with low-voltage control signals. Optionally, an on-chip load resistor is added to avoid a floating rail and for output quick discharge when the switch is turned off.
An integrated load switch offers:

- a low R_{DSON},
- a low threshold control inputs,
- an ultra-small packaging,
- a quiescent and shutdown current below 1 μA,
- an internally controlled slew rate.

Nevertheless an even simpler solution is possible as detailed below.

3.1.4. Pin "Enable" or execution of a priority interrupt
When the load to be powered exhibits an *Enable*-type input command, the simplest implementation of an UVLO procedure lies in the use of this command [**4**], provided first that the command logic outputs are compatible, and then that the corresponding quiescent current is low enough. If the load is a microcontroller, an input command triggering a priority interrupt execution deactivating peripherals and switching to a deep sleep mode is also an option (as an example, we measured 2μA in deep sleep with the Jennic 5148).

4. Control logic
With respect to the control logic, it has to monitor V_{SC} and to determine the logic command to be applied to the switch or circuit inputs. A simple practical solution is a hysteresis comparator using two thresholds V_{IH} and V_{IL}, with $V_{IH} = V_{MIN} + \Delta V$ and $V_{IL} = V_{MIN}$.[**5**]
ΔV is a voltage (and therefore an energy) margin allowing extra power consumption during start-up preventing V_{SC} from falling rapidly below V_{MIN} before the end of the transient associated with the start-up itself. Worth to be mentioned is the voltage drop associated with the series parasitic resistors of the supercapacitors, adding an extra component to V_{SC} drop for high currents.

4.1. Hysteresis comparator with discrete elements
The driving circuit on figure 8 is realized with discrete elements using an ultralow power comparator. It includes a voltage reference and is supplied by the supercapacitor voltage.

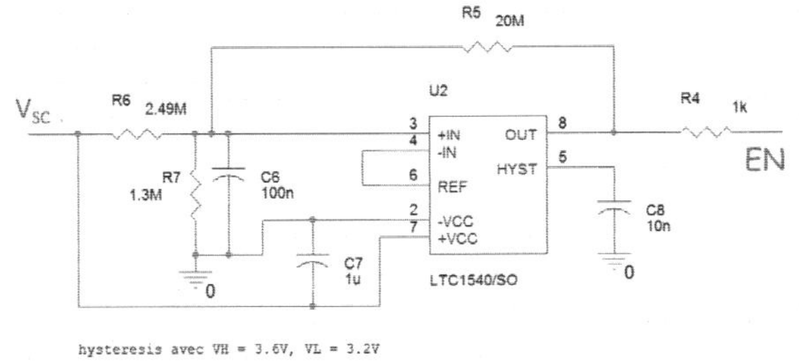

Figure 8. *Comparator with hysteresis.*

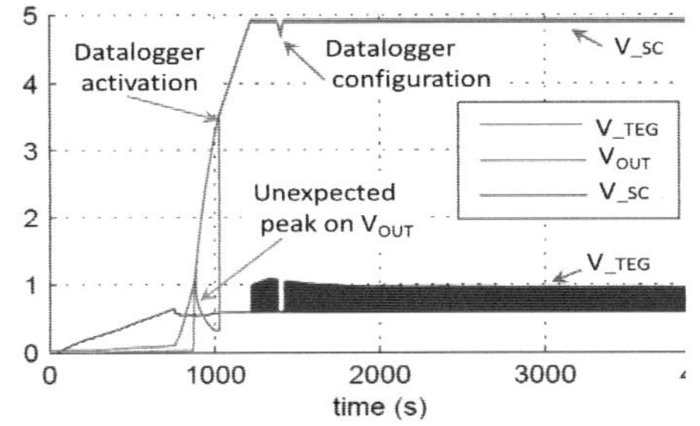

Figure 9. *Start-up sequence.*

Passive component values are chosen to minimize energy losses. Total consumption of the circuit of figure 8 is about 2μA. It has been implemented for the topology depicted in figures 6 and 7. In this case the power source is a thermoelectric generator connected to a DC/DC converter, whereas the load is a commercial datalogger. Corresponding experimental results are presented in figure 9. It can be seen that, even though the datalogger is actually turned on as soon as V_{SC} reaches the high voltage level V_H=3.6V, an unexpected voltage peak takes place on the UVLO output when V_{SC} = 1V. Such a parasitic behavior is caused by the hysteresis comparator, which is transiently unable to maintain its output low as its supply voltage is increasing, yet below its minimum value (2V). This can be avoided by selecting an NMOS with a threshold voltage greater than the comparator's minimum supply voltage. Fortunately, in this configuration the peak is too small to turn on the datalogger, and disappears as soon as the comparator is properly supplied.

4.2. Integrated hysteresis comparator

The desired feature is often included on commercial energy harvesting components. As an illustration, Texas Instruments BQ25504, which is a boost charger IC for ultra-low power energy harvesting, provides a V_{BAT_OK} output command, and three external resistors can adjust thresholds V_{IH} and V_{IL}. If energy source disappear, this energy harvesting device stays active and takes its supply power from the supercapacitors and drains 30 μA. This value should be compared with the 2 μA consumption only of the hysteresis comparator on Fig. 8.

5. Conclusion

Solving the start-up issue of a battery-free autonomous system is essential to optimize both its power consumption and its fast and effective start-up, and thus allowing a normal device operation. We have presented the reasons and main implementations of an Under Voltage Lock-Out (UVLO) strategy when intermediate storage in supercapacitors is used. Solutions using integrated circuits are often the more attractive in terms of simplicity and cost. Discrete implementations are the more appropriate while in a design and prototyping phase and can be used as a reference vs. power consumption.

6. References

[1] Meekhun, D. 2011 Buck converter design for Photovoltaic generators with supercapacitor energy storage" , *Renewable Energy & Power Quality Journal*, N°9, ISSN: 2172-038X

[2] Huang Xiaozong et al. 2012 A proposed under-voltage lockout of compensated temperature coefficient threshold voltage without comparator, *J. of Theoretical and Applied Information Technology*, 15th Oct. 2012. Vol. 44 No.1, pp.117-120.

[3] http://low-powerdesign.com/article_TI_pichot.htm

[4] http://www.monolithicpower.com/DesignNoteDoc/DN0002_r1.0.pdf

[5] http://www.diodes.com/_files/products_appnote_pdfs/zetex/an65.pdf

An AlN cantilever for a wake-up switch triggered by air pressure change

Y. Kaiho[1,2], H. Takahashi[3], Y. Tomimatsu[2], T. Kobayashi[4], K. Matsumoto[3], I. Shimoyama[3], T. Itoh[1,4] and R. Maeda[1,4]
[1]NMEMS Technology Research Organization, JAPAN
[2]Seiko Instruments Inc., JAPAN
[3]The University of Tokyo, JAPAN
[4]National Institute of Advanced Industrial Science and Technology (AIST, JAPAN)

E-mail:ykaihou@nmems.or.jp

Abstract. This research reports an AlN cantilever with an air chamber for a wake-up switch triggered by air pressure change. The proposed sensor is designed to fulfil both high sensitivity and low power consumption. By combining an air chamber to the one side of the AlN cantilever surface, the barometric pressure change generates a piezoelectric voltage. Thus, a wake-up switch triggered by air pressure change can be achieved using an AlN cantilever. The size of the fabricated AlN cantilever was 2000 μm × 1000 μm × 2 μm. The sensitivity to static differential pressure was 11.5 mV/Pa at the range of −20 Pa to 20 Pa. We evaluated the response of the sensor, which was composed of the AlN cantilever and the chamber of 60 ml in volume, when air pressure change was applied. The output voltage increased with increasing the applied air pressure change. It was observed that the maximum output voltage of 50 mV was generated when the air pressure change was 13 Pa.

1. Introduction

In the wireless sensor network (WSN) system, the monitoring objects, such as temperature, humidity, and the amount of particle, are related to air pressure change due to human moving. Thus, information of the change of air pressure around the sensor nodes will manage WSN system efficiently [1,2]. A barometric pressure sensor with low power consumption is suitable for the application as the trigger switch of WSN systems due to limited battery power. However, it was difficult to monitor air pressure change with high sensitivity and low power consumption by the conventional methods.

Previously, we presented a differential pressure sensor using a cantilever with piezoelectric element which can achieve both high sensitivity and low power consumption due to the deformable structure and sensing element characteristics [3,4]. By combining an unsealed air chamber to the one side of the piezoelectric cantilever surface, the air pressure change around the other side generates a piezoelectric voltage due to the cantilever deformation as shown in Figure 1(a) [5,6].

In this research, a piezoelectric cantilever with an air chamber was designed and fabricated for a wake-up switch triggered by air pressure change. Aluminum nitride (AlN) was used as a piezoelectric element due to its excellent chemical stability and a small dielectric constant [7,8]. The response of the fabricated sensor to static differential pressure was measured. We also evaluated the sensor response to instantaneous pressure change.

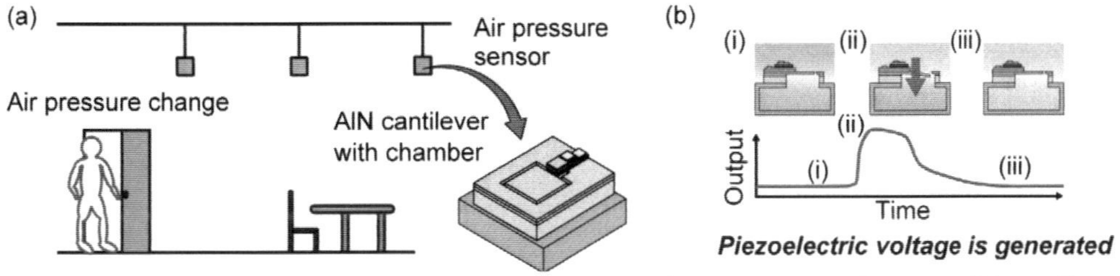

Figure 1 Concept (a) and principle (b) of an air pressure change sensor using an AlN cantilever with an air chamber.

2. Design and Fabrication

2.1. The principal of pressure change detection

The principle of pressure change detection using a piezoelectric cantilever is shown in Figure 1(b). The sensor consists of a AlN thin cantilever with small gaps around the cantilever fixed on an air chamber. When pressure difference is applied between the upper surface and the lower surface of the cantilever, the cantilever is bent. Then, the piezoelectric voltage is generated between the upper and the lower cantilever surfaces.

Pressure in the chambers is the normally same as atmospheric pressure as shown in Figure 1(b)(i). When the air pressure around the sensor changes, the pressure difference between upper and lower surfaces occurs because the pressure in the chamber is still atmospheric pressure as shown in Figure 1(b)(ii). Therefore, the cantilever can detect barometric pressure change. In a few moments both pressures in the upper and lower surfaces become same pressure because of the air leak through the gap around the cantilever as shown in Figure 1(b)(iii).

In cases where the sensor nodes were deployed in an industrial clean room, for example, pressure difference between inside and outside of the room is several tens of pascals. Thus, door opening and closing will generate air pressure difference of several tens of pascals. For this reason, the proposed sensor is required to be able to detect air pressure change of ten pascals.

The generated piezoelectric voltage becomes a trigger signal due to barometric pressure change. Because the sensing element needs no power consumption in principle, the proposed device can perform as a barometric pressure change trigger without power consumption.

2.2. Design and Fabrication process

A design of the AlN cantilever is shown in detail in Figure 2(a). The cantilever is 2000 μm × 1000 μm × 2 μm in size. The front edge of the cantilever is wide to increase the area where pressure acts. The base of the cantilever is narrow to have stress concentration and low capacitance [3,4]. The base area consists of Pt/Ti/AlN/Pt/Ti/SiO_2/Si multi-layer. The cantilever gap is designed to be 5 μm in order to prevent air leaks through the gap [3-6]. The fabrication process was shown in Figure 2(b). The AlN cantilever was fabricated from the multi-layer of Pt/Ti/AlN/Pt/Ti/SiO_2 on a silicon on insulator (SOI) wafer as shown in Figure 2(b)(i). Firstly, the Pt/Ti/AlN/Pt/Ti/SiO_2 multi-layer was etched as shown in Figure 2(b)(i). An Au layer was deposited and patterned for pads. Secondly, the device Si/SiO_2 layers were etched to form the cantilever shape as shown in Figure 2(b)(iii). Finally, the handle Si and SiO_2 layers were etched as shown in Figure 2(b)(iv).

Figure 2 (c)(d) shows a photograph of the fabricated sensor chip and an SEM image of the cross-sectional surface of the cantilever. The sensor chip is 3.0 mm × 5.0 mm × 0.4 mm in size. It is observed that the Pt/Ti/AlN/Pt/Ti/SiO_2 multi-layer was formed on the Si layer. The capacitance of the fabricated AlN cantilever was 68 pF.

We measured the response of the fabricated AlN cantilever when static differential pressure was applied between upper and lower surfaces of the cantilever. The cantilever chip was mounted on a substrate that separated two chambers. The differential pressure was provided by applying air pressure

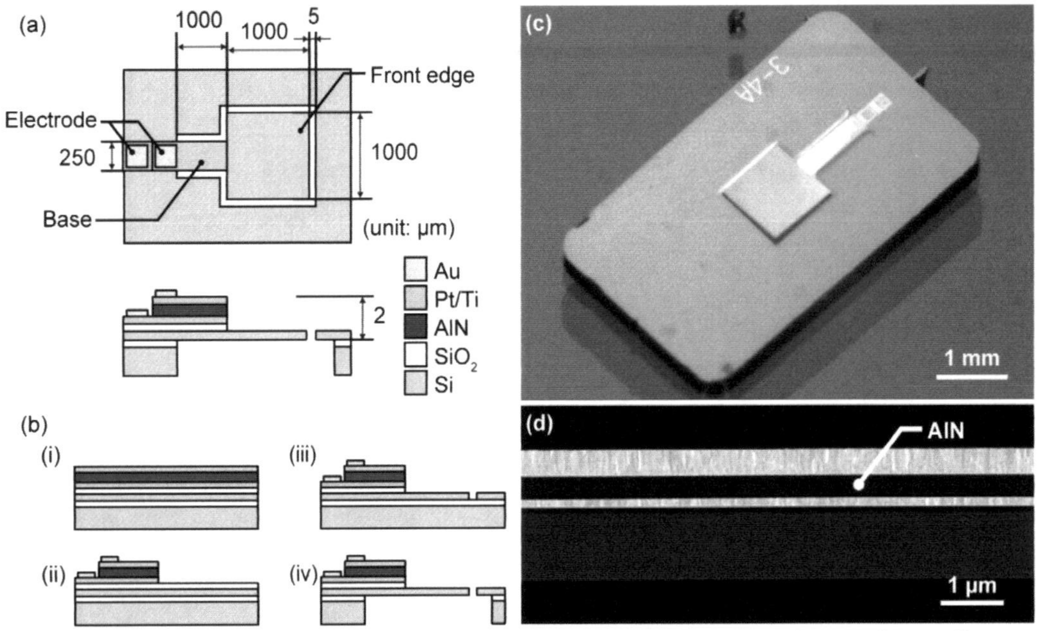

Figure 2 (a) Design of the AlN cantilever. (b) Fabrication process of the AlN cantilever. (i) An AlN cantilever is fabricated from the multi-layer of Pt/Ti/AlN/Pt/Ti/SiO_2 on an SOI wafer. (ii) The Pt/Ti/AlN/Pt/Ti/SiO_2 layers are etched. Au layer is deposited and patterned. (iii) The device Si layer and buried oxide (BOX) are etched. (iv) The handle Si layer and BOX are etched. (c)(d) Photograph and SEM image of the fabricated device.

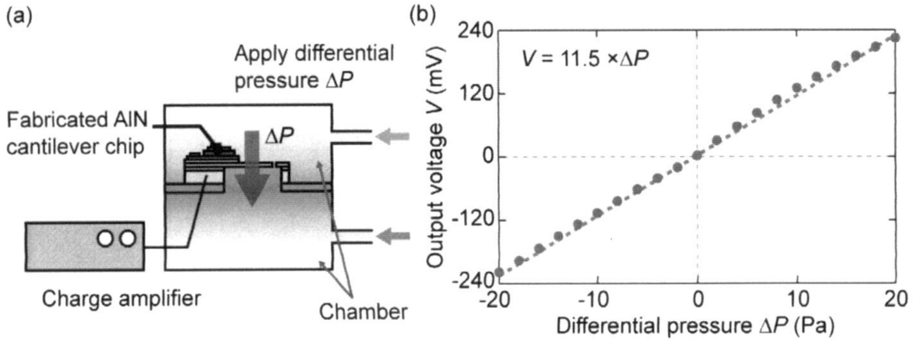

Figure 3 (a) Schematic image of the experimental setup to measure the response to static differential pressure change. (b) Relationship between the differential pressure and the output voltage.

to two chambers through a pressure generator as shown in Figure 3(a). Then, the piezoelectric voltage was measured through a charge amplifier. Figure 3(b) shows the relationship between the output voltage and the static differential pressure. The output voltage was proportional to the applied differential pressure. The sensitivity to the differential pressure was calculated to be 11.5 mV/Pa at the range of −20 Pa to 20 Pa.

3. Experiment and result

We measured the response of the sensor, which was composed of the AlN cantilever and an air chamber, when air pressure change was applied as shown in Figure 4(a)(b). The volume of the sensor chamber is 60 ml. Chamber 1 and Chamber 2 were connected to the sensor though valves. The volume

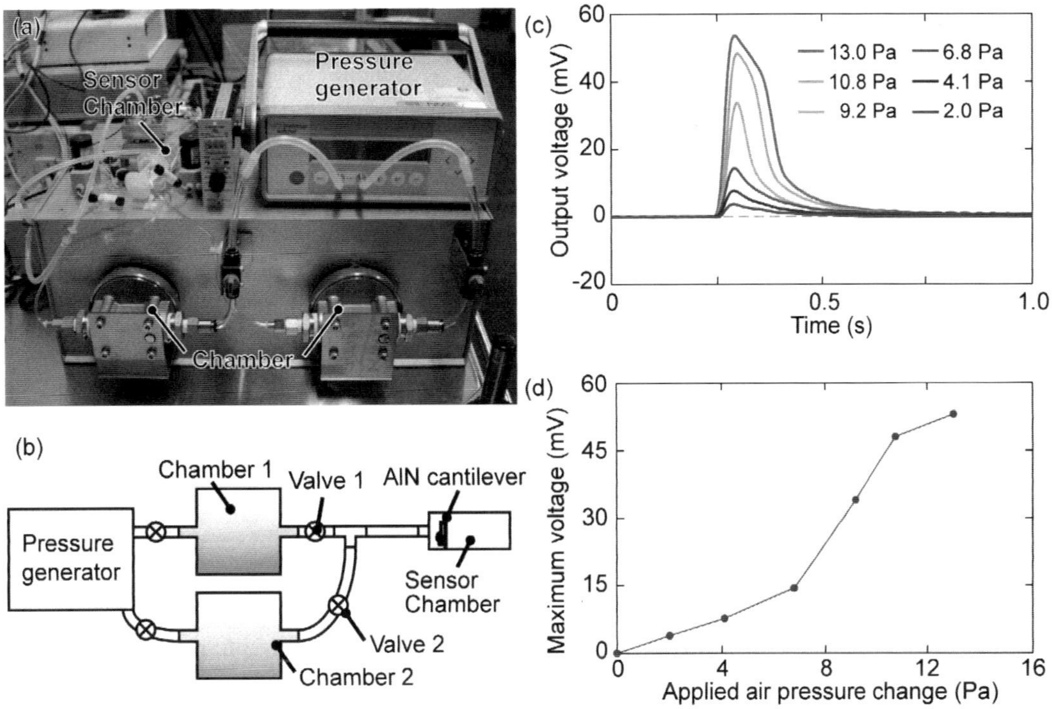

Figure 4 (a)(b) Photograph and schematic image of the experimental setup to measure the response to barometric pressure change. (c) Real-time response to pressure change. (d) Relationship between the pressure change and the maximum output voltage.

of each chamber was 1000 ml. The chambers were also connected to a pressure generator, which applied differential pressure between the two chambers. At first, the valve 1 and valve 2 were opened and closed, respectively. Thus, the air pressures of the upper and lower surfaces of the cantilever were equal to that of the chamber 1. When the states of the valves were switched, the barometric pressure change was applied to the sensor. Then, the sensor generated piezoelectric voltage according to the pressure difference between the chamber 1 and chamber 2. We applied the barometric pressure change of 2.0, 4.1, 6.8, 9.2, 10.8 and 13.0 Pa.

Figure 4(c) shows the real-time response of the sensor to barometric pressure change. It was observed that there was the voltage peak shortly after changing the state of the valves in each pressure change. Then, we obtained the relationship between the maximum output voltage of the sensor and the magnitude of the applied air pressure change as shown in Figure 4(d). As shown in Figure 4(c) and (d), the output voltage increased when the applied air pressure change was increased. The maximum output voltage was approximately 50 mV when applied pressure was 13 Pa. The generated voltage was thought to be high enough to switching an electrical circuit. Thus, using the proposed sensor, a high sensitive switch with a low consumption triggered by air pressure change would be achieved.

The generated output voltage corresponded to differential pressure change of 4.3 Pa. It was thought that the differential pressure between the chamber 2 and the sensor chamber did not reach to 13.0 Pa because of the air leak through the gap. The waveform of the output voltage was different between over 10 Pa and under 10 Pa. It was also considered that there was the different behaviour of the cantilever bending due to the air leak in a boundary 10 Pa. This phenomenon would be related to the cantilever and the sensor chamber sizes. By improving the sensor design, the output voltage would become larger. We will research the relationship of the cantilever and the chamber sizes and the applied air pressure change in future work.

4. Conclusion

The AlN cantilever with the air chamber for a barometric pressure change trigger was designed and fabricated. The fabricated AlN cantilever was 2000 μm × 1000 μm × 2 μm in size. In the static differential pressure detection, the fabricated AlN cantilever had a good linearity between -20 Pa and 20 Pa. The measured sensitivity was 11.5 mV/Pa. As a performance of a barometric pressure sensor, the output voltage of 50 mV was generated when the air pressure change of 13.0 Pa was applied.

Acknowledgments

A part of this work was supported by New Energy and Industrial Technology Development Organization (NEDO). The photolithography masks were fabricated using the EB lithography apparatus of the VLSI Design and Education Center (VDEC) of the University of Tokyo.

Reference

[1] Anastasi G, Conti M, Di Francesco M, and Passarella A 2009 Energy conservation in wireless sensor networks: A survey *Ad Hoc Networks* **7 no. 3** 537–568

[2] Akyildiz I F, Suv W, Sankarasubramaniamv Y, and Cayirci E 2002 Wireless sensor networks: a survey *Computer Networks* **38 no. 4** 393–422

[3] Tomimatsu Y, Takahashi H, Kobayashi T, Matsumoto K, Shimoyama I, Itoh T and Maeda R 2013 A wake-up switch using a piezoelectric differential pressure sensor *Proc. of ISSNIP2013* 23-26

[4] Tomimatsu Y, Takahashi H, Kuwana K, Kobayashi T, Matsumoto K, Shimoyama I, Itoh T and Maeda R 2013 A piezoelectric flow sensor for use as a wake-up switch for a wireless sensor network node *Mechatronics* **02** 003.

[5] Nguyen M-D, Takahashi H, Matsumoto K and Shimoyama I 2011 Barometric pressure change measurement *Proc. of Transducers '11* 898-901

[6] Takahashi H, Nguyen M-D, Matsumoto K, and Shimoyama I 2012 Differential pressure sensor using a piezoresistive cantilever *Journal of Micromechanics and Microengineering* **22 no. 5** 055015

[7] Cimalla V, Pezoldt J, and Ambacher O 2007 Group III nitride and SiC based MEMS and NEMS: materials properties, technology and applications *Journal of Physics D: Applied Physics* **40 no. 20** 6386–6434

[8] Dubois M-A and Muralt P 1999 Properties of aluminum nitride thin films for piezoelectric transducers and microwave filter applications *Applied Physics Letters* **74 no. 20** 3032

Low Frequency Vibration Energy Harvester Using Spherical Permanent Magnet with Non-uniform Mass Distribution

Y Choi[1], S Ju[1], S H Chae[1], S Jun[1], S M Park[1], S Lee[1], H W Lee[2], and C-H Ji[1]

[1] Department of Electronics Engineering, Ewha Womans University, Seoul, Korea
[2] Department of Neurology, Ewha Womans University School of Medicine, Seoul, Korea

cji@ewha.ac.kr

Abstract. We present a non-resonant vibration energy harvesting device using springless spherical permanent magnet with non-uniform mass distribution as a proof mass. The magnet has its center-of-mass below the geometrical center, which generates a roly-poly-like motion in response to external vibrations. Two different types of magnet assemblies with different center-of-mass position have been fabricated and tested. Using the roly-poly-like magnets, proof-of-concept electromagnetic energy harvesters have been fabricated and tested. Moreover, effect of ferrofluid as a lubricant has been tested with the fabricated energy harvester. Maximum open-circuit voltage of 154.4mV and output power of 4.53μW have been obtained at 3g vibration at 12Hz with the fabricated device.

1. Introduction

Energy harvester is one of the promising research topics in the area of micro power generation due to the abundance of energy sources in environment. Most widely pursued target applications of energy harvesting include wireless sensor nodes and low power electronics. Among the various sources of energy, environmental vibration is an excellent candidate for a variety of applications due to ease of utilization and abundance in nature. In general, there are three ways to convert vibration into electrical energy which are piezoelectric, electromagnetic and electrostatic transduction mechanisms [1].

Vibration from human motion is three-dimensional and can be characterized with its random nature, relatively high acceleration, and very low frequency range [2]. These factors make it hard to develop a vibration energy harvester for human-body-induced motion using conventional spring-mass-damper-based resonant systems. To overcome these issues, energy harvesters with springless proof mass have been researched as alternatives to the proliferating spring-mass-damper-based harvesting devices. Various approaches including an impact based piezoelectric energy harvester, an electromagnetic harvester using random motion of a spherical magnet in the cavity wrapped with coil, and electrostatic harvester utilizing rolling magnet on variable capacitor array have been demonstrated [3-5].

In this research, we have proposed a non-resonant vibration energy harvesting device using springless spherical permanent magnet which generates a spontaneous wobbling motion to restore the upright position when an external vibration is applied. While utilizing the mass-controlled spherical magnet as a source for time-varying magnetic field, series connected windings have been wrapped on top and bottom of the cavity for power generation. Fabricated device has been characterized at low frequency vibrations. Effect of the center-of-mass position on magnet behavior and effectiveness of ferrofluid as a lubricant have also been investigated.

2. Device design

2.1. Magnet design
To analyze the effect of center-of-mass position, magnet assemblies with two different center-of-mass positions have been fabricated and tested. Magnet #1 consists of NdFeB ball magnet covered with two hemispherical shells made of Teflon and SUS as lighter and heavier shells, respectively (figure 1(a)). Magnet #2 has been designed to have its center-of-mass closer to the geometrical center by reducing the height of the SUS shell (h) by half that of magnet #1. Distance between the center-of-mass and geometrical center (d) for magnet #1 and #2 are 0.57mm and 0.39mm, respectively.

2.2. Harvester design
As shown in figure 1(b), designed harvester consists of copper windings on top and bottom of the magnet housing. The magnet assembly has been inserted in a cylindrical cavity formed inside the housing. Diameter of the magnet assembly is 7mm. Outer diameter and height of the housing measure 17mm and 23mm, respectively. Inner cavity measures 9mm in diameter with 8mm-height. Bobbins are formed on top and bottom parts of the housing for copper windings. Effect of ferrofluid as a lubricant for the magnet assembly has been analyzed by comparing the harvester output.

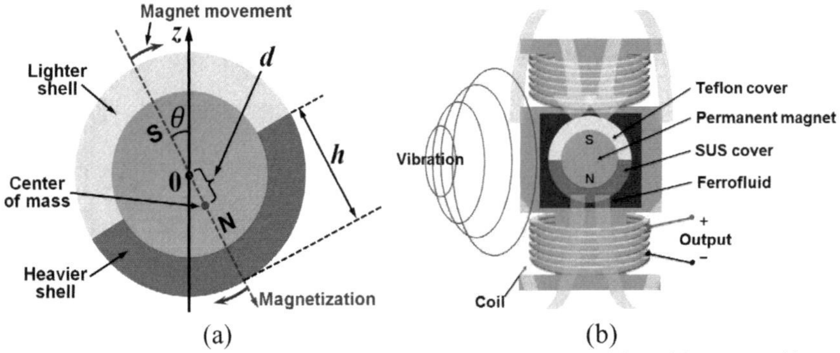

Figure 1. Schematics of the designed device: (a) magnet assembly with non-uniform mass distribution (θ= tilt angle), (b) electromagnetic energy harvester

3. Fabrication
A proof-of-concept energy harvester has been fabricated by enclosing a roly-poly-like magnet assembly inside Teflon housing with cylindrical cavity. As shown in figure 2, roly-poly-like magnet has been prepared by attaching 1mm-thick shells made of Teflon and SUS on a 5mm-diameter NdFeB ball magnet with epoxy adhesive. Top and bottom parts of the Teflon housing have been joined using miniature screws and 0.1mm-diameter copper coil has been wound around the 5mm-diameter bobbin formed on top and bottom (figure 3). Number of winding turns measures 700 turns each on top and bottom. Minimum gap between the winding and cavity is 1mm.

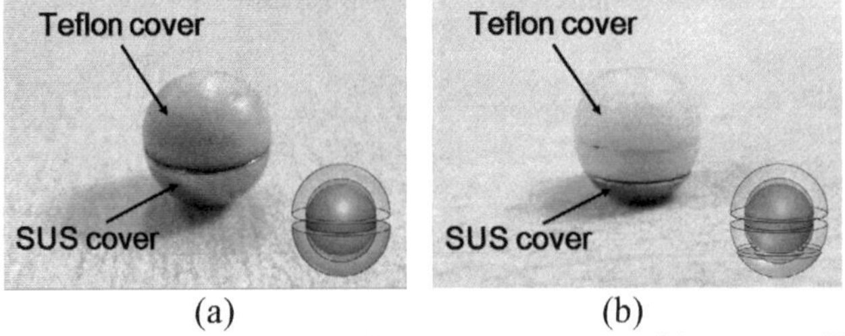

Figure 2. Magnet assemblies: (a) magnet #1, (b) magnet #2

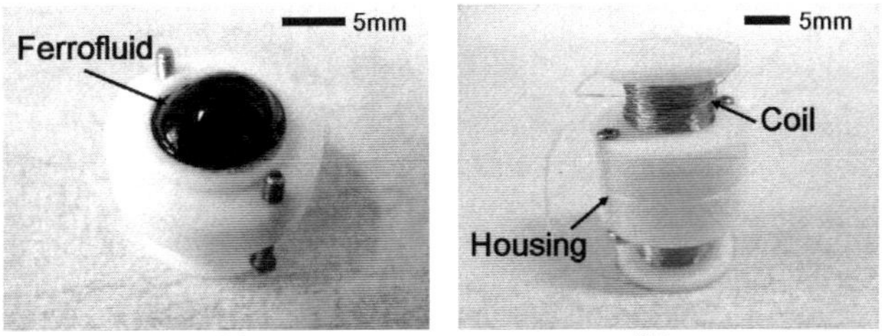

(a) (b)

Figure 3. (a) magnet assembly in Teflon housing covered with ferrofluid, (b) assembled electromagnetic harvesting device

4. Results and discussion

4.1. Characterization of magnet assemblies

For a better understanding of the device architecture, resonant characteristics of two roly-poly-like magnets have been tested. Initial tilt angle was set to 30 degrees and time responses of the roly-poly-like magnets were recorded using 30fps video camera and analyzed. Although pure rotational vibration along specific rotational axis without wobbling motion could not be achieved, relatively clear time responses have been obtained (figure 4). Time constants for exponential decay were 2sec and 0.67sec for magnet #1 and #2, respectively. Estimated resonant frequencies of magnet #1 and #2 were 4.74Hz and 4.52Hz, respectively.

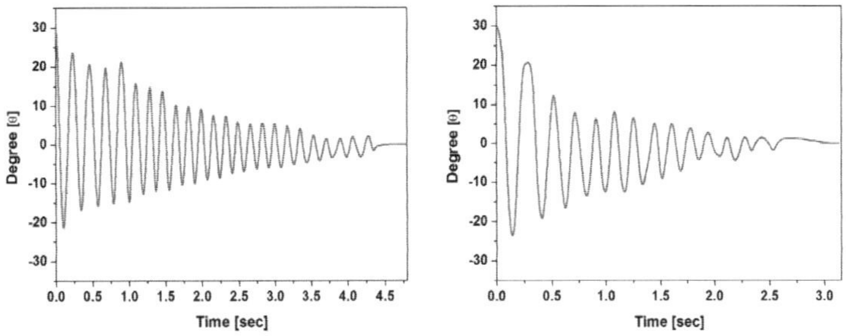

(a) (b)

Figure 4. Time response of the roly-poly-like magnet: (a) magnet #1(b) magnet #2

4.2. Vibration exciter test results

Fabricated device has been mounted on a vibration exciter and sinusoidal horizontal vibration has been applied at frequencies ranging from 7 to 20Hz with 1Hz step while increasing the acceleration from 0.5 to 3g with 0.5g step at each frequency. Vibration at frequencies under 6Hz and acceleration over 3g could not be applied due to the limitation of the vibration exciter. Harvesting device #1 and #2 had magnet #1 and #2 integrated inside, respectively.

Firstly, open-circuit voltages of the harvesters with magnet assemblies having different center-of-mass have been measured and compared without ferrofluid. As shown in figure 5, both devices showed increasing trend in output voltage as the input acceleration was increased with relatively flat frequency response. Open-circuit voltage was higher for device #2 at accelerations lower than 2g, but the increase in output in response to increased acceleration started to saturate at a faster pace than device #1 at accelerations higher than 2g. Maximum open-circuit voltage of 106.4mV has been obtained with device #1 at 12Hz and 80.4mV has been obtained with device #2 at 15Hz, both at 3g vibrations.

Open-circuit voltages of the harvesters with ferrofluid have also been analyzed. In contrast to relatively flat frequency response shown in figure 5, a non-linear resonant behavior has been observed due to addition of ferrofluid with decreasing resonant frequency at higher accelerations. Open-circuit voltage increase was pronounced near the peak points but saturating trend has been observed afterwards. As shown in figure 6, maximum open-circuit voltage of 140.8mV has been obtained with device #1 at 13Hz and 154.4mV has been obtained with device #2 at 12Hz at 3g vibrations. Despite the initial differences in open-circuit voltage trend at various frequencies and accelerations, both devices showed relatively similar tendency after ferrofluid had been added.

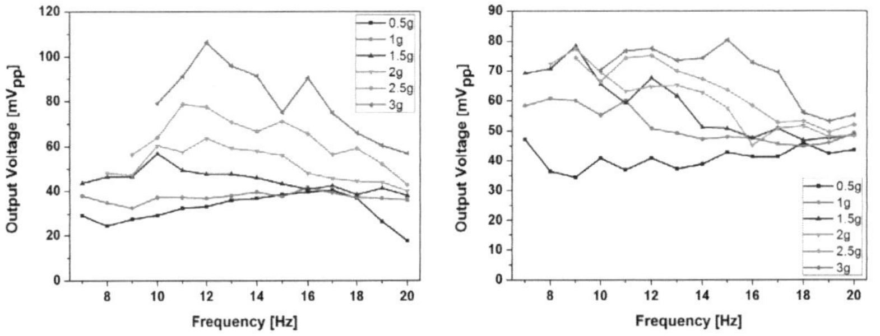

(a) (b)

Figure 5. Open circuit voltage at various input frequencies and accelerations without ferrofluid: (a) device #1 (b) device #2

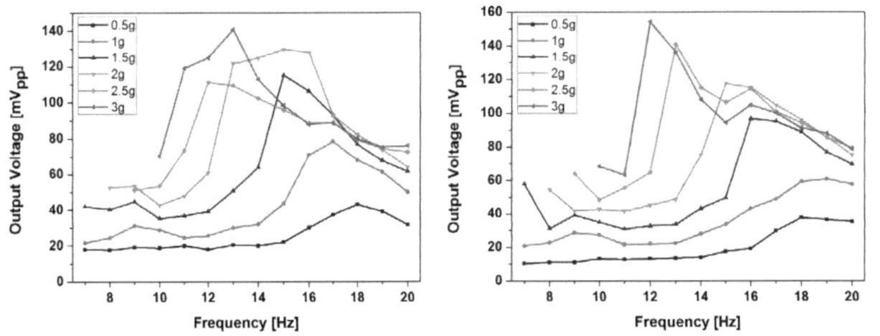

(a) (b)

Figure 6. Open circuit voltage at various input frequencies and accelerations with ferrofluid: (a) device #1 (b) device #2

Figure 7 shows the open circuit voltage waveforms of device #2 at 3g vibration at 12Hz, with and without the ferrofluid. In additions to the increase in amplitude, much smoother waveform has been observed which can be ascribed to the lubrication of magnets by ferrofluid.

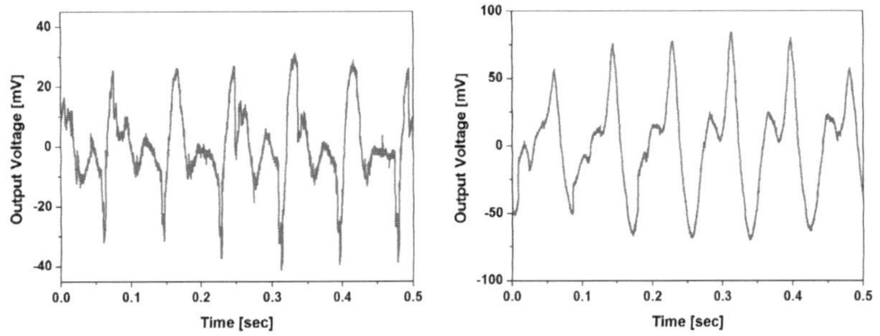

(a) (b)

Figure 7. Output voltage waveform at 3g acceleration at 12Hz (device #2): (a) without ferrofluid (b) with ferrofluid

Output voltage and power have been analyzed with varying load resistance at conditions where the fabricated harvesters with ferrofluid showed maximum open-circuit voltage. Output from device #1 and #2 has been measured at 13Hz and 12Hz, respectively, at 3g acceleration. Maximum output power reached 3.72μW and 4.53μW across load resistance of 70Ω for device #1 and #2, respectively.

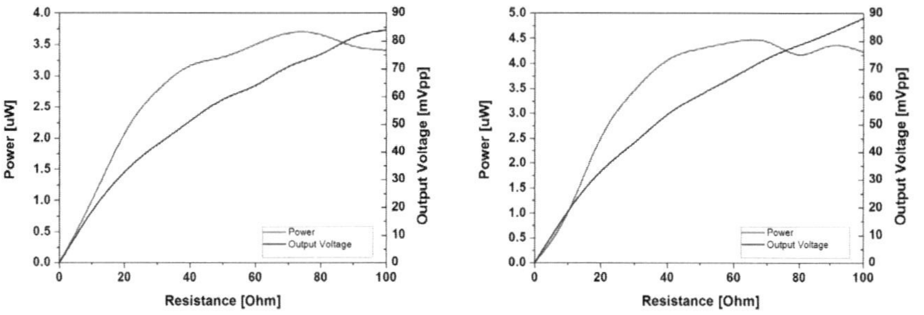

(a) (b)

Figure 8. Output power and output voltage vs. load resistance: (a) 3g vibration at 13Hz (device #1) (b) 3g vibration at 12Hz (device #2)

Without the ferrofluid, device with magnet assembly having smaller time constant (magnet #2) generated relatively higher open-circuit voltage at smaller input accelerations but the trend for accelerations higher than 2g was the opposite. Use of ferrofluid as a lubricant increased the output voltage at the expense of flat frequency response. Although effects of center-of-mass position and ferrofluid as a lubricant require further investigation, the concept of utilizing roly-poly-like magnet in an electromagnetic vibration energy harvester has been tested successfully.

5. Conclusion

We have developed a non-resonant electromagnetic vibration energy harvester using roly-poly-like magnet. A proof-of-concept vibration energy harvester has been designed, fabricated, and tested. Maximum open-circuit voltage of 154.4mV has been obtained at 3g vibration at 12Hz and maximum output power of 4.53μW for a 70Ω load have been achieved at the same condition. Although fabricated device requires further characterization and optimization, we have successfully verified the feasibility of utilizing unique magnet design in electromagnetic harvesting device.

6. Acknowledgement

This work was supported by the Ewha Global Top 5 Grant 2011 of Ewha Womans University and by the Pioneer Research Center Program through the National Research Foundation of Korea funded by the Ministry of Science, ICT & Future Planning (2010-0019453).

References

[1] Roundy S, Wright P K, Rabaey J 2003 A study of low level vibrations as a power source for wireless sensor nodes *Comput. Commun.* **26** 1131–1144

[2] Büren T, Mitcheson P D, Green T C, Yeatman E M, Holmes A S and Tröster G 2006 Optimization of inertial micropower generators for human walking motion *IEEE Sensor J.* **6** 28–38

[3] Renaud M, Fiorini P, Van Schaijk R, and Van Hoof C 2009 An impact based piezoelectric harvester adapted to low frequency environmental vibrations *Proc. Transducers 2009* (Denver, 21-25 June 2009) 2094-2097

[4] Bowers B J and Arnol D P 2009 Spherical, rolling magnetgenerators for passive energy harvesting from human motion *J. Micromech. Microeng.* **19** 094008

[5] Galchev T, Raz R, and Paul O 2012 New multi-dimentional low-frequency springless proximity inertial harvester for converting human and environmental motion *Proc. PowerMEMS 2012,* pp. 117-120

Fabrication of ethanol blended hydrogen peroxide 50 mN class MEMS thruster

Jeongmoo Huh, Jeongsub Lee, Daeban Seo, Shinjae Kang and Sejin Kwon
Department of Aerospace Engineering, Korea Advanced Institute of Science and Technology, 373-1 Guseong-dong, Yuseong-gu, Daejeon, 305-701, Republic of Korea

E-mail: trumpet@kaist.ac.kr

Abstract. MEMS thruster with blended propellant was fabricated and experimentally tested in order to improve specific impulse of micro scale monopropellant thruster and to improve stability of thrust to be better. 90 wt. % H_2O_2 blended with 25 O/F ratio ethanol was used as propellant of thruster and platinum on alumina support was used as catalyst for decomposition of propellant. Thruster was made by five layers of photosensitive glasses. Four layers were integrated by thermal bonding method and catalyst was directly inserted into chamber before UV bonding process for the last layer bonding. Results of experimental tests showed ethanol blended hydrogen peroxide had higher specific impulse than unblended hydrogen peroxide. Expected improvement of thrust stability due to the blended propellant was found only in the transient state of thrust. Also, unlike the thrust instability of vertical type thruster of previous research, improvement of thrust stability was found owe to horizontal type thruster pattern on glass, despite aspect ratio limitation of glass fabrication with wet etching process. During the experimental test, combustion phenomena of ethanol with decomposed hydrogen peroxide were observed through glass layer and it made fracture on structure of thruster.

1. Introduction

Microelectromechanical Systems technology has made conventional machining technology be in a completely different way suggesting revolutionary fabrication technology for micro scale product with low cost and mass product. MEMS technology has been applied to space technology and it is desirable because cheaper, lighter and faster are significant in space technology. One example of applications of MEMS for space technology is micro and nano satellite. Micro and nano satellites, which has mass of 10~100 kg and 1~10 kg respectively, has great advantages at cost of launch vehicle owe to its low mass. Moreover, group operation of distributed small satellites flying can make diverse operation possible, improve revisit time, and make high system reliability than that of classical satellites. For attitude control and orbit maneuver of miniaturized spacecraft, small size novel propulsion system is needed. MEMS thruster fabricated by MEMS technology has dimensions of micro scale and can make mili-newton class thrust. Many research groups have developed miniaturized thruster.

In early 2000, research of Hitt[1] suggested the possibility of micro scale hydrogen peroxide monopropellant thruster. Result of the research showed it was not enough to decompose propellant with coated silver catalyst on diamond shape pillar within the chamber, because of the low durability of silicon pillar in high pressure and high temperature of chamber. Takahashi[2] deposited platinum catalyst on surface of porous silicon within chamber, and he found porous structure was robust against fluidic motion but catalyst particles were went away. Lee[3] made platinum catalyst coated on alumina support to ensure large surface area for catalyst reaction and inserted it directly into chamber made by photosensitive glass. Because glass has lower thermal conductivity than that of silicon, it is desirable to make thruster with glass to reduce heat loss. But there is limitation of aspect ratio of patterning and thickness of glass. As a result, He found thrust instability with his vertical type thruster. Result of the research said instability occurred because of not enough injector thickness and pressure drop within injector.

For improvements of previous research, ethanol blending hydrogen peroxide MEMS thruster is suggested in this article. Blending propellant is one way to improve specific impulse of thruster[4]. If blending propellant is used in monopropellant micro scale thruster, the thruster would have not only system simplicity as monopropellant type but also improved specific impulse. It is also expected that higher chamber pressure occurred by blending propellant can improve the instability problem of previous work. With these advantages, to cope with difficulties of small patterning on relatively thick glass, the new thruster in planar type was fabricated and tested.

2. Catalyst Fabrication

Most case of monopropellant micro scale thruster use micro fabrication technology to maximize surface of strucuture for catalyst coating. However, structural coating catalyst showed not enough decomposition efficiency for propellant. But, previous research[3] showed use of porous support of catalyst and direct insertion into structrure of thruster had high decomposition efficiency. Generally, for experimental test in which blended propellant is used, catalyst should be chosen to be durable in high temperature and high pressure of chamber environment. And it is recommended that ideal support material should be physically, thermally, and chemically robust with high surface to volume ratio[5]. Therefore, in previous research[4], barium hexaaluminate was used which had approximately six times higher than that of alumina at temperature of 1200 °C, because alumina turned its phase and lost high surface to mass ratio at the temperature. However, in this micro scale reserach, high surface to volume ratio of micro scale thruster has high thermal energy loss, so that the support which has highly porous but phase turnable in high temperature could be chosen in this research.

As catalyst, platinum was chosen because it has good performance for propallant decompositon, high temperature melting point as 2041 K, and durability in high temperatrue and high pressure[6,7]. As support, γ-alumina was chosen because γ-alumina has high surface to mass ratio as 255 m^2/g, strong adhesion with metal, and thermally physically robust[7-9]. Pellet type γ-alumina was grinded and was strained through sieve.

For fabrication of catalyst, alumina granules were washed with water before heated in oven at 300 °C, 24 hours. After eliminiate the water from alumina, alumina was in H_2PtCl_6 solution for wetness impregnation, followed by 90 °C, 12 hours process in oven. After the solution were completely evaporated, alumina was heated in furnace with 500 °C, 3 hours for calcination process. Finally, reduction process was completed in condition of 500 °C, 3 hours with hydrogen gas 200cc/min volume flow. Fabricated catalyst is in figure 1 and EDS result show element of catalyst in figure 2 and table 1.

Table 1. Element weight percent and atomic percent of catalyst.

Element	Wt %	At %
O	26.22	47.07
Al	45.2	48.12
Pt	27.67	4.07
Cl	0.91	0.74
Total	100	100

Figure 1. Pt/Al_2O_3 catalyst with size of 40-45mesh, 355-425μm.

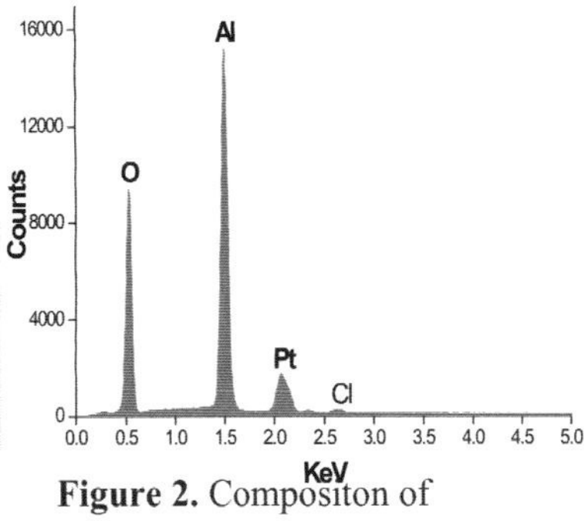

Figure 2. Compositon of catalyst.

3. Design and fabrication of thruster

To get objective thrust, 50 mN, specific impulse of blended propellant was calculated theoretically. With objective thrust, Isp, and gravitational acceleration, propellant mass flow rate was calculated. From designed chamber pressure 5 bar and ambient pressure 1 bar, Mach number was calculated. And after c^* was calculated with chamber temperature, throat area was determined and lastly expansion ratio was designed. Designed expansion ratio was 1.67 and expansion angle of conical nozzle was 12 °.

Table 3. Specification of designed thruster.

Thruster specification	
Thrust	50 mN
Specific impulse	117 s
Chamber pressure	5 bar
Propellant mass flow	0.034 g/s
Catalyst capacity	2 g/s/cm^3
Catalyst volume	0.0196 cm^3
Catalyst/supprot	Pt/Al_2O_3
Support size	40-45 mesh
Chamber aspect ratio	1.5

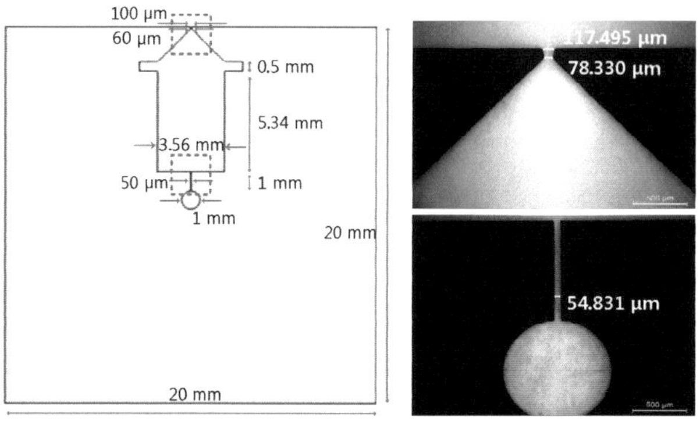

Figure 3. Drawing of designed thruster(left), micro nozzle and injector magnified by optical microscope(right).

Designed thruster profile was patterned on photosensitive glass by HF wet etching. Figure 3 shows profile of thruster in the left and magnified nozzle and injector by microscope in the right. 4 layers were combined with thermal bonding and the last layer was bonded by UV after Pt/Al_2O_3 catalyst was directly inserted into chamber. Nickel mesh was used as catalyst holder. By using two type of bonding method, unlike previous work[3] which needed additional reduction process for oxidized catalyst due to high temperature of thermal bonding, there was no need to second reduction process for catalyst in this case. With planar type thruster, despite limitation of aspect ratio of glass patterning, design length of injector was flexible and elongated injector was fabricated for appropriate pressure distribution at injector, chamber and nozzle. Figure 4 shows drawing of five layer of thruster and figure 5 shows image of 4 layers thermally bonded in the left side and integrated micro thruster in the right side.

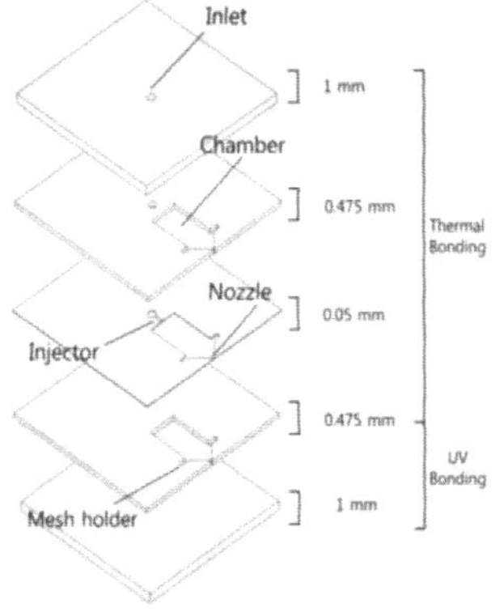

Figure 4. Drawing of five layers of thruster component.

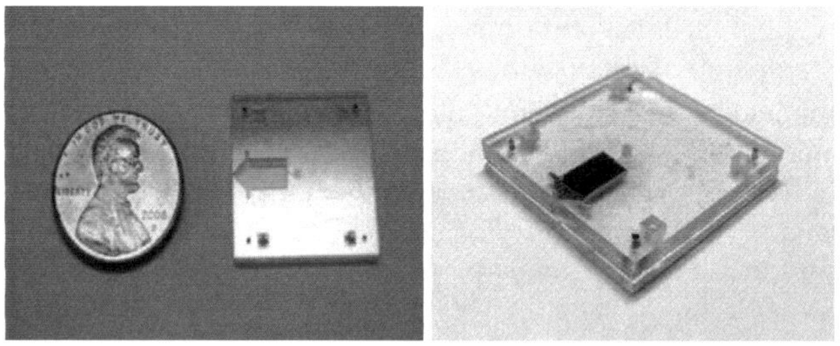

Figure 5. After 4 layers thermally bonded (left), integrated micro thruster(right).

4. Experimental test and discussion

4.1 Experimental setup

In order to measure the thrust, micro force sensor with sensor plate, thrust stand and feeding line were setup. Solenoid valve and syringe pump were used to feed propellant. After propellant was fed to thruster, thrust acted to downward direction and the sensor measured magnitude of thrust.

Figure 6 shows drawing and picture of experimental setup.

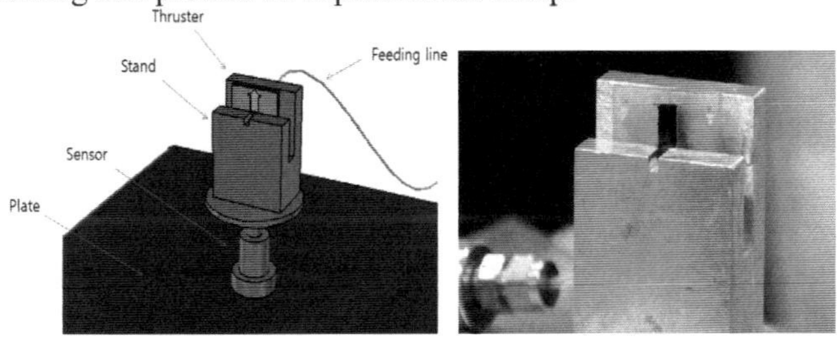

Figure 6. Experimental setup for performance test.

4.2 Result of experiment and discussion

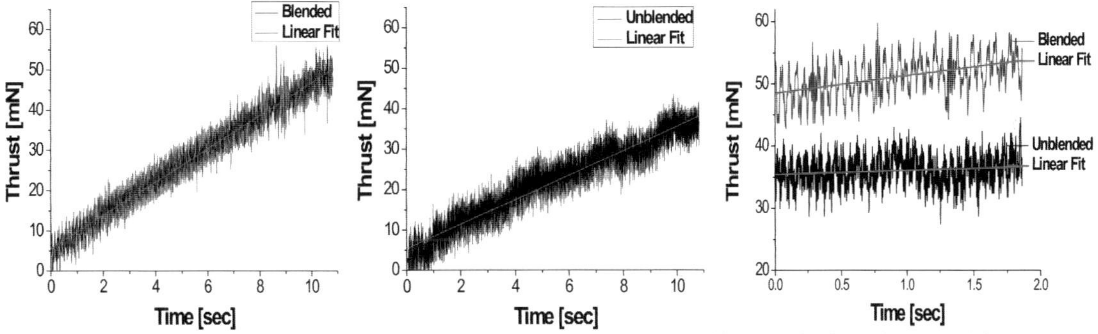

Figure 7. Experimental results of thrust of transient state(left, middle) and of steady state(right).

The experimental results said that thruster worked by ethanol blended hydrogen peroxide propellant got higher thrust than thruster without blending under same mass flow. Figure 7 shows result of performance tests, left and middle figures are about transient thrust performance and right about steady state thrust. It took about 10 second to get target thrust. In experiment, two phase gas and liquid flow was observed at former part of transient section and only gas phase flow was observed latter part of transient section. It is estimated that because micro fluid flow of propellant was handled, even small quantity of water which was product of decomposed hydrogen peroxide, would made catalyst wetted in cold start and reduced catalyst decomposition efficiency. As more propellant was decomposed, temperature of chamber rose and it made water of decomposed hydrogen peroxide turn to vapor, catalyst dried, and catalyst have higher decomposition efficiency at high temperature. This phenomenon would be much effective in case of blended propellant thruster because of more enthalpy fed. It is observed in the left two plot of figure 7. Standard deviation of error between thrust acquired and curve fitting data was 2.8 for blended propellant and 3.2 for unblended, so expected small perturbation was identified. However, no small perturbation was found in case of steady state. Generally, it is considered that high chamber pressure makes low perturbations of thrust[10], but this was not prominent in this micro scale thruster.

Steady state thrust of figure 7 shows ethanol blended hydrogen peroxide propellant got higher thruster than that of hydrogen peroxide without blending. Experimented propellant flow was 3ml/min, which was larger than designed flow rate 1.7ml/min. Average magnitude of thrust is 51 mN for blended propellant and 36 mN for unblended propellant. Assuming mass flow rate was accurate as 3ml/min, specific impulse is approximately 74 sec and 52 sec for blended and unblended respectively. Thrust magnitude fluctuation was approximately ± 9.1% and ± 8.8% for ethanol blended hydrogen peroxide and unblended hydrogen peroxide respectively. Standard deviation was 2.8 for blended propellant and 3.2 for unblended.

Despite no stability improvement in steady state of two second, the improvement found in case of transient state during about ten seconds shows stabilization ability of blended propellant. Considering possibility of thrust stabilization and increase of specific impulse of this experiment, blended propellant have advantages for micro scale monopropellant thruster.

Figure 9 shows combustion phenomenon of blended ethanol with decomposed hydrogen peroxide.

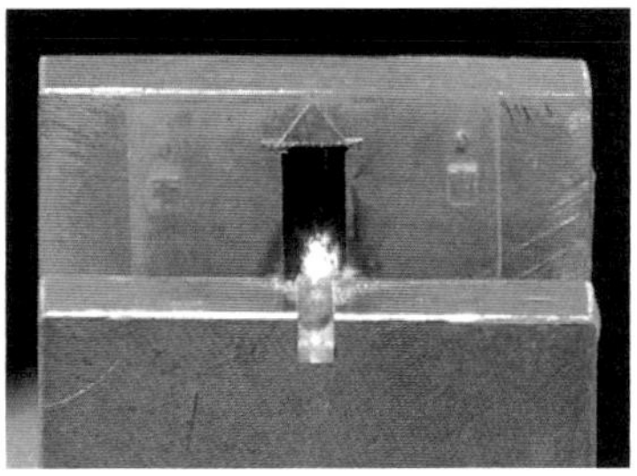

Figure 9. Auto-ignition of blended ethanol by decomposed hydrogen peroxide.

5. Conclusion

To improve performance of stability of micro thruster, plannar type catalyst direct insertion MEMS monopropellant thruster with blending propellant was fabricated and evaluated.

Under the limitation of aspect ratio of glass pattern with wet etching, horizontal type thruster had more chances to improve stability of thrust than vertical type thruster.

The effectiveness of high enthalpy with blended propellant for thruster performance was prominent in the transient state rather in steady state, so that the expected improvement of stability of blended propellant thruster was valid only in the case of transient section not in steady section. Blended propellant suggests the possibility of thrust stabilization.

By blended propellant, micro scale monopropellant thruster can have not only system simplicity as monopropellant thruster, but also improved specific impulse and also possibility of stability improvement.

Acknowledgment

This work was supported by the National Research Foundation of Korea(NRF) grant funded by the Korea government(MEST) (No. 2012R1A2A1A05026398).

References

[1] Hitt D L, Zakrzwski C M and Thomas M K 2001 *J. Smart Mater. Struct.* **10** 1163-75.

[2] K Takahashi, T Ikuta, Y Dan, K Nagayama and M Kishida 2006 *Proc. Sens. Symp.* 513-6

[3] J Lee, S Kim, S Kwon and M Yu 2012 *J. Propul. Power.* **28** 396-404.

[4] J Lee and S Kwon 2013 Evaluation of ethanol-blended hydrogen peroxide monopropellant on a 10 N class thruster *J. Propul. Power.* AIAA Early Edition

[5] Satterfield C N 1993 *Heterogeneous catalysis in industrial practice 2nd edition* McGraw-Hill Singapore

[6] Lim H, An S, Kwon S, and Rang S 2007 *J. Propul. Power* **23** 1147-59

[7] Sungyong An and Sejin Kwon 2009 Scaling and Evaluation of Pt/Al2O3 Catalytic Reactor for Hydrogen Peroxdie Monopropellant Thruster *J. Propul. Power* **25** 1041-5

[8] Tian H, Zhang T, Sun X, Liang D and Lin L 2001 Performance and deactivation of Ir/ γ -Al2O3 catalyst in the hydrogen peroxide monopropellant thruster *Applied Catalysis A* **210** 55-62

[9] Sungyong An, Jeongsub Lee, Rachid Brahmi, Charles Kappenstein, Sejin Kwon 2010 Comparison of catalyst support between monolith and pellet in hydrogen peroxide thrusters *J. Propul. Power* **26** 439-45

[10] Sungyong An, Jungkun Jin, Jeongsub Lee, Sungkwon Jo, Daejong Park and Sejin Kwon 2011 Chugging instability of H2O2 monopropellant thrusters with reactor aspect ratio and Pressures *J. Propul. Power* **27** 422-7

Radioistopes to Solar to High Energy Accelerators - Chip-Scale Energy Sources

Amit Lal [1]
SonicMEMS Laboratory, School of Electrical and Computer Engineering, Cornell University

E-mail: amit.lal@cornell.edu

Abstract. This talk will present MEMS based power sources that utilize radioisotopes, solar energy, and potentially nuclear energy through advancements in integration of new structures and materials within MEMS. Micro power harvesters can harness power from vibration, radioisotopes, light, sound, and biology may provide pathways to minimize or even eliminate batteries in sensor nodes. In this talk work on radioisotope thin films for MEMS will be include the self-reciprocating cantilever, betavoltaic cells, and high DC voltages. The self-reciprocating cantilever energy harvester allows small commercially viable amounts of radioisotopes to generate mW to Watts of power so that very reliable power sources that last 100s of years are possible. The tradeoffs between reliability and potential stigma with radioisotopes allow one to span a useful design space with reliability as a key parameter. These power sources provide pulsed power at three different time scales using mechanical, RF, and static extraction of energy from collected charge. Multi-use capability, both harvesting radioisotope power and local vibration energy extends the reliability of micro-power sources further.

1. Introduction

In addition to providing electrical power, our work with radioisotope thin films has led to self-powered vacuum pumps, vacuum gauges, counting clocks, light sources, and surface acoustic wave transponders. In addition to providing energy, we have shown that radioisotopes can be used for self-powered electron lithography (SPEL). The electrons emitted from radioisotope thin films can also be used to radiate electron beam resists for less than 5nm structures, while we have demonstrated structures as small as 30nm. This approach to electron lithography eliminates the use of electron beams, vacuum, and may enable nanolithography over extremely large and curved surfaces. We have used this lithography to create nano-wire arrays that can be themselves used to harvest energy and enable new sensors and actuators based on thermo-optical actuation. For example, we have used self-powered electron lithography for fabricating nanowire solar cell fabrication which yield high efficiency and large solar collection angles. We have also used these devices for femto-molar DNA and protein concentration sensing. With SPEL one can also envision the development of massively parallel micro-mirror arrays for concentrating solar power for applications such as planar heliostats. The development of planar heliostats will be described not only at the microscale, but also at the mesoscale for energy harvesting applications.

Lastly the talk will present recent development of chip and wafer scale electrostatic guidance of charged particles (electrons and ions) that could lead to very high energy beams for active methods for energy generation, and reduction of energy usage in applications such as medical imaging.

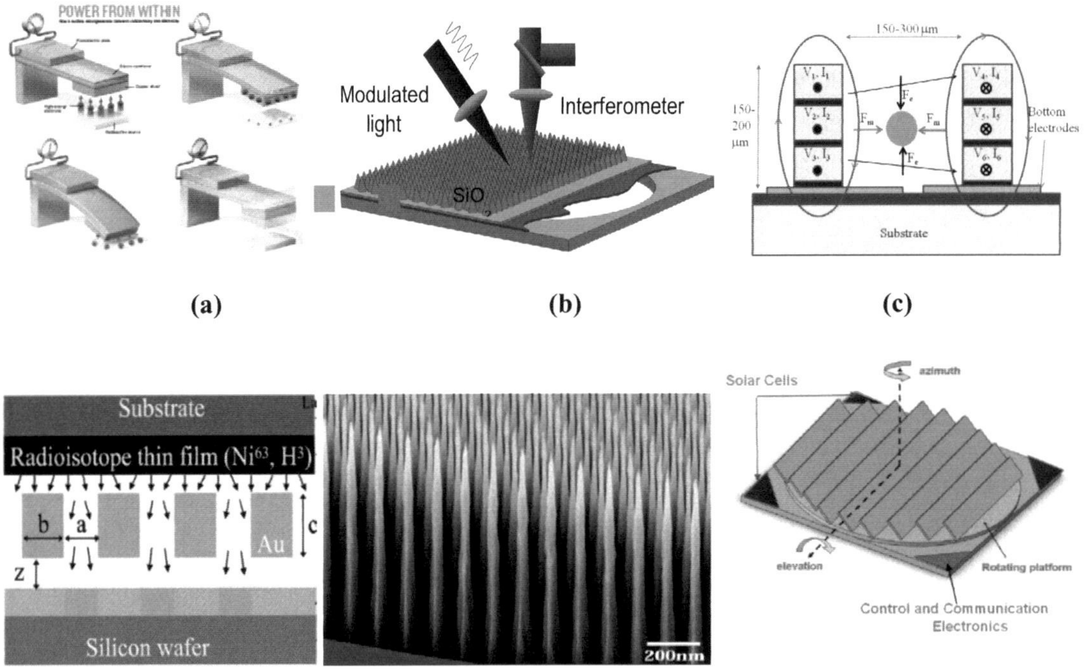

(a) (b) (c)

(d) (e) (f)

Figure 1. Chip scale power source

2. About The Speaker

Amit Lal is a Professor in the School of Electrical and Computer Engineering at Cornell University, Ithaca, NY. He received his B.S. degree (1990) in Electrical Engineering from the California Institute of Technology, and his Ph.D. degree (1996) in Electrical Engineering from the University of California, Berkeley. His technical interests and activities are in the areas of MEMS, ultrasonics, optics, micromachining, piezoelectric systems, design and analysis of integrated circuits, and applications of radioactivity in microsystems. From 2005-2009, Prof. Lal served as a Program Manager in the Microsystems Technology Office of the Defense Advanced Research Projects Agency (DARPA), where he developed and managed programs in the areas of navigation, low-energy computation, bio-robotics, and atomic microsystems. He holds 18 patents and has published more than 145 research papers in the area of microsystem engineering. He is the recipient of the NSF CAREER award and, with his students, several best paper awards at the IEEE Ultrasonics and Frequency Control Symposium. He is also a recipient of the Department of Defense Exceptional Service Award, and a Best Program Manager Award for his work at DARPA.

Multi-frequency Operation of a MEMS Vibration Energy Harvester by Accessing Five Orders of Parametric Resonance

Y Jia, J Yan, K Soga and A A Seshia

Department of Engineering, University of Cambridge, Cambridge CB2 1PZ, UK

E-mail: yj252@cam.ac.uk

Abstract. The mechanical amplification effect of parametric resonance has the potential to outperform direct resonance by over an order of magnitude in terms of power output. However, the excitation must first overcome the damping-dependent initiation threshold amplitude prior to accessing this more profitable region. In addition to activating the principal (1st order) parametric resonance at twice the natural frequency ω_0, higher orders of parametric resonance may be accessed when the excitation frequency is in the vicinity of $2\omega_0/n$ for integer n. Together with the passive design approaches previously developed to reduce the initiation threshold to access the principal parametric resonance, vacuum packaging (< 10 torr) is employed to further reduce the threshold and unveil the higher orders. A vacuum packaged MEMS electrostatic harvester (0.278 mm^3) exhibited 4 and 5 parametric resonance peaks at room pressure and vacuum respectively when scanned up to 10 g. At 5.1 ms^{-2}, a peak power output of 20.8 nW and 166 nW is recorded for direct and principal parametric resonance respectively at atmospheric pressure; while a peak power output of 60.9 nW and 324 nW is observed for the respective resonant peaks in vacuum. Additionally, unlike direct resonance, the operational frequency bandwidth of parametric resonance broadens with lower damping.

1. Introduction

In addition to the conventional use of direct resonance for resonant-based vibration energy harvesting, parametric resonance has been previously demonstrated as a potential superior resonant mechanism in terms of peak power attainable at certain conditions [1]. Instead of yielding a forced response, parametric resonant build-up arises from a periodic modulation in one of the system parameters. The principal (1st order) parametric resonance can be activated around twice the natural frequency. Moreover, the same system is theoretically predicted to possess several higher orders of parametric resonance at $2\omega_0$/n, where ω_0 is the natural frequency and n is a positive integer representing the order number of parametric resonance. The unveiling of these higher orders introduces additional operational frequency bands across the power spectrum. However, the onset of the higher parametric orders is less trivial due the rapid narrowing of the instability regions [2] as can be seen from the Strutt diagram of the damped Mathieu equation (Equation 1) shown in Figure 1.

$$\ddot{x} + c\dot{x} + (\delta + 2\varepsilon\cos(2t))x = 0 \tag{1}$$

where x is displacement, c is damping, δ is squared of natural frequency, ε is excitation amplitude, t is time while assuming unit mass. The shaded unstable areas in the Strutt diagram represent the parametric regimes. Upon activation, the oscillatory amplitude growth is not bounded by linear damping and only saturates due to nonlinearities at high amplitudes.

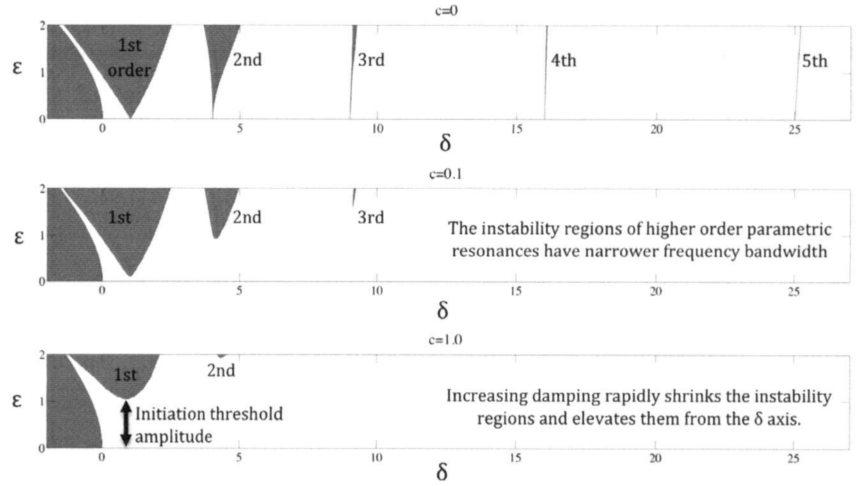

Figure 1. δ-ε plane stability chart for the Mathieu equation (Strutt diagram) showing up to 5 orders. Shaded instability areas represent achievement of parametric resonance.

The system response for all orders of parametric resonance is always near the natural frequency despite the distinct respective excitation frequencies. However, the accessibility of the higher orders is increasingly difficult with higher damping as a result of both the rise of the damping-dependent initiation threshold amplitude that the excitation must minimally attain and the ever-narrowing frequency bandwidth. Therefore, the higher order instability regions rapidly shy away from the base axis with increasing damping as seen in figure 1. Damping due to air drag, which typically dominates the damping in MEMS devices at atmospheric pressure, is directly related to the cross-sectional area of the shuttle and the body-shape-dependent drag coefficient. Air drag for comb finger structures is relatively smaller than that for gap-closing plate configurations and the relative accessibility of higher order instability regions in MEMS parametric resonators has been previously demonstrated [3].

2. Apparatus

The electrostatic harvester employed for this investigation can be seen in figures 2 and 3. The shuttle consists of a cantilever beam with an array of capacitive comb fingers running off its sides. The supposedly anchored end of the cantilever is coupled to the centre of an orthogonal clamped-clamped beam, which acts as the initial spring for the amplification of the base excitation that is fed into the parametric resonator (the cantilever). This threshold-aid design has been previously shown to decrease the initiation threshold of an otherwise sole cantilever structure by over an order of magnitude [1].

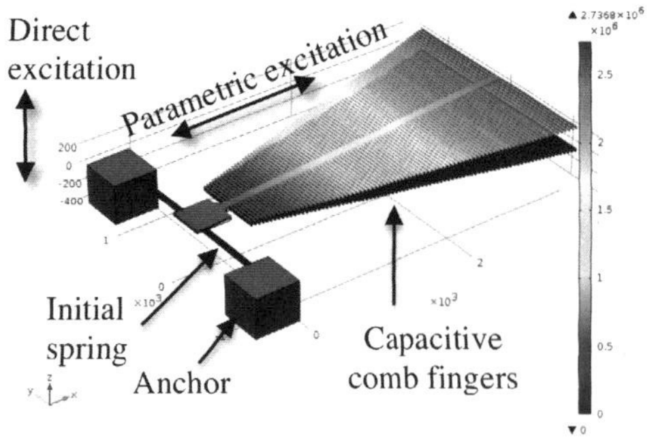

Figure 2. Harvester design with initial spring to reduce initiation threshold.

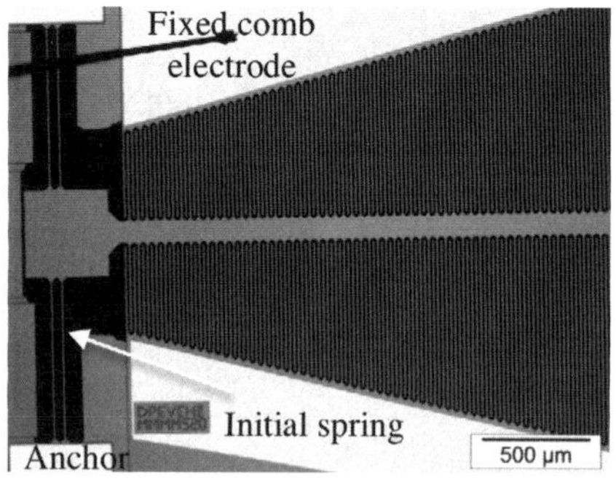

Figure 3. Optical micrograph of the MEMS harvester prototype.

The active device, with 0.278 mm^3 device and anchor volume, 25 μm thick device silicon, is driven on a mechanical shaker with the free end of the resonator positioned upright to help overcome the non-zero initial displacement criterion for activating parametric resonance from an orthogonal excitation. An acceleration scan of up to 10 g was carried out for the predicated frequency vicinities of the fundamental mode of direct resonance as well as the various orders of parametric resonance.

In addition to operation in atmospheric pressure conditions, the device was also vacuum packaged (pressure < 10 torr) with an in-house procedure to reduce damping as shown in figure 4. The vacuum pressure is a conservative estimate extrapolated from the quality factor of a MEMS double-ended tuning fork located on the same chip. Figure 5 shows the laser vibrometer scan of the direct resonant response of the device around the natural frequency (COMSOL® eigenfrequency simulated to be ~*303.9* Hz). The out-of-plane direct resonant peak showed an increase in quality factor from ~1.5 in room pressure to ~9 with the decreased damping inside the vacuum seal.

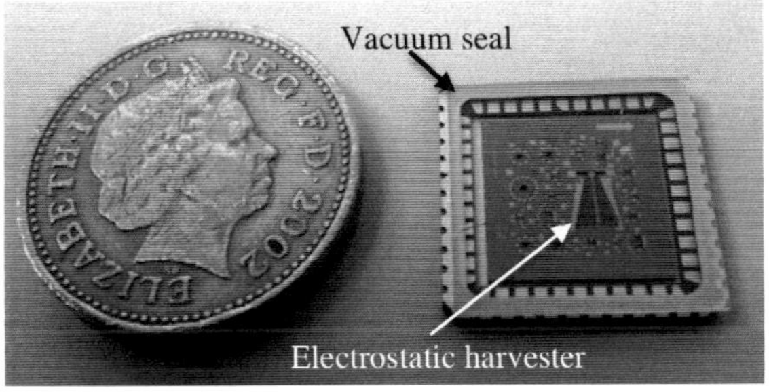

Figure 4. Vacuum packaged device.

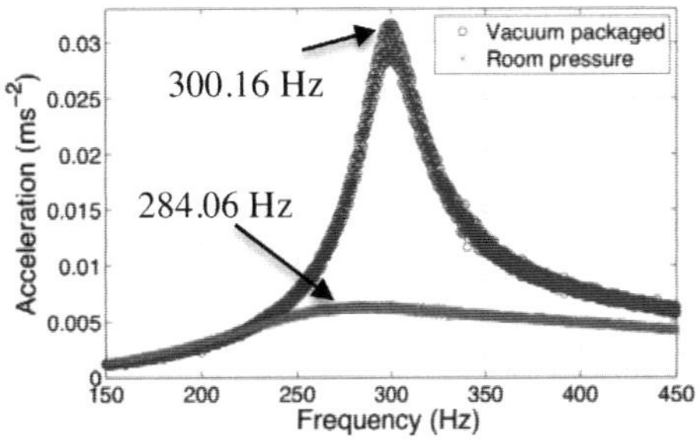

Figure 5. Vibrometer scan ~ f_n.

3. Result

3.1. Initiation threshold amplitude

The initiation threshold for various resonant peaks and their power response per excitation acceleration is given by table 1 and figure 6. At room pressure, the first four orders of parametric resonance onset at 1.57 ms^{-2}, 2.45 ms^{-2}, 3.24 ms^{-2} and 4.22 ms^{-2} respectively; while in vacuum, five orders of parametric resonance were uncovered at noticeably lower initiation thresholds of 0.98 ms^{-2}, 1.64 ms^{-2}, 2.62 ms^{-2}, 3.92 ms^{-2} and 13.73 ms^{-2} respectively. No other higher order resonant peaks were recorded up to 10 g of acceleration. Upon activating a parametric resonant regime, the power output rapidly outraces the direct resonant response with increasing excitation.

Table 1. Peak power output, half power bandwidth (HPB) and initiation threshold amplitude (ITA) of the fundamental direct mode and first 5 parametric orders of the prototype in room and vacuum pressure conditions; 'n/r' denotes not recorded and 'n/a' denotes not applicable.

Resonant peaks	Room pressure			Vacuum packaged		
	Driven at 5.1 ms^{-2}			Driven at 5.1 ms^{-2}		
	Power (nW)	HPB (Hz)	ITA (ms^{-2})	Power (nW)	HPB (Hz)	ITA (ms^{-2})
1st mode	20.8	40	n/a	60.9	11	n/a
1st order	166	80	1.57	324	160	0.98
2nd order	105	21	2.45	259	35	1.64
3rd order	67	7	3.24	175	9.8	2.62
4th order	22	2.7	4.22	84	3.5	3.92
5th order	n/r	n/r	n/r	0	0	13.73

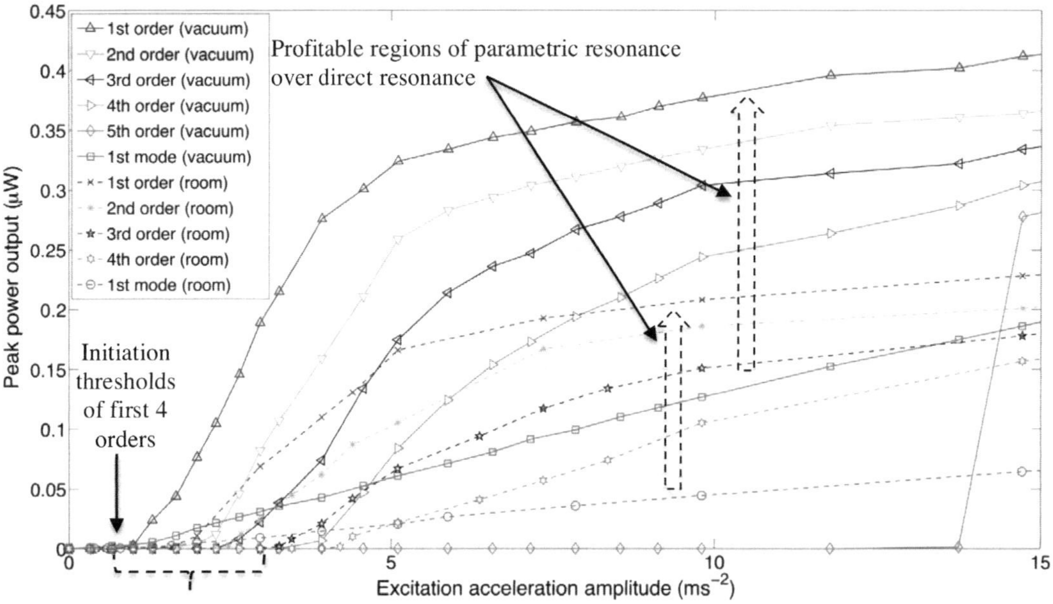

Figure 6. Experimental power response versus excitation acceleration. The initiation thresholds of parametric resonance decrease with lower damping in vacuum.

Although supplementary operational frequency regions from the higher orders were observed, only the principal order parametric resonance was relatively easy to access along with the direct resonant response. The second order overlaps with the fundamental mode and a timed modulation between the linear harmonic and parametric motion takes place, which usually returns to the direct resonant regime. The third and higher orders have significantly narrower frequency bandwidth and longer transient build-up time (> 1 minute). Therefore, the higher orders are only practical for very specific and non-time variant excitation frequencies.

The onset of the additional fifth parametric order for the vacuum packaged device was at significantly higher amplitude than that of the first four orders and the power response rapidly attained a high level upon accessing the parametric regime. This implies a lower true initiation threshold exists for the fifth order and the potential presence of further higher orders for both atmospheric pressure operation and vacuum packaged devices within the scanned acceleration range; however, these additional instability regions have been experimentally elusive due to the impractically narrow operational frequency bandwidth for the frequency sweeps to reveal. To accommodate the prolonged transient build-up time of the parametric regimes, manual frequency sweeps with frequency holds for each individual frequency point were carried out. Therefore, the tedious experimental procedure could have missed these narrow peaks, if they existed.

3.2. Power and frequency

Table 1 and figure 7 present the power peaks and frequency bandwidth for an input acceleration level of 5.1 ms^{-2}. The power peaks and half power bands at atmospheric pressure for direct resonance and principal parametric resonance were 20.8 nW and 40 Hz, and 166 nW and 80 Hz respectively; while that for vacuum were 60.9 nW and 11 Hz and 324 nW and 160 Hz respectively. The higher parametric orders exhibited rapidly-narrowing bandwidth as can be seen from the logarithmic frequency scale of figure 1. The lower attainable peaks for the higher orders are a result of the system operating at relatively less dominant parts of the instability regions for the given boundary conditions.

For direct resonance, despite the increase of power output in vacuum from the lower damping, the frequency bandwidth is sacrificed as expected. This narrowing in operational frequency with increasing sensitivity has become synonymous with higher quality factor (ratio of energy stored and energy dissipated). In the case of parametric resonance however, as damping decreases, the frequency bandwidth broadens along with the increase in power peak and quality factor. This is a result of the unstable tongues (Figure 1) approaching the horizontal base axis and the system moving deeper into the dominant parts of the instability region.

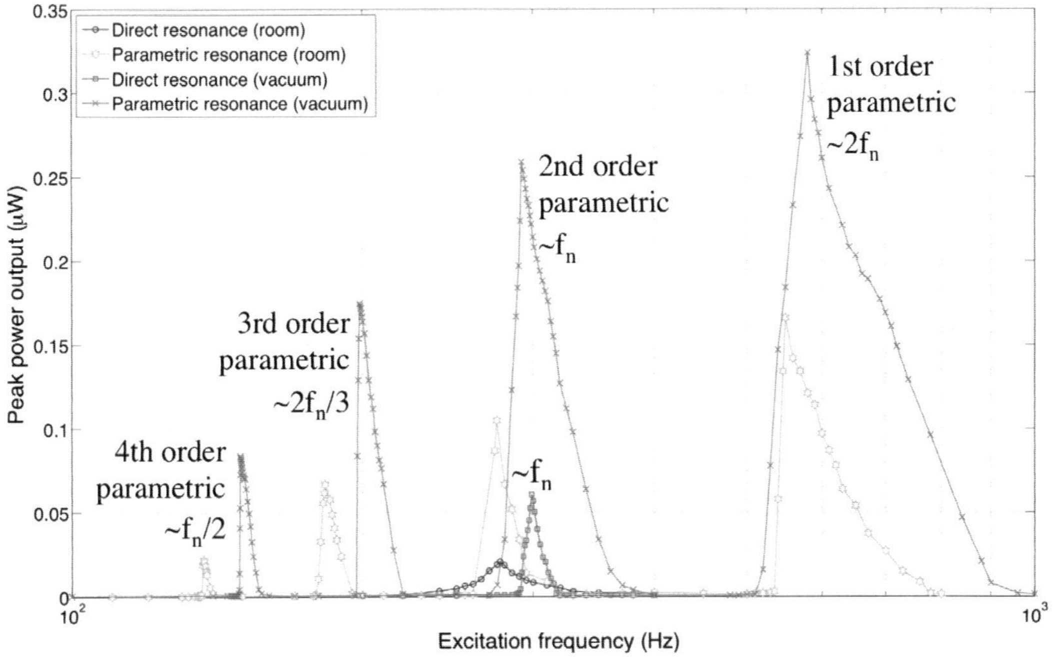

Figure 7. Power spectrum for acceleration of 5.1 ms^{-2}. Note the logarithmic scale for frequency.

This also implies lower initiation threshold and onset of the profitable regions at lower excitation levels as shown in figure 6. Lower initiation threshold also yields easier accessibility for the higher orders of parametric resonance. Nonetheless, the fast narrowing of the frequency bands at higher orders still provides an experimental and practical challenge for accessibility despite their theoretical presence. The drift of the resonant frequency with time further adds to the difficulty of rediscovering these hard to locate frequency bands with their strict combination of boundary conditions.

Conclusion

A MEMS electrostatic harvester was driven into direct resonance and various orders of parametric resonance. The device in room pressure experimentally revealed 4 orders of parametric resonance while the vacuum packaged scenario uncovered the 5th order as well. With lower damping in vacuum, the parametric peaks experienced both power increase and frequency broadening. Coupled with the initial spring structure previously investigated, the vacuum packaged parametrically excited harvester demonstrated the onset of the profitable regions over direct resonance for an input acceleration of 1 ms^{-2}. Future work involves design of structures with minimal air drag and higher quality vacuum packaging to further improve the accessibility of the various orders of parametric resonance.

References

[1] Jia Y, Yan J, Soga K and Seshia A A 2012 *Proc. PowerMEMS 2012* (Atlanta) pp. 215-218
[2] Levy D M and Keller J B 1963 *Commun. Pure Appl. Math.* **16** pp. 469-476
[3] Turner K L, Miller S A, Hartwell P G, MacDonald N C, Strogatz S H and Adams S G 1998 *Lett. Nature* **396** pp. 149-152

Wideband Electrostatic Vibration Energy Harvester (e-VEH) Having a Low Start-Up Voltage Employing a High-Voltage Integrated Interface

A. Dudka[1*], P. Basset[2], F. Cottone[2], E. Blokhina[3], D. Galayko[1]
[1]LIP6-UPMC – Sorbonne, France,
[2]Université Paris-Est, ESYCOM, ESIEE Paris, France
[3]UCD, Ireland

E-mail: *andrii.dudka@lip6.fr

Abstract. This paper reports on an electrostatic Vibration Energy Harvester (e-VEH) system, for which the energy conversion process is initiated with a low bias voltage and is compatible with wideband stochastic external vibrations. The system employs the auto-synchronous conditioning circuit topology with the use of a novel dedicated integrated low-power high-voltage switch that is needed to connect the charge pump and flyback - two main parts of the used conditioning circuit. The proposed switch is designed and implemented in AMS035HV CMOS technology. Thanks to the proposed switch device, which is driven with a low-voltage ground-referenced logic, the e-VEH system may operate within a large voltage range, from a pre-charge low voltage up to several tens volts. With such a high-voltage e-VEH operation, it is possible to obtain a strong mechanical coupling and a high rate of vibration energy conversion. The used transducer/resonator device is fabricated with a batch-processed MEMS technology. When excited with stochastic vibrations having an acceleration level of 0.8 g rms distributed in the band 110-170 Hz, up to 0.75 μW of net electrical power has been harvested with our system. This work presents an important milestone in the challenge of designing a fully integrated smart conditioning interface for the capacitive e-VEHs.

1. Introduction

The kinetic vibrations present in many environments are potential sources for the supply of autonomous microsystems. The Vibration Energy Harvesters based on electrostatic transducers (e-VEHs) are particularly suitable for miniature systems based on integrated CMOS and MEMS technologies. This, in turn, enables the integration of the transducer/resonator on the unique silicon substrate together with power processing and communication electronics.

To achieve a mechanical energy conversion, a conditioning circuit charges and discharges the transducer capacitance synchronously with the transducer capacitance variation [1]. Because of complexity and precision of the required control on the transducer electrical state, design of the conditioning circuit for electrostatic transducers is the main bottleneck associated with e-VEH implementation. Moreover, high bias voltage is necessary to achieve strong mechanical coupling with an electrostatic transducer. This fact, combined with ultra low power requirement for the control circuitry, makes the design of the conditioning circuit very challenging. The low power requirement is related with the fact the energy harvesting is self-supplying, and the power available from microscale e-VEH is of order of few microwatts.

Several architectures of conditioning circuit exist [1, 2]. Our study focuses on the architecture based on a charge pump with an inductive flyback (Fig. 1a), [3, 4]. In this circuit, the

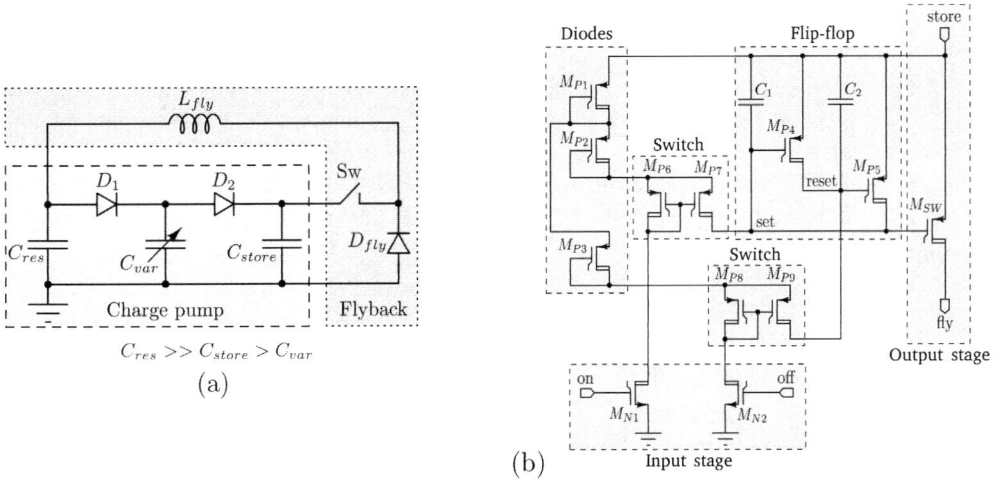

Figure 1. a) Energy harvesting system employing the auto-synchronous conditioning circuit initially proposed by B. Yen [4]; b) Proposed dynamic flip-flop level shifter driving the flyback switch M_{SW}, implemented in AMS035HV CMOS process.

synchronization of the charge flow is done internally thanks to the switching operation of the diodes D_1 and D_2. The unique switch Sw requiring an external control operates rarely (once during tens of vibration cycles), and a synchronization with the mass vibrations is not needed. A significant advantage of this circuit over charge-constant [1] or voltage-constant [2] architectures is its suitability for operation with a resonator excited by broadband stochastic vibrations. This is experimentally proven in our work.

This paper presents the first integrated on silicon high-side switch Sw commuting high voltage (tens of volts) between the charge pump and the flyback in the conditioning circuit of Fig. 1a. The high-side Sw requires a level shifting driver, which allows an interface with a ground-referenced low voltage control signal. To comply with the speed and power consumption requirements, we designed a novel ultra-low power level shifter with zero static power consumption. This driver can be used to implement a smart control algorithm so to adapt the conditioning circuit operation to different application contexts [5]. The implemented switch driver was tested in the circuit of Fig. 1 with a real MEMS resonator presented in [6] excited with stochastic vibrations.

In contrast with previous integrated circuits for capacitive transducer conditioning [2, 7], we target the level of operation voltages which exceed the gate drive voltages of the used technology (e.g., for the AMS035HV process the gate voltage is 3.3 V, and the maximal drain-source voltage is 50 V). That is because the full potential of electrostatic transducers for energy conversion may only be explored with high-voltage operation.

2. System of e-VEH: structure and operation

This section describes the operation of the circuit of Fig. 1a [4]. The role of the charge pump is to pump the charges from a large capacitor C_{res} (≈1-10 μF) toward a smaller capacitor C_{store} (≈1-5 nF), making use of the variation of the transducer capacitance C_{var} (≈10-100 pF) which varies between its extrema values C_{max} and C_{min}. As the charge pump operates, the voltage across C_{store} increases, while V_{res} remains virtually constant (since $C_{res} >> C_{store}$). The energy for the C_{store} voltage elevation comes from the mechanical domain. The converted energy is stored in the voltage difference $V_{store} - V_{res}$. When V_{store} becomes significantly above V_{res} and approaches the saturation value $V_{res}C_{max}/C_{min}$, the charge pump starts to deliver less and less energy to C_{store}, and the energy conversion rate (the power) decreases.

To avoid this decrease of the convered power, the voltage difference between V_{store} and V_{res} must be reduced. This is achieved using a flyback return circuit with an inductive BUCK DC-DC converter topology. When the switch Sw is on, the charges from C_{store} are transferred to C_{res} via the inductor L_{fly} and freewheeling diode D_{fly}, hence putting the total (injected and converted) energy on C_{res}.

Our theoretical analysis identified the optimal V_{store} voltage range (V_1, V_2), which corresponds to the maximal harvested power [8]. This result provides an indication of the optimal Sw switching sequence: the architecture implementing an optimal scenario is discussed in [5]. The bottleneck in implementation of this optimal operation is related to design of electrical interface between the control circuitry and the switch Sw. The main issue lies in the driving of the gate of a high-side MOS transistor with a low-voltage signal referenced to the ground. Therefore, an appropriate level shifter with very stringent power consumption specifications is needed. In the next section we present the designed architecture of a level shifter satisfying these requirements.

3. Switch driver architecture

The proposed level shifter is based on a zero static power topology based on an analog flip-flop [9], Fig. 1b. The flip-flop is composed of state-holding capacitors C_1, C_2 and high-voltage (HV) PMOS pull-up transistors M_{P4} and M_{P5}, as shown in Fig. 1b. The input stage consists of two HV NMOS transistors drawing the current from V_{store} when the pulses *on* or *off* are high.

Initially both inputs *on* and *off* are low, and the state of the flip-flop is so that *set* node voltage equals V_{store}, and *reset* node voltage is $V_{store} - 3V$. When *on* goes high, C_1 is charged to 3 V so that the *set* voltage become $V_{store} - 3V$; this turns M_{P4} on, further pulling up *reset* to V_{store} and discharging C_2. Consequently, M_{P2} turns off. In that way the state of the flip-flop changes to the opposite. The reset of the flip-flop occurs similarly when *off* goes high.

MOS diodes $M_{P1} - M_{P3}$ are used to limit the voltage drop to 3 V across the capacitors. The basic level shifter topology of [9] was supplemented with two switches M_{P6}, M_{P7} and M_{P8}, M_{P9}. They are used to electrically isolate the capacitors from discharge through the parasitic junction capacitances of the diodes, and thus to ensure stable states of the flip-flop during a long period without a need of refreshing.

The proposed level shifter and the high-side power switch have been fabricated in AMS035HV CMOS technology and successfully characterized. The measurement results have a good matching with simulations, as compared in Table 1.

Table 1. Level shifter and switch characteristics: measured versus simulated.

Characteristic	Measured	Simulated
Voltage range (V)	4–31	4–50
Switch ON-resistance (Ω)	32–46	39
Consumed static power (nW) (depending on V_{store} voltage)	0.8–44.8	0.44–111.7
Dynamic energy consumed per *on*/*off* cycle (nJ) (depending on V_{store} voltage)	0.065–1.38	0.05–2.92
Switching speed (ns)	73	40
Max. duration of *on*-state without refresh (s)	5.8	11.2

4. Experiment

The goal of the experiment focuses on testing the operation of the e-VEH presented in (Fig. 1a), with use of the designed and fabricated switch/driver block. The energy conversion process is characterized by the charges accumulated on the reservoir capacitor C_{res}, which can be visualized by the voltage measurement on this capacitor.

The schematic of the test setup of the e-VEH system is given in Fig. 2. The circuit is initially energized with an external DC voltage source, which was connected for a short time to C_{res} and

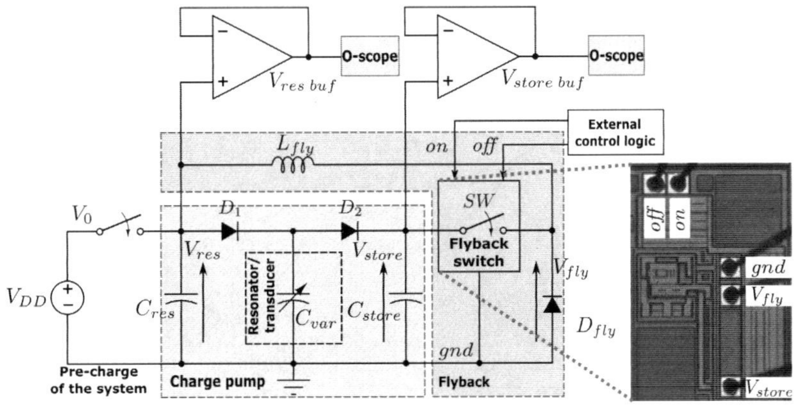

Figure 2. Schematic of a studied vibration energy harvester and micro-photograph of the implemented block.

so pre-charging it to a low initial voltage $V_0 = 5.5V$. The implemented flyback switch whose micro-photograph is illustrated in Fig. 2, is driven by the external control logic that generates periodically a sequence of *on* and *off* signals. A high input impedance and low offset operational amplifiers are used for the voltage probing across C_{res} and C_{store}.

The MEMS transducer/resonator device employed in the experiment was fabricated and fully characterized [6]. The resonance frequency of the resonator is 162 Hz. In large amplitude mode (1g of external vibration acceleration), it has half-power bandwidth of more than 30 % of the central frequency. The transducer capacitance varies between 45 pF and 74 pF.

When the external voltage source V_{DD} is disconnected from C_{res}, the e-VEH system becomes electrically autonomous, with the exception of the power source for the external control logic. The mechanical resonator is submitted to stochastic external vibrations whose energy is distributed in the frequency band (100-180) Hz.

5. Measurement results

The experiments described above were repeated with three sets of acceleration amplitude and bandwidth of vibrations. For the first, second and third experiments vibration parameters are 0.7 g rms @ 100-180 Hz, 0.8 g rms @ 110-170 Hz and 1 g rms @ 90-175 Hz, respectively. All three tests started with C_{res} pre-charged to 5.5 V. The flyback activation frequency was 5 Hz.

Fig. 3 presents the evolution in time of C_{res} voltage (upper plot) and energy (bottom plot). After 3000-4000 seconds of autonomous system operation, the voltage across C_{res} rises up to 15-20 volts. Sudden drops of the V_{res} voltage can be observed. This occurs as V_{res} voltage reaches a level at which a dynamic pull-in occurs. The V_{res} drops are due to a short circuit between the plates of the variable capacitor C_{var}: when it happens, C_{res} discharges very fast through the diode D_1 until the voltage at which the pull-in effect disappears. This is an undesirable effect related with the physics of the MEMS device presented in [6]. The voltage of dynamic pull-in depends on the acceleration amplitude of the external vibrations: at lower acceleration amplitudes, the dynamic pull-in voltage is higher. As a result, the higher levels of V_{res} can be achieved at lower acceleration amplitude, as shown by the measurement. The short circuit "emulates" a load consuming the energy of the storage capacitor, and it can be seen that the energy production on C_{res} is stable on the observed time interval (4000-7000s)

During three experiments the maximum achievable V_{res} voltage was 21 V, 21.5 V and 17 V, this corresponds to the energy on C_{res} of 220 μJ, 230 μJ and 145 μJ, respectively (to be compared with the initial energy of 15 μJ corresponding to $V_{res} = V_0$). The maximal average energy conversion rate (power) measured as the slope of the straight segments of the energy curve is

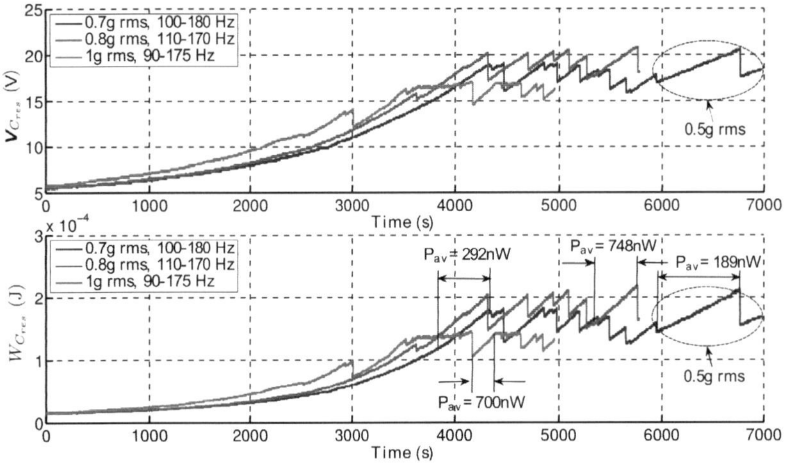

Figure 3. Evolution of the state of the studied system in autonomous mode submitted to stochastic external vibrations, with three different stimuli. Upper plot: the voltage across the reservoir capacitor $C_{res} = 1\mu$F, lower plot: the energy of C_{res}.

between 190 nW and 750 nW (cf. Fig. 3, lower plot). These figures give the net energy and power available for the load supply, accounting for the consumption of the flyback switch, losses in reactive elements and in the diodes.

6. Conclusions

We presented an investigation in the design of e-VEH system based on the auto-synchronous architecture. The work addressed a novel integrated ultra-low power high-voltage switch employed in the conditioning circuit. With the use of the implemented switch we tested the autonomous operation of the MEMS capacitive energy harvester excited with stochastic vibrations in a large bandwidth 100-180 Hz. Initially biased to 5.5 V with 15 μJ, the system increased its voltage up to 21.5 V with the accumulated energy 230 μJ. The maximal measured net power is 748 nW without considering supply of the generator of switching events. The design of a latter block is a subject of ongoing work.

References

[1] S. Meninger, J. Mur-Miranda, R. Amirtharajah, A. Chandrakasan, and J. Lang, "Vibration-to-electric energy conversion," *IEEE Trans. on VLSI Systems*, vol. 9, no. 1, pp. 64–76, Feb. 2001.

[2] E. O. Torres and G. A. Rincón-Mora, "Electrostatic energy-harvesting and battery-charging CMOS system prototype," *IEEE Trans. on Circuits and Systems I: Regular Papers*, vol. 56, no. 9, pp. 1938–1948, 2009.

[3] S. Roundy, P. K. Wright, and J. Rabaey, "A study of low level vibrations as a power source for wireless sensor nodes," *Computer Communications*, vol. 26, no. 11, pp. 1131–1144, Jul. 2003.

[4] B. C. Yen and J. H. Lang, "A Variable-Capacitance Vibration-to-Electric Energy Harvester," *IEEE Transaction on Circuits and Systems–I: Regular papers IEEE*, vol. 53, no. 2, pp. 288–295, 2006.

[5] A. Dudka, D. Galayko, and P. Basset, "Design of integrated smart adaptive conditioning circuit for capacitive vibration energy harvester," in *Workshop of Energy and Wireless Sensors*, Besançon, France, 2012.

[6] R. Guillemet, P. Basset, D. Galayko, F. Cottone, F. Marty, and T. Bourouina, "Wideband MEMS electrostatic vibration energy harvesters based on gap-closing interdigited combs with a trapezoidal cross section," in *Micro Electro Mechanical Systems, 2013 IEEE 26th International Conference on*, 2013, pp. 817–820.

[7] A. Kempitiya, D.-A. Borca-Tasciuc, and M. M. Hella, "Low-Power Interface IC for Triplate Electrostatic Energy Converters," *IEEE Transactions on Power Electronics*, vol. 28, no. 2, pp. 609–614, Feb. 2013.

[8] D. Galayko, R. Pizarro, P. Basset, A. M. Paracha, and G. Amendola, "AMS modeling of controlled switch for design optimization of capacitive vibration energy harvester," *2007 IEEE International Behavioral Modeling and Simulation Workshop*, pp. 115–120, Sep. 2007.

[9] P. Basset, "Conception, réalisation et test d'un micro système de transport alimenté à distance," Ph.D. dissertation, Univesité de Lille1, 2003.

Journal of Physics: Conference Series **476** (2013) 012128 doi:10.1088/1742-6596/476/1/012128

A Wideband Triboelectric Energy Harvester

Lokesh Dhakar [1, 2], Huicong Liu[1, 3], F. E. H. Tay[2, 4] and Chengkuo Lee[1]

[1] Department of Electrical and Computer Engineering, National University of Singapore, Singapore
[2] NUS Graduate School for Integrative Sciences and Engineering, Singapore
[3]Robotics and Microsystems Center, Soochow University, P. R. China
[4] Department of Mechanical Engineering, National University of Singapore, Singapore

E-mail: elelc@nus.edu.sg

Abstract. Contact electrification or triboelectric charging has mostly been seen as a negative phenomenon in various engineering applications. In electronic applications, static charge build-up may even destroy some of the components. In this paper, we demonstrate a cantilever based design using triboelectric charging mechanism for energy harvesting applications. The voltage output of triboelectric energy harvester (TEH) continuously increases with increased mass applied force and proof mass loading on the cantilever. The voltage output increases from 0.4V to 1.8V as the excitation acceleration increases from 0.4g to 1.8g. The design has been shown to demonstrate wide bandwidth characteristics. The FWHM increases from 3.76 to 12.64 Hz as the acceleration increases from 0.4g to 1.8g. The peak power output is found to be 0.18μW at 1.4g.

1. Introduction

Green energy is gaining increased importance in today's world when earnest efforts are being made by society and governments to decrease the dependency on non-renewable energy resources. Energy harvesting devices have emerged as one of the most promising solutions to serve smaller energy needs e.g. powering low power electronic devices and wireless sensors [1]. Vibrations in environment provide an important source of energy which can be tapped using energy harvesting devices. Traditionally, three approaches have been used in energy harvesting devices: electrostatic, electromagnetic and piezoelectric [2][3][4]. Recently, tribolelectricity has been demonstrated as a potential mechanism for energy harvesting [5]. But the area of utilizing triboelectric mechanism for vibration based energy harvesting is mostly unexplored. This paper presents a triboelectric energy harvester (TEH) design working on the principle of contact electrification. According to contact electrification theory, two dissimilar materials get charged when they are contacted and separated. The triboelectric charging process for different materials is well studied for different metals and polymers [6]. Different materials have different tendency to attract electrons which is a property of material also known as electronegativity. According to the tendency of a material to attract electrons, the materials can be arranged in a 'triboelectric series'[6] which lists different materials in the order of relative polarity of contact charge acquired. The materials with a tendency of acquiring a positive charge are placed at top of the series whereas materials with a tendency to acquire negative charges are placed at lower positions in the series. There have been some published works in past couple of years demonstrating the principle of triboelectric energy harvesting [7][8].

Traditional resonant energy harvesting devices based on electrostatic, electromagnetic and piezoelectric mechanisms have confronted problem of narrow operational bandwidth. This paper proposes an energy harvesting design based on triboelectric contact and separation method which solves the narrow bandwidth problem. Wide operation bandwidth(BW) is obtained by use of incorporating spring hardening effect in the device [9][10].

2. Design and fabrication of triboelectric energy harvester

The schematic of the proposed device using triboelectric mechanism is shown Figure 1. The fabricated top part is assembled on a cantilever end which vibrates when mechanically excited by vibrations. The source of vibration can be human motion , wind flow or machine vibrations[11]. The fabricated bottom part is assembled right below the top part. The top part makes contact with the patterned bottom part and then releases due to vibrational motion of cantilever. As making contact and releasing is essential part in process of triboelectric mechanism, the proposed device is designed to take advantage of BW widening phenomenon using a stopper [9]. In the presented design, the bottom part serves as a mechanical stopper for the cantilever which significantly increases the operating BW of TEH.

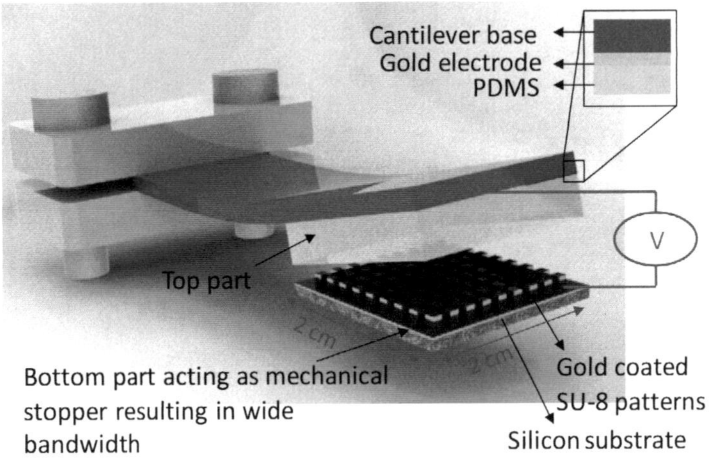

Figure 1. Schematic of the cantilever based triboelectric energy harvester.

The process flow for the fabrication of top and bottom parts as shown in Figure 1 is described in Figure 2. For the bottom part fabrication, pillar arrays are formed on Si wafer using negative photoresist SU-8 by photolithography. After the formation of pillar arrays on Si wafer, 100 nm gold layer is deposited on top of SU-8 pillar arrays using thermal evaporation technique which serves as the bottom electrode. Contacts are made from the bottom electrode using conductive silver paste and heating it at 120°C for 20 minutes so that solvent evaporates and hardens the silver paste.

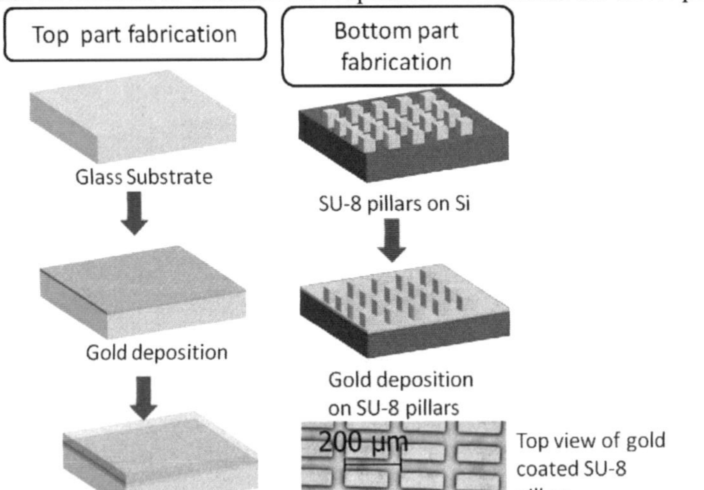

Figure 2. Process flow for the fabrication of top and bottom parts.

For the top part fabrication, 100 nm gold layer is deposited on glass which serves as the top electrode for TEH. Thereafter, a PDMS layer is spin coated using SYLGARD® 184 silicone elastomer kit. PDMS layer is then cured at 80°C for 2 hours. This top part is then assembled on an Aluminum cantilever which can vibrate to provide the contact and separation motion for top and bottom parts. As shown in Figure 1. The top part moves with the cantilever and the bottom part remains fixed on a support.

3. Experiments and results

TEH was characterized by mounting it on an electromagnetic shaker and exciting at various frequencies and accelerations with different mass loadings. The output voltages and accelerations are recorded using a data acquisition system.

3.1. Characterization at different accelerations

TEH was characterized at different excitation accelerations to study the effect on output voltage. The acceleration was increased in steps of 0.2g from 0.4g to 1.8g. The peak output voltage continuously increased from 450 mV at 0.4g to 1845 mV at 1.8g. During the testing, distance between the top part and bottom part is kept constant. Due to constant distance, the impact force of top part to bottom part increases as the acceleration is increased. Therefore as the impact force increases, the output voltage of TEH continuously increases as shown in Figure 3.

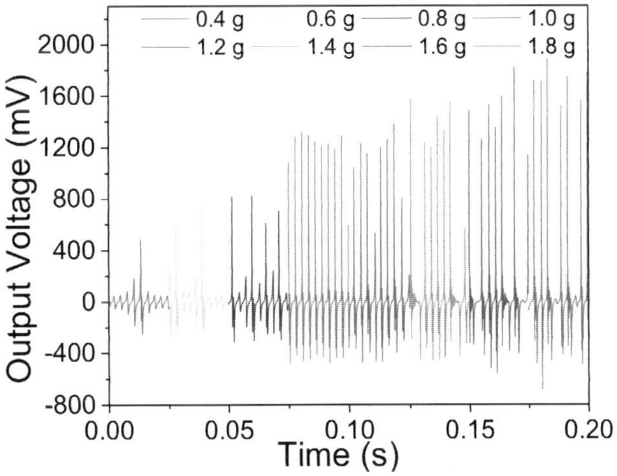

Figure 3. Output voltages at various acceleration levels

3.2. Effect of force and mass loading

The top part which is attached to an aluminum cantilever is loaded with extra proof mass to observe the effect on output voltage. Three different mass loading were used: 1) without any extra proof mass (mass loading = 0 gram) 2) mass loading = 3.2 gram and 3) mass loading = 3.8 gram for which peak voltages were observed to be 1002 mV, 1157 mV and 1570 mV respectively. As the mass loading is increased on the cantilever , the peak output voltage increases as shown in Figure 4.

To quantify the applied force values on TEH, the device was tested using a load cell. In the experimental setup, the top part is attached to a human finger as shown in inset image in Figure 5 and the bottom part is placed on a load cell. The top and bottom part were contacted and separated using the tapping motion of finger. As the top part contacted bottom part, the applied force readings were recorded using the load cell. Corresponding to force values from load cell, the output voltage values from TEH are acquired using LabView DAQ system. The average values of force and peak voltages are plotted in Figure 5. The average peak output voltage continuously increases as the force applied using finger is increased. This result is in accordance to the results obtained by changing the proof mass loadings in cantilever design.

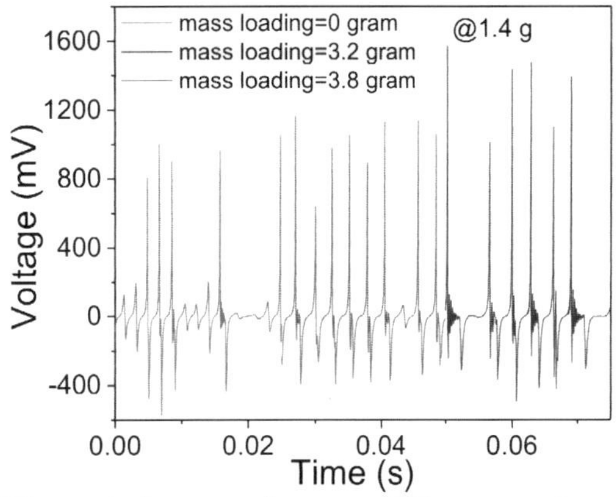

Figure 4. Output voltages at different proof mass loadings

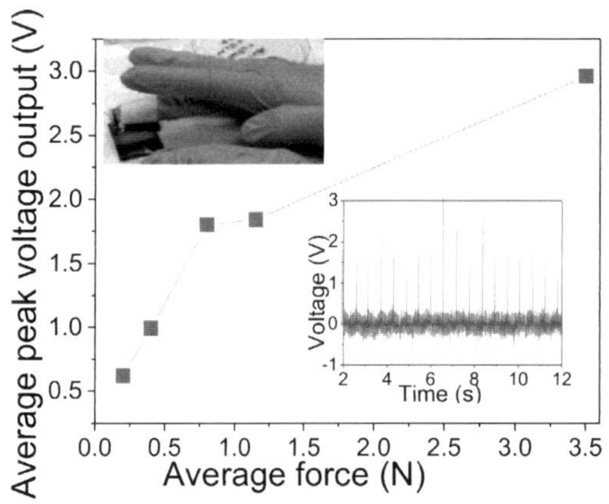

Figure 5. Variation of the peak voltage output with force in finger tapping experiment. The inset graph shows voltage output in time domain.

3.3. Wide bandwidth characteristics

One of the main advantages of using cantilever design for triboelectric energy harvesting mechanism is wide BW phenomenon. The bottom part acts as the stopper which obstructs the amplitude of top part attached to cantilever. This results into stiffening of cantilever which means the spring constant of the cantilever increases after contacting the bottom part. The stiffening of cantilever is responsible for widening of the operating BW of TEH as shown in Figure 6. In the experiments, the acceleration is increased in steps of 0.2g from 0.4g to 1.8g. For a constant level of acceleration, the frequency is swept from 10Hz to 40Hz which covers the operating frequency of TEH. The full width at half maximum (FWHM) after Gauss fit of the signals was observed to be increasing from 3.76 Hz to 12.64 Hz for an acceleration level of 0.4g and 1.8g respectively. Therefore, in conclusion the operating BW of TEH increases by more than 300% as the acceleration increases from 0.4g to 1.8g.

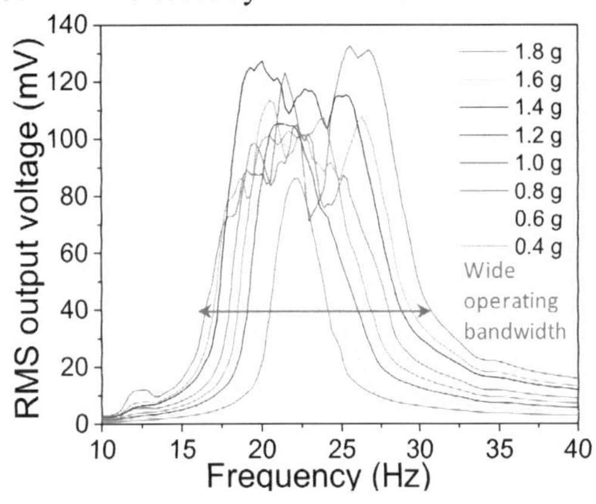

Figure 6. Wide operating bandwidth characteristics of TEH

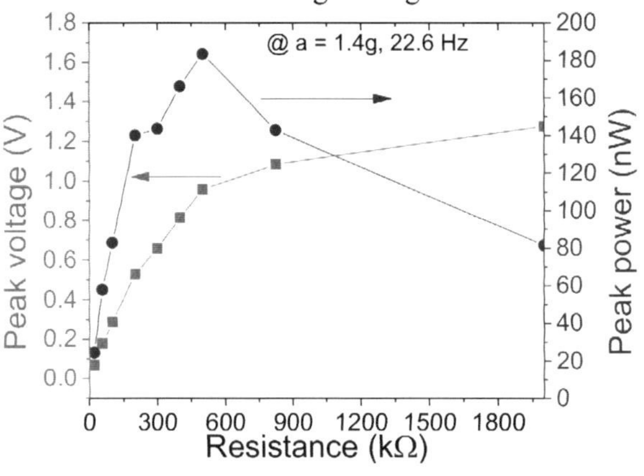

Figure 7. Peak output voltage and power generated by TEH

3.4. Power characteristics

To obtain the power characteristics of TEH, the device was excited at a constant acceleration of 1.4g and frequency of 22.6 Hz. Thereafter, the top and bottom electrodes were connected to a load resistor which is varied from 20 kΩ to 2000 kΩ. The peak output voltage and corresponding power values were measured for different values of load resistances as shown in Figure 7. For the given experimental parameters, the maximum power was obtained to be nearly 180 nW at a load resistance value of 500 kΩ.

4. Conclusion

A novel design for contact electrification (triboelectric mechanism) based energy harvester is proposed and fabricated. TEH is characterized at with three different parameters: acceleration, mass loading and force. The output voltage was observed to be continuously increasing with each of these three parameters. TEH also demonstrated increasingly widened BW characteristics as the excitation acceleration level increased. The maximum FWHM was found to be 12.64 Hz at an acceleration of 1.8g. The peak power was obtained to be 180nW at 1.4g and 22.6 Hz with a load resistance of 500 kΩ. The power output of the presented device can be further improved by optimizing the dimensions of SU-8 pillars in the bottom part.

Acknowledgements

This work is partially supported by the Faculty Research Committee (FRC) Grant (No. R-263-000-692-112) at the National University of Singapore, and the NRF2011 NRF-CRP001-057 Program 'Self-powered body sensor for disease management and prevention-orientated healthcare' (R-263-000-A27-281) from the National Research Foundation (NRF), Singapore.

References

[1] S. Roundy and P. K. Wright 2004 *Smart Mater. Struct.* **13** 1131
[2] S. Meninger, J. O. Mur-Miranda, R. Amirtharajah, A. Chandrakasan, and J. H. Lang 2001 *IEEE Trans. Very Large Scale Integr. Syst.* **9** 64–76
[3] H. Liu, B. W. Soon, N. Wang, C. J. Tay, C. Quan, and C. Lee 2012 *J. Micromechanics Microengineering* **22** 125020
[4] S. P. B. and M. J. T. and N. M. White 2006 *Meas. Sci. Technol.* **17** R175
[5] G. Zhu, Z.-H. Lin, Q. Jing, P. Bai, C. Pan, Y. Yang, Y. Zhou, and Z. L. Wang 2013 *Nano Lett.* **13**847–53
[6] J. Henniker 1962 *Nature* **196** 474
[7] F.-R. Fan, Z.-Q. Tian, and Z. Lin Wang *Nano Energy* **1** 328–334
[8] P. Bai, G. Zhu, Z.-H. Lin, Q. Jing, J. Chen, G. Zhang, J. Ma, and Z. L. Wang 2013 *ACS Nano* **7** 3713-3719
[9] M. S. M. Soliman, E. M. Abdel-Rahman, E. F. El-Saadany, and R. R. Mansour 2008 *J. Micromechanics Microengineering* **18** 115021
[10] H. Liu, C. J. Tay, C. Quan, T. Kobayashi, and C. Lee 2011 *J. Microelectromechanical Syst.* **20** 1131–1142
[11] E. K. Reilly, L. M. Miller, R. Fain, P. Wright, M. Engineering, and C. Berkeley 2009 *Proceedings of PowerMEMS 2009* 312–315.

A New Wideband Electromagnetic Vibration Energy Harvester with Chaotic Oscillation

Takahiro Sato[1], **Hajime Igarashi**
Graduate School of information science and technology, Hokkaido University, Kita 14, Nishi 9, Kita-ku, Sapporo, 060-0814, JAPAN.

E-mail: tsato@em-si.eng.hokudai.ac.jp

Abstract. This paper presents a new electromagnetic vibration energy harvester (VEH) which harvests the electric power in wide frequency range. The present VEH contains the ferrite core embedded in the axis hole of the coil bobbin which generates the magnetic force acting on the oscillator. The performance of the present VEH is estimated using the coupled analysis method for VEH in which motion, Ampere, and circuit equations are alternatively solved until convergence. Using the coupled analysis method, the output power characteristic in the present VEH against input frequency is analyzed. It is shown from the analysis results that the present VEH has chaotic oscillation and wide frequency range for power generation. Moreover, the condition that the present VEH has nonlinear oscillation is discussed.

1. Introduction

The electromagnetic vibration energy harvesters (VEHs) [1]-[4] have received much attention as new small power source for wireless sensor network systems. The sensor nodes with VEHs can harvest the electric power from their environment. Using VEHs, it is possible to realize wireless sensor systems which are free from the battery replacements. For the developments of practical VEH systems, the output power in VEHs must be improved. In addition, the VEHs are required to harvest power in wide frequency range. However, when the VEH has linear oscillation, output can only be obtained near to the resonant frequency. It has been indicated that the nonlinearity in the VEHs can expand the bandwidth [5], [6]. For this reason, the nonlinear VEHs have been developed recently [7], [8]. Because the oscillators in the VEHs with nonlinearity have complicated motion, the conventional spring-damper theory [9] is not applicable to estimate their performance. This is one of the reasons that it is difficult to design the nonlinear VEHs.

In this paper, we present a nonlinear VEH which has wide operation range. The present VEH contains ferrite core embedded in the axis hole of the coil bobbin which generates the magnetic force acting on the oscillator. Because of the magnetic force, the oscillator in the present VEH has complicated motion. To estimate the performance of the present VEH, the coupled analysis method of VEH [10] is employed in which motion, Ampere, and circuit equations are alternatively solved until convergence. It is numerically shown from the analysis results that the present VEH has chaotic oscillation. It is also shown that the preset VEH has wide frequency range for power generation. Moreover, the reason that the present VEH has nonlinear oscillation is discussed.

[1] Corresponding author: Takahiro Sato

2. Coupled analysis method for VEH [10]

The performance of the conventional linear VEH can be estimated from the spring-damper model [8], while, that of the nonlinear VEHs cannot be estimated using it. To estimate the nonlinear VEHs, we have presented the coupled analysis method for VEH, which numerically solves the following equations:

$$m(\ddot{p}-\ddot{q})+c_{\mathrm{m}}(\dot{p}-\dot{q})+k_{\mathrm{m}}(p-q)-F(\boldsymbol{A})=0, \tag{1}$$

$$\nabla\times\nu(\nabla\times\boldsymbol{A})=\boldsymbol{J}+\nu_0\nabla\times\boldsymbol{M}, \tag{2}$$

$$\mathrm{Z}(\boldsymbol{i})\boldsymbol{i}=\boldsymbol{V}(\boldsymbol{A}), \tag{3}$$

where m is mass of objects on the tip of the oscillator, c_{m} and k_{m} are effective damping and stiffness coefficients of the oscillator, p and q are the oscillator and base displacements, respectively. In addition, $\boldsymbol{A}$, $\boldsymbol{M}$, ν and ν_0 are vector potential, magnetization vector, reluctivity of magnetic material and vacuum, respectively. Moreover, $F(\boldsymbol{A})$ and $\boldsymbol{J}$ are the magnetic force acting on the oscillator and the current density in the coil, respectively, $\mathrm{Z}(\boldsymbol{i})$ is the impedance matrix for a circuit, and $\boldsymbol{V}(\boldsymbol{A})$ is the voltage vector corresponding to the electromotive force (EMF) induced in the coils which is calculated as follows:

$$(\boldsymbol{V}(\boldsymbol{A}))_i=\int_{V_i^C}\frac{\mathrm{D}\boldsymbol{A}}{\mathrm{D}t}\cdot\boldsymbol{l}_i\,\mathrm{d}V, \tag{4}$$

where D/Dt is the Lagrange derivative to consider EMF due to the coil movement, and V_i^C, $\boldsymbol{l}_i$ are the volume of the coil and the tangential vector of the current in the coil for circuit loop i, respectively. Note that, if no coil is included the loop i, $(\boldsymbol{V}(\boldsymbol{A}))_i$ is zero.

In the coupled analysis, the governing equations are discretized with respect to time and alternatively solved until convergence at each time step. This iteration is here called sub-cycle. When the all solutions converge in the sub-cycle, that is,

$$\left|p_n^k-p_n^{k-1}\right|<\varepsilon_p,\ \left\|\boldsymbol{A}_n^k-\boldsymbol{A}_n^{k-1}\right\|<\varepsilon_A,\ \left\|\boldsymbol{i}_n^k-\boldsymbol{i}_n^{k-1}\right\|<\varepsilon_I, \tag{5}$$

then the time is incremented, and new sub-cycle starts, where n and k is the iteration number of time and sub-cycle, respectively. Here, (1) and (3) are solved using the Runge-kutta and Newton-Raphon methods, respectively. In addition, (2) is solved by the finite element method for electromagnetic field in which, to easily treat the moving objects, the nonconforming technique is employed. The flow of the coupled analysis is shown in figure 1.

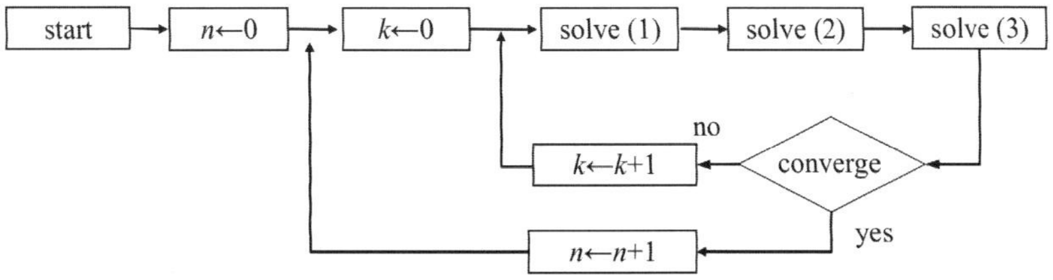

Figure 1. Flowchart of coupled analysis method.

3. Chaotic vibration energy harvester

We consider here the nonlinear VEH shown in figure 2 which contains the ferrite core embedded in the axis hole of the coil bobbin. The ferrite core forms nearly closed flux path with the magnets to increase the flux across the coil. In addition, the magnetic force acting on the magnets is generated, which makes the cantilever motion complicated and gives rise to nonlinear oscillation in the present VEH. When the oscillation becomes chaotic due to the nonlinearity, the frequency range in this system would become wider.

In this work, the performance of the present VEH is analyzed using the coupled analysis method. The mechanical parameters of the cantilever are summarized in Table 1. The full-wave rectifier and the storage capacitor whose capacitance is 1mF are connected to the present VEH as shown in figure 3, where threshold voltage of diodes and winding resistance are set to 0.2V and 100Ω, respectively. The output characteristic against the input frequency, *f*, is analyzed where the input amplitude is set to 0.1mm.

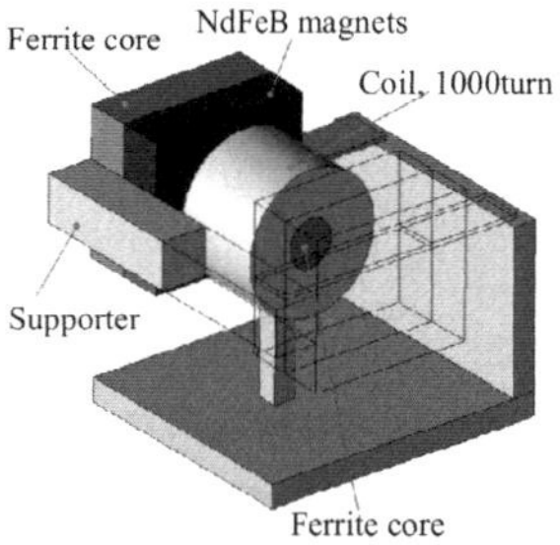

Figure 2(a). Overview of present VEH.

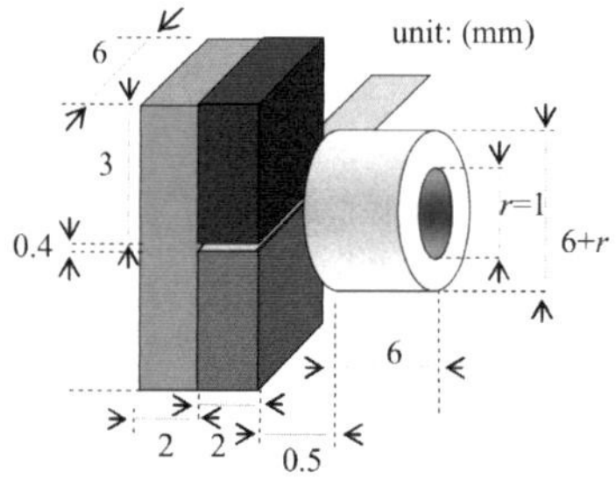

Figure 2(b). Sizes specification.

Table 1. Mechanical parameters in present VEH.

k_m	c_m	m
600 N/m	0.05 Ns/m	3.5 g

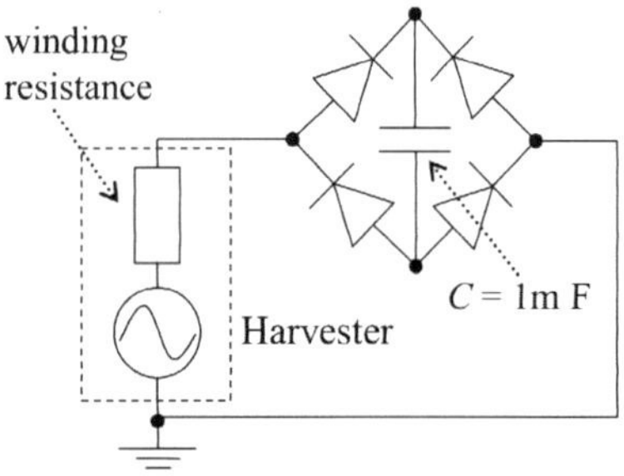

Figure 3. Full-wave rectifier circuit.

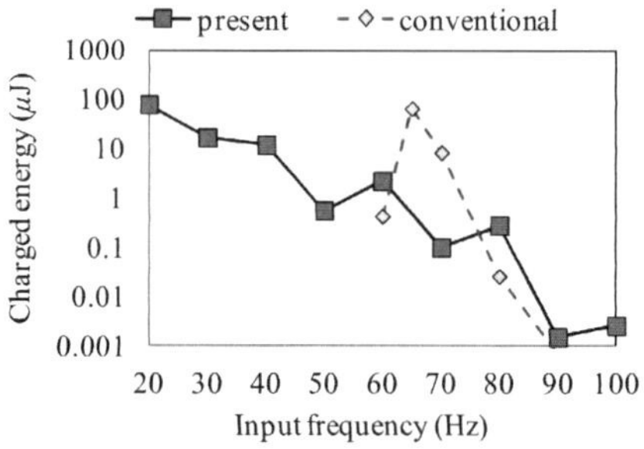

Figure 4. Charged energy against frequency.

The electric energy charged in the capacitor over 20 input cycles is shown in figure 4, in which the energy obtained by the VEH without the ferrite core (conventional VEH) is also shown to clarify the effect of the introduction of the ferrite core. Note that the energy under 0.001μJ is not plotted. It can be seen from figure 4 that we can have wider frequency range for power generation in comparison with the conventional VEH. The trajectories in the *z-v* plane are shown in figure 5, where $z=p-q$ and $v=\dot{p}-\dot{q}$ represent relative displacement and velocity, respectively. The transitions of z and induced voltage, V, are shown in figure 6. We can see that the cantilever oscillation has higher harmonics especially when f is 50Hz and 80Hz. To analyze the complicity in the oscillation, the Lyapunov exponent, λ, is computed using the Rosenstein's method [11], where positive value corresponds to chaotic motion. The Lyapunov exponents for all frequency are shown in figure 7, from which we can see that the cantilever oscillations are chaotic. From these results, it can be said that the chaotic oscillation can expand the operation frequency range for power generation.

To reveal the condition that the present VEH has chaotic motion, the distribution of λ in *r-f* plane is plotted in figure 8, where r denotes the diameter of the ferrite core embedded in the coil bobbin. In figure 8, the VEH has no oscillation when λ is 0. We can see that λ is positive in wide frequency range when r is small, while λ is negative at high frequencies when r is large. This result suggests that there is a lower limit for r to have chaotic motion in the VEH. Furthermore, linear oscillation is observed when r is large.

Here, we discuss the reason why the present VEH has chaotic and linear oscillations when r is small and large, respectively. Figure 9 shows the magnetic force acting on the magnets when r is 1mm and 4mm, respectively, in addition to the spring force. We can see from figure 9 that the magnetic force is has the same direction to the spring force when r=4mm, while its direction is reverse when r=1mm. This is because the flux path completely changes depending on r as shown in figure 10. The total force of magnetic and spring force is shown in figure 11 from which it can be seen that the total force when r=1mm is clearly nonlinear with respect to the relative displacement. As a result, the nonlinearity is given rise to the present VEH. On the other hand, when r=4mm, the magnetic force enhances spring force, and the initial total force is linear with respect to the relative displacement. Thus, the VEH has linear oscillation at high frequencies when r is large. The reason why the VEH has no oscillation at low frequencies is that the resonant frequency gets larger in comparison with the original one of the VEH, because the effective spring constant becomes larger due to the magnetic force.

From these discussions, it can be said that one can design the nonlinear VEH by creating the flux path using the magnetic material to generate the magnetic force whose direction is opposite to the spring force.

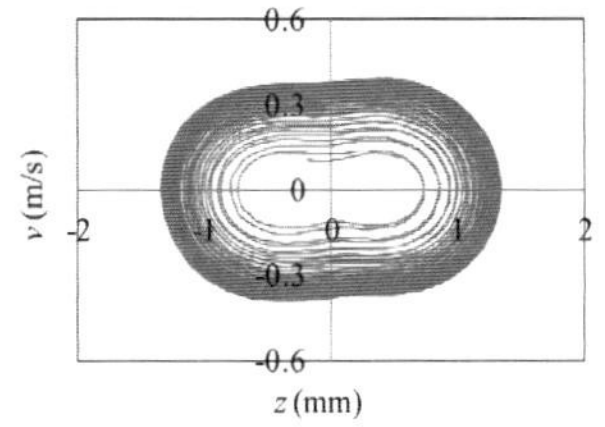

Figure 5(a). Trajectory for 20Hz.

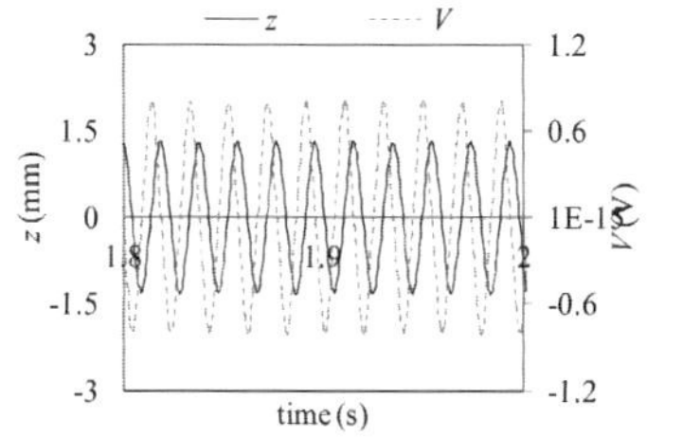

Figure 6(a). Transitions for 20Hz.

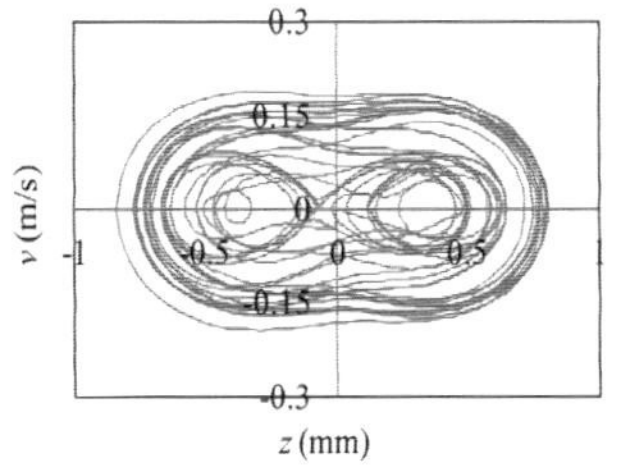

Figure 5(b). Trajectory for 50Hz.

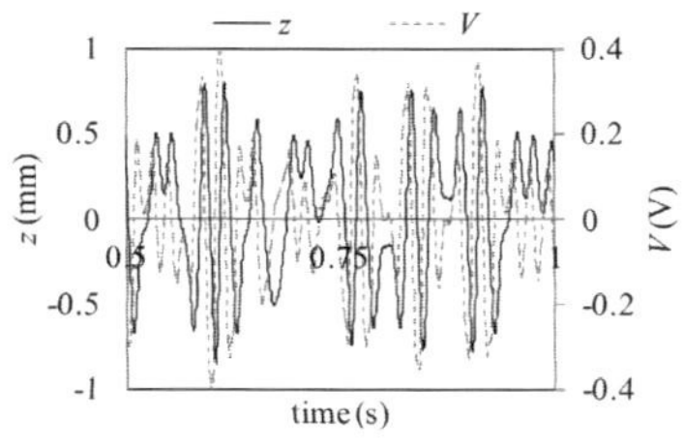

Figure 6(b). Transitions for 50Hz.

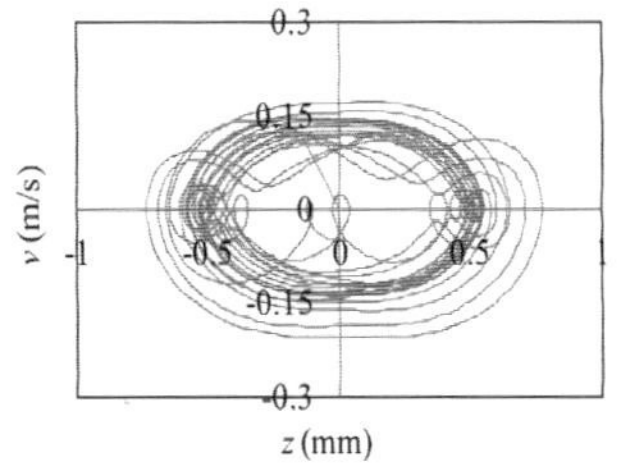

Figure 5(c). Trajectory for 80Hz.

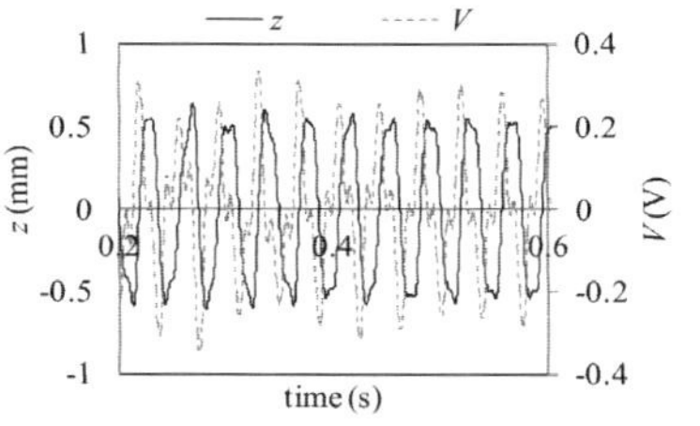

Figure 6(c). Transitions for 80Hz.

4. Conclusion

The new nonlinear VEH which contains ferrite core embedded in the axis hole of the coil bobbin has been presented. The performance of the present VEH is analyzed using the coupled analysis method. It is shown that the present VEH has chaotic oscillation. It is also found from the simulation that the present VEH has wide frequency range for power generation in comparison with conventional linear VEH. Moreover, the condition that the present VEH has chaotic oscillation is shown based on the distribution of the Lyapunov exponent in *z*-*f* plane. The numerical result has suggested that *r* must be smaller than a curtain threshold to have the chaotic motion.

For the future works, the analysis results of the present VEH will be compared with measurement data.

5. Acknowledgment

This study was supported by research grants from Japan Power Academy and a Grant-in-Aid for Scientific Research (24310117).

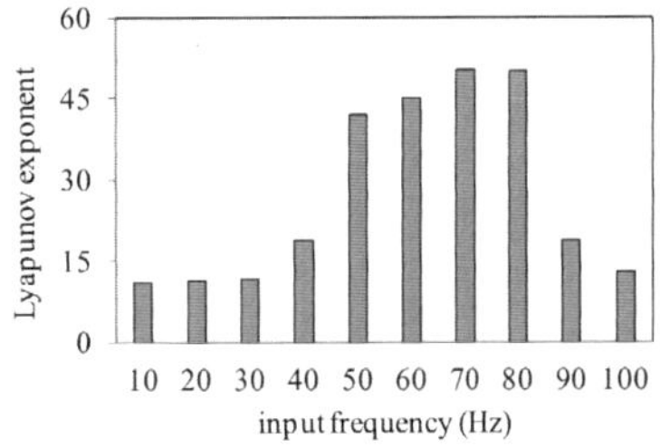

Figure 7. Lyapunov exponents for *z*.

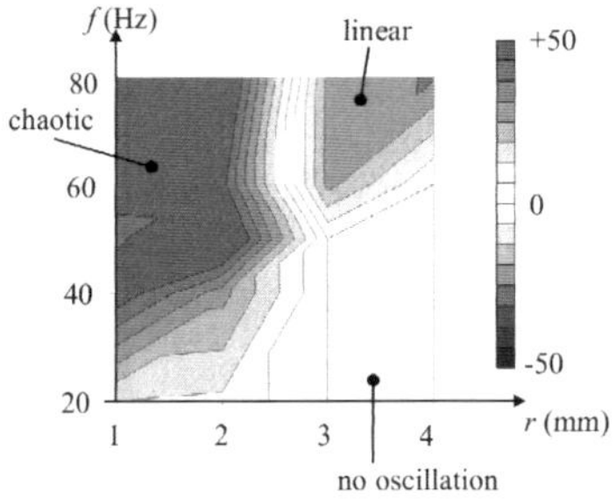

Figure 8. Distribution of Lyapunov exponents.

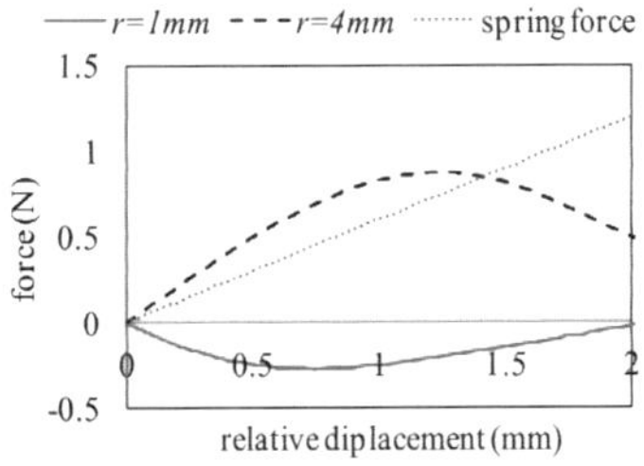

Figure 9. Magnetic and spring force.

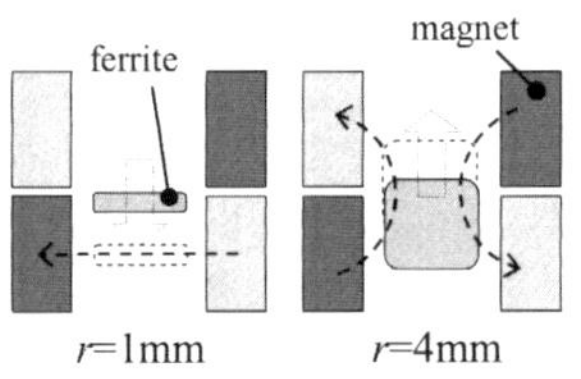

Figure 10. Flux path in VEH.

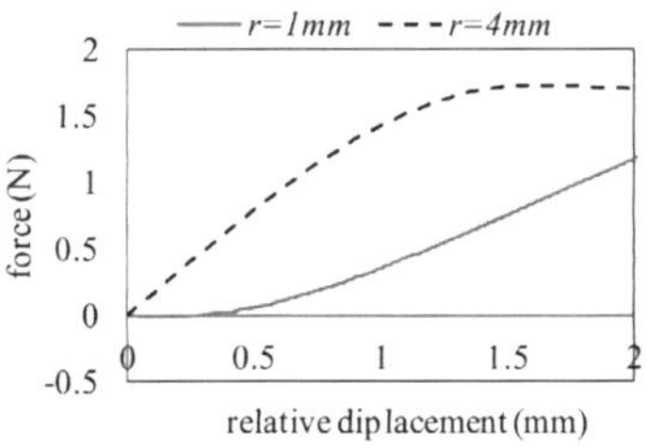

Figure 11. Total force.

6. References

[1] S. P. Beeby, R. N. Torah, M. J. Tudor, P. Glynne-Jones, T. O'Donnell, C. R. Saha and S. Roy, *Micromech. Microeng* 2007, vol. 17, pp 1257-1265.

[2] Dibin Zhu, Stephen Roberts, Thomas Mouille, Michael J Tudor, and Stephen P Beeby, *Smart Mater. Struct.*, 2012, vol. 21, pp. 1-14.

[3] Yingjun Sang, Xueliang Huang, Hexiang Liu, and Ping Jin, *IEEE Trans. Magn.*, 2012, vol. 48, No. 11, pp.4495-4498.

[4] Yang Zhu and Jean W. Zu, *IEEE Trans. Magn.*, 2012, vol 48, No. 11, pp. 3344-3347.

[5] Lars Geir Whist Tvedt, Duy Son Nguyen, and Einar Halvorsen, *Journal of Microelectromechanical systems*, 2010, vol. 19, No. 2, pp. 305-316.

[6] M. Ferrari, V. Ferrari, M. Guizzetti, B. Andò, S. Baglio, C. Trigona, *Sensors and actuators A*, 2010, Vol. 162, pp. 425-431.

[7] A. Erturk, D.J.Inman, *Journal of Sound and Vibration*, 2011, vol. 330, pp 2339–2353.

[8] Louis Van Blarigan, Per Danzl, Jeff Moehlis, *Appl. Phys. Lett.*, 2012, 100, 253904.

[9] N.G. Stephen, *Journal of Sound and Vibration*, 2006, vol. 293, pp. 409–425.

[10] T. Sato, K. Watanabe, H. Igarashi, *IEEE Trans. Magn.*, 2014 (to be published)

[11] Michael T. Rosenstein, James J. Collins, Carlo J. De Luca, 1993, *Physica D*, vol 65, pp. 117-134.

A Hair-cell Structure based Piezoelectric Energy Harvester Operating under Three Dimensional Arbitrary Vibrations

H Park, Y Na, J Park, and J Y Park[1]

Micro/Nano Device and Packaging Lab., Department of Electronic Engineering, Kwangwoon University, 20, Gwangun-ro, Nowon-gu, Seoul, 139-701, South Korea

E-mail: jaepark@kw.ac.kr

Abstract. A hair-cell structure based piezoelectric energy harvester was newly developed to effectively scavenge three-dimensional vibrations. The cantilever of the proposed energy harvester, called a hair-cell structure, is deliberately elongated and curled so that it oscillates with decent displacement under not only vertically induced vibrations, but also under longitudinally and horizontally induced vibrations. The proposed energy harvester is comprised of an elongated and curled piezoelectric cantilever and a proof mass with high aspect ratio at the free end of the cantilever. The fabricated device generated the peak output voltage of 15 mV under vertically induced vibrations with an acceleration of 50 m/s^2 at its resonance frequency of 116 Hz. Furthermore, it also generated the peak output voltage of 33 mV and 10 mV under longitudinally and horizontally induced vibrations, respectively.

1. Introduction

The development of low-power wireless sensor networks has recently opened up a possibility that energy harvesters can be used to power self-sustainable autonomous sensor nodes instead of traditional batteries [1-4]. Micro electromechanical system (MEMS) energy harvesters have been actively developed to reduce size and weight and to integrate with other integrated circuits (ICs) and MEMS components. The reported MEMS energy harvesters generated several micro- watts of output power from induced vibrations and are based on the cantilever structure [4-7]. The cantilever structure is comprised of a piezoelectric thin film attached on a spring structure and proof mass at the free end of the cantilever and it is based on the spring-mass-damper system. MEMS energy harvesters that employ this cantilever structure produce their maximum output power when the frequency of the induced vibrations matches the resonance frequency of the cantilever. Furthermore, most of this cantilever structure utilized for MEMS energy harvesters is flat. Therefore, they produce their maximum output powers when the normal direction of the cantilever plane matches the direction of induced vibration, otherwise their output powers dramatically decrease [4-8]. Therefore, these MEMS energy harvesters need to be individually installed on the vibration sources to maximize their output powers.

A curled cantilever structure based MEMS energy harvester was developed for two-dimensional operation [9]. The curled cantilever structure could be interpreted as the series connections of short flat cantilevers and each short flat cantilever had each own tangential component to induced vibrations

[1] To whom any correspondence should be addressed.

along the cantilever curve, which caused maximum bending momentum. Therefore, this curled cantilever successfully oscillated under both vertically and longitudinally induced vibration and generated uniform output under two-dimensionally induced vibrations. However, it is prohibited to scavenge horizontally induced vibrations.

In this paper, a hair-cell structure based piezoelectric energy harvester was newly proposed to effectively scavenge quasi three-dimensional ambient vibrations. The proposed cantilever structure, called a hair-cell structure, is based on a curled cantilever structure and deliberately elongated so that it oscillates under not only vertically induced vibrations but also under longitudinally and horizontally induced vibrations. The proposed device was successfully micro-fabricated and generated electrical power under vertically, horizontally, and longitudinally induced vibrations.

2. Design and fabrication

2.1. Design

The conventional flat cantilever structure is restricted to oscillating in the tangential direction over the neutral surface under vertically induced vibrations [4-8]. On the other hand, a curled cantilever structure was previously reported to scavenge two-dimensional vibration and its feasibility was verified [9]. A curled cantilever can be interpreted as the series connections of short flat cantilevers and each short flat cantilever has each own tangential component along the cantilever curve, which causes maximum bending momentum. The tangentially induced force at the specific point of curled cantilever is the projection of arbitrary vibrations on the tangential direction of spring component and leads to oscillate vertically and longitudinally.

In addition, a new cantilever structure that is deliberately elongated and curled around the longitudinal direction of the cantilever, called a hair-cell structure, was proposed to scavenge quasi three-dimensional ambient vibrations. The cantilever structure obtains an additional degree of freedom when the ratio of the length to the width of the cantilever is high enough. Therefore, the hair-cell structure based energy harvester can generate electrical energy from quasi three-dimensional ambient vibrations via both the typical fundamental mode and harmonic mode of torsional operation, as shown in Figure 1 (a) and (b), respectively. The proposed energy harvester is comprised of a high aspect ratio proof mass and elongated piezoelectric cantilever curled around the longitudinal direction.

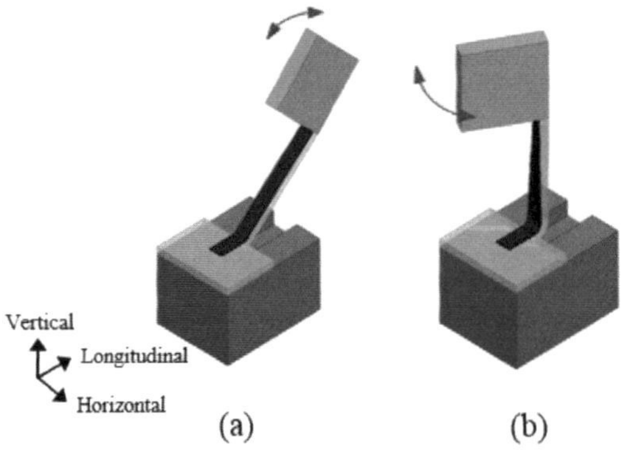

Figure 1. Operation principle of the proposed hair-cell structure based energy harvester: (a) typical fundamental mode for vertical and longitudinal vibration energy harvesting and (b) harmonic mode torsional operation for horizontal vibration energy harvesting.

2.2. Fabrication

The fabrication sequences of the proposed hair-cell structure based energy harvester are schematically shown in Figure 2. The fabrication started with the deposition of LPCVD SiNx (0.8 μm) and LTO (0.2 μm) for an elastic layer on a Si wafer. After sputtering a Ti/Pt (300 Å/1200 Å) thin film for the formation of the bottom electrode, a $Pb(Zr_{0.52}Ti_{0.48})O_3$ (3500 Å) thin film was sol-gel spin coated. A Pt thin film was sputtered for the formation of the top electrode and then Ti/Pt/PZT/Pt layers were patterned by using the inductively coupled plasma (ICP) dry etching technique. A Si_3N_4 (2000 Å) thin film was deposited by using a PECVD and patterned by the RIE technique for the passivation between the PZT layer and NiFe proof mass. A Ti/Au (300/3500 Å) was deposited by the E-beam evaporation technique and patterned by the lift-off technique to form electrode bonding pads. The SiNx/LTO layer on the Si substrate was then patterned to form the cantilever structure. A photoresist layer was coated as a sacrificial layer and Ti/Cu (300/1500 Å) thin film was then sputtered as a seed layer for electroplating. After forming a thick photoresist mold, a $Ni_{0.8}Fe_{0.2}$ (22 μm) alloy was electroplated to form a proof mass. Finally, the XeF_2 etching technique was used to release the cantilever and the released cantilever was curled up due to the intrinsic bending deformation of multi-stacked layer.

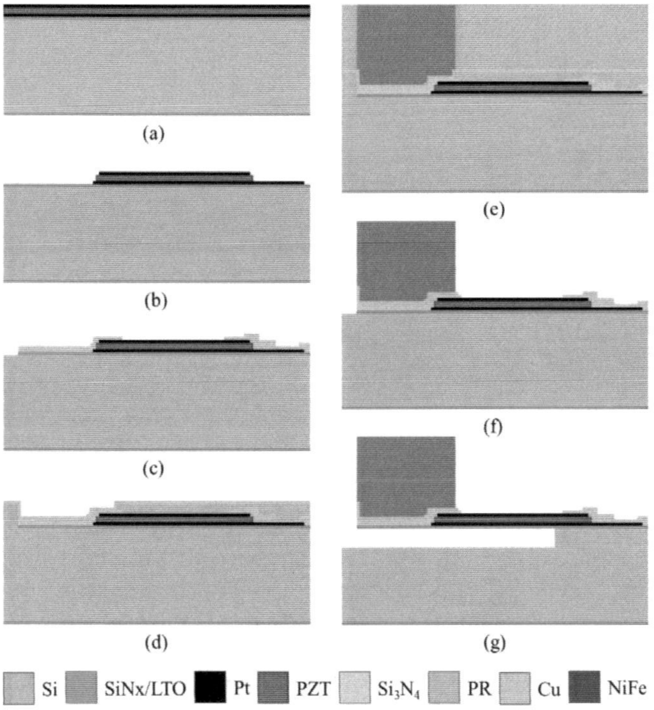

Figure 2. Fabrication sequences of the proposed hair-cell structure based piezoelectric energy harvester: (a) deposition of SiNx/LTO/Ti/Pt/PZT/Pt on a Si wafer, (b) patterning of Ti/Pt/PZT/Pt (c) deposition and patterning of Si_3N_4 as a passivation layer and Ti/Au as an interconnection layer, (d) deposition of photoresist as a sacrificial layer and Ti/Cu as a seed layer for electroplating, (e) formation of photoresist mold and electroplating of NiFe as a proof mass, (f) removal of mold, seed, and sacrificial layers, (g) XeF_2 etching of Si selectively for releasing the cantilever.

Figure 3 shows the photograph and SEM image of fabricated MEMS piezoelectric energy harvester. The ratio of length to width of the cantilever was 67:1. The PZT cantilever was curled up to the longitudinal direction of the cantilever due to the intrinsic bending deformation and the NiFe proof mass was located at the free end of the cantilever. The volumes of fabricated piezoelectric cantilever and NiFe proof mass are 1000×15×1.7 (height) μm^3 and 500×15×22 μm^3, respectively.

(a) (b)

Figure 3. (a) Photograph and (b) SEM image of the micro-fabricated hair-cell structure based piezoelectric energy harvester.

3. Experimental results

The PZT thin film of the fabricated hair-cell cantilever structure was exposed to the electric field ranging from -10 V to 10 V and its polarization curve was measured by the Radiant Precision LC system. The fabricated PZT thin film had a saturation polarization, remnant polarization and coercive field of approximately P_{MAX} = 23.15 μC/cm^2, 2P_r = 6.22 μC/cm^2 and 6.64 kV/cm, respectively. Figure 4 shows the experimental set-up for the fabricated device. A rotational jig between the accelerometer and fabricated device was used to apply three-dimensional vibrations to the fabricated device in order to verify the main concept that the fabricated device generated electrical energy under three-dimensional ambient vibrations.

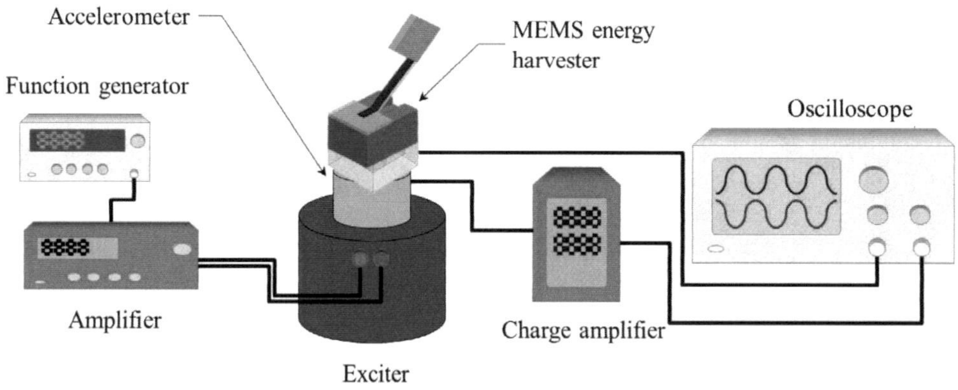

Figure 4. Experimental set-up with a rotational jig for the fabricated device.

The fabricated device was located under vertically induced vibrations at various frequencies to find its fundamental resonance frequency. The output voltage of the fabricated device was maximized to 15 mV at its resonance frequency of 116 Hz under vertically induced vibrations with frequency swept from 95 to 135 Hz and an acceleration of 50 m/s^2, as shown in Figure 5. Figure 6 shows the waveform of the generated output voltage from induced vibrations. In order to verify three-dimensional vibration harvesting, the fabricated hair-cell structure based energy harvester was placed under longitudinally and horizontally induced vibrations. As shown in Figure 7, the fabricated device successfully generated the output voltage of 33 mV and 10 mV under longitudinally and horizontally induced vibrations with a resonance frequency of 116 Hz and acceleration of 50 m/s^2, respectively. Even though the output voltage of proposed energy harvesting device was smaller than the expected value, the performance was not affected by installation angle with respect to vibration.

4. Conclusion

In this paper, a piezoelectric energy harvester for three-dimensional vibration scavenging has been presented. The cantilever of the proposed energy harvester, called a hair-cell structure, was deliberately elongated and curled so that it has a quasi three-dimensional operation with decent displacement under both the typical fundamental mode and harmonic mode of torsional operation. The output voltage needs to be improved for the practical use as a power source and it is expected to generate higher output voltage if the proof mass and PZT thin film becomes heavier and thicker, respectively.

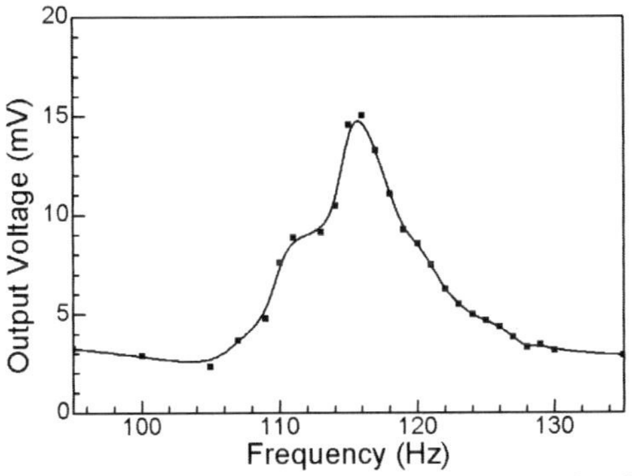

Figure 5. Frequency response of the fabricated hair-cell structure based energy harvester under vertically induced vibrations with an acceleration of 50 m/s^2.

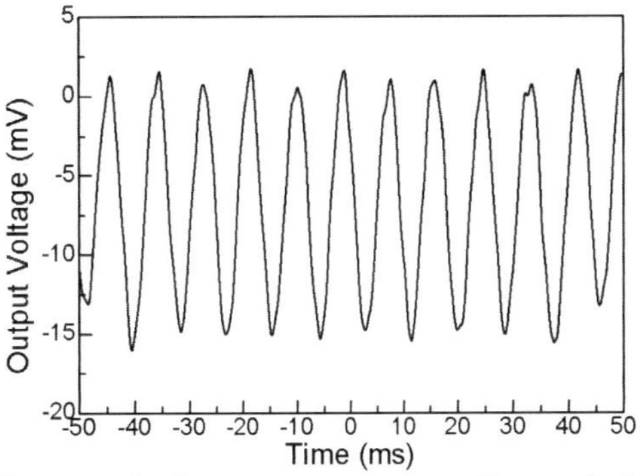

Figure 6. Output voltage waveform of the fabricated energy harvester under vertically induced vibrations with an acceleration of 50 m/s^2 at its resonance frequency of 116 Hz.

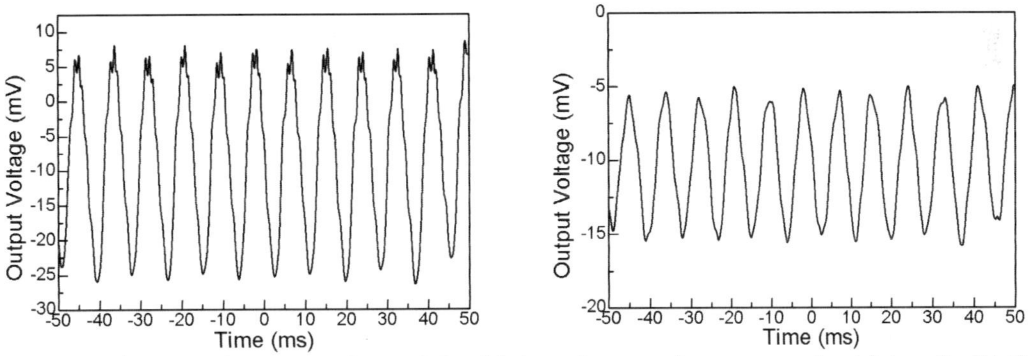

Figure 7. Output voltage waveform of the fabricated energy harvester under (a) longitudinally and (b) horizontally induced vibrations with an acceleration of 50 m/s^2 at its resonance frequency of 116 Hz.

Acknowledgements
The authors are grateful to acknowledge the Basic Science Research Program (2010-0024618) and the Pioneer Research Centre Program (2010-0019313) through the National Research Foundation of Korea (NRF) funded by the Ministry of Science, ICT and Future Planning, Korea.

References
[1] Beeby S P, Tudor M J and White N M 2006 *Meas. Sci. Technol.* **17** R175
[2] Kim H S, Kim J and Kim J 2011 *Int. J. Precis. Eng. Manuf.* **12** 1129
[3] Roundy S, Wright P K and Rabaey J 2003 *Meas. Sci. Technol.* **17** R175.
[4] Park J C, Park J Y and Lee Y 2010 *J. Microelectromech. Syst.* **19** 1215
[5] Fang H B, Liu J Q, Xu Z Y, Dong L, Wang L, Chen D, Cai B C and Liu Y 2006 *Microelectron. J.* **37** 1280
[6] Elfrink R, Kamel T M, Goedbloed M, Matova S, Hohlfeld D, Andel Y V and Schaijk R V 2009 *J. Micromech. Microeng.* **19** 094005
[7] Isarakorn D, Briand D, Janphuang P, Sambri A, Gariglio S, Triscone J, Guy F, Reiner J W Ahn C H and Rooij N F 2011 *Smart Mater. Struct.* **20** 025015
[8] Liu H, Tay C J, Quan C, Kobayashi T and Lee C 2011 *J. Microelectromech. Syst.* **20** 1131
[9] Park J C, Khym S and Park J Y 2013 *Appl. Phys. Lett.* **102** 043901

Passivation of Zinc Oxide Nanowires for Improved Piezoelectric Energy Harvesting Devices

Nimra Jalali[1], Joe Briscoe[1], Peter Woolliams[2], Mark Stewart[2], Paul M. Weaver[2], Markys Cain[2] and Steve Dunn[1]
[1]Queen Mary University of London, [2]National Physical Laboratory

Email: s.c.dunn@qmul.ac.uk

Abstract. This paper evaluates the improvement in performance of ZnO nanowires energy harvesters using p-type copper thiocyanate (CuSCN) passivation. Two types of p-n junction based devices: ZnO/PEDOT:PSS (poly(3,4-ethylene-dioxythiophene) poly(styrenesulfonate)) and ZnO/CuSCN/PEDOT:PSS were fabricated. It was observed that, the passivation of nanowires using CuSCN improved the performance four times, yielding a peak power density of 303 μWcm^{-2} for a load of 3.54 kΩ. The results were supported by impedance analysis of each device and it was observed that the piezoelectric voltage output of the device depends on its RC time constant.

1. Introduction

Energy harvesting from ambient sources is a potential substitute to conventional power generation methods. In order to deal with energy supply and storage issues, research work has developed promising macro-power and micro-power energy harvesting systems. In case of micro-power energy harvesters, piezoelectric devices combine the ability of being self-powered with size compactness[1]. MEMS (Micro-Electro-Mechanical Systems) piezoelectric systems are commercially applied as sensors and detectors for navigation, automotive and smartphones[2][3]. They have gained wide attention in wireless sensing and monitoring for remote operations of systems on which wired connections are impractical[4][5]. The most commonly used piezoelectric material is lead zirconate (PZT). PZT has k_{33} electromechanical coupling coefficient of $\approx$0.67[6] but, being brittle in nature, it has limitations in applications[7].

The piezoelectric voltage measurement of ZnO nanowires was demonstrated in 2006[8]. Since then, it became an attractive material for nanostructure-based piezoelectric energy harvesters. The Young's Modulus of ZnO nanowires, has been reported to be up to 100 GPa[9], which can make it withstand loadings higher than PZT. ZnO nanogenerator energy harvesters have been fabricated with Schottky contacts[10], insulator contacts[11] and p-n junctions[12]. ZnO is an n-type semiconductor and allows mobile charges to flow under piezoelectric voltage. These charges internally screen the polarization field[12]. In addition, external screening of the field occurs through contact of ZnO with metals and metal oxides[12]. ZnO has intrinsic surface donor defects such as oxygen vacancies and zinc interstitials[13][11]. The interaction of the material surface with moisture creates hydroxyl OH^- ions which were also identified as donor species[14]. These defect states reduce the piezoelectric voltage by screening the polarization charges[11][15]. Therefore, surface modification techniques have been applied previously[11][15] to suppress their effect. In this paper, passivation technique using p-type CuSCN is reported. The observations are supported by impedance analysis results. Two devices comprising of ZnO/PEDOT:PSS and ZnO/CuSCN/PEDOT:PSS were fabricated. The effect of screening rate on the voltage output was studied using the RC time constant of the devices.

2. Method

ZnO nanowires were grown on flexible indium-tin oxide (ITO)-coated polyethylene terephthalate (PET) substrates (2x1 cm^2) using aqueous solutions of 0.025 M $Zn(NO_3)_2.6H_2O$ and 0.025 M hexamethylenetetramine at 90°C[12]. For the ZnO/CuSCN/PEDOT:PSS device, the nanowires were spray-deposited with 5 ml of 0.15 M CuSCN solution in dipropyl sulphide using a method described elsewhere [16]. Both the CuSCN-passivated and non-CuSCN-passivated nanorods were thermally treated at 50°C for 44 hours. For passivated nanowire, thermal treatment evaporated dipropyl sulphide. Whereas, for the non-CuSCN passivated nanowires, it was performed to analyze the output difference from our previously reported PEDOT:PSS device[17]. Two layers of PEDOT:PSS were spin-coated at 2000 RPM for 30 seconds onto the nanowires (Figure 1(d)). After deposition, each layer was dried at 100°C. A gold electrode was sputtered on top of PEDOT:PSS. Copper wires were connected as extensions to ITO and gold electrodes. The 2x1 cm^2 device was mounted on a 500 μm thick plastic substrate. One end of the substrate was displaced to ≈6 mm using a cam connected to rotating motor shaft. The other end was fixed to a sample holder[17]. The generated voltage was measured under open-circuited conditions and across load resistances using NI-PXIe 1062Q and Meatest M602 programmable decade box. Short-circuit current was measured using Low-Noise Current Preamplifier SR570. Impedance analysis was carried out using 4294a Agilent Impedance Analyzer.

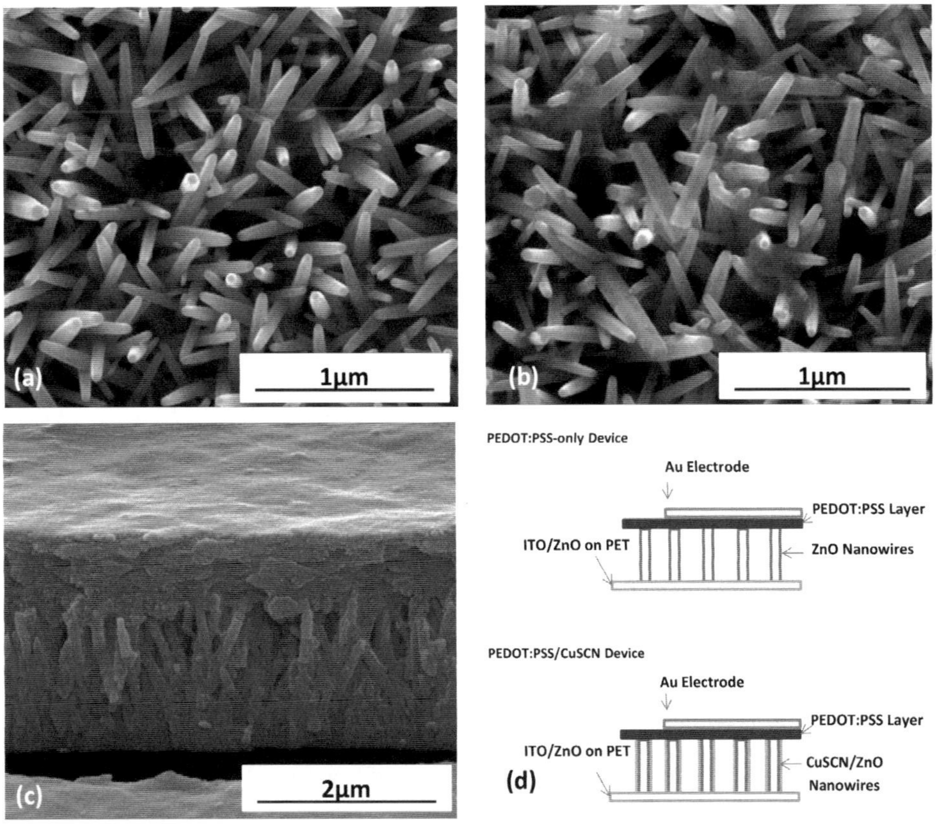

Figure 1. SEM Images: ZnO nanowires (a) ZnO nanowires with CuSCN deposition (b) PEDOT:PSS layer on top of ZnO nanowires (c). Schematic of ZnO/PEDOT:PSS and ZnO/CuSCN/PEDOT:PSS devices (d).

3. Results and Discussion

Vertical c-axis oriented ZnO nanorods having ≈70 nm thickness and ≈2 µm length are shown in Figure 1(a). The nanowires were not grown in closely packed fashion and were tilted due to seed layer texturing. As shown in Figure 1(b), the nanowire arrays were coated with CuSCN along the length. The spray coating technique was used to control the amount of deposition. Figure 1(c) shows, ZnO/PEDOT:PSS junction with PEDOT:PSS layer settled on top of the nanowires.

A ZnO/PEDOT:PSS device with thermal treatment generated an open-circuit voltage of ≈225 mV (Figure 1(a)) which was twice the voltage of our previously reported unannealed ZnO based PEDOT:PSS device[17]. This performance improvement could be related to the removal of surface adsorbed OH^- ions and chemical precursor material. ZnO has six times improved voltage output when annealed at 350°C[15], which may also be due to an annealing-induced reduction in oxygen vacancies[18]. The peak voltage output of the ZnO/CuSCN/PEDOT:PSS device was ≈673 mV (Figure 1(c)). This is a combined effect of thermal and CuSCN passivation of the device defect states on nanowire surface. ZnO has a number of surface defects. They include oxygen defects, which have been identified as the cause of the green emissions in the nanorods[13], and the moisture-generated OH^- ions[14]. They are the main donor species found on ZnO surface[14] and therefore they tend to increase the conductivity of ZnO[15]. This leads to increase in internal screening of polarization charges. Surface passivation reduces the parasitic activity of surface defects which causes a decrease in defect-induced carrier density in ZnO[11]. Therefore, the rate of internal screening reduces and voltage across piezoelectric ZnO improves.

The power density of ZnO/CuSCN/PEDOT:PSS was obtained as 303 μWcm^{-2} across an optimal load resistor of 3.54 kΩ (Figure 2(e))which was approximately four times higher than the ZnO/PEDOT:PSS device. The Nyquist impedance representation is indicative that, CuSCN passivation improved the internal resistance. The real axis semi-circle diameter of the Nyquist plot for ZnO/CuSCN/PEDOT:PSS was ≈10 kΩ as compared to ≈2.5 kΩ for ZnO/PEDOT:PSS (Figure 3). This correlates with a higher optimum load point for the ZnO/CuSCN/PEDOT:PSS device (Figure 2(e)). As observed from the impedance response, the devices can be modeled as RC circuits with their charge and discharge time constants determined by $\tau=RC$ ($\tau=f_c^{-1}$)[19] (Figure 3). Evidently, for the ZnO/PEDOT:PSS device the time constant was 0.55 ms which increased to 2.28 ms with the deposition of CuSCN. This indicated the increase in charge storage in the piezoelectric device and decrease in screening rate. This agrees with the proposed reduction in screening rate achieved by CuSCN passivation of the ZnO surface, which could account for the higher peak voltage and power output.

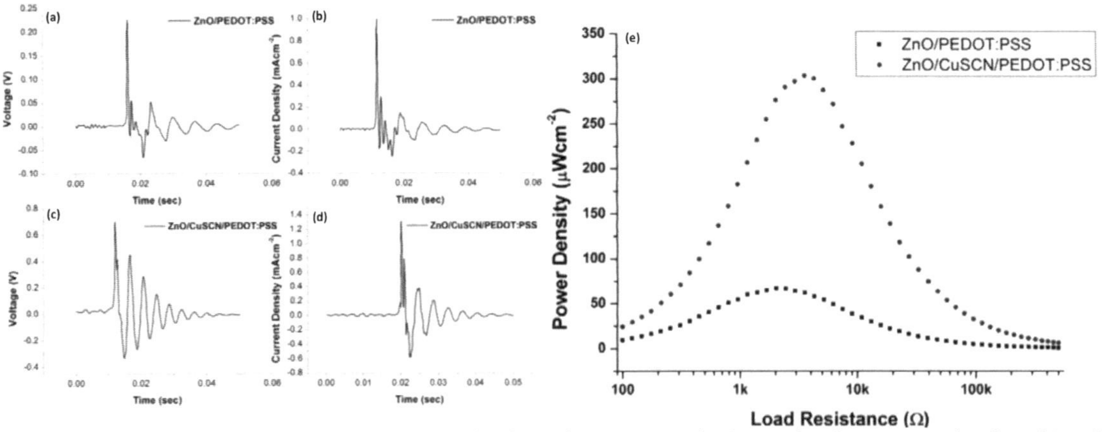

Figure 2. Performance measurement: open-circuit voltage (a) and short circuit current density (b) of ZnO/PEDOT:PSS device, open-circuit voltage (c) and short-circuit current density (d) of ZnO/CuSCN/PEDOT:PSS device, (e) load curves of ZnO/PEDOT:PSS and ZnO/CuSCN/PEDOT:PSS device showing power density at optimum load resistance.

Table 1. Performance parameters of ZnO/PEDOT:PSS and ZnO/CuSCN/PEDOT:PSS devices.

Device	Open Circuit Voltage	Short Circuit Current	Load Resistance	Voltage On Load Resistance	Current Density On Load Resistance	Power Density
	mV	$mAcm^{-2}$	kΩ	mV	$mAcm^{-2}$	μWcm^{-2}
ZnO/ PEDOT:PSS	225	1.00	2.00	114	0.58	66.80
ZnO/CuSCN/ PEDOT:PSS	673	1.30	3.54	389	0.78	303.39

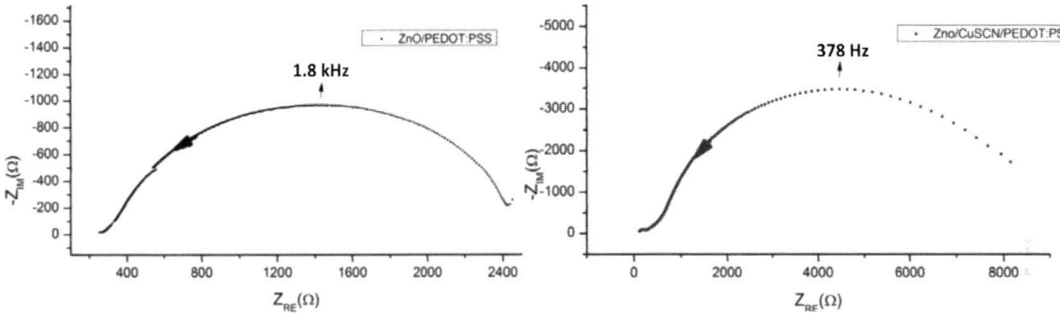

Figure 3. Nyquist plot obtained from impedance analysis (40 Hz-110 MHz) of ZnO/PEDOT:PSS (a) and ZnO/CuSCN/PEDOT:PSS (b) devices.

4. Conclusion

A suitable passivation technique was adopted which improved the energy harvester's power density from ≈66 to ≈303 μWcm^{-2} across an optimum load resistor.

This work demonstrates the importance of impedance analysis for energy harvester optimization. It unfolded important relation between circuit behaviour and rate of screening. The voltage build-up across ZnO nanowires was found to be related to the RC time constant of the devices. This supports the explanation that surface defects and impurities increase the screening effect in piezoelectric ZnO. Passivation reduced the parasitic effects and improved circuit time constant from ≈0.55 to ≈2.28 ms.

References

[1] Uchino K 2008 Piezoelectric Motors and Transformers *Piezoelectricity SE - 11* vol 114 (Springer Berlin Heidelberg) pp 257–77
[2] Marek J 2011 Automotive MEMS sensors — Trends and applications *Proceedings of 2011 International Symposium on VLSI Technology, Systems and Applications* (IEEE) pp 1–2
[3] Lim H J, Singh R P, Chuan K C T, Nuttman D and Je M 2011 Double regulated voltage supply for high precision MEMS accelerometers *2011 International Symposium on Integrated Circuits* (IEEE) pp 571–4
[4] James E P, Tudor M J, Beeby S P, Harris N R, Glynne-Jones P, Ross J N and White N M 2004 An investigation of self-powered systems for condition monitoring applications *Sensors Actuators A Phys.* **110** 171–6
[5] Lallart M, Guyomar D, Jayet Y, Petit L, Lefeuvre E, Monnier T, Guy P and Richard C 2008 Synchronized switch harvesting applied to self-powered smart systems: Piezoactive microgenerators for autonomous wireless receivers *Sensors Actuators A Phys.* **147** 263–72
[6] Priya S and Nahm S 2012 *Lead-Free Piezoelectrics* (New York, NY: Springer New York)
[7] Anton S R and Sodano H A 2007 A review of power harvesting using piezoelectric materials (2003–2006) *Smart Mater. Struct.* **16** R1–R21

[8] Wang Z L and Song J 2006 Piezoelectric nanogenerators based on zinc oxide nanowire arrays. *Science* **312** 242–6
[9] Hoffmann S, Östlund F, Michler J, Fan H J, Zacharias M, Christiansen S H and Ballif C 2007 Fracture strength and Young's modulus of ZnO nanowires *Nanotechnology* **18** 205503
[10] Xu S, Qin Y, Xu C, Wei Y, Yang R and Wang Z L 2010 Self-powered nanowire devices. *Nat. Nanotechnol.* **5** 366–73
[11] Hu Y, Lin L, Zhang Y and Wang Z L 2012 Replacing a battery by a nanogenerator with 20 V output. *Adv. Mater.* **24** 110–4
[12] Briscoe J, Stewart M, Vopson M, Cain M, Weaver P M and Dunn S 2012 Nanostructured p-n Junctions for Kinetic-to-Electrical Energy Conversion *Adv. Energy Mater.* **2** 1261–8
[13] Lv Y, Yao W, Ma X, Pan C, Zong R and Zhu Y 2013 The surface oxygen vacancy induced visible activity and enhanced UV activity of a ZnO1−x photocatalyst *Catal. Sci. Technol.*
[14] Liu W Z, Xu H Y, Ma J G, Liu C Y, Liu Y X and Liu Y C 2012 Effect of oxygen-related surface adsorption on the efficiency and stability of ZnO nanorod array ultraviolet light-emitting diodes *Appl. Phys. Lett.* **100** 203101
[15] Pham T T, Lee K Y, Lee J-H, Kim K-H, Shin K-S, Gupta M K, Kumar B and Kim S-W 2013 Reliable operation of a nanogenerator under ultraviolet light via engineering piezoelectric potential *Energy Environ. Sci.* **6** 841
[16] Hatch S M, Briscoe J, Sapelkin A, Gillin W P, Gilchrist J B, Ryan M P, Heutz S and Dunn S 2013 Influence of anneal atmosphere on ZnO-nanorod photoluminescent and morphological properties with self-powered photodetector performance *J. Appl. Phys.* **113** 204501
[17] Briscoe J, Jalali N, Woolliams P, Stewart M, Weaver P M, Cain M and Dunn S 2013 Measurement techniques for piezoelectric nanogenerators *Energy Environ. Sci.* **6** 3035
[18] Kushwaha A, Tyagi H and Aslam M 2013 Role of defect states in magnetic and electrical properties of ZnO nanowires *AIP Adv.* **3** 042110
[19] Keawboonchuay C and Engel T G 2003 Electrical power generation characteristics of piezoelectric generator under quasi-static and dynamic stress conditions *IEEE Trans. Ultrason. Ferroelectr. Freq. Control* **50** 1377–82

Influence of Bipolar Pulse Poling Technique for Piezoelectric Vibration Energy Harvesters using $Pb(Zr,Ti)O_3$ Films on 200 mm SOI Wafers

N. Moriwaki [1, 2], T. Kobayashi [2, 3], Y. Suzuki [4], N. Makimoto [3], K. Fujimoto [1, 2], K. Suzuki [1], T. Itoh [2, 3] and R. Maeda [2, 3]

[1] Dai Nippon Printing Co., Ltd, 250-1 Wakashiba, Kashiwa, Chiba 277-0871, Japan
[2] NMEMS Technology Research Organization, 1-2-1 Namiki, Tsukuba, Ibaraki 305-8564, Japan
[3] National Institute of Advanced Industrial Science and Technology, 1-2-1 Namiki, Tsukuba, Ibaraki 305-8564, Japan
[4] Ibaragi University, 4-12-1 Nakanarusawa, Hitachi, Ibaraki 316-8511, Japan

E-mail: nmoriwaki@nmems.or.jp

Abstract. Piezoelectric vibration energy harvester arrays using $Pb(Zr,Ti)O_3$ thin films on 200 mm SOI wafers were fabricated. In-plane distribution of influence of bipolar pulse poling technique on direct current (DC) power output from the harvesters was investigated. The results indicate that combination poling treatment of DC and bipolar pulse poling increases a piezoelectric property and reduces a dielectric constant. It means that this poling technique improves the figure of merit of sensors and harvesters. Maximum DC power from a harvester treated by DC poling after bipolar pulse poling is about five times larger than a one treated by DC poling only.

1. Introduction

Recently, there have been growing research and development activities of wireless sensor network using MEMS based wireless sensor nodes. Wireless sensor network systems need a large amount of autonomous sensor nodes. Piezoelectric thin films are widely applied to sensor node elements of low power consumption sensors like piezoelectric trigger switches [1,2] and power sources like piezoelectric vibration energy harvesters. High power piezoelectric devices using $Pb(Zr,Ti)O_3$ (PZT) thin films are in demand, but there are few reports on a mass-production technology including a PZT film deposition, fabrication, and poling treatment. In the present study, piezoelectric vibration energy harvester arrays using a PZT film on a 200 mm SOI wafer were fabricated, and in-plane characterization of influence of bipolar pulse poling (BPP) technique [3,4] on piezoelectric and ferroelectric properties was carried out.

2. Experimental procedure

The fabrication process flow is shown in figure 1. $Pb(Zr_{0.52},Ti_{0.48})O_3$ thin films (2 μm, (100)/(001) – oriented) were deposited by an automated sol-gel deposition system, as reported in our previous study [5]. A multilayer of 175 nm Pt/ 15 nm Ti/ 2 μm PZT/ 175 nm Pt/ 15 nm Ti/ 1 μm SiO_2 on a 200 mm SOI wafer was fabricated into piezoelectric vibration energy harvesters through the following process.

First, the top Pt/Ti layer was removed by the reactive ion etching (RIE), and the PZT film was etched by the mixed acid etchant (STEP 1). Next, the bottom Pt/Ti layer and barrier SiO_2 layer were also etched by the RIE (STEP 2). Thirdly, the Au contact pad was deposited by a sputtering and patterned using Au etchant (STEP 3). Then, the patterning of the structural Si layer to the cantilever shape was carried out using a BOSCH etch process (DRIE), and the BOX layer was removed by RIE (STEP 4). Finally, the Si substrate was also removed by DRIE, the BOX layer under the cantilever was etched by RIE from the backside to release the cantilever chips (STEP 5). The fabricated 200 mm SOI wafer is shown in figure 2. Straight cantilevers with a seismic mass were obtained as shown in figure 3.

The waveform for bipolar pulse poling (BPP) is shown in figure 4. Bipolar pulse poling was treated to PZT thin films using a ferroelectric test system (FCE-1, TOYO Corp.). Direct current (DC) poling was applied by constant voltage power equipment (PMC 18-1A, KIKUSUI). When the voltage of the top electrode is higher than that of the bottom electrode, the voltage is defined as positive and vice versa.

3. Results and discussion

3.1. Influence of microfabrications on ferroelectric properties

Influence of process damages on ferroelectric properties was characterized by the double beam laser

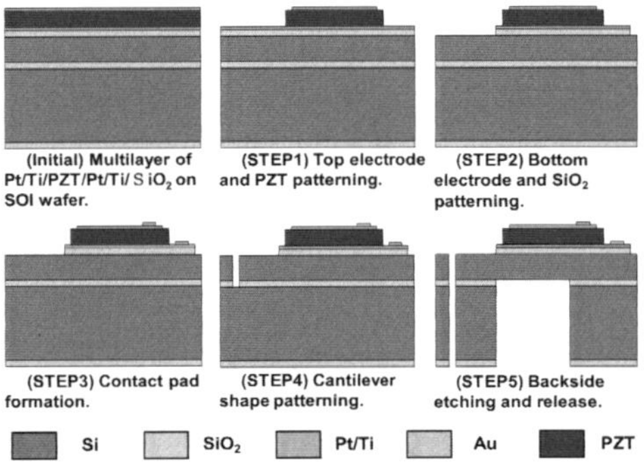

Figure 1. Schematics of the fabrication process for piezoelectric vibration energy harvesters.

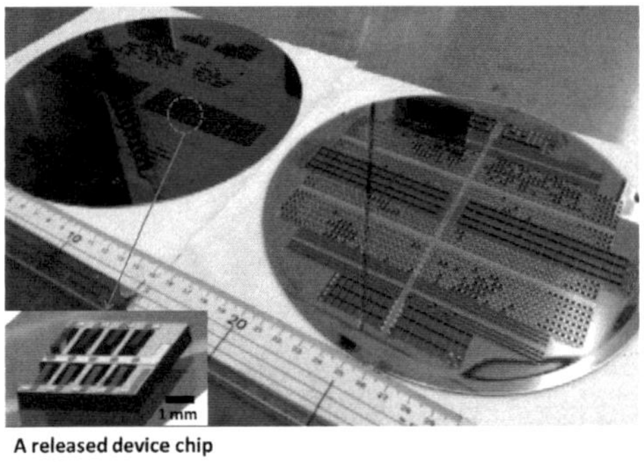

Figure 2. Overall view of the fabricated PZT thin film devices on a 200 mm SOI wafer (right in the figure) and the released device chips on a dummy wafer (left).

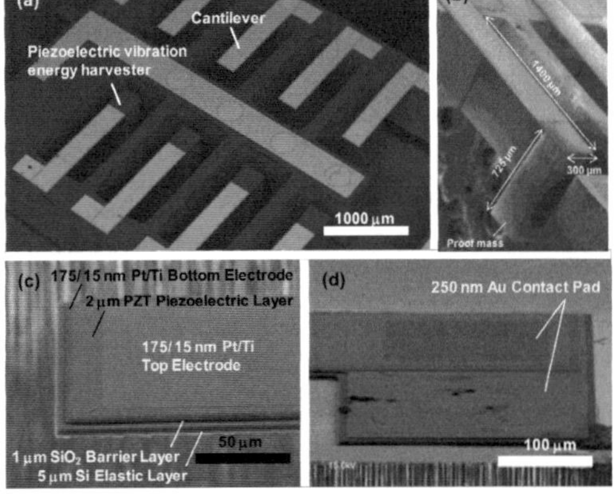

Figure 3. SEM images of the (a) fabricated piezoelectric thin film device array, (b) harvester, (c) tip and (d) base of a cantilever.

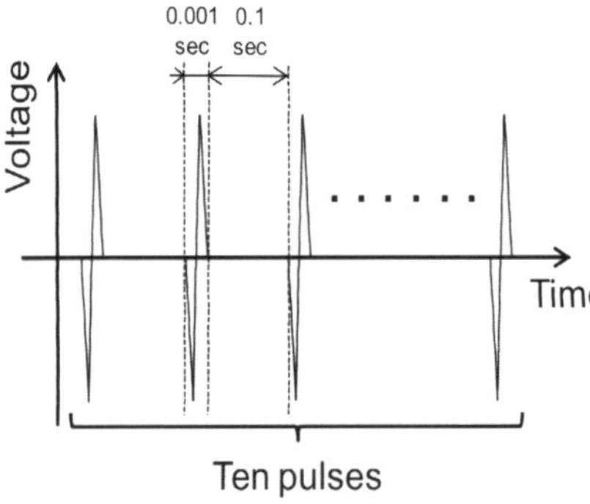

Figure 4. The waveform for bipolar pulse poling.

interferometer system with the ferroelectricity measurement module (TF 2000E, aixACCT SYSTEMS GMBH). Figure 5 shows the *P-V* hysteresis curve measured at 40 V and 1 kHz after STEP 1 and 5. Figure 6 shows the variation in positive and negative remanent polarizations with the progress of the fabrication process. Bipolar pulse poling amplitude was 90 V. Significant degradation was not observed after the fabrication, but the unbiased estimate of the variance of properties was increased after BPP from ±0.5 to ±3.1 μC/cm^2. It appears that any chemical reactions occurred at an interface between the Pt/Ti electrode and PZT film during plasma processes such as STEP 1 and 2 [6].

3.2. Influence of poling conditions on piezoelectric properties

Figure 7 shows the transverse piezoelectric constant $-d_{31}$ and dielectric constant of as a function of DC poling time. A harvester treated by 110 V BPP before DC poling was compared with an unpoled one. The transverse piezoelectric constant was estimated by the calculation using the value of the measured tip displacement of the microcantilever actuated by small AC voltages (±1 V) [7]. The result indicates that a combination poling treatment of DC and bipolar pulse poling increased the $-d_{31}$ from 45 to 60 pmV^{-1}, and reduced a dielectric constant ε from 1400 to 900. It means that this poling technique doubled the figure of merit of sensors and harvesters like a g-constant ($= -d_{31}\varepsilon^{-1}$). Figure 8 shows the estimated $-d_{31}$ before and after 110 V BPP as an in-plane distribution on a 200 mm wafer. The $-d_{31}$ was increased from 33±13 to 65±8 pmV^{-1} after the combination poling.

3.3. Direct current output power from piezoelectric vibration energy harvesters treated by bipolar pulse poling

Direct current output power from a harvester in the mechanical resonance at the vibration of 9.8 ms^{-2} was measured using a shaker, an AC-DC conversion circuit and an oscilloscope shown in figure 9. Harvesters were treated in three different ways, (1) DC poling only (+15 V, 10 sec), (2) 90 V BPP only, and (3) DC poling (+15 V, 2 sec) after 90 V BPP. The result shown in figure 10 indicates that the effect of poling treatment showed different dependency on the position in a wafer. The difference of dependency appears associated with the distribution of plasma damages. Direct current output power from a harvester located between the radius from 60 to 90 mm treated by BPP and DC poling is about five times larger than a one treated by DC poling only. It seems that the effect of the increase of the g-constant increased electro-mechanical conversion efficiency above.

4. Conclusions

Piezoelectric thin film devices using a PZT film on 200 mm SOI wafers were fabricated, and influence of the bipolar pulse poling on piezoelectric and ferroelectric properties of the devices was

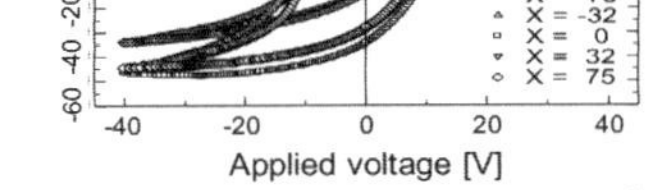

Figure 5. The variation in *P-V* hysteresis curves of five points within a range of the 150 mm diameter after fabrication (a) step 1 and (b) step 5. X in the figure means distance [mm] from the wafer center.

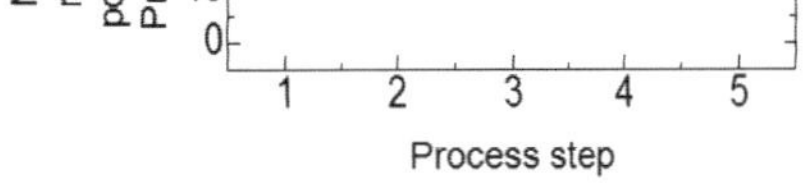

Figure 6. The variation in remnant polarization on thirty four points within a range of the 150 mm diameter after each fabrication step. Error bars signify the unbiased estimate of the variance.

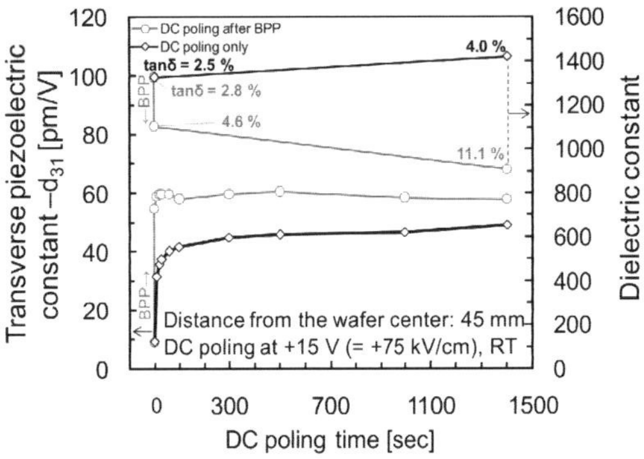

Figure 7. The transverse piezoelectric constant $-d_{31}$ and dielectric constant as a function of DC poling time.

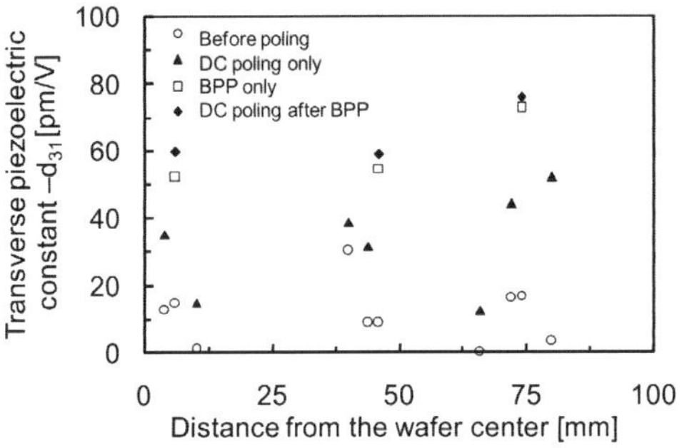

Figure 8. The result of characterization of (a) transverse piezoelectric constant $-d_{31}$ as in-plane distribution.

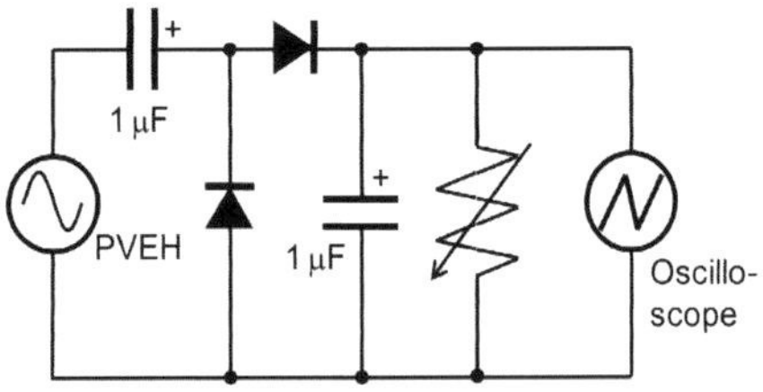

Figure 9. The schematic diagram of direct current output power measurement circuit.

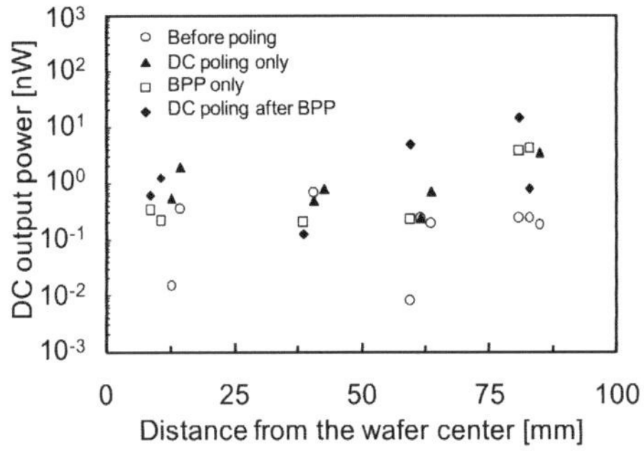

Figure 10. DC output power of a harvester in the mechanical resonance.

characterized. Combination poling treatment of DC and bipolar pulse poling increased the $-d_{31}$ from 33 ± 13 to 65 ± 8 pmV^{-1}. Maximum DC power from a harvester treated by DC poling after bipolar pulse poling is about five times larger than a one treated by DC poling only. The results indicate that the combination poling is a suitable poling method for mass-production of PZT thin film devices compared with direct current poling.

Acknowledgements
A part of this work was supported by New Energy and Industrial Technology Development Organization (NEDO).

References
[1] Kobayashi T, Okada H, Gang V Z, Maeda R, Masuda T and Itoh T 2011 *Tech. Digest of Transducers 2011* 1014
[2] Tomimatsu T, Takahashi H, Kobayashi T, Matsumoto K, Shimoyama I, Itoh T and Maeda R 2012 *Proc. PowerMEMS 2012* 295
[3] Kobayashi T, Makimoto N, Oikawa T, Wada A, Funakubo H and Maeda R 2013 *Proc. MEMS 2013* 413
[4] Kobayashi T, Makimoto N, Suzuki Y, Funakubo H, Oikawa T, Wada A, Maeda R 2013 *Jpn. J. Appl. Phys.* **52** 09KA01
[5] Uryu T, Kobayashi T, Makimoto N, Yoshimi S, Fujimoto K, Suzuki K, Itoh T and Maeda R, 2012 *Proc. PowerMEMS 2012* 323
[6] Shimamoto Y, Kushida-Abdelghafar K, Miki H and Fujisaki Y 1997 *Appl. Phys. Lett.* **70** 3096
[7] Smits J G and Choi W S 1991 *IEEE Trans. Ultrason. Ferroelectr. Freq. Control.* **38** 256

Highly Efficient Low-frequency Energy Harvester Using Bulk Piezoelectric Ceramics

M. Colin[1], S. Basrour[1], L. Rufer[1], C. Bantignies[2], A. Nguyen-Dinh[2]

[1] Univ. Grenoble Alpes, TIMA, F-38031 Grenoble

[1] CNRS, TIMA, F-38031 Grenoble

[2] Vermon, 180 rue du General-Renault, B.P. 93813, 37038 Tours Cedex, France

mikael.colin@imag.fr

Abstract. This paper describes a new way of manufacturing efficient vibration energy harvesters using thick films of piezoelectrics. The presented fabrication process is based on the thinning of high-density bulk Lead Zirconate Titanate (PZT) ceramic substrates, which enables the realization of thick layers (10-100 μm). Using this fabrication approach, we prepared two types of cantilever-based vibration energy scavengers (unimorph and bimorph) operating at very low frequency (~15 Hz) with a 50 μm PZT final thickness. Given that under a harmonic 10 mg vibration the harvested mean power was 1.3 μW and 3 μW respectively, these devices rank among the best ever-reported vibration energy scavengers according to commonly accepted figures of merit. The presented fabrication approach is therefore believed to be a good candidate for the manufacturing of highly efficient piezoelectric energy scavengers operating at very low frequency.

1. Introduction

Most of the applications that could benefit from energy harvesters, such as energy-autonomous body implants for instance; require devices that can scavenge vibration energy at low frequency (10-100 Hz) within a limited volume. In combination with the minimum level of energy to be delivered, this set of constraints lead to the need of enhanced energy density. Regarding resonant piezoelectric energy harvesters, this can be achieved either by improving the piezoelectric material properties or by maximizing the amount (i.e. the thickness) of electro-active material integrated onto the harvesting structure. In this work, we focused on the latter approach.

Thick-film technologies were proposed by different teams to fill the existing gap between existing thicknesses provided with bulk devices (>100 μm) and thin-film devices (<10 μm). Approaches as screen-printing [1], sol-gel process [2], electrophoretic deposition [3], tape-casting [4], or aerosol deposition [5] have shown the feasibility of thick piezoelectric layers with thicknesses ranging from 10 to 100 μm. Some of these fabrication techniques were originally applied in high-frequency ultrasonic transducers but can be successfully transferred to energy harvesting applications. Main challenges in piezoelectric thick-film fabrication are associated with the thickness uniformity, crack-free material, high mechanical density, reproducibility, and high piezoelectric performance. These challenges are difficult to master as thick-film technologies involve processes, which are hostile and destructive to ceramic free-standing structures. For instance, the screen-printing suffers from high processing temperature and low volume density that both degrade resulting mechanical and piezoelectric performances of fabricated structures.

Another way offering to conserve superior piezoelectric behaviour of a bulk piezoceramic material consists in a thinning of such a bulk material bonded on a substrate. This technique, applied in [6], resulted in a harvesting device delivering a power of 10,2 μW as a response to an input acceleration of 2 g at 252 Hz. Our aim is to push the same approach, based on the bulk piezoelectric material thinning, to fabricate harvesting devices working in a frequency band from 10 to 20 Hz. Furthermore, a high density PZT (HD PZT) was selected for this work. Indeed, preliminary thinning experiments of standard PZT ceramics have shown adhesion issues when the targeted thickness is in the range of few grains size (Figure 1).

(a)

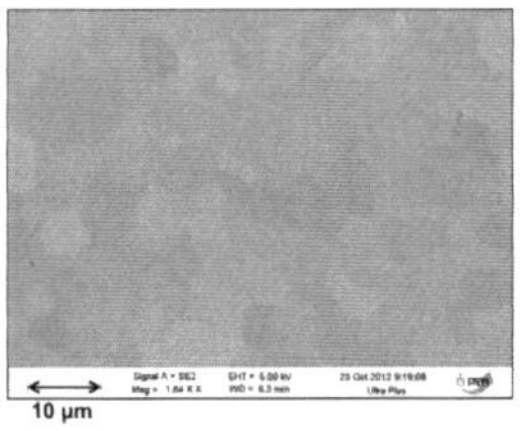

(b)

Figure 1. SEM micrograph of PZT layer surfaces after grinding and polishing steps. (a) Low surface quality is observed with standard PZT, (b) excellent surface quality is obtained with HD PZT.

2. Fabrication

The proposed fabrication process combines the bonding and the grinding/polishing of electrode plated HD PZT substrates (40 x 40 x 0.4 mm^3) onto metallic plates which act as electrodes. In the case of the monomorph scavengers, the ceramic substrate is firstly bonded on a 75 μm thick steel foil, while two ceramic substrates are bonded on both sides of a stainless steel foil (12 μm) for the fabrication of the bimorph scavengers. The bonding operation was done using low temperature non-conductive glue. The specificity of the proposed fabrication process is that it takes advantage of the roughness of PZT substrates to make the electrical contact between PZT inner electrodes and the metal foil. Doing so, usual thermal mismatch issues between PZT and the substrate, present in other high temperature bonding processes, are avoided. It should be noted that though such a surface roughness on the outer PZT faces may cause cracks during the bending, the mechanical integrity of the structures is not put at risk since only low strain is developed in this inner region close to the cantilever neutral plane. Before the grinding/polishing process was performed, the samples were diced at the final cantilever dimensions (5 x 40 mm^2). Then, during the grinding/polishing process, thickness tolerances were accurately controlled for a final PZT layer thickness of 50 μm +/- 2 μm (Figure 2).

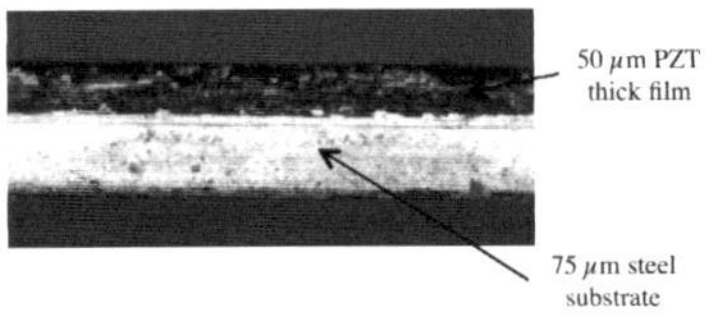

(a)

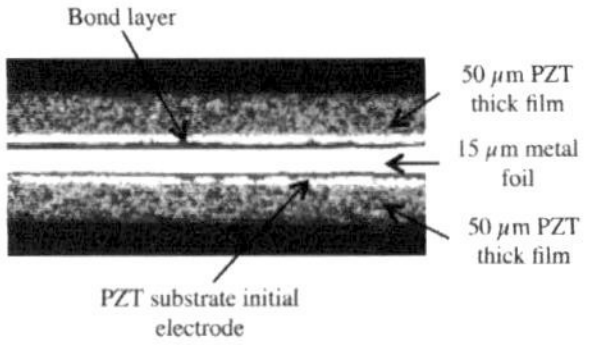

(b)

Figure 2. Cross section view of fabricated (a) single-layer and (b) double-layers piezoelectric scavengers.

Next step was the electrode plating of the external surfaces of the ceramic layers; chromium / gold electrode was used to ensure good conductivity and immunity to air oxidation (Figure 3). The poling process was finally applied to the device in order to maximize piezoelectric properties. In the case of the bimorph scavengers, PZT layers were poled in an anti-parallel configuration by applying the poling voltage between the inner metallic foil and the shunted outer electrodes.

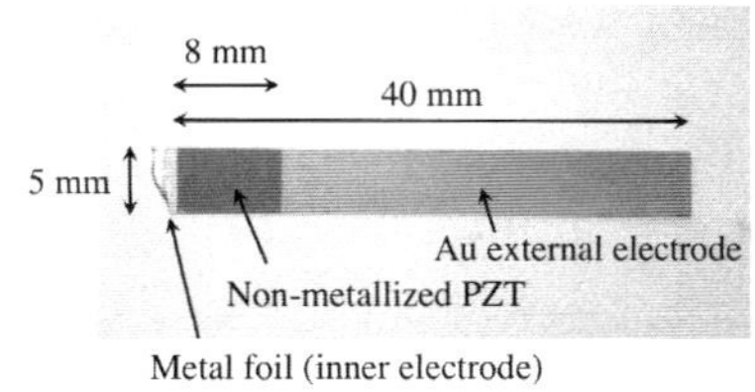

(a)

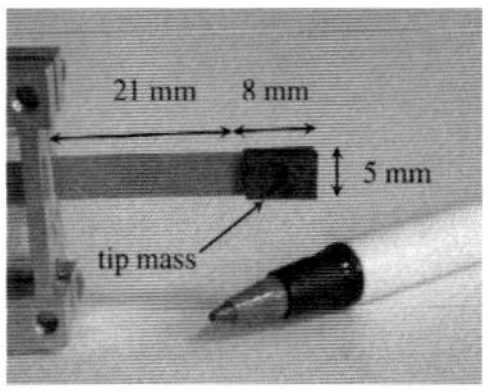

(b)

Figure 3. Pictures of the fabricated bimorph scavenger: (a) as obtained at the end of the fabrication process (b) mounted in the test clamping element and with the tip mass.

3. Experimental results and discussion

For the test of the harvesters under vibrations, we used a shaker (GW-V20/PA100E from Data Physics) in combination with an Agilent 33220A wave generator. The acceleration level on the vibrating base was monitored using a highly sensitive accelerometer (355B04 from PCB Piezotronics, 1V/g). The test set-up was driven in "closed-loop" in order to overcome the non-linear response of the vibrating chain. The device output electrodes were connected to a variable resistive load (from 200 Ω to 20 MΩ) and the generated mean power was then calculated as $P_{mean} = V_{RMS}^2 / R$. Finally, the velocity of the cantilever tip was simultaneously measured using a laser interferometer from Polytec.

3.1. Unimorph energy harvester

We firstly tested the monomorph energy scavenger (Figure 2 (a)). As previously mentioned, the metallic substrate is used as a bottom electrode. We added a 5.4 x 6.15 x 5 mm^3 proof mass made of tungsten (2.36 g) in order to decrease the resonance frequency of the device down to 15 Hz. The total length was 31 mm. Under a harmonic excitation of 10 mg only, the mean power transferred to a 150 kΩ resistance was 1.3 μW (V_{RMS} = 0.43 V) and the transverse displacement amplitude is 300 μm.

3.2. Bimorph energy harvester

3.2.1. Electrical impedance measurement. The impedance of the double-layers scavenger (Figure 2 (b)) was first measured using an Agilent 5061B Network Analyser in order to assess the quality of the piezoelectric coupling.

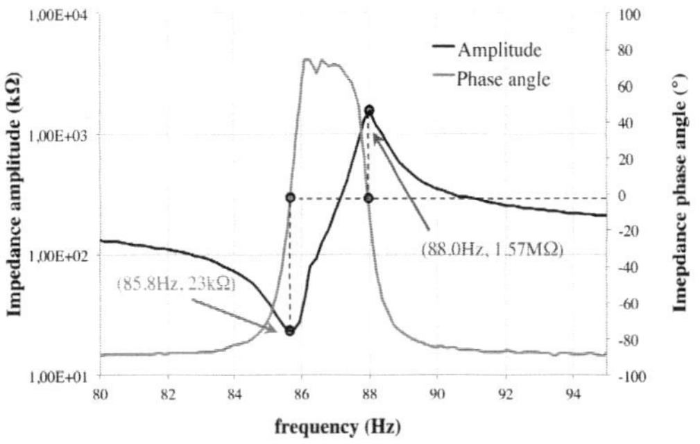

Figure 4. Impedance amplitude and phase angle of the double-layers scavenger with no tip mass.

However, it must be noted at this point that it was not possible, due to the limitation of the Agilent 5061B in Gain-Phase Series mode, to measure the high value of the impedance at low frequency for the device with the tip mass. We therefore tested the scavenger without the tip mass since it presents lower impedance than in the previous case as this one can roughly be considered as inversely proportional to the frequency - if the motional contribution is omitted. The measured impedance of the device with no tip mass is illustrated in Figure 4. It can be noted that since the impedance phase angle spans up to positive values, there are two frequencies with a 0° phase angle. These two specific frequencies correspond to the resonance f_r and the anti-resonance f_a frequencies of the device. Such an impedance phase angle curve illustrates a high piezoelectric coupling, which is also commonly given by the coupling coefficient $k = (f_a^2/f_r^2 - 1)^{1/2}$. The resonance point is characterized by the frequency of 85.8 Hz and by the real impedance of 23 kΩ. The anti-resonance point corresponds to the frequency of 88.0 Hz. The measured anti-resonance impedance is beyond the accuracy limit of the Agilent 5061B.

3.2.2. Harmonic characterization. The bimorph scavenger (with no tip mass) was then characterized in sensor mode. Since the PZT layers are polarized in opposite directions (anti-parallel configuration), the voltage output was measured between the top and bottom electrode.

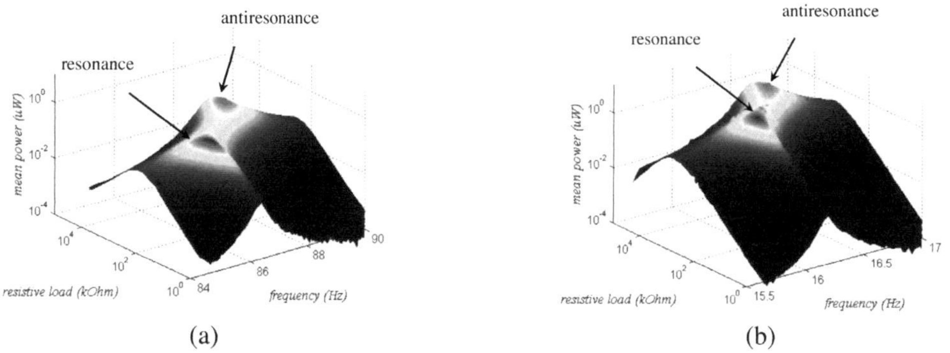

Figure 5. Mean harvested power from a harmonic vibration signal by the bimorph scavenger (a) with no tip mass, 40 mg acceleration (b) with a 1.5g tip mass, 10mg acceleration.

Figure 5 (a) illustrates the measured mean power with 40 mg input acceleration. As anticipated through the impedance analysis, the measurement results show a piezoelectric coupling that is high enough to allow the existence of an antiresonance power peak. A quality factor of $Q \sim 95$ of the structure is obtained from the tip displacement response of the structure. In combination with the coupling coefficient $k = 0.23$ obtained from the impedance measurement, we can assess the piezoelectric coupling figure of merit of the structure $k^2Q = 4.95$. The mean power is found to be 0.65 μW with an RMS voltage of 0.12 V across an optimal resistive load of 20 kΩ, which corresponds to the device impedance at the resonance frequency. In that case, the displacement amplitude of the cantilever tip is 150 μm.

Next, we added a 1.5 g tungsten seismic mass at the tip of the cantilever (Figure 3 (b)) in order to decrease the natural frequency of the structure. The test results at 10 mg acceleration are presented on Figure 5 (b). The mean harvested power is measured at 3 μW at the resonance frequency (16 Hz), and a corresponding RMS voltage of about 0.66 V across a 150 kΩ optimal resistance. The main cause of the increase of this optimal resistance is the lower resonance frequency. Here again, the antiresonance power peak can be observed, but it appears to be closer to the resonance power peak. Though the antiresonance has the same normalized frequency (1.025) as in the previous case, and the coupling coefficient remains unchanged ($k^2 = 0.05$), the "valley" between the two peaks is less pronounced than previously. This effect is linked with a lower quality factor $Q = 64$. The coupling figure of merit k^2Q is now 3.5. It is believed that the lower quality factor is due to the larger tip displacement (590 μm) induced by the lower natural frequency. Additional damping mechanisms are likely to appear because of this larger displacement.

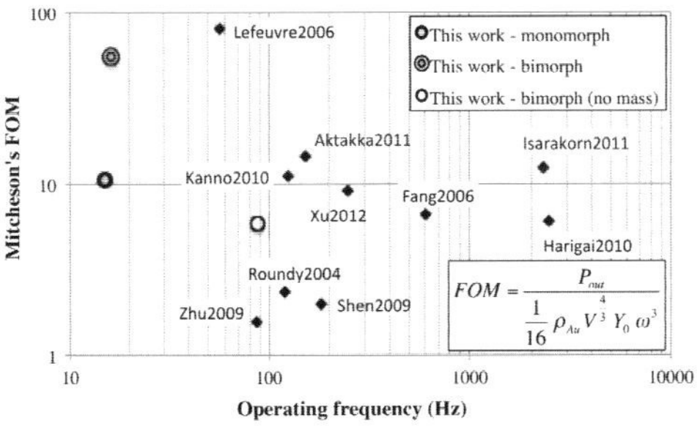

(a)

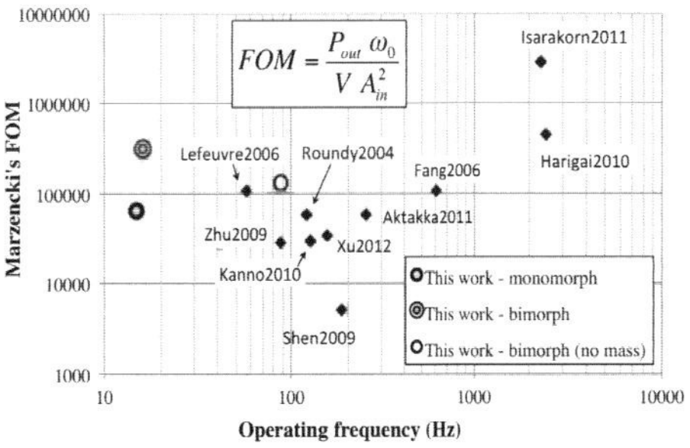

(b)

Figure 6. Comparison of different PZT-based scavengers: Mitcheson's FoM (a), Marzencki's FoM (b).

4. Conclusion

We have presented the fabrication process based on a thinning of the bulk piezoceramic material bonded on a substrate. We have applied such a process for the fabrication of three different piezoelectric energy scavengers operating at very low frequency (~15 Hz): unimorph, bimorph with no tip mass, and bimorph with a tip mass. Based on the test results presented in the previous section, we have evaluated the corresponding figures of merit (FoM) according to Mitcheson and Marzencki. The Mitcheson's FoM for unimorph, bimorph with no tip mass, and bimorph with a tip mass scavengers are 21%, 5%, and 55%, respectively (Figure 6 (a)). The Marzencki's FoM corresponding to the same structures are 64500, 136000, and 315000, respectively (Figure 6 (b)). As can be seen from Figure 6, comparing different PZT-based scavengers developed by other teams, FoM of our devices are situated among the highest values, especially taking into account the low frequency band of operation.

Acknowledgement
This work was supported by the French OSEO project HBS in the framework of FUI 1009044V.

References

[1] N. M. White, P. Glynne-Jones, and S. P. Beeby, "A novel thick-film piezoelectric micro-generator," *Smart Materials and Structures*, Vol. 10, No. 4, p. 850-2, 2001.
[2] K. L. Gentry, J. M. Zara, S. D. Bu, C. B. Eom, and S. W. Smith, "Thick Film Sol Gel PZT Transducer Using Dip Coating", *IEEE Ultrasonics Symposium*, pp. 977-980, 2000.
[3] G. Feuillard, D. Kuscer, L. P. Tran-Huu-Hue, E. Le Clezio, M. Kosec, M. Lethiecq, "Electroacoustic performance of high frequency PZT based transducer fabricated by electrophoretic deposition: comparison with screen printing technique", *IEEE International Ultrasonics Symposium Proceedings*, pp. 1138-1141, 2009.
[4] L. Qin, Y. Sun, Q. M. Wang, Y. Zhong, M. Ou, Z. Jiang, and W. Tian, "Fabrication and Characterization of Thick-Film Piezoelectric Lead Zirconate Titanate Ceramic Resonators by Tape-Casting", *IEEE Trans. On Ultrasonics, Ferroelectrics, and Frequency Control*, Vol. 59, No. 12, pp. 2803-12, 2012.
[5] S. Baba, H. Tsuda, and J. Akedo, "Thickness Dependence of Electrical Properties of PZT Films Deposited on Metal Substrates by Laser-Assisted Aerosol Deposition", *IEEE International Ultrasonics Symposium Proceedings*, Vol. 55, No. 5, pp. 1009-16, 2008.
[6] E. E. Aktakka, R. L. Peterson, K. Najafi, "A CMOS Compatible Piezoelectric Vibration Energy Scavenger Based on the Integration of Bulk PZT Films on Silicon," *IEEE Int. Electron Devices Meeting (IEDM)*, San Francisco, USA, pp. 31.5.1-31.5.4, 2010.

A novel piezoelectric energy harvester designed for single-supply pre-biasing circuit

N Mohammad pour[1 2], D Zhu[1*], R N Torah[1], A D T Elliot[3], P D Mitcheson[3] and S P Beeby[1]

[1]Electronics and Computer Science, University of Southampton, Southampton, UK
[2]Department of Mechanical and Aerospace Engineering, Politecnico di Torino, Turin, Italy
[3]Electrical and Electronic Engineering, Imperial College, London, UK

E-mail: dz@ecs.soton.ac.uk

Abstract. In this paper, the design and test of a novel screen printed piezoelectric energy harvester for the single-supply pre-biasing (SSPB) circuit is presented. It was demonstrated previously that by using the SSPB circuit, power delivered to the load was over three times greater than that in the case of using a bridge rectifier circuit. For maximum power extraction from energy harvesters using the SSPB circuit, the SSPB switches must be triggered when the piezoelectric beam reaches its maximum point of displacement. These points coincide with the peaks and troughs of the voltage across the piezoelectric material, thus accurate peak detection is required. A new design of piezoelectric energy harvester was presented to integrate a sensing piezoelectric component with the main piezoelectric layer for energy harvesting. Maximum power generation by SSPB requires the sensing voltage to be in phase with the generating voltage in order to correctly detect the peaks. However, the difference in areas between the sensing piezoelectric component and the energy harvesting piezoelectric component leads to a difference in capacitance that causes phase shift between the two signals. To overcome this issue, impedance matching was performed. Simulations and experimental rests were presented.

1. Introduction

Energy harvesting is the process of capturing and accumulating energy from ambient energy sources. Vibration energy harvesting is receiving increasing interest as an alternative power source to batteries for autonomously operating devices dues to their longer lifetime, lower cost and lower environmental impact.

The three main transduction methods used to convert mechanical vibration energy into electrical energy are electromagnetic, electrostatic or piezoelectric transducers [1][2]. Over the past decade, the last has received special attention due to their simplicity in structure, which makes them easily to be integrated into self-powered systems.

Power extraction circuits for piezoelectric energy harvesters were compared in [3], which showed that the single-supply pre-biasing (SSPB) circuit is the most suitable for low amplitude energy harvesting applications. In [4], a multilayer piezoelectric harvester with a SSPB circuit was demonstrated to deliver more than three times the power to the load than a bridge rectifier circuit.

The SSPB circuit discharges and pre-biases the piezoelectric beam using switches in an H-bridge configuration [5]. To achieve maximum power extraction from the energy harvester, the switches must be triggered when the piezoelectric beam reaches its maximum point of displacement, which coincides with the peaks and troughs of the voltage generated across piezoelectric material. Triggering early or late causes a drop of energy output [4], thus accurate peak detection is vital to achieving maximum power extraction. One can use an external sensor to detect peaks. However, it increases complexity in fabrication of energy harvesters. A piezoelectric sensing layer can be added to the harvester, however it must be electrically isolated from the generation layer, otherwise the sensing signal will be adversely effected by the pre-biasing circuit operation [4].

This paper presents a new way to generate the piezoelectric sense signal by integrating an electrically isolated piezoelectric segment into the design of the piezoelectric harvester. The complete design is presented and its performance is evaluated through simulations and experiments.

2. Design of the piezoelectric generator with integrated sensing area

Figure 1 and Figure 2 show the single layer harvester with two separate PZT (lead zirconate titanate) areas, respectively. The larger area is for power generation while the smaller area is dedicated to generating the sensing voltage required by the SSPB circuit. The performance of the piezoelectric generator simulated using ANSYS simulation with direct coupled field analysis and a coupled physics circuit ANSYS simulation. A 3D structure was designed on SolidWorks and imported in ANSYS Workbench to improve the accuracy of the results. The RMS amplitude of the sensing signal was chosen as 1.8V based on the experimental set-up in [4]. Figure 2 shows the sensing area designed in a corner near the clamping area. This simplifies the wire connections to the bottom and top electrodes around the sensing PZT and to impose greater stress to the sensing area and generate a higher voltage.

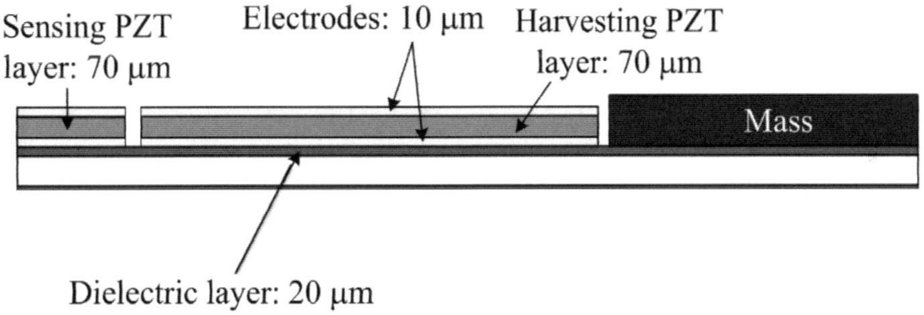

Figure 1. Cross section illustration of the piezoelectric harvester

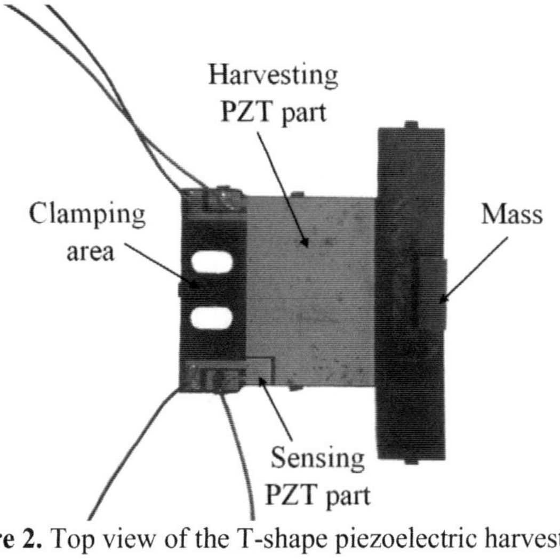

Figure 2. Top view of the T-shape piezoelectric harvester.

Three different structures were designed using a fixed PZT area is 27 × 20 mm^2 and mass area is 47 × 10 mm^2. Each structure has a different area for the sensing PZT component (3 × 3 mm^2, 4 × 4 mm^2, and 5 × 5 mm^2). The simulation results showed that the 3 × 3 mm^2 area would be great enough to generate the open circuit voltage requested for accurate peak detection, which is 1.8V_{pk}.

For all layers, materials are defined and connections between them are set with the Normal Lagrange formulation. The structure is fixed at one end and an acceleration of 1.96 m·s^{-2} is applied to the structure to simulate the vibration. To model the piezoelectric circuit, the CIRCU94 element was added to the ANSYS model in Workbench, where the simulation of the open circuit voltage is performed by connecting a high value load between the bottom and top electrodes. The electrodes of the sensing PZT are electrically isolated from the electrodes of generating PZT. The modal analysis is performed to find the natural frequency of the structure and then harmonic analysis is used to estimate the output voltage of the harvester at the resonant frequency found during the modal analysis.

Results of the simulation show that the natural frequency is at 48 Hz while the open circuit voltage generated by the sensing PZT and harvesting PZT components are 3.2 V and 4.4 V, respectively.

Screen printing technology was used to fabricate the harvester. The main advantages of this technology are the low cost of fabrication and its simplicity. Detailed fabrication process can be found in [6].

3. Experimental results

The SSPB circuit requires the sensing voltage to be in phase with the generating voltage to accurately detect the peaks. The difference in areas of the two parts however, results in a difference in capacitance. As the harvesting and sensing section do not share a common ground, the difference in capacitance leads to a phase shift between the two voltages. Simulation and experimental results show that the phase difference can be cancelled if the impedance of the sensing part is matched to the impedance of the harvesting part.

Furthermore, the harvester was tested on a shaker (Labworks ET-126) with a programmable resistance box and a computer with LabVIEW software collecting the data under the sinusoid vibration with the amplitude of 1.96 m·s^{-2}. As expected, both results show phase shift between the sensing voltage and the harvesting voltage because of the difference in capacitance of the two areas. The small PZT sensing section has a capacitance of 0.47 nF while the larger harvesting section has a capacitance of 24.88 nF, resulting in a phase shift of 66.24°. Figure 3 illustrates the phase shift between the two signals measured experimentally.

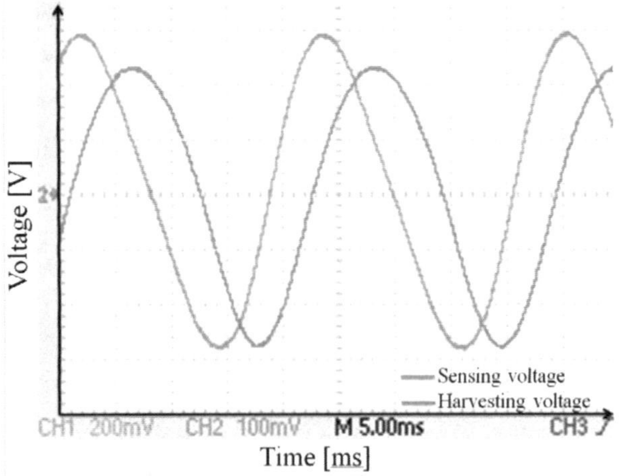

Figure 3. Experimentally measured waveforms showing piezoelectric open circuit voltage.

It was simulated and experimentally verified that sharing a common ground between generation and sense elements was able to cancel the phase shift. However, this arrangement adversely effects the operation of the SSPB circuit as shown in previous study [4]. Therefore, impedance matching was used to avoid the phase shift by connecting a 25 nF capacitor in parallel with the sensing voltage. Figure 4 shows the two signals when this capacitor was connected.

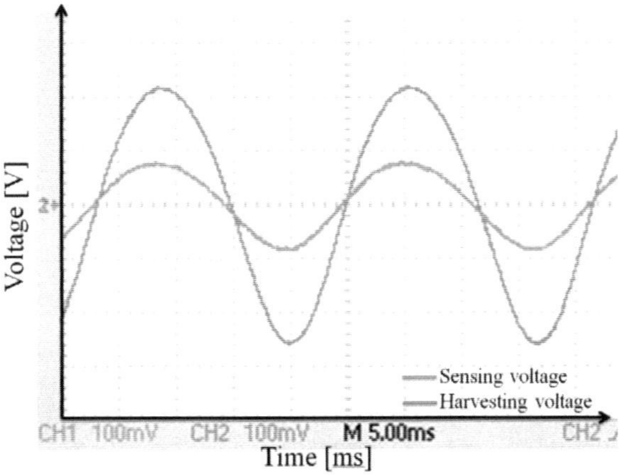

Figure 4. Experimentally measured waveforms showing piezoelectric open circuit voltage with a capacitance of 25 nF connected in parallel to the sensing PZT.

The open circuit voltages measured experimentally of both the harvesting part and the sensing part are shown in Figure 5. The output power measured experimentally with respect to the frequency at its optimum load resistance (96 kΩ) is shown in Figure 6.

Difference between experimental results and simulation results presented in Section 2 is caused by the different target and actual printed thicknesses of PZT layer. Actual thickness of PZT was measured to be between 60 to 64 μm compared to the target thickness of 70 μm, which causes a lower actual resonant frequency and lower open circuit voltages.

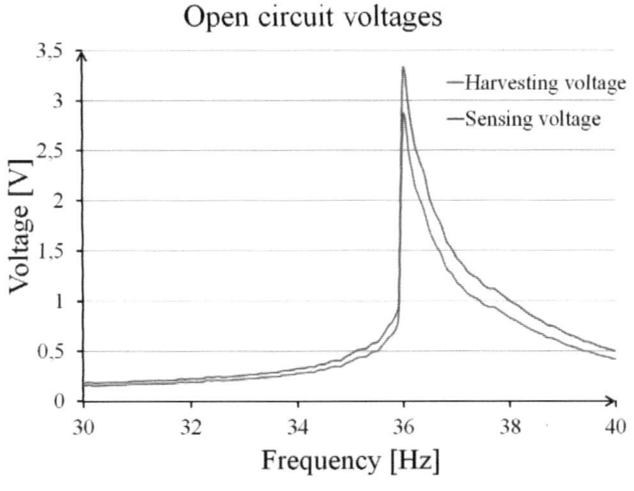

Figure 5. Open circuit voltages of the piezoelectric energy harvester.

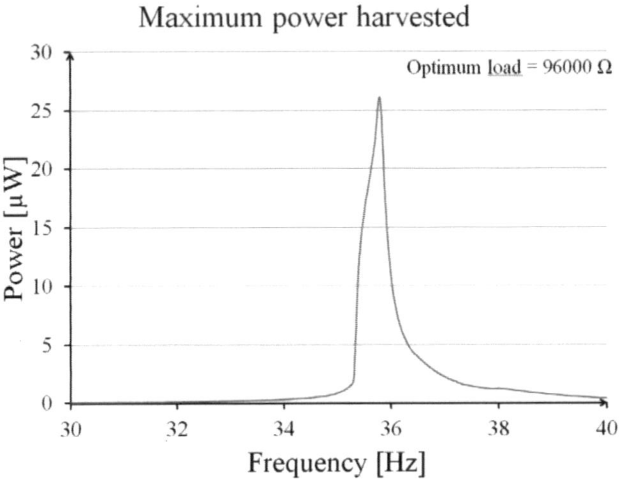

Figure 6. Maximum output power of the piezoelectric energy harvester.

4. Conclusions

In this paper, design and test of a novel screen printed piezoelectric energy harvester for the single-supply pre-biasing circuit is presented. The SSPB circuit needs a sensing signal to detect the timing of the maximum deflection of the piezoelectric beam which is unaffected by the SSPB circuit operation. The new energy harvester design integrated peak sensors with the energy harvester in the same fabrication processes, i.e. a larger area used for energy harvesting while a smaller area generating an electrically isolated sensing voltage.

The difference in areas of these two parts resulted in a difference in capacitance. Since the harvesting and sensing areas do not share a common ground, the difference in capacitance leads to a phase shift between the two voltages. The phase difference can successfully be cancelled by matching the impedance of the sensing part to the impedance of the harvesting part. Although the amplitude of the sensing signal can also be reduced by the impedance matching, it can be amplified in the peak detection circuit.

Future work will consist of power generation improvement achieved by this new harvester design when connected to a SSPB circuit. The expected improvement in peak detection should result in a large improvement in the power generation.

Acknowledgements

This work was supported by the contribution of European Union, in the framework of POR-FESR 2007/2013 within a technology transfer project of Autonomous Region of Aosta Valley and the Ministry of Labour and Social Policy.

References

[1] Beeby S P, Tudor M J and White N M 2006 *Measurement Science and Technology* **17** 175-195
[2] Priya S and Inman D 2008 Energy Harvesting Technologies (Springer)
[3] Dicken J, Mitcheson P, Stoianov I and Yeatman E 2012 IEEE Transactions on Power Electronics 27 4514-4529
[4] Elliott A, Zhu D, Beeby S and Mitcheson P 2012 *PowerMEMS 2012, Atlanta, USA*
[5] Dicken J, Mitcheson P D, Elliott A D T and Yeatman E M 2011 *PowerMEMS 2011, Seoul, South Korea*
[6] Zhu D, Beeby S P, Tudor M J and Harris N 2011 *Sensors and Actuators A* **169** 317-325

Compact and Low-Frequency Vibration Energy Scavenger using the longitudinal excitation of a piezoelectric bar

M. Colin[1], Q. Mortier[1], S. Basrour[1], N. Bencheikh[2]

[1] Univ. Grenoble Alpes, TIMA, F-38031 Grenoble

[1] CNRS, TIMA, F-38031 Grenoble

[2] Cedrat Technologies S.A., 38246 Meylan Cedex, France

mikael.colin@imag.fr

Abstract. This paper introduces an innovative architecture of a piezoelectric harvester, which enables harvesting vibration energy at low frequency using the {33}-transduction mode of a piezoelectric element. Unlike cantilevers integrating ferroelectric material combined with interdigitated electrodes, the concept that we propose is based on the elongation/compression excitation of a piezoelectric bar.

1. Introduction

The conversion of mechanical energy from the environment into useable electrical energy using piezoelectric materials is a well-known technique, which has been widely investigated during the last decade. Most of the time, this vibration energy is located at low frequency (typically from 10-100 Hz) and the corresponding level of acceleration is low. Most of the presented works focus on compliant structures that use flexion mode (cantilevers, clamped-clamped beams, spirals, membranes...) in order to combine low resonance frequency of the structure and actuation of stiff piezoelectric material. The energy is then converted through the {31} piezoelectric transduction mode (Figure 1 a). Also, several works have been presented where the energy is rather converted using the {33}-transduction mode (Figure 1 b) using interdigitated electrode [1,2]. This transduction mode offers higher coupling coefficients but requires an additional poling step since the polarisation vector must be orientated in the layer plane. Furthermore, only a part of the piezoelectric material is actually actuated. The goal of this paper is to propose a compact, low frequency device using the {33}-transduction mode excited by elongation/ compression of a piezoelectric bar (Figure 2). Though the excitation of the piezoelectric elements via direct excitation as illustrated in Figure 2 (a) is not suitable for scavenging low frequency vibration energy because of the very high corresponding resonant frequency, it becomes possible with our structure.

The first part of the paper focuses with the description of the mechanical converter able to work at low frequency and providing an amplification of the applied force on the piezoelectric ceramic bar. The second part is dedicated to the electromechanical simulation of the whole device including the piezoelectric transducer. Moreover, for a given electrical power, we can adapt the voltage output by changing the number of layers in the piezoelectric stack.

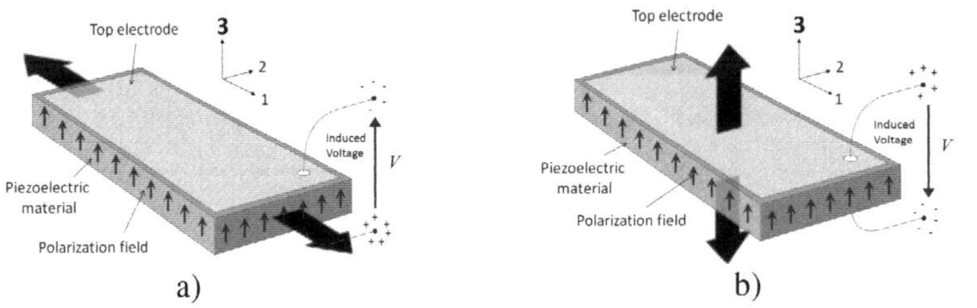

Figure 1. Piezoelectric linear transduction modes: a) {31} mode and b) {33} mode

2. Design of the mechanical amplifier

The {33} transduction is very attractive since it is characterized by three times higher conversion efficiency than the transverse {31}-transduction. However, the stiffness of most of the piezoelectric materials makes the {33}-mode very unpractical for energy scavenging at low frequencies - where most of the ambient vibration energy is localized. Based on previous works on piezoelectric micro-actuators [3-5], we propose a device architecture that converts a large-displacement/low-force excitation into a small-displacement/high-force excitation applied on a piezoelectric element (Figure 2). This structure is therefore termed "force amplifier" or simply "amplifier". With such a device, the use of the elongation/compression of a piezoelectric bar for vibration energy scavenging at low frequencies becomes possible.

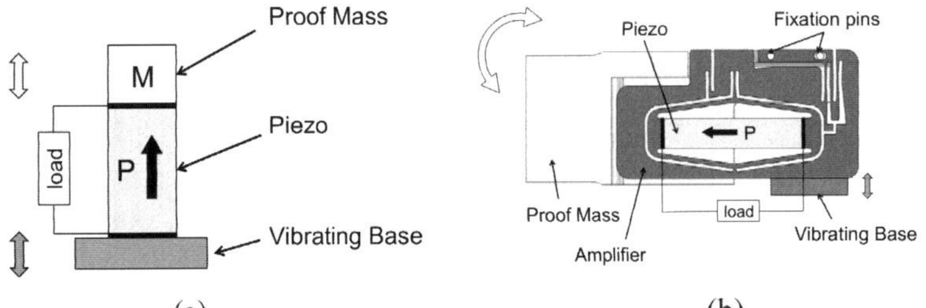

Figure 2. a) Schematic of a basic vibration energy harvester operating in the {33} mode at high frequency. b) Proposed concept for a low frequency vibration energy scavenger using longitudinal excitation of a piezoelectric bar

2.1. Description of the mechanical transducer

The proposed amplifier is used in combination with a proof mass, which is distributed on both parts of the device and tightly attached to the input arm with fixation pins (Figure 3 a). When acceleration is applied on the structure, the displacement of a proof mass creates a rotational moment on the input arm. This input moment is then converted into a pure axial force and transferred to the piezoelectric element by a series of two lever stages (Figure 3 b and c). These are created from the combination of rigid parts and flexural hinges which width is set such as to match the desired resonant frequency (Figure 5). The complexity of the amplifier architecture proposed in this study results from a design effort to optimize the generator compactness.

Besides the ability to excite the piezoelectric bar at low frequency, another main advantage of the proposed concept is that it allows the mechanical excitation of the piezoelectric material in the compressive domain only. During the assembly, a pre-stress shim is placed between the piezoelectric bar and the amplifier to achieve the stress bias. This represents a great advantage in terms of device reliability since no tensile force is applied on the ceramics.

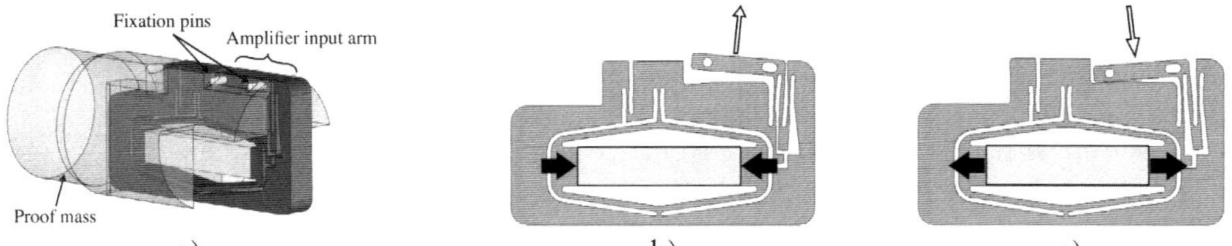

Figure 3. a) Displacement of the proof mass is transferred to the amplifier through the input arm, b) negative stress is applied on the piezoelectric bar when mass is moved upward, and c) positive stress is applied on the piezoelectric bar when mass is moved downward.

2.2. Prototyping
In the frame of this study, we have fabricated a prototype (Figure 4). It is composed of an amplifier made of aluminum alloy, which was fabricated by wire-cut EDM (Electric Discharge Machining), a 3.3 g proof mass mounted on the input arm, and a piezoelectric bar made of PZT NCE51 (from Noliac). The piezoelectric element is 5.7 mm long and its section area is 1.2 x 1.2 mm². The hinges are 50 μm wide. The complete device has a length of 13.5 mm, and the mass diameter is 5 mm. Finite Element Modelling (FEM) simulation returns a device resonance frequency at 25 Hz.

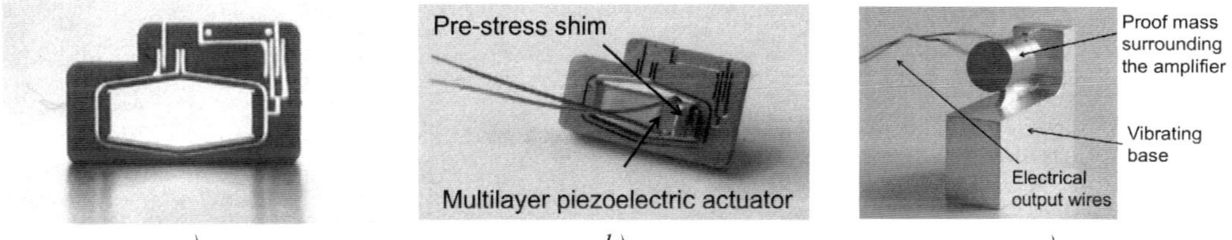

Figure 4. a) Naked 10mm long amplifier, b) amplifier mounted with a commercially available multi-layered piezoelectric actuator and pre-stress shim, c) final assembly with proof mass.

2.3. Mechanical coupling
The simulated load characteristic of the amplifier is illustrated on Figure 6. Under a 40 mN input force (1 mm displacement of the mass), the resulting blocked force is 2.8 N. The gain relative to the force is then calculated as 70. The stroke of the amplifier is about 1 μm. When the amplifier is associated with a piezoelectric element which stiffness is 50 GPa, the corresponding applied force and deformation are 2.3 N and 200 nm respectively. The operation point is clearly not optimal and the mechanical coupling is about 1% only. Further design optimization is still required (hinges, lever lengths...) in order to increase this value.

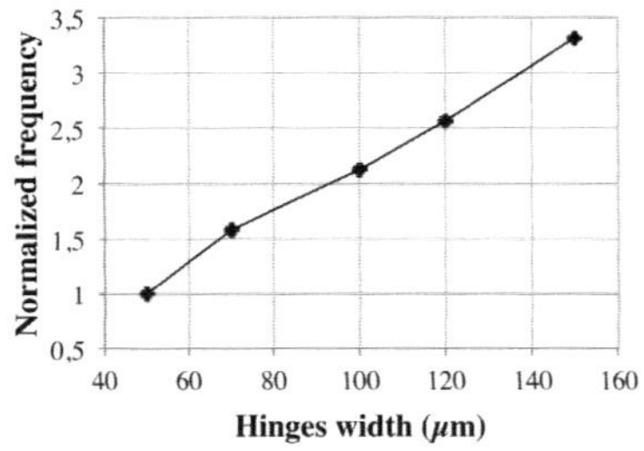

Figure 5. Variation of the resonance frequency of a device made of Aluminium alloy with a piezoelectric element which stiffness was arbitrarily set at 200GPa. These data result from a Finite Element Modelling.

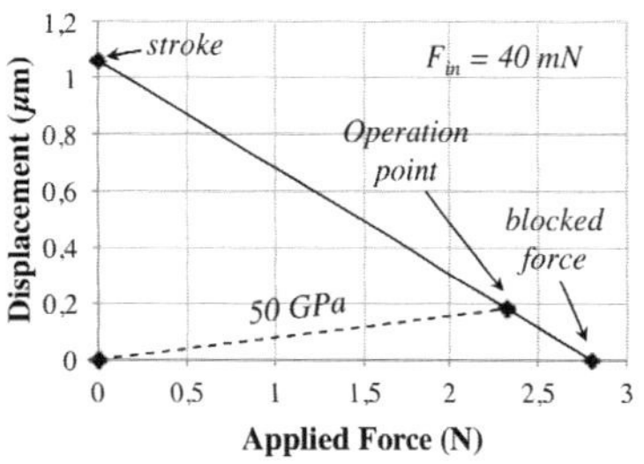

Figure 6. Load characteristics of the amplifier. The blocked force is 2.8 N and the stroke is 1 μm. With a piezoelectric element which stiffness is 50 GPa, the applied force is 2.3 N and corresponding displacement is 200 nm.

Journal of Physics: Conference Series **476** (2013) 012135 doi:10.1088/1742-6596/476/1/012135

3. Electromechanical modelling

Though the mechanical coupling of the structure is not optimized yet, the above-described piezoelectric scavenger has been fully characterized using FEM software with the aim of assessing the power that could be scavenged.

3.1. Monolithic piezoelectric bar

In this case, we assume a monolithic piezoelectric element mounted together with the amplifier i.e. there are only two electrodes located on both ends of the bar. With an input acceleration of 0.1g, the generated mean power transferred to a 1.1 GΩ optimal resistive load is 9 μW. The corresponding RMS voltage is 99 V (Figure 7). This high value of the output voltage comes from the piezoelectric element design, which induces a high electrical impedance.

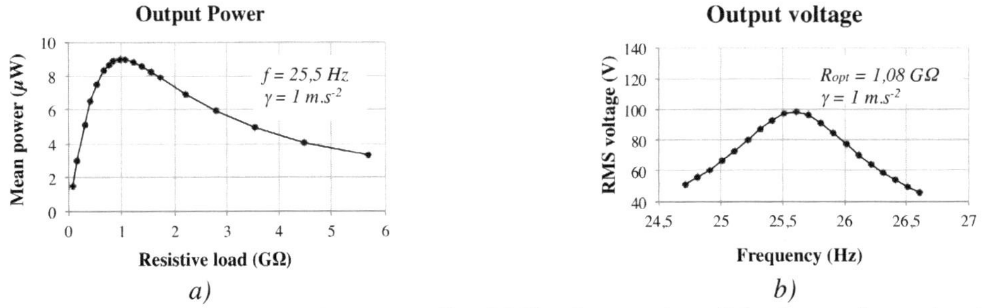

Figure 7. Scavenged mean power and corresponding RMS voltage under a 0.1 g harmonic acceleration. The device resonant frequency is about 25 Hz.

3.2. Multi-layered piezoelectric bar

In order to get a voltage that fit the requirements regarding the design of the AC/DC and DC/DC converter, we propose to use low-impedance configuration: a multilayered piezoelectric bar with the elementary piezoelectric elements connected in parallel (Figure 8 a). Because the amount of piezoelectric material and the applied force remain unchanged, this configuration is expected to return the same power as the monolithic configuration. Based on the fact that 3V DC voltage is usually required for powering rechargeable batteries for instance, we adapt the number of piezoelectric layers such as to set the output voltage to 3V. Using, for instance the PZT NCE51, this number of layers is set to 30 (Figure 8).

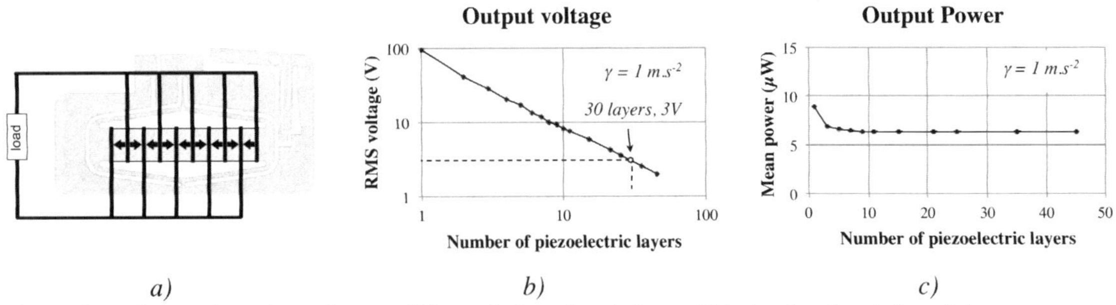

Figure 8. a) Connection scheme for a multilayered piezoelectric bar and b) simulated variation of the output voltage versus the number of layers. For instance, a 3V output voltage is obtained with 30 piezoelectric layers c) mean output power as a function of number of layers.

4. Conclusion

In this paper, we presented an original piezoelectric energy scavenger architecture that allows combination of elongation/compression excitation of a piezoelectric bar with operation at low frequency. This electromechanical transduction mode is known to be more efficient than the traditional {31} mode, which is commonly implemented in bending structures. Furthermore, it has been presented how this approach allows shifting the stress applied to the piezoelectric element in the compressive domain only using a pre-stress shim. Finally, it was shown that by setting the number of piezoelectric layers, the output voltage can be set independently to the output power.

Acknowledgements

This work was supported by the French OSEO project HBS in the framework of FUI 1009044V.

References

[1] Q.Q. Zhang et al., " Lead zirconate titanate films for d33 mode cantilever actuators", Sensor and Actuators A 105 (2003) pp. 91-97.

[2] P. Muralt et al., "Vibration energy harvesting with PZT micro device", Procedia Chemistry, Proceedings of the Eurosensors XXIII conference, vol.1, no.1, (2009), pp.1191-1194.

[3] S. Basrour et al., "Mechanical characterization of microgrippers realized by LIGA technique", International Conference on Solid-State Sensors and Actuators, Transducers '97 (1997), pp. 599-602.

[4] F. Claeyssen, R. Le Letty, N. Lhermet, "Actionneur piezoactif amplifié à raideur élevée", patent FR 95 12598 (1997).

[5] http://www.dynamic-structures.com .

Split-electrode piezoelectric scavengers for harvesting energy from torsional motions

Jie Mei[1], **Lijie Li**[1, 2]

[1]Multidisciplinary Nanotechnology Centre, College of Engineering, Swansea University, SA2 8PP, UK; [2]Wuhan University of Technology, P.R. China

E-mail: L.Li@swansea.ac.uk

Abstract. As most of cantilever scavengers are limited in working in bending mode, it motivates the idea to develop a device that can work with torsional motions. From the finite element analysis, it is found that the torsional modal frequency is near the bending mode frequency, which makes it possible to widen the working frequency range.By using the coupled piezoelectric-circuit finite element method, the harmonic performances of bulk and split electrode devices are quantitatively characterised. The maximum output voltages and power for split-electrode devices are calculated to be $26.7V$, $29.35V$ and $0.023W$, which present much more advantages over the bulk-electrode device($1.21V$ and $43.3\mu W$ correspondingly). In this analysis, it is deduced that the cancelling effect occurs across the surfaces of bulk electrodes. For further validation, transient analysis is conducted and it proved that average output power of split-electrode device is almost 25.35% higher than that in bulk electrode device.

1. Introduction

One common form of energy harvesting devices is the cantilever structure as it is easily matched to ambient vibration frequency for maximum output power. The coupled piezoelectric-circuit finite element model (CPC-FEM) in ANSYS software to develop a sandwiched bimorph cantilever structure was reported, which could calculate the output power directly[1]. A plucked piezoelectric bimorph for knee-joint energy harvesting has also been developed. [2] The device is able to vibrate at resonance even it is excited at much lower frequency by the frequency up-conversion mechanism. However, these devices are still limited in just working in bending mode. Cheng et al proposed a circular cylindrical ceramic shell for harvesting power from angular motions based on torsional mode.[3] Abdelkefi studied the energy harvesting mechanism from a tuned bending-torsion system by adjusting the asymmetry of the tip masses. [4] Recently, Cha et al. [5] analysed underwater energy harvesting from torsional excitation which utilized similar mechanism with our previous work[6].

A model of the split electrode energy harvester is built in this paper. As shown in figure 1, the steel substrate is sandwiched by PZT-5H piezo layers, with electrodes separated into two equal sizes on both top and bottom surfaces. One end of the cantilever beam is fixed to a rotator and the other end is bonded with a steel proof mass. When rotator starts rotating, the cantilever will be twisted by the inertial proof mass. Then the corresponding torsional deformation will generate opposite charges in the piezo layer surfaces according to the direct piezoelectric converting effect. Since the charges across the pieozoelectric layer surface are electrically seperated by the gap in the middle of the electrode, the cancelling effect could be avoided. As the torsional modes

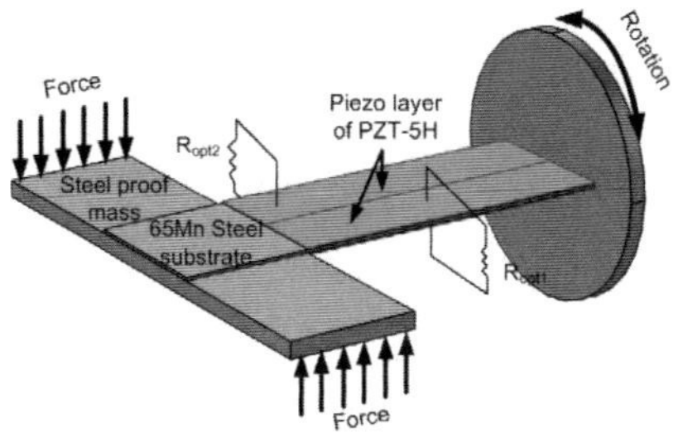

Figure 1. Schematic of split-electrode piezoelectric bimorph scavengers

normally happened at the higher working frequencies, this structure could make it possible to work in a wider frequency range by harvesting both bending and torsional mechanical energy.

2. Modelling

2.1. Modal analysis

In this paper, the presented model is developed in $ANSYS13.0$. The structure of the split-electrode energy harvesting device is made of substrate layer($length \times width \times thickness = 28 \times 9 \times 0.2mm$), piezoelectric layer ($length \times width \times thickness = 20 \times 9 \times 0.1mm$) and proof mass ($length \times width \times thickness = 8 \times 30 \times 1mm$) shown as Figure 1. The materials chosen for these three parts are 65Mn steel, PZT-5H and 45 steel. Then density, poisson ratio and Young's modulus of 65Mn steel are $7850kg/m^3$, 0.23 and $198.6GPa$ respectively. And those parameters of 45 steel are $7850kg/m^3$, 0.28 and $205GPa$. The density of PZT-5H is $7500kg/m^3$ The quality factor and dielectric loss are 80, and 0.014.The elasticity stiffness, piezoelectric stress coefficients and dielectric matrix for PZT-5H are shown as following,

$$[c^E_{ansys}] = \begin{bmatrix} 12.6 & 8.41 & 7.95 & 0 & 0 & 0 \\ 8.41 & 11.7 & 8.41 & 0 & 0 & 0 \\ 7.95 & 8.41 & 12.6 & 0 & 0 & 0 \\ 0 & 0 & 0 & 2.3 & 0 & 0 \\ 0 & 0 & 0 & 0 & 2.3 & 0 \\ 0 & 0 & 0 & 0 & 0 & 2.33 \end{bmatrix}, [e_{ansys}] = \begin{bmatrix} 0 & -6.5 & 0 \\ 0 & 23.3 & 0 \\ 0 & -6.5 & 0 \\ 17 & 0 & 0 \\ 0 & 0 & 17 \\ 0 & 0 & 0 \end{bmatrix}, [\varepsilon^S] = \begin{bmatrix} 3200 & 0 & 0 \\ 0 & 3200 & 0 \\ 0 & 0 & 3200 \end{bmatrix}.$$

In the analysis, 3-D 20-node coupled-field solid element 226 is used to model the piezoelectric layer, and corresponding structure element is solid 186. As previously illustrated, the steel substrate is sandwiched by two PZT-5H layers, which are set in opposite polarities. The bulk electrode layer is separately modelled using coupled degrees of freedom command. One end of the cantilever is fixed, and the other end is set free. In order to do coupled piezoelectric-circuit analysis, the external load resistance is modelled using resistor option of CIRCU94. Then the load resistor is serially connected to the corresponding top and bottom electrodes. The first and second resonant frequencies are calculated to be 105.051 Hz and 252.803 Hz respectively in the modal analysis. As the second resonant frequency is comparable to that of first bending mode, the torsional mode is supposed to be utilised for the environmental vibration energy harvesting. However, In order to prove the efficiency of harvesting energy in torsional mode, both of the bulk-electrode and split-electrode devices are compared in the following analysis.

2.2. Optimal load resistance

Since the load resistance is serially connected with top and bottom surfaces of the scavenger, it is necessary to determine the optimal load resistance for maximized harvesting energy according to equation (1),

$$R_{opt} = \frac{1}{\omega_n C_s} \tag{1}$$

Where R_{opt} is the optimal resistance, ω_n is the resonant frequency, C_s is the serial capacitance which could be obtained by the static analysis in ANSYS software by setting the voltage across the top and bottom surfaces of piezoelectric layer to be $1V$. Then the value of capacitance is equal to the value of surface charge. Table 1 is the calculated capacitance and optimal resistances corresponding to each mode.

Table 1. Calculated capacitance and optimal resistances corresponding to each mode of the bulk and split-electrode device.

Split 1($C_1 = 1.963 \times 10^{-8}F$)		Split 2($C_2 = 1.963 \times 10^{-8}F$)		Bulk($C_b = 3.945 \times 10^{-8}F$)	
$1^{st}mode$	$2^{nd}mode$	$1^{st}mode$	$2^{nd}mode$	$1^{st}mode$	$2^{nd}mode$
$77.176k\Omega$	$32.07k\Omega$	$77.174k\Omega$	$32.069k\Omega$	$38.392k\Omega$	$15.954k\Omega$

2.3. Harmonic analysis

Although the optimal resistance can be theoretically determined by equation (1), it is still necessary to conduct the harmonic analysis with different load resistances to validate the result. The load resistances are set to be $16.035k\Omega$, $32.07k\Omega$, $64.14k\Omega$ and open circuit respectively in the split electrode device. For bulk-electrode device, they are set to be $7.977k\Omega$, $15.954k\Omega$, $31.908k\Omega$ and open circuit. And the damping ratios of the two devices are both set to be 0.015625. Since the electrode is modeled using coupled degrees of freedom command, it will not change the resonant frequency for the split-electrode device. At the same time, in order to model the torsional mode in harmonic analysis, the proof mass is excited by external forces(with amplititude of $0.16N$) along the top surface edges in parallel but opposite directions that shown as Figure 1. Figure 2 is the displacement along the cantilever length at different position in width direction at resonant frequency. It is found in figure 2 that the deformation is symmetric about its central face(x, y plane in ANSYS software), which results in torsional deformation. And the maximum displacement is shown to be 1.29 mm at the corner points of the free end. Figures 3(a)- 3(c) show the output voltages versus frequencies at torsional mode. The peak voltages for the second mode occur at around $252.503Hz$. The output voltages across the load resistor increased with load resistance. Moreover, the output power has optimal values at $32.07K\Omega$ and $15.954k\Omega$ for the split and bulk-electrode devices respectively shown in Figures 4(a) and (b), which proves the results calculated by equation (1) to be right. In addition to this, the maximum output voltages and harvested power of split-electrode devices are calculated to be $26.7V$, $29.35V$ and $0.023W$, which are much higher than that of bulk one($1.21V$ and $43.3\mu W$ correspondingly), and therefore it demonstrate a clear cancelling effect as reported in reference [5] and [6].

2.4. Transient analysis

In real operation, as energy harvesting devices usually work in dynamic vibrations rather than harmonic vibrations, it is necessary to conduct transient analysis to characterise the performance

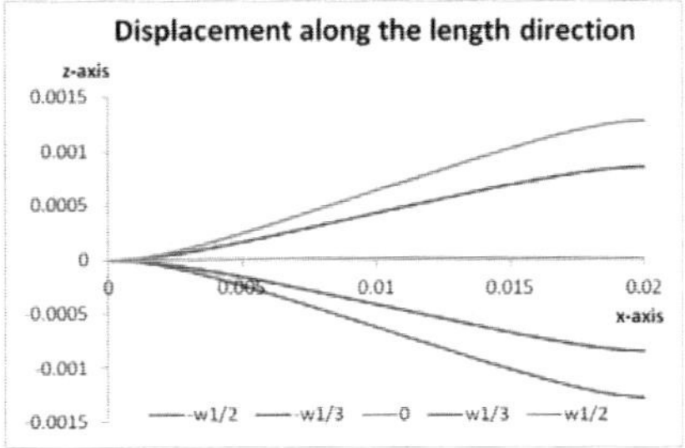

Figure 2. Simulated displacement of the cantilever under torsional loads. Coloured lines represent displacement on cross sectional lines along the length. These cross sectional lines are drown at different position in width direction.

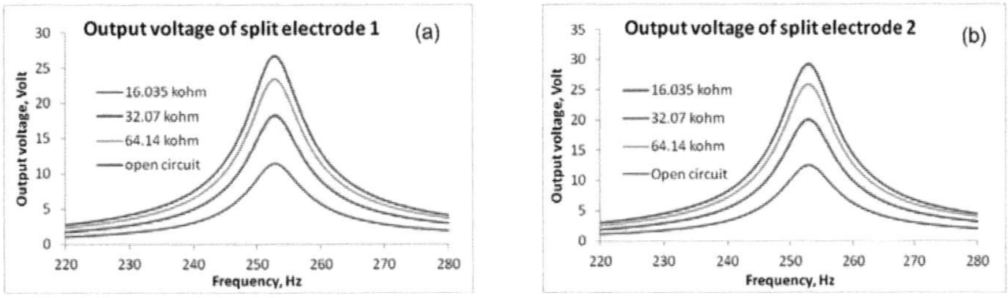

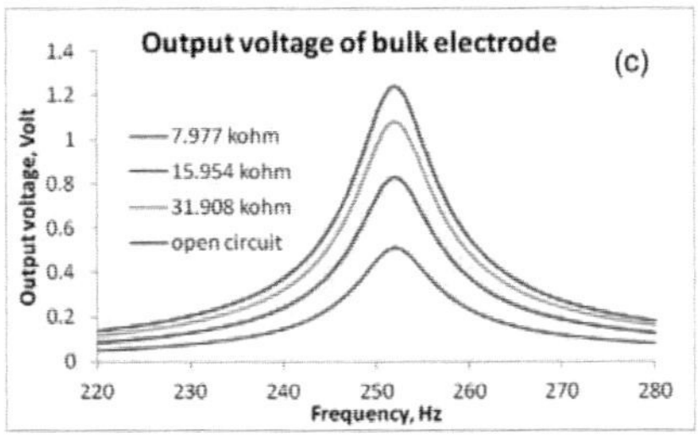

Figure 3. (a) output voltage at electrode 1 of the split-electrode device corresponding different load resistances; (b) output voltage at electrode 2 of the split-electrode device corresponding different load resistances; (c)output voltage of bulk electrode device.

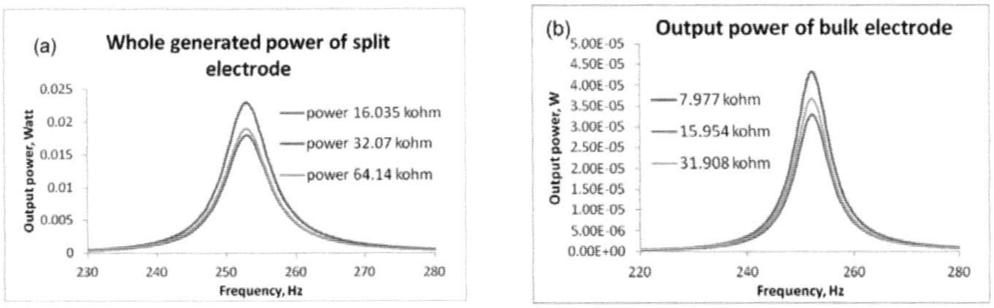

Figure 4. (a)Whole generated power of split-electrode piezoelectric bimorph scavengers; (b) Output power of bulk-electrode piezoelectric bimorph scavengers.

of energy harvesting devices in a loading cycle. For simplicity, the device is actuated by displacements rather than loading forces in the simulation. The displacement is set to be $1.6mm$ equivalent to the loading force($0.16N$) in opposite directions at two corners of the surface of piezo layers(shown in Figure 2). And the stiffness matrix multiplier is set to be 7.87×10^{-6}. There are three steps in the loading phases, which are ramp-type loading step with initial displacement being zero, loading step with constant deformation after phase one and release step followed by structural dynamic vibration. The time is set to be $0.3ms$ in phase one where the displacement is linearly increased from zero to be $1.6mm$, and in phase two the time is set to be $97ms$ where the displacement being $1.6mm$ constantly. After phase two, the external load is released and the structure freely vibrates until it reaches stable after $1\ s$. The displacement response at corner points of the top surface of free tip end of the piezo layer in the transient analysis is shown as Figure 5. Since the instantaneous power is calculated by, [2]

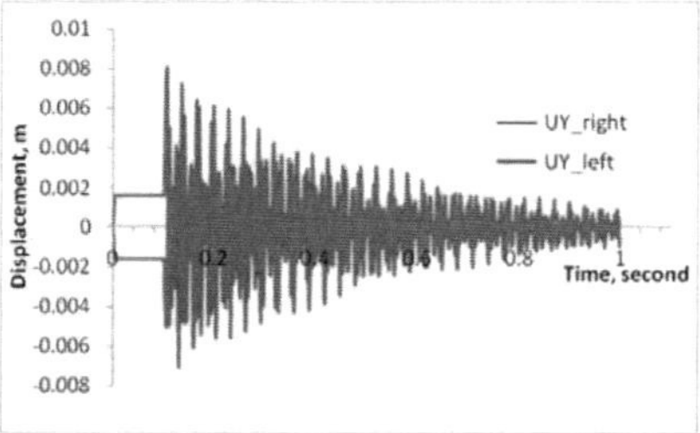

Figure 5. The displacement at corner of the free end of the cantilever in a transient analysis.

$$P_{(t_i)} = \frac{V^2(t_i)}{R} \tag{2}$$

Where $V(t_i)$ is the voltage across the resistor at the corresponding sub step, the instantaneous energy and average power are deduced as,

$$E(t_n) = \sum_{i=0}^{n} P(t_i)\Delta t_i \tag{3}$$

$$P_{av}(t_n) = \frac{1}{t_n} E(t_n) \tag{4}$$

where n is the number of time step, Δt_i is the duration of every time step, t_n is the length of time

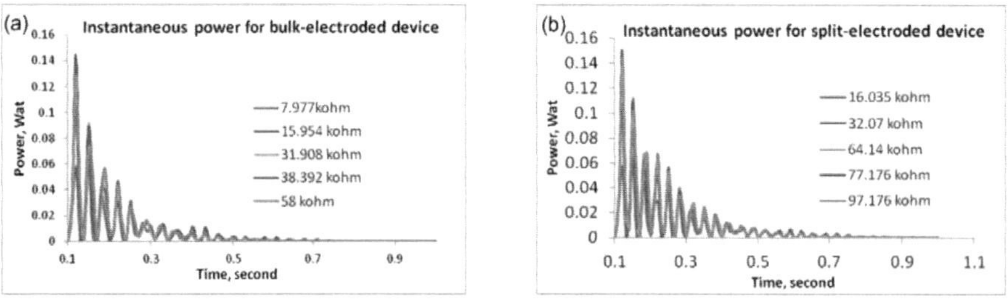

Figure 6. (a)Instantaneous power for bulk-electrode device; (b)Instantaneous power for split-electrode device.

when average power P_{av} produced. Since the generated energy is much higher in release step than that in loading step, it is only focused on the results in the release step. The instantaneous power for bulk-electrode device and split-electrode device are shown as Figures 6(a) and 6(b), where the ripples of the generated power curves are due to natural vibration modes in the release loading step. In Figure 7, the average power for the two kinds of devices are also presented, where the

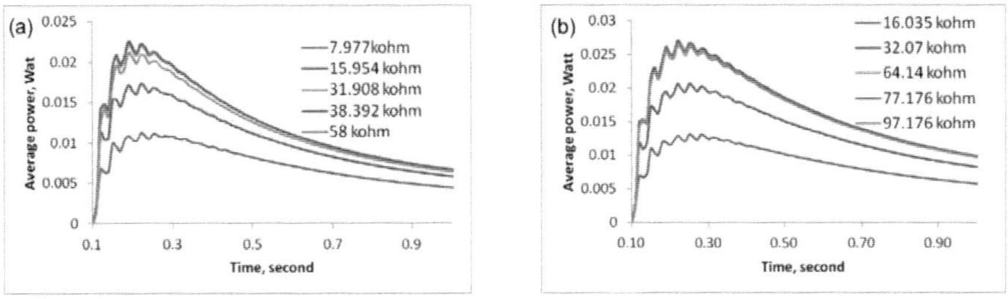

Figure 7. (a) Average output power of the bulk electrode device in the release step; (b) Average output power of the split-electrode device in the release step.

load resistances are are set to be $7.7977k\Omega$, $15.954k\Omega$, $31.908k\Omega$, $38.392k\Omega$ and $58k\Omega$ respectively for bulk-electrode device, and split-electrode device being $16.035k\Omega$, $32.07k\Omega$, $64.14k\Omega$, $77.176k\Omega$ and $97.176k\Omega$. It is found that optimal load resistances are $38.392k\Omega$ and $77.176k\Omega$ rather than $15.954k\Omega$ and $32.07k\Omega$ in the harmonic analysis step. The reason for this phenomenon is because the bending mode is still in a predominant situation in the release step. However, the torsional mode still plays an important role in the whole generated power in the split device, since the maximum average output power of split-electrode device is $0.0054W$ higher than bulk device, which is almost 25.35% higher than that in bulk electrode device. Furthermore, the maximum output average power occurred at $0.25s$ and it keeps stable around $9.85mW$ at $1s$. If the loading cycle is repeated, it could make the split-electrode scavenger much more efficient than bulk-electrode device.

3. Conclusion

Model of split-electrode piezoelectric bimorph scavengers for harvesting energy from torsional motions has been presented. The torsional modal frequency($252.803Hz$) is calculated to be comparable with the bending mode($105.051Hz$) based on the theoretical analysis. It is found that in the torsional mode, the harvested power for the optimal external load resistor($32.07Kohm$) is $0.023W$, which is much higher than that of bulk-electrode device. Moreover, the transient analysis is conducted to characterise performance of the split-electrode device. Harmonic and transient analysis for both split and bulk electrode have been conducted using ANSYS software. The simulation results have shown that the split electrode devices generate much more power bulk electrode devices for any type of vibrational excitations.

References

[1] M.L. Zhu, E. Worthington, and J. Njuguna, IEEE Trans. Ultroson. Ferroelectr. Freq. Control, 56(2009), pp. 1309-1318.
[2] M. Pozzi, and M.L. Zhu, Smart Mater. Struct., 20 (2011), pp. 055007(10pp).
[3] Z. Chen, Y.T. Hu and J.S. Yang, Applied Mathematics Mechanis(English Edition), 28 (2007), pp. 779-784.
[4] A. Abdelkefi, A.H. Nayfeh, M.R. Hajj and F. Najar, Smart Mater. Struct., 21(2012), pp. 075029(9pp).
[5] Y.S. Cha, L.F. Shen, M. Porfiri, Smart Mater. Struct., 22(2013), pp. 055027(13pp).
[6] Z.M. Liu, L.J. Li, 2011 Annual Conf. on IEEE Industrial Electronics Society., (2011), pp. 4059-4063.
[7] C.Q. Chen, X.M. Wang, and Y.P. Shen, Computer and Structures, 60(1996), pp. 505-512.
[8] S. Lee, B.D. Youn, IEEE Transaction Ultrasonics, Ferroelectrics, And Frequency Control, 58(2011), pp. 629-643.

On the experimental determination of the efficiency of piezoelectric impact-type energy harvesters using a rotational flywheel

P. Janphuang[1], R. Lockhart[1], S. Henein[2], D. Briand[1], and N.F. de Rooij[1]

École Polytechnique Fédérale de Lausanne (EPFL),
Institute of Microengineering (IMT),
[1]The Sensors, Actuators and Microsystems Laboratory (SAMLAB),
Rue Jaquet-Droz 1, Case postale 526, CH-2002 Neuchâtel, Switzerland
[2]Patek Philippe Chair in Micromechanical Design (INSTANT-LAB), Microcity, Rue de la Maladière 71b, 2000 Neuchâtel, Switzerland

E-mail: pattanaphong.janphuang@epfl.ch

Abstract. This paper demonstrates a novel methodology using a rotational flywheel to determine the energy conversion efficiency of the impact based piezoelectric energy harvesters. The influence of the impact speed and additional proof mass on the efficiency is presented here. In order to convert low frequency mechanical oscillations into usable electrical energy, a piezoelectric harvester is coupled to a rotating gear wheel driven by flywheel. The efficiency is determined from the ratio of the electrical energy generated by the harvester to the mechanical energy dissipated by the flywheel. The experimental results reveal that free vibrations of the harvester after plucking contribute significantly to the efficiency. The efficiency and output energy can be greatly improved by adding a proof mass to the harvester. Under certain conditions, the piezoelectric harvesters have an impact energy conversion efficiency of 1.2%.

1. Introduction

Converting low frequency vibrations (< 30 Hz) into usable electrical energy using an impact based approach, in which environmental motion is coupled to an inertial object through the physical impacts, has received significant interest. Several configurations used to couple the low frequency excitation into a piezoelectric transducer have been proposed and demonstrated [1-12]. An approach referred to as plucking (or impulse excitation) based frequency up-conversion has received significant interest as it is said to improve the electromechanical coupling and efficiency of energy harvester [7-10]. Using this approach, the piezoelectric cantilever is plucked by an inertial object, which is excited at low frequency before it is left to freely oscillate at its resonant frequency.

Theoretical attempts to determine the conversion efficiency of the impact-type harvesters have been proposed by calculating the mechanical work done on the harvester [7] or using the potential energy of the inertial object [4] as the input energy. Calculating the ratio of electrical damping to the total damping of the harvester can be also used to define the conversion efficiency [5, 11]. With these methods, however, the interaction energy (e.g. friction) between the harvester and the inertial mass is ignored.

We have previously presented a novel and compact approach for extracting the energy from a rotating gear using an AFM-like piezoelectric MEMS harvester [12]. Here, we focus on experimentally evaluating the efficiency of this approach by introducing a flywheel to quantify the mechanical input energy that is provided to the harvester, including the interaction energy between the cantilever and the inertial mass. With this method, the configuration efficiency of the impact-type harvester can be studied in terms of the plucking mechanism, the coupling geometry between the harvester and the gear, the impact speed, and the harvester stiffness.

2. Theory

A flywheel is an inertial energy storage device consisting of a heavy wheel (mass) mounted on a rotating shaft (figure 1(a)) [13]. Kinetic energy is transferred to and stored in the mass as rotational energy, where the magnitude depends on the inertia and the speed of the rotation (as indicated in Eq. 1). Flywheels are typically designed to have a significant moment of inertia to allow it to resist changes to the rotational speed (angular velocity). By applying a torque, its rotational speed is increased and kinetic energy is transferred and stored in the flywheel. As the flywheel turns, the energy stored in the mass is dissipated due to friction from the air and the bearing, as well as the transfer of energy to a coupled mechanical load, thereby reducing its rotational speed.

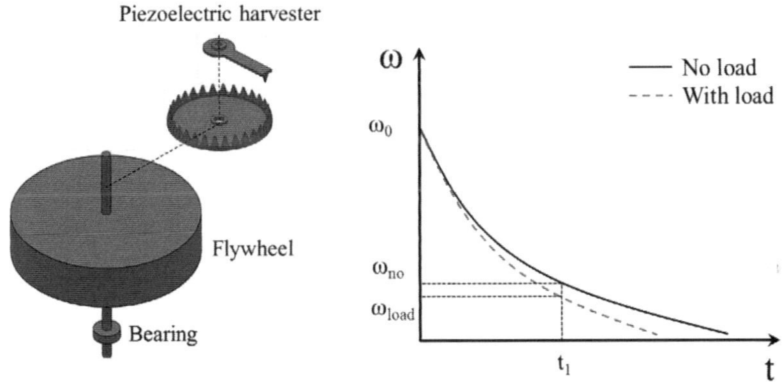

(a) (b)

Figure 1. (a) Schematic diagram of the proposed concept (b) Deceleration of the rotational speed of the flywheel as a function of time for different loading conditions.

The concept proposed for measuring the efficiency of impact energy harvesters is illustrated in figure 1(a). A piezoelectric cantilever is placed directly above a rotating crown gear coupled to the rotating flywheel that is suspended by bearings. The tip at the end of the cantilever extends down between the gear teeth and is plucked as each tooth passes. The gear, which is driven by the flywheel, rotates in-plane. Exploiting the cyclical nature of the flywheel, this method can be used to determine the mechanical energy that is dissipated at each impact. The kinetic energy stored in the flywheel (E_k) can be calculated by

$$E_k = \frac{1}{2} I \omega^2 \quad (1)$$

where I is the moment of inertia, and ω is the angular frequency of the flywheel. The moment of inertia of a solid cylinder is given by

$$I = \frac{1}{2}mr^2 \tag{2}$$

where m is the mass and r is the radius of the cylinder.

Figure 1(b) presents an example of the flywheel operation over time. The flywheel starts with an angular velocity of ω_0. If the flywheel is not subjected to mechanical loading, the reduction of its angular velocity is mainly due to the friction losses from air and the bearing. The change in kinetic energy of a system at a time, t_1, is given as

$$E_1 = \frac{1}{2}I(\omega_0^2 - \omega_{no}^2) \quad \text{(no load condition)} \tag{3}$$

$$E_2 = \frac{1}{2}I(\omega_0^2 - \omega_{load}^2) \quad \text{(load condition)} \tag{4}$$

Therefore, the mechanical input energy can be calculated by comparing the rate of deceleration in each case,

$$E_2 - E_1 = \frac{1}{2}I(\omega_{no}^2 - \omega_{load}^2) \tag{5}$$

Using expression in (5), the mechanical input energy used deflect the harvester in the range of rotational speeds of interest can be quantified.

3. Results and discussion

The experimental setup consists of an aluminium flywheel with a diameter of 50 mm and a thickness of 10 mm. The flywheel is coupled to a crown gear with a single tooth as illustrated in figure 2. An AFM-like MEMS piezoelectric cantilever, consisting of 80 μm-thick PZT layer bonded to a 150 μm-thick silicon layer, is placed above the rotating gear such that the cantilever is plucked once per rotation. More details on the harvester dimension and fabrication process can be found in [12]. A mass weighing approximately 8 mg is added to the top of the cantilever to evaluate its effect on the performance of the harvester (figure 2(c)).

(a) (b) (c)

Figure 2. (a) The efficiency measurement setup showing a harvester above a crown gear which is connected to the flywheel. The laser and optical detector for detecting the velocity of the flywheel are also visible. (b) A closer view of a harvester positioned above the crown gear with a single tooth. (c) The cantilever tip / gear tooth coupling for mechanical plucking. The 8 mg proof mass is also visible.

The flywheel was charged using an electric motor to an angular frequency of 125 rad/s (~20 rps). The velocity of the flywheel was monitored by optical detection. The overlap between the tip of cantilever and the gear tooth was controlled using a micrometric stage. In the experiment, harvesters with and without a proof mass were deflected 30 μm by adjusting the tip depth into the gear wheel in order to compare the performances and efficiency of both devices. The displacement of the cantilevers was observed by the laser Doppler vibrometer (LDV).

Figure 3(a) illustrates the evolution of the angular velocity of the flywheel for different loading conditions. By adding a proof mass to the cantilever, the mechanical load on the system is increased leading to larger frictional force. Thus, a greater portion of the rotational energy stored in the flywheel is required to displace the cantilever causing the angular frequency to drop faster than when a cantilever without mass is used. However, as a result of the proof mass, more energy per pluck is transferred to the load resistance due to a decrease in the mechanical damping as illustrated in figure 3(c). The energy dissipated in the load resistance can be calculated using the following expression

$$U_E=\sum P\Delta t=\sum \left(\frac{V^2}{R}\right)\Delta t \tag{6}$$

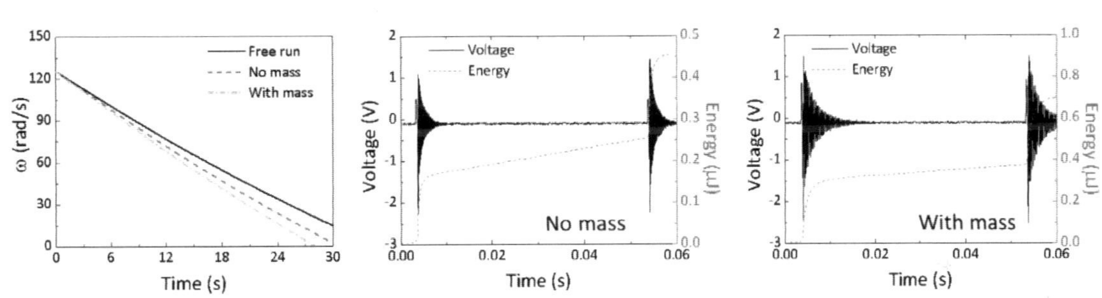

(a) (b) (c)

Figure 3. (a) Angular velocity of the flywheel as a function of time. Output voltage and energy dissipated in the respective optimal load at an angular velocity of 125 rad/s (~20 rps) (b) without a proof mass, and (c) with a proof mass.

Figure 3(b) shows the output voltage and the energy generated from the harvester without a proof mass at its optimal load resistance (R_L) of 5.6 kΩ. The harvester produces 0.25 μJ of electric energy per pluck at a rotational speed of 125 rad/s. After adding the proof mass to the free end of the cantilever, the resonant frequency of the harvester was reduced from 6.5 to 3.2 kHz. With the proof mass, the harvester is capable of producing an output energy of 0.4 μJ at an optimal load of 7.8 kΩ. The electrical output energy increases by more than 70 % with the addition of an 8 mg proof mass. The optimal load (R_L) also increases as a result of a decrease in the resonant frequency since $R_L=1/\omega C_p$ where ω is the resonant frequency of the harvester and C_p is the capacitance of the piezoelectric layer.

By comparing the mechanical input energy to the electrical energy generated by the harvester, the efficiency of the harvester as a function of load resistance, and rotational speed are illustrated in figure 4. The efficiency was improved by adding a proof mass to the cantilever since the increase in the electrical output energy is greater than additional mechanical input energy required to deflect the proof mass as shown in figure 4(a). The efficiency was measured to be 1.2 % for rotational speeds between 113 to 125 rad/s. However, the efficiency is reduced as the gear speed falls as shown in figure 4(b). At slow rotational velocities, the harvester is not able to freely oscillate since the tip of the cantilever simply follows the contour of the gear tooth. This occurs when the movement of the gear tooth is slower than the natural frequency of the harvester. In this situation, less energy is generated by the harvester, and therefore, the efficiency of the system is reduced.

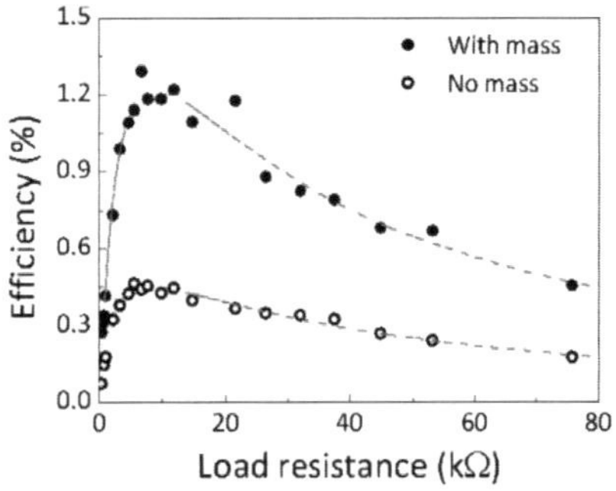

(a)

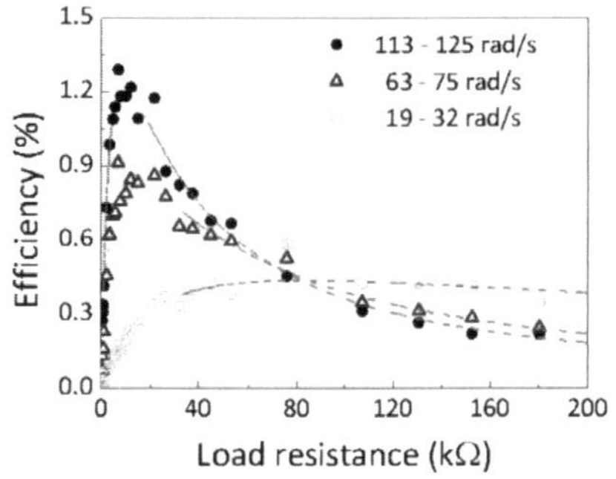

(b)

Figure 4. Measured efficiency (a) comparing cantilevers with and without a proof mass, and (b) as a function of rotational speed.

4. Conclusion

The experimental setup presented here is a novel approach for evaluating the efficiency of impact harvesters by implementing a rotational flywheel to quantify the mechanical input energy. Using this method, the interaction forces between the harvesters and the inertial mass, including friction, for example, are accounted for. The experimental results indicate that the efficiency can be improved by minimizing the mechanical damping ratio while increasing the rotational speed. Experiments are currently being conducted with this novel setup using various harvester configurations in order to optimize the efficiency of our harvesting system.

References

[1] Umeda M, Nakmura K and Uesha s 1996 *Japan. J. Appl. Phys.* **35** 3267-73
[2] Priya S 2005 *Appl. Phys. Lett.* **87** 184101
[3] Cavallier B, Nouira H, Foltete E, Hirsinger L and Ballandras S 2005 *Proc. Symp. On Design, Test, Integration and Packaging of MEMS/MOEMS DTIP05 (Montreux, Switzerland)* pp 391-3
[4] Renaud M, Fiorini P, van Schaijk R and van Hoof C 2009 *Smart. Mater. Struct.* **18** 035001
[5] Gu L 2011 *Microelectronics Jouranal* **42** 277-82
[6] Gu L and Livermore C 2011 *Smart. Mater. Struct.* **20** 045004
[7] Pozzi M and Zhu M 2011 *Smart. Mater. Struct.* **20** 055007
[8] Pozzi M, Aung M S H, Zhu M, Jones R K and Goulermas J Y 2012 *Smart. Mater. Struct.* **21** 075023
[9] Pillatch P, Yeatman E M and Holmes A S 2011 *Proc. 11thInt. Workshop on Micro and Nanotechnology for Power Generation and Energy Conversion Applications, PowerMEMS 2011 (Seoul, Republic of Korea)* pp. 245-48
[10] Pillatch P, Yeatman E M and Holmes A S 2013 *Proc. 17thInt. Conf. on Solid-State Sensors and Actuators, Transducers 2013 (Barcelona, Spain)* pp. 1364-67
[11] Shu Y C and Lien I C 2006 *J. Micromech. Microeng.* **16** 2429-38
[12] Janphuang P, Isarakorn D, Briand D and de Rooij N F 2011 *Proc. 16thInt. Conf. on Solid-State Sensors and Actuators, Transducers 2013 (Beijing, China)* pp. 735-38
[13] Bolund B, Bernhoff H and Leijon M 2007 *Renewable and Sustainable Energy Reviews* **11** 235-58

Elastocaloric cooling using shape memory alloy films

H Ossmer[1], C Chluba[2], B Krevet[1], E Quandt[2], M Rohde[3] and M Kohl[1]

[1] Karlsruhe Institute of Technology, IMT, P.O. Box 3640, 76021 Karlsruhe, Germany
[2] Christian Albrecht University Kiel, Institute for Material Science, 24143 Kiel, Germany
[3] Karlsruhe Institute of Technology, IAM, P.O. Box 3640, 76021 Karlsruhe, Germany

E-mail: hinnerk.ossmer@kit.edu

Abstract. The elastocaloric effect in magnetron-sputtered $Ni_{50.4}Ti_{49.6}$ films of 20 µm thickness is studied by means of uniaxial tensile tests and infrared thermography. For the investigated films, the usable quantity of latent heat is about 7.2 J/g. When relieving the stress after tensile loading and subsequent temperature equalization at strain rates larger than dε/dt = 0.2 s^{-1}, a maximum temperature change of ΔT = -16 K is observed as expected for adiabatic conditions. Compared to bulk specimens, the heat transfer times are reduced to about 850 ms due to the larger surface-to-volume ratio, which is attractive for rapid cooling.

1. Introduction

Shape memory alloys (SMA) absorb a large quantity of latent heat during reverse martensitic transformation [1]. Compared to magnetocaloric materials that have already been in the focus of research for several years, elastocaloric materials provide higher adiabatic temperature changes and do not require a large magnetic field [2,3]. Therefore, solid-state cooling by pseudoelastic SMAs is receiving large interest in recent years.

Pseudoelastic SMAs like the binary NiTi alloy undergo a diffusionless first-order phase transformation from austenite (A) to martensite (M) phase [4]. The stress-induced martensite (SIM) transformation is exothermic resulting in a temperature increase. When releasing a stressed sample after temperature equalization with the environment, the endothermic reverse transformation takes place resulting in a temperature decrease below ambient temperature. The theoretical maximum temperature change $\Delta T_{ad} = L/c_p$ that can be obtained under adiabatic conditions, depends only on the latent heat L of the transformation and the material's heat capacity c_p. In the case of NiTi, values of latent heat reported in literature show a large scatter with maximum values as high as 35 J/g that strongly depend on stoichiometry, carbon and oxygen impurity concentrations [5]. From an engineering point of view, impurity concentrations have to be kept low to avoid brittleness. Chemical compositions have to be adjusted so that the transition temperatures meet the requirements of the application. Thermal treatments and functional fatigue need also to be considered as they significantly affect the latent heat [6]. Due to these reasons, values of latent heat being accessible for elastocaloric cooling applications may be considerably smaller.

Until now, research concerning the elastocaloric effect is mainly focused on bulk materials. Tensile test experiments on Cu-based alloys reveal a maximum temperature decrease during reverse martensitic transformation of 15 K [7]. A maximum temperature decrease by 17 K has been reported for a NiTi wire with a diameter of 3 mm [8]. The highest value of temperature decrease by about 28 K has been observed for NiTi belt-type specimens [9]. However, due to the rather large dimensions of the investigated specimens, heat transfer times are well above 10 s. Considerable improvements are expected for pseudoelastic SMA films as their large surface-to-volume ratio may enhance heat transfer and thus, enables the increase of cycling frequencies and cooling power. Therefore, it is advantageous to combine many SMA films to distributed film arrays instead of using a single bulk specimen for elastocaloric cooling.

This work presents a series of tensile tests and in-situ infrared (IR) thermography experiments on magnetron-sputtered NiTi films of 20 µm thickness to investigate the spatially and time-resolved course of phase transformation, the heat transfer dynamics, as well as the corresponding maximum temperature changes.

2. Experimental

NiTi films are produced by dc magnetron sputtering onto a cooled glass substrate with a thin Cu layer at a power of 300 W and Ar atmosphere of 10^{-7} mbar. The deposition rate is 3.8 nm s^{-1}. Details on film deposition and structuring can be found in [10]. Investigated films have a thickness of 20 µm. For tensile testing, the films are structured to stripes of 1.75 mm width and 35 mm length by optical lithography and sacrificial layer etching. Since as-deposited films are amorphous, a heat treatment procedure is established that comprises two-step rapid thermal annealing at 650 °C for 10 min and at 450 °C for another 10 min.

The transition temperatures of the material are determined from electrical resistance measurements as shown in figure 1. The material exhibits a two-stage phase transformation from austenite to martensite via an intermediate R-phase. As the austenite finish temperature is below 20 °C, the material is pseudoelastic at room temperature. The specific heat capacity for austenite is determined to be 0.45 $Jg^{-1}K^{-1}$ by differential scanning calorimetry (DSC) using a synthetic sapphire standard (results not shown here).

A schematic setup of the experiment is depicted in figure 2. Uniaxial tensile tests (Zwick/Roell, Germany) are carried out with a 50 N force sensor. Samples are bonded to glass plates on both ends with a two-component adhesive. The plates are fixed by screws in the clamping of the tensile test machine. The reference clamping length is 20 mm for each sample. Tensile tests are performed at different strain rates between 10^{-4} s^{-1} and 20 s^{-1} in strain-control mode. A pre-load of 1 N is applied in order to avoid buckling of the NiTi film. The sample elongation ('engineering strain' ε) is periodically varied between 0% and 5%. At the end positions of each semi-cycle at $\varepsilon = 0$ and 5%, the strain is held constant for 10 s in order to let the sample temperature equilibrate with the environment. All experiments are conducted at room temperature in air. Before recording experimental data, ten cycles are performed at low strain rate in order to obtain reproducible results.

During tensile tests, samples are monitored by an IR camera (FLIR, USA) with a spatial resolution of 25 µm to obtain time-resolved local temperature distributions. Since sputtered NiTi samples have a very flat and reflective surface, they are covered with a thin layer of graphite spray to increase emissivity.

3. Results and Discussion

A typical stress-strain curve for a pseudoelastic film sample is shown in figure 3. At small strain rate of 10^{-4} s^{-1} corresponding to quasi-stationary conditions, the stress-strain characteristics exhibit pronounced stress plateaus during forward and reverse transformation. In this case, heat exchange with the environment is faster than the release of latent heat. Stress-induced martensitic transformation starts at a critical strain of 1.3% and proceeds at a constant stress level of about 500 MPa. Upon unloading, the martensitic reverse transformation proceeds at a lower stress level of about 350 MPa resulting in a pronounced hysteresis.

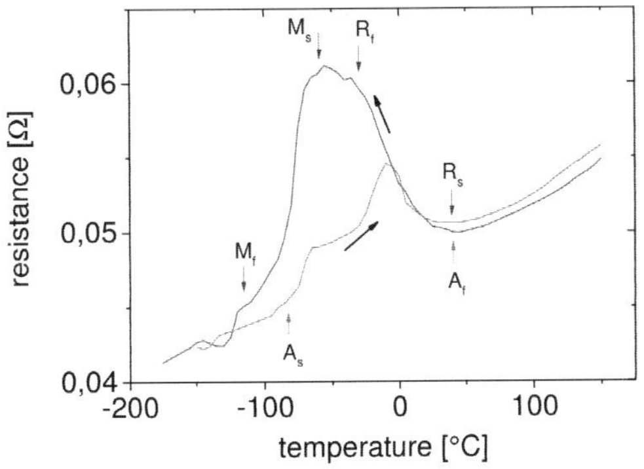

Figure 1. Electrical resistance characteristics upon cooling and heating in the temperature range of phase transformation. The start (s) and finish (f) temperatures of phase transformations are indicated by arrows. Legend: M – martensite; A – austenite; R – rhombohedral phase.

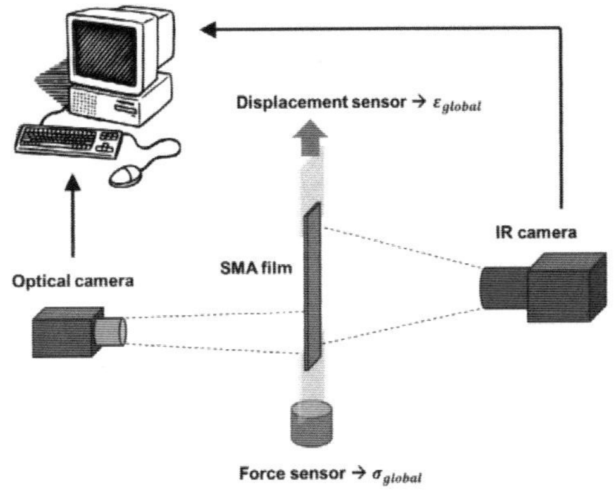

Figure 2. Schematic of tensile test setup for measurement of engineering stress and strain. An infrared camera monitors the evolution of local temperature distribution.

With increasing strain rate, the stress-strain characteristics become steeper and the onset of the transformation is less sharp. Whilst part of the work applied during mechanical loading can be regained during unloading, the area enclosed by the stress-strain-curve of a full load-unload cycle corresponds to work which is dissipated. Irrespective of the direction of the martensitic transformation, dissipation always leads to a heating and therefore counteracts elastocaloric cooling. Based on the stress-strain characteristics recorded at different strain rates, we determine the work per unit volume for loading and unloading by numerical integration. The difference $\Delta W = W_{load} - W_{unload}$, that is plotted versus strain rate in figure 4, corresponds to the total dissipated work. A pronounced maximum is observed at a strain rate of $d\varepsilon/dt = 0.2\ s^{-1}$.

The shape of the stress-strain characteristics is also affected by functional fatigue, which is discussed elsewhere [6].

During forward and reverse martensitic transformation, the release and absorption of latent heat leads to local heating and cooling of the sample, respectively. Figure 5 shows two series of time-dependent IR profiles recorded during loading and unloading at a strain rate of $0.1\ s^{-1}$. Temperature fronts start from the clamping, where the film is interconnected to the sample holder that remains at room temperature. Heat conduction causes broadening of the heated regions. Lüders-like bands are observed similar to bulk specimens [9], but are clearly visible only in the absence of graphite coating, which causes blurring due to rapid heat conduction. Upon loading, the temperature rises to a maximum in a time interval that is proportional to the load rate and then decays due to heat transfer to the environment by air convection and heat conduction via the clamping. After stationary conditions are reached, unloading results in a strong undercooling below ambient temperature, followed by a more gradual increase of temperature due to heat transfer.

Figure 6 shows a typical time-resolved temperature characteristic in the center of the sample during reverse martensitic transformation at a large strain rate of $1.0\ s^{-1}$. Temperature equilibration with the environment can be described by a single exponential decay. The corresponding time constant is determined by least-squares fit to be about 850 ms. The maximum temperature decrease is about $\Delta T = -16$ K. This result is in line with an estimate of the adiabatic temperature change according to $\Delta T_{ad} = L/c_p$ using a latent heat of about 7.2 Jg^{-1} and a specific heat capacity c_p of 0.45 $Jg^{-1}K^{-1}$. The value of 7.2 Jg^{-1} is lower than typical values for a complete martensitic transformation in NiTi alloys indicating that only a part of the two-stage transformation is accessible in the present case. Further adjustment of phase transformation temperatures by tailoring of stoichiometry and thermo-mechanical treatment could be a means to increase the accessible latent heat and thus to further raise the temperature decrease.

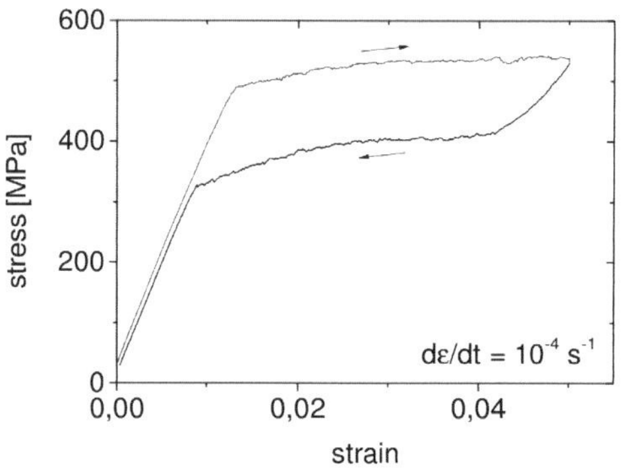

Figure 3. Stress-strain characteristics upon loading and unloading of a NiTi sample for a strain rate of 10^{-4} s^{-1}.

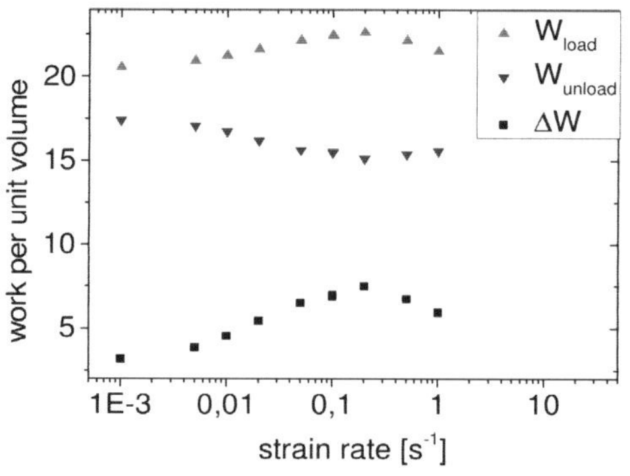

Figure 4. Strain rate dependence of mechanical work during loading and unloading and corresponding mechanical work difference W_{load} - W_{unload} of a complete loading cycle.

The cooling efficiency is described by the coefficient of performance (*COP*), given by the ratio of absorbed heat and mechanical work. Optimum results are obtained by assuming that the unloading work is recovered in a closed cooling cycle. Based on our results for the total dissipated mechanical work during loading and unloading at a strain rate of 1 s^{-1}, $\Delta W = 6$ MPa = 0.93 Jg^{-1} (see figure 4) and the latent heat estimated from the nearly-adiabatic temperature change, $L = 7.2$ Jg^{-1}, we obtain a *COP* of 7.7 in this case. Theoretical estimates on the available COP of NiTi films indicate that this value could be further increased by a factor of two [11].

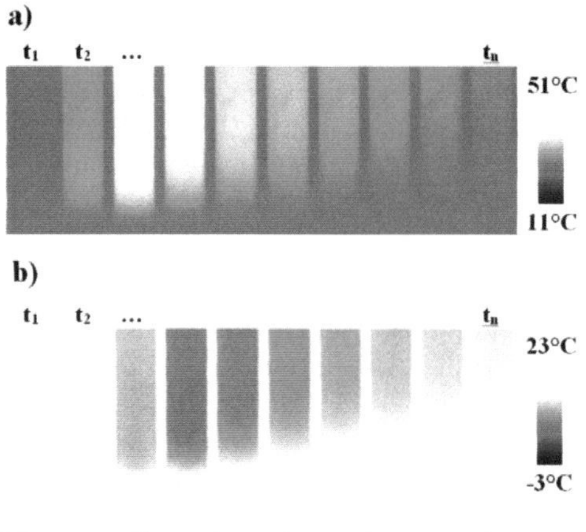

Figure 5. Time-dependent series of thermograms of a NiTi tensile test sample during loading (a) and unloading (b) at a strain rate of 0.1 s^{-1}. The time between two images is $\Delta t = 0.28$ s. The lower edge corresponds to the clamping.

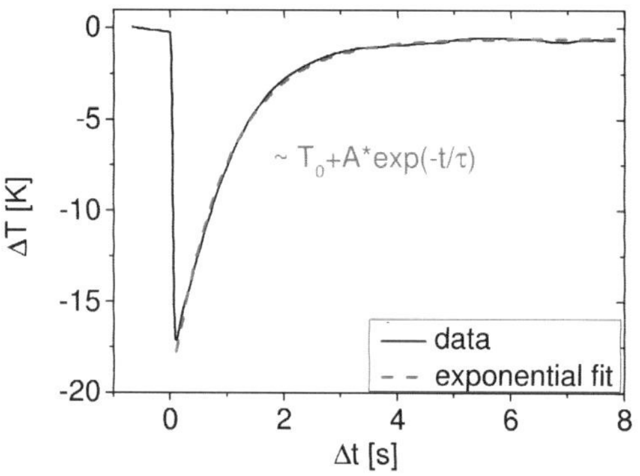

Figure 6. Temperature evolution of a reference point in the center of a NiTi tensile test sample. The sample is released after pre-straining by 5% at a strain rate of 1 s^{-1}. After temperature decrease ΔT of -16 K, the temperature exponentially increases with a time constant τ of about 850 ms.

4. Conclusion

In this paper, we investigate the elastocaloric effect in magnetron-sputtered $Ni_{50.4}Ti_{49.6}$ film samples of 20 μm thickness showing pseudoelastic stress-strain behavior at room temperature. Because of the small cross-section of the test samples with respect to their surface area, localized phenomena like the propagation of Lüders-like bands play a less pronounced role than in macroscopic wire and belt-like samples. The observed temperature fronts have a similar spatial extent as the sample itself.

While bulk samples show heating times in the order of 10 s [8], thermal equilibration within the film sample occurs within milliseconds. Hence, the cooling time is essentially determined by the strain rate. Adiabatic conditions are reached at strain rates larger than 0.2 s^{-1}. Rapid heat exchange with the environment occurs by air convection and heat conduction through the clamping within about 850 ms. These results are important for the engineering of elastocaloric cooling devices based on film samples. In this case, the film would be brought into physical contact with either a solid heat source or a liquid heat transfer medium. Both methods of heat exchange are expected to have shorter time constants than air convection, presumably enabling cycling rates of several Hertz.

We observe a maximum temperature decrease ΔT of -16 K during the reverse martensitic transformation at a strain release rate of 1 s^{-1}. This result is in line with reported values for NiTi wires. However, only a reduced quantity of latent heat of about 7.2 J/g has been accessible in the investigated test samples at room temperature. Therefore, we expect that the effect size can be further increased by tailoring the material properties. Regarding the maximum possible latent heat observed for NiTi-base alloys in the order of 20 J/g, an increase of effect size by a factor of two seems to be possible. The large effect size and rapid heat transfer being accessible in tailored NiTi-base alloys are highly attractive for solid state cooling and heat pumping applications.

5. Acknowledgements

The authors gratefully acknowledge funding by the Deutsche Forschungsgemeinschaft (DFG) within the priority program SPP1599 (www.ferroiccooling.de).

References

[1] Bonnot E, Romero R, Mañosa L, Vives E and Planes A 2008 *Phys. Rev. Lett.* **100** 125901
[2] Fähler S, Rößler UK, Kastner O, Eckert J, Eggeler G, Emmerich H, Entel P, Müller S, Quandt E and Albe K 2012 *Adv. Eng. Mater.* **14** 10-19
[3] Mañosa L, Planes A and Acet M 2013 *J. Mater. Chem.* A **1** 4925-4936
[4] Otsuka K and Wayman CM (ed) 1998 *Shape Memory Materials* (Cambridge, UK: Cambridge University Press)
[5] Otubo J, Rigo OD, Coelho AA, Neto CM and Mei PR 2008, *Mater. Sci. Eng.* A **481-482** 639-642
[6] Bechtold C, Chluba C, Lima de Miranda R and Quandt E 2012 *Appl. Phys. Lett.* **101** 091903
[7] Mañosa L, Planes A, Vives E, Bonnot E and Romero R 2009 *Funct. Mat. Lett.* **2** 73-78
[8] Cui J, Wu Y, Muehlbauer J, Hwang Y, Radermacher R, Fackler S, Wuttig M and Takeuchi I 2012 *Appl. Phys. Lett.* **101** 073904
[9] Pieczyska EA, Gadaj SP, Nowacki WK and Tobushi H 2006 *Exp. Mech.* **46** 531-542
[10] Lima de Miranda R, Zamponi C and Quandt E 2013 *Adv. Eng. Mater.* **15** 66-69
[11] Krevet B, Pinneker V, Rhode M, Bechthold C, Quandt E and Kohl M 2013 *Mater. Sci. Forum* **738-739** 287-291

Soft ferrite cores characterization for integrated micro-inductors

Yen Mai Nguyen[1,3], Thomas Lopez[1], Jean-Pierre Laur[1,3], David Bourrier[1,3], Samuel Charlot[1,3], Zarel Valdez-Nava[2,3], Vincent Bley[2,3], Céline Combettes[2,3], Magali Brunet[1,3]
[1]CNRS, LAAS, 7 avenue colonel Roche, F-31400 Toulouse, France
[2]CNRS, LAPLACE, 118 route de Narbonne, F-31062 Toulouse, France
[3]Université de Toulouse; UPS, INPT; LAAS, LAPLACE, Toulouse, France

Email: tymnguye@laas.fr

Abstract. Ferrite-based micro-inductors are proposed for hybrid integration on silicon for low-power medium frequency DC-DC converters. Due to their small coercive field and their high resistivity, soft ferrites are good candidates for a magnetic core working at moderate frequencies in the range of 5-10 MHz. We have studied several soft ferrites including commercial ferrite film and U70 and U200 homemade ferrites. The inductors are fabricated at wafer level using micromachining and assembling techniques. The proposed process is based on a sintered ferrite core placed in between thick electroplated copper windings. The low profile ferrite cores of 1.2 × 2.6 × 0.2 mm^3 are produced by two methods from green tape-casted films and ferrite powder. This paper presents the magnetic characterization of the sintered ferrite films cut and printed in rectangular shape and sintered at different temperatures. The comparison is made in order to find out the best material for the core that can reach the required inductance (470 nH at 6 MHz) under 0.6A current DC bias and that generate the smallest losses. An inductance density of 285 nH/ mm^2 up to 6 MHz was obtained for ESL 40011 cores that is much higher than the previously reported devices. The small size of our devices is also a prominent point.

1. Introduction

Nowadays, the trend toward miniaturization of mobile electronic products, with the demand of improving functionality and performance confronts challenges on power management. The effort to fabricate more compact electronic devices, having more features will require higher power density systems such as the DC-DC converters. Up to now, the control circuits and the power semiconductor devices can be fully integrated but the passive components especially inductors and transformers, which are bulky and off-chip components, are still an obstacle for further reducing the size of DC-DC converters. Plenty of researches have been carried out to achieve inductor integration on-chip or in-package with different technologies like sputtering, electroplating and screen printing. Some demonstrators of air-core inductors, thin film magnetic inductors, and ferrite based ones have shown that each method has its own merit and difficulties [1-3].

The approaches to the fabrication of micro-inductors involving sputter deposition and electro-deposition of magnetic cores have been demonstrated over last thirty years. Sputtering can be used to deposit a wide range of magnetic materials including alloys and oxides with high resistivity which is useful in limiting eddy current losses such as Co-Zr based, Co-Hf-Ta-Pd, Co-Fe-Hf-O, $Ni_{80}Fe_{20}$, Fe-B-N, Fe-Co-B-C amorphous thin films. That shows promising characteristics at frequency from several MHz to several GHz [4]. However, the problem arises when the film is thicker than several micro-meters due to the process of etching thick metallic films. Dry etching of thick films usually involves long etch time resulting in heating of the substrate and degradation of the magnetic film properties while wet etch of thick films results in severe undercutting. Electroplating is a less costly method than sputtering and more suitable for micro-inductor cores with large cross-sectional areas. The films deposited by electroplating have the thickness of several micro-meters to several tens of micro-meter including alloys of NiFe with different composition such as $Ni_{80}Fe_{20}$, $Ni_{45}Fe_{55}$, $Ni_{50}Fe_{50}$ and others like CoFeNiC, CoPFe, CoNiFe, CoFeCu, NiFeMo [5-7]. A problem is that only conductive materials can be electroplated; hence, eddy currents can be considerable and should be controlled by lamination, especially for thick films. Screen printing of ferrite materials or magnetic composites is an alternative approach compared to the two above-mentioned methods. It has strong advantages of easy deposition and much larger achievable thickness up to several hundreds of micro-meter. By cons, the magnetic composite needs to be sintered at high temperature to achieve desired magnetic properties. Besides the technique, the choice of magnetic materials and the design play an important role in the inductors final performance. This article presents a possible solution for integrating very small size (< 3mm^2) micro-inductor in a PwrSiP (Power system in Package) approach based on ferrite films and ferrite powder.

2. Specifications and design

In this work, the target is to integrate micro inductors in silicon wafers with a good compromise between performance and cost. In order to achieve an integration of the inductor with high specific inductance, it is necessary to work at high switching frequencies. NiZn ferrites are good candidates for a magnetic core working at moderate frequencies (5-10 MHz). The targeted specifications of our inductors are 470 nH of inductance, equivalent serial resistance < 0.1Ω at bias current of 0.6 A and operating at 6 MHz. The inductors are constituted by a thin film ferrite in a rectangular shape wrapped by multi-level metal windings as shown in figure 1.

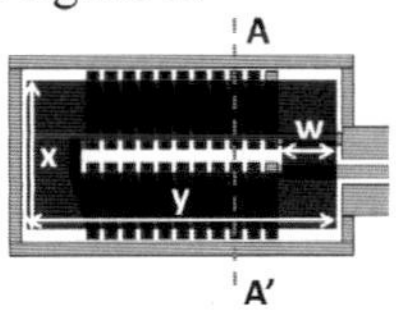

Figure 1. Schematic of micro-inductor

Based on target specifications of the inductor, the design is made by adjusting the parameters of core's properties and winding dimensions. For a given footprint area (3 mm^2), the geometrical input parameters are: core width (x), core length (y), magnetic core width (w), magnetic core thickness (t), thickness and width of copper wire and number of turns. The output parameters are the DC and AC resistance, the total inductance and maximum magnetic induction. After taking into account these parameters, in order to meet specifications of inductance and resistance and to avoid saturation in the core at the given DC current, the copper windings should have a cross section of at least 50 x 50 μm^2 and the desired magnetic core thickness should be 150 μm.

3. Component realization process

Figure 2 presents the fabrication steps for the ferrite-based inductors. 50 μm thick copper bottom tracks are deposited on $Si/SiO_2/Ti/Cu$ wafer by photolithography and electroplating. Then, the copper tracks are planarized by spin coating with a thin layer of SU8. Ferrite cores are placed on Cu tracks and covered by another thin layer of SU8. Vias positions are opened by process of photolithography and a seed layer of Ti/Au is deposited on the wafer and then, coated with a layer of BPN photo-resist [8]. Vias and 50 μm thick copper top tracts are deposited by photolithography and electroplating technique. Finally, residues of BPN and seed layer are removed.

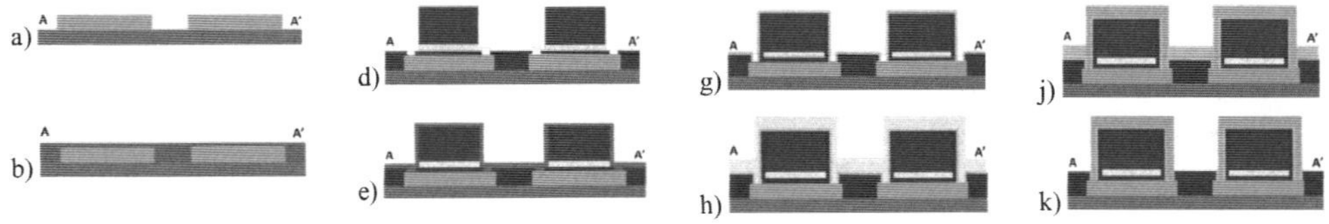

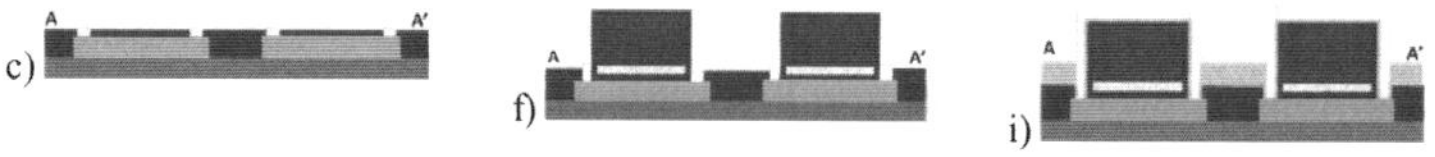

Figure 2. Scheme of fabrication steps in cross section a) Electroplating of bottom Cu tracks on Si wafer b) Planarization of Cu tracks c) Open vias d) Adjunction of thin film ferrite core on Cu tracks by photo-sensitive glue e) Cover ferrite cores with SU8 f) Open vias g) Sputter a seed layer Ti/Au on the wafer h) Cover a layer of BPN on the wafer i) Open vias and top Cu tracks j) Electroplating of top Cu tracks and vias k) Remove residues of BPN and Ti/Au seed layer

4. Experiments and results

There are two proposed approaches to fabricate ferrite cores.

- From commercial ferrite film: Two layers of thin film ferrite (ESL 40010® and ESL 40011®), about 70 μm each, are needed. The layers are bonded together by a wafer bonding machine. The double-layer films (140 μm thick) are then cut by a micro-milling machine into the expected and compensated dimension.
- From home-made ferrite powder: Synthesized NiZn ferrite powder is mixed with organics to form a magnetic slurry or paste. The thick photo-resist mold is formed by photolithography technique. The ferrite paste is then filled into the mould by screen printing technique followed by thirty minutes vacuum treatment to degas the paste. The ferrite composite is dried at 110°C for ten minutes and collected. Thicknesses of printed cores are 90-105 μm; the width and length are similar as mill cores.

The dimension of the fabricated magnetic core is 1.2 x 2.6 mm^2 with the magnetic core width of 350 μm. Then, the milled cores are sintered at 885°C and 950°C during 3 and 2 hours respectively; the printed cores are sintered at 980°C during 2 hours. Microstructures of cores are observed by scanning electron microscopy. Compositions of different ferrites are estimated by energy dispersive analysis (EDS). The magnetic properties of thin film ferrites are characterized by a vibrating sample magnetometer. To extract the complex permeability (μ' and μ'') of the used ferrite materials, tore-shape samples are also prepared and measured by a permeameter (Agilent 16454A). For electrical characterization, sintered cores are placed on gold/copper tracks on silicon wafer; wires are bonded to make the winding around the cores (see figure 3). Bonded cores are measured by an impedance analyzer (Agilent 4294) to determine the inductance versus frequency and versus DC bias and on a LCR meter (Agilent 4284A) to extract core losses versus AC current.

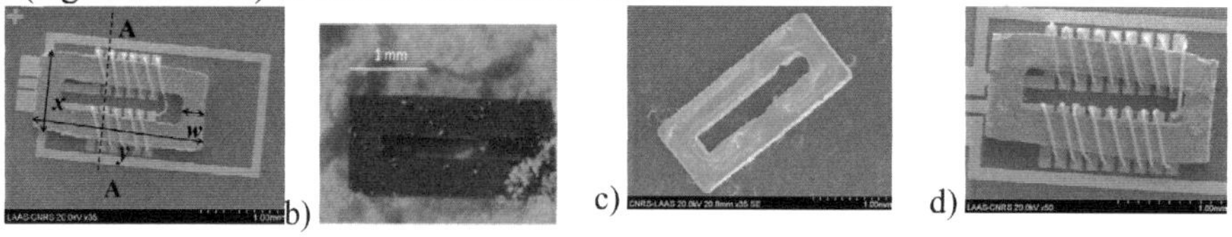

Figure 3. a) Top view of the rectangular test inductor and images of thin film ferrite cores after sintering b) ESL milled core and printed cores c) front side and d) back side

4.1. Magnetic material characterization

Figure 4 shows the microstructure of different ferrites after sintering. The grain sizes of thin film ESL 40010 and 40011 have a bimodal distribution, with large grains of 5-8 μm, covering around 50%vol, and small grains of 2-3 μm in diameter. Grain sizes of printed ferrite U70 and U200 are in the range of 1.5-6 μm. The microstructure of printed ferrites can be sharpened if one presses the ferrite paste after they are printed in the mould. For all cores, the shrinkage was measured to be 15-20%. Sintering temperature of each ferrite is determined after the shrinkage peak identified by the dilatometry characterization. The compositions of these ferrites are identified by EDS as listed in table 1.

The B-H characteristics of ferrites are presented in figure 5 for sintering temperatures 950°C. The induction saturation is in the range of 0.22-0.30 T, with a coercive field of 0.66-5.8 Oe (or 53-462 A/m). Magnetic remanences are 0.8-8%.

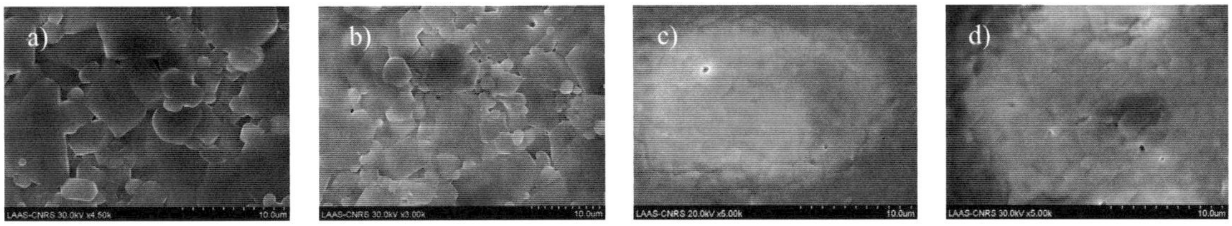

Figure 4. Microstructure a) b) ESL 40010 and 40011 sintered at 950°C/2hours, c) d) U70 and U200 sintered at 980°C/2hours

Table 1. Composition of different ferrite estimated by EDS

Ferrite	Composition
ESL 40010 (Ni/Zn = 1.49)	$Ni_{0.49}Zn_{0.33}Cu_{0.18}Fe_2O_4$
ESL 40011(Ni/Zn = 0.47)	$Ni_{0.28}Zn_{0.60}Cu_{0.12}Fe_2O_4$
U70 (Ni/Zn = 0.54, Co = 0.035)	$Ni_{0.30}Zn_{0.55}Cu_{0.15}Co_{0.035}Fe_2O_4$
U200 (Ni/Zn = 0.43, Co = 0.035)	$(Ni_{0.24}Zn_{0.56}Cu_{0.20})_{0.965}Co_{0.035}Fe_2O_4$

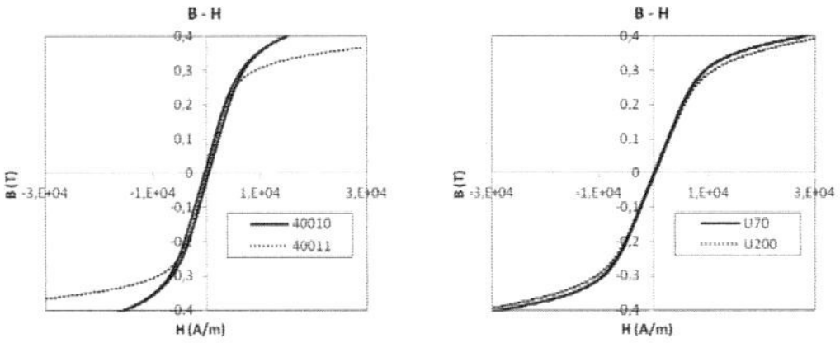

Figure 5. B-H curves of thin film ferrite. Samples were sintered at 950°C/2h with temperature ramp 3.3°C/min

4.2. *Electrical characterization*

The complex permeability of each material (ESL 40010, 40011, U70 and U200) is extracted as a function of frequency (as shown in figure 6). U200 and ESL 40011 films show a high primary permeability: 190 and 300 respectively up to 10 MHz. However, the secondary permeability of ESL 40011 increases quickly from 1-3 MHz while that of U200 remains small up to 10 MHz, which means that the core losses of ESL 40011 are more important than that of U200.

The complex permeability of the magnetic films was extracted from impedance of the test fixture with the tore mounted Z^*_m compared to the impedance of the text fixture without the tore Z^*_{sm} using the following formula:

$$\mu_r^* = \frac{2\pi(Z_m^* - Z_{sm}^*)}{j\omega\mu_0 h \ln\frac{c}{b}} + 1$$

in which ω is measurement pulsation $\omega = 2\pi f$, μ_0 is permeability of free space, h is thickness of the tore; c and b are external and internal diameter of the tore respectively.

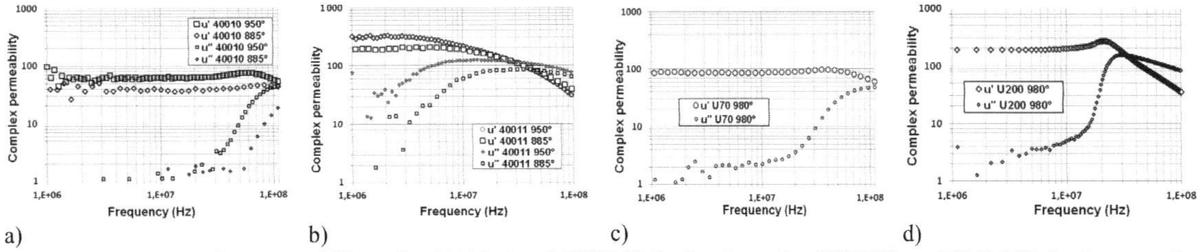

Figure 6. Complex permeability: ESL 40010 a) and 40011 b) ferrite sintered at 950°C/2h and 885°C/3h (b=4 mm, c=6 mm and h= 108 μm), U70 c) and U200 d) home-made ferrite sintered at 980°/2h (b=5 mm, c=14 mm and h=1.5 mm)

Electrical characterizations were performed for different magnetic cores with 21 turns coil. Figure 7 a) shows the inductance of the test inductors as a function of frequency for cores sintered at 885° and 950°C. An inductance as high as 860 nH or 285 nH/mm^2 was obtained for the ESL 40011 core and 287 nH or 95 nH/mm^2 for ESL 40010 core at 6 MHz while the air core inductor has only an inductance of 18 nH. These high values of inductance density are an advantage of our small rectangular inductors comparing to other reported ferrite inductors [9-11]. However, ESL 40011 shows a big decrease of permeability as superimposing DC bias current; this is a weak point of these materials for high DC current applications as compared to other magnetic composites [9-11]. Meanwhile, ESL 40010 cores have permeability more stable with DC bias current, yet they have lower permeability (see figure 6 and figure 7 b)). The core losses per unit volume were extracted from the under-probe measurements of test inductors with varying sine-wave amplitudes. ESL 40010 cores sintered at 885°C show the smallest losses at 1 MHz that are in the range of 100mW/cm^3-300mW/cm^3 at a current of 100mA-150mA (see figure 7 c)). In term of magnetic induction, the loss is about 100 mW/cm^3 at 10mT and 1MHz, which is comparable with the reported value 100mW/cm^3 at 1.5 MHz and 10mT by Mu[12].

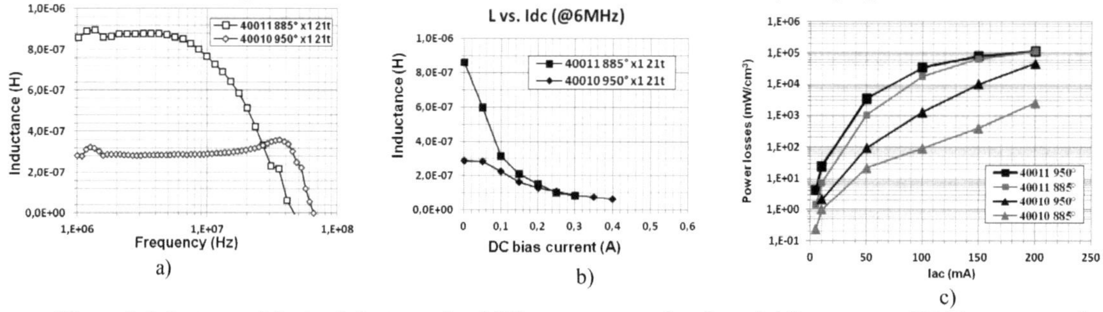

Figure 7. Inductance of the test inductor made of different cores as a function of a) frequency, and b) bias current at 6 MHz, c) Magnetic losses (mW/cm^3) of ESL 40010 and 40011 cores as a function of amplitude of I_{AC} at 1 MHz

5. Conclusions and perspectives

Ferrite rectangular cores were fabricated and sintered at different temperatures. The higher sintering temperature brings the material a higher permeability but also higher losses. These ferrites are very interesting in term of permeability and our inductors achieve a very high inductance density comparing to other reported micro ferrite inductors. For the moment, at 6 MHz, 285 nH/mm^2 of inductance is recorded for ESL 40011 cores and 95 nH/mm^2 for ESL 40010 cores. However, as superimposing a DC bias current all cores of four materials present a big decrease of permeability, which is a disadvantage of soft ferrite cores for high DC current application. It might be necessary to increase the core's volume or introduce air gaps in order to increase the saturation current. In term of losses, ESL 40010 shows the smallest losses beside its lowest permeability. A thorough consideration is needed to find out the material in a suitable design to achieve the target specification of inductor.

Acknowledgement

This work was carried out at LAAS, CNRS and LAPLACE (Toulouse, France) and funded by PRIIM project using micro fabrication platform at LAAS and 3DPhi platform at LAPLACE.

References

[1] T. Sato, H. Tomita, A. Sawabe, T. Inoue, T. Mizoguchi, M. Sahashi, IEEE Transactions on Magnetics 30 (1994).
[2] Y. Fukuda, T. Inoue, T. Mizoguchi, S. Yatabe, Y. Tachi, IEEE Transactions on Magnetics 39 (2003).
[3] N. Wang, E.J. O'Sullivan, P. Herget, B. Rajendran, L.E. Krupp, L.T. Romankiw, B.C. Webb, R. Fontana, E.A. Duch, E.A. Joseph, S.L. Brown, X. Hu, G.M. Decad, N. Sturcken, K.L. Shepard, W.J. Gallagher, Journal of Applied Physics 111 (2012).
[4] K.H. Kim, J. Kim, H.J. Kim, S.H. Han, IEEE Transactions on Magnetics 38 (2002).
[5] E.J. Brandon, E. Wesseling, V. White, C. Ramsey, L. Del Castillo, U. Lieneweg, IEEE Transactions on Magnetics 39 (2003) 2049.
[6] T. El Mastouli, J.P. Laur, J.L. Sanchez, M. Brunet, D. Bourrier, M. Dilhan, in: M.A. Maher, J.C. Chiao, P.J. Resnick (Eds.), Micromachining and Microfabrication Process Technology Xiii, 2008, p. A8820.
[7] M. Brunet, T. O'Donnell, L. Baud, N.N. Wang, J. O'Brien, P. McCloskey, S.C. O'Mathuna, IEEE Transactions on Magnetics 38 (2002) 3174.
[8] D. Bourrier, M. Dilhan, A. Ghannam, H. Granier, Comparisons of the new thick negative resist to Su8 resist, Conference on Advances in Resist Materials and Processing Technology XXVIII, San Jose, CA, 2011.
[9] D.H. Bang, J.Y. Park, IEEE Transactions on Magnetics 45 (2009) 2762.
[10] M.L. Wang, J.P. Li, K.D.T. Ngo, H.K. Xie, IEEE Transactions on Power Electronics 26 (2011) 1310.
[11] X. Fang, R. Wu, L. Peng, J.K.O. Sin, IEEE Electron Device Letters 34 (2013) 292.
[12] M. Mu, Y. Su, Q. Li, F.C. Lee, IEEE Energy Conversion Congress and Exposition (Ecce) (2011).

Low-cost, high-efficiency organic/inorganic hetero-junction hybrid solar cells for next generation photovoltaic device

P R Pudasaini[1, 2] and A A Ayon[1]
[1]MEMS Research Laboratory, Department of Physics and Astronomy, The University of Texas at San Antonio
One UTSA Circle, San Antonio, Texas, 78249, USA

[2]E-mail: pbz620@my.utsa.edu

Abstract. Organic/inorganic hybrid structures are considered innovative alternatives for the next generation of low-cost photovoltaic devices because they combine advantages of the purely organic and inorganic versions. Here, we report an efficient hybrid solar cell based on sub-wavelength silicon nanotexturization in combination with the spin-coating of poly (3,4-ethylene-dioxythiophene):polystyrenesulfonate (PEDOT:PSS). The described devices were analyzed by collecting current-voltage and capacitance-voltage measurements in order to explore the organic/inorganic heterojunction properties. ALD deposited ultrathin aluminium oxide was used as a junction passivation layer between the nanotextured silicon surface and the organic polymer. The measured interface defect density of the device was observed to decrease with the inclusion of an ultrathin Al_2O_3 passivation layer leading to an improved electrical performance. This effect is thought to be ascribed to the suppression of charge recombination at the organic/inorganic interface. A maximum power conversion efficiency in excess of 10% has been achieved for the optimized geometry of the device, in spite of lacking an antireflection layer or back surface field enhancement schemes.

1. Introduction

Sun light is an abundant and renewable energy resource, and converting sunlight into electricity has been regarded as one of the most promising approaches to provide clean energy. Conventional silicon solar cells usually require high purity silicon materials and relatively complex fabrication methods, therefore, they are not necessarily cost effective for a wide range of applications. Organic photovoltaic solar cells (OPV) on the other hand, have become a promising candidate for achieving low cost solar energy conversion devices, since they can be fabricated by relatively simpler and cost-effective processing schemes, such as screen printing, ink jet printing, dip coating and roll to roll processing. Additionally, they also have the advantage of being lightweight and mechanically flexible. However, as of today, state-of-the-art organic solar cells exhibit a relatively low power conversion efficiency (*PCE*) compared to their single crystal silicon counterparts. One of the greatest challenges of OPV is the relative paucity of electron accepting materials which may be paired with hole conducting polymers to induce exciton dissociation at the interface. An interesting approach to this problem is to use semiconducting nanostructures as an electron accepting phase to form organic/inorganic hybrid solar cells [1-3]. Hybrid solar cells comprised of p-type organic polymer thin film and n-type silicon substrate have been extensively investigated in the recent development of high efficiency and cost effective solar cells device. To date, a *PCE* value as high as ~10% has been reported for hybrid heterojunction devices, fabricated by spin coating the polymer (3,4-ethylene-dioxythiophene):polystyrenesulfonate (PEDOT:PSS) on a nanotextured silicon surface [4-7], however, it should be noted that there is a noticeable widespread in PCE values extending to the relatively low value of 1% or less [8,9]. Additionally, vertically aligned silicon nanowires (SiNWs) have been reported to exhibit excellent light absorption characteristics both theoretically and experimentally [10-12]. The quality of the interface between PEDOT:PSS and nanotextured silicon surface is believed to contribute greatly to the large discrepancy observed in the reported *PCE* values which is predicted to have a strong dependence in several variables, namely, the conductivity and wettability of PEDOT:PSS, the spin coating parameters (spin casting speed and time) employed to control the thickness of the PEDOT:PSS layer on the textured silicon surface, and the annealing temperature after the spin coating process, that could all influence the interface properties of a hybrid solar cell. A thin dielectric interface passivation layer between silicon and PEDOT:PSS layer could also have a significant effect on the electrical performance of the proposed hybrid devices. Different groups have employed a variety of dielectric materials as a junction passivation layer (i.e., recombination barrier) for the proposed PEDOT:PSS/Silicon hybrid solar cells and reported the enhanced electrical performance of the device. He *et al.* [13] and Zhang *et al.* [14] independently reported that a thin native oxide (SiOx) is crucial for the enhanced photovoltaic performance of the device compared to the H-terminated silicon surface. This can be attributed to the formation of a favorable internal electric field at the interface for an efficacious carrier extraction. In this letter, we report an efficient hybrid solar cell composed of vertically aligned silicon nanostructured arrays and PEDOT:PSS organic polymer employing an ultra-thin ALD deposited Al_2O_3 junction passivation layer.

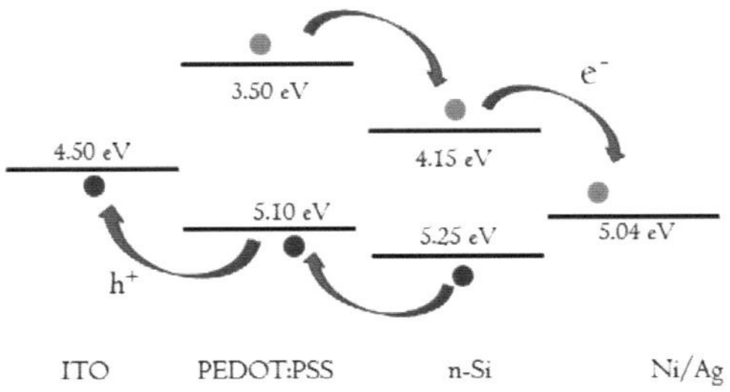

Figure 1. Schematic illustration of the fabricated Silicon/PEDOT:PSS solar cell (left), energy band diagram of the described hybrid solar cell (right).

2. Experimental Section

Highly ordered wafer-scale silicon nanopillar (SiNPs) array textured surfaces were fabricated by metal assisted electroless chemical etching in combination with nanosphere lithography. The detailed fabrication procedure can be found elsewhere [15-17]. The ALD of Al_2O_3 was carried out with a Cambridge Nanotech Savanah 200 system. The platen temperature was maintained at 110°C during the deposition process. The precursors, trimethylaluminium (TMA) and water (H_2O) were kept at room temperature. Nitrogen (N_2) was employed as a carrier and purging gas. ALD is a deposition process that employs cyclical self-limiting gas-surface reactions, and this property is important for a conformal deposition on high aspect ratio structures due to the time required for the reactant gases to fully diffuse into the narrow channel. Prior to device fabrication, the produced textured silicon samples were cleaned with a modified RCA procedure. After native-oxide removal the samples were transferred to the ALD chamber for a ~7A° Al_2O_3 deposition. Subsequently, 10 nm of nickel and 400 nm of silver were deposited on the back side and the samples were annealed at 425°C for 30 min to form a rear ohmic contact. Highly conductive PEDOT:PSS (Sigma-Aldrich) mixed with dimethyl sulphoxide and Triton X-100 (surfactant) solution was spin cast at 300 rpm for 10 s and 2000 rpm for 60 s to form a core-shell radial junction. The samples were then annealed on a hotplate at 140°C for 10 min to remove the solvent. Finally, an ITO coated glass with a resistivity of 8-12 Ω/sq (Sigma-Aldrich) was employed as a front contact to complete the solar cell. Each device had an active area of 1 cm^2. Figure 1 depicts the schematic illustration of the fabrication of a SiNPs/PEDOT:PSS solar cell. The morphology of the samples was collected by using high-resolution scanning electron microscopy (SEM) with a Field emission gun Hitachi S-5500. The PV measurement was performed using a solar simulator (Newport Sol2A) under AM 1.5G illumination (1000W/m^2) at standard testing conditions. Prior to each sample measurement, the simulator intensity was calibrated with a reference solar cell from Newport (Irvine CA, USA) to ensure that the irradiation variation was within 3%.

3. Results and discussion

Figure 2 includes a large SEM micrograph of the vertically aligned nanopillar array (top view) fabricated by the aforementioned metal assisted electroless chemical etching method. The same Figure 2 includes an inset in the upper right corner with higher magnification SEM micrographs of the same sample. Nanosphere lithography was employed to control the dimension of the SiNPs (diameter and inter-pillar spacing). The distance between the centers of any two pillars was fixed at 650 nm, and this value was determined by using Polystyrene nanoparticles with that particular diameter. The density of the SiNPs array was measured to be $\sim 2.7\times10^6$ pillars/mm^2. The inset in the lower left corner in Figure 2 shows the top view of PEDOT:PSS coated SiNPs array textured surface. The thickness of the PEDOT:PSS layer was approximately 40 nm. SiNPs arrays with average heights of 200 nm, 400 nm, 800 nm, and 1200 nm were fabricated by etching the sample for 30 s, 60 s, 120 s, and 180 s, respectively. The height of the SiNPs plays an important role in device performance. The current-voltage characteristics of the SiNPs/PEDOT:PSS hybrid solar cells having different nanopillar heights were measured under 100 mW/cm^2 illumination. Figure 3 depicts the average value of short circuit current density (J_{SC}), open circuit voltage (V_{OC}), fill factor (FF) and power conversion efficiency (PCE) of the fabricated devices as a function of SiNPs height. The measured value of short circuit current density for the fabricated SiNPs/PEDOT:PSS solar cells, increases as the nanopillar height increases, reaches a maximum of 29.5 mA/cm^2 for a SiNPs height of 0.4 μm (see Figure 3a), beyond this particular point, it decreases quasi-linearly to 21.2 mA/cm^2 as the height is further increased to 1.2 μm. Similar effects have also been reported elsewhere [5,18]. The remarkable antireflection property of the SiNPs array textured surface is directly reflected in the measured value of J_{SC} of the SiNPs/PEDOT:PSS solar cell. The maximum J_{SC} of 29.5 mA/cm^2 for a SiNPs array textured cell with SiNPs height of 0.4 μm is almost 36.6% greater than a planar/PEDOT:PSS cell. On the other hand, the measured open circuit voltage of the SiNPs/PEDOT:PSS solar cells was observed to decrease continuously from a maximum value of 538 mV to a minimum of 490 mV as the SiNPs height varied from 0.2 μm to 1.2 μm (see Figure 3a). This can be attributed to an increased junction recombination with increases in surface area. The power conversion efficiency of the SiNPs/PEDOT:PSS solar cells produced, reached a maximum value of 9.65%, for a nanopillar height of 0.4 μm, which compares favorably to the 7.02% observed for a planar/PEDOT:PSS cell. This is mainly due to the increase in short circuit current density and fill factor for the device, despite the slight decrease in V_{OC}. A promising fill factor value of 62 was achieved for a SiNPs/PEDOT:PSS hybrid cell which also compares favorably to FF of 60 for a planar/PEDOT:PSS solar cell. This could be attributed to an increased carrier separation due to the increased junction area. Furthermore, the SiNPs/PEDOT:PSS heterojunction provides a shorter pathway for the minority charge carriers towards the respective electrode. One factor limiting the efficiency of SiNPs/PEDOT:PSS hybrid cells is the low carrier collection efficiency due to increased surface recombination, in spite of having better light absorption

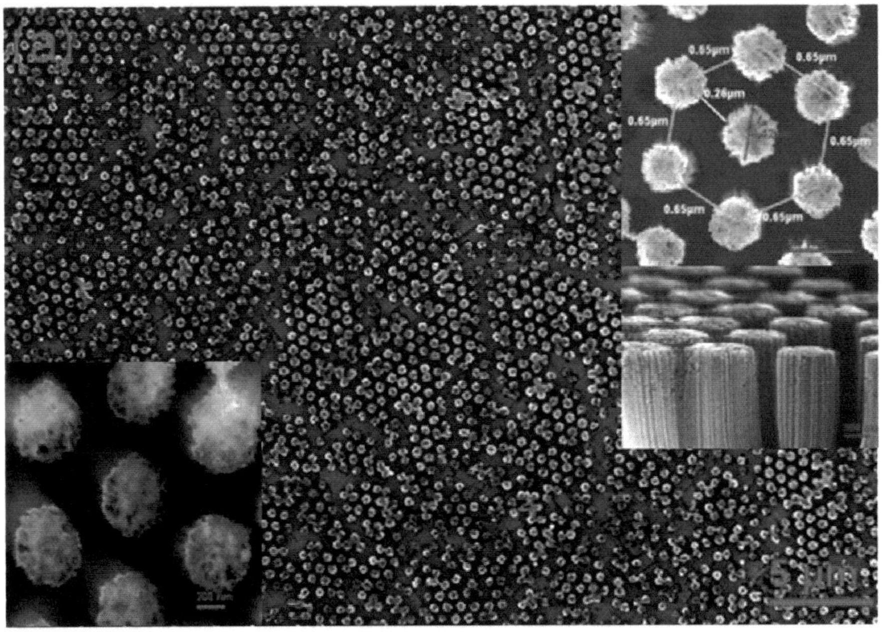

Figure 2. SEM micrographs of top view of a SiNP array textured silicon surface; the inset on the upper right corner includes higher magnification images of the same sample, where it is possible to discern the hexagonal order of the SiNPs array. The inset on the lower left corner shows a high magnification image of the SiNPs array, coated with the transparent and conductive polymer PEDOT:PSS by spin casting at 2000 rpm.

characteristics. The open circuit voltage of a SiNPs/PEDOT:PSS solar cells, irrespective of pillar height, is observed to be smaller compared to their planar/PEDOT:PSS counterparts.To reduce the recombination at the nanotextured silicon surface with a SiNPs/PEDOT:PSS structure we employed an ultrathin (~0.7 nm) ALD Al_2O_3 as an interface passivation layer. ALD Al_2O_3 was chosen as a passivation layer due to its unique chemical and field effect passivation characteristics, as well as the ability of ALD to deposit high quality, conformal films on high aspect ratio features, with an angstrom-level control of film thickness at low temperatures. The height of the SiNPs was fixed at 0.4 μm. It was found that the power conversion efficiency of the devices increased with the utilization of Al_2O_3 interface layer, with a maximum value of 10.56%, while the *PCE* of the same device without Al_2O_3 was 9.65%. This can be attributed mainly to the increased open circuit voltage of the device. The maximum open circuit voltage of 578 mV was achieved with an interfacial ALD Al_2O_3, which is ~9.1% higher compared to the same device without Al_2O_3 interface layer. The maximum short circuit current density of the cell with the Al_2O_3 barrier layer was 30.1 mA/cm^2 compared to 29.5 mA/cm^2 for a cell without Al_2O_3 despite the slight decrease in *FF*. To investigate the underlying reasons for the observed voltage gain, we measured and graphed the dark current density versus voltage characteristics of the hybrid SiNPs/PEDOT:PSS solar cells with and without Al_2O_3 interface layer. The observations indicate that the dark current density is suppressed significantly after employing the ultrathin Al_2O_3 barrier layer (see figure 4 a) (see Figure 4(a)). From the best fitting of the dark *J-V* characteristic curves at the forward bias condition, we extracted the values of saturation current density (J_o) and diode ideality factor (n) of the described device. We obtained the value of J_o to be 0.48 μA/cm^2 for a SiNPs/PEDOT:PSS hybrid cell with an Al_2O_3 barrier layer, which was lower than that of a cell without a barrier layer, namely, 2.19 μA/cm^2.

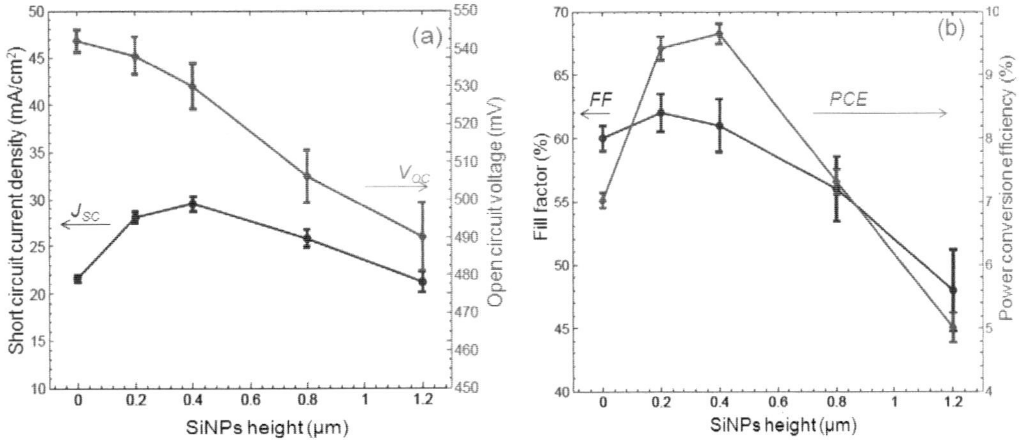

Figure 3. Average photovoltaic performance parameters of (a) *Jsc* and *Voc* and (b) *FF* and *PCE* of the SiNPs/PEDOT:PSS solar cell as a function of SiNPs height.

The ideality factors of 2.06 and 2.21, respectively, were obtained for the SiNPs/PEDOT:PSS cells with and without the Al_2O_3 barrier layer.

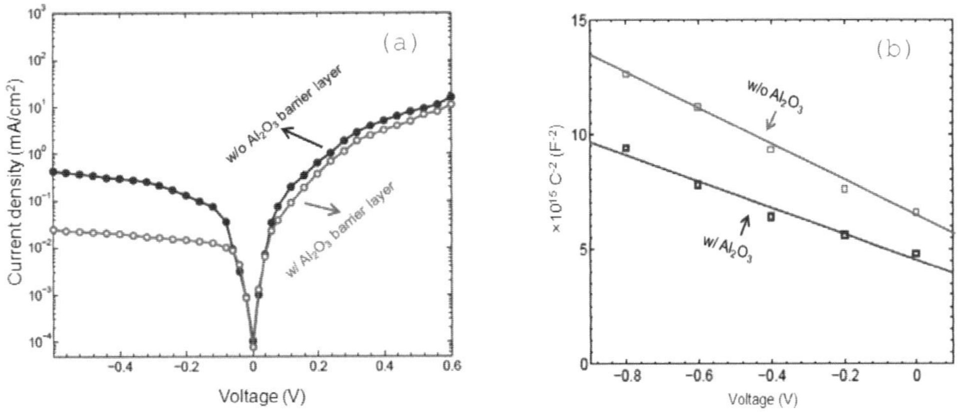

Figure 4. The dark *J-V* characteristics curve (a), C^{-2}-*V* plots (b), for SiNPs/PEDOT:PSS solar cells with and without Al_2O_3 passivation layer.

We also performed the capacitance voltage (*C-V*) measurements for the devices with and without Al_2O_3 passivation layers. Neglecting the series resistance of the devices, the interfacial defect density arising from the presence of chemical defects, such as dangling bonds, nonconformal coating of organic polymer etc., can be calculated by using the equations below.

$$\left(\frac{1}{C^2}\right) = \frac{2}{c_2{}^2 q\varepsilon N_A}\left(\Phi_B - c_2 V - v_p\right) \tag{1}$$

$$c_2 = \frac{\varepsilon_0}{\varepsilon_0 + q^2 \delta D_{it}} \tag{2}$$

where, Φ_B is the barrier height, V is the applied voltage, v_p is the difference between the Si Fermi level and valence band, N_A is the doping concentration, δ is the interface thickness, D_{it} is the interfacial defect density and ε_0 and ε are the permittivity of vacuum and silicon respectively. Figure 4(b) depicts the C^{-2} versus V plot for the optimized SiNPs/PEODT:PSS solar cells with and without interfacial Al_2O_3 passivation layer. Ostensibly, the C^{-2}-V plot is linear and we extracted the value of interfacial defect density of the devices from the slope of the corresponding C-^{2}-V plot. For the device with the Al_2O_3 passivation layer, the D_{it} value of 2.80×10^{12} $cm^{-2}eV^{-1}$ was achieved while for the device without passivation layer a significantly higher value of 4.16×10^{12} $cm^{-2}eV^{-1}$ was obtained. The reduced interfacial defect density was responsible for the lower saturation current density and hence the improved open circuit voltage and power conversion efficiency of the device.

4. Conclusions

We have employed a relatively simple and wafer scalable process to produce and characterize a heterojunction PEDOT:PSSs/SiNP hybrid solar cell that could prove useful in the quest of low-cost yet efficient photovoltaic devices. The effect of SiNP height on solar cell performance was investigated. The *PCE* of a SiNPs/PEDOT:PSS hybrid cell with an optimized SiNPs height of 0.4 μm was observed to be 9.65%. With the utilization of an ultrathin ALD deposited Al_2O_3 junction passivation layer, we observed a short circuit current density and an open circuit voltage as high as 30.1 mA/cm^2 and 578 mV respectively, which led to a *PCE* value in excess of 10.56%. The described hybrid device was formed by spin coating the organic polymer PEDOT:PSS on SiNPs arrays fabricated by relatively simple, low-cost, room temperature methods, which could lead to the proliferation of low cost photovoltaic devices in the future.

Acknowledgement

We thank Dr. Mike Gerhold, Technical Manager of the U.S. Army Research Office, for the financial support provided for this project (ARO grant number W911NF-13-1-0110).

References

[1] Lin C Y, Holman Z C and Kortshagen U R 2009 *Nano Letts.* **9** 449
[2] He L, Jiang C, Rusli, Lai D and Wang H 2011 *Appl. Phys. Letts.* **99** 021104
[3] Huynh W U, Dittmer J J and Alivisatos A P 2002 *Science* **295** 2425
[4] Avasthi S, Lee S, Loo Y L and Sturm J C 2011 *Adv. mater.* **23** 5762
[5] He L, Jiang C, Wang H, Lai D and Rusli 2012 *ACS appl. mater. interfaces* **4** 1704
[6] Jeong S, Garnett E C, Wang S, Yu Z, Fan S, Brongersma M L, McGehee M D and Cui Y 2012 *Nano letts.* **12** 2971
[7] Khatri I, Tang Z, Liu Q, Ishikawa R, Ueno K and Shirai H 2013 *Appl. Phys. Letts.* **102** 063508
[8] Lu W, Wang C, Yue W and Chen L 2011 *Nanoscale* **3** 3631
[9] Moiz S A, Nahhas A M, Um H D, Jee S W, Cho H K, Kim S W and Lee J H 2012 *Nanotechnol.* **23** 145401
[10] Han S E and Chen G 2010 *Nano letts.* **10** 1012
[11] Pudasaini P R, Elam D and Ayon A A 2013 *J. Phys. D: Appl. Phys.* **46** 235104
[12] Pudasaini P R, and Ayon A A 2013 *Microsyst. Technol.* **19** 871
[13] He L, Jiang C, Wang H, Lai D and Rusli 2012 *Appl. Phys. Letts.* **100** 073503
[14] Zhang F, Sun B, Song T, Zhu X and Lee S 2011 *Chem. Mater.* **23** 2084
[15] Mikhael B, Elise B, Xavier M, Sebastian S, Johann M and Laetitia P 2011 *ACS appl. mater. interfaces* **3** 3866
[16] Pudasaini P R and Ayon A A 2013 *Microelect. Eng.* **110** 126
[17] Huang Z, Geyer N, Werner P, Boor J and Gosele U 2011 *Adv. mater.* **23** 285
[18] Syu H J, Shiu S C and Lin C F 2012 *Sol. Energy Mater. Sol. Cells* **98** 267

Comparative Study of Bio-implantable Acoustic Generator Architectures

D Christensen, S Roundy

University of Utah, Mechanical Engineering, 50 S. Central Campus Drive, Salt Lake City, UT, USA

E-mail: dave.christensen@utah.edu

Abstract. This paper is a comparative study of the design spaces of two bio-implantable acoustically excited generator architectures: the thickness-stretch-mode circular piezoelectric plate and the bending-mode unimorph piezoelectric diaphragm. The generators are part of an acoustic power transfer system for implanted sensors and medical devices such as glucose monitors, metabolic monitors, drug delivery systems, etc. Our studies indicate that at small sizes the diaphragm architecture outperforms the plate architecture. This paper will present the results of simulation studies and initial experiments that explore the characteristics of the two architectures and compare their performance.

1. Introduction

Advances in sensing and CMOS technology have opened up the possibility of mm scale bio-implantable sensors. However, the size reduction of power and communications technologies (i.e. antennas) has not kept pace with sensing and computation. Therefore, the size of implantable sensors is primarily driven by the power and communications sub-systems. Denisov et. al.[1] have shown that at small sizes and large implant depths acoustic power transmission outperforms electromagnetic power transfer and thus can enable smaller bio-implant sizes.

Two piezoelectric generator architectures that may be employed for acoustic energy transfer are the standard thickness-stretch-mode circular plate (or disk) and the bending-mode unimorph diaphragm. The plate is the more widely used architecture in ultrasonic applications such as imaging and therapy, and seems the more likely candidate for a bio-implantable acoustic generator. On the other hand, the diaphragm seems a mediocre candidate for two reasons. First, it operates in the 3-1 mode which has a much lower coupling coefficient than the 3-3 mode utilized by stretch mode plates. Secondly, diaphragms use significantly less piezoelectric material than plates of the same size and are thus usually outperformed by plates. Prior published works [2][3] have focused on either the plate or diaphragm generator architecture and optimized the generator for a specific application.

2. Model & Experimental Setup

The scaling results of this paper are based on simulation models that include experimental data to evaluate the accuracy of results. The simulation model and experimental setup used are as follows:

2.1. *Simulation Model*

The simulation models for both architectures are shown in Figure 1. The simulations utilize Mason's network model [4] for the plate and the impedances of the medium. A slightly modified equivalent network model was used for the diaphragm based on the model proposed by Prasad [2]. In prior published works, the transmitter dynamics have played an important role in the power and efficiency results of the generator [4]. The transmitter dynamics are included in the experimental validation of the model, but are excluded in the scaling simulation. In place of the transmitter is substituted mechanical power into the tissue at 720 mW/cm^2, the maximum safe power input allowed by the United States FDA [5]. This substitution leads to a better comparison and more insightful results.

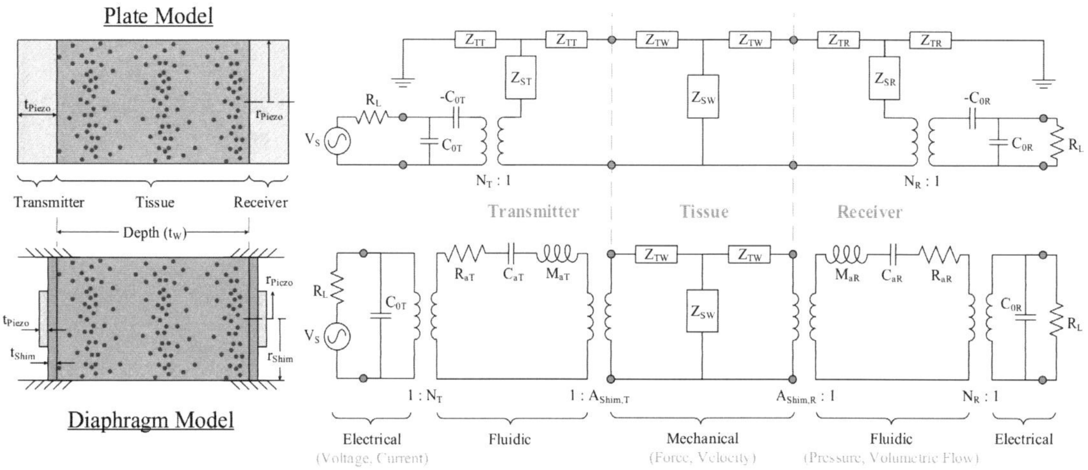

Figure 1. Diagram and simulation model of the plate and diaphragm models.

2.2. Experimental Setup
A tank with adjustable arms was constructed to facilitate testing. Plates (APC Inc., 850 material, 3.43 mm thick, 12.7 mm diameter) and diaphragms (Steminc, SMPD11D11T10F95) were mounted to ABS tubing (McMaster Carr, 1839T371) with cyanoacrylate. All tests were run with a sinusoidal input from a function generator at 2.5 volts amplitude in series with a 100 ohm resistor (R_L = 100 ohms in Figure 1). Plate configurations were given a frequency sweep between 1 Hz and 1 MHz. Diaphragm configurations were given a frequency sweep between 1 Hz and 10 kHz.

3. Results
Experiments with macro-scale devices were run to validate the simulation models. Then scaling simulations were performed to explore the behaviour of the two architectures as dimensions are scaled down to MEMS sizes. First, each transducer was modelled as a stand-alone transmitter. Source current and surface velocity were measured. Damping ratios were experimentally determined. Second, the tested transducers were coupled together and placed in water. A plate transmitter was coupled with a plate receiver and a diaphragm transmitter was coupled with a diaphragm receiver. Transmitter source and receiver generated currents were measured. Lastly, simulations were run to compare the plate and diaphragm architectures at frequencies and diameters of interest. The simulations replace the transmitter with a source intensity and replace water with tissue.

3.1. Transducer Validation

3.1.1. Plate. Figure 2 shows a frequency sweep of the simulated and measured current through the plate transducer. The experimental data shows some lower frequency peaks that are not present in the simulations. These peaks are from the radial vibration mode and harmonics which are not included in the simulation model. The magnitude of the simulated current of the primary extension mode is about 2.2 times greater than the experimentally measured current of the same mode. While the simulated and experimental currents do not match, the simulated and measured velocities of the plate face do match.

This discrepancy of a factor of 2.2 in the source current leads to a mismatch between simulated and measured power output of about a 5 times.

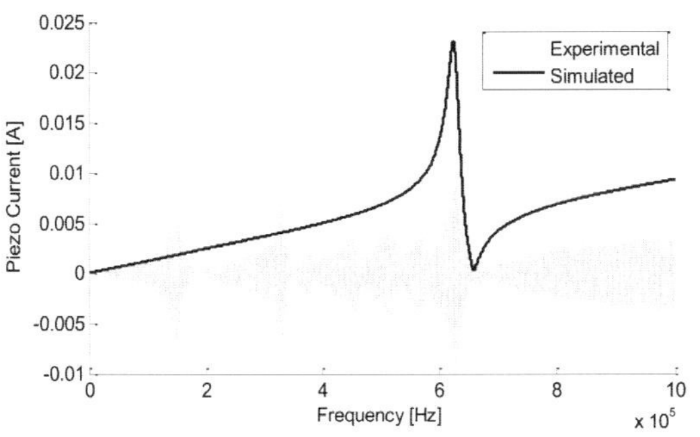

Figure 2. Plate transducer tested as a stand-alone transmitter in air. Simulation results are plotted over experimental data.

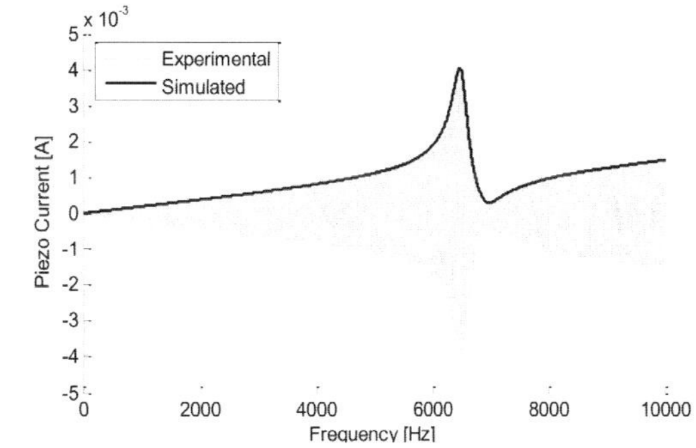

Figure 3. Diaphragm transducer tested as stand-along transmitter in air. Simulation results are plotted over experimental data.

3.1.2. Diaphragm. The diaphragm simulation results closely match the experimental data as shown in Figure 3. While not shown in Figure 3, the velocities measured by the laser vibrometer also indicate a good match. Experimentally measured values of the damping ratio were incorporated into the model.

3.2. Transducers + Medium Validation

3.2.1. Plate. Figure 4 shows simulation and experimental results of the full power transmission model for two plate transducers. The simulation results match the experimental results in expected shape but not magnitude. The magnitudes of both the transmitter and receiver simulation peaks exhibit the 2.2 gain discussed previously.

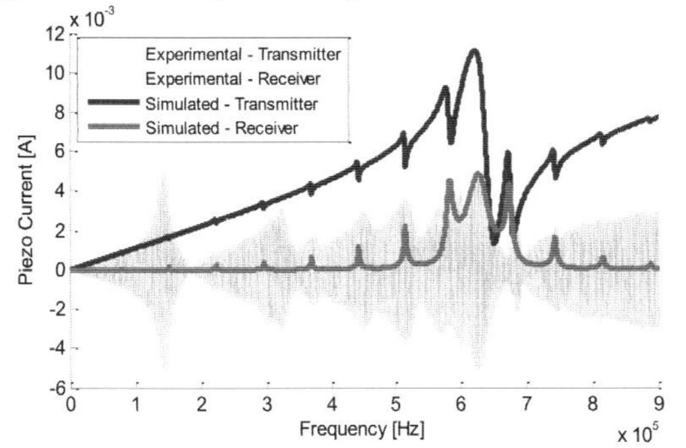

Figure 4. Simulated and measured transmitter and generator current. Plate transmitter and plate receiver separated by 1 cm of water.

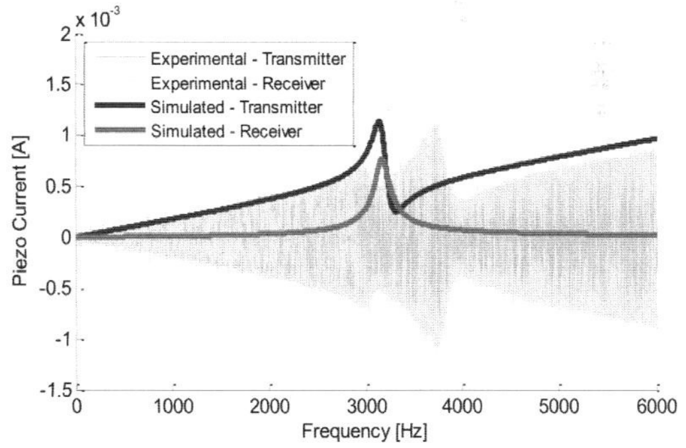

Figure 5. Simulated and measured currents for diaphgram. Diaphragm transmitter and receiver separated by 1 cm of water.

3.2.2. Diaphragm. Figure 5 shows simulation and experimental results of the full power transmission model for two diaphragm transducers. The damping ratio of the diaphragms was found to be much higher in water than in air. Note that the water shifts the resonance of the plate significantly from that measured in air as shown in Figure 5. The models over-estimate this shift.

3.3. Scaled Simulation Comparison

3.3.1. Plate. The scaling simulation of the plate shown in Figure 6 exhibits power loss due to increased absorption at higher frequencies. The diameter to thickness ratio (aspect ratio) is an important consideration for the model. At aspect ratios lower than 1, the plate becomes a cylinder and the radial mode vibrations play a more significant role in the transducer dynamics. Also, the practical design space for a completed plate device would at least have an aspect ratio of 2 or greater to account for space needed for power electronics, matching layer, and housing. As is clear in Figure 6, it is not practical to design a very small plate generator for low (i.e. 100's of kHz) frequencies where the attenuation in human tissue is lower.

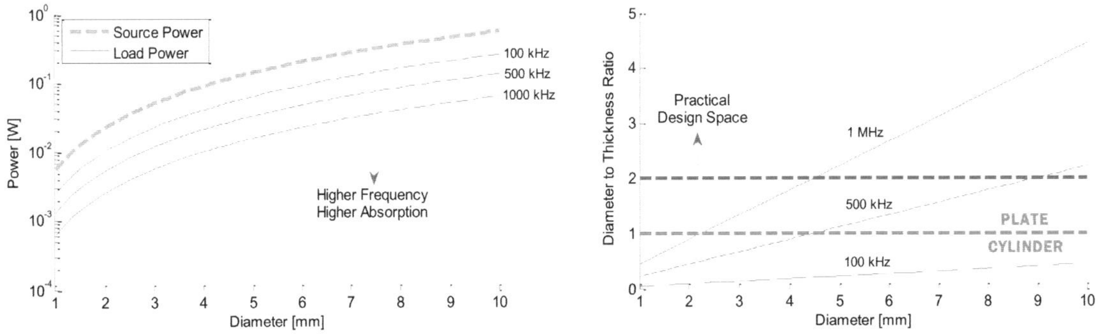

Figure 6. Scaling simulation of the plate architecture at source intensity of 720 mW/cm^2. The optimal plate thickness is ½ the acoustic wavelength, which is determined by the frequency.

3.3.2. Diaphragm. The scaling simulation of the diaphragm shown in Figure 7 exhibits power loss due to increased absorption at higher frequencies like the plate. In the case of the diaphragm, the aspect ratio is important because at low aspect ratios the diaphragm starts to behave more like a plate. At aspect ratios above 10, bending is the significantly dominant strain mode in the diaphragm. At aspect ratios lower than 5, bending strain is insignificant to thickness-stretch strain and a different model would have to be used. Figure 7 indicates that the practical design space for diaphragms is at low ultrasonic frequencies where attenuation is lower and at small diameters which are more desirable for bio-implantable sensors.

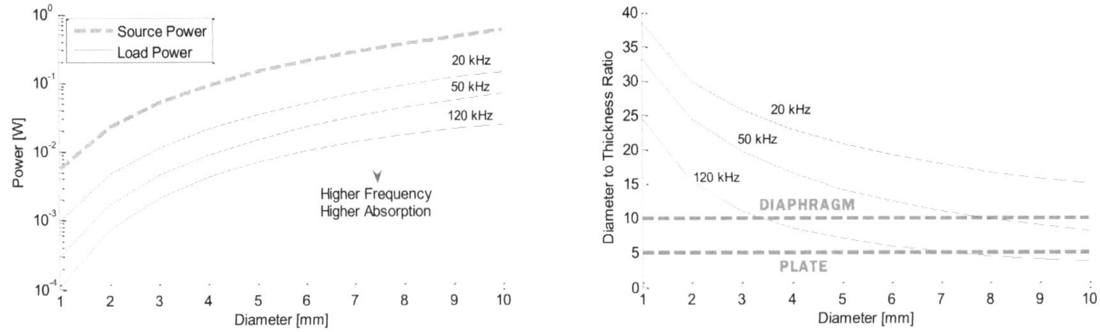

Figure 7. Scaling simulation of the diaphragm architecture at source intensity of 720 mW/cm^2.

3.3.3. Performance Comparison. Combining data from Figure 6 and Figure 7, a direct comparison of the plate and diaphragm is made and shown in Figure 8. The power curves reflect plate architectures with aspect ratios of 2 or greater and diaphragm architectures of 10 or greater. It can be seen that the diaphragm architecture is able to be utilized at lower frequencies with similar efficiency to a larger plate architecture at higher frequencies. The power curves also lead to the not so apparent conclusion that the diaphragm architecture, not the plate, will produce more power for generators of sizes below roughly 5 mm in diameter. Thus, it would be more appropriate for MEMS implementations of ultrasonic bio-implantable sensors.

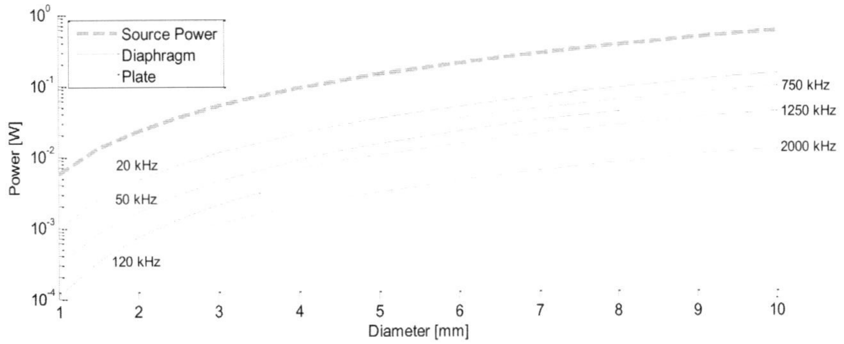

Figure 8. Comparison of plate and diaphragm generators versus diameter for several frequencies at source intensity of 720 mW/cm^2.

4. Discussion

The simulation output shows that the diaphragm will generate more power than the plate at very small sizes (i.e. diameters of 5 mm or less). It should be noted that while the diaphragm model matched the experimental results in terms of power output well, the plate model over-predicted the power output by a factor of approximately 5. Therefore, the plate curves shown in Figure 8 may actually over-predict the power output. This would indicate that at a given size (even up to 9 mm), the comparison may be even more favorable to the diaphragm architecture. However, at larger sizes and lower frequencies, diaphragm displacements may be quite large which could limit their applicability to biological implants.

The ultimate goal of this work is the implementation of a mm scale acoustic power generator capable of producing enough power to operate implantable sensors and communications. On the pathway to this goal, model discrepancies still need to be identified and corrected. This will be followed by ex-vivo tests in tissue or tissue phantoms to demonstrate the power transmission channel and MEMS scale generators.

5. Conclusions

This paper has explored the design space for an acoustically excited generator to power bio-implantable sensors. Two architectures, the standard thickness-stretch-mode plate and a unimorph piezoelectric diaphragm, were evaluated and compared. Simulations and preliminary experimental results indicate that at sizes below roughly 5 mm in diameter the diaphragm architecture is capable of producing more power than the plate.

References

[1] Denisov A and Yeatman E 2010 Ultrasonic vs. Inductive Power Delivery for Miniature Biomedical Implants *2010 International Conference on Body Sensor Networks* (IEEE) pp 84–9

[2] Prasad S A N, Gallas Q, Horowitz S B, Homeijer B D, Sankar B V., Cattafesta L N and Sheplak M 2006 Analytical Electroacoustic Model of a Piezoelectric Composite Circular Plate *AIAA J.* **44** 2311–8

[3] Ozeri S and Shmilovitz D 2010 Ultrasonic transcutaneous energy transfer for powering implanted devices. *Ultrasonics* **50** 556–66

[4] Sherrit S, Badescu M, Bao X, Bar-cohen Y and Chang Z 2005 Efficient Electromechanical Network Models for Wireless Acoustic- Electric Feed-throughs *Proc. SPIE Smart Struct. Conf. San Diego, CA* **5758**

[5] U.S. Department of Health and Human Services 2008 *Guidance for Industry and FDA Staff: Information for Manufacturers Seeking Marketing Clearance of Diagnostic Ultrasound Systems and Transducers*

AUTHOR INDEX

AUTHOR INDEX

AUTHOR INDEX

AUTHOR INDEX

AUTHOR INDEX